Encyclopedia of Materials, Parts, and Finishes

Second Edition

Mel Schwartz

CRC PRESS

Boca Raton   London   New York   Washington, D.C.

Library of Congress Cataloging-in-Publication Data

Schwartz, Mel M.
　　Encyclopedia of materials, parts, and finishes / by Mel Schwartz.—2nd ed.
　　　p. cm.
　　ISBN 1-56676-661-3
　　1. Smart materials—Encyclopedia. I. Title.

TA418.9.S62 S39 2002
620.1'18—dc21   2002019220

This book contains information obtained from authentic and highly regarded sources. Reprinted material is quoted with permission, and sources are indicated. A wide variety of references are listed. Reasonable efforts have been made to publish reliable data and information, but the author and the publisher cannot assume responsibility for the validity of all materials or for the consequences of their use.

Neither this book nor any part may be reproduced or transmitted in any form or by any means, electronic or mechanical, including photocopying, microfilming, and recording, or by any information storage or retrieval system, without prior permission in writing from the publisher.

The consent of CRC Press LLC does not extend to copying for general distribution, for promotion, for creating new works, or for resale. Specific permission must be obtained in writing from CRC Press LLC for such copying.

Direct all inquiries to CRC Press LLC, 2000 N.W. Corporate Blvd., Boca Raton, Florida 33431.

Trademark Notice: Product or corporate names may be trademarks or registered trademarks, and are used only for identification and explanation, without intent to infringe.

Visit the CRC Press Web site at www.crcpress.com

© 2002 by CRC Press LLC

No claim to original U.S. Government works
International Standard Book Number 1-56676-661-3
Library of Congress Card Number 2002019220
Printed in the United States of America 1 2 3 4 5 6 7 8 9 0
Printed on acid-free paper

Preface

This encyclopedia represents an update of existing materials and presents new materials that have been invented or changed, either by new processes or by an innovative technique. The encyclopedia covers basic materials such as rubber and wood.

This two-volumes-in-one includes two decades of the process of materials; the process/fabrication selection has been hindered by new and unusual demands from all quarters. No change in this trend is expected in the foreseeable future.

This trend has been visible in several industries — aerospace, automotive, electronic, space, computers, chemical, and oil — and in many other commercial endeavors. Metals (wrought, cast, forged, powder), plastics (thermoplastics/thermosets), composites, structural ceramics, and coatings are continually finding new applications in the above industries.

The trend toward combining high strength and light weight is evident in fiber/particle/whisker-reinforced composites. This encyclopedia/handbook covers not only these matrix composites (metallic, plastic, ceramic, and intermetallic), but also other materials of the future — nano and functionally graded structures, fullarenes, plastics (PEEK, PES, etc.), smart piezoelectric materials, shape memory alloys, and ceramics.

Higher processing temperatures as well as more resistant and effective high-temperature materials have attracted the attention of engineers, scientists, and materials workers in many industries. Engines now operate more efficiently at temperatures higher than those attainable with the materials of the past. For example, interest in 2000°F (1093°C) turbine engines has brought more high-temperature, high-strength ceramics into development and use.

The use of a vacuum environment has improved many materials not only in their initial production and processing, i.e., steels, but also eventually in their fabrication. For example, a vacuum environment in brazing and welding and in hot isostatic pressing removes voids and consolidates material structures.

New environmental regulations by government agencies (the Environmental Protection Agency, the Occupational Safety and Health Administration, etc.) have sent the technologist back to the drawing board and laboratory to design and develop new and better materials and processes that are not potential health hazards, and many of these new material substitutes are included in this revised edition.

Finally, political diplomacy, rather than economics and regulation, could well be the most important factor in materials supply in the near future. The major supply of many critical raw materials and supplies for the processes needed to sustain the future economies of many nations lies in the hands of a few small nations. Consequently, there is no guarantee of a steady supply of these strategic materials, and we must continually innovate and explore new sources of materials development (ocean floor and space).

Editor

Mel M. Schwartz is a consultant to the vast field of materials and processes. He is editor of the *Journal of Advanced Materials* and editor-in-chief of the *Smart Materials Encyclopedia*. Schwartz received his bachelor of arts degree from Temple University, his master's degree from Drexel University, and is currently working in the doctorate program at the University of Sarasota. His professional experience includes a career in metallurgy, manufacturing research, and development and metals processing at the U.S. Bureau of Mines, U.S. Chemical Corp., Martin-Marietta Corp., Rohr Industries, and Sikorsky Aircraft, from which he retired in 1999.

Awards and honors include Inventor of the Year for Martin-Marietta, the Jud Hall Award (Society of Manufacturing Engineers), the first G. Lubin Award (Society for the Advancement of Materials and Processing Engineers), and Engineer of the Year in Connecticut (1973). He is an elected Fellow for the Society for the Advancement of Materials and Processing Engineers and American Society for Materials International, and sits on several peer-review committees; as well, he is a member of numerous national and international societies. Schwartz has written 14 books and over 100 technical papers and articles and has given company in-house courses and numerous seminars around the world.

Dedication

To Carolyn, Anne-Marie, and Perry whose enormous courage, will, and determined spirit are overwhelming.

Mel Schwartz

A

ABRASIVE

An abrasive is defined as a material of extreme hardness that is used to shape other materials by a grinding or abrading action. Abrasive materials may be used as loose grains, as grinding wheels, or as coatings on cloth or paper. They may be formed into ceramic cutting tools that are used for machining metal in the same way that ordinary machine tools are used. Because of their superior hardness and refractory properties, they have advantages in speed of operation, depth of cut, and smoothness of finish.

Abrasive products are used for cleaning and machining all types of metal, for grinding and polishing glass, for grinding logs to paper pulp, for cutting metals, glass, and cement, and for manufacturing many miscellaneous products such as brake linings and nonslip floor tile.

Abrasive Materials

These may be classified in two groups, the natural and the synthetic (manufactured). The latter are by far the more extensively used, but in some specific applications natural materials still dominate.

The important natural abrasives are diamond (the hardest known material), corundum (a relatively pure, natural aluminum oxide, Al_2O_3), and emery (a less-pure Al_2O_3 with considerable amounts of iron). Other natural abrasives are garnet, an aluminosilicate mineral; feldspar, used in household cleansers; calcined clay; lime; chalk; and silica, SiO_2, in its many forms — sandstone, sand (for grinding plate glass), flint, and diatomite.

The synthetic abrasive materials are silicon carbide SiC, aluminum oxide Al_2O_3, titanium carbide TiC, and boron carbide B_4C. The synthesis of diamond puts this material in the category of manufactured abrasives. There are other carbides, as well as nitrides and cermets, which can be classified as abrasives but their use is special and limited.

Various grades of each type of synthetic abrasive are distinguishable by properties such as color, toughness, and friability. These differences are caused by variation in purity of materials and methods of processing.

The sized abrasive may be used as loose grains, as coatings on paper or cloth to make sandpaper and emery cloth, or as grains for bonding into wheels.

Abrasive Wheels

A variety of bonds is used in making abrasive wheels: vitrified or ceramic, essentially a glass or glass-plus crystals; sodium silicate; rubber; resinoid; shellac; and oxychloride. Each type of bond has its advantages. The more rigid ceramic bond is better for precision-grinding operations, and the tougher, resilient bonds, such as resinoid or rubber, are better for snagging and cutting operations.

Ceramic-bonded wheels are made by mixing the graded abrasive and binder, pressing to general size and shape, firing, and truing or finishing by grinding to exact dimensions.

Grinding wheels are specified by abrasive type, grain size (grit), grade or hardness, and bond type. The term *hardness* as applied to a wheel refers to its behavior in use and not to the hardness of the abrasive material itself.

Literally thousands of types of wheels are made with different combinations of characteristics, not to mention the multitude of sizes and shapes available; therefore, selecting the best grinding wheel for a given job is not simple.

ABS PLASTICS

ABS plastics are a family of opaque thermoplastic resins formed by copolymerizing acrylonitrile, butadiene, and styrene (ABS) monomers. ABS plastics are primarily notable for especially high impact strengths coupled with high rigidity or modulus. Consisting of particles of a rubberlike toughener suspended in a continuous phase of styreneacrylonitrile (SAN) copolymer, ABS resins are hard, rigid, and tough, even at low temperatures. Various grades of these amorphous, medium-priced thermoplastics are available offering different levels of impact strength, heat resistance, flame retardance, and platability.

Most natural ABS resins are translucent to opaque, but they are also produced in transparent grades, and they can be pigmented to almost any color. Grades are available for injection molding, extrusion, blow molding, foam molding, and thermoforming. Molding and extrusion grades provide surface finishes ranging from satin to high gloss. Some ABS grades are designed specifically for electroplating. Their molecular structure is such that the plating process is rapid, easily controlled, and economical.

Compounding of some ABS grades with other resins produces special properties. For example, ABS is alloyed with polycarbonate to provide a better balance of heat resistance and impact properties at an intermediate cost. Deflection temperature is improved by the polycarbonate, molding ease by the ABS. Other ABS resins are used to modify rigid polyvinyl chloride (PVC) for use in pipe, sheeting, and molded parts. Reinforced grades containing glass fibers, to 40%, are also available.

Related to ABS is SAN, a copolymer of styrene and acrylonitrile (no butadiene) that is hard, rigid, transparent, and characterized by excellent chemical resistance, dimensional stability, and ease of processing. SAN resins are usually processed by injection molding, but extrusion, injection-blow molding, and compression molding are also used. They can also be thermoformed, provided that no post-mold trimming is necessary.

The triangle in Figure A.1 illustrates the properties and characteristics that each constituent acrylonitrile, butadiene, and styrene

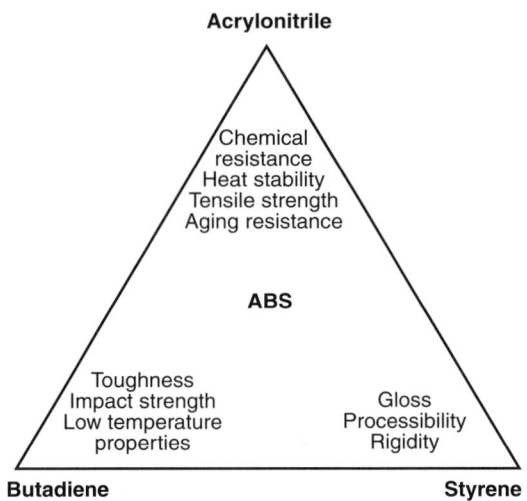

FIGURE A.1 Properties and characteristics of acrylonitrile, butadiene, and styrene.

contributes to the final product. Polymerization of these materials produces the ABS terpolymer, a two-phase system consisting of a continuous matrix of styrene-acrylonitrile copolymer and a dispersed phase of butadiene rubber particles. Properties are varied principally by adjusting the proportions in which the materials are combined and by altering the molecular weight of the SAN.

PROPERTIES

The unique combinations of excellent impact strength with high mechanical strength and rigidity plus good long-term, load-carrying ability or creep resistance are characteristic of the ABS plastics family. In addition, all types of ABS plastics exhibit outstanding dimensional stability, good chemical and heat resistance, surface hardness, and light weight (low specific gravity), Table A.1.

These materials yield plastically at high stresses, so ultimate elongation is seldom significant in design; a part usually can be bent beyond its elastic limit without breaking, although it does stress-whiten. Although not generally considered flexible, ABS parts have enough spring to accommodate snap-fit assembly requirements.

The individual commercially available ABS polymers span a wide range of mechanical properties. Most suppliers differentiate types on the

ABS PLASTICS

TABLE A.1
Properties of ABS and SAN

ASTM or UL Test	Property	Standard ABS Grades			Special-Purpose ABS Grades					SAN Grades
		High Impact	Superhigh Impact	Medium Impact	High Heat	Flame Retardant	Clear	Expandable	Plating	
		Physical								
D792	Specific gravity	1.01–1.05	1.02–1.05	1.04–1.06	1.04–1.06	1.19–1.22	1.05	0.55–0.85	1.05–1.07	1.07–1.08
D792	Specific volume (in.³/lb)	27	27	28	28	—	26	—	26	26
		Mechanical								
D638	Tensile strength (psi)	6,000	5,000–6,300	6,000–7,500	6,000–7,500	5,500–10,000	5,800–6,300	3,000–4,000	5,500–6,600	9,000–12,000
D638	Elongation (%)	5–20	5–70	5–25	3–20	5–25	25–75	—	—	1–4
D638	Tensile modulus (10⁵ psi)	3.3	2.0–3.4	3.6–3.8	3.0–4.0	3.2–3.7	3.0–3.3	1.0–2.5	3–3.8	4.5–5.6
D790	Flexural strength (psi)	10,500	6,000–11,500	11,500	10,000–13,000	9,000–12,250	10,500	3,000–8,000	8,700–11,500	14,000–17,000
D790	Flexural modulus (10⁵ psi)	3.4	2.0–3.5	3.6–4.0	3.1–4.0	3.0–3.4	3.4–3.9	1.4–2.8	3.0–3.8	5.5
D256	Impact strength, Izod (ft-lb/in. of notch)	6.5	7.0–8.0	4.0–5.5	2.3–6.0	4.0–13.0	2.5–4.0	—	5.0–7.0	0.35–0.50
D785	Hardness, Rockwell R	103	69–105	107	111	90–117	100–105	60–70[a]	103–109	M85
		Thermal								
D696	Coefficient of thermal expansion (10⁻⁵) in./in.-°F	5.3	5.6	4.6	3.9–5.1	3.7–4.6	4.6	4.9	4.7–5.3	3.0
D648	Deflection temperature[b] (°F)									
	At 264 psi	188	192	184	220–240	180–220	168	160	189	210
	At 66 psi	203	208	201	230–245	198–238	180–185	185	214	—
UL94	Flammability rating	HB	HB	HB	HB	V-0 to V-1[c]	HB	HB–V-0	HB	HB
		Electrical								
D149	Dielectric strength (V/mil) Short time, 1/8-in. thk	400	350–500	350–500	350–500	400+	400	—	—	—
D495	Arc resistance (s)	89	50–85	50–85	50–85	20–60	120–130	—	—	—

[a] Density has a marked effect.
[b] Unannealed.
[c] 0.060-in.-thick samples.

ASTM = American Society for Testing and Materials; UL = Underwriters' Laboratories.

Source: Mach. Design Basics Eng. Design, June, p. 674, 1993. With permission.

basis of impact strength and fabrication method (extrusion or molding). Some compounds feature one particularly exceptional property, such as high heat deflection temperature, abrasion resistance, or dimensional stability.

Impact properties of ABS are exceptionally good at room temperature and, with special grades, at temperatures as low as −40°C. Because of its plastic yield at high strain rates, impact failure of ABS is ductile rather than brittle. Also, the skin effect, which in other thermoplastics accounts for a lower impact resistance in thick sections than in thin ones, is not pronounced in ABS materials. A long-term tensile design stress of 6.8 to 10.3 MPa (at 22.8°C) is recommended for most grades.

General-purpose ABS grades are adequate for some outdoor applications, but prolonged exposure to sunlight causes color change and reduces surface gloss, impact strength, and ductility. Less affected are tensile strength, flexural strength, hardness, and elastic modulus. Pigmenting the resins black, laminating with opaque acrylic sheet, and applying certain coating systems provide weathering resistance. For maximum color and gloss retention, a compatible coating of opaque, weather-resistant polyurethane can be used on molded parts. For weatherable sheet applications, ABS resins can be coextruded with a compatible weather-resistant polymer on the outside surface.

ABS resins are stable in warm environments and can be decorated with durable coatings that require baking at temperatures to 71°C for 30–60 min. Heat-resistant grades can be used for short periods at temperatures to 110°C in light load applications. Low moisture absorption contributes to the dimensional stability of molded ABS parts.

Molded ABS parts are almost completely unaffected by water, salts, most inorganic acids, food acids, and alkalies, but much depends on time, temperature, and especially stress level. Food and Drug Administration (FDA) acceptance depends to some extent on the pigmentation system used. The resins are soluble in esters and ketones, and they soften or swell in some chlorinated hydrocarbons, aromatics, and aldehydes.

Properties of SAN resins are controlled primarily through acrylonitrile content and by adjusting the molecular weight of the copolymer. Increasing both improves physical properties, at a slight penalty in processing ease. Properties of the resins can also be enhanced by controlling orientation during molding. Tensile and impact strength, barrier properties, and solvent resistance are improved by this control.

Special grades of SAN are available with improved ultraviolet (UV) stability, vapor-barrier characteristics, and weatherability. The barrier resins — designed for the blown-bottle market — are also tougher and have greater solvent resistance than the standard grades.

Fabrication and Forms

ABS plastics are readily formed by the various methods of fabricating thermoplastic materials extrusion, injection molding, blow molding, calendering, and vacuum forming. Molded products may be machined, riveted, punched, sheared, cemented, laminated, embossed, or painted. Although the ABS plastics process easily and exhibit excellent moldability, they are generally more difficult flowing than the modified styrenes and higher processing temperatures are used. The surface appearance of molded articles is excellent and buffing may not be necessary.

Moldings

The need for impact resistance and high mechanical properties in injection-molded parts has created a large use for ABS materials. Advances in resin technology coupled with improved machinery and molding techniques have opened the door to ABS resins. Large complex shapes can be readily molded in ABS today.

Pipe

The ABS plastics as a whole are popular for extrusion and they offer a great deal for this type of forming. The outstanding contribution is their ability to be formed easily and to hold dimension and shape. In addition, very good extrusion rates are obtainable. Because ABS materials are processed at stock temperatures of 400 to 500°F, it is generally necessary to preheat and dry the material prior to extrusion.

The largest single ABS end product is plastic pipe, where the advantages of high long-term mechanical strength, toughness, wide service temperature range, chemical resistance, and ease of joining by solvent welding are used.

Sheet

ABS sheet is manufactured by calendering or extrusion and molded articles are subsequently vacuum-formed. The hot strength of the ABS materials coupled with the ability to be drawn excessively without forming thin spots or losing embossing have made them popular with fabricators. The excellent mechanical strengths, formability, and chemical resistance, particularly to fluorocarbons, are largely responsible for the rapid increase in the use of ABS.

APPLICATIONS

Molded ABS products are used in both protective and decorative applications. Examples include safety helmets, camper tops, automotive instrument panels and other interior components, pipe fittings, home-security devices and housings for small appliances, communications equipment, and business machines. Chrome-plated ABS has replaced die-cast metals in plumbing hardware and automobile grilles, wheel covers, and mirror housings.

Typical products vacuum-formed from extruded ABS sheet are refrigerator liners, luggage shells, tote trays, mower shrouds, boat hulls, and large components for recreational vehicles. Extruded shapes include weather seals, glass beading, refrigerator breaker strips, conduit, and pipe for drainwaste-vent (DWV) systems. Pipe and fittings comprise one of the largest single application areas for ABS.

Typical applications for molded SAN copolymers include instrument lenses, vacuum-cleaner and humidifier parts, medical syringes, battery cases, refrigerator compartments, food-mixer bowls, computer reels, chair shells, and dishwasher-safe houseware products.

ACETAL PLASTICS

Acetals are independent structural units or a part of certain biological and commercial polymers, and acetal resins are highly crystalline plastics based on formaldehyde polymerization technology. These engineering resins are strong, rigid, and have good moisture, heat, and solvent resistance.

Acetals were specially developed to compete with zinc and aluminum castings. The natural acetal resin is translucent white and can be readily colored with a high sparkle and brilliance. There are two basic types — homopolymer (Delrin) and copolymer (Celcon). In general, the homopolymers are harder, more rigid, have higher tensile flexural and fatigue strength, but lower elongation; however, they have higher melting points. Some high-molecular-weight homopolymer grades are extremely tough and have higher elongation than the copolymers. Homopolymer grades are available that are modified for improved hydrolysis resistance to 82°C, similar to copolymer materials.

The copolymers remain stable in long-term, high-temperature service and offer exceptional resistance to the effects of immersion in water at high temperatures. Neither type resists strong acids, and the copolymer is virtually unaffected by strong bases. Both types are available in a wide range of melt-flow grades, but the copolymers process more easily and faster than the conventional homopolymer grades.

Both the homopolymers and copolymers are available in several unmodified and glass-fiber-reinforced injection-molding grades. Both are available in polytetrafluoroethylene (PTFE) or silicone-filled grades, and the homopolymer is available in chemically lubricated low-friction formulations.

The acetals are also available in extruded rod and slab form for machined parts. Property data listed in Table A.2 apply to the general-purpose injection-molding and extrusion grade of Delrin 500 and to Celcon M90.

Acetals are among the strongest and stiffest of the thermoplastics. Their tensile strength ranges from 54.4 to 92.5 MPa, tensile modulus is about 3400 MPa, and fatigue strength at room temperature is about 34 MPa. Acetals are also among the best in creep resistance. This combined with low moisture absorption (less than 0.4%) gives them excellent dimensional stability. They are useful for continuous service up to about 104°C.

TABLE A.2
Properties of Acetals

ASTM or UL Test	Property	Copolymer	Homopolymer
	Physical		
D792	Specific gravity	1.41	1.42
D792	Specific volume (in.^{3}/lb)	19.7	19.5
D570	Water absorption, 24 h, 1/8-in. thk (%)	0.22	0.25
	Mechanical		
D638	*Tensile strength (psi)		
	At 73°F	8,800	10,000
	At 160°F	5,000	6,900
D638	*Elongation (%)	60	40
D638	*Tensile modulus (10^5 psi)	4.1	5.2
D790	Flexural strength (psi)	13,000	14,100
D790	Flexural modulus (10^5 psi)		
	At 73°F	3.75	4.10
	At 160°F	1.80	2.30
D256	Impact strength, Izod (ft-lb/in.)		
	Notched	1.3	1.4
	Unnotched	20	24
D671	Fatigue endurance limit, 10^7 cycles (psi)	3,300	4,300
D785	Hardness, Rockwell M	80	94
	Thermal		
C177	Thermal conductivity		
	(10^{-4} cal-cm/s-cm^2-°C)	5.5	8.9
	Btu-in./hr-ft^2-°F)	1.6	2.6
D696	Coefficient of thermal expansion		
	−40 to +185°C (10^{-5} in./in.·°C)	8.5	10.0
D648	Deflection temp (°F)		
	At 264 psi	230	277
	At 66 psi	316	342
UL94	Flammability rating	HB	HB
	Electrical		
D149	Dielectric strength		
	Short time (V/mil)		
	5 mils	2,100	3,000
	20 mils	—	2,000
	90 mils	500	500
D150	Dielectric constant		
	At 1 kHz	3.7	3.7
	At 1 MHz	3.7	3.7
D150	Dissipation factor		
	At 1 kHz	0.001	0.001
	At 1 MHz	0.006	0.005
D257	Volume resistivity (ohm-cm)		
	At 73°F, 50% RH	10^{14}	10^{15}
D495	Arc resistance (s) 120 mils	240 (burns)	220 (burns, no tracking)
	Frictional		
—	Coefficient of friction		
	Self	0.35	0.3
	Against steel	0.15	0.15

* At 0.2 in./min loading rate.

Source: Mach. Design Basics Eng. Design, June, p. 676, 1993. With permission.

Injection-molding powders and extrusion powders are the most frequently used forms of the material. Sheets, rods, tubes, and pipe are also available. Colorability is excellent.

The range of desirable design properties and processing techniques provides outstanding design freedom in the areas (1) style (color, shape, surface texture and decoration), (2) weight reduction, (3) assembly techniques, and (4) one-piece multifunctional parts (e.g., combined gear, cam, bearing, and shaft).

ACETAL HOMOPOLYMERS

The homopolymers are available in several viscosity ranges that meet a variety of processing and end-use needs. The higher-viscosity materials are generally used for extrusions and for molded parts requiring maximum toughness; the lower-viscosity grades are used for injection molding. Elastomer-modified grades offer greatly improved toughness.

Properties

Acetal homopolymer resins have high tensile strength, stiffness, resilience, fatigue endurance, and moderate toughness under repeated impact. Some tough grades can deliver up to 7 times greater toughness than unmodified acetal in Izod impact tests and up to 30 times greater toughness as measured by Gardner impact tests (Table A.2).

Homopolymer acetals have high resistance to organic solvents, excellent dimensional stability, a low coefficient of friction, and outstanding abrasion resistance among thermoplastics. The general-purpose resins can be used over a wide range of environmental conditions; special, UV-stabilized grades are recommended for applications requiring long-term exposure to weathering. However, prolonged exposure to strong acids and bases outside the range of pH 4 to 9 is not recommended.

Acetal homopolymer has the highest fatigue endurance of any unfilled commercial thermoplastic. Under completely reversed tensile and compressive stress, and with 100% relative humidity (at 73°F), fatigue endurance limit is 30.9 MPa at 10^6 cycles. Resistance to creep is excellent. Moisture, lubricants, and solvents including gasoline and gasohol have little effect on this property, which is important in parts incorporating self-threading screws or interference fits.

The low friction and good wear resistance of acetals against metals make these resins suitable for use in cams and gears having internal bearings. The coefficient of friction (nonlubricated) on steel, in a rotating thrust washer test, is 0.1 to 0.3, depending on pressure; little variation occurs from 22.8 to 121°C. For even lower friction and wear, PTFE-fiber-filled and chemically lubricated formulations are available.

Properties of low moisture absorption, excellent creep resistance, and high deflection temperature suit acetal homopolymer for close-tolerance, high-performance parts.

Applications

Automotive applications of acetal homopolymer resins include fuel-system and seat-belt components, steering columns, window-support brackets, and handles. Typical plumbing applications that have replaced brass or zinc components are showerheads, ball cocks, faucet cartridges, and various fittings. Consumer items include quality toys, garden sprayers, stereo cassette parts, butane lighter bodies, zippers, and telephone components. Industrial applications of acetal homopolymer include couplings, pump impellers, conveyor plates, gears, sprockets, and springs.

ACETAL COPOLYMERS

The copolymers have an excellent balance of properties and processing characteristics. Melt temperature can range from 182 to 232°C with little effect on part strength. UV-resistant grades (also available in colors), glass-reinforced grades, low-wear grades, and impact-modified grades are standard. Also available are electroplatable and dimensionally stable, low-warpage grades.

Properties

Acetal copolymers have high tensile and flexural strength, fatigue resistance, and hardness. Lubricity is excellent. They retain much of their toughness through a broad temperature range and are among the most creep resistant of the crystalline thermoplastics. Moisture absorption is low, permitting molded parts to serve reliably in environments involving humidity changes.

Good electrical properties, combined with high mechanical strength and an Underwriters' Laboratories (UL) electrical rating of 100°C, qualify these materials for electrical applications requiring long-term stability.

Acetal copolymers have excellent resistance to chemicals and solvents. For example, specimens immersed for 12 months at room temperature in various inorganic solutions were unaffected except by strong mineral acids — sulfuric, nitric, and hydrochloric. Continuous contact is not recommended with strong oxidizing agents such as aqueous solutions containing high concentrations of hypochlorite ions. Solutions of 10% ammonium hydroxide and 10% sodium chloride discolor samples in prolonged immersion, but physical and mechanical properties are not significantly changed. Most organic reagents tested have no effect, nor do mineral oil, motor oil, or brake fluids. Resistance to strong alkalies is exceptionally good; specimens immersed in boiling 50% sodium hydroxide solution and other strong bases for many months show no property changes.

Strength of acetal copolymer is only slightly reduced after aging for 1 year in air at 116°C. Impact strength holds constant for the first 6 months, and falls off about one-third during the next 6-month period. Aging in air at 82°C for 2 years has little or no effect on properties, and immersion for 1 year in 82°C water leaves most properties virtually unchanged. Samples tested in boiling water retain nearly original tensile strength after 9 months.

The creep–modulus curve of acetal copolymer under load shows a linear decrease on a log-log scale, typical of many plastics. Acetal springs lose over 50% of spring force after 1000 h and 60% in 10,000 h. The same spring loses 66% of its force after 100,000 h (about 11 years) under load.

Plastic springs are best used in applications where they generate a force at a specified deflection for limited time but otherwise remain relaxed. Ideally, springs should undergo occasional deflections where they have time to recover, at less than 50% design strain. Recovery time should be at least equal to time under load.

Applications

Industrial and automotive applications of acetal copolymer include gears, cams, bushings, clips, lugs, door handles, window cranks, housings, and seat-belt components. Plumbing products such as valves, valve stems, pumps, faucets, and impellers utilize the lubricity and corrosion and hot water resistance of the copolymer. Mechanical components that require dimensional stability, such as watch gears, conveyor links, aerosols, and mechanical pen and pencil parts, are other uses. Applications for the FDA-approved grades include milk pumps, coffee spigots, filter housings, and food conveyors. Parts that require greater load-bearing stability at elevated temperatures, such as cams, gears, television tuner arms, and automotive underhood components, are molded from glass-fiber-reinforced grades.

More costly acetal copolymer has excellent load-bearing characteristics for long-lasting plastic springs. To boost resin performance, engineers use fillers, reinforcing fibers, and additives. Although there are automotive uses for large fiber-reinforced composite leaf springs, unfilled resins are the better candidates for small springs. Glass fibers increase stiffness and strength, but they also limit deflection. And impact modifiers reduce modulus and make plastics more flexible but decrease creep resistance.

ACETAL RESINS

Processing Acetals

Acetal resin can be molded in standard injection molding equipment at conventional production rates. The processing temperature is around 204°C. Satisfactory performance has been demonstrated in full-automatic injection machines using multicavity molds. Successful commercial moldings point up the ability of the material to be molded to form large-area parts with thin sections, heavy parts with thick sections, parts requiring glossy surfaces or different surface textures, parts requiring close tolerances, parts with undercuts for snap fits, parts requiring metal inserts, and parts requiring no flash. It can also be extruded as rod, tubing, sheeting, jacketing, wire coating, or shapes on standard commercial equipment. Extrusion temperatures are in the range of 199 to 204°C.

Generally the same equipment and techniques for blow molding other thermoplastics work with acetal resin. Both thin-walled and thick-walled containers (aerosol type) can be produced in many shapes and surface textures.

Various sheet-forming techniques including vacuum, pressure, and matched-mold have been successfully used with acetal resins.

Fabrication

Acetal resin is easy to machine (equal to or better than free-cutting brass) on standard production machine shop equipment. It can be sawed, drilled, turned, milled, shaped, reamed, threaded and tapped, blanked and punched, filed, sanded, and polished.

The material is easy to join and offers wide latitude in the choice of fast, economical methods of assembly. Integral bonds of acetal-to-acetal can be formed by welding with a heated metal surface, hot gas, hot wire, or spin-welding techniques. High-strength joints result from standard mechanical joining methods such as snap fits, interference or press fits, rivets, nailing, heading, threads, or self-tapping screws. Where low joint strengths are acceptable, several commercial adhesives can be used for bonding acetal to itself and other substrates.

Acetal resin can be painted successfully with certain commercial paints and lacquers, using ordinary spraying equipment and a special surface treatment or followed by a baked top coat. Successful first-surface metallizing has been accomplished with conventional equipment and standard techniques for application of such coatings. Direct printing, process printing, and roll-leaf stamping (hot stamping) can be used for printing on acetal resin. Baking at elevated temperatures is required for good adhesion of the ink in direct and screen-process printing. In hot stamping, the heated die provides the elevated temperature. Printing produced by these processes resists abrasion and lifting by cellophane adhesive tape.

ACETYLENE

Acetylene is a colorless, flammable gas with a garlic-like odor. Under compressed conditions, it is highly explosive; however, it can be safely compressed and stored in high-pressure cylinders if the cylinders are lined with absorbent material soaked with acetone. Users are cautioned not to discharge acetylene at pressures exceeding 15 psig (103 kPa), as noted by the red line on acetylene pressure gauges.

With its intense heat and controllability, the oxyacetylene flame can be used for many different welding and cutting operations including hardfacing, brazing, beveling, gouging, and scarfing. The heating capability of acetylene also can be utilized in the bending, straightening, forming, hardening, softening, and strengthening of metals.

ACRYLIC PLASTICS

The most widely used acrylic plastics are based on polymers of methyl methacrylate. This primary constituent may be modified by copolymerizing or blending with other acrylic monomers or modifiers to obtain a variety of properties. Although acrylic polymers based on monomers other than methyl methacrylate have been investigated, they are not as important as commercial plastics and are generally confined to uses in fibers, rubbers, motor oil additives, and other special products.

STANDARD ACRYLICS

Poly(methyl methacrylate), the polymerized methyl ester of methacrylic acid, is thermoplastic. The method of polymerization may be varied to achieve specific physical properties, or the monomer may be combined with other components. Sheet materials may be prepared by casting the monomer in bulk. Suspension polymerization of the monomeric ester may be used to prepare molding powders.

Conventional poly(methyl methacrylate) is amorphous; however, reports have been published of methyl methacrylate polymers of regular configuration, which are susceptible to crystallization. Both the amorphous and crystalline forms of such crystallization-susceptible polymers possess physical properties that are different from those of the conventional polymer, and suggest new applications.

Service Properties

Acrylic thermoplastics are known for their outstanding weatherability. They are available in cast sheet, rod, and tube; extruded sheet and film; and compounds for injection molding and extrusion. They are also characterized by good impact strength, formability, and excellent resistance to sunlight, weather, and most chemicals. Maximum service temperature of heat-resistant grades is about 200°F. Standard grades are rated as slow burning, but a special self-extinguishing grade of sheet is available. Although acrylic plastic weighs less than half as much as glass, it has many times greater impact resistance. As a thermal insulator, it is approximately 20% better than glass. It is tasteless and odorless.

When poly(methyl methacrylate) is manufactured with scrupulous care, excellent optical properties are obtained. Light transmission is 92%; colorants produce a full spectrum of transparent, translucent, or opaque colors. Most colors can be formulated for long-term outdoor durability. Acrylics are normally formulated to filter UV energy in the 360-nm and lower band. Other formulations are opaque to UV light or provide reduced UV transmission; infrared light transmission is 92% at wavelengths up to 1100 millimicrons, failing irregularly to 0% at 2200 millimicrons; scattering effect is practically nil; refractive index is 1.49 to 1.50; critical angle is 42°; dispersion 0.008. Because of its excellent transparency and favorable index of refraction, acrylic plastic is often used in the manufacture of optical lenses. Superior dimensional stability makes it practicable to produce precision lenses by injection molding techniques.

In chemical resistance, poly(methyl methacrylate) is virtually unaffected by water, alkalies, weak acids, most inorganic solutions, mineral and animal oils, and low concentrations of alcohol. Oxidizing acids affect the material only in high concentrations. It is also virtually unaffected by paraffinic and olefinic hydrocarbons, amines, alkyl monohalides, and esters containing more than ten carbon atoms. Lower esters, aromatic hydrocarbons, phenols, aryl halides, aliphatic acids, and alkyl polyhalides usually have a solvent action. Acrylic sheet and moldings are attacked, however, by chlorinated and aromatic hydrocarbons, esters, and ketones.

Mechanical properties of acrylics are high for short-term loading. However, for long-term service, tensile stresses must be limited to 1500 psi to avoid crazing or surface cracking.

The moderate impact resistance of standard formulations is maintained even under conditions of extreme cold. High-impact grades have considerably higher impact strength than standard grades at room temperature, but impact strength decreases as temperature drops. Special formulations ensure compliance with UL standards for bullet resistance.

Although acrylic plastics are among the most scratch resistant of the thermoplastics, normal maintenance and cleaning operations can scratch and abrade them. Special abrasion-resistant sheet is available that has the same optical and impact properties as standard grades.

Toughness of acrylic sheet, as measured by resistance to crack propagation, can be improved severalfold by inducing molecular orientation during forming. Jet aircraft cabin windows, for example, are made from oriented acrylic sheet.

Transparency, gloss, and dimensional stability of acrylics are virtually unaffected by years of exposure to the elements, salt spray, or corrosive atmospheres. These materials withstand exposure to light from fluorescent lamps without darkening or deteriorating. They ultimately discolor, however, when exposed to high-intensity UV light below 265 nm. Special formulations resist UV emission from light sources such as mercury-vapor and sodium-vapor lamps.

Product Forms

Cell-cast sheet is produced in several sizes and thicknesses. The largest sheets available are 120×144 in., in thicknesses from 0.030 to 4.25 in. Continuous-cast material is supplied as flat sheet to $1/2$ in. thick, in widths to 9 ft. Acrylic sheet cast by the continuous process (between stainless steel belts) is more uniform in thickness than cell-cast sheet. Cell-cast sheet, on the other hand, which is cast between glass plates, has superior optical properties and surface quality. Also, cell-cast sheet is available in a greater variety of colors and compositions. Cast acrylic sheet is supplied in general-purpose grades and in UV-absorbing, mirrored, super-thermoformable, and cementable grades, and with various surface finishes. Sheets are available in transparent, translucent, and opaque colors.

Acrylic film is available in 2-, 3-, and 6-mil thicknesses, in clear form and in colors. It is supplied in rolls to 60 in. wide, principally for use as a protective laminated cover over other plastic materials.

Injection-molding and extrusion compounds are available in both standard and high-molecular-weight grades. Property differences between the two formulations are principally in flow and heat resistance. Higher-molecular-weight resins have lower melt-flow rates and greater hot strength during processing. Lower-molecular-weight grades flow more readily and are designed for making complex parts in hard-to-fill molds; see Table A.3.

Fabrication Characteristics

When heated to a pliable state, acrylic sheet can be formed to almost any shape. The forming operation is usually carried out at about 290 to 340°F. Aircraft canopies, for example, are usually made by differential air pressure, either with or without molds. Such canopies have been made from (1) monolithic sheet stock, (2) laminates of two layers of acrylic, bonded by a layer of polyvinyl butyral, and (3) stretched monolithic sheet. Irregular shapes, such as sign faces, lighting fixtures, or boxes, can be made by positive pressure-forming, using molds.

Residual strains caused by forming are minimized by annealing, which also brings cemented joints to full strength. Cementing can be readily accomplished by using either solvent or polymerizable cements.

Acrylic plastic can be sawed, drilled, and machined like wood or soft metals. Saws should be hollow ground or have set teeth. Slow feed and coolant will prevent overheating. Drilling can be done with conventional metal-cutting drills. Routing requires high-speed cutters to prevent chipping. Finished parts can be sanded, and sanded surfaces can be polished with a high-speed buffing wheel. Cleaning should be by soap or detergent and water, not by solvent-type cleaners.

Acrylic molding powder may be used for injection, extrusion, or compression molding. The material is available in several grades, with a varying balance of flow characteristics and heat resistance. Acrylics give molded parts of excellent dimensional stability. Precise contours and sharp angles, important in such applications as lenses, are achieved without difficulty, and this accuracy of molding can be maintained throughout large production runs.

Since dirt, lint, and dust detract from the excellent clarity of acrylics, careful handling and storage of the molding powder are extremely important.

Applications

In merchandising, acrylic sheet has become the major sign material for internally lighted faces and letters, particularly for outdoor use where resistance to sunlight and weathering is important. In addition, acrylics are used for counter dividers, display fixtures and cases, transparent demonstration models of household appliances and industrial machines, and vending machine cases.

The ability of acrylics to resist breakage and corrosion, and to transmit and diffuse light efficiently has led to many industrial and architectural applications. Industrial window glazing, safety shields, inspection windows, machine covers, and pump components are some of the uses commonly found in plants and factories. Acrylics are employed to good advantage as the diffusing medium in lighting fixtures and large luminous ceiling areas. Dome skylights formed from acrylic sheet are an increasingly popular means of admitting daylight to industrial, commercial, and public buildings and even to private homes.

Shower enclosures and deeply formed components such as tub–shower units, which are subsequently backed with glass-fiber-reinforced polyester and decorated partitions, are other typical applications. A large volume of the material is used for curved and flat windshields on pleasure boats, both inboard and outboard types.

TABLE A.3
Properties of Acrylics

ASTM Test	Property	Cast Sheet	Molding Grade Standard	High Impact
		Physical		
D792	Specific gravity	1.19	1.19	1.15–1.17
D792	Specific volume (in.^{3}/lb)	23.3	23.3	24.1
D570	Water absorption, 24 h, 1/8-in. thk (%)	0.2	0.3	0.3
		Mechanical		
D638	Tensile strength (psi)	10,500	10,500	5,400–7,000
D638	Elongation (%)	5	5	50
D638	Tensile modulus (10^5 psi)	4.5	4.3	2.2–8.2
D790	Flexural strength (psi)	16,500	16,000	7,000–10,500
D790	Flexural modulus (10^5 psi)	4.5	4.5	0.65–2.5
D256	Impact strength, Izod (ft-lb/in. of notch)	0.4	0.4	0.6–1.2
D785	Hardness, Rockwell	M100–102	M95	R99–M68
		Thermal		
D696	Coefficient of thermal expansion (10^{-5}) in./in.-°C	3.9	3.6	3.8
D648	Deflection temperature (°F)			
	At 264 psi	200–215	198	170–190
	At 66 psi	225	214	187
		Electrical		
D149	Dielectric strength (V/mil) Short time, 1/8-in. thk	500	500	383–450
D150	Dielectric constant			
	At 1 kHz	3.3	3.3	3.9
	At 1 MHz	2.5	2.3	2.5–3.0
D150	Dissipation factor			
	At 1 kHz	0.04	0.04	—
	At 1 MHz	0.02–0.03	0.02–0.03	0.01–0.02
D257	Volume resistivity (ohm-cm) At 73°F, 50% RH	>10^{17}	>10^{17}	>10^{15}
D495	Arc resistance (s)	No track	No track	No track
		Optical		
D542	Refractive index	1.49	1.49	1.49
D1003	Transmittance (%)	92	92	90

Source: Mach. Design Basics Eng. Design, June, p. 678, 1993. With permission.

Acrylic sheet is the standard transparent material for aircraft canopies, windows, instrument panels, and searchlight and landing light covers. To meet the increasingly severe service requirements of pressurized jet aircraft, new grades of acrylic have been developed that have improved resistance to heat and crazing. The stretching technique has made possible enhanced resistance to both crazing and shattering. Large sheets, edge-lighted, are used as radar plotting boards in shipboard and ground-control stations.

In molded form, acrylics are used extensively for automotive parts, such as taillight and stoplight lenses, medallions, dials, instrument panels, and signal lights. The beauty and durability of molded acrylic products have led to their wide use for nameplates, control knobs, dials, and handles on all types of home appliances. Acrylic molding powder is also used for the manufacture of pen and pencil barrels, hairbrush backs, watch and jewelry cases, and other accessories. Large-section moldings, such as covers for fluorescent street lights, coin-operated phonograph panels, and fruit juice dispenser bowls, are being molded from acrylic powder. The extrusion of acrylic sheet from molding powder is particularly effective in the production of thin sheeting for use in such applications as signs, lighting, glazing, and partitions.

The transparency, strength, light weight, and edge-lighting characteristics of acrylics have led to applications in the fields of hospital equipment, medical examination instruments, and orthopedic devices. The use of acrylic polymers in the preparation of dentures is an established practice. Contact lenses are also made of acrylics. The embedment of normal and pathological tissues in acrylic for preservation and instructional use is an accepted technique. This has been extended to include embedment of industrial machine parts, as sales aids, and the preparation of various types of home decorative articles.

HIGH-IMPACT ACRYLICS

High-impact acrylic molding powder is used in large-volume, general use. It is used where toughness greater than that found in the standard acrylics is desired. Other advantages include resistance to staining, high surface gloss, dimensional stability, chemical resistance, and stiffness, and they provide the same transparency and weatherability as the conventional acrylics.

High-impact acrylic is off-white and nearly opaque in its natural state and can be produced in a wide range of opaque colors. Several grades are available to meet requirements for different combinations of properties. Various members of the family have Izod impact strengths from about 0.5 to as high as 4 ft-lb/in. notch. Other mechanical properties are similar to those of conventional acrylics.

High-impact acrylics are used for hard service applications, such as women's thin-style shoe heels and housings, ranging from electric razors to outboard motors, piano and organ keys, and beverage vending machine housings and canisters — in short, applications where toughness, chemical resistance, dimensional stability, stiffness, resistance to staining, lack of unpleasant odor or taste, and high surface gloss are required.

ADHESIVES

These are materials capable of fastening two other materials together by means of surface attachment. The words *glue*, *mucilage*, *mastic*, and *cement* are synonymous with adhesive. In a generic sense, the word *adhesive* implies any material capable of fastening by surface attachment, and thus will include inorganic materials such as portland cement and solders such as Wood's metal. In a practical sense, however, adhesive implies the broad set of materials composed of organic compounds, mainly polymeric, that can be used to fasten two materials together. The materials being fastened together by the adhesive are the adherends, and an adhesive joint or adhesive bond is the resulting assembly. Adhesion is the physical attraction of the surface of one material for the surface of another.

From an industrial manufacturing standpoint, the advent of the stealth aircraft and all the structural adhesive bonding it entails has drawn widespread attention to the real capabilities of adhesives. Structural bonding uses adhesives to join load-bearing assemblies. Most often, the assemblies are also subject to severe service conditions. Such adhesives, regardless of chemistry, generally have the following properties:

- Tensile strengths in the 1500 to 4500 psi range
- Very high impact and peel strength
- Service temperature ranges of about −65 to 3500°F

If these types of working conditions are expected, then one should give special consideration to proper adhesive selection and durability testing.

THEORIES

The phenomenon of adhesion has been described by many theories. The most widely accepted and investigated is the wettability–adsorption theory. Basically, this theory states that for maximum adhesion the adhesive must come into complete intimate contact with the surface of the adherend. That is, the adhesive must completely wet the adherend. This wetting is considered to be maximized when the intermolecular forces are the same forces as are normally considered in intermolecular interactions such as the van der Waals, dipole–dipole, dipole–induced dipole, and electrostatic interactions. Of these, the van der Waals force is considered the most important. The formation of chemical bonds at the interface is not considered to be of primary importance for achieving maximum wetting, but in many cases it is considered important in achieving durable adhesive bonds.

If the situation is such that the adhesive completely wets the adherend, the strength of the adhesive joint depends on the design of the joint, the physical properties of the adherends, and, most importantly, the physical properties of the adhesive.

PARAMETERS

Innumerable adhesives and adhesive formulations are available today. The selection of the proper type for a specific application can only be made after a complete evaluation of the design, the service requirements, production feasibility, and cost considerations. Usually such selection is best left up to adhesive suppliers. Once they have been given the complete details of the application they are in the best position to select both the type and specific adhesive formulation.

Types and Forms

Adhesives have been in use since ancient times and are even mentioned in the Bible. The first adhesives were of natural origin; for example, bitumen, fish oil, and tree resins. In more modern times, adhesives were still derived from natural products but were processed before use. These modern natural adhesives include animal-derived (such as blood, gelatin, and casein), vegetable-derived (such as soybean oil and wheat flour), and forest-derived (pine resins and cellulose derivatives) products.

Forms include liquid, paste, powder, and dry film. The commercial adhesives include pastes, glues, pyroxylin cements, rubber cements, latex cement, special cements of chlorinated rubber, synthetic rubbers, or synthetic resins, and the natural mucilages.

Characteristics

Adhesives are characterized by degree of tack (or stickiness), by strength of bond after setting or drying, by rapidity of bonding, and by durability. The strength of bond is inherent in the character of the adhesive itself, particularly in its ability to adhere intimately to the surface to be bonded. Adhesives prepared from organic products are in general subject to disintegration on exposure. The life of an adhesive usually depends on the stability of the ingredient that gives the holding power, although otherwise good cements of synthetic materials may disintegrate by the oxidation of fillers or materials used to increase tack. Plasticizers usually reduce adhesion. Some fillers such as mineral fibers or walnut-shell flour increase the thixotropy and the strength, while some such as starch increase the tack but also increase the tendency to disintegrate.

CLASSIFICATION

Adhesives can be grouped into five classifications based on chemical composition. These are summarized in Table A.4.

Natural

These include vegetable- and animal-based adhesives and natural gums. They are inexpensive, easy to apply, and have a long shelf life. They develop tack quickly, but provide only low-strength joints. Most are water soluble. They are supplied as liquids or as dry powders to be mixed with water.

ADHESIVES

TABLE A.4
Adhesives Classified by Chemical Composition

	Natural	Thermoplastic	Thermosetting	Elastomeric	Alloys[a]
Types within group	Casein, blood albumin, hide, bone, fish, starch (plain and modified); rosin, shellac, asphalt; inorganic (sodium silicate, litharge-glycerin)	Polyvinyl acetate, polyvinyl alcohol, acrylic, cellulose nitrate, asphalt, oleo-resin	Phenolic, resorcinol, phenol-resorcinol, epoxy, epoxy-phenolic, urea, melamine, alkyd	Natural rubber, reclaim rubber, butadiene-styrene (GR-S), neoprene, acrylonitrile-butadiene (Buna-N), silicone	Phenolic-polyvinyl butyral, phenolic-polyvinyl formal, phenolic-neoprene rubber, phenolic-nitrile rubber, modified epoxy
Most used form	Liquid, powder	Liquid, some dry film	Liquid, but all forms common	Liquid, some film	Liquid, paste, film
Common further classifications	By vehicle (water emulsion is most common but many types are solvent dispersions)	By vehicle (most are solvent dispersions or water emulsions)	By cure requirements (heat and/or pressure most common but some are catalyst types)	By cure requirements (all are common); also by vehicle (most are solvent dispersions or water emulsions)	By cure requirements (usually heat and pressure except some epoxy types); by vehicle (most are solvent dispersions or 100% solids); and by type of adherends or end-service conditions
Bond characteristics	Wide range, but generally low strength; good resistance to heat, chemicals; generally poor moisture resistance	Good to 150–200°F; poor creep strength; fair peel strength	Good to 200–500°F; good creep strength; fair peel strength	Good to 150–400°F; never melt completely; low strength; high flexibility	Balanced combination of properties of other chemical groups depending on formulation; generally higher strength over wider temp range
Major type of use[b]	Household, general purpose, quick set, long shelf life	Unstressed joints; designs with caps, overlaps, stiffeners	Stressed joints at slightly elevated temp	Unstressed joints on lightweight materials; joints in flexure	Where highest and strictest end-service conditions must be met; sometimes regardless of cost, as military uses
Materials most commonly bonded	Wood (furniture), paper, cork, liners, packaging (food), textiles, some metals and plastics; industrial uses giving way to other groups	Formulation range covers all materials, but emphasis on nonmetallics—esp wood, leather, cork, paper, etc.	Epoxy-phenolics for structural uses of most materials; others mainly for wood; alkyds for laminations; most epoxies are modified (alloys)	Few used "straight" for rubber, fabric, foil, paper, leather, plastics, films; also as tapes; most modified with synthetic resins	Metals, ceramics, glass, thermosetting plastics; nature of adherends often not as vital as design or end-service conditions (i.e., high strength, temp)

[a] "Alloy," as used here, refers to formulations containing resins from two or more *different* chemical groups. There are also formulations that benefit from compounding two resin types from the same chemical group (e.g., epoxy-phenolic).
[b] Although some uses of the "nonalloyed" adhesives absorb a large percentage of the quantity of adhesives sold, the uses are narrow in scope; from the standpoint of diversified applications, by far the most important use of any group is the forming of adhesive alloys.

Casein-latex type is an exception. It consists of combinations of casein with either natural or synthetic rubber latex. It is used to bond metal to wood for panel construction and to join laminated plastics and linoleum to wood and metal. Except for this type, most natural adhesives are used for bonding paper, cardboard, foil, and light wood.

Synthetic Polymer. The greatest growth in the development and use of organic compound-based adhesives came with the application of synthetically derived organic polymers. Broadly, these materials can be divided into two types: thermoplastics and thermosets. Thermoplastic adhesives become soft or liquid upon heating and are also soluble. Thermoset adhesives cure upon heating and then become solid and insoluble. Those adhesives that cure under ambient conditions by appropriate choice of chemistry are also considered thermosets.

An example of a thermoplastic adhesive is a hot-melt adhesive. A well-known hot-melt adhesive in use since the Middle Ages is sealing wax. Modern hot-melt adhesives are composed of polymers such as polyamides, polyesters, ethylene-vinyl acetate copolymers, and polyethylene. Modern hot melts are heavily compounded with wax and other materials. Another widely used thermoplastic adhesive is polyvinyl acetate, which is applied from an emulsion.

Thermoplastic Adhesives

They can be softened or melted by heating and hardened by cooling. They are based on thermoplastic resins (including asphalt and oleoresin adhesives) dissolved in solvent or emulsified in water. Most of them become brittle at subzero temperatures and may not be used under stress at temperatures much above 150°F. As they are relatively soft materials, thermoplastic adhesives have poor creep strength. Although lower in strength than all but natural adhesives and suitable only for noncritical service, they are also lower in cost than most adhesives. They are also odorless and tasteless and can be made fungus resistant.

Pressure Sensitive. Pressure-sensitive adhesives are mostly thermoplastic in nature and exhibit an important property known as tack. That is, pressure-sensitive adhesives exhibit a measurable adhesive strength with only a mild applied pressure. Pressure-sensitive adhesives are derived from elastomeric materials, such as polybutadiene or polyisoprene.

Thermosetting Adhesives

Based on thermosetting resins, they soften with heat only long enough for the cure to initiate. Once cured, they become relatively infusible up to their decomposition temperature. Although most such adhesives do not decompose at temperatures below 500°F, some are useful only to about 150°F. Different chemical types have different curing requirements. Some are supplied as two-part adhesives and mixed before use at room temperature; some require heat or pressure to bond.

As a group, these adhesives provide stronger bonds than natural, thermoplastic, or elastomeric adhesives. Creep strength is good and peel strength is fair. Generally, bonds are brittle and have little resilience and low impact strength.

Elastomeric Adhesives

Based on natural and synthetic rubbers, elastomeric adhesives are available as solvent dispersions, latexes, or water dispersions. They are primarily used as compounds that have been modified with resins to form some of the adhesive "alloys" discussed below. They are similar to thermoplastics in that they soften with heat, but never melt completely. They generally provide high flexibility and low strength, and without resin modifiers, are used to bond paper and similar materials.

Alloy Adhesives

This term refers to adhesives compounded from resins of two or more different chemical families, e.g., thermosetting and thermoplastic, or thermosetting and elastomeric. In such adhesives the performance benefits of two or more types of resins can be combined. For example, thermosetting resins are plasticized by a second resin resulting in improved toughness, flexibility, and impact resistance.

Structural Adhesives

Structural adhesives are, in general, of the alloy or thermosetting type and have the property of fastening adherends that are structural materials (such as metals and wood) for long periods of time even when the adhesive joint is under load. Phenolic-based structural adhesives were among the first structural adhesives to be developed and used.

The most widely used structural adhesives are based on epoxy resins. Epoxy resin structural adhesives will cure at ambient or elevated temperatures, depending on the type of curative. Urethanes, generated by isocyanate-diol reactions, are also used as structural adhesives. Acrylic monomers have also been utilized as structural adhesives. These acrylic adhesives use an ambient-temperature surface-activated free radical cure. A special type of acrylic adhesive, based on cyanoacrylates (so-called superglue), is a structural adhesive that utilizes an anionic polymerization for its cure. Acrylic adhesives are known for their high strength and extremely rapid cure. Structural adhesives with resistance to high temperature (in excess of 390°F, or 200°C) for long times can be generated from ladder polymers such as polyimides and polyphenyl quinoxalines.

Three of the most commonly used adhesives are the modified epoxies, neoprene-phenolics, and vinyl formal-phenolics. Modified epoxy adhesives are thermosetting and may be of either the room-temperature-curing type, which cure by addition of a chemical activator, or the heat-curing type. They have high strength and resist temperature up to nearly 500°F (260°C).

A primary advantage of the epoxies is that they are 100% solids, and there is no problem of solvent evaporation after joining impervious surfaces. Other advantages include high shear strengths, rigidity, excellent self-filleting characteristics, and excellent wetting of metal and glass surfaces. Disadvantages include low peel strength, lack of flexibility, and inability to withstand high impact.

Neoprene-phenolic adhesives are alloys characterized by excellent peel strength, but lower shear strength than modified epoxies. They are moderately priced, offer good flexibility and vibration absorption, and have good adhesion to most metals and plastics.

Neoprene-phenolics are solvent types, but special two-part chemically curing types are sometimes used to obtain specific properties.

Vinyl formal-phenolic adhesives are alloys whose properties fall between those of modified epoxies and the thermoset-elastomer types. Vinyl formal-phenolics have good shear, peel, fatigue, and creep strengths and good resistance to heat, although they soften somewhat at elevated temperatures.

They are supplied as solvent dispersions in solution or in film form. In the film form the adhesive is coated on both sides of a reinforcing fabric. Sometimes it is prepared by mixing a liquid phenolic resin with vinyl formal powder just prior to use.

Other Adhesives/Cements

Paste adhesives are usually water solutions of starches or dextrins, sometimes mixed with gums, resins, or glue to add strength, and containing antioxidants. They are the cheapest of the adhesives, but deteriorate on exposure unless made with chemically altered starches. They are widely employed for the adhesion of paper and paperboard. Much of the so-called vegetable glue is tapioca paste. It is used for the cheaper plywoods, postage stamps, envelopes, and labeling. It has a quick tack, and is valued for pastes for automatic box-making machines. Latex pastes of the rub-off type are used for such purposes as photographic mounting, as they do not shrink the paper as do the starch pastes. Glues are usually water solutions of animal gelatin, and the only difference between animal glues and edible gelatin is in the degree of purity. Hide and bone glues are marketed as dry flake, but fish glue is liquid. Mucilages are light vegetable glues, generally from water-soluble gums.

Rubber cements for paper bonding are simple solutions of rubber in a chemical solvent. They are like the latex pastes in that the excess can be rubbed off the paper. Stronger rubber cements are usually compounded with resins, gums, or synthetics. An infinite variety of these cements is possible, and they are all waterproof with good initial bond, but they are subject to deterioration on exposure, as the rubber is uncured. This type of cement is also made from synthetic rubbers that are self-curing. Curing cements are rubber compounds to be cured by heat and pressure or by chemical curing agents. When cured, they are stronger, give better adhesion to metal surfaces, and have longer life. Latex cements are solvent solutions of rubber latex. They provide excellent tack and give strong bonds to paper, leather, and fabric, but they are subject to rapid disintegration unless cured.

In general, natural rubber has the highest cohesive strength of the rubbers, with rapid initial tack and high bond strength. It also is odorless. Neoprene has the highest cohesive strength of the synthetic rubbers, but it requires tackifiers. Graphite–sulfur rubber (styrene–butadiene) is high in specific adhesion for quick bonding, but has low strength. Reclaimed rubber may be used in cements, but it has low initial tack and needs tackifiers.

Pyroxylin cements may be merely solutions of nitrocellulose in chemical solvents, or they may be compounded with resins, or plasticized with gums or synthetics. They dry by the evaporation of the solvent and have little initial tack, but because of their ability to adhere to almost any type of surface they are called household cements. Cellulose acetate may also be used. These cements are used for bonding the soles of women's shoes. The bonding strength is about 10 lb/in.^{2} (0.07 MPa), or equivalent to the adhesive strength of the outer fibers of the leather to be bonded. For hot-press lamination of wood the plastic cement is sometimes marketed in the form of thin sheet.

Polyvinyl acetate-crotonic acid copolymer resin is used as a hot-dip adhesive for book and magazine binding. It is soluble in alkali solutions, and thus the trim is reusable. Polyvinyl alcohol, with fillers of clay and starch, is used for paperboard containers. Vinyl emulsions are much used as adhesives for laminates.

Epoxy resin cements give good adhesion to almost any material and are heat-resistant to about 400°F (204°C). An epoxy resin will give a steel-to-steel bond of 3100 lb/in.^{2} (22 MPa), and an aluminum-to-aluminum bond to 3800 lb/in.^{2} (26 MPa).

Some pressure-sensitive adhesives are mixtures of a phenolic resin and a nitrile rubber in a solvent, but adhesive tapes are made with a wide variety of rubber or resin compounds.

Furan cements, usually made with furfural-alcohol resins, are strong and highly resistant to chemicals. They are valued for bonding acid-resistant brick and tile.

Acrylic adhesives are solutions of rubber-based polymers in methacrylate monomers. They are two-component systems and have characteristics similar to those of epoxy and urethane adhesives. They bond rapidly at room temperature, and adhesion is not greatly affected by oily or poorly prepared surfaces. Other advantages are low shrinkage during cure, high peel and shear strength, excellent impact resistance, and good elevated temperature properties. They can be used to bond a great variety of materials, such as wood, glass, aluminum, brass, copper, steel, most plastics, and dissimilar metals.

Ultraviolet cure adhesives are anaerobic structural adhesives formulated specifically for glass bonding applications. The adhesive remains liquid after application until ultraviolet light triggers the curing mechanism.

A ceramic adhesive developed by the Air Force for bonding stainless steel to resist heat to 1500°F (816°C) is made with a porcelain enamel frit, iron oxide, and stainless steel powder. It is applied to both parts and fired at 1750°F (954°C), giving a shear strength of 1500 lb/in.^{2} (10 Mpa) in the bond. But ceramic cements that require firing are generally classed with ordinary adhesives. Wash-away adhesives are used for holding lenses, electronic crystal wafers, or other small parts for grinding and polishing operations. They are based on acrylic or other low-melting thermoplastic resins. They can be removed with a solvent or by heating.

Electrically conductive adhesives are made by adding metallic fillers, such as gold, silver, nickel, copper, or carbon powder. Most conductive adhesives are epoxy-based systems, because of their excellent adhesion to metallic and nonmetallic surfaces. Silicones and polyimides are also frequently the base in adhesives used in bonding conductive gaskets to housings for electromagnetic and radio-frequency interference applications.

Properties

An important property for a structural adhesive is resistance to fracture (toughness). Thermoplastics, because they are not cured, can deform under load and exhibit resistance to fracture. As a class, thermosets are quite brittle, and thermoset adhesives are modified by elastomers to increase their resistance to fracture.

Applications

Hot-melt adhesives are used for the manufacture of corrugated paper, in packaging, in carpeting, in bookbinding, and in shoe manufacture. Pressure-sensitive adhesives are most widely used in the form of coatings on tapes. These pressure-sensitive adhesive tapes have numerous applications, from electrical tape to surgical tape. Structural adhesives are applied in the form of liquids, pastes, or 100% adhesive films. Epoxy liquids and pastes are very widely used adhesive materials, having application in many assembly operations ranging from general industrial to automotive to aerospace vehicle construction. Solid-film structural adhesives are used widely in aircraft construction. Acrylic adhesives are used in thread-locking operations and in small-assembly operations such as electronics manufacture, which require rapid cure times. The largest-volume use of adhesives is in plywood and other timber products manufacture. Adhesives for wood bonding range from the natural products (such as blood or casein) to the very durable phenolic-based adhesives.

ALKYDS

Several types of alkyds exist.

Alkyd coatings are used for such diverse applications as air-drying water emulsion wall paints and baked enamels for automobiles and appliances. The properties of oil-modified alkyd coatings depend on the specific oil used as well as the percentage of oil in the composition. In general, they are comparatively low in cost and have excellent color retention, durability, and flexibility, but only fair drying speed, chemical resistance, heat resistance, and salt spray resistance. The oil-modified alkyds can be further modified with other resins to produce resin-modified alkyds.

Alkyd resins are a group of thermosetting synthetic resins known chemically as hydroxycarboxylic resins, of which the one produced from phthalic anhydride and glycerol is representative. They are made by the esterification of a polybasic acid with a polyhydric alcohol, and have the characteristics of homogeneity and solubility that make them especially suitable for coatings and finishes, plastic molding compounds, calking compounds, adhesives, and plasticizers for other resins. The resins have high adhesion to metals; are transparent, easily colored, tough, flexible, and heat and chemical resistant; and have good dielectric strength.

Alkyd plastic molding compounds are composed of a polyester resin and usually a diallyl phthalate monomer plus various inorganic fillers, depending on the desired properties. The raw material is produced in three forms — granular, putty, and glass-fiber-reinforced. As a class, the alkyds have excellent heat resistance up to about 150°C, high stiffness, and moderate tensile and impact strength. Their low moisture absorption combined with good dielectric strength makes them particularly suitable for electronic and electrical hardware, such as switch-gear, insulators, and parts for motor controllers and automotive ignition systems. They are easily molded at low pressures and cure rapidly.

Alkyds are part of the group of materials that includes bulk-molding compounds (BMCs) and sheet-molding compounds (SMCs). They are processed by compression, transfer, or injection molding. Fast molding cycles at low pressure make alkyds easier to mold than many other thermosets. They represent the introduction to the thermosetting plastics industry of the concept of low-pressure, high-speed molding. Typical properties are shown in Table A.5.

Alkyds are furnished in granular compounds, extruded ropes or logs, bulk-molding compound, flake, and putty-like sheets. Except for the putty grades, which may be used for encapsulation, these compounds contain fibrous reinforcement. Generally, the fiber reinforcement in rope and logs is longer than that in granular compounds and shorter than that in flake compounds. Thus, strength of

TABLE A.5
Properties of Alkyds

ASTM Test	Property	Filler	
		Mineral	Glass
	Physical		
D792	Specific gravity	1.60–2.30	2.0–2.3
D570	Water absorption, 24 h, 1/8-in. thk (%)	0.05–0.50	0.03–0.5
	Mechanical		
D638	Tensile strength (psi)	3,000–9,000	4,000–9,500
D638	Tensile modulus (10^5 psi)	5–30	20–28
D790	Flexural strength (psi)	6,000–17,000	8,500–26,000
D790	Flexural modulus (10^5 psi)	20	20
D256	Impact strength, Izod (ft-lb/in. of notch)	0.3–0.5	0.5–16
D785	Hardness, Rockwell	E98	E95
	Thermal		
C177	Thermal conductivity (10^{-4} cal-cm/s-cm²-°C)	12.2–25	15–25
D696	Coefficient of thermal expansion (10^{-5} in./in.-°C)	2–5	1.5–3.3
D648	Deflection temperature (°F)		
	At 264 psi	350–500	400–500
	Electrical		
D149	Dielectric strength, (V/mil)		
	Short time, 1/8-in. thk	350–450	250–530
D150	Dielectric constant		
	At 1 kHz	5.5–6.0	—
D150	Dissipation factor		
	At 1 kHz	0.02–0.04	—
D257	Volume resistivity (ohm-cm)		
	At 73°F, 50% RH	10^{13}–10^{15}	—
D495	Arc resistance (s)	180+	180+

Source: Mach. Design Basics Eng. Design, June, p. 680, 1993. With permission.

these materials is between those of granular and flake compounds. Because the fillers are opaque and the resins are amber, translucent colors are not possible. Opaque, light shades can be produced in most colors, however.

Molded alkyd parts resist weak acids, organic solvents, and hydrocarbons such as alcohol and fatty acids; they are attacked by alkalies.

Depending on the properties desired in the finished compound, the fillers used are clay, asbestos, fibrous glass, or combinations of these materials. The resulting alkyd compounds are characterized in their molding behavior by the following significant features: (1) no liberation of volatiles during the cure, (2) extremely soft flow, and (3) fast cure at molding temperatures.

Although the general characteristics of fast cure and low-pressure requirements are common to all alkyd compounds, they may be divided into three different groups that are easily discernible by the physical form in which they are manufactured.

1. Granular types, which have mineral or modified mineral filters, providing superior dielectric properties and heat resistance

2. Putty types, which are quite soft and particularly well suited for low-pressure molding
3. Glass fiber-reinforced types, which have superior mechanical strengths

For each of these distinct types a more detailed description follows.

GRANULAR TYPES

The physical form of materials in this group is that of a free-flowing powder. Thus, these materials readily lend themselves to conventional molding practices such as volumetric loading, preforming, and high-speed automatic operations. The outstanding properties of parts molded from this group of compounds are high dielectric strength at elevated temperatures, high arc resistance, excellent dimensional stability, and high heat resistance. Compounds are available within this group that are self-extinguishing and certain recently developed types display exceptional retention of insulating properties under high humidity conditions.

These materials have found extensive use as high-grade electrical insulation, especially in the electronics field. One of the major electronic applications for alkyd compounds is in the construction of vacuum tube bases, where the high dry insulation resistance of the material is particularly useful in keeping the electrical leakage between pins to a minimum. In the television industry, tuner segments are frequently molded from granular alkyd compound since electrical and dimensional stability is necessary to prevent calibration shift in the tuner circuits. Also, the granular alkyds have received considerable usage in automotive ignition systems where retention of good dielectric characteristics at elevated temperatures is vitally important.

PUTTY TYPES

This group contains materials that are furnished in soft, puttylike sheets. They are characterized by very low pressure molding requirements (less than 800 psi), and are used in molding around delicate inserts and in solving special loading problems. Molders customarily extrude these materials into a ribbon of a specific size, which is then cut into preforms before molding. Whereas granular alkyds are rather diversified in their various applications, putty has found widespread use in one major application: molded encapsulation of small electronic components, such as mica, polyester film, and paper capacitors; deposited carbon resistors; small coils; and transformers.

The purpose here is to insulate the components electrically, as well as to seal out moisture. Use of alkyds has become especially popular because of their excellent electrical and thermal properties, which result in high functional efficiency of the unit in a minimum space, coupled with low-pressure molding requirements, which prevent distortion of the subassembly during molding.

GLASS-FIBER-REINFORCED TYPES

This type of alkyd molding compound is used in a large number of applications requiring high mechanical strength as well as electrical insulating properties. Glass-fiber-reinforced alkyds can be either compression or plunger molded permitting a wide variety of types of applications, ranging from large circuit breaker housings to extremely delicate electronic components.

OTHER TYPES OF ALKYD MOLDING COMPOUNDS

Halogen and/or phosphorus-bearing alkyd molding compounds with antimony trioxide added provide improved flame resistance. Other flame-resistant compounds are available that do not contain halogenated resins. Many grades are UL-rated at 94V-0 in sections under $1/16$ in. Flammability ratings depend on specific formulations, however, and can vary from 94HB to V-0. Flammability ratings also vary with section thickness.

Glass- and asbestos-filled compounds have better heat resistance than the cellulose-modified types. Depending on type, alkyds can be used continuously to 350°F and, for short periods, to 450°F. Alkyd molding compounds retain their dimensional stability and electrical and mechanical properties over a wide temperature range.

Molding Characteristics

Although full realization of the advantages of molding alkyds is best attained through the use of high-speed, lightweight equipment, nearly all modern compression presses are suitable for use with these materials. Since these compounds are quite fast curing, the press utilized in molding them should be capable of applying full pressure within approximately 6 to 8 s after the mold has been charged. In selecting a press to operate a specific mold for alkyds, the following rule should prove useful: for average draws, the press should furnish about 1500 psi over the projected area of the cavity and lands for molding granular alkyds; about 800 psi for alkyd putty; and about 2000 psi for glass-reinforced alkyd.

Alkyd parts are in successful production in positive, semipositive, and flash molds. In general, the positive and semipositive types are recommended to obtain uniformly dense parts with lowest shrinkage. However, flash molds are frequently used with alkyd putty because of its low bulk factor. In any case, hardened, chromium-plated steel molds are recommended.

The resin characteristics of alkyd molding compounds are such that the material goes through a very low viscosity phase momentarily when heat and pressure are applied. This low viscosity phase makes possible the complete filling of the mold at pressures much lower than those required for other thermosets. Under ordinary conditions, alkyd materials have good release characteristics, and no lubrication is necessary to ensure ejection from the mold.

Applications

High-impact grades of alkyd compounds (with high glass content) are used in military switchgear, electrical terminal strips, and relay and transformer housings and bases. Mineral-filled grades, which can be modified with cellulose to reduce specific gravity and cost, are used in automotive ignition parts, radio and television components, switch-gear, and small appliance housings. Alkyds with all-mineral fillers have high moisture resistance and are particularly suited for electronic components. Grades are available that can withstand the temperatures of vapor-phase soldering.

ALLOY

An alloy is a metal product containing two or more elements as a solid solution, as an intermetallic compound, or as a mixture of metallic phases. Except for native copper and gold, the first metals of technological importance were alloys. Bronze, an alloy of copper and tin, is appreciably harder than copper. This quality made bronze so important an alloy that it left a permanent imprint on the civilization of several millennia ago now known as the Bronze Age.

Alloys are used because they have specific properties or production characteristics that are more attractive than those of the pure, elemental metals. For example, some alloys possess high strength, others have low melting points, others are refractory with high melting temperatures, some are especially resistant to corrosion, and others have desirable magnetic, thermal, or electrical properties. These characteristics arise from both the internal and the electronic structure of the alloy. In recent years, the term *plastic alloy* also has been applied to plastics.

Metal alloys are more specifically described with reference to the major element by weight, which is also called the base metal or parent metal. Thus, the terms *aluminum alloy*, *copper alloy*, etc. Elements present in lesser quantities are called alloying elements. When one or more alloying elements are present in substantial quantity or, regardless of their amount, have a pronounced effect on the alloy, they, too, may be reflected in generic designations.

Metal alloys are also often designated by trade names or by trade association or society designations. Among the more common of the latter are the three-digit designations for the major families of stainless steels and the four-digit ones for aluminum alloys.

Structurally there are two kinds of metal alloys — single phase and multiphase. Single-phase alloys are composed of crystals with the same type of structure. They are formed by "dissolving" together different elements to produce a solid solution. The crystal structure of a solid solution is normally that of the base metal.

In contrast to single-phase alloys, multiphase alloys are mixtures rather than solid solutions. They are composed of aggregates of two or more different phases. The individual phases making up the alloy are different from one another in their composition or structure. Solder, in which the metals lead and tin are present as a mechanical mixture of two separate phases, is an example of the simplest kind of multiphase alloy. In contrast, steel is a complex alloy composed of different phases, some of which are solid solutions. Multiphase alloys far outnumber single-phase alloys in the industrial material field, chiefly because they provide greater property flexibility. Thus, properties of multiphase alloys are dependent upon many factors, including the composition of the individual phases, the relative amounts of the different phases, and the positions of the various phases relative to one another.

When two different thermoplastic resins are blended, a plastic alloy is obtained. Alloying permits resin polymers to be blended that cannot be polymerized. Not all plastics are amenable to alloying. Only resins that are compatible with each other — those that have similar melt traits — can be successfully blended.

TYPES OF ALLOYS

Bearing Alloys

These alloys are used for metals that encounter sliding contact under pressure with another surface; the steel of a rotating shaft is a common example. Most bearing alloys contain particles of a hard intermetallic compound that resists wear. These particles, however, are embedded in a matrix of softer material that adjusts to the hard particles so that the shaft is uniformly loaded over the total surface. The most familiar bearing alloy is babbitt metal, which contains 83 to 91% tin (Sn); the remainder is made up of equal parts of antimony (Sb) and copper (Cu), which form hard particles of the compounds SbSn and CuSn in a soft tin matrix. Other bearing alloys are based on cadmium (Cd), copper, or silver (Ag). For example, an alloy of 70% copper and 30% lead (Pb) is used extensively for heavily loaded bearings. Bearings made by powder metallurgy techniques are widely used. These techniques are valuable because they permit the combination of materials that are incompatible as liquids, for example, bronze and graphite. Powder techniques also permit controlled porosity within the bearings so that they can be saturated with oil before being used, the so-called oilless bearings.

Corrosion-Resisting Alloys

Certain alloys resist corrosion because they are noble metals. Among these alloys are the precious metal alloys, which will be discussed separately. Other alloys resist corrosion because a protective film develops on the metal surface. This passive film is an oxide that separates the metal from the corrosive environment. Stainless steels and aluminum alloys exemplify metals with this type of protection. Stainless steels are iron alloys containing more than 12% chromium (Cr). Steels with 18% Cr and 8% nickel (Ni) are the best known and possess a high degree of resistance to many corrosive environments. Aluminum (Al) alloys gain their corrosion-deterring characteristics by the formation of a very thin surface layer of aluminum oxide (Al_2O_3), which is inert to many environmental liquids. This layer is intentionally thickened in commercial anodizing processes to give a more permanent Al_2O_3 coating. Monel, an alloy of approximately 70% nickel and 30% copper, is a well-known corrosion-resisting alloy that also has high strength. Another nickel-base alloy is Inconel, which contains 14% chromium and 6% iron (Fe). The bronzes, alloys of copper and tin, also may be considered to be corrosion resisting.

Dental Alloys

Amalgams are predominantly alloys of silver and mercury, but they may contain minor amounts of tin, copper, and zinc for hardening purposes, for example, 33% silver, 52% mercury, 12% tin, 2% copper, and less than 1% zinc. Liquid mercury is added to a powder of a precursor alloy of the other metals. After compaction, the mercury diffuses into the silver-base metal to give a completely solid alloy. Gold-base dental alloys are preferred over pure gold because gold is relatively soft. The most common dental gold alloy contains gold (80 to 90%), silver (3 to 12%), and copper (2 to 4%). For higher strengths and hardnesses, palladium and platinum (up to 3%) are added, and the copper and silver are increased so that the gold content drops to 60 to 70%. Vitallium, an alloy of cobalt (65%), chromium (5%), molybdenum (3%), and nickel (3%), and other corrosion-resistant alloys are used for bridgework and special applications.

Die-Casting Alloys

These alloys have melting temperatures low enough so that in the liquid form they can be injected under pressure into steel dies. Such castings are used for automotive parts and for office and household appliances that have moderately complex shapes. This processing procedure eliminates the need for expensive machining and forming operations. Most die castings are made from zinc-base or aluminum-base alloys. Magnesium-base alloys also find some application when weight reduction is paramount. Low-melting alloys of lead and tin are not common because they lack the necessary strength for the above applications. A common zinc-base alloy contains approximately 4% aluminum and up to 1% copper. These additions provide a second phase in the metal to give added strength. The alloy must be free of even minor amounts (less than 100 ppm) of impurities such as lead, cadmium, or tin, because impurities increase the rate of corrosion. Common aluminum-base alloys contain 5 to 12% silicon, which introduces hard-silicon particles into the tough aluminum matrix. Unlike zinc-base alloys, aluminum-base alloys cannot be electroplated; however, they may be burnished or coated with enamel or lacquer.

Advances in high-temperature die-mold materials have focused attention on the die-casting of copper-base and iron-base alloys. However, the high casting temperatures introduce costly production requirements, which must be justified on the basis of reduced machining costs.

Eutectic Alloys

In certain alloy systems a liquid of a fixed composition freezes to form a mixture of two basically different solids or phases. An alloy that undergoes this type of solidification process is called a eutectic alloy. A typical eutectic alloy is formed by combining 28.1% of copper with 71.9% of silver. A homogeneous liquid of this composition on slow cooling freezes to form a mixture of particles of nearly pure copper embedded in a matrix (background) of nearly pure silver.

The advantageous mechanical properties inherent in composite materials such as plywood composed of sheets or lamellae of wood bonded together and fiberglass in which glass fibers are used to reinforce a plastic matrix have been known for many years. Attention is being given to eutectic alloys because they are basically natural composite materials. This is particularly true when they are directionally solidified to yield structures with parallel plates of the two phases (lamellar structure) or long fibers of one phase embedded in the other phase (fibrous structure). Directionally solidified eutectic alloys are being given serious consideration for use in fabricating jet engine turbine blades. For this purpose eutectic alloys that freeze to form tantalum carbide (TaC) fibers in a matrix of a cobalt-rich alloy have been heavily studied.

Fusible Alloys

These alloys generally have melting temperatures below that of tin (450°F, or 232°C), and in some cases as low as 120°F (50°C). Using eutectic compositions of metals such as lead, cadmium, bismuth, tin, antimony, and indium achieves these low melting temperatures. These alloys are used for many purposes, for example, in fusible elements in automatic sprinklers, forming and stretching dies, filler for thin-walled tubing that is being bent, and anchoring dies, punches, and parts being machined. Alloys rich in bismuth were formerly used for type metal because these low-melting metals exhibited a slight expansion on solidification, thus replicating the font perfectly for printing and publication.

High-Temperature Alloys

Energy conversion is more efficient at high temperatures than at low; thus the need in power-generating plants, jet engines, and gas turbines for metals that have high strengths at high temperatures is obvious. In addition to having strength, these alloys must resist oxidation by fuel–air mixtures and by steam vapor. At temperatures up to about 1380°F (750°C), the austenitic stainless steels (18% Cr–8% Ni) serve well. An additional 180°F (100°C) may be realized if the steels also contain 3% molybdenum. Both nickel-base and copper-base alloys, commonly categorized as superalloys, may serve useful functions up to 2000°F (1100°C). Nichrome, a nickel-base alloy containing 12 to 15% chromium and 25% iron, is a fairly simple superalloy. More sophisticated alloys invariably contain five, six, or more components; for example, an alloy called René-41 contains approximately 19% Cr, 1.5% Al, 3% Ti, 11% Co, 10% Mo, 3% Fe, 0.1% C, 0.005% B, and the balance Ni. Other alloys are equally complex. The major contributor to strength in these alloys is the solution-precipitate phase of Ni_3(TiAl). It provides strength because it is coherent with the nickel-rich phase. Cobalt-base superalloy may be even more complex and generally contain carbon, which combines with the tungsten (W) and chromium to produce carbides that serve as the strengthening agent. In general, the cobalt-base superalloys are more resistant to oxidation than the nickel-base alloys are, but they are not as strong. Molybdenum-base alloys have exceptionally high strength at high temperatures, but their brittleness at lower temperatures and their poor oxidation resistance at high temperatures have limited their use. However, coatings permit the use of such alloys in an oxidizing atmosphere, and they are finding increased application. A group of materials called cermets, which are mixtures of metals and compounds such as oxides and carbides, have high strength at high temperatures, and although their ductility is low, they have been found to be usable. One of the better-known cermets consists of a mixture of TiC and nickel, the nickel acting as a binder or cement for the carbide.

Joining Alloys

Metals are bonded by three principal procedures: welding, brazing, and soldering. Welded joints melt the contact region of the adjacent metal; thus, the filler material is chosen to approximate the composition of the parts being joined. Brazing and soldering alloys are chosen to provide filler metal with an appreciably lower melting point than that of the joined parts. Typically, brazing alloys melt above 750°F (400°C) whereas solders melt at lower temperatures. A 57% Cu–42% Zn–1% Sn brass is a general-purpose alloy for brazing steel and many nonferrous metals. A Si–Al eutectic alloy is used for brazing aluminum, and an aluminum-containing magnesium eutectic alloy brazes magnesium parts. The most common solders are based on Pb–Sn alloys. The prevalent 60% Sn–40% Pb solder is eutectic in composition and is used extensively for electrical circuit production, in which temperature limitations are critical. A 35% Sn–65% Pb alloy has a range of solidification and is thus preferred as a wiping solder by plumbers.

Light-Metal Alloys

Aluminum and magnesium, with densities of 2.7 and 1.75 g/cm^3), respectively, are the bases for most of the light-metal alloys. Titanium (4.5 g/cm^3) may also be regarded as a light-metal alloy if comparisons are made with metals such as steel and copper. Aluminum and magnesium must be hardened to receive extensive application. Age-hardening processes are used for this purpose. Typical alloys are 90% Al-10% Mg, 95% Al-5% Cu, and 90% Mg-10% Al. Ternary (three element) and more complex alloys are very important light-metal alloys because of their better properties. The Al–Zn–Mg system of alloys, used extensively in aircraft applications, is a prime example of one such alloy system.

Low-Expansion Alloys

This group of alloys includes Invar (64% Fe–36% Ni), the dimensions of which do not vary over the atmospheric temperature range. It has special applications in watches and other temperature-sensitive devices. Glass-to-metal seals for electronic and related devices require a matching of the thermal-expansion characteristics of the two materials. Kovar (54% Fe–29% Ni–17% Co) is widely used because its expansion is low enough to match that of glass.

Magnetic Alloys

Soft and hard magnetic materials involve two distinct categories of alloys. The former consists of materials used for magnetic cores of transformers and motors, and must be magnetized and demagnetized easily. For AC applications, silicon–ferrite is commonly used. This is an alloy of iron containing as much as 5% silicon. The silicon has little influence on the magnetic properties of the iron, but it increases the electric resistance appreciably and thereby decreases the core loss by induced currents. A higher magnetic permeability, and therefore greater transformer efficiency, is achieved if these silicon steels are grain-oriented so that the crystal axes are closely aligned with the magnetic field. Permalloy (78.5% Ni–21.5% Fe) and some comparable cobalt-base alloys have very high permeabilities at low field strengths, and thus are used in the communications industry. Ceramic ferrites, although not strictly alloys, are widely used in high-frequency applications because of their low electrical conductivity and negligible induced energy losses in the magnetic field. Permanent or hard magnets may be made from steels that are mechanically hardened, either by deformation or by quenching. Some precipitation-hardening, iron-base alloys are widely used for magnets. Typical of these are the Alnicos, for example, Alnico-4 (55% Fe–28% Ni–12% Al–5% Co). Since these alloys cannot be forged, they must be produced in the form of castings. Hard magnets are being produced from alloys of cobalt and the rare earth type of metals. The compound RCo_5, where R is samarium (Sm), lanthanum (La), cerium (Ce), and so on, has extremely high coercivity.

Precious-Metal Alloys

In addition to their use in coins and jewelry, precious metals such as silver, gold, and the heavier platinum (Pt) metals are used extensively in electrical devices in which contact resistances must remain low, in catalytic applications to aid chemical reactions, and in temperature-measuring devices such as resistance thermometers and thermocouples. The unit of alloy impurity is commonly expressed in karats, when each karat is $1/24$ part. The most common precious-metal alloy is sterling silver (92.5% Ag, with the remainder being unspecified, but usually copper). The copper is very beneficial in that it makes the alloy harder and stronger than pure silver. Yellow gold is an Au–Ag–Cu alloy with approximately a 2:1:1 ratio. White gold is an alloy that ranges from 10 to 18 karats, the remainder being additions of nickel, silver, or zinc, which change the color from yellow to white. The alloy 87% platinum–13% rhodium (Rh), when joined with pure platinum, provides a widely used thermocouple for temperature measurements in the 1830 to 3000°F (1000 to 1650°C) temperature range.

Shape Memory Alloys

These alloys have a very interesting and desirable property. In a typical case, a metallic object of a given shape is cooled from a given temperature T_1, to a lower temperature T_2, where it is deformed to change its shape. Upon reheating from T_2 to T_1, the shape change accomplished at T_2 is recovered so that the object returns to its original configuration. This thermoelastic property of the shape memory alloys is associated with the fact that they undergo a martensitic phase transformation (that is, a reversible change in crystal structure that does not involve diffusion) when they are cooled or heated between T_1 and T_2.

For a number of years the shape memory materials were essentially scientific curiosities. Among the first alloys shown to possess these properties was one of gold alloyed with 47.5% cadmium. Considerable attention has been given to an alloy of nickel and titanium known as Nitinol. The interest in shape memory alloys has increased because it has been realized that these alloys are capable of being employed in a number of useful applications. One example is for thermostats; another is for couplings on hydraulic lines or electrical circuits. The thermoelastic properties can also be used, at least in principle, to construct heat engines that will operate over a small temperature differential and will thus be of interest in the area of energy conversion.

Thermocouple Alloys

These include Chromel, containing 90% Ni and 10% Cr, and Alumel, containing 94% Ni, 2% Al, 3% Cr, and 1% Si. These two alloys together form the widely used Chromel–Alumel thermocouple, which can measure temperatures up to 2200°F (1204°C). Another common thermocouple alloy is Constantan, consisting of 45% Ni and 55% Cu. It is used to form iron-Constantan and copper-Constantan couples, used at lower temperatures. For precise temperature measurements and for measuring temperatures up to 3000°F (1650°C), thermocouples are used in which one metal is platinum and the other metal is platinum plus either 10 or 13% rhodium.

Prosthetic Alloys

Prosthetic alloys are alloys used in internal prostheses, that is, surgical implants such as artificial hips and knees. External prostheses are devices that are worn by patients outside the body; alloy selection criteria are different from those for internal prostheses. In the United States, surgeons use about 250,000 artificial hips and knees and about 30,000 dental implants per year.

Alloy selection criteria for surgical implants can be stringent primarily because of biomechanical and chemical aspects of the service environment. Mechanically, the properties and shape of an implant must meet anticipated functional demands; for example, hip joint replacements are routinely subjected to cyclic forces that can be several times body weight. Therefore, intrinsic mechanical properties of an alloy, for example, elastic modulus, yield strength, fatigue strength, ultimate tensile strength, and wear resistance, must all be considered. Similarly, because the pH and ionic conditions within a living organism define a relatively hostile corrosion environment for metals, corrosion properties are an important consideration. Corrosion must be avoided not only because of alloy deterioration but also because of the possible physiological effects of harmful or even cytotoxic corrosion products that may be released into the body. (Study of the biological effects of biomaterials is a broad subject in itself, often referred to as biocompatibility.) The corrosion resistance of all modern alloys stems primarily from strongly adherent and passivating surface oxides, such as TiO_2 on titanium-based alloys and Cr_2O_3 on cobalt-base alloys.

The most widely used prosthetic alloys therefore include high-strength, corrosion-resistant ferrous, cobalt-base, or titanium-base alloys. Examples include cold-worked stainless steel; cast Vitallium, a wrought alloy of cobalt, nickel, chromium, molybdenum, and titanium; titanium alloyed with aluminum and vanadium; and commercial-purity titanium. Specifications for nominal alloy compositions are designated by the American Society for Testing and Materials (ASTM).

Prosthetic alloys have a range of properties. Some are easier than others to fabricate into the complicated shapes dictated by anatomical constraints. Fabrication techniques include investment casting (solidifying molten metal in a mold), forging (forming metal by deformation), machining (forming by machine-shop processes, including computer-aided design and manufacturing), and hot isostatic pressing (compacting fine powders of alloy into desired shapes under heat and pressure). Cobalt-base alloys are difficult to machine and are therefore usually made by casting or hot isostatic pressing. Some newer implant designs are porous coated; that is, they are made from the standard ASTM alloys but are coated with alloy beads or mesh applied to the surface by sintering or other methods. The rationale for such coatings is implant fixation by bone ingrowth.

Some alloys are modified by nitriding or ion-implantation of surface layers of enhanced surface properties. A key point is that prosthetic alloys of identical composition can differ substantially in terms of structure and properties, depending on fabrication history. For example, the fatigue strength approximately triples for hot isostatically pressing vs. as-cast Co–Cr–Mo alloy, primarily because of a much smaller grain size in the microstructure of the former.

No single alloy is vastly superior to all others; existing prosthetic alloys have all been used in successful and, indeed, unsuccessful implant designs. Alloy selection is only one determinant of performance of the implanted device.

Superconducting Alloys

Superconductors are materials that have zero resistance to the flow of electric current at low temperatures. There are more than 6000 elements, alloys, and compounds that are known superconductors. This remarkable property of zero resistance offers unprecedented technological advances such as the generation of intense magnetic fields. Realization of these new technologies requires development of specifically designed superconducting alloys and composite conductors. An alloy of niobium and titanium (NbTi) has a great number of applications in superconductivity; it becomes superconducting at 9.5 K (critical superconducting temperature, T_c). This alloy is preferred because of its ductility and its ability to carry large amounts of current at high magnetic fields, represented by $J_c(H)$ (where J_c is the critical current and H is a given magnetic field), and still retain its superconducting properties. Brittle compounds with intrinsically superior superconducting properties are also being developed for magnet applications. The most promising of these are compounds of niobium and strontium (Nb_3Sn), vanadium and gallium (V_3Ga), niobium and germanium (Nb_3Ge), and niobium and aluminum (Nb_3Al), which have higher T_c (15 to 23 K) and higher $J_c(H)$ than NbTi.

Superconducting materials possess other unique properties such as magnetic flux quantization and magnetic-field-modulated supercurrent flow between two slightly separated superconductors.

These properties form the basis for electronic applications of superconductivity such as high-speed computers or ultrasensitive magnetometers. Development of these applications began using lead or niobium (T_c of 7 and 9 K) in bulk form, but the emphasis then was transferred to materials deposited in thin-film form. PbIn and PbAu alloys are more desirable than pure lead films, as they are more stable. Improved vacuum deposition systems eventually led to the use of pure niobium films as they, in turn, were more stable than lead alloy films. Advances in thin-film synthesis techniques led to the use of the refractory compound niobium nitride (NbN) in electronic applications. This compound is very stable and possesses a higher T_c (15 K) than either lead or niobium.

Novel high-temperature superconducting materials have revolutionary impact on superconductivity and its applications. These materials are ceramic, copper-oxide-based materials that contain at least four and as many as six elements. Typical examples are yttrium–barium–copper–oxygen (T_c 93 K), bismuth–strontium–calcium–copper–oxygen (T_c 110K), and thallium–barium–calcium–copper (T_c 125 K). These materials become superconducting at such high temperatures that refrigeration is simpler, more dependable, and less expensive. Much research and development has been done to improve the technologically important properties such as $J_c(H)$, chemical and mechanical stability, and device-compatible processing procedures. It is anticipated that the new compounds will have a significant impact in the growing field of superconductivity.

ALLOY STRUCTURES

Metals in actual commercial use are almost exclusively alloys, and not pure metals, since it is possible for the designer to realize an extensive variety of physical properties in the product by varying the metallic composition of the alloy. As a case in point, commercially pure or cast iron is very brittle because of the small amount of carbon impurity always present, whereas the steels are much more ductile, with greater strength and better corrosion properties. In general, the highly purified single crystal of a metal is very soft and malleable, with high electrical conductivity, whereas the alloy is usually harder and may have a much lower conductivity. The conductivity will vary with the degree of order of the alloy, and the hardness will vary with the particular heat treatment used.

The basic knowledge of structural properties of alloys is still in large part empirical, and indeed, it will probably never be possible to derive formulas that will predict which metals to mix in a certain proportion and with a certain heat treatment to yield a specified property or set of properties. However, a set of rules exists that describes the qualitative behavior of certain groups of alloys. These rules are statements concerning the relative sizes of constituent atoms, for alloy formation, and concerning what kinds of phases to expect in terms of the valence of the constituent atoms. The rules were discovered in a strictly empirical way, and for the most part, the present theoretical understanding of alloys consists of rudimentary theories that describe how the rules arise from the basic principles of physics. These rules were proposed by W. Hume-Rothery concerning the binary substitutional alloys and phase diagrams.

ALLYLICS (DIALLYL PHTHALATE PLASTICS)

Allylics are thermosetting materials developed since World War II. The most important of these are diallyl phthalate (DAP) and diallyl isophthalate (DAIP), which are currently available in the form of monomers and prepolymers (resins). Both DAP and DAIP are readily converted to thermoset molding compounds and resins for preimpregnated glass cloth and paper. Allyls are also used as cross-linking agents for unsaturated polyesters.

DAP resin is the first all-allylic polymer commercially available as a dry, free-flowing white powder. Chemically, DAP is a relatively linear partially polymerized resin that softens and flows under heat and pressure (as in molding and laminating), and cross-links to a three-dimensional insoluble thermoset resin during curing.

Properties

In preparing the resin, DAP is polymerized to a point where almost all the change in specific gravity has taken place. Final cure, therefore, produces very little additional shrinkage. In fact, DAP is cured by polymerization without water formation. The molded material, depending on the filler, has a tensile strength from 30 to 48 MPa, a compressive strength up to 210 MPa, a Rockwell hardness to M108, dielectric strength to 16.9×10^6 V/m, and heat resistance to 232°C.

Allylic resins enjoy certain specific advantages over other plastics, which make them of interest in various special applications. Allylics exhibit superior electrical properties under severe temperature and humidity conditions. These good electrical properties (insulation resistance, low loss factor, arc resistance, etc.) are retained despite repeated exposure to high heat and humidity. DAP resin is resistant to 155 to 180°C temperatures, and the DAIP resin is good for continuous exposures up to 206 to 232°C temperatures. Allylic resins exhibit excellent post-mold dimensional stability, low moisture absorption, good resistance to solvents, acids, alkalis, weathering, and wet and dry abrasion. They are chemically stable, have good surface finish, mold well around metal inserts and can be formulated in pastel colors with excellent color retention at high temperatures.

DAP resin currently finds major use in (1) molding and (2) industrial and decorative laminates. Both applications utilize the desirable combination of low shrinkage, absence of volatiles, and superior electrical and physical properties common to DAP.

Molding Compounds

Compounds based on allyl prepolymers are reinforced with fibers (glass, polyester, or acrylic) and filled with particulate materials to improve properties. Glass fiber imparts maximum mechanical properties, acrylic fiber provides the best electrical properties, and polyester fiber improves impact resistance and strength in thin sections. Compounds can be made in a wide range of colors because the resin is essentially colorless; see Table A.6.

Prepregs (preimpregnated glass cloth) based on allyl prepolymers can be formulated for short cure cycles. They contain no toxic additives, and they offer long storage stability and ease of handling and fabrication. Properties such as flame resistance can be incorporated. The allyl prepolymers contribute excellent chemical resistance and good electrical properties.

Other molding powders are compounded of DAP resin, DAP monomer, and various fillers like asbestos, Orlon, Dacron, cellulose, glass, and other fibers. Inert fillers used include ground quartz and clays, calcium carbonate, and talc.

Allyl moldings have low mold shrinkage and post-mold shrinkage — attributed to their

TABLE A.6
Properties of DAP Molding Compounds

ASTM Test	Property	Filler			Arc-Track Resistant
		Polyester	Long Glass	Short Glass	
	Physical				
D792	Specific gravity	1.39–1.42	1.70–1.90	1.6–1.8	1.87
D792	Specific volume (in.^{3}/lb)	19.96–19.54	17.90–16.32	17.34–15.42	14.84
D570	Water absorption, 24 h, 1/8-in. thk (%)	0.2	0.05–0.2	0.05–0.2	0.14
	Mechanical				
D638	Tensile strength (psi)	5,000	9,000	7,000	7,000–10,000
D790	Flexural strength (psi)	11,500–12,500	18,000	16,000	24,000
D790	Flexural modulus (10^5 psi)	6.4	16	17	19
D256	Impact strength, Izod (ft-lb/in. of notch)	4.5–12	6.0	0.8	3.6
D785	Hardness, Rockwell M	108	105–110	105–110	112
	Thermal				
C177	Thermal conductivity (10^{-4} cal-cm/s-cm^2-°C)	—	14–16	14–15	15–17
D696	Coefficient of thermal expansion (10^{-5} in./in.-°C)	—	2.0–3.0	2.0–3.0	23–27
D648	Deflection temperature (°F) At 264 psi	290	450	420	>572
	Electrical				
D149	Dielectric strength, (V/mil) Step by step, 1/8-in. thk	400	385	400	400
D150	Dielectric constant At 1 kHz	0.008	0.004–0.006	0.006	0.003–0.008
D150	Dissipation factor At 1 kHz	3.6	4.2	4.4	4.1–4.5
D257	Volume resistivity (ohm-cm) At 73°F, 50% RH	2–3×10^{15}	2–3×10^{15}	2–3×10^{15}	10^{16}
D495	Arc resistance (s)	125	140	135	125–180
	Frictional				
—	Coefficient of friction			Stat/Dyn	
	Self	—	—	0.14/0.13	—
	Against steel	—	—	0.20/0.19	—

Source: Mach. Design Basics Eng. Design, June, p. 680, 1993. With permission.

nearly complete addition reaction in the mold — and have excellent stability under prolonged or cyclic heat exposure. Advantages of allyl systems over polyesters are freedom from styrene odor low toxicity, low evaporation losses during evacuation cycles, no subsequent oozing or bleed-out, and long-term retention of electrical-insulation characteristics.

Applications

Uses of such DAP molding compounds are largely for electrical and electronic parts, connectors, resistors, panels, switches, and insulators. Other applications for molding compounds include appliance handles, control knobs, dinnerware, and cooking equipment.

Decorative laminates containing DAP resin can be made from glass cloth (or other woven and nonwoven materials), glass mat, or paper. Such laminates may be bonded directly to a variety of rigid surfaces at lower pressures (50 to 300 psi) than generally required for other plastic laminates. A short hot-to-hot cycle is employed, and press platens are always held at curing temperatures. DAP laminates can, therefore, be used to give a permanent finish to high-grade wood veneers (with a clear overlay sheet) or to upgrade low-cost core materials (by means of a patterned sheet).

Allyl prepolymers are particularly suited for critical electronic components that serve in severe environmental conditions. Chemical inertness qualifies the resins for molded pump impellers and other chemical-processing equipment. Their ability to withstand steam environments permits uses in sterilizing and hot-water equipment. Because of their excellent flow characteristics, DAP compounds are used for parts requiring extreme dimensional accuracy. Modified resin systems are used for encapsulation of electronic devices such as semiconductors and as sealants for metal castings.

A major application area for allyl compounds is electrical connectors, used in communications, computer, and aerospace systems. The high thermal resistance of these materials permits their use in vapor-phase soldering operations. Uses for prepolymers include arc-track-resistant compounds for switchgear and television components. Other representative uses are for insulators, encapsulating shells, potentiometer components, circuit boards, junction boxes, and housings.

DAP and DAIP prepregs are used to make lightweight, intricate parts such as radomes, printed-circuit boards, tubing, ducting, and aircraft parts. Another use is in copper-clad laminates for high-performance printed-circuit boards.

ALUMINA

The oxide of aluminum is Al_2O_3. The natural crystalline mineral is called corundum, but the synthetic crystals used for abrasives are designated usually as aluminum oxide or marketed under trade names. For other uses and as a powder it is generally called alumina. It is widely distributed in nature in combination with silica and other minerals, and is an important constituent of the clays for making porcelain, bricks, pottery, and refractories.

The crushed and graded crystals of alumina when pure are nearly colorless, but the fine powder is white. Off colors are due to impurities. American aluminum oxide used for abrasives is at least 99.5% pure, in nearly colorless crystals melting at 2050°C. The chief uses for alumina are for the production of aluminum metal and for abrasives, but it is also used for ceramics, refractories, pigments, catalyst carriers, and in chemicals.

Aluminum oxide crystals are normally hexagonal, and are minute in size. For abrasives, the grain sizes are usually from 100 to 600 mesh. The larger grain sizes are made up of many crystals, unlike the single-crystal large grains of SiC. The specific gravity is about 3.95, and the hardness is up to 2000 Knoop.

There are two kinds of ultrafine alumina abrasive powder. Type A is alpha alumina with hexagonal crystals with particle size of 0.3 μm, density 4.0, and hardness 9 Mohs, and Type B is gamma alumina with cubic crystals with particle size under 0.1 μm, specific gravity of 3.6, and a hardness 8. Type A cuts faster, but Type B gives a finer finish. At high temperatures gamma alumina transforms to the alpha crystal. The aluminum oxide most frequently used for refractories is the beta alumina in hexagonal crystals heat-stabilized with sodium.

Activated alumina is partly dehydrated alumina trihydrate, which has a strong affinity for moisture or gases and is used for dehydrating organic solvents, and hydrated alumina is alumina trihydrate.

$Al_2O_3 \cdot 3H_2O$ is used as a catalyst carrier.

Activated alumina F-1 is a porous form of alumina, Al_2O_3, used for drying gases or liquids and is also used as a catalyst for many chemical processes.

Alumina ceramics are the most widely used oxide-type ceramic, chiefly because Al_2O_3 is plentiful, relatively low in cost, and equal to or better than most oxides in mechanical properties. Density can be varied over a wide range, as can purity — down to about 90% Al_2O_3 — to meet specific application requirements.

Al_2O_3 ceramics are the hardest, strongest, and stiffest of the oxides. They are also outstanding in electrical resistivity and dielectric strength, are resistant to a wide variety of chemicals, and are unaffected by air, water vapor, and sulfurous atmospheres. However, with a melting point of only 2037°C, they are relatively low in refractoriness, and at 1371°C retain only about 10% of room-temperature strength. Besides wide use as electrical insulators and chemical and aerospace applications, the high hardness and close dimensional tolerance capability of alumina make this ceramic suitable for such abrasion-resistant parts as textile guides, pump plungers, chute linings, discharge orifices, dies, and bearings.

Alumina Al-200, which is used for high-frequency insulators, gives a molded product with a tensile strength of 172 MPa, compressive strength of 2000 MPa, and specific gravity of 3.36. The coefficient of thermal expansion is half that of steel, and the hardness about that of sapphire. Alumina AD-995 is a dense vacuum-tight ceramic for high-temperature electronic use. It is 99.5% Al_2O_3 with no SiO_2. The hardness is Rockwell N80, and dielectric constant 9.27. The maximum working temperature is 1760°C, and at 1093°C it has a flexural strength of 200 MPa.

Other alumina products have found their way in the casting of hollow jet engine cores. These cores are then incorporated in molds into which eutectic superalloys are poured to form the turbine blades.

Alumina balls are available in sizes from 0.6 to 1.9 cm for reactor and catalytic beds. They are usually 99% alumina, with high resistance to heat and chemicals. Alumina fibers in the form of short linear crystals, called sapphire whiskers, have high strength up to 1375 MPa for use as a filler in plastics to increase heat resistance and dielectric properties. Continuous single-crystal sapphire (alumina filaments) have unusual physical properties: high tensile strength (over 2069 MPa) and modulus of elasticity of 448.2 to 482.7 GPa. The filaments are especially needed for use in metal composites at elevated temperatures and in highly corrosive environments. An unusual method for producing single-crystal fibers in lieu of a crystal growing machine is the floating zone fiber-drawing process. The fibers are produced directly from a molten ceramic without using a crucible.

FP, a polycrystalline alumina (Al_2O_3) fiber, has been developed. The material has greater than 99% purity, and a melting point of 2045°C, which makes it attractive for use with high-temperature metal-matrix composite (MMC) processing techniques. Thanks to a mechanism, currently not explainable by the developer of FP fibers (Du Pont), a silica coating results in an increase in the tensile strength of the filaments to 1896 MPa even though the coating is approximately 0.25 μm thick and the modulus does not change. Fiber FP has been demonstrated as a reinforcement in magnesium, aluminum, lead, copper, and zinc, with emphasis to date on aluminum and magnesium materials.

Fumed alumina powder of submicrometer size is made by flame reduction of aluminum chloride. It is used in coatings and for plastics reinforcement and in the production of ferrite ceramic magnets.

Aluminum oxide film, or alumina film, used as a supporting material in ionizing tubes, is a strong, transparent sheet made by oxidizing aluminum foil, rubbing off the oxide on one side, and dissolving the foil in an acid solution to leave the oxide film on the other side. It is transparent to electrons. Alumina bubble brick is a lightweight refractory brick for kiln lining, made by passing molten alumina in front of an airjet, producing small hollow bubbles which are then pressed into bricks and shapes.

The foam has a density of 448.5 kg/m³ and porosity of 85%. The thermal conductivity at 1093°C is 0.002 W/(cm²)(°C).

ALUMINIDES

True metals include the alkali and alkaline earth metals, beryllium, magnesium, copper, silver, gold, and the transition elements. These metals exhibit those characteristics generally associated with the metallic state.

The B subgroups comprise the remaining metallic elements. These elements exhibit complex structures and significant departures from typically metallic properties. Aluminum, although considered under the B subgroup metals, is somewhat anomalous in that it exhibits many characteristics of a true metal.

The alloys of a true metal and a B subgroup element are very complex, because their components differ electrochemically. This difference gives rise to a stronger tendency toward definite chemical combination than to solid solution. Discrete geometrically ordered structures usually result. Such alloys are also termed *electron compounds*. The aluminides are phases in such alloys or compounds. A substantial number of beta, gamma, and epsilon phases have been observed in electron compounds, but few have been isolated and evaluated.

The development of intermetallic alloys into useful and practical structural materials remains, despite recent successes, a major scientific and engineering challenge. As with many new and advanced materials, hope and the promise of major breakthroughs in the near future have kept a very active and resilient fraction of the metallurgical community focused on intermetallic alloys.

Compared to conventional aerospace materials, aluminides of titanium, nickel, iron, niobium, etc., with various compositions offer attractive properties for potential structural applications. The combination of good high-temperature strength and creep capability, improved high-temperature environmental resistance, and relatively low density makes this general class of materials good candidates to replace more conventional titanium alloys and, in some instances, nickel-base superalloys. Moreover, titanium aluminide matrix composites appear to have the potential to surpass the monolithic titanium aluminides in a number of important property areas, and fabrication into composite form may be a partial solution to some of the current shortcomings attributed to monolithic titanium aluminides.

The material classes include both monolithic and continuous fiber composite materials based on the intermetallic composition Ti_3Al (α_2-phase) and monolithic alloys based on the intermetallic composition TiAl (γ-phase). In their monolithic form, and as a matrix material for continuous fiber composites, titanium aluminides are important candidates to fill a need in the intermediate-temperature regime of 600 to 1000°C. Before these materials can become flightworthy, however, they must demonstrate reliable mechanical behavior over the range of anticipated service conditions.

The β and γ phases that are found to exist in the Mo–Al alloy system are generally considered to correspond to the compositions $MoAl_5$ and $MoAl_2$, respectively.

Powder metallurgy techniques have proved feasible for the production of alloys of molybdenum and aluminum, provided care is taken to employ raw materials of high purity (99% +). As the temperature of the compact is raised, a strong exothermic reaction occurs at about 640°C causing a rapid rise in temperature to above 960°C in a matter of seconds. Bloating occurs, transforming the compact into a porous mass. Complete alloying, however, is accomplished. This porous, friable mass can be subsequently finely comminuted, repressed, and sintered (or hot-pressed) to form a useful body quite uniform in composition. Vacuum sintering at 1300°C for 1 h at 0.04 μm produces clean, oxide-free metal throughout. Wet comminution prevents caking of the powder, and a pyrophoric powder can be produced by prolonged milling.

Hot pressing is a highly successful means of forming bodies of molybdenum and aluminum previously reacted as mentioned above. Graphite dies are employed to which resistance heating techniques are applicable. A parting compound is required since aluminum is highly reactive with carbon causing sticking to the die walls.

Hot-pressed small bars exhibit modulus of rupture strengths ranging from 40,000 to 50,000 psi at room temperature, decreasing to 38,000 to 40,000 psi at 1040°C. Room temperature resistance to fuming nitric acids is excellent.

As has been recognized for some time, ordered intermetallic compounds have a number of properties that make them intrinsically more appealing than other metallic systems for high-temperature use. The primary requirements for high-temperature structural intermetallics, as with any high-temperature structural material, are that they (1) have a high melting point, (2) possess some degree of resistance to environmental degradation, (3) maintain structural and chemical stability at high temperatures, and (4) retain high specific mechanical properties at elevated temperatures whether they are intended as monolithic components or as reinforcing fibers or matrix in composite structures.

Melting point is a useful first approximation of the high-temperature performance of a material, as various high-temperature mechanical properties (e.g., strength and creep resistance) are limited by thermally assisted or diffusional processes and thus tend to scale with the melting point of the material. Therefore, the intermetallics can be crudely ranked in terms of their melting points to indicate their future applicability as high-temperature structural materials.

As may be seen in Figure A.2, metallic materials (intermetallics or otherwise) that are currently in use or being studied melt at temperatures much lower than 1650°C. If these materials are discounted from consideration, the remaining intermetallics in Figure A.2 may be roughly divided into two groups: those that fall in the temperature range just above 1650°C and those whose melting points extend to much higher temperatures.

This second group of intermetallic compounds (IMCs) belongs to a group of intermetallics that are predicted on the basis of the Engel-Brewer phase stability theory.

There are several techniques that have been developed and used to improve the toughness of intermetallics as well as intermetallic compounds:

- Crystal structure modification (macroalloying)
- Microalloying
- Control of grain size or shape
- Reinforcement by ductile fibers or particles
- Control of substructure

Table A.7 includes the above major categories, however, the use of hydrostatic pressure and suppression of environment should also be cited.

Additions of chromium and manganese have induced appreciable compressive ductility and modest improvements in bend ductility of Al_3Ti, but significant tensile ductility remains unattainable.

The fracture toughness of Ti_3Al alloys also can be markedly improved by a control of composition, microstructure, and processing techniques. However, the maximum benefits are obtained at about 400°C.

Microstructural control has proved to be a particularly effective means of ductilizing TiAl and Ti_3Al. It is now generally accepted that lamellar microstructures in TiAl, consisting of alternating γ- and α_2-plates, provide the highest ductility.

The interest in aluminides has covered the high-melting-point phases in metallic systems with aluminum.

Ordered intermetallics constitute a unique class of metallic materials that form long-range-ordered crystal structures (Figure A.3) below a critical temperature that is generally referred to as the critical ordering temperature (T_c). These intermetallics usually exist in relatively narrow compositional ranges around simple stoichiometric ratios.

The search for new high-temperature structural materials has stimulated much interest in ordered intermetallics. Recent interest has been

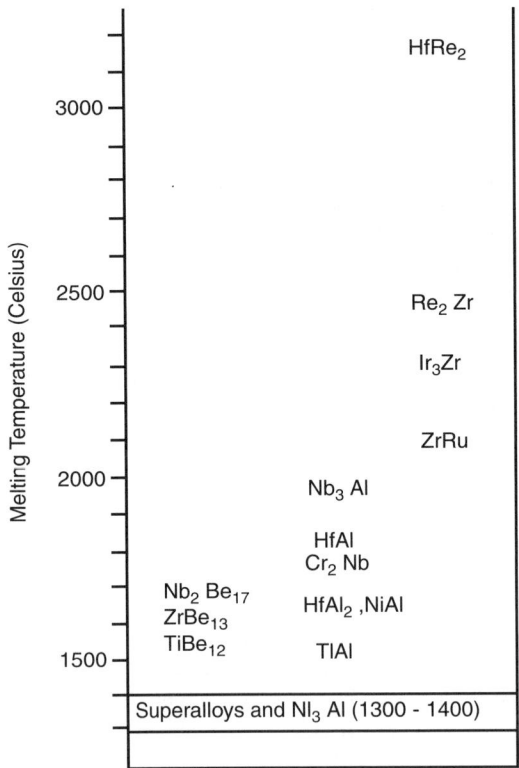

FIGURE A.2 Melting points of various intermetallic compounds relative to superalloys. (From Schwartz, M., *Emerging Technology*, Technomics, 19. With permission.)

ALUMINIDES

**TABLE A.7
Toughness and Ductility Improvements**

Microalloying
  B in Ni_3Al, Ni_3Si, PdIn
  Be in Ni_3Al
  Fe, Mo, Ga in NiAl
  Ag in Ni_3Al
Macroalloying
  Fe in Co_3V
  Mn, V, Cr in TiAl
  Nb in Ti_3Al
  Mn, Cr in Al_3Ti
  Pd in Ni_3Al
Grain Size Refinement
  NiAl
Hydrostatic Pressure
  Ni_3Al
Martensite Transformation
  Fe in NiAl
Composites (Fibers or Tubes)
  NiAl/304SS
  Al_3Ta/Al_2O_3
  $MoSi_2$/Nb-1Zr
Composites (Ductile Particles)
  Nb in TiAl
  Fe, Mn in Ni_3Al
  Nb in $MoSi_2$

Source: Schwartz, M., *Emerging Technology*, Technomics. With permission.

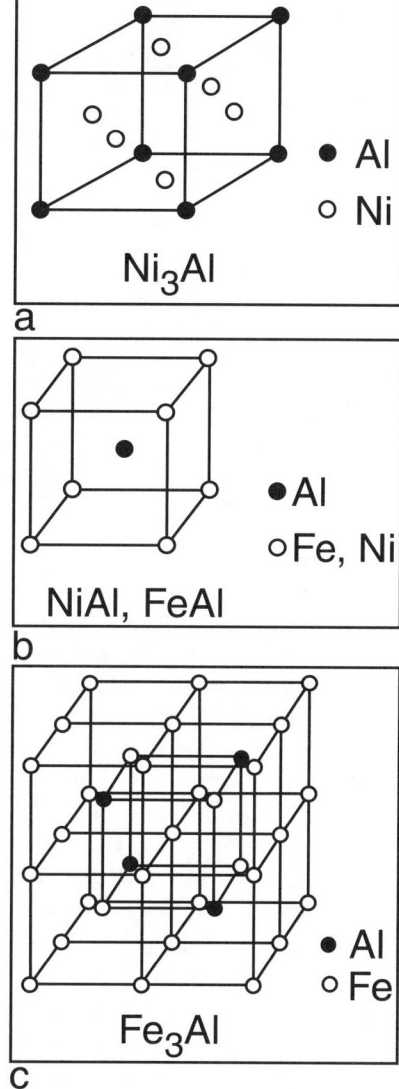

FIGURE A.3 The crystal structure of nickel and iron aluminides: (a) LI_2, (b) B_2, (c) DO_3. (From Schwartz, M., *Emerging Technology*, Technomics, 19. With permission.)

focused on aluminides of nickel, iron, and titanium. These aluminides possess many attributes that make them attractive for high-temperature structural applications. They contain enough aluminum to form, in oxidizing environments, thin films of aluminide oxides that often are compact and protective. They have low densities, relatively high melting points, and good high-temperature strength properties. For example, Ni_3Al shows an increase rather than a decrease in yield strength with increasing temperatures. The aluminides of interest are described in Table A.8.

In the range of 14 to 34% aluminum by weight occur the two intermetallic phases Ni_3Al and NiAl. The alloys are prepared by powder metallurgy and by casting techniques. Compacts of NiAl + 5% Ni, produced by powder techniques, exhibit room-temperature modulus of rupture values of 144,000 psi, and heat shock resistance is considered excellent. Good resistance to red and white fuming nitric acids is obtained.

The cast alloys of nickel and aluminum exhibit an increasing exothermic character with increasing aluminum content. Alloys with the exception of the 34% aluminum alloy and NiAl (31.5% Al) produce sound castings free of excessive porosity. A room-temperature tensile strength of approximately 49,000 psi is exhibited by the Ni_3Al compound with about 5%

TABLE A.8
Properties of Nickel and Iron Aluminides

Alloy	Crystal Structure[a]	Critical Ordering Temp. (°C)	Melting Point (°C)	Material Density (g/cm³)	Young's Modulus (Gpa)
Ni$_3$Al	L1$_2$ (ordered fcc)	1390	1390	7.50	178
NiAl	B2 (ordered bcc)	1640	1640	5.86	294
Fe$_3$Al	D0$_3$ (ordered bcc)	540	1540	6.72	141
	B2 (ordered bcc)	760	1540		
FeAl	B2 (ordered bcc)	1250	1250	5.56	260

[a] fcc = face-centered cubic; bcc = body-centered cubic.

Source: Schwartz, M., *Emerging Technology,* Technomics. With permission.

elongation. The 25% Al (NiAl) alloy has a room temperature tensile strength of 24,000 psi.

The 17.5% aluminum alloy that contained a mixture of the phases Ni$_3$Al and NiAl exhibits the best strength properties. Room-temperature tensile strength is approximately 80,000 psi with about 2% elongation, while at 815°C the tensile strength is about 50,000 psi. This alloy can be rolled at 1315°C, and possesses good thermal shock and oxidation resistance. Impact resistance is fair. The 100-h stress-to-rupture strength at 734°C is 14,000 psi, but, it should be noted, creep rates are high.

Ni$_3$Al is an intermetallic phase that forms at the nickel-rich end of the Ni–Al system and has a crystal structure (Figure A.3). Ni$_3$Al is the most important strengthening constituent of nickel-base superalloys. This is because the aluminide possesses an excellent elevated-temperature strength in addition to good oxidation resistance. Unlike conventional materials, Ni$_3$Al and its alloys show a yield anomaly; that is, its yield strength increases rather than decreases with increasing temperature.

Boron has been found to be most effective in improving the tensile ductility of Ni$_3$Al (<25% aluminum) tested in air at room temperature. Other tests indicate that boron-doped Ni$_3$Al is not sensitive to test environment at room temperature.

Boron-free Ni$_3$Al (24% Al) showed a ductility of only 8.2% in dry O$_2$ while boron-doped Ni$_3$Al exhibited 42.8%. This suggests that boron segregation also enhances the grain-boundary cohesion in Ni$_3$Al. In addition to boron, iron, chromium, and zirconium appear to enhance the grain-boundary properties and improve the tensile ductility of Ni$_3$Al at room temperature.

The mechanical properties of Ni$_3$Al strongly depend on deviation from alloy stoichiometry (i.e., aluminum/nickel ratio) and ternary alloying additions. The aluminide is capable of dissolving substantial alloying additions without forming second phases; as a result, its deformation and fracture behavior can be strongly altered by alloying.

Mechanical and metallurgical properties of Ni$_3$Al can be improved by alloying additions. Recent alloy design efforts have led to the development of Ni$_3$Al-base alloys with the following composition range (in atomic percent) for structural use at elevated temperatures in hostile environments: Ni–(14 to 18%) Al–(6 to 9%) Cr–(1 to 4%) Mo–(0.01 to 1.5%) Zr/Hf–(0.01 to 0.20%) boron.

NiAl containing more than about 41 at% forms a single-phase ordered B2 structure based on the body centered cubic (bcc) lattice (Figure A.3). In terms of thermophysical properties, B2 NiAl offers more potential for high-temperature structural applications than Ll NiAl. It has a higher melting point (1638°C), a substantially lower density (5.86 g/cm²), and a distinctly higher thermal conductivity, 76 W/(m × K), at ambient temperatures. In addition, NiAl has excellent oxidation resistance at high temperatures.

The structural use of NiAl suffers from two major drawbacks: poor fracture resistance at ambient temperatures and low strength and creep resistance at elevated temperatures. Single crystals of NiAl are quite ductile in compression, but both single-crystal and polycrystalline NiAl appear to be brittle in tension at ambient temperatures.

Because of the potential use of NiAl at high temperatures, considerable effort has been devoted to understanding brittle fracture and improving mechanical properties of NiAl during the past years.

The room-temperature tensile ductility of NiAl is increased from 1% to as high as 6% by adding about 0.2 at% iron. The ductility decreases sharply when the iron content is beyond 0.4 at%. A similar effect has been observed for molybdenum and gallium.

In addition to its widespread popularity as a model system for the basic understanding of a broad range of physical and mechanical properties, NiAl is a material with several potential engineering applications. Coatings currently comprise the principal engineering use of NiAl and NiAl-base alloys, and several other potential applications are in different stages of research and development.

APPLICATIONS

Coatings

NiAl is the basis of a family of oxidation- and corrosion-resistant coatings that have been used on nickel-base and cobalt-base superalloys over the past 30 years. NiAl coatings 25 to 100 mm thick are typically applied by a pack cementation process.

Both inward diffusion and outward diffusion coatings have been applied. The mechanisms of coating degradation include a depletion due to spallation and reaction with the substrate. A variety of alloying additions, most notably chromium and yttrium, have been applied to improve performance in thermal cycling and hot corrosion.

Turbine Blades and Vanes

The advantages associated with NiAl as a candidate structural material in advanced gas turbine engines include a 30% reduction in density over current nickel-base superalloys, good intrinsic environmental resistance, a thermal conductivity three to eight times larger than nickel-base superalloys (for improved cooling efficiency), and a melting temperature that is approximately 128°C higher than superalloys (for higher operating temperatures).

The processing, fabrication, and engineering design of NiAl, along with an understanding and improvement in impact properties, will also provide significant technical challenges that must be overcome before NiAl can be successfully employed as a structural material that may compete with current nickel-base superalloys. It is important to point out that a single crystal approach is crucial to obtaining a balance of both room-temperature ductility and high-temperature strength. Although NiAl-base composites may provide an alternative avenue to balanced properties, this approach is still in its infancy.

Electronic Applications

NiAl is an attractive material for metallizations on III–V semiconductors, and is also a critical component in complex "metal"/ III–V semiconductor heterostructures, such as enhanced-barrier Schottky contacts and semiconductor-clad metallic quantum wells. The characteristics that make NiAl an attractive candidate include a good lattice parameter match with several semiconductors (for epitaxy), chemical stability, good electrical conductivity, and a stable native oxide that allows patterning in air without destroying electrical continuity. An NiAl layer 3.3 mm thick is currently the thinnest independently contacted electrode in any solid-state device.

Among the alloys currently being considered as replacements for superalloys in high-temperature applications, some of the most promising are the intermetallic compounds in the Nb–Al system. The three compounds $NbAl_3$, Nb_2Al and Nb_3Al have melting temperatures in excess of many candidate replacement materials and densities that are superior to or competitive with those materials. Furthermore, the crystal structures, although complex, are reasonably well understood. In addition to these physical properties, there is a distinct processing consideration that leads to heightened interest in this system. The compounds of the Nb–Al system represent the highest melting temperature alloys that can be processed using currently active superalloy techniques, i.e., the various vacuum and hearth melting techniques.

The intermetallic compounds of the Nb–Al system, when systematically viewed as a whole, demonstrate both an orderly progression of high-temperature properties and a great deal of promise for future applications. Vickers microhardness as a function of temperature proceeds from $NbAl_3$, which is inferior to Ni_3Al, to Nb_3Al, which is the highest-melting-temperature compound with a temperature superiority of about 600°C over Ni_3Al.

Iron aluminides based on Fe_3Al and FeAl (Figure A.3) have excellent oxidation and corrosion resistance because they are capable of forming protective oxide scales at elevated temperatures in hostile environments. These aluminides exhibit corrosion rates lower than those of promising iron-based alloys (including coating material) by a couple orders of magnitude when tested in a severe sulfidizing environment at 800°C. The major drawbacks of the aluminides are their poor ductility and brittle fracture at ambient temperatures and poor strength and creep resistance above 600°C.

Iron aluminides have been known to be brittle at room temperature for more than 50 years; however, the major cause of their low ductility and brittle fracture was not identified until recently. The study of the environmental effect on tensile properties indicates that the poor ductility commonly observed in air tests is caused mainly by an extrinsic effect — environmental embrittlement.

The understanding of the cause of brittleness in iron aluminides has led to new directions in the design of ductile iron aluminide alloys. The schemes used to improve the ductility of Fe_3Al–FeAl alloys include control of surface conditions, reduction of hydrogen solubility and diffusion by alloy additions, refinement of grain structure by thermomechanical treatment, enhancement of grain-boundary cohesion by microalloying, and control of grain shape and recrystallization condition. The ductile Fe_3Al alloys show a distinctly high yield strength and a high ductility of 13 to 16% when tested in air at room temperature. Similarly, FeAl (35.8 at% Al) alloys show an increase in room-temperature ductility from 2 to 11% as a result of alloying additions (zirconium, molybdenum, and boron) and grain-structure refinement. This ductility further increased to 14% by formation of protective oxide scales on specimen surfaces through preoxidation at 700°C. The room-temperature ductility of Fe_3Al and FeAl can be significantly improved by refining grain structure via powder extrusion with or without second-phase particles (such as TiB_2). The strength of Fe_3Al and FeAl alloys at elevated temperatures can be improved by alloying with molybdenum, zirconium, niobium, titanium, and TiB_2.

Gamma alloys of titanium aluminide can be grouped into single-phase (γ)-alloys and two-phase ($\alpha_2 + \gamma$)-alloys (Table A.9). Single-phase alloys attracted attention initially because of their excellent resistance to environmental attack (e.g., oxidation and hydrogen absorption). In spite of a lack of progress toward overcoming their poor ductility and fracture toughness, however, interest in these engineering alloys has not diminished. Gamma titanium aluminide alloys of importance are thus two-phase alloys based on Ti -(45 to 49%) Al with appropriate combinations of alloying elements from these groups designated X_1, X_2, and X_3. (The footnotes in Table A.9 list the symbols of elements comprising the three groups.) The group X_1 elements increase ductility in two-phase alloys. For example, an addition of 1 to 3 at% V, Cr, Cr + Mn, or V + Cr to Ti–48Al nearly

TABLE A.9
Compositions of Multicomponent Gamma Titanium Aluminide Alloys

Alloy Class	Compositions (at%)
Single phase[a]	Ti-(50–52)Al-(1–2)X_2
Two phase[b]	Ti-(44–49)Al-(113)X_1-(1–4)X_2-(0.1–1)X_3

[a] X_2 = W, Nb, Ta.
[b] X_1 = V, Mn, Cr; X_2 = Nb, Ta, W, Mo; X_3 = Si, C, B, N, P, Se, Te, Ni, Mo, Fe.

Source: Schwartz, M., *Emerging Technology*, Technomics. With permission.

doubles the room-temperature ductility of the material.

Additions from either group X_1 or X_2 elements strengthen the alloys by solid solution strengthening, with chromium appearing to be the most effective and manganese the least.

The group X_1 elements are known to reduce oxidation resistance of gamma alloys. The X_2 elements do not increase the ductility of the Ti–48Al-base alloys; however, they are very effective in improving oxidation resistance. Such improvements in both oxidation resistance and solid solution strengthening are two reasons almost all multicomponent alloys developed to date contain at least 2 at% niobium. Small additions of the group X_3 elements have various effects wherein carbon and N_2 are known to improve creep resistance, silicon, boron, nickel, and iron decrease the melt viscosity, and silicon may yield some improvements in oxidation resistance and room-temperature ductility. Minor additions of phosphorus, selenium, or terbium have recently been shown to result in remarkably increased oxidation resistance.

Processing of Gamma Alloys

Gamma alloys may be processed by:

- Casting
- Ingot metallurgy (IM)
- Powder metallurgy (P/M)
- Sheet-forming

The XD™ process for producing gamma alloys containing TiB_{2p} can be classified separately, although the XD alloys are produced by either the casting process or the ingot metallurgy process. Each process can involve one or more routes. Figure 1.4 is a flowchart for typical processing methods and their specific routes.

The conventional method of producing foils from brittle materials was chemical milling. This approach involves suspending hot-rolled and surface-ground 0.050-mm-thick plates of the material in a pickling bath, and dissolving the metal until the plates are about 0.015 mm thick. For γ-TiAl, the chemical milling technique yields a poor surface finish on the foil.

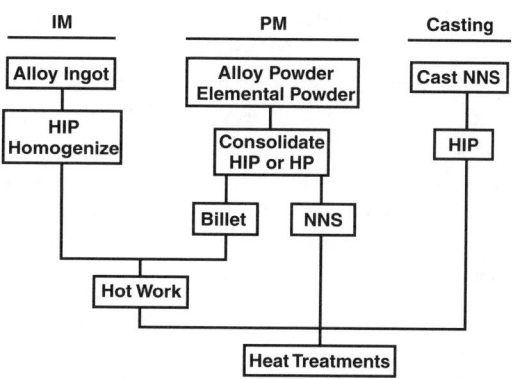

FIGURE A.4 Typical processing routes for gamma titanium aluminite alloys. HIP = hot isostatic pressing, HP = hot pressing, and NNS = near-net-shape. (From Schwartz, M., *Emerging Technology*, Technomics, 19. With permission.)

Engineers knew that the γ-TiAl, while brittle in tension, accepts high compression strains without failing. So they designed a new "isobaric" rolling mill to maintain triaxial compression on the material as it passes through the rolls. This proprietary process increases yield from the original ingot to around 50%, and provides thinner (0.03 to 0.05 mm) foils with smaller grains. As a result, isobaric cold rolling is a breakthrough technology. A cold-rolled and annealed γ-Ti-48-2-2 strip has a hardness of 260 H_v (Vickers).

Other processing methods include the self-propagating high-temperature synthesis (SHS) process, which starts with a structure of alternating layers of pure metal sheets or foils (e.g., iron, nickel, or titanium) and aluminum sheets or foils. The ductile metal layers are formed into the desired shape, either before or during the SHS reaction. The SHS reaction is initiated by heating in a vacuum. When the system reaches the initiation temperature (typically less than 660°C, the melting point of aluminum), an SHS reaction begins at the metal interface, i.e., between the aluminum and the iron, nickel, or titanium metals. The heat generated from the reaction melts the aluminum allowing rapid reaction with adjacent metal layers. That is, the SHS reaction occurs between the elemental metals producing well-bonded layers of unreacted elemental metal and metal–aluminum intermetallics. In this technique, the aluminum completely reacts, forming a titanium–aluminide–titanium, a nickel–aluminide–nickel, or an iron–aluminide–iron layered composite structure.

A second technique takes elemental powder mixtures that are placed between layers of an elemental metal foil or sheet, after which the entire layup is heated under pressure in a vacuum. Starting powders have been iron, nickel, or titanium mixed with aluminum. Starting elemental metal sheets have been titanium, nickel, or iron.

Combining P/M (Extrusion), Rolling, and SPF (Superplastic Forming)

A mixture of titanium and 34 wt% aluminum, i.e., 48 at%, was prepared by blending commercially available elemental powders with a maximum grain size of 100 mm compacted at room temperature in a uniaxial pressing machine to billets of 50 mm in diameter.

Titanium aluminide foils were produced using a combination of a standard P/M technique (extrusion), aluminum technology (rolling), and advanced titanium technology (SPF device). Claims have been made that since these techniques were available on a large scale, it should be possible to produce titanium aluminide foils in production.

A series of low-pressure turbine blades fabricated from titanium aluminide was successfully tested and this development could eventually result in reducing the weight of aviation gas turbines by hundreds of pounds. 49% Ti–47% Al–2% Cr–2% Nb blades, which have about half the weight of comparable components made from conventional nickel-base alloys, were run in the fifth low-pressure turbine stage of a CF6-80C2 power plant and during the tests the blades went through 1000 flight cycles. Compared with traditional nickel-base alloys, the material has half the density and is comparable in strength up to about 760°C. The titanium aluminide also is about 50% stiffer than conventional titanium alloys. If used in the low-pressure section of new aircraft engines as blade material, the titanium aluminide could cut engine weight by more than 136.01 kg.

The 98 blades used in the tests weighed 217 g. Comparable nickel alloy components weigh 383 g. In an ideal situation, the lighter blades would also allow turbine wheels to be lighter and less robust because the reduced-weight blades create lower stresses during operation. The titanium aluminide alloy used to fabricate the blades were solid airfoils and were cast using conventional foundry techniques.

Future IM Developments

New titanium aluminide alloys, stronger and tougher than conventional α_2(Ti$_3$Al) materials, are based on the ordered orthorhombic-phase Ti$_3$AlNb (Ti–25 at% Al-25 at% Nb).

The materials are potential weight-saving alternatives to nickel-base superalloys in relatively low temperature (650°C) aircraft gas-turbine engine applications. Several examples include exhaust-nozzle structures, compressor casings, and various compressor components; while the density of Ti$_3$AlNb-base alloys (and other titanium aluminides) is less than two thirds that of nickel-base superalloys, such as Alloy 718, the lower weight can be translated into higher thrust-to-weight ratios or gains in fuel efficiency.

Of two new alloys, the Ti–22% Al–27% Nb has the best combination of high-temperature strength and room-temperature ductility and fracture toughness. The alloy is much stronger than conventional α_2 titanium aluminides such as Ti–24% Al–11% Nb and Ti–25% Al–10% Nb–3% V–1 % Mo. It also has a higher strength-to-weight ratio, even though it is slightly denser. For example, Ti–22% Al–27% Nb has a factor-of-two advantage in strength-to-weight ratio over Ti–25% Al–10% Nb–3% V–1% Mo, an α_2-alloy, from room temperature to 650°C. Fracture toughness (K_{Ic}) of Ti–22% Al–27% Nb averages about 28 MPa m$^{1/2}$, which is higher than those of the α_2-alloys Ti–25% Al–10% Nb–3% V–1% Mo (K_{Ic} = 14 MPa m$^{1/2}$) and Ti–24% Al–17% Nb–1% Mo (19 MPa m$^{1/2}$). Creep behavior of the Ti$_3$AlNb-base alloy is competitive with lower toughness titanium aluminides that have been optimized for creep resistance.

ALUMINUM

Called aluminium in England, aluminum is a white metal with a bluish tinge (symbol Al, atomic weight 26.97), obtained chiefly from bauxite. It is the most widely distributed of the elements next to O_2 and silicon, occurring in all common clays. Aluminum metal is produced by first extracting alumina (aluminum oxide) from the bauxite by a chemical process. The alumina is then dissolved in a molten electrolyte, and an electric current is passed through it, causing the metallic aluminum to be deposited on the cathode. The metal was discovered in 1727, but was obtained only in small amounts until it was reduced electrolytically in 1885.

Pure (99.99%) aluminum has a specific gravity of 2.70 or a density of 2685 kg/m^3, a melting point of 660°C, electrical and thermal conductivities about two thirds that of copper, and a tensile modulus of elasticity of 62,000 MPa. The metal is nonmagnetic, highly reflective, and has a face-centered cubic (fcc) crystal structure. Soft and ductile in the annealed condition, it is readily cold-worked to moderate strength. It resists corrosion in many environments as the result of the presence of a thin aluminum oxide film.

Iron, silicon, and copper are the principal impurities in commercially pure aluminum, wrought products containing at least 99% aluminum and a foil product containing at least 99.99%. Such unalloyed aluminum is available in a wide variety of mill forms and constitutes the aluminum 1XXX series in the designation system for wrought aluminum and aluminum alloys. Annealed sheet is quite ductile — 35 to 45% tensile elongation — but weak, having a tensile yield strength of 27 to 35 MPa. Cold reduction of 75% increases yield strength to 124 to 166 MPa. Thus, the unalloyed metal is used far more for its electrical, thermal, corrosion-resistant, and cosmetic characteristics than for its mechanical properties. Applications include electrical and thermal conductors, capacitor electrodes, heat exchangers, chemical equipment, packaging foil, heat and light reflectors, and decorative trim.

Aluminum flake enhances the reflectance and durability of paints. Aluminum powder is used for powder-metal parts. Aluminum powder and aluminum paste are used in catalysts, soaps, explosives, fuels, and thermite welding. Aluminum shot is used to deoxidize steel, and aluminum foam, made by foaming the metal with zirconium hydride or other hydrides, is an effective core material for lightweight structures.

Anodized aluminum is aluminum with a hard aluminum oxide surface imparted electrolytically using the metal as the anode. The coating, which is much thicker than the naturally formed aluminum oxide film, provides additional corrosion and weathering resistance in certain environments, but is generally no more protective in acidic or alkaline solutions outside the 4 to 8.5 pH range. The coating is nonconductive, wear resistant, and can be produced in various colors, thus enhancing decorative appeal. Whereas the naturally formed oxide film is less than a millionth of an inch thick, anodized coatings range from 0.005 to 0.008 mm for bright auto trim to as much as 0.030 mm for architectural applications.

ALUMINUM ALLOYS

Alloying aluminum with various elements markedly improves mechanical properties, strength primarily, at only a slight sacrifice in density, thus increasing specific strength, or strength-to-weight ratio. Traditionally, wrought alloys have been produced by thermomechanically processing cast ingot into mill products such as billet, bar, plate, sheet, extrusions, and wire. For some alloys, however, such mill products are now made by similarly processing "ingot" consolidated from powder. Such alloys are called PM (powder metal) wrought alloys or simply PM alloys. To distinguish the traditional type from these, they are now sometimes referred to as ingot-metallurgy (IM) alloys or ingot-cast alloys. Another class of PM alloys are those used to make PM parts by pressing and sintering the powder to near-net shape. There are also many cast alloys. All told, there are about 100 commercial aluminum alloys.

There are two principal kinds of wrought alloys: (1) heat-treatable alloys — those strengthened primarily by solution heat treatment or solution heat treatment and artificially aging (precipitation hardening), and (2) non-heat-treatable alloys — those that depend primarily on cold work for strengthening. Alloy designations are a continuation of the four-digit system noted for aluminum, followed with a letter to designate the temper or condition of the alloy: F (as-fabricated condition), O (annealed), H (strain-hardened), W (solution heat-treated and unstable; that is, the alloy is prone to natural aging in air at room temperature), and T (heat-treated to a stable condition). Numerals following T and H designations further distinguish between tempers or conditions. T3, for example, refers to alloys that have been solution heat-treated, cold-worked, and naturally aged to a substantially stable condition. T6 denotes alloys that have been solution heat-treated and artificially aged. H designations are followed by two or three digits. The first (1, 2, or 3) indicates a specific sequence of operations applied. The second (1 to 8) refers to the degree of strain hardening (the higher the number, the greater the amount of strain, or hardening, 8 corresponding to the amount induced by a cold reduction of about 75%). The third (1 to 9) further distinguishes between mill treatments (Table A.10).

The aluminum alloy 2XXX series is characterized by copper (2.3 to 6.3%) as the principal alloying element. Most of these alloys also contain lesser amounts of magnesium and manganese and some may contain small amounts of other ingredients, such as iron, nickel, titanium, vanadium, zinc, and zirconium. 2XXX alloys are strengthened mainly by solution heat treatment, sometimes by solution heat treatment and artificial aging. Among the more common, especially for structural aircraft applications, are aluminum alloys 2014, 2024, and 2219, which can be heat-treated to tensile yield strengths in the range of 276 to 414 MPa. A recently developed alloy for auto body panels is aluminum alloy 2036, which in the T4 temper provides a tensile yield strength of about 195 MPa. Aluminum alloy 2XXX series is not as corrosion resistant as other aluminum alloys and thus is often clad with a thin layer of essentially pure aluminum or a more corrosion-resistant aluminum alloy, especially for aircraft applications.

Manganese (0.5 to 1.2%) is the distinguishing alloying element in the aluminum alloy 3XXX series. Aluminum alloy 3003 also contains 0.12% copper. Aluminum alloys 3004 and 3105 also contain 1 and 0.5% magnesium, respectively. Strengthened by strain hardening, 3XXX alloys provide maximum tensile yield strengths in the range of 186 to 248 MPa and are used for chemical equipment, storage tanks, cooking utensils, furniture, builders' hardware, and residential siding.

The aluminum alloy 4XXX series is characterized by the addition of silicon: about 12%, for example, in aluminum alloy 4032 and 5% in aluminum alloy 4043. Aluminum alloy 4032, which also contains 1% magnesium and almost as much copper and nickel, is heat-treatable, providing a tensile yield strength of about 315 MPa in the T6 temper. This, combined with its high wear resistance and low thermal expansivity, has made it popular for forged engine pistons. Aluminum alloy 4043, which is alloyed only with silicon, is a strain-hardenable alloy used for welding rod and wire. Some Al–Si alloys are also used for brazing, others for architectural applications. Their appeal for architectural use stems from the dark gray color they develop in anodizing.

The principal alloying element in the aluminum alloy 5XXX series is magnesium, which may range from about 1 to 5%, and is often combined with lesser amounts of manganese and/or chromium. Like 3XXX alloys, the 5XXX are hardenable only by strain hardening and all are available in a wide variety of H tempers. Tensile yield strengths range from less than 69 MPa in the annealed condition to more than 276 MPa in highly strained conditions. Aluminum alloys 5083, 5154, 5454, and 5456 are widely used for welded structures, pressure vessels, and storage tanks, others for more general applications, such as appliances, cooking utensils, builders' hardware, residential siding, auto panels and trim, cable sheathing, and hydraulic tubing.

The aluminum alloy 6XXX series is characterized by modest additions (0.4 to 1.4%) of silicon and magnesium and can be strengthened by heat treatment. Except for auto sheet aluminum alloys 6009 and 6010, which are typically supplied and used in the T4 temper, the alloys are strengthened by solution heat treatment and artificial aging. As a class, these alloys are intermediate in strength to aluminum alloy 2XXX and aluminum alloy 7XXX but provide good overall fabricability. The auto sheet alloys are relatively new. More traditional among the

TABLE A.10
Aluminum Wrought Alloys

Designation	1100		2011		2014		2024		2036	2219		3003		3004		5052		5083		6061		6066		6262	7050	7075
Temper	O	H18	T3	T8	O	T6	O	T4	T4	O	T62	O	H18	O	H38	O	H38	O	H321	O	T6	O	T6	T9	T74	T6
Yield strength (10³ psi)	5	22	43	45	14	60	11	47	28	11	42	6	27	10	36	13	37	21	33	8	40	12	52	55	70	73
Tensile strength (10³ psi)	13	24	55	59	27	70	27	68	49	25	60	16	29	26	41	28	42	42	46	18	45	22	57	58	81	83
Shear strength (10³ psi)	9	13	32	35	18	42	18	41	30	—	—	11	16	16	21	18	24	25	—	12	30	14	34	35	45	48
Fatigue limit (10³ psi)	5	9	18	18	13	18	13	20	18	—	15	7	10	14	16	16	20	—	23	9	14	—	16	13	21	23
Elongation in 2 in. (%)	45	15	15	12	18	13	22	19	24	—	—	40	10	25	6	30	8	22	16	30	17	18	12	10	10	11
Modulus of elasticity (10⁶ psi)	10.0	10.0	10.2	10.2	10.6	10.6	10.6	10.3	10.6	10.6	—	10.0	10.0	10.0	10.0	10.2	10.2	10.3	10.3	10.0	10.0	10.0	10.0	10.0	10.3	10.4
Melting temperature (°F)	1,190–1,215	1,190–1,215	1,005–1,190	1,005–1,190	945–1,080	945–1,080	935–1,180	935–1,180	1,030–1,200	1,010–1,190	1,010–1,190	1,190–1,210	1,190–1,210	1,165–1,210	1,165–1,210	1,125–1,200	1,125–1,200	1,095–1,180	1,095–1,180	1,080–1,205	1,080–1,205	1,045–1,195	1,045–1,195	1,080–1,205	910–1,165	890–1,175
Coefficient of thermal expansion (10⁻⁶ in./in.-°F)	13.1	13.1	12.7	12.7	12.8	12.8	12.9	12.9	13.0	12.4	12.4	12.9	12.9	13.3	13.3	13.2	13.2	13.2	13.2	13.1	13.1	12.9	12.9	13.0	12.8	13.1
Thermal conductivity (Btu-in./h-ft²-°F)	1,540	1,510	1,050	1,190	1,340	1,070	1,340	840	1,100	1,190	840	1,340	1,070	1,130	1,130	960	960	810	—	1,250	1,160	1,070	1,020	1,190	1,092	900
Electrical resistivity (Ohm-cir mil/ft)	18	18	27	23	21	26	21	35	25	24	35	21	26	25	25	30	30	36	—	22	24	26	28	24	35	31

Source: *Mach. Design Basics Eng. Design*, June, p. 54–55, 1993. With permission.

dozen or so alloys of this kind are aluminum alloy 6061, which is used for truck, marine, and railroad-car structures, pipelines, and furniture, and aluminum alloy 6063 for furniture, railings, and architectural applications. Other applications include complex forgings, high-strength conductors, and screw-machine products.

Zinc is the major alloying element in the aluminum alloy 7XXX series, and it is usually combined with magnesium for strengthening by heat treatment. An exception is aluminum alloy 7072, which is alloyed only with 1% zinc, is hardenable by strain hardening, and is used for fin stock or as a clad for other aluminum alloys. The other alloys, such as aluminum alloys 7005, 7049, 7050, 7072, 7075, 7175, 7178, and 7475, contain 4.5 to 7.6% Zn, 1.4 to 2.7 Mg, and, in some cases, may also include copper, magnesium, silicon, titanium, or zirconium. These heat-treatable alloys are the strongest of aluminum alloys, with tensile yield strengths of some exceeding 483 MPa in the T6 temper. They are widely used for high-strength structures, primarily in aircraft.

CAST ALUMINUM ALLOYS

Of some 40 or so standard casting alloys, 10 are aluminum die-casting alloys. The others are aluminum sand-casting alloys and/or aluminum permanent-mold-casting alloys. Some of the latter also are used for plaster-mold casting, investment casting, and centrifugal casting. Although the die-casting alloys are not normally heat-treated, those for sand and/or permanent-mold casting often are, usually by solution heat treatment and artificial aging. Both heat-treatable and non-heat-treatable alloys are available. The heat-treatable alloys, which can be solution heat-treated and aged similar to wrought heat-treatable grades, carry the temper designations T2, T4, T5, T6, or T7. Die castings are seldom solution heat-treated because of the danger of blistering.

Table A.11 shows the nominal compositions of several important casting alloys. The major alloying addition is silicon, which improves the castability of aluminum and provides moderate strength. Other elements are added primarily to increase the tensile strength. Most die castings are made of alloy 413.0 or 380.0. Alloy 443.0 has been very popular in architectural work, and 355.0 and 356.0 are the principal alloys for sand casting. Number 390.0 is employed for die-cast automotive engine cylinder blocks, and alloy F332.0 is used for pistons for internal combustion engines.

Casting alloys are significant users of secondary metal (recovered from scrap for reuse). Thus, casting alloys usually contain minor amounts of a variety of elements; these do no

TABLE A.11
Nominal Composition and Casting Procedure for Common Aluminum Casting Alloys

Alloy	Form[a]	Si	Cu	Mg
413.0	D	12	—	—
B 443.0[b]	B, C	5.3	0.15 max	—
F 332.0[b]	C	9.5	3.0	1.0
355.0	B, C	5.0	1.3	0.5
356.0	B, C	7.0	—	0.3
380.0	D	8.5	3.5	—
390.0	C, D	17	4.5	0.55

[a] B = sand casting; C = permanent mold casting; and D = die casting.
[b] The letter indicates modifications of alloys of the same general composition or differences in impurity limits, from alloys having the same four-digit numerical designations.

Source: McGraw-Hill Encyclopedia of Science and Technology, 8th ed., Vol. 1, McGraw-Hill, New York, 532. With permission.

harm as long as they are kept within certain limits. The use of secondary metal is also of increasing importance in wrought alloy manufacturing as producers take steps to reduce the energy that is required in producing fabricated aluminum products.

Aluminum alloy 380.0 and its modifications constitute the bulk of die-casting applications. Containing 8.5% Si, 3.5% Cu, and as much as 2% Fe, it provides a tensile yield strength of about 165 MPa, good corrosion resistance, and is quite fluid and free from hot shortness, and thus is readily castable. Engine cylinder heads, typewriter frames, and various housings are among its applications. Aluminum alloy 390.0, a high-silicon (17%) Cu–Mg–Zn alloy, is the strongest of the die-casting alloys, providing a tensile yield strength of about 240 MPa as cast and 260 MPa in the T5 temper. Like high-silicon wrought alloys, it also features low thermal expansivity and excellent wear resistance. Typical applications include auto engine cylinder blocks, brake shoes, compressors, and pumps requiring abrasion resistance. Aluminum alloys 383.0, 384.0, 413.0, and A413.0 provide the best die-filling capacity, and have excellent resistance to hot cracking and die sticking. Aluminum alloy 518.0 provides the best corrosion resistance, machinability, and polishability, but is more susceptible to hot cracking and die sticking than the other alloys.

Sand and/or permanent-mold casting alloys are encompassed in each of the aluminum alloy 2XX.X to 8XX.X series. As-cast, tensile yield strengths range from about 97 MPa for 208.0 to 200 MPa for aluminum alloy A390.0. In various solution-treated and aged conditions, several alloys can provide tensile yield strengths exceeding 280 MPa. Strongest of the alloys — 415 MPa in the T7 temper and 435 MPa in the T6 — is aluminum alloy 201.0, which contains 0.7% Ag and 0.25% Ti in addition to 4.6% Cu and small amounts of magnesium and manganese. The T7 temper is suggested for applications requiring stress-corrosion resistance. Applications for 201.0 include aircraft and ordnance fittings and housings; engine cylinder heads; and pistons, pumps, and impellers. Other more commonly used high-strength alloys are aluminum alloys 354.0, 355.0, and 356.0, which are alloyed primarily with silicon, copper, magnesium, manganese, iron, and zinc, and which are used for auto and aircraft components. Premium-quality aluminum-alloy castings are those guaranteed to meet minimum tensile properties throughout the casting or in specifically designated areas.

Aluminum–Lithium Alloys

Aluminum–lithium (Al–Li) alloys are a significant recent development in high-strength wrought aluminum alloys. Lithium is the lightest metal in existence. For each weight percent of lithium added to aluminum, there is a corresponding decrease of 3% in the weight of the alloy. Therefore, Al–Li alloys offer the attractive property of low density. Because of the very low density of lithium, every 1% of this alkali metal can provide a 3% reduction in density and a 10% increase in stiffness-to-density ratio relative to conventional 2XXX and 7XXX alloys. Thus, these alloys, such as aluminum alloys 2090 and 2091 (there are several others, including proprietary ones), are extremely promising for future aircraft applications, see Table A.12. These alloys are comparable in strength with some of the strongest traditional alloys. Some are ingot-cast products, others PM-wrought alloys.

A second beneficial effect of lithium additions is the increase in elastic modulus (stiffness). Also, as the amount of lithium is increased, there is a corresponding increase in strength due to the presence of very small precipitates that act as strengthening agents to the aluminum. As the precipitates grow during heat treatment, the strength increases to a limit, then begins to decrease. Al–Li alloys therefore come under the classification of precipitation-strengthening alloys.

With the exception of the PM alloy AA 5091 in Table A.12, all the current commercially available Al–Li alloys are produced by direct-chill casting, and require a precipitation-aging heat treatment to achieve the required properties. In Al–Li alloys containing greater than 1.3% (by weight) of lithium, the intermetallic phase δ-Al_3Li precipitates upon natural or artificial aging, but the associated strengthening effect is insufficient to meet the medium or high

TABLE A.12
Aluminum–Lithium Alloys

Product Form	Alloy	Temper of End Use Condition and Definition	Property Category
Sheet	AA8090	T81-Solution treated, quenched, stretched, and precipitation heat-treated to an underaged condition	Damage tolerant: low strength but enhanced toughness and impact resistance
	AA8090	T621-Re-solution treated by the user, quenched and precipitation heat-treated to an underaged condition (same aging time and temperature as for T81)	Re-solution treatment by the user to promote enhanced formability
	AA8090	T8-Solution treated, quenched, stretched, and precipitation heat-treated to a near-peak aged condition	Medium strength where some reduction in toughness and impact resistance compared to T81 and T621 can be tolerated
Extruded sections (solid and hollow)	AA8090	T8511-Solution treated, quenched, controlled-stretched, and precipitation heat-treated to a near-peak aged condition	Medium/high strength
Die forgings	AA8090	T852-Solution treated, quenched, cold compressed, and precipitation heat-treated to a near-peak aged condition	Medium/high strength
	AA5091	H112—As forged	Medium/high strength but where cold compression is impractical or insufficient to achieve properties if made from AA8090

Source: Adv. Mater. Proc., October, pp. 41–43, 1992. With permission.

strength levels usually required (the damage tolerant temper in AA 8090 is an exception).

Cold-working operation is a prime requirement for IM alloys containing relatively high levels of lithium, such as AA 8090 or AA 2090, if optimum medium-strength or high-strength properties are to be achieved. Cold working is readily applied to sheet, plate, and extrusions by stretching, although it must be recognized that in most cases, this can only be carried out by the metal manufacturer.

Alloy AA 5091 is made by the PM process of mechanical alloying (MA). This technology avoids the limitations arising from the need to cold-compress alloys AA 8090 and AA 2090 as forgings, because it is dispersion-strengthened during manufacture. Through careful selection of the Al–Li–Mg base composition, a single-phase solid solution is achieved, and the need for heat treatment is eliminated.

The MA process is inherently more expensive than direct-chill-casting and is not a practical route for producing large rectangular-section ingots for sheet or plate rolling. However, it is useful for certain parts in which the configuration of the forged components would not allow adequate cold compression to be achieved in alloys AA 8090 or AA 2090.

POWDER METALLURGY (PM) ALLOYS

Elements such as iron, cobalt, and nickel have been added to aluminum in large quantities to produce alloys that have relatively stable structures at elevated temperatures and higher elastic moduli than conventional aluminum alloys.

The PM process is a rapid solidification process that involves the transformation of finely divided liquid either in the form of droplets or thin sheets that become solid at high solidification rates (on the order of 10^4 K/s). Solidification occurs as heat is removed from the molten metal.

TABLE A.13
Properties of Selected Beryllium and Aluminum Aerospace Alloys

Alloy	Elastic Modulus, Gpa (Msi)	Density, g/cm^3 (lb/in.^{3})	Thermal Conductivity, W/m-K	CTE, ppm/K
AlBeMet 162	200 (29)	2.1 (0.076)	210	13.9
Al 6061	69 (10)	2.8 (0.101)	170	23.6
Beryllium	300 (44)	1.8 (0.067)	210	11.5
Aluminum–lithium	90 (13)	2.5 (0.092)	120	23.6

Source: Adv. Mater. Proc., October, pp. 37–39, 1997. With permission.

Rapid-solidification-process aluminum alloys containing iron and cerium appear to result in alloys that have refined particle-size distribution and improved high-temperature stability. These PM alloys appear competitive with titanium up to 191°C.

Aluminum PM parts are made mainly from aluminum alloys 201 AB and 601 AB. Both are copper, silicon, and magnesium compositions that can be heat-treated to the T4 or T6 tempers after sintering. 201 AB is the stronger, having 324 MPa tensile yield strength for 95% dense material in the T6 temper as compared with about 241 MPa for 601 AB for these conditions.

MECHANICAL ALLOYING (MA)

This process circumvents the limitations of conventional ingot casting. Blends of powders are mixed in a ball mill. A drum is mounted horizontally and half-filled with steel balls and blends of elemental metal powders. As the drum rotates, the balls drop on the metal powder. The degree of homogeneity of the powder mixture is determined by the size of the particles in the mixture.

ALUMINUM–BERYLLIUM ALLOYS (ALBEMET)

AlBeMetAE is an alloy of aluminum and beryllium. The Al–62Be alloy has the modulus of steel at 200 GPa, and density of 2.1 g/cm^3, approximately 20% that of steel, which gives a specific modulus of 950×10^6 cm. Table A.13 shows a comparison of the properties of aluminum, beryllium, Al–Li and AlBeMet.

Powder processing for AlBeMet is well established. First, inert-gas atomization produces prealloyed powder. The powder is cold isostatically pressed to a density of about 80% of theoretical, then consolidated by extrusion.

The alloy varies from 60–65% Be and 35–40% Al. The alloy is formed by melting spherical beryllium and aluminum powder and then is atomized and the metals are mixed. As the metal mixture cools, an alloy with a matrix-like structure is formed. The resulting material is extruded and can be rolled or formed with hot isostatic pressing. AlBeMet also can be dip-brazed or welded.

The extruded bar may be fabricated directly into parts, or cut for cross-rolling into sheet. Sheet and plate thicknesses of 7.6 mm to 1.6 mm are produced by rolling, followed by grinding to final gauge. This sheet has essentially isotropic properties in the sheet plane.

A range of applications is made economically possible by the extrusion process, which has produced extrusions with diameters up to 25 cm. For example, profiled AlBeMet extrusions have been fabricated into computer hard disk drive arms, with over 1 million parts machined from extruded bar stock.

AlBeMet is also produced as hot isostatically pressed block. This is quite useful for prototype parts, and for components in which strength and ductility are of secondary importance to stiffness, density, modulus, thermal conductivity, and thermal expansion. The input powder for hot isostatic pressing is the same as that for wrought product manufacture. The process is carried out after degassing to remove adsorbed species such as water vapor.

The physical properties of AlBeMet 162 are largely independent of processing route. However, they are critically dependent on the beryllium content of the alloy. Although AlBeMet 162 with 62 wt% Be is the most common composition, others can be quite useful. The most striking physical properties are modulus and density. The modulus is substantially higher, and the density is significantly lower than conventional materials such as Al–Li, which is typically 2.6 g/cm^3.

Unlike its physical properties, the mechanical properties of AlBeMet 162 are strongly dependent upon the processing route.

To meet many of the economic requirements imposed by modern designs, net shape or near-net shape processes are required. Investment casting has long been recognized as a process capable of producing parts to near net shape. However, only recently has the capability of investment casting been developed for Al–Be alloy parts. AlBeCast 910 is the first of a family of Al–Be investment casting alloys.

The properties of AlBeCast 910 include 0.2% yield strength of 124 MPa, ultimate tensile strength of 180 MPa, and elongation of 3%.

In addition to conventional processes such as casting and mechanical working, significant progress has been made in adapting new technologies to AlBeMet. The semisolid processing of Al–Be alloys enables the net shape and high production capability of permanent mold die casting for a material whose high melting point and reactivity preclude conventional die casting. As a prototype, a sabot from a 30-mm projectile was semisolid-formed, and the uniformity and repeatability of the process was proved out in a semiproduction run of approximately 2500 parts.

The high modulus-to-density ratio, 3.8 times that of aluminum or steel, minimizes flexure and reduces the chance of mechanically induced failure. Therefore, AlBeMet plays a dual role in its avionics applications.

The stiffness, low density, and thermal characteristics of Al–Be alloys combine to produce the highest performance brake components for the demanding Formula I race car circuit, specifically the brake caliper from extrusion. It is the stiffness that is particularly important in such parts, because excessive bending of the caliper leads to reduced braking efficiency. In addition, the thermal management characteristics allow higher and more consistent braking performance than possible with conventional materials.

As a structural material, AlBeMet 162 sheet is being qualified as the rudder material for advanced versions of a U.S. Air Force tactical fighter. In addition, as an essentially isotropic material, it allows simplified design, reduced weight, and increased performance.

Space structures also benefit from AlBeMet, including the OrbComm vehicle. On this satellite, the face sheets for the honeycomb bus (the circular portion of the structure), the brackets that hold the bus to the payload, and the long boom are all fabricated AlBeMet structures.

Aluminum–Scandium Alloys (AlSc)

An AlSc-base alloy, Sc7000, has been developed for use as bicycle frame tubing, shock absorbers, and handlebars. The alloy is 50% stronger than aluminum bicycle alloys and reduces the weight in the bicycle frame by 10 to 12%.

New and Special Alloys

Aluminum alloy 7033-T6 is 50% stronger than AA 6061-T6 and provides higher fatigue performance. Developed to provide improved fracture toughness for forgings, the alloy offers corrosion and stress-corrosion properties nearly equal to those of AA 6061-T6, and superior to those of AA 2014-T6.

Ultimate tensile strength of AA 7033 in the longitudinal direction is 517 MPa, vs. 338 MPa for AA 6061. Fracture toughness in the L-S orientation is 66 MPa m$^{1/2}$, and in the S-L orientation is 41 MPa m$^{1/2}$. This compares with 36 MPa m$^{1/2}$ in the L-S direction and 36 MPa m$^{1/2}$ in the S-L direction for 6061-T6. Close control of alloy chemistry and processing enables products to develop consistent microstructures with typical grain size of 100 µm or less.

Aluminum Powder

Aluminum powder is produced by drawing a stream of the molten metal through an atomizing nozzle and impinging that stream with compressed air or inert gas, solidifying and disintegrating the metal into small particles, which are then drawn into a collection system and screened, graded, and packaged. Particle sizes range from −325 mesh (fine) to +200 mesh (granules). In the process, the metal reacts with oxygen in the air and moisture, causing a thin film of Al_2O_3 to form on the surface of the particles. Oxide content, which increases with decreasing particle size, ranges from 0.1 to 1.0% by weight.

First used to produce aluminum flake by ball milling for paint pigments, aluminum powder now finds many other uses: ferrous and nonferrous metals production, P/M parts, P/M-wrought aluminum alloy mill products, coatings for steel, asphalt-roof products, spray coatings, and vacuum metallizing. Other applications include rocket fuels and explosives, incendiary bombs, pyrotechnics, signal flares, and heat and magnetic shields. It is also used in permanent magnets, high-temperature lubricants, industrial cements, chalking compounds, printing inks, and cosmetic and medical products.

AMALGAM

An amalgam is a combination of a metal with mercury (Hg). The amalgams have the characteristic that when slightly heated they are soft and easily workable, and they become very hard when set. They are used for filling where it is not possible to employ high temperatures.

Dental amalgams are prepared by mixing mercury with finely divided alloys composed of varying proportions of silver, tin, and copper. Amalgams used in dental work require the following composition: Ag, 65% minimum, Cu, 6% maximum, Zn, 2% maximum, and Sn, 1% minimum.

AMINE

An amine is a member of a group of organic compounds that can be considered as derived from ammonia by replacement of one or more hydrogens by organic radicals. Generally, amines are bases of widely varying strengths, but a few that are acidic are known.

Amines constitute one of the most important classes of organic compounds. The lone pair of electrons on the amine nitrogen enables amines to participate in a large variety of reactions as a base or a nucleophile. Amines play prominent roles in vitamins, for example, epinephrine (adrenaline), thiamin or vitamin B_1, and Novocaine.

Amines are used to manufacture many medicinal chemicals, such as sulfa drugs and anesthetics. The important fiber nylon is an amine derivative.

Aramids are synthetic fibers produced from long-chain polyamides (nylons) in which 85% of the amide linkages are attached directly to two aromatic rings. The fibers are exceptionally stable and have good strength, toughness, and stiffness, which is retained well above 150°C. Two aramids are Nomex and Kevlar from Du Pont Co., and Twaron from Akzo NV. They have high strength, intermediate stiffness, and are suitable for cables, ropes, webbings, and tapes. Kevlar 49, with high strength and stiffness, is used for reinforcing plastics. Nomex, best known for its excellent flame and abrasion resistance, is used for protective clothing, air filtration bags, and electrical insulation.

AMINO

Melamine and urea are the principal commercial thermosetting polymers called aminos. The amino resins are formed by an addition reaction of formaldehyde and compounds containing NH_2 amino groups. They are supplied as liquid or dry resins and filled molding compounds. Applying heat in the presence of a catalyst converts the materials into strong, hard products. Aminos are used as molding compounds, laminating resins, wood adhesives, coatings, wet-strength paper resins, and textile-treating resins.

Moldings made from amino compounds are hard, rigid, abrasion resistant, and have high resistance to deformation under load. They have excellent electrical insulation characteristics. Melamines are superior to ureas in resistance to acids, alkalies, heat, boiling water, and for applications involving wet/dry cycling.

Melamines and ureas are not resistant to strong oxidizing acids or strong alkalies, but they can be used safely with conventional household chemicals such as naphtha and detergents. They are unaffected by organic solvents such as acetone, carbon tetrachloride, ethyl alcohol, heptane, and isopropyl alcohol. Petroleum, paraffin hydrocarbons, gasoline, kerosene, motor oil, aromatic hydrocarbons, and fluorinated hydrocarbons (Freon) have no apparent effect on urea and melamine moldings. Dimensional stability is good, but moldings do swell and shrink slightly in varying moisture conditions.

AMORPHOUS METALS

Also known as metallic glasses, amorphous metals are produced by rapid quenching of molten metal–metalloid alloys, resulting in a noncrystalline grain-free structure in the form of ribbon or narrow strips. They are extremely strong and hard yet reasonably ductile and quite corrosion resistant. Perhaps of greatest interest is the magnetic performance of several complex iron-base compositions, called Metglas. Because of their very low hysteresis and power losses, they can markedly reduce the size of transformer cores traditionally made of silicon steels. Several nickel-alloy compositions also are available for use as brazing foils.

ANODIC COATINGS

Anodic oxidation or anodizing is the common commercial term used to designate the electrolytic treatment of metals in which stable oxide films or coatings are formed on surfaces. Aluminum and magnesium are anodized to the greatest extent on a commercial basis. Some other metals such as zinc, beryllium, titanium, zirconium and thorium can also be anodized to form films of varying thicknesses but they are not used to any large extent commercially.

Aluminum

It is a well-known fact that a thin oxide film forms on aluminum when it is exposed to the atmosphere. This thin, tenacious film provides excellent resistance to corrosion. The ability of aluminum to form an adherent oxide film led to the development of electrochemical processes to produce thicker and more effective protective and decorative coatings.

Anodic coatings can be formed on aluminum in a wide variety of electrolytes utilizing either AC or DC current or combinations of both. Electrolytes of sulfuric, oxalic, and chromic acids are considered to be the most important commercially. Other electrolytes such as borates, citrates, chromates, bisulfates, phosphates, and carbonates can also be used for specialized applications.

Anodic coatings produced in a sulfuric acid type of electrolyte are generally translucent, can be produced with a wide range of properties by varying operating techniques, and have a greater number of pores that are smaller in diameter than other anodic coatings.

Coatings obtained by the chromic acid anodic process are much thinner than those produced by the sulfuric acid process based upon the same time of anodic oxidation. Although the coatings are thin, they provide high resistance to corrosion because of the presence of chromium compounds in combination with their relatively thick barrier layer. Because the conventional anodic coatings produced in a chromic acid electrolyte are thin, they have low resistance to abrasion but a high degree of flexibility.

The chromic acid anodic process is critical with respect to alloy composition. In general, the process is not recommended for aluminum alloys containing more than 7% Si or 4.5% Cu. Alloys containing over 7% total alloying elements are also not recommended. It is difficult to form films on these alloys, particularly on casting alloys. Anodic solution (surface attack) will generally occur unless the processing conditions are controlled very carefully.

On most aluminum alloys, anodic coatings of this type do not require sealing. However, sealing in a boiling dichromate or a dilute chromic acid solution produces coatings with the best resistance to salt spray corrosion on AA2014-T6 and AA7075-T6 alloys.

Anodic oxidation in dilute oxalic acid solutions produces coatings that are essentially transparent. Their color varies from a light yellow to bronze. Such coatings are dense and have little absorptive capacity but possess high resistance to abrasion.

Hard Coatings

Procedures have been developed that produce finishes of greater thickness and density than the conventional anodic coatings. They have high resistance to abrasion, erosion, and corrosion. The coatings have thicknesses in the range of 0.1 to 5 mils, depending upon the application.

Hard coatings are popular for applications requiring light weight in combination with high resistance to wear, erosion, and corrosion. These applications include helicopter rotor blade surfaces, pistons, pinions, gears, cams, cylinders, impellers, turbines, and many others. Also, due to their attractive gray color, hard anodic coatings are now being used for architectural applications.

The processing conditions for obtaining hard coatings are such that thick coatings with maximum density can be obtained on most aluminum alloys. The selection of alloy is of utmost importance.

Variations of the anode oxidation process that produce conventional hard anodic coatings also form thick, dense, colored anodic coatings for architectural applications. Attractive bronze, gray, or black coatings are obtained by utilizing certain organic acids as electrolytes.

Other Coatings

Utilizing oxalic acid in combination with titanium, zirconium and thorium salts for the electrolyte produces a dense oxide coating that has an opaque, light-gray appearance.

Anodic oxidation of aluminum in sulfamic acid produces coatings that are denser than those produced by the sulfuric acid process. This anodic process is expensive owing to the high cost of sulfamic acid.

Alternating current may be used to form anodic coatings on aluminum and alloys with all of the electrolytes previously mentioned, but the aluminum surface is anodic only half of the cycle so that an oxide coating is formed only at half the rate of coatings formed with direct current.

Superimposed AC–DC has also been used for anodizing and has produced hard, thick, anodic coatings.

Alloy Selection

The previous discussion on the anodic oxidation characteristics of various electrolytes was based upon the use of relatively pure aluminum. Alloy selection is important. Even aluminum of 99.3% minimum purity, such as the AA1100 alloy, is in a sense an aluminum alloy from the standpoint of anodic oxidation, because the other elements present have an effect on the characteristics of the coating.

The response of the different constituents of aluminum alloys varies considerably and is an important factor. Some constituents will be dissolved by the anodic reaction, whereas others may be unaffected. For alloys where the constituents are dissolved during the anodic treatment, the coatings will have voids that decrease the density of the coating and also lower its resistance to corrosive action and abrasion. Al–Cu alloys are an example of this type.

Silicon in Al–Si alloys is an example of a constituent that is unaffected by conventional anodic oxidation. The silicon particles remain unchanged in the coating in their original position.

Some constituents of aluminum alloys will themselves oxidize under anodic oxidation and the oxidation products will color the coating. For example, constituents such as manganese will produce a brownish opaque appearance due to the presence MnO_5. Also, chromium as a constituent will give a yellowish tint to the coating from oxidation products of chromium.

The lower the concentration of constituents present or the purer the aluminum, the more continuous and transparent will be the oxide coating. The so-called superpurity (99.99%) aluminum produces the most transparent oxide coating.

Properties and Applications

The oxide coatings produced by anodizing have many properties that make them commercially important. Anodic coatings are essentially Al_2O_3, which is a very hard substance. Because the Al_2O_3 is integral with the surface it will not chip or peel from the surface; this outstanding characteristic is useful for architectural applications. The combination of high wear resistance and attractive satin sheen of the finish makes it a logical choice for aluminum hardware, handrails, moldings, and numerous other architectural components. Because the anodic finish reproduces the texture of the surface from which it is formed, a wide variety of attractive effects are possible by variation in the surface preparatory procedures. Many commercial architectural applications can use a natural aluminum finish, but for applications where it is desirable to preserve initial appearance, the anodic finish will require less maintenance.

Because the oxide coating is brittle compared with the aluminum underneath, it will crack if the coated article is bent. It is possible, however, to produce oxide coatings that are relatively flexible. In general, the thicker coatings formed in sulfuric and oxalic acid electrolytes will crack or craze to a much greater extent than oxide coatings formed in a chromic acid electrolyte. For many applications this cracking may not be objectionable, because it is usually difficult to detect by visual observation. These fine cracks have an adverse effect on the bending properties of the metal, however, and may sometimes cause fracture if the bends are severe. For this reason, it is generally recommended that the finish be applied after forming.

If fatigue is a critical factor, then proper allowance must be made for the reduction in endurance limits produced by relatively thick oxide coatings. Fatigue tests indicate that a coating 0.1 mil thick on smooth surfaces will have little effect on fatigue strength. Thicker coatings in the range of 0.3 to 0.5 mil on smooth surfaces have a slight detrimental effect at high stresses.

Anodic coatings also provide substantial protection against corrosion. There are many factors that must be considered in this connection such as the continuity of the coating and the choice of alloy. Since continuity is dependent upon constituents present in the alloy, anodic coatings on high-purity aluminum are the most resistant to corrosion. On the other hand, anodic coatings formed from Al–Cu alloys have much lower resistance to corrosion.

Sealing of anodic coatings in dichromate solutions results in an appreciable improvement in resistance to corrosion, particularly by chlorides. The results of atmospheric exposure tests indicate that anodic coatings with a thickness of 0.4 mil or greater will provide greatly increased resistance to weathering.

The ability of anodic coatings to absorb coloring substances such as dyestuffs and pigments makes it possible to obtain finishes in a complete range of colors including black. The colors are unique because the luster of the underlying metals gives them a metallic sheen that is particularly attractive for applications that simulate metals such as gold, copper, bronze, and brass. Colorants are available that, when used to color anodic coatings on the proper alloys, will reproduce the natural colors of the metals listed above. Colored finishes can be used in a wide range of applications for nameplates, panels, appliance trim, optical goods, cameras, fishing reels, instruments, giftware, and jewelry.

Anodic coatings have good electrical insulating properties. Anodic films produced in boric acid electrolytes are used commercially on aluminum foil for electrical capacitors. The voltage necessary to break down the anodic coatings is generally proportional to the thickness.

Anodic oxidation in oxalic acid and phosphoric acid electrolytes produces coatings that have been successfully used as preparatory treatments for electroplating. Copper, nickel, cadmium, silver and iron have been successfully deposited over oxide coatings. Plating solutions that are highly alkaline should be avoided as they attack the coating and destroy the porous structure necessary for the best adhesion.

A "Krome-Alume" process utilizes an oxalic acid electrolyte to form the oxide coating, and subsequent modification of the coating with hydrofluoric acid produces a structure satisfactory for electroplating. Furthermore, the anodic oxidation process utilizing a phosphoric acid electrolyte produces a structure that requires no further modification to condition it for electroplating. The alloy has an important effect on coating structure and, in general, the phosphoric anodic process is not recommended for preparing high-purity aluminum, Al–Mg wrought alloys, and most die-casting alloys for electroplating.

ANODIC COATINGS

Production Methods

Anodic coatings are applied to aluminum and its alloys by a variety of methods, including batch, bulk, continuous conveyor, and continuous strip.

The batch method of applying anodic coatings is similar to that used in electroplating except that the parts are anodic instead of cathodic. The continuous conveyor method is also similar to the conventional plating method.

The bulk method for applying anodic coatings to small parts such as rivets, washers, and screws is radically different from bulk electroplating methods. The barrel-finishing method is not suitable for applying anodic coatings because the initial flow of current forms an anodic coating on the parts and even if they contact each other during the rotation of the barrel, no current will flow because the coating is an insulator. The bulk methods employed for anodic coatings utilize special perforated nonmetallic cylindrical containers. Pressure, applied to the parts in the container through a threaded center contact post, maintains the initial contact between the surface of the parts.

The continuous strip process has been used commercially to apply anodic coatings to aluminum sheet that is subsequently roll-formed into weather strip. In Europe, this same process is used to apply anodic coatings to aluminum sheet that is formed into food containers such as sardine cans.

Magnesium

Although all of the magnesium alloys in commercial use today have good resistance to corrosion, many parts are provided with maximum resistance to corrosion and abrasion through electrolytic treatments based on anodic treatments. The anodic treatments produce relatively thick and dense coatings with excellent adhesion and high resistance to corrosion and abrasion. As in the case of aluminum, anodic coatings on magnesium alloys are an excellent base for lacquers and enamels.

It is well known that magnesium alloys are attacked by most inorganic and organic acids. However, because magnesium alloys are resistant to alkalies, fluorides, borates, and chromates, the electrolytes for anodizing are generally based upon these chemicals.

The simplest electrolyte for anodizing magnesium alloys is a 5% caustic soda solution. This electrolyte is used in a temperature range of 60 to 70°C. A voltage of 5 to 6 is satisfactory; anodic oxidation time is generally 30 min. All magnesium alloys will respond to this treatment. The coatings produced are approximately 0.3 mil thick and are essentially crystalline magnesium hydroxide. They have relatively high resistance to abrasion and are gray or tan in color in the as-formed condition. The coatings may be colored for decorative applications by immersion in water-soluble dyestuffs in much the same way as the coloring procedures used for anodic coatings on aluminum alloys. If maximum resistance to corrosion is required, immersion (sealing) in a 5% sodium chromate solution at 77 to 82°C is recommended.

Another anodizing treatment for magnesium alloys capable of producing a coating with many desirable properties is the "H.A.E." process. This process is also based upon an alkaline electrolyte. All magnesium alloys, both wrought and cast, respond to this process to give coatings with excellent resistance to corrosion, high dielectric strength, and high hardness.

Other Metals

Thin oxide films may be formed on beryllium by anodic oxidation in an electrolyte composed of 10% nitric acid containing 200 g/l of chromic acid. The anodic oxidation is carried out at approximately 25 A/ft^2. Such films retard high temperature oxidation and corrosion.

Thin oxide films can also be formed on zirconium, titanium and thorium by anodizing in an electrolyte composed of 70% glacial acetic and 30% nitric acid. The corrosion resistance of titanium is substantially improved by anodizing for 10 to 15 min in a 15 to 22% (by weight) sulfuric acid solution (room temperature) at approximately 18 V DC. Anodic coatings on titanium are also used as a base for lubricants.

ZINC

Much of the work on anodic coatings for zinc has been conducted in alkaline electrolytes consisting of a two-stage process. The electrolyte for the first oxidation treatment is conducted in an alkali–carbonate solution followed by anodizing in a silicate solution.

ANTIMONY

Antimony is a bluish-white metal, symbol Sb, with a crystalline scalelike structure that exhibits poor electrical and heat conductivity. It is brittle and easily reduced to powder. It is neither malleable nor ductile and is used only in alloys or in its chemical compounds. Like arsenic and bismuth, it is sometimes referred to as a metalloid, but in mineralogy it is called a semimetal. The element is available commercially in 99.999+% purity and is finding increasing use in semiconductor technology.

Antimony is produced either by roasting the sulfide with iron, or by roasting the sulfide and reducing the sublimate of Sb_4O_6 thus produced with carbon; high-purity antimony is produced by electrolytic refining. Antimony is one of the few elements that exhibits the unique property of expanding on solidification. Antimony is ordinarily stable and not readily attacked by air or moisture. Under controlled conditions it will react with O_2 to form oxides. The chief uses of antimony are in alloys, particularly for hardening lead-base alloys.

Antimony imparts hardness and a smooth surface to soft-metal alloys, and alloys containing antimony expand on cooling, thus reproducing the fine details of the mold. This property makes it valuable for type metals. When alloyed with lead, tin, and copper, it forms the babbitt metals used for machinery bearings. It is also much used in white alloys for pewter utensils. Its compounds are used widely for pigments.

ARAMID (AROMATIC POLYAMIDE)

Aramid fibers are characterized by excellent environmental and thermal stability, static and dynamic fatigue resistance, and impact resistance. These fibers have the highest specific tensile strength (strength/density ratio) of any commercially available continuous-filament yarn. Aramid-reinforced thermoplastic composites have excellent wear resistance and near-isotropic properties — characteristics not available with glass or carbon-reinforced composites.

Aramid fiber, trade-named Kevlar, is available in several grades and property levels for specific applications. The grade designated simply as Kevlar is made specifically to reinforce tires, hoses, and belting, such as V-belts and conveyor belts.

Kevlar 29 is similar to the basic Kevlar in properties but is designated specifically for use in ropes and cables, protective apparel, and as the substrate for coated fabrics. In short fiber or pulp form, Kevlar 29 can substitute for asbestos in friction products or gaskets. Fabrics of Kevlar 29 can be made into bullet-resistant vests. Clothing made from Kevlar 29 can be as heat resistant as that made from asbestos and also be extremely cut resistant.

Kevlar 49 has half the elongation (2.5%) and twice the modulus (117,885 MPa) of Kevlar 29. Applications are principally in reinforcing plastic compounds used in lightweight aircraft boat hulls and sports equipment. Composites containing Kevlar are also used as interior panels and secondary structural parts, such as fairings and doors on commercial aircraft.

Kevlar 149 is a highly crystalline aramid that has a modulus of elasticity 40% greater than that of Kevlar 49 and a specific modulus nearly equal to that of high-tenacity graphite fibers. It is used to reinforce composites for aircraft components.

Nomex aramid fiber is characterized by excellent high-temperature durability with low shrinkage. It will self-extinguish and does not melt, retaining a high percentage of its initial strength at elevated temperatures. It is available as continuous-filament yarn, staple, and tow. Nomex is used in military and civilian protective apparel, dry gas filtration, rubber reinforcement, and industrial fabrics. Nomex aramid fibers are also available as a paper for use in high-temperature electrical insulation and in resilient, corrosion-proof honeycomb core for aerospace and other transportation applications.

ARC DISCHARGE

Arc discharge is a type of electrical conduction in gases characterized by high current density and low potential drop. A typical arc runs at a voltage drop of 100 V with a current drain of 10 A. The arc has negative resistance — the voltage drop decreases as the current increases — so a stabilizing resistor or inductor in series is required to maintain it. The high-temperature gas rises like a hot-air balloon while it remains anchored to the current-feeding electrodes at its ends. It thereby acquires an upward-curving shape, which accounts for its being called an arc.

There are many applications of such an intensely hot object. The brilliant arc and the incandescent carbon adjacent to it form the standard light source for movie theater projectors. The electronic flash-gun in a camera uses an intense pulsed arc in xenon gas, simulating sunlight. Since no solid-state material can withstand this temperature for long, the arc is used industrially for welding steel and other metals. Alternatively, it can be used for cutting metal very rapidly. Electric arcs form automatically when the contacts in electrical switches in power networks are opened, and much effort goes into controlling and extinguishing them. Lightning is an example of a naturally occurring electric arc.

ARGON

A chemical element (symbol Ar), argon is the third member of the gaseous elements called the noble, inert, or rare gases, although argon is not actually rare. The earth's atmosphere is the only commercial argon source; however, traces of the gas are found in minerals and meteorites.

Argon is colorless, odorless, and tasteless. The element is a gas under ordinary conditions, but it can be liquefied, solidified readily, and is a major industrial gas.

Argon does not form any chemical compounds in the ordinary sense of the word, although it does form some weakly bonded clathrate compounds with water, hydroquinone, and phenol. There is one atom in each molecule of gaseous argon.

Most argon is produced in air-separation plants. Air is liquefied and subjected to fractional distillation. Because the boiling point of argon is between that of N_2 and O_2, an argon-rich mixture can be taken from a tray near the center of the upper distillation column. The argon-rich mixture is further distilled and then warmed and catalytically burned with H_2 to remove O_2. A final distillation removes H_2 and N_2, yielding a very high purity.

The oldest large-scale use for argon is in filling electric lightbulbs. Welding and cutting metal consumes the largest amount of argon. Metallurgical processing constitutes an important application.

Argon and Ar–Kr mixtures are used, along with a little mercury vapor, to fill fluorescent lamps. The inert gases make the lamps easier to start, help to regulate the voltage, and supplement the radiation produced by the excited mercury vapor.

Argon mixed with a little neon is used to fill luminous electric-discharge tubes employed in advertising signs (similar to neon signs) when a blue or green color is desired instead of the red color of neon.

Argon is used to fill the space between the panes of higher-quality double-pane windows, reducing heat transfer by gaseous conduction by about 30% compared to air filling.

Argon is also used in gas-filled thyratrons. Geiger–Müller radiation counters, ionization chambers that measure cosmic radiation, and electron tubes of various kinds. Argon atmospheres are used in dry boxes during manipulation of very reactive chemicals in the laboratory and in sealed-package shipments of such materials. In high-energy physics research, a tank of liquid argon can form a calorimeter to detect certain subatomic particles.

AROMATIC HYDROCARBON

A hydrocarbon with a chemistry similar to that of benzene. Aromatic hydrocarbons are either benzenoid or nonbenzenoid. Benzenoid aromatic hydrocarbons contain one or more benzene rings and are by far the more common and the more important commercially.

Nonbenzenoid aromatic hydrocarbons have carbon rings that are either smaller or larger than the six-membered benzene ring. Their importance arises mainly from a theoretical interest in understanding those structural features that impart the property of aromaticity.

Benzenoid aromatic hydrocarbons are also called arenes. Benzene itself is the prototypical arene.

ARSENIC

Arsenic (symbol As) is a soft, brittle, poisonous element of steel-gray color and metallic luster. In atomic structure it is a semimetal, lacking plasticity, and is used only in alloys and in compounds. The bulk of the arsenic used is employed in insecticides, rat poisons, and weed killers, but it has many industrial uses, especially in pigments. It is also used in poison gases for chemical warfare.

Metallic arsenic is stable in dry air. When exposed to humid or moistened air, the surface first becomes coated with a superficial golden bronze tarnish, which on further exposure turns black. On heating in air arsenic will vaporize and burn to As_2O_3.

Dermatitis can result from handling arsenical compounds; hence, it is desirable to use impervious gloves or protective creams. The best preventative for dermatitis is strict personal hygiene. In areas where arsenical dusts and fumes are present, effective exhaust ventilation is necessary, or when impractical, a respirator should be used.

In the field of electronics, high-purity arsenic is finding an important use as a constituent of the III–V semiconductor compounds. The compounds of most interest are indium arsenide, which may be used for Hall effect devices, and gallium arsenide used for making diodes, transistors, and solar cells.

ARTIFICIAL INTELLIGENCE

Artificial intelligence is the subfield of computer science concerned with understanding the nature of intelligence and constructing computer systems capable of intelligent action, abbreviated AI. It embodies the dual means of furthering basic scientific understanding and of making computers more sophisticated in the service of humanity.

AI is primarily concerned with symbolic representations of knowledge and heuristic methods of reasoning, that is, using common assumptions and rules of thumb. Two examples of problems studied in AI are planning how a robot, or person, might assemble a complicated device, or move from one place to another; and diagnosing the nature of a person's disease, or of a machine's malfunction, from the observable manifestations of the problem. In both cases, reasoning with symbolic descriptions predominates over calculating.

ARTIFICIALLY LAYERED STRUCTURES

Manufactured, reproducibly layered structures with layer thicknesses approaching interatomic distances. Modern thin-film techniques are at a stage at which it is possible to fabricate these structures, also known as artificial crystals or superlattices, opening up the possibility of engineering new desirable properties into materials.

These structures serve as model systems and as a testing ground for theoretical models and for other naturally occurring materials that have similar structures. For example, ceramic superconductors consist of a variable number of conducting CuO_2 layers intercalated by various other oxide layers, and therefore, artificially layered structures may be used to study predictions of the behavior of suitably manufactured materials of this class.

ASBESTOS

Any of six naturally occurring minerals characterized by being extremely fibrous (asbestiform), strong, and incombustible. They are utilized in commerce for fire protection; as reinforcing material for tiles, plastics, and cements; for friction materials; and for thousands of other uses. Because of a great concern over the health effects of asbestos, many countries have promulgated strict regulations for its use. The six minerals designated as asbestos also occur in a nonfibrous form. In addition, there are many other minerals that morphologically mimic asbestos because of their fibrous nature.

The six naturally occurring minerals exploited commercially for their desirable physical properties, which are in part derived from their asbestiform habit, are chrysotile asbestos, a member of the serpentine mineral group, and grunerite asbestos, riebeckite asbestos, anthophyllite asbestos, tremolite asbestos, and actinolite asbestos, all members of the amphibole mineral group. Individual mineral particles, however processed and regardless of their mineral name, are not demonstrated to be asbestos if the average length-to-width ratio is less than 20:1.

Chrysotile is a hydrated magnesium silicate from the serpentine mineral group. Amosite and erocidolite are iron silicates from the amphibole mineral group. Anthophyllite is a magnesium silicate of the amphibole group with the magnesium isomorphically replaced by varying amounts of iron and aluminum.

The important characteristics of the asbestos minerals that make them unique are their fibrous form; high strength and surface area; resistance to heat, acids, moisture, and weathering; and good bonding characteristics with most binders such as resins and cement.

Asbestos is used for many types of products because of its chemical and thermal stability, high tensile strength, flexibility, low electrical conductivity, and large surface area. Past uses of asbestos, such as sprayed-on insulation, where the fibers may become easily airborne, have been generally abandoned. Asbestos is used predominantly for the construction industry in the form of cement sheets, coatings, pipes, and roofing products. Additional important uses are for reinforcing plastics and tiles, for friction materials, and packings and gaskets.

- *Asbestos fabrics* are often woven mixed with some cotton. For brake linings and clutch facings the asbestos is woven with fine metallic wire.
- *Asbestos shingles*, for fireproof roofing and siding of houses, are normally made of asbestos fibers and portland cement formed under hydraulic pressure.
- *Asbestos board* is a construction or insulating material in sheets made of asbestos fibers and portland cement molded under hydraulic pressure. Ordinary board for the siding and partitions of warehouses and utility buildings is in the natural mottled-gray color, but pigmented boards are also marketed in various colors.
- *Asbestos lumber* is asbestos-cement board molded in the form of boards for flooring and partitions, usually with imitation wood grain molded into the surface. Asbestos siding, for house construction, is grained to imitate cypress or other wood and is pigmented with titanium oxide to give a clear white color. Asbestos roofing materials may also be made with asphalt or other binder instead of cement.
- *Asbestos paper* is a thin asbestos sheeting made of asbestos fibers bonded usually with a solution of sodium silicate. It is strong, flexible, and white in color, and is fireproof and a good heat insulator. For covering steam pipes and for insulating walls, it is made in sheets of two or three plies. For wall insulation it is also made double with one-corrugated sheet to form air pockets when in place.

ASPHALT

Asphalt refers to varieties of naturally occurring bitumen. Asphalt is also produced as a petroleum byproduct. Both substances are black and largely soluble in carbon disulfide. Asphalts are of variable consistency, ranging from a highly viscous fluid to a solid.

Asphalt is derived from petroleum in commercial quantities by removal of volatile components. It is an inexpensive construction material used primarily as a cementing and waterproofing agent.

Asphalt is composed of hydrocarbons and heterocyclic compounds containing N_2, sulfur, O_2; its components vary in molecular weight from about 400 to 5000. It is thermoplastic and viscoelastic; at high temperatures or over long loading times it behaves as a viscous fluid; at low temperatures or short loading times as an elastic body.

The three distinct types of asphalt made from petroleum residues and their uses are described in Table A.14.

TABLE A.14
Asphalts and Their Uses

Asphalt Type and % of Production	Manufacturing Process	Properties	Uses
Straight-run, 70–75%	Distillation or solvent precipitation	Nearly viscous flow	Roads, airport runways, hydraulic works
Air-blown, 25–30%	Reacting with air at 204–316°C	Resilient: viscosity less susceptible to temperature change than straight-run	Roofing, pipe coating, paints, underbody coatings, paper laminates
Cracked, less than 5%	Heating to 427–538°C	Nearly viscous flow; viscosity more susceptible to temperature change than that of straight-run asphalt	Insulation board saturant, dust laying

Source: McGraw-Hill Encyclopedia of Science and Technology; 8th ed., Vol. 2, McGraw-Hill, New York, 176. With permission.

B

BABBITT METAL

The original name for tin–antimony–copper (Sn–Sb–Cu) white alloys used for machinery bearings, the term now applies to almost any white bearing alloy with either tin or lead (Pb) base. The alloy consists of 88.9% tin, 7.4% antimony, and 3.7% copper. This alloy melts at 239°C. It has a Brinell hardness of 35 at 21°C, and 15 at 100°C. As a general-utility bearing metal, the original alloy has never been improved greatly, and makers frequently designate the tin-base alloys close to this composition as genuine babbitt.

Commercial white bearing metals now known as babbitt are of three general classes: tin-base, with more than 50% tin hardened with antimony and copper, and used for heavy-duty service; intermediate, with 20 to 50% tin, with lower compressive strength and more sluggish behavior as a bearing; and lead-base, made usually with antimonial lead with smaller amounts of tin together with other elements to hold the lead in solution.

Copper hardens and toughens the alloy and raises the melting point. Lead increases fluidity and raises antifriction qualities, but softens the alloy and decreases its compressive strength. Antimony hardens the metal and forms hard crystals in the soft matrix, which improve the alloy as a bearing metal.

Alloys containing up to 1% arsenic (As) are harder at high temperatures and are fine-grained, but arsenic is used chiefly for holding lead in suspension. Zinc (Zn) increases hardness but decreases frictional qualities, and with much zinc the bearings are inclined to stick. Even minute quantities of iron (Fe) harden the alloys, and iron is not used except when zinc is present. Bismuth (Bi) reduces shrinkage and refines the grain, but lowers the melting point and lowers the strength at elevated temperatures. Cadmium (Cd) increases the strength and fatigue resistance, but any considerable amount lowers the frictional qualities, lowers the strength at higher temperatures, and causes corrosion. Nickel (Ni) is used to increase strength but raises the melting point. The normal amount of copper in babbitts is 3 or 4%, at which point the maximum fatigue-resisting properties are obtained with about 7% antimony. More than 4% copper tends to weaken the alloy, and raises the melting point. When the copper is very high, tin–copper crystals are formed and the alloy is more a bronze than a babbitt.

Because of increased speeds and pressures in bearings and the trend to lighter weights, heavy-cast babbitt bearings are now little used despite their low cost and the ease of casting the alloys. The alloys are used mostly as antifriction metals in thin facings on steel backings, with the facing usually less than 0.03 cm thick, in order to increase their ability to sustain higher loads and dissipate heat.

BARIUM

Barium (symbol Ba) is a metallic element that occurs in combination in the minerals witherite and barite, which are widely distributed. The metal is silvery white in color and can be obtained by electrolysis from the chloride, but it oxidizes so easily that it is difficult to obtain in the metallic state. Its melting point is 850°C, and its specific gravity 3.78. The most extensive use of barium is in the form of its compounds. The salts that are soluble, such as sulfide and chloride, are toxic. An insoluble, nontoxic barium sulfate salt is used in radiography. Barium compounds are used as pigments, in chemical manufacturing, and in deoxidizing alloys of tin, copper, lead, and zinc. Barium is introduced into lead-bearing metals by electrolysis to harden the lead.

Barium is also a key ingredient in ceramic superconductors.

BEARING MATERIALS

A large variety of metals and nonmetallic materials in monolithic and composite (laminate) form are used for bearings. Monolithic ferrous bearings are made of gray cast iron, pressed and sintered iron and steel powder, and many wrought steels, including low-carbon and high-carbon plain-carbon steels, low-alloy steels, alloy steels, stainless steels, and tool steels. Most cast-iron bearings are made of gray iron because it combines strength with the lubricity of graphitic carbon. Pressed and sintered bearings can be made to controlled porosity and impregnated with oil for lubricity. Because of its wide use in ball and roller bearings, one of the best-known bearing steels is AISI (American Iron and Steel Institute) 52100 steel, a through-hardening 1% carbon–1.3 to 1.6% chromium (Cr) alloy steel. Many steels, however, are simply surface-hardened for bearing applications. In recent years, the performance of bearing steels has been markedly improved by special melting practices that reduce the presence of nonmetallic inclusions.

Monolithic nonferrous bearings include copper (90%)–zinc (10) bronze, leaded bronzes, and unleaded bronzes, and an aluminum–tin (Al–Sr) alloy, containing about 6% tin as the principal alloying element. The bronze and aluminum alloy provide similar load-bearing capacity and fatigue resistance, but the bronze is somewhat better in resistance to corrosion by fatty acids that can form with petroleum-base oils. It is also less prone to seizure and abrasion from mating shafts; more able to embed foreign matter and, thus, prevent shaft wear; and more tolerant of shaft misalignment.

Applications include auto engine starter-motor bearings, or bushings, for the copper-zinc bronze; auto engine connecting-rod bearings for the aluminum alloy; and various bearings in motors, machine tools, and earth-moving equipment for the tin bronzes.

Monolithic bearings are also made of cemented tungsten (W) and chromium carbides, plastics, carbon-graphite, wood, and rubber. Plastics provide good combinations of inherent lubricity, corrosion resistance, and adequate strength at room to moderately elevated temperatures. Thermal conductivity and other performance features that may be required can be provided by metal and other fillers. Plastic bearings can be made of acetal, nylon, polyester, polytetrafluoroethylene, polysulfone, polyphenylene sulfide, polyimide, and polyamide-imide. Carbon-graphite bearings are more heat resistant but rather brittle, thus limited to nonimpact applications. Wood bearings are made of maple and the hard lignum vitae. Rubber bearings, usually steel-backed, are used for bearings requiring resilience.

Nonferrous metals are widely used in dual- or tri-metal systems. Dual-metal bearings comprise a soft, thin, inner liner metallurgically bonded to stronger backing metal. Steel lined with bronze containing 4 to 10% lead provides the highest load-bearing capacity, 55 MPa, or about twice that of the bronze alone, and fatigue strength. However, the aluminum alloy with a steel backing provides the best corrosion resistance and only moderately less load-bearing capacity. Tin and lead babbitt linings excel in surface qualities conducive to free-sliding conditions and are used with either steel, bronze, or aluminum alloy backings.

Dual-metal systems cover a gamut of bearings for motors, pumps, piston pins, camshafts, and connecting rods.

Tri-metal bearings, all with steel backings, have an inner liner of tin or lead babbitt and an intermediate layer of a more-fatigue-resistant metal, such as leaded bronze, copper–lead, aluminum–tin, tin-free aluminum alloys, silver (Ag), or silver–lead. Load-bearing capacity ranges from 10 to 83 MPa. The silver bearing systems provide the best combination of load-bearing capacity, fatigue and corrosion resistance, and compatibility to mating materials, but a lead babbitt, medium-lead bronze, and steel system is a close second, sacrificing only a moderate reduction in corrosion resistance but at a reduction in cost. Applications include connecting-rod, camshaft, and main bearings in auto engines and reciprocating aircraft engines.

BENZENE

A colorless, liquid, inflammable, aromatic hydrocarbon which boils at 80.1°C and freezes at 5.4 to 5.5°C. Benzene is used as a solvent and particularly in Europe as a constituent of motor fuel. In the United States the largest uses of benzene are for the manufacture of styrene and phenol. Other important outlets are in the production of chlorinated benzenes (used in DDT and moth flakes), and benzene hexachloride, an insecticide.

BERYL

At least 50 different beryllium-bearing minerals are known, but in only about 30 is it a regular constituent. The only beryllium-bearing mineral of industrial importance is beryl, a hexagonal beryllium–aluminum silicate with the ideal composition $Be_3Al_2Si_6O_{18}$, equivalent to 5% beryllium or 14% beryllium oxide, 19% Al_2O_3, and 67% SiO_2. The precious forms of beryl, emerald and aquamarine, approach the ideal composition. The emerald is a transparent, intensely green variety of beryl.

BERYLLIUM

Among structural metals, beryllium (symbol Be) has a unique combination of properties. It has low density (two thirds that of aluminum), high modulus per weight (five times that of ultrastrength steels), high specific heat, high strength per density, excellent dimensional stability, and transparency to x-rays. Beryllium is expensive, however, and its impact strength is low compared to values for most other metals.

Beryllium is a steel-gray lightweight metal, used mainly for its excellent physical properties rather than its mechanical properties. Except for magnesium (Mg), it is the lightest in weight of common metals, with a density of 1855 kg/m³. It also has the highest specific heat (1833 J/kg K) and a melting point of 1290°C. It is nonmagnetic, has about 40% the electrical conductivity of copper, a thermal conductivity of 190 W/m K, high permeability to x-rays, and the lowest neutron cross section of any metal having a melting point above 500°C. Also, its tensile modulus (28.9×10^4 MPa) is far greater than that of almost all metals.

Ultimate tensile strength ranges from 228 to 690 MPa and tensile elongation from 1 to 40%, depending on mill form. Thus, because of its low density, beryllium excels in specific strength, and especially in specific stiffness. However, tensile properties, especially elongation, are extremely dependent on grain size and orientation, and are highly anisotropic so that results based on uniaxial tensile tests have little significance in terms of useful ductility in fabrication or fracture toughness in structural applications. From these standpoints, the metal is considered to be quite brittle. Ductility, as measured by elongation in tensile tests, increases with increasing temperature to about 400°C, then decreases above about 500°C. Although resistant to atmospheric corrosion under normal conditions, beryllium is attacked by O_2 and N_2 at elevated temperatures and certain acids, depending on concentration, at room temperature.

Available forms include block, rod, sheet, plate, foil, extrusions, and wire. Machining blanks, which are machined from large vacuum hot pressings, make up the majority of beryllium purchases. However, shapes can also be produced directly from powder by processes such as cold-press/sinter/coin, CIP/HIP, CIP/sinter, CIP/hot-press, and plasma spray/sinter. (CIP is cold isostatic press, and HIP is hot isostatic press.) Mechanical properties depend on powder characteristics, chemistry, consolidation process, and thermal treatment. Wrought forms, produced by hot working, have high strength in the working direction, but properties are usually anisotropic.

Beryllium parts can be hot-formed from cross-rolled sheet and plate as well as plate-machined from hot-pressed block. Forming rates are slower than for titanium, for example, but tooling and forming costs for production items are comparable.

Structural assemblies of beryllium components can be joined by most techniques such as mechanical fasteners, rivets, adhesive bonding, brazing, and diffusion bonding. Fusion-welding processes are generally avoided because they cause excessive grain growth and reduced mechanical properties.

Beryllium behaves like other light metals when exposed to air by forming a tenacious protective oxide film that provides corrosion protection. However, the bare metal corrodes readily when exposed for prolonged periods to tap water or seawater or to a corrosive environment that includes high humidity. The corrosion resistance of beryllium in both aqueous and gaseous environments can be improved by applying chemical conversion, metallic, or nonmetallic coatings. Beryllium can be electroless nickel-plated, and flame or plasma sprayed.

All conventional machining operations are possible with beryllium, including EDM and ECM. However, beryllium powder is toxic if inhaled. Because airborne beryllium particles and beryllium salts present a health hazard, the metal must be machined in specially equipped facilities for safety.

Machining damages the surface of beryllium parts. Strength is reduced by the formation of microcracks and "twinning." The depth of the damage can be limited during finish machining by taking several light machining cuts and sharpening cutting tools frequently or by using nonconventional metal-removal processes. For highly stressed structural parts, 0.05 to 0.10 mm should be removed from each surface by chemical etching or milling after machining. This process removes cracks and other surface damage caused by machining, thereby preventing premature failure. Precision parts should be machined with a sequence of light cuts and intermediate thermal stress reliefs to provide the greatest dimensional stability.

Beryllium is toxic if inhaled or ingested, necessitating special precaution in handling. Most applications are quite specialized and stem largely from the good thermal and electrical properties of the metal. Uses include precision mirrors and instruments, radiation detectors, x-ray windows, neutron sources, nuclear reactor reflectors, aircraft brakes, and rocket nozzles.

Beryllium typically appears in military aircraft and space-shuttle brake systems, in missile reentry body structures, missile guidance systems, and satellite structures. The modulus-to-density ratio is higher than that of unidirectionally reinforced, "high-modulus" boron, carbon, and graphite fiber composites. Beryllium has an additional advantage because its modulus of elasticity is isotropic.

The largest-volume uses of beryllium metal are in the manufacture of beryllium–copper alloys and in the development of beryllium-containing moderator and reflector materials for nuclear reactors. Addition of 2% beryllium to copper forms a nonmagnetic alloy that is six times stronger than copper. These beryllium–copper alloys find numerous applications in industry as nonsparking tools, as critical moving parts in aircraft engines, and in the key components of precision instruments, mechanical computers, electrical relays, and camera shutters. Beryllium–copper hammers, wrenches, and other tools are employed in petroleum refineries and other plants in which a spark from steel against steel might lead to an explosion or fire.

One of the largest uses of beryllium metal is in nuclear reactors as a moderator to lessen the speed of fission neutrons and as a reflector to reduce leakage of neutrons from the reactor core. Beryllium is useful in nuclear applications because of its relatively high neutron-scattering cross section, low neutron-absorption cross section, and low atomic weight.

Another large-scale use of beryllium is in the manufacture of beryllium bronze, which has high tensile strength and a capacity for being hardened by heat treatment. Beryllium–copper molds are used in manufacturing plastic furniture with the appearance of wood-grain surfaces.

A small but important use of beryllium is in sheet or foil form as window material in x-ray tubes. Beryllium transmits x-rays 17 times better than an equivalent thickness of aluminum and six to ten times better than Lindemann glass. This, together with its high melting point, makes possible the use of x-ray beams of greater intensity.

Beryllium oxide is used in the manufacture of high-temperature refractory material and high-quality electrical porcelains, such as aircraft spark plugs and ultrahigh-frequency radar insulators. The high thermal conductivity of beryllium oxide and its good high-frequency electrical insulating properties find application in electrical and electronic fields.

Another use of beryllium oxide is as a slurry for coating of graphite crucibles to insulate the graphite and to avoid contamination of melted alloys with carbon. Beryllium oxide crucibles are used where exceptionally high purity or reactive metals are being melted. In the field of beryllium-oxide ceramics, a type of beryllia has been developed that can be formed into custom shapes for electronic and microelectronic circuits. Beryllium oxide has a high thermal conductivity, equal to that of aluminum, and excellent insulating properties, which permits closer packing of semiconductor functions in silicon (Si) integrated circuits.

The lightweight, very high elastic modulus, and heat stability of beryllium make it an attractive material for use as construction material in aircraft and missiles. However, its lack of ductility is a drawback. Were it not for its toxicity and scarcity, beryllium would find use as a rocket fuel because it produces more heat energy per unit volume than any other element. In multistage missiles a small weight reduction in the final stage, such as might be achieved by using beryllium in place of steel, permits a much larger weight reduction in the earlier stages in terms of fuel and structure. Research in the utilization of beryllium metal and beryllium-containing materials for aircraft and missiles is carried out very actively. These and other still-developing applications together with the continuing uses of beryllium in nuclear technologies sustain the ever-mounting production levels of beryllium.

BERYLLIUM ALLOYS

Dilute alloys of base metals that contain a few percent of beryllium in a precipitation-hardening system are the principal useful beryllium alloys manufactured today. Although beryllium has some solid solubility in copper, silver, gold, nickel, colbalt, platinum, palladium, and iron and forms precipitation-hardening alloys with these metals, the copper–beryllium system and, to a considerably lesser degree, the nickel–beryllium alloys are the only ones used commercially.

Other than the precipitation-hardening systems, small amounts of beryllium are used in alloys of the dispersion type wherein there is little solid solubility (aluminum and magnesium). Various amounts of beryllium combine with most elements to form intermetallic compounds. Development of beryllium-rich alloys have been chiefly confined to the ductile matrix beryllium–aluminum, beryllium–copper solid solution alloy with up to 4% copper, and dispersed-phase type alloys having relatively small amounts of compounds (0.25 to 6%), chiefly as beryllium oxide or intermetallics, for dimensional stability, elevated temperature strength, and elastic limit control.

BERYLLIUM–COPPER ALLOYS

Commercial alloys of the beryllium–copper group are divided into cast or wrought types, usually in ternary combination such as copper–beryllium–cobalt (see Table B.1). Alloys with higher beryllium content have high strength while those with low beryllium have high electrical and thermal conductivity.

Age-hardenable copper–beryllium–cobalt alloys offer a wide range of properties, because they are extremely ductile and workable in the annealed condition and are strong and hard after precipitation or aging treatment. Cobalt is added to inhibit grain growth and provide uniform heat-treatment response. These alloys also have inherent characteristics of substantial electrical and thermal conductivity and resistance to corrosion and wear, being protected by beryllium oxide films which impart this property to all materials containing beryllium. These age-hardenable alloys resist dimensional change and fatigue.

**TABLE B.1
Composition of Principal Beryllium–Copper Alloys**

Alloy Grade	Beryllium, %	Other, %
25	1.80–2.05	0.20–0.35 cobalt
165	1.60–1.79	0.20–0.35 cobalt
10	0.40–0.70	2.35–2.70 cobalt
50	0.25–0.50	1.40–1.70 cobalt
		0.90–1.10 silver
275C	2.60–2.85	0.35–0.65 cobalt
245C	2.30–2.55	0.35–0.65 cobalt
20C	2.00–2.25	0.35–0.65 cobalt
165C	1.60–1.85	0.20–0.65 cobalt
10C	0.55–0.75	2.35–2.70 cobalt
50C	0.40–0.65	1.40–1.70 cobalt
		1.00–1.15 silver

Primary applications are found in the electronics, automotive, appliance, instrument, and temperature-control industries for electric-current-carrying springs, diaphragms, electrical switch blades, contacts, connectors, terminals, fuse clips, and bellows (foil, strip, and wire), as well as resistance-welding dies, electrodes, clutch rings, brake drums, and switch gear (rod, plate, and forgings). With 1.5% beryllium or more, the melting point of copper is severely depressed and a high degree of fluidity is encountered, allowing casting of intricate shapes with very fine detail. The characteristic is important for plastic injection molds.

For special applications specific alloys have been developed. Free machining and nonmagnetic alloys have been made, as well as high-purity materials. A precipitation-hardening beryllium–Monel for oceanographic application, containing about 30% nickel, 0.5% beryllium, 0.5% silicon, and the remainder copper, illustrates one of a series of alloys having strength, corrosion resistance to seawater, and antifouling characteristics.

New applications in structural, aerospace, and nuclear fields are submarine repeater cable housings for transoceanic cable systems, wind tunnel throats, liners for magnetohydrodynamic generators for gas ionization, and scavenger tanks for propane-Freon bubble chambers in high-energy physics research. Important developing applications for beryllium–copper are trunnions and pivot bearing sleeves for the landing gear of heavy, cargo-carrying aircraft, because these alloys allow the highest stress of any sleeve bearing material.

Beryllium–copper master alloys are produced by direct reduction of beryllium oxide with carbon in the presence of copper in an arc furnace. Because Be_2C forms readily, the master alloy is usually limited to about 4.0 to 4.25% beryllium. The master alloy is remelted with additional copper and other elements to form commercial alloys. Rolling billets up to 680 kg have been made by continuous-casting techniques.

Beryllium–copper alloys can be fabricated by all the industrial metalworking techniques to produce principally strip, foil, bar, wire, and forgings. They can be readily machined and can be joined by brazing, soldering, and welding. Annealing, to provide high plasticity at room temperature, is accomplished by heat-treating (from 790 to 802°C for the high-beryllium alloys or 900 to 930°C for the low-beryllium alloys) and water quenching. Precipitation-hardening is accomplished by heating to (400 to 480°C; low beryllium) and (290 to 340°C; high beryllium).

ALLOYS WITH NICKEL AND IRON

Nickel containing 2% beryllium can be heat-treated to develop a tensile strength of 1700 MPa with 3 to 4% elongation. Little commercial use is made of the hard nickel alloys although they have been employed, principally as precision castings, for aircraft fuel pumps, surgical instruments, and matrices in diamond drill bits. Another nickel alloy with 2.0 to 2.3% beryllium, 0.5 to 0.75% carbon, and the balance of nickel and refractory metals has been used for mold components and forming tools for glass pressware of optical and container quality. Thermal conductivity, wear resistance, and strength, coupled with unusual machinability for a nickel–beryllium alloy, make this alloy particularly advantageous for glassware tooling.

Wrought beryllium–nickel contains about 2% beryllium, 0.5% titanium, and the balance nickel. Casting alloys contain a bit more beryllium (2 to 3%) and, in one alloy, 0.4% carbon. Arsenic in the case of beryllium–copper alloys, mechanical properties vary widely, depending on temper condition — from 310 to 1586 MPa in tensile yield strength and Rb 70 to Rc 55 in hardness at room temperature. The alloys retain considerable yield strength at high temperature: 896 to 1172 MPa at 538°C. They also have good corrosion resistance in general atmospheres and reducing media. Because beryllium is toxic, special precautions are required in many fabricating operations. The wrought alloy is used for springs, bellows, electrical contacts, and feather valves, and the casting alloys for molding plastics and glass, pump parts, seal plates, and metal-forming tools.

Attempts have been made to add beryllium to a number of ferrous alloys. Small amounts are used to refine grains and deoxidize ferritic steels in Japan, and promising properties have been developed for austenitic and martensitic steels. Stainless steels (iron–nickel–chromium) may be made maraging by adding 0.15 to 0.9% beryllium, developing strengths as high as 1800 MPa as cast or 2300 MPa as rolled while retaining their oxidation- and corrosion-resistant characteristics. Amounts of 0.04 to 1.5% beryllium have been added to various iron alloys containing nickel, chromium, cobalt, molybdenum (Mo), and tungsten for special applications such as watch springs.

Beryllium-Base Alloys

Three types of beryllium-base alloys are of interest. These consist of dispersed-phase types containing up to 4% beryllium oxide; ductile-phase or duplex alloys of beryllium and aluminum, particularly 38% aluminum in beryllium; and solid solution alloys of up to 4% copper in beryllium.

Dispersed-phase alloys containing oxides, carbides, nitrides, borides, and intermetallic compounds in a beryllium matrix are chiefly characterized by increased strength and resistance to creep at elevated temperatures. Major commercial alloys in this series are of the fine-grain, high-beryllium oxide (4.25 to 6%), hot-pressed types such as materials used for inertial guidance instruments characterized by high dimensional stability, high-precision elastic limit (55 to 103 MPa), and good machinability.

The 62% beryllium, 38% aluminum alloy previously discussed under aluminum alloys was developed as a structural aerospace alloy to combine high modulus and low density with the machining characteristics of the more common magnesium-base alloys. This alloy in sheet form has at room temperature about 344 MPa ultimate strength, 324 MPa yield strength, and about 8% elongation. It is also produced as extrusions. It has a duplex-type microstructure characterized by a semicontinuous aluminum phase. Other alloys of the beryllium–aluminum type have been reported, but the 62% beryllium, 38% aluminum alloy is the most used.

Intermetallic Compounds

Beryllium, combined with most other elements, forms intermetallic compounds with high strength at high temperature (up to 552 MPa modulus of rupture at 1260°C), good thermal conductivity, high specific heat, and good oxidation resistance. The beryllium oxide film formed by surface oxidation is protective to volatile oxides (molybdenum) and to elements of high reactivity (zirconium and titanium) at temperatures 1100 to 1500°C and for short times up to 1650°C.

The beryllides are of interest to the nuclear field, to power generation, and to aerospace applications. Evaluation of the intermetallics as refractory coatings, reactor hardware, fuel elements, turbine buckets, and high-temperature bearings has been carried out.

BERYLLIUM OXIDE

A colorless to white crystalline powder of the composition beryllium oxide, also called beryllia. It has a specific gravity of 3.025, a high melting point, about 2585°C, and a Knoop hardness of 2000. It is used for polishing hard metals and for making hot-pressed ceramic parts. Its high heat resistance and thermal conductivity make it useful for crucibles, and its high dielectric strength makes it suitable for high-frequency insulators. Single-crystal beryllia fibers, or whiskers, have a tensile strength above 6800 MPa.

Ceramic parts with beryllia as the major constituent are noted for their high thermal conductivity, which is about three times that of steel, and second only to that of the high-conductivity metals (silver, gold, and copper). They also have high strength and good dielectric properties. Properties of typical grades of beryllia ceramics are tensile strength, 96 MPa; compressive strength, 2068 MPa; hardness (micro), 1300 Knoop; maximum service temperature, 2400°C; dielectric strength, 5.8. Beryllia ceramics are costly and difficult to work with. Above 1650°C they react with water to form a volatile hydroxide. Also, because beryllia dust and particles are toxic, special handling precautions are required. Beryllia parts are used in electronic, aircraft, and missile equipment. A more recent application has been the use of beryllia as thermocouple insulators in vacuum furnace equipment operating below 1650°C.

Beryllium oxide powder is available in three particle size ranges: (1) submicron to 1 to 2 mm, used for fabricating both ceramic components and BeO–UO$_2$ nuclear fuel elements, (2) 2 to 8 mm, used primarily for fabricating beryllia bodies of 96 to 99.5% purity, and (3) ultrahigh-density grains of specific size distribution for admixing with resins and other organics to provide very high thermal conductivity coatings and potting compounds.

Beryllia ceramics have these characteristics: outstanding resistance to wetting and corrosion by many metals and nonmetals; mechanical properties only slightly less than those of 96% Al$_2$O$_3$ ceramics; valuable nuclear properties, including an exceptionally low, thermal neutron absorption cross section; and ready availability in a wide variety of shapes and sizes. Like Al$_2$O$_3$ and some other ceramics, beryllia is readily metallized by a variety of thick- and thin-film techniques.

Major markets for beryllium oxide ceramics are: microwave tube parts such as cathode supports, envelopes, spacers, helix supports, collector isolators, heat sinks, and windows; substrates, mounting pads, heat sinks, and packages for solid-state electronic devices; and bores or plasma envelopes for gas lasers.

Other uses include klystron and ceramic electron tube parts, radiation and antenna windows, and radar antennae. The exceptional resistance of beryllia to wetting (and thus corrosion) by many molten metals and slags makes it suitable for crucibles for melting uranium (U), thorium (Th), and beryllium.

The high general corrosion resistance of beryllia has helped it capture new applications in the chemical and mechanical fields. And other uses in aircraft, rockets, and missiles are predicted.

Beryllium oxide is tapped for nuclear reactor service because of its refractoriness, high thermal conductivity, and ability to moderate (slow down) fast neutrons. The "thermal" neutrons that result are more efficient in causing fusion of uranium-235. Nuclear industry uses for beryllia include reflectors and the matrix material for fuel elements. When mixed with suitable nuclear poisons, beryllium oxide may be a new candidate for shielding and control rod assembly applications.

The market for electrically insulating, heat-conductive encapsulants based on beryllia grain–polymer mixtures is both small and restricted. Although these composites have thermal conductivities 10 to 20 times higher than those of other filled plastics, the handling restrictions necessitated by the presence of beryllia limit their use.

BISMUTH

Bismuth (symbol Bi) is a brittle, crystalline metal with a high metallic luster with a distinctive pinkish tinge. The metal is easily cast but not readily formed by working. Within a narrow range of temperature, around 225°C, it can be extruded. Its crystal structure is rhombohedral.

It is one of the few metals that expand on solidification; the expansion is 3.3%. The thermal conductivity of bismuth is lower than that of any metal, with the exception of mercury.

Bismuth is the most diamagnetic of all metals (mass susceptibility of -1.35×10^6). It shows the greatest Hall effect (increase in resistance under influence of a magnetic field). It also has a low capture cross section for thermal neutrons (0.034 barn).

Bismuth is inert in dry air at room temperature, although it oxidizes slightly in moist air. It rapidly forms an oxide film at temperatures above its melting point, and it burns at red heat, forming the yellow oxide, Bi$_2$O$_3$. The metal combines directly with halogens and with sulfur, selenium, and tellurium; however, it does not combine directly with nitrogen or phosphorus.

Bismuth is not attacked at ordinary temperatures by air-free water, but it is slowly oxidized at red heat by water vapor. Bismuth does not dissolve in nonoxidizing acids, but does dissolve in HNO$_3$ and in hot concentrated H$_2$SO$_4$. The formation of intermetallic compounds involves mainly the strongly electropositive metals.

Bismuth combined with a number of metallic elements forms a group of interesting and useful low-melting alloys. Some of the lowest melting of these are as follows:

	Melting Point
49.5 bismuth, 10.1 cadmium, 13.1 tin, 27.3 lead	70°C
49.0 bismuth, 12.0 tin, 18 lead, 21 indium	57°C
44.7 bismuth, 5.3 cadmium, 8.3 tin, 22.6 lead, 19.1 indium	47°C

The fusible alloys are used in many ingenious ways, e.g., sprinkler-system triggering devices, bending pipes, anchoring tools during machining, accurate die patterns, etc.

Bismuth metal (0.1%) is also added to cast iron and steel to improve machinability and mechanical properties. An alloy of 50% bismuth and 50% lead is added to aluminum for screw machine stock, to increase machinability.

A permanent magnet (bismanol) with excellent resistance to demagnetization is produced from manganese and bismuth.

The development of refrigerating systems depending on the Peltier effect for cooling uses a bismuth–tellurium or selenium alloy for thermocouples. Bismuth telluride is used extensively for thermoelectric cooling and for low-temperature thermoelectric power production.

Bismuth is playing an important role in nuclear research. Its high density gives it excellent shielding properties for gamma rays while its low thermal neutron capture cross section allows the neutrons to pass through. For investigations in which it is desired to irradiate objects, i.e., animals, with neutrons but protect them from gamma rays, castings of bismuth are used as neutron windows in nuclear reactors.

Bismuth has been proposed as a solvent-coolant system for nuclear power reactors. The bismuth dissolves sufficient uranium so that, when the solvent and solute are pumped through a moderator (graphite), criticality is reached and fission takes place. The heat generated from the fission reaction raises the temperature of the bismuth. The heated bismuth is then pumped to conventional heat exchangers producing the steam power required for eventual conversion to electricity.

The advantages of such a reactor are that (1) it has potential for producing low-cost power, (2) it has an integrated fuel processing system, and (3) it converts thorium to fissionable uranium.

Another important use of bismuth is in the manufacture of pharmaceutical compounds. Various bismuth preparations have been employed in the treatment of skin injuries, alimentary diseases, such as diarrhea and ulcers, and syphilis. The oxide and basic nitrate are perhaps the most widely used compounds of bismuth. The trioxide is used in the manufacture of glass and ceramic products, and the basic nitrate is used in the porcelain painting to fire on gilt decoration.

BITUMINOUS COATINGS

Bitumens have been defined as mixtures of hydrocarbons of natural or pyrogenous origin or combinations of both (frequently accompanied by their nonmetallic derivatives), which can be gaseous, liquid, semisolid, or solid and which are completely soluble in carbon disulfide.

Bitumens used in the manufacture of coatings are of the semisolid and solid variety and are derived from three sources:

1. Asphalt produced by the distillation of petroleum
2. Naturally occurring asphalts and asphaltites
3. Coal tar produced by the destructive distillation of coal

It is customary to classify bituminous coatings by their application characteristics as well as by their generic composition. All of the coatings can be divided into two classes depending on whether or not they require heating prior to application.

1. *Hot-applied coatings.* These are either 100% bitumen or bitumen blended with selected fillers. A common loading for coatings employing fillers is 10 to 20% filler. Hot-applied coatings are brought to the desired application viscosity by heating. The majority of buried pipelines are coated with this type of bituminous coating.
2. *Cold-applied coatings.* These employ both solvents and water to attain the desired application viscosity. A wide range of solvents is used and the choice depends mainly on the drying characteristics desired and the solvent power required to dissolve the particular bitumen being used. Various fillers are also used in cold-applied coatings to obtain specific applications and end-use properties.

Bituminous coatings can be formulated from many combinations of bitumens, solvents, or dispersing agents and fillers. This makes possible a great variety of end products to meet application and service requirements. The coatings that can be produced range from thin-film (3-mil) coatings to protect machined parts in storage, up to thick (100-mil), tough coatings to protect buried pipelines.

As with all other coatings, the conditions of the surface to which a bituminous coating is applied is an important, life-determining factor. A good sandblast is preferred, especially if the surface is badly corroded and the exposure is severe. In any case the surface should be free of moisture, grease, dust, salts, loose rust, and poorly adherent scale. A thin, penetrating bituminous primer can be beneficial on rusty surfaces that can only be cleaned by wire brushing and scraping.

End-Use Requirements

The application will dictate which performance properties of a coating should be given greatest consideration.

Service Temperature

Many applications of bituminous coating require a moderate service temperature range, often no greater than that caused by weather changes. However, other applications, such as coatings for chemical processing vessels, may require much wider service temperature ranges. In any event, it is possible to obtain good performance with bituminous coatings over a range of −73 to 163°C.

Thermal and Electrical Insulation

Bitumens themselves are relatively poor thermal insulators. However, by using low-density fillers, coatings with good insulation properties can be formulated. These coatings both protect from corrosion and provide thermal insulation. An added advantage of the coatings is that the insulating material (low-density filler) is completely surrounded by bitumen and is permanently protected from moisture. Thus, they are not subject to a loss in efficiency (as are conventional insulating systems) from damage or failure of the protecting vapor barrier.

Bitumens are naturally good electrical insulators. This is an important consideration in systems using cathodic protection as a complementary corrosion-prevention device.

Abrasion Resistance

Many bituminous coatings need to have high abrasion resistance. Automotive undercoatings, for example, need high abrasion resistance because they are continually buffeted by gravel and debris thrown up by the wheels of the vehicle.

Abrasion-resistant coatings are also used to protect interior surfaces of railroad cars or other vessels handling chemical solids or abrasive slurries.

Weathering Resistance

Asphalt-base bituminous coatings generally weather better than coal-tar-base coatings. Also, there is quite a wide difference in the performance of asphalt coatings derived from different petroleum crudes. By critically selecting the crude and the processing method, asphalt coatings can be formulated that will weather for many years.

In industrial areas, corrosive solids, solutions, and vapors will affect weathering performance. In general, the resistance of bituminous coatings to corrosive media is equal to that of the best organic coatings. Bituminous coatings have good resistance to dilute hydrochloric, sulfuric, and phosphoric acids as well as to sodium hydroxide. They also have good resistance to solutions of ammonium nitrate and ammonium sulfate. However, they have poor resistance to dilute nitric acid, and most coatings are not resistant to oils, greases, and petroleum solvents.

Mechanical Impact and Thermal Shock

Bituminous coatings generally have good adhesion when subjected to mechanical impact. However, where severe mechanical impact is expected, the coatings should first

BITUMINOUS COATINGS

be field-tested or tested in the laboratory under simulated service conditions.

Resistance to thermal shock is an important consideration in coatings used on some types of processing equipment. Laboratory tests in which a coated panel is transferred back and forth between hot and cold chambers can be used to predict field behavior. Bituminous coatings are available that will withstand thermal shock over a temperature range of −51 to 60°C.

Types of Coatings

Because bitumens are very dark in color, heavy loadings of pigments are required to produce colors other than black. However, if color is an important consideration, certain colored paints, including whites, can be used as topcoats. Granules can also be blown into the coating before it has completely cured to produce a wide range of decorative effects.

Thin-Film Coatings

Coatings less than 6 mils thick are arbitrarily included in this group. For the most part, these are solvent cutback coatings. They are black except in the case of aluminum paints employing bituminous bases. The coatings are inexpensive and can be used to give good protection from corrosion when color is not important.

Asphalt coatings of this type are used extensively to protect machined parts in storage. Because the coatings retain their solubility in low-cost petroleum solvents even after long weathering, kerosene or similar solvents can be used to remove a coating from the protected part just prior to placing it in service. Coal-tar-base coatings are much more difficult to dissolve and cannot be used for this purpose. However, this property makes coal tar useful for protecting crude-oil tank bottoms and in other applications requiring resistance to petroleum fractions.

Industrial Coatings

Heavy-bodied industrial coatings incorporate low-density and fibrous mineral fillers. Coatings can be formulated that will not slump or flow on vertical surfaces when applied as thick as 250 mils. However, they are usually used in thicknesses from 6 to 120 mils.

The coatings are used extensively in industrial plants to protect tanks and structural steel from such corrosive environments as acids, alkalis, salt solutions, ammonia, sulfur dioxide, and hydrogen sulfide gases.

Industrial coatings are also used in large volume by the railroad industry. Complete exteriors of tank cars carrying corrosive liquids are often coated and provide good protection of the saddle area where spillage is likely to occur. Coatings on the exteriors and interiors of hopper cars in dry chemical service also provide protection from both corrosive action and abrasive wear.

Railroad car bituminous cements are used to seal sills and joints in boxcars. An application of this material followed by an overcoating of granules makes an excellent roofing system for railroad cars.

Coatings for Use over Insulation

Practically all industrial insulating materials must be protected from the weather and moisture; otherwise they would lose their efficiency. Bituminous coatings formulated to have low rates of moisture vapor transmission give best results on installations operating at low (−73°C) to moderate (82°C) temperatures. Coatings that allow a higher rate of moisture transmission (breathing type) are needed to protect the insulation on systems operating at 82°C and above. This is necessary so that moisture trapped beneath the coating can escape when the unit is brought to operating temperatures.

Thermal Insulating Coatings

Low-density fillers can be employed in bituminous mastics to produce coatings with relatively good insulating values; a k value of 0.6 Btu/ft^2/h/°F is typical. Insulating coatings are usually applied somewhat more thickly than conventional mastics to obtain the insulating value desired. They are commonly used in thicknesses of 250 to 375 mils and because of their thickness and resiliency they have excellent resistance to mechanical damage.

Automotive Underbody Coatings

These are mastic type coatings containing fibrous and other fillers. They are used to coat the undersides of floor panels, fenders, gasoline tanks, and frames to protect against corrosion and provide sound deadening and joint sealing.

The coatings have high resistance to deicing salts, moisture, and water. They also have sound-deadening properties that noticeably reduce the noise level inside an automobile. This provides for a more pleasant and less-fatiguing ride. The sealing and bridging action of the coatings is also especially effective in reducing drafts and dust infiltration.

Sound-Deadening Coatings

High-efficiency, sound-deadening coatings can be formulated from selected resinous bases and high-density fillers. They have better sound-deadening properties than automotive underbody coatings and are used on the wall, roof, and door panels of automotive equipment where sound deadening is the primary need, rather than abrasion resistance or protection from corrosion. They are also used on railroad passenger cars, house trailers, stamped bathtubs, kitchen sinks, air-conditioning cabinets, and ventilation ducts.

Pipe Coatings

Industrial coatings are excellent for protection of pipe above ground. However, the environment of underground exposure and the complementary use of cathodic protection make it necessary to use specially designed coatings. The stresses created by shrinking and expanding soil require that the coating be very tough. Rocks and other sharp objects can be expected to cause high localized pressures on the coating surface. A coating must have good cold-flow properties to resist penetration by objects, which can cause localized pressures as high as 690 MPa.

Cathodic protection (an impressed negative electrical potential) is widely used to prevent the corrosion processes from occurring at flaws in the pipe coatings. Both asphalts and coal tars are good electrical insulators and make excellent coatings for cathodic protection applications.

Coatings for pipe are usually of the hot application type. Application can be made at the mill, at a special pipe-coating yard, or over the ditch, depending on the terrain, size of pipe, and other factors. The coating may be given added strength by embedding it with a glass fabric while it is still hot. Outer wrappings of rag, asbestos, and glass felts are sometimes used to give added resistance to damage by soil stresses.

APPLICATION

Bituminous coatings that are cut back with solvents or emulsified with water can be produced to consistencies suitable for application by dipping, brushing, spraying, or troweling at ambient temperatures.

- Dipping is usually used to coat small parts. As a rule, coating viscosity is adjusted to produce a thickness of 1 to 6 mils.
- Brushing is used on areas that cannot be reached by spraying and on jobs that do not warrant setting up spray equipment. Coating thicknesses can range from 1 to 65 mils.
- Spraying is the most popular method for applying cold coatings. The thickness required in one application determines the consistency of the formulation, and thicknesses of 1 to 250 mils are obtainable by spraying. Conventional paint-spray equipment can be used for coatings up to 6 mils thick. Heavier coatings require the use of mastic spray guns fed from pressure pots or heavy-duty pumps. Heated vessels and feed lines can also be used to decrease viscosity and permit faster application and the buildup of thicker films in one application.
- Troweling is usually used in inaccessible areas or where it is necessary to produce a very heavy coating in one application. Trowel coats are usually applied in thicknesses above 250 mils.

Bituminous coatings can also be applied hot without the need for any diluents. Such coatings are widely used on piping in thicknesses of about 95 mils. They are heated to 177 to 288°C and then pumped into a special apparatus that surrounds and travels along the pipe.

BLOW MOLDING

Essentially, blow molding involves trapping a hollow tube of thermoplastic material in a mold. Air pressure applied to the inside of the heated tube blows the tube out to take the shape of the mold. There are many variations on the basic technique.

In short, the process is an economical high-speed, high-production-rate method of forming thermoplastic parts of hollow shape, or parts that can be simply made from a hollow shape.

Uses include the container and toy field, where bottles and toys of many different shapes are formed in large quantities at low cost. The most commonly used material is polyethylene (PE).

Although any thermoplastic resin can be considered a candidate for blow molding, PE was the first used when blow molding started with low-density PE for blow-molded squeeze bottles. Now, low-, intermediate-, and high-density PE resins are used, as well as special ethylene copolymers designed to provide greatly improved stress cracking resistance compared with PE homopolymers, needed for detergent containers.

One of the main criteria of selection of a PE resin for blow molding is the proper balance of physical properties required for the specific use.

With the extension of blow molding into broader use in industrial products, the need for engineering properties other than those of PE has stimulated interest in other thermoplastics. The main plastics available for blow molding, other than PE, include cellulosics, polyamides (nylons), polyacetals, polycarbonates, polypropylene, and vinyls.

The cellulosic family of plastics includes acetate, butyrate, propionate, and ethyl cellulose. For blow molding, the cellulosics offer strength, stiffness, transparency, and high surface gloss. They have unlimited color possibilities. Chemical resistance and availability of nontoxic resins make them potentially suitable for medicine and food packaging. Their strength, stiffness, and transparency make them suitable for industrial parts, toys, and numerous decorative and novelty items.

Polyamides or nylons, although relatively high cost materials, offer potential benefits in industrial parts and special containers, such as aerosols. Special developments have resulted in formulations tailored to special viscosity requirements for blow molding.

Polyacetal resins for blow molding offer toughness, rigidity, abrasion resistance, high heat distortion temperatures, and excellent resistance to organic solvents. Also, they are resistant to aliphatic and aromatic hydrocarbons, alcohols, ketones, strong detergents, weak organic acids, and to some weak inorganic bases. Aerosol containers are another application for blow molding.

Polycarbonates have found their place in blow-molded industrial parts. Primarily they offer high toughness, strength, and heat resistance. They are transparent with almost unlimited colorability and are self-extinguishing.

Polypropylenes, somewhat similar to higher-density PEs, but with lower specific gravity, higher rigidity, strength and heat resistance, and lower permeability, offer interesting properties at low cost. Because of their lower permeability they are used in containers where PE is unsuited. They also have excellent stress-crack resistance.

PVC (polyvinyl chloride) for blow-molded parts offers benefits in terms of variability of engineering properties. PVCs are available with properties ranging all the way from high rigidity in the unplasticized grades to highly flexible plasticized PVCs. The variability in performance resulting from the many possible formulations means that engineers must consult with the materials supplier in attempting to obtain a formulation with the proper performance and processing characteristics to meet their needs.

The scope of blow-molded design has already broadened beyond that of round hollow objects. Production of such parts as housings by blowing a unit and sawing the item along the parting line to produce two housings is already a reality. As many as ten cavities have been incorporated into a mold, using a wide tube and allowing air to pass through a hollow sprue or runner system between parts.

The future design possibilities of blow molding appear bright. The constant improvements in equipment design continue to add flexibility to the blow-molding process.

Secondary Operations

The equipment used will determine to a great extent the amount of finishing required on the parts. Parts must usually be trimmed and often decorated. Trimming may be by hand, or may involve highly automated trimming, reaming, and cutting.

Decorating of parts depends on the shape of the part and the type of material used. But generally, a variety of techniques is available, including labeling, hot stamping, silk screening, and offset printing.

By far the largest market for blow-molding applications is in the container field, ranging from containers for food, drugs, and cosmetics to household and industrial chemicals. Toys and housewares represent sizable markets.

The industrial product area represents one of the biggest potential uses for the process. Present products include controls for television sets, rollers for lawn mowers, oil dispensers, toilet floats, molds for epoxy potting, and auto ducting for air conditioning.

BONDING

A method of holding the parts of an object together with or without the aid of an adhesive such as an epoxy or a glue. Composite materials such as fiber-reinforced plastics require strong interfacial bonding between the reinforcement and the matrix. In the case of atomic or optical contact bonding, interatomic forces hold the parts together. In optical contact bonding, surface flatness and cleanliness between the mating parts determine the bonding strength, and the atoms at the surface provide the necessary forces. The number of valence electrons in the atoms of a material determines the bonding strength between the group of atoms that constitutes a molecule. In these cases the term *chemical bonding* is used.

Wire bonding is an interconnect technique widely used in microchip manufacturing to provide electrical continuity between the metal pads of an integrated circuit chip and the electrical leads of the package housing the chip. The two common methods of wire bonding are thermocompression and ultrasonic bonding. In these, a fine aluminum or gold wire is bonded at one end to the metal pad of the integrated circuit chip, and at the other to the electrical lead of the package. There are three types of thermocompression bonds: wedge, stitch, and ball. In thermocompression bonding, a molecular metallurgical bond is formed at the two metal junctions — bond wire and IC metal pad, and bond wire and package lead metal — by applying heat and pressure without melting. In ultrasonic bonding, the molecular metallurgical bond is achieved through a combination of ultrasonic energy and pressure. The bonding operation is done under pressure to break the few surface layers of the material and form the bond between the contamination-free surfaces. Thermocompression bonding has higher throughput and speed than ultrasonic bonding. The bonding wire is usually aluminum, which does not introduce any intermetallic problems.

BONDING MATERIALS

These are matrix materials in glass-fiber-reinforced plastics whereby the fiber glass reinforcement has been previously sized and supplied with these polymeric materials, which are catalysts, accelerators, fillers, inhibitors, and plasticizers.

- *Acrylic polymers*. Limited use as a laminating resin in the form of thermosetting acrylic syrup to make glazing panels. The monomer is used in conjunction with styrene as a cross-linking agent in polyesters to improve weathering resistance.
- *Alkyds*. Reaction products of polymerization of polyhydroxy alcohols and polycarboxylic acids, such as glycerol and phthalic anhydride. Modified grades are reinforced by glass fibers in chopped strand form.
- *Epoxy resins*. Reaction products of bisphenol A and epichlorohydrin. Epoxies are cross-linked by the addition of acids or amines to form laminating resins of exceptional bonding

properties. These boast excellent electrical properties and high resistance to acids and alkalies and are used with fiberglass cloth and strands.
- *Melamine formaldehyde.* Melamine reacts with 1 to 6 mol of formaldehyde to form various compounds with good electrical characteristics. Used in flat electrical laminates, reinforced by glass cloth, these resins also are used for bonding glass fibers in thermal insulation.
- *Phenol formaldehyde resins.* Condensation polymerization products of phenol and formaldehyde are used primarily for bonding fiberglass insulation products. Also used in flat fiberglass cloth laminates for electrical applications
- *Polyesters.* Unsaturated polyester resins are made by condensation polymerization of polyfunctional acids such as maleic acid and phthalic anhydride with glycols and subsequent addition of cross-linking agents such as styrene. These represent the largest family of resins used in laminating and molding fiberglass cloth and strands.

BORON

Boron (symbol B) is a metallic element closely resembling silicon. Boron has a specific gravity of 2.31, a melting point of about 2200°C, and a Knoop hardness of 2700 to 3200, equal to a Mohs hardness of about 9.3. At 600°C, boron ignites and burns with a brilliant green flame. Minute quantities of boron are used in steels for case hardening by the nitriding process to form a boron nitride, and in other steels to increase hardenability, or depth of hardness. In these boron steels, as little as 0.003% is beneficial, forming an iron boride, but with larger amounts the steel becomes brittle and susceptible to hot-short unless it contains titanium or some other element to stabilize the carbon . In cast iron, boron inhibits graphitization and also serves as a deoxidizer. It is added to iron and steel in the form of ferroboron.

The free element is prepared in crystalline or amorphous form. The crystalline form is an extremely hard, brittle solid. It is of jet-black to silvery-gray color with a metallic luster. One form of crystalline boron is bright red. The amorphous form is less dense than the crystalline and is a dark-brown to black powder.

Crystalline boron, although relatively brittle compared to diamond, is second only to diamond in hardness.

The chemical properties of boron depend largely on the physical form as well as on the purity of samples. Amorphous boron oxidizes slowly in the air even at room temperature, and is spontaneously flammable at about 860°C.

Boron is not affected by either hydrochloric or hydrofluoric acids, even on prolonged boiling. Crystalline boron is quite stable to heat and oxidation even at relatively high temperatures. It is slowly attacked and oxidized by hot concentrated nitric acid and by mixtures of sodium dichromate and sulfuric acid. Hydrogen peroxide and ammonium persulfate also slowly oxidize crystalline boron.

All varieties of boron are completely oxidized by molten mixtures of alkali carbonates and hydroxides.

Boron reacts vigorously with sulfur at about 600°C to form a mixture of boron sulfides. Boron nitride is formed when boron is heated in an N_2 or NH_3 atmosphere above 1000°C. At high temperatures, boron combines with phosphorus and with arsenic to form a phosphide, BP, and an arsenide, BAs.

It reacts with silicon to form silicon borides at temperatures above 2000°C.

Boron reacts with a majority of metals and metallic oxides at high temperatures to form metallic borides. Boron does not conform to the usual rules of valence in forming these compounds. The borides, along with the carbides and the nitrides, are called interstitial compounds. These compounds have crystal structures and properties very similar to those of the original metal.

Metallic borides are usually prepared by hot sintering of the elements. They may also be prepared by carbothermic and aluminothermic reduction of metal oxide–boron oxide mixtures. A number of metallic borides of high purity have been prepared on a semicommercial scale by vapor deposition methods, in which the volatile halides of the elements are deposited on a hot substrate, such as tungsten or tantalum wire, in an atmosphere of H_2. Borides have much in common with true metals. They have a high electrical conductivity, high melting points, and extreme hardness. Many metal borides are used as components of cermet compositions.

Boron fibers for reinforced structural composites are continuous fine filaments that are, themselves, composites. They are produced by vapor deposition of boron on a tungsten or carbon substrate. Their specific gravity is about 2.6, and they range from 0.10 to 0.15 mm in diameter. They have tensile strengths around 3450 MPa, and a modulus of elasticity of nearly 0.5 million MPa. The fibers are used chiefly in aluminum or epoxy matrixes. Unidirectional boron–aluminum composites have tensile strengths ranging from 758 to over 1378 MPa. Their strength-to-weight ratio is about three times greater than that of high-strength aluminum alloys.

Boron compounds are employed for fluxes and deoxidizing agents in melting metals, and for making special glasses. Boron, like silicon and carbon, has an immense capacity for forming compounds, although it has a different valence. The boron atom appears to have a lenticular shape, and two boron atoms can make a strong electromagnetic bond, with the boron acting like carbon but with a double ring.

Boron Nitride

Boron nitride (BN) has many potential commercial applications. It is a white, fluffy powder with a greasy feel. It has an x-ray diffraction pattern almost identical with that of graphite, indicating a close similarity in structure. It is called white graphite. BN has a very low coefficient of friction, but unlike carbon, it is a nonconductor of electricity, and it is attacked by nitric acid. It sublimes at 3000°C. It reacts with carbon at about 2000°C to form boron carbide (B_4C). It is used for heat-resistant parts by molding and pressing the powder without a binder to a specific gravity of 2.1 to 2.25.

BN may be prepared in a variety of ways, for example, by the reaction of boron oxide with ammonia, alkali cyanides, and ammonium chloride, or of boron halides and ammonia. The usually high chemical and thermal stability, combined with the high electrical resistance of BN, suggests numerous uses for this compound in the field of high-temperature technology. BN can be hot-pressed into molds and worked into desired shapes.

BN powders can be used as mold-release agents, high-temperature lubricants, and additives in oils, rubbers, and epoxies to improve thermal conductance of dielectric compounds. Powders also are used in metal- and ceramic-matrix composites (MMC and CMC) to improve thermal shock and to modify wetting characteristics.

Hexagonal BN may be hot-pressed into soft (Mohs 2) and easily machinable white or ivory billets having densities 90 to 95% of theoretical (2.25 g/cm^3). Thermal conductivities of 17 to 58 W/mK and coefficients of thermal expansion of 0.4 to 5 $\times$ 10^{-6}/C are obtained, depending on density, orientation with respect to pressing direction, and amount of boric oxide binder phase. Because of its porosity and relatively low elastic modulus (50 to 75 GPa), hot-pressed BN has outstanding thermal shock resistance and fair toughness. Pyrolytic BN, produced by chemical vapor deposition (CVD) on heated substrates, also is hexagonal; the process is used to produce coatings and shapes with thin cross sections.

Uses for hexagonal BN shapes include crucibles, parts for chemical and vacuum equipment, metal casting fixtures, boron sources for semiconductor processing, and transistor mounts.

A cubic form of BN, called borazon, has been prepared at pressures near 6500 MPa and temperatures near 1500°C. It is comparable to diamond in hardness and apparently has properties superior to those of diamond with regard to oxidation, electrical resistance, and thermal stability. This material will probably take its place with diamond for industrial grinding. Borazon is stable up to about 1925°C. This molded material is resistant to molten aluminum, is not wetted by molten silicon or glass, and is used for crucibles. Cubic BN tooling typically outlasts Al_2O_3 and carbide tooling and is preferred in applications where diamond is not appropriate, such as grinding of ferrous metals.

BN fibers are produced in diameters as small as 5 to 7 μm and in lengths to 0.38 m.

The fibers have a tensile strength of 1378 MPa. They are used for filters for hot chemicals, and as reinforcement to plastic lamination. BN HCJ is a fine powder, 99% pure, used as a filler in encapsulating and potting compounds to add thermal and electric conductivity. BN is basically the same as B_4C and can only be densified by hot pressing. Hot-pressed hexagonal BN is easy to machine and has a low density but must be kept free of boron oxide (or the boron oxide stabilized with CaO-containing additives), so BN is not destroyed by hydration during heating.

A tough, coherent, hard abrasive compact, consisting of a cubic form of BN bonded with B_4C, has been developed.

Boron Carbide

Boron carbide (B_4C) is produced by the high-temperature (about 1371 to 2482°C) interaction of boric oxide, B_2O_3, and carbon in an electrical resistance-type furnace. It is a black, lustrous solid. It is used extensively as an abrasive, because its hardness approaches that of the diamond. It is also used as an alloying agent, particularly in molybdenum steels.

Additionally, it is used in drawing dies and gauges, or into heat-resistant parts such as nozzles. The composition is either B_6C or B_4C; the former is the harder but usually contains an excess of graphite difficult to separate in the powder. It can be used thus as a deoxidizing agent for casting copper, and also for lapping, since the graphite acts as a lubricant. Boroflux is B_4C with flake graphite, used as a casting flux.

B_4C parts are fabricated by hot pressing, sintering, and sinter-HIPing (HIP = hot-isostatic press). Industrially, densification is carried out by hot pressing (2100 to 2200°C, 20 to 40 MPa) in argon. The best properties are obtained when pure fine powder is densified without additives. Pressureless sintering to high density is possible using ultrafine powder, with additives (notably carbon). Less expensive than hot pressing, sintering also can be used for more complex shapes.

Special part formulations include bonding B_4C with fused sodium silicate, borate frits, glasses, plastics, or rubbers to lend strength, hardness, or abrasion resistance. B_4C-based cermets and MMC (especially Al/B_4C, Mg/B_4C, Ti/B_4C), and CMCs (e.g., TiB_2/B_4C) have unique properties, including superior ballistic performance, that make these materials suitable for highly specialized applications. High-temperature strength, light weight, corrosion resistance, and hardness make these composites especially attractive. B_4C shapes can be reaction-bonded using SiC as the bonding phase. B_4C–C mixtures are formed, then reacted with silicon to create the SiC bond. SiC also can be used as a sintering aid for B_4C, and vice versa.

As an abrasive, B_4C is used for fine polishing and ultrasonic grinding and drilling, either as a loose powder or as a slurry. Tendency to oxidize at workpiece temperatures precludes its use in bonded abrasive wheels. Abrasion-resistant parts made from B_4C include spray and blasting nozzles, bearing liners, and furnace parts. The refractory properties of B_4C, in addition to its abrasion resistance, are of value in the latter application.

B_4C is chemically inert, although it reacts with O_2 at elevated temperatures and with white hot or molten metals of the iron group, and certain transition metals. B_4C reacts with halogens to form boron halides — precursors for the manufacture of most non-oxide boron chemicals. B_4C also is used in some reaction schemes to produce transition metal borides. Boronizing packings containing B_4C are used to form hard boride surface layers on metal parts.

B_4C and elemental boron are used for nuclear reactor control elements, radiation shields, and moderators.

B_4C platelets are single crystals of B_4C or $B_{13}C_2$ that are of 1 to 5 mm average diameter and about 0.2 mm thick. The tiny B_4C platelets have proved to be highly effective in reinforcing ceramic materials. Toughness values of 8.1 MPa.m$^{1/2}$ have been reported in B_4C platelet-reinforced Al_2O_3 with corresponding four-point bed strengths of more than 650 MPa.

B_4C Whiskers

These are single crystals of B_4C or $B_{13}C_2$ that range in diameter from 1 to 10 mm and an average length of 30 to 300 mm. B_4C whiskers have been utilized in matrices of Al_2O_3, SiC, and Si_3N_4. Toughness values exceeding 9.0 MPa.m$^{1/2}$ have been reported in a matrix of Al_2O_3.

B_4C whiskers have been tested in metal-matrix composites with high values reported in ductility.

Another boron compound used for the production of high-temperature ceramic parts by pressing and sintering is boron silicide (B_4Si). It is a black, free-flowing crystalline powder. The powder is microcrystalline, with particles about 75 µm in diameter, and the free silicon is less than 0.15%. This compound normally reacts at 1200°C to form B_6Si and silicon, but when compacted and sintered, the ceramic forms a boron silicate oxygen protective coating, and the parts have a serviceable life in air at temperatures to 1400°C. Molded parts have high thermal shock resistance, and can be water-quenched from 1093°C without shattering.

In combination with plastics or aluminum, boron provides an effective lightweight neutron-shielding material. Boron-containing shields are valuable because of their satisfactory mechanical properties and because boron absorbs neutrons without producing high-energy gamma rays. Rods and strips of boron steel have been used extensively as control rods in atomic reactors.

The physical properties that make boron attractive as a construction material in missile and rocket technology are its low density (15% lighter than aluminum), extreme hardness, high melting point, and remarkable tensile strength in filament form. Production of boron fibers by vapor deposition methods has been developed on a commercial scale. These fibers are being used in an epoxy (or other plastic) carrier material or matrix. The resulting composite is stronger and stiffer than steel and 25% lighter than aluminum. The balance of strength and stiffness of the composite makes it ideal for aircraft applications, where high performance is of primary importance. Another development in this area is the incorporation of boron filaments in metal matrices.

The use of boron has found its way in such diversified applications as in motor-starting devices, phonograph needles, lightning arresters, thermal cutouts for transformers, igniters in rectifier and control tubes, alloys resistant to high-temperature abrasion and scaling, constant-potential controllers, thermoelectric couples, and resistance thermometers.

Certain compounds of boron, such as borax and boric acid, have been known and used for a long time in glass, enamel, ceramic, and mining industries. Refined borax, $Na_2B_4O_7 \cdot H_2O$, is an important ingredient of a variety of detergents, soaps, water-softening compounds, laundry starches, adhesives, toiletry preparations, cosmetics, talcum powder, and glazed paper. It is also used in fireproofing, disinfecting of fruit and lumber, weed control, and insecticides, as well as in the manufacture of leather, paper, and plastics.

BRASS

Brass is an alloy of copper and zinc. In manufacture, lump zinc is added to molten copper, and the mixture is poured either into castings ready for use or into billets for further working by rolling, extruding, forging, or a similar process. Brasses containing 75 to 85% copper are red-gold and malleable; those containing 60 to 70% are yellow and also malleable; and those containing 50% or less copper are white, brittle, and not malleable.

Copper–zinc alloys whose zinc content ranges up to 40% with the copper crystal structure face-centered cubic (fcc) are considered brass. This solid solution alpha brass has good mechanical properties, combining strength with ductility. Corrosion resistance is very good, but electrical conductivity is considerably lower than for copper.

Beta brass contains nearly equal proportions of copper and zinc. Specific brasses are designated as follows: gilding (95% copper; 5% zinc), red (85:15), low (80:20), and admiralty (70:29, with balance of tin). Naval brass is 59 to 62% copper, with about 1% tin, less than that of lead and iron, and the remainder zinc. With small amounts of other metals, other names are used. Leaded brass is used for castings. These alloys have essentially the same range of zinc content as the straight brasses. Lead is present, ranging from less than 1 to 3.25%, to improve machinability and related operations. Lead also improves anti-friction and bearing properties.

Another group, the tin brasses, are copper–zinc alloys with small amounts of tin. The tin improves corrosion resistance. Pleasing colors are also obtained when tin is added to the low brasses. Tin brasses in sheet and strip form, with 80% or more copper, are used widely as low-cost spring materials.

The mechanical properties of brasses vary widely. Strength and hardness depend on alloying and/or cold work. Tensile strengths of annealed grades are as low as 206 MPa, although some hard tempers approach 620 MPa.

Brass is annealed for drawing and bending by quenching in water from a temperature of about 538°C. Simple copper–zinc brasses are made in standard degrees of temper, or hardness. This hardness is obtained by cold-rolling after the first anneal, and the degree of hardness depends on the percentage of cold reduction. When the thickness is reduced about 10.9%, the resulting sheet is $1/4$-hard. The other grades are $1/2$-hard, hard, extra hard, spring, and finally extra spring, which corresponds to a reduction of about 68.7% without intermediate annealing. Degrees of softness in annealed brass are measured by the grain size, and annealed brass is furnished in grain sizes from 0.010 to 0.150 mm.

Even slight additions of other elements to brass alter characteristics drastically. Slight additions of tin change the structure, increasing the hardness but reducing the ductility. Iron hardens the alloy and reduces the grain size, making it more suitable for forging, but making it difficult to machine. Manganese increases strength, increases the solubility of iron in the alloy, and promotes stabilization of aluminum, but makes the brass extremely hard. Slight additions of silicon increase strength, but large amounts promote brittleness, loss of strength, and danger of oxide inclusion. Nickel increases strength and toughness, but when any silicon is present, the brass becomes extremely hard and more a bronze than a brass.

Brass is widely used in cartridge cases, plumbing fixtures, valves and pipes, screws, clocks, and musical instruments.

BRAZING

Brazing constitutes a group of welding processes in which coalescence is produced by heating to suitable temperatures above 450°C and by using a filler metal that must have a liquidus temperature above 450°C and below the solidus temperature of the base metal. The filler metal is distributed between the closely fitted surfaces of the joint by capillary attraction. Brazing is distinguished from soldering in that the latter employs a filler metal having a liquidus below 450°C.

Brazing provides advantages over other welding processes, especially because it permits joining of dissimilar metals that, because of metallurgical incompatibilities, cannot be joined by traditional fusion processes. Since base metals do not have to be melted to be joined, it does not matter that they have widely different melting points. Therefore, steel can be brazed to copper as easily as steel to steel.

Brazing also generally produces less thermally induced distortion (warping) than fusion welding does, because an entire part can be brought up to the same brazing temperature, thereby preventing the kind of localized heating that can cause distortion in welding.

Finally, and perhaps most important to the manufacturing specialist, brazing readily lends itself to mass-production techniques.

The important elements of the brazing process are filler-metal flow, base-metal and filler-metal characteristics, surface preparation, joint design and clearance, and temperature, time, rate, and source of heating.

A myriad of applications of brazing occur in many fields, including automotive and aircraft designs; engines and engine components; electron tubes; vacuum equipment and nuclear components, such as production of reliable ceramic-to-metal joints; and miscellaneous applications such as fabrication of food-service dispensers (scoops) used for ice cream and corrosion-resistant and leak-proof joints in stainless steel blood-cell washers.

BRICK

Brick is a construction material usually made of clay and extruded or molded as a rectangular block. Three types of clay are used in the manufacture of bricks: surface clay, fire clay, and shale. Adobe brick is a sun-dried molded mix of clay, straw, and water, manufactured mainly in Mexico and some southern regions of the United States.

The first step in manufacture is crushing the clay. The clay is then ground, mixed with water, and shaped. Then the bricks are fired in a kiln at approximately 1093°C.

Substances in the clay such as ferrous, magnesium, and calcium oxides impart color to the bricks during the firing process. The color may be uniform throughout the bricks, or the bricks may be manufactured with a coated face. The latter are classified as glazed, claycoat, or engobe. Engobes are coatings, also called slurries, that are applied to plastic or dry body brick units to develop the desired color and texture. Claycoat is a type of engobe that is sprayed on as a coating of liquid clay and pigments.

Clay bricks are manufactured for various applications. The most commonly used brick product is known as facing brick. In addition to standard bricks, decorative bricks molded in special shapes are available in both standard and custom sizes. They are used to form certain architectural details such as water tables, arches, copings, and corners. Bricks are also used to create sculptures and murals.

BRONZE

The term bronze is generally applied to any copper alloy that has as the principal alloying element a metal other than zinc or nickel. Originally the term was used to identify copper–tin alloys that had tin as the only, or principal, alloying element. Some brasses are called bronzes because of their color, or because they contain some tin. Most commercial copper–tin bronzes are now modified with zinc, lead, silver, or other elements.

Bronze is used in bearings, bushings, gears, valves, and other fittings both for water and steam. Tin bronze, including statuary bronze, contains 2–20% tin; bell metal 15–25%; and speculum metal up to 33%. Gun metal contains 8–10% tin plus 2–4% zinc.

The properties of bronze depend on its composition and working. Phosphor bronze is tin bronze hardened and strengthened with traces of phosphorus; it is used for fine tubing, wire springs, and machine parts. Lead bronze may contain up to 30% lead; it is used for cast parts such as low-pressure valves and fittings. Maganese bronze with 0.5 to 5% manganese plus other metals, but often no tin, has high strength. Aluminum bronze also contains no tin; its mechanical properties are superior to those of tin bronze, with up to 3% silicon, casts well, and can be worked hot or cold by rolling, forging, and similar methods. Beryllium bronze (also called beryllium copper) has about 2% beryllium and no tin. The alloy is hard and strong and can be further hardened and strengthened by precipitation hardening; it is one of the few copper alloys that responds to heat treatment, approaching three times the strength of structural steel.

Tin bronze is harder, stronger in compression, and more resistant to corrosion than the brasses. Bronze that will not be exposed to extremes of weather can be protected from corrosion by warming it to slightly over 100°C in an oxygen atmosphere. A thin layer of oxide or patina forms to prevent further oxidation. A patina may be formed on art objects by exposure first to acid fumes and then drying as above. While still warm, the object can be further protected by a spray of wax in a solvent.

For bearings, sintered bronze is compacted from 10% tin, up to 2% graphite, and the balance by weight of copper.

Gear bronze may be any bronze used for casting gears and worm wheels, but usually means a tin bronze of good strength deoxidized with phosphorus and containing some lead to make it easy to machine and to lower the coefficient of friction. A typical gear bronze contains 88.5% copper, 11% tin, 0.25% lead, and 0.25% phosphorus.

Bronzes containing more than 90% copper are reddish; below 90% the color changes to orange-yellow, which is the typical bronze color. Ductility rapidly decreases with increasing tin content. Above 20% tin the alloy rapidly becomes white in color and loses the characteristics of bronze. A 90% copper and 10% tin bronze weighs 8774 kg/m^3; an 80–20 bronze weighs 8719 kg/m^3. The 80–20 bronze melts at 1020°C, and a 95–5 bronze melts at 1360°C.

BUCKYBALLS

Buckyballs, the soccer-shaped molecules discovered over a decade ago, are made entirely of carbon atoms linked with unusual chemical bonds.

Buckybowls, large fragments of buckyballs, have been synthesized. The molecules have the ability to take up electrons and give them back later, under the right conditions, in higher concentrations than buckyballs.

The only practical application thus far has been the computer industry's use of another relative, the buckytube, as an atomic-scale probe. Some researchers believe that the buckybowls may facilitate the development of plastic batteries that would be lighter, smaller, and more environmentally friendly than the rechargeable batteries now used to power cellular phones and laptop computers.

The shape of the molecule may also allow it to bond with other molecules. It fits over the buckyball much like a contact lens, and may be able to serve as a medium that links other substances to the balls.

BUTADIENE–STYRENE THERMOSETTING RESINS

Butadiene–styrene resins are low-molecular-weight, all-hydrocarbon, thermosetting copolymers designed primarily for surface coatings. They form tough, inert films by numerous and different mechanisms: oxidation, polymerization, solvent evaporation, and through the use of cross-linking agents.

Proper selection of curing conditions and/or modifiers provides conversions ranging from seconds to hours at temperatures from 21.1 to 593°C.

Butadiene–styrene resins are prepared in a variety of forms. The basic resin is a viscous, clear material with a high degree of unsaturation. It is supplied in a solvent-free state for use in special applications such as can linings and thin film metal-strip coatings. It is readily soluble in hydrocarbon thinners and chlorinated hydrocarbons. Another resin is based on butadiene alone. It is similar in molecular weight and physical properties to the basic resin; however, it possesses certain metal wetting and fabrication characteristics that make it superior for such uses as tinplate coatings. Other resins are produced by modifying the base resin to introduce such polar groups as hydroxyls, carbonyls, and carboxyls. As a result, they have a much more active chemical nature and are more compatible with other coating resins than either of the previous two.

A wide range of film properties can be obtained by varying cure conditions or by modification with other resins and cross-linking agents. Basically, films have excellent hardness and flexibility as well as abrasion resistance equal or superior to other high-quality industrial coating resins.

Adhesion. Films provide excellent adhesion to steel, galvanized tinplate, aluminum, brass, and die cast zinc. A wide variety of plastics, wood, glass, untreated concrete, and most other surfaces can also be coated successfully.

Chemical resistance. Butadiene–styrene films have excellent resistance to corrosive atmospheres, particularly salt-spray and detergent solutions. In chemical resistance, the films are generally equivalent or superior to other chemically resistant coating systems. Typical properties of baked butadiene–styrene films include extended resistance to water, salt water, acids, ketones, alcohols, aromatics, and other solvents.

Some butadiene–styrene resins are frequently combined with other resins to obtain specific coating properties. They are used as modifiers for such leading conventional coating resins as alkyd, vinyl, acrylic, and epoxy. They can also be modified with other resins, primarily ureas, melamine, and nitrocellulose.

Butadiene–styrene resins can be cured by several systems — air-dry, low-temperature bake, high-temperature bake, and instant curing (2 s at 427°C).

Four means of instant cure have been found suitable: flame curing, high-density infrared heating, flame spraying, and induction heating. The principle is simply to supply a great deal of heat to the coating as rapidly as possible.

With the addition of metallic driers, butadiene-styrene resins will air-dry in 8 to 24 h. Curing time is rapid, however, when the resin is modified with nitrocellulose. This rapid, lacquer-type drying composition is particularly suitable for wood furniture and hardboard finishes.

The resins best suited for use in special thin-film coatings are up to 1.2 mils in thickness and are used in applications such as can linings, flat steel primers, beverage-can base coats, as well as sanitary enamels. Additionally, automobiles appliances, metal wall partitions, and preprimed steel sheeting are among other uses for the resins in the primer field.

In the lacquer field, a resin modified with nitrocellulose produces nonpenetrating, high-build, high-solids, chemically resistant finishes for such uses as wood furniture, metal cabinets, and plastics. This resin system is also suitable as the primer-surfacer for hardboard and wall paneling.

Other uses of butadiene–styrene resins include metal strip coatings and pipe coatings produced by instant cure methods and coatings for drum and tank liners.

BUTYL RUBBER

Butyl is a general-purpose synthetic rubber. There are two types of rubber in this category and both are based on crude oil. The first is polyisobutylene with an occasional isoprene unit inserted in the polymer chain to enhance vulcanization characteristics. The second is the same, except that chlorine is added (approximately 1.2% by weight), resulting in greater vulcanization flexibility and cure compatibility with general-purpose rubbers.

Butyl rubbers have outstanding impermeability to gases and excellent oxidation and ozone resistance. The chemical inertness is further reflected in lack of molecular-weight breakdown during processing, thus permitting the use of hot-mixing techniques for better polymer/filler interaction.

Flex, tear, and abrasion resistance approach those of natural rubber, and moderate-strength (14.3 MPa) unreinforced compounds can be made at a competitive cost. Butyls lack the toughness and durability, however, of some of the general-purpose rubbers.

Butyl is superior to natural rubber and styrene–butadiene resin (SBR) in resistance to aging and weathering, sunlight, heat, air, ozone and O_2, chemicals, flexing and cut-growth, and in impermeability to gases and moisture. Because of their resistance to deterioration, butyl vulcanizates retain their original properties in actual use longer than either natural rubber or SBR vulcanizates. On heating in an oxidative atmosphere, natural rubber and SBR tend to become hard and brittle, whereas butyl remains unchanged or becomes slightly softer. Butyl can be compounded to have abrasion resistance equal to that of natural rubber and SBR. Butyl also has excellent dielectric properties. In many of its performance properties butyl corresponds to the more expensive synthetics such as the polychloroprenes (neoprene), the polysulfides, and the various nitrile copolymers, although like natural rubber and SBR, butyl alone is neither oil and grease nor flame resistant. In common with other synthetics, butyl has advantages of uniformity and freedom from foreign matter, as compared with natural rubber.

Butyl has relatively high hysteresis loss, and thus unique dynamic properties that make it the preferred rubber in applications requiring shock, vibration, and sound absorption.

Butyl has found a great many applications in the automotive transport, mechanical goods, electrical, chemical, proffed-fabric, building, and consumer-goods fields.

The attribute responsible for the high-volume use of butyl rubber in automotive inner tubes and tubeless tire interliners is its excellent permeability to air.

Butyl passenger car tires have been successful because the unique dynamic properties of butyl give a remarkably soft, cushioned ride, no squeal on braking or rounding corners, and higher coefficients of friction between tires and road, which result in shorter stopping distances. The abrasion resistance of butyl tires is comparable to that of other commercial tires.

The resistance of butyl to weathering and to cutting and chipping has been demonstrated in off-the-road machinery and farm tractor tires. Its resistance weathering, sunlight, and ozone has opened many applications in automotive weatherstrips, windshield gaskets, curtain-wall gaskets and scalers, caulking compounds, and the like. Butyl-coated fabrics are extensively used as convertible tops, tarpaulins, outdoor furniture covers, etc. Such fabrics have been used as liners for irrigation ditches, with adjacent widths of fabric cemented together with butyl cement at the site to make linings wide enough for large ditches.

Resistance to heat is essential in many butyl applications, such as high-pressure steam hose. Butyl has become the preferred material for curing bags and bladders used in the vulcanization of tires.

Butyl is also used as the inner liner in many tubeless tires, particularly truck tires, because butyl liners remain impermeable even after exposure to the heat of several successive recapping operations.

The outstanding electrical properties of butyl, coupled with its age, ozone, and moisture resistance, have made it useful in many electrical applications. Its resistance to corona and tracking makes it a preferred insulation material for power cable, and because of its heat resistance, butyl-insulated cable can be used to carry more current than cable of equal diameter insulated with SBR or other rubbers. Butyl is also used in electrical encapsulation compounds, as bus-way, factory wire, and communications wire insulation, and in other miscellaneous electrical applications.

Many automotive and mechanical applications take advantage of the ability of butyl to absorb shock, vibration, and noise. These include axle and body bumpers, truck load cushions, boat dock bumpers, bowling alley bumpers, etc. Motor and machinery mounts, bridge pads, drive-shaft insulators, and gasketing applications also benefit from the good resistance of butyl to permanent set from prolonged compression and vibration. Although not usually considered oil resistant, butyl has given years of satisfactory under-the-hood service in spark plug nipples, motor mounts, grommets, radiator hose, and other parts occasionally exposed to oil and grease.

Chemical and heat resistance requirements are satisfied by butyl in applications such as chemical tank linings, gaskets, hose, diaphragms, etc. Butyl is the preferred gasketing material for use in contact with ester-type hydraulic fluids of low flammability. Its acid resistance is used to advantage in dairy hose, and in large rubber containers for the bulk transportation of foodstuffs. Chemical, heat, and abrasion resistance are essential in butyl conveyor belts used to transport hot granular solids in chemical plants. Resistance to chemicals and compression set and impermeability to gases are important in butyl food-jar seals and medicine bottle stoppers.

C

CADMIUM

Cadmium (symbol Cd) is a silvery-white crystalline metal that has a specific gravity of 8.6, is very ductile, and can be rolled or beaten into thin sheets. It resembles tin and gives the same characteristic cry when bent, but is harder than tin. A small addition of zinc makes it very brittle. It melts at 320°C and boils at 765°C. Cadmium is employed as an alloying element in soft solders and in fusible alloys, for hardening copper, as a white corrosion-resistant plating metal, and in its compounds for pigments and chemicals. It is also used for Ni–Cd batteries and to shield against neutrons in atomic equipment; but gamma rays are emitted when the neutrons are absorbed, and these rays require an additional shielding of lead.

Most of the consumption of cadmium is for electroplating. For a corrosion-resistant coating for iron or steel a cadmium plate of 0.008 mm is equal in effect to a zinc coat of 0.025 mm. The plated metal has a silvery-white color with a bluish tinge, is denser than zinc, and harder than tin, but electroplated coatings are subject to H_2 embrittlement, and aircraft parts are usually coated by the vacuum process. Cadmium plating is not normally used on copper or brass since copper is electronegative to it, but when these metals are employed next to cadmium-plated steel a plate of cadmium may be used on the copper to lessen deterioration.

Cadmium oxide is used extensively in plastics. In conjunction with barium it forms a compound used to stabilize the color of finished plastics. It is also a major constituent of phosphors used in television tubes.

Cadmium is mutually soluble in a number of other metallic elements. It is combined with several of these elements to form a number of commercial alloys with special properties.

The alloys of cadmium with lead, tin, bismuth, and indium are unique because of their low melting points. For example:

Alloy	Melting Point, °C
33% Cd–67% Sn	176
18% Cd–51.2% Sn–30.6%, Pb	145
40% Cd–60% Bi	144
20.2% Cd–25.9% Sn–53.9% Bi	103
10.1% Cd–13.1% Sn–49.5% Bi, 27.3% Pb	70
5.3% Cd–12.6% Sn–47.5% Bi, 25.4% Pb, 19.1% In	47

These fusible alloys are used in applications ranging from fire-detection apparatus to accurate proof casting.

Cadmium has limited use in soft solders combined with tin, lead, and zinc but its major application in the field of joining is for joints requiring higher-temperature strength than can be obtained with the soft solders. Cadmium combined with silver, copper, and zinc forms several brazing alloys. A typical alloy contains 35% silver, 26% copper, 21% zinc, 18% cadmium and has a melting point of 608°C. Cadmium (0.7 to 1%) is added to copper to make a strong ductile metal that has a high annealing temperature but no serious loss of electrical conductivity. (Trolley wire is an example of its use.)

The importance of cadmium in the nuclear field depends on its high thermal neutron capture cross section. By absorbing neutrons, cadmium is employed to control the rate of fission in a nuclear reaction.

Cadmium sulfide exhibits both photosensitivity and electroluminescence; i.e., it can convert light to electricity and electricity to light.

Finally, the fumes of cadmium, its compounds, and solutions of its compounds are very toxic, and cadmium-plated articles should not be used in food, nor should cadmium-coated articles be welded or used in ovens.

CALCIUM

Calcium (symbol Ca) is a metallic element belonging to the group of alkaline earths. It is one of the most abundant materials, occurring in combination in limestones and calcareous clays. The metal is obtained 98.6% pure by electrolysis of the fused anhydrous chloride. By further subliming, it is obtained 99.5% pure. Calcium metal is yellowish white in color. It oxidizes easily and, when heated in air, burns with a brilliant white light. It has a density of 1.55 g/cm^3, a melting point of 838°C, and a boiling point of 1440°C. Its strong affinity for O_2 and sulfur is utilized as a cleanser for nonferrous alloys. As a deoxidizer and desulfurizer it is employed in the form of lumps or sticks of calcium metal or in ferroalloys and Ca–Cu.

Many compounds of calcium are employed industrially, in fertilizers, foodstuffs, and medicine. It is an essential element in the formation of bones, teeth, shells, and plants. Oyster shells form an important commercial source of calcium for animal feeds. They are crushed, and the fine flour is marketed for stock feeds and the coarse for poultry feeds. The shell is calcium carbonate.

Edible calcium, for adding calcium to food products, is calcium lactate, a white powder of the composition $Ca(C_3H_5O_3)_2 \cdot 5\ H_2O$, derived from milk. Calcium lactobionate is a white powder that readily forms chlorides and other double salts, and is used as a suspending agent in pharmaceuticals. It contains 4.94% available calcium. Calcium phosphate, used in the foodstuffs industry and in medicine, is marketed in several forms. Calcium diphosphate, known as phosphate of lime, is $CaHPO_4 \cdot 2H_2O$, or in anhydrous form. It is soluble in dilute citric acid solutions, and is used to add calcium and phosphorus to foods, and as a polishing agent in toothpastes.

Calcium oxide is made by the thermal decomposition of carbonate minerals in tall kilns using a continuous-feed process. Care must be taken during the heating to decompose the limestone at a low enough temperature so that the oxide will slake freely with water. If too high a temperature is used, so-called dead-burnt lime is formed. The oxide is used in high-intensity arc lights (limelights) because of its unusual spectral features and as an industrial dehydrating agent. At high temperatures, lime combines with sand and other siliceous material to form fusible slags; hence, the metallurgical industry makes wide use of it during the reduction of ferrous alloys.

Because of the low cost of calcium hydroxide, it is used in many applications where hydroxide ion is needed. Slaked lime is an excellent absorbent for carbon dioxide (CO_2) to produce the carbonate. Because of the great insolubility of the carbonate, gases are easily tested qualitatively for CO_2 by passing them through a saturated lime–water solution and watching for a carbonate cloudiness. The hydroxide is also used in the formation of mortar, which is composed of slaked lime (1 volume), sand (3 to 4 volumes), and enough water to make a thick paste. The mortar gradually hardens because of the evaporation of the water and the cementing action of the deposition of calcium hydroxide, and because of the absorption of CO_2.

Calcium silicide, CaSi, is an electric-furnace product made from lime, silica, and a carbonaceous reducing agent. This material is useful as a steel deoxidizer because of its ability to form calcium silicate, which has a low melting point.

Calcium carbide is a hard, crystalline substance of grayish-black color, used chiefly for the production of acetylene gas for welding and cutting torches and for lighting. It is made by reducing lime with coke in the electric furnace, at 2000 to 2200°C. It can also be made by heating crushed limestone to a temperature of about 1000°C, flowing a high-methane natural gas through it, and then heating to 1700°C. The composition is CaC_2, and the specific gravity is 2.26. It contains theoretically 37.5% carbon. When water is added to calcium carbide, acetylene gas is formed, leaving a residue of slaked lime.

Calcium sulfate dihydrate is called gypsum. It constitutes the major portion of portland cement, and has been used to help reduce soil alkalinity. A hemihydrate of calcium sulfate, produced by heating gypsum at elevated temperatures, is known under the commercial name plaster of paris. When mixed with water, the hemihydrate reforms the dihydrate, evolving considerable heat and expanding in the process, so that a very sharp imprint of the mold is formed. Thus, plaster of paris finds use in the casting of small art objects and mold testing.

Calcium metal is employed as an alloying agent for aluminum-bearing metal, as an aid in removing bismuth from lead, and as a controller for graphitic carbon in cast iron. It is also used as a deoxidizer in the manufacture of many steels, as a reducing agent in preparation of such metals as chromium, thorium, zirconium, and uranium, and as a separating material for gaseous mixtures of N_2 and argon. When added to magnesium alloys (0.25%), it refines the grain structure, reduces their tendency to take fire, and modifies the strengthening heat treatment. It finds use also in the precipitation-hardening Pb–Ca alloys.

Calcium in the biosphere is an invariable constituent of all plants because it is essential for their growth. It is contained both as a structural constituent and as a physiological ion. The calcium ion is able to counteract the toxic effects of potassium, sodium, and magnesium ions. Calcium may also affect the growth of plants because its presence in soil affects the alkalinity of the latter.

Calcium is found in all animals in the soft tissues, in tissue fluid, and in the skeletal structures. The bones of vertebrates contain calcium as calcium fluoride, calcium carbonate, and calcium phosphate. In some lower animals, magnesium replaces either totally or partially the skeletal calcium. The importance of calcium in animals as a structural constituent is based on its abundance and on the low solubility of the three calcium salts just listed. Calcium is also essential in many biological functions of the vertebrates.

CALENDERED SHEET

Calendering is the process of forming a continuous sheet of controlled size by squeezing a softened thermoplastic material between two or more horizontal rolls. Along with extrusion and casting, calendering is one of the major techniques used to process thermoplastics into film and sheeting. It also is used in the manufacture of flooring and to apply a plastic coating on paper, textiles, and other supporting materials.

The calendering process is used in the plastics, rubber, linoleum, paper, and metals industries for various roll-forming operations in the manufacture of sheeted materials. This coverage will be concerned only with calendering of thermoplastic materials (plastics and elastomers, the latter of which are calendered in a thermoplastic state prior to vulcanization).

The process consists of five steps: preblending, fluxing, calendering, cooling, and wind-up. Blending of the resin powder with plasticizers, stabilizers, lubricants, colorants, and fillers is usually done in large ribbon blenders. The compound is fluxed, i.e., heated and worked until it reaches a molten or doughlike consistency.

When a Banbury mill is used, the molten material from it is discharged to a two-roll horizontal mill and thence to the calender either in a continuous strip or in batches. When an extruder is used for fluxing, the extrudate is fed directly to the calender. Alternatively, the preblend can in some cases be fed directly to the calender.

After passage through the calender, the continuous sheet of hot plastic is stripped off the last calender roll with a small, higher-speed stripping roll. The hot sheet is cooled as it travels over a series of cooling drums. The film or sheeting is finally automatically cut into individual sheets or wound up in a continuous roll.

The desired surface finish on the calendered film or sheeting is imparted by the last pair of calender rolls and may range from a high gloss to a heavy matte finish. Extra-high gloss or special engraved patterns can be made either by having a polished or engraved roll impinge on the last calender roll (contact embossing) or by passing the hot sheeting directly from the calender between an engraved metal roll and a rubber backup roll (in-line embossing). Many attractive patterns are made by these techniques.

The major groups of thermoplastics and elastomers that are calendered into film or sheeting are polyvinyl chloride (PVC) (plasticized and unplasticized) and natural and synthetic rubbers. ABS (acrylonitrile, butadiene, and styrene) polymers, polyolefins, and silicones are also calendered, but in much smaller quantities.

Calendered vinyl is used extensively for floor tile and continuous flooring, rainwear, shower curtains, table covers, pressure-sensitive tape, automotive and furniture upholstery, wall coverings, luminous ceilings, signs and displays, credit cards, etc.

In contrast, calendered rubber — except for some fabric coating — is mainly an intermediate product used in the manufacture of a multitude of articles such as automobile tires, footwear, molded mechanical goods, etc.

In applications that involve the use of calenders for paper and cloth coating, the paper or fabric is fed into the last calender nip so that the plastic or rubber coating is formed on top of the material (calender coating). Frictioning is a variation of this technique by which a thin layer of rubber is squeezed or rubbed into the fabric itself, a technique widely used in the rubber industry for the manufacture of friction tapes and cord fabric for tires.

Production speeds vary greatly with materials, from 10 to 300 fpm (feet per minute).

Calenders are heavy, high-precision machines that are made in a variety of designs and with elaborate control equipment that automatically monitors film variations and readjusts the machine to produce the desired thickness.

CAPACITOR

A capacitor is an electrical device capable of storing electrical energy. In general, a capacitor consists of two metal plates insulated from each other by a dielectric. The capacitance of a capacitor depends primarily upon its shape and size and upon the relative permittivity ε_r of the medium between the plates. In vacuum, in air, and in most gases, ε_r ranges from one to several hundred.

One classification of capacitors comes from the physical state of their dielectrics, which may be gas (or vacuum), liquid, solid, or a combination of these. Each of these classifications may be subdivided according to the specific dielectric used. Capacitors may be further classified by their ability to be used in AC or DC circuits with various current levels.

Capacitors are also classified as fixed, adjustable, or variable. Capacitors made with air, gas, or vacuum as the dielectric between plates are constructed with flat parallel metallic plates (or cylindrical concentric metallic plates).

Solid-dielectric types of capacitors use one of several dielectrics such as a ceramic, mica, glass, or plastic film. Alternate plates of metal, or metallic foil, are stacked with the dielectric, or the dielectric may be metal-plated on both sides.

Plastic-film types are capacitors that use dielectrics such as polypropylene, polyester, polycarbonate, or polysulfone with a relative permittivity ranging from 2.2 to 3.2. This plastic film may be used alone or in combination with Kraft paper. The most common electrodes are aluminum or zinc vacuum-deposited on the film, although aluminum foil is also used.

Other major types of capacitors can be seen in Table C.1.

CARBIDES

Of the several classes of carbides, only two types need be considered for engineering applications: the covalent carbides of silicon and boron, and the interstitial carbides of the transition metals — titanium, zirconium, vanadium, niobium (columbium), tantalum, chromium, molybdenum, and tungsten. All of these may be characterized as hard, refractory materials of extreme chemical inertness. Other carbides such as those of aluminum, iron, and manganese are too reactive to be considered for engineering applications. Chemically, all the inert carbides, with the exception of SiC, are unusual in that they exist in their typical form over a range of composition. In this respect they are more similar to alloys than true chemical compounds.

Structurally, interstitial carbides may be described as metal lattices into which the small carbon atoms have been inserted. In the ideal form this leads to an fcc (face-centered cubic) structure for all of the metal carbides (MC) types except tungsten carbide (WC), which like the M_2C types has a simple hexagonal structure. B_4C is rhombohedral, and consists of a distorted

TABLE C.1
Major Types of Capacitors

Type	Capacitance	Voltage (working voltage, DC), V[a]	Applications
Monolithic ceramics	1 pF–2.2 μF	50–200	Ultrahigh frequency, rf coupling, computers
Disk and tube ceramics	1pF–1 μF	50–1000	General, very high frequency
Paper	0.001–1 μF	200–1600	Motors, power supplies
Film			
Polypropylene	0.0001–0.47 μF	400–1600	Television vertical circuits, rf circuits
Polyester	0.001–4.7 μF	50–600	Entertainment electronics
Polystyrene	0.001–1 μF	100–200	General, high stability
Polycarbonate	0.01–18 μF	50–200	General
Metallized polypropylene	4–60 μF	400[a]	Alternating-current motors
Metallized polyester	0.001–22 μF	100–1000	Coupling, rf filtering
Electrolytic			
Aluminum	100,00 μF	5–500	Power supplies, filters
Tantalum	0.1–2200 μF	3–150	Small space requirement, low leakage
Gold	0.022–10 μF	2.5–5.5	Memory backup
Nonpolarized (either aluminum or tantalum)	0.47–1000 μF	10–200	Loudspeaker crossovers
Mica	330 pF–0.05 μF	50–1000	High frequency
Silver-mica	5–820 pF	50–500	High frequency
Variable			
Ceramic	1–5 to 16–100 pF	200	Radio, television, communications
Film	0.8–5 to 1.2–30 pF	50	Oscillators, antenna, rf circuits
Air	10–365 pF	50	Broadcast receiver
Poly(tetrafluoroethylene)	0.25–1.5 pF	2000	Very high frequency, ultrahigh frequency

[a] Alternating-current voltage at 60 Hz.

Source: McGraw-Hill Encyclopedia of Science and Technology, 8th ed., Vol. 3, McGraw-Hill, New York, 215. With permission.

B lattice in which part of the boron atoms have been replaced by carbon atoms. SiC exists in a number of crystalline polytypes, both cubic (diamond or zinc blende structures) and hexagonal (wurtzite type). All these structures may be described on the basis that every silicon atom is surrounded tetrahedrally by four carbon atoms and that every carbon atom is surrounded tetrahedrally by four silicon atoms.

Properties of carbides that make them unique are their extreme hardness, exceptional corrosion resistance, extreme refractoriness, high Young's modulus of elasticity, and high-temperature strength.

The simplest method to produce carbides is a direct combination of the metal with carbon, used exclusively in the preparation of carbides where purity is more important than cost considerations. Most carbides are, however, prepared by reduction of metal oxides with carbon.

With the exception of SiC, all carbide bodies are fabricated using powder metallurgical techniques, i.e., sintering and hot pressing. SiC bodies are formed by infiltrating a preformed carbonaceous body with elemental silicon, forming SiC in the shape desired.

Complete pump assemblies have been built to transfer molten metals (corrosion resistance); pipes for heat exchangers (high thermal conductivity); cyclone separators and sandblast nozzles (abrasion resistance); rocket nozzles (refractoriness and erosion resistance); electric-light sources (high melting temperature); and suction box covers for papermaking machines (ability to retain smooth finish without wear).

Several metal carbides mentioned above qualify as engineering ceramics. Most commonly used are B_4C and SiC.

B_4C is noted for its very high hardness and low density — unusual qualities for a brittle ceramic — which qualify this ceramic for lightweight, bulletproof armor plate. The material has the best abrasion resistance of any ceramic, so it is also specified for pressure-blasting nozzles and similar high-wear applications. A limitation of B_4C is its low strength at high temperatures.

Despite its higher cost, SiC is challenging Al_2O_3, particularly for the more critical applications. SiC is one of the high-temperature, high-strength "superstars" of the engineering ceramics. It is one of the strongest structural ceramics for high-temperature oxidation-resistant service. However, SiC does not easily self-bond. Consequently, many processing variations have been devised to fabricate parts from this material, creating a number of trade-offs in cost, fabricability, and properties. The ceramic can be consolidated by hot pressing. Under the combination of high temperature and pressure — with, in some cases, additives that act as bond-forming catalysts — fully dense material can be formed.

The hot-pressed ceramic is extremely strong and tough at high temperatures, but the manufacturing process is limited to simple shapes, bars, or billets. Complex parts made by hot pressing must be machined to shape — a slow and costly process of ultrasonic machining, EDM (if possible), or diamond grinding.

On the other hand, SiC particles can be bonded without pressure by a number of processes, variously called reaction bonding, recrystallization (for SiC), or reaction sintering. With these processes, "green" parts can be dry or isostatically pressed, extruded, slip-cast, or, in some cases, formed by conventional plastic molding techniques such as injection molding, then sintered. Complex shapes, close to finished size, can be produced by these techniques, but the ceramic is only about 80% as dense as the hot-pressed counterpart and has lower strength and poorer thermal shock resistance. SiC — either hot pressed or reaction bonded — is not as strong as Si_3N_4 up to about 1427°C; Si_3N_4 grain boundaries soften, or creep, and strength drops. Above 1427°C, SiC is the stronger ceramic. At 1316°C, however, strength of hot-pressed ceramics nearly equals that of reaction-bonded ceramics.

Hot-pressed SiC is harder and more difficult to EDM than Si_3N_4, which has lower thermal expansion and better thermal shock resistance than SiC. Electrical resistivity of SiC is low at low frequencies and high at high frequencies — an unusual characteristic that qualifies this material as a semiconductor.

REFRACTORY HARD METALS

Refractory hard metals (RHMs) are a ceramic-like class of materials made from metal-carbide particles bonded together by a metal matrix. Often classified as ceramics and sometimes called cemented or sintered carbides, these metals were developed for extreme hardness and wear resistance.

The RHMs are more ductile and have better thermal shock resistance and impact resistance than ceramics, but they have lower compressive strength at high temperatures and lower operating temperatures than most ceramics. Generally, properties of RHMs are between those of conventional metals and ceramics. Parts are made by conventional powder-metallurgy compacting and sintering methods. Many metal carbides such as SiC and B_4C are not RHMs but are true ceramics. The fine distinction is in particle binding: RHMs are always bonded together by a metal matrix, whereas ceramic particles are self-bonded. Some ceramics have a second metal phase, but the metal is not used primarily for bonding.

Four RHM systems are used for structural applications and, in most cases, several grades are available within each system. WC with a 3 to 20% matrix of cobalt is the most common structural RHM. The low-cobalt grades are used for applications requiring wear resistance; the high-cobalt grades serve where impact resistance is required.

TaC and WC combined in a matrix of nickel, cobalt, and/or chromium provide an RHM formulation especially suited for a combination of corrosion and wear resistance. Some grades are almost as corrosion resistant as platinum. Nozzles, orifice plates, and valve components are typical uses.

TiC in a molybdenum and nickel matrix is formulated for high-temperature service. Tensile and compressive strengths, hardness, and oxidation resistance are high as 1093°C. Critical parts for welding and thermal metalworking tools, valves, seals, and high-temperature gauging equipment are made from grades of this RHM.

Tungsten-titanium carbide ($WTiC_2$) in cobalt is used primarily for metal-forming applications such as draw dies, tube-sizing mandrels, burnishing rolls, and flaring tools. The $WTiC_2$ is a gall-resistant phase in the RHM containing WC as well as cobalt.

CARBON

Carbon is a nonmetallic element, symbol C, existing naturally in several allotropic forms, and in combination as one of the most widely distributed of all the elements. It is quadrivalent, and has the property of forming chain and ring compounds, and there are more varied and useful compounds of carbon than of all the other elements. The black amorphous carbon has a specific gravity of 1.88; the black crystalline carbon known as graphite has a specific gravity of 2.25; the transparent crystalline carbon, as in the diamond, has a specific gravity of 3.51. Amorphous carbon is not soluble in any known solvent. It is infusible, but sublimes at 3500°C, and is stable and chemically inactive at ordinary temperatures. At high temperatures it burns and absorbs O_2, forming the simple oxides CO and CO_2; the latter is the stable oxide present in the atmosphere and a natural plant food. Carbon dissolves easily in some molten metals, notably iron, exerting great influence on them. Steel, with small amounts of chemically combined carbon, and cast iron, with both combined carbon and graphitic carbon, are examples of this.

Carbon occurs as hydrocarbons in petroleum, and as carbohydrates in coal and plant life, and from these natural basic groupings an infinite number of carbon compounds can be made synthetically. Carbon for chemical, metallurgical, or industrial use is marketed in the form of compounds in a large number of different grades, sizes, and shapes; or in master alloys containing high percentages of carbon; or as activated carbons, charcoal, graphite, carbon black, coal-tar carbon, petroleum coke; or as pressed and molded bricks or formed parts with or without binders or metallic inclusions. Natural deposits of graphite, coal tar, and petroleum coke are important sources of elemental carbon.

Combined with carbonaceous binders, such as tars, pitches, and resins, the carbon is compacted by molding or extrusion, and baked at between 816 and 1649°C to produce what is known as industrial carbon or baked carbon. Conventional industrial graphite (Gr) is made by mixing mined, natural graphite with carbon to produce in effect a C–Gr composite, or baked carbon can be heat treated at about 2985°C, at which temperature the carbon graphitizes.

Manufactured or artificial carbon has a two-phase structure consisting of carbon particles (or grains) in a matrix of binder carbon. Both phases consist essentially of disordered, or uncrystallized, carbon surrounding embryonic carbon crystallites.

Graphites, except for the pyrolytic types, have a two-phase structure similar to that of carbon, but as the result of high-temperature processing contain well-developed graphite crystallites in both phases. These multicrystalline graphites exhibit many of the properties of single-crystal graphite, such as high electrical conductivity, lubricity, and anisotropy. Compared to carbon, graphite has higher electrical and heat conductivity, better lubricity, and is easier to machine. Because of their more favorable properties, graphites have broader application as engineering materials than do carbons. Further discussion on graphite will be found under **Graphite**.

TYPES AND FORMS OF CARBON

Elemental carbon exists in two well-defined crystalline allotropic forms, diamond and graphite. Other forms, which are poorly developed in crystallinity, are charcoal, coke, and carbon black.

Charcoal is prepared by the ignition of wood, sugar, blood, and other carbon-containing compounds in the absence of air. X-ray diffraction studies reveal that it has a graphite structure but is not very well developed in crystallinity. The lack of crystallinity is the result of defects in the crystal structure and the high surface area. In the activated state, charcoal adsorbs gases, liquids, and solids.

Coke, another form of amorphous carbon, is prepared by heating coal in the absence of air. It is used primarily for the reduction of metal oxides to the free metals.

Chemically pure carbon is prepared by the thermal decomposition of sugar (sucrose) in the absence of air. Impurities in the carbon are removed by treatment with chlorine (Cl) gas at red heat. The substance is then washed with water and the low residual chlorine is removed by heating in an atmosphere of H_2 gas.

Carbon-13 is one of the isotopes of carbon, used as a tracer in biologic research where its heavy weight makes it distinguished from other carbon. Carbon-14, or radioactive carbon, has a longer life. It exists in the air, formed by the bombardment of N_2 by cosmic rays at high altitudes, and enters into the growth of plants. The half-life is about 6000 years. It is made from N_2 in a cyclotron.

CARBON FIBERS/YARN/FABRIC/WOOL

These forms are made by pyrolysis of organic precursor fibers in an inert atmosphere. Pyrolysis temperatures can range from 1000 to 3000°C; higher process temperatures generally lead to higher-modulus fibers. Only three precursor materials, rayon, polyacrylonitrile (PAN), and pitch have achieved significance in commercial production of carbon fibers (see Table C.2). The first high-strength and high-modulus carbon fibers were based on a rayon precursor. These fibers were obtained by being stretched to several times their original length at temperatures

TABLE C.2
Carbon Fibers

	Nominal Fiber Modulus (10^6 psi)							
	Pan Fibers					Pitch Fibers		
	30[a]	40[b]	40[c]	50[d]	70[e]	50[f]	75[g]	100+[h]
Fiber strength (10^3 psi)	500	820	650	350	300	275	300	360
Tensile modulus (10^6 psi)	33	41	42	57	75	55	75	110
Composite strength (10^3 psi)								
Longitudinal	250	470	360	205	110	135	140	180
Transverse	10	10	10	5	4	5	5	3
Composite strength (10^6 psi)								
Longitudinal	20	25	25	35	44	35	49	72
Transverse	2	2	1.5	1	4	1	1	1

[a] Thornel 300 (Amoco).
[b] Thornel T-40.
[c] Thornel T650/42.
[d] Thornel 50 PAN.
[e] Celanese GY-70.
[f] Thornel P-55S.
[g] Thornel P-75S.
[h] Thornel P-100S.

Source: Mach. Design Basics Eng. Design, June, p. 730, 1993. With permission.

above 2800°C. The second generation of carbon fibers is based on a PAN precursor and has achieved market dominance. In their most common form, these carbon fibers have a tensile strength ranging from 2413–3102 MPa, a modulus of 0.2–0.5 GPa, and a shear strength of 90–117 MPa. This last property controls the traverse strength of composite materials. The high-modulus fibers are highly graphitic in crystalline structure after being processed from PAN at temperatures in excess of 1982°C. Higher-strength fibers obtained at lower temperatures from rayon feature a higher carbon crystalline content. There are also carbon and graphite fibers of intermediate strength and modulus. The third generation of carbon fibers is based on pitch as a precursor. Ordinary pitch is an isotropic mixture of largely aromatic compounds. Fibers spun from this pitch have little or no preferred orientation and hence low strength and modulus. Pitch is a very inexpensive precursor compared with rayon and PAN. High-strength and high-modulus carbon fibers are obtained from a pitch that has first been converted into a mesophase (liquid crystal). These fibers have a tensile strength of more than 2069 MPa, and a Young's modulus ranging from 0.38 to 0.52 GPa.

The average filament diameter of continuous yarn is 0.008 mm. Pitch-based carbon and graphite fibers are expected to see essentially the same applications as the more costly PAN and rayon-derived fibers, e.g., ablative, insulation, and friction materials and in metals and resin matrices.

Carbon fibers added to thermoplastic resins provide the highest strength, modulus, heat-deflection temperature, creep, and fatigue — endurance values commercially available in composites. These property improvements, coupled with greatly increased thermal conductivity and low friction coefficients, make carbon fibers ideal for wear and frictional applications where the higher cost can be tolerated. In applications where the abrasive nature of glass fibers wears the mating surface, the softer carbon fibers can be substituted to reduce the wear rate. Carbon fibers can also be used in conjunction with internal lubricants to further improve surface characteristics of most thermoplastic resin system.

Another useful property of carbon-fiber-reinforced thermoplastics is their low volume and surface resistivities. Most resin systems reinforced with 15% or more carbon fibers can effectively dissipate static charge, which is a problem common to gears, slides, and bearings used in business machine, textile, electrical, and conveying equipment.

Thornel is a yarn made from these filaments for high-temperature fabrics. It retains its strength to temperatures above 1538°C. Carbon yarn is 99.5% pure carbon. It comes in plies from 2 to 30, with each ply composed of 720 continuous filaments of 0.0076-mm diameter. Each ply has a breaking strength of 0.91 kg. The fiber has the flexibility of wool and maintains dimensional stability to 3150°C. Ucar is a conductive carbon fabric made from carbon yarns woven with insulating glass yarns with resistivities from 0.2 to 30Ω for operating temperatures to 288°C. Carbon wool, for filtering and insulation, is composed of pure carbon fibers made by carbonizing rayon. The fibers, 5 to 50 μm in diameter, are hard and strong, and can be made into rope and yarn, or the mat can be activated for filter use.

Carbon
Brushes/Brick/Paper/Electrodes

Carbon brushes for electric motors and generators, and carbon electrodes, are made of carbon in the form of graphite, petroleum coke, lampblack, or other nearly pure carbon, sometimes mixed with copper powder to increase the electric conductivity, and then pressed into blocks or shapes and sintered.

Carbon brick, used as a lining in the chemical-processing industries, is carbon compressed with a bituminous binder and then carburized by sintering. If the binder is capable of being completely carbonized, the bricks are impervious and dense.

There is a carbon filter paper for filtering hot gases and liquids made from carbon fibers pressed into a paperlike mat, 0.18 to 0.127 mm thick, and impregnated with activated carbon.

Diamond

Diamond is the cubic crystalline form of carbon. When pure, diamond is water clear, but impurities add shades of opaqueness including black. It is the hardest natural material with a hardness on the Knoop scale ranging from 5500 to 7000. It will scratch and be scratched by the hardest anthropogenic material Borozon. It has a specific gravity of 3.5. Diamond has a melting point of around 3871°C, at which point it will graphitize and then vaporize. Diamonds are generally electrical insulators and nonmagnetic. Synthetic diamonds are produced from graphite at extremely high pressures (5444 to 12,359.9 MPa) and temperatures from 1204 to 2427°C. They are up to 0.01 carat in size and are comparable to the quality of industrial diamonds. In powder form they are used in cutting wheels. Of all diamonds mined, about 80% by weight are used in industry. Roughly 45% of the total industrial use is in grinding wheels. Tests have shown that under many conditions synthetic diamonds are better than mined diamonds in this application.

Carbides

Carbon forms binary compounds known as carbides with elements less electronegative than carbon (C–H_2 compounds are excluded). Effectively then, carbides are composed of metal–carbon compounds if boron and silicon are included among the normal metals. Essentially no volatile compounds (except AlC) are known, because decomposition sets in at higher temperatures before volatization of the carbide as such can occur.

Most carbides can be prepared by heating a mixture of the powdered metal and carbon, usually to high temperatures, but not necessarily as high as the melting point. Generally, the same result is possible by heating a mixture of the oxide of the metal with carbon.

It is useful to classify carbides as ionic (salt-like), metallic (interstitial), and covalent. The more electropositive elements (groups I, II, and III, and to some extent, members of the lanthanide and actinide series of the periodic system) form ionic carbides with transparent (or saltlike) crystals.

Highly refractory covalent carbides are formed with silicon and boron. SiC (diamond structure) is formed from a mixture of SiO_2 and coke. The very hard B_4C can be made similarly from its oxide; it is unusual both structurally and in having a fairly high electrical conductivity.

The carbides of chromium, manganese, iron, cobalt, and nickel are intermediate between the interstitial and covalent types, but are much nearer the former in properties. The presence of Fe_3C in iron is an important factor in the properties of steel.

Mechanical/Thermal/Electrical/Other Properties

Compared to metals and most polymers, room-temperature tensile strengths of conventional carbon and graphite are low, ranging from 6.8 to 9.16 MPa. Compressive strengths range from 20.6 to 54.9 MPa. These are generally with-grain values (i.e., specimen length is parallel to the grain).

Carbon and graphite do not have a true melting point. They sublime at 4399°C. Conventional graphites have exceptionally high thermal conductivity at room temperature, whereas carbon has only fair conductivity. Conductivity with the grain in graphite is comparable to that of aluminum; across the grain it is about the same as brass. Conductivity increases with temperature up to about 0°C; it then remains relatively high, but decreases slowly over a broad temperature range, before it drops sharply. In pyrolitic graphite, thermal conductivity with the grain approaches that of copper; across grain, it serves as a thermal insulator and is comparable to that of ceramics.

Thermal expansion of carbon and graphite is quite low (1 to 1.5×10^{-6}/°F)—less than one third that of many metals. Expansion of graphite across grain increases with increasing density, whereas along the grain it decreases with increasing density. But expansion increases in both directions with increasing temperatures.

The electrical characteristics associated with carbon and graphite use as electrodes or anodes are relatively well known. Carbon and graphite are actually semiconductors, with their electrical resistivity, or conductivity, falling between those of common metals and common semiconductors. At temperatures approaching absolute zero, carbon and graphite have few conducting electrons, the number increasing with increasing temperature. Thus, electrical resistivity decreases with increasing temperature. On the other hand, although increasing electron density tends to reduce resistivity as temperature rises, scattering effects may become dominant at certain temperatures in the range of 982°C and thus modify or even reverse this trend. Pyrolytic graphite with its higher density has improved electrical conductivity (along the grain). Further, its high degree of anisotropy results in a high degree of electrical resistivity across the grain.

Other properties of graphite are excellent lubricity and relatively low surface hardness; carbon has fair lubricity and relatively high surface hardness. Further, certain types of carbon graphitize relatively easily; others do not. Consequently, a wide variety of carbon, C–Gr, and graphite materials are available, each designed to provide specific types of surface characteristics for such uses as bearings and seals. Grades are also available impregnated with a wide variety of substances, from synthetic resin or oil to a bearing metal.

The nuclear grades of carbon and graphite are of exceptionally high purity. As a moderator and reflector in nuclear reactors, they have no equal because of their low thermal neutron absorption cross section and high scattering cross section coupled with high strength at elevated temperatures and thermal stability in nonoxidizing environments. In general, the properties of carbon and graphite are improved by exposure to nuclear radiation. Hardness and strength increase while thermal and electrical conductivity decrease.

Applications

The free element of carbon has many uses, ranging from ornamental applications of the diamond in jewelry to the black-colored pigment of carbon black in automobile tires and printing inks. Another form of carbon, graphite, is used for high-temperature crucibles, arc-light and dry-cell electrodes, lead pencils, and as a lubricant. Charcoal, an amorphous form of carbon, is used as an absorbent for gases and as a decolorizing agent.

The compounds of carbon find many uses. CO_2 is used for the carbonation of beverages, for fire extinguishers, and in the solid state as a refrigerant. CO finds use as a reducing agent for many metallurgical processes. Carbon tetrachloride and carbon disulfide are important solvents for industrial uses. Gaseous dichlorodifluoromethane, commonly known as Freon, is used in refrigeration devices. Calcium carbide is used to prepare acetylene, which is used for the welding and cutting of metals as well as for the preparation of other organic compounds. Other metal carbides find important uses as refractories and metal cutters.

Carbon nanotubes may one day be used to construct extremely small-scale circuits. Scientists have observed ballistic conductance — electrons passing through a conductor without heating it — at room temperature in 5-µm carbon nanotubes. Because the nanotubes remain cool, extremely large current densities can flow through them.

The most common use of manufactured carbon is as sliding elements in mechanical devices. It is used as the primary rubbing face in most mechanical seals. Used as a brush, it transfers electrical current to the rotating commutator on small electric motors. Carbon vanes, piston rings, or cylinder liners are used in most small air pumps and drink-dispenser pumps. Carbon is also used for pistons in chemical-metering pumps and for metering valves in gasoline pumps. All these applications require the carbon to slide on metal with a coefficient of friction below 0.2 and a wear rate below 0.001 in. per million inches rubbed.

Bearings are another significant application area for manufactured carbon. In many cases, characteristics of the carbon bearing are tailored to satisfy a wide range of requirements. This is done by impregnating the porous, as-baked or as-graphitized carbon with various materials — for example, resin, babbitt, copper, or glass — or by combining impregnation with chemical conversion of the carbon surface (to hard SiC).

These self-lubricating materials are particularly suited for environments containing dust or lint, repeated steam cleaning, solvents or corrosive fluids, low or high temperatures, high static loads, or hard vacuums.

CARBON–CARBON COMPOSITES

Some of these conditions require impregnants in the carbon material. Other applications using manufactured carbon bearings are those inaccessible for lubrication or where product contamination (from lubricants) cannot be tolerated.

CARBON–CARBON COMPOSITES

Carbon–carbon composites (C–C) provide high strength, light weight, and resistance to high temperature and corrosion for pistons in both stationary and mobile engines.

Advantages in using carbon–carbon for both gasoline and diesel engines include increased performance (higher power output and faster engine acceleration); higher temperature operation for improved fuel economy; reduced levels of air pollutant emissions; and reduced noise, vibration, and harshness (NVH) levels. These benefits also could provide lower operational costs and reduced environmental impact.

Another carbon–carbon originally developed for use as rocket nozzles is now a replacement for graphite, quartz, and ceramic furnace components used in Czochralski crystal pulling, a processing step in semiconductor manufacturing. The material has a high strength-to-weight ratio, maintains its strength and rigidity despite high temperatures, resists wear, and is relatively chemically inert. When used in Czochralski furnaces, the thermal properties of the material lead to a 10% reduction in energy use and a 50% cut in cooling times. Because of the purity of the material, chipmakers achieve a 7% increase in wafer yields compared to furnaces with graphite components. More information can be found under **Composite Materials**.

CARBON MONOXIDE

Carbon monoxide (CO) is a product of incomplete combustion, and is very reactive. It is one of the desirable products in synthesis gas for making chemicals; the synthesis gas made from coal contains at least 37% CO. It is also recovered from top-blown O_2 furnaces in steel mills. It reacts with H_2 to form methanol, which is then catalyzed by zeolites into gasoline. Acetic acid is made by methanol carbonylation, and acrylic acid results from the reaction of CO, acetylene, and methanol. CO forms a host of neutral, anionic, and cationic carbonyls, with such metals as iron, cobalt, nickel, molybdenum, chromium, rhodium, and ruthenium. These metals are from groups I, II, VI, VII, and VIII of the periodic table. The metal carbonyls can be prepared by the direct combination of the metal with CO, although several of the compounds require fairly high pressures. The metal carbonyls react with the halogens to produce metal carbonyl halides. With H_2, a similar reaction takes place to form metal carbonyl hydrides.

Nickel carbonyl finds application in purification and separation of nickel from other metals. Iron carbonyl has been used in antiknock gasoline preparations and to prepare high-purity iron metal.

CO is an intense poison when inhaled and is extremely toxic even in the small amounts from the exhausts of internal-combustion engines.

With chlorine, in the presence of sunlight, CO forms highly poisonous phosgene, $COCl_2$; with sulfur, carbonyl sulfide, COS, is obtained.

CARBON NITRIDE

Traditional materials synthesis techniques, which require high-temperature conditions to facilitate diffusion of atoms in the solid state, generally preclude the rational assembly of atoms since the products are restricted to those phases stable at high temperature. For example, heating graphite in molecular N_2 does not provide carbon nitride (C–N) solids. However, binary C–N solids are desirable materials to prepare and study because they may possess extreme hardness and high thermal conductivity. Extreme hardness is important in thin-film coatings of cutting tools used for machining and in bearing surfaces used in a variety of high-performance mechanical devices, whereas high thermal conductivity is important to the fabrication of advanced microelectronics devices. To overcome the limitations imposed by classical methodologies of solid-state chemistry and provide access to these potentially exciting materials requires better control of the reactants and reactions that might lead to C–N solids. A new experimental approach that meets these requirements and has provided access to C–N materials utilizes pulsed-laser evaporation of graphite to generate reactive carbon fragments and an atomic N_2 beam as a source of N_2 that can readily react with the carbon fragments.

The atomic N_2 beam is generated by using a radio-frequency discharge within an Al_2O_3 nozzle through which N_2 seeded in helium flows. This process produces a very high flux of atomic N_2. Furthermore, by varying the N_2:He ratio and the radio-frequency power, it is possible to control both the flux and energy of this critical reactant. Hence, this synthetic approach provides a ready means of producing reactants that are free from impurities and that have controllable energies.

STRUCTURE

The structural properties of the C–N materials produced by the laser ablation approach have been investigated and show that carbon and N_2 are bound covalently within this solid but cannot provide information about the three-dimensional structure. Diffraction ring patterns suggest that a single crystalline C–N phase was obtained by using the new synthetic strategy.

OTHER KEY PROPERTIES

The new C–N materials also exhibit interesting physical properties that may be attractive for high-performance engineering applications. Qualitative scratch tests indicate that C–N materials produced by the new laser ablation technique are hard. For example, rubbing C–N and hard amorphous carbon surfaces against one another produces damage in the carbon but not the C–N material.

Studies have shown that C–N is an excellent electrical insulator, and that these electrical properties are stable to thermal cycling. Because C–N is also expected to exhibit good thermal conductivity, it could be an attractive candidate for the dielectric in advanced microelectronic devices where thermally conducting electrical insulators are needed to enable further miniaturization of devices.

CARBON STEEL

Carbon (C) steel, also called plain carbon steel, is a malleable, iron-based metal containing carbon, small amounts of manganese, and other elements that are inherently present. The old shop names, machine steel and machinery steel, are still used to mean any easily worked low-carbon steel. By definition, plain carbon steels are those that contain up to about 1% carbon, not more than 1.65% manganese, 0.60% silicon, and 0.60% copper, and only residual amounts of other elements, such as sulfur (0.05% max) and phosphorus (0.04% max). They are identified by means of a four-digit numerical system established by the American Iron and Steel Institute (AISI). The digits are preceded by either "AISI" or "SAE." The first digit is the number 1 for all carbon steels. A 0 after the 1 indicates nonresulfurized grades, a 1 for the second digit indicates resulfurized grades, and the number 2 for the second digit indicates resulfurized and rephosphorized grades. The last two digits give the nominal (middle of the range) carbon content, in hundredths of a percent. For example, for grade 1040, the 40 represents a carbon range of 0.37 to 0.44%. If no prefix letter is included in the designation, the steel was made by the basic open-hearth, basic O_2, or electric furnace process. The prefix B stands for the acid Bessemer process, which is obsolete, and the prefix M designates merchant quality. The letter L between the second and third digits identifies leaded steels, and the suffix H indicates that the steel was produced to hardenability limits.

For all plain carbon steels, carbon is the principal determinant of many performance properties. Carbon has a strengthening and hardening effect. At the same time, it lowers ductility, as evidenced by a decrease in elongation and reduction of area. In addition, increasing carbon content decreases machinability and weldability, but improves wear resistance. The amount of carbon present also affects physical properties and corrosion resistance. With an increase in carbon content, thermal and electric conductivity decline, magnetic permeability decreases drastically, and corrosion resistance is less.

Carbon steels may be specified by chemical composition, mechanical properties, method of oxidation, or thermal treatment (and the resulting microstructure). Carbon steels are available in most wrought mill forms, including bar, sheet, plate, pipe, and tubing. Sheet is primarily a low-carbon-steel product, but virtually all grades are available in bar and plate. Plate, usually a low-carbon or medium-carbon product, is used mainly in the hot-finished condition, although it also can be supplied heat-treated. Bar products, such as rounds, squares, hexagonals, and flats (rectangular cross sections), are also mainly low-carbon and medium-carbon products and are supplied hot-rolled or hot-rolled and cold-finished. Cold finishing may be by drawing (cold-drawn bars are the most widely used); turning (machining) and polishing; drawing, grinding, and polishing; or turning, grinding, and polishing. Bar products are also available in various quality designations, such as merchant quality (M), cold-forging quality, cold-heading quality, and several others. Sheet products also have quality designations as noted in low-carbon steels, which follow. Plain carbon steels are commonly divided into three groups, according to carbon content: low carbon, up to 0.30%; medium carbon, 0.31 to 0.55%; and high carbon, 0.56 to 1%.

Low-carbon steels are the grades AISI 1005 to 1030. Sometimes referred to as mild steels, they are characterized by low strength and high ductility, and are nonhardenable by heat treatment except by surface-hardening processes. Because of their good ductility, low-carbon steels are readily formed into intricate shapes. These steels are also readily welded without danger of hardening and embrittlement in the weld zone. Although low-carbon steels cannot be through-hardened, they are frequently surface-hardened by various methods (carburizing, carbonitriding, and cyaniding, for example) that diffuse carbon into the surface. Upon quenching, a hard, wear-resistant surface is obtained.

Low-carbon sheet and strip steels (1008 to 1012) are widely used in cars, trucks, appliances, and many other applications. Hot-rolled products are usually produced on continuous hot strip mills. Cold-rolled products are then made from the hot-rolled products, reducing thickness and enhancing surface quality. Unless the fully work-hardened product is desired, it is then annealed to improve formability and temper-rolled to further enhance surface quality. Hot-rolled sheet and strip and cold-rolled sheet are designated commercial quality (CQ), drawing quality (DQ), drawing quality special killed (DQSK), and structural quality (SQ). The first three designations refer, respectively, to steels of increasing formability and mechanical property uniformity. SQ, which refers to steels produced to specified ranges of mechanical properties and/or bendability values, does not pertain to cold-rolled strip, which is produced to several tempers related to hardness and bendability. Typically, the hardness of CQ hot-rolled sheet ranges from Rb 40 to 75 and tensile properties range from ultimate strengths of 276 to 469 MPa, yield strengths of 193 to 331 MPa, and elongations of 14 to 43%. For DQ hot-rolled sheet: Rb 40 to 72, 276 to 414 MPa, 186 to 310 MPa, and 28 to 48%, respectively. For CQ cold-rolled sheet: Rb 35 to 60, 290 to 393 MPa, 159 to 262 MPa, and 30 to 45%. And for DQ cold-rolled sheet: Rb 32 to 52, 262 to 345 MPa, 138 to 234 MPa, and 34 to 46%.

Low-carbon steels 1018 to 1025 in cold-drawn bar (16 to 22 mm thick) have minimum tensile properties of about 483 MPa ultimate strength, 413 MPa yield strength, and 18% elongation. Properties decrease somewhat with increasing section size, to, for example, 379 MPa, 310 MPa, and 15%, respectively, for 50- to 76-mm cross sections.

Medium-carbon steels are the grades AISI 1030 to 1055. They usually are produced as killed, semikilled, or capped steels, and are hardenable by heat treatment. However, hardenability is limited to thin sections or to the thin outer layer on thick parts. Medium-carbon steels in the quenched and tempered condition provide a good balance of strength and ductility. Strength can be further increased by cold work. The highest hardness practical for medium-carbon steels is about 550 Bhn (Rockwell C55). Because of the good combination of properties, they are the most widely used steels for structural applications, where moderate mechanical properties are required. Quenched and tempered, their tensile strengths range from about 517 to over 1034 MPa.

Medium-carbon steel 1035 in cold-drawn bar 16 to 22 mm thick has minimum tensile properties of about 586 MPa ultimate strength, 517 MPa yield strength, and 13% elongation. Strength increases and ductility decreases with increasing carbon content, to, for example, 689 MPa, 621 MPa, and 11%, respectively, for medium-carbon steel 1050. Properties decrease somewhat with increasing section size, to, for example, 483 MPa, 414 MPa, and 10%, respectively, for 1035 steel 50- to 76-mm cross sections.

High-carbon steels are the grades AISI 1060 to 1095. They are, of course, hardenable with a maximum surface hardness of about Rockwell C64 achieved in the 1095 grade. These steels are thus suitable for wear-resistant parts. So-called spring steels are high-carbon steels available in annealed and pretempered strip and wire. In addition to their spring applications, these steels are used for such items as piano wire and saw blades. Quenched and tempered, high-carbon steels approach tensile strengths of 1378 MPa.

Free-machining carbon steels are low-carbon and medium-carbon grades with additions usually of sulfur (0.08 to 0.13%), S–P combinations, and/or lead to improve machinability. They are AISI 1108 to 1151 for sulfur grades, and AISI 1211 to 1215 for phosphorus and sulfur grades. The latter may also contain bismuth (Bi) and be lead free. The presence of relatively large amounts of sulfur and phosphorus can reduce ductility, cold formability, forgeability, weldability, as well as toughness and fatigue strength. Calcium deoxidized steels (carbon and alloy) have good machinability, and are used for carburized or through-hardened gears, worms, and pinions.

Low-temperature carbon steels have been developed chiefly for use in low-temperature equipment and especially for welded pressure vessels. They are low-carbon to medium-carbon (0.20 to 0.30%), high-manganese (0.70 to 1.60%), silicon (0.15 to 0.60%) steels, which have a fine-grain structure with uniform carbide dispersion. They feature moderate strength with toughness down to –46°C.

For grain refinement and to improve formability and weldability, carbon steels may contain 0.01 to 0.04% columbium. Called columbium steels, they are used for shafts, forgings, gears, machine parts, and dies and gauges. Up to 0.15% sulfur, or 0.045 phosphorus, makes them free-machining, but reduces strength.

Rail steel, for railway rails, is characterized by an increase of carbon with the weight of the rail. Railway engineering standards call for 0.50 to 0.63% carbon and 0.60% manganese in a 27-kg rail, and 0.69 to 0.82% carbon and 0.70 to 1.0 manganese in a 64-kg rail. Rail steels are produced under rigid control conditions from deoxidized steels, with phosphorus kept below 0.04%, and silicon 0.10 to 0.23%. Guaranteed minimum tensile strength of 551 MPa is specified, but it is usually much higher.

Sometimes a machinery steel may be required with a small amount of alloying element to give a particular characteristic and still not be marketed as an alloy steel, although trade names are usually applied to such steels. Superplastic steels, developed at Stanford University, with 1.3 to 1.9% carbon, fall between high-carbon steels and cast irons. They have elongations approaching 500% at warm working temperatures of 538 to 650°C, and 4 to 15% elongation at room temperature. Tensile strengths range from 1034 to over 1378 MPa. The extra-high ductility is a result of a fine, equiaxed grain structure obtained by special thermomechanical processing.

Production Types

Steelmaking processes and methods used to produce mill products, such as plate, sheet, and bars, have an important effect on the properties and characteristics of a steel.

Deoxidation Practice

Steels are often identified in terms of the degree of deoxidation resulting during steel production. Killed steels, because they are strongly deoxidized, are characterized by high composition and property uniformity. They are used for forging, carburizing, and heat-treating applications. Semikilled steels have variable degrees of uniformity, intermediate between those of killed and rimmed steels. They are used for plate, structural sections, and galvanized sheets and strip. Rimmed steels are deoxidized only slightly during solidification. Carbon is highest at the center of the ingot. Because the outer layer of the ingot is relatively ductile, these steels are ideal for rolling. Sheet and strip made from rimmed steels have excellent surface quality and cold-forming characteristics. Capped steels have a thin low-carbon rim, which gives them surface qualities similar to rimmed steels. Their cross-section uniformity approaches that of semikilled steels.

Melting Practice

Steels are also classified as air melted, vacuum melted, or vacuum degassed. Air-melted steels are produced by conventional melting methods, such as open hearth, basic O_2, and electric furnace. Vacuum-melted steels encompass those produced by induction vacuum melting and consumable electrode vacuum melting. Vacuum-degassed steels are air-melted steels that are vacuum processed before solidification. Compared with air-melted steels, those produced by vacuum-melting processes have lower gas content, fewer nonmetallic inclusions, and less center porosity and segregation. They are more costly, but have better mechanical properties, such as ductility and impact and fatigue strengths.

Rolling Practice

Steel mill products are produced from various primary forms such as heated blooms, billets, and slabs. These primary forms are first reduced to finished or semi-finished shape by hot-working operations. If the final shape is produced by hot-working processes, the steel is known as hot rolled. If it is finally shaped cold, the steel is known as cold finished, or more specifically as cold rolled or cold drawn. Hot-rolled mill products are usually limited to low and medium, non-heat-treated, carbon steel grades. They are the most economical steels, have good formability and weldability, and are used widely for large structural shapes. Cold-finished shapes, compared with hot-rolled products, have higher strength and hardness and better surface finish, but are lower in ductility.

CASE-HARDENING MATERIALS

Case-hardening materials are those for adding carbon or other elements to the surface of low-carbon or medium-carbon steels or to iron so that upon quenching a hardened case is obtained, the center of the steel remaining soft and ductile. The material may be plain charcoal, raw bone, or mixtures marketed as carburizing compounds. A common mixture is about 60% charcoal and 40% barium carbonate. The latter decomposes, yielding CO_2, which is reduced to CO in contact with the hot charcoal. If charcoal is used alone, the action is slow and spotty. Coal or coke can be used, but the action is slow, and the sulfur in these materials is detrimental. Salt is sometimes added to aid the carburizing action. By proper selection of the carburizing material, the carbon content may be varied in the steel from 0.80 to 1.20%. The carburizing temperature for carbon steels typically ranges from 850 to 950°C but may be as low as 790°C or as high as 1095°C. The articles to be carburized for case hardening are packed in metallic boxes for heating in a furnace, and the process is called pack hardening, as distinct from the older method of burying the red-hot metal in charcoal.

Steels are also case-hardened by the diffusion of carbon and N_2, called carbonitriding, or N_2 alone, called nitriding. Carbonitriding, also known as dry cyaniding, gas cyaniding, liquid cyaniding, nicarbing, and nitrocarburizing, involves the diffusion of carbon and N_2 into the case. Nitriding also may be done by gas or liquid methods. In carbonitriding, the steel may be exposed to a carrier gas containing carbon and as much as 10% ammonia (NH_3), the N_2 source, or a molten cyanide salt, which provides both elements. NH_3, from gaseous or liquid salts, is also the N_2 source for nitriding. Although low- and medium-carbon steels are commonly used for carburizing and carbonitriding, nitriding is usually applied only to alloy steels containing nitride-forming elements, such as aluminum, chromium, molybdenum, or vanadium. In ion nitriding, or glow-discharge nitriding, electric current is used to ionize low-pressure N_2 gas. The ions are accelerated to the workpiece by the electric potential, and the workpiece is heated by the impinging ions, obviating an additional heat source. Of the three principal case-hardening methods, all provide a hard wear-resistant case. Carburizing, however, which gives the greater case depth, provides the best contact-load capacity. Nitriding provides the best dimensional control and carbonitriding is intermediate in this respect.

The principal liquid-carburizing material is sodium cyanide, which is melted in a pot that the articles are dipped in, or the cyanide is rubbed on the hot steel. Cyanide hardening gives an extremely hard but superficial case. N_2 as well as carbon is added to the steel by this process. Gases rich in carbon, such as methane, may also be used for carburizing, by passing the gas through the box in the furnace. When NH_3 gas is used to impart N_2 to the steel, the process is not called carburizing but is referred to as nitriding.

Chromized steel is steel surface-alloyed with chromium by diffusion from a chromium salt at high temperature. The reaction of the salt produces an alloyed surface containing about 40% chromium.

Metalliding is a diffusion coating process involving an electrolytic technique similar to electroplating, but done at higher temperatures (816 to 1093°C). The process uses a molten fluoride salt bath to diffuse metals and metalloids into the surface of other metals and alloys. As many as 25 different metals have been used as diffusing metals, and more than 40 as substrates. For example, boride coatings are applied to steels, nickel-base alloys, and refractory metals. Beryllide coatings can also be applied to many different metals by this process. The coatings are pore-free and can be controlled to a tolerance of 0.025 mm.

CASTINGS

Strength and performance of a cast part do not depend solely on part geometry. Proper alloy selection is crucial to a cost-effective casting design and trouble-free engineering and manufacturing. Materials selection software has been developed to help meet the need for information about alloys. Because the packages are in a database format, users can search through the data to choose those materials that meet their requirements. Once the basic material selection has been made, three-dimensional (3-D) modeling, casting simulation, and finite-element analysis (FEA) software can be used to confirm the behavior of that material/part combination during manufacture and in service.

RAPID PROTOTYPING

Computer technologies known as rapid prototyping allow manufacturers to fabricate 3-D models, prototypes, patterns, tooling, and production parts directly from computer-aided design (CAD) data in a fraction of the total time and cost of conventional methods. Several practical rapid prototyping systems are commercially available, and their use by OEM product parts designers as well as producers of cast metal parts and tooling shops are becoming widespread.

Each technology shares the same basic approach: A computer analyzes a 3-D CAD file that defines the object to be fabricated and "slices" the object into thin cross sections. The cross sections are then systematically recreated and combined to form a 3-D object.

Here's a thumbnail sketch of how they work: Stereolithography recreates the object by sequentially solidifying layers of photoactive liquid polymer by exposing the liquid to ultraviolet light.

Ultimately, scientists hope to develop numerical simulations that provide enough information to optimize all the variables in a casting operation. To improve understanding of the process and to improve gating design, researchers use an x-ray system that makes images of the molten metal as it fills the mold. Computer codes that predict mechanical properties are then compared with experimental results and modified to match the behavior of specific alloys.

METAL-CASTING PROCESSES

Expendable molds are for use only once (sand castings); other molds (or dies) are made of metal (permanent molding and diecasting) and can be used repeatedly. The pattern must be removable from the mold without damage, and the casting must be removable from the mold or die without damage to either the die or the casting.

Sand Casting Processes

More than 80% of all castings made in the United States are produced by green sand molding. (The term *green sand* does not refer to color, but to the fact that a raw sand and binder mixture has been tempered with water.) Sand molding is a versatile metal-forming process that provides freedom of design with respect to size, shape, and product quality.

Permanent Molding

In permanent mold casting, which also is referred to as gravity diecasting, a metal mold (or die) consisting of at least two parts is used repeatedly, usually for components that require relatively high production. Molds usually are made of cast iron, although steel, graphite, copper, and aluminum have been used as mold materials with varying degrees of success.

When molten metal is poured into a permanent mold, it cools more rapidly than in a sand mold and produces a finer-grained structure, a sounder and denser casting, and enhanced mechanical properties.

Diecasting Process

The diecasting process is used widely for high production of zinc, lead, tin, aluminum, copper, and magnesium cast components of intricate design. Molten alloy is poured manually or automatically into a shot well and injected into the die under pressure. An important factor in diecasting machine operation is the locking force (in tons), which keeps the die halves firmly closed against the injection pressure exerted by the plunger as it injects the molten metal.

There are two basic types of diecasting machines — hot chamber and cold chamber. The hot-chamber machine makes shots automatically and is used for low-melting-point materials, such as zinc alloys. The cold-chamber method, for higher-melting-point materials, such as aluminum and magnesium, holds molten metal at a constant temperature in a holding furnace of the bailout type. Metal is poured into the shot well either by hand or by automatic devices.

Vacuum diecasting sometimes is used to evacuate the die cavity. Its objectives are reduction of porosity, assisting metal flow in thin sections, and improving surface finish while at the same time permitting the use of injection pressures lower than those normally applied.

It should be noted that the meanings of the term *diecasting* in U.S. and in European usage are different. Diecasting in Europe is any casting made in a metal mold. Pressure diecasting in Europe is a casting made in a metal mold in which the metal is injected under high pressure. In the U.S., this is simply "diecasting." Gravity diecasting in Europe is a casting poured in a metal mold by gravity, with no application of pressure. In the U.S., this is "permanent molding."

Investment Casting

In investment casting, a ceramic slurry is poured around a disposable pattern (normally of modified paraffin waxes, but also of plastics) and allowed to harden to form a disposable mold. The pattern is destroyed when it melts out during the firing of the ceramic mold. Later, molten metal is poured into the ceramic mold, and after the metal solidifies, the mold is broken up to remove the casting.

Two processes are used to produce investment casting molds — the solid mold and the ceramic shell methods. The ceramic shell method has become the predominant production technique today. The solid investment process is used primarily to produce dental and jewelry castings. Ceramic shell molds are used primarily for the investment casting of carbon and alloy steels, stainless steels, heat-resistant alloys, and other alloys with melting points above 1093°C. The process can be mechanized.

Almost any degree of external and internal complexity can be accommodated; the only limitation is the state of the art in ceramic core manufacturing. Many problems inherent in producing a component by forging, machining, or multiple-piece fabricated assembly can be solved by utilizing the investment casting process. Sheet metal components, assembled by riveting, brazing, soldering, or welding, have been investment-cast as a single unit. Advantages realized include weight savings and better soundness.

Shell Molding Process

The essential feature of the shell, or Croning, process is the use of thin-walled molds and cores. Thermosetting resin-bonded silica sand is placed on a heated pattern for a predetermined length of time. Heating cures the resin, causing the sand grains to adhere to each other to form a sturdy shell that constitutes half the mold. Because of pattern costs, this method is best suited to volume production of cast metal components.

Shell molds are similar to typical green sand molds in that a set of plate-mounted patterns is fitted with integral gates and feeders. The basic production sequence in shell molding is depicted in an accompanying line drawing. The complete assembly is preheated to 177 to 204°C and the surface treated with a parting compound (silicone emulsion). The cured resin binders are nonhygroscopic, permitting prolonged storage of shell molds and allowing flexibility for production scheduling.

Castings made by the shell molding process may be more accurate dimensionally than conventional sand castings. A high degree of reproducibility as well as dimensional accuracy can be achieved with a minimum of dependence on the craftsmanship that sometimes is required with other molding processes. Only metal patterns and metal coreboxes can be used in the shell process.

Lost Foam Casting

This process is also referred to as expanded polystyrene (EPS) molding, expendable pattern casting, evaporative foam casting, the full mold process, the cavityless casting process, and the cavityless EPS casting process. The process is an economical method of producing complex, close-tolerance castings, and uses unbonded sand; the pattern material is EPS.

The process involves attaching patterns to gating systems also made of EPS, then applying a refractory coating to the total assembly. Molten metal poured into the down-sprue vaporizes the polystyrene instantly and reproduces the pattern exactly. Gases formed from the vaporized pattern escape through the pattern coating, the sand, and the flask vents. A separate pattern is required for each casting.

To the designer, a major advantage of the process is that no cores are required. Cast-in features and reduced finishing stock usually are benefits of using the lost foam process. Inserts can be cast into the metal, and bimetallic castings can be made commercially.

Vacuum Molding

The vacuum molding process, popularly known as the V-process, is a sand molding process in which no binders are used to retain the shape of the mold cavity. Instead, unbonded sand is positioned between two sheets of thin plastic that are held in place by the application of a vacuum.

Replicast Process

The replicast process is said to overcome the shortcomings of another process that is prone to cause the formation of lustrous carbon defects in steel castings, as well as undesirable carbon pickup. Outstanding features of the replicast process include surface finish comparable to that obtainable on investment castings, elimination of cores through the use of core inserts in pattern-making tooling, improved casting yields because of absence of sprues and runners, and high quality levels with regard to casting integrity and dimensional accuracy.

Other Casting Processes

Other molding systems and casting processes are used to make metal castings. For example, certain types of castings are produced in centrifugal casting machines.

Plaster mold casting is another specialized casting process used to produce nonferrous castings that is said to offer certain advantages over other processes.

Certain techniques are used to make castings in ceramic molds that are different from ceramic shell investment molding. The main difference is that the ceramic molds consist of a cope and a drag or, if the casting shape permits, a drag only.

Squeeze casting, also known as liquid-metal forging, is a process by which molten metal (ferrous or nonferrous) solidifies under pressure within closed dies positioned between the plates of a hydraulic press. The applied pressure and the instant contact of the molten metal with the die surface produce a rapid heat transfer condition that reportedly yields a pore-free, fine-grain casting with good mechanical properties.

CAST IRON

Cast iron is a generic term for a group of metals that basically are alloys of carbon and silicon with iron. Cast iron compositions are shown below:

Metal	Carbon	Silicon	Manganese	Phosphorus	Sulfur
		Typical Composition, %			
Cast steel	0.5-0.9	0.2-0.7	0.5-1.0	0.05	0.05
White iron	1.8-3.6	0.5-2.0	0.2-0.8	0.18	0.10
Malleable iron	2.0-3.0	0.6-1.3	0.2-0.6	0.15	0.10
Gray iron	2.5-3.8	1.1-2.8	0.4-1.0	0.15	0.10
Ductile iron	3.2-4.2	1.1-3.5	0.3-0.8	0.08	0.02

In addition to the major elements given above, all of these metals may contain other alloys added to modify their properties.

Although cast iron is often considered a simple metal to produce and to specify, the metallurgy of cast iron is more complex than that of steel and most other metals.

Steels and cast irons are both primarily iron with carbon as the main alloying element. Steels contain less than 2% and usually less than 1% carbon; all cast irons contain more than 2% carbon. Because 2% is about the maximum carbon content at which iron can solidify as a single-phase alloy with all the carbon solution in austenite, the cast irons, by definition, solidify as heterogeneous alloys and always have more than one constituent in their microstructure. In addition to carbon, cast irons must also contain silicon, usually from 1 to 3%; thus, they are actually Fe–C–Si alloys.

High carbon content and silicon in cast irons give them excellent castability. Their melting temperatures are appreciably lower than those of steel. Molten iron is more fluid than molten steel and less reactive with molding materials. Formation of lower-density graphite during solidification makes production of complex shapes possible. Cast irons, however, do not have sufficient ductility to be rolled or forged.

The carbon content of iron is the key to its distinctive properties. The precipitation of carbon (as graphite) during solidification counteracts the normal shrinkage of the solidifying metal, producing sound sections. Graphite also provides excellent machinability (even at wear-resisting hardness levels), damps vibration, and aids lubrication on wearing surfaces (even under borderline lubrication conditions). When most of the carbon remains combined with the iron (as in white iron), the presence of hard iron carbides provides good abrasion resistance.

In some cases, iron microstructure may be all ferrite — the same constituent that makes low-carbon steels soft and easily machined. But the ferrite of iron is different because it contains sufficient dissolved silicon to eliminate the characteristic gummy nature of low-carbon steel. Thus, cast irons containing ferrite do not require sulfur or lead additions in order to be free machining.

Because the size and shape of a casting control its solidification rate and strength, design of the casting and the casting process involved must be considered in selecting the type of iron to be specified. Although most other metals are specified by a standard chemical analysis, a single analysis of cast iron can produce several entirely different types of iron, depending upon foundry practice and shape and size of the casting, all of which influence cooling rate. Thus, iron is usually specified by mechanical properties. For applications involving high temperatures or requiring specific corrosion resistance, however, some analysis requirements may also be specified.

Pattern making is no longer a necessary step in manufacturing cast iron parts. Many gray, ductile, and alloy–iron components can be machined directly from bar that is continuously cast to near-net shape. Not only does this "parts-without-patterns" method save the time and expense of pattern making, continuous-cast iron also provides a uniformly dense, fine-grained structure, essentially free from porosity, sand, or other inclusions. Keys to the uniform microstructure of the metal are the ferrostatic pressure and the temperature-controlled solidification that are unique to the process.

For each basic type of cast iron, there are a number of grades with widely differing mechanical properties. These variations are caused by differences in the microstructure of the metal that surrounds the graphite (or iron carbides). Two different structures can exist in the same casting. The microstructure of cast iron can be controlled by heat treatment, but once graphite is formed, it remains.

Pearlitic cast iron grades consist of alternating layers of soft ferrite and hard iron carbide. This laminated structure — called pearlite — is strong and wear resistant, but still quite machinable. As laminations become finer, hardness and strength of the iron increase. Lamination size can be controlled by heat treatment or cooling rate.

Cast irons that are flame-hardened, induction-hardened, or furnace-heated and subsequently oil-quenched contain a martensite structure. When tempered, this structure provides machinability with maximum strength and good wear resistance.

Gray Iron

This is a supersaturated solution of carbon in an iron matrix. The excess carbon precipitates out in the form of graphite flakes. Gray iron is specified by a two-digit designation; Class 20, for example, specifies a minimum tensile strength of 138 MPa. In addition, gray iron is specified by the cross section and minimum strength of a special test bar. Usually, the test bar cross section matches or is related to a particularly critical section of the casting. This second specification is necessary because the strength of gray iron is highly sensitive to cross section (the smaller the cross section, the faster the cooling rate and the higher the strength).

Impact strength of gray iron is lower than that of most other cast ferrous metals. In addition, gray iron does not have a distinct yield point (as defined by classical formulas) and should not be used when permanent, plastic deformation is preferred to fracture. Another important characteristic of gray iron — particularly for precision machinery — is its ability to damp vibration. Damping capacity is determined principally by the amount and type of graphite flakes. As graphite decreases, damping capacity also decreases; see Table B.1.

The high compressive strength of gray iron — three to five times tensile strength — can be used to advantage in certain situations. For example, placing ribs on the compression side of a plate instead of the tension side produces a stronger, lighter component.

Gray irons have excellent wear resistance. Even the softer grades perform well under certain borderline lubrication conditions (as in the upper cylinder walls of internal-combustion engines, for example).

To increase the hardness of gray iron for abrasive-wear applications, alloying elements can be added, special foundry techniques can be used, or the iron can be heat-treated. Gray iron can be hardened by flame or induction methods, or the foundry can use a chill in the mold to produce hardened, "white-iron" surfaces.

Typical applications of gray iron include automotive engine blocks, gears, flywheels, brake disks and drums, and machine bases. Gray iron serves well in machinery applications because of its good fatigue resistance.

Ductile Iron

Ductile, or nodular, iron contains trace amounts of magnesium which, by reacting with the sulfur and O_2 in the molten iron, precipitates out carbon in the form of small spheres. These spheres improve the stiffness, strength, and shock resistance of ductile iron over gray iron. Different grades are produced by controlling the matrix structure around the graphite, either as-cast or by subsequent heat treatment.

A three-part designation system is used to specify ductile iron. The designation of a typical alloy, 60-40-18, for example, specifies a minimum tensile strength of 414 MPa, a minimum yield strength of 276 MPa, and 18% elongation in 5.08 mm.

Ductile iron is used in applications such as crankshafts because of its good machinability, fatigue strength, and high modulus of elasticity; in heavy-duty gears because of its high yield strength and wear resistance; and in automobile door hinges because of its ductility. Because it contains magnesium as an additional alloying element, ductile iron is stronger and more shock resistant than gray iron. But although ductile Fe also has a higher modulus of elasticity, its damping capacity and thermal conductivity are lower than those of gray iron.

By weight, ductile iron castings are more expensive than gray iron. Because they offer higher strength and provide better impact resistance, however, overall part costs may be about the same.

Although it is not a new treatment for ductile iron, austempering has become increasingly known to the engineering community in the past 5 to 10 years. Austempering does not produce the same type of structure as it does in steel because of the high carbon and silicon content of iron. The matrix structure of austempered ductile iron (ADI) sets it apart from other cast irons, making it truly a separate class of engineering materials.

In terms of properties, the ADI matrix almost doubles the strength of conventional ductile iron while retaining its excellent toughness. Like ductile iron, ADI is not a single material; rather, it is a family of materials having various combinations of strength, toughness, and wear resistance. Unfortunately, the absence of a standard specification for the materials has restricted its widespread acceptance and use. To help eliminate this problem, the Ductile Iron Society has proposed property specifications for four grades of austempered ductile iron; see Table C.3.

Most current applications for ADI are in transportation equipment — automobiles, trucks, and railroad and military vehicles. The same improved performance and cost savings are expected to make these materials attractive in equipment for other industries such as mining, earthmoving, agriculture, construction, and machine tools.

White Iron

White iron is produced by "chilling" selected areas of a casting in the mold, which prevents graphitic carbon from precipitating out. Both gray and ductile iron can be chilled to produce a surface of white iron, consisting of iron carbide, or cementite, which is hard and brittle. In castings that are white iron throughout, however, the composition of iron is selected according to part size to ensure that the volume of metal involved can solidify rapidly enough to produce the white-iron structure.

The principal disadvantage of white iron is its brittleness. This can be reduced somewhat by reducing the carbon content or by thoroughly stress-relieving the casting to spheroidize the carbides in the matrix. However, these measures increase cost and reduce hardness.

TABLE C.3
Proposed Specification for Four ADI Grades

ADI Grade	Tensile Strength, min (10^3 psi)	Yield Strength, min (10^3 psi)	Elongation, min (%)	Hardness (Bhn)	Impact Strength, unnotched, min (ft-lb)
1	125	85	10	269–331	80
2	150	100	7	302–363	65
3	175	120	4	341–401	45
4	200	140	2	375–461	30

Source: Mach. Design Basics Eng. Design, June, p. 781, 1993. With permission.

Chilling should not be confused with heat-treat hardening, which involves an entirely different metallurgical mechanism. White iron, so called because of its very white fracture, can be formed only during solidification. It will not soften except by extended annealing, and it retains its hardness even above 538°C.

White irons are used primarily for applications requiring wear and abrasion resistance such as mill liners and shot-blasting nozzles. Other uses include railroad brake shoes, rolling-mill rolls, clay-mixing and brickmaking equipment, and crushers and pulverizers. Generally, plain (unalloyed) white iron costs less than other cast irons.

COMPACTED GRAPHITE IRON

Until recently, compacted graphite iron (CGI), also known as vermicular iron, has been primarily a laboratory curiosity. Long known as an intermediate between gray and ductile iron, it possesses many of the favorable properties of each. However, because of process-control difficulties and the necessity of keeping alloy additions within very tight limits, CGI has been extremely difficult to produce successfully on a commercial scale. For example, if the magnesium addition varied by as little as 0.005%, results would be unsatisfactory.

Processing problems have been solved by an alloy-addition package that provides the essential alloying ingredients — magnesium, titanium, and rare earths — in exactly the right proportions.

Strength of CGI parts approaches that of ductile cast iron. CGI also offers high thermal conductivity, and its damping capacity is almost as good as that of gray iron; fatigue resistance and ductility are similar to those properties in ductile iron. Machinability is superior to that of ductile iron, and casting yields are high because shrinkage and feeding characteristics are more like gray iron.

The combination of high strength and high thermal conductivity suggests the use of CGI in engine blocks, brake drums, and exhaust manifolds of vehicles. CGI gear plates have replaced aluminum in high-pressure gear pumps because of the ability of the iron to maintain dimensional stability at pressures above 10.3 MPa.

MALLEABLE IRON

Malleable iron is white iron that has been converted by a two-stage heat treatment to a condition having most of its carbon content in the form of irregularly shaped nodules of graphite, called temper carbon. Resulting properties are opposite from those of the white iron from which it is derived. Rather than being hard and brittle, it is malleable and easily machined. Malleable iron castings generally cost slightly less than ductile iron castings.

The three basic types of malleable iron are ferritic, pearlitic, and martensitic. Ferritic grades are more machinable and ductile, whereas the pearlitic grades are stronger and harder. Generally, the martensitic grades are grouped with the pearlitic materials; they might be thought of as extensions (at the higher strength end of the range) of pearlitic malleable iron.

In sharp contrast to ferritic malleable iron, whose microstructure is free from combined carbon, pearlitic malleable Fe contains from 0.3 to 0.9% carbon in the combined form. Since this constituent can be transformed readily into the hardest form of combined carbon by a simple heating and quenching treatment, pearlitic malleable iron castings can be selectively hardened. Depth of hardening is controlled by the rate of heat input, time at temperature, and quenching rate. Heat treating can produce surface hardness to about Rockwell C60.

Carbon in malleable irons helps retain and store lubricants. In extreme-wear service, the pearlitic malleable iron surface wears away in harmless, micron-size particles, which are less damaging than other types of iron particles. The porous malleable iron surface traps abrasive debris that accumulates between bearing surfaces. Gall streaks can form on malleable iron, but galling does not usually progress.

Malleable iron castings are often used for heavy-duty bearing surfaces in automobiles, trucks, railroad rolling stock, and farm and construction machinery. Pearlitic grades are highly wear resistant, with hardnesses ranging from 152 to over 300 Bhn. Applications are limited, however, to relatively thin-sectioned castings because of the high shrinkage rate and the need for rapid cooling to produce white iron.

HIGH-ALLOY IRONS

High-alloy irons are ductile, gray, or white irons that contain 3% to more than 30% alloy content. Properties are significantly different from those of unalloyed irons. These irons are usually specified by chemical composition as well as by various mechanical properties.

White high-alloy irons containing nickel and chromium develop a microstructure with a martensite matrix around primary chromium carbides. This structure provides a high hardness with extreme wear and abrasion resistance. High-chromium irons (typically, about 16%) combine wear and oxidation resistance with toughness. Irons containing from 14 to 24% nickel are austenitic; they provide excellent corrosion resistance for nonmagnetic applications. The 35% nickel irons have an extremely low coefficient of thermal expansion and are also nonmagnetic and corrosion resistant.

CAST NONFERROUS ALLOYS

Nonferrous metals and alloys can be categorized as follows:

1. Aluminum-base
2. Copper-base (brasses and bronzes)
3. Lead-base
4. Magnesium-base
5. Nickel-base
6. Tin-base
7. Zinc-base
8. Titanium-base

Another common way of grouping nonferrous alloys is to divide them into heavy metals (copper, zinc, lead, and nickel base) and light metals (aluminum, magnesium, and titanium base).

ALUMINUM AND ITS ALLOYS

Aluminum and its alloys continue to grow in acceptance, particularly in the automotive industry. The electronics industry is another major user of cast aluminum for chassis, enclosures, terminals, etc.

Most metallic elements can be alloyed readily with aluminum, but only a few are commercially important — among them copper, lithium, magnesium, manganese, silicon, and zinc. Many other elements serve as supplementary alloying additions to improve properties and metallurgical characteristics.

Aluminum can be cast by virtually all of the common casting processes — particularly diecasting, sand mold, permanent mold, expendable pattern (lost foam) casting, etc.

Aluminum alloy castings are produced by virtually all commercial processes in a range of compositions possessing a wide variety of useful engineering properties.

- Al–Mg alloys offer excellent corrosion resistance, good machinability, and an attractive appearance when anodized. Careful gating and risering are required.
- Binary Al–Si alloys exhibit good weldability, high corrosion resistance, and low specific gravity.
- Al–Zn alloys have good machinability characteristics and age at room temperature to moderately high strengths in a relatively short period of time without solution heat treatment.
- Al–Sn alloys were developed for bearings and bushings with high load-carrying capacity and fatigue strength. Corrosion resistance is superior, but there is a susceptibility to hot cracking,
- Al–Li alloys are of recent commercial significance because of their good mechanical properties. They are believed to have many potential aerospace applications, and alloy development is widespread.

COPPER CASTING ALLOYS

These alloys are grouped according to composition by these general categories: pure copper, high-copper alloys, brasses, leaded brasses, bronzes, aluminum bronzes, silicon bronzes,

Cu–Ni alloys, and Cu–Ni–Zn alloys known as "nickel silvers." The UNS designations for copper-base casting alloys range from C80000 through C99900. In brasses, zinc is the principal alloying element. For cast brass, there are Cu–Zn–Sn alloys (red, semi-red, and yellow brasses), leaded and unleaded manganese bronze alloys (high-strength yellow brasses), and Cu–Zn–Si alloys.

Tin is the principal alloying element in cast bronze alloys, which consist of four families: tin bronzes, leaded and highly leaded tin bronzes, Ni–Sn bronzes, and aluminum bronzes.

ZINC ALLOYS

Zinc has a low melting temperature and is readily and economically diecast. In recent years, zinc "foundry alloys" known as ZA–8, ZA–12, and ZA–27 (containing 8, 12, and 27% aluminum, respectively) have been developed. They are finding wide application as gravity-cast sand and permanent mold castings as well as diecastings.

Zinc alloys can be cast in thin sections and with tight dimensional control. The principal alloys used for diecastings contain low percentages of magnesium, 3.9 to 4.3% aluminum, and small controlled quantities of impurities such as tin, lead, and cadmium. Copper, and nickel are significant alloying additions.

MAGNESIUM ALLOYS

The usefulness of magnesium, lightest of the commercial metals, is enhanced considerably by alloying. Magnesium can be used for sand and permanent mold castings and diecastings. Hot-chamber diecasting is a more recent development that is finding wider application each year. With the help of different heat treatments, alloy tensile strengths range from 136 to 324 MPa, yield strengths from 69 to 262 MPa, and elongations from 1 to 15%.

Magnesium foundry technology has advanced significantly in recent years. Complex components are successfully made as sand castings, with wall thicknesses down to 3.5 mm and tolerances to ±0.6 mm for automotive, aircraft, and electronics applications.

TITANIUM CASTINGS

Applications are broadening. Titanium is particularly suitable for withstanding corrosive environments or applications that take advantage of its light weight, high strength-to-weight ratio, and nonmagnetic properties. Long applied in military aircraft, titanium alloys are now solving problems in nonmilitary equipment and jewelry.

METAL MATRIX COMPOSITES (MMC)

MMC are an important new technology. These composites consist of an inorganic reinforcement-particle filament, or whisker, in a metal matrix. Nearly all metals can be used as a matrix, but current applications center around aluminum, magnesium, and titanium. The lighter metals offer the best weight-to-strength ratio. MMCs also show improved wear resistance and allow adjustment of the coefficient of thermal expansion to meet varying requirements.

SUPERALLOYS

Superalloys are nickel, Fe–Ni, and cobalt-base alloys generally used at temperatures above about 538°C. The Fe–Ni-base superalloys are an extension of stainless steel technology and generally are wrought, whereas cobalt-base and nickel-base superalloys can be wrought or cast.

The more highly alloyed compositions normally are processed as castings. Fabricated cast structures can be built up by welding or brazing, but many highly alloyed compositions containing a high amount of hardening phase are difficult to weld. Properties can be controlled by adjustments in composition and by processing including heat treatment.

CAST PLASTICS

Plastics casting materials can be generally classified in two groups: (1) those resins, usually thermosetting, that are cast as liquids and cured by chemical cross-linking either at room temperature or elevated temperatures, and (2) thermoplastics that are supplied essentially in suspension or monomeric form and fused or polymerized at elevated temperatures.

MATERIALS

Each casting resin has a unique combination of properties, such as heat resistance, strength, electrical properties, chemical resistance, cost, and shrinkage, which dictate their use for specific applications.

The most commonly used casting resins are phenolics, polyesters, and epoxies. Others are vinyls, acrylics, and urethane elastomers. Following is a brief discussion of each major type.

Phenolic Resins

Phenolic resins are used for low-cost parts requiring good electrical insulating properties, heat resistance, or chemical resistance. The average shelf life of this resin is about 1 month at 21.1°C. This can be extended by storing it in a refrigerator at 1.6 to 10°C. Varying the catalyst (according to the thickness of the cast) and raising the cure temperature to 93°C will alter the cure time from as long as 8 h to as short as 15 min.

Some shrinkage occurs in the finished casting (0.012 to 0.6 mm/mm), depending on the quantity of filler, amount of catalyst, and the rate of cure. Faster cure cycles produce a higher rate of shrinkage. Since the cure cycle can be accelerated, phenolics are used in short-run casting operations.

Cast phenolic parts are easily removed from the mold if the parting agents recommended by the supplier are used. Posteuring improves the basic properties of the finished casting.

Polyester Resins

Polyester resins are primarily used in large castings such as those required in the motion picture industry or for large sculptures for museums, parks, and display purposes. Since polyester shrinkage is about 0.024 to 0.032 mm/mm, castings are usually reinforced with glass cloth or mat and are generally cast in a flexible mold. Catalysts used to initiate the cure are peroxides or hydroperoxides and activators are cobalt naphthanate, alkyl mereaptans, or dialkyl aromatic amines. Recently, isophthalic polyesters have been introduced containing isophthalic acid, which provides improved heat, chemical, and impact resistance.

Clear polyester castings can be made using diallyl or triallyl cyanurate type polyesters. Triallyl cyanurate polyesters are used in casting clear sheets because they have excellent scratch and heat resistance.

Acrylic Resins

The process involved in casting acrylic resins is complex and forms a specialized field. The methyl-methacrylate monomer contains inhibitors that must be removed before adding the catalyst. The resin must be cured under very accurately controlled conditions.

The primary use of cast acrylics is in optically clear sheet, rod, or tube stock and in the embedment of specimens for museums and display of industrial parts, as well as for the embedment of decorative motifs in the jewelry industry.

Vinyls

Plasticized vinyls (polyvinyl chloride, or PVC) are used in industry in a variety of plastisol processing techniques, e.g., slush casting and rotational casting. Electroformed molds are commonly used for this purpose. Since the conversion of the semiliquid vinyl plastisol to a solid consists of fusing the suspended vinyl particles to each other, it is only necessary to raise the temperature of the mass to 82 to 177°C according to the formulation. Cast-vinyl prototype parts can be produced that are comparable with molded parts.

Epoxy Resins

Most cast epoxy resins, other than those used in plastic tooling, are used in encapsulating electrical components and in casting prototype parts. The variations in properties possible make them very versatile materials. They can be made to have almost infinite shelf life, can be varied from a liquid to a thixotropic gel, and can be highly flexibilized by the addition of polysulfides and polyamides. They can be formulated to provide heat resistance up to 260°C.

Cure cycles may range from 1 to 16 h, at which time the casting is removed from the mold. Epoxy resins for casting are available in transparent water-white, semitransparent, and

CAST STEELS

TABLE C.4
Minimum Mechanical Properties of Cast Low-Alloy Steels

	ASTM Classification							
	A352	A217	A148	A148	A148	A148	A148	A148
				GRADE				
	LCl Normalized & Tempered	WC4 Normalized & Tempered	80–50 Normalized & Tempered	90–60 Normalized & Tempered	105–85 Quenched & Tempered	115–95 Quenched & Tempered	150–135 Quenched & Tempered	180–145 Quenched & Tempered
Yield strength (10^3 psi)	35	40	50	60	85	95	135	145
Tensile strength (10^3 psi)	65	70	80	90	105	115	150	160
Impact strength, Charpy (ft-lb)	60	55	48	40	58	45	30	24
Fatigue endurance limit, polished specimen (10^3 psi)	20	23	25	31	34	37	44	48
Elongation in 2 in. (%)	4	20	22	20	17	14	7	6
Hardness, Bhn	—	—	163	187	217	248	311	363

Source: Mach. Design Basics Eng. Design, June, p. 787, 1993. With permission.

opaque formulations. Room-temperature cures are effected when aliphatic amines are added to the resin in exact amounts. Heat resistance of such systems is about 82°C. Cure is relatively rapid; therefore, exothermic reaction produces relatively high temperatures. Thus, casting must generally be limited in thickness.

Proprietary amine hardeners and epoxy resin systems are available whose heat resistance is about 121 to 177°C. The use of liquid anhydrides yields castings with heat resistance above 204°C. Such systems permit casting in large masses because the exothermic reaction and the curing temperatures are low (about 121°C). The pot life is several days, because elevated tempertures are required to initiate the reaction.

Several facts are basic in the use of these systems. Because it is necessary to use acidic catalysts in order to achieve heat resistance, castings tend to be more brittle; also, more difficulty is encountered in releasing the cast from the mold. The use of the proper release agent with a given resin system is necessary to overcome this tendency.

CAST STEELS

The general nature and characteristics of cast steels are, in most respects, closely comparable to wrought steels. Cast and wrought steels of equivalent composition respond similarly to heat treatment and have fairly similar properties. A major difference between them is that cast steel is more isotropic in structure. Therefore, properties tend to be more uniform in all directions in contrast to wrought steel, whose properties generally vary, depending on the direction of hot or cold working.

Five basic steel groups are available:

1. Carbon steels
2. Low-alloy steels
3. High-alloy steels
4. Tool steels
5. Stainless steels

The most common types of steels used in castings are the carbon steels, which contain only carbon as their principal alloying element. Other elements are present in small quantities, including those added for deoxidation. Silicon and manganese in cast carbon steels typically range from 0.25 to about 0.8% silicon and 0.50 to 1% manganese, respectively.

Cast plain carbon steels can be divided into three groups — low, medium, and high carbon. However, cast steel is usually specified by mechanical properties, primarily tensile strength, rather than composition. Standard classes are 414, 483, 586, and 690 MPa. Low-carbon grades, used mainly in the annealed or normalized conditions, have tensile strength ranging from 380 to 448 MPa. Medium-carbon grades, annealed and normalized, range from 483 to 690 MPa. When quenched and tempered, strength exceeds 690 MPa.

Ductility and impact properties of cast steels are comparable, on average, to those of wrought carbon steel. However, the longitudinal properties of rolled and forged steels are higher than those of cast steel. Endurance limit strength ranges between 40 and 50% of ultimate tensile strength.

By definition, low-alloy steels contain alloying elements, in addition to carbon, up to a total content of 8%. A casting containing more than 8% alloy content is classified as a high-alloy steel. Technically, tool steels and stainless steels are high-alloy steels but are normally classified separately.

Small quantities of titanium and aluminum are also used for grain refinement; see Table C.4 (Minimum Mechanical Properties of Cast Low-Alloy Steels).

Stainless steels are grouped in three classes: martensitic, ferritic, and austenitic. They are all more resistant to corrosion than plain carbon steels or lower alloy steels, and they contain either significant amounts of chromium or chromium and nickel. Cast stainless steel grades are designated in general as either heat resistant or corrosion resistant. Stainless steel castings are specified by ACI designations.

The C series of ACI stainless grades designates corrosion-resistant steels; the H series designates heat-resistant steels that are suitable for service temperatures in the 649 to 1204°C range. Typical casting applications for C-series grades are valves, pumps, and fittings. H-series grades are used for furnace parts, turbine engine components, and other high-temperature requirements.

Interest has renewed in cast duplex stainless steels because the 50% ferritic and 50% austenitic metallographic structure of these grades makes them extremely resistant to stress corrosion cracking. In addition, they have twice the yield strength of austenitic grades and may cost less.

Although the basic properties of cast duplex alloys are determined primarily by the 50/50 mixture of ferrite and austenite, the metallurgy can be complex because of the numerous carbides, nitrides, and intermetallic phases or compounds that can form. Ferrite formation is a function of cooling rate and chemistry, and austenizing temperature affects strength.

Hardenability of cast steels does not vary significantly from that of wrought steels of similar composition. The one principal difference between wrought and cast steels is the effect of the casting surface. It may contain scale and oxides and may not be chemically or structurally equivalent to the base metal; see Table C.5 (Cast Stainless Steels).

A few of the industries and some of the specific products that are being made from cast steel are: automotive (frames, wheels, gears); electrical manufacturing (rotors, bases, housings, frames, shafts); transportation (couplings, draw bars, brake shoes, wheel truck frames); marine (rudders, stems, anchor chain, ornamental fittings, capstans); off-the-road equipment (crawler side frames, levers, shafts, tread links, turntables, buckets, dipper teeth); municipal (fire hydrants, catch basins, manhole frames and covers); miscellaneous (ingot and pig molds, rolling mill rolls, blast-furnace ingot buggies, engine housings, cylinder blocks and heads, crankshafts, flanges and valves).

CELLULOSE

Cellulose is the main constituent of the structure of plants (natural polymer) that, when extracted, is employed for making paper, plastics, and in many combinations. Cellulose is made up of long-chain molecules in which the complex unit $C_6H_{10}O_5$ is repeated as many as 2000 times. It consists of glucose

TABLE C.5
Cast Stainless Steels

	Corrosion-Resistant Grades (ACI Designation)						Heat-Resistant Grades (ACI Designation)[f]						
	CA–15[a]	CB–30[b]	CC–50[c]	CE–30[d]	CF–8[e]	CH–20[d]	HC	HE	HF	HH	HT	HX	CG–6 MMN[h]	CF–10 MnN[h]
									Equivalent AISI Grades[g]				UNS	
	410	442	446	312	304	309							S20910	S21800
Yield strength (10^3 psi)														
Grades	100	60	65	63	35–40	50	—	—					50	47
HR grades	—	—	75	45	45	50	40	25					40	—
Tensile strength (10^3 psi)	115	95	70–110	87–92	75–92	80–88	69–70	56–75					94	96
Impact strength, Charpy, 70°F (ft-lb)	35	2	45	10	70	15	4	—					71	134
Creep strength, 0.001%/h (10^3 psi)														
HR grades only, 1400°F	—	—	1.3	3.5	6	7.0	8.0	6.4					—	—
1800°F	—	—	0.336	1.0	3.2	2.1	2.0	1.6					—	—
Elongation (%)														
R grades	29	—	29	35	28	28	27	25					28	26
Hardness, Bhn														
CR grades	225	195	210	190	140	190	—	—					210	195
HR grades	—	—	223	200	165	185	180	176					—	—
Melting temperature (°F)	2700–2790	2700–2750	2650–2750	2600–2700	2550–2600	2500–2600	2400–2450	2350					—	—
Coefficient of thermal expansion (10^{-6} in./in.-°F)														
70–212°F	5.5	5.7	5.9	—	9.0	8.3	—	—					9.0	8.8
70–1000°F	6.4	6.5	6.4	9.6	10.9	9.6	9.8	9.5					10.2	10.0
Thermal conductivity (Btu-ft/h-ft²-°F)	14.5	12.8	12.6	10.0	9.0	8.2	7.7	11.1					9.0	—
Density (lb/in.³)	0.275	0.272–0.274	0.272–0.277	0.276	0.280	0.279	0.286	0.294					0.285	0.276
Electrical resistivity, CR grades (microhm-cm)	78.0	76.0	77.0	85.0	76.2	84.0	—	—					82.0	98.2
Magnetic	Yes	Yes	Yes	Partially	CF, partially, HF, no	Partially	Partially	—					Slightly	Slightly

Note: CR = corrosion resistant; HR = heat resistant.

[a] 1800°F air cooled, 1200°F tempered.
[b] 1450°F air cooled.
[c] 1900°F air cooled.
[d] 2000°F air water quenched.
[e] >1900°F water quenched.
[f] As cast.
[g] Equivalent wrought grades are given for comparison only; the ACI designations, generally included in ASTM A743 and A297, are used to specify the cast stainless steel grades.
[h] Annealed.

Source: Mach. Design Basics Eng. Design, June, p. 790, 1993. With permission.

molecules with three hydroxyl groups for each glucose unit.

Cellulose is the most abundant of the nonprotein natural organic products. It is highly resistant to attack by the common microorganisms. However, the enzyme cellulase digests it easily, and this substance is used for making paper pulp, for clarifying beer and citrus juices, and for the production of citric acid and other chemicals from cellulose. Cellulose is a white powder insoluble in water, sodium hydroxide, or alcohol, but it is dissolved by sulfuric acid.

One of the simplest forms of cellulose used industrially is regenerated cellulose, in which the chemical composition of the finished product is similar to that of the original cellulose. It is made from wood or cotton pulp digested in a caustic solution. Cellophane is a regenerated cellulose in thin sheets for wrapping and other special uses include windings on wire and cable.

CELLULOSE PLASTICS

For plastics, pure cellulose from wood pulp or cotton linters (pieces too short for textile use) is reacted with acids or alkalis and alkyl halides to produce a basic flake. Depending upon the reactants, any one of four esters of cellulose (acetate, propionate, acetate butyrate, or nitrate) or a cellulose ether (ethyl cellulose) may result. The basic flake is used for producing both solvent cast films and molding powders.

Ethyl cellulose plastics are thermoplastic and are noted for their ease of molding, light weight, and good dielectric strength, 15 to 20.5 $\times 10^6$ V/m, and retention of flexibility over a wide range of temperature from −57 to 66°C, the softening point. They are the toughest, the lightest, and have the lowest water absorption of the cellulosic plastics. But they are softer and lower in strength than cellulose-acetate plastics. Typical ethyl cellulose applications include football helmets, equipment housings, refrigerator parts, and luggage.

For molding powders, the flake is then compounded with plasticizers, pigments, and sometimes other additives. At this stage of manufacture, the plastics producer is able to adjust hardness, toughness, flow, and other processing characteristics and properties. In general, these qualities are spoken of together as flow grades. The flow of a cellulose plastic is determined by the temperature at which a specific amount of the material will flow through a standard orifice under a specified pressure. Manufacturers offer cellulosic molding materials in a large number of standard flow grades, and, for an application requiring a nonstandard combination of properties, are often able to tailor a compound to fit. Cellulose can be made into a film (cellophane) or into a fiber (rayon), but it must be chemically modified to produce a thermoplastic material.

Cellulosics are synthethic plastics, but they are not synthethic polymers; see Table C.6.

Because the cellulosics can be compounded with many different plasticizers in widely varying concentrations, property ranges are broad. These materials are normally specified by flow, defined in American Society for Testing and Materials (ASTM) D569, which is controlled by plasticizer content. Hard flows (low plasticizer content) are relatively hard, rigid, and strong. Soft flows (higher plasticizer content) are tough, but less hard, less rigid, and less strong. They also process at lower temperatures. Thus, within available property ranges listed, no one formulation can provide all properties to the maximum degree. Most commonly used formulations are in the middle flow ranges.

Molded cellulosic parts can be used in service over broad temperature ranges and are particularly tough at very low temperatures. Ethyl cellulose is outstanding in this respect. These materials have low specific heat and low thermal conductivity — characteristics that give them a pleasant feel.

Dimensional stability of butyrate, propionate, and ethyl cellulose is excellent. Plasticizers used in these materials do not evaporate significantly and are virtually immune to extraction by water. Water absorption (which causes dimensional change) is also low, with that of ethyl cellulose the lowest. The plasticizers in acetate are not as permanent as those in other plastics, however, and water absorption of this material is slightly higher.

Butyrate and propionate are highly resistant to water and most aqueous solutions except strong acids and strong bases. They

TABLE C.6
Properties of Cellulosics

ASTM or UL Test	Property	Cellulose Acetate	Cellulose Propionate	Cellulose Acetate Butyrate	Ethyl Cellulose
	Physical				
D792	Specific gravity	1.22–1.34	1.16–1.24	1.15–1.22	1.09–1.17
D792	Specific volume (in.^{3}/lb)	22.7–20.6	23.4–22.4	24.1–22.7	25.5–23.6
D570	Water absorption, 24 h, 1/8-in. thk (%)	1.7–4.5	1.2–2.8	0.9–2.2	0.8–1.8
	Mechanical				
D638	Tensile strength (psi)	2200–6900	1400–7200	1400–6200	3000–4800
D638	Tensile modulus (10^5 psi)	0.65–4.0	0.6–2.15	0.5–2.0	2.2–2.5
D790	Flexural strength (psi)	2500–10,400	1700–10,600	1800–9250	4700–6800
D790	Flexural modulus (10^5 psi)	1.2–3.6	1.15–3.7	0.9–3.0	—
D256	Impact strength, Izod (ft-lb/in. of notch)	1.0–7.3	1.0–10.3	1.1–9.1	3.0–8.0
D785	Hardness, Rockwell R	To 122	To 115	To 112	79–106
	Thermal				
C177	Thermal conductivity (10^{-4} cal-cm/s-cm^2-°C)	4–8	4–8	4–8	3.8–7.0
D696	Coefficient of thermal expansion (10^{-5} in./in.-°C)	8–16	11–17	11–17	10–20
D648	Deflection temperature (°F)				
	At 264 psi	111–195	111–228	113–202	115–190
	At 66 psi	120–209	147–250	130–227	170–180
UL94	Flammability rating	V–2, HB	HB	HB	—
	Electrical				
D149	Dielectric strength[a] (V/mil)				
	Short time, 1/8-in. thk	250–600	300–500	250–400	350–500
D150	Dielectric constant				
	At 1 kHz	3.2–7.0	3.3–4.0	3.4–6.4	3.0–4.1
D150	Dissipation factor				
	At 1 kHz	0.01–0.10	0.01–0.05	0.01–0.04	0.002–0.020
D257	Volume resistivity (Ω-cm)				
	At 73°F, 50% RH	10^{10}–10^{14}	10^{12}–10^{16}	10^{11}–10^{15}	10^{12}–10^{14}
D495	Arc resistance (s)	50–310	175–190	—	60–80
	Optical				
D542	Refractive Index	1.46–1.50	1.46–1.49	1.46–1.49	—
D1003	Transmittance[b] (%)	80–92	80–92	80–92	—

[a] At 500V/s rate of rise.
[b] For 1/8-in. thick specimen.

Source: Mach. Design Basics Eng. Design, June, p. 684, 1993. With permission.

resist nonpolar materials such as aliphatic hydrocarbons and ethers, but they swell or dissolve in low-molecular-weight polar compounds such as alcohols, esters, and ketones, as well as in aromatic and chlorinated hydrocarbons. Acetate is slightly less resistant than butyrate and propionate to water and aqueous solutions, and slightly more resistant to organic materials. Ethyl cellulose dissolves in all the common solvents for this polymer, as well as in such solvents as cyclohexane and diethyl ether. Like the cellulose esters, ethyl cellulose is highly resistant to water.

Although unprotected cellulosics are generally not suitable for continuous outdoor use, special formulations of butyrate and propionate are available for such service. Acetate and ethyl cellulose are not recommended for outdoor use.

APPLICATIONS

Acetate applications include extruded and cast film and sheet for packaging and thermoforming.

Cellulose Acetate

Cellulose acetate is an amber-colored, transparent material made by the reaction of cellulose and acetic acid or acetic anhydride in the presence of sulfuric acid.

It is thermoplastic and easily molded. The molded parts or sheets are tough, easily machined, and resistant to oils and many chemicals. In coatings and lacquers the material is adhesive, tough, and resilient, and does not discolor easily. Cellulose acetate fiber for rayons can be made in fine filaments that are strong and flexible, nonflammable, mildew proof, and easily dyed. Standard cellulose acetate for molding is marketed in flake form.

In practical use, cellulose acetate moldings exhibit toughness superior to most other general-purpose plastics. Flame-resistant formulations are currently specified for small appliance housings and for other uses requiring this property. Uses for cellulose acetate molding materials include toys, buttons, knobs, and other parts where the combination of toughness and clear transparency is a requirement.

Extruded film and sheet of cellulose acetate packaging materials maintain their properties over long periods. Here also the toughness of the material is advantageously used in blister packages, skin packs, window boxes, and overwraps. It is a breathing wrap and is solvent and heat sealable.

Large end uses for cellulose acetate films and sheets include photographic film base, protective cover sheets for notebook pages and documents, index tabs, sound recording tape, as well as the laminating of book covers. The grease resistance of cellulose acetate sheet allows its use in packaging industrial parts with enclosed oil for protection.

For eyeglass frames, cellulose acetate is the material in widest current use. Because fashion requires varied and sometimes novel effects, sheets of clear, pearlescent, and colored cellulose acetate are laminated to make special sheets from which optical frames are fabricated.

The electrical properties of cellulosic films combined with their easy bonding, good aging, and available flame resistance bring about their specification for a broad range of electrical applications. Among these are as insulations for capacitors; communications cable; oil windings; in miniaturized components (where circuits may be vacuum metallized); and as fuse windows.

Cellulose triacetate is widely used as a solvent cast film of excellent physical properties and good dimensional stability. Used as photographic film base and for other critical dimensional work such as graphic arts, cellulose triacetate is not moldable.

Cellulose Propionate

Cellulose propionate, commonly called "CP" or propionate, is made by the same general method as cellulose acetate, but propionic acid is used in the reaction. Propionate offers several advantages over cellulose acetate for many applications. Because it is "internally" plasticized by the longer-chain propionate radical, it requires less plasticizer than is required for cellulose acetate of equivalent toughness.

Cellulose propionate absorbs much less moisture from the air and is thus more dimensionally stable than cellulose acetate. Because of better dimensional stability, cellulose propionate is often selected where metal inserts and close tolerances are specified.

Largest-volume uses for cellulose propionate are as industrial parts (automotive steering wheels, armrests, and knobs, etc.), telephones, toys, findings, ladies' shoe heels, pen and pencil barrels, and toothbrushes.

Cellulose Acetate Butyrate

Commonly called butyrate or CAB, it is somewhat tougher and has lower moisture absorption and a higher softening point than acetate. CAB is made by the esterification of cellulose with acetic acid and butyric acid in the presense of a catalyst. It is particularly valued for coatings, insulating types, varnishes, and lacquers.

Special formulations with good weathering characteristics plus transparency are used for outdoor applications such as signs, light globes, and lawn sprinklers. Clear sheets of butyrate are available for vacuum-forming applications. Other typical uses include transparent dial covers, television screen shields, tool handles, and typewriter keys. Extruded pipe is used for electric conduits, pneumatic tubing, and low-pressure waste lines. Cellulose acetate butyrate also is used for cable coverings and coatings. It is more soluble than cellulose acetate and more miscible with gums. It forms durable and flexible films. A liquid cellulose acetate butyrate is used for glossy lacquers, chemical-resistant fabric coatings, and wire-screen windows. It transmits ultraviolet light without yellowing or hazing and is weather-resistant.

Cellulose Acetate Propionate

This substance is similar to butyrate in both cost and properties. Some grades have slightly higher strength and modulus of elasticity. Propionate has better molding characteristics, but lower weatherability than butyrate. Molded parts include steering wheels, fuel filter bowls, and appliance housings. Transparent sheeting is used for blister packaging and food containers.

Cellulose Nitrate

Cellulose nitrates are materials made by treating cellulose with a mixture of nitric and sulfuric acids, washing free of acid, bleaching, stabilizing, and dehydrating. For sheets, rods, and tubes it is mixed with plasticizers and pigments and rolled or drawn to the shape desired. The lower nitrates are very inflammable, but they do not explode like the high nitrates, and they are the ones used for plastics, rayons, and lacquers, although their use for clothing fabrics is restricted by law. The names *cellulose nitrate* and *pyroxylin* are used for the compounds of lower nitration, and the term *nitrocellulose* is used for the explosives.

Cellulose nitrate is the toughest of the thermoplastics. It has a specific gravity of 1.35 to 1.45, tensile strength of 41 to 52 MPa, elongation 30 to 50%, compressive strength 137 to 206 MPa, Brinell hardness 8 to 11, and dielectric strength 9.9 to 21.7×10^6 V/m. The softening point is 71°C, and it is easy to mold and easy to machine. It also is readily dyed to any color. It is not light stable, and is therefore no longer used for laminated glass. It is resistant to many chemicals, but has the disadvantage that it is inflammable. The molding is limited to pressing from flat shapes.

Among thermoplastics, it is remarkable for toughness. For many applications today, however, cellulose nitrate is not practical because of serious property shortcomings: heat sensitivity, poor outdoor aging, and very rapid burning.

Cellulose nitrate cannot be injection-molded or extruded by the nonsolvent process because it is unable to withstand the temperatures these processes require. It is sold as films, sheets, rods, or tubes, from which end products may then be fabricated.

Cellulose nitrate yellows with age; if continuously exposed to direct sunlight, it yellows faster and the surface cracks. Its rapid burning must be considered for each potential application to avoid unnecessary hazard.

The outstanding toughness properties of cellulose nitrate lead to its continuing use in such applications as optical frames, shoe eyelets, ping pong balls, and pen barrels.

CENTRIFUGAL CASTINGS

Centrifugal castings can be produced economically and with excellent soundness. They are used in the automotive, aviation, chemical, and process industries for a variety of parts having a hollow, cylindrical form or for sections or segments obtainable from such a form.

There are three modifications of centrifugal casting: (1) true centrifugal casting, (2) semi-centrifugal casting, and (3) centrifuging.

1. True centrifugal casting is used for the production of cylindrical parts. The mold is rotated, usually in a horizontal plane, and the molten metal is held against the wall by centrifugal force until it solidifies.
2. Semicentrifugal casting is used for disk- and wheel-shaped parts. The mold is spun on a vertical axis, the metal is introduced at the center of the mold, and centrifugal force throws the metal to the periphery.
3. Centrifuging is used to produce irregular-shaped pieces. The method differs from static casting only in that the mold is rotated. Mold cavities are fastened at the periphery of a revolving turntable, the metal is introduced at the center, and thrown into the molds through radial ingates.

The nature of the centrifugal casting process assures a dense, homogeneous cast structure free from porosity. Because the metal solidifies in a spinning mold under centrifugal force, it tends to be forced against the mold wall while impurities, such as sand, slag, and gases, are forced toward the inside of the tube. Another advantage of centrifugal casting is that recovery can run as high as 90% of the metal poured.

Certain types of castings are produced in centrifugal casting machines. There are essentially two types of those machines — the horizontal type that rotates about a horizonal axis and the vertical type that rotates about a vertical axis. In general, horizontal machines are used to make pipe, tubes, bushings, cylinder sleeves, and other cylindrical or tubular castings that are simple in shape. Castings that are not cylindrical, or even symmetrical, can be made using vertical centrifugal casting machines.

Ferrous Castings

Centrifugal castings can be made of many of the ferrous metals — cast irons, carbon and low-alloy steels, and duplex metals.

Mechanical Properties

Regardless of alloy content, the tensile properties of irons cast centrifugally are reported to be higher than those of static castings produced from the same heat. Hydrostatic tests of cylinder liners produced by both methods show that centrifugally cast liners withstand about 20% more pressure than statically cast liners.

Freedom from directionality is one of the advantages that centrifugal castings have over forgings. Properties of longitudinal and tangential specimens of several stainless grades are substantially equal.

Shapes, Sizes, Tolerances

The external contours of centrifugal castings are not limited to circular forms. The contours can be elliptical, hexagonal, or fluted, for example. However, the nature of the true centrifugal casting process limits the bore to a circular cross section.

Iron and steel centrifugally cast tubes and cylinders are produced commercially with diameters ranging from 28.6 to 1500 mm, wall thickness of 0.25 to 102 mm, and in lengths up to 14.30 m. Generally it is impractical to produce castings with the a ratio of the outside diameter to the inside diameter greater than about 4 to 1. The upper limit in size is governed by the cost of the massive equipment required to produce heavy castings.

As-cast tolerances for centrifugal castings are about the same as those for static castings. For example, tolerances on the outside diameter of centrifugally cast gray iron pipe range from 0.3 mm for 76 mm diameter to ±0.6 mm. for 1.2 m diameter. Inside-diameter tolerances are greater, because they depend not only on the mold diameter, but also on the quantity of metal cast; the latter varies from one casting to another. These tolerances are generally about 50% greater than those on outside diameters. Casting tolerances depend to some extent also on the shrinkage allowance for the metal being cast.

The figures given above apply to castings to be used in the unmachined state. For castings requiring machining, it is customary to allow 2.35 to 3.2 mm on small castings and up to 6.4 mm on larger castings. If the end use requires a sliding fit, broader tolerances are generally specified to permit additional machining on the inside surface.

Cast Irons

Large tonnages of gray iron are cast centrifugally. The relatively low pouring temperatures and good fluidity of the common grades make them readily adaptable to the process. Various alloy grades that yield pearlitic, acicular, and chill irons are also used. In addition, specialty iron alloys such as "Ni-Hard" and "Ni-Resist," have been cast successfully.

Carbon and Low-Alloy Steels

Centrifugal castings are produced from carbon steels having carbon contents ranging from 0.05 to 0.90%. Practically all of the AISI (American Iron and Steel Institute) standard low-alloy grades have also been cast.

Small-diameter centrifugally cast tubing in the usual carbon steel grades is not competitive in price with mechanical tubing having normal wall thicknesses. However, centrifugally cast tubing is less expensive than statically cast material.

High-Alloy Steels

Most of the AISI stainless and heat-resisting grades can be cast centrifugally. A particular advantage of the process is its use in producing tubes and cylinders from alloy compositions that are difficult to pierce and to forge or roll.

The excellent ductility resulting in the stainless alloys from centrifugal casting makes it possible to reduce the rough cast tubes to smaller-diameter tubing by hot- or cold-working methods. For example, billets of 18-8 stainless steel, 114.5 mm outside diameter by 16 mm wall, have been redated to 27-gauge capillary tubing without difficulty.

Duplex Metals

Centrifugal castings with one metal on the outside and another on the inside are also in commercial production. Combinations of hard and soft cast iron, carbon steel, and stainless steel have been produced successfully.

Duplex metal parts have been centrifugally cast by two methods. In one, the internal member of the pair is cast within a shell of the other. This method has been used to produce aircraft brake drums by centrifugally casting an iron liner into a steel shell.

In the second method, both sections of the casting are produced centrifugally; the metal that is to form the outer portion of the combination is poured into the mold and solidified and the second metal is introduced before the first has cooled. The major limitation of this method is that the solidification temperature of the second metal poured must be the same or lower than that of the first. This method is said to form a strongly bonded duplex casting.

The possibilities of this duplex method for producing tubing for corrosion-resistant applications and chemical pressure service have been developed.

Nonferrous Castings

Nonferrous centrifugal castings are produced from copper alloys, nickel alloys, and tin- and lead-base bearing metals. Only limited application of the process is made to light metals because it is questionable whether any property improvement is achieved; for example, differences in density between aluminum and its normal impurities are smaller than in the heavy metals and consequently separation of the oxides, a major advantage of the process, is not so successful.

Shapes, Sizes, Tolerances

As with ferrous alloys, the external shapes of nonferrous centrifugal castings can be elliptical, hexagonal, or fluted, as well as round. However, the greatest overall tonnage of nonferrous castings is produced in plain or semiplain cylinders. The inside diameter of the casting is limited to a straight bore or one that can be machined to the required contour with minimum machining cost.

Nonferrous castings are produced commercially in outside diameter ranging from about 25.4 mm to 1.8 m and in lengths up to 8.1 m. Weights of individual castings range from 0.2268 to 27300 kg.

Although tolerances on as-cast parts are about the same as those for sand castings, most centrifugal castings are finished by machining. An advantage of centrifugal casting is that normally only a small machining allowance is required; this allowance varies from as little as 1.53 mm on small castings to 6.4 mm on the outside diameter of large-diameter castings. A slightly larger machining allowance is required on the bore to permit removal of dross and other impurities that segregate in this area.

Copper Alloys

A wide range of copper casting alloys is used in the production of centrifugal castings. The alloys include the plain brasses, leaded brasses and bronzes, tin bronzes, aluminum bronzes, silicon bronzes, manganese bronzes, nickel silvers, and beryllium copper. The ASTM lists 32 copper alloys for centrifugal casting; in addition, there are a number of proprietary compositions that are regularly produced by centrifugal casting.

Most of these alloys can be cast without difficulty. Some trouble with segregation has been reported in casting the high leaded (over 10% lead) alloys. However, alloys containing up to 20% lead are being cast by some foundries; the requirements are (1) rapid chilling to prevent excessive lead segregation and (2) close control of speed.

The mechanical properties of centrifugally cast copper alloys vary with the composition and are affected by the mold material used. Centrifugal castings produced in chill molds have higher mechanical properties than those obtained by casting in sand molds. However, centrifugal castings made in sand molds have properties about 10% higher than those obtained on equivalent sections of castings produced in static sand molds. (Castings produced in centrifugal chill molds have properties 20 to 40% higher than those produced in static sand molds.)

Nickel Alloys

Centrifugal castings of nickel 210, 213, and 305; "Monel" alloys 410, 505, and 506; and "Inconel" alloys 610 and 705 are commercially available in cylindrical tubes. Centrifugal castings are also produced from the heat-resisting alloys 60% nickel–12% chromium and 66% nickel–17% chromium. These alloys should behave like other materials and show improved density with accompanying improvement in mechanical properties. The nickel alloys are employed for service under severe corrosion, abrasion, and galling conditions.

Bearing Metals

Centrifugal casting is a standard method of producing lined bearings. Steel cylinders, after being cleaned, pickled, and tinned, are rotated while tin- or lead-base bearing alloys are cast into them. The composite cylinder is then cut lengthwise, machined, and finished into split bearings.

CERAMIC FIBERS

Alumina-silica (Al_2O_3–SiO_2) fibers, frequently referred to as ceramic fibers, are formed by subjecting a molten stream to a fiberizing force. Such force may be developed by high-velocity gas jets or rotors or intricate combinations of these. The molten stream is produced by melting high-purity Al_2O_3 and SiO_2, plus suitable fluxing agents, and then pouring this melt through an orifice. The jet or rotor atomizes the molten stream and attenuates the small particles into fine fibers as supercooling occurs.

The resulting fibrous material is a versatile high-temperature insulation for continuous service in the 538 to 1260°C range. It thus bridges the gap between conventional inorganic fiber insulating materials (e.g., asbestos, mineral wool, and glass) and insulating refractories.

Al_2O_3–SiO_2 fibers have a maximum continuous use temperature of 1093 to 1260°C, and a melting point of over 1760°C. If the fiber is exposed to temperature in excess of 1093°C for extended periods of time, a phenomenon called devitrification occurs. This is a change in the orientation of the molecular structure of the material from the amorphous state (random orientation) to the crystalline state (definitely arranged pattern). Insulating properties are not affected by this phase change but the material becomes more brittle.

Most ceramic fibers have an Al_2O_3 content from 40 to 60%, and an SiO_2 content from 40 to 60%. Also contained in the fibers are from 1.5 to 7% oxides of sodium, boron, magnesium, calcium, titanium, zirconium, and iron.

Fibers as formed resemble a cottonlike mass with individual fiber length varying from short to 254 mm, and diameters from less than 1 to 10 µm. Larger-diameter fibers are produced for specific applications. In all processes, some unfiberized particles are formed that have diameters up to 40 µm.

Low density, excellent thermal shock resistance, and very low thermal conductivity are the properties of Al_2O_3–SiO_2 fibers that make them an excellent high-temperature insulating material. Available in a variety of forms, ceramic fiber is in ever-increasing demand due to higher and higher temperatures now found in industrial and research processes.

Applications

Ceramic fibers were originally developed for application in insulating jet engines. Now, this is only one of numerous uses for this material. It can be found in aircraft and missile applications where a high-temperature insulating medium is necessary to withstand the searing heat developed by rockets and supersonic aircraft. Employed as a thermal-balance and pressure-distribution material, ceramic fiber in the form of paper has made possible the efficient brazing of metallic honeycomb-sandwich structures.

Successful trials have been conducted in aluminum processing where this versatile product in paper or molded form has been used to transport molten metal with very little heat loss. Such fibrous bodies are particularly useful in these applications because they are not readily wet by molten aluminum.

Industrial furnace manufacturers utilize lightweight ceramic fiber insulation between firebrick and the furnace shell. It is also used for "hot topping," heating element cushions, and as expansion joint packing to reduce heat loss and maintain uniform furnace temperatures.

Use of this new fiber as combustion chamber liners in oil-fired home heating units has materially improved heat-transfer efficiencies. The low heat capacity and light weight, compared to previously used firebrick, improve furnace performance and offer both customer and manufacturer many benefits.

SiC Fibers

These fibers, capable of withstanding temperatures to about 1200°C, are manufactured from a polymer precursor. The polymer is spun into a fine thread, then pyrolyzed to form a 15-µm ceramic fiber consisting of fine SiC crystallites and an amorphous phase. An advantage of the process is that it uses technology developed for commercial fiber products such as nylon and polyester. Two commercial SiC fiber products are the Ube Industries Tyranno fiber and the Nippon Carbon Nicalon fiber.

CERAMIC-MATRIX COMPOSITE

The class of materials known as ceramic-matrix composites, or CMCs, shows considerable promise for providing fracture-toughness values similar to those for metals such as cast iron. Two kinds of damage-tolerant ceramic–ceramic composites have been developed. One incorporates a continuous reinforcing phase, such as a fiber; the other, a discontinuous reinforcement, such as whiskers. The major difference between the two is in their failure behavior. Continuous-fiber-reinforced materials do not fail catastrophically. After matrix failure, the fiber can still support a load. A fibrous failure is similar to that which occurs in wood.

Incorporating whiskers into a ceramic matrix improves resistance to crack growth, making the composite less sensitive to flaws. These materials are commonly described as being flaw tolerant. However, once a crack begins to propagate, failure is catastrophic.

Of particular importance to the technology of toughened ceramics has been the development of high-temperature SiC reinforcements.

SiC Filaments

SiC filaments are prepared by chemical vapor disposition. A thick layer of SiC is deposited on a thin fiber substrate of tungsten or carbon. Diameter of the final product is about 140 μm.

Although developed initially to reinforce aluminum and titanium matrices, SiC filaments have since been used as reinforcement in Si_3N_4.

SiC Whiskers

SiC whiskers consist of a fine (0.5–5 μm-diameter) single-crystal structure in lengths to 100 μm. The material is strong (to 15.9 GPa) and is stable at temperatures to 1800°C. Whiskers can be produced by heating SiO_2 and carbon sources with a metal catalyst in the proper environments.

Although these materials are relatively new, at least one successful commercial product is already being marketed. An SiC-whisker-reinforced Al_2O_3 cutting-tool material is being used to machine nickel-based superalloys. In addition, considerable interest has been generated in reinforcing other matrices such as mullite, SiC, and Si_3N_4 for possible applications in automotive and aerospace industries.

DIMOX Process

CMCs are steadily moving from the laboratory to initial commercial applications. For example, engineers are currently evaluating these materials for use in the hot gas zones of gas turbine engines, because ceramics are known for their strength and favorable creep behavior at high temperatures. Advanced ceramics, for example, can potentially be used at temperatures 204 to 482°C above the maximum operating temperature for superalloys.

Until recently, however, there has been more evaluation than implementation of advanced ceramics for various reasons. Monolithic or single-component ceramics, for example, lack the required damage tolerance and toughness. Engine designers are put off by the potential of ceramic material for catastrophic, brittle failures. Although many CMCs have greater toughness, they are also difficult to process by traditional methods, and may not have the needed long-term high-temperature resistance.

A relatively new method for producing CMCs developed by Lanxide Corp. promises to overcome the limitations of other ceramic technologies. Called the DIMOX directed metal oxidation process, it is based on the reaction of a molten metal with an oxidant, usually a gas, to form the ceramic matrix. Unlike the sintering process, in which ceramic powders and fillers are consolidated under heat, directed metal oxidation grows the ceramic matrix material around the reinforcements.

Examples of ceramic matrices that can be produced by the DIMOX directed metal oxidation process include Al_2O_3, Al_2Ti_5, AlN, TiN, ZrN, TiC, and ZrC. Filler materials can be anything chemically compatible with the ceramic, parent metal, and growth atmosphere.

The first step in the process involves making a shaped preform of the filler material. Preforms consisting of particles are fabricated with traditional ceramic-forming techniques, while fiber preforms are made by weaving, braiding, or laying up woven cloth. Next, the preform is put in contact with the parent metal alloy. A gas-permeable growth barrier is applied to the surfaces of this assembly to limit its shape and size.

The assembly, supported in a suitable refractory container, is then heated in a furnace. For aluminum systems, temperatures typically range from 899 to 1149°C. The parent metal reacts with the surrounding gas atmosphere to grow the ceramic reaction product through and around the filler to form a CMC.

Capillary action within the growing ceramic matrix continues to supply molten alloy to the growth front. There, the reaction continues until the growing matrix reaches the barrier. At this point, growth stops, and the part is cooled to ambient temperature. To recover the part, the growth barrier and any residual parent metal are removed. However, some of the parent metal (5 to 15% by volume) remains within the final composite in micron-sized interconnected channels.

Traditional ceramic processes use sintering or hot pressing to make a solid CMC out of ceramic powders and filler. Part size and shapes are limited by equipment size and the shrinkage that occurs during densification of the powders can make sintering unfeasible. Larger parts pose the biggest shrinkage problem. Advantages of the directed metal oxidation process include no shrinkage because matrix formation occurs by a growth process. As a result, tolerance control and large part fabrication can be easier with directed metal oxidation.

In addition, the growth process forms a matrix whose grain boundaries are free of impurities or sintering aids. Traditional methods often incorporate these additives, which reduce high-temperature properties. And cost comparisons show the newer process is a promising replacement for traditional methods.

CERAMICS

Ceramics are inorganic, nonmetallic materials processed or consolidated at high temperature. Ceramics, one of the three major material families, are crystalline compounds of metallic and nonmetallic elements. The ceramic family is large and varied, including such materials as refractories, glass, brick, cement and plaster, abrasives, sanitaryware, dinnerware, artware, porcelain enamel, ferroelectrics, ferrites, and dielectric insulators. There are other materials that, strictly speaking, are not ceramics, but that nevertheless are often included in this family. These are carbon and graphite, mica, and asbestos. Also, intermetallic compounds, such as aluminides and beryllides, which are classified as metals, and cermets, which are mixtures of metals and ceramics, are usually thought of as ceramic materials because of similar physical characteristics to certain ceramics.

Ceramic materials can be subdivided into traditional and advanced ceramics. Traditional ceramics include clay-base materials such as brick, tile, sanitaryware, dinnerware, clay pipe, and electrical porcelain. Common-usage glass, cement, abrasives, and refractories are also important classes of traditional ceramics.

Advanced materials technology is often cited as an "enabling" technology, enabling engineers to design and build advanced systems for applications in fields such as aerospace, automotive, and electronics. Advanced ceramics are tailored to have premium properties through application of advanced materials science and technology to control composition and internal structure. Examples of advanced ceramic materials are Si_3N_4, SiC, toughened ZrO_2, ZrO_2-toughened Al_2O_3, AlN_3, PbMg niobate, PbLa titanate, SiC-whisker-reinforced Al_2O_3, carbon-fiber-reinforced glass ceramic, SiC-fiber-reinforced SiC, and high-temperature superconductors. Advanced ceramics can be viewed as a class of the broader field of advanced materials, which can be divided into ceramics, metals, polymers, composites, and electronic materials. There is considerable overlap among these classes of materials.

Advanced ceramics can be subdivided into structural and electronic ceramics based on primary function or application. Optical and magnetic materials are usually included in the electronic classification. Structural applications include engine components, cutting tools, bearings, valves, wear- and corrosion-resistant parts, heat exchangers, fibers and whiskers, and biological implants. The electronic-magnetic-optic functions include electronic substrates, electronic packages, capacitors, transducers, magnets, waveguides, lamp envelopes, displays, sensors, and ceramic superconductors. Thermal insulation, membranes, and filters are important advanced ceramic product areas that do not fit well into either the structural or the electronic class of advanced ceramics.

Advanced ceramics are differentiated from traditional ceramics such as brick and porcelain by their higher strength, higher operating temperatures, improved toughness, and tailorable properties. Also known as engineered ceramics, these materials are replacing metals in applications where reduced density and higher melting points can increase efficiency and speed of operation. The nature of the bond between ceramic particles helps differentiate engineering ceramics from conventional ceramics. Most particles within an engineering ceramic are self-bonded, that is, joined at grain boundaries by the same energy-equilibrium mechanism that bonds metal grains together. In contrast, most nonengineering ceramic particles are joined by a so-called ceramic bond, which is a weaker, mechanical linking or interlocking of particles. Generally, impurities in nonengineering ceramics prevent the particles from self-bonding.

A broad range of metallic and nonmetallic elements are the primary ingredients in ceramic materials. Some of the common metals are aluminum, silicon, magnesium, beryllium, titanium, and boron. Nonmetallic elements with which they are commonly combined are O_2, carbon, or N_2. Ceramics can be either simple, one-phase materials composed of one compound, or multiphase, consisting of a combination of two or more compounds. Two of the most common are single oxide ceramics, such as alumina (Al_2O_3) and magnesia (MgO), and mixed oxide ceramics, such as cordierite (magnesia alumina silica) and forsterite (magnesia silica). Other newer ceramic compounds include borides, nitrides, carbides, and silicides. Macrostructurally, there are essentially three types of ceramics: crystalline bodies with a glassy matrix; crystalline bodies, sometimes referred to as holocrystalline; and glasses.

The specific gravities of ceramics range roughly from 2 to 3. As a class, ceramics are low-tensile-strength, relatively brittle materials. A few have strengths above 172 MPa, but most have less than that. Ceramics are notable for the wide difference between their tensile and compressive strengths. They are normally much stronger under compressive loading than in tension. It is not unusual for a compressive strength to be five to ten times that of the tensile strength. Tensile strength varies considerably depending on composition and porosity.

One of the major distinguishing characteristics of ceramics, as compared to metals, is their almost total absence of ductility. They fail in a brittle fashion. Lack of ductility is also reflected in low impact strength, although impact strength depends to a large extent on the shape of the part. Parts with thin or sharp edges or curves and with notches have considerably lower impact resistance than those with thick edges and gently curving contours.

Ceramics are the most rigid of all materials. A majority of them are stiffer than most metals, and the modulus of elasticity in tension of a number of types runs as high as 0.3 to 0.4 million MPa compared with 0.2 million MPa for steel. In general, they are considerably harder than most other materials, making them especially useful as wear-resistant parts and for abrasives and cutting tools.

Ceramics have the highest known melting points of materials. Hafnium and TaC, for example, have melting points slightly above 3870°C, compared to 3424°C for tungsten. The more conventional ceramic types, such as Al_2O_3, melt at temperatures above 1927°C, which is still considerably higher than the melting point of all commonly used metals. Thermal conductivities of ceramic materials fall between those of metals and polymers. However, thermal conductivity varies widely among ceramics. A two-order magnitude of variation is possible between different types, or even between different grades of the same ceramic. Compared to metals and plastics, the thermal expansion of ceramics is relatively low, although like thermal conductivity it varies widely between different types and grades. Because the compressive strengths of ceramic materials are five to ten times greater than tensile strength, and because of relatively low heat conductivity, ceramics have fairly low thermal-shock resistance. However, in a number of ceramics, the low thermal expansion coefficient succeeds in counteracting to a considerable degree the effects of thermal conductivity and tensile-compressive-strength differences.

Practically all ceramic materials have excellent chemical resistance, and are relatively inert to all chemicals except hydrofluoric acid and, to some extent, hot caustic solutions. Organic solvents do not affect them. Their high surface hardness tends to prevent breakdown by abrasion, thereby retarding chemical attack. All technical ceramics will withstand prolonged heating at a minimum of 999°C. Therefore, atmospheres, gases, and chemicals cannot penetrate the material surface and produce internal reactions that are normally accelerated by heat.

Unlike metals, ceramics have relatively few free electrons and therefore are essentially nonconductive and considered to be dielectric. In general, dielectrical strengths, which range between 7.8×10^6 and 13.8×10^6 V/m, are lower than those of plastics. Electrical resistivity of many ceramics decreases rather than increases with an increase in impurities, and is markedly affected by temperature.

Fabrication Processes

A wide variety of processes are used to fabricate ceramics. The process chosen for a particular product is based on the material, shape, complexity, property requirements, and cost. Ceramic fabrication processes can be divided into four generic categories: powder, vapor, chemical, and melt processes.

Powder Processes

Traditional clay-base ceramics and most refractories are fabricated by powder processes as are the majority of advanced ceramics. Powder processing involves a number of sequential steps. These are preparation of the starting powders, forming the desired shape (green forming), removal of water and organics, heating with or without application of pressure to densify the powder, and finishing.

Vapor Processes

The primary vapor processes used to fabricate ceramics are chemical vapor deposition (CVD) and sputtering. Vapor processes have been finding an increasing number of applications. CVD involves bringing gases containing the atoms to make up the ceramic into contact with a heated surface, where the gases react to form a coating. This process is used to apply ceramic coatings to metal and tungsten carbide (WC) cutting tools as well as to apply a wide variety of other coatings for wear, electronic, and corrosion applications. CVD can also be used to form monolithic ceramics by building up thick coatings. A form of CVD known as chemical vapor infiltration (CVI) has been developed to infiltrate and coat the surfaces of fibers in woven preforms.

Several variations of sputtering and other vacuum-coating processes can be used to form coatings of ceramic materials. The most common process is reactive sputtering, used to form coatings such as TiN on tool steel.

Chemical Processes

A number of different chemical processes are used to fabricate advanced ceramics. The CVD process described above as a vapor process is also a chemical process. Two other chemical processes finding increasing application in advanced ceramics are polymer pyrolysis and sol-gel technology.

Melt Processes

These are used to manufacture glass, to fuse-cast refractories for use in furnace linings, and to grow single crystals. Thermal spraying can also be classified as a melt process. In this process a plasma-spray gun is used to apply ceramic coatings by melting and spraying powders onto a substrate.

Metal Oxide Ceramics

Although most metals form at least one chemical compound with O_2, only a few oxides are useful as the principal constituent of a ceramic. And of these, only three are used in their fairly pure form as engineering ceramics: Al_2O_3, BeO, and ZrO_2.

The natural alloying element in the Al_2O_3 system is SiO_2. However, Al_2O_3s can be alloyed with chromium (which forms a second phase with the Al_2O_3 and strengthens the ceramic) or with various oxides of silicon, magnesium, or calcium.

Al_2O_3s serve well at temperatures as high as 1925°C provided they are not exposed to thermal shock, impact, or highly corrosive atmospheres. Above 2038°C, strength of Al_2O_3 drops. Consequently, many applications are in steady-state, high-temperature environments, but not where abrupt temperature changes would cause failure from thermal shock. Al_2O_3s have good creep resistance up to about 816°C above which other ceramics perform better. In addition, Al_2O_3s are susceptible to corrosion from strong acids, steam, and sodium. See **Aluminum**.

BeO ceramics are efficient heat dissipaters and excellent electrical insulators. They are used in electrical and electronics applications, such as microelectric substrates, transistor bases, and resistor cores. BeO has excellent thermal shock resistance (some grades can withstand 816°C/s changes), a very low coefficient of thermal expansion (CTE), and a high thermal conductivity. It is expensive, however, and is an allergen to which some persons are sensitive. See **Beryllium**.

ZrO_2 is used primarily for its extreme inertness to most metals. ZrO_2 ceramics retain strength nearly up to their melting point — well over 2205°C, the highest of all ceramics. Applications for fused or sintered ZrO_2 include crucibles and furnace bricks. See **Zirconium**.

Transformation-toughened ZrO_2 ceramics are among the strongest and toughest ceramics made. These materials are of three main types: Mg-PSZ (ZrO_2 partially stabilized with MgO), Y-TZP (Y_2O_3 stabilized tetragonal ZrO_2 polycrystals), and ZTA (ZrO_2-toughened Al_2O_3).

Applications of Mg-PSZ ceramics are principally in low- and moderate-temperature abrasive and corrosive environments — pump and valve parts, seals, bushings, impellers, and knife blades. Y-TZP ceramics (stronger than Mg-PSZ but less flaw tolerant) are used for pump and valve components requiring wear and corrosion resistance in room-temperature service. ZTA ceramics, which have lower density, better thermal shock resistance, and lower cost than the other two, are used in transportation equipment where they need to withstand corrosion, erosion, abrasion, and thermal shock.

Many engineering ceramics have multioxide crystalline phases. An especially useful one is cordierite ($MgO-Al_2O_3-SiO_3$), which is used in cellular ceramic form as a support for a washcoat and catalyst in catalytic converters in automobile emissions systems. Its low CTE is a necessary property for resistance to thermal fracture.

Glass Ceramics

Glass ceramics are formed from molten glass and subsequently crystallized by heat treatment. They are composed of several oxides that form complex, multiphase microstructures. Glass ceramics do not have the strength-limiting porosity of conventional sintered ceramics. Properties can be tailored by control of the crystalline structure in the host glass matrix. Major applications are cooking vessels, tableware, smooth cooktops, and various technical products such as radomes.

The three common glass ceramics, $Li-Al-SiO_3$ (LAS, or beta spodumene), $Mg-Al-SiO_3$ (MAS, or cordierite), and $Al-SiO_3$ (AS, or aluminous keatite), are stable at high temperatures, have near-zero CTEs, and resist various forms of high-temperature corrosion, especially oxidation. LAS and AS have essentially no measurable thermal expansion up to 427°C. The high SiO_2 content of LAS is responsible for the low thermal expansion, but the SiO_2 also decreases strength. LAS is attacked by sulfur and sodium.

MAS is stronger and more corrosion resistant than LAS. A multiphased version of this material, MAS with $AlTiO_3$, has good corrosion resistance up to 1093°C.

AS, produced by leaching lithium out of LAS particles prior to forming, is both strong and corrosion resistant. It has been used, for example, in an experimental rotating regenerator for a turbine engine.

A proprietary ceramic (Macor) called machinable glass ceramic (MGC), is about as strong as Al_2O_3. It also has many of the high-temperature and electrical properties of the glass ceramics. The main virtue of this material is that it can be machined with conventional tools. It is available in bars, or it can be rough-formed, then finish-machined. Machined parts do not require firing.

A similar glass ceramic is based on chemically machinable glass that, in its initial state, is photosensitive. After the glass is sensitized by light to create a pattern, it is chemically machined (etched) to form the desired article. The part can then be used in its glassy state, or it can be fired to convert it to a glass ceramic. This material/process combination is used where precision tolerances are required and where a close match to thermal expansion characteristics of metals is needed. Typical applications are sliders for disk-memory read/write heads, wire guides for dot-matrix printers, cell sheets for gas-discharge displays, and substrates for thick-film and thin-film metallization.

Another ceramic-like material, glass-bonded mica, the moldable/machinable ceramic, is also called a "ceramoplastic" because its properties are similar to those of ceramics, but it can be machined and molded like a plastic material. A glass/mica mixture is pressed into a preform, heated to make the glass flow, then transfer- or compression-molded to the desired shape. The material is also formed into sheets and rods that can be machined with conventional carbide tooling. No firing is required after machining.

The thermal-expansion coefficient of glass-bonded mica is close to that of many metals. This property, along with its extremely low shrinkage during molding, allows metal inserts to be molded into the material and also ensures close dimensional tolerances. Molding tolerances as close as ±0.01 mm can be held. Continuous service temperatures for glass-bonded mica range from cryogenic to 371 or 704°C depending upon material grade.

CERIUM

A chemical element, cerium (Ce) is the most abundant metallic element of the rare earth group in the periodic table. Cerium occurs mixed with other rare earths in many minerals, particularly monazite and blastnasite, and is found among the products of the fission of uranium, thorium, plutonium.

Ceric oxide, CeO_2, is the oxide usually obtained when cerium salts of volatile acids are heated. CeO_2 is an almost white powder that is insoluble in most acids, although it can be dissolved in H_2SO_4 or other acids when a reducing agent is present. The metal is an iron-gray color and it oxidizes readily in air, forming a gray crust of oxide. Misch metal, an alloy of cerium, is used in the manufacture of lighter flints. Cerium has the interesting property that, at very low temperatures or when subjected to high pressures, it exhibits a face-centered cubic form, which is diamagnetic and 18% denser than the common form.

CERMETS

A composite material made up of ceramic particles (or grains) dispersed in a metal matrix. Particle size is greater than 1 μm, and the volume fraction is over 25% and can go as high as 90%. Bonding between the constituents results from a small amount of mutual or partial solubility. Some systems, however, such as the metal oxides, exhibit poor bonding between phases and require additions to serve as bonding agents. Cermet parts are produced by powder-metallurgy (P/M) techniques. They have a wide range of properties, depending on the composition and relative volumes of the metal and ceramic constituents. Some cermets are also produced by impregnating a porous ceramic structure with a metallic matrix binder. Cermets can also be used in powder form as coatings. The powdered mixture is sprayed through an acetylene flame, and it fuses to the base material; see Table C.7.

Although a great variety of cermets have been produced on a small scale, only a few types have significant commercial use. These fall into two main groups: oxide-base and carbide-base cermets. Other important types include the TiC-base cermets, Al_2O_3-base cermets, and UO_2 cermets specially developed for nuclear reactors.

TABLE C.7
Representative Components of Cermets

Class	Ceramic	Metal Addition
Oxides	Al_2O_3	Al, Be, Co, Co–Cr, Fe, stainless steel
	Cr_2O_3	Cr
	MgO	Al, Be, Co, Fe, Mg
	SiO_2	Cr, Si
	ZrO_2	Zr
	UO_2	Zr, Al, stainless steel
Carbides	SiC	Ag, Si, Co, Cr
	TiC	Mo, W, Fe, Ni, Co, Inconel, Hastelloy, stainless steel, Vitallium
	WC	Co
	Cr_3C_2	Ni, Si
Borides	Cr_3B_2	Ni
	TiB_2	Fe, Ni, Co
	ZrB_2	Ni
Silicides	$MoSi_2$	Ni, Co, Pt, Fe, Cr
Nitrides	TiN	Ni

Source: McGraw-Hill Encyclopedia of Science and Technology, 8th ed., Vol. 3, McGraw-Hill, New York, 483. With permission.

The most common type of oxide-base cermets contains Al_2O_3 ceramic particles (ranging from 30 to 70% volume fraction) and a chromium or chromium-alloy matrix. In general, oxide-base cermets have specific gravities between 4.5 and 9.0, and tensile strengths ranging from 144 to 268 MPa. Their modulus of elasticity ranges between 0.25 and 0.34 million MPa, and their hardness range is A70 to 90 on the Rockwell scale. The outstanding characteristic of oxide-base cermets is that the metal or ceramic can be either the particle or the matrix constituent. The 6MgO–94Cr cermets reverse the roles of the oxide and chromium, i.e., the magnesium is added to improve the fabrication and performance of the chromium. Chromium is not ductile at room temperature. Adding MgO not only permits press-forging at room temperature but also increases oxidation resistance to five times that of pure chromium. Of the cermets, the oxide-base alloys are probably the simplest to fabricate. Normal P/M or ceramic techniques can be used to form shapes, but these materials can also be machined or forged. The oxide-base cermets are used as a tool material for high-speed cutting of difficult-to-machine materials. Other uses include thermocouple-protection tubes, molten-metal-processing equipment parts, mechanical seals, gas-turbine flame holders (resistance to flame erosion), and flow control pins (because of $Cr–Al_2O_3$'s resistance to wetting and erosion by many molten metals and to thermal shock).

There are three major groups of carbide-base cermets: tungsten, chromium, and titanium. Each of these groups is made up of a variety of compositional types or grades. WC cermets contain up to about 30% cobalt as the matrix binder. They are the heaviest type of cermet (specific gravity is 11 to 15). Their outstanding properties include high rigidity, compressive strength, hardness, and abrasion resistance. Their modulus of elasticity ranges between 0.45 and 0.65 million MPa, and they have a Rockwell hardness of about A90. Structural uses of WC–Co cermets include wire-drawing dies, precision rolls, gauges, and valve parts. Higher-impact grades can be applied where die steels were formerly needed to withstand inpact loading. Combined with superior abrasion resistance, the higher impact strength results in die-life improvements as high, in some cases, as 5000 to 7000%. Most TiC cermets have nickel or nickel alloys as the metallic matrix, which results in high-temperature resistance. They have relatively low density combined with high stiffness and strength at temperatures above 1204°C. Typical properties are specific gravity, 5.5 to 7.3; tensile strength, 517 to 1068 MPa; modulus of elasticity, 0.25 to 0.38 million MPa; and Rockwell hardness, A70 to 90. Typical uses are integral turbine wheels, hot-upsetting anvils, hot-spinning tools, thermocouple protection tubes, gas-turbine nozzle vanes and buckets, torch tips, hot-mill-roll guides, valves, and valve seats. CrC cermets contain from 80 to 90% CrC, with the balance being either nickel or nickel alloys. Their tensile strength is about 241 MPa, and they have a tensile modulus of about 0.34 to 0.39 million MPa. Their Rockwell hardness is about A88. They have superior resistance to oxidation, excellent corrosion resistance, and relatively low density (specific gravity is 7.0). Their high rigidity and abrasion resistance make them suitable for gauges, oil-well check valves, valve liners, spray nozzles, bearing seal rings, and pump rotors.

Other cermets are berium–carbonate–nickel and tungsten-thoria, which are used in higher-power pulse magnetrons. Some proprietary compositions are used as friction materials. In brake applications, they combine the thermal conductivity and toughness of metals with the hardness and refractory properties of ceramics. UO_2 cermets have been developed for use in nuclear reactors. Cermets play an important role in sandwich-plate fuel elements, and the finished element is a siliconized SiC with a core containing UO_2. Control rods have been fabricated from B_4C–stainless steel and rare earth oxides–stainless steel. Other cermets developed for use in nuclear equipment include $Cr–Al_2O_3$ cermets, Ni–MgO cermets, and Fe–ZrC cermets. Nonmagnetic compositions can be formulated for use where magnetic materials cannot be tolerated.

Interactions

The reactions taking place between the metallic and ceramic components during fabrication of cermets may be briefly classified and described as follows:

1. Heterogeneous mixtures with no chemical reaction between the components, characterized by a mechanical interlocking of the components without formation of a new phase, no penetration of the metallic component into the ceramic component, and vice versa, and no alteration of either component (example, MgO–Ni).
2. Surface reaction resulting in the formation of a new phase as an interfacial layer that is not soluble in the component materials. The thickness of this layer depends on the diffusion rate, temperature, and time of the reaction (example, Al_2O_3–Be).
3. Complete reaction between the components, resulting in the formation of a solid solution characterized by a polyatomic structure of the ceramic and the metallic component (example, TiC–Ni).
4. Penetration along grain boundaries without the formation of interfacial layers (example, Al_2O_3–Mo).

Bonding Behavior

One important factor in the selection of metallic and ceramic components in cermets is their bonding behavior. Bonding may be by surface interaction or by bulk interaction. In cermets of the oxide-metal type, for example, investigators differentiate among three forms of surface interaction: macrowetting, solid wetting, and wetting assisted by direct lattice fit.

Combinations

One distinguishes basically between four different combinations of metal and ceramic components: (1) the formation of continuous interlocking phases of the metallic and ceramic components, (2) the dispersion of the metallic component in the ceramic matrix, (3) the dispersion of the ceramic component in the metallic matrix, and (4) the interaction between the metallic and ceramic components.

Applications

Aside from the high-temperature applications in turbine buckets, nozzle vanes, and impellers for auxiliary power turbines, there is a wide variety of applications for cermets based on various other properties. One of the most successful applications for the TiC-base cermets is in elements of temperature sensing and controlling thermostats where their oxidation resistance together with their low coefficient of thermal expansion as compared with nickel-base alloys are the important properties. Their ability to be welded directly to the alloys is also important.

The TiC-base cermets are also used for bearings and thrust runners in liquid metal pumps, hot flash trimming and hot spinning tools, hot rod mill guides, antifriction and sleeve-type bearings, hot glass pinch jaws, rotary seals for hot gases, oil well valve balls, etc.

CESIUM

A chemical element, cesium (symbol Cs) is the heaviest of the alkali metals in group I. It is a soft, light, very low melting temperature metal. It is the most reactive of the alkali metals and indeed is the most electropositive and the most reactive of all the elements.

Cesium oxidizes easily in the air, ignites at ordinary temperatures, and decomposes water with explosive violence. It can be contained in vacuum, inert gas, or anhydrous liquid hydrocarbons protected from O_2 and air. The specific gravity is 1.903, melting point 28.5°C, and boiling point 670°C. It is used in low-voltage tubes to scavenge the last traces of air. It is usually marketed in the form of its compounds such as cesium nitrate, $CsNO_3$, cesium fluoride, CsF, or cesium carbonate, Cs_2CO_3. In the form of cesium chloride, CsCl, it is used on the filaments of radio tubes to increase sensitivity. It interacts with the thorium of the filament to produce positive tons. In photoelectric cells CsCl is used for a photosensitive deposit on the cathode, since cesium releases its outer electron under the action of ordinary light, and its color sensitivity is higher than that of other alkali metals. The high-voltage rectifying tube for changing AC to DC has cesium metal coated on the nickel cathode, and has cesium vapor for current carrying. The cesium metal gives off a copious flow of electrons and is continuously renewed from the vapor. Cesium vapor is also used in the infrared signaling lamp as it produces infrared waves without visible light. Cesium salts have been used medicinally as antishock agents after administration of arsenic drugs.

Cesium metal is generaly made by thermochemical processes. The carbonate can be reduced by metallic magnesium, or the chloride can be reduced by CaC. Metallic cesium volatilizes from the reaction mixture and is collected by cooling the vapor.

CHEMICAL MILLED PARTS

Chemical milling is the process of producing metal parts to predetermined dimensions by chemical removal of metal from the surface.

Acid or alkaline, pickling, or etching baths have been formulated to remove metal uniformly from surfaces without excessive etching, roughening, or grain boundary attack. Simple immersion of a metal part will result in uniform removal from all surfaces exposed to the chemical solution. Selective milling is accomplished by use of a mask to protect the areas where no metal is to be removed. By such means optimum strength per unit of construction weight is achieved. Nonuniform milling can be done by the protective masking procedure or by programmed withdrawal of the part from the milling bath. Complex milling is done by multiple masking and milling or withdrawal steps.

Versatility Offered

The aircraft industry, as an example, utilizes production chemical milling for weight reduction of large parts by means of precise etching. The process is the most economical means of removal of metal from large areas, nonplaner surfaces, or complex shapes. A further advantage is that metal is just as easily removed from fully hardened as from annealed parts. The advantages of chemical milling result from the fact that metal removal takes place on all surfaces contacted by the etching solution. The solution will easily mill inside and reentrant surfaces as well as thin metal parts or parts that are multiple racked. The method does not require elaborate fixturing or precision setups, and parts are just as easily milled after forming as in the flat. Job lots and salvage are treated, as well as production runs.

Maximum weight reduction is possible through a process of masking, milling, measuring, and remilling with steps repeated as necessary. Planned processing is the key to production of integrally stiffened structures milled so that optimum support of stresses is attained without the use of stiffening by attachment, welding, or riveting.

A level of ability comparable to that required for electroplating is necessary to produce chemically milled parts. Planned processing, solution control, and developed skill in masking and handling of the work are requisite to success. Periods to train personnel, however, are relatively short as compared to training for other precision metal-removal processes.

Tooling requirements are simple. Chemicals, tanks, racks, templates, a hoist, hangers, and a few special hand and measuring tools are required.

Although chemical milling skill can be acquired without extensive training, it is not feasible to expect to produce the extremes of complexity and precision without an accumulation of considerable experience. However, a number of organizations are available that will produce engineering quality parts on a job shop basis. The processes are well established, commercially, either in or out of the plant.

Specific Etchants Needed

It is anticipated that any metal or alloy can be chemically milled. On the other hand, it does take time to develop a specific process and only those metals can be milled for which an etchant has been developed, tested, and made available. Aluminum alloys have been milled for many years. Steel, stainless steel, nickel alloys, titanium alloys, and superalloys have been milled commercially and a great number of other metals and alloys have been milled experimentally or on a small commercial scale.

It is advantageous to be able to mill a metal without changing the heat-treated condition or temper, as can be done chemically. Defective or nonhomogeneous metal, however, can respond unfavorably. Porous castings will develop holes during milling and mechanically or thermally stressed parts will change in shape as stressed metal is removed. Good-quality metal and controlled heat treating, tempering, and stress relieving are essential to uniformity and reproducibility.

Process Characteristics

Almost any metal size or shape can be milled; limitations are imposed only by extreme conditions such as complex shapes with inverted pockets that will trap gases released during milling, or very thin metal foil that is too flimsy to handle. Shapes can be milled that are completely impractical to machine. For example, the inside of a bent pipe could easily be reduced in section by chemical removal of metal. This possibility is used to advantage to reduce weight on many difficult-to-machine areas such as webs of forgings or walls of tubing. Thin sections are produced by milling when alternate machining methods are excessively costly and the optimum in design demands thin metal shapes that are beyond commercial casting, drawing, or extruding capabilities.

Surface roughness is often reduced during milling from a rough-machined, cast, or forged surface to a semimatte finish. The milled finish may vary from about 30 to 250 μin., depending on the original finish, the alloy, and the etchant. In some instances the production of an attractive finish reduces finishing steps and is a cost advantage. So-called etching that takes place during milling often causes a brightened finish and etchants have been developed that do not result in a loss of mechanical properties.

Complement Machining

Chemical milling has flourished in the aircraft industry where paring away of every ounce of weight is important. It has spread to instrument industries where weight or balance of working parts is important to the forces required to initiate and sustain motion. It has also become a factor in the design of modern weight-limited portable equipment.

A realistic appraisal of the limitations and advantages of the process is essential to optimum designs. The best designs result from complementing mechanical, thermal, and chemical processes. Chemical milling is not a substitute for mechanical methods but, rather, is more likely as an alternative where machining is difficult or economically unfeasible. It does not compete with low-cost, mass production mechanical methods but, rather, is successful where other methods are limited due to the configuration of the part.

Tolerances

It is good design practice to allow a complex shape to be manufactured by the most economical combinations of mechanical and chemical means. To allow this, print tolerances must reflect allowances that are necessary to apply chemical milling. Chemical milling will produce less well-defined cuts, radii, and surface finishes. The tolerance of a milled cut will vary with the depth of the cut. For 2.5-mm cut, a tolerance in depth of cut of ±0.10 m is commercial. This must be allowed in addition to the original sheet tolerance. Line definition (deviation from a straight line) is usually ±0.8 mm. Unmilled lands between two milled areas should be 0.004 mm minimum. Greater precision can be had at a premium price.

In general, milling rates are about 0.0004 mm/s and depth of cut is controlled by the immersion time. Cuts up to 12.7 mm are not unrealistic although costs should be investigated before designs are made that are dependent on deep cuts.

Limitations

There are limitations to the process. Deep cuts on opposite sides of a part should not be taken simultaneously. One side can be milled at a time but it is less costly to design for one cut rather than two. Complex parts can be made by step milling or by programmed removal of parts to produce tapers. In general, step milling is less expensive and more reliable. Chemical milling engineers should be consulted relative to the feasibility and cost of complex design. Very close tolerance parts can be produced by milling, checking, masking, and remilling but such a multiple-step process could be more costly than machining.

CHLORINATED POLYETHER

Chlorinated polyether is a thermoplastic resin used in the manufacture of process equipment. Chemically, it is a chlorinated polyether of high molecular weight, crystalline in character, and is extremely resistant to thermal degradation at molding and extrusion temperatures. It possesses a unique combination of mechanical, electrical, and chemical-resistant properties, and can be molded in conventional injection and extrusion equipment.

PROPERTIES

Chlorinated polyether provides a balance of properties to meet severe operating requirements. It is second only to the fluorocarbons in chemical and heat resistance and is suitable for high-temperature corrosion service.

Mechanical Properties

A major difference between chlorinated polyether and other thermoplastics is its ability to maintain its mechanical strength properties at elevated temperatures. Heat distortion temperatures are above those usually found in thermoplastics and dimensional stability is exceptional even under the adverse conditions found in chemical plant operations. Resistance to creep is significantly high and in sharp contrast to the lower values of other corrosion-resistant thermoplastics. Water absorption is negligible, assuring no change in molded shapes between wet and dry environments.

Chemical Properties

Chlorinated polyether offers resistance to more than 300 chemicals and chemical reagents, at temperatures up to 121°C and higher, depending on environmental conditions. It has a spectrum of corrosion resistance second only to certain of the fluorocarbons.

In steel construction of chemical processing equipment chlorinated polyether liners or coatings on steel substrates provide the combination of protection against corrosion plus structural strength of metal.

Electrical Properties

Along with the mechanical capabilities and chemical resistance, chlorinated polyether has good dielectric properties. Loss factors are somewhat higher than those of polystyrenes, fluorocarbons, and polyethylenes, but are lower than many other thermoplastics. Dielectric strength is high and electrical values show a high degree of consistency over a range of frequencies and temperatures.

FABRICATION

The material is available as a molding powder for injection-molding and extrusion applications. It can also be obtained in stock shapes such as sheet, rods, tubes, or pipe, and blocks for use in lining tanks and other equipment, and for machining gears, plugs, etc. In the form of a finely divided powder it is used in a variety of different coating processes.

The material can be injection-molded by conventional procedures and equipment. Molding cycles are comparable to those of other thermoplastics. Rods, sheet, tubes, pipe, blocks, and wire coatings can be readily extruded on conventional equipment and by normal production techniques. Parts can be machined from blocks, rods, and tubes on conventional metalworking equipment.

Sheet can be used to convert carbon steel tanks into vessels capable of handling highly corrosive liquids at elevated temperatures. Using a conventional adhesive system and hot gas welding, sheet can be adhered to sandblasted metal surfaces.

Coatings of finely divided powder can be applied by several coating processes and offer chemical processors an effective and economical means for corrosion control. Using the fluidized bed process, pretreated, preheated metal parts are dipped in an air-suspended bed of finely divided powder to produce coatings, which after baking are tough, pinhole free, and highly resistant to abrasion and chemical attack. Parts clad by this process are protected against corrosion both internally and externally.

Uses

Complete anticorrosive systems are available with chlorinated polyether, and lined or coated components, including pipe and fittings, tanks and processing vessels, valves, pumps, and meters.

Rigid uniform pipe extruded from solid material is available in sizes ranging from 12.7 to 50.8 mm in either Schedule 40 or 80, and in lengths up to 6 m. This pipe can be used with injection-molded fittings with socket or threaded connections.

Lined tanks and vessels are useful in obtaining maximum corrosion and abrasion resistance in a broad range of chemical exposure conditions. Storage tanks, as well as processing vessels protected with this impervious barrier, offer a reasonably priced solution to many processing requirements.

A number of valve constructions can be readily obtained from leading valve manufacturers. Solid injection-molded ball valves, coated diaphragm, and plug valves are among the variety available. Also available are diaphragm valves with solid chlorinated polyether bodies.

CHLORINATED POLYETHYLENE

This family of elastomers is produced by the random chlorination of high-density polyethylene. Because of the high degree of chemical saturation of the polymer chain, the most desirable properties are obtained by cross-linking with the use of peroxides or by radiation. Sulfur donor cure systems are available that produce vulcanizates with only minor performance losses compared to that of peroxide cures. However, the free radical cross-linking by means of peroxides is most commonly used and permits easy and safe processing, with outstanding shelf stability and optimum cured properties.

Chlorinated polyethylene elastomers are used in automotive hose applications, premium hydraulic hose, chemical hose, tubing, belting, sheet packing, foams, wire and cable, and in a variety of molded products. Properties include excellent ozone and weather resistance, heat resistance to 149°C (to 177°C in many types of oil), dynamic flexing resistance, and good abrasion resistance.

CHLOROSULFONATED POLYETHYLENE

This material, more commonly known as Hypalon, can be compounded to have an excellent combination of properties including virtually total resistance to ozone and excellent resistance to abrasion, weather, heat, flame, oxidizing chemicals, and crack growth. In addition, the material has low moisture absorption, good dielectric properties, and can be made in a wide range of colors because it does not require carbon black for reinforcement. Resistance to oil is similar to that of neoprene. Low-temperature flexibility is fair at –40°C.

The material is made by reacting polyethylene with chlorine and SO_2 to yield chlorosulfonated polyethylene. The reaction changes the thermoplastic polyethylene into a synthetic elastomer that can be compounded and vulcanized. The basic polyethylene contributes chemical inertness, resistance to damage by moisture, and good dielectric strength. Inclusion of chlorine in the polymer increases its resistance to flame (makes it self-extinguishing) and contributes to its oil and weather resistance.

Selection

Hypalon is a special-purpose rubber, not particularly recommended for dynamic applications. The elastomer is produced in various types, with generally similar properties. The design engineer can best rely on the rubber formulator to select the appropriate type for a given application, based on the nature of the part, the properties required, the exposure, and the performance necessary for successful use.

In combination with properly selected compounding ingredients, the polymer can be extruded, molded, or calendered. In addition, it can be dissolved to form solutions suitable for protective or decorative coatings.

Initially used in pump and tank linings, tubing, and comparable applications where chemical resistance was of prime importance, this synthetic rubber is now finding many uses where its weatherability, its colorability, its heat, ozone, and abrasion resistance, and its electrical properties are of importance. Included are jacketing and insulation for utility distribution cable, control cable for atomic reactors, automotive primary and ignition wire, and linemen's blankets. Among heavy-duty applications are conveyor belts for high-temperature use and industrial rolls exposed to heat, chemicals, or abrasion.

Interior, exterior, and underhood parts for cars and commercial vehicles are an increasingly important area of use. Representative automotive applications are headliners, window seals, spark plug boots, and tractor seat coverings.

Chlorosulfonated polyethylene is used in a variety of mechanical goods, such as V-belts, motor mounts, O-rings, seals, and gaskets, as well as in consumer products like shoe soles and garden hose. It is also used in white sidewalls on automobile tires. In solution, it is used for fluid-applied roofing systems and pool liners, for masonry coatings, and various protective-coating applications. It can also be extruded as a protective and decorative veneer for such products as sealing and glazing strips.

CHROMIUM

An elementary metal, chromium (symbol Cr) is used in stainless steels, heat-resistant alloys, high-strength alloy steels, electrical-resistance alloys, wear-resistant and decorative electroplating, and, in its compounds, for pigments, chemicals, and refractories. The specific gravity is 6.92, melting point 1510°C, and boiling point 2200°C. The color is silvery white with a bluish tinge. It is an extremely hard metal; the electrodeposited plates have a hardness of 9 Mohs. It is resistant to oxidation, is inert to HNO_3, but dissolves in HCl and slowly in H_2SO_4. At temperatures above 816°C, it is subject to intergranular corrosion.

Chromium occurs in nature only in combination. Its chief ore is chromite, from which it is obtained by reduction and electrolysis. It is marketed for use principally in the form of master alloys with iron or copper.

Most pure chromium is used for alloying purposes such as the production of Ni–Cr or other nonferrous alloys where the use of the cheaper ferrochrome grades of metal is not possible. In metallurgical operations such as the production of low-alloy and stainless steels, the chromium is added in the form of ferrochrome, an electric-arc furnace product that is the form in which most chromium is consumed.

Uses

Its bright color and resistance to corrosion make chromium highly desirable for plating plumbing fixtures, automobile radiators and bumpers, and other decorative pieces. Unfortunately, chrome plating is difficult and expensive. It must be done by electrolytic reduction of dichromate in H_2SO_4 solution. It is customary, therefore, to first plate the object with copper, then with nickel, and finally, with chromium.

Alloys

In alloys with iron, nickel, and other metals, chromium has many desirable properties. Chrome steel is hard and strong and resists corrosion to a marked degree. Stainless steel contains roughly 18% chromium and 8% nickel. Some chrome steels can be hardened by heat treatment and find use in cutlery; still others are used in jet engines. Nichrome and chromel consist largely of nickel and chromium; they have low electrical conductivity and resist corrosion, even at red heat, so they are used for heating coils in space heaters, toasters, and similar devices. Other important alloys are Hastelloy C — chromium, molybdenum, tungsten, iron, nickel — used in chemical equipment that is in contact with HCl, oxidizing acids, and hypochlorite. Stellite — cobalt, chromium, nickel, carbon, tungsten (or molybdenum) — noted for its hardness and abrasion resistance at high temperatures, is used for lathes and engine valves, and Inconel — chromium, iron, nickel — is used in heat treating and in corrosion-resistant equipment in the chemical industry.

Biological Aspects

Chromium is essential to life. A deficiency (in rats and monkeys) has been shown to impair glucose tolerance, decrease glycogen reserve, and inhibit the utilization of amino acids. It has also been found that inclusion of chromium in the diet of humans sometimes, but not always, improves glucose tolerance.

On the other hand, chromates and dichromates are severe irritants to the skin and mucous membranes, so workers who handle large amounts of these materials must be protected against dusts and mists. Continued breathing of the dusts finally leads to ulceration and perforation of the nasal septum. Contact of cuts or abrasions with chromate may lead to serious ulceration. Even on normal skin, dermatitis fequently results.

CHROMIUM ALLOYS AND STEELS

Chromium Copper

A name applied to master alloys of copper with chromium used in the foundry for introducing chromium into nonferrous alloys or to Cu–Cr alloys, or chromium copper alloys, which are high-copper alloys. A Cr–Cu master alloy. Electromet chromium copper, contains 8 to 11% chromium, 88 to 90% copper, and a maximum of 1% iron and 0.50% silicon.

Wrought chromium copper alloys are designated C18200, C18400, and C18500, and contain 0.4 to 1.0% chromium. C18200 also contains as much as 0.10% iron, 0.10% silicon, and 0.05% lead. C18400 contains as much as 0.15% iron and 0.10% silicon, and several other elements in small quantities. C18500 is iron-free, and contains as much as 0.015% lead, and several other elements in small quantities. Soft, thus ductile, in the solution-treated condition, these alloys are readily cold-worked and can be subsequently precipitation-hardened. Depending on such treatments, tensile properties range from 241 to 482 MPa ultimate strength, 103 to 427 MPa yield strength, and 15 to 42% in elongation. Electrical conductivity ranges from 40 to 85% that of copper. Chromium copper alloys are used for resistance-welding electrodes, cable connectors, and electrical parts.

Cr–Mo Steel

This is any alloy steel containing chromium and molybdenum as key alloying elements. However, the term usually refers specifically to steels in the American Iron and Steel Institute (AISI) 41XX series, which contain only 0.030 to 1.20% chromium and 0.08 to 0.35% molybdenum. Chromium imparts oxidation and corrosion resistance, hardenability, and high-temperature strength. Molybdenum also increases strength, controls hardenability, and reduces the tendency to temper embrittlement. AISI 4130 steel, which contains 0.30% carbon, and 4140 (0.40%) are probably the most common and can provide tensile strengths well above 1379 MPa. Many other steels have greater chromium and/or molybdenum content, including high-pressure boiler steels, most tool steels, and stainless steels. Croloy 2, which is used for boiler tubes for high-pressure superheated steam, contains 2% chromium and 0.50% max molybdenum, and is for temperatures to 621°C, and Croloy 5, which has 5% chromium and 0.50% max molybdenum, is for temperatures to 649°C and higher pressures as well as Croloy 7, which has 7% chromium and 0.50% molybdenum.

Chromium Steels

Any steel containing chromium as the predominating alloying element may be termed chromium steel, but the name usually refers to the hard, wear-resisting steels that derive the property chiefly from the chromium content. Straight chromium steels refer to low-alloy steels in the AISI 50XX, 51XX, and 61XX series. Chromium combines with the carbon of steel to form a hard CrC, and it restricts graphitization. When other carbide-forming elements are present, double or complex carbides are formed. Chromium refines the structure, provides deep-hardening, increases the elastic limit, and gives a slight red-hardness so that the steels retain their hardness at more elevated temperatures. Chromium steels have great resistance to wear. They also withstand quenching in oil or water without much deformation. Up to about 2% chromium may be included in tool steels to add hardness, wear resistance, and nondeforming qualities. When the chromium is high, the carbon may be much higher than in ordinary steels without making the steel brittle. Steels with 12 to 17% chromium and about 2.5% carbon have remarkable wear-resisting qualities and are used for cold-forming dies for hard metals, for broaches, and for rolls. However, chromium narrows the hardening range of steels unless balanced with nickel. Such steels also work-harden rapidly unless modified with other elements. The high-chromium steels are corrosion resistant and heat resistant but are not to be confused with the high-chromium stainless steels that are low in carbon, although the non-nickel 4XX stainless steels are very definitely chromium steels. Thus, the term is indefinite but may be restricted to the high-chromium steels used for dies, and to those with lower chromium used for wear-resistant parts such as ball bearings.

Chromium steels are not especially corrosion resistant unless the chromium content is at least 4%. Plain chromium steels with more than 10% chromium are corrosion resistant even at elevated temperatures and are in the class of stainless steels, but are difficult to weld because of the formation of hard brittle martensite along the weld.

Chromium steels with about 1% chromium are used for gears, shafts, and bearings. One of the most widely used bearing steels is AISI 52100, which contains 1.3 to 1.6% chromium. Many other chromium steels have greater chromium content and, often, appreciable amounts of other alloying elements. They are used mainly for applications requiring corrosion, heat, and/or wear resistance.

Cr–V Steels

Alloy steel containing a small amount of chromium and vanadium, the latter having the effect of intensifying the action of the chromium and the manganese in the steel and controlling grain growth. It also aids in formation of carbides, hardening the alloy, and in increasing ductility by the deoxidizing effect.

The amount of vanadium is usually 0.15 to 0.25%. These steels are valued where a combination of strength and ductility is desired. They resemble those with chromium alone, with the advantage of the homogenizing influence of the vanadium. A Cr–V steel with 0.92% chromium, 0.20% vanadium, and 0.25% carbon has a tensile strength up to 689 MPa, and when heat-treated has a strength up to 1034 MPa and elongation 16%. Cr–V steels are used for such parts as crankshafts, propeller shafts, and locomotive frames. High-carbon Cr–V steels are the mild-alloy tool steels of high strength, toughness, and fatigue resistance. The chromium content is usually about 0.80%, with 0.20% vanadium, and with carbon up to 1%.

CLAD METALS

Cladding means the strong, permanent bonding of two or more metals or alloys metallurgically bonded together to combine the characteristic properties of each in composite form. Copper-clad steel, for example, is used to combine the electrical and thermal characteristics of copper with the strength of steel. A great variety of metals and alloys can be combined in two or more layers, and they are available in many forms, including sheet, strip, plate, tubing, wire, and rivets for application in electrical and electronic products, chemical-processing equipment, and decorative trim, including auto trim; see Figure C.1.

Cladding Processes

In the process a clad metal sheet is made by bonding or welding a thick facing to a slab of base metal; the composite plate is then rolled to the desired thickness. The relative thickness of the layers does not change during rolling. Cladding thickness is usually specified as a percentage of the total thickness, commonly 10%.

Other cladding techniques, including a vacuum brazing process, have been developed. The pack rolling process is still the most widely used, however.

CLAD METALS

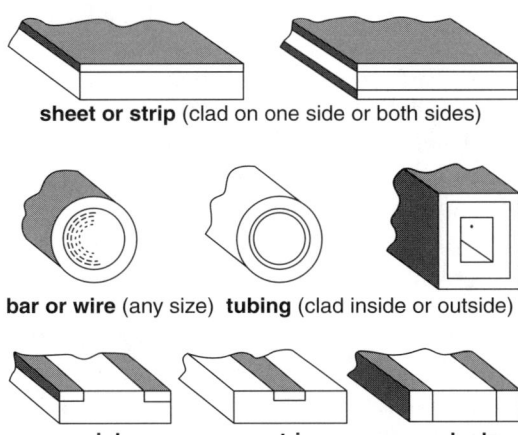

FIGURE C.1 Types of cladding. (From *McGraw-Hill Encyclopedia of Science and Technology*, 8th ed., Vol. 3, McGraw-Hill, New York, 738. With permission.)

ALLOYS

Generally speaking, the choice of alloys used in cladding is dictated by end use requirements such as corrosion, abrasion, or strength.

Cladding supplies a combination of desired properties not found in any one metal. A base metal can be selected for cost or structural properties, and another metal added for surface protection or some special property such as electrical conductivity. Thickness of the cladding can be made much heavier and more durable than obtainable by electroplating.

Combinations

The following clad materials are in common use:

- Stainless steel on steel. Provides corrosion resistance and attractive surface at low cost for food display cases, chemical-processing equipment, sterilizers, and decorative trim.
- Stainless steel on copper. Combines surface protection and high thermal conductivity for pots and pans, and for heat exchangers for chemical processes.
- Copper on aluminum. Reduces cost of electrical conductors and saves copper on appliance wiring.
- Copper on steel. Adds electrical conductivity and corrosion resistance needed in immersion heaters and electrical switch parts; facilitates soldering.
- Nickel or Monel on steel. Provides resistance to corrosion and erosion for furnace parts, blowers, chemical equipment, toys, brush ferrules, and many mechanical parts in industrial and business machines; more durable than electroplating.
- Titanium on steel. Supplies high-temperature corrosion resistance. Bonding requires a thin sheet of vanadium between titanium and steel.
- Bronze on copper. Usually clad on both sides, for current-carrying springs and switch blades; combines good electrical conductivity and good spring properties.
- Silver on copper. Provides oxidation resistance to surface of conductors, for high-frequency electrical coils, conductors, and braiding.
- Silver on bronze or nickel. Adds current-carrying capacity to low-conductivity spring material; cladding sometimes is in form of stripes or inlays with silver areas serving as built-in electrical contacts.
- Gold on copper. Supplies chemical resistance to a low-cost base metal for chemical processing equipment.
- Gold on nickel or brass. Adds chemical resistance to a stronger base metal than copper; also used for jewelry, wristbands, and watchcases.

APPLICATIONS

Gold-filled jewelry has long been made by the cladding process: the surface is gold, the base metal bronze or brass with the cladding thickness usually 5%. The process is used to add corrosion resistance to steel and to add electrical or thermal conductivity, or good bearing properties, to strong metals. One of the first industrial applications was the use of a nickel-clad steel plate for a railroad tank car to transport caustic soda; stainless clad steels are used for food and pharmaceutical equipment. Corrosion-resistant pure aluminum is clad to a strong duralumin base, and many other combinations of metals are widely used in cladding; there is also a technique for cladding titanium to steel for jet engine parts.

Today's coinage uses clad metals as a replacement for rare silver. Dimes and quarters have been minted from composite sheet consisting of a copper core with Cu–Ni facing. The proportion of core and facing used duplicates the weight and electrical conductivity of silver so the composite coins are acceptable in vending machines.

A three-metal composite sole plate for domestic steam irons provides a thin layer of stainless steel on the outside to resist wear and corrosion. A thick core of aluminum contributes thermal conductivity and reduces weight, and a thin zinc layer on the inside aids in bonding the sole plate to the body of the iron during casting.

Clad metals have been applied in nuclear power reactor pressure vessels in submarines as well as in civilian power plants.

Other applications where use of clad is increasing are in such fields as fertilizer, chemicals, mining, food processing, and even seagoing wine tankers.

Producers see a market for clad metal curtain wall building panels, and even stainless clad bus and automobile bumpers.

CLAY

Clay is composed of naturally occurring sediments that are produced by chemical actions resulting during the weathering of rocks. Often clay is the general term used to identify all earths that form a paste with water and harden when heated. The primary clays are those located in their place of formation. Secondary clays are those that have been moved after formation to other locations by natural forces, such as water, wind, and ice. Most clays are composed chiefly of SiO_2 and Al_2O_3. Clays are used for making pottery, tiles, brick, and pipes, but more particularly the better grade of clays are used for pottery and molded articles not including the fireclays and fine porcelain clays. Kaolins are the purest forms of clay.

The fineness of the grain of a clay influences not only its plasticity but also such properties as drying performance, drying shrinkage, warping, and tensile, transverse, and bonding strength. For example, the greater the proportion of fine material, the slower the drying rate, the greater the shrinkage, and the greater the tendency to warping and cracking during this stage. Clays with a high fines content usually are mixed with coarser materials to avoid these problems.

For two clays with different degrees of plasticity, the more plastic one will require more water to make it workable, and water loss during drying will be more gradual because of its more extensive capillary system. The high-plasticity clay also will shrink more and will be more likely to crack.

The most important clays in the pottery industry are the ball clays and china clays (kaolin).

COMMERCIAL CLAY

Commercial clays, or clays utilized as raw material in manufacturing, are among the most important nonmetallic mineral resources. The value of clays is related to their mineralogical and chemical composition, particularly the clay mineral constituents kaolinite, montmorillonite, illite, chlorite, and attapulgite. The presence of minor amounts of mineral or soluble salt impurities in clays can restrict their use. The more common mineral impurities are quartz, mica, carbonates, iron oxides and sulfides, and feldspar. In addition, many clays contain some organic material.

MINING AND PROCESSING

Almost all the commercial clays are mined by open-pit methods, with overburden-to-clay ratios ranging as high as 10 to 1. The overburden is removed by motorized scrapers, bulldozers, shovels, or draglines. The clay is won with draglines, shovels, or bucket loaders, and transported to the processing plants by truck, rail, aerial trainways, or belt conveyors, or as slurry in pipelines.

The clay is processed dry or, in some cases, wet. The dry process usually consists of crushing, drying, and pulverizing. The clay is crushed to egg or fist size or smaller and dried usually in rotary driers. After drying, it is pulverized to a specified mesh size such as 90% retained on a 200-mesh screen with the largest particle passing a 30-mesh screen. In other cases, the material may have to be pulverized to 99.9% finer than 325 mesh. The material is shipped in bulk or in bags. All clays are produced by this method.

PROPERTIES

Most clays become plastic when mixed with varying proportions of water. Plasticity of a materal can be defined as the ability of the material to undergo permanent deformation in any direction without rupture under a stress beyond that of elastic yielding. Clays range from those that are very plastic, called fat clay, to those that are barely plastic, called lean clay. The type of clay mineral, particle size and shape, organic matter, soluble salts, adsorbed ions, and the amount and type of nonclay minerals all affect the plastic properties of a clay.

Strength

Green strength and dry strength properties are very important because most structural clay products are handled at least once and must be strong enough to maintain shape. Green strength is the strength of the clay material in the wet, plastic state. Dry strength is the strength of the clay after it has been dried.

Shrinkage

Both drying and firing shrinkages are important properties of clay used for structural clay products. Shrinkage is the loss in volume of a clay when it dries or when it is fired. Drying shrinkage is dependent on the water content, the character of the clay minerals, and the particle size of the constituents. Drying shrinkage is high in most very plastic clays and tends to produce cracking and warping. It is low in sandy clays or clays of low plasticity and tends to produce a weak, porous body.

Color

Color is important in most structural clay products, particularly the maintenance of uniform color. The color of a product is influenced by the state of oxidation of iron, the state of division of the Fe minerals, the firing temperature and degree of vitrification, the proportion of Al_2O_3, lime, and MgO in the clay material, and the composition of the fire gases during the burning operation.

Uses

All types of clay and shale are used in the structural products industry but, in general, the clays that are used are considered to be relatively low grade. Clays that are used for conduit tile, glazed tile, and sewer pipe are underclays and shales that contain large proportions of kaolinite and illite.

Clays used for brick and drain tile must be plastic enough to be shaped. In addition, color and vitrification range are very important. For common brick, drain tile, and terra-cotta, shales and surface clays are usually suitable, but for high-quality face bricks, shales and underclays are used.

COATINGS

Plastic, metal, or ceramic coatings can be applied to the surface of a material in a variety of ways to achieve desired properties.

Coatings improve appearance, corrosion resistance, abrasion resistance, and electrical or optical properties. They can be applied by wet or dry techniques, with simple or complex equipment. The choices are almost limitless because almost any coating material offers some degree of protection as long as it retains its integrity. If it provides a continuous barrier between the substrate and the environment, even a thin, decorative coating can do the job in a relatively dry and mild environment.

METAL COATINGS

Many new materials have been developed, but steel remains the principal construction material for automobiles, appliances, and industrial machinery. Because of the vulnerability of steel to attack by aggressive chemical environments or even from simple atmospheric oxidation, coatings are necessary to provide various degrees of protection. They range from hot-dipped and electroplated metals to tough polymers and flame-sprayed ceramics.

In general, corrosive environments contain more than one active material, and the coating must resist penetration by a combination of oxidizers, solvents, or both. Thus, the best barrier is one that resists "broadband" corrosion.

Physical integrity of the coating is as important as its chemical barrier properties in many applications. For instance, coatings on impellers that mix abrasive slurries can be abraded quickly; coatings on pipe joints will cold-flow away from a loaded area if the creep rate is not low; and coatings on flanges and support brackets can be chipped or penetrated during assembly if impact strength is inadequate. Selecting the best coating for an application requires evaluating all effects of the specific environment, including thermal and mechanical conditions.

Zinc

One of the most common and inexpensive protection methods for steel is provided by zinc. Zinc-coated, or galvanized, steel is produced by various hot-dipping techniques, but more steel companies have moved into electrogalvanizing so they can provide both.

Oxidation protection of steel by zinc operates in two ways — first as a barrier coating, then as a sacrificial coating. If the zinc coating is scratched or penetrated, it continues to provide protection by galvanic action until the zinc layer is depleted. This sacrificial action also prevents corrosion around punched holes and at cut edges.

The grades of zinc-coated steel commercialized in recent years have been designed to overcome the drawbacks of traditional galvanized steel, which has been difficult to weld and to paint to a smooth finish. The newer materials are intended specifically for stamped automotive components, which are usually joined by spot welding and which require a smooth, Class A painted finish.

Aluminum

Two types of aluminium-coated steel are produced, each a different kind of corrosion protection. Type I has a hot-dipped Al–Si coating to provide resistance to both heat and corrosion. Type 2 has a hot-dipped coating of commercially pure aluminium, which provides excellent durability and protection from atmospheric corrosion. Both grades are usually used unpainted.

Type 1 aluminium-coated steel resists heat scaling to 677°C and has excellent heat reflectivity to 482°C. Nominal aluminium-alloy coating is about 1 mil on each side. The sheet is supplied with a soft, satiny finish. Typical applications include reflectors and housings for industrial heater panels, interior panels and heat exchangers for residential furnaces, microwave ovens, automobile and truck muffler systems, heat shields for catalytic converters, and pollution-control equipment.

Type 2 aluminized steel, with an aluminum coating of about 1.5 mil on each side, resists atmospheric corrosion and is claimed to outlast zinc-coated sheet in industrial environments by as much as five to one. Typical applications are industrial and commercial roofing and siding, drying ovens, silo roofs, and housings for outdoor lighting fixtures and air conditioners.

Electroplating

Use of protective electroplated metals has changed in recent years, mainly because of rulings by the Environmental Protection Agency (EPA). Cyanide plating solutions and cadmium and lead-bearing finishes are severely restricted or banned entirely. Chromium and nickel platings are much in use, however, applied both by conventional electroplating techniques and by new, more efficient methods such as fast rate electrodeposition (FRED), which has also been used successfully to deposit stainless steel on ferrous substrates.

Functional chromium, or "hard chrome," plating is used for antigalling and low-friction characteristics as well as for corrosion protection. These platings are usually applied without copper or nickel underplates in thicknesses from about 0.3 to 2 mil. Hard-chrome plating is recommended for use in saline environments to protect ferrous components.

Nickel platings, in thickness from 0.12 to 3 mil, are used in food-handling equipment, on wear surfaces in packaging machinery, and for cladding in reaction vessels.

Electroless nickel plating, in contrast to conventional electroplating, operates chemically instead of using an electric current to deposit metal. The electroless process deposits a uniform coating regardless of substrate shape, overcoming a major drawback of electroplating — the difficulty of uniformly plating irregularly shaped components. Conforming anodes and complex fixturing are unnecessary in the electroless process. Deposit thickness is controlled simply by controlling immersion time. The deposition process is autocatalytic, producing thicknesses from 0.1 to 5 mil.

Proprietary electroless-plating systems contain, in addition to nickel, elements such as phosphorus, boron, and/or thallium. A relatively new composition, called the polyalloy, features three or four elements in the bath. These products are claimed to provide superior wear resistance, hardness, and other properties, compared with those of generic electroless-plating methods.

One polyalloy contains nickel, thallium, and boron. Originally developed for aircraft gas turbine engines, it offers excellent wear resistance. Comparative tests show that relative wear for a polyalloy-coated part is significantly less than that for hard chromium and Ni–P coatings.

In general, Ni–B coatings are nodular. As coating thickness increases, nodule size also increases. Because the columnar structure of the coating flexes as the substrate moves, Ni–B resists chipping and wear.

Adhesion quality depends on factors such as substrate material, part preparation, and contamination. Although it is excellent for tool steels, stainless steel, high-performance nickel- and cobalt-base alloys, and titanium, a few metal substrates are not compatible. These include metals with high zinc or molybdenum content, aluminum, magnesium, and tungsten carbide (WC). Modifications can, however, eliminate this incompatibility.

Another trend in composite electroless plating appears to be toward codeposition of particulate matter within a metal matrix. These coatings are commercially available with just a few types of particulates — diamond, SiC, Al_2O_3, and polytetrafluoroethylene (PTFE) — with diamond heading the list in popularity.

These coatings can be applied to most metals, including iron, carbon steel, cast iron, aluminum alloys, copper, brass, bronze, stainless steel, and high-alloy steels.

Conversion Coatings

Electroless platings are more accurately described as conversion coatings, because they produce a protective layer or film on the metal surface by means of a chemical reaction. Another conversion process, the black oxide finish, has been making progress in applications ranging from fasteners to aerospace. Black oxide is gaining in popularity because it provides corrosion resistance and aesthetic appeal without changing part dimensions.

On a chemical level, black oxiding occurs when the Fe within the surface of the steel reacts to form magnetite (Fe_3O_4). Processors use inorganic blackening solutions to produce the reaction. Oxidizing salts are first dissolved in water, then boiled and held at 138 to 140°C. The product surface is cleaned in an alkaline soak and then rinsed before immersion in the blackening solution. After a second rinse, the finish is sealed with rust preventatives, which can produce finishes that vary from slightly oily to hard and dry.

Black oxiding produces a microporous surface that readily bonds with a topcoat. For example, a supplemental oil topcoat can be added to boost salt-spray resistance to the same level as that of zinc plate with a clear chrome coating (100 to 200 h).

Black oxide can be used with mild steel, stainless steel, brass, bronze, and copper. As long as parts are scale free and do not require pickling, the finish will not produce H_2 embrittlement or change part dimensions. Operating temperatures range from cryogenic to 538°C.

Chromate conversion coatings are formed by the chemical reaction that takes place when certain metals are brought in contact with acidified aqueous solutions containing basically water-soluble chromium compounds in addition to other active radicals. Although the majority of the coatings are formed by simple immersion, a similar type of coating can be formed by an electrolytic method.

Protective chromate conversion coatings are available for zinc and zinc alloys, cadmium, aluminum and aluminum alloys, copper and copper alloys, silver, magnesium and magnesium alloys. The appearance and protective value of the coatings depends on the base metal and on the treatment used.

Chromate conversion coatings both protect metals against corrosion and provide decorative appeal. They also have the characteristics of low electrical resistance, excellent bonding characteristics with organic finishes, and can be applied easily and economically. For these reasons the coatings have developed rapidly, and they are now one of the most commonly used finishing systems. They are particularly applicable where metal is subjected to storage environments such as high humidity, salt, and marine conditions.

The greatest majority of chromate conversion coatings are supplied as proprietary materials and processes. These are available usually as liquid concentrates or powdered compounds that are mixed with water. In the case of the powdered compounds, they are often adjusted with additions of acid for normal operation.

Chromate conversion coatings are formed immersing the metal in an aqueous acidified chromate solution consisting substantially of chromic acid or water-soluble salts of H_2CrO_3 together with various catalysts or activators. The chromate solutions, which contain either organic or inorganic active radicals or both, must be acid and must be operated within a prescribed pH range.

Maximum corrosion protection is obtained by using drab or dark bronze coatings on zinc and cadmium surfaces, and yellow to brown-colored coatings on the other metals. Lighter iridescent yellow type coatings generally provide medium protection, and the clear-bright type coatings, produced either in one dip or by leaching, provide the least protection.

Chromate conversion coatings provide maximum corrosion protection in salt spray or marine types of environment, and in high humidity such as encountered in storage, particularly where stale air with entrapped water may be present. They also provide excellent protection against tarnishing, staining, and finger marking, or other conditions that normally produce surface oxidation.

Olive drab type coatings are widely used on military equipment because of their high degree of corrosion protection coupled with a nonreflective surface. Iridescent yellow coatings are widely used for corrosion protection where appearance is not a deciding factor. The clear-bright chemically polishing type coatings for zinc and cadmium have been widely used to simulate nickel and chromium electroplate and are primarily used for decorative appeal rather than corrosion protection. Where additional corrosion protection or abrasion resistance is desired, these clear coatings act as an excellent base for a subsequent clear organic finish.

Heavy olive drab and yellow coatings for zinc, cadmium, and aluminum can be dyed various colors. Generally speaking, the dyed colors are used for identification purposes only since they are not lightfast and will fade upon exposure to direct sunlight or other sources of ultraviolet.

Because of their low electrical resistance, chromate conversion coatings are widely used for electronics equipment. Surface resistance depends on the type and thickness of the film deposited, the pressure exerted at the contact, and the nature of the contact. Low-resistance coatings are particularly important on aluminum, silver, magnesium, and copper surfaces.

Chromate conversion coatings can also be soldered and welded. A chromate coating on aluminum, for example, facilitates heliarc welding. Because of the slight increase in electrical resistance, an adjustment in current (depending upon the thickness of the coating) must be made to satisfactorily spot-weld. Soldering, using rosin fluxes, can be performed on cadmium-plated surfaces that have been treated with clear bright chromate conversion coatings. Clear, bright coatings on zinc-plate surfaces and colored coatings on both zinc and cadmium necessitate the use of an acid flux or removal of the film by an increase in soldering iron temperature, which burns through the coating, or by mechanical abrasion, which removes the film and provides a clean metal surface for the soldered joint.

Most chromate conversion treatments are applied by simple immersion in an acidified chromate solution. Because no electrical contacts need be made during immersion, the coatings can be applied by rack, bulk, or strip line operation. Under special situations, swabbing or brush coating can be used where small areas must be coated, as in a touch-up operation.

Chromate conversion coatings can also be applied by an electrolytic method in which the electrolyte is composed essentially of water-soluble chromium compounds and other radicals operated at neutral or slightly alkaline pH. This type of application is limited primarily to rack-type operation.

In general, processing can be placed in two categories: (1) over freshly electroplated surfaces; and (2) over electroplated surfaces that have been aged or oxidized, or other metal surfaces such as zinc die castings, wrought metals, or hot-dipped surfaces.

Sputtering

Formerly used primarily to produce integrated-circuit components, sputtering has moved on to large, production-line jobs such as "plating" of automotive trim parts. The process deposits thin, adherent films, usually of metal, in a plasma environment on virtually any substrate.

Sputtering offers several advantages to automotive manufacturers for an economical replacement for conventional chrome plating. Sputtering lines are less expensive to set up and operate than plating systems. And because sputtered coatings are uniform as well as thin, less coating material is required to produce an acceptable finish. Also, pollution controls are unnecessary because the process does not produce any effluents. Finally, sputtering requires less energy than conventional plating systems.

Chrome coating of plastics and metals is only one application for sputtering. The technique is not limited to depositing metal films. PTFE has successfully been sputtered on metal, glass, paper, and wood surfaces. In another application, cattle bone was sputtered on metallic prosthetic devices for use as hip-bone replacements. The sputtered bone film promotes bone growth and attachment to living bone.

Sputtering is the only deposition method that does not depend on melting points and vapor pressures of refractory compounds such as carbides, nitrides, silicides, and borides. As a result, films of these materials can be sputtered directly onto surfaces without altering substrate properties.

Much sputtering has been aimed at producing solid-film lubricants and hard, wear-resistant refractory compounds. NASA is interested in these tribological applications because coatings can be sputter-deposited without a binder, with strong adherence, and with controlled thickness on curved and complex-shaped surfaces such as gears and bearing retainers, races, and balls. Also, because sputtering is not limited by thermodynamic criteria (unlike most conventional processes that involve heat input), film properties can be tailored in ways not available with other deposition methods.

Most research on sputtered solid-lubricant films has been done with MoS_2. Other films that have been sputtered are WC, TiN, PbO_2, gold, silver, tin, lead, indium, cadmium, PTFE, and polyimide (PI). Of these coatings, the gold-colored TiN coatings are most prominent.

TiN coatings are changing both the appearance and performance of high-speed steel metal-cutting tools. Life of TiN-coated tools, according to producers' claims, increases by as much as tenfold, metal-removal rates can be doubled, and more regrinds are possible before a tool is discarded or rebuilt.

Sputter Coating Process

The SCX™ sputter coating process, a proprietary, computer-aided process developed by Engelhard-CLAL, Carteret, NJ, a producer of high-purity materials, enables the coating of base or refractory metals with precious metals. The source of the coating material can be almost any metallic composition. A major benefit of sputtering is the ability to deposit alloys or compounds that cannot be mechanically worked or alloyed as is required in the cladding process. By fabricating a segmented target comprising of two or more individual elements, a deposition can be made that is a uniformly dispersed "alloy" of the constituents.

SCX sputtering is conducted at low temperatures (<300°C), permitting deposition on plastics and other temperature-sensitive materials in addition to metals. Conducted at reduced pressures of inert gases, entrapped gases are kept to a minimum. Finally, by replacing the inert gas with a reactive gas such as H_2, N_2, or O_2, a compound formed by the gas can be deposited. This permits the reaction of very interesting coatings, such as nitrides, hydrides, and oxides. The unique sputter coating process makes it possible to attain very thin as well as relatively thick coatings equally as well in the range of $1/2$ to over 6 µm.

Typical substrate dimensions are wire: 0.08 to 1 mm diameter, continuous lengths up to 3000 m at 0.08 mm diameter, and 300 m at 0.89 mm diameter; ribbon: 0.017 to 0.50 mm thick and widths from 0.25 to 3.2 mm, continuous lengths from 120 to 500 m; rods: 3.2 mm diameter by 508 mm long; metallic foil: 0.05 mm thick and up, to 102×508 mm window dimension; polymeric: 0.25 mm thick and up, to 102×508 mm window; rigid metallic and nonmetallic: up to 127 mm thick and up to 102×508 mm window dimension. Flexible and discrete parts can be coated selectively on one or both sides.

Application to Power Tube Grids

Power tube grids control the flow of tube current by providing a bias between the cathode and anode. Semiconductor devices have replaced the bulk of electron tube usage, especially in receiving applications. However, in extremely high power applications, the Triode style thermal emissions tube still finds global use. The electron tube is expected to provide long-lasting, high-quality performance throughout the typical frequency range of 20 kHz to over 20 GHz, with grid temperatures from 600 to 1300°C, depending on the application. Secondary electron emission is of major concern to tube designers. Without controls or limits, a tube could easily become unstable and quickly self-destruct.

Platinum-coated molybdenum and tungsten are traditional materials for grid construction. Traditional platinum-clad molybdenum grid wire is produced with very thick precious metal coatings because it is difficult to produce claddings without base material breakthrough, and the molybdenum tends to diffuse through the platinum, embrittling the grid as well as causing an increase in emission. The SCX-PC sputter coating process accomplishes the same function as cladding, but with a precious metal savings of 15 to 30%. This is achieved by introducing a diffusion barrier into the coating during the manufacturing process, which effectively prohibits the interdiffusion of the core and the coating.

Other unique coatings include SCXPZC, which includes zirconium in the deposition process to permit higher-temperature grid usage with closer cathode spacing, and SCX-TH, which produces titanium-hydride (TiH_2) coatings to control primary and secondary emissions and enables the grid to act as a getter for nascent gas molecules.

Ion Plating

The basic difference between sputtering and ion plating is that sputtered material is generated by impact evaporation and transferred by a momentum transfer process. In ion plating, the evaporant is generated by thermal evaporation. Ion plating combines the high throwing power of electroplating, the high deposition rates of thermal evaporation, and the high energy impingement of ions and energetic atoms of sputtering and ion-implantation processes.

The excellent film adherence of ion-plated films is attributed to the formation of a graded interface between the film and substrate, even where the two materials are incompatible. The graded interface also strengthens the surface and subsurface zones and increases fatigue life.

The high throwing power and excellent adherence makes possible the plating of complex three-dimensional configurations such as internal and external tubing, gear teeth, ball bearings, and fasteners. Gears for space applications, for example, have been ion plated with 0.12 to 0.2 µm of gold for lubrication and to prevent cold welding of the gear pitch line. Ion plating has also been used, on a production basis, to plate aluminum on aircraft landing-gear components for corrosion protection.

Ion plating is also one of the two methods used to deposit diamond-like coatings (DLCs). A relative newcomer to the coatings field, DLCs are commonly made from hydrocarbon (often methane) and H_2 gases heated to 2000°C. The carbon coatings are prized for their wear resistance, as well as electrical and optical properties. Although they represent a huge potential, present DLCs are at the earliest stages of commercialization. However, their wide range of properties, along with their relatively low cost, leads many to predict huge growth in DLCs.

Researchers have proposed that the coatings be used to improve wear resistance in tool bits, as electronic heat sinks, and to boost wear and corrosion resistance in optical materials.

Chemical vapor deposition (CVD) is the method most often used to deposit DLCs. Adjusting deposition conditions allows the processor to change the coating from graphite to diamond-like. One process used at Battelle deposits the DLC in a gas atmosphere at reduced pressure without a fixed target. This plasma-assisted CVD allows large workpieces to be coated on all sides without turning. However, substrates must be heated to roughly 800°C when using CVD.

Reduced substrate temperatures are offered by dual ion-beam-enhanced deposition. Substrate temperature reaches only 66°C, and the dual ion-beam process does not rely on epitaxial growth for its formation as CVD does. Epitaxial growth requires a crystalline substrate; because dual ion-beam processing is free of this need, it enables amorphous materials to be coated as well.

Materials that are compatible with the Diond process include ferrous and nonferrous metals, glasses, ceramics, plastics, and composites. In addition to the Diond coating, dual ion-beam enhanced deposition can apply metallic coatings to fiber-reinforced carbon–carbon materials.

The basic ion-implantation process sends beams of elemental atoms (produced in a particle accelerator) into the surface of the target component. With dual ion-beam enhanced deposition, two simultaneous beams are used. One beam continuously sputters carbon onto the surface, providing the carbon material necessary to grow a diamond film. A second beam, consisting of inert gas at higher energy, drives some of the diamond layer into the interface zone. Then, the energy of the second beam is reduced to allow diamond growth. Implanting diamond material within the interface zone optimizes adhesion.

Technologies developed for electronic and optical thin films are often transferred into the engineering coatings sector, leading to a wider use of ion- and plasma-based techniques.

A novel nonequilibrium plasma treatment method with unusual characteristics is now being developed by EA Technology Ltd. of Capenhurst, near Chester, Great Britain. The process acts like a low-pressure, nonthermal glow discharge. Although it appears to provide many of the conditions needed for plasma surface engineering, the process runs at atmospheric pressure. Atmospheric deposition is generally much simpler than traditional vacuum plasma deposition, and its higher reactant concentrations make it much faster and, therefore, more cost-effective.

The equipment needed for this new approach is little more than a modified commercial microwave oven in which the plasma is sustained within a flask by microwave energy. Processing can be done either within the plasma or in a downstream gas that flows through the flask. In-plasma treatment is more energetic. Those materials that can withstand high temperatures can be coated within the flask. Downstream processing is easier and particularly effective at coating epoxies and polymers such as polymethyl methacrylate with materials such as TiO_2. This technique improves the weather resistance of the surface. The plasma can even be used to break down noxious gases such as volatile organics with more than 97% efficiency.

Vacuum Plasma Processing

This is the heart of advanced physical vapor deposition (PVD) coatings used on tools such as molds, dies, drills, and cutters. Although TiN is the most widely used of these coatings, many demanding applications now use TiAlN. High-quality PVD coatings are produced when electron beams evaporate the coating material while N_2-rich plasma bombards the substrates.

Electron beams cannot be used to evaporate alloys such as TiAl, since the vapor pressure of aluminum is 100 times higher than that of titanium. However, to overcome this problem, a new control technology has been developed for electron-beam deposition systems. This technology measures optical emissions from elements in the plasma, picking up characteristic titanium and aluminum emission lines. These are used to control individual electron-beam sources for alloy elements.

As a result, users can control the composition of the alloy coating — they can even modify the gradation of the coating chemistry, a technique not possible with alloy arc and sputter sources.

C-coated components, when compared to nitrided, nickel, and chromium coatings, exhibit improved wear-resisting performance. In particular, Balinit® C WC/C coatings are claimed to offer the proper combination of low coefficient of friction and high hardness needed by highly loaded automotive and machine parts.

Balinit C WC/C coatings are made of hard WC particles in a soft amorphous carbon matrix. Ion bombardment of a WC target removes coating material for deposition onto component surfaces under controlled conditions. Several applications demonstrate the ability of the material to solve wear problems:

- A coating of WC/C, specifically developed for highly stressed machine components operating under less-than-optimum lubricating conditions, has given design engineers a way out of such predicaments. Produced by PVD, WC/C is said to improve seizure resistance and reduce failure due to particle-contaminated hydraulic oils.
- Pump pistons coated with Balinit C operate longer than nitrided pistons or nickel- and chromium-coated pistons. Replacement of sliding shoes made from bronze with Balinit C coated steel also cuts down on wear.
- Application results in a low coefficient of friction and "smoothing" of the surface of the part. Hardness measures 1000 VHN (25 g); thickness is approximately 3 μm.
- Balinit C WC/C coatings, with a maximum working temperature of 300°C, are at present used in racing, as well as industrial gear systems.

In tribological tests conducted on CrN, TiAlN, TiAlCN, TiCN, TiCN+C, TiN+C, TiB_2, WC, and molybdenum coatings for load capacitance, adhesion power, abrasion force, hardness, and fatigue strength showed TiAlN, the very hard, metallic coating used on cutting tool inserts, gave the best results. Unlike cutting tools, however, bearing components cannot be coated in a standard CVD process at temperatures over 500°C. Even with a speciality developed PVD process, which deposits hard coatings with high adhesion power at a temperature level of only 160°C, bearing rings must undergo special annealing after final grinding at a temperature of approximately 240°C to achieve the roundness deviation after coating within the normal manufacturing tolerance range.

Thermal Spraying

Arc spraying, a form of thermal spraying of metals, is done on a prepared (usually grit-blasted) metal surface, using a wire-arc gun. The coating metal is in the form of two wires that are fed at rates that maintain a constant distance between their tips. An electric arc liquefies the metal, and an air spray propels it onto the substrate. Because particle velocity can be varied considerably, the process can produce a range of coating finishes from a fine to a coarse texture.

Arc-sprayed coatings are somewhat porous, as they are composed of many overlapping platelets. Used in applications where appearance is important, thermally sprayed coatings can be sealed with pigmented vinyl copolymers or paints, which usually increase the life of the metal coating. Arc-sprayed coatings are thicker than those applied by hot dipping, ranging from 3 to 5 mil for light-duty, low-temperature applications to 7 to 12 mil for severe service.

Because zinc and aluminum are, under most conditions, more corrosion resistant than steel, they are the most widely used spray-coating metals. In addition, since both metals are anodic to steel, they act galvanically to protect ferrous substrates.

In general, aluminum is more durable in acidic environments, and zinc performs better in alkaline conditions. For protecting steel in gas or chemical plants, where temperatures might reach 204°C, aluminum is recommended. Zinc is preferred for protecting steel in fresh, cold waters; in aqueous solutions above 66°C, aluminum is the usual choice.

For service to 538°C, a thermally sprayed aluminum coating should be sealed with a silicone–aluminum paint. Between 538°C and 899°C, the aluminum coating fuses and reacts with the steel base metal, forming a coating that, without being sealed, protects the structure from an oxidizing environment. And, for continuous service to 982°C, a nickel–chrome alloy is used, sometimes followed by aluminum.

In Europe, where thermally sprayed metal coatings for corrosion protection have been far more widely used than in the United States, many structures such as bridges are still in good condition after as long as 40 years, with minimum maintenance. Other applications include exhaust-gas stacks, boat hulls, masts, and many outdoor structures.

Thermal spraying has become much more than a process for rebuilding worn metal surfaces. Thanks to sophisticated equipment and precision control, it is now factored into the design process, producing uniform coatings of metals and ceramics. With some of the processes, even gradated coatings can be applied. This is done by coating the substrate with a material that provides a good bond and that has compatible expansion characteristics, then switching gradually to a second material to produce the required surface quality such as wear resistance, solderability, or thermal-barrier characteristics.

Plasma Spraying

Plasma-spray coating relies on a hot, high-speed plasma flame (N_2, H_2, or argon) to melt a powdered material and spray it onto the substrate. A DC arc is maintained to excite gases into the plasma state.

The high-heat plasma (in excess of 7075°C) enables this process to handle a variety of coating materials — most metals, ceramics, carbides, and plastics. Although most coating materials are heated to well beyond their melting points, substrate temperatures commonly remain below 121°C.

This process has found wide acceptance in the aircraft industry. Plasma-sprayed metallic coatings protect turbine blades from corrosion, and sprayed ceramics provide thermal-barrier protection for other engine parts.

Proprietary refinements in plasma-spray technology include a wear-resistance coating material that lends itself to forming amorphous/microcrystalline phases when plasma sprayed. The resultant coating provides excellent corrosion resistance with minimal oxidation at higher temperatures. This promises to eliminate problems of work-hardened crystalline coatings that chip or delaminate in response to stress, which have previously been taken care of by expensive alloying elements.

Another amorphous alloy development involves a crystalline material that, upon abrasive wear, transforms to an amorphous hard-phase alloy. The top layer, 3 to 5 µm thick, results in hardness levels over 1300 Vickers. Wear tests have indicated this material is superior to more expensive WC coatings.

Detonation gun coatings considered by many to be an industry standard, use a detonation wave to heat and accelerate powdered material to 732 mm/s. In the line-of-sight process, each individual detonation deposits a circle of coating with a 2.54-cm diameter and 2-µm thickness. Coatings, thus, consist of multiple layers of densely packed lenticular particles tightly bonded to the surface.

The Super D-Gun has been developed to increase particle velocities. New coatings (the UCAR 2000 Series) applied with the gun offer improved wear resistance without affecting fatigue performance. The system has been targeted for fatigue-sensitive aircraft components.

Other low-pressure plasma-spray (LPPS) coatings protect turbine vanes by improving the sulfidation and oxidation resistance of complex components. Inert-atmosphere LPPS systems are an effective means for applying complex corrosion-resistant coatings such as NiCoCrAlY to high-temperature engine components.

PS300 is a self-lubricating solid coating material for use in sliding contacts at temperatures up to 800°C. PS300 is a composite of metal-bonded CrO_2 with $BaFl_2/CaFl_2$ eutectic and silver as solid lubricant additives. The "PS" in the name of this and other self-lubricating, high-temperature composite materials signifies that the material is applied to a substrate by plasma spraying of a powder blend of its constituents.

Spray Coatings

Stabilized Zirconia (ZrO_2)

Yttria-stabilized zirconia (YZP) represents the bulk of all sprayed ceramics. This material is used primarily for thermal barrier coatings (TBCs) in aircraft, rocket, and reciprocating engines. TBCs are applied to engine components to lower substrate temperatures so that combustion gas temperatures can be higher, thereby increasing engine power and efficiency and lowering emissions. Stabilized ZrO_2 is unique for its high CTE and low thermal conductivity. The high CTE correlates well with the base metals to which ZrO_2 is commonly applied, reducing stresses that are induced by differential expansion. Plasma-sprayed YZP is also reasonably resistant to thermal fatigue and chemical attack.

Other elements used to stabilize ZrO_2 include the oxides of magnesium, calcium, and cerium. Phase stabilization is used to mitigate the large volume change that ZrO_2 undergoes during heating to and cooling from service temperatures. The phase transformation from low-temperature monoclinic to high-temperature tetragonal can be arrested by the inclusion of stabilizing components such as YO_2. Fully stabilized ZrO_2 maintains a cubic structure throughout heating. Partially stabilized ZrO_2, which has both cubic and tetragonal phases, is reported to be tougher and to have a better match of CTE with engine materials.

MCRALYS

Metallic coatings are used between the ceramic coating and substrate both to enhance bonding and to provide a barrier that prevents substrate oxidation and corrosion. As a class, these materials are denoted by the term MCRALY, which is derived from the components: a base metal (M), chromium (CR), aluminum (AL), and yttria (Y). The M component is iron (FE), cobalt (CO), or nickel (NI), singly or in combination. The coatings are then called FECRALY, COCRALY, NICRALY, and so forth.

Recent advances in thermal spraying have focused on controls. Process control includes barfeedstock, materials, and processing parameters. Historically, thermal spray coatings have been applied at a confidence level of around 70.

Applications

Common use of thermally sprayed YZP involves net-shape manufacturing. O_2 sensors for automobile emission control systems are manufactured by applying the coatings of YZP to remove mandrels. When the mandrel is removed, a free-standing shape is left.

Net- and near-net-shape (NNS) techniques facilitate the fabrication of parts that is not practicable by other means. Freestanding net-shape ion engines have been manufactured from tungsten by using load chamber, plasma-arc spray. Plasma spraying in an argon atmosphere eliminates oxidation of reaction materials, such as tungsten. Precision Al_2O_3 tubes (0.75 mm) wall thickness, 75 mm diameter, and 1.2 m length would be difficult, if not impossible, to fabricate by casting and grinding; however, they have been successfully fabricated by thermal-spray net-shape techniques. On the other extreme, multilayer ceramic tubes with 1-mm inside diameter have been made to join blood vessels.

Ceramic coatings are applied to medical instruments used for endoscopic and other forms of minimal invasive surgery.

POLYMER COATINGS

Polymeric coatings designed for corrosion protection are usually tougher and are applied in heavier films than are appearance coatings. Requirements of such coatings are much more stringent: They must adhere well to the substrate and must not chip easily or degrade from heat, moisture, salt, or chemicals.

Environmental factors also drive the technology behind polymer coatings replacing chromium and cadmium coatings. This is partly due to increasing concern about heavy metals. Also, automakers must now contend with acid rain in addition to salt spray, and polymers surpass chromium and cadmium in acid rain resistance.

Acrylics and alkyds are widely used for farm equipment and industrial products requiring good corrosion protection at a moderate cost. Alkyd resins, particularly, play a major role in maintenance painting because of their good weathering characteristics and ease of application with low-cost, low-toxicity solvents. Alkyd paints are also relatively high in solids, permitting good buildup of a paint film with a minimum number of coats.

Silicone modification of organic resins improves overall weatherability and durability. Compared to organic coatings in general, silicones have greater heat stability, longer life, better resistance to deterioration from sunlight and moisture, and greater biological and chemical inertness.

For optimum weatherability, silicone content should be 25 to 30%. Performance of the waterborne formulations is proving to be almost identical to that of the solvent-based coatings.

For coatings requiring higher heat resistance, silicone resins can be used alone for the paint vehicles, or they can be blended with various organic resins. These finishes are used on space heaters, clothes dryers, and barbecue grills. Similar formulations are used on smokestacks, incinerators, boilers, and jet engines. Performance of formulations containing ceramic frits approaches that of ceramic materials.

Polyurethane enamels are characterized by excellent toughness, durability, and corrosion resistance. These thermosetting materials, available in both one- and two-part formulations, cost more than the alkyds and acrylics.

Urethane chemistry is versatile enough to provide a hard, durable, environmentally resistant film, a tough, elastomeric coating, or a surface somewhere between. Urethanes have traditionally been available as solvent-based coatings containing 25 to 45% solids, but environmental concerns have prompted manufacturers to also supply them in high solids, 100% solids, and waterborne formulations.

Coating thickness of polyurethanes ranges from about 2 mil for average requirements to as much as 30 mil for applications requiring impact and/or abrasion resistance as well as corrosion resistance. Typical uses are on conveyor equipment, aircraft radomes, tugboats, road-building machinery, and motorcycle parts. Abrasion-resistant coatings of urethanes are applied on railroad hopper cars, and linings are used in sandblasting cabinets and slurry pipes.

Epoxy finishes have better adhesion to metal substrates than do most other organic materials. Epoxies are attractive economically because they are effective against corrosion in thinner films than are most other finishing materials. They are often used as primers under other materials that have good barrier properties but marginal adhesive characteristics.

Coating thickness can vary from 1 mil for light-duty protection to as much as 20 mil for service involving the handling of corrosive chemicals or abrasive materials. Performance of epoxies is limited in the heavier thicknesses, however, because they are more brittle than other organic materials.

Nylon 11 coatings provide attractive appearance as well as protection from chemicals, abrasion, and impact. Applied by electrostatic spray in thicknesses from 2.5 to 8 mil, nylon coatings are used on office and outdoor furniture, hospital beds, vending-machine parts, and building railings. Heavier coatings — to 50 mil — are applied by the fluidized-bed method and are used to protect dishwasher baskets, food-processing machinery, farm and material-handling equipment, and industrial equipment such as pipe, fittings, and valves.

Fluorocarbons are more nearly inert to chemicals and solvents than all other polymers. The most effective barriers among the fluorocarbons for a variety of corrosive conditions are PFA, PTFE, ECTFE, FEP, and PVDF.

Ethylene-chlorotrifluoroethylene (ECTFE) is a member of the fluoropolymer family of resins. This high-temperature coating provides corrosion resistance and mechanical and electrical qualities up to 149°C. It is easy to apply and has excellent release and low-friction properties. Multiple layers of ECTFE can be applied up to 100 mils. Although more expensive than other powders, its performance often justifies a higher cost. ECTFE has a smooth surface as applied and is therefore used in water handling systems to minimize bacterial buildup.

Fluorinated ethylene propylene (FEP) is a soft fluoropolymer coating similar to PTFE and PFA. It has the best release properties among powder coatings. The corrosion-resistant qualities of FEP are better than PTFE, but FEP does not stand up to high temperatures as well.

Perfluoroalkoxy (PFA) is another member of the fluoropolymer family that resists corrosion and has better release and nonwetting qualities than PTFE. With wide temperature limits, PFA can be used in applications ranging from cryogenic levels to 260°C. It is typically used for coating molding cavities, and this food-grade-quality powder can also be used for baking surfaces.

Polytetrafluoroethylene (PTFE) is a soft and waxy material that has release properties similar to PFA. The high-temperature coating protects against corrosion in environments reaching 260°C. However, its softness limits it to non-abrasive and moderately abrasive applications.

Polyvinylidenefluoride (PVDF) is the hardest powder coating with qualities similar to PPS, making it ideal for high-load and high-modulus applications. The coating works well in corrosive environments for components in pulp mills, waste treatment plants, and petrochemical facilities. Once applied, PVDF coatings can be removed only with heat. The coating has a service temperature limit of 130°C and is not commonly used for release and low-friction applications.

For impact service, PVDF and ECTFE coatings are recommended, in that order. PTFE, FEP, and PFA are also suitable, but they have a greater tendency to creep under load. For abrasive conditions, PVDF is outstanding among the fluorocarbons. Recommended for high service temperatures — drying ovens and steam-handling equipment, for example — are PFA and PVDF. These materials are also used on engine components and welders. PVDF also has the highest compressive strength of the fluorocarbons. PTFE has the highest allowable service temperature (316°C) of the fluorocarbons.

Coatings based on PTFE are being used to reduce wear in the U.S. automotive industry. Fluoropolymer coatings prevent binding and galling in disk brake systems at temperatures over 100°C. PTFE is also used as a dry lubricant. In addition, PTFE can be used as a coating on automotive fasteners, and a new process uses PTFE to prevent seizure in valve springs. This process is FluoroPlate impingement, a process whereby a mixture of inorganic and organic particles bombard the spring surface, thus relieving internal stresses and reducing surface flaws. The coating also helps the springs to repel oil.

A new class of coating — an alloy of fluoropolymer and other resins — has a different viscosity behavior than that of the earlier organics. Viscosity of "fastener-class" coating resins decreases sharply as film shear increases (as in application by the dip/spin process). Then, when the spinning basket stops, viscosity returns almost instantaneously to its original value. Thus, when applied to the dip/spin process, the coating clings to sharp edges, threads, and points.

Film thickness typically ranges from 0.5 to 0.7 mil, but formulations can be adjusted to provide films of 0.3 to 0.4 mil for parts with fine threads or other intricate features. Not only do these extremely tough coatings provide a more uniform barrier to corrosives, but they are also based on polymers that are inherently stable in the presence of a wide spectrum of acids, bases, and aqueous solutions.

Combination coatings blend the advantages of anodizing or hard-coat platings with the controlled infusion of low-friction polymers and/or dry lubricants. The coatings become an integral part of the top layers of metal substrates, providing increased hardness and other surface properties.

These coatings are different for each class of metals. For example, a Tufram coating for aluminum combines the hardness of Al_2O_3 and the protection of a fluorocarbon topcoat to impart increased hardness, wear and corrosion resistance, and permanent lubricity.

In the multistep process, the surface is first converted to Al_2O_3. Submicron particles of PTFE are then fused into the porous anodized surface, forming a continuous plastic/ceramic surface that does not chip, peel, or delaminate. The coating is claimed to have greater abrasion resistance than case-hardened steel or hard chromium plate.

Another proprietary coating that penetrates PTFE into precision hardcoat anodizing is Nituff. The coating achieves a self-curing, self-lubricating surface with low friction, high corrosion resistance, and dielectric properties superior to ordinary hardcoat anodizing. It is used extensively in aerospace, textile, food processing, packaging, and other industries, where it allows manufacturers to benefit from the light weight and easy machinability of aluminum enhanced by the durability, cleanliness, and dry lubrication of the Nituff surface.

Other proprietary combination coatings have been developed for steel, stainless steel, copper, magnesium, and titanium that provide similar surface improvement. Coatings are also available that enhance specific properties such as lubricity, corrosion resistance, or wear resistance.

Powder Coatings

Powder coatings are generally much thicker than fluids, typically greater than 5 mils compared to 1 or 2 mils for fluids. Powders commonly protect substrates from corrosion and erosion wear, or provide release or aesthetic qualities. Fluid coatings, in contrast, are preferred for friction, abrasion, and spalling wear — corrosion applications that require low friction and release.

In addition to performance attributes, powder coatings also provide processing advantages over fluids. Powders are environmentally friendly because they do not contain volatile organic compounds (VOCs) that attack the ozone layer.

The powder coating combines properties of both plastics and paints. The coatings are manufactured using typical plastics-industry equipment. They are first sent through a melt-mix extruder and then ground. When applied as a coating, however, the powder becomes a coating film that is exactly like paint.

These coatings have been developed in response to pressures to reduce VOC emissions, which have increased over the past few years. Overspray from liquid paints contains solvents that are released into the atmosphere. Even with recovery systems, some volatile components escape. Powder coatings, on the other hand, are completely recyclable. Overspray can be collected easily and reused. If a small amount becomes too contaminated for recycling, safe disposal techniques are available.

Powder coatings also show promise as a substitute for clear coats in the automotive industry, Present solvent-based paints could be replaced by a clear powder coating that cures at roughly the same temperature as conventional paints. Powder coatings may also replace the baked-on porcelain enamel used for appliance parts. Washer and dryer lids are now powder-coated in industry.

Applications are not all that is new, however. Materials have changed. The majority of powder coatings have relied on either an epoxy or polyester resin base. Acrylics, however, are becoming more important, and other possible bases include nylon, vinyl, and various fluoropolymers.

Two processes for applying coatings have undergone refinements. With electrostatic spraying, the most popular method, powder is given a charge and sprayed onto electrically grounded parts. Baking them completes the cure. Nonconductive parts must be primed or heated to provide them with more electrostatic attraction.

In the fluid-bed process, air passes through a porous membrane at the bottom of a tank and aerates the powder so that it swirls around in the tank. A part is then heated and dipped into the tank, so that the powder melts on the surface. This process is used for thick-film protection coatings, and is suitable only for metal parts that can retain heat long enough to be coated.

Liquid Layers

Despite strict governmental restrictions that continue to thwart the amount of VOCs released into the atmosphere, the use of high-performance fluid coatings remains widespread. Unlike powder coatings, fluids can be applied in films as thin as 0.2 mils without affecting their integrity. Powder coatings, in contrast, require more clearance between components to accommodate thicknesses greater than 5 mils.

Another advantage of fluid coatings is that manufacturers can apply them by conventional and electrostatic spray processes, as well as by brush and dip methods. This is particularly important for components with deep recesses that are difficult to coat with powders using electrostatic means because the charged particles adhere to the outer edge of the recess. Probe-spray systems, in turn, are much more efficient for covering recesses with fluid coatings.

Following is a summary of the fluid coatings commonly used to protect engineering materials; see Figure C.2.

Matrix coatings have the highest mechanical properties, such as tensile strength and wear resistance, among fluids coatings. They consist of one or more polymer binders (PPS, polyimide) combined with a dry lubricant such as PTFE, PFA, MoS_2, or graphite. When applied, the dispersion of this composite mixture consists of lubricant evenly distributed within the binder. Typical applications include large-thread fasteners, actuators, pistons, bearings, impellers for mixers and superchargers, and rotating and sliding powder-metal components.

Stratified coatings also consist of binders and lubricants; however, their formulation keeps most of the low-friction, high-release agent on the bearing surface of the coating. With lubricant segregated to the outer surface, the top layer of stratified coatings is softer and has lower mechanical properties than matrix coatings. However, this concentration of lubricant improves release qualities, and with certain fluoropolymers, the coatings protect substrates from corrosion. Typical applications include components used in photocopiers, valves, and fuel-handling systems.

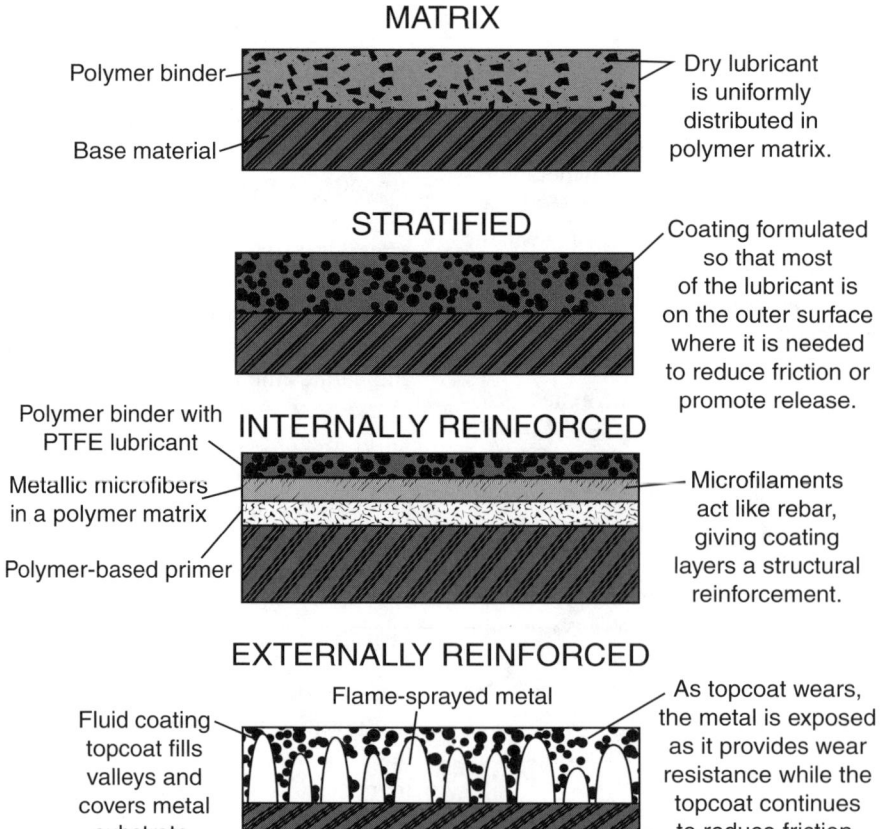

FIGURE C.2 A closer look at coatings. (From *Machine Designs*, September 10, p. 103, 1998. With permission.)

Hybrids are a new class of coatings that have greater strength than both powders and matrix-fluid coatings. To boost strength, the films are formulated with either internal or external reinforcements.

Internally reinforced coatings use microfilaments to provide a mechanical structure inside the film, acting in much the same way that rebar strengthens concrete structures. The overlapping filaments produce a film that has a wear limit that exceeds conventional measurement methods. Engineers typically use these coatings in applications with high loads such as cutting blades.

Externally reinforced coatings combine a rigid substrate such as stainless steel with a conventional high-release fluid coating containing a low-friction release agent such as high percentages of PTFE. The films consist of two layers. The first is a continuous layer of flame-sprayed stainless steel. Manufacturers then cover the stainless steel with a thick topcoat of low-friction fluid to create a total film thickness of 1 to 1.4 mils. The topcoat fills the depressions and smoothes over rough spots on the flame-sprayed metal coat.

Under high loads or extreme wear, the topcoat may wear away, exposing the substrate material. However, because the fluid coat remains in the asperities of the base metal, the composite coating maintains a low-friction surface while the steel helps resist wear. Typical applications are large locking and fastening mechanisms, tumblers, agitators, and parts subjected to high abrasive wear.

Conformal Coatings

Printed circuit board (PCB) assemblies used in avionics, marine, automotive, and military applications generally perform in environments affected by heat, moisture, industrial pollutants, manual handling, and process residues, which typically magnifies such effects.

The PCB designer must determine if and how the assembly must be protected with a worst-case environment in mind. Consequences ranging from malfunction to product failure will devolve from this decision. Other factors include feasibility and cost of implementation of such protection, but the designer will have several options: reducing circuit sensitivity through designed signal characteristics, increasing conductor spacing, and using buried vias. Other protective options include sealing the circuit in a box pressurized with inert gas or in a polymer via potting or molding, or employing an electrically insulating barrier between conductors and the ambient environment.

Using a polymeric-film barrier for board protection is the most common and practical approach. There are, however, limitations and peculiarities accompanying the two categories of coatings:

1. Bare-board coatings, or "permanent soldermasks," are applied during the board fabrication process as liquids or dry films over conductive finishes or bare copper, and polymerized (cured) either by heat or ultraviolet light.
2. Assembly coatings offer protective coverage against water in the form of atmospheric humidity and condensation. They are commonly referred to as "conformal" coatings, implying compliance of the film to the contour line of the assembly. However, this term may be a misnomer, since newer application techniques permit only specific areas be coated to eliminate the cost of masking.

The use of conformal coating materials is provided in four basic material types (acrylic, epoxy, silicone, and urethane resins) for liquid-applied coatings, and one (paraxylylene) for vapor deposition. As a result, a coating must be selected on the basis of electrical, thermal, mechanical, and other pertinent properties as dictated by requirements for circuit performance and characteristics, type and degree of environmental exposure, and consequences of failure; see Table C.8.

TABLE C.8
Five Basic Prerequisites for Good Coating Performance

1. Surfaces to be coated must be free of contaminants and volatiles.
2. Electrical contact areas are adequately masked to be free of coating.
3. Coating film should be capable of developing an adequate bond to the applied surfaces.
4. The coating film should be completely cured and cover all intended areas at a specific thickness.
5. The assembly coating, substrate, and, when applicable, the bare-board coating, should yield a dielectric system that complies with design requirements.

COBALT

Cobalt (symbol Co) is a lustrous, silvery-blue metallic chemical element, resembling nickel but with a bluish tinge instead of the yellow of nickel. It is rarer and costlier than nickel and its price has varied widely in recent years. Although allied to nickel, it has distinctive differences. It is more active chemically than nickel. It is dissolved by dilute H_2SO_4, HNO_3, or HCl acids, and is attacked slowly by alkalis. The oxidation rate of pure cobalt is 25 times that of nickel. Its power of whitening copper alloys is inferior to that of nickel, but small amounts in Ni–Cu alloys will neutralize the yellowish tinge of the nickel and make them whiter. The metal is diamagnetic like nickel, but has three times the maximum permeability. Like tungsten, it imparts red-hardness to tool steels. It also hardens alloys to a greater extent than nickel, especially in the presence of carbon, and can form more chemical compounds in alloys than nickel.

Its chemical properties resemble, in part, those of both nickel and iron. Cobalt is the metal with the highest Curie temperature (1121°C) and the lowest allotropic transformation temperature (399°C). Below 421°C, cobalt is close-packed hexagonal; above, it is face-centered cubic.

PROPERTIES

Cobalt has a specific gravity of 8.756, a melting point of 1495°C, 85 Brinell hardness, and an electrical conductivity about 16% that of copper. The ultimate tensile strength of pure cast cobalt is 234 MPa, but with 0.25% carbon it is increased to 427 MPa. Strength can be increased slightly by annealing and appreciably by swaging or zone refining. The metal is used in tool-steel cutters, in magnet alloys, in high-permeability alloys, and as a catalyst, and its compounds are used as pigments and for producing many chemicals.

The natural cobalt is cobalt–59, which is stable and nonradioactive, but the other isotopes from 54 to 64 are all radioactive, emitting beta and gamma rays. Most have very short life except cobalt–57, which has a half-life of 270 days, cobalt–56 with a half-life of 80 days, and cobalt–58 with a half-life of 72 days. Cobalt–60, with a half-life of 5.3 years, is used for radiographic inspection. It is also used for irradiating plastics, and as a catalyst for the sulfonation of paraffin oils because the gamma rays cause the reaction of SO_2 and liquid paraffin. Co 60 emits gamma rays of 1.1- to 1.3-MeV energy, which gives high penetration for irradiation. The decay loss in a year is about 12%, the cobalt changing to nickel.

The best known cobalt alloys are the cobalt-base superalloys used for aircraft turbine parts. The desirable high-temperature properties of low creep, high stress-rupture strength, and high thermal-shock resistance are attributed to the allotropic change of cobalt to a face-centered cubic structure at high temperatures. Besides containing 36 to 65% cobalt, usually more than 50%, most of these alloys also contain about 20% chromium for oxidation resistance and substantial amounts of nickel, tungsten, tantalum, molybdenum, iron, and/or aluminum, and small amounts of still other ingredients. Carbon content is in the 0.05 to 1% range. These alloys include L–605; S–816; V–36; WI–52; X–40; J–1650; Haynes 21 and 151; AiResist 13, 213, and 215; and MAR-M 302, 322, and 918. Their 1000-h stress-rupture strengths range from about 276 to 483 MPa at 649°C and from about 28 to 103 MPa at 982°C. Cobalt is also an important alloying element in some nickel-base superalloys, other high-temperature alloys, and alloy steels. Besides tool steels, the maraging steels are a good example. Although cobalt-free grades have been developed, due to the scarcity of this metal at times, most maraging steels contain cobalt, as much as 12%. Cobalt is also a key element in magnet steels, increasing residual magnetism and coercive force, and in non-ferrous-base magnetic alloys.

An important group of cobalt alloys is the stellites. They contain chromium and various other elements such as tungsten, molybdenum, and silicon. The extremely hard alloy carbides in a fairly hard matrix give excellent abrasion and wear resistance and are used as hard-facing alloys and for aircraft jet engine parts.

The interesting properties of cobalt-containing permanent, soft, and constant-permeability magnets are a result of the electronic configuration of cobalt and its high curie temperature. In addition, cobalt in well-known Alnico magnet alloys decreases grain size and increases coercive force and residual magnetism.

Cobalt is a significant element in many glass-to-metal sealing alloys and low-expansion alloys. One iron-base alloy containing 31% nickel and 5% cobalt provides a lower CTE than the iron–36% nickel alloy called Invar, and is less sensitive to variations in heat treatment. Co–Cr alloys are used in dental and surgical applications because they are not attacked by body fluids. Alloys named Vitallium are used as bone replacements and are ductile enough to permit anchoring of dentures on neighboring teeth. They contain about 65% cobalt.

Cobalt is a necessary material in human and animal metabolism, and is used in fertilizers.

Applications

The major uses of cobalt are in cobalt-base and cobalt-containing materials for high-temperature alloys, permanent magnets, and steels. In addition, cemented carbides, which are considered cutting tools, are used in balls for ballpoint pens and high-temperature ball bearings.

The hard-facing alloys are useful because of their resistance to corrosion, abrasion, and oxidation at high temperatures for plowshares, oil bits, crushing equipment, tractor treads, rolling mill guides, knives, punches, shears, billet scrapers, valves for high-pressure steam, oil refineries, and diesel and auto engines.

The superalloys have found use as searchlight reflectors, and are also useful under the severe operating conditions of high-temperature nuclear reactors. Their superior elevated-temperature properties compensate, to some extent, for the high thermal-neutron absorption for cobalt. They are used in reactors in certain wear-resistant components as guides for control rods. In nuclear submarines, these alloys are used where severe wear in contact with seawater is encountered.

Although the main application of Alnico permanent magnets is in motors, generators, regulating devices, instruments, radar, and loudspeakers, some are used in games, novelties, and door latches.

Cobalt-containing tool and high-speed steels are also used for dies.

COLD-MOLDED PLASTICS

Cold-molded plastics are one of the oldest of the so-called plastic materials; they were introduced in the United States in 1908. For the first time they provided the electrical engineer with materials that could be molded into more complicated shapes than could porcelain or hard rubber, providing better heat resistance than hard rubber, and better impact strength than porcelain. They could also incorporate metal inserts.

General Nature and Properties

So-called cold-molded plastics are formulated and mixed by the molder (usually in a proprietary formulation). The materials fall into two general categories: inorganic or refractory materials, and organic or nonrefractory materials.

Inorganic cold-molded plastics consist of asbestos fiber filler and either an SiO_2-lime cement or portland cement binder. Clay is sometimes added to improve plasticity. The SiO_2-lime materials are easier to mold although they are lower in strength than the portland cement types.

In general, advantages of these materials include high arc resistance, heat resistance, good dielectric properties, comparatively low cost, rapid molding cycles, high production with single-cavity molds (thus low tool cost), and no need for heating of molds. On the other hand, they are relatively heavy, cannot be produced to highly accurate dimensions, are limited in color, and can be produced only with a relatively dull finish. They have been used generally for arc chutes, arc barriers, supports for heating coils, underground fuse shells, and similar applications.

Organic cold-molded plastics consist of asbestos fiber filler materials bound with bituminous (asphalt, pitches, and oils), phenolic, or melamine binders. The binder materials are mixed with solvents to obtain proper viscosities, then thoroughly mixed with the asbestos, ground and screened to form molding compounds. The bituminous-bound compounds are lowest in cost and can be molded more rapidly than the inorganic compounds; the phenolic and melamine-bound compounds have better mechanical and electrical properties than the bituminous compounds and have better surfaces as well as being lighter in color. Like the inorganic compounds, organic compounds are cold-molded, followed by oven curing.

Compounds with melamine binders are similar to the phenolics, except that melamines have greater arc resistance, lower water absorption, are nontracking, and have higher dielectric strength.

Major disadvantages of these materials, again, are relatively high specific gravity, limited colors, and inability to be molded to accurate dimensions. Also, they can be produced only with a relatively dull finish.

Compounds with bituminous binders are used for switch bases, wiring devices, connector plugs, handles, knobs, and fuse cores. Phenolic and melamine compounds are used for similar applications where better strength and electrical properties are required.

An important benefit of cold-molded plastics is the relatively low tooling cost usually involved for short-run production. Most molding is done in single-cavity molds, in conventional compression-molding presses equipped for manual, semiautomatic, or fully automatic operation.

The water-fillable plastics used to replace wood or plaster of paris for ornamental articles, such as plaques, statuary, and lamp stands, and for model making, are thermoplastic resins that cure to closed-cell lattices that entrap water. The resin powders are mixed with water and a catalyst and poured into a mold without pressure. They give finer detail than plasters, do not crack or chip, are light in weight, and the cured material can be nailed and finished like wood. Water content can be varied from 50 to 80%.

Design Considerations

Cross sections are generally heavier than hot molded materials to provide durability in handling. Taper is not usually necessary on the part, except on projecting barriers or bosses, as well as on sides of recesses or depressions. Generous fillets should always be provided. Undercuts and reentrant angles should be avoided as they will increase mold cost and reduce production rate.

In molding, a variation of ±0.038 mm must be allowed in thickness of part. Also, because parts are cured out of the mold, dimensional tolerances cannot be held very closely.

Lettering, figures, and simple designs can be molded on surfaces; marking is usually of the raised type and is placed on recessed surfaces to prevent rubbing off.

COLD-ROLLED STEEL

Cold-rolled steels are flat steel products produced by cold-rolling hot-rolled products. The hot-rolled product is cleaned of oxide scale by pickling and passed through a cold-reduction mill to reduce and more uniformly control thickness and to enhance surface finish. Cold rolling also increases hardness, reducing ductility. Although the steel is sometimes used as rolled, it is often subsequently annealed to improve formability and then temper-rolled or roller-leveled for flatness. Cold-rolled steels are available in carbon and alloy grades as well as high-alloy grades, such as stainless steels. For plain carbon steels, carbon content is usually 0.25% maximum, often less. Quality designations include commercial quality (CQ) steel, which is produced from rimmed, capped, or semikilled steel; drawing quality (DQ), which is made from specially processed steel and is more ductile and uniform in forming characteristics; and drawing-quality special-killed (DQSK) steel, which is still more ductile and more uniform in forming characteristics. Cold-rolled structural-quality (SQ) steel refers to cold-rolled steel produced to specific mechanical properties. Bar and rod products are often cold-drawn through dies and called cold-drawn bar steel, or cold-finished in other ways and called cold-finished bar steel.

A series of SQ ultrahigh-strength steels featuring minimum tensile strength levels of between 1000 and 1400 MPa combine superior performance with low weight. Named Docol UHS, the new steels are particularly applicable to the automotive industry. Their high-energy-absorbing properties, for example, make them useful as structural members and for components used in a car's crumple zone.

Since the steel is hardened prior to leaving the factory, industries using these steels no longer require their own warm-up plants and hardening furnaces. Cutting, shaping, and welding are achieved with traditional methods. The Docol UHS series consists of three standard steels: Docol 1000 DP, Docol 1200 DP, and Docol 1400 DP. Numbers relate to maximum loads measured in megapascals, MPa.

COLUMBIUM AND ALLOYS

One of the basic elements, columbium (Cb) is also known as niobium and occurs in the minerals columbite and tantalite. A refractory metal, it closely resembles tantalum, is yellowish-white in color, has a specific gravity of 8.57, a melting point of 2468°C, and an electrical conductivity 13.2% relative to copper. It is quite ductile when pure or essentially free of interstitials and impurities, notably N_2, O_2, and H_2, which are limited to very small amounts. Tensile properties depend largely on purity, and columbium, with a total interstitial content of 100 to 200 ppm (parts per million), provides about 276 MPa ultimate strength, 207 MPa yield strength, 30% elongation, and 105,000 MPa elastic modulus. Drawn wire having an ultimate tensile strength of 896 MPa has been produced. The metal is corrosion resistant to many aqueous media, including dilute mineral and organic acids, and to some liquid metals, notably lithium, sodium, and NaK. It is strongly attacked, however, by strong dilute alkalis, hot concentrated mineral acids, and HFl acid. At elevated temperatures, gaseous atmospheres attack the metal primarily by oxidation even if O_2 content is low, with the attack especially severe at 399°C and higher temperatures, necessitating the use of protective coatings. Columbium tends to gall and seize easily in fabrication. Sulfonated tallow and various waxes are the preferred lubricants in forming, and carbon tetrachloride in machining. Ferrocolumbium is used to add the metal to steel. Columbium is also an important alloying element in nonferrous alloys.

SECONDARY FABRICATION

Pure columbium is considered one of the most workable of the refractory metals, and commercially fabricated columbium can be forged, rolled, swaged, drawn, and stamped by existing commercial techniques. In the primary or mill fabrication, an ingot is hot-worked by forging or extruding, following which the surface is conditioned to remove the contaminated layer, annealed in vacuum or an inert atmosphere to obtain a recrystallized structure, and then cold-worked (with intermediate anneals, if required) by any desired technique to final shape and size. Columbium metal containing less than a total of 0.12% combined O_2, N_2, and carbon can be given a cold reduction of over 90% in cross-sectional area. Secondary fabrication is done cold to avoid O_2 contamination, and lubrication is used to minimize galling or sizing on the working tools.

Vapor degreasing is an effective way to remove oils or grease from columbium parts, while immersion in various hot acids can be used for surface cleaning.

COLUMBIUM ALLOYS

These alloys are noted mainly for their heat resistance at temperatures far greater than those that can be sustained by most metals, but protective coatings are required for oxidation resistance. Thus, they find use for aircraft turbine components and in rocket engines, aerospace reentry vehicles, and for thermal and radiation shields. Table C.9 lists the composition of many of the early columbium alloys as well as the properties of those in use today. Cb–Sn and

TABLE C.9
Composition and Density of Some Columbium-Base Alloys

Alloy	Zr	V	Ti	Hf	Mo	W	Ta	Density lb/cu in.
1% Zr	0.6–1.2	—	—	—	—	—	—	0.31
D31	—	—	10	—	10	—	—	0.292
D41	—	—	10	—	6	20	—	0.31
FS-82	1	—	—	—	—	—	33	0.368
F48	1	—	—	—	5	15	—	0.34
F50	1	—	5	—	5	15	—	0.33
Cb-65	1	—	8	—	—	—	—	—
Cb-752	5	—	—	—	—	10	—	0.32
B33	—	4	—	—	—	—	—	0.306
B66	1	5	—	—	5	—	—	0.305
B77	1	5	—	—	—	10	—	0.319
FS-85	0.5	—	—	—	—	12	27	0.39
D14	5	—	—	—	—	—	—	0.31
D36	5	—	10	—	—	—	—	0.252
C103	—	—	1	10	—	—	—	0.32
C120	1	—	—	—	5	15	—	—
SCb-291	—	—	—	—	—	10	10	0.362

Cb–Ti alloys have found use as superconductors, and Cb–1Zr, a columbium–1% zirconium alloy, has been used for high-temperature components, liquid-metal containers, sodium or magnesium vapor-lamp parts, and nuclear applications. It has a tensile yield strength of about 255 MPa at 21°C and 165 MPa at 1093°C. Thin cold-rolled sheet of columbium alloy C-103, which contains 10% hafnium and 1% titanium, has a tensile yield strength of 648 MPa at 21°C and 172 MPa at 1093°C. After recrystallization at 1315°C, however, yield strength drops to 345 MPa at 21°C and 124 MPa at 1093°C. The room-temperature tensile properties of the 10% tungsten, 10% hafnium, 0.1% yttrium columbium alloy, known as columbium alloy C-129, are 620 MPa ultimate strength, 517 MPa yield strength, 25% elongation, and 110,000 MPa elastic modulus. Its strength falls rapidly with increasing temperatures; tensile yield strength declining to about 234 MPa at 1000°C. Other columbium alloys and their principal alloying elements are Cb-752 (10% tungsten, 2.5% zirconium), B-66 (5% molybdenum, 5% vanadium, 1% zirconium), Cb-132M (20% tantalum, 15% tungsten, 5% molybdenum, 1.5% zirconium, 0.12% carbon), FS-85 (28% tantalum, 10% tungsten, 1% zirconium), and SCb-291 (10% tantalum, 10% tungsten). Typical tensile properties of columbium alloy B-66 at room temperature and 1093°C, respectively, are 882 and 448 MPa ultimate strength, 745 and 400 MPa yield strength, 12 and 28% elongation, and 105,000 and 82,700 MPa elastic modulus. Additionally, $CbSe_2$, is more electrically conductive than graphite and forms an adhesive lubricating film. It is used in powder form with silver, copper, or other metal powders for self-lubricating bearings and gears.

The secondary fabrication of columbium-base alloys creates a problem because of their greater strength at high temperatures and as a result are more difficult to work than the unalloyed metal. Procedures have been developed for the primary and secondary fabrication. Breakdown of the initial ingot requires higher temperatures and finish cold rolling involves more frequent annealing or, in some case, hot rolling.

COMPOSITE MATERIAL

Composite materials are based on the controlled distribution of one or more reinforcement materials in a continuous matrix. Plastics are the most common matrix materials,

TABLE C.10
Commercially Available Cb Alloys for High-Temperature Service

Alloy	Composition, wt.%	Thermal Conductivity at 800°C (1470°F), W/m·°C	Thermal Conductivity at 1200°C (2190°F), W/m·°C	Total Emissivity at 800°C (1470°F)	Total Emissivity at 1200°C (2190°F)	Density, g/cm³ (lb/in.3)	Melting Point, °C (°F)	Coefficient of Thermal Expansion at 20°C, × 10⁻⁶/°C
Pure niobium	Nb	—	—	—	—	8.57	2468	7.1
C-103	Nb–10Hf–1Ti	37.4	42.4	0.28	0.40 0.70–0.82 (silicide coated)	8.85 (0.320)	2350±50 (4260±90)	8.73
Nb–1Zr	Nb–1Zr	59.0	63.1	0.14	0.18	8.57 (0.310)	2410±10 (4365±15)	6.8
PWC-11	Nb–1Zr–0.1C	—	—	—	—	8.57 (0.310)	—	6.8
WC-3009	Nb–30Hf–9W	—	—	—	—	10.1 (0.365)	—	7.5
FS-85	Nb–28Ta–10W–1Zr	52.8	56.7	—	—	10.6 (0.383)	—	7.1

Source: Adv. Mater. Proc., 156(6), 126–127, 1999. With permission.

TABLE C.11
Typical Room-Temperature Tensile Properties of Columbium Alloys

Alloy	Yield Strength, Mpa (ksi)	Ultimate Tensile Strength, Mpa (ksi)	Elongation, %	Elastic Modulus at 20°C, Gpa (Msi)	Elastic Modulus at 1200°C, Gpa (Msi)
C-103	296 (42.93)	420 (60.91)	26	90 (13.1)	64 (9.3)
Nb–1Zr	150 (21.75)	275 (39.88)	40	80 (11.7)	28 (4.1)
PWC-11	175 (25.38)	320 (46.41)	26	80 (11.7)	28 (4.1)
WC-3009	752 (109.06)	862 (125.02)	24	123 (17.9)	NA
FS-85	462 (67.00)	570 (82.67)	23	140 (20.4)	110 (16.0)

Source: Adv. Mater. Proc., 156(6), 126–127, 1999. With permission.

although metals, ceramics, and intermetallics are also used. Reinforcements include ceramics, glass, polymers, carbon, and metals. They can be in the shape of filaments, spheres, irregularly shaped particles, short fibers known as whiskers, or flat particles known as flakes.

Composites are also found in nature. Wood is a composite of cellulose fibers bonded by a matrix of natural polymers, mainly lignin. Egyptians reinforced mud with straw to make bricks. Concrete can be classified as a ceramic composite in which stones are dispersed among cement. And in the 1940s, short glass fibers impregnated with thermosetting resins, known as fiberglass, became the first composite with a plastic matrix.

In a properly designed composite, the reinforcement compensates for low properties of the matrix. Furthermore, in many cases synergism enables the reinforcing material to improve properties in the matrix. Composites also offer the capability of placing specific properties where they are needed on the part.

All these developments mean a larger and more-complicated choice of materials. This diversity has made plastics applicable to a broad range of consumer, industrial, automotive, and aerospace products. It has also made the job of selecting the best materials from such a huge array of candidates quite challenging.

DEFINITION

Composite materials are macroscopic combinations of two or more distinct materials with a discrete and recognizable interface separating them.

CONSTITUENTS AND CONSTRUCTION

In principle, composites can be constructed of any combination of two or more materials — metallic, organic, or inorganic; but the constituent forms are more restricted. The matrix is the body constituent serving to enclose the composite and give it bulk form. Major structural constituents are fibers, particles, laminae or layers, flakes, fillers, and matrices. They determine the internal structure of the composite. Usually, they are the additive phase.

Because the different constituents are intermixed or combined, there is always a contiguous region. It may simply be an interface, that is, the surface forming the common boundary of the constituents. An interface is in some ways analogous to the grain boundaries in monolithic materials. In some cases, however, the contiguous region is a distinct added phase, called an interphase. Examples are the coating on the glass fibers in reinforced plastics and the adhesive that bonds the layers of a laminate together. When such an interphase is present, there are two interfaces, one between the matrix and the interphase and one between the fiber and the interface.

Interfaces are among the most important yet least understood components of a composite material. In particular, there is a lack of understanding of processes occurring at the atomic level of interfaces, and how these processes influence the global material behavior. There is

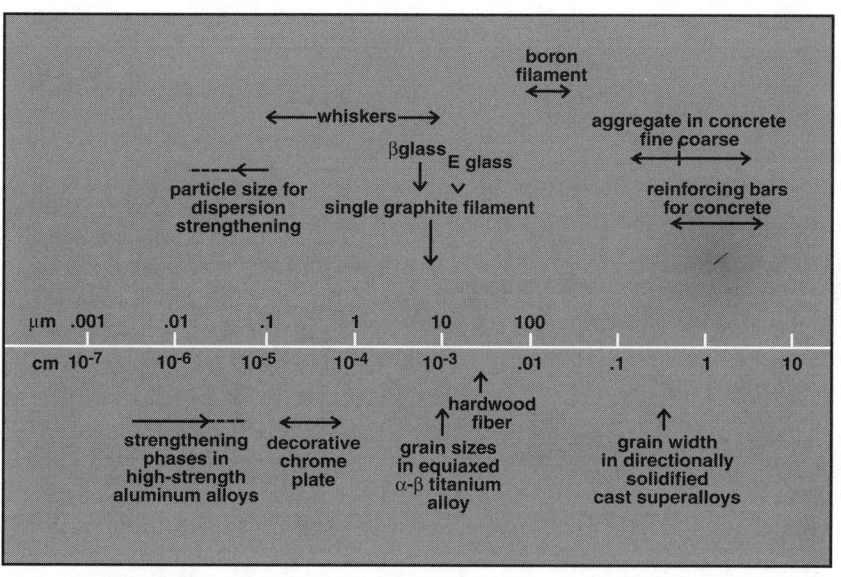

FIGURE C.3 Dimensional range of microstructural features in composite and conventional materials. Filament and fiber dimensions are diameters, 1 cm = 0.39 in. (From *McGraw-Hill Encyclopedia of Science and Technology*, 8th ed., Vol. 4, McGraw-Hill, New York, 252. With permission.)

a close relationship between processes that occur on the atomic, microscopic, and macroscopic levels. In fact, knowledge of the sequence of events occurring on these different levels is important in understanding the nature of interfacial phenomena. Interfaces in composites, often considered as surfaces, are in fact zones of compositional, structural, and property gradients, typically varying in width from a single atom layer to micrometers. Characterization of the mechanical properties of interfacial zones is necessary for understanding mechanical behavior.

Nature and Performance

Several classification systems for composites have been developed, including classification by (1) basic material combinations, for example, metal–organic or metal–inorganic; (2) bulk-form characteristics, such as matrix systems or laminates; (3) distribution of the constituents, that is, continuous or discontinuous; and (4) function, for example, electrical or structural.

There are five classes under the classification by basic material combinations: (1) fiber composites, composed of fibers with or without a matrix; (2) flake composites, composed of flat flakes with or without a matrix; (3) particulate composites, composed of particles with or without a matrix; (4) filled (or skeletal) composites, composed of a continuous skeletal matrix filled by a second material; and (5) laminar composites, composed of layer or laminar constituents.

There is also a classification based on dimensions. The dimensions of some of the components of composite materials vary widely and overlap the dimensions of the microstructural features of common conventional materials (Figure C.3). They range from extremely small particles or fine whiskers to the large aggregate particles or rods in reinforced concrete.

The behavior and properties of composites are determined by the composition, form and arrangements, and interaction between the constituents. The intrinsic properties of the materials of which the constituents are composed largely determine the general order or range of properties of the composite. Structural and geometrical characteristics — that is, the shape and size of the individual constituents, their structural arrangement and distribution, and the relative amount of each — contribute to overall performance. Of far-reaching importance are the effects produced by the combination and interaction of the constituents. The basic principle is that by using different constituents it is possible to obtain combinations of properties and property values that are different from those of the individual constituents.

A performance index is a property or group of properties that measures the effectiveness of a material in performing a given function. The values of performance indices for a composite differ from those of the constituents.

Fiber–Matrix Composites

Fiber–matrix composites have two constituents and usually a bonding phase as well.

Fibers

The performance of a fiber-matrix composite depends on orientation, length, shape, and composition of the fibers; mechanical properties of the matrix; and integrity of the bond between fibers and matrix. Of these, orientation of the fibers is perhaps most important.

Fiber orientation determines the mechanical strength of the composite and the direction of greatest strength. Fiber orientation can be one dimensional, planar (two-dimensional), or three-dimensional. The one-dimensional type has maximum composite strength and modulus in the direction of the fiber axis. The planar type exhibits different strengths in each direction of fiber orientation; and the three-dimensional type is isotropic but has greatly decreased reinforcing values. The mechanicl properties in any one direction are proportional to the amount of fiber by volume oriented in that direction. As fiber orientation becomes more random, the mechanical properties in any one direction become lower.

Fiber length also impacts mechanical properties. Fibers in the matrix can be either continuous or short. Composites made from short fibers, if they could be properly oriented, could have substantially greater strengths than those made from continuous fibers. This is particularly true of whiskers, which have uniform high tensile strengths. Both short and long fibers are also called chopped fibers. Fiber length also has a bearing on the processibility of the composite. In general, continuous fibers are easier to handle but have more design limitations than short fibers.

Bonding

Fiber composites are able to withstand higher stresses than their individual constituents because the fibers and matrix interact, resulting in redistribution of the stresses. The ability of constituents to exchange stresses depends on the effectiveness of the coupling or bonding between them. Bonding can sometimes be achieved by direct contact of the two phases, but usually a specially treated fiber must be used to ensure a receptive adherent surface. This requirement has led to the development of fiber finishes, known as coupling agents. Both chemical and mechanical bonding interactions occur for coupling agents.

Voids (air pockets) in the matrix are one cause of failure. A fiber passing through the void is not supported by resin. Under load, the fiber may buckle and transfer stress to the resin, which readily cracks. Another cause of early failure is weak or incomplete bonding. The fiber–matrix bond is often in a state of shear when the material is under load. When this bond is broken, the fiber separates from the matrix and leaves discontinuities that may cause failure. Coupling agents can be used to strengthen these bonds against shear forces.

Reinforced Plastics

Probably the greatest potential for lightweight high-strength composites is represented by the inorganic fiber–organic matrix composites, and no composite of this type has proved as successful as glass-fiber-reinforced composites. As a group, glass-fiber–plastic composites have the advantages of good physical properties, including strength, elasticity, impact resistance, and dimensional stability; high strength-to-weight ratio; good electrical properties; resistance to chemical attack and outdoor weathering; and resistance to moderately high temperatures (about 260°C).

A critical factor in reinforced plastics is the strength of the bond between the fiber and the polymer matrix; weak bonding causes fiber pullout and delamination of the structure, particularly under adverse environmental conditions. Bonding can be improved by coatings and the use of coupling agents. Glass fibers, for example, are treated with silane (SiH_4) for improved wetting and bonding between the fiber and the matrix.

Generally, the greatest stiffness and strength in reinforced plastics are obtained when the fibers are aligned in the direction of the tension force. Other properties of the composite, such as creep resistance, thermal and electrical conductivity, and thermal expansion, are anisotropic. The transverse properties of such a unidirectionally reinforced structure are much lower than the longitudinal. Seven mechanical and thermal properties are of direct interest in assessing the potential of a new composite: density, modulus, strength, toughness, thermal conductivity, expansion coefficient, and heat capacity; others, such as fracture toughness and thermal diffusivity, are calculated from them.

Tailoring Properties

The ideal way to develop a product made of composites is to model and analyze it extensively by computer before a prototype is built. But this is difficult because most computer programs were developed for metals and do not work well with composites.

Many new applications for composites are structural. Since the objective of structural parts generally is to maximize strength-to-weight ratios, a key design objective is to optimize configurations as well as materials.

After a design is defined, manufacturing is the next challenge. Building a single part normally is not technically taxing. The trick comes in fabricating composite parts reliably in mass production. Manufacturing operations tend to be expensive because fabrication is labor intensive, and the labor must be skilled.

The processes for fabricating composites also may produce built-in defects. For this reason, provisions for nondestructive testing should go hand-in-hand with fabrication. Unfortunately, available methods for nondestructive testing often leave a lot to be desired.

All these problems are being combated. Better guidelines are being developed to help designers select a composite and define its shape. Software is being developed to cope with the analytical complexities posed by composites and to help with the optimization process. Finally, major efforts are being exerted to automate fabrication processes and refine nondestructive testing operations.

Thermoplastic Composites

No longer is product design constrained to the property limits and performance characteristics of unmodified grades of resins. Thermoplastics that are reinforced with high-strength, high-modulus fibers provide dramatic increases in strength and stiffness, toughness, and dimensional stability. The performance gain of these composites usually more than compensates for their higher cost. Processing usually involves the same methods used for unreinforced resins.

Glass and Mineral Fibers

Glass fibers used in reinforced compounds are high-strength, textile-type fibers, coated with a binder and coupling agent to improve compatibility with the resin and a lubricant to minimize abrasion between filaments. Glass-reinforced thermoplastics are usually supplied as ready-to-mold compounds. Molded products may contain as little as 5% and as much as 60% glass by weight. Pultruded shapes (usually using a polyester matrix) sometimes have higher glass contents. Most molding compounds, for best cost/performance ratios, contain 20 to 40% glass.

Practically all thermoplastic resins are available in glass-reinforced compounds. Those used in largest volumes are nylon, polypropylene, polystyrene, ABS, and SAN, probably because most experience with reinforced thermoplastics has been based on these resins. The higher-performance resins — PES, PEI, PPS, PEEK, and PEK, for example — are also available in glass-fiber-reinforced composites, and some with carbon or aramid fibers as well.

Glass-fiber reinforcement improves most mechanical properties of plastics by a factor of two or more. Tensile strength of nylon, for example, can be increased from about 70 MPa to over 210 MPa, and deflection temperature to almost 260°C, from 77°C. A 40% glass-fortified acetal has a flexural modulus of 1.89 MPa, a tensile strength of 150.5 MPa, and a deflection temperature of 168°C. Reinforced polyester has double the tensile and impact strength and four times the flexural modulus of the unreinforced resin.

Also improved in reinforced compounds are tensile modulus, dimensional stability, hydrolytic stability, and fatigue endurance.

Fiber reinforcement of a resin always changes its impact behavior and notch sensitivity. The change may be in either direction, depending on the specific resin involved. But even when the change is an improvement, these properties may not be high enough for certain demanding applications. This need has led to the development of impact-modified compounds — specifically, nylon 6 and 6/6 alloys, a nylon 6/6 copolymer, and a polypropylene copolymer — with up to 50% improvement over reinforced unmodified compounds. Although the impact properties of a glass-reinforced compound are not always superior to those of the unreinforced compound, the reinforced modified compounds are always superior to the reinforced unmodified grades.

Applications

Molded glass-reinforced and mineral-reinforced plastics are used in a broad range of structural and mechanical parts. For example, glass-reinforced nylon, because of its strength and stiffness, is used in gears and automotive under-the-hood omponents, whereas mineral-reinforced nylon is used in housings and body parts because it is tougher and has low warpage characteristics. Polypropylene applications include automotive air-cleaner housings and dishwasher tubs and inner doors. Polycarbonate is used in housings for water meters and power tools. Polyester applications include motor components — brush holders and fans — high-voltage enclosures, television tuner gears, electrical connectors, and automobile exterior panels.

ADVANCED COMPOSITES

Advanced composites comprise structural materials that have been developed for high-technology applications, such as airframe structures, for which other materials are not sufficiently stiff. In these materials, extremely stiff and strong continuous or discontinuous fibers, whiskers, or small particles are dispersed in the matrix. A number of matrix materials are available including carbon, glass, ceramics, metals, and polymers. Advanced composites possess enhanced stiffness and lower density compared to fiberglass and conventional monolithic materials. Although composite strength is primarily a function of the reinforcement, the ability of the matrix to support the fibers or particles and to transfer load to the reinforcement is equally important. Also, the matrix frequently dictates service conditions, for example, the upper temperature limit of the composite.

Reinforcements

Continuous filamentary materials that are used as reinforcing constituents in advanced composites are carbonaceous fibers, organic fibers, inorganic fibers, ceramic fibers, and metal wires. Reinforcing inorganic materials are used in the form of discontinuous fibers and whiskers.

Carbon and graphite fibers offer high modulus and the highest strength of all reinforcing fibers. These fibers are produced in a pyrolysis chamber from three different precursor materials — rayon, polyacrylonitrile (PAN), and pitch. High-modulus carbon fibers are available in an array of yarns and bundles of continuous filaments (tows) with differing moduli, strengths, cross-sectional areas, twists, and plies.

Almost any polymer fiber can be used in a composite structure, but the first one with high-enough tensile modulus and strength to be used as a reinforcement in advanced composites was an aramid, or aromatic polyamide, fiber. Aramid fibers have been the predominant organic reinforcing fiber; graphite is a close second. See **Aramid**.

The most important inorganic continuous fibers for reinforcement of advanced composites are boron and SiC, both of which exhibit high stiffness, high strength, and low density. Continuous fibers are made by chemical vapor deposition (CVD) processes. Other inorganic compounds that provide stiff, strong discontinuous fibers that predominate as reinforcements for metal-matrix composites (MMC) are SiC, Al_2O_3, graphite, Si_3N_4, TiC, and carbon carbide. See **Boron**.

Polycrystalline Al_2O_3 is a commercial continuous fiber that exhibits high stiffness, high strength, high melting point, and exceptional resistance to corrosive environments. One method to produce the fibers is dry spinning followed by heat treatment. See **Ceramics**.

Whiskers are single crystals that exhibit fibrous characteristics. Compared to continuous or discontinuous polycrystalline fibers, they exhibit exceptionally high strength and stiffness. SiC whiskers are prepared by chemical processes or by pyrolysis of rice hulls. Whiskers made of Al_2O_3 and Si_3N_4 are also available. Particulates vary widely in size, characteristics, and function, and since particulate composites are usually isotropic, their distribution is usually random rather than controlled.

Organic-Matrix Composites

In many advanced composites the matrix is organic, but metal matrices are also used. Organic matrix materials are lighter than metals, adhere better to the fibers, and offer more flexibility in shaping and forming. Ceramic-matrix composites (CMC), carbon–carbon composites (C–C), and intermetallic-matrix composites (IMC) have applications where organic or metal matrix systems are unsuitable.

Materials

Epoxy resins have been used extensively as the matrix material. However, bismaleimide (BMI) resins and polyimide (PI) resins have been developed to enhance in-service temperatures. Thermoplastic resins, PEK, and polyphenylene sulfide (PPS) are in limited use.

The continuous reinforcing fibers for organic matrices are available in the forms of monofilaments, multifilament fiber bundles, unidirectional ribbons, roving (slightly twisted fiber), and single-layer and multilayer fabric mats. Frequently, the continuous reinforcing fibers and matrix resins are combined into a nonfinal form known as a prepreg.

Fabrication

Many processes are available for the fabrication of organic matrix composites. The first process is contact molding in order to orient the unidirectional layers at discrete angles to one another. Contact molding is a wet method, in which the reinforcement is impregnated with the resin at the time of molding. The simplest method is hand layup, whereby the materials are placed and formed in the mold by hand and the squeezing action expels any trapped air and compacts the part.

Molding may also be done by spraying, but these processes are relatively slow and labor costs are high, even though they can be automated. Many types of boats, as well as buckets for power-line servicing equipment, are made by this process.

Another process is vacuum-bag molding, where prepregs are laid in mold to form the desired shape. In this case, the pressure required to form the shape and achieve good bonding is obtained by covering the layup with a plastic bag and creating a vacuum. If additional heat and pressure are desired, the entire assembly is put into an autoclave. To prevent the resin from sticking to the vacuum bag and to facilitate removal of excess resin, various materials are placed on top of the prepreg sheets. The molds can be made of metal, usually aluminum, but more often are made from the same resin (with reinforcement) as the material to be cured. This eliminates any problem with differential thermal expansion between the mold and the part.

In filament winding, the resin and fibers are combined at the time of curing. Axisymmetric parts, such as pipes and storage tanks, are produced on a rotating mandrel. The reinforcing filament, tape, or roving is wrapped continuously around the form. The reinforcements are impregnated by passing them through a polymer bath. However, the process can be modified by wrapping the mandrel with prepreg material. The products made by filament winding are very strong because of their highly reinforced structure. For example, filament winding can be used directly over solid-rocket-propellant forms.

Pultrusion is a process used to produce long shapes with constant profiles, such as rods or tubing, similar to extruded metal products. Individual fibers are often combined into a tow, yarns, or roving, which consists of a number of tows or yarns collected into a parallel bundle without twisting (or only slightly so). Filaments can also be arranged in a parallel array called a tape and held together by a binder. Yarns or tows are often processed further by weaving, braiding, and knitting or by forming them into a sheetlike mat consisting of randomly oriented chopped fibers or swirled continuous fibers held together by a binder.

Weaving to produce a fabric is a very effective means of introducing fibers into a composite. There are five commonly used patterns, which include box or plain, basket, crowfoot, long-shaft, and leno weave. Although weaving is usually thought of as a two-dimensional process, three-dimensional weaving is often employed.

Knitting is a process of interlooping chains of tow or yarn. Advantages of this process are that the tow or yarn is not crimped as happens in weaving, and higher mechanical properties are often observed in the reinforced product. Also, knitted fabrics are easy to handle and can be cut without falling apart.

In braiding, layers of helically wound yarn or tow are interlaced in a cylindrical shape, and interlocks can be produced at every intersection of fibers. During the process, a mandrel is fed through the center of a braiding machine at a uniform rate, and the yarn or tow from carriers is braided around the mandrel at a controlled angle. The machine operates like a maypole, the carriers working in pairs to accomplish the over-and-under sequencing. The braiding process is most effective for cylindrical geometries. It is used for missile heat shields, lightweight ducts, fluid-sealing components such as packings and sleevings, and tubes for insulation.

Metal-Matrix Composites (MMCs)

MMCs are usually made with alloys of aluminum, magnesium, or titanium, and the reinforcement is typically a ceramic in the form of particulates, platelets, whiskers, or fibers, although other systems may be used. MMCs are often classified as discontinuous or continuous, depending on the geometry of the reinforcement. Particulates, platelets, and whiskers are in the discontinuous category, whereas the continuous category is reserved for fibers and wires. The type of reinforcement is important in the selection of an MMC because it determines virtually every aspect of the product, including mechanical properties, cost, and processing method. The primary methods for processing of discontinuous MMCs are powder metallurgy, liquid metal infiltration, squeeze or pressure casting, and conventional casting; however, most of these methods do not result in finished parts. Therefore, most discontinuously reinforced MMCs require secondary processing, which includes conventional wrought metallurgy operations such as extrusion, forging, and rolling; standard and nonstandard machining operations; and joining techniques such as brazing and welding.

Ceramic-Matrix Composites (CMCs)

One type of CMC incorporates a continuous fiber, and another type a discontinuous reinforcement such as whiskers. Both approaches enhance fracture resistance, but the mechanism is substantially different. Continuous-fiber-reinforced ceramics resist catastrophic failure because, after the matrix fails, the fiber supports the load. When whiskers are used as reinforcements, the resistance to crack propagation is enhanced and hence the composite is less sensitive to flaws. However, once a crack begins to propagate, the failure will be catastrophic.

Carbon–Carbon Composites (C–C)

A carbon–carbon composite is a specialized material made by reinforcing a carbon matrix with continuous carbon fiber. This type of composite has outstanding properties over a wide range of temperatures in both vacuum and inert atmospheres. It will even perform well at elevated temperatures in an oxidizing environment for short times. It has high strength, modulus, and toughness up to 2000°C; high thermal conductivity; and a low coefficient of thermal expansion. A material with such properties is excellent for rocket motor nozzles and exit cones, which require high-temperature strength as well as resistance to thermal shock. Carbon–carbon composites are also used for aircraft and other high-performance brake applications that take advantage of the fact that C–C composites have the highest energy capability of any known material. If a carbon–carbon composite is exposed to an O_2-containing atmosphere above 600°C for an appreciable time, it oxidizes, and therefore it must be protected by coatings.

APPLICATIONS

The use of fiber-reinforced materials in engineering applications has grown rapidly. Selection of composites rather than monolithic materials is dictated by the choice of properties. The high values of specific stiffness and specific strength may be the determining factor, but in some applications wear resistance or strength retention at elevated temperatures is more important. A composite must be selected by more than one criterion, although one may dominate.

Components fabricated from advanced organic-matrix-fiber-reinforced composites are used extensively on commercial aircraft as well as for military transports, fighters, and bombers. The propulsion system, which includes engines and fuel, makes up a significant fraction of aircraft weight (frequently 50%) and must provide a good thrust-to-weight ratio and efficient fuel consumption. The primary means of improving engine efficiency are to take advantage of the high specific stiffness and strength of composites for weight reduction, especially in rotating components, where material density directly affects both stress levels and critical dynamic characteristics, such as natural frequency and flutter speed.

Composites consisting of resin matrices reinforced with discontinuous glass fibers and continuous glass-fiber mats are widely used in truck and automobile components bearing light loads, such as interior and exterior panels, pistons for diesel engines, drive shafts, rotors, brakes, leaf springs, wheels, and clutch plates.

The excellent electrical insulation, formability, and low cost of glass-fiber-reinforced plastics have led to their widespread use in electrical and electronic applications ranging from motors and generators to antennas and printed circuit boards.

Composites are also used for leisure and sporting products such as the frames of rackets, fishing rods, skis, golf club shafts, archery bows and arrows, sailboats, racing cars, and bicycles.

Advanced composites are used in a variety of other applications, including cutting tools for machining of superalloys and cast iron and laser mirrors for outer-space applications. They have made it possible to mimic the properties of human bone, leading to development of biocompatible prostheses for bone replacements and joint implants. In engineering, composites are used as replacements for fiber-reinforced cements and cables for suspension bridges.

COMPRESSION AND TRANSFER MOLDED PLASTICS

Compression and transfer molding techniques are the most commonly used methods of molding thermosetting molding compounds as well as rubber parts. They may also be used for forming thermoplastic materials (e.g., compression molding of vinyl phonograph records), but usually other methods are more economical for molding thermoplastics.

The two processes are somewhat similar in terms of sizes and shapes produced. The major difference lies in the greater control of material flow permitted in transfer molding, allowing use of more delicate inserts and production of somewhat more complicated shapes.

COMPRESSION MOLDING

This method involves forming a part by placing the material into an open heated mold, shaping the part by closing the mold, and subsequently curing or hardening the part in the closed mold under pressure; see Figure C.4.

The materials to be molded are generally softened by preheating in conventional ovens or, more frequently, in a dielectric preheating unit prior to placement in the mold.

Compression molding techniques are used most extensively for the manufacture of products made from thermosetting plastic materials and rubbers. These materials require the relatively high pressures and temperatures afforded

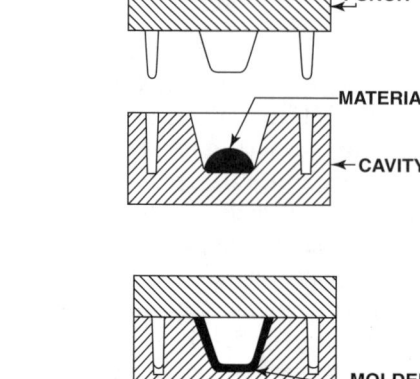

FIGURE C.4 Compression molding.

by the compression molding process. Such materials include the phenolics, the melamines, the ureas, and the polyester resins. Under special circumstances, thermoplastics are compression-molded, but the injection molding process is usually more economical for the production of thermoplastic parts.

The process is ideal for the production of such items as thermosetting radio cabinets, television cabinets, trays, and other products that require resistance to heat. The size of the compression molded articles is generally limited only by the platen size and tonnage capacity of the presses used. Through the use of multicavity molds, small electrical components such as switch plates and terminal blocks may be produced economically. Compression molding is also used to produce extremely large parts such as fiberglass boat hulls and the complete fuselage for radio-controlled target aircraft.

Compression molding is not practical for many intricate products where complicated molds are required. Thermosetting materials are extremely stiff or viscous plastic masses during the mold-closing period. The internal pressures developed by these materials tend to distort or break delicate core pins and other small mold components. The process also may be unsatisfactory for the production of extremely close tolerance articles, particularly where critical dimensions are influenced by the mold parting line. Flash thickness produced at the mold parting line tends to vary from cycle to cycle, thus changing the dimensions in the direction of the stroke of the press.

Molding pressures range from as low as 0.35 MPa for certain polyester compounds to as high as 2.3 kg/mm^2 for stiff high-impact phenolic materials. Process temperatures range from 121 to 177°C depending upon the material used.

Thickness of the molded part influences production rates. A general rule of thumb allows a minute of mold close-time for each 3.2 mm of wall thickness. Standard dimensional tolerances are usually figured at ±0.05/mm, although closer tolerances can be held under special circumstances. Where phenolic and urea materials are to be molded, hardened steel is used for the mold construction. Such molds are usually carburized and hardened to about 50 Rockwell C. For use with polyester materials, where pressure requirements are not high, pre-hardened steel at about 38 Rockwell C is usually satisfactory.

Transfer Molding

Transfer molding is best described as a closed-mold technique wherein the material is injected or transferred into a closed mold through a gate or runner system; see Figure C.5. Essentially, transfer molding is a one-shot injection molding technique. The process is particularly adaptable again to thermosetting materials, since these

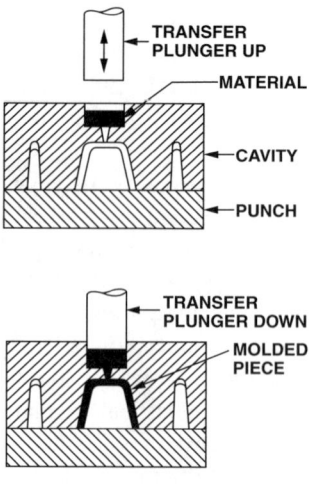

FIGURE C.5 Transfer molding.

materials retain their plastic condition after preheating for only a short length of time. The technique is also used extensively for the molding of unplasticized polyvinyl chloride, a material that tends to degrade when held at plasticizing temperatures for any length of time.

Generally, the comments made regarding compression molding apply to this process as well. The distinct advantage in the transfer molding process lies in the fact that the mold is completely closed and under clamping pressure before the material is injected into the mold cavity, This results in little or no flash and accurate control of dimensions.

A preweighed material charge is plasticized generally in a dielectric preheating unit. The charge is then placed in a pot that is usually positioned above the closed mold. A ram enters the pot and forces the material through an orifice into the closed mold. The transfer plunger and mold are kept under pressure for a predetermined time to allow the chemical/heat hardening process to proceed. When the mold is opened, the small amount of material remaining in the transfer pot and that which has filled the orifice is removed as a cull and discarded.

Transfer molding generally operates at faster cycles than compression molding. Because of the highly plastic condition of the material, complex part designs involving cores, undercuts, and moving die parts are best adapted to this process. Molded-in inserts can be held in position more easily in the transfer molding operation. Since the mold is closed during the entire molding operation, control of dimensions is more satisfactory.

With a properly designed mold, transfer molded articles require fewer finishing operations with a resultant lower net cost.

The process results in higher material costs due to the loss of material in the transfer pot and runner system. High-impact phenolic materials have generally lower physical properties when transfer molded compared with those obtained by compression molding.

Transfer molding techniques are used in the manufacture of a wide range of product shapes and sizes. Small, complex electrical components with molded-in terminals are made by this process. Radio and television cabinets weighing up to 1.7 kg have been transfer molded from phenolic materials.

CONCRETE

Concrete is a construction material composed of portland cement and water combined with sand, gravel, crushed stone, or other inert material such as expanded slag or vermiculite. The cement and water form a paste that hardens by chemical reaction into a strong stonelike mass. The inert materials are called aggregates, and for economy no more cement paste is used than is necessary to coat all the aggregate surfaces and fill all the voids. The concrete paste is plastic and easily molded into any form or troweled to produce a smooth surface. Hardening begins immediately, but precautions are taken, usually by covering, to avoid rapid loss of moisture because the presence of water is necessary to continue the chemical reaction and increase the strength. Too much water, however, produces a concrete that is more porous and weaker. The quality of the paste formed by the cement and water largely determines the character of the concrete.

Proportioning of the ingredients of concrete is referred to as designing the mixture, and for most structural work the concrete is designed to give compressive strengths of 16 to 34 MPa. A rich mixture for columns may be in the proportion of 1 volume of cement to 1 of sand and 3 of stone, whereas a lean mixture for foundations may be in the proportion of 1:3:6. Concrete may be produced as a dense mass that is practically artificial rock, and chemicals may be added to make it waterproof, or it can be made porous and highly permeable for such use as filter beds. An air-entraining chemical may be added to produce minute bubbles for porosity or light weight. Normally, the full hardening period of concrete is at least 7 days. The gradual increase in strength is due to the hydration of the tricalcium aluminates and silicates. Sand used in concrete was originally specified as roughly angular, but rounded grains are now preferred. The stone is usually sharply broken. The weight of concrete varies with the type and amount of rock and sand. A concrete with traprock may weigh 2483 kg/m^3. Concrete is stronger in compression than in tension, and steel bars or mesh are embedded in structural members to increase the tensile and flexural strengths. In addition to the structural uses, concrete is widely used in precast units such as block, tile, sewer and water pipe, and ornamental products.

Concretes are similar in composition to mortars, which are used to bond unit masonry. Mortars, however, are normally made with sand as the sole aggregate, whereas concretes contain much larger aggregates and thus usually have greater strength. As a result, concretes have a much wider range of structural applications, including pavements, footings, pipes, unit masonry, floor slabs, beams, columns, walls, dams, and tanks.

CLASSIFICATION

Concretes may be classified as flexible or rigid. These characteristics are determined mainly by the cementitious materials used to bond the aggregates. Flexible concretes tend to deform plastically under heavy loads or when heated. Rigid concretes are considerably stronger in compression than in tension and tend to be brittle. To overcome this deficiency, strong reinforcement may be incorporated in the concrete, or prestress may be applied to keep the concrete under compression.

Flexible Concretes

Usually, bituminous or asphaltic concretes are used when a flexible concrete is desired. The main use of such concretes is for pavements.

The aggregates generally used are sand, gravel or crushed stone, and mineral dust; the binder is asphalt cement, an asphalt specially refined for the purpose. A semisolid at normal temperatures, the asphalt cement may be heated until liquefied for bonding the aggregates. Ingredients usually are mixed mechanically in a pug mill, which has pairs of blades revolving in opposite directions. While the mix is still hot and plastic, it can be spread to a specified thickness and shaped with a paving machine and compacted with a roller or by tamping to a desired density. When the mix cools, it hardens sufficiently to withstand heavy loads.

Sulfur, rubber, or hydrated lime may be added to an asphalt–concrete mix to improve the performance of the product.

Rigid Concretes

Ordinary rigid concretes are made with portland cement, sand, and stone, or crushed gravel. The mixes incorporate water to hydrate the cement to bond the aggregates into a solid mass. Admixtures may be added to the mix to impart specific properties to it or to the hardened concrete.

Other types of rigid concretes include nailable concretes; insulating concretes; heavyweight concretes; lightweight concretes; fiber-reinforced concretes, in which short steel or glass fibers are embedded for resistance to tensile stresses; polymer and pozzolan concretes, which exhibit improvement in several properties; and silica-fume concretes, which possess high strength. Air-entrained concrete formulations, in which tiny air bubbles have been incorporated, may be considered variations of ordinary concrete.

STRESS AND REINFORCEMENT

Because ordinary concrete is much weaker in tension than in compression, it is usually reinforced or prestressed with a much stronger material, such as steel, to resist tension. Use of plain, or unreinforced, concrete is restricted to structures in which tensile stresses will be small, such as massive dams, heavy foundations, and unit-masonry walls. For reinforcement of other types of structures, steel bars or structural steel shapes may be incorporated in the concrete.

CEMENTITIOUS MATERIALS

These are for general-purpose cements, cements modified to achieve low heat of hydration and to produce a concrete resistant to sulfate attack, high-early-strength cement, and air-entraining cement.

Several other types of cement are sometimes used instead of portland cement for specific applications — for example, hydraulic cements, which can set and harden underwater. Included in this category, in addition to portland cement, are aluminous, natural, and white portland cements, and blends, such as portland blast-furnace slag, portland-pozzolan, and slag cements.

Aggregates for Ordinary Concrete

Aggregates should be inert, dimensionally stable particles, preferably hard, tough, and round or cubical. They should be free of clay, silt, organic matter, and salts.

Polymer Concretes

Polymers are used in several ways to improve concrete properties; they are impregnated in hardened concrete, are incorporated in a mix, or replace portland cement.

Impregnation is sometimes used for concrete road surfaces. It can more than double compressive strength and elastic modulus, decrease creep, and improve resistance to freezing and thawing.

Monomers and polymers added as admixtures are used for restoring and resurfacing deteriorated roads. The concrete hardens more rapidly than ordinary concrete, enabling faster return of roads to service.

Polymer concretes in which a polymer replaces portland cement possess strength and other properties similar to those of impregnated concrete. After curing for a relatively short time, for example, overnight at room temperatures, polymer concretes are ready for use; in comparison, ordinary concrete may have to cure for a week or more before exposure to service loads.

Casting

There are various methods employed for casting ordinary concrete. For very small projects, sacks of prepared mixes may be purchased and mixed on the site with water, usually in a drum-type, portable, mechanical mixer. For larger projects, mix ingredients are weighed separately and deposited in a stationary batch mixer, a truck mixer, or a continuous mixer. Concrete mixed or agitated in a truck is called ready-mixed concrete. Paving mixers are self-propelled, track-mounted machines that move at about 0.45 m/s while mixing.

At the site, concrete may be conveyed to the forms in wheelbarrows, carts, trucks, chutes, conveyor belts, buckets, or pipelines. Where chutes or belts are used, the flow of material is maintained as continuous as possible. On large projects, concrete may be delivered to the formwork in dump buckets by cableways or by cranes. In many instances, direct pumping through pipelines is used to convey concrete from the mixer or hopper to the point of placement.

CONDUCTIVE (ELECTRICAL) POLYMERS AND ELASTOMERS

These are polymers made electrically conductive by the addition of carbon black, nickel, silver, or other metals. Volume resistivities of plastics and rubbers, which normally are in excess of 10^8 Ω/cm, can be lowered to between 10^{-1} and 10^6 Ω/cm by addition of conductive materials. Carbon black is the most widely used filler. The relationship of carbon black loading and volume resistivity is not proportioned. Up to a 25% loading, conductivity significantly increases, but it falls off sharply thereafter. Generally, the addition of carbon black lowers the mechanical properties of the polymer. However, the use of carbon fibers to enhance conductivity improves mechanical properties.

Polyethylene and polyvinyl chloride resins loaded with carbon black are perhaps the most widely used conductive plastics. Plastics often made conductive by adding up to 30% carbon fiber are polysulfone, polyester, polyphenylene sulfide, nylon 6/6, ethylene tetrafluoroethylene, and vinylidene fluoride-polytetrafluoroethylene.

Although silicone is the most widely used base polymer for conductive rubber, other rubbers frequently used in compounding conductive elastomers include SBR, EPDM, TPR, and neoprene.

Another type of electrically conductive polymer includes materials that are doped with either electron acceptors, such as alkali metal ions and iodine, or electron donors, such as arsenic trifluoride. Also referred to as organic conductors, their conductivity is about one-hundreth that of copper. The most widely used are polyacetylene, polyparaphenylene, and polyparaphenylene sulfide.

CONDUCTORS

This term is usually applied to materials, generally metals, used to conduct electric current, although heat conductors and sound conductors have important uses. Good conductors of electricity tend to be good conductors of heat, as well. Silver is the best conductor of electricity, but copper is the most commonly used. The conductivity of pure copper is 97.6% that of silver. The electric conductivity of metals is often expressed as a percentage of the electric conductivity of copper, which is arbitrarily set at 100%. Tough-pitch copper is the standard conductivity metal, and it is designated as the International Annealed Copper Standard (IACS).

Because of the low conductivity of zinc, the brasses have low current-carrying capacity, but are widely used for electric connections and parts because of their workability and strength. The electric conductivity of aluminum is only 63% that of copper, but it is higher than that of most brasses. Copper wire for electric conductors in high-temperature environments has a plating of heat-resistant metal. Aluminum wire with a steel core is used for power transmission because of the long spans possible. Steel has a conductivity only about 12% that of copper, but the current in a wire tends to travel near the surface, and the small steel core does not reduce greatly the current-carrying capacity. Aluminum is now widely used to replace brass in switches and other parts. Aluminum wire for electric equipment is usually commercially pure aluminum with small amounts of alloying elements, such as magnesium, which give strength without appreciably reducing the conductivity. Plastics, glass, and other nonconductors are given conductive capacity with coatings of transparent lacquer containing metal powder, but conductive glass usually is made by spraying on at high temperature an extremely thin invisible coating of SnO. Coated glass panels are available with various degrees of resistivity.

COPOLYESTERS

These thermoplastic elastomers are generally tougher over a broader temperature range than the urethanes. Also, they are easier and more forgiving in processing. Several grades include Hytrel, Riteflex, and Ecdel, ranging in hardness from 35 to 72 Shore D. These materials can be processed by injection molding, extrusion, rotational molding, flow molding, thermoforming, and melt casting. Powders are also available.

Copolyesters, which along with the urethanes are high-priced elastoplastics, have excellent dynamic properties, high modulus, good elongation and tear strength, and good resistance to flex fatigue at both low and high temperatures. Brittle temperature is below –68°C, and modulus at –40°C is only slightly higher than at room temperature. Heat resistance to 149°C is good.

Resistance of the copolyesters to nonoxidizing acids, some aliphatic hydrocarbons, aromatic fuels, sour gases, alkaline solutions, hydraulic fluids, and hot oils is good to excellent. Thus, they compete with rubbers such as nitriles, epichlorohydrins, and polyacrylates. However, hot polar materials, strong mineral acids and bases, chlorinated solvents, phenols, and cresols degrade the polyesters. Weathering resistance is low but can be improved considerably by compounding ultraviolet stabilizers or carbon blacks with the resin.

Copolyester elastomers are not direct substitutes for rubber in existing designs. Rather, such parts must be redesigned to use the higher strength and modulus, and to operate within the elastic limit. Thinner sections can usually be used — typically one-half to one-sixth that of a rubber part.

Applications of copolyester elastomers include hydraulic hose, fire hose, power-transmission belts, flexible couplings, diaphragms, gears, protective boots, seals, oil-field parts, sports-shoe soles, wire and cable insulation, fiber-optic jacketing, electrical connectors, fasteners, knobs, and bushings.

A copolyester-based thermoplastic elastomer, Lomod, was introduced as a general-purpose, flame-retardant, and high-heat grade, developed for airdams, fascias, and filler panels with excellent impact resistance down to

−40°C and capable of withstanding on-line painting. Lomod thermoplastic elastomers are also used in connectors, wire, cable, hose, tubing, and other applications.

COPOLYMER

A copolymer is a macromolecule in which two or more different species of monomer are incorporated into a polymer chain. The properties of copolymers depend on both the nature of the monomers and their distribution in the polymer.

Copolymers can be prepared by all the known methods of polymerization.

RANDOM COPOLYMERS

Styrene and butadiene, polymerized by free radicals in emulsion, form a random copolymer. Polyvinyl chloride is a very useful plastic, but it is very difficult to process because it decomposes near its molding temperature. A copolymer containing a small amount of vinyl acetate is more easily fabricated; yet it contains most of the good properties of the homopolymer. Major uses of poly(vinyl chloride) are in floor coverings, film, and phonograph records.

BLOCK COPOLYMERS

Although styrene–butadiene random copolymer is an important sulfur-vulcanizable rubber, a styrene–butadiene–styrene block copolymer (SBS) has very different properties. This copolymer is important because it need not be vulcanized to make a useful rubber. At elevated temperature a uniform melt is formed, and the material can be readily extruded or injection-molded, processes that are much more economical than those needed to form sulfur-vulcanizable rubbers. As the material cools, however, the polystyrene blocks solidify and form a separate phase. These tiny balls of plastic suspended in a rubbery matrix serve the same function as the sulfur cross-links in a vulcanized rubber. This inexpensive fabrication method has led to very rapid growth in areas where high temperatures are not encountered, such as in shoe soles. For some purposes isoprene is substituted for butadiene, and hydrogenation of the double bonds leads to much better stability.

GRAFT COPOLYMERS

Acrylonitrile–butadiene–styreneterpolymers (ABS resins) are important thermoplastic structural plastics. The terpolymer is prepared by copolymerizing acrylonitrile and styrene dissolved in polybutadiene rubber. The chains of copolymer grow attached to the rubber by chemical bonds.

COPPER AND ALLOYS

Copper (symbol Cu) is one of the most useful of the metals, and probably the one first used by humans. It is found native and in a large number of ores. Its apparent plentifulness is only because it is easy to separate from its ores and is often a by-product from silver and other mining. Copper has a face-centered cubic crystal structure. It is yellowish red in color, tough, ductile, and malleable; gives a brilliant luster when polished; and has a disagreeable taste and a peculiar odor. It is the best conductor of electricity next to silver, with a conductivity 97% that of silver. Copper refers to the metal at least 99.3% pure. Standard wrought grades number more than 50, many of which are more than 99.7% pure. They are represented by the C10XXX to C15XXX series of copper and copper Alloy Numbers of the Copper Development Association. These include O_2-free coppers, O_2-free-with-silver coppers, and O_2-bearing coppers (C10100 to C10940); electrolytic-tough-pitch coppers and tough-pitch-with-silver coppers (C11100 to C11907); phosphorus-deoxidized coppers, fire-refined tough-pitch coppers, and fire-refined tough-pitch-with-silver coppers (C12000 to C13000); and certain coppers distinguished by very small amounts of specific ingredients such as cadmium copper (not to be confused with the high-copper alloys having a greater cadmium content), tellurium-bearing copper, sulfur-bearing copper, zirconium copper, and aluminum-bearing coppers (C14XXX to C15XXX). The highest-purity grade, O_2-free-electronic copper, is at least 99.99% pure. There are seven standard cast coppers (C80XXX to C81100), and their minimum purity ranges from 99.95% (C80100) to 99.70% (C81100).

O_2-free coppers (C10100 and C10200) have a melting point of 1082°C, a density of

8.94 mg/m³, an electrical conductivity of 101% — or slightly greater than the 100% for electrolytic-tough-pitch copper (C11100) used as the International Annealed Copper Standard (IACS) for electrical conductivity — a thermal conductivity of 391 W/m∗K, and a specific heat of 385 J/kg∗K. Typical tensile properties of thin flat products and small-diameter rod and wire having an average grain size of 0.050 mm are 220 MPa ultimate strength, 69 MPa yield strength, 45 to 50% elongation, and 117,000 MPa elastic modulus. Hardness is about Rf 40. These properties are fairly typical of other wrought coppers as well. Strength increases appreciably with cold work, yield strengths reaching 345 MPa in the spring and hard-drawn conditions. Zirconium copper, which may be heat-treated after cold working, can provide yield strengths of 345 to 483 MPa in rod and wire forms and retains considerable strength at temperatures to 426°C. Cast coppers are suitable for sand, plaster, permanent-mold, investment, and centrifugal castings as well as for continuous casting. Regardless of grade, typical tensile properties are 172 MPa ultimate strength, 48 MPa yield strength, and 40% elongation. Hardness is typically Brinell 44.

Copper and copper-base alloys are unique in their desirable combination of physical and mechanical properties. The following properties are found or may be developed in these materials:

1. Moderate to high strength and hardness
2. Excellent corrosion resistance
3. Ease of workability both in primary and secondary fabricating operations
4. Pleasing color and wide color range
5. High electrical and thermal conductivity
6. Nonmagnetic properties
7. Superior properties at subnormal temperatures
8. Ease of finishing by polishing, plating
9. Good-to-excellent machinability
10. Excellent resistance to fatigue, abrasion, and wear
11. Relative ease of joining by soldering, brazing, and welding
12. Moderate cost
13. Availability in a wide variety of forms and tempers

The importance of copper and copper-base alloys is in no small part because many other metals, whether singly or in combination, may be alloyed with copper. There are many binary alloys, and ternary and quaternary alloy systems are quite common. The copper content of alloys in commercial use ranges from about 58 to about 100%. Alloys of 63% to about 100% copper are of the alpha (single-phase) solid solution type, have excellent ductility, and are ideally suited for cold-working processes such as rolling, drawing, and press working. Copper alloys in the lower range, 62% and less, are of the alpha plus beta (two-phase) type and are particularly suited to hot-working processes such as hot rolling, hot extrusion, and hot forging.

The chief elements alloyed with copper are zinc, tin, lead, nickel, silicon, and aluminum, and to a lesser extent, beryllium, phosphorus, cobalt, boron, arsenic, zirconium, antimony, cadmium, iron, manganese, chromium, and very recently, mercury.

Cast Products

Foundry alloys are used in the form in which they are originally molded. In general, cast alloys are not easily workable and subsequent operations are limited to such treatment as machining, electroplating, soldering, and brazing. The use of alloys in cast form permits the production of intricate and irregular shapes impractical or impossible to make by other means. The use of cores permits the making of hollow shapes. The alloying metals are generally greater in amount than for wrought alloys. Types of castings in commercial use include sand castings, die castings, permanent mold, plaster mold, shell mold, and centrifugal and investment castings. The alloy involved, the properties desired, and the quantity to be made govern the choice of casting process. Where regularity of shape permits, a number of foundry alloys are available in the continuous-cast condition and may be either solid or hollow shapes. The classification of copper-base foundry alloys generally follows that used for wrought alloys. As in the wrought-alloy field, the number of cast alloys in commercial use runs well into the hundreds; many are modifications of standard alloys.

Wrought Products

Wrought products are those originally cast into starting shapes such as billets, cakes (slabs), and wire bar, which are further fabricated into useful forms such as sheet, strip, plate, rod, bar, seamless tubes, extruded shapes, and forgings by hot rolling, cold rolling, cold drawing, hot piercing, hot extrusion, or combinations of these primary fabricating processes. Secondary fabricating processes include such operations as machining, bending, spinning, cold heading, hot press forging, soldering, brazing, electroplating, cold extrusion, and operations normally associated with pressworking (blanking, deep drawing, coining, staking, embossing, drifting).

The various types of copper and the copper-base alloys are not fabricated with equal facility into usable forms or by secondary fabricating processes. This fact, together with the wide range and combination of physical and mechanical properties desirable for end use, accounts for the large number of commercial wrought copper-base alloys. Design considerations usually dictate a compromise between service properties desired, secondary fabricating properties, forms and sizes available, and cost.

The Alloys

In the copper industry, and in the brass mill section of that industry in particular, the alloys of copper have been designated by a sort of "word-terminology" that is often confusing and misleading. Names assigned to various alloys are not always descriptive or indicative of the composition. Several names may be in common use for the same composition. Numerous proprietary names are also used and, in some cases, several names in this category are in general use for essentially the same composition or type of alloy.

Since the more important alloying elements are used effectively in varying amounts, a convenient and logical classification is made possible by separating the alloys into well-recognized groups according to composition. Literally hundreds of wrought copper alloys are used. Those listed are standard alloys that account for the large majority of tonnage used. Alloys not listed are, in general, modifications of standard alloys often used for very special purposes.

Certain terms require clarification. Brass is an alloy of copper with zinc as the principal alloying element, used with or without additional elements. High brasses are Cu–Zn alloys containing more than 20% zinc, more properly called yellow brasses. Low brasses are alloys containing 20% zinc or less and low brass is a specific alloy, 80% copper and 20% zinc. Rich low brass is a term used for brass containing nominally 85% copper and 15% zinc, more properly called red brass. Straight brasses are those alloys containing copper and zinc only.

The term *bronze* in modern usage properly requires a modifying adjective. Early bronzes were alloys of copper with tin as the only, or principal, alloying metal, and thus the term *bronze* has been traditionally associated with alloys of copper and tin. The widespread use of numerous other alloy systems in recent times requires that a bronze be defined as an alloy of copper with some element, or elements, other than zinc as the chief alloying metal. Thus, "tin-bronze" is the proper name for alloys of copper and tin, silicon-bronze for alloys of copper and silicon, aluminum-silicon-bronze for an alloy of Cu–Al–Si. Tin-bronzes are also commonly called phosphor-bronzes because of the long-established practice of deoxidizing these alloys with phosphorus.

The "coppers" include pure copper and its modified forms, and also certain materials that might more properly be classed as alloys but that are commonly called "coppers" where the alloying element is less than about 1%, e.g., "TeCu" — 0.5% tellurium, the balance, copper.

Refractory copper-base alloys are those that, because of their hardness or abrasiveness, require greater dimensional tolerances than those established for nonrefractory alloys. Examples of refractory alloys are the phosphor bronzes, silicon bronzes, aluminum bronzes, and nickel silvers.

Coppers

Many types and variations of commercial copper are available; the differences are slight or important depending upon the method of production or the intentional minor additions of other metals. Two types refer to the method of refining. Fire-refined copper is that finished by furnace refining. Electrolytically refined copper is that finished by electrolytic deposition where the original ore requires such refining because of certain undesirable impurities that can be satisfactorily removed only by this method, or where the original ore contains sufficient quantities of silver and gold, recoverable in electrolytic refining, to make the electrolytic process profitable. Over 85% of all copper is electrolytically refined. Three major types of commercial copper with respect to composition and method of casting are tough pitch, O_2-free, and phosphorus deoxidized. Tough-pitch copper contains a small but controlled amount of O_2 (about 0.04%) necessary to obtain a level set, i.e., the correct pitch, on the refinery casting (wire bar, cake, billet); the level set is necessary to provide adequate ductility for hot or cold primary fabricating operations, i.e., sufficient "toughness." Electrolytic tough-pitch copper is the standard copper of industry for electrical and many other purposes. Its low electrical resistance (0.15328 Ω/g m^2 at 20°C) equivalent to 100% IACS conductivity is the standard with which other metals are compared. Copper has the highest conductivity of any base metal and is exceeded only by that of Silver.

Tough-pitch copper has one distinct disadvantage. When heated above about 399°C in atmospheres containing reducing gases, particularly H_2, the CuO_2 is reduced to copper and water vapor, the water changing to steam at the temperatures involved, thus "gassing" or embrittling the copper and destroying its ductility. Hence, the term H_2 *embrittlement*. Under circumstances where H_2 embrittlement is likely to occur (brazing, welding, annealing), O_2-free copper may be used without sacrifice of conductivity. In cases where some conductivity loss may be tolerated, various types of deoxidized copper may be used. Tough-pitch copper, while ductile enough for most fabricating work, is somewhat lacking in the ductility required for difficult forming operations (severe drawing, edgewise heading) and O_2-free copper is often used in these instances. Also, tough-pitch copper is not as free machining as the varieties made for that purpose, i.e., leaded copper, TeCu, sulfur copper. Silver is added to copper (without sacrifice of conductivity) for the purpose of increasing the recrystallization temperature, thus permitting soldering operations without reduction of strength and hardness. The amount of silver added varies with the application. Lake copper is a general term for silver-bearing copper, having varying but controlled amounts of silver up to about 30 oz/ton.

Phosphorized copper, with excellent hot and cold workability, is the second largest type in use with respect to tonnage and is widely used for pipe and tube for domestic water service and refrigeration equipment.

O_2-free copper is made either by melting and casting copper in the presence of carbon or carbonaceous gases or by extruding compacted, specially prepared cathode copper under a protective atmosphere, so that no O_2 is absorbed. No deoxidizing agent is required; therefore, no residual deoxidants are present, and optimum electrical conductivity is maintained. O_2-free copper is used extensively in the electrical and electronics industries, e.g., metal-to-glass seals.

Wrought Cu Alloys

Copper and zinc melted together in various proportions produce one of the most useful groups of copper alloys, known as the brasses. Six different phases are formed in the complete range of possible compositions. The relationship between composition and phases alpha, beta, gamma, delta, epsilon, and eta are graphically shown in the well-established constitution diagram for the Cu-Zn system. Brasses containing 5 to 40% zinc constitute the largest volume of copper alloys. One important alloy, cartridge brass (70% copper, 30% zinc), has innumerable uses, including cartridge cases, automotive radiator cores and tanks, lighting fixtures, eyelets, rivets, screws, springs, and plumbing products. Tensile strength ranges from 310 MPa as annealed to 900 MPa for spring temper wire.

Lead is added to both copper and the brasses, forming an insoluble phase that improves machinability of the material. Free cutting brass (61% copper, 3% lead, 36% zinc) is the most important alloy in the group. In rod form it has a strength of 340 to 480 MPa, depending on temper and size. It is machined into parts on high-speed (10,000 rpm) automatic screw machines for a multiplicity of uses.

Increased strength and corrosion resistance are obtained by adding up to 2% tin or aluminum to various brasses. Admiralty brass (70% copper, 2% aluminum, 0.023% arsenic, 21.97% zinc) are two useful condenser-tube alloys. The presence of phosphorus, antimony, or arsenic effectively inhibits these alloys from dezincification corrosion.

Alloys of copper, nickel, and zinc are called nickel silvers. Typical alloys contain 65% copper, 10 to 18% nickel, and the remainder zinc. Nickel is added to the Cu–Zn alloys primarily because of its influence on the color of the resulting alloys; color ranges from yellowish-white, to white with a yellowish tinge, to white. Because of their tarnish resistance, these alloys are used for table flatware, zippers, camera parts, costume jewelry, nameplates, and some electrical switch gear.

Copper forms a continuous series of alloys with nickel in all concentrations. The constitution diagram is a simple all alpha-phase system.

Nickel slightly hardens copper, increasing its strength without reducing its ductility. Copper with 10% nickel makes an alloy with a pink cast. More nickel makes the alloy appear white.

Three copper-base alloys, containing 10, 20, and 30% nickel, with small amounts of manganese and iron added to enhance casting qualities and corrosion resistance, are important commercially. These alloys are known as cupronickels and are well suited for application in industrial and marine installations as condenser and heat-exchanger tubing because of their high corrosion resistance and particular resistance to impingement attack. Heat-exchanger tubes in desalinization plants use the cupronickel 10% alloy.

Cu–Sn alloys (3 to 10% tin), deoxidized with phosphorus, form an important group known as phosphorus bronzes. Tin increases strength, hardness, and corrosion resistance, but at the expense of some workability. These alloys are widely used for springs and screens in papermaking machines.

Silicon (1.5 to 3.0%), plus smaller amounts of other elements, such as tin, iron, or zinc, increases the strength of copper, making alloys useful for hardware, screws, bolts, and welding rods.

Sulfur (0.35%) and tellurium (0.50%) form insoluble compounds when alloyed with copper resulting in increased ease of machining.

Alloys of copper that can be precipitation-hardened have the common characteristic of a decreasing solid solubility of some phase or constituent with decreasing temperature. Precipitation is a decomposition of a solid solution leading to a new phase of different composition to be found in the matrix. In such alloy systems, cooling at the appropriate rapid rate (quenching) from an equilibrium temperature well within the all-alpha field will preserve the alloy as a single solid solution possessing relatively low hardness, strength, and electrical conductivity.

A second heat treatment (aging) at a lower temperature will cause precipitation of the unstable phase. The process is usually accompanied by an increase in hardness, strength, and electrical conductivity. Some 19 elements form copper-base binary alloys that can be age- or precipitation-hardened.

Two commercially important precipitation-hardenable alloys are Be–Cu and Cr–Cu. Be–Cu (2.0 to 2.5% beryllium plus cobalt or nickel) can have a strength of 1400 MPa and an electrical conductivity of less than 50% of IACS. Cobalt adds high-temperature stability and nickel acts as a grain refiner. These alloys have found use as springs, diaphragms, and bearing plates and in other applications requiring high strength and resistance to shock and fatigue.

CuCr (1% chromium) can have a strength of 550 MPa and an electrical conductivity of 80%. Cu–Cr alloys are used to make resistance-welding electrodes, structural members for electrical switch gear, current-carrying members, and springs.

Cu–Ni, with silicon or phosphorus added, forms another series of precipitation-hardenable alloys. Typical composition is 2% nickel, 0.6% silicon. Strength of 830 MPa can be obtained with high ductility and electrical conductivity of 32% IACS.

Zr–Cu is included in this group because it responds to heat treatment, although its strength is primarily developed through application of cold deformation or work. Heat treatment restores high electrical conductivity and ductility and increases surface hardness. Tensile strength of 500 MPa coupled with an electrical conductivity of 88% can be developed. Uses are resistance welding wheels and tips, stud bases for transistors and rectifiers, commutators, and electrical switch gear.

As little as 2% beryllium content, combined with the thermal conductivity of copper, has improved wear resistance to extend mold life and accelerated mold cycle times. Advancements in electrode materials are directly responsible for use of alloys like Be–Cu in mold making.

Conventional EDM performance in Be–Cu alloys with graphite electrodes is similar to that of other copper work metals. However, differences in EDM performance for various grades of Be–Cu alloys are evident at all power levels.

Be–Cu alloys with beryllium contents of 0.2 to 0.6% and 1.6 to 2.0% have shown that the higher-content group exhibited improved conventional EDM performance over the lower-content group. Testing of EDM with Be–Cu must still be performed relative to the possible health hazards involving fumes or gases, which must be recognized. Adequate EDM tank ventilation and dust collection for machining particulates could eliminate the hazards of inhaling any free beryllium or copper.

Cast Copper Alloys

Copper alloy castings of irregular and complex external and internal shapes can be produced by various casting methods, making possible the use of shapes for superior corrosion resistance, electrical conductivity, good bearing quality, and other attractive properties. High-copper alloys with varying amounts of tin, lead, and zinc account for a large percentage of all copper alloys used in the cast form. Tensile strength ranges from 250 to 330 MPa depending on composition and size. Leaded tin-bronzes with 6 to 10% tin, about 1% lead, and 4% zinc are used for high-grade pressure castings, valve bodies, gears, and ornamental work. Bronzes high in lead and tin (5 to 25% lead and 5 to 10% zinc) are mostly used for bearings. High tin content is preferred for heavy pressures or shock loading, but lower tin and higher lead is preferred for lighter loads, higher speeds, or where lubrication is less certain. A leaded red brass containing 85% copper and 5% each of tin, lead, and zinc is a popular alloy for general use.

High-strength brasses containing 57 to 63% copper, small percentages of aluminum and iron, and the balance zinc have tensile strengths from 490 to 830 MPa, high hardness, good ductility, and resistance to corrosion. They are used for valve stems and machinery parts requiring high strength and toughness.

Copper alloys containing less than 2% alloying elements are used when relatively high electrical conductivity is needed. Strength of these alloys is usually notably less than that of other cast alloys.

Aluminum bronzes containing 5 to 10.5% aluminum, small amoumts of iron, and the balance copper have high strengths even at elevated temperature, high ductility, and excellent corrosion resistance. The higher-aluminum-content castings can be heat-treated, increasing their strength and hardness. These alloys are used for acid-resisting pump parts, pickling baskets, valve seats and guides, and marine propellers.

Additives impart special characteristics. Manganese is added as an alloying element for high-strength brasses where it forms intermetallic compounds with other elements, such as iron and aluminum. Nickel additions refine cast structures and add toughness, strength, and corrosion resistance. Silicon added to copper forms Cu–Si alloys of high strength and high corrosion resistance in acid media. Beryllium or chromium added to copper forms a series of age- or precipitation-hardenable alloys. Cu–Be alloys are among the strongest of the copper-base cast materials.

Copper Alloy Designations

Copper serves as the base metal for a great variety of wrought and cast alloys and the major wrought alloys and their designations, or alloy numbers, are high-copper alloys (C16200 to C19750), which include CdCu, BeCu, and CrCu; CuZn brasses (C20500 to C28580); Cu–Zn–Pb leaded brasses (C31200 to C38590); Cu–Zn–Sn alloy or tin brasses (C40400 to C49080); Cu–Sn phosphor bronzes (C50100 to C52400); Cu–Sn–Pb bronze alloys or leaded phosphor bronzes (C53200 to C54800); Cu–P alloys and Cu–Ag–P alloys (C55180 to C55284); Cu–Al alloys or aluminum bronzes (C60600 to C64400); Cu–Si alloys or silicon bronzes (C64700 to C66100); miscellaneous Cu–Zn alloys (C66400 to C69950); and Cu–Ni–Zn alloys or nickel silvers (C73150 to C79900). All told, there are about 300 standard wrought alloys and many have cast counterparts (C81300 to C99750). There are about 140 standard cast alloys. Many copper alloys are also available in powder form for powder-metal parts.

Cu–Ni wrought alloys are designated C70100 to C72950; cast alloys, C96200 to C96800. The alloys also have been referred to as cupronickels, Cu–Ni 20% (or whatever the percent nickel), also 80–20 (or whatever the percent copper and nickel), nickel content may be as low as 2 to 3% (C70200) or as high as 43 to 46% (C72150), but intermediate amounts, nominally 10% (C70600 and C96200), 20% (C71000 and C96300), and 30% (C71500 and C96400), are the most common. Most of the alloys also contain small to moderate amounts of iron, zinc, manganese, and other alloying ingredients, and some contain substantial amounts of tin (1.8–2.8% in C72500; 7.5–8.5% in C72800).

THE BRASSES/BRONZES/SILVERS/CUPRONICKELS

The straight brasses are without question the most useful and widely used of all copper-base alloys. Zinc is used in substantial portions in five of the alloy groups listed. Most of the brasses are within the alpha field of the phase diagrams, thus maintaining good cold-working properties. With a higher zinc content of about 40%, the beta phase facilitates hot working. The addition of zinc decreases the melting point, density, electrical and thermal conductivities, and modulus of elasticity. It increases the coefficient of expansion, strength, and hardness. Work hardening increases with zinc content but 70 to 30 brass has the best combination of strength and ductility.

The low zinc brasses (20% zinc or less) have excellent corrosion resistance, are highly resistant to stress-corrosion cracking (SCC), and cannot dezincify. All of the low brasses are extensively used where these properties are required, along with good forming characteristics, and moderate strength. Commercial bronze is the standard alloy for domestic screen cloth, rotating bands on shells, and weatherstripping. Red brass is highly corrosion resistant, more so than copper in many cases, and is used for condenser tubes and process piping. The brasses have a pleasing color range varying from the red of copper through bronze and gold colors to the yellow of the high zinc brasses. The alloy 87.5% copper, 12.5% zinc very closely matches the color of 14-karat gold, and the low brasses in particular are used in inexpensive jewelry, for closures, and in other decorative items.

The high brasses, cartridge brass and yellow brass, have excellent ductility and high strength and have innumerable applications for structural and decorative parts that are to be fabricated by drawing, stamping, cold heading, spinning, and etching. Cartridge brass is able to withstand severe cold working in practically all fabricating operations, and derives its name from the deep-drawing operations necessary on small-arms munition cases, for which it is ideally suited. Muntz metal is primarily a hot-working alloy and is cold-formed with difficulty. It is used in applications where cold spinning, drawing, or upsetting are not necessary. A typical use of Muntz metal is for condenser tube plates.

Silicon, Aluminum, and Manganese Brass

Silicon red brass has the corrosion resistance of the low brasses, with higher electrical resistance than that normally inherent in those brasses, and is especially suited to applications involving resistance welding. Aluminum brass is a moderate-cost alloy primarily made in tube form for use in condensers and heat exchangers, where its improved resistance to the corrosive and erosive action of high-velocity seawater is desired.

Manganese brass serves the same purpose as silicon brass.

Leaded Brasses

The primary purpose in adding lead to any copper alloy is to improve machinability and related operations such as blanking or shearing. Lead also improves antifriction properties and is sometimes useful in that respect. Lead has an adverse effect in that both hot- and cold-working properties are hindered by reduced ductility. Optimum machinability is reached in free-cutting brass rod with 3.25% lead, ideally suited for automatic screw-machine work. Other leaded brasses in rod and other forms, with lead content ranging from 0.5% to over 2%, and with varying copper content, are available to permit varying combinations of corrosion resistance, strength, and hardness, and equally important, secondary fabricating processes (flaring, bending, drawing, stamping, thread rolling). Applications of all these alloys are found throughout industry.

Tin Brasses

Tin improves the corrosion resistance and strength of Cu–Zn alloys. Pleasing colors are obtainable when tin is alloyed with the low brasses. The tin brasses in sheet and strip form and with copper content of 80% or greater comprise a group of materials used as low-cost spring materials, for hardware, or for color effect alone, as in inexpensive jewelry, closures, and decorative items. Admiralty brass is the standard alloy for heat exchanger and condenser tubes, with good corrosion resistance to both seawater and domestic-water supplies. The addition of a small amount of antimony, phosphorus, or arsenic inhibits dezincification, a type of corrosion to which the high-zinc alloys are susceptible. Naval brass and manganese bronze are widely used for applications requiring good corrosion resistance and high strength, particularly in marine equipment for shafting and fastenings.

Phosphor Bronzes

As a group, these range in tin content from 1.25 to 10% nominal tin content. They have excellent cold-working characteristics, high strength, hardness, and endurance properties, low coefficient of friction, and excellent general corrosion resistance, and hence find wide use as springs, diaphragms, bearing plates, Fourdrinier wire, bellows, and fastenings.

Aluminum Bronzes

The alpha aluminum bronzes (containing less than about 8% aluminum have good hot- and cold-working properties, while the alpha–beta alloys (8 to 12% aluminum, with nickel, iron, silicon, and manganese) are readily hot-worked. All have excellent corrosion resistance, particularly to acids, and high strength and wear resistance. Certain alloys in this category may be hardened by a heat-treatment process similar to that used for steels. Certain alloys in this group are used for spark-resistant tools.

Cupronickels

These are straight Cu–Ni alloys with nickel content ranging from 2.5 to 30% nickel and have increased corrosion resistance with increasing nickel content. Cupronickels are markedly superior in their resistance to the corrosive and erosive effects of high-velocity seawater. They are moderately hard but quite tough and ductile, so that they are particularly suited to the manufacture of tubes for condenser and heat-exchanger use, and have flaring, rolling, and bending characteristics necessary for installation.

Cadmium Bronzes

Alloys of copper and cadmium, with or without small additions of tin, are primarily used for electrical conductors where high electrical conductivity and improved properties are found.

Nickel Silvers

Actually nickel brasses, these alloys contain varying amounts of copper, zinc, and nickel. They have a color ranging from ivory white in the lower nickel alloys to silver white in the higher nickel alloys. Their pleasing silver color, coupled with excellent corrosion resistance and cold formability, makes them ideally suited as a base for silver-plated flatware and hollowware, and structural and decorative parts in optical goods, cameras, and costume jewelry. The low-copper 18% nickel alloy is used for springs and other applications where high fatigue strength is desired.

Silicon Bronzes

The silicon bronzes are an extremely versatile series of alloys, having high strength, exceptional corrosion resistance, and excellent weldability, coupled with excellent hot and cold workability. The low silicon bronzes (1.5% silicon) are widely used for electrical hardware. The high silicon bronzes are used for structural parts throughout industry. Al–Si bronze is an important modification with exceptional strength and corrosion resistance, particularly suited to hot working and with better machinability than the regular silicon bronzes.

Copper Alloy Powders

Besides copper, there are a rather large range of compositions of copper alloy powders available, including the brasses, bronzes, and Cu–Ni. Brass powders are the most widely used for powder-metallurgy (P/M) structural parts. Conventional grades are available with zinc contents from around 10 to 30%. Sintered brass parts have tensile strengths up to around 245 and 280 MPa and elongations of from 15 to around 40% depending on composition, design, and processing. In machinability they are comparable to cost and wrought brass stock of the same composition. Brass P/M parts are well suited for applications requiring good corrosion resistance, and where free machining properties are desirable.

Cu–Ni, or Ni–Ag, powders contain 10 or 18% nickel. Their mechanical properties are rather similar to the brasses, with slightly higher hardness and corrosion resistance. Because they are easily polished, they find considerable use in decorative applications.

Copper and bronze powders are used for filters, bearings, and electrical and friction products. Bronze powders are relatively hard to press to densities to give satisfactory strength for structural parts. Probably the most commonly used bronze contains 10% tin. The strength properties are considerably lower than iron-base and brass powders, usually below 140 MPa.

COPPER OXIDE

Cuprous oxide (Cu_2O) and cupric oxide or black CuO are used in ceramics. Cupric oxide is generally preferred in glazes and cuprous oxide in glasses.

Cupric oxide, when used in glazes, has a wide range of color. It may be used either as the raw oxide in a raw glaze, as the raw oxide in a fritted glaze, or as part of the frit itself.

CORROSION-RESISTANT ALLOYS

There are now several corrosion-resistant alloys available (Table C.12) that are capable of meeting most difficult design requirements.

Cast Alloys

In general, these are the cast counterparts to 3XX and 4XX wrought stainless steels and, thus, are also referred to as cast stainless steels. Designations of the Alloy Casting Institute of the Steel Founders Society of America and the wrought designations to which they roughly correspond (compositions are not identical) include CA-15 (410), CA-40 (420), CB-30 (431), CC-50 (446), CE-30 (312), CF-3 (304L), CF-3M (316L), CF-8 (304), CF-8C (347), CF-8M (316), CF-12M (316), CF-16F (303), CF-20 (302), CG-8M (317), CH-20 (309), and CK-20 (310). There are also other alloys that do not correspond to wrought grades. The cast alloys corresponding to 3XX wrought grades have chromium contents in the range of 17 to 30% and nickel contents in the range of 8 to 22%. Silicon content is usually 2.00% maximum (1.50 for CE-8M), Manganese 1.50 maximum, and carbon 0.08 to 0.30 maximum, depending on the alloy. Other common alloying elements include copper and molybdenum. Those corresponding to 4XX grades may contain as much chromium but much less nickel: 1 to 5.5%, depending on alloy. Manganese and silicon contents are also generally less and carbon may be 0.15 to 0.50%, depending on the

TABLE C.12
Corrosion-Resistant Alloys

Alloy	UNS	Form	Specification ASME	Specification ASTM
AL-6XN®	N08367	Plate, sheet, strip	SB-688	B-688
		Welded pipe	SB-675	B-675
		Welded tube	SB-676	B-676
		Rod, bar, wire	SB-691	B-691
Carpenter 20Cb-3® Stainless	N08020	Plate, sheet, strip	SB-463	B-463
		Welded pipe	SB-464	B-464
		Bar, wire	SB-473	B-473
		Wrought, welded fittings	SB-366	B-366
RA825	N08825	Plate, sheet	SB-424	B-424
		Seamless pipe	SB-423	B-423
		Welded pipe	SB-705	B-705
RA2205	S32205 & S31803	Plate, sheet, strip	SA-240	A-240
		Seamless, welded Pipe	SA-790	A-790
		Bar	SA-276	A-276

Note: AL-6XN® is a registered trademark of Allegheny Ludlum Corporation. 20Cb-3® is a registered trademark of Carpenter Technology Corporation.

Source: Ind. Heating, May, p. 80, 1998. With permission.

alloy. All the alloys are Fe–Cr–Ni alloys and the widest used are CF-8 and CF-8M, which limit carbon content to 0.08%. CN-7M and CN-7MS contain more nickel than chromium and, thus, are referred to as Fe–Ni–Cr alloys.

The alloys are noted primarily for their outstanding corrosion resistance in aqueous solutions and hot gaseous and oxidizing environments. Oxidation resistance stems largely from the chromium. Nickel improves toughness and corrosion resistance in neutral chloride solutions and weak oxidizing acids. Molybdenum enhances resistance to pitting in chloride solutions. Copper increases strength, and permits precipitation hardening to still greater strength. After a 482°C age, for example, the room-temperature tensile properties of CB-7Cu are 1290 MPa ultimate strength, 1100 MPa yield strength, 10% elongation, and 196,500 MPa elastic modulus. Hardness is Brinell 412 and impact strength (Charpy V-notch) 9.5 J. At 426°C, yield strength approaches 827 MPa. Higher aging temperatures, to 621°C, decrease strength somewhat but markedly increase impact strength. The alloys are widely used for pumps, impellers, housings, and valve bodies in the power-transmission, marine, and petroleum industries, and for chemical, food, pulp and paper, beverage, brewing, and mining equipment.

CORUNDUM

Corundum is a very hard crystalline mineral used chiefly as an abrasive, especially for grinding and polishing optical glass. It is in the alpha, or hexagonal, crystal form, usually containing some lime and other impurities. The physical properties are theoretically the same as for synthetic alpha Al_2O_3, but they are not uniform. The melting point and the hardness are generally lower because of impurities, and the crystal structure also varies.

Pure corundum is transparent and colorless, but most specimens contain some transition elements substituting for aluminum, resulting in the presence of color. Substitution of chromium results in a deep red color; such red corundum is known as ruby. The term *sapphire* is used in both a restricted sense for the "cornflower blue" variety containing iron and titanium, and in a general sense for gem-quality corundums of any color other than red.

Corundum is synthesized by a variety of techniques for use as synthetic gems of a variety of colors and as a laser source (ruby). The most important technique is the Verneuil flame-fusion method, but others include the flux-fusion and Czochralski "crystal-pulling" techniques.

CRYOGENICS

Today's limited acceptance and use of deep-cryogenic treatment at liquid N_2 temperature (–196°C) is usually attributed to a lack of understanding of the technology, as well as to the absence of generally acceptable practices for deploying it.

When the above two factors are resolved the "science" of deep cryogenics and its potential contributions will be shown. For example, a deep-cryogenic treatment at –196°C between quenching and tempering optimizes the mechanical properties of AISI T15 high-speed steel and dramatically improves its wear resistance, compared with standard heat treatments that do not incorporate a cryogenic step.

When the above is used, deep cryogenics does indeed become a science — results are predictable and the mechanical properties of the heat-treated parts are optimized. This is why it is believed that the use of the process during heat treating should now be standard rather than optional.

In future heat-treating guidelines and practices and procedures it is recommended that when using deep cryogenics as part of the heat treatment, the temperature of the subsequent temper be noted for the specific steel in the hardness vs. tempering temperature.

CUTTING TOOLS

Polycrystalline diamond (PCD) and polycrystalline cubic boron nitride (PCBN) inserts could be the best means of increasing productivity despite their high cost. In the past, PCD and PCBN cutting tools were difficult to cost-justify, unless they were essential for the machining job.

Today, improvements in quality and reliability make these tools, although still costly, competitive in many machining applications in the automotive, aerospace, and medical equipment industries. More rigid machines and tooling setups enable manufacturers to take full advantage of the potential for improved productivity offered by PCD and PCBN inserts. Also, having more cutting-tool options gives manufacturing engineers an opportunity for cost-effective productivity improvement in various machining applications.

PCBN

New, thicker PCBN solid inserts with larger grain size possess improved wear and impact resistance — the key to effective machining of materials like cast irons containing less than 10% ferrite content. Such performance improvement is especially important when machining alloy cast irons for automotive applications. Previously, manufacturers had to grind these castings.

In roughing operations on alloy cast irons, tools must withstand interrupted cuts due to surface cracks, sand inclusions, and other surface discontinuities inherent to the casting process. Good wear resistance comes into play in finishing operations, where parts containing 28 to 30% chromium have a hardness between Rc 68 and Rc 72.

Solid PCBN inserts provide multiple cutting edges on two sides, reducing insert cost per part produced. PCBN inserts also come in full-face and tipped types. The full-face type has a complete PCBN top face sintered onto a carbide substrate, and provides multiple cutting edges on one side only. These inserts are less expensive than solid PCBN inserts. The tipped style contains a small PCBN segment brazed onto one corner of a carbide insert, providing either a single cutting edge or double cutting edges. Most PCBN inserts used today are tipped. Both full-face and tipped PCBN inserts come in industry-standard sizes and, like the solid insert, can be used in the insert pockets of standard tool-holders and milling cutters.

Inserts made of PCBN work best in hard-part machining applications. In practice, the low end for part hardness falls at about Rc 45.

Machining softer parts using PCBN produces insert cratering.

In roughing operations, maximum depths of cut using solid-style PCBN inserts range from about 4.76 mm for white iron and other hard, high-chromium irons to about 6.4 mm for unalloyed, "clean" cast irons. Finishing speeds range from 107 to 122 m/min on high-chromium irons to as high as 2134 m/min on gray cast irons. All PCBN operations require the use of very rigid tooling, work fixturing, machine spindles, and machine tools.

Appropriate PCBN finishing speeds vary dramatically depending on the work material, material hardness, and part size and shape. For example, one can achieve a cutting speed of 183 m/min at a 0.51-mm depth of cut on hardened steels in the range of Rc 60 to 62. A feed rate of 0.05 to 0.1 mm/rev typically produces a surface finish of about 8 rms. Harder materials require faster speeds, but speeds greater than about 198 m/min result in excessive wear. One can attain much higher speeds on plain cast irons.

Currently, the largest growth area for PCBN inserts is in hard turning/finish turning of alloy steel automotive engine components such as gears, shafts, and bearings with hardnesses between Rc 60 and 65. Traditionally, manufacturers ground these parts to obtain very tight dimensional tolerances and a fine surface finish. Hard turning gives the same results using a CNC lathe.

Although hard enough to resist deformation from the heat generated during machining, PCBN is brittle and can be cracked by the thermal shock produced because of exposure to a coolant. Therefore, machining operations generally should run dry, especially in the case of interrupted cuts. Never use a coolant when running them.

Insert edge preparation also strongly influences the success of PCBN machining. To promote good tool life, a reinforced cutting edge with the proper edge preparation is a must. Edge preparation ranges from a small hone for finish-machining cast irons to a T-land measuring 3 to 8 mm wide by 15° for heavy roughing of white iron. One can combine lands and hones. Adding a land to the insert increases and strengthens the cutting edge. For example, a 20° land added to a 90° corner provides a 110° corner; the larger the corner angle, the stronger the cutting edge.

PCD

PCD inserts are cost-effective because they dramatically outperform carbide in most nonferrous applications, not because of any developments in the material itself.

Most PCD inserts used today are tipped. They employ a small segment of PCD brazed onto one corner of a carbide insert. PCD inserts come in industry-standard sizes for use in the insert pockets of standard tool-holders and milling cutters. Unlike carbide inserts, however, PCD-tipped inserts are not indexable, and provide only a single cutting edge.

Standard-size, full-face-style PCD inserts provide a complete PCD top face sintered onto a carbide substrate.

More secure insert retention — especially in rotating tools — has played a key role in increasing the use of PCD inserts. Today's combinations of wedges and screws, as well as tapered inserts and wedges, enhance insert retention. Conventional screw or clamp designs, and cutters using direct-mount cartridges where the PCD segment is brazed directly onto the cartridge body, also provide the necessary rigidity. Rigid tooling setups provide more reliable results with PCD.

More secure part fixturing and more rigid machine tools and spindles also improve performance. Today's CNC lathes and machining centers provide all the rigidity required by PCD or PCBN cutting tools. And as companies install new equipment, the machine shop environment becomes more suited to high-performance PCD and PCBN cutting tools.

Cutters used for finish milling of aluminum automotive manifolds often combine PCD-tipped and plain carbide inserts to cut through the casting flash encountered at the insert depth-of-cut line.

Another major use of PCD-tipped inserts is roughing automotive parts in a process called "cubing." Foundries economize on shipping costs and recover aluminum chips for recycling by rough machining castings such as aluminum cylinder heads. They then semifinish the castings (top, bottom, side, and edge) at very high metal-removal rates using a PCD cutting tool. Machined parts are easier to handle and pack because of their rectangular shape, and shipping costs are lower for the smaller parts because they take up less space.

PCD-tipped inserts are preferred over carbide in applications where manufacturers want higher production rates, consistent part finish, and longer tool life (eliminating frequent tool changes). These inserts excel in machining abrasive high-silicon aluminum automotive parts like cylinder heads, engine blocks, manifolds, transmission cases, and wheels.

When using PCD inserts, good chip evacuation is essential. Because the inserts produce chips so rapidly, the system must remove them from the work zone quickly and continuously. Coolant, air, mist, refrigerated air, or combinations of all these methods will work. PCD cutting speeds as high as 3048 m/min are possible with effective chip evacuation and a rigid machine and operating setup. Today's CNC machining and turning centers easily achieve these speeds.

D

DENTAL MATERIALS

The field of dental materials is an interdisciplinary area that applies biology, chemistry, and physics to the development, understanding, and evaluation of materials used in the practice of dentistry. It is principally involved in restorative dentistry, prosthodontics, pedodontics, and orthodontics, and to a smaller extent in the other areas of practice. The field of dental materials has advanced rapidly with the application of technologies such as genetic engineering of filling materials, computer designing and machining of restorations, adhesive bonding to teeth, dental implants, and new types of restorative materials.

RESTORATIVE MATERIALS

The restoration of missing tooth structure is one of the most challenging areas for the use of dental materials. The enamel that covers the crown of the tooth is the hardest substance in the human body and is very resistant to physical and chemical attack. Direct restorations are put in place and harden by a setting reaction, whereas indirect restorations require a replica of the tooth, which has to be fabricated.

Gold

Pure gold or gold foil has long been used as a restorative material. Pellets of pure gold are formed from very thin gold sheets or foils. The gold is compacted with compressed air or manually, and contoured and polished to the desired shape. These restorations are long-lasting and biologically inert. They are placed in low-stress areas because gold is soft.

Amalgam

Dental amalgam is the most frequently used direct restorative material for posterior (molar and premolar) teeth. It is formed by mixing mercury with an alloy powder of silver, tin, and copper. After mixing, there is time to insert and shape the material before it hardens. Ag–Cu compounds and elements such as indium improve the alloy. Its advantages are ease of application, durability, and low cost; the disadvantages are chiefly aesthetic.

Resins

Composite resins are tooth-colored filling materials that were first developed for restoring anterior teeth (incisors and canines) but have found increased use as an attractive alternative to dental amalgam in posterior teeth. A composite resin consists of an organic polymer matrix reinforced with up to 80% of an inorganic filler such as SiO_2 or glass. Composite resin from a single paste is placed on the tooth and then polymerized with the help of a fiber-optic visible light source. This allows more time for contouring and gives a restoration with improved color stability. Composite resin restorations are strengthened by etching the enamel with 37% orthophosphoric acid prior to placement of the resin to create a strong attachment to the enamel.

Gl Ionomer

The only dental restorative material that forms a durable chemical bond to dentin is formed by the reaction of aluminosilicate glass with polyacrylic acid. Glass ionomer is particularly useful for restoring eroded or carious areas on exposed root surfaces with little or no enamel for bonding with composite resin. This material slowly releases fluoride to provide some additional protection from new decay. It is brittle and not used in areas of heavy stress.

Metal Castings

Crowns and inlays are made of metal castings and used when there is insufficient tooth structure to support filling. A negative of the tooth is prepared with synthetic rubber or agar, into which high-strength gypsum (plaster) is poured. This forms the replica of the tooth, to which wax is applied. An indirect lost-wax process is used in fabricating dental castings. Alloy ingots used for the castings cover a range of compositions from 80% of noble metals — gold and the platinum group — combined with copper and other minor elements to base metal alloys of chromium, cobalt, and nickel. Traditionally, only noble alloys were used, but as the cost of gold increased, other alloys were developed.

Porcelain

Porcelain jacket crowns and metal crowns to which porcelain is bonded are used in visible areas. In one type an inner core of high-strength aluminous (40 to 50% Al_2O_3) porcelain is covered by feldspathic porcelain. The latter is applied to a refractory gypsum model covered with platinum foil and sintered under a vacuum at a high temperature. The porcelain powder contains metal oxides to color the crown. The other type is made from a glass ceramic of SiO_2 that is cast by the lost-wax process. After casting and contouring, the crown is heat-treated to form an opaque crystalline material, mica. Finally, color is applied by firing oxide stains on the surface. Porcelain and cast ceramics are used for a cosmetic veneer of the front surface of anterior teeth. The thin veneers are bonded to etched enamel with a composite resin cement. The third type of crown has a cast metal substructure for strength, onto which feldspathic porcelain is bonded by the same process described above.

Provisional Restoration

For provisional restorations different materials are frequently used. While a permanent crown is being fabricated, a temporary or provisional crown protects the tooth. For best appearance, a custom-formed poly(methyl methacrylate) or polycarbonate crown is lined with acrylic resin; aluminum crowns lined with acrylic resin can be used with posterior teeth along with the temporary crowns. These crowns are easily removed, because a cement is used that sets when pastes containing the active ingredients Zn_2O and eugenol are combined. A Zn_2O-eugenol cement reinforced with poly(methyl methacrylate) beads is also used as a temporary direct filling material for patients with deep decay.

Prosthodontic Dental Materials

The materials used with fixed prosthodontics are similar to those described above under **Metal Castings** and **Porcelain**.

Orthodontic Materials

The movement and stabilization of teeth can be controlled by several unique dental materials. Stainless steel alloys are commonly used for brackets and bands on teeth, but plastic brackets are also being used to improve appearance. Brackets are bonded to etched enamel with the help of composite resin without inorganic filler. Bands encircling posterior teeth are made of stainless steel and are generally secured with a cement formed by combining Zn_2O and HPO_4. Wires in different configurations are adapted to the dental arch to create the desired static forces for controlled tooth movement. The wires are commonly formed from wrought stainless steel, but alloys of Ni–Ti, Ti–Mo, and Co–Cr–Ni are also used.

DIAMOND

Diamond is a highly transparent and exceedingly hard crystalline stone of almost pure carbon. When pure, it is colorless, but it often shows tints of white, gray, blue, yellow, or green. It is the hardest known substance, and is placed as 10 on the Mohs hardness scale. But the Mohs scale is only an approximation, and the hardness of the diamond ranges from 5500 to 7000 on the Knoop scale, compared with 2670 to 2940 for B_4C, which is designated as 9 on the Mohs scale. The diamond always occurs in crystals in the cubical system, and has a specific gravity of 3.521 and a refractive index of 2.417.

In addition, diamonds are generally electrical insulators and nonmagnetic. They are quite transparent to x-rays, and thus make useful pressure cell windows for x-ray measurements. Nearly all diamonds show residual stresses in polarized light.

The melting point of diamond is given in terms of temperature and pressure because the phase diagram of carbon is such that unless pressure is maintained on diamond when at a temperature greater than 4000°C, the diamond will graphitize and then vaporize.

The diamond has been valued since ancient times as a gemstone, but it is used extensively as an abrasive, for cutting tools, and for dies for drawing wire. These industrial diamonds are diamonds that are too hard or too radial-grained for good jewel cutting.

Synthetic Diamonds

These diamonds are produced from graphite at pressures from 5512 to 12,402 MPa and temperatures from 1204 to 2427°C. A molten metal catalyst of chromium, cobalt, nickel, or other metal is used, which forms a thin film between the graphite and the growing diamond crystal. Without the catalyst much higher pressures and temperatures are needed. The shape of the crystal is controllable by the temperature. At the lower temperatures cubes predominate, and at the upper limits octahedra predominate; at the lower temperatures the diamonds tend to be black, whereas at higher temperatures they are yellow to white. The synthetic diamonds produced by the General Electric Co. are up to 0.01 carat in size, and are of industrial quality comparable with natural diamond powders.

CVD Diamonds

A newer production method used chemical vapor deposition (CVD) to produce a sheet of diamonds consisting of countless diamonds, each only a few micrometers across. The CVD diamonds have similar properties to those made by the older method. However, they conduct heat better than any known material — five times better than copper — making them useful as heat sinks to conduct heat away from electrical components.

The CVD process has been used to coat the cutting surfaces of tungsten carbide rotating tools such as routers, end mills, and drills. The technique should bring substantial productivity improvement to the machining of composites, carbon fiber, and glass/fiberglass-reinforced composites, epoxy resins, and graphite.

Tests show diamond-coated tools last up to 50 times longer than tungsten carbide ones and more than twice as long as polycrystalline diamond tools in aerospace applications. In addition to greater longevity, longer unattended runs and fewer tool changes, diamond-coated tools also can run at higher speeds and feeds than conventional tools.

The CVD process allows round tools of virtually any shape or size to be diamond-coated with no subsequent brazing or grinding.

Applications

Several types of grinding wheels (usually distinguished by the bond material) are made that carry diamonds as the abrasive material. Wheel bonds may be resinoid, vitrified, metal, etc. Diamond grit sizes in these wheels may vary from 0.0254 to 0.762 mm. In general, the diamond-carrying material is placed in a rather thin layer on the working surface of the wheel.

Wheel surface speeds of 22.5 to 33 mps are recommended. Most grinding is done wet. By far the greatest use for diamond wheels is in shaping and sharpening tungsten carbide tools. In addition, diamond wheels are used for edging and beveling glass and for lens grinding. Also, artificial sapphire is sliced and formed with diamond grinding wheels.

Diamond dust is a powder obtained by crushing the fragments of bort, or from refuse from the cutting of gem diamonds. It is used as an abrasive for hard steels, for cutting other stones, and for making diamond wheels for grinding. Grit sizes for grinding wheels are 80 to 400. The coarsest, No. 80, for fast cutting, has about 1400 particles per square inch in the face of the bonded wheel.

Diamond drills range from small, diamond-tipped dental drills up to oil-well core bits containing 1000 carats. A 10.16-cm core bit may be set with 150 to 300 diamonds varying in size from $1/10$ to $1/30$ carat. As much as 3048 cm of limestone can be drilled before the diamonds require resetting in the hard metal matrix.

Diamond wire drawing dies are required for tungsten filament wire and preferred for many other wire materials. The low friction coefficient may contribute to the success of diamond in this application. Stones for wire drawing are large (up to 8 carats for 2.54-mm copper wire) and free from flaws and inclusions.

Single-point diamond lathe tools are made by setting a stone in the end of a steel shank suitable for mounting in a machine tool such as a lathe. The stone is mounted with consideration for its shape and crystallographic axis. After mounting, the tool is ground to the required shape on diamond grit wheels and polished on a scaife.

Single-point diamond tools are designed with angles between faces of 80° or greater to avoid the breakage that would occur if slender points were used. Diamond tools are not suitable for heavy cuts particularly on ferrous material. Accurate, fine finish cuts may be made on nonferrous metals and plastics. Tool wear is slow, so accuracy is maintained for many workpieces. A diamond tool maintains its cutting edge about 25 times as long as a high-speed steel tool and 3 to 4 times as long as a carbide tool.

DIAMOND FILMS

A new method has been developed for creating amorphous diamond films that are harder than any other coating, except for crystalline diamond. The new coatings are also smooth and can be created at room temperature, whereas crystalline diamond coatings need high temperatures and have rough surfaces.

This newly developed method uses a pulsed laser hitting a graphite target to deposit a carbon film containing a high percentage of diamond-like bonds. Heating the film eliminates internal stresses while allowing the film to retain diamond-like properties, including 90% of the stiffness and hardness of crystalline diamond. An added benefit is that researchers can control the level of stress. In fact, scientists have made stress-free films more than 7 μm thick and applied the coatings to plastic substrates as well as free-standing membranes more than an inch in diameter and less than 600 Å thick, and tested to 800°C. Besides being incredibly hard, the films resist wear with low coefficients of friction and are almost inert to all chemicals. The films could find use as coatings for metal tools, auto parts, and biomedical devices.

Other methods have been developed that offer a lower-cost alternative by coating and chemically bonding an inexpensive substrate with a thin film of diamond-like carbon (DLC).

These diamond films have great potential as chemically inert protective coatings that make machine tools and parts last ten times longer. Other applications include optical instruments, medical equipment, watch crystals, and eyeglasses.

This method is called direct ion-beam deposition for applying DLCs to a substrate. An ion generator creates a stream of ions from a hydrocarbon gas source; the C ions impinge directly on the target substrate and "grow" into a thin DLC film. This low-pressure, low-temperature process allows use of plastics and other substrates that cannot withstand the high pressures and temperatures normally used to synthesize diamonds.

Other commercial DLC applications include coatings for magnetic data-storage disks, surgical needles, and a diamond-coated ball for an artificial hip joint.

DIE CASTINGS

Die castings are made by forcing molten metal under high pressure into a steel die containing an impression of the part to be made, which is called the die cavity. The process is most useful for the high production manufacture of small to medium-size castings. Economies of the process are developed through:

1. High speed of production. The output from a die is many times more than for other casting processes.
2. Strategic use of metal. Through casting thinner walls, by coring holes and passages, and by contouring the die

halves carefully, material can be saved and the casting made lighter.

3. Dimensional accuracy. The reproducibility is very good and dimensional tolerances can be held close enough to eliminate many machining operations.

PRODUCTION

Die castings are made in a special machine containing a clamping mechanism to open and close the die and to hold the die halves together, and an injection mechanism to force the metal into the die. The injection mechanisms can be of two types. The first system, the immersed plunger type, is suitable for alloys such as tin, lead, zinc, and sometimes magnesium where attack of the molten metal on the parts of the mechanism is insignificant. The die containing the cavity is closed and held tightly in the clamping mechanism while the metal is injected by the downward motion of the plunger, which is operated by a hydraulic cylinder. The filling of the die is accomplished in less than $1/10$ s. Dies can be opened in a few seconds, the casting removed by ejector pins, and the cycle repeated. Casting rates vary from 50/h for heavy parts to 15,000/h for small parts made on automatic machines.

Metals such as aluminum, brass, and magnesium, which because of either their solvent action or high melting temperatures are not adaptable to the immersed plunger system, can be cast by the second type, the cold chamber system, where the metal is melted in a separate unit and is hand-ladled or metered into a shot cylinder securely fastened to the die. The injection force is supplied by a plunger that in turn is operated by a hydraulic cylinder. Higher pressures (200 to 1000 atm) are used in a cold chamber machine permitting improved density and mechanical properties of the die casting.

CHARACTERISTICS

Die castings have smooth, dense surfaces, which require a minimum amount of finishing. A major advantage of die casting is the close tolerances that can be held. These substantially reduce machining operations and are largely responsible for the economy of the process.

Intricate casting shapes can be readily made through the use of cores and slides, ribs, and bosses. Special properties can be obtained by casting in inserts such as bronze sleeves, laminated pole shoes, magnets, and hardened steel plates.

One limitation of the process is the expense of the dies, which must be amortized over the expected production. However, most automotive, appliance, or business machine production is large enough that the tool and die costs become a small portion of the total casting cost and they are completely offset by manufacturing economies due to elimination of machining of parts after processing.

Average wall thickness for medium-size parts in aluminum and magnesium is 2.2 mm and for large parts is 2.5 to 3.9 mm. Wall thickness for zinc will run approximately 75% of aluminum and magnesium because of the ease of casting of zinc in comparison to the light metals. Minimum wall thicknesses vary with the size of the part — 1 mm is generally accepted for zinc, 1.3 mm for aluminum and magnesium.

APPLICATIONS

The largest single consumer of die castings is the automotive industry. Zinc is used for ornamentation and hardware such as radio grilles, instrument clusters and housings, as well as for functional parts such as carburetors, fuel pumps, and speedometer housings. Zinc is used in appliances such as washing machines, dryers, electric ranges, and refrigerators for decorative and functional parts. It is also used for camera cases and projector housings and for toys. Military uses include precision fuze parts.

Because aluminum and magnesium can be used almost interchangeably, relative cost is often the governing selection factor. Automotive uses of aluminum die castings include transmission cases, torque converter housings and extensions, valve bodies, and carburetors. In the business machine field, typewriters are the largest application for aluminum die castings. The optical industry also uses aluminum die castings for binoculars, camera cases, projector frames, and other parts. Military uses are varied and include ordnance parts such as fuzes, windshields, aircraft engine accessory parts, and airframe parts, as well as some small missile airframe parts. The electrical industry uses aluminum and magnesium for instrument housings and frames and for brush holders.

Materials

Dies are generally made of alloy steels and are relatively expensive. Dies for zinc can be made from prehardened steels of the SAE 4140 type. Dies for aluminum and magnesium are made from the Cr–Mo H-11 or H-13 steels heat-treated to C46 to C48 Rockwell.

Zinc castings are made from high-purity Zn–Al alloys containing 4% aluminum with or without copper and containing approximately 0.03% magnesium. Impurities such as tin, lead, and cadmium must be kept at very low values, on the order of 0.005% or less, to avoid deterioration of the alloy in humid atmospheres. Zinc alloys have high tensile strength (292.4 MPa), and high impact strength (54.2 J). These properties change with decreasing temperature so that the useful temperature range for zinc is only recommended between 0 and 93°C.

Aluminum die-casting alloys are made from the Al–Si, Al–Si–Cu, or Al–Mg series. These alloys have the best castability and physical properties for the die-casting process. Mechanical properties will vary between 276 to 325 MPa tensile strength, 135 to 172 MPa yield strength, and 3 to 5% elongation. Die castings of aluminum have high fatigue strength (135 to 142 MPa), which makes them particularly adaptable to structural uses where cyclic stresses are involved.

Magnesium alloys are of the Al–Zn type having tensile strength of 234 MPa, yield strength of 142 MPa, and an elongation of 3%.

Copper alloys used are low-melting silicon brasses with tensile strength of 345 to 584 MPa, yield strength of 241 to 345 MPa, and an elongation of 15 to 25%.

DIE STEELS

Die steels include any of the various types of tool steels used for cold- and hot-forming dies, including forging, casting, and extrusion dies; stamping and trim dies; piercing tools and punches; molds; and mandrels. In general, all of the major families of tool steels except the high-speed types are used for dies, including the hot-work, cold-work, shock-resisting, mold, special-purpose, and water-hardening types. The high-speed types, however, which are typically used for cutters, are also used for punches.

DIELECTRIC MATERIALS

These are materials that are electrical insulators or in which an electric field can be sustained with a minimum dissipation of power. A solid is a dielectric if its valence band is full and is separated from the conduction band by at least 3 eV.

Dielectrics are employed as insulation for wire cables and electrical equipment, as polarizable media for capacitors, in apparatus used for the propagation or reflection of electromagnetic waves, and for a variety of artifacts, such as rectifiers and semiconductor devices, piezoelectric transducers, dielectric amplifiers, and memory elements.

The ideal dielectric material does not exhibit electrical conductivity when an electric field is applied. The term *dielectric*, although it may be used for all phases of matter, is usually applied to solids and liquids. In practice, all dielectrics have some conductivity, which generally increases with increase in temperature and applied field. For a good dielectric, such as polytetrafluoroethylene, the low-field DC conductivity at room temperature may be lower than 10^{-16} $\Omega \cdot m$, whereas the corresponding figure for some specimens of plasticized polyvinyl chloride (PVC) may be as high as 10^{-4} $\Omega \cdot m$, similar to that of the low-conductivity semiconductors.

Breakdown

If the applied field is increased to some critical magnitude, the material abruptly becomes conducting, a large current flows (often accompanied by a visible spark), and local destruction occurs to an extent dependent upon the amount of energy that the source supplies to the low-conductivity path. This critical field depends on the geometry of the specimen, the shape and material of the electrodes, the nature of the medium surrounding the dielectric, the time variation of the applied field, and other factors. Temperature instability can occur because of the heat generated through conductivity or dielectric losses, causing thermal breakdown.

INDUSTRIAL DIELECTRIC MATERIALS

Many of the traditional materials are still in common use, and they compete well in some applications with newer materials regarding their electrical and mechanical properties, reliability, and cost. For example, oil-impregnated paper is still used for high-voltage cables, usually paper and various types of pressboard and mica, often as components of composite materials, are also in use. Elastomers and press-molded resins are also of considerable industrial significance. However, synthetic polymers such as polyethylene, polypropylene, polystyrene, polytetrafluoroethylene, PVC, polymethyl methacrylate, polyamide, and polyimide have become important, as has polycarbonate because it can be fabricated into very thin films.

PROPERTIES OF POLYMERS

Generally, these have crystalline and amorphous regions, increasing crystallinity causing increased density, hardness, and resistance to chemical attack, but often producing brittleness. Many commercial plastics are amorphous copolymers, and often additives are incorporated in polymers to achieve certain characteristics or to improve their workability. Additives include inorganic fillers, antioxidants, stabilizers, and pigments.

Polyethylene

Polyethylene ($C_n H_{2n+2}$) is the most common, most investigated polymer. It is relatively inexpensive and is easily worked. Polyethylene readily oxidizes, and so its working temperature is normally not above 70°C; although it may be worked at higher temperatures (for example, up to 90°C) in suitable circumstances it normally contains antioxidants.

Polyethylene Terephthalate

Polyethylene terephthalate is a useful material that has good resistance to acids and alkalis provided that they are diluted. It is insoluble in many common solvents, although it does dissolve in aromatic and chlorinated hydrocarbons, especially when warm. It softens above 250°C.

Polypropylene

Polypropylene is similar in structure to polyethylene, but every other carbon atom has one of its H_2 atoms replaced by a CH_2 group. Although electrically similar to polyethylene, polypropylene can be made in thinner films, say 5 µm as against about 25 µm for polyethylene. These films replace paper for impregnated capacitors, with reduced loss.

Polystyrene

Polystyrene is brittle at room temperature, becomes soft at 80°C, and is often modified by copolymerization. Traditionally, it is used in film form for capacitors, and it remains competitive for this application. Polystyrene is also used for coaxial-cable insulation, but in wound strip or bead form, because the solid is not very flexible.

Polytetrafluoroethylene

Polytetrafluoroethylene is similar to polyethylene, with all the H_2 atoms replaced by fluorine atoms. Highly crystalline (about 95%), it resists twisting and bending and is mechanically tough, with a very low coefficient of friction, 0.06. It is highly resistant to chemical attack and is useful for hostile environmental conditions.

Polyvinyl Chloride

PVC has a structure similar to that of polyethylene, but every other carbon atom has one of its H_2 atoms replaced by a chlorine atom. It is only about 10% crystalline and is compatible with a large number of other polymers. It is easily decomposed by heat and is not suitable for continuous use at 70°C. PVC is not well characterized, and depending on admixtures, may be molded; its main electrical use is coating wires.

Polymethyl Methacrylate

Polymethyl methacrylate is a clear solid, hard but easily scratched. Its dielectric properties are only moderate, and its use is restricted to undemanding conditions at normal ambient temperatures, usually in a molded form, where appearance is important.

Polyamide

Polyamide may be in linear form, known as nylon, or in aromatic form. Neither of these is of much value as a dielectric, but nylons, because of their toughness and resistance to solvents, are used to form coatings to protect insulation. The aromatic form can also make yarns, but it is most often used in impregnated board from which it gives low-voltage insulation capable of being used continuously at temperatures of 150°C, or even higher.

Polyimide

Polyimide is related to aromatic polyamide, but with additional aromatic groups. Expensive to produce, it can be made in films down to 25 μm thick, and in varnish form for wire coating. It is very tough, does not burn but chars above 800°C, and remains flexible at liquid helium temperatures; it can be used continuously up to 250°C.

Resins

The resins are important members of the family of cross-linked polymers. Epoxy resins have a high mechanical strength, absorb very little water, and bond easily to most materials but not to polyethylene. They are used for bonding and encapsulation, and their properties can be modified by the curing process used and by the use of fillers and hardeners.

DIFFUSION COATINGS

A large number of elements can be diffused into the surface of metals to improve their hardness and resistance to wear, corrosion, and oxidation. Diffusion coatings (sometimes called cementation coatings) are applied by heating the base metal in an atmosphere of the coating material, which diffuses into the metal.

CALORIZED

Aluminum (calorized) coatings are applied diffusion to carbon and alloy steels to improve their resistance to high-temperature oxidation. They can be applied by treating the metal in powdered aluminum compound or in $AlCl_3$ vapor, or by spraying the aluminum on and subsequently heat-treating it. The alloy coating formed (about 25% aluminum) protects the metal by sealing it from the surrounding air. The coatings range in depth from 5 to 40 mils and permit parts to remain serviceable for many years at temperatures up to 760°C. They have also been used for intermittent exposure as high as 927°C. Typical high-temperature uses are chemical and metal processing pots, bolts, air heater tubes, and parts for furnaces, steam superheaters, and oil and gas polymerizers.

CARBURIZED

Carburizing allows steels to retain high internal strength and toughness and at the same time have high surface hardness. The hardened surface is produced by introducing carbon into a steel surface by heating the metal above the transformation temperature while it is in contact with carbonaceous material, which may be a solid, gas, or liquid.

In general, carburizing is limited to steels low enough in carbon (below about 0.45%) to take up that element readily. Plain carbon steels are generally used if surface hardness is the principal requirement and core properties are not too critical. Alloy steels must usually be used if high strength and toughness are needed in the core. Typical applications are gears, cams, pawls, racks, and shafts.

CHROMIZED

Chromizing is the process of diffusing chromium into ferrous metals to improve their resistance to corrosion, heat, and wear. Typical of the chromizing methods that have been developed is one in which the parts to be treated are packed in a proprietary powdered chromium compound and heated to 816 to 1038°C. This method produces high Cr–Fe alloy on ferrous metals with a low carbon content. The case (3 mils thick) exhibits good resistance to sealing and corrosion at high temperatures. A CrC case is produced on high-carbon materials such as cast iron, iron powder, tool steel, and plain carbon steels containing over 0.40% carbon. This case ($1/2$ to 2 mils thick) has a hardness of 1600 to 1800 vpn.

Cyanided, Carbonitrided

Both cyaniding and carbonitriding produce a hard and wear-resistant surface on low-carbon steels. Both methods cause carbon and N_2 to diffuse into the surface of the base metal. The case developed has high hardness after quenching. The methods differ in that a liquid bath is used in cyaniding, whereas a gas atmosphere is used in carbonitriding.

In general, cyaniding and carbonitriding are used with the same base metals and for the same applications as carburizing. Warpage is usually less serious than in carburizing. Quenching is usually required for full hardness, but file hardness can be obtained without quenching.

Ni–P Coated

With some exceptions, Ni–P coatings can be roughly classified as diffusion coatings. The coatings are prepared from NiO_2 dibasic NH_3PO_4, and water, and are applied to ferrous surfaces just like a paint. Subsequent heat treatment in a controlled atmosphere produces coatings with a degree of corrosion resistance approaching that of stainless steel and the high-nickel alloys. The coatings have little porosity and high resistance to heat and abrasion.

Nitrided

Nitriding is a means of improving wear resistance. In the most widely used process, steel is exposed to gaseous NH_4 at a temperature (about 538°C) suitable for the formation of metallic nitrides. The hardest cases are obtained with aluminum-bearing steels such as the Nitralloys. Where lower hardness is acceptable, steels containing no aluminum, such as medium-carbon steels containing chromium and molybdenum, can be used.

Stainless steels can also be case-hardened by nitriding. Straight chromium steels are more readily nitrided than Ni–Cr steels, although both are used. Tool steels can also be given a thin hard case.

Nitriding produces minimum distortion. Some growth occurs, but this can be allowed for. In general, nitriding is used for the same applications as carburizing.

Siliconized

Substantial improvements in the wear resistance and hardness of steel and iron parts can be obtained by impregnating with silicon (about 14%). The most wear-resistant cases are formed on low-carbon, low-sulfur steels. High-carbon, low-sulfur steels can also be impregnated, although treatment time is longer. White and malleable iron can also be siliconized; siliconizing of gray irons is not recommended.

The case of a siliconized surface (about 5 to 10 mils) is rather brittle, hardness varying from Rockwell B80 to B85. Siliconized surfaces are virtually nongalling and are especially effective in resisting combined wear and corrosion.

DIRECTED METAL OXIDATION PROCESS (DIMOX)

This process is based on the reaction of a molten metal with an oxidant, usually a gas, to form a ceramic matrix. Unlike the sintering process, in which the ceramic powders and fillers are consolidated under heat, directed metal oxidation grows the ceramic matrix material around the reinforcements.

Examples of ceramic matrices that can be produced by the DIMOX-directed metal oxidation process include Al_2O_3, Al_2Ti_5, AlN, TiN, ZrN, TiC, and ZrC. Filler materials can be anything chemically compatible with the caramic, parent metal, and growth atmosphere.

The first step in the process involves making a shaped preform of the filler material. Preforms consisting of particles are fabricated with traditional ceramic-forming techniques, while fiber preforms are made by weaving, braiding, or laying up woven cloth. Next, the preform is put in contact with the parent metal alloy. A gas-permeable growth barrier is applied to the surfaces of this assembly to limit its shape and size.

The assembly, supported in a suitable refractory container, is then heated in a furnace. For aluminum systems, temperatures typically range form 899 to 1149°C. The parent metal reacts with the surrounding gas atmosphere to grow the ceramic reaction product through and around the filler to form a ceramic-matrix composite (CMC).

Capillary action within the growing ceramic matrix continues to supply molten alloy to the growth front. There, the reaction continues until the growing matrix reaches the barrier. At this point, growth stops, and the part is cooled to ambient temperature. To recover the part, the growth barrier and any residual parent metal are removed. However, some of the parent metal (5 to 15% by volume) remains within the final composite in micron-sized interconnected channels.

Traditional ceramic processes use sintering or hot pressing to make a solid CMC out of ceramic powders and filler. Part size and shapes are limited by equipment size and the shrinkage that occurs during densification of the powders can make sintering unfeasible. Larger parts pose the biggest shrinkage problem. Advantages of the DIMOX process include no shrinkage because matrix formation occurs by a growth process. As a result, tolerance control and large part fabrication can be easier with directed metal oxidation.

In addition, the growth process forms a matrix whose grain boundaries are free of impurities or sintering aids. Traditional methods often incorporate these additives, which reduce high-temperature properties. And cost comparisons show the newer process is a promising replacement for traditional methods.

DISPERSION-STRENGTHENED METALS

Particulate composites in which a stable material, usually an oxide, is dispersed throughout a metal matrix. The particles are less than 1 μm in size, and the particle volume fraction ranges from only 2 to 15%. The matrix is the primary load bearer while the particles serve to block dislocation movement and cracking in the matrix. Therefore, for a given matrix material, the principal factors that affect mechanical properties are the particle size, the interparticle spacing, and the volume fraction of the particle phase. In general, strength, especially at high temperatures, improves as interparticle spacing decreases. Depending on the materials involved, dispersion-hardened alloys are produced by either powder-metallurgy, liquefied-metal, or colloidal techniques. They differ from precipitation-hardened alloys in that the particle is usually added to the matrix by nonchemical means. Precipitation-hardened alloys derive their properties from compounds that are precipitated from the matrix through heat treatment.

There is a rather wide range of dispersion-hardened alloy systems. Those of aluminum, nickel, and tungsten, in particular, are commercially significant. Tungsten thoria, a lamp-filament material, has been in use for more than 30 years. Dispersion-hardened aluminum alloys, know as SAP alloys, are composed of aluminum and Al_2O_3 and have good oxidation and corrosion resistance plus high-temperature stability and strength considerably greater than that of conventional high-strength aluminum alloys. Another dispersion-hardened metal, TD nickel, has dispersion of thoria in a nickel matrix. It is three to four times stronger than pure nickel at 871 to 1316°C. TD–Ni–Cr has also been produced for increased resistance to high-temperature oxidation. Other metals that have been dispersion-strengthened include copper, lead, zinc, titanium, iron, and tungsten alloys. The copper is used for resistance-welding (spot-welding) tips.

DOLOMITE

Dolomite is the carbonate mineral $CaMg(CO_3)_2$, often with small amounts of iron, manganese, or excess calcium, which replace some of the magnesium; cobalt, zinc, lead, and barium are more rarely found.

Dolomite is a very common mineral, occurring in a variety of geologic settings. It is often found in ultrabasic igneous rocks, notably in carbonates and serpentinites, in metamorphosed carbonate sediments, where it may recrystallize to form dolomite marbles, and in hydrothermal veins. The primary occurrence of dolomite is in sedimentary deposits, where it constitutes the major component of dolomite rock and is often present in limestones.

Dolomite is normally white or colorless with a specific gravity of 2.9 and a hardness of 3.5-4 on Mohs scale.

DRY ICE

Dry ice is a solid form of carbon dioxide, CO_2, which finds its largest application as a cooling agent in the transportation of perishables. It is nontoxic and noncorrosive and sublimes directly from a solid to a gas, leaving no residue. At atmospheric pressure it sublimes at −78.7°C, absorbing its latent heat of 573.1 kJ/kg. Including sensible heat absorption, the cooling effect per pound (kilogram) of dry ice is approximately 628.0 kJ at storage temperatures above −9°C and 581.5 kJ at lower temperatures. Slabs of dry ice can easily be cut and used in shipping containers for frozen foods, in refrigerated trucks, and as a supplemental cooling agent in refrigerator cars.

The manufacture of CO_2 gas is a chemical process. The gas is liquefied by compressing it to 6.2 to 6.9 MPa gauge in three stages of reciprocating compressors and then condensing it in water-cooled condensers. The liquid is expanded to atmospheric pressure where the temperature is below the triple point — 56.6°C.

DUCTILE IRON

Ductile iron, also called spheroidal graphite iron, S.G. iron, or nodular iron, is a graphite containing ferrous metal in which the graphite appears as rounded particles or spheroids rather than the usual plates, flakes, or clumps typical of other graphite ferrous metals.

In spherical form, the graphite exerts a minor influence on the properties of the steel-like matrix because in that form the graphite-to-metal contact area is at a minimum. It is relatively economical to produce and finds application in many industries.

Ductile iron is similar to other high-carbon, high-silicon alloy with respect to corrosion resistance, sliding wear, and machinability except insofar as the shape of the graphite alters these characteristics. The spheroidal form of the graphite, with lower surface area, alters corrosion resistance where the graphite is cathodic to the surrounding material. In comparison, with flake graphite iron, greater power will be required to remove metal in machining, while in metal-to-metal contact the greater toughness resists removal of metal particles and therefore increases resistance to certain types of wear.

The spheroidal graphite irons have very good machining characteristics by virtue of their contained graphite and compare favorably with other materials at equivalent hardnesses and strengths. The soft annealed grade provides a strong, tough material that can be turned at very high speeds and feeds. Maximum turning speeds of well over 400 sfpm are attainable in this grade. The cutting speed is, however, determined by the microstructure with maximum turning speed of the pearlitic grade reduced to about 150 sfpm and of the hardest oil-quenched grade to 100 sfpm. The chip produced in machining operations tends to be long and continuous, which complicates such operations as deep-hold drilling and internal broaching and reaming. For fast metal removal in these operations, special tool configurations are required to provide access for chip removal. Cutting fluids are recommended for all machining operations and, of course, production machining requires the use of carbide-tipped tools.

WELDING

It is possible and practical to weld the spheroidal graphite irons by several welding processes. Most commonly used are the metallic arc and oxacetylene processes. Due to the high carbon content and consequent high hardenability, the material must be handled very carefully to prevent excessive FeC from forming at the welding temperature. If maximum ductility and machinability are desired in the heat-affected zone, the welding operation should be followed by annealing. Filler rods, which undergo no phase changes, such as the 60% Ni–40% Fe rods, are the easiest and most practical to use. The operation requires the lowest practical input of amperage and the use of welding techniques that prevent overheating of the base metal in any location. The annealed grade, containing little or no combined carbon (to harden on cooling), is the grade on which welding is most easily performed.

Brazing

The joining of ductile iron castings to similar or dissimilar metals can be accomplished with silver or copper brazing filler metals, usually without resorting to special techniques in preparing the surfaces to be brazed. Overlaying with hard surfacing alloys also presents no special problems.

Available Forms

Because its normal composition makes it eminently suited, ductile iron is essentially a casting alloy. In this condition a wide variety of control is available on its properties through composition and heat treatment. Ferritic castings may be processed further by hot bending to shapes too difficult to cast conveniently. Tubular shapes may be formed by centrifugal casting and by the hot extruding of billets. The forging, rolling, extruding, and stamping processes may be used to produce bars, shafts, pipe, plate, angles, beams, gear blanks, etc., in order to utilize the good machinability and corrosion resistance of this high-silicon, high-carbon material. Certain shapes have been produced by explosive forming.

Applications

The incorporation into one material of high strength in all section sizes, excellent castability, wear resistance, machinability, and relatively simple control of properties suggests that such material would find application in many different industries and applications. It is lighter than nongraphic ferrous metals and, with its high strength and good castability, permits the design of thin-section, lighter-weight machine components. The largest tonnages are used in automobile and diesel engines for crankshafts, rocker arms, and other engine and clutch components; in farm implements and tractors for gears, sprockets, brackets, transmission casings, housings, and many structural components; in earthmoving machinery for wheels, gears, rope drums; and in heavy machinery for load-carrying components requiring good wear, easy machining, high strength, and antideflection properties. Other applications include fluid-handling devices such as valves and fittings; paper-mill machinery — rolls, roll heads, gears; steel mill rolls and mill equipment; as well as many others in a variety of industries.

ELASTOMERS

Elastomers and rubber are differentiated from polymers by the mechanical property of returning to their original shape after being stretched to several times their length. The rubber industry differentiates between the terms *elastomer* and *rubber* on the basis of how long a deformed material sample requires to return to its approximate original size after a deforming force is removed, and on its extent of recovery. The American Society for Testing and Materials (ASTM) defines an elastomer as "a polymeric material which at room temperature can be stretched to at least twice its original length and upon immediate release of the stress will return quickly to approximately its original length." ASTM D1566 is more specific and quantitative in defining rubber as "a material that is capable of recovering from large deformations quickly and forcibly... (and which), in its modified state, free of diluents, retracts within one minute to less than 1.5 times its original length after being stretched at room temperature to twice its length and held for one minute before release."

The major distinguishing characteristic of elastomers is their great extensibility and high-energy storing capacity. Unlike many metals, for example, which cannot be strained more than a fraction of 1% without exceeding their elastic limit, elastomers have usable elongations up to several hundred percent. Also, because of their capacity for storing energy, even after they are strained several hundred percent, virtually complete recovery is achieved.

Before World War II, almost all rubber was natural. During the war, synthetic rubbers began to replace the scarce natural rubber, and since that time production of synthetics has increased until now their use far surpasses that of natural rubber. There are thousands of different elastomer compounds. Not only are there many different classes of elastomers, but individual types can be modified with a variety of additives, fillers, and reinforcements. In addition, curing temperatures, pressures, and processing methods can be varied to produce elastomers tailored to the needs of specific applications.

Today, the term *rubber* means any material capable of extreme deformability, with more or less complete recovery upon removal of the deforming force. Synthetic materials such as neoprene, nitrile, styrene–butadiene (SBR), sometimes also called Buna S, and GR-S and butadiene rubber are now grouped with natural rubber. These materials serve engineering needs in fields dealing with shock absorption, noise and vibration control, sealing, corrosion protection, abrasion protection, friction production, electrical and thermal insulation, waterproofing, confining other materials, and load bearing.

Of all the available choices, SBR dominates the field, accounting for approximately one-half of all rubber — natural and synthetic — used in the United States. The demand for SBR has been responsible for the building of a massive production capability for this material. More than half of SBR production goes into passenger-car tires in the United States. Natural rubber is used almost exclusively in more demanding areas such as truck, bus, aircraft, and off-highway tires.

In the raw-material or crude stage, elastomers are thermoplastic. There are roughly 20 major classes of elastomers. For example, hard rubbers, which have the highest cross-linking of the elastomers, in many respects are similar to phenolics. In the unstretched state, elastomers are essentially amorphous because the polymers are randomly entangled and there is no special preferred geometrical pattern present. However, when stretched, the polymer chains tend to straighten and become aligned,

TABLE E.1
Properties of Thermoplastic Elastomers

ASTM Test	Property	Polyurethane	Copolyester	Styrene Block Copolymer	Olefin
	Physical				
D792	Specific gravity	1.10–1.24	1.15–1.25	0.93–1.10	0.88–1.00
D792	Specific volume (in.^{3}/lb)	26.5–22.0	—	37.4–27.5	31.5–27.7
D570	Water absorption, 24 h, $1/8$-in. thk (%)	0.1–0.3	0.3–1.6	0.19–0.39	0.01–0.1
	Mechanical				
D638	Tensile strength (psi)	4,000–9,000	4,490–7,600	300–5,000	650–4,000
D638	Elongation (%)	225–570	250–800	250–1,350	180–600
D638	Tensile modulus (10^3 psi)	0.7–35	7–75	0.8–50	0.8–34
D790	Flexural strength (psi)	600–1,000	—	—	—
D790	Flexural modulus (10^3 psi)	—	5–75	4–150	1.5–20
D256	Impact strength, Izod (ft-lb/in. of notch)	No break	No break	No break	No break
D785	Hardness, Shore	65A–80D	35–72D	28–95A	60A–60D
	Thermal				
C177	Thermal conductivity (10^{-4} cal-cm/s-cm^2-°C)	5	3.6–4.5	3.6	4.5–5
D696	Coefficient of thermal expansion (10^{-5} in./in.-°F)	5.6–11	2.0–2.8	7.2–7.6	7.2–9.4
D648	Deflection temperature (°F)				
	At 264 psi	90	115–122	<75	85–90
	At 66 psi	145	129–284	95	120–180
	Electrical				
D149	Dielectric strength (V/mil)				
	Short time, $1/8$-in. thk	450–500	400–460	420–520	400–800
D150	Dielectric constant				
	At 1 kHz	6.7–7.5	3.9–5.1	2.5–3.4	2.2–3.1
D150	Dissipation factor				
	At 1 kHz	0.050–0.060	0.008–0.02	0.001–0.003	0.0006–0.003
D257	Volume resistivity (Ω-cm)				
	At 73°F, 50% RH	2×10^{11}	1.1×10^{12} 1.8×10^4	2.5×10^{16}	$>10^{14}$
D495	Arc resistance (s)	122	—	95	120

Source: *Mach. Design Basics Eng. Design*, June, p. 741, 1993. With permission.

thus increasing in crystallinity. This tendency to crystallize when stretched is related to the strength of an elastomer. Thus, as crystallinity increases, strength also tends to increase (Table E.1).

Neoprene, also known as chloroprene, has the distinction of being the first commercial synthetic rubber. It is chemically and structurally similar to natural rubber, and its mechanical properties are also similar. Its resistance to oils, chemicals, sunlight, weathering, aging, and ozone is outstanding. Also, it retains its properties at temperatures up to 121°C, and is one of the few elastomers that does not support combustion, although it is consumed by fire. In addition, it has excellent resistance to permeability by gases, having about one fourth to one tenth the permeability of natural rubber, depending on the gas. Although it is slightly inferior to rubber in most mechanical properties, neoprene has superior resistance to compression set, particularly at elevated temperatures. It can be used for low-voltage insulation, but is relatively low in dielectric strength. Typical products made of chloroprene elastomers are heavy-duty conveyor belts, V belts, hose covers, footwear, brake diaphragms, motor mounts, rolls, and gaskets. Butyl rubbers, also referred to as isobutylene–isoprene elastomers, are copolymers of isobutylene and 1 to 3% isoprene. They are similar in many ways to natural rubber and are one of the lowest-priced synthetics. They have excellent resistance to abrasion, tearing, and flexing. They are noted for low gas and air permeability (about 10 times better than natural rubber), and for this reason make a good material for tire inner tubes, hose, tubing, and diaphragms. Although butyls are non-oil-resistant, they have excellent resistance to sunlight and weathering, and generally have good chemical resistance. They also have good low-temperature flexibility and heat resistance up to around 149°C; however, they are not flame resistant. They generally have lower mechanical properties such as tensile strength, resilience, abrasion resistance, and compression set, than the other elastomers. Because of their excellent dielectric strength, they are widely used for cable insulation, encapsulating compounds, and a variety of electrical applications. Other typical uses include weather stripping, coated fabrics, curtain wall gaskets, high-pressure steam hoses, machinery mounts, and seals for food jars and medicine bottles.

Isoprene is synthetic natural rubber. It is processed like natural rubber and its properties are quite similar, although isoprene has somewhat higher extensibility. Like natural rubber, its notable characteristics are low hysteresis, low heat buildup, and high tear resistance. It also has excellent flow characteristics, and is easily injection molded. Its uses complement those of natural rubber, and its good electrical properties plus low moisture absorption make it suitable for electrical insulation. Polyacrylate elastomers are based on polymers of butyl or ethyl acrylate. They are low-volume use, specialty elastomers, chiefly used in parts involving oils (especially sulfur-bearing) at elevated temperatures of 149°C and even as high as 204°C. A major use is for automobile transmission seals. Other oil-resistant uses are gaskets and O rings. Mechanical properties such as tensile strength and resilience are low. Further, except for recent new formulations, they lose much of their flexibility below –23°C. The new grades extend low-temperature service to –40°C. Polyacrylates have only fair dielectric strength, which improves, however, at elevated temperatures.

Nitrile elastomers, or NBR rubbers, known originally as Buna N, are copolymers of acrylonitrile and butadiene. They are principally known for their outstanding resistance to oil and fuels at both normal and elevated temperatures. Their properties can be altered by varying the ratio of the two monomers.

Most commercial grades range from 20 to 50% acrylonitrile. Those at the high end of the range are used where maximum resistance to fuels and oils is required, such as in oil well parts and fuel hose. Low-acrylonitrile grades are used where good flexibility at low temperatures is of primary importance. Medium-range types, which are the most widely used, find applications between these extremes. Typical products are flexible couplings, printing blankets, rubber rollers, and washing machine parts. Nitriles as a group are low in most mechanical properties. Because they do not crystallize appreciably when stretched, their tensile strength is low, and resilience is roughly one third to one half that of natural rubber. Depending on acrylonitrile content, low-temperature brittleness occurs at from –26 to –60°C. Their electrical insulation quality varies from fair to poor. Polybutadiene elastomers are notable for their low-temperature performance. With the exception of silicone, they have the lowest brittle or glass transition temperature, –73°C, of all the elastomers. They are also one of the most resilient, and have excellent abrasion resistance. However, resistance to chemicals, sunlight, weathering, and permeability by gases is poor. Some uses are shoe heels, soles, gaskets, and belting. They are also often used in blends with other rubbers to provide improvements in resilience, abrasion resistance, and low-temperature flexibility.

Polysulfide elastomer is rated highest in resistance to oil and gasoline. It also has excellent solvent resistance, extremely low gas permeability, and good aging characteristics. Thus, it is used for such products as oil and gasoline hoses, gaskets, washers, and diaphragms. Its major use is for equipment and parts in the coating production and application field. It is also widely applied in liquid form in sealants for the aircraft and marine industries. Its mechanical properties, including strength, compression set, and resilience, are poor. Although it is poor in flame resistance, it can be used in temperatures up to 121°C. Ethylene–propylene elastomers, or EPR rubber, are available as copolymers and terpolymers. They offer good resilience, flexing characteristics, compression-set resistance, and hysteresis resistance, along with excellent resistance to weathering, oxidation, and sunlight. Although fair to poor in oil resistance, their resistance to chemicals is good. Their maximum continuous service temperature is around 177°C. Typical applications are electrical insulation, footwear, auto hose, and belts. Urethane elastomers are copolymers of diisocyanate with their polyester or polyether. Both are produced in solid gum form and viscous liquid. With tensile strengths above 34 MPa and some grades approaching 49 MPa, urethanes are the strongest available elastomers. They are also the hardest, and have extremely good abrasion resistance. Other notable properties are low compression set, and good aging characteristics and oil and fuel resistance. The maximum temperature for continuous use is under 93°C, and their brittle point ranges from –51 to –68°C. Their largest field of application is for parts requiring high wear resistance or strength. Typical products are forklift truck wheels, airplane tail wheels, shoe heels, bumpers on earth-moving machinery, and typewriter damping pads. Chlorosulfonated polyethylene elastomer, commonly known as Hypalon, contains about one third chlorine and 1 to 2% sulfur. It can be used by itself or blended with other elastomers. Hypalon is noted for its excellent resistance to oxidation, sunlight, weathering, ozone, and many chemicals. Some grades are satisfactory for continuous service at temperatures up to 177°C. It has moderate oil resistance. It also has unlimited colorability. Its mechanical properties are good but not outstanding, although abrasion resistance is excellent. Hypalon is frequently used in blends to improve oxidation and ozone resistance. Typical uses are tank linings, high-temperature conveyor belts, shoe soles and heels, seals, gaskets, and spark plug boots. Epichlorohydrin elastomers are noted for their good resistance to oils, and excellent resistance to ozone, weathering, and intermediate heat. The homopolymer has extremely low permeability to gases. The copolymer has excellent resilience at low temperatures. Both have low heat buildup, making them attractive for parts subjected to repeated shocks and vibrations. Fluorocarbon elastomers, fluorine-containing elastomers, like their plastic counterparts, are highest of all the elastomers in resistance to oxidation, chemicals, oils, solvents, and heat, and they are also the highest in price. They can be used continuously at a temperature 127°C and do not support combustion. Their brittle temperature, however, is only –23°C. Their mechanical and electrical properties are only moderate. Unreinforced types have tensile strengths of less than 13 MPa and only fair resilience. Typical applications are brake seals, O rings, diaphragms, and hose. The phosphonitrile plastics and elastomers have high elasticity and high temperature resistance. They are derived from chlorophosphonitrile, or phosphonitrilic chloride, $P_3N_3Cl_3$, are highly resistant to oils and solvents, and remain flexible and serviceable at temperatures from –57 to 177°C.

Viton is a vinylidene fluoride hexafluoropropylene tetrafluoroethylene copolymer as well as Fluorel with extra high resistance to solvents, hydrocarbons, steam, and water.

Silicone elastomers are polymers composed basically of silicon and O_2 atoms. There are four major elastomer composition groups. In terms of application, silicone elastomers can be divided roughly into the following types: general-purpose, low-temperature, high-temperature, low-compression-set, high-tensile, high-tear, fluid-resistant, and room-temperature vulcanizing. All silicone elastomers are high-performance, high-price materials. The general-purpose grades, however, are competitive with some of the other specialty rubbers, and are less costly than the fluorocarbon elastomers. Silicone elastomers are the most stable group of all the elastomers. They are outstanding in resistance to high and low temperatures, oils, and chemicals. High-temperature grades have maximum continuous service temperatures up to 316°C; low-temperature grades have glass transition temperatures of −118°C. Electrical properties, which are comparable to the best of the other elastomers, are maintained over a temperature range from −73°C to over 260°C. However, most grades have relatively poor mechanical properties. Tensile strength runs only around 8 MPa. However, grades have been developed with much improved strength, tear resistance, and compression set. Fluorosilicone elastomers have been developed that combine the outstanding characteristics of the fluorocarbons and silicones. However, they are expensive and require special precautions during processing. A unique characteristic of one of these elastomers is its relatively uniform modulus of elasticity over a wide temperature range and under a variety of conditions. Silicone elastomers are used extensively in products and components where high performance is required. Typical uses are seals, gaskets, O rings, insulation for wire and cable, and encapsulation of electronic components.

ELASTOMERIC LININGS

Elastomeric lining materials are available in natural rubber, GR-S, butyl, neoprene, nitrile, and Hypalon synthetic rubbers as well as in polyvinyl chloride (PVC) and polyvinylidene chloride (saran) plastics and fluorocarbon rubbers. These basic materials, when formulated and processed into linings, are normally sold and applied as heavy sheets at 4.7 to 6.4 mm thick.

Elastomeric linings are manufactured by plying up thin calendered films of 0.038 to 0.076 mm thickness to the full thickness of the lining.

They are applied in the soft and plastic uncured state to the base metal and are cured in place.

NATURAL RUBBER

About 70% of the applied linings are estimated to be based on natural rubber. This is due to the fact that natural rubber can be formulated in an extremely wide range of physical properties. Linings from natural rubber are classified as soft (sulfur content at 2 to 3%), semihard (sulfur content at 15 to 30%), and hard (sulfur content at 30 to 45%). As the sulfur content is increased, general chemical resistance increases from good to excellent, but at the same time the lining also becomes harder and less flexible.

Linings from natural rubber are among the lowest in cost, and all natural rubber linings can be applied without difficulty. Even hard rubber can be used satisfactorily on large vessels if the vessel is designed in sections that are then bolted together and the joints covered with lap straps. With proper usage and suitable repair when damaged, they can be expected to give service for 12 to 15 years or more, which makes for a very low overall cost.

Natural rubber linings give excellent service for many water solutions of chemicals and can be used for practically all plating solutions (except chromium). Hard rubber linings will also withstand paraffenic oils, greases, and fats at moderate temperatures. The linings will not withstand the solvating action of aniline, benzen, carbon disulfide, carbon tetrachloride, and other chlorinated or halogenated hydrocarbons. They are not resistant to the chemical attack of the oxidizing acids, HNO_3 above 10% or H_2SO_4 acid above 50% at 71°C.

GR-S

With the exception of gum (nonfilled) rubber linings, GR-S can be used to duplicate practically all of the linings that are now made from natural rubber and for the same type of service.

BUTYL RUBBER

Butyl rubber linings are available in physical characteristics approaching those of the soft natural rubbers.

Butyl linings have a higher degree of general chemical resistance than natural rubber, and are suitable for weak oxidizing acids and for organic chemical that deteriorate soft rubber linings over a long period of time (e.g., fats, greases, and soaps).

Butyl linings are suitable for use above 71°C and for continued use at 149°C as a lining material. These characteristics allow butyl use in tank trucks and railroad tank cars where hot solutions are to be transported, and where the empty car can be returned without fear of rupture of the lining by shock in cold climates. Also, butyl is used where the continual vibration and shock encountered in railroad hauling eliminates the semihard and hard rubbers that would normally be used for such solutions. HF_3 and ethylene chlorohydrin can be handled with butyl linings.

Neoprene

Neoprene linings have gained acclaim for their resistance to caustic solutions and are recommended for use with fluorides and phosphates. Flexible neoprene linings also have good resistance to a large group of chemicals such as all fatty acids, oils, greases, and aliphatic hydrocarbons; they are superior to butyl rubber in this respect. Such resistance enables neoprene linings to give good service in chemical processes where mixtures of acids with kerosene, oils, or other organic materials are involved.

Because neoprene is more expensive than natural rubber, it is normally used only where it gives service superior to natural rubber. Neoprene linings are specifically used where flexibility is needed above 71°C.

Nitrile Rubber

Nitrile rubber linings, because of their relatively high cost, are used only with specific organic solvents, where organic solvents are used in conjunction with acids and other water-based corrosive systems and where superior oil resistance is needed. The general chemical resistance of nitrile rubbers is only fair, but their resistance to oils is good.

Hypalon

Hypalon linings are flexible and are useful because of their resistance to heat as well as to oxidizing chemicals. Hypalon linings are especially suitable for use with hypochlorites and H_2SO_4; cold solutions of H_2SO_4 up to 90% exhibit very little attack upon Hypalon. It probably is the best lining available for hot solutions of H_2SO_4 above 50% concentration.

Polyvinyl Chloride (PVC)

PVC linings are available both in the flexible and rigid state. Flexible PVC is used almost exclusively where H_2CrO_3 solutions are involved in chromium plating, and where HNO_3 or mixtures of HNO_3 and HF acids are involved in the electropickling of stainless steels. No other linings are suitable for these operations. PVC linings are also suitable where mixtures of water solutions of corrosive chemicals with oils or specific organic solvents are involved.

Similar to PVC are the saran (polyvinylidene chloride) linings, which look like and are applied in much the same manner as PVC. They are good for use with HCl.

Application

Elastomeric linings are applied in much the same manner as wallpaper. In the application of rubber linings, it is important that the lining be applied when the rubber is uncured and is plastic and workable.

PVC and polyvinylidene chloride linings are applied similarly to rubber linings but they are not cured. They are applied with butt joints covered with a narrow sealing strip.

Uses

Elastomeric linings can be compounded to provide specific corrosion resistance for many applications and can be applied to almost any shape of equipment or surface under almost any set of conditions and at any location.

Those industries using the highest volume of linings include:

1. Chemical process industry, where practically every type of lining is utilized for a wide variety of chemicals
2. The steel and aluminum industries, where the linings are used in pickling and anodizing operations
3. In the automotive and appliance industries, for plating
4. In ordnance, where missile fuels are involved
5. The food processing industry, where specifically compounded linings are used in smaller volume

ELECTRICAL CERAMICS

Electrical ceramics, like all other ceramics, consist of randomly oriented small crystallites, bonded together by either a glassy matrix or by close interlocking of one or several crystalline phases. The atomic structures of the crystalline and glassy phases determine the physical properties of the finished ceramic.

RAW MATERIALS

Conventional raw materials for ceramic production are earthy, natural minerals, used in finely-ground form, such as clays, talc, feldspar, and flint. These are used for the production of electrical insulators, often carefully selected and refined from gross impurities by flotation or filtering processes. In many instances, these naturally occurring minerals are replaced by synthesized inorganic compounds, which are prepared from various oxides. This may be done by melting, recrystallizing, and grinding, or more frequently by high-temperature solid-state reactions.

These compounds may be composed of oxides of practically any metal, singly or in combination, or they may be refractory compounds such as carbides and silicides. The use of these inorganic synthetic raw materials has opened the field of likely ceramic compositions to practically infinite numbers, and the possibility of variation in physical and electrical properties is therefore equally staggering.

FABRICATION TECHNIQUES

The compositions, which may consist of either finely-ground minerals or synthetic compounds or combinations of both, are blended into so-called bodies. These are then shaped into desired forms and fired. The resultant products are hard, dense, and brittle. Further shaping is possible only by grinding. Other compositions can be formed by glass-shaping techniques, such as casting in the molten state or hot pressing.

The demand for complex shapes, close dimensional tolerances, and controlled physical and chemical properties has resulted in new developments in forming techniques. The older methods of slip casting and jiggering are still used for forming electrical porcelain, which contains a considerable amount of plastic clays.

Other methods of forming, particularly adaptable to nonplastic compositions and to line production techniques, are automatic dry pressing, extruding, and injection or compression molding. Film casting and stamping of ceramic sheets is another method used for the production of accurately formed thin ceramic shapes, such as capacitor dielectrics or vacuum tube spacers.

MATERIALS

Electrical Porcelain

Porcelain is the outstanding ceramic insulator, because it combines mechanical stability and strength with heat and arc resistance and the ability to resist the passage of an electric current.

Conventional electrical porcelain is made of the natural minerals clay, flint, and feldspar and is very similar in composition to porcelain used for high-quality vitrified dinnerware. After forming, the ceramic is fired to complete vitrification to develop maximum mechanical and dielectric strength. For outdoor use and at locations exposed to humidity, it is desirable to use insulators with glazed surface. The glaze consists essentially of the same components as the porcelain itself, but in different proportions.

Electrical porcelain is a satisfactory insulator for low-tension electrical wiring systems, for outlets and switches, lamp sockets, and electrical appliances. It is unsurpassed for insulation of outdoor high-voltage power transmission and distribution systems, and is used extensively for suspension and pin-type insulators. Porcelain high-tension insulators also find numerous applications in transformers and as lead-in insulators.

High-Frequency Insulation

A dielectric material for use at high frequencies must have the additional characteristic of low dielectric loss, and its properties must not be affected by changes over the required temperature range. Porcelain has a rather high dielectric loss under high frequency conditions, but a number of low-loss ceramic materials have been developed especially for high-frequency insulation. It is generally agreed that a low-loss ceramic body should consist chiefly of uniformly small crystallites bonded with only a very small amount of glassy matrix. Typical special ceramics are frequently named according to the predominant crystalline phase in the ceramic structure, which is the chief contributor to the specific properties of the ceramic product.

Among the best-known high-frequency ceramics are *Steatite* ceramics ($MgO \cdot SiO_2$). They are based on the mineral talc or steatite, a hydrous magnesium silicate. Other low-loss ceramic compositions are *Forsterite* ($2MgO \cdot SiO_2$), *Wollastonite* ($CaO \cdot SiO_2$), *Zircon* ($ZrO_2 \cdot SiO_2$), *Mullite* ($3Al_2O_3 \cdot 2SiO_2$), and *Spinel* ($MgO-Al_2O_3$).

Sintered Al_2O_3

Highest physical and dielectric strength and low dielectric loss over a wide temperature range is found in a sintered aluminum oxide or Al_2O_3 ceramic. The main crystalline phase is corundum Al_2O_3.

Sintered Al_2O_3 ceramics are impervious to gases and meet the very exacting requirements of spark plug insulation, of envelopes for ultra-high-frequency receiving and power vacuum tubes. Paper-thin wafers of sintered Al_2O_3 are used as wire supports in vacuum tubes, as a substitute for stamped natural mica spacers.

Thermal Shock-Resistant Electrical Ceramics

Numerous applications in the low- and high-frequency field demand ceramics that will withstand sudden heat shock. For example, insulation for electrical appliances such as toasters, ovens, also electric arc chambers and switch gears, thermocouple insulation, and supports for electrical resistors, are subjected to sudden thermal changes. Low thermal expansion, combined with high heat conductivity, is desirable in ceramics for such applications. The outstanding material in this group is *Cordierite* ($2MgO \cdot 2Al_2O_3 \cdot 5SiO_2$). Other ceramics of low thermal expansion are based on the mineral β-Spodume, a complex lithium aluminosilicate. Both groups of ceramics have lower mechanical strength than steatite and sintered Al_2O_3, but their excellent thermal shock resistance makes these materials very useful for the applications just mentioned.

Glass

A unique combination of properties has made glass an indispensable material of construction in the electrical industry. Among its most important properties are transmission of light, imperviousness to gases, ability to seal readily to metals, good dielectric properties, and ease of fabrication into many shapes. The chemical compositions of glasses vary widely, but in general it can be stated that silica (SiO_2) is the foundation of most commercial glasses, fused with such metallic oxides as soda, potash, Ca, Pb, Ba, Mg, and B.

The properties of glasses are primarily determined by their composition. The most widely used glasses for electrical insulation are of the soda-lime-silicate type, the lead glasses, and the boron alumino silicate (Pyrex) glasses. The major characteristics of the latter type are good mechanical strength, low thermal expansion, good weathering stability, and good dielectric properties.

Glass insulators are used extensively as line insulators for open wire lines, radio antennas, and for envelopes of incandescent lamps, vacuum tubes, mercury switches, and other devices. In the electronic field, glass serves as an insulation basis for electronic components, such as resistors and inductors and as a dielectric for capacitors.

Certain glasses can be converted to finely crystalline bodies bonded together with a vitreous matrix or by fusion of the crystallites at their grain boundaries (Pyroceram). Articles to be converted from glass to a crystalline ceramic are first fabricated in the glassy state with special nucleating agents added to the glass composition to form crystallites in the body. After forming, crystallization is effected by either exposure to ultraviolet radiation and temperature treatment, or by heat treatment alone. The outstanding advantage of this new group of material is that they can be fabricated into a variety of shapes by conventional glass-forming methods and have the higher strength and stability of crystalline ceramics.

Fibrous glass is an important insulating material for electrical equipment. It may be used in the form of yarn, tapes, sleeving, or for reinforcement of organic plastics. On the other hand, plastics improve abrasion resistance and dielectric strength of fibrous glass constructions.

Ceramic Papers

There are a number of ceramic papers on the market that serve as high-temperature electrical insulation in a similar way as fibrous glass. These papers are made from reconstituted mica flakes (Samica, Mica Mat), clays, or glass and mineral (mullite) fibers.

Glass-Bonded Mica

As the name suggests, this group of materials consists of either natural or synthetic mica flakes that are bonded under heat and pressure by a glassy matrix. Vitrified ceramics can only be lapped or ground with the hardest abrasives, such as SiC, corundum, or diamond powder. In contrast, glass-bonded mica can be subjected to all normal machining operations. The material is impervious to moisture, has high dielectric strength, and can be molded to accurate dimensions. Metal inserts can be molded directly into mica-bonded articles. This is a distinct advantage over other ceramics, which require assembly or attachment of metal parts after processing.

Electronic Ceramics

Great progress has been made during the past decades in the area of solid-state research. Solid-state components are finding many applications in electronics. Some of these, such as diodes, rectifiers, or transistors, are based on single crystals, germanium and silicon, grown from a melt and cut into required shape. These are not considered ceramics, but others based on polycrystalline systems and fabricated by ceramic processes are considered among electric ceramics.

Ferroelectric Ceramics

Ferroelectricity, the spontaneous alignment of electric dipoles under the influence of an electric field, is an outstanding property of certain ceramic materials, especially those based on barium titanate. Besides a barium titanate, additional ferroelectric materials were discovered in the field of niobates, tantalates, and zirconates. Ferroelectric ceramics exhibit a high dielectric constant, which makes them attractive for dielectrics in capacitors. The dielectric constant of these ceramics can be varied through compositional changes; values as high as 10,000 at room temperature can be obtained. Very high dielectric constant ceramics have rather steep negative temperature coefficients of capacity. The capacitance of special barium titanate capacitors is sensitive to applied voltage, which suggests their use in dielectric amplifiers. Nonlinear capacitors of this type have also been employed for frequency modulation and remote tuning devices. Ferroelectric ceramics can be made to show piezoelectric effects, i.e., the ability to convert mechanical strain into electric charges and, conversely, the ability to transform a voltage into mechanical force. This is produced by exposing the material to an orienting electric field during the cooling period after firing. Piezoelectric ceramics of the barium titanate type have found applications for photograph pickups, ultrasonic thickness gauges, accelerometers, ultrasonic cutting tools, and even in electromedical instruments for measuring heart conditions.

Ferromagnetic Ceramics

Ferromagnetic ceramics, also known as ferrites, are compounds of various metal oxides and have the general formula $MO \cdot Fe_2O_3$, where M stands for a bivalent metal ion, such as zinc, nickel, magnesium, and others. They are ceramic materials with a crystalline structure of the spinel ($MgO \cdot Al_2O_3$) type. The mineral magnetite ($FeO \cdot Fe_2O_3$) is the only naturally occurring mineral of this type and has been well known and used for ages as lodestone for its magnetic properties. In contrast to metallic magnetic materials, ceramic ferrites have high-volume resistivity and high permeability. Their specific gravity is between 4 and 5, considerably less than iron (8). They can be made both into "soft" and "hard" permanent magnetic materials. The properties of soft ferrites can be varied over a wide range to meet specific application requirements. For cores of radio and television loop antennas they are made with emphasis on a high quality (Q) factor, to attain optimum in reception quality and selectivity. For memory cores in electron computers, they are so compounded that they exhibit a square hysteresis loop of magnetization and have an extremely low switching time between magnetic saturation and demagnetization. Ceramic magnets are about 35% lower in density than metallic magnets, an important factor for military airborne applications.

Resistor Ceramics

Certain nonmetallics pass limited amounts of electricity and therefore are useful as electrical resistors. Some of these have advantages over metals in that they are more resistant to oxidation and are therefore useful at higher temperatures.

SiC, either self-bonded or bonded with silicon in form of rods or tubes, is used for resistor heating elements in electric furnaces. These can be heated to 1400°C in air for indefinite periods of time and may be heated as high as 1600°C for shorter periods.

SiC as a resistor element shows nonlinear characteristics and is used where a nonohmic variation of current is desired (Thyrite resistors). Nonlinear resistors of this type are chiefly used for voltage regulation and current suppression, such as in lightning arrestors.

Many oxides become electrically conductive at elevated temperatures. The electrical conductivity of oxides is governed to a large degree by the amount of impurities present. For example, the electrical resistance of thoria is greatly reduced by the presence of such oxides as ceria, yttria, erbia, etc. Thoria and ZrO_2 resistors have been used as electric furnace heating elements up to 2000°C. These oxides are nonlinear with temperature change and therefore have found applications as thermistors. These components are used as temperature sensors in pyrometers, temperature bridges, and microwave power meters.

Other Semiconductor Ceramics

Semiconductive ceramics of silicide, telluride, and oxide compositions have been used for thermoelectric devices, to convert heat directly into electrical energy (Seebeck effect) and for refrigeration (Peltier effect).

ELECTRICAL-CONTACT MATERIALS

These are materials used to make or break electrical contact, thus make-and-break electric circuits, or to provide sliding or constant contact. Both require high electrical conductivity to ease current flow, high thermal conductivity to dissipate heat, high melting point or range to inhibit arc erosion and prevent sticking, corrosion and oxidation resistance to prevent formation of films that impede current flow, high hardness for wear resistance, and amenability to welding, brazing, or other means of joining. The sliding contact types also require low friction, and a lubricant is always required between the sliding materials to prevent seizing and galling.

The materials used range from pure metals and alloys to composites, including those made by powder metallurgy (P/M) methods. Copper is widely used but requires protection from oxidation and corrosion, such as by immersion in oil, coating, or vacuum sealing. Cu–W alloys or mixtures of Cu–Gr increase resistance to arcing and sticking and some copper alloys provide greater hardness, thus greater wear resistance and better spring characteristics. Silver is more oxidation resistant in air and, pure or alloyed, is the most widely used metal for make-and-break contacts for application at currents to 600 A. Ag–Cu alloys provide greater hardness but less conductivity and oxidation resistance; Ag–Cd alloys increase resistance to arc erosion and sticking; and Ag–Pt alloys, Ag–Pd alloys and Ag–Au alloys increase hardness, wear resistance, and oxidation resistance. All alloying elements, however, decrease conductivity. Gold has outstanding oxidation and sulfidation resistance but, being soft and prone to wear and arc erosion, is limited to low-current (0.5 A maximum) applications. To enhance these properties, gold alloys, such as Au–Ag, Au–Cu, Au–Ag–Pt, Au–Ag–Ni, and Au–Cu–Pt–Ag are more commonly used. Platinum and palladium are also used for contacts, but, again, in alloy form more than as pure metals. Among the most common ones are Pt–Ir, Pt–Ru, Pt–Pd–Ru, Pd–Ru, Pd–Cu, and Pd–Ag. A Pd–Ag–Pt–Au–Ag alloy for brushes and sliding contacts is noted for its exceptional modulus of elasticity and high proportional limit. Aluminum, tungsten, and molybdenum are also used for electrical contacts but mainly in composite form. Al used for contacts provides an electrical conductivity of about 60% that of copper, but is prone to oxidation and, thus, is clad or plated with silver, tin, or copper. The refractory metals, although providing excellent resistance to wear and arc erosion, are poor conductors and oxidize readily.

The principal metals made in composite form by P/M methods are the refractory metals, including those in carbide form, and copper-base and silver-base metals. The refractory metals, notably tungsten and molybdenum or their carbides, usually serve as a base for infiltrating with copper or silver, thus combining electrical conductivity and resistance to wear and arc erosion in a single material. Many such composites are common, including W–Cu, W–Ag, WC–Ag, WC–Cu, W–Gr–Ag, and Mo–Ag. The amount of conductive metal may exceed or be less than that of the refractory metal or refractory-metal carbide. A common silver composite is Ag–CdO, which, for a given amount of silver, provides greater conductivity than Ag–Cd alloys as well as greater hardness and resistance to sticking. Others include Ag–Gr, Ag–Ni, and Ag–Fe. The Ag–Gr composites are used mainly for sliding or brush contacts.

ELECTRICAL-RESISTANCE METALS AND ALLOYS

This major family of metals, including alloys as well as pure metals, includes alloys used in controls and instruments to measure or regulate electrical performance, heating alloys used to generate heat, and thermostat metals used to convert heat to mechanical energy. There are seven types of electrical-resistance alloys: (1) radio alloys, which contain 78 to 98% Cu with the balance Ni; (2) manganins, 87% Cu and 13% Mn or 83 to 85% Cu, 10 to 13% Mn, and 4% Ni; (3) constantans, 55 to 57% Cu and 43 to 45% Ni; (4) Ni–Cr–Al alloys, 72 to 75% Ni, 20% Cr, 3% Al, and either 5% Mn or 2% Cu, Fe, or Mn; (5) Fe–Cr–Al alloys, 73 to 81% Fe, 15 to 22% Cr, 4 to 5.5% Al; (6) various other alloys, mostly nickel-base alloys, some of which contain substantial amounts of chromium or iron, chromium and iron, chromium and silicon, and, in some cases, magnesium and aluminum, and (7) pure metals, notably aluminum, copper, iron, nickel, precious metals, and refractory metals.

Key characteristics of resistance alloys are uniform resistivity, stable resistance, reproducible temperature coefficients of resistance, and low thermoelectric potential vs. copper. Less critical, but also important, are coefficient of thermal expansion (CTE); strength and ductility; corrosion resistance; and joinability to dissimilar metals by welding, brazing, or soldering. Heating alloys require high heat resistance, including resistance to oxidation and creep in particular environments, such as furnaces, in which they are widely used; high electrical resistivity; and reproducible temperature coefficients of resistance. Also desirable are high emissivity, low CTEs, and low modulus to minimize thermal fatigue, strength, and resistance to thermal shock and ductility for fabricability. Thermostat metals, two or more bonded materials of which one may be nonmetallic, are chosen based on different electrical resistivities and thermal expansivities so that applied heat can be converted to mechanical energy.

The electrical resistivity in ohm·circular nohm·m for the alloys encompassed by the seven groups range from 16 for silver to 1450 for an alloy made up of 72.5% Fe, and 5.5% Al. Iron has the highest resistivity, 970, of the pure metals, followed by Ta, 135; Pt, 106; Ni, 80; and W, 55. The radio alloys are in the 50 to 300 range, the manganins 380 to 480, the constantans 490 to 500, and most of the Ni–Cr–Al, Fe–Cr–Al, and various other alloys are in the 1015 to 1450 range. Temperature coefficients of resistance in parts per million per degree Celsius range from ±10 to ±15 at 15 to 45°C for the manganins to +6000 at 20 to 35°C for pure nickel. Thermoelectric potentials vs. copper in the µV/°C range from –43 at 25 to 105°C for the 57% Cu–43% Ni constantan to +12.2 at 0 to 100°C for pure Fe. The refractory metals have the lowest CTEs –4.3 µm/m·°C while W–Al has the highest (23.9 µm/m·°C), and most of the alloys are intermediate in this respect –11 to 19 µm/m·°C. Tensile strength and ductility also range widely depending on the alloy or metal. Maximum operating temperatures in air for the commonly used resistance-heating alloys range from 927°C for 43.5% Fe–35% Ni–20% Cr–1.5% Si alloy to 1374°C for 72.5% Fe–22% Cr–5.5% Al. For platinum, this temperature is 1510°C. The refractory metals are suitable for still higher temperatures in vacuum and, in the case of molybdenum and tungsten, in select environments.

Electrical-resistance alloys are mainly wire products, and the alloys have been known by a multitude of trade names. The standard alloy for electrical-resistance wire for heaters and electrical appliances in Ni–Cr, but Ni–Mn and other alloys are used. For high-temperature furnaces, tungsten, molybdenum, and alloys of the more expensive high-melting metals are employed. The much-used alloy with 80% Ni and 20% Cr resists scaling and oxidation to 1177°C, but it is subject to an intergranular corrosion, known as green rot, which may occur in chromium above 816°C unless modified with other elements. The 60 to 80% Ni–Cr family of alloys are used in heater elements, resistors, rheostats, resistance thermometers, and in potentiometers.

The Cr–Al–Fe alloys have high resistivity and high oxidation resistance, but have a tendency to become brittle. The Kanthal alloys have 20 to 25% chromium with some cobalt and aluminum, and the balance iron. Kanthal A, with 5% aluminum, will withstand temperatures to 1299°C and is resistant to H_2SO_4. The tensile strength is 813 MPa with elongation of 12 to 16%. Kanthal A-1, for furnaces, has an operating temperature to 1373°C.

Cu–Mn alloys have high resistivity and alloys with 96 to 98% manganese, although they may be brittle and difficult to make into wire. Addition of nickel makes them ductile, but lowers resistivity. A typical alloy contains 35% Mn, 35% Ni, and 30% Cu.

Other alloys are used for rheostats, electrical heaters for 1093°C temperature operations, and thermocouples up to 1250°C.

ELECTROCHEMICAL PROCESS

The principles of electrochemistry may be adapted for use in the preparation of commercially important quantities of certain substances, both inorganic and organic in nature.

INORGANIC PROCESSES

Inorganic chemical processes can be classified as electrolytic, electrothermic, and miscellaneous processes including electric discharge through gases and separation by electrical means. In electrolytic processes, chemical and electrical energy are interchanged. Current passed through an electrolytic cell causes chemical reactions at the electrodes. Voltaic cells convert chemicals into electricity. Electrothermic processes use electricity to attain the necessary temperature for reaction.

Voltaic cells are used for the intermittent production of small amounts of electricity. When the chemicals involved are exhausted and must be replaced, the unit is called a primary cell. A special case of the primary cell is the fuel cell in which the fuel and anoxidizer are fed continuously to the cell, converted to electricity, and the products removed. If exhausted components can be revived by passing electricity backward through the unit, it is called a secondary cell, storage battery, or accumulator. Cells may be connected in parallel or in series to form a battery.

ORGANIC PROCESSES

Organic electrochemistry was once regarded as a tantalizing area with many important laboratory achievements but few successes in commercial practice. This situation has changed, however, in that electroorganic processes are commercially advantageous if they can fulfill either of two conditions: (1) performance under conditions of voltage corresponding thermodynamically to the conversion of an organic group to a reduced or oxidized group, with the cell products relatively easy to isolate and purify; (2) performance of a highly selective, specific technique to make an addition at a double bond, or to split a particular bond (for example, between carbon atoms 17 and 18 of a complex molecule having 25 carbon atoms).

Selectivity and specificity are highly important in electroorganic processes for the manufacture of complicated molecules of vitamins and hormones—as well as for the medicinal products whose action on pathogenic organisms is a function of their spiral arrangement, steric forms, and resonance.

The electrolytic oxidation and reduction of organic compounds differ from the corresponding and more familiar inorganic reactions only in that organic reactions tend to be complex and have low yields. The electrochemical principles are precisely those of inorganic reactions, while the procedures for handling the chemicals are precisely those of organic chemistry.

Oxidations

Commercial success in inorganic electrochemistry has come about by well-engineered combinations of organic and inorganic techniques in areas where strictly chemical methods are either impossible or inefficient, for example, in catalytic hydrogenation or oxidation.

Reductions

Substances that are easy to reduce may be acted on, at the interface of cathodes, with low H_2 overvoltage. Hard-to-reduce materials may require much higher overvoltages, which are reached through either the cathode composition or the current density.

ELECTROFORMED PARTS

Electroforming is essentially an electrolytic plating process for manufacturing metal parts. In general, it is best to consider electroforming for applications where a part is impossible or difficult to make by any other standard method or if the tooling required by another method such as forging or die-casting is extremely expensive. Electroforming is not generally used for large-quantity production because it is a batch operation. It is, however, valuable for short-run, simple parts or long-run, complicated parts because of its low tooling costs.

Electroforming is the technology of creating exact (mirror image) copies of uniquely shaped objects by electrodepositing a layer of heavy metal onto an original and subsequently separating the two.

Self-supporting parts can be made from such metals as Ni, Cu, Fe, Ag, Cr, and Au. In all instances, the pure metal has been found to be the only practical deposit obtainable. There are, however, a number of bimetallic deposits such as cobalt and/or nickel and tungsten.

Nickel is neither the toughest nor the most corrosion-resistant metal, and yet it is becoming more popular with plastic molders. Uniquely suited for one application in particular—mold insert electroforming—nickel makes possible the cost-effective mass production of many of today's advanced products. Today the process is used to manufacture so-called Fresnel lenses, used in overhead projectors, as well as many other plastic optical products. A single precision-machined mandrel generates numerous inserts, assuring that each molded optical component exhibits the same geometric characteristics and surface finish essential for consistent optical performance. Nickel molds electroformed have high dimensional stability (meaning they display minimal warping), have an optical-quality surface finish, and exhibit the durability necessary to produce thousands of precision lenses using only a few electroformed mold inserts.

Among the easiest metals to electroplate out of environmentally benign aqueous solutions, nickel is the logical choice for electroformed tooling. Nickel is also sufficiently strong, hard, and tarnish-resistant to withstand molding conditions encountered in the processing of many popular plastics. Moreover, it is also easily machinable, brazeable, and weldable.

ADVANTAGES

Following are the chief advantages of electroforming:

1. Initial tooling costs are extremely low. This has several advantages. Short-run parts can be produced to determine market acceptability and, as is the case in missile and aircraft applications, small numbers of parts may be produced economically. The low initial tooling cost makes it economically possible to try various shapes and configurations without undue tooling charges.
2. Because this is an electrochemical process, the surface finish of the

mandrel is duplicated exactly, thereby providing a means of producing parts to any surface finish—from a high polish to a surface resembling gravel. Many examples of these are found in industry. Among them are molds to make records, plastic tile resembling different fabric textures, and surface finish standards.
3. Complex shapes such as ducting and tubing may be made in one piece, thereby eliminating welding and soldering.
4. Parts requiring extremely close tolerances are easily made by the electroforming process, inasmuch as the mandrel or shape upon which the metal is deposited may be machined or produced to extremely close tolerances, inspected easily, and thereby duplicated exactly. Examples of extremely close tolerance parts that can be made are radar plumbing, hot-air ducting and tubing, wind-tunnel test nozzles and liners, reflectors, collectors, mirrors, nose cones, etc.

SIZES

Parts produced by electroforming range in size from miniature to extremely large. Typical miniature parts such as small electronic devices are 0.50 to 0.76 mm in diameter and 3.2 mm long with a wall thickness of 0.03 ± 0.01 mm. Large sizes may be as big as 4.8 m in diameter. Thicknesses also may vary from 0.03 to 25.4 mm or even thicker.

Parts such as wind-tunnel nozzles 4.8 m long, 762 mm in diameter, and varying in thickness, have been fabricated from 1.5 to 25.4 mm. Tolerances on a part like this are extremely close and are usually ±0.03 mm in the throat area and ±0.05 mm in the downstream sections. Tolerances on waveguide plumbing may be as close as ±0.01 mm.

The range of mechanical properties of the materials that may be deposited by electroforming are shown below:

Material	Brinell Hardness, Bhn	Ultimate Tensile Strength, MPa	Yield Strength, MPa	Elongation, % in 2 in.
Nickel	140–500	379–1552	276–862	2–20
Copper	51–170	248–552	83–276	21–39

ELECTROLESS PLATING

This is a chemical reduction processing that, once initiated, as autocatalytic. The process is similar to electroplating except that no outside current is needed. The metal ions are reduced by chemical agents in the plating solutions, and deposit on the substrate. Electroless plating is used for coating nonmetallic parts. Decorative electroless plates are usually further coated with electrodeposited nickel and chromium. There are also applications for electroless deposits on metallic substrates, especially when irregularly shaped objects require a uniform coating. Electroless copper is used extensively for printed circuits, which are produced either by coating the nonmetallic substrate with a very thin layer of electroless copper and electroplating to the desired thickness or by using the electroless process only. Electroless iron and cobalt have limited uses. Electroless gold is used for microcircuits and connections to solid-state components. Deeply recessed areas that are difficult to plate can be coated by the electroless process.

Nonmetallic surfaces and some metallic surfaces must be activated before electroless deposition can be initiated. Activation on nonmetals consists of the application of stannous and palladium chloride solutions. Once electroless plating is begun, it will continue to a desired thickness; that is, it is autocatalytic. The process thus differs from a displacement reaction, in which a more noble metal is deposited while a less noble one goes into solution; this ceases when the more noble deposit, if pore-free, covers the less noble substrate.

ELECTROLYTE

This is a material that conducts an electric current when it is fused or dissolved in a solvent, usually H_2O. Electrolytes are composed of positively charged species, call cations, and negatively charged species, called anions. For example, NaCl is an electrolyte composed of sodium cations (Na^+) and chlorine anions (Cl^-). The ratio of cations to anions is always such that the substance is electrically neutral. If two wires connected to a lightbulb and to a power source are placed in a beaker of H_2O, the lightbulb will not glow. If an electrolyte, such as NaCl, is dissolved in the H_2O, the lightbulb will glow because the solution can now conduct electricity. The amount of electric current that can be carried by an electrolyte solution is proportional to the number of ions dissolved. Thus, the bulb will glow more brightly if the amount of NaCl in the solution is increased.

HYDRATION

H_2O is a special solvent because its structure has two different sides. On one side is the O_2 atom, and on the other are two H_2 atoms. Covalent molecules, such as H_2O, are held together by covalent bonds, which are formed when two atoms share a single pair of electrons. However, when two different atoms form a covalent bond, the sharing of electrons is not always equal. An electron in a covalent bond between two different atoms might spend more time near one atom or the other. In H_2O, the electrons from the O–H bonds spend more time near the O_2 atom than near the H_2 atoms. As a result, the O_2 has somewhat more negative charge than the H_2 atoms. This phenomenon is not the same as ionization, where the electron is completely transferred from one atom to the other, so to indicate the difference; the O_2 is said to have a partial negative charge (δ^-) and the H_2 to have a partial positive charge (δ^+).

When H_2O molecules surround dissolved ions, the ions are said to be hydrated. The electrostatic attraction associated with hydration provides the energetic driving force for dissolving ions.

ELECTROPLATED COATINGS

Electroplating may be defined as the electrodeposition of an adherent metallic coating upon an electrode for the purpose of securing a surface with properties or dimensions different from those of the base metal. It must not be confused with electroforming or "electroless plating." "Brush plating," on the other hand, is a special method of electroplating.

In brush plating the anode may be soluble or insoluble. It is covered with a cloth or similar spongelike material and moistened with the plating solution while it is gently moved back and forth across the surface to be plated. It is usually used for specialized applications.

Electroplating is extensively used to produce printed circuit boards. Its main advantage is that the circuit can be produced directly rather than having to be etched out of a piece of copper sheet. Electroplating is also widely used to impart corrosion resistance. Most parts of automobile bodies are zinc plated for corrosion resistance. Because zinc is more readily attacked by most corrosive agents that automobiles encounter than steel, it provides galvanic or sacrificial protection. An electrolytic cell is formed in which zinc, the less noble metal, is the anode and the steel, the more noble one, is the cathode. The anode corrodes and the cathode is protected. Zinc also provides a good base for paint. If a metal is more noble than the one on which it is electroplated, it provides protection against corrosion only if it is completely continuous. If a small area of the substrate is exposed such as under a pinhole, corrosion occurs there, rapidly forming a pit.

An example of an electroplated coating applied primarily for wear resistance is hard chromium on a rotating shaft. Electroplating is also used to build up worn or undersized parts. Gold-plated jewelry is an example of a decorative application. Gold and also palladium are electrodeposited on electrical contacts. Here the absence of an oxide film avoids the rise in the electrical resistance of the contact. Nickel and aluminum are plated for some decorative applications; however, their wide use in the automotive industry has diminished considerably, primarily because of the associated environmental problems. Magnetic components made of such alloys as permalloy can be manufactured by electroplating.

PROCESS

The electroplating process consists essentially of connecting the parts to be plated to the negative terminal of a DC source and another piece of metal to the positive pole, and immersing both in a solution containing ions of the metal to be deposited (see Figure E.1). The part connected to the negative terminal becomes the cathode, and the other piece is the anode. In general, the anode is a piece of the same metal that is to be plated. Metal dissolves at the anode and is plated at the cathode. If the current is used only to dissolve and deposit the metal to be plated, the process is 100% efficient. Often, fractions of the applied current are diverted to other reactions such as the evolution of H_2 at the cathode; this usage results in lower efficiencies as well as changes in the acidity (pH) of the plating solution. In some processes, such as chromium plating, a piece of metal that is essentially insoluble in the plating solution is the anode. When such insoluble anodes are used, metal ions in the form of soluble compounds must also be added periodically to the plating solution. The anode area is generally about the same as that of the cathode; in some applications it is larger.

Most plating solutions are of the aqueous type. There is a limited use of fused salts or organic liquids as solvents. Nonaqueous solutions are employed for the deposition of metals such as aluminum that have overvoltages lower than H_2. Such metals cannot be plated in the presence of H_2O, as H_2 would be preferentially reduced.

In addition to metal ions, plating solutions contain relatively large quantities of various substances used to increase the electrical conductivity, to buffer, and in some instances to form complexes with the metal ions. Relatively small amounts of other substances, which are called addition agents, are also present in plating solutions to level and brighten the deposit, to reduce internal stress, to improve the mechanical properties, and to reduce the size of the metal crystals or grains or to change their orientation.

The quantity of metal deposited, that is, the thickness, depends on the current density (A/m), the plating time, and the cathode efficiency. The current is determined by the applied voltage, the electrical conductivity of the plating solution, the distance between anode and cathode, and polarization. Polarization potentials develop because of the various reactions and processes that occur at the anode and cathode, and depend on the rates of these reactions, that is, the current density. If the distance between anode and cathode varies because the part to be plated is irregular in shape, the thickness of the deposit may vary. A quantity called the throwing power represents the degree to which a uniform deposit thickness is attained on areas of the cathode at varying distances from the anode. Good throwing power results if the plating efficiency is low because of polarization where the current density is high.

PLATING OF SPECIFIC METALS

Most metals can be electroplated from either aqueous or fused-salt solutions. The more important metals plated from aqueous baths are Cr, Cu, Au, Ni, Ag, Sn, and Zn. Alloys can also be electroplated. Electrodeposited Cu–Zn and Pb–Sn alloys are used extensively.

Chromium

Electroplated chromium is used primarily to produce wear and corrosion-resistant coatings. Chromium is not deposited for decorative

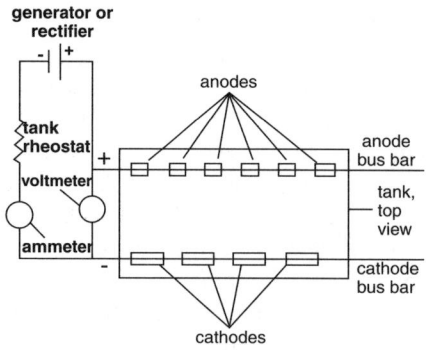

FIGURE E.1 Typical connections for a simple electroplating process. (From *McGraw-Hill Encyclopedia of Science and Technology*, 8th ed., Vol. 6, McGraw-Hill, New York, 1997, 299. With permission.)

purposes as extensively as in the past because of the associated pollution problem from the discharge of hexavalent chromium. Chromium plating solutions consist primarily of Cr_2O_3, H_2SO_4, and H_2O.

Copper

Electroplated copper is used extensively in the manufacture of printed circuit boards. Because copper cannot be directly plated on the insulating material substrates, they must be rendered electrically conductive first. The main advantage of using electrolytically deposited copper to produce printed circuits is that the areas of the board that are made conductive can be controlled. The actual circuit can be produced by selective etching or selective plating involving techniques such as photosensitizing, photoresist, and etch resist. Areas exposed through a mask can become conductive by coating them with electroless copper after activation with a solution of stannous and palladium chloride. Suitable organic materials are also used to render the board conductive. The areas made conductive serve as the substrate for copper electrodeposition. Some circuits are produced only with electroless plating of copper. Only the through holes are plated, generally with electroplated copper over electroless plated copper when the boards are made by laminating copper sheet to the plastic substrate. The copper segments of the printed circuit may be coated with an electroplated Sn–Pb alloy to facilitate subsequent soldering and also to protect them from oxidation. Gold deposits are also used for this purpose. There are other uses of electroplated copper in the electronic industry, such as in the production of microchips. Electroplated copper is also the undercoat for decorative Ni–Cr deposits.

Gold

Gold plating is used for electrical contacts that must remain free of oxides, connections for microcircuits, certain information-storage devices, solid-state components, and jewelry. The use of gold for electrical and electronic application exceeds that for decorative purposes. Electroplated palladium is being substituted for gold in some applications.

Nickel

Nickel coatings covered with chromium provide corrosion-resistant and decorative finishes for steel, brass, and zinc-base die castings. The most widely used plating solution is the Watts bath, which contains nickel sulfate, nickel chloride, and boric acid. All-chloride, sulfamate, and fluoroborate plating solutions are also used. The sulfamate solution is used for low-stress applications. There are a number of compounds, mostly organic, which can be added to nickel-plating baths.

Silver

The principal use of silver electrodeposits is for tableware because of their corrosion resistance (except to sulfur-containing foods) and pleasing appearance. Other important uses are for bearings and electrical circuits, waveguides, and hot-gas seals. The plating solutions are of the cyanide type and generally contain additives that produce bright deposits.

Tin

Tin is used in electrodeposition as a component of solders. Solder is electroplated over copper to protect it from oxidation and to facilitate subsequent joining operations. The advantage of electroplating the solder is that it can be applied only where it is needed. Because of the desirability of eliminating lead in solders, those with higher tin contents are used.

Tin-plated steel for cans in which food is preserved has limited use, because lacquers are preferred to prevent contact between steel and food. Tin plating is employed for refrigerator coils and bearing surfaces. Bright pore-free surfaces can be produced by melting the electroplated tin and allowing it to "reflow."

Stannous chloride and stannous sulfate are the main components of acid tin-plating solutions.

Zinc

The sacrificial protection of steel against corrosion is the main reason for plating zinc. It is used extensively in the automobile industry for this purpose. Screws, bolts, and washers are barrel-plated with zinc also for corrosion protection. Continuous zinc electroplating of wire and strip is another application. The advantage of electrodepositing zinc over hot dipping is the ability to apply thinner coatings and higher purity.

Properties of Electrodeposits

In the various applications of electrodeposits, certain properties must be controlled and, therefore, must be measured. The properties of electroplated metal that should be considered, depending on the use of the deposit, are thickness, adhesion to substrate, brightness, corrosion resistance, wear resistance, the mechanical properties of yield strength, tensile strength, ductility, and hardness, internal stress, solderability, density, electrical conductivity, and the magnetic characteristics.

ENGINEERING FILMS

Specific properties that separate engineering films from their commodity counterparts include greater tensile and impact strength; improved moisture and gas barrier characteristics; good heat resistance and weatherability; better bonding and lamination; and improved electrical ratings. One or more of these properties can be obtained by choosing from a number of different polymer films.

Several melt-processible engineering thermoplastic films such as oriented polyester, oriented nylon, and unoriented nylon, exhibit high strength, especially at high temperatures. In addition, they provide toughness at low temperatures, stiffness and abrasion resistance, and good chemical resistance. Polyester film is made from the PET polymer. The monomer is polymerized, extruded, cast into a web, and biaxially oriented, forming a drawn polyester film.

Oriented polypropylene, a thermoplastic with low specific gravity, has excellent resistance, relatively high melting point, and good strength. Polycarbonate film is specified for its toughness, clarity, and high heat-deflection temperature.

Polyimides, both thermoplastics and thermosets, retain their principal properties over a wide temperature range. They have useful mechanical properties, even at cryogenic temperatures. At −162°C, the film can be bent around a 6.4-mm mandrel without breaking and, at 500°C, its tensile strength is 31 MPa. Room-temperature mechanical properties are comparable to those of polyester film.

To minimize transmission of moisture vapors, fluoropolymers are the best choice. This family of material has a general paraffinic structure with some or all of the hydrogen atoms replaced by fluorine.

Polyetheretherketone (PEEK), a high-performance thermoplastic, offers outstanding thermal properties as well as resistance to many solvents and proprietary fluids. This film can be used for interbonding or cladding in PEEK structural components. Thermoplastic and thermoset acrylics are noted for exceptional clarity and weatherability, and also offer favorable stiffness, density, chemical resistance, and toughness.

Film produced from PEEK resins can be laminated to itself or to other substrates. Bond strength depends on surface preparation and adhesive type. The film is available in a transparent, thermoformable grade and in a higher-temperature, heat-stabilized version, which is more crystalline and less transparent (also thermoformable).

Plastic film can be manufactured from almost any resin, but not every resin produces an engineering film. Generally, the properties of a resin-based film are related to the chemistry of the basic polymer; however, properties may be further affected by subsequent processing techniques. Manufacturers can choose from, and end users can specify, a range of process treatments that significantly enhance heat stability, mechanical properties, electrical characteristics, barrier properties, and bondability.

Coatings are typically applied by emulsion, solvent, and dry methods. Results vary according to the formula used: PVDC coatings improve barrier properties; polyurethane improves abrasion resistance; and aluminum coatings alter electrical characteristics.

Some processors have developed proprietary antistatic coatings that are cured by electron-beam radiation. Metal coatings produce conductive capabilities and also enhance barrier properties. Often, metallization is used to improve moisture-barrier properties for biaxially oriented nylon and polypropylene.

Surface treatment, which removes low-molecular-weight residue, improves adhesion and appearance. Several methods may be used. Corona discharge techniques position the film between an electrically grounded roller and a high-voltage electrode. A continuous-art discharge (corona) is generated to clean and activate the film surface.

In gas-plasma surface treatments, film is placed in a reaction chamber. After evacuation, the chamber is charged with O_2, A, helium, or N_2, while a radiofrequency field ionizes the gas. A resultant glow discharge creates free radicals on the surface, in adhesion.

EPOXY RESINS

Epoxy (Ep) resins comprise an extremely broad and diverse family of materials. They are used in the form of protective coatings, adhesives, reinforced plastics, molding compounds, casting and potting compounds, and foams.

Epoxies, perhaps best known as adhesives, are premium thermosetting plastics, and are generally employed in high-performance uses where their high cost is justified. They are available in a wide variety of forms, both liquid and solid, and are cured into the finished plastic by a catalyst or with hardeners containing active H_2. Depending on the type, they are cured at either room temperature or at elevated temperatures.

Liquid epoxies are used for casting, for potting or encapsulation, and for laminating. They are used unfilled or with any of a number of different mineral or metallic powers. Molding compounds are available as liquids, and also as powders with various types of fillers and reinforcements.

PROPERTIES

The epoxy resins have very high adhesion to metals and nonmetals, heat resistance from 177 to 260°C, dielectric strength to 22 V/m, and hardness to Rockwell M110. The tensile strength may be up to 82 MPa, with elongation to 2 to 5%, but some resilient encapsulating resins are made with elongation to 150% with lower tensile strengths. The resins have high resistance to common solvents, oils, and chemicals.

An unlimited variety of epoxy resins is possible by varying the basic reactions with different chemicals or different catalysts, or both, by combination with other resins, or by cross-linking with organic acids, amines, and other agents. To reduce cost when used as laminating adhesives, they may be blended with furfural resins, giving adhesives of high strength and high chemical resistance. Blends with polyamides have high dielectric strength, mold well, and are used for encapsulating electrical components. By using a polyamide curing agent, an epoxy can be made water emulsifiable for use in water-based paints. An epoxy resin with 19% bromine in the molecule is flame-resistant. Another grade, with 49% bromine, is a semisolid, used for heat-resistant adhesives and coatings.

The epoxies have excellent electrical, thermal, and chemical resistance. Their strength can be further increased with fibrous reinforcement or mineral fillers. The variety of combinations of epoxy resins and reinforcements provides a wide latitude in properties obtainable in molded parts.

Molded epoxy parts are hard, rigid, relatively brittle, and have excellent dimensional stability over a broad temperature range. Some fiber-reinforced formulations can withstand service temperatures above 260°C for brief periods. Their excellent electrical properties, in combination with high mechanical strength, qualify them for electrostructural applications. Resins based on bisphenol–A are adequate for most services. However, cycloaliphatics are recommended for parts subjected to arcing conditions or those requiring outdoor weatherability; see Table E.2.

Excellent adhesion in structural applications is another outstanding property of epoxy systems. Epoxy adhesives for bonding many

TABLE E.2
Properties of Epoxies

| | | Molding Compounds | | Casting Resins | | |
| | | | | Bisphenol A | | |
ASTM Test	Property	Glass-Fiber Filler	Mineral Filler	No Filler	Silica Filler	Cycloaliphatic
		Physical				
D792	Specific gravity	1.6–2.0	1.6–3.0	1.11–1.40	1.6–2.0	1.16–1.21
D792	Specific volume (in.^{3}/lb)	15.4–14.0	14.2–13.4	26.9–20.0	17.3–13.9	—
D570	Water absorption, 24 h, $1/8$-in. thk (%)	0.05–0.20	0.04	0.08–0.15	0.04–0.10	0.08–0.15
		Mechanical				
D638	Tensile strength (psi)	10,000–20,000	5,000–10,000	4,000–13,000	7,000–13,000	10,000–20,000
D638	Elongation (%)	4	—	3–6	1.3	2–10
D638	Tensile modulus (10^5 psi)	30.4	—	3.5	—	5–9
D790	Flexural strength (psi)	10,000–60,000	8,000–15,000	13,300–21,000	8,000–14,000	15,000–32,000
D256	Impact strength, Izod (ft-lb/in. of notch)	2–30	0.3–0.4	0.2–1.0	0.3–0.45	0.2–1.0
D785	Hardness, Rockwell M	100–110	100–110	80–110	85–120	85–120
		Thermal				
C177	Thermal conductivity (10^{-4} cal-cm/s-cm^2-°C)	4–10	4–30	4.0–5.0	10–20	—
D696	Coefficient of thermal expansion (10^{-5}in./in.-°C)	1.1–3.5	2.0–5.0	4.5–6.0	2.0–4.0	—
D648	Deflection temperature (°F) At 264 psi	250–500	250–500	115–550	160–550	200–450
		Electrical				
D149	Dielectric strength (V/mil) Short time, $1/8$-in. thk	300–400	300–400	300–500	400–550	420–440
D150	Dielectric constant At 1 kHz	3.5–5.0	3.5–5.0	3.5–4.5	3.2–4.0	3.9–4.5
D150	Dissipation factor At 1 kHz	0.01	0.01	0.002–0.02	0.008–0.03	—
D257	Volume resistivity (Ω-cm) At 73°F, 50% RH	>10^{14}	>10^{14}	10^{12}–10^{17}	10^{13}–10^{16}	10^{15}
D495	Arc resistance (s)	120–180	150–190	45–120	150–300	150–180

Source: Mach. Design Basics Eng. Design, June, p. 688, 1993. With permission.

dissimilar materials can be supplied either as one- or two-part systems. One-part systems require heat for curing; two-part systems usually cure at room temperature, but properties are improved when the materials are heat-cured. Some epoxy adhesive systems can withstand temperatures to 232°C, although properties at such temperatures are considerably lower than at room temperature.

Applications

- Casting resins are primarily used for potting or encapsulating electrical or electronic equipment. The excellent adhesion and extremely low shrinkage of epoxies, coupled with their high dielectric properties, provide a

- well-sealed, voidless, well-insulated component.
- Foaming resins have also been used for electronic potting.
- Liquid resins are also formulated with a variety of fillers, such as metal powder, to provide effective patching and repair putties or pastes. Such compounds can be used to patch both metal and plastic surfaces.
- Liquid resin systems are used to produce low-pressure reinforced laminates and moldings, high-pressure industrial thermosetting laminates (NEMA Grade G-10 glass cloth-base, as well as paper-based laminates for electrical uses), and filament wound shapes.
- Epoxy laminates are also widely used in plastic tools. They are used for drilling, checking, and locating fixtures, where dimensional accuracy is critical. They are also used to provide durable surfaces for metal-forming tools, such as draw dies for short-run production.
- Filament wound structures are used for such applications as rocket-motor booster cases, pressure vessels, and chemical tanks and pipe.
- Molding compounds provide the performance characteristics of epoxy resins with the automated speed and economy of compression and transfer molding. They are being used primarily for electrical components.

Handling and Resin Selection

Because the cure of an epoxy is brought about by a chemical reaction, and not by a simple process of solvent evaporation, it is quite important that users recognize at the outset that handling (particularly for the two-component, low-temperature curing compounds) requires more care than that of older materials. However, because this is, and probably will be, a problem for some time to come, procedures have been developed that keep difficulties in handling to a minimum. One simple and inexpensive way of eliminating frequent weighing is to use calibrated mixing containers where it is necessary only to fill to the first calibration with the base, to the second with the hardener, mix, and apply.

Formulators meet this problem by supplying the compounds packaged in preweighed containers, while equipment manufacturers use automatic mixers and dispensers. It is possible through the use of these machines to eject preweighed and premixed shots of compound as the operator requires.

One-component systems, as the name indicates, eliminate problems of weighing, mixing, and pot life, and are subdivided into solids, most often "B" staged epoxy powders, and liquids containing latent hardeners such as boron trifloride, or anhydride/solvent solutions.

Powders have been used to some extent as adhesives and in potting, but are employed most widely for the fluidized-bed coating method or in compression and transfer molding. The epoxy fluidized-bed approach provides a relatively easy way of applying encapsulant coatings in uniform thickness over contours but is not suitable for impregnation, or for coating units that cannot be heated.

In resin selection, epoxy compounds can be categorized (although there are exceptions) as (1) two-component liquids that will cure at temperatures from 21.1 to 60°C, but that have short pot lives; (2) two-component liquid systems requiring cure at temperatures up to 177°C, but that offer longer pot lives and improved operational characteristics; and (3) one-component liquids and powders that reduce handling problems, offer optimum operating characteristics, but that require higher curing temperatures (in the range of 177 to 204°C).

With respect to possible methods of application, with the exception of the powders, epoxies are supplied for use by spray, brush, spatula, roller coat, knife coat, dipping, filament winding, laminating, and casting. All these except casting are self-descriptive and this is simply the method of producing a defined shape by pouring a compound into a mold where it cures and from whence it can later be removed. On the other hand, if an object has been placed in the mold and epoxy cast around it, the process is called potting.

ETCHING MATERIALS

These are chemicals, usually acids, employed for cutting into, or etching, the surface of metals, glass, or other material. In the metal industries they are called etchants. The usual method of etching is to coat the surface with a wax, asphalt, or other substance not acted upon by the acid; cut the design through with a sharp instrument; and then allow the acid to corrode or dissolve the exposed parts. For etching steel, a 25% solution of H_2SO_4 in water or an $FeCl_3$ solution may be used. For etching stainless steels a solution of $FeCl_3$ and HCl in water is used. For high-speed steels, brass, or nickel, a mixture of HNO_3 and HCl acids in water solution is used, or nickel may be etched with a 45% solution of H_2SO_4. Copper may be etched with a solution of $HcrO_3$ acid. Brass and nickel may be etched with an acid solution of $FeCl_3$ and $KClO_3$. For red brasses, deep etching is done with concentrated HNO_3 acid mixed with 10% HCl acid, with the latter added to keep the SnO_2 in solution and thus retain a surface exposed to the action of the acid. For etching aluminum a 9% solution of $CuCl_3$ in 1% acetic acid, or a 20% solution of $FeCl_3$ may be used, followed by a wash with strong HNO_3 acid. NaOH, NH_4OH, or any alkaline solutions are also used for etching aluminum. Zinc is preferably etched with weak HNO_3 acid, but requires a frequent renewal of the acid. Strong acid is not used because of the heat generated, which destroys the wax coating. A 5% solution of HNO_3 acid will remove 0.005 cm of zinc/min, compared with the removal of over 0.013 cm/min in most metal-etching processes. Glass is etched with HF acid or with white acid. White acid is a mixture of HF acid and ammonium bifluoride, a white crystalline material of the composition $(NH_4)FHF$.

The process in which the metal is removed chemically to give the desired finish as a substitute for mechanical machining is called chemical machining.

ETHER

Ether is the common name for ethyl ether, or diethyl ether, a highly volatile, colorless liquid of the composition $(C_2H_5)_2O$ made from ethyl alcohol. It is used as a solvent for fats, greases, resins, and nitrocellulose, and in medicine as an anesthetic. The specific gravity is 0.720, boiling point 34.2°C, and freezing point –116°C. Its vapor is heavier than air and is explosive.

Recently, the promotion for the production and use of cleaner-burning fuels was announced. As a result, methyl tertiary butyl ether (MTBF) became a very important petrochemical. MTBF is the most widely produced ether for oxygenates. It is commonly produced by the dehydrogenation of isobutane and the subsequent reaction of isobutylene with methanol.

ETHYLENE

Ethylene, also called ethane, is a colorless, inflammable gas, $CH_2:CH_2$, produced in the cracking of petroleum. Ethylene liquefies at –68.2°C. Ethylene is the largest-volume organic chemical produced today, and is the basic building block of the petrochemical industry. Polymerization of ethylene is its largest use. When ethylene is reacted in the presence of transition metal catalysts, such as Mo_2O_5 or Cr_2O_3, at high pressures, it forms low-density polyethylene (LDPE); at lower pressures, high-density polyethylene (HDPE) is produced. Recently, low pressures have been employed for making a new variant, linear low-density polyethylene (LLDPE). Ethylene is now used to produce ethyl alcohol, acrylic acid, and styrene, and it is the basis for many types of reactive chemicals.

Trichloroethylene is a colorless liquid of pleasant odor of the composition $CHCl:CCl_2$, also known as westrosol. Its boiling point is 87°C and its specific gravity 1.471. It is insoluble in water and is unattacked by dilute acids and alkalis. It is non flammable and is less toxic than tetrachlorethane. Trichloroethylene is a powerful solvent for fats, waxes, resins, rubber, and other organic substances, and is employed for the extraction of oils and fats, for cleaning fabrics, and for degreasing metals preparatory to plating. The freezing point is –88°C, and it is also used as a refrigerant. It is also used in soaps employed in the textile industry for degreasing.

ETHYLENE GLYCOL

This substance, also known as glycol and ethylene alcohol, is a colorless syrupy liquid CH_2OHCH_2OH, with a sweetish taste, very soluble in water. It has a low freezing point, $-25°C$, and is much used as an antifreeze in automobiles. A 25% solution has a freezing point of $-20.5°C$, without appreciably lowering the boiling point of the water. It has the advantage over alcohol that it does not boil away easily, and permits the operation of the engines at much higher temperatures than with water, giving greater fuel efficiency. Ethylene glycol is also used for the manufacture of acrylonitrile fibers, and as a solvent for nitrocellulose. It is highly toxic in contact with the skin.

ETHYLENE-PROPYLENE ELASTOMER

Ethylene-propylene elastomer is a completely saturated copolymer made by solution polymerization. The remarkable properties of the material include exceptional ozone resistance, excellent electrical properties, good high- (149 to 163°C) and low-temperature properties, good stress–strain characteristics, and resistance to chemicals, light, and other types of aging.

Ethylene-propylene rubber has required a peroxide or peroxide-sulfur modified curing system. Sulfur improves the peroxide curing efficiency and assists in chemical cross-linking of the polymer chains, thereby imparting better physical properties to the vulcanizate.

Some plasticizers used in other rubbers are not suitable for ethylene-propylene rubber. Most acceptable plasticizers are saturated materials of relatively low polarity such as paraffinic hydrocarbon oils and waxes.

EXTRUDED METALS

Extrusion is the forcing of solid metal through a suitably shaped orifice under compressive forces. Extrusion is somewhat analogous to squeezing toothpaste through a tube, although some cold extrusion processes more nearly resemble forging, which also deforms metals by application of compressive forces. Most metals can be extruded, although the process may not be economically feasible for high-strength alloys.

HOT EXTRUSION

The most widely used method for producing extruded shapes is the direct, hot extrusion process. In this process, a heated billet of metal is placed in a cylindrical chamber and then compressed by a hydraulically operated ram; see Figure E.2. The opposite end of the cylinder contains a die having an orifice of the desired shape; as this die opening is the path of least resistance for the billet under pressure, the metal "squirts" out of the opening as a continuous bar with the same cross-sectional shape as the opening. By using two sets of dies, stepped extrusions can be made. The small section is extruded to the desired length, the small split die is replaced by the large die, and the large section is then extruded.

The most outstanding feature of the extrusion process is its ability to produce a wide variety of section configurations. Structural shapes can be extruded that have complex nonuniform and nonsymmetrical sections that would be difficult or impossible to roll.

In many instances, extrusions can replace bulky assemblies made up by joining, welding, or riveting rolled structural shapes, or sections previously machined from bar, plate, or pipe.

An extrusion die is relatively simple to make and inexpensive when compared to a pair of rolls or a set of forging dies. The low cost of dies and the short lead time for die changes

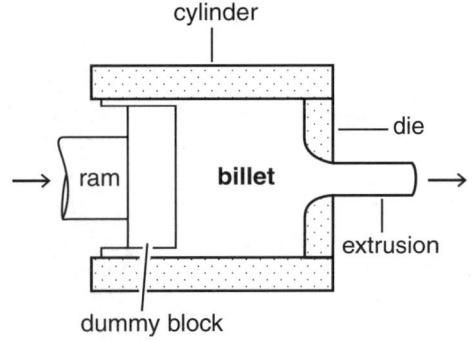

FIGURE E.2 Schematic representation of the direct extrusion process (hot). (From *McGraw-Hill Encyclopedia of Science and Technology*, 8th ed., Vol. 7, McGraw-Hill, New York, 1997, 651. With permission.)

make it possible to extrude small quantities more economically than by most other methods.

Lubricants are used to minimize friction and protect the die surfaces. Graphite is a common lubricant for nonferrous alloys, whereas for hot extrusion of steel, glass is an excellent lubricant.

Indirect, or inverted, extrusion was developed to overcome such difficulties as surface friction and entrainment of surface oxide of direct extrusion. In the indirect process the ram is hollow, the die opening is in the dummy block, and the opposite end of the cylinder is closed. As the ram advances, the billet does not move as in the case of direct extrusion, and the metal is extruded backward through the die and the hollow ram. However, the process is not very popular because the hollow ram is weaker, resulting in lower machine capacity; trouble-free operation requires that the extruded product be straight and not hit the inside of the ram.

Cold Extrusion

The extrusion of cold metal is variously termed *cold pressing*, *cold forging*, *cold extrusion forging*, *extrusion pressing*, and *impact extrusion*. The term *cold extrusion* has become popular in the steel fabrication industry, while *impact extrusion* is more widely used in the nonferrous field.

The original process (identified as impact extrusion) consists of a punch (generally moving at high velocity) striking a blank (or slug) of the metal to be extruded, which has been placed in the cavity of a die. Clearance is left between the punch and the die walls; as the punch comes in contact with the blank, the metal has nowhere to go except through the annular opening between punch and die. The punch moves a distance that is controlled by a press setting. This distance determines the base thickness of the finished part. The process is particularly adaptable to the production of thin-walled, tubular-shaped parts with thick bottoms, such as toothpaste tubes.

Advantages of cold extrusion are high strength because of severe strain-hardening, good finish and dimensional accuracy, and economy due to fewer operations and minimum of machining required.

Metals Extruded

Extrusion can be used to fabricate practically all structural metals and alloys. Among the more common metals extruded on a commercial or semi-commercial basis are alloys in the following metal systems:

Magnesium	Carbon and alloy steels
Aluminum	Stainless steel
Brass	Iron superalloys
Copper	Nickel superalloys
Titanium	Columbium or Niobium
Zirconium	Molybdenum
Beryllium	Tantalum
Nickel	Tungsten

These materials span a range of working temperatures from about 316 to 2205°C, in approximately the order shown above.

The wide range of extrusion temperatures gives rise to the major differences in processing that center around such variables as extrusion lubricants, die materials, die design, billet preparation, and extrusion speed. Present tool materials are capable of maintaining adequate strength and wear resistance for extrusion at temperatures only slightly higher than 538°C. At higher temperatures, lubricants are necessary not only to reduce friction but to insulate and protect the tooling surface from overheating. Also, the speed of extrusion must be more rapid to avoid prolonged contact between the tools and the hot billet. Thus, the extrusion method for magnesium and aluminum is quite different from that for the other metal systems.

The Light Alloys

Magnesium and aluminum are extruded at temperatures below 538°C with no lubrication and flat, sharp-cornered dies. Deformation of the billet occurs by shear flow, which is from within the billet so that the surface skin of the billet is retained in the container as discard. This type of turbulent flow is possible because of the ability of these materials to form sound welds when severely deformed, but requires comparatively slow pressing speeds, often less than 1.66 m/min. With clean tools and no lubricants there are no contaminants present to cause internal defects or laminations, and several sections can be extruded at one time by using multihole dies. Precise dimensional control is attained with ordinary hot-work tool-steel dies, which last for hundreds of extrusions.

Other Metals

With higher-temperature materials, e.g., titanium, steels, refractory metals, it is necessary to use lubricants and die designs so that deformation occurs by uniform flow. In this case, the surface of the billet becomes the surface of the extrusion; otherwise, laminations and inclusions could occur. Graphitic lubricants are suitable for producing relatively short lengths at temperatures up to about 1093°C if the operation is performed at high speeds. The Ugine–Sejournet process in which molten glass serves as a lubricant is most widely used for high-temperature extrusion. Because of the insulating as well as lubricating properties of glass, overheating of tools does not occur and die life is increased. For titanium and steels, dies are usually made of tungsten hot-work tool steels. Ceramic coatings (Al_2O_3 or ZrO_2) on the dies are necessary at the temperatures required for refractory metals. Pressing speeds are usually in the range of 8.33 to 33.32 m/min.

SHAPE, SURFACE, AND TOLERANCE LIMITATIONS

Extruded shapes are generally classified by configuration and include rod, bar, tube, and hollow, semihollow, and solid shape.

Although many asymmetrical shapes can be produced, probably the most important factor in the extrudability of a shape is symmetry. Hollow and semihollow shapes cost more than solid shapes and usually cannot be extruded with as thin sections. Semihollow shapes with long thin voids should be avoided. For best extrudability the length-to-width ratio of partially enclosed voids, channels, or grooves should not exceed 3:1 for aluminum and magnesium, 2:1 for brass, or 1:1 for copper, titanium, and steels. Wall thickness surrounding the voids should be as uniform as possible.

The size and weight of extruded shapes are limited both by the section configuration and by the material properties. The maximum size that can be extruded on a press of given capacity is determined by the circumscribing circle, which is the smallest circle that will enclose the shape. The circumscribing circle size controls the die size, which in turn is limited by the press size.

Thickness limitations are related to the size of the cross section as well as the type of material. As a rule, thicker sections are required with increased section size.

Sharp corners and edges are usually possible with aluminum and magnesium alloys, but 0.38-mm corner and fillet radii are preferred. Minimum fillet radii of 3.2 mm for steel and 4.5 mm for titanium are suggested by most extruders. Typical minimum corner radii are 0.8 mm for steel and 1.5 mm for titanium.

Smooth surfaces with finishes better than 30 μin. rms are readily attainable in aluminum and magnesium alloys. High-temperature alloys are characteristically rougher; an extruded finish of 125 μin. rms is generally considered acceptable for most steels and titanium alloys. Improved surface finishes can be produced by a cold-draw finishing operation.

Although extruded shapes minimize and often eliminate the need for machining, they do not possess the dimensional accuracy of machined parts. The tolerances of any given dimension vary somewhat depending on the size and type of shape, and the relative location of the dimension. Detailed standard tolerances covering straightness, flatness, twist, and cross-sectional dimensions such as section thickness, angles, contours, and corner and fillet radii have been established for magnesium, aluminum, copper, and brass by most extruders and are published in handbooks. Standard tolerances also have been established for steels and titanium alloys in simple sections, but in many instances these are subject to mill inquiry.

EXTRUDED PLASTICS

Extrusion is a process for making articles of constant cross section, called "continuous shapes," by forcing softened material through a hole approximating the desired shape. With plastics the process is carried out by one of two methods: ram extrusion or screw extrusion.

In ram extrusion, the softened mass fills a cylinder to which the die—the shaped hole—is attached at one end. A closely fitting piston, the ram, enters the cylinder and pushes the mass through the die at pressures ranging up to 68 MPa. The product, or extrudate, is cooled or otherwise hardened shortly after leaving the die. Subsequent handling depends on the material and the shape. Ram extrusion is used chiefly for extruding TFE fluorocarbon (tetrafluoroethylene) resin (Teflon), which is damaged by the shearing action of screw extrusion, and for cellulose nitrate, whose extreme heat sensitivity and inflammability make screw extrusion dangerous.

Screw extrusion, by far the more economical and commercially important process, centers around the screw extruder. This consists of a heavy cylindrical barrel inside which turns a motor-driven screw, or worm. The screw is essentially a thick shaft with a helical blade, or flight, wrapped around it. At the rear end of the barrel, a feed hopper admits cold plastic particles that normally fall into the screw channel by gravity. As the screw rotates, the particles are dragged forward by frictional action between screw, plastic, and barrel. Electric band heaters on the outside of the barrel heat the plastic, which is further heated by the frictional action of the screw. Soon, the particles coalesce into a voidless mass that softens further to become a melt. This plastic melt is very viscous (a million times as viscous as water), so considerable pressure, on the order of 3.4 to 68 MPa, must be developed to force it through the die at the front of the extruder at economical rates.

Extrudable Materials

All thermoplastics can be extruded by either ram or screw extrusion. Today, high-viscosity grades of type 66 nylon and other nylons are available and their extrusion presents no special difficulties. In extruding rigid polyvinyl chloride (PVC), extreme care must be taken not to overheat the resin since thermal decomposition, once started, snowballs. This simply means being careful to avoid extreme temperatures everywhere and to streamline meticulously all passages through which the melt must pass. To a lesser degree, CFE fluorocarbon (trifluorochloroethylene) resin (Kel-F et al.), cellulosics, nylons, and acetal (polyoxymethylene) are similarly heat sensitive.

Some thermosets can be extruded provided they are formulated to flow at temperatures safely below the curing temperatures. The process has been used to make pipe and structural shapes, to coat wire with thermosetting compositions, and to prepare "rope" and pellets for compression molding. The extrusion of rubbers closely resembles plastics extrusion.

F

FABRICS, NONWOVEN BONDED

Although there are several types of fabrics that are not woven, the term *nonwoven fabrics* is recognized in the textile trade as applying to those materials composed of a fibrous web held together with a bonding agent to obtain fabric-like qualities. These fabrics may be of a uniform, close-bonded fibrous structure or of a foraminous unitary construction.

They may be formed by processing on modifications either of textile type machines or papermaking equipment. In either case, the fibers as laid up in the basic web prior to bonding or to postforming may be oriented in one or more prescribed directions, or may be distributed in a completely random fashion. They are secured in place by suitable adhesives incorporated in the web. The application of these adhesives may be controlled to coat and bond the fibers completely, or to bond them only in selected areas, or at points of individual fiber contact.

Nonwovens may be thick or thin and of either low or high density. The conditions under which nonwoven fabrics are manufactured and the possible combinations of fibers and adhesives permit the production of structures offering a wide range of physical and chemical properties.

Production Methods

Production of a nonwoven fabric may be divided into two basic steps: (1) formation of the web and (2) bonding the web. The most widely used means for forming the web is a series of cotton cards feeding to a common conveyor belt to build up a unidirectional composite web of the desired weight. The number of cards per line will depend on the maximum weight of the product to be produced. Each card in the line may be geared to produce webs ranging from some 35 grains up to 100 grains per yard at speeds of 66.66 m/min down to 15 m/min for the heavier material. Material from these lines is usually limited to 101.6-cm widths, with strength favoring the machine direction over the cross machine direction in ratios from 3:1 to as much as 20:1. Where wider material of heavier weight, or when the strength balanced in the machine and cross machine directions is required, a web production line consisting of a single breaker and a finisher garnett equipped with a cross lapper may be used to advantage.

The most versatile type of web for nonwoven fabrics is that produced by the air disposition of precarded fibers that are collected with a minimum of orientation as a uniform mat.

In contrast to nonwoven fabrics made from webs that have been dry-processed on modified textile equipment are those produced from wet-layup webs using papermaking machines. Such webs usually depend on adhesive additives or postbonding to impart the necessary physical properties.

Once the web has been formed, by either the dry or wet layup process, it may be further modified by techniques such as needle punching, aeration, or impingement with gaseous, liquid, or other means, to produce a patterned configuration of desired characteristics.

Nonwovens of such postformed webs may be characterized by added resistance to delamination, superior drape, flexibility, porosity, abrasion, and flame resistance or other desirable properties.

Types of Fibers

In the production of nonwoven fabrics most every type of natural and anthropogenic fiber can be used. Price, equipment, and quality, as well as chemical and physical requirements of the product, govern the particular fiber used as well as the bonding agent. The use of more virgin first-quality fiber, especially the anthropogenic cellulosics, is preferred, in everything from diapers to casket liners.

Rayon is the predominant fiber used for both utility as well as aesthetic appeal. It is made in a wide variety of descriptions to fit the different manufacturing methods and end-use requirements. The finer deniers give the best tensile, tear, and bursting strength values. For special applications calling for particular chemical or electrical resistance, more use of the expensive synthetic fibers such as Acrilan, nylon, or Dacron may be warranted.

Bonding Agents

Properties of nonwoven fabrics are as dependent on the bonding agent as they are on the fiber that forms the foundation of the material. Both are selected with the end use in mind, and each must be compatible with the other.

Bonding agents may be grouped into three broad classifications: (1) liquid dispersions, (2) powdered adhesives, and (3) thermoplastic fibers.

Liquid dispersions. Liquid dispersions are the most extensive type used. Among these are polyvinyl alcohol, generally used as a preliminary binder or where high strength and permanence are not essential; polyvinyl acetate for good strength and flexibility where freedom from odor and taste are important; polyvinyl chloride (PVC) for good wet and dry strength, and toughness; synthetic latices of butadiene–acrylonitrile or butadiene–styrene for good adhesive and elastic properties where strength and a high degree of permanence are more important than color, stability, and odor; the acrylics for good strength, soft "hand," color stability, and permanence. These dispersions are applied (1) by spraying, generally used for low-density materials, (2) by saturator, for denser, more durable material, and (3) by printing, usually for selective bonding of localized areas in soft absorbent products.

Powdered adhesives. These are usually of thermosetting or thermoplastic resin types and are sifted into the fiber web as formed. They are used especially in the low-density, high-bulk nonwovens where wetting by the binder or the application of pressure might cause excessive matting and compression of the material. Bonding is effected by heating either with or without the use of pressure.

Thermoplastic fibers. The thermoplastic fiber binders have the advantage of constituting an integrated structural part of the fiber web that forms the fabric. To bond the web they may be activated by solvents or by heat and pressure. By regulation of the amount of heat and pressure as well as the amount of thermoplastic fiber present, a wide variety of characteristics may be built into these nonwoven fabrics.

Applications

Construction and performance of nonwoven fabrics have not been standardized. They are usually constructed to fit a particular end-use requirement or are built around particular specifications.

Industrial products for which nonwoven fabrics have been used include acoustical curtains; artificial leather and chamois; automotive plumpers; backing for adhesive tapes; base for vinyl and rubber coatings; bagging; buffing wheels; cable and wire wrappings; electrical tapes; filters for air, gases, and liquids; insulation; laminate reinforcements; polishing and wiping cloths; and wall coverings.

FABRICS, WOVEN

By far the greatest volume of textile materials is used in consumer textiles, such as apparel. But textiles are extremely versatile materials, which have been applied to a large number of engineered uses, e.g., thermal, acoustical, and electrical insulation; padding and packaging; barrier applications; filtration, both dry and wet; upholstery and seating; reinforcing for plastics or rubber; and various mechanical uses such as fire hose jackets, tenting, tarpaulins, parachutes, and marine lines.

Textiles are highly complex materials. Their properties depend not only on the fiber but on the form in which it is used — whether the form be a felt, a bonded fabric, a woven or knit fabric, or cordage. Properties such as heat, chemical, and weather resistance depend primarily on the type of fiber used; properties such as mechanical strength, thermal transmission, and air or liquid permeability depend both on the fiber and the textile form.

The versatility of textiles stems from (1) the wide range of fibers that can be used, and (2) the range of complicated textile structures that can be formed from the fibers.

There are two important factors that should be considered in discussing textile needs with textile suppliers:

1. The types of finishes that can be applied to the finished textile product can substantially alter or modify the stability, "hand," and/or durability of the textile.
2. Combining of textiles with other materials, such as resins or rubber, either by impregnation or by coating, will substantially alter performance characteristics of the final composite.

Textile Constructions

Textile engineering materials can be classified generally as (1) nonwoven fabrics, including both felts and bonded fabrics; (2) woven or knit fabrics, and (3) cordage. Nonwoven fabrics are discussed in a separate article.

Woven and Knit Fabrics

Woven fabrics and knit fabrics are composed of webs of fiber yarns. The yarns may be of either filament (continuous) or staple (short) fibers. In knit fabrics, the yarns are fastened to each other by interlocking loops to form the web. In woven fabrics, the yarns are interlaced at right angles to each other to produce the web. The lengthwise yarns are called the warp, and the crosswise ones are the filling (or woof) yarns.

The many variations of woven fabrics can be grouped into four basic weaves. In the plain weave fabric, each filling yarn alternates up and under successive warp yarns. With a plain weave, the most yarn interlacings per square inch can be obtained for maximum density, "cover," and impermeability. The tightness or openness of the weave, of course, can be varied to any desired degree. In twill weave fabrics, a sharp diagonal line is produced by the warp yarn crossing over two or more filling yarns. Satin weave fabrics are characterized by regularly spaced interlacings at wide intervals. This weave produces a porous fabric with a smooth surface. Satins woven of cotton are called sateen. In the leno weave fabrics, the warp yarns are twisted and the filling yarns are threaded through the twist openings. This weave is used for meshed fabrics and nets.

Cordage

The term *cordage* includes all types of threads, twine, rope, and hawser. Essentially all cordage consists of fibers twisted together, plied, and in many cases cabled to produce essentially continuous strands of desired cross section and strength.

In addition to the type of fiber used, the most important determinants of the end properties of cordage are the type and degree of twist employed. The two major types of twist are (1) cable twist, in which the direction of twisting is alternated in each successive operation, i.e., singles may be "S" twisted, plies "Z" twisted, and cables "S" twisted (a yarn or cord has an "S" twist if, when held in a vertical position, the spirals conform in direction of slope to the central portion of the letter "S," and a "Z" twist if the spirals conform in direction of slope to the central portion of the letter "Z"), and (2) hawser twist, in which the singles, plies, and cables are twisted "SSZ" or "ZZS." Hawser twist generally provides higher strength and resilience.

Specifications

Textile specifications contain two important types of information: (1) descriptive information and (2) service property requirements.

Specifications that physically describe the textile fabric usually include (1) width, in inches, (2) weight, usually in ounces per square yard, (3) type of weave, such as twill, broken twill, leno, or satin, (4) thread count, both in warp and filling (e.g., 68 × 44 denotes 68 warp yarns/in. and 44 filling yarns/in.), (5) type of fiber and whether the yarn is to be filament or staple, (6) crimp, in percent, (7) twist per inch, and (8) yarn number both for warp and fill.

Yarn number designations are somewhat complex, as they have been developed in a relatively unorganized fashion over the years, and different systems are used in different types of fibers. (Filament yarns are usually stated simply in denier, which is the weight in grams of 9000 m of yarn.)

Essentially, yarn numbers provide a measure of weight per unit length, or length per unit weight. A typical yarn designation on a specification may appear as "210 (denier)/1 × 20/2 (cotton system)." This means that (1) the warp yarn is 210-denier single yarn, and (2) the filling yarn contains 2 plies, each of which is a 20 singles yarn (determined by the cotton numbering system).

A number of fabric-designation systems have been formalized by tradition. For example, sheetings, drills, twills, jeans, broken twills, and sateens are designated only by width in inches, number of linear yards per pound, and number of warp and filling threads per inch. "Specs" for equivalent synthetic fabrics also include fiber type, whether staple or filament.

FASTENERS

Mechanical fasteners are among the most common components in fabricated products. Thus, fasteners are extremely important from both a manufacturing and a product standpoint. The proper selection of fasteners is necessary to provide the most value for the manufacturer as well as for the consumer.

The most common types of fasteners can be grouped into the following categories: bolts/screws, studs, pins, nuts, rivets, and holes with or without threads tapped into them.

Fastener Types and Materials Selection

Bolts and screws, two names for externally threaded fasteners, are certainly well-recognized fastener types. Like all fasteners, they are also a highly engineered method of joining. Figure F.1 is a schematic drawing of the significant features of a bolt, showing the tensile

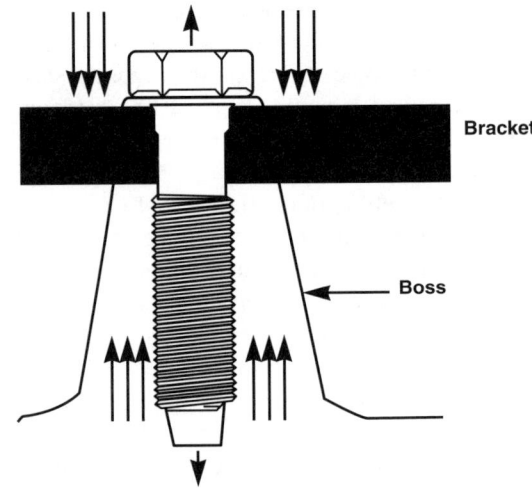

FIGURE 6.1 Features of a bolt. (From *Adv. Mater. Proc.*, 154(4), 49, 1998. with permission.)

force on the bolt and the compressive loading of the joint members.

The ability of a bolt to translate rotational motion into a clamping force has made it the standard fastener. The common bolt is also an ideal fastener from a reusability standpoint. Its helical thread form enables it to attach, detach, then reattach joint members an almost infinite number of times. The simple tools required are another reason that bolts have become so common. Bolts provide an excellent ratio of performance to cost, but during both the design phase and installation, they require more consideration than other fastener types to achieve an optimized joint.

Many different materials types are available for fasteners. Standard metric bolts and nuts are made of steels with varying chemical compositions.

A rivet is a non-reusable fastener that is deformed to provide a mechanical clinching of joint members. Both aluminum and steel may be fabricated into rivets. In fact, the stem and body of a rivet can be easily made with two different materials. Low-carbon steel alloys in the AISI/SAE 1XXX series may be selected when cost is the most important factor, or when a higher-strength rivet is needed. Typical tensile strength values for joints made with steel rivets are 5 to 10 kN, which is about a third to half the value of equivalently sized bolts. An aluminum rivet usually has about half to two thirds the strength of steel rivets.

However, the shear resistance of both types of rivets can equal their tensile resistance, and can even exceed the value for a similar sized bolt. The typical aluminum alloys for rivets in automotive applications are AA 5052, 5056, 7075, 7178, and 2014. These alloys provide good strength, high corrosion resistance (without the need for additional corrosion protection), and light weight.

Some rivets are made of thermoplastics, namely, polyamide 6 (PA6), polyamide 66 (PA66), and polyoxymethylene (POM). These are semicrystalline polymers with high strength, low mass, and resistance to automotive fluids and salt environments. As a result, they are also ideal for the myriad pins and clips that connect plastic components.

Polyamides are also coated onto bolts and nuts and the plastic is added to reduce the allowance between the mating metallic threads, making them more resistant to vibrational loosening.

Fluoropolymers are commonly added to the threaded region of weld nuts prior to installation, because of their low coefficient of friction.

Finally, polymer adhesives are often added to bolts. These bolts resist vibrational loosening through mechanical means, such as thermoplastic additions or locally deformed features. In fact, the application of adhesives is the only method that provides long-term resistance to loosening.

A final material being considered for fasteners is titanium. The mechanical properties of titanium are rather high. If low density is also required, as it is for automotive structures, then titanium appears to be the only current material that can challenge traditional steel alloys. The workhorse alloy titanium, 6% aluminum, 4% vanadium (Ti–6Al–4V) could definitely be chosen for fastener applications.

The advantages of titanium are its extremely high strength-to-weight ratio and its almost complete imperviousness to corrosive attack by chloride ions. This means that a lower-mass fastener can be made, one that does not require the expensive and time-consuming application of corrosion-resistant coatings. Many mechanical joints in aerospace structures currently rely on titanium alloys, including Ti–6Al–4V.

If the raw material costs of titanium can be reduced, it will become a viable alternative material for automotive applications, with fasteners at the top of the list.

FELDSPAR

Feldspars (Al_2O_3 and SiO_4 tetrahedra) constitute 60% of the outer 13 to 17 km of the earth's crust and are the most common mineral in crystalline rocks. Feldspars are aluminum silicates of potassium, sodium, and calcium.

The importance of the many feldspars that occur so widely in igneous, metamorphic, and some sedimentary rocks cannot be underestimated, especially from the viewpoint of a petrologist attempting to unravel Earth history.

With weathering, feldspars form commercially important clay materials. Economically. feldspars are valued as raw material for the ceramic and glass industries, as fluxes in iron smelting, and as constituents of scouring powders. Occasionally, their luster or colors qualify them as semiprecious gemstones. Some decorative building and monument stones are predominantly composed of weather-resistant feldspars.

Knowledge of the composition of a feldspar and its crystal structure is indispensable to an understanding of its properties. However, it is the distribution of the aluminum and silicon atoms among the available tetrahedral sites in each chemical species that is essential to a complete classification scheme, and is of great importance in unraveling clues to the crystallization and thermal history of many igneous and metamorphic rocks.

FELTS

Classes of felts in use as engineering materials are wool and synthetic fiber felts.

Wool felt is a fabric obtained as a result of the interlocking of wool fibers under suitable combinations of mechanical work, chemical action, moisture, and heat, alone or in combination with other fibers.

Synthetic fiber felt is a fabric obtained as a result of interlocking of synthetic fibers by mechanical action.

PROPERTIES

Neither wool nor synthetic fiber felts require binders and exist as 100% fibrous materials.

Felts generally exhibit the same chemical properties as do the fibers of which the felt is composed. Wool felts are characterized by excellent resistance to acids but are damaged by exposure to strong alkalies. They exhibit remarkable resistance to atmospheric aging and are as a class probably the most inert of all nonmetallic engineering materials with respect to nonaqueous liquids, oils, or solvents. Wool felts are not generally recommended for stressed dry uses at temperatures in excess of 82°C, because of changes in physical properties, but are used as dry spacers and gaskets in many applications up to 149°C, and the use of the materials as oil wicks and lubricating system components at ambient temperatures up to 149°C is common.

Synthetic fiber felts are available in virtually all classes of fiber composition including regenerated cellulose, cellulose acetate, cellulose triacetate; polyamide, polyester, acrylic, modacrylic, olefin, and TFE fluorocarbon (tetrafluoroethylene). The variety of types available provides a virtually infinite range of physical and chemical properties for application beyond the natural versatility of wool fiber felts. Outstanding among the properties of this class of engineering material are chemical, solvent, thermal, and biological stability as well as low moisture absorption, quick drying after aqueous wetting, abrasion resistance, and frictional and dielectrical features.

FORMS

Wool felts are produced as "sheet" stock in a standard 91.4 × 91.4 cm size in thickness ranging from 1.6 to 76 mm. There are a number of density classifications based on the weight for a 914 × 914-mm sheet in 25.4-mm thickness from 5.4 to 14.4 kg with the weight for any given thickness in the density class proportioned to the weight per square meter at the 25.4-mm thickness. "Roll" felts are produced in either 1522- or 1830-mm widths in lengths up to 160 m and standard thicknesses range from 0.8 to 25.4 mm.

Synthetic fiber felt widths are available from 600 to 1830 mm in thicknesses from 0.8 to 19 mm.

Both wool and synthetic fiber felts are nonforming and nonraveling and thus provide considerable ease of cutting and fabricating. In addition, wool felt lends itself to most grinding, cutting, shaping, extruding, and other machining operations so that special shaped parts such as polishing laps, round wicking, ink rollers, and others are produced.

APPLICATIONS

The structural elasticity of felts as a class makes these materials suitable for molding and forming and parts of these types are produced to provide special shaped gaskets, seals, fillers, instrument covers, and the like. In many cases these shaped parts are stabilized through the use of resinous or rubber impregnants and this modified class of felt is finding ever-increasing use as flat stock for special gasketing, sealing, and other applications. Combinations of felt and plastic and elastomer sheet materials are also produced for application where resilience plus nonpermeability is desired as in sealing and frictional uses.

Use of felt as an engineering material covers a broad spectrum of application. Major contributing properties include resilience, mechanical, thermal, and acoustic energy absorption; high porosity-to-weight ratio; resistance to aging; thermal and chemical stability; high effective surface area per unit volume; and solvent resistance.

Applied uses include wet and dry filtration, thermal and acoustical insulation, vibration isolation, impact absorption, cushioning and packaging, polishing, frictional surfacing, liquid absorption and reservoirs, wicking, gasketing, sealing, and percussion mechanical dampening.

FERRITE

A ferrite is any of the class of magnetic oxides. Typically, the ferrites have a crystal structure that has more than one type of site for the cations. Usually the magnetic moments of the metal ions on sites of one type are parallel to each other, and antiparallel to the moments on at least one site of another type. Thus, ferrites exhibit ferromagnetism.

COMMERCIAL TYPES

There are three important classes of commercial ferrites. One class has the spinal structure. The second class of commercially important ferrites has the garnet structure. Yttrium-based garnets are used in microwave devices. Thin monocrystalline films of complex garnets have been developed for bubble domain memory devices. The third class of ferrites has a hexagonal structure of the magnetoplumbite type. Because of their large magnetocrystalline anisotropy, the hexagonal ferrites develop high coercivity and are an important member of the permanent magnet family.

PROPERTIES

The important intrinsic parameters of a ferrite are the saturation magnetization, Curie temperature, and magnetocrystalline (K_1) and magnetostrictive (Ω_x) anisotropies. These properties are determined by the choice of the cations and their distribution in the various sites.

In addition to the intrinsic magnetic parameters, microstructure plays an equally important role in determining device properties. Thus, grain size, porosity, chemical homogeneity, and foreign inclusions dictate in part such technical properties as permeability, line width, remanence, and coercivity in polycrystalline ceramics. In garnet films for bubble domain device applications, the film must essentially be free of all defects such as inclusions, growth pits, and dislocations.

PREPARATION

Polycrystalline ferrites are most economically prepared by ceramic techniques. Component oxides or carbonates are mixed, calcined at elevated temperatures for partial compound formation, and then granulated by ball milling. Dispersants, plasticizers, and lubricants are added, and the resultant slurry is spray-dried, followed by pressing to desired shape and sintering. The last step completes the chemical reaction to the desired magnetic structure and effects homogenization, densification, and grain growth of the compact. It is perhaps the most critical step in optimizing the magnetic properties of commercial ferrites.

FERRITE DEVICES

These are electrical devices whose principle of operation is based on the use and properties of ferrites, which are magnetic oxides. Ferrite devices are divided into two categories, depending on whether the ferrite is magnetically soft (low coercivity) or hard (high coercivity). Soft ferrites are used primarily as transformers, inductors, and recording heads, and in microwave devices. Since the electrical resistivity of soft ferrites is typically 10^6 to 10^{11} times that of metals, ferrite components have much lower eddy current losses and hence are used at frequencies generally above about 10 kHz. Hard ferrites are used in permanent-magnet motors, loudspeakers, and holding devices, and as storage media in magnetic recording devices.

CHEMISTRY AND CRYSTAL STRUCTURE

Soft ferrite devices are spinels with the general formula of MFe_2O_4, in which M is a divalent metal ion. The commercially practical ferrites are those in which the divalent ion represents one or more magnesium, manganese, iron, cobalt, nickel, copper, zinc, and cadmium ions. The trivalent iron ion may also be substituted by other trivalent ions such as aluminum. The compositions are carefully adjusted to optimize the device requirements, such as permeability, loss, ferromagnetic resonance line width, and so forth.

The ferromagnetic garnets have the general formula of $M_3Fe_5O_{12}$, in which M is a rare earth or ytrrium ion. Single-crystal garnet films form the basis of bubble domain device technology. Bulk garnets have applications in microwave devices.

Hard ferrites for permanent-magnetic device applications have the hexagonal magneto-plumbite structure, with the general formula $MFe_{12}O_{19}$, where M is usually barium or strontium.

The material of choice for magnetic recording is δ–Fe_2O_3, which has a spinel structure.

With the exception of some single crystals used in recording heads and special microwave applications, and particulates used as storage media in magnetic recording, all ferrites are prepared in polycrystalline form by ceramic techniques.

APPLICATIONS

Applications of ferrites may be divided into nonmicrowave, microwave, and magnetic recording applications. Further, the nonmicrowave applications may be divided into categories determined by the magnetic properties based on the B–H behavior, that is, the variation of the magnetic induction or flux density B with magnetic field strength H. The categories are linear B–H, with low flux density, and nonlinear B–H, with medium to high flux density. The highly nonlinear B–H, with a square or rectangular hysteresis loop, was once exploited in computer memory cores.

Linear B–H Devices

In the linear region, the most important devices are high-quality inductors, particularly those used in filters in frequency-division multiplex telecommunications systems and low-power wideband and pulse transformers. Virtually all such devices are made of either MnZn ferrite or NiZn ferrite, although predominantly the former.

Nonlinear B–H Devices

The largest usage of ferrite measured in terms of material weight is in the nonlinear B–H range, and is found in the form of deflecting yokes and flyback transformers for television receivers. Again, MnZn and NiZn ferrites dominate the use in these devices.

A rapidly growing use of ferrites is in the power area, where ferrite transformers are extensively used in switched mode (AC to DC) and converter mode (DC to DC) power supplies. Such power supplies are widely used in various computer peripheral equipment and private exchange telephone systems.

Microwave devices. Microwave devices make use of the reciprocal propagation characteristics of ferrites close to or at a gyromagnetic resonance frequency in the range of 1 to 100 GHz. The most important of such devices are isolators and circulators. The garnets have highly desirable, small, ferromagnetic-resonance linewidths, particularly in single-crystal form.

Magnetic recording devices. Vast amounts of audio and video information and digital data from computers are stored in magnetic tapes and disks. Here magnetic recording materials function as hard magnetic materials. The most widely used particles in magnetic recording are δ-Fe_2O_3 and co-modified δ-Fe_2O_3. The basic δ-Fe_2O_3 is generally used in audio recording, while the higher-coercivity Co-δ-Fe_2O_3 dominates video recording.

FERROALLOYS

Ferroalloys compose an important group of metallic raw materials required for the steel industry. Ferroalloys are the principal source of such additions as silicon and manganese, which are required for even the simplest plain-carbon steels, and chromium, vanadium, tungsten, To, and molybdenum, which are used in both low- and high-alloy steels. Also included are many other more complex alloys. Ferroalloys are unique in that they are brittle and otherwise unsuited for any service application, but they are important as the most economical source of these elements for use in the manufacture of the engineering alloys. These same elements can also be obtained, at much greater cost in most cases, as essentially pure metals. The ferroalloys contain significant amounts of iron and usually have a lower melting range than the pure metals and are therefore dissolved by the molten steel more readily than the pure metal. In other cases, the other elements in the ferroalloy serve to protect the critical element against oxidation during solution and thereby give higher recoveries. Ferroalloys are used both as deoxidizers and as a specified addition to give particular properties to the steel.

Many ferroalloys contain combinations of two or more desirable alloy additions, and well over 100 commercial grades and combinations

TABLE F.1
Analysis of Typical Ferroalloys (wt%)

Type of Ferroalloy[a]	Mn	Si	C	Cr	Mo	Al	Ti	V
Ferromanganese								
Standard	78–82	1.25	7.5					
Medium carbon	80–85	1.25–2.5	1–3[b]					
Low carbon	80–85	1.25–7.0	0.75					
Ferrosilicon								
50% regular	—	47–52	0.15					
75% regular	—	73–78	0.15					
Ferrochromium								
High carbon	—	1–2	4.5–6.0[b]	67–70				
Low carbon		0.3–1.0	0.03–2.0	68–71				
SM low carbon	4–6	4–6	1.25	62–65				
Ferromolybdenum								
High carbon	—	1.5	2.5	—	55–70			
Ferrovanadium								
High carbon	—	13.0	3.5	—	—	1.5	—	30–40
Ferrotitanium								
Low carbon	—	3–5	0.1	—	—	6–10	38–43	

[a] In all cases the balance is iron, with the exception of minor impurities. The latter are usually specified, such as 0.10% max P.
[b] In several specified grades within this range.

Source: *McGraw-Hill Encyclopedia of Science and Technology*, 8th ed., Vol. 7, McGraw-Hill, New York, 62. With permission.

are available. Although of less general importance, other sources of these elements for steelmaking are metallic nickel, nickel, SiC, Mo_2O_5, and even misch metal (a mixture of rare earths).

Analyses of a few typical ferroalloys are given in Table F.1.

The three ferroalloys that account for the major tonnage in this class are the various grades of silicon, manganese, and chromium. For example, 5.9 kg of manganese is used on the average in the United States for every ton of open-hearth steel produced. Elements supplied as ferroalloys are among the most difficult metals to reduce from ore.

FERROCHROMIUM

A high-chromium iron master alloy used for adding chromium to irons and steel, ferrochromium is also called ferrochrome. It is made from chromite ore by smelting with lime, silica, or fluorspar in an electric furnace. High-carbon-ferrochrome is used for making tool steels, ball-bearing steels, and other alloy steels. It melts at about 1250°C. Low-carbon-ferrochrome is used for making stainless steels and acid-resistant steels. Simplex ferrochrome comes in pellet form and is used for making low-carbon stainless steels. Low-carbon ferrochrome is also preferred for alloy steel mixtures where much scrap is used because it keeps down the carbon and inhibits the formation of hard chromium carbides. The various grades of ferrochromium are also marketed as high-N_2 ferrochrome, and for use in making high-chromium cast steels where the N_2 refines the grain and increases the strength. Foundry-grade ferrochrome is used for making cast irons, as well as for ladle additions to cast iron to give uniform structure and increase the strength and hardness.

FERROMANGANESE

This is a master alloy of manganese and iron used for deoxidizing steels, and for adding manganese to iron and steel alloys and bronzes. Manganese is the common deoxidizer and cleanser of steel, forming oxides and sulfides that are carried off in the slag. Ferromanganese is made from the ores in either the blast furnace or the electric furnace. Spiegeleisen is a form of low-manganese ferromanganese and the German name, meaning mirror iron, is derived from the fact that the crystals of the fractured face shine like mirrors. Spiegeleisen has the advantage that it can be made from low-grade manganese ores, but the quantity needed to obtain the required proportion of manganese in the steel is so great that it must be premelted before adding to the steel. It was used for making irons and steels by the Bessemer process.

FERROPHOSPHORUS

This substance is an iron containing a high percentage of phosphorus, used for adding phosphorus to steels. Small amounts of phosphorus are used in open-hearth steels to make them free-cutting, and phosphorus is also employed in tinplate steels to prevent the sheets from sticking together in annealing. Ferrophosphorus is made by melting phosphate rock together with the ore in making pig iron. There is also a master alloy, ferroselenium, for adding selenium to steels, especially stainless steels, to give free-machining qualities.

FERROSILICON

This is a high-silicon master alloy used for making silicon steels, and for adding silicon to transformer irons and steels. It is made in the electric furnace by fusing quartz or silica with iron turnings and carbon. Silicon is often added to steels in combination alloys with deoxidizers or other alloying elements. Ferrosilicon aluminum is a more effective deoxidizer for steel than aluminum alone. It is also used for adding silicon to aluminum casting alloys. The alloy serves as a deoxidizer, fluxes the slag inclusions, and also controls the grain size of the steel.

FERROTITANIUM

A master alloy of titanium with iron, ferrotitanium is used as a purifying agent for irons and steel owing to the great affinity of titanium for O_2 and N_2 at temperatures above 800°C. The value of the alloy is as a cleanser, and little or no titanium remains in the steel unless the percentage is gauged to leave a residue. The ferrocarbon titanium is made from ilmenite in the electric furnace, and the carbon-free alloy is made by reduction of the ore with aluminum. Ferrotitanium comes in lumps, crushed, or screened, and it is used for ladle additions for cleansing steel. Low-carbon ferrotitanium is used as a deoxidizer and as a carbide stabilizer in high-chromium steels. Graphidox improves the fluidity of steel, increases machinability, and adds a small amount of titanium to increase the yield strength, and the Grainal alloys which are used to control alloy steels and have various compositions.

MnTi is used as a deoxidizer for high-grade steels and for nonferrous alloys.

NiTi is used for hard nonferrous alloys and columbium is used for adding columbium to steel. It has an exothermic reaction that prevents chilling of the molten metal.

FERROUS P/M

Specifying powder-metallurgy (P/M) parts and their consolidation process used to be a simple process: Design the part, select the metal powders and lubricants that provide the required properties, compact the powders into a briquette, and sinter the briquette into its finished form. Through this procedure, millions of parts have been produced for applications ranging from automobiles to appliances and from business to farm and garden machines.

However, the needs of industries have changed significantly. Removing weight from all products has risen to primary importance. Energy, tooling, and materials costs now figure prominently in parts design, and productivity has emerged as the new watchword

With these changes have come changes in P/M technology. Through the many manufacturing processes, improvements have been made in the powders themselves — improvements such as lower levels of inclusions and higher compressibility. In addition to conventional iron and steel metals, the list of available powders has been expanded to include new classes of tool steel, as well as materials such as cermets and alloys of titanium, nickel, and aluminum.

Accompanying these developments has been the growth of new consolidation technologies. As a result, design engineers need current information on which P/M technologies are viable, cost-effective, and production-effective, and which have potentially wide application.

Although P/M is used to fabricate parts from just about any metal, the most commonly used metals are the iron-base alloys. Low-density iron P/M parts (5.6 to 6.0 g/cm^3), with a typical tensile strength of 108.8 MPa are usually used in bearing applications. Copper is commonly added to improve both strength and bearing properties. Alloy-steel powders are sometimes hot-forged to high or nearly theoretical density to form parts with improved mechanical properties that, when heat-treated, may have tensile strengths to 1156 MPa. Powder forging (P/F) is now established as a serious contender for parts formerly made as wrought forgings or machined from mill forms.

Iron P/M or sintered Fe–Cu alloy strength can be varied by adjusting density, carbon content (up to 15%), or all three to satisfy specific design requirements. The mechanical properties of ferrous powder parts can be considerably improved by impregnating or infiltrating them with any one of a number of different materials, both metallic and nonmetallic, such as oil, wax, resins, copper, lead, and babbit.

Low-density P/M parts are used in bearing applications because they provide porosity for oil storage. Impregnating sintered-metal bearings with oil usually eliminates the need for relubrication.

For higher-strength needs, alloyed (frequently prealloyed nickel–molybdenum–iron) iron, compacted to a higher density, is used. When carbon or other alloying elements are mixed with the iron powders and densities exceed 6.2 g/cm^3, the parts are considered to be steel rather than iron. As carbon content is increased up to 1%, the strength of steel P/M parts increases, just as the strength of wrought steel increases with higher carbon content.

Ferrous-base P/M parts can range in size from about 2.5 mm thick and 3.2 mm in diameter to 50.8 mm thick and over 612 mm in diameter. Because they can be mass produced at relatively low cost, iron-base P/M parts find a wide variety of uses in such high-volume products as appliances, business machines, power tools, and automobiles. Typical parts are gears, bearings, rotors, valves, valve plates, cams, levers, ratchets, and sprockets.

Additional applications can be accommodated by sealing the pores in iron P/M parts. The sealing materials used are copper, polyesters, and anaerobics; each requires a different processing system to impregnate the parts. Impregnation of sintered P/M parts is done for any of several reasons:

- To serve in pressure-tight applications
- To improve surface finish (impregnated parts are platable)
- To improve machinability
- To improve corrosion resistance

Although high precision has been achieved in P/M parts for many years, their application was once restricted because of mechanical property limitations. Now, however, mechanical properties can be increased in steel P/M parts by hot forging in closed dies. Properties of P/M parts forged to 100% theoretical density in production conditions are claimed to be equal, and sometimes superior, to those of wrought steels of similar composition.

FIBER-REINFORCED GLASS

Fiber-reinforced glass (FRG) composites are glass or glass ceramic matrices reinforced with long fibers of carbon or SiC. These composites are lighter than steel but just as strong as many steel grades, and can resist higher temperatures. They also have outstanding resistance to impact, thermal shock, and wear, and can be formulated to control thermal and electrical conductivity. With proper tooling, operations such as drilling, grinding, and turning can be completed in half the time required for nonreinforced glass.

Currently, FRG components are primarily used for handling hot glass or molten aluminum during manufacturing operations. FRG is also under test as an engineering material in a variety of markets, including the aerospace, automotive, and semiconductor industries.

PROPERTIES

Glass and glass ceramics are versatile materials because of such characteristics as high chemical resistance and special electrical behavior, but fragility under tensile stress has limited their structural applications. However, the addition of continuous fibers of carbon and SiC into glass produces a material that withstands very high mechanical stresses and loads. As a result, it is now possible to manufacture glass matrix composites that are capable of structural roles (Table F.2).

The reinforcement process provides an increase in both admissible maximum stress and ultimate elongation: brittle fracture is replaced by an almost ductile behavior. These properties, together with others, make FRG composites advantageous as a material in the machine-building industry. For example, low density and high modulus of elasticity allow extremely strong and stiff structures.

Another important characteristic of long-fiber-reinforced glasses is that their mechanical properties are largely independent of the condition at the surface. This means that they can be drilled, even near edges, and joined with other parts by means of screws and bolts.

In addition, FRG composites have very high resistance to temperature changes, low coefficient of thermal expansion, high specific heat capacity, and good chemical resistance.

They also exhibit good tribological and wear properties. However, because of the great variety of tribological and tribo-mechanical processes and applications, each service environment must be evaluated before selecting an FRG composite.

Fiber Effects

Performance of FRG composites under various operating conditions depends primarily on the type of fiber, the amount of fiber, and its orientation within the matrix. It also depends on the operating environment and the duration or cycle of the specific application.

Fiber Types

Carbon fibers are stable up to >2000°C under inert atmospheres, but few applications require such performance. Moreover, at those temperature levels, the matrix itself becomes soft. In air, carbon fibers remain stable to about 450°C. This temperature limit also applies to composites containing carbon fibers, unless the fibers have been completely sealed.

By contrast, SiC fibers are stable in air up to about 1200°C. In this case, the heat resistance of the matrix is the limiting factor. Therefore, SiC FRG products can be considered stable to 500°C if the matrix consists of the alkali–borosilicate–glass Duran.

On the other hand, if an alkali-free aluminosilicate glass is the matrix, the SiC-fiber-containing composite is stable up to 750°C.

When quenched from 350 to 20°C, most non-reinforced glasses break. However, SiC FRG is capable of withstanding a 60-fold quench (thermal shock) from 550°C to ambient temperatures. Furthermore, the product has exhibited good fracture toughness at temperatures as low as –200°C.

Amount of Fiber

Performance at high temperatures and resistance to thermal shock can be improved by increasing the percentage of fiber within the composite. However, fiber is the chief cost factor in producing FRG. Therefore, performance advantages must continue to outweigh costs as fiber count is increased.

Fiber Orientation

Depending on the arrangement of the fibers, the properties of a FRG composite can be either isotropic or anisotropic. For example, fibers may be arranged so that heat conductivity is low in one direction, but high in the perpendicular direction. Absolute values of heat conductivity range from 1.7 to ~25 W/m·K or half the value of steel. As in high-temperature performance, heat conductivity depends on the direction, type, and amount of fiber in the composite.

Relatively high conductivity values result when the composite contains carbon fibers aligned longitudinally in the direction of heat travel; otherwise, the material may work as a heat barrier.

TABLE F.2
Properties of FRG Composites vs. Unreinforced Glass and Various Steel Grades

Property	SiC-Reinforced Duran Glass, Fibers Oriented 0/90	Carbon-Reinforced Duran Glass, Fibers Oriented 0/90	SiC-Reinforced 8252 Aluminosilicate	SiC-Reinforced Machinable Glass Ceramic, Fibers Oriented 0/90	Unreinforced Duran Glass	Stainless Steel 1.4301/1.4304[a]	Carbon Steel
Density, g/cm^3	2.5	1.9–2.2	2.5	2.5	2.2	8	1.8–2.5
Bending strength, MPa (3-point bending)	450	500	450	400–500	30–50	Mean 600	40
Young's modulus, Gpa	110	130	110	100	63	190–210	10
Max. strain, %	0.5–1	0.5–1	0.5–1	1	0.1	15	0.1
Work of fracture, J/m^2	3.5×10^4	4×10^4	3.5×10^4	[b]	1.2×10^2	[c]	[c]
Coefficient of thermal expansion, ppm/K	2–4	0–5 at 0° / 5–15 at 90°	2–4	2–4	3.3	15–18	4–8
Thermal conductivity, W/(K·m)	1.5–3	1.5 at 0° / 1.5 at 90°	1.5–3	1.5–3	1.1	20	50–150
Thermal shock resistance, ΔK	>530	>450	>730	>900	>100	[c]	[c]
Max. application temperature, °C							
Short time (seconds)	600	500	800	1000	600	650	550
Long time (hours)	550	450	750	900	530	550	450

[a] 1.4301 (Fe/Cr18/Ni10) and 1.4304 (Fe/Cr18/Ni10/Mo3) are material numbers for two types of stainless steel, used in, i.e., medical or vacuum applications.
[b] Values not measured until now.
[c] Values for those properties are not known (here) or not meaningful for those materials.

Similarly, the electrical resistance of composites has been measured at room temperature in a composite 40% fiber by volume. In the fiber direction, specific electrical resistance for the SiC product was 10 $\Omega \cdot$cm, while for the carbon product, it was 0.01 $\Omega \cdot$cm. Normal to the fibers, the specific electrical resistance may be several orders of magnitude higher.

POTENTIAL APPLICATIONS

FRG components are used as replacements for asbestos and other materials for handling hot glass during glass manufacturing operations. These include pushers, grips, transporters, and takeout pads as well as other parts in contact with hot glowing glass. It is an application characterized by extreme conditions in terms of heat and thermal shock, in which FRG composites have a great advantage over traditional materials.

Within the metals industry, FRG pads act as thermal buffers in the handling of molten aluminum and its alloys. In this application, they form insulation pads between the melt vessel and supporting structural work, thereby minimizing heat conductivity away from the melt.

Other applications being exploited are the replacement for fragile glass substrates that hold silicon wafers during chemical vapor deposition (CVD) coating processes. Required properties include high fracture toughness for higher yields, a low coefficient of thermal expansion, and high stiffness.

Automotive applications include piston inserts, valve-control components, and other parts subjected to thermal and mechanical shock. FRG may also be applied in products for the protection of areas that must be insulated from heat, or where fracture toughness is needed.

The tribological and wear properties of FRG suggest applications as bearings and seals in pump manufacturing, and for off-road industrial vehicle and equipment brakes. Other potential applications include aerospace components with glass ceramic matrices for temperatures above 1000°C; scanning-mirror substrates for space and missile systems; and protective inserts for parts requiring high impact resistance. As FRG manufacturing technology advances and costs are reduced, the composites would be suitable for the construction of safes and strong rooms, and as armor in vehicles.

FRG components are being produced in several configurations, which include plates, disks, or rings that have a maximum dimension (length, width, or diameter) of 400 mm, and thickness of 50 mm. The minimum thickness of unprocessed forms is 0.5 mm. Semifinished material can be reprocessed into end products with complex geometric forms.

FIBER-REINFORCED PLASTICS

Fiber-reinforced plastics (FRPs) comprise a broad group of composite materials composed of fibers embedded in a plastic resin matrix. In general, they have relatively high strength-to-weight ratios and excellent corrosion resistance compared to metals. They can be formed economically into virtually any shape and size. In size, FRP products range from tiny electronic components to large boat hulls. Between these extremes, there are a wide variety of FRP gears, bearings, housings, and parts used in all product industries.

FRPs are composed of three major components — matrix, fiber, and bonding agent. The plastic resin serves as the matrix in which are embedded the fibers. Adherence between matrix and fibers is achieved by a bonding agent or binder, sometimes called a coupling agent. Most plastics, both thermosets and thermoplastics, can be the matrix material. In addition to these three major components, a wide variety of additives — fillers, catalysts, inhibitors, stabilizers, pigments, and fire retardants — can be used to fit specific application needs.

FIBERS

Glass is by far the most-used fiber in FRPs. Glass-fiber-reinforced plastics are often referred to as GFRP or GRP. Asbestos fiber has some use, but is largely limited in applications where maximum thermal insulation or fire resistance is required. Other fibrous materials used as reinforcements are paper, sisal, cotton, nylon, and Kevlar. For high-performance parts and components, more costly fibers, such as boron, carbon, and graphite, can be specified.

FIBER-REINFORCED PLASTICS

The standard glass fiber used in GRP is a borosilicate type, known as E-glass (E-Gl). The fibers are spun as single glass filaments with diameters ranging from 0.01 to 0.03 mm. These filaments, collected into strands, usually around 200 per strand, are manufactured into many forms of reinforcement. The E-Gl fibers have a tensile strength of 3447 MPa. Another glass fiber, known as S-G1, is higher in strength, but because of its higher cost its use is limited to advanced, high-performance applications. In general, in reinforced thermoplastics, glass content runs between 20 and 40%; with thermosets it runs as high as 80% in the case of filament-wound structures.

There are a number of standard forms in which glass fiber is produced and applied in GRP.

1. Continuous strands of glass supplied either as twisted, single-end strands (yarn) or as untwisted multistrands (continuous) roving
2. Fabrics woven from yarns in a variety of type, weights, and widths
3. Woven rovings — continuous rovings worked into a coarse, heavy, drapable fabric
4. Chopped strands made from either continuous or spun roving cut into 3.2- to 12.7-mm lengths
5. Reinforcing mats made of either chopped at random or continuous strand laid down in a random pattern
6. Surfacing mats composed of continuous glass filaments in random patterns

Resins

Although a number of different plastic resins are used as the matrix for reinforced plastics, thermosetting polyester resins are the most common. The combination of polyester and glass provides a good balance of mechanical properties as well as corrosion resistance, low cost, and good dimensional stability. In addition, curing can be done at room temperature without pressure, thus making for low processing-equipment costs.

Polyesters are also available as casting resins, both in water-extended formulations for low-cost castings, and in compounds filled with ground wood or pecan-shell flour for furniture components.

Low-profile molding resins are mixtures of polyester resins, thermoplastic polymers, and glass-fiber reinforcements. These are used to mold parts with smooth surfaces that can be painted without the need for prior sanding.

For high-volume production, special sheet-molding compounds (SMC) are available in continuous-sheet form. Resin mixtures of thermoplastics with polyesters have been developed to produce high-quality surfaces in the finished molding.

These polyester resins are available for use as coatings for curing by ultraviolet radiation. These 100%-solids materials cure in a matter of seconds and release no solvents. Although some styrene may be lost upon exposure to ultraviolet radiation, the amount is small.

Prepregs are partially cured thermoset resin-coated reinforcing fabrics in roll or sheet. The prepregs (short for preimpregnated) can be laid or wrapped in place and then fully cured by heat.

Other glass-reinforced thermosets include phenolics and epoxies. GR (glass-reinforced) phenolics are noted for their low cost and good overall performance in low-strength applications. Because of their good electrical resistivity and low water absorption, they are widely used for electrical housings, circuit boards, and gears. Since epoxies are more expensive than polyesters and phenolics, GR epoxies are limited to high-performance parts where their excellent strength, thermal stability, chemical resistance, and dielectric strength are required.

Initially, GRP materials were largely limited to thermosetting plastics. Today, however, more than 1000 different types and grades of reinforced thermoplastics or GR+P are commercially available. Leaders in volume are nylon and the styrenes. Unlike thermosetting resins, Gr+P parts can be made in standard injection-molding machines. The resin can be supplied as pellets containing chopped glass fibers. As a general rule, a GR+P with chopped fibers at least doubles the tensile strength and stiffness of the plastic. Glass-reinforced thermoplastics are also produced as sheet materials for forming on metal-stamping equipment and compression-molding machines.

Processing Methods

Matched Metal Die Molding

This is the most efficient and economical method for mass-producing high-strength parts. Parts are press-molded in matched male and female molds at pressures of 1380 to 2068 MPa and at heats of 113 to 127°C.

Four main forms of thermosetting resin reinforcement are used:

1. Chopped fiber preforms, shaped like the part, are saturated with resin at the mold. They are best for deep-draw, compound curvature parts.
2. Flat mat, saturated with resin at the mold, is used for shallow parts with simple curvature.
3. SMC, a preimpregnated material, has advantages for parts with varying thickness. SMCs consist of polyester resin, long glass fibers (to 5.08 cm), a catalyst, and other additives. They are supplied in rolls, sandwiched between polyethylene carrier films. Uniformity of SMC materials is closely controlled, making these materials especially suited for automated production. Structural SMCs contain up to 65% glass in continuous, as well as random, fiber orientation. This compares with 20 to 35% glass, in random orientation only, of conventional SMC.
4. Bulk molding compound (BMC), a premix of polyester resin, short glass fibers (3.2 to 6.4 mm long), filler, catalyst, and other additives for specific properties. BMC is supplied in bulk form or as extruded rope for ease of handling and it is used for parts similar to castings.

Injection Molding

In this high-volume process, a mix of short fibers and resin is forced by a screw or plunger through an orifice into the heated cavity of a closed matched metal mold. It is the major method for forming reinforced thermoplastics and is used for thermoplastic-modified thermosetting BMCs.

Hand Layup

This is the simplest of all methods of forming thermosetting composites. It is best employed for quantities under 1000, for prototypes and sample runs, for extremely large parts, and for larger volume where model changes are frequent, as in boats. In hand layup, only one mold is used, usually female, which can be made of low-cost wood or plaster. Duplicate molds are inexpensive. The reinforcing mat or fabric is cut to fit, laid in the mold, and saturated with resin by hand, using a brush, roller, or spray gun. Layers are built up to the required thickness; then the laminate is cured to permanent hardness, generally at room temperature.

Spray-Up

Like hand layup, the spray-up method uses a single mold, but it can introduce a degree of automation. This method is good for complex thermoset moldings, and its portable equipment eases on-site fabrication and repair. Short lengths of reinforcement and resin are projected by a specially designed spray gun so they are deposited simultaneously on the surface of the mold. Cure is usually accomplished by a catalyst in the resin at room temperature.

Filament Winding

This method produces moldings with the highest strength-to-weight ratio of any reinforced thermoset because of its high glass-to-resin ratio. It is generally limited to surfaces of revolution — round, oval, tapered, or rectangular — but it can achieve a high degree of automation. Continuous fiber strands are wound on a suitably shaped mandrel or core and precisely positioned in predetermined patterns. The mandrel may be left in place permanently or removed after cure. The strands may be preimpregnated or the resin may be applied during or after winding. Heat is used to effect final cure.

Centrifugal Casting

This is another method of producing round, oval, tapered, or rectangular parts. It offers low labor and tooling costs, uniform wall thicknesses, and good inner and outer surfaces. Chopped fibers and resin are placed inside a mandrel and uniformly distributed as the mandrel is rotated inside an oven.

Continuous Laminating

Continuous laminating is the most economical method of producing flat and corrugated panels, glazing, and similar products in large volume. Reinforcing mat or fabric is impregnated with resin, run through laminating rolls between cellophane sheets to control thickness and resin content, and then cured in a heating zone.

Pultrusion

Pultrusion produces shapes with high unidirectional strength such as "I" beams, flat stock for building siding, fishing rods, and shafts for golf clubs. Continuous fiber strands, combined with mat or woven fibers for cross strength, are impregnated with resin and pulled through a long heated steel die. The die shapes the product and controls resin content.

PROPERTIES

The mechanical properties of fiber composites are dependent on a number of complex factors. Two of the dominating ones in GRPs are the length of fibers and the glass content by weight. In general, strength increases with fiber length. For example, reinforcing a thermoplastic with chopped glass fibers at least doubles the strength of the plastic, whereas long-glass fiber reinforced the plastics thermoplastics exhibit increases of 300 and 400%. Also heat distortion temperatures usually increase by about 37.8°C, and impact strengths are appreciably increased. Similarly, as a general rule, an increase in glass content results in the following property changes:

- Tensile and impact strength increase.
- Modulus of elasticity increases.
- Heat deflection temperature increases, sometimes as much as 149°C.
- Creep decreases and dimensional stability increases.
- Thermal expansion decreases.

Properties of thermoset polyesters are so dependent on type, compounding, and processing method that a complete listing covering all combinations would be almost impossible. However, typical strength ranges obtainable in parts fabricated from various forms of polyester/glass compounds and processed by several methods are listed in Table F.3.

FIBERS

By definition, a fiber has a length at least 100 times its diameter or width, and its length must be at least 0.5 cm. Length also determines whether a fiber is classified as staple or filament. Filaments are long and/or continuous fibers. Staple fibers are relatively short, and, in practical applications, range from under 2.5 to 15.2 cm long (except for rope, where the fibers can run to several centimeters). Of the natural fibers, only silk exists in filament form; synthetics are produced as both staple and filaments.

The internal, microscopic structure of fibers is basically no different from that of other polymeric materials. Each fiber is composed of an aggregate of thousands of polymer molecules. However, in contrast to bulk plastic forms, the polymers in fibers are generally longer and aligned linearly, more or less parallel to the fiber axis. Thus, fibers are generally more crystalline than are bulk forms.

Also in contrast to bulk forms, fibers are not used alone, but either in assemblies or aggregates such as yarn or textiles or as a constituent with other materials, such as in composites. Also, compared with other materials, the properties and behavior of both fibers and textile forms are more critically dependent on their geometry. Hence, fibers are sometimes characterized as tiny microscopic beams, and, as such, their structural properties are dependent on such factors as cross-sectional area and shape, and length. The cross-sectional shape and diameter of fibers vary widely. Glass,

TABLE F.3
Properties of FRP Parts

Process and Reinforcement	Glass Fiber (% by wt)	Specific Gravity	Density (g/cm³)	Tensile Strength (MPa)	Flexural Strength (MPa)	Flexural Modulus (MPa)
Contact Molding						
Layup, mat	30–45	1.4–1.6	1.35–1.56	62–126	110–193	56
Layup, woven roving	30–65	1.5–1.7	1.45–1.64	207–385	207–455	69–77
Spray-up	30–50	1.4–1.6	1.35–1.56	62–126	110–193	51
Compression Molding						
Preform, mat	35–50	1.5–1.7	1.45–1.64	241–207	69–276	91–126
BMC	15–35	1.8–21	1.75–2.02	27–69	60–140	98–140
SMC	15–30	1.7–2.1	1.64–2.02	56–140	126–207	98–140
Cold Press Molding						
Preform, mat	20–30	1.5–1.7	1.45–1.64	83–140	152–265	91–97

Source: Mach. Design Basics Eng. Design, June, p. 713, 1993. With permission.

nylon, Dynel, and Dacron, for example, are essentially circular. Some other synthetics are oval, and others are irregular and serrated round. Cotton fibers are round tubes, and silk is triangular.

Fiber diameters range from about 0.01 to 0.04 mm. Because of the irregular cross section of many fibers, it is common practice to specify diameter or cross-sectional area in terms of fineness, which is defined as a weight-to-length or linear density relationship. One exception is wool, which is graded in micrometers. The common measure of linear density is the denier, which is the weight in grams of a 9000-m length of fiber. Another measure is the tex, which is defined as grams per 1 km. A millitex is the number of grams per 1000 km.

Of course, the linear density, or denier, is also directly related to fiber density. This is expressed as the denier/density value, commonly referred to as denier per unit density, which represents the equivalent denier for a fiber with the same cross-sectional area and a density of 1.

The cross-sectional diameter or area generally has a major influence on fiber and textile properties. It affects, for example, yarn packing, weave tightness, fabric stiffness, fabric thickness and weight, and cost relationships. Similarly, the cross-sectional shape affects yarn packing, stiffness, and twisting characteristics. It also affects the surface area, which in turn determines the fiber contact area, air permeability, and other properties.

For efficient production, a number of filaments are pulled simultaneously from several orifices in the bushing. These filaments (usually numbering about 204) are collected into a bundle, called a "strand," at a gathering device, where a "size" is applied to the filament surfaces. The strand is then wound into a forming package called a "cake." From this cake, shippable forms of fibrous glass are produced.

FIBROUS GLASS

The primary engineering benefits of glass fibers are their (1) inorganic nature, which makes them highly inert; (2) high strength-to-weight ratio; (3) nonflammability; and (4) resistance to heat, fungi, and rotting.

Glass fibers are produced in both filament and staple form. Their major engineering uses are (1) thermal and/or acoustical insulation and (2) as reinforcements, primarily for plastics.

TYPES

The largest volume of glass fibers used for engineering applications are so-called "E" type, made from a lime-Al_2O_3 borosilicate glass that is relatively soda-free. Although its initial strength at the bushing may be about 2758 to 3447 MPa, surface damage to fibers (both mechanical damage in handling and effects of moisture) reduces usable strength to 1034 to 1380 MPa. But at 1380 MPa tensile strength, the relatively low density of glass (0.092 lb/ft^3) produces a strength-to-weight ratio of about 5,511,800 cm, superior to that of a 3060-MPa tensile strength steel. Modulus of E-glass fibers is about 68,666 MPa. Although essentially unaffected by low temperatures, E-glass is limited to a maximum continuous operating temperature of about 316°C.

Other specialized types of glass (primarily used in specialty reinforced plastics applications) include:

1. High silica, leached glass fiber — Fibers with silica content of 96 to 99% are produced by leaching glass fibers. Such fibers provide excellent heat resistance, but relatively low strengths. They are usually used in short-fiber form for molding compounds.
2. Silica or quartz fiber — Fibers of pure silica provide optimum heat resistance (to about 93°C), although strength is somewhat lower than that of conventional E-glass.
3. High modulus fibers — Fibers of a beryllia-containing glass have been developed (primarily for filament winding use) with modulus of about 109,866 to 123,599 MPA.

Production Methods

Most fibrous glass is produced either by air, steam, or flame blowing or by mechanical pulling or drawing. Blowing produces relatively short staple fibers; mechanical drawing produces continuous monofilaments.

In blowing, steam or air jets impinge upon and break up molten streams of glass, forming fibers. The type of fiber produced depends on the pressure of the steam or air and the temperature and viscosity of the molten glass.

In the mechanical drawing process, the molten glass is fed into a "bushing," which contains a number of orifices through which the glass flows. Continuous filaments are then drawn from the molten stream. During the early stages of cooling, the stream is attenuated into filaments by being pulled at very high speeds — usually ranging from 25 to 50 m/s.

For efficient production, a number of filaments are pulled simultaneously from several orifices in the bushing. These filaments (usually numbering about 204) are collected into a bundle, called a "strand," at a gathering device where a "size" is applied to the filament surfaces. The strand is then wound into a forming package called a "cake." From this cake, shippable forms of fibrous glass are produced.

Insulation

In general, fibrous glass insulation is available in densities ranging from 0.5 to 12 lb/ft^3. Maximum operating temperature is about 316 to 1093°C, depending on type of glass. It provides high sound absorption, relatively high tensile strength, and resistance to moisture, fire, rotting, and fungi and bacteria growth. It is available in either flexible or rigid form.

The excellent insulating properties of fibrous glass are due to the large pockets of air between the fibers. These air pockets take up considerable volume.

Fibrous glass is not affected by low temperatures and has been used satisfactorily at temperatures as low as –212°C. Heat resistance depends on type of glass: borosilicate glass is generally limited to operating temperatures of 316 to 538°C; high silica glasses are capable of operating at 999°C; silica (quartz) fibers are usable up to 1093°C. The heat resistance of bonded insulations is normally limited by the heat resistance of the binder (maximum, about 232 to 316°C).

Forms

Following are the various forms in which fibrous glass is used in reinforced plastics:

1. Rovings consist of a number of strands (usually 60) gathered together from cake packages and wound on a tube to form a cylindrical package. Rovings have very little or

no twist. They are used either to provide completely unidirectional strength characteristics, such as in filament winding, or are chopped into predetermined lengths for preform-matched metal or spray molding.
2. Chopped strand consists of strands that have been cut into short lengths (usually 12.7 to 50.8 mm) in a manner similar to chopped roving, for use in preform-matched metal or spray molding, or to make molding compounds. It is the least expensive form of fibrous-glass reinforcement.
3. Milled fibers are produced from continuous strands that are hammer-milled into small modules of filamented glass (nominal lengths of 0.8 to 3.2 mm). Largely used for filler reinforcement in casting resins and in resin adhesives, they provide greater body and dimensional stability.
4. Yarns are twisted from either filaments or staple fibers on standard textile equipment. Although primarily an intermediate form from which woven fabrics are made, yarns are used for making rod stock, and for some very high strength, unidirectionally reinforced shapes. A common form in which yarn is available is the "warp beam," where many parallel yarns are wrapped on a mandrel.
5. Nonwoven mats are available both as reinforcing mats and as surfacing or overlay mats. Reinforcing mats are made of either chopped strands or swirled continuous strands laid down in a random pattern. Strands are held together by resinous binders. In laminates, mats provide relatively low strength levels, but strengths are isotropic. Surfacing or overlay mats are both thin mats of staple monofilaments. They provide practically no reinforcing, but serve to stabilize the surface resin coat, providing better appearance.
6. Woven fabrics and rovings provide the highest strength characteristics to reinforced plastic laminates (except for filament-wound structures), although strengths are orthotropic. A wide variety of fabrics and weaves are available both in woven yarns and woven rovings. Probably the most common types used are plain, basket, crowfoot satin, long shaft satin, unidirectional, and leno weaves.

FILAMENT-WOUND REINFORCED PLASTICS

The true fiberglass filament-wound structure may be more appropriately termed a resin-bonded filament-wound structure because it comprises approximately 80% glass fibers by weight and 20% bonding resin. Fibers are generally oriented to resist the principal stresses and the resin protects while secondarily supporting the fiber system. Filament winding is well adapted to the fabrication of internal pressure vessels; it has also performed well under external pressure and can be designed to function efficiently as a column or beam. This structure has the tensile strength of moderately heat-treated alloy steel and one quarter of the weight.

THE WINDING PROCESS

Bands of parallel glass filaments (usually in the form of roving) are wound over a mandrel following a precise pattern in such a way that subsequent bands lie adjacent, progressively covering the mandrel in successive layers, thus generating a shell structure. Liquid resin is simultaneously applied, generally by passing the filament band through a bath of catalyzed resin.

Tension generates a running load between the curved work surface and filament band which forces out air and excess resin and allows each successive layer ultimately to rest on solid material while the remaining interstices are filled with resin. Precision of filament placement plus tension and viscosity control are primary controlling factors in the attainment of high fiber content, which is generally desired for high strength.

Preimpregnation of the fiber strands is also used as a means of applying the resin binder. Such prepregs must fuse on contact with the work to accomplish a bond so the fundamental relationships remain the same. The fiber bands are parallel strands only, because a cross weave or other structural filler would not bear primary loads and would preclude the true maintenance of equal tension on fibers in the band or roving when winding over crowned surfaces.

Tension serves only the purpose of accomplishing high fiber content and cannot be considered as accomplishing any prestressing. This is primarily true because the dry strength of glass fibers, as wound, is only about one quarter of the ultimate resin consolidated strength. Also, unless some structural component is to remain within the wound shell (rather than a removable mandrel), there is no member against which a prestress can be maintained.

Range and Accuracy

Large or small structures are easily fabricated by adhering to the basic principles. Winding precision is important as is the relation of filament tension, resin viscosity, and radius of curvature of the filament path on the work surface. Small tubes have been made down to 0.65 cm in diameter, and there appear to be no fixed limitations in either direction. Wall thicknesses may be several centimeters or more because the normal bonding resins contain no volatile components and the glass content is so high that there is little danger of the exothermic heat becoming excessive.

Dimensional control in winding depends upon mandrel accuracy as well as both material and process control. The bands of filaments are generally 0.13 mm thick, and a full layer requires coverage by both right- and left-hand helices making the layer thickness 0.25 mm. Glass fiber thickness is subject to some variation, and resin content variation will also affect thickness. Wall thickness can generally be held to ±5%. Length and diameter are easily held to $^1/_{10}$ of 1% and less, because there is little resin shrinkage or wound-in strain due to winding tension.

Machining may be accomplished by carbide tools or grinding techniques. Tolerances can be held as closely as in metals. The inner surface as-wound against a good steel mandrel can have a finish of approximately 30 µin., and normal machining or grinding will produce a 40- to 60-µin. finish. Cutting of surface fibers in machining does not weaken the structure.

Component Materials

There are three primary materials in this composite: the glass fiber, fiber finish, and the bonding resin. Glass fibers are continuous and each "end" contains 204 monofilaments approximately 0.01 mm in diameter. These ends are plied together without twist to form a strand of "roving"; 12-end, equally tensioned, roving is generally used for the winding of high-performance structures.

The fiber finish generally includes compounds with a chemical affinity for the glass surface and the bonding resin. These are called coupling agents. Other functional components are lubricants and "film formers" for the generation of strand integrity. Both improve handling properties but do not contribute to the performance of filament structures.

The resin binder materials are generally liquid at room temperatures or are wound hot for liquid integration of the system. Best results are obtained with strong tough resins such as the epoxies. Polyesters have also been extensively used (primarily for radomes where electrical properties are critical) and any bonding resin should be free of polymerization products of a volatile nature, which would have to escape through the cured structure.

Chemical resistance and electrical properties of the constituents are usually critical in selecting the proper resin, as applications of the filament-wound structure are found in both the electrical and chemical industries.

FLAME-SPRAYED COATINGS

Processes

Flame-spraying methods are used to produce coatings of a wide variety of heat-fusible materials, including metals, metallic compounds, and ceramics. There are three basic flame-spray processes.

Wire-Type Guns

These guns are used to produce coatings of metals, alloys, and, in some cases, ceramics.

With this type of equipment wire or rod from 0.76 to 4.7 mm in diameter is fed axially through the center of a fuel gas/O_2 flame at a controlled rate. The flame is surrounded by an annular blast of air, which imparts high velocity to the burning gases and provided the kinetic energy needed to atomize the metal as it melts. Acetylene is the most commonly used fuel gas, although propane, natural gas, and H_2 are also used.

Wire-type guns are the most widely used, because of their versatility and ease of operation.

Handheld guns are used to coat large areas such as bridges, ship hulls, and tanks. Machine-mounted guns are used principally for machine element applications, as in surfacing rolls or salvaging journals by building up worn areas. In many production applications one or more electronically controlled guns are operated and cycled automatically by a central console.

Powder-Type Guns

A variety of materials that cannot be readily produced in the form of wire or rod utilize this type of gun. The guns are also used for spraying low-melting metals that are readily available in powder form.

With the powder gun, metal or ceramic powders are fed axially through a fuel gas/O_2 flame at a controlled rate. The powders are entrained in a carrier gas, which may be air, acetylene, or O_2. Because the powders are finely divided and dispersed as they enter the flame, further atomization is not required and annular air blast is not needed.

The powder supply for guns of this type may be a reservoir connected to the gun by a powder tube or hose, or it may be a cannister attached directly to the gun.

Powder guns are used for flame-spraying a wide variety of ceramics, and for coating with "self-fluxing" alloys, which are subsequently fused to the base material. They have also been used for many years to apply zinc and aluminum coatings for corrosion prevention.

Plasma Guns

The plasma flame is produced by passing suitable gases through a confined arc, where dissociation and ionization occur. The ionized gases form a conductive path within a water-cooled nozzle, so that an arc of considerable length is maintained. The gases most commonly used are N_2, H_2, and argon.

Temperature of the plasma flame depends on the type and volume of gas used, the size of the nozzle, and the amount of current used. For flame-spraying purposes temperature ranges of 5482.4 to 8232.4°C are generally employed, although much higher temperatures may be attained if desired. Plasma flame-spray guns usually operate at 20 to 40 kW, using 47 to 141 l/min of gas.

In addition to their high-temperature capabilities, plasma guns have other advantages. Extremely high velocities are possible, and favorable environmental conditions can be obtained by proper selection of gases. Spraying within a controlled atmosphere chamber permits the production of oxide-free coatings.

With the plasma gun, metal or ceramic powders are fed into the flame at a point downstream from the actual arc path. Current, gas flow, and powder flow must be adjusted for different coating materials, many of which would otherwise be completely vaporized.

Surface Preparation

Regardless of the flame-spray method used or the type of coating material being applied, some sort of surface preparation is usually required. Bond to the base is often largely mechanical, and, in general, the greater the degree of surface roughness, the better the bond. Thin coatings require less elaborate preparation than thick coatings, and some coating materials require much more thorough preparation than other materials.

Bonding methods used include abrasive blasting, rough threading, molybdenum bonding coats, heating, or combinations of these steps. Abrasive blasting is probably the most widely used.

Finishing Methods

Flame-sprayed deposits may be finished by machining or by grinding, depending on the hardness of the particular coating material. Sintered carbide tools are generally used for machining because even the softer materials contain some amount of abrasive oxides that may cause rapid wear of tool steel. For those materials that must be finished by grinding, specific wheel recommendations are available from flame-spray equipment manufacturers.

Flame-sprayed coatings may require sealing, depending on the type of coating and service requirements. A wide variety of impregnants are used to reduce porosity, enhance physical properties, or to improve friction characteristics.

FLINT

Flint is SiO_2 and is a black, gray, or brown cryptocrystalline variety of quartz. In the United States, ceramists often employ the term *flint* to include other siliceous minerals in addition to true flint.

Calcined and ground flint is used in pottery to reduce shrinkage in drying and firing and to give the body a certain rigidity. Flint is employed in the manufacture of whiteware, such as fine earthenware, bone china, and porcelain.

FLUIDIZED-BED COATINGS

The fluidized-bed process is used to apply organic coatings to parts by first preheating the parts and then immersing them in a tank of finely divided plastic powders which are held in a suspended state by a rising current of air. In this suspended state, the powders behave and feel like a fluid. The method produces an extremely uniform, high-quality, fusion bond coating that offers many technical and economic advantages.

Fusion bond coatings are generally applied to metal parts, although other substrates have been successfully coated. The major fields of application are electrical, chemical, and mechanical equipment as well as household appliances. The process is now being used in every major industrial country in the world for applying many different plastics to a variety of parts.

COATING APPLICATION

Objects to be coated are preheated in an oven to a temperature above the melting point of the plastic coating material. The preheat temperature depends on the type of plastic used, the thickness of the coating to be applied, and the mass of the article to be coated.

After preheating, the parts are immersed with suitable motion in the fluidized bed. Air used to fluidize the plastic powders enters the tank through a specially designed porous plate located at the bottom of the unit.

When the powder particles contact the heated part, they fuse and adhere to the surface, forming a continuous and extremely uniform coating. In many cases, the part is postheated to coalesce the coating completely and improve its appearance. Thickness of the coating depends on the temperature of the part surface and how long it is immersed in the fluidized bed.

COATING TYPES

Cellulosic

Cellulosic fluidized-bed coatings are noted for their all-around combination of properties and are especially popular for decorative/protective applications. They combine good impact and abrasion resistance with outstanding electrical insulation. The coatings have excellent weathering properties, salt spray resistance, high gloss, and can be made in an almost unlimited range of colors. They can be solvent-etched to provide a satin finish and heat-embossed for additional effects.

The economy of cellulosic coatings combined with their excellent appearance and durability are particularly useful for such applications as indoor and outdoor furniture, kitchen fixtures, home and marine hardware, metal stampings, fan guards, and sporting goods. Major uses of cellulosic fusion bond finishes are coated transformer tanks and covers, reclosure tanks and covers, outdoor electrical equipment housings, and many pole line hardware parts.

Vinyl

Vinyl fluidized-bed coatings have a good combination of chemical resistance, decorative appeal, flexibility, toughness, and low-frequency insulating properties. They have excellent salt-spray resistance, outstanding outdoor weathering characteristics, and can be used for general-purpose electrical insulation. The vinyl fusion bond coatings are claimed to have better uniformity and edge coverage than plasticol coatings.

Fusion bond vinyl coatings have been especially successful on wire goods for applications such as dishwasher racks, washing-machine parts, and refrigerator shelves. They are also being used on wire furniture and hardware, and in industrial applications such as bus bars, pump impellers, transformer tanks and covers, auto battery brackets, conveyor rollers, and other material-handling equipment. Cast iron, die castings, and expanded metal parts can be readily coated.

Epoxy

Both thermoplastic and thermosetting materials can be applied. Epoxy coatings have a smooth, hard surface and exceptionally good electrical-insulation properties over a wide temperature range. They are available in rigid and semiflexible variations with different combinations of electrical, physical, and chemical properties. Epoxy coatings on electrical motor laminations provide good dielectric strength and uniform coverage over sharp edges. When properly applied, epoxy coatings have good impact resistance and do not sacrifice toughness for surface hardness. Other electrical applications include torroidal cores, wound coils, encapsulated printed circuit boards, bus bars, watt hour meter coils, resistors, and capacitors.

Nylon

The combination of properties that have made the polyamide plastics unique for molded and extruded parts are obtainable in coatings, and nylon used as a solution coating technique applied by the fluidized-bed process offers many advantages to the design engineer.

Nylon fusion bond finishes combine a decorative, smooth, and glossy appearance with low surface friction and excellent bearing and wear properties. They minimize scratching and cut down undesirable noise. The frictional heat developed on nylon is dissipated more rapidly when used as a coating over a conductive metal surface than when used as a solid plastics member. Coated metal parts offer increased dimensional stability. Because of the unique properties of nylon-coated metal parts, many users of the process are able to reduce the number of metal component parts at substantial savings.

By an additional immersion of the heated and coated part into a fluid bed of whirling molybdenum sulfide or graphite, an impregnated nylon surface can be produced with unusual bearing, frictional, and wear characteristics.

Nylon fusion bonds are effectively used in machine shop fixtures, modern indoor furniture to simulate a wrought iron finish, aircraft instrument panels, ball-joint suspensions, collars, guards, and slide valves used in textile and farm equipment, knitting machine parts, switch box cover panels, tractor control handles, and radar and calculator component parts.

Polyethylene

Polyethylene coatings combine low water absorption and excellent chemical resistance with good electrical insulation properties. They can be applied successfully in thicknesses of 10 to 60 mils by the fluidized-bed process. The primary uses for polyethylene coatings are for protecting chemical processing equipment and on food-handling equipment. Typical chemical applications include pipe and fittings, pump and motor housings, valves, battery hold-downs, fans, and electroplating jigs.

Chlorinated Polyether

This fluidized-bed coating has an excellent combination of mechanical, chemical, thermal, and electrical properties. It has good resistance to wear and abrasion. It provides good electrical insulation even under high humidity and high temperature conditions, and has very low moisture absorption. It can be used continuously at 121°C and even up to 149°C.

Chlorinated polyether coatings have excellent chemical resistance and are widely used to coat equipment for the chemical industry such as valves, pipe and pipe fittings, pump housings and impellers, electroplating jigs and fixtures, cams, and bushings. However, chlorinated polyether coatings should not be used in contact with some chlorinated organic solvents, or with fuming sulfuric and nitric acids.

FLUOROCARBONS

These are any of the organic compounds in which all of the hydrogen atoms attached to a carbon atom have been replaced by fluorine. Fluorocarbons are usually gases or liquids at room temperature, depending on the number of carbon atoms in the molecule. A major use of gaseous fluorocarbons is in radiation-induced etching processes for the microelectronics industry; the most common one is tetrafluoromethane. Liquid fluorocarbons possess a unique combination of properties that has led to their use as inert fluids for cooling of electronic devices and soldering. Solubility of gases in fluorocarbons has also been used to advantage. For example, they have been used in biological cultures requiring O_2, and as liquid barrier filters for purifying air.

Fluorocarbons may be made part hydrocarbon and part fluorocarbon, or may contain chlorine. The fluorocarbons used as plastic resins may contain as much as 65% fluorine and also chlorine, but are very stable. Liquid fluorocarbons are used as heat-transfer agents, hydraulic fluids, and fire extinguishers. Benzene-base fluorocarbons are used for solvents, dielectric fluids, lubricants, and for making dyes, germicides, and drugs. Synthetic lubricants of the fluorine type consist of solid particles of a fluorine polymer in a high-molecular-weight fluorocarbon liquid. Chlorine reacts with fluorocarbons to form chlorofluorocarbons, commonly referred to as CFCs. CFC 11 is used as a foam-blowing agent, and CFC 12 is employed as a refrigerant. CFC 113 is a degreasant in semiconductor manufacturing. Because they are strong depletants of stratospheric ozone, the use of CFCs as aerosol propellants has been banned in the United States since 1978, and is being phased out in Europe. Alternatives to CFCs are being sought for other applications by partially substituting the chlorine with other elements. CFC 22, which has 95% less ozone-depleting capacity than CFC 12, is a potential candidate to replace CFC 12.

FLUOROPLASTICS

Also termed fluoropolymers, fluorocarbon resins, and fluorine plastics, fluoroplastics are a group of high-performance, high-price engineering plastics. They are composed basically of linear polymers in which some or all of the hydrogen atoms are replaced with fluorine, and are characterized by relatively high crystallinity and molecular weight. All fluoroplastics are natural white and have a waxy feel. They range from semirigid to flexible. As a class, they rank among the best of the plastics in chemical resistance and elevated-temperature performance. Their maximum service temperature ranges up to about 260°C. They also have excellent frictional properties and cannot be wet by many liquids. Their dielectric strength is high and is relatively insensitive to temperature and power frequency. Mechanical properties, including tensile creep and fatigue strength, are only fair, although impact strength is relatively high.

PTFE, FEP, AND PFA

There are three major classes of fluoroplastics. In order of decreasing fluorine replacement of hydrogen, they are fluorocarbons, chlorotrifluoroethylene, and fluorohydrocarbons. There are two fluorocarbon types: tetrafluoroethylene (PTFE or TFE) and fluorinated ethylene propylene (FEP). PTFE is the most widely used fluoroplastic. It has the highest useful service temperature, 260°C, and chemical resistance.

Their high melt viscosity prevents PTFE resins from being processed by conventional extrusion and molding techniques. Instead, molding resins are processed by press-and-sinter methods similar to those of powder metallurgy or by lubricated extrusion and sintering. All other fluoroplastics are melt processible by techniques commonly used with other thermoplastics.

PTFE resins are opaque, crystalline, and malleable. When heated above 341°C, however, they are transparent, amorphous, relatively intractable, and they fracture if severely deformed. They return to their original state when cooled.

The chief advantage of FEP is its low melt viscosity, which permits it to be conventionally molded. FEP resins offer nearly all of the desirable properties of PTFE, except thermal stability. Maximum recommended service temperature for these resins is lower by about 37.8°C. Perfluoroalkoxy (PFA) fluorocarbon resins are easier to process than FEP and have higher mechanical properties at elevated temperatures. Service temperature capabilities are the same as those of PTFE.

PTFE resins are supplied as granular molding powders for compression molding or ram extrusion, as powders for lubricated extrusion, and as aqueous dispersions for dip coating and impregnating. FEP and PFA resins are supplied in pellet form for melt extrusion and molding. FEP resin is also available as an aqueous dispersion.

Teflon is a tetrafluoroethylene of specific gravity up to 2.3. The tensile strength is up to 23.5 MPa, elongation 250 to 350%, dielectric strength 39.4×10^6 V/m, and melting point 312°C. It is water resistant and highly chemical resistant. Teflon S is a liquid resin of 22% solids, sprayed by conventional methods and curable at low temperatures. It gives a hard, abrasion-resistant coating for such uses as conveyors and chutes. Its temperature service range is up to 204°C. Teflon fiber is the plastic in extruded monofilament, down to 0.03 cm in diameter, oriented to give high strength. It is used for heat- and chemical-resistant filters. Teflon tubing is also made in fine sizes down to 0.25 cm in diameter with wall thickness of 0.03 cm. Teflon 41-X is a colloidal water dispersion of negatively charged particles of Teflon, used for coating metal parts by electrodeposition. Teflon FEP is fluorinated ethylenepropylene in thin film, down to 0.001 cm thick, for capacitors and coil insulation. The 0.003-cm film has a dielectric strength of 126×10^6 V/m, tensile strength of 20 MPa, and elongation of 250%.

Properties

Outstanding characteristics of the fluoroplastics are chemical inertness, high- and low-temperature stability, excellent electrical properties, and low friction. However, the resins are fairly soft and resistance to wear and creep is low. These characteristics are improved by compounding the resins with inorganic fibers or particulate materials. For example, the poor wear resistance of PTFE as a bearing material is overcome by adding glass fiber, carbon, bronze, or metallic oxide. Wear resistance is improved by as much as 1000 times, and the friction coefficient increases only slightly. As a result, the wear resistance of filled PTFE is superior, in its operating range, to that of any other plastic bearing material and is equaled only by some forms of carbon.

The static coefficient of friction for PTFE resins decreases with increasing load. Thus, PTFE bearing surfaces do not seize, even under extremely high loads. Sliding speed has a marked effect on friction characteristics of unreinforced PTFE resins; temperature has very little effect.

PTFE resins have an unusual thermal expansion characteristic. A transition at 18°C produces a volume increase of over 1%. Thus, a machined part, produced within tolerances at a temperature on either side of this transition zone, will change dimensionally if heated or cooled through the zone.

Electrical properties of PTFE, FEP, and FPA are excellent, and they remain stable over a wide range of frequency and environmental conditions. Dielectric constant, for example, is 2.1 from 60 to 10^9 Hz. Heat-aging tests at 300°C for 6 months show no change in this value. Dissipation factor of PTFE remains below 0.0003 up to 10^8 Hz. The factor for FEP and PFA resins is below 0.001 over the same range. Dielectric strength and surface arc resistance of fluorocarbon resins are high and do not vary with temperature or thermal aging (Table F.4).

CTFE or CFE

Chlorotrifluoroethylene (CTFE or CFE) is stronger and stiffer than the fluorocarbons and

TABLE F.4
Properties of Fluoroplastics

ASTM or UL Test	Property	PTFE	FEP	PFA	PVDF	CTFE[a]	Modified ETFE
	Physical						
D792	Specific gravity	2.13–2.24	2.12–2.17	2.12–2.17	1.75–1.78	2.13	1.70
D792	Specific volume (in.^{3}/lb)	13–12.3	13.0–12.7	13.0–12.7	15.7–15.6	—	16.3
D570	Water absorption, 24 h, 1/8-in. thk (%)	<0.01	<0.01	0.03	0.04	0.01	<0.03
	Mechanical						
D638	Tensile strength (psi)	3,350	3,000	4,000	5,200–7,400	5,430	6,500
D638	Elongation (%)	300	300	300	100–300	125	275
D638	Tensile modulus (10^5 psi)	0.5	—	—	1.6	1.86	1.2
D790	Flexural strength (psi)	No break	No break	No break	No break	10,700	No break
D790	Flexural modulus (10^5 psi)	0.5–0.9	0.95	0.95	2.0	2.54	2.0
D256	Impact strength, Izod (ft-lb/in. of notch)	3.5	No break	No break	3–4	3.1	No break
D785	Hardness, Rockwell	—	—	—	—	S 85	R50
	Shore D	50–65	55	60	80	79	—
	Thermal						
C177	Thermal conductivity (Btu-in./hr-ft^2-°F)	1.7	1.4	1.8	0.7–0.9	1.83	1.65
D696	Coefficient of thermal expansion (10^{-5} in./in.-°F)	5.5–8.4	4.6–5.8	6.7	8.0–8.5	4.8–15	5.2
D648	Deflection temperature (°F)						
	At 264 psi	132	24	118	195	167	165
	At 66 psi	250	158	164	300	265	220
UL94	Flammability rating	V-0	V-0	V-0	V-0	V-0	V-0
	Electrical						
D149	Dielectric strength (V/mil)						
	Short time, 1/8-in. thk	500–600	500–600	500–600	260	490	400–500
D150	Dielectric constant						
	At 1 kHz	2.1	2.1	2.1	7.5	2.45	2.6
D150	Dissipation factor						
	At 1 kHz	0.00005	0.00005	0.0003	0.019	0.0247	0.0008
D257	Volume resistivity (Ω-cm)						
	At 73°F, 50% RH	>10^{18}	>10^{18}	>10^{18}	2×10^{14}	2.5×10^{16}	10^{16}
D495	Arc resistance (s)	>300	>180	>180	50–70	3360	75
	Optical						
D542	Refractive Index	1.350	1.344	1.350	1.42	1.435	1.403
D1003	Transmittance (%)						
	1-mil film	—	>95	>95	>90	>90	—
	Frictional						
—	Coefficient of friction						
	Against steel (100 psi, 10 fpm)	0.050	0.330	0.214	0.14	—	0.400

[a] Crystalline compound. Below and above 135°F.

Source: Mach. Design Basics Eng. Design, June, p. 205, 1993. With permission.

has better creep resistance. Like FEP and unlike PTFE, it can be molded by conventional methods.

Sensitivity to processing conditions is greater in CTFE resins than in most polymers. Molding and extruding operations require accurate temperature control, flow channel streamlining, and high pressure because of the high melt viscosity of these materials. With too little heat, the plastic is unworkable; too much heat degrades the polymer. Degradation begins at about 274°C. Because of the lower temperatures involved in compression molding, this process produces CTFE parts with the best properties.

Thin parts such as films and coil forms must be made from partially degraded resin. The degree of degradation is directly related to the reduction in viscosity necessary to process a part. Although normal, partial degradation does not greatly affect properties, seriously degraded CTFE becomes highly crystalline, and physical properties are reduced. Extended usage above 121°C also increases crystallinity.

CTFE plastic is often compounded with various fillers. When plasticized with low-molecular-weight CTFE oils, it becomes a soft, extensible, easily shaped material. Filled with glass fiber, CTFE is harder, more brittle, and has better high-temperature properties.

Properties

CTFE plastics are characterized by chemical inertness, thermal stability, and good electrical properties, and are usable from 400 to –400°C. Nothing adheres readily to these materials, and they absorb practically no moisture. CTFE components do not carbonize or support combustion. Up to thicknesses of about 3.2 mm, CTFE plastics can be made optically clear. Ultraviolet absorption is very low, which contributes to its good weatherability.

Compared with PTFE, FEP, and PFA fluorocarbon resins, CTFE materials are harder, more resistant to creep, and less permeable; they have lower melting points, higher coefficients of friction, and are less resistant to swelling by solvents than the other fluorocarbons.

Tensile strength of CTFE moldings is moderate, compressive strength is high, and the material has good resistance to abrasion and cold flow. CTFE plastic has the lowest permeability to moisture vapor of any plastic. It is also impermeable to many liquids and gases, particularly in thin sections.

PVF_2 AND PVF

The fluorohydrocarbons are of two kinds: polyvinylidene fluoride (PVF_2) and polyvinyl fluoride (PVF). Although similar to the other fluoroplastics, they have somewhat lower heat resistance and considerably higher tensile and compressive strength.

Except for PTFE, the fluoroplastics can be formed by molding, extruding, and other conventional methods. However, processing must be carefully controlled. Because PTFE cannot exist in a true molten state, it cannot be conventionally molded. The common method of fabrication is by compacting the resin in powder form and then sintering.

PVF_2, the toughest of the fluoroplastic resins, is available as pellets for extrusion and molding and as powders and dispersions for corrosion-resistant coatings. This high-molecular-weight homopolymer has excellent resistance to stress fatigue, abrasion, and to cold flow. Although insulating properties and chemical inertness of PVDF are not as good as those of the fully fluorinated polymers, PTFE and FEP, the balance of properties available in PVDF qualifies this resin for many engineering applications. It can be used over the temperature range from –73 to 149°C and has excellent resistance to abrasion.

PVDF can be used with halogens, acids, bases, and strong oxidizing agents, but it is not recommended for use in contact with ketones, esters, amines, and some organic acids.

Although electrical properties of PVDF are not as good as those of other fluoroplastics, it is widely used to insulate wire and cable in computer and other electrical and electronic equipment. Heat-shrinkable tubing of PVDF is used as a protective cover on resistors and diodes, as an encapsulant over soldered joints.

Valves, piping, and other solid and lined components are typical applications of PVDF in chemical-processing equipment. It is the only fluoroplastic available in rigid pipe form.

Woven cloth made from PVDF monofilament is used for chemical filtration applications.

A significant application area for PVDF materials is as a protective coating for metal panels used in outdoor service. Blended with pigments, the resin is applied, usually by coil-coating equipment, to aluminum or galvanized steel. The coil is subsequently formed into panels for industrial and commercial buildings.

A recently developed capability of PVDF film is based on the unique piezoelectric characteristics of the film in its so-called beta phase. Beta-phase PVDF is produced from ultrapure film by stretching it as it emerges from the extruder. Both surfaces are then metallized, and the material is subjected to a high voltage to polarize the atomic structure.

When compressed or stretched, polarized PVDF generates a voltage from one metallized surface to the other, proportional to the induced strain. Infrared light on one of the surfaces has the same effect. Conversely, a voltage applied between metallized surfaces expands or contracts the material, depending on the polarity of the voltage.

PFA, ECTFE, AND ETFE

The following three fluoroplastics are melt processible. Perfluoroalkoxy (PFA) can be injection-molded, extruded, and rotationally molded. Compared to FEP, PFA has slightly greater mechanical properties at temperatures over 150°C and can be used up to 260°C.

Ethylene-chlorotrifluoroethylene (ECTFE) copolymer resins also are melt processible with a melting point of 240°C. Their mechanical properties — strength, wear resistance, and creep resistance, in particular — are much greater than those of PTFE, FEP, and PFA, but their upper temperature limit is about 165°C. ECTFE also has excellent property retention at cryogenic temperatures.

Ethylene-tetrafluoroethylene (ETFE) copolymer resin is another melt-processible fluoroplastic with a melting point of 270°C. It is an impact-resistant, tough material that can be used at temperatures ranging from cryogenic up to about 179°C.

One of the advantages of the copolymers of ethylene and TFE — called modified ETFE — and of ethylene and CTFE — called ECTFE — compared with PTFE and CTFE is their ease of processing. Unlike their predecessors, they can be processed by conventional thermoplastic techniques. Various grades can be made into film or sheet, into a monofilament, or used as a powder coating; all grades can be heat-sealed or welded.

Although these resins have lower heat resistance than PTFE or CTFE, they offer a combination of properties and processability that is unattainable in the predecessor resins. Maximum service temperature for no-load applications is in the range of 149 to 199°C for ETFE and ECTFE, compared with 199°C for CTFE and 288°C for PTFE. Glass reinforcement increases these values by 10°C.

Both tensile strength and toughness of these resins are higher than those of the other fluoropolymers; they are rated "no break" in notched Izod tests. The modulus of ECTFE is higher than that of ETFE up to about 100°C; above 150°C, ETFE has a higher modulus. Deflection temperature of both resins is similar, with ECTFE slightly higher (116°C, compared with 104°C, at 0.44 MPa, and 77°C compared with 71°C at 1820 MPa). Hardness of ETFE is Rockwell R50; that of ECTFE is R93; see Table F.4. The limiting oxygen index (LOI) of ETFE is 31; that of ETCFE is 60. (LOIs of PTFE, FEP, and CTFE are over 95.)

As with other fluoroplastics, these resins are compatible with most chemicals, even at high temperatures. ETFE is not attacked by most solvents to temperatures as high as 199°C. ECTFE is similar to 121°C, but is attacked by chlorinated solvents at higher temperatures. ETFE has better chemical stress-crack resistance.

Applications for these resins include wire and cable insulation, chemical-resistant linings and molded parts, laboratoryware, and molded electrostructural parts.

FLUORSPAR

Also called fluorite, fluorspar is a crystalline or massive granular mineral of the composition CaF_2, used as a flux in the making of steel, for making hydrofluoric acid, in opalescent glass, in ceramic enamels, for snaking artificial cryolite, as a binder for vitreous abrasive wheels, and in the production of white cement. It is a better flux for steel than limestone, making a fluid slag, and freeing the iron of sulfur and phosphorus. About 2.5 kg of fluorspar is used per ton of basic open-hearth steel.

Acid spar is a grade used in making hydrofluoric acid. It is also used for making refrigerants, plastics, and chemicals, and for aluminum reduction. Optical fluorspar is the highest grade but is not common. Fluoride crystals for optical lenses are grown artificially from acid-grade fluorspar. Pure calcium fluoride, Ca_2F_6, is a colorless crystalline powder used for etching glass, in enamels, and for reducing friction in machine bearings. It is also used for ceramic parts resistant to hydrofluoric acid and most other acids. Calcium fluorite has silicon in the molecule and is a crystalline powder used for enamels. The clear rhombic fluoride crystals used for transforming electric energy into light are lead fluoride, PbF_2.

FLUX

Flux is a substance added to a refractory material to aid in its fusion. A secondary action of a flux, which may also be a primary reason for its use, is as a reducing agent to deoxidize impurities.

Any material that lowers the melting temperature of another material or mixture of materials is a flux. Fluxing substances may occur as natural impurities in a raw material. Thus, the alkali content of a clay will flux the clay. In other cases, fluxes are separate raw materials. Example: use of feldspar to flux a mixture of clays and flint.

An auxiliary flux is a third component that may make the primary flux more effective. Thus, addition of 2% dolomite, talc, or fluorspar to a whiteware mixture that contains 25% feldspar will produce a substantial decrease in vitrification temperature. The auxiliary constituent may be incapable of producing the same result (or too expensive to use) as the sole flux.

Compounds of alkali metals (sodium, potassium, and lithium) are popular fluxes for clay bodies. Compounds of alkaline earth metals (calcium, magnesium, and, to a lesser extent, barium and strontium) are common auxiliary fluxes. However, they also may be primary fluxes for such products as low-loss dielectrics. Lead and boron compounds are important fluxes for glasses, glazes, and enamels. And premelted glasses or frits may be used to flux clay or other bodies.

The term *flux* also may be used to specify a low-melting glass used in decorating glass products or an overglaze for clayware. Pigments are mixed with the powdered glass flux and then applied to the object to produce a vitrifiable coating at temperatures <650°C.

FOAM MATERIALS

These are materials with a spongelike, cellular structure. They include the well-known sponge rubber, plastic foams, glass foams, refractory foams, and a few metal foams. Ordinary chemically blown sponge rubber is made up of interconnecting cells in a labyrinth-like formation. When made by heating latex, it may show spherical cells with the porous walls perforated by the evaporation of moisture. It is also called foam rubber. Special processes are used to produce cell-tight and gas-tight cellular rubber, which is nonabsorbent.

Cellular rubber comes in sheets of any thickness for gaskets, seals, weather stripping, vibration insulation, and refrigerator insulation. Most of the so-called sponge rubbers are not made of natural rubber but are produced from synthetic rubbers or plastics, and may be called by a type classification, such as urethane foam.

Others include phenolic foam, which is made by incorporating sodium bicarbonate and an acid catalyst into liquid phenol resin. The reaction liberates CO_2 gas, expanding the plastic.

Cellular cellulose acetate is expanded with air-filled cells for use as insulation and as a buoyancy material for floats. It is tough and resilient. Strux is cellulose acetate foam, made by extruding the plastic mixed with barium sulfate in an alcohol-acetone solvent. When the pressure is removed, it expands into a light, cellular structure. Polystyrene foam is widely used for packaging and for building insulation. It is available as prefoamed board or sheet, or as beads that expand when heated.

Styrofoam is polystyrene expanded into a multicellular mass 42 times the original size. It has only one sixth the weight of cork, but will withstand hot water or temperatures above

77°C, as it is thermoplastic. It is used for cold-storage insulation, and is resistant to mold. Polyethylene foam and polypropylene foam are also available. Compared with expanded polystyrene, expanded polyolefins have greater toughness and can be molded more easily. Current uses include automotive bumper cores, sunvisor cores, and electronic packaging.

Other foams include foamed vinyl plastisols and expanded polyvinyl chloride. Polyester foam is odorless, flame-resistant, and resistant to oils and solvents. It has only half the weight of foamed rubber with greater strength and high resistance to oxidation. It is used for upholstery and insulation.

Silicone foam, used for insulation, is silicone rubber foamed into a uniform unicellular structure. Vinyl foams are widely used and they are made from various types of vinyl resins with the general physical properties of the resin used. Epoxy foams come as a powder consisting of an epoxy resin mixed with diaminodiphenyl sulfone. Glass foam is used as thermal insulation for buildings, industrial equipment, and piping. Ceramic foams of alumina, silica, and mullite are used principally for high-temperature insulation. Aluminum foam is a metal foam that has found appreciable industrial use as a core material in sandwich composites. Foamed zinc is a lightweight structural metal with equal strength in all directions, made by foaming with an inert gas into a closed cell structure. It is used particularly for shock and vibration insulation. Foaming agents for metals are essentially the same as those used for plastics. They are chemical additives that release a gas to expand the material by forming closed bubbles. Or they may be used to cause froth as in detergents or firefighting foams.

FOAM POLYMERS

Most polymers, thermoplastics, or thermosets can be expanded into a foam by the addition of a foaming or blowing agent. They can be foamed to desired density, ranging from near the weight of solid resin, at over 64.08×10^{-2} g/cm^3, to extremely light, at 1.602×10^{-2} g/cm^3. The foamed product will maintain the general characteristics of the base resin as its mechanical properties decrease along with its density.

Thermoplastic Foams

Thermoplastic foams can be extruded by heating the resin and blowing agent to a threshhold temperature, causing the blowing agent to react and expand into a gas. The gas combines with the resin, forming a frothy mass that is extruded through a shaping die. The resulting buns, logs, or sheets can then be cut into shapes for inclusion in composite panels and other applications.

Bead foaming is accomplished by heating resin beads that include the blowing agent in the formulation. A mold cavity is partially filled with beads and heated to the temperature at which the blowing agent gives off gas. This causes the softened beads to expand, filling the mold and welding them together. The process commonly used to make polystyrene foam creates slab stock or shapes molded to specific dimensions. Foam stock forms can be mechanically knife-cut or band-sawed into sheets, blocks, or other desired shapes.

Thermoset Foams

Some type of liquid foaming is necessary to manufacture thermoset foams, used primarily for the rigid structural foam-type applications mentioned earlier: The liquid formulation for these foams contains not only the base resin and blowing agent, but also the catalyst required for curing in thermoset chemistry. Liquid foaming techniques, which include spraying, can be employed for open- or closed-mold processes and often include a reinforcement or filler material.

Polyurethane

Polyurethane, long considered a workhorse of low-cost foams, can be formed by first reacting two liquid components together. Water is the principle blowing agent (fluorocarbon is also used). This reaction produces CO_2, which is trapped in closed cells, giving good thermal insulation to the foam. Depending on the reaction and the chemicals or additives used, the foams can either be soft and flexible or tough, hard, and rigid.

Cell Structure and Density

Polymer foams have either open- or closed-cell structures, depending primarily on the type of resin and foaming process employed. Closed-cell foams, which dominate the marine industry, are impermeable, preventing water ingress. Depending on the resin chemistry used, closed-cell foams may also be impervious to corrosive liquid chemicals and solvents, making them useful as an insulating material for industrial tanks.

Most low-density (1.05 to $3.20 \times 10^{-2} \text{g/cm}^3$) rigid urethane foams have a closed-cell content of 90% or greater in slab form. Under controlled conditions, the closed-cell content can be increased to 99%.

Medium- to high-density polyurethane foams (4.81 to 64.08×10^{-2} g/cm^3) can be formulated with various characteristics and diverse high-performance capabilities that range from insulating and isolating toxic industrial wastes and radioactive nuclear materials to sandwich panels requiring higher modulus and high-temperature resistance for use in aircraft.

Higher-density, heavier foams, sometimes formulated with phenolic resins, are more dimensionally stable at elevated temperatures. Low-density foams may swell when exposed to elevated temperatures because of gas expansion and increased pressure from inner cell air and moisture. If this happens, foam cells will then erupt and collapse, since internal pressure is necessary to maintain cell structure. Contraction can continue until the foam becomes so dense that it resembles a solid more than a foamed plastic.

Foam Additives and Reinforcements

A variety of additives can be included in foam formulations that greatly impact the density, cost, manufacturability, and performance characteristics of the finished product. Randomly distributed short-fiber reinforcements can be blown into foam during spray-up, which become an integral part of the cell walls. Fiber in mat, fabric, or long-fiber form can also be introduced into the mold. Mechanical properties of the finished product will increase dramatically with the introduction of fiber reinforcements. From low-cost mineral fillers, to microspheres and oriented fiber webs, new reinforced foam products are being developed that offer elegant, low-cost solutions to esoteric application requirements.

Structural Foam

Structural foam, SRIM-type constructions, are similar in configuration to laminated sandwich panels. They have full-density outer skins and lightweight cellular cores, and the relative benefits inherent in this type of configuration, such as high stiffness-to-weight and strength-to-weight ratios, are similar to sandwich panels as well. These rigid foam products, usually made with polyurethane-based chemistry, are often reinforced, enhancing their mechanical properties.

Structural foams are made using both high- and low-pressure processes. When products are made using high pressure, the mold is solidly filled under pressure with unfoamed resin; then it is expanded (or a core removed) and the foaming agent is activated or injected to form the core layer. High-pressure processes are able to provide dense, smooth skins and achieve high levels of surface detail. Tooling for these processes is usually relatively expensive.

In low-pressure methods, the mold is partially filled with molten resin that expands to fill the mold and forms a skin upon contact with the walls. In the past, there has been some concern about surface quality with self-skinning open-molded structural foams but recent advances in resin chemistry and molding techniques are producing class A finishes for demanding automotive applications under low pressures.

Applications

Sandwich Panel Construction

One of the oldest and best-understood uses for foam core materials is in the construction of structural sandwich panels. The typical sandwich panel consists of strong outer laminate skins made with oriented glass, carbon, or aramid fiber and a lightweight inner core.

Tailoring foam core selection to the specific application is especially important. When failure of a sandwich panel occurs, it is usually in the core, because plastic foams have low shear rigidity compared to the skins. The core material must be strong enough to stabilize the sandwich structure, providing a shear load path between the laminate skins to prevent buckling.

An inexpensive foam product, a lightweight, non-chlorofluorocarbon-blown polyisocyanurate, ELFOAM is a product for insulation applications, where nonwoven fiberglass incorporates webs within the foam to carry the structural load in the design for structural sandwich panel construction. The foam is co-cured by resin infusion or pressurized processes and the dry fiber provides channels through which resin is drawn into the core. This offers high strength-to-weight and stiffness-to-weight ratios, creating tough, thick, lightweight structures ideal for industrial or construction applications.

Boatbuilding

Commercial boatbuilding has used composite materials over the years. Tough, durable thermoplastic PVC foam formulations are the most commonly used marine panel applications. In fact, the educated consumer has grown to appreciate and demand the lightweight, self-insulating, low-condensation improvements provided by two fiberglass skins over a foam core.

Another resin option designed for boat hull applications that must withstand repeated exposure to dynamic loads is linear structural foam. The new chemistry addresses the discrepancy between a high-modulus core with no flexure and a flexural core with low modulus. Core-Cell is a linear formulation that offers the stiffness of cross-linked PVCs, while maintaining its plastic range for high impact and damage tolerance. Core-Cell has heat distortion temperature comparable to cross-linked PVC foams, making it suitable for deck and superstructure laminates.

There is also a spray-applied foam for marine applications that is based on polyester/polyurethane hybrid chemistry. For use in open-mold processes only, the unsaturated polyester provides stiffness and hardness, and the polyurethane enables a fast cure and internal toughness.

The spray-applied foam provides high flexural modulus and flexural strength. It can be sprayed at a 0.38 to 12.7 mm thickness to provide structural strength at less weight than fiber-reinforced plastics/polyester. Ultracore foams are bondable with epoxy, vinyl ester, and polyurethane resins.

High-Performance Foams

High-performance foam core materials are typically striving for exceptionally high-temperature resistances or strength-to-weight ratios, or for other esoteric properties. These are generally derived from state-of-the-art resin chemistry or additives. Cores made from resins with exotic chemistries like polymethacrylimide (PMI) or bismaleimide (BMI) lend themselves to use in aerospace applications that demand dimensional stability at temperatures up to 149°C. For example, Rohacell PMI rigid foams have been selected for the interstage and payload fairings of the Delta II-IV rockets. These components are co-cured in thermoformed-to-shape sandwich cores. Rohacell XT is the newest, closed-cell isotropic foam with an extended temperature grade available that is suitable for use with BMI prepreg skins, co-curing at temperatures up to 190°C with a 232°C postcure.

Syntactic foams are an entirely different concept in foam technology. Cell structure is created by incorporating hollow microspheres into a compound (usually a liquid resin), eliminating the need for a blowing agent. The microspheres can be made from a variety of materials such as glass, diatomaceous earth, or plastic, with foam density controlled by how much resin is used to hold the microspheres together. Deriving properties primarily from the microsphere material, syntactic foams are highly customizable for esoteric applications.

Foam-Filled Honeycomb

Honeycomb cores, filled with foam and sandwiched between laminates, offer both high static-load and dynamic-load properties, as well as high fatigue endurance, strength and stiffness, and resistance to localized compression of random loads. Combining honeycomb and foam technologies in a single product raises the mechanical properties significantly. Additionally, the foam technology provides flexibility, memory, good elongation properties, and shear strength of 1.7 MPa and above.

FOIL

Foil is very thin sheet metal used chiefly for wrapping, laminating, packaging, insulation, and electrical applications. Tinfoil is higher in cost than some other foils, but is valued for wrapping food products because it is not poisonous; it has now been replaced by other foils such as aluminum.

Lead foil, used for wrapping tobacco and other nonedible products, is rolled to the same thickness as tinfoil, but because of its higher specific gravity gives less coverage. Stainless steel foil is produced in thicknesses from 0.005 to 0.038 cm for laminating and for pressure-sensing bellows and diaphragms.

Aluminum foil has high luster, but is not as silvery as tinfoil. The thin foil usually has a bright side and a matte side because two sheets are rolled at one time. Aluminum foil also comes with a satin finish, or in colors or embossed designs. For electrical use the foil is 99.999% pure aluminum, but foil for rigid containers is usually aluminum alloy 3003, with 1 to 1.5% manganese, and most other foil is of aluminum alloy 1145, with 99.45% aluminum. The tear resistance of thin aluminum foil is low, and it is often laminated with paper for food packaging. Trifoil is aluminum foil coated on one side with Teflon and on the other with an adhesive. It is used as coatings for tables and conveyors or liners in chemical and food plants. Since polished aluminum reflects 96% of radiant heat waves, this foil is applied to building boards or used in crumpled form in walls for insulation.

Aluminum yarn, for weaving ribbons, draperies, and dress goods, is made from aluminum foil, 0.003 to 0.008 cm thick, by gang-slitting to widths from 0.0317 to 0.3175 cm and winding the thread on spools. Gold foil is called gold leaf and is not normally classed as foil. It is used for architectural coverings and for hot-embossed printing on leather. It is made by hammering in books, and can be made as thin as 0.0000083 cm, a gram of gold covering 3.4 m^2. Usually, gold leaf contains 2% silver, and copper for hardening.

Metal film, or metal foil, for overlays for plastics and for special surfacing on metals or composites, comes in thicknesses from flexible foils as thin as 0.005 cm to more rigid sheets for blanking and forming casings for intricately shaped parts. There are many types of hot-stamping foils, including metallic pigment foils, printed foils, and vacuum metallized foils. Composite metal films come in almost any metal or alloy such as film of tungsten carbide in a matrix of nickel alloy for wear-resistant overlays.

FORGINGS

Forging is a process of plastic deformation, usually at elevated temperatures, of ingot, bar, or billet, mill product, or metal powder to produce a desired shape and mechanical properties. Forging develops a metallurgically sound, uniform, and stable material that will have optimum properties in the operating component after being completely processed and assembled.

During the forging process, metal is distributed within the required shape as it is needed according to function and stress requirements for purposes of improved design, material utilization, producibility, and cost. Forged integral components permit reduction in the number of mechanical and welded joints and elimination of welds in critical areas, with resultant reduction in stress concentrations, increased reliability and weight savings.

Forging processes are usually classified either by the type of equipment used or by the geometry of the end product. The simplest forging operation is upsetting, which is carried out by compressing the metal between two flat parallel platens. From this simple operation, the process can be developed into more-complicated geometries with the use of dies. A number of variables are involved in forging; among major ones are properties of the workpiece and die materials, temperature, friction, speed of deformation, die geometry, and dimensions of the workpiece. One basic principle in forging is the fact that the material flows in the direction of least resistance.

Forgeability

In practice, forgeability is related to the strength, ductility, and friction of the material. Because of the great number of factors involved, no standard forgeability test has been devised. For steels, which constitute the majority of forgings, torsion tests at elevated temperatures are the most predictable; the greater the number of twists of a round rod before failure, the greater is its ability to be forged. A number of other tests, such as simple upsetting, tension, bending, and impact tests, have also been used.

In terms of factors such as ductility, strength, temperature, friction, and quality of forging, various engineering materials can be listed as follows in order of decreasing forgeability: aluminum alloys, magnesium alloys, copper alloys, carbon and low-alloy steels, stainless steels, titanium alloys, iron-base superalloys, cobalt-base superalloys, columbium alloys, tantalum alloys, molybdenum alloys, nickel-base alloys, tungsten alloys, and beryllium.

Types of Forging Methods

All methods of forging are basically related to hammering or pressing; the main difference between the two is the speed of pressure application. At times, the two methods may be interchangeable, depending on the availability of equipment and forging characteristics of the alloy. Certain metals and alloys resist rapid deformation and require the normally slower pressing operation.

Hammers are energized by gravity, air, or steam, and repeated blows of a vertically guided ram on metal (usually heated) resting on the anvil cause the metal to change shape. The steam hammer is generally used in the modern forge shop because of its flexibility of operation.

Pressing causes metal to move as the result of a slowly applied force. Presses operate by hydraulic, air, or steam action or by mechanical means such as crank or screw. Hydraulic presses have slower action than hammers, and their pressure application may be more closely controlled, allowing the maintenance of sustained pressures. Mechanical action presses, particularly of the crank type, may closely approach the speed of hammer blows and cannot maintain sustained pressures.

Although presses are usually assumed to operate in a vertical plane, upsetters, which fall into the category of mechanical presses, operate in a horizontal plane with side or gripper dies moving in coordination and at 90° to the basic press movement. Ring rolling machines come in various designs, but fundamentally operate as presses, with moving rolls forcing the ring to the desired shape and size. Extrusion presses may operate either vertically or horizontally.

Swaging or, as it is also called, rotary swaging consists of two or four dies which are activated radially by blocks in contact with a series of rollers. The rotation of the block die assembly causes the curved ends of the blocks (either circular or modified sine curve) to be in contact with the rollers; relative motion between the blocks and the roller housing then gives a reciprocating motion to the dies. The hammering action on the outer surface of the stock reduces its diameter; the stock is generally prevented from rotating.

Die Types

In forging, force is transmitted to the workpiece through dies generally made of chromium–molybdenum–vanadium steels, sometimes modified by addition of nickel or tungsten. In open dies, the primary force (compression) is applied locally, and different parts of the forging are progressively worked. In closed dies, the primary force is applied on the entire surface and the metal is forced into a cavity for forging to shape.

Open die (or hand) forgings are produced either in a hammer or a press, using a minimum of tooling. This method offers low tool cost and fast initial delivery, but relatively poor utilization of material and slow production rates. Closed die forgings involve higher tool cost for small quantities of finished parts, but offer relatively good utilization of material, generally better properties, close tolerances, good production rates, and good reproducibility.

Flat die forgings are produced on the anvil and offer simple configurations. Blocker-type forgings have fairly generous design tolerances and are usually made in one set of closed dies. Normal (or commercial) dies impart a more exact configuration to the workpiece that may be machined to required tolerances. No-draft close-tolerance dies impart a finished shape and size to the forging, which requires little or no machining. Generally, precision is inversely proportional to size, melting point of the metal, forging temperature, and the reactive tendencies of the metal surface with the atmosphere and lubricants at the forging temperature. Precision is directly proportional to the number of dies used.

With most configurations, therefore, economics dictate that it is cheaper to machine a commercial forging than to continue to approach the finished tolerances with die forging alone.

Advantages of Forgings

Orientation of crystal structure of the base metal and the flow pattern distribution of secondary phases (nonmetallics and alloy segregation) aligned in the direction of working is called grain flow. Metals in the solid state are a crystalline aggregate and are, therefore, anisotropic to a degree in properties such as strength, ductility, and impact and fatigue resistance. This anisotropy or directionality can be employed to the desired extent by orienting the metal during forging (through die design) in the direction requiring maximum strength.

It is possible to develop the maximum strength potential of a particular alloy in the forging process. Only quality-controlled rolled bar or cast ingot stock is used. Quality is determined by chemical analysis, macrostructure, microstructure, ultrasonics, and mechanical testing. Further improvement comes during the forging process, because work on the material itself achieves recrystallization and grain refinement to produce material for optimum heat-treatment response.

Forgings are better than cast material for many applications because of their greater strength and ductility in a particular alloy as well as greater soundness, uniformity in chemistry, and finer grain size. There is no drastic change of state or volume in forging as there is in castings during solidification.

Forgings are highly reproducible because of this, and because of the use of carefully controlled material, controlled working temperatures, and controlled metal flow in specially designed permanent die cavities. Forged components are also stronger than welded fabrications because weld efficiencies rarely equal 100%. Ordinary sintered products do not develop the full strength potential of an alloy because of porosity in the sintered part. As a result, a larger section is required for equivalent strength, but even here, ductility will be appreciably lower than in a part forged from material of the same chemistry.

Forgings, designed to approach the desired finished part configuration, make better utilization of material than parts machined from plate or bar stock. The closeness of this approach will be determined by the amount of money desired to be spent on forge tools rather than machining capacity. Also, bar or plate stock has only one direction of grain flow. As a result, changes in section size cut across flow lines and render the material more notch, fatigue, and stress-corrosion sensitive.

Other Comparisons

Forging vs. Stamping

Stampings are suitable at most levels of production, but are most economical where annual production is high. Press productivity and die costs logically depend on the number of dies and presses required. The list price of sheet stock is competitive with forging stock, but stamping is not usually as material efficient. Energy consumption, however, is low since the stock is not heated.

Six factors give forging advantages over sheet metal stamping.

1. The engineered scrap rate for some types of stampings that are alternatives to forging may be as high as 50%.
2. Most stampings are made in stages requiring separate dies. Processing

costs climb as the number of dies increases.
3. Many applications require several stampings to be separately formed and joined in an assembly. Production costs climb with the amount of tooling, fixturing, and joining required. In some cases forgings have been chosen over one-piece stampings to achieve weight advantages or to gain secondary benefits from shapes that cannot be stamped.
4. Stampings are usually made from stock of uniform thickness. The stock thickness of a stamping is driven by the mechanical requirements of one critical feature. Alternatives, such as added reinforcements or tailored blanks, require separate parts to be processed or joined. Forging allows designers to tailor feature thickness to functional requirements, reducing overall component weight.
5. Stamping assemblies require features, such as flanges, to facilitate joining. These features usually increase the amount of purchased stock and the weight of the end product.
6. Stamping processes work-harden metal to some degree, increasing strength and hardness and decreasing ductility in some areas of the product. These increases are driven by the process, and usually cannot be optimized to the application as can be done in forging. In some cases, work hardening requires intermediate annealing.

Forgings vs. Weldments

Weldments are generally made from bar, tubing, and plate. Part shapes are made by burning, laser cutting, shearing, or sawing, depending on complexity and thickness. Tooling cost is very low but cutting and welding can be labor intensive.

Weldments may offer an advantage over forgings in low production quantities, but this economic advantage decreases as quantities increase. Applications where production requirements start low can often be introduced as weldments and converted to forgings as production grows.

Forgings can also become part of a weldment. For example, when special features are added to a forging it is sometimes more economical to weld two forgings together than to forge the entire part. A good example is friction welding a bar of steel to a flange to form a long axle shaft.

Forging vs. Foundry Casting

Forgings offer significant advantages over castings in applications where reliability, high tensile strength, or fatigue strength are required. Forgings are free from porosity, which is difficult to eliminate in castings. This is particularly true in areas where geometric transitions occur, which are typically areas of stress concentration.

The superior fracture toughness of forgings often must be considered when designing equivalent castings by applying a "casting factor." This casting factor imposes a weight penalty that is often enough to make forgings the more economical choice.

Forging vs. Investment Casting

Investment casting, sometimes known as the "lost-wax" process, is used with a wide range of alloys, including carbon and alloy steels, stainless steels, titanium, nickel, cobalt, and aluminum alloys. Tools consist of aluminum molds for injection molding the wax patterns. They are relatively inexpensive and require very little maintenance.

The investment casting process is more labor-intensive than forging, and is more suited to lower production quantities. Investment casting is most advantageous for small and medium-size castings of highly complex shapes that require extremely tight dimensional precision and good surface quality.

Forging vs. CNC Machining

There is virtually no limit to the shapes that can be produced by CNC machines. In most cases, standard cutting tools are used, eliminating the need to purchase special tooling. Processing and material costs are generally high because much of the purchased stock is removed. Hogouts are useful for complex or precision components in very low quantities. They can also be used on a limited basis for prototype forgings. Forgings generally exhibit superior directional properties and fatigue performance due to grain flow.

Forging vs. Powder Metallurgy

Conventional powder metallurgy (P/M), metal injection molding (MIM), and powder forging (P/F) are the three most commonly specified powder metallurgy processes. MIM is currently limited to very small components, up to approximately 100 g of complex configuration. It is rarely an alternative to forging. Conventional P/M and P/F may be alternatives in some applications.

Conventional P/M produces very close dimensional precision, but the process is characterized by porosity, which reduces mechanical properties. Tensile and yield strengths decline approximately in proportion to the level of porosity. Ductility and dynamic properties, such as impact toughness, fracture toughness, and fatigue strength, are usually much lower than for forgings. Additional processes, such as infiltration and special sintering procedures, can improve these properties. However, the metallurgical properties of such products are not equivalent to forgings made from similar alloys.

Material properties of P/M parts approach those of forgings only when the porosity is reduced to 0.5% or less. Generally, this can only be achieved by P/F. P/F is an alternative to impression die forging for small and medium-sized components with a high degree of symmetry and high production volumes, such as automotive connecting rods.

Forgings vs. Reinforced Plastics and Composites

Reinforced plastics and composites generally utilize thermoset plastics, and occasionally thermoplastics, as a matrix. Reinforcing fibers of glass, mineral, carbon, and aramid are added to increase strength and stiffness. These materials are well established in applications where low weight is essential and increased cost can be tolerated, such as aerospace applications. Regardless of the reinforcing fibers used, the operating temperature range of reinforced plastics and composites is limited by the polymer matrix materials.

Forgings offer advantages of lower cost and higher production rates. Forging materials outperform composites in almost all physical and mechanical properties, especially impact toughness, fracture toughness, and compression strength.

FUEL CELLS

Solid Oxide Fuel Cells

Solid oxide fuel cells (SOFC) can be classified in terms of their structures into tubular, planar, and honeycomb types. Among them, a planar SOFC will be the most suitable for large power plant utilization because the higher power density can be expected. Furthermore, well-established ceramic processing methods such as the tape-casting method can be applied to fabricate planar SOFC components cheaply and easily.

SOFC are high-temperature (900 to 1000°C), ceramic, electrochemical reactors that directly convert chemical into electrical energy. The basic unit of a cell consists of two porous gas-diffusion electrodes separated by a gastight (i.e., dense), oxygen-ion-conducting electrolyte. Yttria-stabilized zirconia (YSZ) has been most popularly used as electrolyte material, because of its relatively high ionic conductivity and high chemical stability both in reducing and oxidizing atmospheres. The perovskite $La_{1-x}Sr_xMnO_{3-y}$ (SLM, with $0 < x < 0.5$) has been used mostly as cathode material and the cermet Ni/8 mol% yttria-stabilized zirconia (8YSZ) has been used as anode material.

Two well-known methods for attaching the LSM to the YSZ electrolyte are plasma spraying and sputtering. Because these are gas-phase processing methods, significantly good contact conditions between the electrode and the electrolyte can be obtained; a thin and homogeneous electrode layer can be produced on the electrolyte.

For optimum SOFC performance, the electrodes require good lateral conductivity, electrochemical activity, and chemical stability toward the electrolyte and gas environment. These factors depend on the composition and microstructure of the electrode, which is determined by the nature of the starting powder and the applied manufacturing technique.

Application

The high-temperature SOFC using zirconia electrolyte are being considered for power generation, because they are expected to provide high energy efficiency, yet produce a low level of pollutant gases. The SOFC plants will be operated at high temperatures near 1000°C, and different cell materials will be exposed to either oxidizing or reducing atmosphere over a long time. Hence, the degradation of the cell performance due to the reaction between electrode materials and the zirconia electrolyte may become an important factor in determining the service life.

OTHER CELLS

An environmentally cleaner way to generate electricity, known as molten carbonate fuel cell technology, could significantly increase the demand for nickel over the next decade.

Stationary plants that will use H_2 and O_2 to generate electrical power, with water and heat as the sole byproducts, could be built. These quiet, nonpolluting plants could become an important part of the power industry in the 21st century.

If the present design concepts hold up, the anodes and cathodes of these massive power plants will be made of porous nickel alloys. Typically, the anode is a Ni–Cr alloy, and the cathode is composed of a lithiated NiO. Nickel catalysts (supported either on MgO or Li-Al-dioxide) are also required to reform hydrocarbon fuel such as natural gas, producing H_2 gas which is needed in the fuel cell.

Fuel cells may be a preferred technology of the future for perfluorinated polymers, which are critical components in membrane fuel cells. Membranes made with perfluorinated polymers act as a separator and electrolyte to allow the fuel cell to run at high current densities and voltages.

Last are the proton exchange membrane (PEM) fuel cells for major office buildings and for the automotive industry.

The fuel cells in a power plant, for example, if installed in a 52-story office building, will convert natural gas to electrical energy through a series of chemical processes. Combined, two cells will generate 400 kW of building power, including external lighting and signage for the building facade and hot water for thermal heat.

Similar to the common battery, the fuel cell uses an electrochemical process to convert chemical energy found in H_2 and O_2 into electricity and water. Fuel cells are more efficient than internal-combustion engines and are virtually pollution- and noise-free. Stationary fuel cells can be powered by gasoline, natural gas, or methanol. Cells designed for cars are powered by gasoline.

The transportation market is the next frontier — to advance PEM technology for so-called electric cars. The resulting fuel cell can produce more than 50 kW of power, enough to power a midsize car. The extraordinary progress being made in using technology to reduce greenhouse gases, improve the air that we breathe, and use our energy more efficiently is being advanced in automotive technology.

FULLERENES

Fullerenes are a family of molecules that contain an even number of carbon atoms in a closed cage. The molecule is a hollow, pure carbon molecule in which the atoms lie at the vertices of a polyhedron with 12 pentagonal faces and any number (other than one) of hexagonal faces. The fullerenes were discovered as a consequence of astrophysically motivated chemical physics experiments that were interpreted by using geodesic architectural concepts. Fullerene chemistry, a field that appears to hold much promise for materials development and other applied areas, was born from pure fundamental science.

Buckminster fullerene (C_{60} or fullerene-60), is the archetypal member of the fullerenes. Other stable members of the fullerene family have similar structures (Figure F.2). The fullerenes can be considered, after graphite and diamond, to be the third well-defined allotrope of carbon.

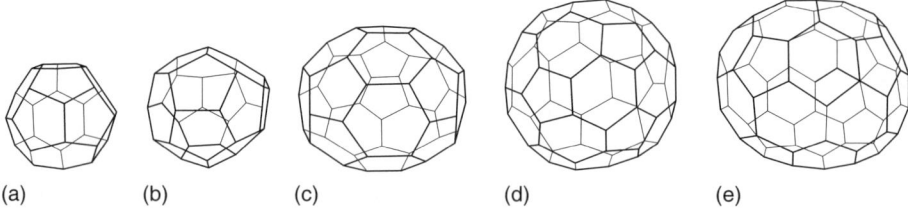

FIGURE F.2 Some of the more stable members of the fullerene family. (a) C_{28}, (b) C_{32}, (c) C_{50}, (d) C_{60}, (e) C_{70}. (From *McGraw-Hill Encyclopedia of Science and Technology*, 8th ed., Vol. 7, McGraw-Hill, New York, 540. With permission.)

The fullerenes promise to have synthetic, pharmaceutical, and industrial applications. Derivatives have been found to exhibit fascinating electrical and magnetic behavior, in particular superconductivity and ferromagnetism.

STRUCTURES

All 60 atoms in fullerene-60 are equivalent and lie on the surface of a sphere distributed with the symmetry of a truncated icosahedron. The molecule was named after R. Buckminster Fuller, the inventor of geodesic domes, which conform to the same underlying structural formula.

CHEMISTRY

Fullerene-60 behaves as a soft electrophile, a molecule that readily accepts electrons during a primary reaction step. It can accept three electrons readily and perhaps even more. The molecule can be multiply hydrogenated, methylated, ammoniated, and fluorinated.

SUPERCONDUCTIVITY

On exposure of C_{60} to certain alkali and alkaline earth metals, exohedrally doped crystalline materials are produced that exhibit superconductivity at relatively high temperatures (10 to 33 K).

Previously only metallic and ceramic materials exhibited superconductivity at temperatures much greater than a few kelvins. This discovery has opened the field of superconductivity to a different range of substances — in this case, molecular superconductors.

NANOPARTICLES AND NANOTUBES

The discovery that graphite networks (single sheets of hexagonally interconnected carbon atoms) can close readily has revolutionized general understanding of certain types of graphitic materials.

Carbon microparticles can spontaneously rearrange at high temperatures to form onion-like structures in which the concentric shells are fullerene or giant fullerenes. This phenomenon reveals the dynamics of carbon "melting."

Most importantly, carbon nanofibers consisting of concentric graphene tubes can form. These structures are essentially elongated giant fullerenes and apparently form quite readily. They will be significant in the production of carbon-fiber composite materials.

APPLICATIONS

The properties of fullerene materials that have been determined suggest that there is likely to be a wide range of areas in which the fullerenes or their derivatives will have uses. The facility for acceptance and release of electrons suggests a possible role as a charge carrier in batteries.

Fullerene nanotubes, tiny, tubular carbon fibers, were recently cut into open-ended pipes for the first time. This allows them to be chemically manipulated for use in nanotechnologies and materials. The attachment of molecules to the ends of the pipes lets them serve as means of binding to other chemical groups or surfaces.

The properties of graphite suggest that lubricative as well as tensile and other mechanical properties of the fullerenes are worthy of investigation. Liquid solutions exhibit excellent properties of optical harmonic generation. The high temperature at which superconducting behavior is observed suggests possible applications in microelectronics devices, as does the detection of ferromagnetism in other fullerene derivatives.

FURFURAL

Also known as furfuraldehyde, furol, and pyromuclealdehyde, furfural is a yellowish liquid with an aromatic odor, soluble in water and in alcohol, but not in petroleum hydrocarbons. On exposure, it darkens and gradually decomposes. Furfural occurs in different forms in various plant life and is obtained from complex carbohydrates known as pentosans, which occur in such agricultural wastes as cornstalks, corncobs, straw, oat husks, peanut shells, bagasse, and rice. Furfural is used for making synthetic plastics, as a plasticizer in other synthetic resins, as a preservative in weed killers, and as a selective solvent especially for removing aromatic and sulfur compounds from lubricating oils. It is also used for the making of butadiene, adiponitrile, and other chemicals.

Various derivatives of furfural are not used, and these, known collectively as furans, are now made synthetically from formaldehyde and acetylene, which react to form butyl nedole.

FUSIBLE ALLOYS

Fusible alloys are those with melting points below the boiling point of water (100°C). They are used as binding plugs in automatic sprinkler systems, for low-temperature boiler plugs, for soldering pewter and other soft metals, for tube bending, and for casting pattern and many ornamental articles and toys. They are also used for holding optical lenses and other parts for grinding and polishing. They consist generally of mixtures of lead, tin, cadmium, and bismuth. The general rule is that an alloy of two metals has a melting point lower than that of either metal alone. By adding still other low-fusing metals to the alloy a metal can be obtained with almost any desired low melting point.

Newton's metal, used as a solder for pewter, contains 50% bismuth, 25% cadmium, and 25% tin. It melts at 95°C, and will dissolve in boiling water. Lipowitz alloy, another early metal, contains 3 parts cadmium, 4 tin, 15 bismuth, and 8 lead. It melts at 70°C, is very ductile, and takes a fine polish. It was employed for casting fine ornaments, but now has many industrial uses. A small amount of indium increases the brilliance and lowers the melting point 1.45°C for each 1% of indium up to a maximum of 18%. Wood's alloy, or Wood's fusible metal, was the first metal used for automatic sprinkler plugs. It contained 7 to 8 parts bismuth, 4 lead, 2 tin, and 1 to 2 cadmium. It melts at 71°C, and this point was adopted as the operating temperature of sprinkler plugs in the United States; in England it is 68°C. The alloy designated as Wood's metal contains 50% bismuth, 25% lead, 2.5% tin, and 12.5% cadmium. It melts at 70°C.

Cerrobend, or Bendalloy, is a fusible alloy for tube bending that melts at 71°C. Cerrocast is a bismuth–tin alloy with pouring range of 138 to 170°C, and shrinkage of only 0.0025 cm/cm, used for making pattern molds. Cerrosafe, or Safalloy, is a fusible metal used for toy-casting sets, as the molten metal will not burn wood or cause fires. Alloys with very low melting points are sometimes used for this reason for pattern and toy casting. A fusible alloy with a melting point at 60°C contains 26.5% lead, 13.5% tin, 50% bismuth, and 10% cadmium. These alloys expand on cooling and make accurate impressions of the molds.

The Tempil pellets are alloy pellets made with melting points in steps of −10.8, 10, and 37.7°C for measuring temperatures from 45 to 1371°C. The Semalloy metals cover a wide range of fusible alloys with various melting points. Semalloy 1010 with a melting point at 47°C can be used where the melting point must be below that of thermoplastics. It contains 45% bismuth, 23% lead, 19% indium, 8% tin, and 5% cadmium. Semalloy 1280, for uses where the desired melting is near the boiling point of water, melts at 96°C. It contains 52% bismuth, 32% lead, and 16% tin.

G

GALLIUM

An elementary metal, symbol Ga, gallium is silvery white, resembling mercury in appearance but having chemical properties more nearly like aluminum. It melts at 30°C and boils at 2403°C, and this wide liquid range makes it useful for high-temperature thermometers. Like bismuth, the metal expands on freezing, the expansion amounting to about 3.8%. Pure gallium is resistant to mineral acids, and dissolves with difficulty in caustic alkali. Commercial gallium has a purity of 99.9%. In the molten state it attacks other metals, and small amounts have been used in Sn–Pb solders to aid wetting and decrease oxidation, but it is expensive for this purpose.

Gallium alloys readily with most metals at elevated temperatures. It alloys with tin, zinc, cadmium, aluminum, silver, magnesium, copper, and others. Tantalum resists attack up to 450°C, and tungsten to 800°C. Gallium does not attack graphite at any temperature and silica-base refractories are satisfactory up to about 1000°C.

APPLICATIONS

Ga–Sn alloy has been used when a low-melting metal was needed. It is also used in rectifiers to operate to 316°C. The material has high electron mobility. This material in single-crystal bars is produced for lasers and modulators.

Ga–As is an interesting new material because both gallium and arsenic are available in the state of extreme purity required for semiconductor applications and because the finished gallium arsenide in the proper state of purity can be used in transistors at high frequencies and high temperatures.

Another device using gallium arsenide is the tunnel diode. Basically, a tunnel diode is a heavily doped junction diode that displays a quantum-mechanical tunneling effect under forward bias. This effect leads to an interesting negative resistance effect. Several applications for tunnel diodes are replacement for phase-locked oscillators; switching circuits; frequency-modulated transmitter circuits; and amplifiers.

Gallium arsenide is also used in increasing quantities for solar cells. A "paddle wheel" satellite in orbit demonstrates a dramatic commercial application of solar batteries. The NASA Deep Space 1 (DS 1) probe has a unique lens system that will enable its solar panels to generate power from a solar-cell area $1/16$th the size of conventional silicon devices. A pair of solar-array wings use refractive Fresnel lenses to concentrate light onto the cells, thus less material is required. The panels are critical to the mission. They supply not only power for electronics but for the electric propulsion system of the vehicle as well. Each of the 5232 × 1600-mm GaAs-based array wings will produce 1.3 kW. The light-concentrating Fresnel lenses are silicon with a thin glass coating. Although more expensive per unit area than conventional silicon arrays, the more efficient (at 23%) multijunction GaAs cells further reduce the required solar-cell area, which, in turn, cuts spacecraft size and mass. Net result: cost is half that of conventional planar panels.

In addition to GaAs, several gallium compounds have found application in the semiconductor field. GaO has been used for vapor-phase doping of other semiconductor materials, and the oxide and halides have application in epitaxial growth of GaAs and GaP. Gallium itself has been used as a dopant for semiconductors. Gallium ammonium chloride has been used in plating baths for the electrodeposition of gallium onto whisker wires used as leads for transistors.

TABLE G.1
Properties of Gallium Semiconductors

Property	GaN	GaP	GaAs	GaSb	GaS
Structure[a]	ZnS hcp	ZnS ccp	ZnS ccp	ZnS ccp	ZnS ccp
Density, g/cm^{-3}	6.10	4.13	5.31	5.61	5.316
Melting point, °C (K)	1500 (1773)	1540(d) (1813)	1245(d) (1518)	712 (985)	962 (1235)
$-\Delta H_f^\circ$ kcal/mol (kJ/mol)	24.9 (104)	17.2 (72)	20 (84)	10.5 (43.9)	23.2
$-\Delta G_f^\circ$ kcal/mol (kJ/mol)	18.6 (77.8)	14.5 (60.7)	18.5 (74.4)	9.0 (38)	
S_{298}°,[b] cal/deg/mol (J/K/mol)	7.1 (29.7)	10.8 (45.2)	15 (63)	18.0 (75.3)	13.8
Energy gap, eV, 27°C (300 K)	3.36	2.26	1.42	0.72	1.35
Mobility, 27°C (300 K), cm^2/V/s					
Electrons	380	110	8500	5000	8800
Holes	—	75	400	850	400
Effective mass/rest mass					
Electrons	0.19	0.82	0.067	0.042	—
Holes	0.60	0.60	0.082	0.40	—

[a] hcp = hexagonal close packed; ccp = cubic close packed.
[b] S_{298}° = absolute entropy at 298 K.

Source: McGraw-Hill Encyclopedia of Science and Technology, 8th ed., Vol. 7, McGraw-Hill, New York, 622. With permission.

GaSe, GaSc, GaI$_3$, and other compounds are also used in electronic applications, see Table G.1.

A new solar cell being developed boasts 50% more power than traditional designs. The new cells, based on the two-junction, Ga–In–P on Ga–As technology, also have a longer life and are more resistant to radiation than silicon-based cells. This makes them well suited for communication satellites. (Silicon solar cells lose half their efficiency after 5 years in space.)

Other gallium alloys are suitable as dental alloys, and gallium is used in gold–platinum–indium alloys for dental restoration. Because of its low vapor pressure, gallium is being used as a sealant for glass joints in laboratory equipment, particularly mass spectrometers. Certain alloys (principally with cadmium and zinc) are used as cathodes in specialized vapor-arc lamps. Hard gallium alloys are used as low-resistance contact electrodes for bonding thermocouples and other wires to ferrites and semiconductors.

A new generation of transistors based on gallium nitride promises to deliver up to a hundred times more power at microwave frequencies than current semiconductors. Tests have shown gallium nitride transistors with an output power of up to 2.2 W/mm at a frequency of 4 GHz. This compares with about 1 W/mm at 10 GHz for GaAs transistors. By combining four devices it may be possible to make transistors with an output power of 12.5 W/mm, with each device 2 mm long, on a monolithic integrated circuit to make a chip with an output power of 100 W at a frequency of 10 GHz.

The crystal from which the chips are made is grown on a heat sink made of either silicon carbide or sapphire (Al$_2$O$_3$). Instead of doping, in which small amounts of another material are added to the crystal, the researchers chose another method. In this technology, a thin layer of gallium aluminum nitride is placed on top of a base of gallium nitride. The bond between the two layers places a strain on the upper layers, which enables free electrons to flow into the gallium nitride layer, producing the holes that make it a semiconductor.

Applications could include military radar, portable satellite phones, and satellite transmitters. The high-power transistors may also save money. Satellites equipped with the devices could operate in higher orbits. This could usher the way for fewer satellites flying higher above the Earth to give the same ground coverage.

GALVANIZING

Galvanizing is the process of coating irons and steels with zinc for corrosion protection. The zinc may be applied by immersing the substrate in a bath of the molten metal (hot-dip galvanizing), by electroplating the metal on the substrate (electrogalvanizing), or by spraying atomized particles of the metal onto otherwise finished parts. The zinc protects the substrate in two ways: (1) as a barrier to atmospheric attack, and (2) galvanically, that is, if the coating is broken, exposing the substrate, the coating will corrode sacrificially, or in preference to the substrate. From the standpoint of barrier protection alone, a coating weight of 400 g/m² on sheet steel will provide a service life of about 30 years in rural atmosphere and about 5 years in severe industrial atmosphere.

Both hot-dip galvanizing and electrogalvanizing are continuous processes applied in the production of galvanized steel, and the coating may be applied on one or both sides of the steel. In the case of hot-dipped galvanized steel, the zinc at the steel face alloys with about 25% iron from the steel. Iron alloying decreases progressively to a region that is 100% zinc. Electrogalvanized steel typically has a more homogeneous but thinner coating of pure zinc and is somewhat more formable than the hot-dipped variety. Hot dipping is widely practiced with mild steel sheet for garbage cans and corrugated sheets for roofing, sheathing, culverts, and iron pipe, and with fencing wire. The electroplating method is also used for wire, as well as for applications requiring deep drawing. An alloy layer does not form; hence, the smooth electroplated coating does not flake in the drawing die.

GARNET

Garnet is a general name for a group of minerals varying in color, hardness, toughness, and method of fracture, used for coating abrasive paper and cloth, for bearing pivots in watches, for electronics, and the finer specimens for gemstones. Garnets are trisilicates of alumina, magnesia, calcia, ferrous oxide, manganese oxide, or chromic oxide. The general formula is $3R''O \cdot R_2'''O_3 \cdot 3SiO_2$, in which R'' is calcium, magnesium, iron, or manganese, and R''' is aluminum, chromium, or iron.

Garnet-coated paper and cloth are preferred to quartz for the woodworking industries, because garnet is harder and gives sharper cutting edges, but Al_2O_3 is often substituted for garnet.

Synthetic garnets for electronic application are usually rare earth garnets with a rare earth metal substituted for the calcium, and iron substituted for the aluminum and the silicon. Yttrium garnet, is thus $Yt_3Fe_2(FeO_4)_3$. Yt–Al garnets of 3-mm diameter are used for lasers. Gadolinium (Gd) garnet has been chosen for microwave use. Gd–Ga garnet made from GdO and GaO is used for computer bubble memories.

GASKET MATERIALS

These are any sheet material used for sealing joints between metal parts to prevent leakage, but gaskets may also be in the form of cordage or molded shapes. The simplest gaskets are waxed paper or thin copper. A usual requirement is the material will not deteriorate by the action of water, oils, or chemicals. Gasket materials are usually marketed under trade names. There are sheets of paper or fiber, 0.025 to 0.318 cm thick, coated to withstand oils and gasoline.

To resist high heat and pressure, there are sheets of metal coated with graphited asbestos, with the sheet metal punched with small tongues to hold the asbestos. There is felt impregnated with zinc chromate to prevent corrosion and electrolysis between dissimilar metal surfaces.

Foamed synthetic rubbers in sheet form, and also plastic impregnants, are widely used for gaskets. Some of the specialty plastics, selected for heat resistance or chemical resistance, are used alone or with fillers, or as binders for fibrous materials.

A gasketing sheet to withstand hot oils and superoctane gasolines is based on Viton, a copolymer of vinylidene fluoride and hexafluoropropylene. It contains about 65% fluorine, has a tensile strength of 13 MPa with elongation of 400%, and will withstand operating temperatures to 204°C with intermittent temperatures to 316°C.

GASOLINE

Gasoline is a colorless liquid hydrocarbon obtained in the fractional distillation of petroleum. It is used chiefly as motor fuel, but also as a solvent. Ordinary gasoline consists of the hydrocarbons between C_6H_{14} and $C_{10}H_{22}$, which distill off between the temperatures 69 and 174°C, usually having the light limit at heptane, C_7H_{16}, or octane, C_8H_{18}. The octane number is the standard of measure of detonation in the engine. Motor fuel, or the general name gasoline, before the wide use of high-octane gasolines obtained by catalytic cracking meant any hydrocarbon mixture that could be used as a fuel in an internal-combustion engine by spark ignition without being sucked in as a liquid and without being so volatile as to cause imperfect combustion and carbon deposition. These included also mixtures of gasoline with alcohol or benzol.

GEL

A gel is a continuous solid network enveloped in a continuous liquid phase; the solid phase typically occupies less than 10 vol% of the gel. Gels can be classified in terms of the network structure. The network may consist of agglomerated particles formed, for example, by destabilization of a colloidal suspension (Figure G.1a); a "house of cards" consisting of plates (as in a clay) (Figure G.1b); fibers polymers joined by small crystalline regions (Figure G.1c); and polymers linked by covalent bonds (Figure G.1d).

In a gel, the liquid phase does not consist of isolated pockets, but is continuous. Consequently, salts can diffuse into the gel almost as fast as they disperse in a dish of free liquid. Thus, the gel seems to resemble a saturated household sponge, but it is distinguished by its colloidal size scale: the dimensions of the open

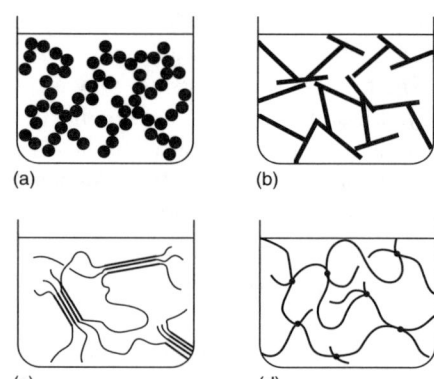

FIGURE G.1 Gel structures. (a) Agglomerated particles, (b) framework of fibers or plates, (c) polymers linked by crystalline junctions, (d) polymers linked by covalent bonds. (From *McGraw-Hill Encyclopedia of Science and Technology*, 8th ed., Vol. 7, McGraw-Hill, New York, 722. With permission.)

spaces and of the solid objects constituting the network are smaller (usually much smaller) than a micrometer. This means that the interface joining the solid and liquid phases has an area on the order of 1000 m/g of solid. As a result, the properties of a gel are controlled by interfacial and short-range forces, such as van der Waals, electrostatic, and hydrogen-bonding forces. Factors that influence these forces, such as introduction of salts or another solvent, application of an electric field, or changes in pH or temperature, affect the interaction between the solid and liquid phases. Variations in these parameters can induce huge changes in volume as the gel imbibes or expels liquid, and this phenomenon is exploited to make mechanical actuators or hosts for controlled release of drugs from gels. For example, a polyacrylamide gel (a polymer linked by covalent bonds) shrinks dramatically when it is transferred from a dish of water (a good solvent) to a dish of acetone (a poor solvent), because the polymer chains tend to favor contact with one another rather than with acetone, so the network collapses onto itself. Conversely, the reason that water cannot be gently squeezed out of such a gel (as from a sponge) is that the network has a strong affinity for the liquid, and virtually all of the molecules of the liquid are close enough to the solid–liquid interface to be influenced by those attractive forces.

Gelation

The most striking feature of a gel, which results from the presence of a continuous solid network, is elasticity: if the surface of a gel is displaced slightly, it springs back to its original position. If the displacement is too large, gels (except those with polymers linked by covalent bonds) may suffer some permanent plastic deformation, because the network is weak. The process of gelation, which transforms a liquid into an elastic gel, occurs with no change in color or transparency and no evolution of heat. It may begin with a change in pH that removes repulsive forces between the particles in a colloidal suspension, or a decrease in temperature that favors crystallization of a solution of polymers or the initiation of a chemical reaction that creates or links polymers.

In many cases, both the liquid and solid phases of a gel are of practical importance (as in timed-release of drugs, or a gelatin dessert). More often (as in preparation of catalyst supports, chromatographic columns, or desiccants), it is only the solid network that is of use. If the liquid is allowed to evaporate from a gel, large capillary tension develops in the liquid, and the suction causes shrinkage of the network. The porous network remaining after evaporation of the liquid is called a xerogel.

To maximize the porosity of the dried product, the gel can be heated to a temperature and pressure greater than the critical point of the liquid phase, where capillary pressures do not exist. The fluid can then be removed with little or no shrinkage occurring. The resulting solid is called an aerogel, and the process is called supercritical drying.

Sol-Gel Processing

Sol-gel processing comprises a variety of techniques for preparing inorganic materials by starting with a sol, then gelling, drying, and (usually) firing. Many inorganic gels can be made from solutions of salts or metallorganic compounds, and this offers several advantages in ceramics processing: the reactants are readily purified; the components can be intimately mixed in the solution or sol stage; the sols can be applied as coatings, drawn into fibers, emulsified or spray-dried to make particles, or molded and gelled into shapes; xerogels can be sintered into dense solids at relatively low temperatures, because of their small pore size.

GERMANIUM

A rare elemental metal, germanium (Ge) has a grayish white crystalline appearance and has great hardness: 6.25 Mohs. Its specific gravity is 5.35, and melting point is 937°C. It is resistant to acids and alkalies. It has metallic-appearing crystals with diamond structure, gives greater hardness and strength to aluminum and magnesium alloys, and as little as 0.35% in tin will double the hardness. It is not used commonly in alloys, however, because of its rarity and great cost. It is used chiefly as metal in rectifiers and transistors. An Au–Ge alloy, with about 12% germanium, has a melting point of 359°C and has been used for soldering jewelry.

Germanium is obtained as a by-product from flue dust of the zinc industry, or it can be obtained by reduction of its oxide from the ores, and is marketed in small irregular lumps. Germanium crystals are grown in rods up to 3.49 cm in diameter for use in making transistor wafers. High-purity crystals are used for both P and N semiconductors. They are easier to purify and have a lower melting point than other semiconductors, specifically silicon.

Alloys

Two alloys of importance to the semiconductor industry are Ge–Al and Ge–Au. At 55 wt% germanium, the Ge–Al system forms a eutectic that melts at 423°C. At 12 wt% germanium, the Ge–Au system forms a eutectic that melts at 356°C. The two alloys are very useful in forming electrical contact systems for germanium transistors, diodes, and rectifiers. The Ge–Au system has been evaluated regarding its use for dental alloys because of its good dimensional stability upon cooling, thus allowing precision castings to be made. The Ge–Si alloy system forms a continuous series of solid solutions. It has been explored quite extensively from the standpoint of its semiconductor characteristics.

Germanium has been investigated as a possible alloying agent for zirconium in the development of corrosion-resistant, high-strength zirconium alloys. Small additions of germanium are known to give increased hardness and strength to copper, aluminum, and magnesium. In addition, small quantities of germanium improve the rolling properties of some alloys and do not create any appreciable increase in production costs. Fundamental studies on the magnetic properties of Fe–Ge and Mn–Ge alloys have been made. These materials are ferromagnetic, as are the alloys UGe_2, PuGe, and CrGe.

Uses

The properties of germanium are such that there are several important applications for this element.

Semiconductor

During World War II, germanium was intensively investigated for its use in the rectification of microwaves for radar applications.

A major development in electronics occurred in 1948 with the invention of the transistor. This solid-state device, which was first made of germanium, had a profound influence on all electronic applications. The transistor captured the hearing aid and radio markets, then moved into industrial applications, such as computers and guidance and control systems for missile and antimissile systems. Transistors cover signal processing from DC to gigahertz frequencies, with power-handling capabilities from microwatts to hundreds of watts.

The development of germanium semiconductor technology quite closely paralleled the development of the germanium transistor. In fabricating a transistor, a need arose for high-purity polycrystalline germanium to be used in the growth of single-crystal germanium, since it was soon discovered that not only purity but also good crystal structure was required for good device performance. The growth processes for single crystals of germanium had to be developed. Several techniques have been extensively studied. Two of the most common are the Czochralski or Teal-Little method and zone leveling. The general method proceeds by allowing germanium to slowly freeze or crystallize onto a single crystal seed, which may be rotated and withdrawn from the melt. The growth (freeze) rate is controlled by a combination of the temperature of the melt and the amount of heat lost from the crystal by conduction up the seed and by radiation from its surfaces. Crystals from 0.16 to 15 cm in diameter have been grown by this process. Growth rates vary with diameter and desired crystal properties but are usually in the inches-per-hour range.

In the zone-leveling technique, or horizontal method, the seed and polycrystalline charge or ingot are usually loaded into a quartz boat. The boat and contents are then placed into the quartz tube of the zone leveler. With this technique, melting occurs on the leading edge of the molten zone, and freezing occurs on the trailing edge as the furnace is moved. Growth rates vary with crystal properties desired but are usually in the inches-per-hour range. The advantage of this process over the Teal-Little for germanium is the extremely uniform resistivity profile of the crystal due to the uniform mass of molten germanium in the growth process. The constant mass of molten germanium enables the ratio of dopant to germanium to remain the same, thus producing uniform resistivity.

In the fabrication of germanium devices, the single-crystal material must be sawed, lapped, or ground and then polished into thin-slice forms with flat, damage-free surfaces. Various techniques have been developed to produce slices with these characteristics. Some of these are slicing with diamond wheels, grinding with diamond, or lapping with an abrasive slurry. These techniques induce damage in the form of microcracks and fissures, which must be removed by chemical, electrochemical, or chemical-mechanical polishing methods.

Two other growth techniques (modifications of melt growth) have been explored. One is dendritic growth, which depends on the growth characteristics of twinned crystals, and the other is shaped crystal growth, which is dependent on a shaped heat zone. Both techniques can produce thin ribbons of germanium with desired characteristics for some type of devices.

Vapor-phase crystal growth at temperatures well below the melting point is used for growing thin films or layers onto slices and is usually referred to as epitaxial growth. The primary advantage of gas-phase growth is that it allows doping impurities to be changed quite rapidly so that thin (micrometer range) layers of quite different resistivities can be grown sequentially.

Optoelectronics

Another use that surfaced as a result of the cost of material such as gallium arsenide (GaAs) and the inability to control its growth is the use of Czochralski- or Teal-Little-grown single-crystal germanium as a substrate for vapor-phase growth of GaAs and gallium arsenide phosphide (GaAsP) thin films used in some light-emitting diodes (LEDs). These devices have been used in digital displays for calculators, watches, and so on.

Infrared Optical Materials

Germanium lenses and filters have been used in instruments that operate in the infrared region of the spectrum. Windows and lenses of germanium are vital components of some laser and infrared guidance or detection systems. Glasses prepared with germanium dioxide have a higher refractivity and dispersion than do comparable silicate glasses and may be used in wide-angle camera lenses and microscopes.

GLASS

Glass, one of the oldest and most extensively used materials, is made from the most abundant of Earth's natural resources — silica sand. For centuries considered as a decorative, fragile material suitable for only glazing and art objects, today glass is produced in thousands of compositions and grades for a wide range of consumer and industrial applications.

COMPOSITION AND STRUCTURE

As just stated, the basic ingredient of glass is silica (silicon dioxide), which is present in various amounts, ranging from about 50 to almost 100%. Other common ingredients are oxides of metals, such as lead, boron, aluminum, sodium, and potassium.

Unlike most other ceramic materials, glass is noncrystalline. In its manufacture, a mixture of silica and other oxides is melted and then cooled to a rigid condition. The glass does not change from a liquid to a solid at a fixed temperature, but remains in a vitreous, noncrystalline state, and is considered as a supercooled liquid. Thus, because the relative positions of the atoms are similar to those in liquids, the structure of glass has short-range order. However, glass has some distinct differences compared to a supercooled liquid. Glass has a three-dimensional framework and the atoms occupy definite positions. There are covalent bonds present the same as those found in many solids. Therefore, there is a tendency toward an ordered structure in that there is present in glass a continuous network of strongly bonded atoms.

PROPERTIES AND PROCESSING

Glass is an amorphous solid made by fusing silica (silicon dioxide) with a basic oxide. Its characteristic properties are its transparency, its hardness and rigidity at ordinary temperatures, its capacity for plastic working at elevated temperatures, and its resistance to weathering and to most chemicals except hydrofluoric acid. The major steps in producing glass products are (1) melting and refining, (2) forming and shaping, (3) heat treating, and (4) finishing. The mixed batch of raw materials, along with broken or reclaimed glass, called cullet, is fed into one end of a continuous-type furnace where it melts and remains molten at around 1499°C. Molten glass is drawn continuously from the furnace and runs in troughs to the working area, where it is drawn off for fabrication at a temperature of about 999°C. When small amounts are involved, glass is melted in pots.

TYPES OF GLASS

There are a number of general families of glasses, some of which have many of hundreds of variations in composition (Table G.2).

Soda-Lime Glasses

The soda-lime family is the oldest, lowest in cost, easiest to work, and most widely used. It accounts for about 90% of the glass used in this

TABLE G.2
Composition of Typical Glasses

Designation	Composition, %								
	SiO_2	Na_2O	K_2O	CaO	MgO	BaO	PbO	B_2O_3	Al_2O_3
Silica glass (fused silica)	99.5+								
96% Silica glass	96.3	<0.2	<0.2					2.9	0.4
Soda-lime (window sheet)	71–73	12–15		8–10	1.5–3.5				0.5–1.5
Soda-lime (plate glass)	71–73	12–14		10–12	1–4				0.5–1.5
Soda-lime (containers)	70–74	13–16	13–16	10–13	10–13	0–0.5			1.5–2.5
Soda-lime (electronic lamp bulbs)	73.6	16	0.6	5.2	3.6				1
Lead-alkali silicate (electrical)	63	7.6	6	0.3	0.2		21	0.2	0.6
Lead-alkali silicate (high-lead)	35		7.2				58		
Borosilicate (low-expansion)	80.5	3.8	0.4					12.9	2.2
Borosilicate (low-electrical loss)	70.0		0.5				1.2	28.0	1.1
Aluminosilicate	57	1.0		5.5	12			4	20.5

country. Soda-lime glasses have only fair to moderate corrosion resistance and are useful up to about 460°C, annealed, and up to 249°C, tempered. Thermal expansion is high and thermal shock resistance is low compared with other glasses. They are the glass of ordinary windows, bottles, and tumblers.

Lead Glasses

Lead or lead alkali glasses are produced with lead contents ranging from low to high. They are relatively inexpensive and are noted for high electrical resistivity and high refractory index. Corrosion resistance varies with lead content, but they are all poor in acid resistance compared with other glasses. Thermal properties also vary with lead content. The coefficient of expansion, for example, increases with lead content. High lead grades are heaviest of the commercial glasses. As a group, lead glasses are the lowest in rigidity. They are used in many optical components, for neon sign tubing, and for electric lightbulb stems.

Borosilicate Glasses

Borosilicate glasses are most versatile and are noted for their excellent chemical durability, for resistance to heat and thermal shock, and for low coefficients of thermal expansion. There are six basic kinds. The low-expansion type is best known as the Pyrex brand ovenware. The low-electrical-loss types have a dielectric loss factor only second to fused silica and some grades of 96% silica glass. Sealing types, including the well-known Kovar, are used in glass-to-metal sealing applications. Optical grades, which are referred to as crowns, are characterized by high light transmission and good corrosion resistance. Ultraviolet-transmitting and laboratory apparatus grades are two other borosilicate-type glasses. Because of this wide range of types and compositions, borosilicate glasses find use in such products as sights and gauges, piping, seals to low-expansion metals, telescope mirrors, electronic tubes, laboratory glassware, ovenware, and pump impellers.

Aluminosilicate Glasses

These glasses are roughly three times more costly than borosilicate types, but are useful at higher temperatures and have greater thermal shock resistance. Maximum service temperature for annealed condition is about 649°C. Corrosion resistance to weathering, water, and chemicals is excellent, although acid resistance is only fair compared with other glasses. Compared to 96% silica glass, which it resembles in some respects, it is more easily worked and is lower in cost. It is used for high-performance power tubes, traveling wave tubes, high-temperature thermometers, combustion tubes, and stovetop cookware.

Fused Silica

Fused silica is 100% silicon dioxide. If naturally occurring, the glass is known as fused quartz. There are many types and grades of both glasses, depending on impurities present and manufacturing method. Because of its high purity level, fused silica is one of the most transparent glasses. It is also the most heat resistant of all glasses and it can be used up to 899°C in continuous service and to 1260°C for short-term exposure. In addition, it has outstanding resistance to thermal shock, maximum transmittance to ultraviolet, and excellent resistance to chemicals. Unlike most glasses, its modulus of elasticity increases with temperature. Because fused silica is high in cost and difficult to shape, its applications are restricted to such specialty applications as laboratory optical systems and instruments and crucibles for crystal growing. Because of a unique ability to transmit ultrasonic elastic waves with little distortion or absorption, fused silica is used in delay lines in radar installations.

96% Silica Glass

These glasses are similar in many ways to fused silica. Although less expensive than fused silica, they are still more costly than most other glasses. Compared to fused silica they are easier to fabricate, have a slightly higher coefficient of expansion, about 30% lower thermal stress resistance, and a lower softening point. They can be used continuously up to 799°C. Uses include chemical glassware and windows and heat shields for space vehicles.

Phosphate Glass

Phosphate glass will resist the action of hydrofluoric acid and fluorine chemicals. It contains no silica, but is composed of P_2O_5 with some alumina and magnesia. It is transparent and can be worked like ordinary glass, but it is not resistant to water.

Sodium-Aluminosilicate Glass

These glasses are chemically strengthened and are used in premium applications, such as aircraft windshields. Molten salt baths are used in the strengthening process.

Industrial Glass

This is a general name usually meaning any glass molded into shapes for product parts. The lime glasses are the most frequently used because of low cost, ease of molding, and adaptability to fired colors. For such uses as light lenses and condenser cases, the borosilicate heat-resistant glasses may be used.

Boric Oxide Glasses

These glasses are transparent to ultraviolet rays. The so-called invisible glass is a borax glass surface treated with a thin film of sodium fluoride. It transmits 99.6% of all visible light rays, thus casting back only slight reflection and giving the impression of invisibility. Ordinary soda and potash glasses will not transmit ultraviolet light. Glass containing 2 to 4% ceric oxide absorbs ultraviolet rays, and is also used for x-ray shields. Glass capable of absorbing high-energy x-rays or gamma rays may contain tungsten phosphate, while the glass used to absorb slow neutrons in atomic-energy work contains cadmium borosilicate with fluorides. The shields for rocket capsule radio antennas are made of 96% silica glass.

Optical Glass

Optical glass is a highly refined glass that is usually a flint glass of special composition, or made from rock crystal, used for lenses and prisms. It is cast, rolled, or pressed. In addition to the regular glassmaking elements, silica and soda, optical glass contains barium, boron, and lead. The high-refracting glasses contain abundant silica or boron oxide. A requirement of optical glass is transparency and freedom from color.

Plate Glass

This is any glass that has been cast or rolled into a sheet, and then ground and polished. However, the good grades of plate glass are, next to optical glass, the most carefully prepared and the closest to perfect of all the commercial glasses. It generally contains slightly less calcium oxide and slightly more sodium oxide than window glass, and small amounts of agents to give special properties may be added, such as agents to absorb ultraviolet or infrared rays, but inclusions that are considered impurities are kept to a minimum. The largest use of plate glass is for storefronts and office partitions. Plate glass is now made on a large scale on continuous machines by pouring on a casting table at a temperature of about 1000°C, smoothing with a roller, annealing, setting rigidly on a grinding table, and grinding to a polished surface.

Conductive Glass

These glasses, which are employed for windshields to prevent icing and for uses where the conductive coating dissipates static charges, are plate glass with a thin coating of stannic oxide produced by spraying glass, at 482 to 704°C, with a solution of stannic chloride. Coating thicknesses are 50 to 550 nm, and the coating will carry current densities of 9.300 W/m^2 indefinitely. The coatings are hard and resistant to solvents. The light transmission is 70 to 88% that of the original glass, and the index of refraction is 2.0, compared with 1.53 for glass.

Transparent Mirrors

These are made by coating plate glass on one side with a thin film of chromium. The glass is a reflecting mirror when the light behind the glass is less than in front, and is transparent when the light intensity is higher behind the glass. Photosensitive glass is made by mixing submicroscopic metallic particles in the glass. When ultraviolet light is passed through the negative on the glass, it precipitates these particles out of solution, and since the shadowed areas of the negative permit deeper penetration into the glass than the high light areas, the picture is in three dimensions and in color. The photograph is developed by heating the glass to 538°C.

Other Glasses

Colored glasses, made by adding small amounts of colorants to glass batches, are used in lamp bulbs, sunglasses, light filters, and signalware.

Opal glasses contain small particles dispersed in transparent glass. The particles disperse the light passing through the glass, producing an opalescent appearance.

Polarized glass for polarizing lenses is made by adding minute crystals of tourmaline or peridote to the molten glass, and stretching the glass while still plastic to bring the axes of the crystals into parallel alignment.

Porous glass is made of silica sand, boric acid, oxides of alkali metals (sodium, potassium, etc.), and a small amount of zirconia. This glass has 500 times better resistance to alkali solution than has otherwise been achieved to date. The glass is made by first melting, heat treatment at 650 to 800°C, then separating into two phases, one composed mainly of silica and the other containing boric acid and alkali metal oxides; this is known as phase splitting. The boric acid–alkali oxides phase elutes in heat treatment, producing small holes that give porous glass its name. The glass is heat-resistant, transmits gas, and permits adhesion of many substances to its surface. It is, therefore, used at high temperatures as a gas-separating membrane, and as a carrier of various substances.

Oxycarbide Glass

This glass has been developed in which substituting carbon for oxygen or even nitrogen can create a whole new category of high-strength glasses. In a 1.0% carbon glass system of Mg–Si–Al–O, Vickers hardness was increased significantly as well as glass transition temperature. The oxycarbide glass was prepared firing SiO_2, Al_2O_3, MgO, and SiC in a molybdenum crucible at 1800°C for 2 h. Oxycarbide glass systems based on Si–metal–O–C and Al–metal–O–C are likely to exist, and could potentially be used to produce refractory glasses. Controlled recrystallization of oxycarbides could lead to stable glass–ceramic matrices for ceramic-reinforced composites.

Products

The most common glass products are containers and flat glass. The latter is principally used in transportation and architecture. Flat glass is produced by the float process, developed in the 1950s and considered one of the important technological achievements of the 20th century.

Containers are largely made on automatic machines; some special pieces may be made by hand blowing into molds, and some very special ware may be made by the free or offhand blowing of a skilled glassblower.

Electric lightbulb envelopes are molded on a special machine that converts a fast-moving ribbon of glass into over 10,000 bulbs/min.

Glass fibers are either continuous for reinforcement or discontinuous for insulation; see **Glass Fiber**.

Glass wool is most often made by spinning, or forcing molten glass centrifugally out of small holes (around 20,000) in the periphery of a rapidly rotating steel spinner, and attenuating the fibers by entrainment with gas burners. As the fibers fall to a conveyor, they are sprayed with a binder that preserves their open structure. The resulting glass wool mat is mainly composed of still air, which accounts for its excellent insulation properties.

GLASS CERAMICS

A family of fine-grained crystalline materials made by a process of controlled crystallization from special glass compositions containing nucleating agents; glass ceramics are sometimes referred to as nucleated glass, devitrified ceramic, or vitro ceramics. Because they are mixed oxides, different degrees of crystallinity can be produced by varying composition and heat treatment. Some of the types produced are cellular foams, coatings, adhesives, and photosensitive compositions.

Glass ceramics are nonporous and generally either opaque white or transparent. Although not ductile, they have much greater impact strength than commercial glasses and ceramics. However, softening temperatures are lower than those for ceramics, and they are generally not useful above 1093°C. Thermal expansion varies from negative to positive values depending on composition. Excellent thermal-shock resistance and good dimensional stability can be obtained if desired. These characteristics are used to advantage in "heatproof" skillets and range tops. Like chemical glasses, these materials have excellent corrosion and oxidation resistance. They are electrical insulators and are suitable for high-temperature, high-frequency applications in the electronics field.

Pyroceram is a hard, strong, opaque-white nucleated glass with a flexural strength to above 206 MPa, a density of 2.4 to 2.62, a softening point at 1349°C, and high thermal shock resistance. It is used for molded mechanical and electrical parts, heat-exchanger tubes, and coatings. Macor is a machinable glass ceramic. Axles for mechanisms that provide power in cardiac pacemakers have been chosen from Macor due to its chemical inertness and light weight. It is also used as welding fixtures and as welding nozzles. Nucerite, which is used for lining tanks, pipes, and valves, is nucleated glass. It has about four times the abrasion resistance of a hard glass, will withstand sudden temperature differences of 649°C, and has high impact resistance. It also has high heat-transfer efficiency.

Applications

The first use of glass ceramics was in radomes for supersonic missiles, where a radar-transmitting material with a combination of strength, hardness, temperature and thermal shock resistance, uniform quality, and precision finishing is required.

The second large-scale application is in "heatproof" skillets and saucepans, again taking advantage of the thermal shock resistance.

One of the products of potential interest to mechanical engineers is bearings for operation at high temperatures without lubrication, or in corrosive liquids. Another is a lightweight, dimensionally stable honeycomb structure, which has promise for use in heat regenerators for gas turbines operating at high temperatures. Still a third possibility is precision gauges, and machine tool parts whose dimensions do not change with temperature.

Bearings

Although metal bearings are perfectly satisfactory for most applications, there are conditions such as high temperature, dry (unlubricated) operation, and presence of corrosive media in which even the best metal bearings may perform poorly. Also, the metals that do perform best are expensive, difficult to fabricate, and heavy.

Since glass ceramic bearings might be expected to be stable at high temperatures, resistant to oxidation and to corrosive conditions, as well as light in weight, a thorough evaluation of their characteristics is being made by a number of laboratories and bearing manufacturers. One of the surprising findings has been that glass ceramic bearings can be finished almost as easily as steel bearings and in the same general type of equipment.

High-Temperature Heat Exchangers

The thermal shock resistance and dimensional stability of the low expansion glass ceramics make them useful in various kinds of heat exchangers. One interesting type has been developed for use in high-temperature turbine engines.

Precision Uses

As requirements for precision instruments and machines become more stringent, the gauges and machine tools required to make them must become still more precise in dimensions. If these are made of relatively high expansion metals such as steel, the dimensions vary with temperature.

GLASS FIBER

Fine flexible fibers made from glass are used for heat and sound insulation, fireproof textiles, acid-resistant fabrics, retainer mats for storage batteries, panel board, filters, and electrical insulating tape, cloth, and rope. Molten glass strings out easily into threadlike strands, and this spun glass was early used for ornamental purposes but the first long fibers of fairly uniform diameter were made in England by spinning ordinary molten glass on revolving drums. The original fiber, about 0.003 cm in diameter, was called glass silk and glass wool, and the loose blankets for insulating purposes were called *navy wool*. The term navy wool is still used for the insulating blankets faced on both sides with flameproofed fabric, employed for duct and pipe insulation and for soundproofing.

Glass fibers are now made by letting the molten glass drop through tiny orifices and blowing with air or steam to attenuate the fibers. The usual composition is that of soda-lime glass, but it may be varied for different purposes. The glasses low in alkali have high electrical resistance, whereas those of higher alkali are more acid-resistant. They have very high tensile strengths, up to about 2757 MPa.

The standard glass fiber used in glass-reinforced plastics is a borosilicate type known as E-glass. The fibers spun as single glass filaments, with diameters ranging from 0.0005 to 0.003 cm are collected into strands that are manufactured into many forms of reinforcement. E-glass fibers have a tensile strength of 3445 MPa. Another type, S-glass, is higher in strength, about one third stronger than E-glass, but because of higher cost, 18 times more costly per pound vs. E-glass, its use is limited to advanced, higher-performance products.

Staple glass fiber is usually from 0.0007 to 0.0009 cm, is very flexible and silky, and can be spun and woven on regular textile machines. Glass fiber yarns are marketed in various sizes and twists, in continuous or staple fiber, and with glass compositions varied to suit chemical or electrical requirements.

Halide glass fibers are composed of compounds containing fluorine and various metals such as barium, zirconium, thorium, and lanthanum. They appear to be promising for fiber-optic communication systems. Their light-transmitting capability is many times better than that of the best silica glasses now being used.

Glass fiber cloth and glass fiber tape are made in satin, broken twill, and plain weaves; satin-wear cloth is 0.02 cm thick weighing 0.24 kg/m². Glass cloth of plain weave of either continuous fiber or staple fiber is much used for laminated plastics. The usual thicknesses are from 0.005 to 0.058 cm in weights from 0.05 to 0.50 kg/m². Cloth woven of monofilament fiber in loose rovings to give better penetration of the impregnating resin is also used. Glass mat, composed of fine fibers felted or intertwined in random orientation, is used to make sheets and boards by impregnation and pressure. Fluffed glass fibers are tough, twisted glass fibers. For filters and insulation the felt withstands temperatures to 538°C. Chopped glass, consisting of glass fiber cut to very short lengths, is used as a filler for molded plastics. Translucent corrugated building sheet is usually made of glass fiber mat with a resin binder. All these products, including chopped fiber, mat, and fabric preimpregnated with resin, and the finished sheet and board, are sold under a wide variety of trade names. Glass fiber bonded with a thermosetting resin can be performed for pipe and other insulation coverings. Glass fiber block is also available to withstand temperatures to 316°C. Glass filter cloth is made in twill and satin weaves in various thicknesses and porosities for chemical filtering. Glass belting, for conveyor belts that handle hot and corrosive materials, is made with various resin impregnations. Many synthetic resins do not adhere well to glass, and the fiber is sized with vinyl chlorosilane or other chemical.

Four major principles should be recognized in using glass fibers as composite reinforcement. Mechanical properties depend on the combined effect of the amount of glass-fiber reinforcement used and its arrangement in the finished composite. The strength of the finished object is directly related to the amount of glass in it. Generally speaking, strength increases directly in relation to the amount of glass. A part containing 80 wt% glass and 20 wt% resin is almost four times stronger than a part containing the opposite amounts of these two materials. Chemical, electrical, and thermal performance are influenced by the resin system used as the matrix. Materials selection, design, and production requirements determine the proper fabrication process to be used. Finally, the cost–performance value achieved in the finished composite depends on good design and judicious selection of raw materials and processes.

GLAZING

Glazing involves the application of finely ground glass, or glass-forming materials, or a mixture of both, to a ceramic body and heating (firing) to a temperature where the material or materials melt, forming a coating of glass on the surface of the ware. Glazes are used to decorate the ware, to protect against moisture absorption, to give an easily cleaned, sanitary surface, and to hide a poor body color.

Glazes are classified and described by the following characteristics: surface — glossy or matte; optical properties — transparent or opaque; method of preparation — fritted or raw; composition — such as lead, tin, or boron; maturing temperatures; and color. Opaque glazes contain small crystals embedded in the glass, but special glazes in which a few crystals grow to recognizable size are called crystalline glazes. A glaze may be applied during the firing; such a glaze is called salt glaze. Common salt, NaCl, or borax, $Na_2B_4O_7 \cdot 10H_2O$, or a mixture of both is introduced into the kiln at the finishing temperature. The salt evaporates and reacts with the hot ware to form the glaze. This type of glaze has been applied to sewer pipe and some fine stoneware.

The most important factor in compounding a glaze, after a suitable maturing temperature has been obtained, is the matching of the coefficient of thermal expansion of the glaze and the body on which it is applied. A slightly lower coefficient for the glaze will place it in compression (the desired condition) when the ware cools. The reverse state (with the glaze in tension because it has a higher coefficient) leads to the formation of fine hairline cracks, a condition known as crazing.

TABLE G.3
Physical Properties of Important Glaze Oxides

Oxide	Molecular Weight	Specific Gravity	Melting Point, °C	Coefficient of Thermal Expansion[a] ($\times 10^{-7}$)
Al_2O_3	102	4	2045	1.5
B_2O_3	69.6	1.8	460	0.6
BaO	153.4	5.7	1923	3.6
CaO	56.1	3.3	2580	4.5
K_2O	94.2	2.3	350	9
Li_2O	29.9	2	445	11.1
MgO	40.3	3.6	2800	0.6
Na_2O	62	2.3	460	11.4
PbO	223.2	9.5	888	2.25
SiO_2	60.08	1.5	1713	0.8
SrO	103.6	4.7	2430	4.2[b]
ZnO	81.4	5.5	1975	3
ZrO_2	123.2	5.6	2715	2.1[c]

[a] Hall factors (JACS, 13(3) 194, 1930).
[b] Unknown source.
[c] M & H Factor.

Source: Handbook of Chemistry and Physics, 55th ed., Robert C. Weast, Ed., CRC Press, Boca Raton, FL, 1974–1975.

GLAZE MATERIALS

As reaction times, melting points, temperatures, and substrates have changed in the whiteware industry, the materials used in glaze formulation have also changed over time. Table G.3 shows the physical properties of important glaze oxides.

Silica is a major glaze component and is added in many forms, such as quartz, feldspar, or wollastonite. Silica acts as a glass former and is used to control thermal expansion and help impart acid resistance to the glaze.

Clay, such as kaolin, ball clay, china clay, or bentonite, continues to be the primary suspending agent used in ceramic glazes. The rheology characteristics required by the application method, as well as physical properties such as glaze drying time or shrinkage characteristics, need to be taken into account when selecting the clay to be used in a ceramic glaze. For example, glazing wet column brick requires glazes with up to 25% clay, while only 5 to 10% clay is needed for glaze suspension.

Feldspathic minerals, such as soda and potash feldspar and nepheline syenite, remain some of the most commonly used raw materials. These materials are a major source of silica and alkali fluxes in a glaze. Feldspar can be used as either a flux or a refractory material in a glaze, depending on the firing temperature.

Alumina is normally added as calcined alumina or alumina trihydrate, although both clay and feldspars are also sources of alumina in the glaze. The alumina is used to improve the scratch resistance or abrasion resistance of the glaze, and also influences the gloss level.

Alkaline earth oxide materials, such as calcium carbonate, wollastonite, and zinc oxide, are also generally added as raw materials. Other alkaline earth oxides, such as lead oxide, strontium oxide, barium oxide, and magnesium oxide, are more typically added in a fritted form. The alkaline earth oxides are beneficial because they provide fluxing action without having a major effect on glaze thermal expansion.

Zirconium silicate is the major opacifier used in ceramic glazes. However, tin oxide is used by some manufacturers, particularly if chrome–tin pigments are being used in the glaze. Using zircon-opacified frits to provide some or all of the zirconium silicate needed in a glaze is also becoming more common. This is especially true in fast-firing cycles, where the use of a high percentage of a refractory material (such as zirconium silicate) is not desirable.

Ceramic frits play a major part in glaze formulation. Frits continue to be a source of highly soluble oxides, such as soda, potassium, or boron. As firing cycles have grown shorter, however, materials such as zircon, calcium, alumina, or barium are commonly added in the fritted form. In addition, frit producers are able to tailor frit formulations for particular uses and processes.

GOLD AND GOLD ALLOYS

Gold (Au) is a soft, ductile, yellow metal, known since ancient times as a precious metal on which all material trade values are based. Commercially pure gold is 99.97% pure, and higher purity material is available.

The outstanding useful property of gold is its oxidation resistance. It is not attacked by the common acids when used singly. It does, however, dissolve in aqua regia (nitric acid plus hydrochloric acid) and cyanide solutions, and is attacked by chlorine above 80°C. It is resistant to dry fluorine up to about 300°C, hydrogen fluoride, dry hydrogen chloride, and dry iodine. It is also resistant to sulfuric acid, sulfur, and sulfur dioxide.

Gold is found widely distributed in all parts of the world. It is used chiefly for coinage, ornaments, jewelry, and gilding.

Gold is the most malleable of metals, and can be beaten into extremely thin sheets. In most uses gold is alloyed to increase its hardness without appreciable loss of oxidation resistance. Copper is a common alloying element along with silver and small amounts of the platinum metals. Some of these alloys can be heat-treated to relatively high strengths. Gold and its alloys are worked into all the usual forms of sheet, wire, ribbon, and tubing.

A gram of gold can be worked into leaf covering 0.6 m^2, and only 0.0000084 cm thick, or into a wire 2.5 km in length. Precision casting is also used to form gold alloys, particularly for jewelry. The expense of gold and its low hardness are often offset by using it as a laminate or plating on base metals. It can also be applied to metals, ceramics, and some plastics by the thermal decomposition of certain gold compounds.

Cast gold has a tensile strength of 137 MPa. The specific gravity is 19.32, and the melting point 1063°C. It is not attacked by nitric, hydrochloric, or sulfuric acid, but is dissolved by aqua regia, or by a solution of azoimide, and is attacked by sodium and potassium cyanide plus oxygen. The metal does not corrode in air, only a transparent oxide film forming on the surface.

Gold alloys (Au–Ag–Cu and Au–Ni–Cu–Zn) can be made in a range of colors from white to many shades of yellow. For this reason, gold is widely used for jewelry and other decorative applications. Similar alloys (Au–Ag–Cu–Pt–Pd) are used in dentistry, the nobility of the alloys and their response to heat-treatment hardening being of concern here. Gold–silver alloys have been used in low-current electrical contacts (under 0.5 A). Gold is often used in electrical and other equipment, which is used for standards, and where stability is of prime concern.

Gold–gallium and gold–antimony alloys for electronic uses come in wire as fine as 0.013 cm in diameter, and in sheet as thin as 0.003 cm. The maximum content of antimony in workable gold alloys is 0.7%. A gold–gallium alloy with 2.5% gallium has a resistivity of 15×10^{-8} Ω · m, and has a tensile strength of 379 MPa and 22% elongation. Gold powder and gold sheet, for soldering semiconductors, are 99.999% pure. The gold wets silicon easily at a temperature of 371°C.

The corrosion resistance and melting point of gold also make it useful as a brazing material. It is used as well in chemical equipment, where its susceptibility to chlorine attack is not a problem. In particular, it is used to line reaction vessels, and as a gasketing material.

Gold may be readily applied by electroplating from cyanide and other solutions. It may also be applied to some metals by simple immersion in special plating solutions. Plating has, of course, many decorative applications.

Gold plating is also used to make reflectors, particularly for the infrared wavelengths. Electrical components are often gold plated, especially for high frequencies, because of the low electrical resistance of gold. Vacuum tube grids may be gold-plated to reduce electron emission and some electrical contacts are gold plated.

Gold may also be applied by using "liquid bright golds." These are varnishlike solutions of gold compounds that may be applied in any suitable manner — brushing, spraying, printing, etc. — to metals, ceramics, and some plastics. After being applied, the material is heated to decompose the compound, depositing a tightly adhering gold layer. Some gold alloys may also be applied in this manner. This method is used for the decoration of china and glassware, as well as for printed circuits and electrical resistance elements.

GRAPHITE

Graphite is a form of carbon. It was formerly known as black lead, and when first used for pencils was called Flanders' stone. It is a natural variety of elemental carbon with a grayish-black color and a metallic tinge.

Carbon and graphite have been used in industry for many years, primarily as electrodes, arc carbons, brush carbons, and bearings. In the last decade or so, development of new types and emergence of graphite fibers as a promising reinforcement for high-performance composites have significantly increased the versatility of this family of materials.

TYPES OF GRAPHITE

Recrystallized graphite is produced by a proprietary hot-working process that yields recrystallized or "densified" graphite with specific gravities in the 1.85 to 2.15 range, as compared with 1.4 to 1.7 for conventional graphites. The major attributes of the material are a high degree of quality reproducibility, improved resistance to creep, a grain orientation that can be controlled from highly anisotropic to relatively isotropic, lower permeability than usual, absence of structural macroflaw, and ability to take a fine surface finish.

Graphite fibers are produced from organic fibers. One line of development used rayon as the precursor, and the other used polyacrylonitrile (PAN). Although the detailed processing conditions for converting cellulose or PAN into carbon and graphite fibers differ in detail, they both consisted fundamentally of a sequence of thermal treatments to convert the precursor into carbon by breaking the organic compound to leave a "carbon polymer." The fibrous carbon formed by the controlled pyrolysis of organic precursor fiber was viscous rayon or acrylonitrile. Carbon fibers produced by the rayon-precursor method have a fine-grained, relatively disordered microstructure, which remains even after treatment at temperatures up to 3000°C. Graphite crystallites with a long-range three-dimensional order do not develop. In both the rayon and PAN processes a high degree of preferred crystal orientation was responsible for the high elastic modulus and tensile strength.

Although the names *carbon* and *graphite* are used interchangeably when related to fibers, there is a difference. Typically, PAN-based carbon fibers are 93 to 95% carbon by elemental analysis, whereas graphite fibers are usually 99+%. The basic difference is the temperature at which the fibers are made or heat-treated. PAN-based carbon is produced at about 1316°C, whereas higher-modulus graphite fibers are graphitized at 1899 to 3010°C. This also applies to carbon and graphite cloths. Unfortunately, with only rare exceptions, none of the carbon fibers is ever converted into classic graphite regardless of the heat treatment.

When used in composites, the fibers are generally made into yarn containing some 10,000 fibers. Depending on the precursor fiber, their tensile strength ranges from 1378 to nearly 3445 MPa, and their modulus of elasticity is from 0.2 million to 0.5 million MPa.

Graphite fiber-reinforced graphite composites can be used to temperatures in excess of 3500°C. No compatibility problems exist because the graphite fiber or filament is in a graphite matrix. This composite system is good in reducing environments; in air or oxidizing atmospheres special protective coatings are sometimes needed.

The graphite matrix is produced by the pyrolytic decomposition of polymeric systems in which the graphite fiber or filaments are originally embedded. Although many of the matrix starting materials are considered proprietary, usable polymer systems include phenolic, furfuryl ester, and epoxy resins. Graphite–graphite composites, even those fabricated with low-modulus materials, are up to 20 times stronger than conventional carbon and graphite materials. At a density of approximately 1384 kg/m^3 they also are about 30% lighter than conventional carbons. They provide a very high strength-to-weight ratio at temperatures to 3300°C and exhibit superior thermal stability. Most Gr–Gr composites are more than 99% carbon (carbon content about 99.5 to 99.9%). This high purity provides good chemical inertness and corrosion resistance. Gr–Gr composites are not wetted by molten metals, which makes them ideally suited to metallurgical applications where high strength, light weight, erosion resistance, and good thermal conductivity are important. Typical properties are tensile strength, 56.5 MPa; flexural strength, 76 MPa; compressive strength, 276 MPa; and modulus, 17.2 GPa.

Graphite fiber–epoxy composites provide exceptionally high strength and stiffness, and because of their light weight are finding increasing use for golf club shafts, tennis racquet frames, and a multitude of sports equipment, as well as extensive use in the aerospace industry (wings, etc.).

PT graphites are graphite fibers impregnated or bonded with an organic resin (such as furfural) and then carbonized. The result is a graphite-reinforced carbonaceous material with a high degree of thermal stability. The composite has a low density (0.93 to 1.2 specific gravity), and what is reported to be the highest strength-to-weight ratio of known material at temperatures in the 2204 to 2760°C range.

Colloidal graphite consists of natural or artificial graphite in very fine particle form, coated with a protective colloid, and dispersed in a liquid. The selection of the liquid — water, oils, or synthetics — is made on the basis of intended use of the product. Significant characteristics of colloidal graphite dispersions are that the graphite particles remain in suspension indefinitely, and the particles "wick" — that is, they are carded by the liquid to most places penetrated by the liquid.

The supergraphite used for rocket casings and other heat-resistant parts is recrystallized molded graphite. It will withstand temperatures to 3038°C. Pyrolytic graphite is an oriented graphite. It has high density, with a specific gravity of 2.22, has exceptionally high heat conductivity along the surface, making it very flame resistant, is impermeable to gases, and will withstand temperatures to 3704°C. It is made by deposition of carbon from a stream of methane on heated graphite, and the growing crystals form with thin planes parallel to the existing surface. The structure consists of close-packed columns of graphite crystals joined to each other by strong bonds along the flat planes, but with weak bonds between layers. This weak and strong electron bonding provides a laminal structure. The material conducts heat and electricity many times faster along the surface than through the material. The flexural strength is 172 MPa compared with less than 55 MPa for the best conventional graphite. At 2760°C the tensile strength is 275 MPa. Sheets as thin as 0.003 cm are impervious to liquids or gases. It is used for nozzle inserts and reentry parts for spacecraft, as well as for atomic shielding with an addition of boron.

Mechanical Properties

The degree of anisotropy in graphites varies, but cross-grain strengths are usually substantially lower. Tensile strengths of the newer engineering graphites are substantially higher, with that of pyrolytic graphite reaching around 95.2 MPa. See **Carbon** and **Carbon Composites**.

GRAY IRONS

Gray iron is characterized by the presence of flakes of graphite supported in a matrix of ferrite, pearlite, austenite, or any other matrix attainable in steel. The major dimension of the flakes may vary from about 0.05 to 1.0 mm. Because of their low density, the graphite flakes occupy about 10% of the metal volume. The flakes interrupt the continuity of the matrix, and have a large effect on the properties of gray iron. In addition, the flakes give a fractured surface that is gray. This is responsible for the name "gray iron."

High carbon content and the flakes of graphite give gray iron unique properties as follows:

1. Lowest melting point of the ferrous alloys, so that low-cost refractories can be used for molds
2. High fluidity in the molten state, so that complex and thin designs can be cast
3. Excellent machinability, better than steel
4. High damping capacity and ability to absorb vibrations
5. High resistance to wear involving sliding
6. Low ductility and low impact strength when compared with steel

Gray iron is by far the most common and widely used cast iron.

Gray iron is encountered almost exclusively as shaped castings used either with or without machining. Typical common applications include:

1. Pipe for underground service for water or gas
2. Ingot molds into which steel and other metals are cast
3. Cylinder blocks and heads for internal combustion engines
4. Frames and end bells for electric motors
5. Bases, frames, and supports for machine tools
6. Sanitaryware such as sinks and bathtubs (usually coated with porcelain enamel)
7. Pumps, car wheels, and transmission cases

The major industries that consume gray iron castings are as follows: automotive, building and construction, utilities, machine tools, architectural, rolling mills (steel plants), general machinery, household appliances, and heating equipment.

For engineering applications where tensile strength is important, gray iron usually is classified on the basis of minimum tensile strength in a specimen machined from a separately cast test bar.

Mechanical and Physical Properties

Compressive strength. Unusually high; at least three times the tensile strength.

Modulus of elasticity. Increases with tensile strength. About 0.87×10^5 MPa for a tensile strength of 136 MPa, and up to about 1.45×10^5 MPa for a tensile strength of 480 MPa.

Endurance limit. About 35 to 50% of tensile strength. Gray iron is relatively insensitive to the effect of notches.

Damping capacity. Very high, especially in irons of high carbon content. Specific damping capacity is about ten times that of steel.

Specific gravity. Varies from about 6.8 for high-carbon, low-strength irons to about 7.6 for low-carbon, high-strength irons.

Coefficient of thermal expansion. Slightly lower than that of steel.

Coefficient of thermal conductivity. About the same as many other ferrous alloys; about 0.11 to 0.14 in CGS units. Can be lowered appreciably by adding alloying elements.

GYPSUM

Gypsum is the most common sulfate mineral, characterized by the chemical formula $CaSO_4 \cdot 2H_2O$; it shows little variation from this composition.

Gypsum is one of the several evaporite minerals. This mineral group includes chlorides, carbonates, borates, nitrates, and sulfates. These minerals precipitate in seas, lakes, caves, and salt flats due to concentration of ions by evaporation. When heated or subjected to solutions with very large salinities, gypsum converts to bassanite ($CaSO_4 \cdot H_2O$) or anhydrite ($CaSO_4$). Under equilibrium conditions, this conversion to anhydrite is direct. The conversion occurs above 42°C in pure water.

Gypsum is used for making building plaster, wallboard tiles, as an absorbent for chemicals, as a paint pigment and extender, and for coating papers. Natural gypsum of California, containing 15 to 20% sulfur, is used for producing ammonium sulfate for fertilizer. Gypsum is also used to make sulfuric acid by heating to 1093°C in an air-limited furnace. The resultant calcium sulfide is reacted to yield lime and sulfuric acid. Raw gypsum is also used to mix with portland cement to retard the set. Compact massive types of the mineral are used as building stones. The color is naturally white, but it may be colored by impurities to gray, brown, or red. The specific gravity is 2.28 to 2.33, and the hardness 1.5 to 2. It dehydrates when heated to about 190°C, forming the hemihydrate $2CaSO_4 \cdot H_2O$, which is the basis of most gypsum plasters. It is called calcined gypsum, or when used for making ornaments or casts is called plaster of Paris. When mixed with water, it again forms the hydrated sulfate that will solidify and set firmly owing to interlocking crystallization. Theoretically, 18% of water is needed for mixing, but actually more is necessary. Insufficient water causes cracking. Water solutions of synthetic resins are mixed with gypsum for casting strong, waterproof articles.

Much calcined gypsum, or plaster of Paris, is used as gypsum plaster for wall finish. For such use it may be mixed in lime water or glue water, and with sand. Because of its solubility in water it cannot be used for outside work.

The presence of halite (NaCl) or other sulfates in the solution lowers this temperature, although metastable gypsum exists at higher temperatures.

Crystals of gypsum are commonly tabular, diamond-shaped, or lenticular; swallow-tailed twins are also common. The mineral is monoclinic with symmetry 2/m. Gypsum is the index mineral chosen for hardness 2 on the Mohs scale. In addition to free crystals, the common forms of gypsum are satin spar (fibrous), alabaster (finely crystalline), and selenite (massive crystalline).

Gypsum is used for a variety of purposes, but chiefly in the manufacture of plaster of Paris, in the production of wallboard, in agriculture to loosen clay rich soils, and in the manufacture of fertilizer. Plaster of Paris is made by heating gypsum to 200°C in air.

Gypsum deposits play an important role in the petroleum industry. The organic material commonly associated with its formation is considered a source of hydrocarbon (oil and gas) generation. In addition, these deposits act as a seal for many petroleum reservoirs, preventing the escape of gas and oil.

H

HAFNIUM AND ALLOYS

Hafnium (Hf), the heaviest of the three metals comprising the Group IV transition metals, is now in production. Because of the startling similarity in their chemical properties, zirconium and hafnium always occur together in nature. In their respective ability to absorb neutrons, however, they differ greatly, and this difference has led to their use in surprisingly different ways in nuclear reactors. Zirconium, with a low neutron-absorption cross section (0.18 barn), is highly desirable as a structural material in water-cooled nuclear reactor cores. Hafnium, on the other hand, because of its high neutron-absorption cross section (105 barns), can be used as a neutron-absorbing control material in the same nuclear reactor cores. Thus, the two elements, which occur together so intimately in nature that they are very difficult to separate, are used as individual and important but contrasting components in the cores of nuclear reactors.

Properties

Pure hafnium is a lustrous, silvery metal that is not so ductile nor so easily worked as zirconium; nevertheless, hafnium can be hot- and cold-rolled on the same equipment and with similar techniques as those used for zirconium. All zirconium chemicals and alloys may contain some hafnium, and hafnium metal usually contains about 2% zirconium. The melting point, 2222°C, is higher than that of zirconium, and heat-resistant parts for special purposes have been made by compacting hafnium powder to a density of 98%. The metal has a close-packed hexagonal structure. The electric conductivity is about 6% that of copper. It has excellent resistance to a wide range of corrosive environments.

Hafnium Alloys and Compounds

Hafnium forms refractory compounds with carbon, nitrogen, boron, and oxygen. Hafnium oxide, or hafnia, HfO_2, is a better refractory ceramic than zirconia, but is costly.

Hafnium carbide, HfC, produced by reacting hafnium oxide and carbon at high temperature, is obtained as a loosely coherent mass of blue-black crystals. The crystals have a hardness of 2910 Vickers, and a melting point of 4160°C. It is thus one of the most refractory materials known. Heat-resistant ceramics are made from hafnium titanate by pressing and sintering the powder. The material has the general composition $_x(TiO_2) \cdot n(HfO_2)$, with varying values of x and n. Parts made with 18% titania and 82% hafnia have a density of 7.197 kg/m^3, a melting point at about 2204°C, a low coefficient of thermal expansion, good shock resistance, and a rupture strength above 68 MPa at 1093°C. Hafnium nitride, with a melting point of 3300°C has the highest melting point of any nitride and hafnium boride, with a melting point of 3260°C has a melting point higher than any other boride. The alloy Ta_4HfC_5 has the highest melting point of any substance known, about 4215°C.

HARD FACINGS

Hard facing is a technique by which a wear-resistant overlay is welded on a softer and usually tougher base metal. The method is versatile and has a number of advantages:

1. Wear resistance can be added exactly where it is needed on the surface.
2. Hard compounds and special alloys are easy to apply.
3. Hard facings can be applied in the field as well as in the plant.

4. Expensive alloying elements can be economically used.
5. Protection can be provided in depth.
6. A unique and useful structure is provided by the hard-surfaced, tough-core composite.

Many of the merits of hard facing stem from the hardness of the special materials used. For example, ordinary weld deposits range in hardness up to about 200 Brinell, hardened steels have a hardness up to 700 to 800 Vickers, and special carbides have hardness up to about 3000 Vickers. However, it is important to note that the hardness of the materials does not always correlate with wear resistance. Thus, special tests should be performed to determine the resistance of the material to impact, gouging abrasion, grinding (high-stress abrasion), erosion (low-stress scratching abrasion), seizing or galling, and hot wear.

Another important point is that durable overlays are not necessarily hard. Most surfacing is used to protect base metals against abrasion, friction, and impact. However, many "hard facings" such as the stainless steels, related nickel-base alloys, and copper alloys are used for corrosion-resistant applications where hardness may not be a factor. Also, the relatively soft leaded bronzes may be used for bearing surfaces. Other facings are also used for heat- and oxidation-resistant applications.

Methods of Application

Hard facings can be applied by:

1. Manual, semiautomatic, and automatic methods using bare or flux coated electrodes
2. Submerged-arc welding
3. Inert-gas shielded arc welding (both consumable and tungsten electrode types)
4. Oxyacetylene and oxyhydrogen gas welding
5. Metal spraying
6. Welded or brazed on inserts

Gas welding and spraying usually provides higher quality and precise placement of surfaces; arc welding is less expensive. Automatic or semiautomatic methods are preferred where large areas are to be covered, or where repetitive operations favor automation.

Surfacing filler metals are available in the form of drawn wire, cast rods, powders, and steel tubes filled with ferroalloys or hard compounds (e.g., tungsten carbide). The electrodes may take the form of filled tubes or alloyed wires, stick types, or coils specially designed for automated operations. The stick-type electrodes may have a simple steel core and a thick coating containing the special alloys. In submerged-arc welding the alloys may be introduced through a special flux blanket. In spray coating, the materials are used in the form of powders or bonded wire.

Sprayed facings are advantageous in producing thin layers and in following surface contours. With this method it is usually necessary to fuse the sprayed layer in place after deposition to obtain good abrasion resistance. However, under boundary lubrication conditions the as-sprayed porosity of the facings may aid against frictional wear.

Hard facings are used in thicknesses from 0.031 to 25.4 cm or more. The thinnest layers are usually deposited by gas welding, usually with low melting alloys that solidify with many free carbide or other hard compound crystals. The thick deposits are usually made from air-hardening or austenitic steels.

Hard overlays are usually strong in compression but weak in tension. Thus, they perform better in pockets, grooves, or low ridges. Edges and corners must be treated cautiously unless the deposit is tough. Brittle overlays should be deposited over a base of sufficient strength to prevent subsurface flow under excessive compression.

Gas welding is a useful method for depositing small, precisely located surfacings in applications where the base metal can withstand the welding temperatures (e.g., steam valve trim and exhaust valve facings). On the other hand, heavy layers and large areas may be impossible to surface without cracks with the harder, more wear-resistant alloys because of the severe thermal stresses that are encountered (e.g., usually in arc welding). Thus, the opposing factors of wear resistance and freedom from cracking frequently require a compromise in process and material selection.

Materials

Basically, hard-facing materials are alloys that lend themselves to weld fusion and provide hardness or other properties without special heat treatment. Thus, for hard surfacing, the steels and the matrices of high-carbon irons may contain enough alloys to cause the hardening transformation during weld cooling, rather than after a quenching treatment.

The properties of the iron-, nickel-, and cobalt-base alloys are strongly affected by carbon content and somewhat by the welding technique used. For example, gas welding usually provides superior abrasion resistance, although carbon pickup may lower corrosion resistance. Arc welding tends to burn out carbon and alloys, thereby lowering abrasion resistance but increasing toughness; high thermal stresses from arc welding may also accentuate cracking tendencies.

The martensitic irons, martensitic steels, and austenitic manganese steels are suited for light, medium, and heavy impact applications, respectively. Gouging abrasion applications usually require an austenitic manganese steel because of the associated heavy impact. Grinding abrasion is well resisted by the martensitic irons and steels. Erosion is most effectively resisted by a good volume of the very hard compounds (e.g., high-chromium irons). Tungsten carbide composites have outstanding resistance to abrasion where heavy impact is not present, but deposits may develop a rough surface.

Selection of materials for hot-wear applications is complicated by oxidation, tempering, softening, and creep factors. Oxidation resistance is provided by using a minimum of 25% chromium. Tempering resistance (up to 593°C) is provided by chromium, molybdenum, tungsten, etc. Creep resistance is provided by the austenitic structure in nickel- or cobalt-bearing alloys. The chromium–cobalt–tungsten grade of materials usually provides a good combination of properties above 649°C.

HARD RUBBER

Hard rubber is a plastic. It is a resinous material mixed with a polymerizing or curing agent and fillers, and can be formed under heat and pressure to practically any desired shape.

The bulk of today's hard rubber is made with SBR synthetic rubber. Other types of synthetic rubbers, such as butyl or nitrite or, in rare cases, silicone or polyacrylic, can also be used.

Once it has gone through the process of heat and pressure, hard rubber cannot be returned to its original state and therefore falls into the class of thermosetting plastics, i.e., those that undergo chemical change under heat and pressure. It differs, however, from other commercial thermosetting plastics such as the phenolics and theureas in that after it has gone through the thermosetting process it will still soften somewhat under heat. In this characteristic it most resembles the thermoplastic acetates, polystyrenes, and vinyls. It differs from all others in that it is available in pliable sheet form before vulcanization and is therefore adaptable to many shapes for which molds and presses are not necessary. Because of this feature and because it can be softened again after vulcanization, it falls into a class by itself in the field of plastics.

The term *hard rubber* is self-descriptive. The hardness is measured on the Shore D scale, which is several orders of magnitude higher than the Shore A scale used for conventional rubbers and elastomers. Similar in composition to soft rubber, it contains a much higher percentage of sulfur, up to a saturation point of 47% of the weight of the rubber in the compound. If sulfur is present in rubber compounds in amounts over 18% of the weight of rubber in the compound when the material is completely vulcanized, the product will be generally known as hard rubber.

Properties and Fabrication

The most important properties of hard rubber are the combination of relatively high tensile strength, low elongation, and extremely low water absorption.

Hard rubber may be compression-, transfer-, or injection-molded. In sheet form it can be hand-fabricated into many shapes. Its machining qualities are comparable to brass, and it may be drilled and tapped. The material lends itself readily to permanent or temporary sealing with hot or cold cements and sealing compounds.

The size and shape of a hard rubber part is dependent only upon the size of press equipment and vulcanizers available.

Uses

Perhaps the largest application for hard rubber is in the manufacture of battery boxes. The water-meter industry is also a large user. Hard rubber linings and coatings either molded or hand laid-up account for large amounts of material. In the electrical industry, hard rubber is used for terminal blocks, insulating materials, and connector protectors. The chemical, electroplating, and photographic industries use large quantities of hard rubber for acid-handling devices.

HEAT-RESISTANT ALLOYS (CAST)

Cast alloys suitable for use at service temperatures to at least 538°C and, for some alloys, to 1093°C are classed as "heat-resistant" or "high-temperature" alloys. They have the characteristic of corroding at very slow rates compared with unalloyed, or low-alloy cast iron or steel in the atmospheres to which they are exposed, and they offer sufficient strength at operating temperature to be useful as load-carrying engineering structures. Iron-base and nickel-base alloys comprise the bulk of production, but cobalt-base, chromium-base, molybdenum-base, and columbium-base alloys are also made.

Although some cast heat-resistant alloys are available in compositions similar to wrought alloys, it is necessary to differentiate between them. Cast alloys are made to somewhat different chemical specifications than wrought alloys; physical and mechanical properties for each group are also somewhat different.

For these reasons, it is advisable to follow the alloys designated as the H-Series by the Alloy Casting Institute of the American Steel Founders Society as well as nickel-base alloys and cobalt-base alloys. Most of the nickel-base and cobalt-base alloys are also known as superalloys because of their exceptional high-temperature stress-rupture strength and creep resistance as well as corrosion and oxidation resistance.

There are, moreover, a number of heat-resistant cast alloys that are not available in wrought form; this is frequently of advantage in meeting special conditions of high-temperature service. In addition to the grades HA to HX discussed below, the industry produces special heat-resistant compositions. Many of these are modifications of the standard types, but some are wholly different and are designed to meet unique service conditions.

Selection of a particular alloy, of course, is dependent upon the application, and in this article composition, structure, and properties of the various cast heat-resistant alloys are discussed from this point of view.

Proper selection of an alloy for a specific high-temperature service involves consideration of some or all of the following factors: (1) required life of the part, (2) range and speed of temperature cycling, (3) the atmosphere and its contaminants, (4) complexity of casting design, and (5) further fabrication of the casting. The criteria that should be used to compare alloys depend on the factors enumerated, and the designer will be aided in the choice by providing the foundry with as much pertinent information as possible on intended operating conditions before reaching a definite decision to use a particular alloy type.

Physical and Mechanical Properties

For high-temperature design purposes a frequently used design stress is 50% of the stress that will produce a creep rate of 0.0001%/h maximum operating temperature. Such a value should be applied only under conditions of direct axial static loading and essentially uniform temperature or slow temperature variation. When impact loading or rapid temperature cycles are involved, a considerably lower percentage of the limiting creep stress should be used. In the selection of design stresses, safety factors should be higher if the parts are inaccessible, nonuniformly loaded, or of complex design; they may be lower if the parts are accessible for replacement, fully supported or rotating, and of simple design with little or no thermal gradient.

H-Series Cast Alloys

The H-Series cast alloys include iron–chromium, iron–chromium–nickel, and iron–nickel–chromium alloys also containing 0.20 to 0.75% carbon, 1 to 2.5% silicon, and 0.35 to 2% manganese. A letter (A to X) following the H is used to distinguish alloy compositions more closely. The iron–chromium cast alloys (HA, HC, and HD) contain as much as 30% chromium and under 7% nickel. The iron–chromium–nickel cast alloys (HE, HF, HH, HI, HK, and HL) contain as much as 32% chromium and 22% nickel. And the iron–nickel–chromium cast alloys (HN, HP, HP-50WZ, HT, HU, HW, and HX) contain as much as 68% nickel (HX) and 32% chromium (HN) so that some of these alloys are actually nickel-base instead of iron-base alloys.

In selecting alloys from this group, consider the following factors:

1. Increasing nickel content increases resistance to carburization, decreases hot strength somewhat, and increases resistance to thermal shock.
2. Increasing chromium content increases resistance to corrosion and oxidation.
3. Increasing carbon content increases hot strength.
4. Increasing silicon content increases resistance to carburization, but decreases hot strength.

All are noted primarily for their oxidation resistance and ability to withstand moderate to severe temperature changes. Most are heat-treatable by aging room-temperature tensile properties in the aged condition ranging from 503 to 793 MPa in terms of ultimate strength, 297 to 552 MPa in yield strength, and 4 to 25% in elongation. Hardness of the aged alloys ranges from Brinell 185 to 270. Applications include heat-treating fixtures, furnace parts, oil-refinery and chemical processing equipment, gas-turbine components, and equipment used in manufacturing steel, glass, and rubber.

Both the nickel-base and cobalt-base alloys are probably best known for their use in aircraft turbine engines for disks, blades, vanes, and other components. The nickel alloys contain 50 to 75% nickel and usually 10 to 20% chromium and substantial amounts of cobalt, molybdenum, aluminum, and titanium, and small amounts of zirconium, boron, and, in some cases, hafnium. Carbon content ranges from less than 0.1 to 0.20%. Because of their complex compositions, they are best known by trade names, such as B-1900; Hastelloy X; IN-100, -738X, -792; Rene 77, 80, 100; Inconel 713C, 713LC, 718, X-750; MAR-M 200, 246, 247; Udimet 500, 700, 710; and Waspaloy. The high-temperature strength of most of these alloys is attributed to the presence of refractory metals, which provide solid-solution strengthening; the presence of grain-boundary-strengthening elements, such as carbon, boron, hafnium, and zirconium; and, because of the presence of aluminum and titanium, strengthening by precipitation of an $Ni_3(Al,Ti)$ compound known as "gamma prime" during age hardening. Many of these alloys provide 1000-h stress-rupture strengths in the range of 690 to 759 MPa at 649°C, and 55 to 124 MPa at 982°C.

The cobalt alloys contain 36 to 65% cobalt, usually more than 50%, and usually about 20% chromium and substantial amounts of nickel, tungsten, tantalum, molybdenum, iron and/or aluminum, and small amounts of still other ingredients. Carbon content is 0.05 to 1%. Although not generally as strong as the nickel alloys, some may provide better corrosion and oxidation resistance at high temperatures. These alloys include L-605; S-816; V-36; WI-52; X-40; J-1650; Haynes 21, 151; AiResist 13, 213, 215; and MAR-M 302, 322, 918. Their 1000-h stress-rupture strengths range from about 276 to 483 MPa at 649°C and from about 28 to 103 MPa at 982°C.

HEAT-RESISTANT PLASTICS (SUPERPOLYMERS)

Several different plastics developed in recent years that maintain mechanical and chemical integrity above 204°C for extended periods are frequently referred to as superpolymers. They are polyimide, polysulfone, polyphenylene sulfide, polyarylsulfone and aromatic polyester.

In addition to their high-temperature resistance, all these materials have in common high strength and modulus, and excellent resistance to solvents, oils, and corrosive environments. They are also among the highest priced plastics, and a major disadvantage is processing difficulty. Molding temperatures and pressures are extremely high compared with conventional plastics. Some of them, including polyimides and aromatic polyester, are not molded conventionally. Because they do not melt, the molding process is more of a sintering operation. Because of their high price, superpolymers are largely used in specialized applications in the aerospace and nuclear energy field.

Indicative of their high-temperature resistance, the superpolymers have a glass transition temperature well over 260°C as compared to less than 177°C for most conventional plastics. In the case of polyimides, the glass temperature is greater than 427°C, and the material decomposes rather than softens when heated excessively.

Polysulfone has the highest service temperature of any melt-processible thermoplastic. Its flexural modulus stays above 2040 MPa at up to 160°C. At such temperatures it does not discolor or degrade.

Aromatic polyester does not melt, but at 427°C can be made to flow in a nonviscous manner similar to metals. Thus, filled and unfilled forms and parts can be made by hot sintering, high-velocity forging, and plasma spraying. Notable properties are high thermal stability, good strength at 316°C, high thermal conductivity, good wear resistance, and extra-high compressive strength.

HEAVY ALLOY

This is a name applied to tungsten–nickel alloy produced by pressing and sintering the metallic powders. It is used for screens for x-ray tubes and radioactivity units, for contact surfaces for circuit breakers, and for balances for high-speed machinery. The original composition was 90% tungsten and 10% nickel, but a proportion of copper is used to lower sintering temperature and give better binding as the copper wets the tungsten. Too large a proportion of copper makes the product porous. In general, the alloys weigh nearly 50% more than lead, permitting space saving in counterweights and balances, and they are more efficient as gamma-ray absorbers than lead. They are highly heat resistant, retain a tensile strength of about 137 MPa at 1093°C, have an electric conductivity about 15% that of copper, and can be machined and brazed with silver solder.

An alloy of 90% tungsten, 7.5% nickel, and 2.5% copper has a tensile strength of 930 MPa, compressive strength of 2757 MPa, elongation 15%, Rockwell hardness C30, and weight of 16,885 kg/m^3. Kenertium has this composition.

HELIUM

Helium is a colorless, odorless, elementary gas, He, with a specific gravity of 0.1368, liquefying at –268.9°C, freezing at –272.2°C. It has a valency of zero and forms no electron-bonded compounds. It has the highest ionization potential of any element. The lifting power of helium is only 92% that of hydrogen, but it is preferred for balloons because it is inert and nonflammable, and is used in weather balloons. It is also used instead of air to inflate large tires for aircraft to save weight. Because of its low density, it is also used for diluting oxygen in the treatment of respiratory diseases. Its heat conductivity is about six times that of air, and it is used as a shielding gas in welding, and in vacuum tube and electric lamps. Because of its inertness helium can also be used to hold free chemical radicals, which, when released, give high energy and thrust for missile propulsion. When an electric current is passed through helium it gives a pinkish-violet light, and is thus used in advertising signs. Helium can be obtained from atmospheric nitrogen, but comes chiefly from natural gas.

Properties

Helium has the lowest solubility in water of any known gas. It is the least reactive element. The density and the viscosity of helium vapor is low. Thermal conductivity and heat content are exceptionally high. Helium can be liquefied, but its condensation temperature is the lowest of any known substance. At pressures below 2.5 MPa helium remains liquid even at absolute zero.

Applications

Gases

Helium was first used as a lifting gas in balloons and dirigibles. This use continues for high-altitude research and for weather balloons.

Welding

The principal use of helium is in inert gas-shielded arc welding. Using helium instead of argon permits a greater heat release, which is useful in welding very heavy sections or in high-speed machine welding of long seams. By mixing helium and argon, the optimum heat release can be obtained for different welding jobs.

Superconductive Devices

The greatest potential for helium use continues to emerge from extremely low temperature applications. Helium is the only refrigerant capable of reaching temperatures below 14 K. In the laboratory many fundamental properties of matter are studied at temperatures near absolute zero with helium refrigeration. Infrared detectors and masers operate with exceptionally low noise distortion at these low temperatures. The chief value of ultralow temperature is the development of the state of superconductivity, in which there is virtually zero resistance to the flow of electricity. Very large currents are carried by even small conductors with little loss of voltage. Electromagnets producing immensely powerful magnetic fields can be made small and light and are energized with modest amounts of electric power through the use of superconducting windings. These magnets are already used in particle accelerators, bubble chambers, and plasma confinement for nuclear physics research. Thermonuclear and magnetohydrodynamic (MHD) power plants are expected to use superconducting magnets. Additional applications are electric motors and generators. Superconductive devices make highly sensitive detectors of electric voltage and frequency, magnetic field strength, and temperature, especially at low temperature levels.

Lasers

Helium is also used in gas-discharge lasers. Energy is transferred by helium to the lasing gas, carbon dioxide or neon, for example.

Rockets

Consumption of helium as a pressurizing gas in liquid-fueled rockets declined with the completion of the Apollo space program. Because it is light, inert, and relatively insoluble in the fuel and oxidizer fluids, helium is an ideal material to fill the tankage as the liquids are consumed.

Breathing Mixtures

Use of helium–oxygen breathing mixtures for divers at great depths is required to eliminate the narcotic effects of nitrogen. The low density and low viscosity of helium also reduce the work of breathing. Similarly, helium–oxygen breathing mixtures promote both intake of oxygen and removal of carbon dioxide for persons whose breathing passages are constricted.

Nuclear Reactors

Inertness and heat-transfer capability make helium an excellent working fluid for gas-cooled nuclear power reactors. Because the reactor core is composed of graphite and ceramic materials, very high temperatures can be attained without damage. Helium-cooled reactors operate with the highest efficiency of all reactor types. In addition to electric power generation, with helium working fluid nuclear reactors can provide the process heat for coal gasification, steel making, and various chemical processes.

Chemical Analysis

Helium is the most frequently used carrier gas for chemical analysis by gas chromatography. It is the most sensitive leak detection fluid and can be used at extremes of high and low temperature.

HIGH-ENERGY RATE FORMING (EXPLOSIVE FORMING)

In high-energy rate forming (HERF), parts are shaped by the extremely rapid application of high pressures. Pressures as high its 13,600 MPa and speeds as high as 914 m/s may be used.

The principal advantages of HERF are as follows:

1. Parts can be formed that cannot be formed by conventional methods.
2. Exotic metals, which do not readily lend themselves to conventional forming processes, may be formed over a wide range of sizes and configurations.
3. The method is excellent for restrike operations.
4. Springback after forming is reduced to a minimum.
5. Dimensional tolerances are generally excellent.
6. Variations from part to part are held to a minimum.
7. Scrap rate is low.
8. Less equipment and fewer dies cut down on production lead time.

Explosive Forming

There are three different explosive forming techniques now being used: free forming, bulkhead forming, and cylinder forming. All are shown schematically in Figure H.1. Both free forming and bulkhead forming allow the workpiece to be heated before forming. Although air can be used as a coupling medium between the explosive and the workpiece, in most cases water is used. Efficiency in air is approximately 4%; in water, 33%.

Changes in mechanical properties caused by explosive forming correlate closely with those obtained in material cold-worked to the same degree.

In HERF of nickel alloys and the 300 series of stainless steel, strength and hardness are increased and, as expected, ductility is decreased.

Studies indicate that explosive impact hardening is useful with materials hardened by cold work, e.g., austenitic stainless steel, Hadfield steel, nickel, molybdenum, etc. An interesting application is the possibility of restoring mechanical properties of parts that have been welded or heat-treated, and thus softened.

Simple forgings can be made by explosive forming techniques. One study showed that aluminum alloys could be explosive-forged if the design had no extreme contours.

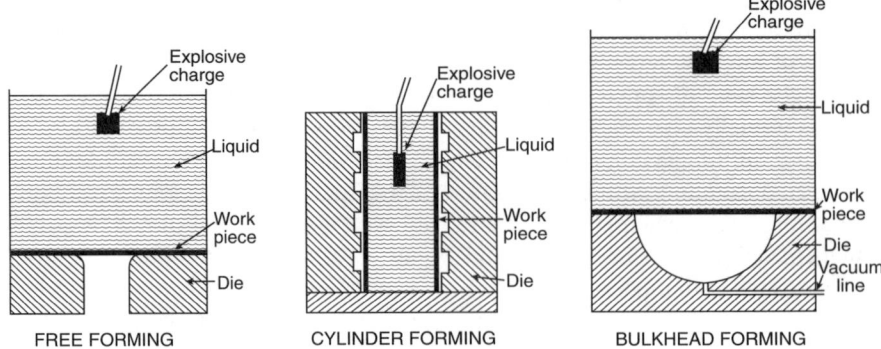

FIGURE H.1 Three general methods for explosive forming using high explosives.

Copper has been welded to copper by the application of explosive force. The joint was a metallurgical, not mechanical.

Expanding Gas

Gases generated by the burning of propellant powders in a closed container produce the pressures required to form metals. Most of the work that uses propellant powder gases as the energy source is classified as bulge-type forming.

Dynapak

Dynapak is another forming method and machine that uses a gas-powered die to form a part and the operation is under close control.

Dynapak can form low-alloy steels such as AISI 4340, austenitic steels of the 200 and 300 series, titanium, and the refractory metals, to name a few examples. It can also be used to compact powders to a density higher than normally obtained with conventional powder-metallurgy processes.

Parts have been extruded with excellent surface finish and close dimensional tolerances. Web thicknesses of 0.25 mm can be obtained and wire 0.50 mm in diameter has been extruded directly from a 25.4-mm billet.

Hot and cold forgings of various materials can be produced with zero draft angles and minimum radii. The smooth, close tolerance surfaces that are produced often minimize finish machining requirements.

Capacitor Discharge Techniques

Explosives and compressed gases are not the only means of achieving high deformation rates. In one type of device, a spark is discharged in a nonconducting liquid medium and generates a shock wave that travels at the speed of sound from the spark source to the workpiece.

This forming technique has several advantages:

1. Explosives, with their potential safety hazard, are eliminated.
2. Parts can be sized into a die by several applications of energy impulses. Since the device is electrical, components of the system do not have to be repositioned after each shot.
3. A standard machine tool, based on this principle, can be constructed for about one tenth the cost of a conventional hydraulic press and occupy only a fraction the floor space.

One problem encountered with capacitor discharge techniques is the containment of high voltages, because stored electrical energy increases with the square of voltage ($E = CV^2$, where C = capacitance and V = voltage). Two other problems: corona and arcing. Normal safety procedures for handling high voltage must be followed.

Magnetic Forming

Use of the pressure generated by a magnetic field permits parts to be formed in 6 s. Of this time, only 10 to 20 μs may be needed for the forming operation; the balance is taken up by setup and removal of the part from the apparatus.

The many possible magnetic coil configurations permit wide variety in forming operations from expanding to compressing to forming flat stock. Coils usually are massively supported since they must be able to withstand the high pressures generated by the magnetic field. For example, at a flux density of 300,000 Gauss, the pressure is approximately 340 MPa. At higher fields (up to a million Gauss) magnetic forming devices may generate pressures exceeding 3400 MPa.

The value of magnetic forming methods lies in the ability to perform quickly and economically many conventional operations such as swaging, bulging, expanding, and assembly.

HIGH-SPEED STEELS

High-speed steels are those alloy steels developed and used primarily for metal-cutting tools. They are characterized by being heat-treatable to very high hardness (usually Rockwell C64 and over) and by retaining their hardness and cutting ability at temperatures as high as 538°C, thus permitting truly high-speed machining. Above 538°C, they rapidly soften and lose cutting ability.

TABLE H.1
High Speed Steel

AISI Type	Austenitizing Temp., °C	Tempering Temp., °C	Resistance to Decarburization[a]	Characteristics
T1	1260–1288	551–580	Excellent	General purpose
T5	1274–1301	551–580	Fair	Extra heat resistance
T15	1218–1245	538–566	Good	Heaviest duty
M1	1177–1218	551–566	Poor	General purpose
M2	1190–1232	551–566	Fair	General purpose
M3	1204–1232	538–551	Fair	Heavy duty, abrasive materials
M7	1190–1227	538–566	Poor	Special applications
M10	1177–1218	538–566	Poor	General purpose
M36	1218–1245	551–580	Fair	Extra heat resistance

[a] During heat treatment.

All high-speed steels are based on either tungsten or molybdenum (or both) as the primary heat-resisting additive, with carbon for high hardness, chromium for ease of heat treating, vanadium for grain refining and, in amounts over 1%, for abrasion resistance, and sometimes cobalt for additional hardness and resistance to heat softening.

Popular compositions of high-speed steels readily available in the United States, the most common characteristics, and recommended temperature ranges are shown in Table H.1. By far the most important of these are M1, M2, and M10. Nearly all twist drills, taps, chasers, reamers, saw blades, and high-speed steel hand tools are made from M1 or M10, whereas the more complex tools such as milling cutters, gear hobs, broaches, and form tools utilize the M2 type.

Special applications, such as machining of hard heat-treated materials, call for a more abrasion-resistant type (such as M3 or T15), and extra heavy-duty cutting involving maximum heat generation leads to the use of the high-cobalt types M36 or T5.

Properties

High-speed steel possesses the highest hardness after heat treating of any well-known ferrous alloy. The value of high-speed steel lies in its ability to retain this hardness under considerable exposure to heat, and to retain a sharp cutting edge when exposed to abrasive wear.

The hardness of high-speed steel when heat-treated is usually Rockwell C64 to 66, equivalent to Brinell 725 to 760. It is brittle at this hardness, particularly in the cobalt-bearing grades, and must be sharpened and handled carefully. This high hardness is obtained by somewhat special heat-treating techniques as compared with lower alloyed steels. Temperatures much in excess of normal steel heat-treating temperatures are employed.

Heat Treatment

In general, for maximum hardness and heat resistance it is necessary to heat-treat at as high a temperature as possible short of the point of initial fusion or grain growth. This would normally result in severe surface damage when done in conventional furnace atmospheres, so surface protection by special gaseous atmospheres or by molten salt (usually $BaCl_2$) is required for production of quality tools.

After quenching in oil, or a salt eutectic ($KCl–NaCl–BaCl_2$) held at about 566-593°C, and air cooling to 52°C, high-speed steel is tempered in the "secondary" hardening range (523 to 566°C) to develop maximum hardness and cutting life. Such tempering is usually done two or three times successively for best results.

Occasionally, a shallow surface treatment (under 0.03 mm) is imparted by nitriding in salt, or by one of several proprietary methods, to elevate surface hardness and reduce friction. These treatments are often very useful in improving cutting life.

USE AND SELECTION

High-speed steels are used for all types of cutting tools, particularly those powered by machines, such as drills, taps, reamers, milling cutters of all types, form cutting tools, shavers, broaches, and lathe, planer, and shaper bits. They have preference over cemented carbide when the tool is difficult to form (high-speed steel is machinable before heat treating), when subject to shock loading or vibration (high-speed steel is tough and resistant to fatigue, considering its hardness level), or when the machining problem is not particularly difficult. High-speed steel is considerably less expensive than carbides and much simpler to form into complex tools, but it does not have the high hardness, abrasion resistance, or tool life in severe high-speed cutting applications associated with cemented carbide. On the other hand, high-speed steel, with good heat resistance, consistently cuts far better than carbon steel, or one of the "fast-finishing" types.

Other uses for high-speed steel are in forming dies, drawing dies, inserted heading dies, knives, chisels, high-temperature bearings, and pump parts. In these applications use is made of the combination of high hardness, heat resistance, and abrasion resistance rather than cutting ability.

Among the types of high-speed steel listed in Table H.1, common mass-produced tools are made from M1 or M10. These grades have the lowest cost, and are easiest to machine, heat-treat, and sharpen. They also are the toughest when hard and thus withstand the abuse often given common tooling drills, taps, threading dies, etc.

More complex, expensive tools are usually made from M2 high-speed steel. It has better abrasion resistance and is easier to heat-treat in complex shapes. Most milling cutters, gear hobs, broaches, and similar multiple-point tools are in this category. M7 is also becoming popular for specialized applications.

Occasionally, extremely difficult machining operations are encountered, such as cutting plastic, synthetic wood, and paperboard products, or hardened alloy steels. Better tool life can then be obtained by use of M3 or T15, the high-vanadium high-speed steels. They are more expensive and considerably more difficult to sharpen and maintain because of resistance to grinding, but these factors are often outweighed by the superior tool life developed.

The high-cobalt high-speed steels T5 and M36 have the best heat resistance, and therefore are particularly suited for tools cutting heavy castings or forgings, where cutting speeds are relatively slow but cuts are deep and the cutting edge gets very hot. T5 and M36 are more expensive than other grades and thus have limited use, but are the most economical for some operations.

FABRICATION

High-speed steels in the annealed condition are machinable by all common techniques. Their machinability rating is about 30% of Bessemer screw stock, and they must be cut slowly and carefully. The recent development of machining high-speed steels has eased this situation, but considerable care is still required.

Ordinarily, tools are machined from bar stock or forgings, either singly for complex tools or in automatic screw machines for mass production items (taps, twist drills, etc.). After finishing almost to final size, the tools are heat-treated to final hardness, then finish-ground. The manufacture of unground tools machined to final size before heat treating is growing because of improvements in heat-treating facilities and better cutting ability of a properly hardened unground surface.

After high-speed steel tools become dull they can easily be resharpened, with some care given to selection of the proper grinding wheel and technique. They are rarely softened by annealing and re-heat-treated, since this may produce a brittle grain structure in the steel unless great care is employed. High-speed steels are never welded after hardening, and tools are seldom repaired by welding because of extreme brittleness in the weld. Often high-speed steel inserts are brazed to alloy steel bodies, or flash-welded to alloy steel shanks (heavy drills, taps, and reamers).

HIGH-STRENGTH HYDROGEN-RESISTANT ALLOY

NASA-23 is a hydrogen-resistant alloy that has been developed for applications in which there are requirements for high strength and high resistance to corrosion.

Adequate resistance to corrosion is necessary for the survival of alloys in hydrogen environments. Alloys for use in such environments typically contain a minimum of 10% chromium. The unique feature of NASA-23 is that it combines high strength with resistance to corrosion, and can readily be made into parts that are suitable for use in hydrogen environments.

The use of high-strength alloys that have low resistance to hydrogen results in frequent replacement of components because hydrogen severely degrades mechanical properties of these alloys. At present, coatings that serve as barriers to hydrogen/metal interactions are often applied to such alloys used in hydrogen. However, the use of coatings does result in higher production costs. Additionally, alloys that have lower resistance to hydrogen can be used by increasing section sizes; however, this method results in an increase in component weight.

NASA-23 can be used for components that encounter temperatures up to 649°C in hydrogen environments.

Innovators of NASA-23 have carefully chosen alloying elements of iron, nickel, cobalt, chromium, niobium, titanium, and aluminum to ensure adequate precipitation hardening for strength and minimum precipitation of detrimental grain boundary precipitates.

HIGH-TEMPERATURE MATERIALS

These are materials that serve above about 540°C. In the broad sense, high-temperature materials can be identified by the following classes of construction solids: stainless steel (limited), austenitic superalloys, refractory metals, ceramics and ceramic composites, metal–matrix composites, and graphitic composites. The first three classes are well proved in industrial use, although stainless steels serve but slightly above 540°C and refractory metals are usually limited to nonoxidizing atmospheric conditions. The other classes are under extensive worldwide research to establish whether they can be utilized to replace and extend the capabilities of austenitic superalloys, which are the mainstay of high-temperature service.

The most-demanding applications for high-temperature materials are found in the aircraft jet engines, industrial gas turbines, and nuclear reactors. However, many furnaces, ductings, and electronic and lighting devices operate at such high temperatures. To perform successfully and economically at high temperatures, a material must have two essential characteristics: it must be strong, because increasing temperature tends to reduce strength; and it must have resistance to its environment, because oxidation and corrosion also increase with temperature.

High-temperature materials have acquired their importance because of the pressing need to provide society with energy and transportation. Machinery that produces electricity or some other form of power from a heat source operates according to a series of thermodynamic cycles, including the basic Carnot cycle and the Brayton cycle, where the efficiency of the device depends on the difference between its highest operating temperature and its lowest temperature. Thus, the greater this difference, the more efficient is the device — a result providing great impetus to the creation of materials that operate at very high temperatures.

METALLIC MATERIALS DESIGNS

Alloys used at high temperatures in heat engines are composed of several elements. High-temperature metallurgists, using both theoretical knowledge and application of empirical experimental techniques, depend upon three principal methods for developing and maintaining strength. The alloys are composed of grains of regular crystalline arrays of atoms and show usable ductility and toughness when one plane of atoms slips controllably over the next (dislocation movement); excessive dislocation movement leads to weak alloys. Thus, in terms of achieving mechanical strength, alloy design attempts to inhibit but not completely block dislocation movement.

The most common technique is solid solution strengthening. Foreign atoms of a size different from that of the parent group cause the crystal lattice to strain. This distortion impedes the tendency of the lattice to slip and thus increases strength.

A second significant mechanism is dispersion strengthening. This is the introduction into the alloy lattice of extremely hard and fine foreign phases such as carbide particles or oxide particles. Carbide dispersions can usually be created by a solid-state chemical reaction within the alloy to precipitate the particles, whereas oxide particles are best added mechanically. These hard panicles impede slippage of the metal lattice simply by intercepting and locking the dislocations in place.

Still another strengthening technique is coherent-phase precipitation; a foreign phase component of similar but modestly differing crystalline structure is introduced into the alloy — always by a chemical solid-state precipitation mechanism. The phase develops significant binding strength with the mother alloy in which it resides (it is "coherent") and, like the other mechanisms, then impedes and controls dislocation flow through the metal lattice. Coherent phase precipitation is a special case within precipitation hardening.

However, certain advances in processing have had significant effects on these classical strengthening approaches in recent years.

METALLIC MATERIALS SYSTEMS

These include superalloys, stainless steel, eutectic alloys, and intermetallic compounds.

Superalloys

High-temperature alloys also must be resistant to chemical attack by the atmosphere in which they exist, more often than not a high-speed, high-pressure gas of highly oxidizing nature. Protection from the atmosphere is achieved by causing the material to develop a tough diffusion-resistant oxide film from various elements in the alloy, or by applying a protective coating to the material.

From all this, a high degree of success has been achieved with the superalloy class of materials.

Superalloys serve under tough conditions of mechanical stressing and atmospheric attack to over 1100°C, which often is within a few hundred degrees of their melting point. These alloys are capable of supporting modern aircraft jet engines.

Undoubtedly, the most complex and sophisticated group of high-temperature alloys is the superalloys. A superalloy is defined as an alloy developed for elevated temperature service, where relatively severe mechanical stressing is encountered and high surface stability is frequently required. Superalloys, classically, are those utilized in the hottest parts of aircraft jet engines and industrial turbines; in fact, the demand of these technically sophisticated applications created the need for superalloys.

The superalloys are strengthened by all of the methods described above, as well as by some even more subtle ones, such as control of the boundaries between the grains of the crystalline metal by minor additives and other little-understood solid-state chemical reactions.

Stainless Alloys

Stainless steels are strengthened principally by carbide precipitations and solid solution strengthening; because they cannot form the γ the coherent precipitate, their use is limited to about 650°C, except where strength may not be needed.

Refractory Alloys

Refractory metal alloys are based on elements that have extremely high melting points — greater than 1650°C. These elements — tungsten, tantalum, molybdenum, and niobium (columbium) — are strengthened principally by a combination of solid solution strengthening, carbide precipitation, and unique metalworking. However, they cannot be used in modern heat engines because no commercially successful method of preventing their extensive reaction with oxidizing environments for broad application has been found.

Eutectic Alloys

These metal-based systems have been under study since the late 1960s, but service applications have not been developed. They are usually similar to or based on superalloys, but they are processed by directional solidification so that a particularly strong stable phase forms as a needlelike structure, giving unusual strength. However, they are expensive to process, and industrial use is questionable.

Intermetallic Compounds

These are metal-based systems centered on the fixed atomic compositions occurring in metallic systems of aluminum with nickel, titanium, and niobium, such as Ni_3Al, Ti_3Al, $TiAl$, and Nb_3Al. Intermetallic compounds are of interest because they often possess a lattice arrangement of atoms that leads to higher melting points and less ease of deformation. Some have shown ductility and toughness potential, and alloying to optimize properties is under significant study. For instance, Ti_3Al is fabricable and ductile, particularly when alloyed with niobium. None is in gas turbine service, but there is a possibility that Ti_3Al will become acceptable if the danger of titanium combustion and hydrogen embrittlement can be avoided.

OXIDATION AND CORROSION

In addition to possessing strength, high-temperature alloys must resist chemical attack from the environment in which they serve. Most commonly this attack is characterized by a simple oxidation of the surface. However, in machines that utilize crude or residual oils or coal or its products for fuel, natural or acquired contaminants can cause severe and complex chemical attack. Involved reactions with sulfur, sodium, potassium, vanadium, and other elements that appear in these fuels can destroy high-temperature metals rapidly — sometimes in a matter of a few hours. This is known as hot corrosion, and it is a major problem facing otherwise well-suited alloys for service in turbines operating in the combustion products of coal.

Some nuclear reactors operate above 540°C, and the working fluid is often not an oxidizing gas. For example, high-temperature gas-cooled reactors in the development stages utilize high-temperature, high-pressure helium as the working fluid. However, the helium inevitably contains very low levels of impurities — oxygen, carbon, hydrogen, and others — and it does not contain enough oxygen to form a protective oxide film on the metals that contain it. Impurities enter these alloys and reduce strength by precipitating excessive amounts of oxide and carbide phases and by other deleterious effects. Refractory metals, nominally of high interest here, have shown long-term degradation by carbide-related mechanisms, and so metallurgists have been working to develop resistant superalloys.

In other reactor applications such as the liquid metal fast-breeder reactor, construction metals must resist the high-temperature liquid metal used to transfer heat from hot uranium–plutonium fuel. The liquid metal is usually sodium or potassium. All classes of high-temperature alloys — stainless steels, superalloys, and refractory metals — have been under evaluation for service. As in helium gas, small amounts of impurities are the critical item. They can react with the metal at one temperature and transfer it to a component operating at a lower temperature. Stainless steels remain the prime construction materials for these breeder reactors.

The most significant problem, however, remains that of resistance to oxidation and high-temperature corrosion in present heat engines. Superalloys contain small-to-moderate amounts of highly reactive elements such as chromium and aluminum, which react easily with oxygen to form a thick tenacious semiplastic oxide surface film. This prevents further reaction of the aggressive environment with the underlying metal. This is the reason stainless steels are "stainless" — all contain a minimum of 10% chromium. Superalloys follow approximately the same rule, but often also contain about 5% aluminum, which further enhances oxidation resistance. It has been found that small (<1.0%) additions of rare earth elements (such as yttrium) further improve oxidation resistance by reducing oxide spalling during the inevitable thermal cycles that gas turbines experience.

The natural protective system works well when oxidation is the only or primary type of attack. However, when the contaminants — vanadium and others — are present, they react in myriad ways in the 540 to 1100°C temperature range to destroy the protective oxide and eventually the alloy. Coating is a method used to combat this problem.

Superalloy Processes

Metallurgists have been seeking ways to increase the capability of superalloys through development of new processes.

Oxide Dispersion Strengthening

One technique to generate improved strength is known as oxide dispersion strengthening. The objective is to distribute very fine, uniformly dispersed nonreactive oxide particles throughout the alloy. The processing usually involves starting with very fine particles of the metal itself, to which the oxide is added by a chemical or mechanical process step. The alloy is then consolidated by a mechanical pressing operation and forged into a final useful shape. Oxide dispersion strengthening materials are characterized by unusual creep resistance at very high temperatures; however, intermediate-temperature creep and all tensile properties are mediocre.

Success has been obtained in combining some of the classic strengthening factors described previously with oxide dispersion strengthening. This gives a balance of property enhancement that can be particularly useful to turbine metallurgists — good high-temperature creep strength from oxide dispersion strengthening and good intermediate-temperature strength from the other mechanisms.

Directional Solidification

A technique adaptable to investment casting is that of directional solidification, which is particularly suitable to the complex shapes of airfoil parts used in gas turbines. It has been found that by commencing the freezing of molten superalloys (held in a ceramic mold of the shape desired) at the bottom, then allowing the freezing process slowly to proceed up through the shape to be cast, the grains of the structure acquire a long slender shape in the direction of freezing. This significantly increases the ability of the structure to withstand mechanical load in the freezing direction, and particularly increases resistance of the superalloy to a complex phenomenon involving stress and temperature cycling or temperature gradients, known as thermal fatigue. Thermal fatigue failures commence at transverse grain boundaries by this process.

Another approach to advanced airfoil materials is to eliminate grain boundaries entirely. In directional solidification, a number of grains nucleate in the first solid to freeze, and the few with the most preferred orientation for growth choke off solidification of the others and grow along the length of the airfoil. By reducing the cross section of the first solid to form, fewer grains are nucleated. If the ceramic mold also has a short length of spiral cavity before the airfoil shape, solidification of metal through this spiral "pigtail" will allow only a single grain to grow into the airfoil section. If a particular crystal orientation with respect to the airfoil is required, a seed crystal is used instead of a pigtail. The seed placed at the bottom of the mold is long enough that the very bottom portion is not melted in the directional solidification furnace. As the mold-containing molten alloy is withdrawn, solidification occurs on the seed, with essentially the same crystal orientation as the seed. The seed can be cut from the solidified airfoil and used to seed additional airfoils. These single crystals, also known as monocrystals, allow elimination of grain boundary strengtheners such as zirconium, carbon, boron, or hafnium. This may allow more flexibility in the remainder of the composition, and higher heat treatment temperatures to optimize alloy mechanical properties and microstructure. Further freezing occurs in the 001 crystallographic direction preferred because it is in the low-elastic modulus direction for the face-centered cubic alloys.

Powder Metallurgy

An old process, powder metallurgy (P/M), has found increased use for high-temperature superalloys, and offers the possibility of added flexibility in alloy chemistry. Alloys originally developed for cast and wrought processing have been modified slightly for use as P/M materials. For large structures such as turbine disks, this has resulted in much more homogeneous chemistry and microstructure and achievement of outstanding thermal fatigue resistance. Metal flow resistance in isothermal forging is reduced because of nearly superplastic grain-boundary sliding.

Coatings

Superalloys and stainless steels must possess a balance of properties for high-temperature service. The most oxidation- and corrosion-resistant alloys do not have acceptable strength for most structural applications. Therefore, a viable solution is to utilize strong alloys for airfoil structures but then coat them to create environmental stability. This is done by adding elements such as chromium and aluminum into the surface of the alloy; these elements react with the aggressive environment to form highly protective oxides. The coating is applied by chemically reacting the turbine part with an atmosphere containing chromium or aluminum halides at very high temperature so that the active elements diffuse into the surface of the alloy to form the protective layer. Thus, ultimately, the coating layer is composed mainly of nickel from the alloy, and aluminum or chromium or both added in the coating process. Other methods develop overlayer coatings, which form not by reaction with the nickel substrate but by deposition from sources of the desired overlayer chemistry that are made by processes such as physical vapor deposition from an electron-beam-heated source, or by low-pressure plasma spraying of powders of desired coating compositions. Coatings formed by any of these methods are bonded to the substrate in the high-temperature heat treatment to cause the necessary interdiffusion. Such aluminum- and chromium-rich coatings can triple the life of industrial gas-turbine parts at temperatures such as 870°C in oxidizing atmospheres.

Eventually, however, the coatings fail by oxidizing away or by further interdiffusion with the superalloy underneath. If a very thin layer of platinum is used, the concentration of aluminum in the surface appears to be enhanced, creating an even greater measure of protection.

Nonmetallic Materials

Considerable activity involving attempts to adapt ceramics to high-temperature applications has occurred since the mid-1960s. These are ceramics of the covalent-bonded type, such as silicon carbide (SiC) and silicon nitride (Si_3N_4). Oxide ceramics, such as aluminum oxide (Al_2O_3) and zirconium oxide (ZrO_2), possess ionically bonded structures, and so tend not to possess usable high-temperature creep resistance. Turbine designers and materials engineers are struggling with the problems of utilizing the covalent ceramics, because they possess great strength. It has also become apparent that these ceramics possess great oxidation and corrosion resistance — features that would be useful in turbine equipment that must handle high-temperature products of combustion from coal. Data have shown that silicon carbide and silicon nitride are attacked to only 0.05 to 0.08 mm in depth after 6000 to 8000 h exposure in corrosive atmospheres; most high-temperature metals or alloys cannot meet this performance level. However, their complete lack of ductility means that new design techniques are required to prevent early and catastrophic failure, and only very small parts have been shown to be useful so far. It is not certain that such ceramics will be usable in heat engines.

To bypass the brittleness problem, the concept of ceramic composites has emerged. In these materials, a ceramic matrix is filled with ceramic fibers, which strengthen the whole body. When subjected to a high mechanical load, the fibers pull against the matrix and slip slightly, giving the material a certain level of tolerance. This is known as fiber pull-out. However, the effect is not elastic, so that it does not approach the deformation tolerance normally seen by metallic alloys that is essential for safe structures in tension.

The use of graphite and graphite/graphite compositions has also been explored and developed. Graphite has a strikingly unique combination of very low density and high elastic modulus, and it demonstrates increasing mechanical strength with increasing temperature. It is capable of mechanical service at temperatures of 2200°C or higher. However, since graphite is a form of carbon, it has virtually no oxidation resistance. Therefore, attempts to utilize graphite must involve truly extraordinary success in protecting it from oxidation. So far, protective coatings have had but limited success, and it appears that use may well be limited to rather low temperatures.

HOT-DIP COATINGS

A hot-dip coating is produced by immersing a base metal in a bath of the molten coating metal. Adhesion results from the tendency of the coating metal to diffuse into the base metal and form an alloy layer. Most hot-dip coatings consist of at least two distinct layers: an alloy layer and a layer of relatively pure coating metal. The alloy layer is usually a brittle intermetallic compound. Hot-dip coatings in which the alloy layer is relatively thick are not readily deformable, but modern techniques make it possible to keep the alloy layer quite thin.

Fairly thick coatings of inexpensive metals can be obtained more cheaply by hot dipping than by electroplating. Except on simple shapes, however, hot-dip coatings are nonuniform and wasteful of material. The nature of the process is such that coating metals are restricted to relatively low-melting metals, and base metals are limited to high-melting metals such as cast iron, steel, and copper.

ZINC

Properties and Uses

Hot-dipped or galvanized zinc coatings have been popular for many years for protecting ferrous products because of their ideal combination of high corrosion protection and low cost. Their corrosion protection stems from three important factors: (1) zinc has a slower rate of corrosion than iron, (2) zinc corrosion products are white and nonstaining, and (3) zinc affords electrolytic protection to iron.

The amount of protection against corrosion depends largely upon coating weight — the heavier the coating, the longer the life of the base metal. For example, a coating 0.04 mm thick is expected to have a life of 25 years in rural atmospheres, whereas a 0.88-mm coating will last 50 years. The life of zinc coatings may be five to ten times greater in rural atmospheres than in industrial atmospheres containing sulfur and acid gases. Nevertheless, the coatings are still popular for industrial use because of their low cost.

Hot dipping is particularly valuable for zinc coating parts that cannot conveniently be made of galvanized sheet. Thus, it is quite popular for structural parts, castings, bolts, nuts, nails, poleline hardware, heater and condenser coils, windlasses, and many other products.

Application Procedures

Hot-dip zinc coatings must be applied to absolutely clean metal. Consequently, surfaces are usually cleaned in a caustic or degreasing medium and then pickled. After rinsing, parts must be fluxed to promote bonding of the zinc coating. The coating itself is applied at 449 to 460°C. A zinc–iron alloy layer is formed between the base metal and coating. Immersion time depends on the thickness of coating desired, with most coatings applied in less than 1 min.

LEAD

Properties and Uses

Hot-dip lead coatings provide many important advantages over ferrous metals. They are relatively inexpensive, provide very good protection against indoor and outdoor atmospheric corrosion, and can be used in contact with many chemicals such as sulfuric, hydrochloric, hydrofluoric, phosphoric, and chromic acids. Their atmospheric corrosion resistance stems from the formation of a superficial oxide film, which is relatively impervious to corrosion.

Because of their softness, lead coatings can withstand severe deformation. Poor adhesion may be a problem since the bond is mechanical, but this problem can be minimized by adding alloying elements. Pinholes formed during application may be potential sources of corrosion but they can be eliminated by slight working or burnishing. Typical successful applications for the coatings are wire, pole-line hardware, nuts, bolts, washers, tanks, barrels, cans, and miscellaneous air ducts.

Application Procedure

Because pure lead will not alloy with ferrous metals it is usually combined with an alloying agent such as tin, which alloys with iron, forming an interface between the base metal and the coating. A typical sequence of operations for coating small parts involves solvent cleaning, electrocleaning, pickling, predipping, fluxing, coating, and quenching.

TIN

Properties and Uses

Hot-dip coatings can be applied to fabricated parts made of mild and alloy steels, cast iron, and copper and copper alloys to improve appearance and corrosion resistance. Like zinc, the coatings consist of two layers — a relatively pure outer layer and an intermediate alloy layer.

An invisible surface film of stannic oxide is formed during exposure, which helps to retard, but does not completely prevent, corrosion. The coatings have good resistance to tarnishing and staining indoors, and in most rural, marine, and industrial atmospheres. They also resist foods. Corrosion resistance in all cases can be markedly improved by increasing thickness and controlling porosity. Typical applications where they can be used are milk cans, condenser and transformer cans, food and beverage containers, and various items of sanitary equipment such as cast iron mincing machines and grinders.

Application Procedure

Steel products first must be thoroughly cleaned and fluxed. They are then immersed in a preliminary tinning pot, followed by immersion in a second pot at lower temperature. Finally, the parts are withdrawn through palm oil or are dipped in a separate oil pot. Small parts are handled in one pot, centrifuged to remove excess tin, and then quenched or air-cooled. Thickness of the coatings varies from 0.3 to 0.5 mm. The above treatments are typical and variations are used for cast iron and copper and copper alloy products.

ALUMINUM

Aluminum hot-dip coatings (usually about 1 mm) are more expensive but much more atmospheric-resistant than zinc. The coatings are also highly heat-reflective and the aluminum–iron alloy layer is highly refractory. Although these coatings seem promising for more general use in outdoor (especially industrial) atmospheres, they are currently used primarily to protect steel from high-temperature oxidation; typical applications include aircraft fire walls, toasters, automobile mufflers, and water heater casings. Hot-dip aluminum-coated (aluminized) steel sheet, strip, and wire are commercially available.

HSLA STEELS

An interesting class of alloys known as high-strength, low-alloy (HSLA) steels has emerged in response to requirements for weight reduction of vehicles. The compositions of many commercial HSLA steels are proprietary and they are specified by mechanical properties rather than composition. A typical example, however, might contain 0.2 wt% carbon and about 1 wt% or less of such elements as manganese, phosphorus, silicon, chromium, nickel, or molybdenum. The high strength of HSLA steels is the result of optimal alloy selection and carefully controlled processing such as hot rolling (deformation at temperatures sufficiently elevated to allow some stress relief).

HYBRID MATERIALS

A hybrid material is a combination of two or more different material systems that results from the attachment of one material to another. For example, a hybrid composite consists of two or more types of reinforcing fibers in one or more types of matrices. By bringing together the different properties of the reinforcements, a hybrid composite can achieve improved performance with a balance in cost. Hybrid composites have been applied successfully in transportation vehicles including race cars, helicopters,

HYBRID MATERIALS

power boats, and in other load-bearing components since the late 1960s. Recently, hybrid composites have been considered in infrastructure applications because of enhanced performance compared to all glass–fiber composites. It has been shown that combining carbon–fiber and glass–fiber composites will result in increased tensile and compressive stiffness, flexural stiffness and strength, fatigue performance, impact properties, and environmental resistance. The tensile failure strain of a glass–carbon hybrid composite is increased compared to all carbon–fiber composites.

Several studies on modeling of hybrid composites have demonstrated that by using an analytical model, carbon–glass hybrid sheet molding compound (SMC) composite can have a Young's modulus up to three times as high and a density up to 15% lower than the glass–SMC. For equal Young's modulus, a carbon–glass hybrid SMC with unfilled matrix cost considerably less than all carbon–fiber SMC.

Because carbon fibers are resistant to water and dilute acids, carbon–glass hybrid composites are more resistant to stress corrosion than all-glass fiber composites. The corrosion resistance of a hybrid composite is also dependent on laminate construction. Additionally, fatigue performance of hybrid composite is enhanced compared to all–glass composites.

POLYMETS

Polymets are a new class of hybrid materials composed of metals and polymers. Recent research has demonstrated the feasibility of extruding metal–polymer composites (polymets) of aluminum with poly(etherether ketone) and of aluminum with a liquid-crystal polymer (Al–LCP). Extrusion through either a high-shear 90° die (Figure H.2) or a converging conical die improves the properties of these aluminum–polymer composites. The improvement in properties is believed to be the result of texture development, that is, the bulk and molecular orientation of the polymer in the extrusion direction. The yield strengths of these metal–polymer composites were found to be 13 to 18% greater than that predicted by the rule of mixtures, and their specific yield strengths 14 to 21% greater than that of the aluminum control specimen.

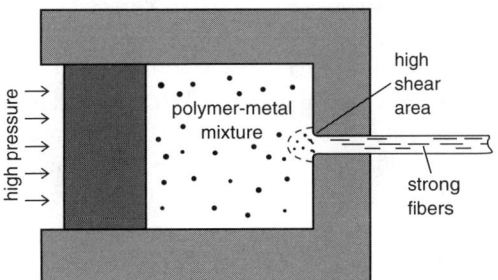

FIGURE H.2 Polymet extrusion process in which fibers are formed in place at the melt temperature. (From *McGraw-Hill 1997 Yearbook of Science and Technology*, McGraw-Hill, New York, 1997, 233. With permission.)

Tensile Properties

The strength of the polymets decreases with increasing polymer content. Ductility also decreases; tensile elongation declines from 1.6 to 1.2%. Thus, the properties of the polymets are affected by the extrusion ratio: high extrusion ratios decrease the yield strength, increase ultimate tensile strength, and improve ductility of the polymets.

Continued research on Al-LCP polymets has yielded the following:

1. Formation of in-place polymer fibrils and films result from extrusion processing.
2. The ultimate tensile strength and ductility of Al-LCP aromatic polyester polymets is improved by increasing the extrusion ratio from 10.7/1 to 114/1.
3. The yield strengths of the polymets decline by increasing the extrusion ratio, but the normalized yield strengths of the polymets increase an average of 26.7%.
4. Increases in the ultimate tensile strengths and normalized yield strengths of the polymets with extrusion ratio are attributable to formation of polymer fibrils and films.
5. The energy required to produce tensile failure is significantly enhanced by increasing the extrusion ratio from 10.7/1 to 114/1.

HYDROCHLORIC ACID

Hydrochloric acid (HCl) is soluble in water and is a strong mineral acid made by the action of sulfuric acid on common salt, or as a byproduct of the chlorination of hydrocarbons such as benzene.

Uses

HCl is used to some extent in pickling of metal prior to porcelain enameling. Pickling solutions generally contain 5 to 10% HCl, and should be contained in steel tanks lined with acid-proof brick. Rubber-lined tanks may be used for small jobs. Pickling temperatures range from room temperature up to –93°C; pickling times range from 2 to 15 min, depending on metal condition, acid strength, and temperature.

HCl will pickle faster than sulfuric acid (H_2SO_4). Also, the metal surface will be cleaner because the iron salts formed during pickling rinse off more readily. The danger of overpickling is greatly reduced by use of HCl, because the chlorine radical does not promote defects as readily as does the sulfate radical of H_2SO_4. HCl requires no steam for heating. Metal sheet should not be immersed in HCl longer than 12 to 15 min.

HYDROGEN

A colorless, odorless, tasteless elementary gas, with an atomic weight of 1.008, hydrogen is the lightest known substance. The specific gravity is 0.0695, and its density ratio in relation to air is 1:14.38. It is liquefied by cooling under pressure, and its boiling point at atmospheric pressure is –252.7°C. Its light weight makes it useful for filling balloons, but, because of its flammable nature, it is normally used only for signal balloons, for which use the hydrogen is produced easily and quickly from hydrides. Hydrogen (H_2) produces high heat, and is used for welding and cutting torches. For this purpose it is used in atomic form rather than the usual H_2 molecular form. Its high thrust value makes it an important rocket fuel. It is also used for the hydrogenation of oil and coal, for the production of ammonia and many other chemicals, and for water gas, a fuel mixture of hydrogen and carbon monoxide made by passing steam through hot coke.

Hydrogen is so easily obtained in quantity by the dissociation of water and as a byproduct in the production of alkalies by the electrolysis of brine solutions that it appears as a superabundant material, but its occurrence in nature is much less than that of many of the other elements. It occurs in the atmosphere to the extent of only about 0.01%, and in the Earth's crust to the extent of about 0.2%, or about half that of the metal titanium. However, it constitutes about one ninth of all water, from which it is easily obtained by high heat or by electrolysis.

Hydrogen has three isotopes. Hydrogen-2, called deuterium, occurs naturally in ordinary hydrogen-1, called protium, to the extent of one part in about 5000. Deuterium has one proton and one neutron in the nucleus, with one orbital electron revolving around. A gamma ray will split off the neutron, leaving the single electron revolving around a single proton. Deuterium is also called double-weight hydrogen. Deuterium oxide is known as heavy water. The formula is H_2O, but with the double-weight hydrogen the molecular weight is 20 instead of the 18 for ordinary water.

Heavy water is used for shielding in atomic reactors because it is more effective than graphite in slowing down fast neutrons. Hydrogen-3 is triple-weight hydrogen, and is called tritium. It has two neutrons and one proton in the nucleus, and is radioactive. It is a by-product of nuclear fission reactors, and most commercial production is from this source. It is a beta emitter with little harmful secondary ray emission, which makes it useful in self-luminous phosphors. It is a solid at very low temperatures. Liquid hydrogen for rocket fuel use is made from ordinary hydrogen.

Hydrides are metals that contain hydrogen in a reduced state and as a solid solution in their lattice. Titanium hydride and zirconium hydride have catalytic activity. Lanthanum and cerium react with hydrogen at room temperature forming hydrides, and are used for storing hydrogen. Sodium borohydride is used commercially as a reducing agent, for removing trace impurities from organic chemicals, in the synthesis of pharmaceuticals, for wood-pulp bleaching, for brightening clay, and for recovering trace metals in effluents. Silicon hydride, also known as silane, is a gas used for manufacturing ultrapure silicon for fabrication into semiconductors.

HYDROGEN FLUORIDE

Hydrogen fluoride is the hydride of fluorine and the first member of the family of halogen acids. Anhydrous hydrogen fluoride is a mobile, colorless liquid that fumes strongly in air. It has the empirical formula HF, melts at −83°C and boils at 19.8°C. The vapor is highly aggregated, and gaseous hydrogen fluoride deviates from perfect gas behavior to a greater extent than any other gaseous substance known.

Hydrogen fluoride is prepared on the large industrial scale by treating fluorspar (calcium fluoride, CaF_2) with concentrated sulfuric acid. The crude product is purified by fractional distillation to yield a product containing more than 99.5% hydrogen fluoride.

Properties

Anhydrous hydrogen fluoride is an extremely powerful acid, exceeded in this respect only by 100% sulfuric acid. Like water, hydrogen fluoride is a liquid of high dielectric constant that undergoes self-ionization and forms conducting solutions with many solutes. Because anhydrous hydrogen fluoride is a superacid, many organic solutes dissolve in it to form stable carbonium ions. Alkali metal fluorides and silver fluoride dissolve readily in hydrogen fluoride to form conducting solutions. The alkali metal fluorides are bases in the hydrogen fluoride system and correspond to solutions of alkali metal hydroxides in water.

Conversely, antimony pentafluoride and boron trifluoride act as acids in hydrogen fluoride and accentuate the already strong acid properties of the solvent.

Uses

Hydrogen fluoride is a widely used industrial chemical. It was formerly used in the petroleum refining industry for the isomerization of aliphatic hydrocarbons to form more desirable automotive fuels, but this application has been superseded by other methods. The largest industrial use of hydrogen fluoride is in making fluorine-containing refrigerants (Freons, Genetrons).

Another important use of hydrogen fluoride is in the preparation of organic fluorocarbon compounds by the Simons electrochemical process. In this procedure, an organic compound is dissolved in hydrogen fluoride, and an electric current is passed through the solution, whereupon the hydrogen atoms in the organic solute are replaced by fluorine. Hydrogen fluoride is employed in the electrochemical preparation of fluorine and for the preparation of inorganic fluorides. Thus, hydrogen fluoride is used for the conversion of uranium dioxide to uranium tetrafluoride, an intermediate in the preparation of uranium metal and uranium hexafluoride. With the great increase in nuclear energy-produced electricity, this represents an important use of hydrogen fluoride.

Both hydrogen fluoride and hydrofluoric acid cause unusually severe burns; appropriate precautions must be taken to prevent any contact of the skin or eyes with either the liquid or the vapor.

IMMERSION AND CHEMICAL COATINGS

Immersion coatings are applied without electricity by immersing parts in a chemical solution or bath containing the metal to be deposited. Deposition can take place either by displacement, when the metal in solution displaces the base metal, or by reduction, where the base metal does not enter into the reaction.

Although many metals can be deposited in immersion baths, comparatively few have proved acceptable for decorative or functional applications. These are nickel, tin, copper, gold, and silver.

Electroless Nickel

Properties

Electroless nickel is generally more expensive than electroplated nickel. For this reason it is used primarily for its functional properties, although a very smooth, bright deposit can be obtained on buffed ferrous and nonferrous metals.

Because of its amorphous structure and phosphorous content (8 to 10%), the coatings are said to have better corrosion resistance than electrolytic or wrought nickel. Hardness of the coatings is relatively high — about 50 R_c — and can be raised to 64 R_c by heat treatment. Thickness of the coatings ranges from 1 to 5 mils, depending on end use.

Applications

The most important uses for electroless nickel are to protect parts from corrosion and to prevent product contamination. The coatings are widely used on tank-car interiors to protect caustic soda, ethylene oxide, tetraethyl lead, tall oil, and many other liquids from contamination. Other similar applications include oil refinery air compressors, missile fuel for plates, gas storage bottles for liquid, and pumps for petroleum and related products.

The hardness of the coatings is particularly valuable in increasing the life of rotating and reciprocating surfaces in gas compressors, pumps, hydraulic cylinders, sheaves, and armatures. The coatings are also used on aluminum electronic devices to facilitate soldering, on stainless steel to facilitate brazing, on moving metal parts to prevent galling, and on stainless steel equipment to prevent stress corrosion cracking.

Tin

Properties

Tin immersion coatings are especially noted for their low cost, bright appearance, good frictional properties, and ease of application to many common metals such as copper, brass, bronze, aluminum, and steel. However, their corrosion resistance is only fair.

As with some other immersion coatings, plating usually stops when the base metal is completely covered. Thus, thickness for common decorative uses is limited to about 0.015 mil. However, thicknesses up to 2 mils for heavy-duty applications have been produced by placing the base metal in contact with a dissimilar metal, thereby generating current and promoting additional plating.

Applications

Tin immersion coatings are popular for decorative finishing of small parts such as safety pins, thimbles, and buckles. They are also applied to copper tubing to prevent discoloration from water, and to aluminum engine pistons to provide lubrication during break-in periods.

COPPER

Properties

The most important characteristics of copper immersion coatings are their high electrical conductivity, good lubrication properties, and unique appearance. In addition to steel, they can be applied to brass and aluminum and to printed circuit boards. Usual thickness range is 0.1 to 1 mil.

Applications

Because of their conductivity, copper immersion coatings have proved particularly useful for printed circuits. They are not especially noted for their decorative appeal, but can be used in applications where a particular appearance is required, e.g., inexpensive, decorative hardware such as casket parts. Because of their good lubrication properties they can also be used on steel wire in die-forming operations.

GOLD

Properties

Gold immersion coatings are relatively inexpensive because of their extreme thinness — about 0.001 mil. The coatings have good electrical conductivity and emissivity characteristics, and a bright, attractive appearance. As deposited, they are not especially resistant to discoloration and abrasion; however, they can be protected with a clear lacquer finish. They are used on a wide variety of ferrous and nonferrous metals, and on copper printed circuit boards.

Applications

Because of their good appearance gold immersion coatings are principally used on costume jewelry, trophies, auto trim, and inexpensive novelties. Their conductivity and solderability are used to advantage in electrical applications such as printed circuits, transistors, and connectors. Also, the unique emissivity properties of the coatings have proved useful in missile applications.

SILVER

Properties

Like gold, silver immersion coatings are relatively inexpensive because of their extreme thinness. The coatings have a bright, attractive appearance when first deposited. Their resistance to tarnishing and abuse is poor; however, they can be protected somewhat with a clear lacquer coating.

Silver immersion coatings can be applied to most base metals except lead, zinc, aluminum, and very active metals. They perform best on copper, nickel, and steel. Usual thickness is about 0.001 mil, but 0.03 mil can be deposited in some cases.

Applications

Because of their poor durability, silver immersion coatings are not too popular; the only applications are cheap decorative products, minor electronic parts, and maintenance plating.

IMPACT EXTRUSIONS

Impact extrusion consists of subjecting metallic materials to very high pressures at room temperature. Under these pressures the metals become "plastic" and assume predetermined shapes. Whereas in coining this process takes place within a closed die, typical impact extrusions allow a portion of the metal to be "squirted" or "squeezed" out of the die cavity, thereby forming an integral part of the desired shape.

From a practical point of view, impact extrusion of suitable parts may result in substantially lower unit manufacturing costs because it permits:

1. High production rates
2. Substantial material savings (up to 75%)
3. Little or no machining
4. Low initial tool costs and long tool life
5. Bright and smooth surface finish ready for decorating

There are two types. Impact extrusion of the slug takes place between punch and die.

Under pressure the metal may be forced to flow counter to the direction of punch travel (backward extrusion) or in the direction of punch travel (forward extrusion). Frequently, parts require a combination of both types.

Starting Material

The raw material for the process is usually referred to as a slug. Slugs may be blanked from sheet or plate, sawed from bar stock, or cast. The cross section of a slug — round, oval, square, or rectangular — fits into the die bottom; its height is determined by the volume of metal required to produce the part.

Equipment Selection

The pressures necessary for impact extrusion are available on mechanical or hydraulic presses. Mechanical presses are usually preferred when higher production rates are required. Both toggle and crank presses are used — depending on load and performance characteristics required. Hydraulic presses have primarily been used for forward extrusions requiring particularly high pressures, and heavy cross sections. Factors bearing on the selection of equipment for impact extrusion are dimensional characteristics of the part to be produced, tonnage, and speed.

Dimensional characteristics of the part (diameter, height, etc.) determine die size of the press and length of stroke required.

Tonnage required for metal flow must be carefully determined in advance. It is predicated on the relative plastic deformation required for the part, i.e., the ratio of the cross-sectional area of the extruded part to the area of the slug. Maximum limits of plastic deformation vary from 90 to 95% for 99.5% aluminum, lead, and tin, to 70 to 80% for mild steel and brasses.

Design Criteria

Impact extrusion should be specified by the designer primarily for:

1. Parts that are essentially hollow shells consisting of a wall and a bottom or partial bottom section. While in drawing the ratio between height and bottom diameter is limited to approximately 2 to 3:1, it is possible to impact-extrude (backwards) up to 8:1. Forward extrusions many times as long as the diameter have also been made.
2. Parts with straight, no-draft walls.
3. Parts requiring high-strength characteristics.
4. Parts with longitudinal ribs, flutes, splines, etc. or with bosses, cavities, etc. in the bottom section.
5. Parts made in large quantities calling for low unit cost.

Applications

It is significant for this process that the flow of metal takes place primarily in a direction parallel to punch travel. Parts produced by impact extrusion are essentially longitudinally oriented, e.g., collapsible tubes, cans, etc.

Originally used only on soft materials (lead, tin, etc.) to make collapsible tubes, the process has, in recent years, found rapidly increasing applications in the field of metallic containers, as well as in the production of a wide range of automotive, electrical, and hardware components. Aluminum and its alloys, copper, high brasses, and mild steel are impact-extruded commercially in large quantities.

Since it is a "chipless" metalworking process, impact extrusion competes in many applications with the automatic screw machine, deep drawing, die casting, hot forging, cold upsetting, and other operations. To achieve optimum results, it is important that likely parts be designed with an understanding of characteristics of metal flow, die design, and proper distribution of pressures in impact extrusion.

IMPREGNATING MATERIALS FOR CASTINGS

Castings are impregnated for several reasons, the most obvious of which is to prevent leakage. Other important reasons for impregnating are to prevent corrosion, improve surface finish, remove sites that may lodge food particles and cause bacterial growth, and to prevent back seepage of occluded fluids. Leakage may be

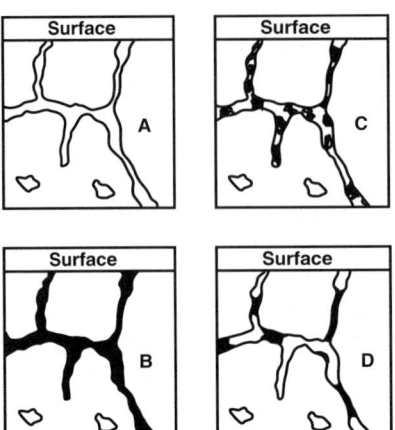

FIGURE I.1 Stages in casting impregnation process.

avoided by blocking the pores at any point along their length, whereas all of the other aims can be met only when the pores are blocked at the surface as well as in depth.

IMPREGNATING PROCESSES

Shown in Figure I.1 are the stages in the impregnation process. Figure I.1A shows a section of a porous casting with different types of porosity (continuous; noncontinuous or blind; and isolated). After evacuation and impregnation under pressure, the condition shown in B (Figure I.1B) is produced. If neither bleeding nor contraction occurs before or during the curing operation, then B will also represent the final condition. However, this ideal condition is rarely attained, and some condition between B and the extreme represented by C (Figure I.1C) is more likely to occur.

Bleeding will lead to incompletely filled pores near the surface, and contraction in curing will lead to incomplete filling along the length of the pore (condition C). The latter condition may arise from loss of solvent from impregnants introduced as a solution; loss of water from vehicles such as sodium silicate; and decomposition of organic impregnants by exposure to excessive temperatures (carbonization). Some compensation for the loss of solvent or water can be obtained if the residue can be made to expand on curing, e.g., by oxidation of metallic particles carried in sodium silicate.

It has been demonstrated that the sealing action of an impregnant generally decreases as the amount of volatile in the formulation increases. One exception to this rule is demonstrated by condition D (Figure I.1D), in which a self-fluxing brazing alloy powder is fused in the pores after evaporation of the vehicle. The molten braze alloy runs into the narrowest part of the pore under capillary action and gives optimum sealing in spite of incomplete filling of the pores.

IMPREGNATING MATERIALS

The oldest method of reducing leakage, particularly in cast iron, is to rust or oxidize the pores, sometimes after "impregnation" with mud. Natural drying oils such as tung and linseed were also among the first impregnant materials to be used. Another early type of impregnant was based on the readily available water glass (sodium silicate).

The contraction of sodium silicate impregnants during drying was reduced by the addition of inert particles such as asbestos, chalk, or oxides. Further development has led to the so-called metallic impregnants in which a large proportion of the solids are metal powders (e.g., copper or iron). Special agents are added to the sodium silicate to reduce sedimentation. During the curing of these "metallic" impregnants, the metal particles oxidize so that a certain amount of expansion occurs to offset the considerable contraction. However, these impregnants have several disadvantages: their resistance to high temperatures is poor (e.g., 149°C max); the impregnants are attacked by steam; and the abrasive oxides make them unsuitable for bearings or machining.

There are several differences between the sodium silicate and plastics impregnants. In general, because plastics or oils do not wet castings readily, a vapor degreasing is usually necessary, followed by pumping under a vacuum of at least 712 mmHg for periods of 30 min (one impregnator has achieved success in difficult cases by holding under a vacuum of 10 μm for 2 h). A second difference results from the higher viscosity of plastics. During the pressure stage of impregnation, a minimum of 30 min under pressure may be necessary to force the plastics to flow deeply into the pores. One solution to the problem of high viscosity is to thin the plastic by solvents, but this leads to condition C.

The cost of impregnating with sodium silicate is usually less (about one quarter less) than with plastics because prior cleaning of castings needs to be less thorough and excess impregnant is washed away with water.

The most important types of plastics impregnants are based on the styrene monomer. A typical composition of this substitute is 80% styrene monomer with 20% linseed oil and a small quantity of organic oxidizer as a catalyst. The instability of the catalyst in the presence of lead, zinc, and copper restricts the use of this type of impregnant primarily to the light metals. The castings are heated for 2 h at 135°C to polymerize the impregnant.

The thermosetting plastics formed by copolymerization of the styrene monomer with polyesters represents an advance on the styrene–linseed oil polymers because it is possible to introduce liquid bath curing. The surfaces are cured rapidly when they make contact with the liquid curing agent, so that subsequent bleeding (which occurs on curing in ovens) is prevented. Other improvements that have been introduced with the styrene–polyester plastics include detergent washing (does not wash the impregnant from pores), freedom from inhibition by copper, and low contraction on curing (e.g., 6 to 7%). A typical curing cycle is 1 h at 135°C. These impregnants will withstand 260°C for short periods and 204°C in continuous service.

Phenolics, epoxy, and furfural plastics can also be used for impregnation. The phenolics are dissolved in a solvent such as alcohol, and require curing in two stages; the first to remove the alcohol (e.g., 1 h at 80°C) and the second to cure the plastic (e.g., 3 h at 190°C). Phenolic resins do not have good adhesion to metal and they seal according to condition C. Therefore, they do not provide the highest-quality seals, although they may be suitable for many types of work. Epoxy resins have good adhesion to metals and resistance to chemicals, but many have to be thinned with solvents to achieve low viscosity so that the impregnation occurs readily. Furfural type plastics have been developed for resistance to alkalies.

Polyester–styrene and sodium silicate are the most widely used impregnants. The plastics give the higher quality seal, but sodium silicate continues to be used because of its low cost.

INDIUM

Indium (symbol In) is a silvery-white metal with a bluish hue, whiter than tin. It has a specific gravity of 7.31, tensile strength of 103 MPa, and elongation 22%. It is very ductile and does not work-harden, because its recrystallization point is below normal room temperature, and it softens during rolling. The metal is not easily oxidized, but above its melting point, 157°C, it oxidizes and burns with a violet flame.

Indium is now obtained as a by-product from a variety of ores. Because of its bright color, light reflectance, and corrosion resistance, it is valued as a plating metal, especially for reflectors. It is softer than lead, but a hard surface is obtained by heating the plated part to diffuse the indium into the base metal. It has high adhesion to other metals. When added to chromium plating baths it reduces brittleness of the chromium.

In spite of its softness, small amounts of indium will harden copper, tin, or lead alloys and increase the strength. About 1% in lead will double the hardness of the lead. In solders and fusible alloys it improves wetting and lowers the melting point. In lead-base alloys a small amount of indium helps to retain oil film and increases the resistance to corrosion from the oil acids. Small amounts may be used in gold and silver dental alloys to increase the hardness, strength, and smoothness. Small amounts are also used in silver-lead and silver–copper aircraft-engine bearing alloys. Lead–indium alloys are highly resistant to corrosion, and are used for chemical-processing equipment parts. Gold-indium alloys have high fluidity, a smooth lustrous color, and good bonding strength. An alloy of 77.5% gold and 22.5 indium, with a working temperature of about 500°C, is used for brazing metal objects with glass inserts. Silver–indium alloys have high hardness and a fine silvery color. A silver–indium alloy used for nuclear control rods contains 80% silver, 15% indium, and 5% cadmium. The melting point is 746°C, tensile strength 289 MPa, elongation 67%, and it retains a strength of 120 MPa at 316°C. It is stable to irradiation, and is corrosion-resistant in high-pressure water up to 360°C. The thermal expansion is about six times that of steel.

MECHANICAL PROPERTIES

Typical properties of annealed indium are as follows: tensile strength, 262 MPa; hardness, 0.9 Bhn; compressive strength, 215 MPa.

Because indium does not work-harden and almost all of the deformation occurring in the tensile test is localized, deformation is very low for such a low-strength material.

REACTIVITY

Indium is stable in dry air at room temperature. The metal boils at 2000°C but sublimes when heated in hydrogen or in vacuum. Because a thin, tenacious oxide film forms on its surface, indium resists oxidation up to, and a little beyond, its melting point. The film, however, dissolves in dilute hydrochloric acid.

Indium can be slowly dissolved in dilute mineral acids and more readily in hot dilute acids. It unites with the halogens directly when warm. Concentrated mineral acids react vigorously with indium but there is no attack by solutions of strong alkalies. The metal dissolves in oxalic acid, but not in acetic acid.

There is no evidence that the metal is toxic and it has no action as a skin irritant.

APPLICATIONS

The three largest uses of indium are in semiconductor devices, bearings, and low melting-point alloys.

Semiconductors

Indium is used to form p-n junctions in germanium. Two characteristics — the fact that indium readily wets the germanium and dissolves it at 500 to 550°C, as well as the fact that after alloying, the indium does not set up contraction stresses in the germanium — make it suitable for this application.

If a piece of n-type germanium is dissolved in indium, germanium containing excess indium recrystallizes after subsequent cooling. The excess indium changes the germanium from n-type to p-type.

Other compounds include InSb where interest is twofold. Its small energy gap (0.18 eV) makes it valuable as a photodetector in wavelengths from the infrared up to ~8 mm. InSb also has a very high electron mobility — up to 80,000 cm^2/Vs at room temperature and up to 800,000 cm^2/Vs at liquid nitrogen temperature. Consequently, the material can be used in devices based on magnetoresistance or the Hall effect (gyrators, switching elements, magnetometers, analog computers, etc.).

InAs is of interest as a semiconductor: energy gap ~0.47 eV; electron mobility ~50,000 cm^2/Vs (room temperature). The energy gap is too small to make InAs practical as a transistor material. It is, however, a candidate for infrared photoconductor applications. The relatively high mobility and small dependence of electrical properties on temperature make InAs of particular interest for Hall effect and magnetoresistance devices.

InO_3 is an n-type semiconductor finding use as a resistive element in integrated circuits, and InP is a semiconductor that is useful for rectifiers and transistors and is a promising material for electronic devices that operate at intermediate temperatures.

Bearings

Impregnating the surface of steel-backed lead-silver bearings increases the strength and hardness, improves resistance to corrosion by acids in the lubricants, and permits better retention of the bearing oil film. Indium-coated bearings can be used for high-duty service such as found in aircraft engines and diesel engines.

Alloying

A common glass sealing alloy contains approximately 50% tin–50% indium. A solder alloy containing 37.5% lead, 37.5% tin and 25% indium has greater resistance to alkalies than the 50% lead–50% tin solder.

Adding indium to gold for use in dental alloys increases the tensile strength and ductility of gold, improves resistance to discoloration, and improves bonding characteristics.

INDUSTRIAL THERMOSETTING LAMINATES

So-called industrial thermosetting laminates are those in which a reinforcing material has been impregnated with thermosetting resins and the laminates have usually been formed at relatively high pressures.

Resins most commonly used are phenolic, polyester, melamine, epoxy, and silicone; reinforcements are usually paper, woven cotton, asbestos, glass cloth, or glass mat.

NEMA (National Electrical Manufacturers' Association) has published standards covering many standard grades of laminates. Each manufacturer, in addition to these, normally provides a range of special grades with altered or modified properties or fabricating characteristics. Emphasis here is on the standard grades available.

Laminates are available in the form of sheet, rod, and rolled or molded tubing. Laminated shapes can also be specified; these are generally custom-molded by the laminate producer.

General Properties

Resin binders generally provide the following characteristics: (1) phenolics are low in cost, have good mechanical and electrical properties, and are somewhat resistant to flame; (2) polyesters used with glass mat provide a low-cost laminate for general-purpose uses; (3) epoxy resins are more expensive, but provide a high degree of resistance to acids, alkalies, and solvents, as well as extremely low moisture absorption, resulting in excellent mechanical and electrical property retention under humid conditions; (4) silicones, high in cost, provide optimum heat resistance; and (5) melamines provide resistance to flame, alkalies, arcing, and tracking, as well as good colorability.

Mechanical Properties

Tensile strengths of paper, asbestos, and cotton fabric-base laminates vary from about 136 to over 552 MPa; flexural strengths are somewhat higher and the moduli of elasticity are about 7000 to 14,000 MPa.

Other Properties

In general, moisture absorption is about 0.3 to 1.3% (24 h, 1.5 mm thickness), depending on resin and reinforcement.

Heat resistance depends on both resin and reinforcements. Most standard laminates are designed for a variety of insulation requirements. General maximum continuous service temperatures are about 121°C for phenolics reinforced with organic reinforcements, 149°C for phenolics, melamines, and epoxies reinforced with inorganic materials, and 260°C for silicones reinforced with inorganic materials.

Flame retardance of most standard grades other than melamines is relatively poor, but special flame-resistant or self-extinguishing grades are available.

Glass cloth reinforcement provides tensile strength values in the 92 to 255 MPa range; modulus of elasticity can approach 20,600 MPa. Typical flatwise compressive strength values fall in the 160 to 340 MPa range.

Electrical Properties

Dielectric strengths (perpendicular to laminates) generally range between 400 and 700 v/mil, and up to 1000 v/mil for paper-based phenolics. Dissipation factor, as-received and at 10^6 cps, ranges from a high of about 0.055 for cotton-based phenolics to a low of about 0.0015 for a glass-reinforced silicone laminate.

INJECTION MOLDING

Injection molding of plastics is analogous to die casting of metals. The plastic is heated to a fluid state in one chamber, then forced at high speed into a relatively cold, closed mold where it cools and solidifies to the desired shape. This method of processing plastics became a commercial reality in the early 1930s. It is the fastest and most economical of all commercial processes for the molding of thermoplastic materials, and is applicable to the production of articles of intricate as well as simple design.

A slightly modified version of injection molding known as "jet molding" is applicable to the molding of thermosetting materials. The principal difference between the two processes is the function of the temperature of the nozzle and mold on the material. Injection molding employs a relatively cold mold to solidify thermoplastics by chilling the mass below the melting point. Jet molding employs a relatively hot nozzle and mold to harden the thermosetting material by completing the cure.

THE PROCESS

The sequence of operation known as the molding cycle is as follows:

1. Two mold halves which, when closed together, combine to form one or more negative forms of the article to be molded, are tightly clamped between the platens of an injection-molding machine.
2. The closed mold is brought into contact with the nozzle orifice of a heating chamber. The heating chamber, known as the plastifying cylinder, is of sufficient size to carry an inventory of material equal to several volumes required to fill the mold. This permits gradual heating of the plastic to fluidity.
3. An automatically weighed or measured quantity of granulated thermoplastic, sufficient to fill the mold cavity, is fed into the rear of the plastifying cylinder.
4. A reciprocating plunger actuated by a hydraulically operated piston forces the material into the plastifying cylinder. An equal quantity of fluid plastic is thus forced out of the front of the cylinder through the nozzle orifice and into the mold.
5. A pressure of several thousand megapascals is maintained on the material within the mold until the plastic cools and solidifies.
6. After the molded item has hardened sufficiently to permit removal from the mold without distortion, the mold is opened and the part ejected.

Injection molding offers several advantages over other methods of molding. Some of the more important of these are as follows:

1. The process lends itself to complete automation for the molding of a great number of parts.
2. Mold parts require little if any post-molding operations.
3. High rates of production are made possible by the high thermal efficiency of the operation and by the short molding cycles possible.
4. There is a low ratio of mold-to-part cost in large volume production.
5. Long tool and machine life requires a minimum of maintenance and relatively low amortization costs.
6. Reuse of material is possible in most applications.

Molding machines vary considerably in design as well as capacity. For example, the clamp mechanism may be hydraulically operated, hydraulic-mechanically operated, or entirely mechanical. The injection of the material into the mold may be accomplished by a rotating screw as well as a plunger.

Machines are rated according to the weight of plastic that can be injected into the mold (the shot) with one stroke of the injection plunger. The capacity of commercially available machines covers a range from a fraction of a gram to 12.5 kg.

LIMITATIONS

The injection molding process is subject to the following limitations:

Material. Any thermoplastic material may be a candidate for injection molding, if upon heating it can be rendered sufficiently fluid to permit injection into a mold and the resulting molded article retains all the desired properties.

Geometry of Part. Any part, regardless of geometry, that can be removed from a mold without damage to the mold or part is moldable.

Weight of Part. The weight of the part must be within the "shot" rating of the particular machine being used.

INJECTION MOLDING

Projected Area and Wall Thickness of Part. The limitation on these dimensions will be governed by several factors including the relative fluidity of the plastic being molded, the pressure necessary to fill the mold, the rigidity of the part to permit ejection from the mold without deformation, and sufficient clamp force available to hold the mold in a closed position when the necessary injection pressure is developed within the cavity.

Production Economy. Economy can be realized only when relatively large quantities are produced. In addition to the material consumed and the cost of the molding operation, the number of pieces produced must also bear the cost of the mold in which they are cast. Depending on the size and complexity of the geometry of the part to be molded, mold costs will vary from a few hundred to many thousands of dollars. Again depending on size and complexity of the part, the rate of production will vary from a few minutes for a single part to several hundred per minute.

In addition to the high production rates made possible by ejection molding, articles of high quality and relatively precise dimensions can be produced. The production of industrial parts held to dimensional tolerances of ±0.002 in./in. is quite common.

The injection-molding process has made possible the development of an extremely large family of plastics. The basic types of thermoplastics and the human-modified compositions available are analogous to metals and alloys. New compositions as well as process variations are constantly being developed to meet the demands of particular situations.

Molds

The molds used in injection-molding machines for producing such large automotive components as body panels can now be made more compact and lighter than ever before. What makes this possible is nickel vapor deposition.

For car manufacturers, this means the capital invested in production molding machines can be significantly reduced, or, in more practical terms, it means smaller equipment is needed to handle the molds.

In the manufacture of shell molds nickel carbonyl vapor is fed into a low-pressure chamber where, at a temperature of 180°C, the vapor decomposes and nickel is deposited, atom by atom, onto a heated metallic master, or mandrel. Carbon monoxide and any unused carbonyl gases are recycled.

The process creates a layer of nickel that is 99.9% pure, at a deposition rate of 0.25 mm/h, faithfully reproducing all surface features. Delivery time for the production of large molds is greatly reduced, compared with conventional mold-making techniques. Uniform wall thicknesses of 1.5 to 25 mm have been achieved using the vapor deposition method.

A surface hardness of up to 42 R_c to a depth of 0.25 mm gives the shell molds excellent wear resistance.

Also, the excellent thermal conductivity of nickel reduces cycle times in the injection molding process because heat is quickly transferred from the shell to water-cooling channels in the mold assembly.

Nickel has other engineering advantages in shell molds, including its outstanding weldability, the uniform wall thicknesses possible, and the absence of residual stresses.

Recently, a 68,080-kg injection mold was created for the side panels of a prototype Composite Concept Vehicle. Larger components, up to 1.2 by 2.5 m, can be easily accommodated.

Updated Processes

Injection-compression molding (ICM) has been around for years — in fact, it is how old vinyl records were made. However, with the recent availability of high-speed microprocessors and advanced software to precisely control the molding cycle, the process now works with long-fiber-reinforced thermoplastics. This offers the prospect of stronger and lighter parts, lower costs, and better part-to-part consistency. While standard injection-molding produces reinforced-thermoplastic parts with good success, ICM can optimize the performance of long-fiber-reinforced thermoplastic materials (see Figure I.2).

The difference between ICM and conventional injection molding is that the shot is injected at low pressure into a partially open tool, as opposed to a closed one. The mold closes to compress and distribute the melt into the far reaches of the cavity, thus completing the filling and packing phase. This eliminates molded-in stresses resulting from high-injection pressures.

There are two basic types of ICM: sequential and simultaneous. In the latter, compression can begin at any point during injection, and cycle times are similar to those for conventional injection molding. With sequential ICM, the injection stroke ends before compression begins, and cycle times are 1 to 2 s longer to accommodate secondary clamping motions.

ICM uses significantly lower injection pressures than standard injection molding so longer fibers remain in the finished part. That translates into better mechanical properties. For example, ICM maintains the 12.7 mm fiber length of commercial composites, resulting in finished parts with higher impact strength and more isotropic mechanical properties. In typical applications, impact strength improves 15 to 20% in 3.2-mm wall thickness and over 50% in 1.5-mm wall thickness.

Thus, one major benefit of ICM is that it can produce thinner-walled, long-fiber-reinforced parts previously unattainable with injection molding. This is an important consideration in automotive applications where companies want to reduce weight and still maintain high stiffness and impact strength.

However, ICM start-up costs may be higher. The tooling, for example, depending on part and size could cost 10 to 15% more. ICM also requires a press with a second-stage compression stroke, as well as precise clamping, accurate shot-size control, and speed control during secondary clamping.

Costs to convert standard injection-molding presses to ICM vary with the type of machine, age, sophistication of the controller, and whether it has precision linear-position encoders to determine clamp and screw locations. In most cases the cost to upgrade a newer injection-molding machine is justified by the reduction in part cost.

Parts can also incorporate ribs, bosses, gussets, and through-holes. ICM is particularly suited to relatively flat parts, such as automobile load floors, sunroof liners, seat backs, and door panels.

Another new process is the plastic injection molding method of forming in which heat-softened material is forced under pressure into a cavity, where it cools and takes the shape of the cavity. The molding operation can be completed in one step because details such as screw threads and ribs may be easily integrated into the mold. High process repeatability is important to economical operation, because it means that few variables must be monitored and adjusted during the process. The key benefit of electric molding technology (EMT) is that it delivers this repeatability, while improving productivity and quality.

Features	Benefits
Preserves fiber length in glass-reinforced materials	Better physical properties and greater toughness
Improves weld-line strength	Maintains physical integrity throughout part
Lower injection pressure	Retains long glass-fiber length
Reduces residual molded-in stresses	Less part warpage and greater dimensional stability over a range of temperatures
Creates biaxial flow patterns and uniform packing forces across entire part	Yields more-isotrpic shrinkage, less shrinkage and warpage, and maximized density, which improves falling-weight impact properties
Allows thinner-walled parts	Lower cost and less weight
One-step molded-in coverstock laminations	Eliminates secondary lamination/adhesion operations, reducing part and labor costs

Source: Mach. Design, March 11, p. 204, 1999. With permission.

FIGURE I.2 Injection-Compression Molding.

Repeatability improvement is the reason that will likely also pull EMT into the mainstream of injection molding. This success is based on the fact that electric machines can control variables much more closely than is possible at comparable cost with hydraulic machinery. This tighter process capability translates into a variety of benefits, including less scrap, lower labor costs, and higher quality.

The essential "repeatability potential" for EMT is inherently higher than that of hydraulic power. The reason for this is that hydraulic drives are typically distributed systems, involving a compressible fluid and a complex network of hoses, tubes, and valves that enable one or two pumps to drive all machine axes. Therefore, any conditions that affect fluid or flow properties also affect positioning of the machine.

By contrast, an electric machine has a motor for each axis. In this case, "axis" may be defined as any motion that is controlled through a feedback loop on the basis of variables such as time, position, pressure, or velocity. On an injection-molding machine, the main linear axes involve clamping, ejection, injection, and sled pull-in. Rotary axes may control extrusion, and in some cases, die-height adjustment.

An all-electric powertrain on one of these axes may consist of a belt, two pulleys, and a ballscrew. With a separate motor for each axis, electric machines have the ability to drive and coordinate all axes of motion simultaneously, a significant advantage over hydraulic machines in some applications. Whatever its configuration, the electromechanical powertrain is rigid, "solid-on-solid."

Currently, EMT is more expensive than hydraulic machines, as is true with most newer technology. However, when comparing prices with hydraulic machines, specifications and capabilities should be balanced. By the time circuits and controls on a hydraulic machine are enhanced to approach the performance of a general-application electric, the cost difference narrows significantly — and hydraulic technology remains less precise.

Because it impacts molding costs in so many ways, hydraulic oil will be seen as a business liability as well as an environmental hazard. Any increase in electricity prices will also drive demand for EMT. In the mold-building industry, a corresponding increase is anticipated in the production of servo-electric systems to actuate core pulls and other functions. The environment of the molding plant will likely change considerably in a relatively few years, along with the standard of quality that is expected from the process.

INORGANIC POLYMER

An inorganic polymer is defined as a giant molecule linked by covalent bonds but with an absence or near-absence of hydrocarbon units in the main molecular backbone; these may be included as pendant side chains. Carbon fibers, graphite, and so forth are considered inorganic polymers. Much of inorganic chemistry is the chemistry of high polymers.

For compounds that do not melt or dissolve without chemical change, both the absence of an equilibrium vapor pressure and the observation of a dissociation pressure resulting from depolymerization bring them into the framework of the definition.

PROPERTIES

Some special characteristics of many inorganic polymers are a higher Young's modulus and a lower failure strain compared with organic polymers. Relatively few inorganic polymers dissolve in the true sense, or alternatively, if they swell, few can revert. Crystallinity and high glass transition temperatures are also much more common than in organic polymers. In highly cross-linked inorganic polymers, stress relaxation frequently involves bond interchange.

The properties of inorganic polymers require a different technology from that of their organic counterparts. Such technology is either completely new (such as reconstructive processing — the spinning of an inorganic compound on an organic support or binder subsequently removed by oxidation/volatilization), or it has been adapted from other fields, for example, glass technology. Thus, reconstructed vermiculite can give flexible sheets. Yarn, paper, woven cloth, and even textiles can be made from alumina and zirconia fibers by the spinning/volatilization process. A mica-forming glass ceramic is resistant to thermal and mechanical shock and can be worked with conventional metalworking tools.

CLASSIFICATION

Inorganic polymers can be classified in a number of ways. Some are based on the composition of the backbone, such as the silicones (Si–O), the phosphazenes (P–N), and polymeric sulfur (S–S). Others are based on their connectivity, that is, the number of network bonds linking the repeating unit into the network. Thus, the silicones based on R_2SiO, the phosphazenes based on NPX_2, and polymeric sulfur each have a connection of two, while boric oxide based on B_2O_3 has a connectivity of three, and amorphous silica based on SiO_2 has a connection of four.

TYPES

The number of inorganic polymers is very large. Sulfur, selenium, and tellurium all form high polymers. Polymers of sulfur are usually elastomeric, and those of selenium and tellurium are generally crystalline. In the melt at 220°C, the molecular weight of the sulfur polymer is about 12,000,000 and that of selenium about 800,000.

Silicones

Perhaps best known of all the synthetic polymers based on inorganic molecular structures are the silicones, which are derived from the basic units.

Chalcogenide Glasses

These are amorphous cross-linked polymers with a connectivity of three. Probably the best known is arsenic sulfide, $(As_2S_3)_n$, which can be used for infrared transparent windows. Threshold and memory switching are also interesting properties of these glasses. Ultraphosphate glasses resemble glassy organic plastics and can be processed by the same methods, such as extrusion and injection molding. They are used for antifouling surfaces for marine applications and in the manufacture of nonmisting spectacle lenses.

Graphite

This is a well-known two-dimensional polymer with lubricating and electrical properties. Intercalation compounds of graphite can have supermetallic anisotropic properties.

Boron Polymers

Structurally related to graphite is hexagonal boron nitride (BN). Like graphite, it has lubricating properties, reflecting the relationship between molecular structure and physical properties, but unlike graphite, it is an electrical insulator. Molybdenum disulfide, $(MoS_2)_n$, with a similar and related structure, is also a solid lubricant. Both graphite and hexagonal boron nitride can be readily machined. Outstanding properties of the latter include high thermal and chemical stability and good dielectric properties. Crucibles and such items as nuts and bolts can be made from this material.

Borate glasses with comparatively low softening points are used as solder and sealing glasses and can be prepared by fusing mixtures of metal oxides with boric oxide, $(B_2O_3)_n$.

Silicate Polymers

The silicates, both crystalline and amorphous, supply a very large number of inorganic polymers. Examples include the naturally occurring fiberlike asbestos and sheetlike mica. The industrially important water-soluble alkali metal silicates can give highly viscous polymeric solutions. Borosilicate glasses form another important group of silicate polymers. The Pyrex type is well known for its resistance to thermal shock; the leached Vycor type is porous and can be used for filtering bacteria and viruses. Asbestos occurs as ladder polymers, of which crocidolite is the most important, and as layer polymer, exemplified by chrysotile. The zeolites, many of which have been found naturally or have been synthesized, are three-dimensional network polymers. Their uses as molecular sieves are well known.

Other Polymers

Silicon nitride, $(Si_3N_4)_n$, is another macromolecule with interesting properties. Prepared by heating of silicon powder in an atmosphere of nitrogen (nitridation) at above 1200°C, the product is a material that can be machined readily and whose good thermal shock resistance and creep resistance at high temperatures, which is further improved by admixture of another inorganic macromolecule silicon carbide, make it useful for applications in gas turbine, diesel engines, thermocouple sheaths, and a variety of components.

Allotropic forms of carbon boron nitride are diamond and cubic boron nitride, both preparable by high-temperature and high-pressure syntheses and characterized by extreme hardness, which make them useful industrially in cutting and grinding tools.

IMPORTANCE

Inorganic polymeric materials are growing in importance as a result of a combination of two major factors: the depletion of the world's fossil fuel reserves (the basis of the petrochemical industry) and the ever-increasing demands of modern technology, coupled with environmental and health regulations, such as flame retardancy and nonflammability.

INSULATORS

These are any materials that retard the flow of electricity and are used to prevent the passage or escape of electric current from conductors. No materials are absolute nonconductors; those rating lowest on the scale of conductivity are therefore the best insulators. An important requirement of a good insulator is that it not absorb moisture, which would lower its resistivity. Glass and porcelain are the most common line insulators because of low cost. Pure silica glass has an average dielectric strength of 20×10^6 V/m, and glass-bonded mica about 17.7×10^6 V/m, while ordinary porcelain may be as low as 8×10^6 V/m, and steatite about 9.4×10^6 V/m. Slate, steatite, and stone slabs are still used for panel boards, but there is now a great variety of insulating boards made by compressing glass fibers, quartz, or minerals with binders, or standard laminated plastics of good dielectric strength may be used.

Synthetic rubbers and plastics have now replaced natural rubber for wire insulation, but some aluminum conductors are insulated only with an anodized coating of aluminum oxide. Wires to be coated with an organic insulator may first be treated with hydrogen fluoride, giving a coating of copper fluoride on copper wire and aluminum fluoride on aluminum wire.

INTEGRATED CIRCUITS

Integrated circuits, based on gallium arsenide (GaAs), have come into increasing use since the late 1970s.

The major advantage of these circuits is their fast switching speed.

GALLIUM ARSENIDE FET

The gallium arsenide field effect transistor (GaAs FET) is a majority carrier device in which the cross-sectional area of the conducting path of the carriers is varied by the potential applied to the gate. Unlike the MOSFET, the gate of the GaAs FET is a Schottky barrier composed of metal and gallium arsenide.

As noted above, the major advantage of gallium arsenide integrated circuits over silicon integrated circuits is the faster switching speed of the logic gate. The reason for the improvement of the switching speed of GaAs FETs with short gate lengths (less than 1 μm) over silicon FETs of comparable size has been the subject of controversy. In essence, the speed or gain-bandwidth product of a FET is determined by the velocity with which the electrons pass under the gate. The saturated drift velocity of electrons in gallium arsenide is twice that of electrons in silicon; therefore, the switching speed of gallium arsenide might be expected to be only twice as fast.

INTERCALATION COMPOUNDS

These are crystalline or partially crystalline solids consisting of a host lattice containing voids into which guest atoms or molecules are inserted. Candidate hosts for intercalation reactions may be classified by the number of directions (0 to 3) along which the lattice is strongly bonded and thus unaffected by the intercalation reaction. Isotropic, three-dimensional lattices (including many oxides and zeolites) contain large voids that can accept multiple guest atoms or molecules. Layer-type, two-dimensional lattices (graphite and clays) swell up perpendicular to the layers when the guest atoms enter (Figure I.3). The chains in one-dimensional structures (polymers such as polyacetylene) rotate cooperatively about their axes during the intercalation reaction to form channels that are occupied by the guest atoms.

Properties

The physical properties of the host are often dramatically altered by intercalation. In carbon-based hosts such as graphite, polyacetylene, and solid C_{60}, the charge donated to the host by the guest is delocalized in at least one direction, leading to large enhancements in electrical conductivity. Some hosts can be intercalated by electron-donating or electron-accepting guests, leading respectively to n-type or p-type synthetic metals, analogous to doped semiconductors. The chemical bonding between guest and host can be exploited to fine-tune the chemical reactivity of the guest. In many cases, superconductors can be created out of nonsuperconducting constituents.

Applications

Many applications of intercalation compounds derive from the reversibility of the intercalation reaction. The best-known example is pottery. Water intercalated between the silicate sheets makes wet clay plastic, while driving the water out during firing results in a dense, hard, durable material. Many intercalation compounds are good ionic conductors and are thus useful as electrodes in batteries and fuel cells. A technology for lightweight rechargeable batteries employs lithium ions, which shuttle back and forth between two different intercalation electrodes as the battery is charged and discharged: vanadium oxide (three-dimensional) and graphite (two-dimensional). Zeolites containing metal atoms remain sufficiently porous to serve as catalysts for gas-phase reactions. Many compounds can be used as convenient storage media, releasing the guest molecules in a controlled manner by mild heating.

INTERFACE OF PHASES

This is the boundary between any two phases. Among the three phases, gas, liquid, and solid, five types of interfaces are possible: gas–liquid, gas–solid, liquid–liquid, liquid–solid, and solid–solid. The abrupt transition from one phase to another at these boundaries, even though subject to the kinetic effects of molecular motion, is statistically a surface only one or two molecules thick.

A unique property of the surfaces of the phases that adjoin at an interface is the surface energy which is the result of unbalanced molecular fields existing at the surfaces of the two phases. Within the bulk of a given phase, the intermolecular forces are uniform because each molecule enjoys a statistically homogeneous field produced by neighboring molecules of the same substance. Molecules in the surface of a phase, however, are bounded on one side by an entirely different environment, with the result that there are intermolecular forces that then tend to pull these surface molecules toward the bulk of the phase. A drop of water, as a result, tends to assume a spherical shape to reduce the surface area of the droplet to a minimum.

INTERMETALLIC COMPOUNDS

Intermetallic compounds are materials composed of two or more types of metal atoms, which exist as homogeneous, composite substances and differ discontinuously in structure from that of the constituent metals. They are also called, preferably, intermetallic phases. Their properties cannot be transformed continuously into those of their constituents by changes of composition alone, and they form distinct crystalline species separated by phase boundaries from their metallic components and mixed crystals of these components. It is generally not possible to establish formulas for intermetallic compounds on the sole basis of analytical data, so formulas are determined in conjunction with crystallographic structural information.

The term *alloy* is generally applied to any homogeneous molten mixture of two or more metals, as well as to the solid material that crystallizes from such a homogeneous liquid phase. Alloys may also be formed from solid-state reactions. In the liquid phase, alloys are essentially solutions of metals in one another, although liquid compounds may also be present. Alloys containing mercury are usually referred to as amalgams. Solid alloys may vary greatly in range of composition, structure, properties, and behavior.

Phase Transformations

Much of the accumulated experimental information about the nature of the interaction and the phase transformations in systems composed of two or more metals is contained in phase or equilibrium diagrams. For example, in the phase relationships in the copper–zinc (brass) system, such phase diagrams, even for binary metal systems, may be of all degrees of complexity, ranging from systems showing the formation of simple solid solutions to dozens of intermetallic phases exhibiting structural (polymorphic), order-disorder, magnetic, bond-type, or deformation-type transformations as a function of composition or temperature, or both. Intermetallic compounds are composed of two or more metals. They may be stable over only a very narrow or over a relatively wide range of composition, which may be stoichiometric, as for the compounds GaAs, $PdCu_{13}$, $Zr_{57}Al_{43}$, and $Nb_{48}Ni_{39}Al_{13}$, or nonstoichiometric, as for the compounds $Co_{1-x}Te_x$ (where x extends continuously from 0.5 to 0.67) and $Al_{-0.08}Ge_{-0.2}Nb_3$ (an important superconducting alloy).

Crystal Structure

The crystal structures found for intermetallic compounds may likewise range from the simple rock-salt structure displayed by BeTe to the extremely complex arrangement found for $NaCd_2$, $Mg_{32}(Al,Zn)_{49}$, and Cu_4Cd_3.

Family Groups

The intermetallic compound families include the borides, hydrides, nitrides, silicides, and aluminides of the transition metal elements of Groups V and VI in the periodic table combined with semimetallic elements of small diameter. More than 1000 possible compounds fit this definition, with approximately 200 of them having melting points above 1500°C.

Methods of preparation for silicides and aluminides are very similar to those used for carbides, nitrides, and borides: (1) synthesis by fusion or sintering, (2) reduction of the metal oxide by silicon or aluminum, (3) reaction of the metal oxide with SiO_2 and carbon, (4) reaction of the metal with silicon halide, or (5) fused salt electrolysis. The simplest preparation method consists of mixing the metal powders in proper ratio, and then heating the mixture in vacuum or inert atmosphere to the temperature where reaction begins. At that point, the exothermic reaction furnishes the heat needed to drive it to completion.

The best forming methods are hot pressing and vacuum sintering. Some intermetallic compounds — ZrB_4 with $MoSi_2$, Mo_2NiB_2, Mo_2CoB, for example — are as hard as tungsten carbide. Borides of zirconium or titanium added in small amounts to Al_2O_3, BeO, ZrO_2, or ThO_2 improve their resistance to heat checking. Nitrides of aluminum, titanium, silicon and boron are used as refractory materials.

Fibers and whiskers of intermetallic materials also are being considered for strength-enhancing fillers in metal- and ceramic-matrix composites.

INTERMETALLIC MATERIALS

Ordered intermetallics bridge the gap between metals and ceramics. They exhibit physical properties characteristic of metals and can be processed by conventional casting techniques, but their mechanical properties resemble those of ceramics. Some intermetallics also have poor resistance to oxidation and sulfidation attack at elevated temperatures.

The primary barrier to their use in load-carrying applications is their brittleness or low crack tolerance. The reasons for this behavior include complex crystal structure and deformation behavior, embrittlement due to defects (interstitials, vacancies) and to hydrogen, and notch sensitivity. The major effort in research and development on intermetallics is concerned with increasing the ductility and toughness while maintaining good high-temperature strength, stiffness, creep resistance, and resistance to oxidation and sulfidation.

ORDERED STRUCTURES

An ordered intermetallic is an alloy of two elements, A and B, in specific ratios (for example, AB, AB_2, AB_3, A_2B, A_3B). In the fully ordered condition, the two species are arranged on specific lattice sites. Depending on their constitution, ordered intermetallics are divided into groups such as aluminides, silicides, and beryllides.

Nickel Aluminides

The two major nickel aluminides are Ni_3Al and NiAl.

In the crystal structure of Ni_3Al, the nickel atoms occupy the face-centered sites and the aluminum atoms occupy the corner sites of a face-centered cubic unit cell. Ni_3Al has excellent elevated-temperature strength and good oxidation resistance; there is an anomaly in the yield strength in that it increases, rather than decreases (as is the case with most other structural materials), with increasing temperature. Single crystals of Ni_3Al are highly ductile, whereas polycrystalline materials are brittle at ambient temperatures because of the brittleness of the grain boundaries. Polycrystalline Ni_3Al is also prone to environmental embrittlement at both ambient and elevated temperatures.

The mechanical properties of Ni_3Al can be improved by alloying additions of boron, chromium, hafnium, molybdenum, and zirconium. Binary Ni_3Al alloys are being examined for alternative applications, such as turbocharger rotors in diesel engine trucks.

The alloy NiAl has two thirds the density of nickel-base superalloys and, in addition, exhibits significantly higher (four to eight times depending on composition and temperature) thermal conductivity and excellent oxidation resistance. Although binary NiAl is brittle even in single-crystal form, minor alloying with iron, molybdenum, and gallium has recently been shown to significantly improve tensile ductility.

Iron Aluminides

The two major iron aluminides are Fe_3Al and FeAl. Both have excellent oxidation resistance and corrosion resistance because they form protective scales at elevated temperatures in hostile environments. They also are low in cost and have low density and good strength up to intermediate temperatures. Their main drawbacks are poor ductility and brittle fracture at ambient temperatures and poor strength and creep resistance above 650°C. The poor ductility at ambient temperatures is caused by the environmental embrittlement due to hydrogen from the moisture in the air. These compounds are being investigated as potential replacements for stainless steel in selected applications in the chemical, petroleum, and coal industries.

Titanium Aluminides

The major titanium aluminides are Ti_3Al and TiAl. They are well suited for aerospace applications because of their low density compared to the standard engine materials, nickel-base superalloys. The density of titanium aluminides is about half that of superalloys. Titanium aluminides also have good high-temperature strength but poor ductility and toughness at ambient temperatures.

Alloys corresponding to Ti_3Al have been developed as two-phase alloys with compositions based on titanium with 23 to 25% aluminum and 10 to 30% niobium (a phase is a constituent of an alloy that is physically distinct and is homogeneous in chemical composition). The oxygen impurity level is a very important consideration, because it has a large effect on the ductility. Ti_3Al has limited ductility at ambient temperatures. The mechanical properties depend on the composition and the microstructure. Increasing the niobium content increases the ductility, but the toughness and creep resistance remain low. Increasing the aluminum content increases the strength. The microstructure can be altered significantly by thermal mechanical processing.

Promising materials corresponding to TiAl have the following compositions: titanium with 46 to 52% aluminum and 1 to 10% M, where M is at least one element from the group of metals chromium, molybdenum, niobium, tantalum, vanadium, and tungsten. The materials are produced as either single-phase or two-phase materials. The two-phase materials contain both the Ti_3Al and the TiAl phases. Additions of niobium or tantalum increase the strength and oxidation resistance of single-phase materials.

The mechanical properties of these alloys are sensitive to these microstructural modifications. Unlike the NiAl alloys, TiAl alloys are produced in the polycrystalline form, and investment casting appears to be the preferred route to producing net-shape components.

A number of gas turbine engine components have been identified for TiAl-base alloy applications: rotational and stationary compressor components such as high-pressure compressor blades; turbine components such as low-pressure turbine blades; combustor components such as diffuser case and swirler; and nozzle components such as flaps and outer skins. A low-pressure turbine wheel with cast TiAl-base alloy blades has passed a rigorous simulated engine test.

Molybdenum Disilicide

High-temperature silicides are a new class of materials with potential applications in the temperature range 1200 to 1600°C. However, there are several issues to be solved before they can be used as structural materials. Such issues include (1) improving the intermediate-temperature oxidation behavior to avoid pesting (the disintegration of a silicide material in an oxygen-bearing environment, generally occurring between 300 and 900°C); (2) increasing the ambient-temperature ductility and toughness; and (3) improving the high-temperature creep resistance. The silicide that seems most promising is molybdenum disilicide ($MoSi_2$) because of its high melting temperature (2200°C) and excellent oxidation resistance. It has a tetragonal crystal structure. A major problem is its absence of ductility at temperatures up to 1000°C. It also has poor high-temperature strength due to the presence of a grain boundary silicon-rich phase that may become viscous at very high temperatures. Recent research has shown that this problem may be solved by the addition of carbon to the material.

PRODUCTION

The primary processing methods for production of intermetallics are melting, casting, ingot processing, and powder metallurgy. The secondary processes include forming, machining, and chemical milling.

Primary Processing

The two major methods of producing intermetallics are by melting and casting or by the production of powders. In the melting of aluminides the large difference between the melting temperature of aluminum and that of iron, nickel, and titanium must be considered. Other pertinent factors are (1) the large amount of aluminum in the intermetallic; (2) the melting temperature of the intermetallic (it may be higher than that of either constituent); and (3) the general reactivity of the elements to be melted. Similar issues exist for $MoSi_2$, except that the difference in the melting temperatures of the elements is smaller.

The melting methods used are air induction melting (AIM), vacuum induction melting (VIM), vacuum arc remelting (VAR), vacuum arc double electrode remelting (VADER), electroslag remelting (ESR), plasma melting, and electron-beam melting. Ni_3Al and the iron aluminides can be melted in air, although the high amount of aluminum promotes the rapid formation of a continuous aluminum-oxide film on the top of the molten materials. For many applications it is best to use processes that do not involve air melting, such as VIM, VAR, VADER, and ESR. The extreme reactivity of molten titanium with moisture and with oxidizing and carburizing environments makes the melting of titanium aluminides in conventional melting crucibles impossible. Titanium aluminides are melted by a process known as the induction skull method (ISM), which combines features of consumable arc melting and conventional ceramic crucible induction melting. Also the VAR process and plasma arc cold hearth melting (PACHM) process have been used to melt titanium aluminides as well as $MoSi_2$. Many applications of intermetallics require cast components. Casting methods include sand, investment, centrifugal, directional solidification, and near-net-shape methods.

Fine-grain wrought products are required in other applications. The commercial use of intermetallics at competitive cost requires fabrication by conventional hot-working operations. The primary processing of cast ingots is feasible, but the requirement for hot working is more stringent than for commercial metallic alloys. The secondary processing of intermetallics is very difficult and varies from intermetallic to intermetallic.

Powder metallurgy offers the most flexibility in producing intermetallics. The problem of using powder metallurgy for this purpose is that these production methods often result in surface contamination of the powders. Each of the powder consolidation methods for producing intermetallics from powders has processing difficulties. These methods are hot pressing, hot isostatic pressing, powder injection molding, extrusion, and explosive compaction. The reaction synthesis process has been uniquely applicable for intermetallics and has been used to produce many different materials.

Secondary Processing

Secondary steps such as machining and joining are critical in using these advanced materials for various applications, and extensive efforts are under way to develop these technologies. Innovative joining techniques such as friction welding, capacitor discharge welding, flash welding, laser welding, welding using the combustion synthesis concept, welding using microwaves and infrared waves, electron-beam welding, brazing, and diffusion bonding are under evaluation. A variety of machining techniques, including electrodischarge machining, water-jet cutting, ultrasonic machining, and laser cutting, are available to precision-machine complex geometries and contours. Conventional grinding, diamond drilling, and boring techniques have seen limited applications in machining TiAl alloys.

INTERMETALLIC MATRIX COMPOSITES (IMC)

IMCs have recently received considerable attention, and a variety of matrices and reinforcements have been examined to date. Reinforcement type, volume fraction, size, shape, and distribution have been shown to affect microstructure and mechanical properties. Several innovative approaches ranging from conventional techniques, such as mechanical alloying, to more exotic techniques, such as reactive consolidation and magnetron sputtering, have been used to produce these composites.

Significant advances in characterization have been made in continuously reinforced Ti_3Al-based alloys (SiC fibers in Ti_3Al + Nb alloys) and particulate-reinforced TiAl alloys (TiAl + TiB_2 particulates), in directionally solidified (DS) eutectics of NiAl, and in discontinuously reinforced $MoSi_2$.

INVESTMENT CASTINGS

The investment-casting process derives its name from one operation of the overall process: that of investing, enclosing, or encasing a disposable pattern or pattern cluster within a refractory slurry that subsequently hardens to form a mold. This method of mold making and casting in a hot mold (871 to 1038°C) distinguishes investment casting from other casting procedures. Other casting processes, such as sand casting, die casting, and shell mold casting, employ a split mold-making technique.

The investment-casting process is also known as the lost-wax, or precision casting process. (Another technique akin to it is the frozen mercury process.) Investment castings are now used in many industries.

GENERAL DESCRIPTION

Investment casting consists of three major operations — pattern production, mold production, and casting.

1. Castings are produced by using disposable patterns. Patterns are formed of wax or plastic by injecting these materials, which are fluid, into a die cavity or cavities. The shape and detail of the part intended for reproduction is cast or machined into the die.

A pattern die is constructed in two parts so that the pattern, or, in the case of a multiple-cavity die, patterns can be removed. Machined-pattern dies may be cut from a solid piece of steel or may be built in components for assembly. Cast-pattern dies require that a metal master pattern first be introduced, which is in turn used to cast a soft metal die cavity or cavities. These dies are usually cast from a low-melting alloy such as tin–bismuth. Steel dies are normally used for injection of plastics. In many instances, a part of the foundry gating system (e.g., feeders and runners) is built into the die cavity to eliminate subsequent assembly work in making up a pattern cluster. A pattern die can be made to produce a single pattern, multiple patterns, or only a portion of a pattern of a given part. In the latter case, more than one die would be used and these components would be put together to make a complete pattern.

After the patterns are produced, the next step is making the pattern assembly or pattern cluster which subsequently forms the mold cavity. The pattern assembly consists of the part pattern or patterns and the gating system, which includes feeders, runners, and metal reservoir.

2. The first operation in mold making is to dip the pattern cluster into a liquid mix of silica flour to wet the patterns and then cover them with a dry mix of flour and sand. This forms the dip coat of the patterns and is the start of the mold buildup. There are now two major methods of completing the molds.

The first method is to seal the pattern cluster to a mold board and place a container on the board that encloses the pattern cluster. A refractory slurry is poured into the container, which invests (encloses) the pattern cluster except at that area at the bottom of the mold board where the pattern cluster is stuck to the board. The fluids in the slurry harden chemically and the result is a very dense, hard, heat-resistant mold with the pattern cluster embedded. The mold board is then removed, exposing the end of wax pattern clusters.

An alternative method of completing the mold after the initial dip coating is to continue with the procedure of alternately dipping in a heavier slurry and coating with dry refractory particles until a ceramic shell is formed around the pattern cluster. Drying is necessary between cycles and the number of cycles depends on the strength requirements to contain the molten metal when poured into the mold.

3. The casting operation includes three steps: (1) removing the pattern cluster from the mold, (2) heating the metal to a molten state and pouring it into the mold, and (3) cooling the mold and removing the cast parts from it. The pattern material (wax or plastic) is removed from the mold by a combination of melting and burning. Approximately an 8-h heating cycle at temperatures of about 871 to 1038°C is used to fire the ceramic mold and remove the pattern material.

This, then, leaves a cavity in the mold. The molds are removed from the furnace shortly before casting so they are near furnace temperature when poured.

Depending on the alloy from which the castings are to be made, various melting–casting techniques are used. Alloys are melted while exposed to normal atmosphere, while covered protectively by an inert gas, or while the melting crucible and mold are in a vacuum chamber. Melting is usually accomplished by indirect arc or induction heating. Molds may be cast by "floor pouring" to fill the mold directly from the furnace or from furnace to ladle to mold. The floor-pouring method is used primarily for large molds. The more prevalent way is to secure the mold on top of the melting furnace and invert both furnace and mold so that the alloy enters directly into the mold from the crucible, normally under slight pressure.

After solidification and cooling have taken place, the mold material is mechanically or hydraulically broken away from the castings. The gates are cut from the casting and the casting is processed through other operations of finishing and inspection as required.

MATERIALS

The majority of the alloys cast by this process are iron-, nickel-, or cobalt-base alloys. The iron-base alloys include the complete range from plain carbon steels to the stainless and high-alloy steels. Many nonferrous alloys can also be used, such as gold, brasses, and bronzes, as well as gold, aluminum, and copper alloys.

The alloy to be cast is generally not considered as a restriction, because the process is sufficiently versatile to accommodate almost all types. However, not all manufacturers cast all alloys. The use of inert gas and vacuum protection for melting and pouring has enabled the industry to cast the ultrahigh-temperature iron- and nickel-base precipitation-hardening alloys.

Sizes and Configurations

The size of investment castings is often thought of in terms of castings weighing less than 2.8 g to castings weighing about 1500 g; this size range includes the majority of the castings poured. However, the capability of the process permits the successful casting of parts over 50,000 g and over 500 mm in diameter with considerable accuracy and detail. Small castings have been produced with edges down to 0.38 mm, holes 3.2 mm in diameter by 152 to 203 mm long, shorter holes to 0.076 mm. The type of alloy selected and the corresponding pouring practice may modify the maximum or minimum sizes and tolerances.

Applications

One of the major uses of investment-cast parts is in gas turbines. Investment-cast parts have proved very successful here because the shapes required (airfoils) are particularly difficult to machine and the high-strength alloy being used is difficult to finish, even for fairly simple operations. Many of the alloy compositions develop their outstanding properties in the cast form and may not be amenable to working in the wrought form.

No single industry, other than the gas-turbine industry, uses a large volume of investment castings. However, practically all industries are using some investment castings in their products or processes.

Advantages of the Process

There are several important advantages in the mass production of metal parts by the investment-casting process, not only by comparison with other casting processes, but also with other production methods such as machining or forging.

Compared with other casting processes (such as sand or shell molding) investment casting provides:

1. An improved casting from the standpoint of cleanliness of the alloy and metallurgical soundness. This is especially true when an inert gas or vacuum is used to protect the alloy.
2. Considerable more versatility in producing intricate shapes and cast detail. This feature opens the possibility of combining two or more components of an assembly into a single casting.
3. Greater dimensional accuracy, which in many cases eliminates or minimizes the need for finishing. This certainly reduces the necessary stock allowance.
4. Better cast surfaces, which are beneficial from the standpoint of quality and appearance. This may further eliminate or minimize finishing operations.

The advantages of investment casting compared with other manufacturing processes are the following:

1. Many alloy compositions develop their outstanding properties in the cast form only. Also, many of the compositions cannot be produced other than in cast form.
2. There has always been controversy whether cast parts could compete with the consistent high quality and uniformity of forged parts. The investment casting industry has proved this can be done, primarily, by consistently producing to the stringent quality requirements of the aircraft industry.
3. There have been many instances on both small and large parts where castings have been very beneficial in eliminating great amounts of machining, either starting with a wrought mill stock or from forged parts. Castings also have greater versatility in accommodating changes in section and shape.

IODINE

Iodine is a purplish-black, crystalline, poisonous elementary solid, chemical symbol I, best known for its use as a strong antiseptic in medicine, but also used in many chemical compounds and war gases. In tablet form it is used for sterilizing drinking water, and has less odor and taste than chlorine for this purpose. It is also used in cattle feeds. Although poisonous in quantity, iodine is essential to proper cell growth in the human body, and is found in every cell in a normal body, with larger concentration in the thyroid gland.

A wide range of compounds are made for electronic and chemical uses. Iodine is also a chemical reagent, used for reducing vanadium pentoxide and zirconium oxide into high-purity metals.

ION IMPLANTATION

Researchers have devised a surface modification technique to develop new materials with unique properties, by forcibly implanting ions into metals and alloys. The new properties may be physical, chemical, electrical, optical, or mechanical. Ion implantation offers broad new areas of applications, including resistance to corrosion.

The alloys ASA M50 and M50NIL are the primary bearing steels used by the Navy in its turboshaft engines. Since the Navy operates over salt water, the environment is very corrosive compared to that experienced by Air Force and commercial aircraft. Refurbishment and replacement of bearings, which cost up to $3000 each, is a significant maintenance expense. Turboshaft bearings must maintain high rolling contact fatigue resistance at relatively high operating temperature; therefore, stainless steel cannot be used. Protection of the bearings with an anticorrosion coating has been unsuccessful due to delamination problems.

Bearings were ion-implanted with chromium ions that produced a 75-nm-thick stainless steel layer on the low-alloy bearing steels ASA M50 and 52100. This dramatically improved the service life and shelf life of the expensive bearings. Results showed that the bearings could be implanted for between $70 and $170 per bearing, and that this cost was more than paid for by the average increase in the bearing service life of 2.5 times.

IRIDIUM

Iridium (symbol Ir) is a grayish-white metal of extreme hardness. It is insoluble in all acids and in aqua regia. The melting point is 2447°C, and the specific gravity is 22.50. It occurs naturally with the metal osmium as an alloy, known as osmiridium, 30 to 60% osmium, used chiefly for making fountain-pen points and instrument pivots.

Iridium is employed as a hardener for platinum, the jewelry alloys usually containing 10%. With 35% iridium the tensile strength of platinum is increased to 965 MPa. Iridium wire is used in spark plugs because it resists attack of leaded aviation fuels. Iridium-tungsten alloys are used for springs operating at temperatures to 800°C. Iridium intermetallic compounds such as Cb_3Ir_2, Ti_3Ir, and $ZrIr_2$ are superconductors.

Upon atmospheric exposure the surface of the metal is covered with a relatively thick layer of iridium dioxide (IrO_2). Only when heated to redness is the metal attacked by oxygen (to give IrO_2) and by the halogens. The vapor pressure of IrO_2 is atm (10^5 Pa) at 1120°C, and consequently pieces of iridium metal lose mass at elevated temperatures in an oxygen atmosphere.

Metallic iridium is available in powder, sponge, wire, and sheet form. Iridium is usually difficult to forge and fabricate, but it is ductile when hot and consequently is worked hot. It remains ductile when cooled as long as it is not annealed. Annealing causes loss of ductility and brittleness. In the absence of oxygen, iridium is not attacked by many molten metals, including lithium, sodium, potassium, gold, lead, and tin.

Uses

Because of its scarcity and high cost, applications of iridium are severely limited. Although iridium metal and many of its complex compounds are good catalysts, no large-scale commercial application for these has been developed. In general, other platinum metals have superior catalytic properties. The high degree of thermal stability of elemental iridium and the stability it imparts to its alloys does give rise to those applications where it has found success. Particularly relevant are its high melting point (2443°C), its oxidation resistance, and the fact that it is the only metal with good mechanical properties that survives atmospheric exposure above 1600°C. Iridium is alloyed with platinum to increase tensile strength, hardness, and corrosion resistance. However, the workability of these alloys is decreased. These alloys find use as electrodes for anodic oxidation, for containing and manipulating corrosive chemicals, for electrical contacts that are exposed to corrosive chemicals, and as primary standards for weight and length. Platinum-iridium alloys are used for electrodes in spark plugs that are unusually resistant to fouling by antiknock lead additives. Iridium–rhodium thermocouples are used for high-temperature applications, where they have unique stability. Very pure iridium crucibles are used for growing single crystals of gadolinium gallium garnet for computer memory devices and of yttrium aluminum garnet for solid-state lasers.

IRON

Iron (symbol Fe) is one of the most common of the commercial metals. It has been in use since the most remote times, but it does not occur native except in the form of meteorites — early tools of Egypt were apparently made from nickel irons from this source. The common iron ores are magnetic pyrites, magnetite, hematite, and carbonates of iron. To obtain the iron the ores are fused to drive off the oxygen, sulfur, and impurities. The melting is done in a blast furnace directly in contact with the fuel and with limestone as a flux. The latter combines with the quartz and clay, forming a slag that is readily removed. The resulting product is crude pig iron, which requires subsequent remelting and refining to obtain commercially pure iron. Sintered iron and steel are also produced without blast-furnace reduction by compressing purified iron oxide in rollers, heating to 1204°C, and hot-strip rolling. The final cold-rolled product is similar to conventional iron and steel.

Originally, all iron was made with charcoal, but because of the relative scarcity of wood and the greater expense, charcoal is now seldom in the blast furnace.

Iron is a grayish metal, which until recently was never used pure. It melts at 1525°C and boils at 2450°C. Even very small additions of carbon reduce the melting point. It has a specific gravity of 7.85. Iron containing more than 0.15% chemically combined carbon is termed steel. When the carbon is increased to above about 0.40%, the metal will harden when cooled suddenly from a red heat. Iron, when pure, is very ductile, but a small amount of sulfur, as little as 0.03%, will make it hot-short, or brittle at red heat. As little as 0.25% of phosphorus will make iron cold-short, or brittle when cold. Iron forms carbonates, chlorides, oxides, sulfides, and other compounds. It oxidizes easily and is also attacked by many acids.

Because pure iron is allotropic, it can exist as a solid in two different crystal forms. From subzero temperatures up to 910°C, it has a body-centered cubic structure and is identified as alpha (α) iron. Between 910 and 1400°C, the crystal structure is face-centered cubic. This form is known as gamma (γ) iron. At 1400°C and up to its melting point of 1540°C, the structure again becomes body-centered cubic. This last form, called delta (δ) iron, has no practical use. The transformation from one allotropic form to another is reversible. Thus, when iron is heated to above 910°C, the alpha body-centered cubic crystal changes into face-centered cubic crystals of gamma iron. When cooled below this temperature, the metal again reverts back to a body-centered cubic structure. These allotropic phase changes inherent in iron make possible the wide variety of properties obtainable in ferrous alloys by various heat-treating processes.

Types

Electrolytic iron is a chemically pure iron produced by the deposition of iron in a manner similar to electroplating. Bars of cast iron are used as anodes and dissolved in an electrolyte of ferrous chloride. The current precipitates almost pure iron on the cathodes, which are hollow steel cylinders. The deposited iron tube is removed by hydraulic pressure or by splitting, and then annealed and rolled into plates. The iron is 99.9% pure, and is used for magnetic cores and where ductility and purity are needed.

Iron powder is made by reducing iron ore by the action of carbon monoxide at a temperature below the melting point of the iron and below the reduction point of the other metallic oxides in the ore. In the United States it is made by the reduction of iron oxide mill scale, by electrolysis of steel borings and turnings in an electrolyte of ferric chloride, or by atomization. Iron powders are widely used for pressed and sintered structural parts, commonly referred to as P/M (powder-metallurgy) parts. Pure iron powders are seldom used alone for such parts. Small additions of carbon in the form of graphite and/or copper are used to improve performance properties.

Iron–copper powders contain from 2 to 11% copper. Small amounts of graphite are sometimes added. Copper increases strength and hardness, and improves corrosion resistance, but lowers ductility somewhat. Tensile strengths range between 206 and 680 MPa, depending on density and heat treatment.

Iron–carbon powders contain up to 1% graphite. When pressed and sintered, internal carburization results and produces a carbon-steel structure, although some free carbon remains. In general, the density of structural or mechanical iron-carbon steel P/M parts is 80 to 95% that of wrought steels, and the greater the density, the greater the strength and other mechanical properties. These carbon-steel P/M parts have higher strength and hardness than those of iron, but they are usually more brittle. As-sintered strengths range from about 241 to 482 MPa depending on density. By heat treatment, strengths up to 861 MPa are achieved. P/M parts can be made to controlled porosity for filters or for filling with oil or other materials for surface lubricity.

IRON ALLOYS

These are solid solutions of metals, where one metal is iron. A great number of commercial alloys have iron as an intentional constituent. Iron is the major constituent of wrought and cast iron and wrought and cast steel. Alloyed with usually large amounts of silicon, manganese, chromium, vanadium, molybdenum, niobium (columbium), selenium, titanium, phosphorus, or other elements, singly or sometimes in combination, iron forms the large group of materials known as ferroalloys that are important as addition agents in steelmaking. Iron is also a major constituent of many special-purpose alloys developed to have exceptional characteristics with respect to magnetic properties, electrical resistance, heat resistance, corrosion resistance, and thermal expansion; Table I.1 lists some of these alloys.

Because of the enormous number of commercially available materials, the listing is limited to the better-known types of alloys. Emphasis is on special-purpose alloys; practically all of these contain relatively large amounts of an alloying element or elements referred to in the classification. Alloys containing less than 50% iron are excluded, with a few exceptions.

Iron–Aluminum Alloys

Although pure iron has ideal magnetic properties in many ways, its low electrical resistivity makes it unsuitable for use in alternating-current (AC) magnetic circuits. Addition of aluminum in fairly large amounts increases the electrical resistivity of iron, making the resulting alloys useful in such circuits.

Three commercial iron–aluminum alloys with moderately high permeability at low field strength and high electrical resistance nominally contain 12% aluminum, 16% aluminum, and 16% aluminum with 3.5% molybdenum, respectively. These three alloys are classified as magnetically soft materials; that is, they become magnetized in a magnetic field but are easily demagnetized when the field is removed.

The addition of more than 8% aluminum to iron results in alloys that are too brittle for many uses because of difficulties in fabrication. However, addition of aluminum to iron markedly increases its resistance to oxidation. One steel containing 6% aluminum possesses good oxidation resistance up to 1300°C.

TABLE I.1
Some Typical Composition Percent Ranges of Iron Alloys Classified by Important Uses[a]

Type	Fe	C	Mn	Si	Cr	Ni	Co	W	Mo	Al	Cu	Ti
Heat-resistant alloy castings	Bal.[b]	0.30–0.50	—	1–2	8–30	0–7	—	—	—	—	—	—
	Bal.	0.20–0.75	—	2–2.5	10–30	8–41	—	—	—	—	—	—
Heat-resistant cast irons	Bal.	1.8–3.0	0.3–1.5	0.5–2.5	15–35	5 max	—	—	—	—	—	—
	Bal.	1.8–3.0	0.4–1.5	1.0–2.75	1.75–5.5	14–30	—	0.5 max	1	—	7	—
Corrosion-resistant alloy castings	Bal.	0.15–0.50	1 max	1	11.5–30	0–4	—	—	—	—	—	—
	Bal.	0.03–0.20	1.5 max	1.5–2.0	18–27	8–31	—	—	—	—	—	—
Corrosion-resistant cast irons	Bal.	1.2–4.0	0.3–1.5	0.5–3.0	12–35	5 max	—	—	4 max	—	3 max	—
	Bal.	1.8–3.0	0.4–1.5	1.0–2.75	1.75–5.5	14–32	—	—	1 max	—	7 max	—
Magnetically soft materials	Bal.	—	—	0.5–4.5	—	—	—	—	—	—	—	—
	Bal.	—	—	—	—	—	—	—	3.5	16	—	—
	Bal.	—	—	—	—	—	—	—	—	16	—	—
	Bal.	—	—	—	—	—	—	—	—	12	—	—
Permanent-magnet materials	Bal.	—	—	—	—	—	12	—	17	—	—	—
	Bal.	—	—	—	—	—	12	—	20	—	—	—
	Bal.	—	—	—	—	20	5	—	—	12	—	—
	Bal.	—	—	—	—	17	12.5	—	—	10	6	—
	Bal.	—	—	—	—	25	5	—	—	12	—	—
	Bal.	—	—	—	—	28	5	—	—	12	—	—
	Bal.	—	—	—	—	14	24	—	—	8	3	—
	Bal.	—	—	—	—	15	24	—	—	8	3	1.25
Low-expansion alloys	Bal.	—	0.15	0.33	—	36	—	—	—	—	—	—
	Bal.	—	0.24	0.03	—	42	—	—	—	—	—	—
	61–53	0.5–2.0	0.5–2.0	0.5–2.0	4–5	33–35	—	1–3	—	—	—	—

[a] This table does not include any AISI standard carbon steels, alloy steels, or stainless and heat-resistant steels or plain or alloy cast iron for ordinary engineering uses; it includes only alloys containing at least 50% iron, with a few exceptions.

[b] Bal. = balance percent of composition.

Source: *McGraw-Hill Encyclopedia of Science and Technology*, 8th ed., Vol. 9, McGraw-Hill, New York, 445–457. With permission.

Iron-Carbon Alloys

The principal iron–carbon alloys are wrought iron, cast iron, and steel.

Wrought iron of good quality is nearly pure iron; its carbon content seldom exceeds 0.035%. In addition, it contains 0.075 to 0.15% silicon, 0.10 to less than 0.25% phosphorus, less than 0.02% sulfur, and 0.06 to 0.10% manganese. Not all of these elements are alloyed with the iron; part of them may be associated with the intermingled slag that is a characteristic of this product. Because of its low carbon content, the properties of wrought iron cannot be altered in any useful way by heat treatment.

Cast iron may contain 2 to 4% carbon and varying amounts of silicon, manganese, phosphorus, and sulfur to obtain a wide range of physical and mechanical properties. Alloying elements (silicon, nickel, chromium, molybdenum, copper, titanium, and so on) may be added in amounts varying from a few tenths to 30% or more. Many of the alloyed cast irons have proprietary compositions.

Steel is a generic name for a large group of iron alloys that include the plain carbon and alloy steels. The plain carbon steels represent the most important group of engineering materials known. Although any iron–carbon alloy containing less than about 2% carbon can be considered a steel, the American Iron and Steel Institute (AISI) standard carbon steels embrace a range of carbon contents from 0.06% maximum to about 1%. In the early days of the American steel industry, hundreds of steels with different chemical compositions were produced to meet individual demands of purchasers. Many of these steels differed only slightly from each other in chemical composition. Studies were undertaken to provide a simplified list of fewer steels that would still serve the various needs of fabricators and users of steel products. The Society of Automotive Engineers (SAE) and the AISI both periodically publish lists of steels, called standard steels, classified by chemical composition. These lists are published in the SAE Handbook and the AISI Steel Products Manuals. The lists are altered periodically to accommodate new steels and to provide for changes in consumer requirements. There are minor differences between some of the steels listed by the AISI and SAE.

A numerical system is used to indicate grades of standard steels. Provision also is made to use certain letters of the alphabet to indicate the steelmaking process, certain special additions, and steels that are tentatively standard. The basic numerals for the AISI classification and the corresponding types of steels in this system include the first digit of the series designation, which indicates the type to which a steel belongs; thus 1 indicates a carbon steel, 2 indicates a nickel steel, and 3 indicates a nickel–chromium steel. In the case of simple alloy steels, the second numeral usually indicates the percentage of the predominating alloying element. Usually, the last two (or three) digits indicate the average carbon content in points, or hundredths of a percent. Thus, 2340 indicates a nickel steel containing about 3% nickel and 0.40% carbon.

All carbon steels contain minor amounts of manganese, silicon, sulfur, phosphorus, and sometimes other elements. At all carbon levels the mechanical properties of carbon steel can be varied to a useful degree by heat treatments that alter its microstructure. Above about 0.25% carbon steel can be hardened by heat treatment. However, most of the carbon steel produced is used without a final heat treatment.

Alloy steels are steels with enhanced properties attributable to the presence of one or more special elements or of larger proportions of manganese or silicon than are present ordinarily in carbon steel. The major classifications of alloy steels are high-strength, low-alloy; AISI alloy; alloy tool; heat-resisting; electrical; and austenitic manganese. Some of these iron alloys are discussed briefly in the following.

Iron-Chromium Alloys

An important class of iron–chromium alloys is exemplified by the wrought stainless and heat-resisting steels of the type 400 series of the AISI standard steels, all of which contain at least 12% chromium, which is about the minimum chromium content that will confer stainlessness. However, considerably less than 12% chromium will improve the oxidation resistance of steel for service up to 650°C, as is true of AISI types

501 and 502 steels that nominally contain about 5% chromium and 0.5% molybdenum. A comparable group of heat- and corrosion-resistant alloys, generally similar to the 400 series of the AISI steels, is covered by the Alloy Casting Institute specifications for cast steels.

Corrosion-resistant cast irons alloyed with chromium contain 12 to 35% of that element and up to 5% nickel. Cast irons classified as heat-resistant contain 15 to 35% chromium and up to 5% nickel.

During World War I a high-carbon steel used for making permanent magnets contained 1 to 6% chromium (usually around 3.5%); it was developed to replace the magnet steels containing tungsten that had been formerly but could not then be made because of a shortage of tungsten.

Iron–Chromium–Nickel Alloys

The wrought stainless and heat-resisting steels represented by the type 200 and the type 300 series of the AISI standard steels are an important class of iron–chromium–nickel alloys. A comparable series of heat- and corrosion-resistant alloys is covered by specifications of the Alloy Casting Institute. Heat- and corrosion-resistant cast irons contain 15 to 35% chromium and up to 5% nickel.

Iron–Chromium–Aluminum Alloys

Electrical-resistance heating elements are made of several iron alloys of this type. Nominal compositions are 72% iron, 23% chromium, 5% aluminum; and 55% iron, 37.5% chromium, 7.5% aluminum. The iron–chromium–aluminum alloys (with or without 0.5 to 2% cobalt) have higher electrical resistivity and lower density than nickel–chromium alloys used for the same purpose. When used as heating elements in furnaces, the iron–chromium–aluminum alloys can be operated at temperatures of 2350°C maximum. These alloys are somewhat brittle after elevated-temperature use and have a tendency to grow or increase in length while at temperature, so that heating elements made from them should have additional mechanical support. Addition of niobium (columbium) reduces the tendency to grow.

Because of its high electrical resistance, the 72% iron, 23% chromium, 5% aluminum alloy (with 0.5% cobalt) can be used for semiprecision resistors in, for example, potentiometers and rheostats.

Iron–Cobalt Alloys

Magnetically soft iron alloys containing up to 65% cobalt have higher saturation values than pure iron. The cost of cobalt limits the use of these alloys to some extent. The alloys also are characterized by low electrical resistivity and high hysteresis loss. Alloys containing more than 30% cobalt are brittle unless modified by additional alloying and special processing. Two commercial alloys with high permeability at high field strengths (in the annealed condition) contain 49% cobalt with 2% vanadium, and 35% cobalt with 1% chromium. The latter alloy can be cold-rolled to a strip that is sufficiently ductile to permit punching and shearing. In the annealed state, these alloys can be used in either AC or DC applications. The alloy of 49% cobalt with 2% vanadium has been used in pole tips, magnet yokes, telephone diaphragms, special transformers, and ultrasonic equipment. The alloy of 35% cobalt with 1% chromium has been used in high-flux-density motors and transformers as well as in some of the applications listed for the higher-cobalt alloy.

Although seldom used now, two high-carbon alloys called cobalt steel were used formerly for making permanent magnets. These were both high-carbon steels. One contained 17% cobalt, 2.5% chromium, 8.25% tungsten; the other contained 36% cobalt, 5.75% chromium, 3.75% tungsten. These are considered magnetically hard materials as compared to the magnetically soft materials.

Iron–Manganese Alloys

The important commercial alloy in this class is an austenitic manganese steel (sometimes called Hadfield manganese steel after its inventor) that nominally contains 1.2% and 12 to 13% manganese. This steel is highly resistant to abrasion, impact, and shock.

Iron–Nickel Alloys

The iron–nickel alloys exhibit a wide range of properties related to their nickel contents.

Nickel content of a group of magnetically soft materials ranges from 40 to 60%; however, the highest saturation value is obtained at about 50%. Alloys with nickel content of 45 to 60% are characterized by high permeability and low magnetic losses. They are used in such applications as audio transformers, magnetic amplifiers, magnetic shields, coils, relays, contact rectifiers, and choke coils. The properties of the alloys can be altered to meet specific requirements by special processing techniques involving annealing in hydrogen to minimize the effects of impurities, grain-orientation treatments, and so on.

Another group of iron–nickel alloys, those containing about 30% nickel, is used for compensating changes that occur in magnetic circuits due to temperature changes. The permeability of the alloys decreases predictably with increasing temperature.

Low-expansion alloys are so called because they have low thermal coefficients of linear expansion. Consequently, they are valuable for use as standards of length, surveyors' rods and tapes, compensating pendulums, balance wheels in timepieces, glass-to-metal seals, thermostats, jet-engine parts, electronic devices, and similar applications.

The first alloy of this type contained 36% nickel with small amounts of carbon, silicon, and manganese (totaling less than 1%). Subsequently, a 39% nickel alloy with a coefficient of expansion equal to that of low-expansion glasses and a 46% nickel alloy with a coefficient equal to that of platinum were developed. Another important alloy is one containing 42% nickel that can be used to replace platinum as lead-in wire in light bulbs and vacuum tubes by first coating the alloy with copper. An alloy containing 36% nickel and 12% chromium has a constant modulus of elasticity and low expansivity over a broad range of temperatures. Substitution of 5% cobalt for 5% nickel in the 36% nickel alloy decreases its expansivity. Small amounts of other elements affect the coefficient linear expansion, as do variations in heat treatment, cold-working, and other processing procedures.

A 9% nickel steel is useful in cryogenic and similar applications because of good mechanical properties at low temperatures. Two steels (one containing 10 to 11% nickel, 3 to 5% chromium, about 3% molybdenum, and lesser amounts of titanium and aluminum and another with 17 to 19% nickel, 8 to 9% cobalt, 3 to 3.5% molybdenum, and small amounts of titanium and aluminum) have exceptional strength in the heat-treated (aged) condition. These are known as maraging steels.

Cast irons containing 14 to 30% nickel and 1.75 to 5.5% chromium possess good resistance to heat and corrosion.

Iron–Silicon Alloys

There are two types of iron–silicon alloys that are commercially important: the magnetically soft materials designated silicon or electrical steel, and the corrosion-resistant, high-silicon cast irons.

Most silicon steels used in magnetic circuits contain 0.5 to 5% silicon. Alloys with these amounts of silicon have high permeability, high electrical resistance, and low hysteresis loss compared with relatively pure iron. Most silicon steel is produced in flat-rolled (sheet) form and is used in transformer cores, stators, and rotors of motors, and so on that are built up in laminated-sheet form to reduce eddy current losses. Silicon–steel electrical sheets, as they are called commercially, are made in two general classifications: grain-oriented and non-oriented.

The grain-oriented steels are rolled and heat-treated in special ways to cause the edges of most of the unit cubes of the metal lattice to align themselves in the preferred direction of optimum magnetic properties. Magnetic cores are designed with the main flux path in the preferred direction, thereby taking advantage of the directional properties. The grain-oriented steels contain about 3.25% silicon, and they are used in the highest-efficiency distribution and power transformers and in large turbine generators.

The non-oriented steels may be subdivided into low-, intermediate-, and high-silicon classes. Low-silicon steels contain about 0.5 to 1.5% silicon and are used principally in rotors and stators of motors and generators; steels containing about 1% silicon are also used for reactors, relays, and small intermittent-duty transformers. Intermediate-silicon steels contain about 2.5 to 3.5% silicon and are used in motors and generators of average to high efficiency and small- to medium-size intermittent-duty transformers, reactors, and motors. High-silicon steels contain about 3.75 to 5% silicon and are used in highest-efficiency motors, generators, and transformers.

High-silicon cast irons containing 14 to 17% silicon and sometimes up to 3.5% molybdenum possess corrosion resistance that makes them useful for acid-handling equipment and for laboratory drain pipes.

Iron–Tungsten Alloys

Although tungsten is used in several types of relatively complex alloy (including high-speed steels), the only commercial alloy made up principally of iron and tungsten was a tungsten steel containing 0.5% chromium in addition to 6% tungsten that was used up to the time of World War I for making permanent magnets.

Hard-Facing Alloys

Hard-facing consists of welding a layer of metal of special composition on a metal surface to impart some special property not possessed by the original surface. The deposited metal may be more resistant to abrasion, corrosion, heat, or erosion than the metal to which it is applied. A considerable number of hard-facing alloys are available commercially. Many of these would not be considered iron alloys by the 50% iron content criterion adopted for the iron alloys. Among the iron alloys are low-alloy facing materials containing chromium as the chief alloying element, with smaller amounts of manganese, silicon, molybdenum, vanadium, tungsten, and in some cases nickel to make a total alloy content of up to 12%, with the balance iron. High-alloy ferrous materials containing a total of 12 to 25% alloying elements form another group of hard-facing alloys; a third group contains 26 to 50% alloying elements. Chromium, molybdenum, and manganese are the principal alloying elements in the 12 to 25% group; smaller amounts of molybdenum, vanadium, nickel, and in some cases titanium are present in various proportions. In the 26 to 50% alloys, chromium (and in some cases tungsten) is the principal alloying element, with manganese, silicon, nickel, molybdenum, vanadium, niobium (columbium), and boron as the elements from which a selection is made to bring the total alloy content within the 26 to 50% range.

Permanent-Magnet Alloys

These are magnetically hard ferrous alloys, many of which are too complex to fit the simple compositional classification used above for other iron alloys. As already mentioned in discussing iron–cobalt and iron–tungsten alloys, the high-carbon steels (with or without alloying elements) are now little used for permanent magnets. These have been supplanted by a group of sometimes complex alloys with much higher retentivities. The ones considered here are all proprietary compositions. Two of the alloys contain 12% cobalt–17% molybdenum and 12% cobalt–20% molybdenum.

Members of a group of six related alloys contain iron, nickel, aluminum, and with one exception cobalt; in addition, three of the cobalt-containing alloys contain copper and one has copper and titanium. Unlike magnet steels, these alloys resist demagnetization by shock, vibration, or temperature variations. They are used in magnets for speakers, watt-hour meters, magnetrons, torque motors, panel and switchboard instruments, and so on, where constancy and degree of magnet strength are important.

ISOPOLYESTER COMPOSITES

An isopolyester composite is used to retrofit concrete causeway support columns. The Snap-Tite composite jacket will use isopolyester made with purified isophthalic acid (PIA). Isopolyesters have a proven record of delivering corrosion resistance and mechanical strength at less cost. The resin encapsulates the reinforcing network, which for most applications is conventional e-glass woven roving and bidirectional fabric. As a result, the compressive strength of the columns will reportedly more

TABLE I.2
Comparison of Isostatic Pressing Methods

Method	Advantages	Limitations	Cycle Time
Cold (room temperature)	Uniform green density	Slower than uniaxial pressing	5–30 min for wet bag
	Waxless, complex shapes	Parts may require post-machining	<1 min for dry bag
	Wet-bag: various shapes per cycle		
	Dry-bag: automated, one part at a time		
Warm (100°C)	Cost-effective for different-shaped parts	Only suitable for specific applications	3–5 min
	Eliminates post-sintering		
Hot (2200°C)	Improves mechanical and physical properties	Cycle times can be slow	10–15 h
	Near net shape		<8 h with uniform rapid cooling
	Full density		

Source: Ceram. Ind., March, p. 33, 1998. With permission.

than double, to 83 MPa or more, and the ductility will improve from 3 to 10%.

ISOSTATIC PRESSING

The isostatic pressing process was pioneered in the mid-1950s and has steadily grown from a research curiosity to a viable production tool. Many industries apply this technique for consolidation of powders or defect healing of castings. The process is used for a range of materials, including ceramics, metals, composites, plastics, and carbon.

Isostatic pressing applies a uniform, equal force over the entire product, regardless of shape or size. It thus offers unique benefits for ceramic and refractory applications. The ability to form product shapes to precise tolerances (reducing costly machining) has been a major driving force for its commercial development.

There are three basic types of isostatic pressing (Table I.2). Cold isostatic pressing (CIP) is applied to consolidate ceramic or refractory powders loaded into elastomeric bags. Warm isostatic pressing (WIP) differs from CIP only in that shapes are pressed at warm temperature to about 100°C. Hot isostatic pressing (HIP) involves both temperature and pressure applied simultaneously to obtain fully dense parts (to 100% theoretical density), and is used mainly for engineered ceramics requiring optimum properties for high-performance applications.

COLD ISOSTATIC PRESSING (CIP)

CIP is mainly a powder-compacting process for obtaining 60 to 80% theoretically dense parts ready for sintering. Because of the good green strength obtained with this forming method, premachining before sintering is feasible without causing breakage.

When comparing uniaxial pressing to isostatic pressing, one can say that uniaxial pressing is more suitable for small shapes at high production rates. Die wall friction may result in non-uniform densities, especially for large aspect ratios (greater than 3:1).

CIP is slower than uniaxial pressing but can be used for small or large, simple or complex shapes. The uniform green density offers more even shrinkage during sintering, which is very important for good shape control and uniform properties. In addition, CIP does not require a wax binder as does uniaxial pressing, thus eliminating dewaxing operations.

Low-cost elastomer tooling is used for isostatic pressing, but close tolerances can only be obtained for surfaces that are pressed against a highly accurate steel mandrel. Surfaces in contact with the elastomer tooling may require post machining when tight tolerances and good surface finishes are specified.

Wet-Bag CIP

Two types of CIP methods have evolved over the years: wet-bag and dry-bag. The so-called wet-bag method (Figure I.4) is used for producing mixed shapes. It is estimated that there are more than 3000 wet-bag presses in use worldwide today, ranging in size from 50 to 2000 mm in diameter.

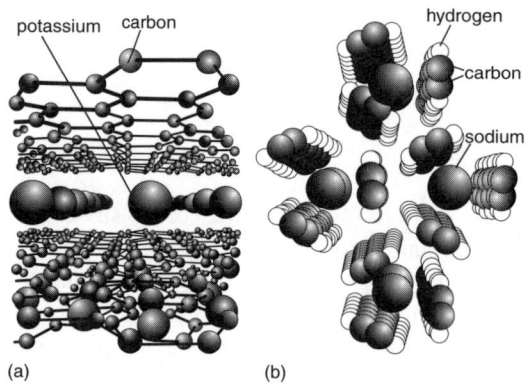

FIGURE I.3 Typical intercalation compounds. (a) Potassium in graphite, KC_8, a prototype layer intercalate, (b) sodium in polyacetylene, $[Na_{0.13}(CH)]_x$, where x denotes infinitely repeating polymer chains. (From *McGraw-Hill Encyclopedia of Science and Technology*, 8th ed., Vol. 9, McGraw-Hill, New York, 313. With permission.)

A typical cycle time for a production press ranges from 5 to 30 min, depending mainly on size, powder volumetric compaction ratio, and pump selected. This speed is rather slow but can be improved by higher-volume pumps, better vessel use, and improved loading mechanisms. A 5-min cycle calculates to 24,000 annual cycles based on a 1-shift, 8-h operation, or 240,000 cycles in 10 years. That should be the minimum design value of a vessel. Therefore, it is very important to select the proper vessel design to meet the end user's specified fatigue requirements as proven by the supplier's theoretical analysis and past performance.

Dry-Bag CIP

The dry-bag process (Figure I.5) is more applicable to producing same-shaped parts. Automated dry-bag isostatic pressing equipment was developed in the 1930s for compacting spark plug insulators, which are exclusively produced that way today for worldwide distribution.

The dry-bag method involves the same flex-bag technology for powder containment as the wet bag except that a stationary polyurethane "master bag," called the membrane, is inserted inside the pressure vessel. The pressure force (water) is transmitted through the membrane to the mold and then to the powder, thus keeping the mold dry. Generally one part is pressed at a time, and loading occurs from the bottom.

A minimum of three cassettes (each cassette typically consisting of the mandrel, mold, and powder) are in motion simultaneously; one

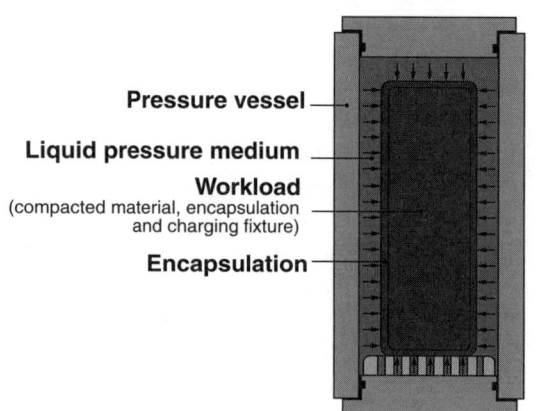

FIGURE I.4 Schematic of wet-bag CIP. (From *Ceram. Ind.*, March, p. 34, 1998. With permission.)

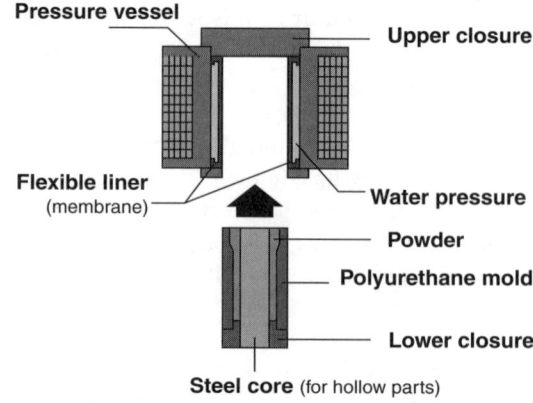

FIGURE I.5 Schematic of dry-bag CIP. (From *Ceram. Ind.*, March, p. 34, 1998. With permission.)

ISOSTATIC PRESSING

is pressed, one is filled with powder, and one is de-bagged. The cycle time is 1 min or less, which calculates to 120,000 cycles/year on a 1-shift, 8-h basis. This rate of cycling places a much higher demand on the pressure vessel fatigue, and proper design is critical to withstand the higher pressures. The dry-bag method is preferred over the wet-bag for automated production of same-size or same-shape parts in lots of about 50 parts per hour or above.

WARM ISOSTATIC PRESSING (WIP)

WIP follows the same path as CIP except the parts are compacted both at pressure and low temperature to 100°C. The pressing fluid water may be substituted with oil. To date, there are a few applications for manufacturers in the electronics industry as a cost-effective means of compacting different shaped parts.

Traditionally, a heated platen press has been applied in these applications. The problem associated with this method is the lack of uniform pressure all around the parts, resulting in dimensional variations from one side to the other. WIP, on the other hand, is a well-suited alternative for applying equal and uniform pressure on all surfaces.

HOT ISOSTATIC PRESSING (HIP)

HIP (Figure I.6) is a densification method for powders, compacts, or castings. It applies a gas pressure of 100 to 200 MPa and temperatures to 2200°C. An inert gas, most commonly argon, is used as the pressing fluid. The goal is to improve the performance of critical parts by eliminating defects and porosity resulting in fully dense compacts. Improvements in mechanical and physical properties, fatigue, surface finish, reliability, and/or rejection rate are possible using HIP.

Two HIP methods are used today for compacting parts: direct HIP, which applies to encapsulated powders, and post-HIP, which applies to pre-sintered compacts without interconnected porosity. (The HIP and CIP processes can also be combined, sometimes called CHIP. In CHIP, loose powder is cold-compacted, then sintered, then post-HIPed to achieve fully dense parts.)

Direct HIP involves a ceramic powder enclosed in a container usually made from an impermeable barrier, such as glass, ceramic, or refractory metal. Glass is the most common barrier today for direct compacting of ceramic powders and serves in two ways: (1) it acts as a barrier for consolidation and (2) it isolates and protects the powder from the processing gas.

Several precompacting methods are available for direct HIP, including injection molding, slip casting, CIP, or dry pressing. After preforming, the parts are glass encapsulated prior to HIP. A major goal is to form parts to near-net configurations to minimize final machining, which usually requires expensive diamond tooling.

Post-HIP is an alternative method and is a widely practiced way to HIP products such as oxide ceramics, tool bits, and ferrites. This method is only valid for materials that are able to sinter without pressure to close surface-connected porosity, equivalent to about 92 to 95% of theoretical density. The powder is shape-formed by either casting or CIP, is then pressureless-sintered, and finally post-HIP is employed for full densification. Post-HIP is also chosen when material decomposition is to be avoided; the post-HIP process allows the pressure fluid to contact the partially dense parts for reactive processing. For example, a mixture of 95% argon and 5% oxygen is preferred for oxide ceramics, whereas nitrogen gas is preferred for nitride ceramics.

The HIP process is rather slow, and a cycle may take 10 to 15 h, depending on part size, material, and furnace design. One of the reasons for the long cycle is that most installations use conventional furnaces that cool not only slowly

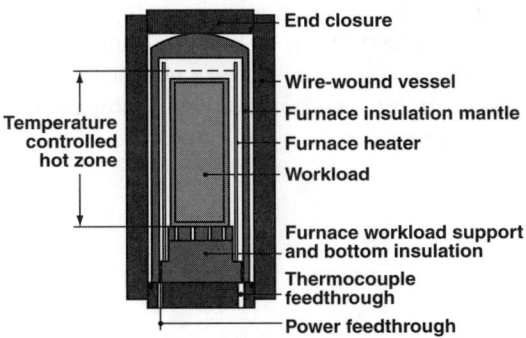

FIGURE I.6 Schematic of conventional CIP. (From *Ceram. Ind.*, March, p. 34, 1998. With permission.)

but also nonuniformly. Advanced furnaces are available with uniform rapid cooling (URC) provisions that have been used for metals since early 1990. Shorter cycle times improve throughput, thus reducing processing costs.

The rate of cooling is programmable from 1 to 50°C/min. A variable-speed fan is used to select the appropriate rate to avoid cracking when processing thermally sensitive ceramic parts. This feature can reduce HIP cycles to less than 8 h but adds only 10 to 15% to the equipment price.

Currently, the largest HIP unit for processing ceramics is 0.64 m in diameter and is rated at 1850°C.

Versatile Process

Isostatic processing has found wide use for many different materials used in a variety of applications. CIP is mainly a powder consolidation process using inexpensive molds as barriers for compacting simple to complex shapes to 60 to 80% densities. The selection of a "wet bag" vs. "dry bag" method depends on the type, mix, and production lots of parts produced.

Warm isostatic pressing has found a niche in certain industries where combined pressure and low temperatures to 100°C are specified.

The HIP process is gaining momentum in the engineered ceramics field for obtaining near-net-shape and fully dense ceramics for high-performance applications. Either direct HIP or post-HIP may be selected, depending on the material or process specified.

ISOTHERMAL FORGING

To meet the increasing forging demands of the aerospace turbine engine industry, a highly integrated, state-of-the-art isothermal (ISO) forging cell has been developed. The cell has the capacity to meet 100% of the current and projected industry need for superalloy, isothermally forged turbine engine components.

Equipment within the isothermal forging cell includes an 8000-ton clearing press, a 1204°C heat treatment furnace, and customized, computer-controlled ultrasonic inspection equipment. The cell can completely produce and process (forge, machine, heat treat, and inspect) isothermally forged engine components within one facility.

Advantages of ISO

The aircraft industry is increasing its use of nickel-based powder alloys. These alloys allow for the manufacture of higher-performing aircraft components that can withstand higher temperatures and greater stress than those manufactured from more traditional wrought alloys. Powder alloys must be extruded or consolidated into a billet, which is then forged and machined to its final net shape.

In addition to these qualitative advantages of utilizing powder alloys, isothermal forging offers a number of performance and production capacity advantages. Isothermal forging is a process in which the billet and the forging dies are heated to the same temperatures. This allows for a nearer-net-shape forging, thus reducing the amount of material necessary to produce the part as well as reducing machining requirements. A reduction in machining requirements reduces materials waste and improves throughput.

The use of an ISO cell has resulted in a significant increase in productivity where the average cycle time has been reduced by 30 to 40%, with machining times reduced by as much as 50%. Forging press setup time has been reduced by up to 50% and downtime is less than 5%. All total, these enhancements to productivity have resulted in significant unit cost reductions.

K

KAOLIN (CHINA CLAY)

The two terms are used interchangeably to describe a type of clay that fires to a white color. The name kaolin comes from the two Chinese words kao-ling, meaning high ridge, and was originally a local term used to describe the region from which the clay was obtained.

Kaolin ($Al_2O_3 2SiO_2 2\ H_2O$) usually contains less than 2% alkalies and smaller quantities of iron, lime, magnesia, and titanium. Because of its purity, kaolin has a high fusion point and is the most refractory of all clays. Lone kaolins are widely used in casting sanitaryware, ceramics, and refractories.

Georgia china clay is one of the most uniform kaolins to be found. Generally speaking, there are two types of Georgia-sourced kaolin, both of which are widely used for casting and other processes. One type imparts unusually high strength and plasticity, and is used for both casting and jiggering where a high degree of workability is required. The other type typically is a fractionated, controlled particle size clay that also behaves well in casting, dries uniformly, and reduces cracking of ware.

Calcined Kaolin

This commercial product is made from a specially prepared kaolin that is low in iron and alkalies. Analyses show the calcined product to be principally mullite ($3Al_2O_3 2SiO_2$) in association with an amorphous siliceous material. In fact, of the alumina present, 96% is converted to mullite. Iron content is not only low but in such a state of oxidation as to facilitate solid solution with the alumina.

Outstanding properties of refractories containing this calcined kaolin are high refractoriness and retention of shape under load; high resistance to corrosion by slags, glasses, and glaze or enamel frits; resistance to thermal shock; and high mechanical strength. It is being used in thermal shock bodies, refractories subjected to reducing atmospheres, kiln furniture compositions, thermal insulation bodies, low expansion bodies, permeable ceramic compositions, high-temperature castables, investment molds for precision casting, as a placing medium, as a kiln wash, as gripping sand for high-tension insulators, and in many other special refractory applications.

KNITTED AND WOVEN METALS

Although they are commonly referred to in same context and are used for some of the same applications, there is a considerable difference between knitted and woven metals. As their name implies, knitted metals are knitted into a mesh structure in much the same way as stockings or sweaters. The structure of woven wire, on the other hand, is usually simpler, consisting of interwoven strands of wire. Whereas weaving usually produces a symmetrical mesh, usually with square openings and parallel wires, knitting produces an asymmetrical mesh of interlocking loops.

Knitted Metals

An important advantage of knitting is that it produces a mesh of interlocking loops each of which acts as a small spring and provides resiliency. Because of this resiliency, knitted metals are usually able to withstand greater loads and deflections without being permanently deformed.

Knitted wire also has both a large surface area and a high percentage of free space. Thus, knitting permits construction of a mesh using wire with a maximum surface area and with interstices of almost any desirable size, regardless of wire diameter. Fine wire, for example, has been knitted into a mesh with as few as three to five openings to the inch.

The free volume of knitted wire can be controlled between 50 and 98%, depending on the interstice size, and regardless of the wire size used. Even when the wires are widely spaced to produce a potential volume of 98%, the structure retains its shape; a similar spacing with woven wire would result in a shape that would be almost impossible to handle.

Another important property of knitted metals is their ability to maintain their dimensions during temperature cycling. When the meshes are slightly stretched in every direction and expansion occurs, the sides of the loops are merely forced closer together without changing overall dimensions or the plane of the surface.

Knitted metals can be produced in wire diameters of 0.01 to 0.6 mm in such materials as steel, copper, brass, aluminum, stainless steel, and various nickel alloys including Monel. In fact, almost any metal that can be drawn into wire can be knitted. Thus, knitted parts can be made in a wide range of strength, corrosion-resistance, wear-resistance, electrical-shielding, and heat-resistance properties.

The asymmetrical mesh of knitted parts is advantageous for electronic shielding because the continuous-loop structure apparently causes induced currents to cancel themselves. Apart from this application, however, the biggest use for knitted wire is to remove one material from another. Thus, it can be used to separate two phases of the same material, to separate two immiscible liquids, or to separate a solid from a liquid or a gas. The degree of separation can be controlled by varying the compression used to form or shape the knitted part, and by controlling the size and shape of the wire and the size of the loops.

A good example of this kind of application is a mist eliminator in which the knitted mesh separates the liquid phase from the gas phase. Gases bearing droplets from such processes as distillation, evaporation, scrubbing, cleaning, or absorption are passed through built-up layers of knitted fabrics up to 152 mm thick. The droplets collect on the loops and slowly run to the bottom where they accumulate. Liquid particles as small as 5 to 20 µm can be separated with efficiencies as high as 98 to 100%. Although small eliminators are usually made in one piece, eliminators can be made up to 8.4 m in diameter by building up layers of crimped mesh.

Knitted metals are also used in many other applications where their special structure and properties are useful. Fuel line filters, for example, of knitted wire are resilient and do not require precise machining for sidewall fit. Gaskets of knitted wire have excellent conductivity and sufficient resiliency to produce tight joints on uneven surface, thereby preventing radio-frequency leakage. Heat dissipation sleeves of knitted metal for subminiature glass tube envelopes provide high cooling efficiency. Knitted metals also provide good shock and vibration control when used as mountings for airborne and industrial equipment.

Woven Metals

Woven wire cloth is used in a wide range of applications for grading materials, filtering, straining, washing, guarding, reinforcing, and decoration. A variety of meshes in different materials and sizes is available to meet these applications.

Like knitted wire, woven wire cloth can be produced in almost any metal that can be drawn into wire, including carbon and stainless steel, copper, brass, bronze, nickel, Monel, Inconel, and aluminum. Plain steel wire cloth is one of the most economical and generally used materials. However, it may require a protective coating to prevent rusting.

Corrosion can be prevented by using tinned or galvanized wire; tinned wire is preferred for handling food products, galvanized for all other applications. Where required, the cloth can also be provided with a protective electroplate of cadmium, chromium, or tin. Phosphate coatings and paints can also be used.

Where severe corrosive conditions are encountered, the weave can be made from such materials as stainless steel, monel, phosphor bronze, and silicon bronze; some of these materials are available only in limited sizes.

Abrasive resistant steels are available for applications where abrasive materials have to be handled. Many grades of stainless can also be used to withstand high temperatures, e.g., 347 stainless (up to 760°C), 309 and 310 (up to 871°C), and 310 stainless as well as Inconel and Nichrome (up to 1003°C).

Woven wire cloth is widely used for filtering and straining, particularly in automotive and aviation applications for carburetors, air screens, and oil and fuel strainers. They are also used to grade materials, and for wire baskets, insect screening, and safety guards.

L

LAMINATED PLASTICS

Laminated plastics are a special form of polymer–matrix composite (PMC) consisting of layers of reinforcing materials that have been impregnated with thermosetting resins, bonded together, and cured under heat and pressure. The cured laminates, called high-pressure laminates, are produced in more than 70 standard grades.

Laminated plastics are available in sheet, tube, and rod shapes that are cut or machined for various end uses. The same base materials are also used in molded-laminated and molded-macerated parts. The molded-laminated method is used to produce shapes that would be uneconomical to machine from flat laminates, where production quantities are sufficient to warrant mold costs.

Strength of a molded shape is higher than that of a machined shape because the reinforcing plies are not cut, as they are in a machined part. The molded-macerated method is used for similar parts that require uniform strength properties in all directions.

Other common forms of laminated plastics are composite sheet laminates that incorporate a third material bonded to one or both surfaces of the laminate. Metals most often used in composites are copper, aluminum, nickel, and steel. Copper-clad sheets (one or both sides) for printed-circuit and multilayer boards comprise the largest volume of metal composite sheet laminates. Nonmetallics include elastomers; vulcanized fiber, and cork. Composite metal/plastic materials are also produced in rods and tubes.

Vulcanized fiber is another product often classified with the laminated plastics because end uses are similar. Vulcanized fiber is made from regenerated cotton cellulose and paper, processed to form a dense material (usually in sheet form) that retains the fibrous structure. The material is tough and has good resistance to abrasion, flame, and impact.

Phenolics are the most widely used resin in laminated plastics. These low-cost resins have good mechanical and electrical properties and resistance to heat, flame, moisture, mild acids, and alkalies. Most paper- and cloth-reinforced laminates are made with phenolics.

LANTHANUM

A chemical element, lanthanum, symbol La, the second most abundant element in the rare earth group, is a metal. The naturally occurring element is made up of the isotopes and is one of the radioactive products of the fission of uranium, thorium, or plutonium. Lanthanum is the most basic of the rare earths and can be separated rapidly from other members of the rare earth series by fractional crystallization. Considerable quantities of it are separated commercially, because it is an important ingredient in glass manufacture. Lanthanum imparts a high refractive index to the glass and is used in the manufacture of expensive lenses. The metal is readily attacked in air and is rapidly converted to a white powder.

Lanthanum becomes a superconductor below about –267°C in both the hexagonal and face-centered crystal forms.

Lanthanum Oxide

Lanthanum oxide (La_2O_3) has a melting point of 2250°C, is soluble in acids, and very slightly soluble in water. This oxide of a rare earth element occurs in monazite and bastnasite. It is marketed as the oxide or as other salts such as the oxalate, nitrate, or hydrate. It quickly absorbs water and CO_2 from the atmosphere.

Its chief use is as an ingredient in nonsilica, rare element optical glass with oxides of tungsten, tantalum, and thorium. Lanthanum increases refractive index, decreases dispersion, and is also used in x-ray image intensifying screens that speed up x-ray exposure as much as two to ten times so that diagnostic dosages may be reduced by as much as 80% with fewer retakes. It is also used in barium titanate capacitors.

LANXIDES

A composite formed in the reaction between a molten metal and oxygen in air to some other vapor-phase oxidant. Normally, such a situation produces an unwelcome scum on the surface of the metal. However, by controlling the temperature of the molten metal and by adding traces of suitable dopant metals, an inch/cm-thick layer of a metal oxide composite can be grown on the surface of the liquid. Composites up to 10 cm thick and weighing up to 18 kg have been grown by this method, with no falling off in growth rate. Under the right conditions, a lanxide composite is considerably stronger than sintered alumina. This makes such a composite potentially useful for armor plating, rocket or jet engines, and other applications.

LASER

Laser Alloying

This is a material-processing method that utilizes the high power density available from focused laser sources to melt metal coatings and a portion of the underlying substrate. Since the melting occurs in a very short time and only at the surface, the bulk of the material remains cool, thus serving as an intimate heat sink. Large temperature gradients exist across the boundary between the melted surface region and the underlying solid substrate. The result is rapid self-quenching and resolidification.

Transitions

The sequence of schematic cross sections in Figure L.1 illustrates the transitions occurring during and following an individual laser exposure. In Figure L.1a, the metal substrate (B) coated with a thin metal film (A) is irradiated with a laser pulse. For metal surfaces and most laser wavelengths, a significant fraction of the incident light will be specularly or diffusely scattered away as shown in the figure. The absorbed energy is "instantaneously" (10^{-12} s) transferred to the lattice. The near-surface region very rapidly reaches the melting point, and a liquid–solid interface starts to move through the film (Figure L.1b). In Figure L.1c, the liquid–solid interface has swept through the original thin-film–substrate interface. Interdiffusion of the film and substrate elements starts. The laser pulse is nearly terminated, and the surface has remained below the vaporization temperature. In Figure L.1d, the maximum melt depth has been reached, and interdiffusion continues. The resolidification interface velocity is momentarily zero and then rapidly increases. In Figure L.1e, the resolidification interface has moved approximately halfway back to the surface from the melt depth. Interdiffusion in the liquid continues, but the resolidified metal behind the liquid–solid interface cools so rapidly that solid-state diffusion may be neglected. In Figure L.1f, the material is completely resolidified, and a "surface alloy" of A in B has been produced.

What makes laser surface alloying both attractive and interesting is the wide variety of chemical and microstructural states that can be retained because of the rapid quench from the liquid phase. These include chemical profiles where the "alloyed" element A is highly concentrated near the atomic surface and decreases in concentration over shallow depths (hundreds of nanometers), and uniform profiles where the concentration of A in B is the same throughout the entire melted region. The types of microstructures observed include extended solid solutions (the concentration of A in B greatly exceeds equilibrium values), metastable crystalline phases (high-temperature phases retained because of the rapid return to room temperature), and metallic glasses.

For a pulsed or Q-switched laser (high-power, short-pulse laser) source, there also is an optical and effective spot size, and the "pulse length" is the time of exposure. The laser pulses are emitted in a train of pulses characterized by a repetition rate. For these lasers, the effective

LASER

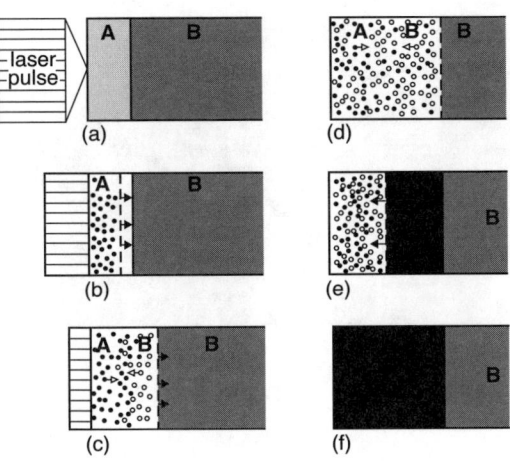

Key:
- • film elements
- ○ substrate elements
- → movement of interface
- ⇢ movement of film/substrate elements

Sequence of schematic cross sections for laser alloying, with time increasing from *a* to *f*. A_xB_{1-x} represents surface alloy (with composition fixed by x) of film elements A in substrate elements B.

FIGURE L.1 Laser alloying. (From *McGraw-Hill Encyclopedia of Science and Technology*, 8th ed., Vol. 9, McGraw-Hill, New York, 659. With permission.)

spot size, degree of overlap, and repetition rate determine the area per unit time that can be produced.

For all the laser sources (continuous-wave pulsed and Q-switched), the exposure time (dwell time or pulse length) strongly influences the depth that will be melted. Longer exposure times result in deeper melting. Because deeper melting means a longer total time in the molten state, that means more time available for diffusion of the one or more alloying elements into the molten portion of the substrate. Deeper melting and longer melt times result in more dilute surface alloys, whereas shallow melting and shorter melt times result in more concentrated surface alloys. It is also evident that in some instances convection, surface tension, and plasma effects can enhance the mixing within the liquid state and drive the melt toward homogenization.

In making laser alloys, many other processing variables need to be considered. In addition to the exposure time just discussed, these include the laser power, the thickness of the film put down prior to laser melting, and in some instances the nature of the gaseous ambient during the laser processing. The processing variables are interrelated, and one variable cannot be freely changed without affecting another. Another consideration is that laser alloying is a liquid state–rapid quenching phenomenon. The near-surface region must be melted and yet vaporization avoided. Different minimum and maximum energy densities are thus defined for each laser exposure time. In addition to these processing constraints, there are certain properties of matter that strongly influence whether or not certain element combinations may be laser-alloyed. For example, it is not possible to laser-alloy a low-melting-point/high-vapor-pressure element like zinc into a high-melting-point/low-vapor-pressure metal substrate such as tungsten. The zinc would vaporize before the underlying tungsten could be melted. Because liquid-state intermixing is required, suitable systems must exhibit miscibility in the molten state. Binary systems like silver–nickel (Ag–Ni), iron–lead (Fe–Pb), copper–molybdenum (Cu–Mo), and aluminum–bismuth (Al–Bi) have miscibility gaps in the liquid state spanning nearly all compositions. Such systems cannot be laser-alloyed.

ADVANTAGES

Surface alloys, particularly laser surface alloys, have advantages that include tailoring surface properties, conservation of materials, and creation of new metal surfaces.

Surface alloying allows alteration of the metal surface to achieve characteristics best suited to the service environment. For example, in an ordinary saw blade the material requirements for the cutting teeth are very different from those of the length of the blade. In many applications requiring corrosion or oxidation resistance, that resistance is needed only on the external surface of the material. It is possible to design the bulk of the material with characteristics most suitable for structural needs or ease of fabrication and to tailor the metal surface for interface requirements. Material conservation is another reason often cited for considering surface alloying. In the case of stainless steels and superalloys, the elements added to impart the special properties are often strategic elements for which there are sparse domestic supplies. For example, chromium and nickel may not be necessary on the inside of knives, forks, and spoons. Further, precious metals are expensive and in relatively short supply. Thus surface alloying can be very cost-effective.

These first two advantages — tailoring for surface properties and material conservation — are generally important considerations in almost any coating technology. However, laser alloying differs from coating technologies in that the near-surface region is a continuous extension of the interior of the metal. There is no interface between bulk and "coating": the laser alloy is a mixture of bulk and surface elements. Also, problems in regard to porosity and adherence do not exist.

Laser alloying involves very large temperature gradients and quenching from the liquid state. In this way it resembles other rapid-solidification technologies.

It is anticipated that laser alloying will produce many new alloys that could not have been produced by conventional methods.

Laser Cooling

Reducing the thermal motion of atoms with the force exerted by a laser beam is laser cooling. Typically, such cooling is used to reduce the temperature of a gas of atoms, or the velocity spread of atoms in an atomic beam.

The keys to using such repeated kicks to reduce the random, thermal motion of a gas of atoms are the monochromatic nature of laser light, the selectivity of absorption of light by atoms, and the Doppler effect.

The viscosity of laser cooling does not stop atoms. Although this might seem to be a fundamental limit, techniques have been developed that go even farther, by arranging that when, by chance, an atom achieves a very low velocity it stops absorbing light and remains cold. In this way, atoms have been cooled below the recoil limit.

Applications

Improving atomic clocks, where the thermal motion of atoms reduces the precision and accuracy, was a major motivation to developing

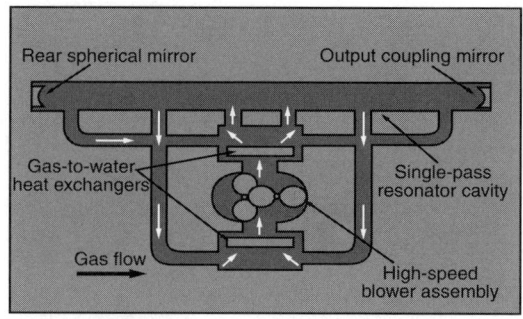

FIGURE L.2 Fast-axial-flow CO_2 laser. (From *Welding Design Fabr.*, December, p. 21, 1998. With permission.)

laser cooling. Clocks using laser-cooled trapped ions or free neutral atoms rival the performance of the best conventional atomic clocks. Laser cooling is also used in atom optics, where well-collimated, monoenergetic atomic beams are more easily and effectively manipulated. In addition, laser cooling has been used to study collisions between very slow atoms.

Laser Cutting

Types of Lasers for Metal Cutting

The continuous-wave (CW) CO_2 laser leads the way for metal cutting. The CW laser beam builds up in small-diameter gas-containing tubes with electrodes at each end. High-voltage direct current flows between electrodes, exciting gas atoms to develop the lasing effect.

The workhorse for metal cutting is the fast-axial-flow (FAF) CO_2 laser. The majority of these units are DC design, an economical laser excitation technique for cutting thin-gauge materials. The FAF laser produces a stable beam for high-quality cuts. Fast-flow technology moves gas rapidly through the discharge area to operate between 500 W and 6 kW. The fast flow efficiently cools the laser gas, making more power available in more compact units (Figure L.2).

In the transverse flow (TF) CO_2 laser, gas flows in a transverse direction to the laser cavity axis. The laser gas circulates at high speed and cools through a heat exchanger. This design results in an output power in the 5- to 45-kW range. The TF laser produces a beam with a mixed mode for high-speed quality cuts. The

LASER

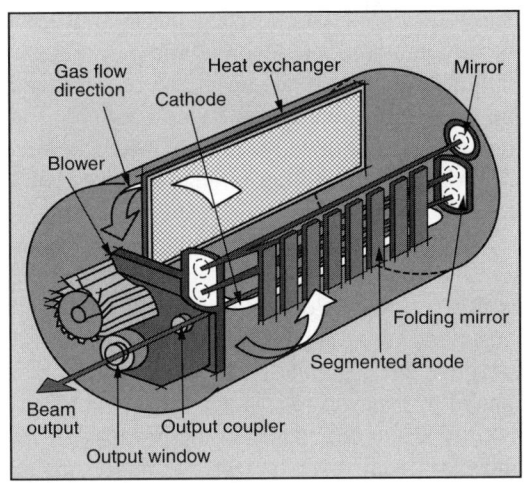

FIGURE L.3 Transverse-flow CO_2 laser. (From *Welding Design Fabr.*, December, p. 21, 1998. With permission.)

compact TF design encloses the laser in the workstation (Figure L.3).

A solid-state laser describes an optically pure material that will lase when doped with specific elements. Neodymium-doped yttrium aluminum garnet (Nd:YAG) rates as the most popular solid-state laser for industrial use in either rod- or rectangular-shaped design.

Types of Cutting Machines

Flying-optics machines are systems that move the beam along the x–y axes and the workpiece is stationary. In the gantry design, a stationary workpiece improves positioning tolerances, especially for heavier workpieces. The space-efficient design and the unobstructed beam motion allow easy integration into production lines. Beam-motion accuracy holds constant over the entire working volume at 0.010 mm with a repeatability of 0.05 mm. Traverse rates can double other machine configurations.

The stationary head, or fixed-beam design, a widely used metal-cutting machine, integrates x and y workpiece motion under a fixed-position laser beam. The beam design requires a minimum number of mirrors to bend the beam and maintain beam characteristics and quality over the travel distance.

The hybrid machine moves the workpiece in the x axis and the optics move in the y axis. To maintain beam accuracy, hybrid units offer shorter beam-delivery paths than full moving-beam systems. Moving the workpiece in only one axis permits more-accurate movement of heavier workpieces than with flying optics and holds position accuracy to ±0.010 mm and repeatability to ±0.05 mm.

The advantages of cutting using robot-beam manipulation, the classic articulated arm with cylindrical or spherical coordinate designs, are obvious. Robots with internal optics for CO_2 laser cutting can cut very complex parts, but require good mechanics because the many joints and mirrors make alignment difficult. Fiber optics used to convey an Nd:YAG beam align more easily, but the minimum bending radius of the fiber can limit mobility. Units can hold 0.05 mm point-to-point repeatability, with dynamic path repeatability of 1 mm at 0.86 mm/s.

LASER INSPECTION

Laser inspection involves an inspection system in which lasers check that the soldering on printed circuit boards has been developed. The 3 Dimensional Solder Shape Inspection Machine detects faulty soldering by the Non-contact 3 Dimensional Shape Measuring System. The system is based on laser measuring technology, and is said to provide accuracy ten times higher than conventional technology in judging faulty soldering. The system reportedly also reduces the time required for pre-operational arrangement to one tenth that of normal.

The system utilizes a thin laser light 5 μm thick, which scans the outline of solder paths on the printed board and compares the shape with a standard. Together with a high-speed camera, the system pictures the solder outline visualized through laser reflection from the solder surface, then processes the picture at a speed of 0.016 s/frame. The system then makes quantitative evaluations based on measured data about the size and the adhesion angle of the solder and the component height.

LASER LITHOGRAPHY

A line narrowed by 193-nm laser radiation has been demonstrated with 21-W output power. This is reportedly the highest ever demonstrated from a single-stage line narrowed argon–fluorine excimer laser oscillator.

The laser was operated with 1-kHz repetition rate and yielded more than 20 mJ/pulse with a bandwidth of less than 0.6 pm, FWHM. The achievement is a result of ongoing studies to improve the efficiency of the line narrowing resonator and laser tube design.

The argon–fluorine excimer lasers will likely be used for the next generation of deep ultraviolet lithography tools for semiconductor production of 1-gigabit memory chips and advanced microprocessors. Development efforts are under way to transfer the new technology into products suitable for 193-nm lithography production, particularly due to the reduced costs operation achievable.

Laser Peening

A laser peening method is being perfected that could significantly improve the durability of critical jet engine components. Although the laser peening concept is not new, a neodymium-doped glass laser with 600 W of average power makes the process faster and economically feasible. The laser is capable of firing 10 pulses/s, compared with only 1 pulse every 2 s from commercial lasers. Laser peening, which uses concentrated light to generate a small, sudden force equivalent to about 30,000 times the pressure of the atmosphere, can leave a compressive residual stress in engine fan blades, disks, rotors, and shafts up to 1 mm deep. That is four times deeper than conventional shot peening. The compression significantly retards metal fatigue and corrosion, and improves damage tolerance.

LAVA

Lava is molten rock material that reaches the Earth's surface through volcanic vents and fissures; also, the igneous rock formed by consolidation of such molten material. Relatively rapid cooling at the Earth's surface may transform fluid lava into a dense-textured volcanic rock composed of tiny crystals or glass or both.

The temperature of liquid lava ranges widely but generally does not exceed 1200°C. Basaltic lavas are usually hotter than rhyolitic ones. The viscosity of lava depends largely upon the temperature, composition, and gas content.

As lava cools, it becomes more viscous and the rate of flow decreases. Rapid cooling, as at the surface of a flow, promotes the formation of glass. Slower cooling as near the center of a flow, favors the growth of crystals.

During many volcanic eruptions the lava is so rapidly ejected that it is blown to bits by the explosive force of expanding gases. The small masses rapidly congeal and settle to the Earth to form thick blankets of volcanic tuff and related pyroclastic rock. Lava flows and volcanic tuffs cover large areas of the Earth's surface and may form more or less alternating layers totaling many thousands of feet in thickness.

Lava is also a name given to ceramic material used for molding gas-burner tips, electrical insulating parts, nozzles, and handles. It may be calcined talc, steatite, or other material. It is molded from magnesium oxide, and it is hardened by heat treatment after shaping and cutting. It is baked at 1093°C. The compressive strength is from 138 to 207 MPa. It will resist moisture and has high dielectric strength. Rods as small as 0.05 cm in diameter can be made. Alsimag is the trade name of lava that is produced from ground talc and sodium silicate.

LEAD AND ALLOYS

A soft, heavy, bluish-gray metal (symbol Pb), lead is obtained chiefly from the mineral galena. It surface-oxidizes easily, but is then very resistant to corrosion. It is soluble in nitric acid but not in sulfuric or hydrochloric, and is one of the most stable of the metals. Its crystal structure is face-centered cubic. It is very malleable, but it becomes hard and brittle on repeated melting because of the formation of oxides. The specific gravity of the cast metal is 11.34, and that of the rolled is 11.37. The melting point is 327°C, and boiling point is 1750°C. The tensile strength is low, that of the rolled metal being about 25 MPa with elongation of 52% at normal temperatures, but at low temperatures the strength is greatly increased. At –40°C it is about 89 MPa, with elongation of 30%. The coefficient of expansion is 0.0000183, and the thermal conductivity is 8.2% that of silver. The electric conductivity is only 7.8% that of

LEAD AND ALLOYS

copper. When used in storage batteries, the metal is largely returned as scrap after a period and is remelted and marketed as secondary lead, as is also that from pipes and cable coverings. Lead is highly toxic and, thus, poses a health hazard. Inhalation of dust and fumes should be avoided and it should not be used in contact with food or drink products.

Not only is lead the most impervious of all common metals to x-rays and gamma radiation, it also resists attack by many corrosive chemicals, most types of soil, and marine and industrial environments. Although lead is one of the heaviest metals, only a few applications are based primarily on its high density. The main reasons for using lead often include low melting temperature, ease of casting and forming, good sound and vibration absorption, and ease of salvaging from scrap.

With its high internal damping characteristics, lead is one of the most efficient sound attenuators for industrial, commercial, and residential applications. Sheet lead, lead-loaded vinyls, lead composites, and lead-containing laminates are used to reduce machinery noise. Lead sheet with asbestos or rubber sandwich pads are commonly used in vibration control.

The natural lubricity and wear resistance of lead make the metal suitable, in alloys, for heavy-duty bearing applications such as railroad-car journal bearings and piston-engine crank bearings. Lead is also widely used as a constituent in solders. Most common solders are the lead–tin alloys; melting temperature can be as low as 183°C.

FORMS

Sheet and Foil

Because of its malleability, lead and its alloys are readily rolled to any desired thickness down to 0.01 mm. Sheets are easily fabricated by burning or soldering. Standard widths run to 2.4 m or more for sheet and sheets may be cut to any desired size. Blanks for impact extrusion, gaskets, washers, or other purposes may be stamped out. Tin-coated lead can be produced by rolling lead and tin together.

Extrusions

Lead is easily extruded in the form of pipe, rod, wire, or any desired cross section like window cames (H-shaped), rounds, hollow stars, rectangular duct. Commercially available extrusions range in size from 612 mm pipe down to solder wire 0.25 mm in diameter. Lead is extruded over paper, rubber, or plastic in making electrical cable and around steel bars. Common flux-cored solder is a lead extrusion; toothpaste tubes are impact extrusions.

Castings

One of the simplest metals to cast, lead is used in tiny die castings and massive cast counterweights. Type metal, renowned for its ability to reproduce minute detail, is a lead alloy. Lead grids for most batteries are die-cast. Casting temperature (usually about 316°C) is moderate. Arsenic, antimony, or tin are frequently alloyed to impart strength or special properties. Small die castings can have wall thicknesses as low as 1.3 mm and "as-cast" dimensions are reproducible to 0.03 mm.

Coatings

Protection of underlying iron and steel is the main objective in most lead coating. In the purely protective class one finds terne-plate for roofing, fireproof frames and doors, automotive parts, and containers for paint and oil. Lubricity imparted by the coating cases drawing and stamping operations and produces an excellent surface for soldering — hence, television chassis and automotive gas tanks are made of terne. Other hot-dip processes as well as electroplating and flame spraying are also used for outdoor hardware automotive mufflers, bearings, bushings, nuts, bolts, and for maintenance as well.

Laminations

Developed originally for x-ray protection, a large family of laminated lead materials now exists. In addition to their original niche these are finding increasing use in sound isolation and noise control. Typical examples include lead–plywood, lead–gypsum board, lead–cinder block, leaded plastic–fabric laminates, leaded plastic, and glass–fiber combinations.

Cladding

Metallic lead in thicknesses from 3.2 to 305 mm or more may be bonded to other metals. Thus, for example, lead and steel may be combined for corrosion resistance and strength or lead and copper for gamma shielding and heat transfer. In many instances a product such as a tank or chemical reactor is completely or partially fabricated in steel and then clad with bonded lead as a unit.

Powder

Spheres, irregular grains, and flakes of lead from 4 µm diameter up find use in special greases, as a constituent of bearings, brake, and clutch facings, in filling plastics and rubber, and in paints and pile-joint compounds. Wire rope is usually treated with such a powder to lubricate it and to fill any nicks in the filter, thus renewing it with self-lubricating surfaces.

Shot

This form of lead is produced in abundance — about $30,360 \times 10^3$ kg go into shotgun ammunition each year. Ammunition sizes range from 1 to 11.3 mm; small shot is made for other uses. Easily handled, it is a preferred form when mass or shielding is required inside an irregular enclosure. It is also used in making free-machining steels.

Wool

By passing molten lead through a fine sieve and allowing it to solidify in the air, a loose rope of filters is produced. Under pressure, usually by being driven into a crevice with a calking iron and hammer, the fibers weld into a homogeneous mass. This permits the forming of a solid metal seal where temperature or explosion hazards prohibit jointing procedures requiring heat. Continuous lead fiber is also produced by being spun on textile machines.

Alloys

In its unalloyed form as 99.85% minimum, lead is soft and weak; it requires support for mechanical applications. This "chemical lead" is used primarily in corrosive chemical-handling applications such as tank linings.

"Hard lead" — lead alloyed with 1 to 13% antimony — has sufficient tensile strength, fatigue resistance, and hardness for many mechanical applications. These alloys can be cast, rolled, or extruded and are especially suited for castings requiring good detail and moderate strength. Rolled antimonial alloys are harder and stronger than the cast alloys. Battery-plate lead contains 7 to 12% antimony.

Calcium (0.03 to 0.12%) forms another series of mechanically suitable alloys with lead. These alloys age-harden naturally at room temperature — usually for 30 to 60 days — after being cast or worked. Properties of wrought Pb–Ca alloys are somewhat directional, being greater in the longitudinal direction. Uses include cable sheathing and grids in storage batteries.

Tin, added to Pb–Ca alloys in amounts to about 1.5%, raises tensile strength and stress-rupture resistance but increases aging time to 180 days. Tin is also used to reduce the coefficient of friction for bearing applications. Higher-tin-bearing alloys are primarily used in solders, which normally contain from 40 to 60% tin.

Lead alloys may exhibit greatly improved mechanical or chemical properties as compared to pure lead. The major alloying additions to lead are antimony and tin. The solubilities of most other elements in lead are small, but even fractional weight percent additions of some of these elements, notably copper and arsenic, can alter properties appreciably.

Cable-Sheathing Alloys

Lead is used as a sheath over the electrical components to protect power and telephone cable from moisture. Alloys containing 1% antimony are used for telephone cable, and lead-arsenical alloys, containing 0.15% arsenic, 0.1% tin, and 0.1% bismuth, for example, are used for power cable. Aluminum and plastic cable sheathing have replaced lead alloy sheathing in many applications.

Battery-Grid Alloys

Lead alloy grids are used in the lead-acid storage battery (the type used in automobiles) to support the active material composing the plates. Lead grid alloys contain 6 to 12% antimony for strength, small amounts of tin to improve castability, and one or more other minor additions to retard dimensional change in service. No lead alloys capable of replacing the lead–antimony alloys in automobile batteries have been developed. An alloy containing 0.03% calcium for use in large stationary batteries has had success.

Chemical-Resistant Alloys

Lead alloys are used extensively in many applications requiring resistance to water, atmosphere, or chemical corrosion. They are noted for their resistance to attack by sulfuric acid. Alloys most commonly used contain 0.06% copper, or 1 to 12% antimony, where greater strength is needed. The presence of antimony lowers corrosion resistance to some degree.

Type Metals

Type metals contain 2.5 to 12% tin and 2.5 to 25% antimony. Antimony increases hardness and reduces shrinkage during solidification. Tin improves fluidity and reproduction of detail. Both elements lower the melting temperature of the alloy. Common type metals melt at 238 to 246°C.

Bearing Metals

Lead bearing metals (babbitt metals) contain 10 to 15% antimony, 5 to 10% tin, and for some applications small amounts of arsenic or copper. Tin and antimony combine to form a compound that provides wear resistance. These alloys find frequent application in cast sleeve bearings, and are used extensively in freight-car journal bearings. In some cast bearing bronzes, the lead content may exceed 25%.

Solders

A large number of lead-base solder compositions have been developed. Most contain large amounts of tin with selected minor additions to provide specific benefits, such as improved wetting characteristics.

Free-Machining Brasses, Bronzes, Steels

Lead is added in amounts from 1 to 25% to brasses and bronzes to improve machining characteristics. Lead remains as discrete particles in these alloys. It is also added to some construction steel products to increase machinability. Only about 0.1% is needed, but the tonnage involved is so large that this forms an important use for lead.

Uses

Lead wool is lead in a shredded form used for calking. Sheet lead is produced by cold-rolling, and is used as a sound barrier in building construction.

Lead has a high capacity for the capture of neutrons and gamma rays and is used for radiation shielding in the form of sheet lead or as metal powder in ceramic mortars and blocks, paints, and in plastic composite structures. DS Lead is a dispersion-strengthened lead containing up to 1.5% lead monoxide evenly distributed through the structure. The oxide combines chemically with the lead, doubling the strength and stiffness of the metal, but increasing its brittleness. It is used for chemical piping and fittings. A neoprene–lead fabric is a neoprene fabric impregnated with lead powder. It has a radiation shielding capacity one-third that of solid lead sheet. It comes in thicknesses of 0.08 to 0.64 cm, and its flexibility makes it suitable for protective clothing and curtains. Shielding cements for x-ray and nuclear installation shielding are metallic mortars containing a high percentage of lead powder with ceramic oxides as binders and other elements for selective shielding. They are mixed with water to form plasters or for casting into sections and blocks. The formulation varies with the intended use for capture, attenuation, or dissipation of neutrons, gamma rays, and other radiation. Shielding paints are blended in the same manner.

Battery-plate lead for the grid plates of storage batteries and Silvium alloy for positive-plate grids are additional applications. Lead-coated copper used for roofing and for acid-resistant tanks as well as frangible bullets, which shatter on striking a target surface and are used for aerial gunner practice, are two other uses.

Antimonial lead is an alloy containing up to 25% antimony with the balance lead, used for storage-battery plates, type metal, bullets, tank linings, pipes, cable coverings, bearing metals, roofing, collapsible tubes, toys, and small cast articles. The alloy is also known as hard lead.

One hard lead has 10% antimony and 90% lead and melts at 252°C. Called cable lead, or sheathing lead, it is used to cover telephone and power cables to protect against moisture and mechanical injury. Terne-plate, as the coated steel is called, is widely used for automobile gasoline tanks and also has been used for roofing on buildings.

LEAD OXIDE

Two types of lead oxide are used in ceramics: litharge, or lead monoxide, and red lead.

- *Litharge:* Lead monoxide (PbO) has a specific gravity of 9.3 to 9.7; a melting point of 888°C. It is insoluble in water but soluble in alkalies, certain acids, and some chloride solutions.
- *Red lead:* Lead oxide (Pb_3O_4) has a specific gravity of 9.0 to 9.2; it decomposes between 500 and 530°C. It is insoluble in water and is decomposed in some acids, leaving insoluble lead peroxide, PbO_2.

Lead oxide is used quite extensively in optical glass, electrical glass, and tableware. It increases the density and refractive index of glass. In addition, it can be cut more easily than other glasses and has superior brilliance, both of which make it good for cut glass.

Lead glasses may be formulated with a wide variety of electrical and acid-resisting characteristics; desirable properties, such as weather resistance, electrical resistivity, etc., will depend upon the total composition of the glass.

Lead has many advantages as a glaze ingredient. The superiority of lead glazes lies in their brilliance, luster, and smoothness, which are due to their lower fusion point and viscosity. Lead glazes are, in general, highly resistant to water solubility and chipping, and have few faults in texture and bond. They have high mobility, refractivity, and elasticity and are softer than leadless glazes.

All investigators agree that the use of fritted glazes, in which all of the lead is fritted, has important health advantages. In this way the raw lead oxide is converted into relatively harmless lead silicates, which are much less soluble in dilute acids or gastric juices. Lead silicates are more slowly absorbed after entering the respiratory system, and can be eliminated with less lead absorption.

Lead oxide also is used in enamels. With an increase in lead and a corresponding decrease in potash, with flint constant, enamels become more fusible, have less tendency to craze, and become more refractive, but are less durable in acid fumes.

LEAD ZIRCONATE TITANATE

Lead zirconate titanate, $Pb(Zr_{0.4}Ti_{0.3})O_3$ to $Pb(Zr_{0.9}Ti_{0.1})O_3$, is also known as PZT ceramics. It is used mainly in the manufacture of piezoelectric ceramic elements.

The wide range of possible compositions provides a wide range of dielectric constant values, piezoelectric activity, and primary transition temperatures. These are normally greater than is possible with barium titanate. Small additions of niobium, strontium, barium, and antimony serve as modifiers.

Two techniques for producing PZT powders are used today. Calcining (CMO) is the more common method of powder production, and the calcine is then formed and fired in its final shape.

PZT is the most widely used polycrystalline piezoelectric material. Its electrical output can measure pressure. It is used in hydrophones, which permit listening to sound transmitted through water.

One variation of the PZT ceramics is PLZT ceramics (lead–lanthanum zirconate titanate). They are a range of ferroelectric, optically active, transparent ceramics based on the $PbZrO_2TiO_2$ system. When hot pressed, the material can reach 99.8% theoretical density. For device applications of PLZT ceramics, the ratio that is important is x:65:35, where x is lanthanum (8 to 10%) and 65:35 is the $PbZrO_3$:$PbTiO_3$ ratio. They have basically the same characteristics as PZT ceramics and are frequently finding use as filters, oscillators, and vibrators in many areas. They are also used for optical shutters.

LEATHER

Leather is made from the hides or skins of animals, birds, reptiles, and fish. Two main steps are involved in this process. First, the hides or skins are cured or dressed to prepare them for tanning. This curing process removes all the flesh, hair, and foreign matter. The hides are then tanned to produce durable, useful leather.

There are two general tanning methods — (1) bark or vegetable and (2) chrome tanning. In chrome tanning, salts of chromium are used to process the hides. Chrome-tanned leather cannot be tooled, but it can be dyed and embossed.

Vegetable tanning makes use of the tannic acid that is found in bark, leaves, and other vegetable products. Leather made by vegetable tanning is toolable and is, in general, the most suitable for leathercraft.

Tanned leather, which is about the color of human skin, is smoother grained, but can be further finished by dyeing, glazing, buffing, graining, or embossing.

While the better leathers are usually left smooth, those having scratches, flaws, or other surface defects after tanning are generally embossed with a grain.

Leather can be purchased in many sizes and thicknesses. The thickness of leather is expressed either in ounces or fractions of an inch. When ounce is used it is really a measure of thickness, and not of weight. Leathers range from about 1 to 8 oz. The weight to use depends on the end-service requirements.

TYPES

Cowhide

This is a strong, tough, durable leather available in a wide variety of grades, types, weights, finishes and colors. Heavy vegetable-tanned cowhide back leather, ranging from 6 to 8 oz, is widely used for tooling, carving, and stamping. Heavy cowhide sides, sometimes called strap leather, are also suitable for tooling deep designs. Shoulder pieces are widely used for belts. Thinner cowhide, both vegetable and chrome tanned, is used in leathercraft kits. Cowhide belly, lowest in cost and quality, can be used where appearance and durability are not important.

Calfskin

Calf has the finest grain of all tooling leathers and is therefore best for tooling. It ranges in weight from 1.5 to 3.5 oz; it is more expensive than cowhide. When moistened and tooled on the grain side, the designs will remain permanently. Calfskin is available in natural finish or in a variety of colors. Chrome-tanned grades are less expensive than the tooling grades and are excellent for lining. (Vegetable-tanned calfskin is known as saddle leather.)

Steerhide

Steerhide has a crinkly surface, ranges in weight from about 2.5 to 4.5 oz, tools easily, and is not readily marred. It is available in many colors and in three-toned combination finish.

Sheepskin

Sheepskin is available in both tooling and nontooling grades. The largest nontooling use is for linings. Several types are available: suede, the best of lining leathers; skivers, thin and lightweight; glazed, with a fine, smooth glossy finish and firm texture. All these are available in many different colors, and are generally inexpensive.

The tooling grade is economical and tools without difficulty. Relatively deep tooling is also possible, but the leather should be dampened only very lightly. Sheepskin is often embossed or dyed to resemble calfskin.

Goatskin or "Morocco"

Although goat is naturally a fine-grained, smooth leather, it is most commonly seen with the pebbly or crinkly "morocco" grain produced by boarding. It is stronger and more attractive than sheepskin of the same finish. In the heavier weights it is suitable for tooling line design and initials. Morocco leather is used extensively for book covers. Glazed goatskin has a high gloss, firm finish, and is excellent for lacing things.

Kidskin

Exceptionally thin, smooth, and strong, kidskin is characterized by tiny holes on the grain side. Excellent for lining, and available in various colors, it is used for lining, handbags, book covers, and wallets.

Pigskin

This leather has a fine grain and smooth surface with visible holes in groups of three. The finest grades are usually imported. Tooling of pigskin is very difficult and not recommended. It is an expensive leather and therefore its use is confined largely to small parts such as billfolds, card holders, and pocket secretaries.

Suede

Suede can be made from many kinds of leather, but most commonly from sheepskin. It has a soft velvet finish, is nontooling, and is suitable as a lining as well as for garments. Velvet Persian is a high-quality suede from skins of Persian sheep.

Alligator

Alligator is very expensive and is therefore restricted to small products. Besides genuine alligator, simulated alligator-grain leather is also available. It is usually made from calfskin or cowhide. It is attractive, strong, nontooling, available in many colors, and relatively inexpensive.

Snake and Lizard

Reptile skins are specialty leathers noted for the beauty of their markings. The skins are durable as well as beautiful. Because they are quite expensive, they are used only for very small items.

LIGHT-EMITTING DIODES (LEDs)

LEDs have been developed for numerous applications and systems for special lighting technology. For example, LEDs developed for experiments to activate light-sensitive, tumor-treating drugs. The light source, consisting of 144 of the tiny diodes, measures only 12.7 mm diameter — about the size of a small human finger. With its cooling system, the entire light source is the size of a medium suitcase. This type of therapy system involves injecting into the patient's bloodstream a drug, which attaches to the unwanted tissues and permeates into them, without affecting the surrounding tissues. A solid-state LED probe is placed near the affected tissue to illuminate the tumor and activate the drug. Once activated by the light, the drug destroys the tumor cells.

LIME

Lime, which is calcium oxide (CaO), has a specific gravity of 3.4, a melting point of 2572°C, boils at 2850°C, and is soluble. It is introduced into ceramic mixtures in several different forms. In pottery bodies and glazes it is bought as whiting (calcium carbonate) or dolomite (calcium carbonate and magnesium carbonate). In glass batches it is introduced by limestone, burned lime (calcined limestone), and dolomite. In the enameling industry it is used in the form of whiting.

Lime as CaO is not found in nature. Calcium carbonate, which is the chief source of lime, is found in the form of the minerals calcite and aragonite.

The chemical requirements of lime used in glass vary with the type of ware produced. The combined CaO and MgO should be at least 89% for bottle glass, 91% for sheet glass, 93% for blown glass, 96% for rolled glass, and 99% for optical glass. The iron oxide should be practically zero for optical glass, whereas in bottle glass as much as 0.5% is permissible, with nearly the same limits for blown or sheet glass. The silica or alumina may run as high as 15% for bottle glass, but should be vented much less for other grades.

LIMESTONE

A number of terms are in general use for the different varieties of limestone based on differences of origin, texture, composition, etc. *Marble* is a limestone that is more or less distinctly crystalline. *Chalk* is a fine-grained aragonite limestone composed of finely divided calcium carbonate usually from marine shell sources.

It may be said, regardless of the impurities that are found in limestone, that lime is in all cases practically the only base found in a pure theoretical limestone. In glass, lime is one of the most important of the common batch ingredients.

Lime gives to glass, when added in proper quantities, stability or permanency, hardness, viscosity, and tenacity, and facilitates melting and refining. Lime decreases the viscosity at high temperatures but increases the rate of setting in working range.

Magnesium lime or dolomitic lime is largely used because of its low iron content. Dolomitic lime seems to have a more powerful fluxing action and a glass using dolomitic lime is said to fine or plain up quicker than one using lime from another source.

Calcium carbonate is not used to the same extent in enamels as it is in glasses and glazes. This is probably because the average burning range of enamels is lower than that of either glasses or glazes, and calcium carbonate exerts strong fluxing action only at high temperatures.

LIQUID CRYSTALS

These are nonisotropic materials — neither crystalline nor liquid — that are composed of long molecules parallel to each other in large clusters and that have properties intermediate between those of crystalline solids and liquids. It is estimated that 1 in every 200 organic compounds has the capability of being produced in the liquid crystal form.

There are three principal types of liquid crystals, based on the arrangement of the molecules. In the smectic type, the molecules are parallel with their ends in line, forming layers that are usually curved or distorted, but are still capable of movement over one another. In the nematic type, the molecules are essentially parallel, but there is no regular alignment of their ends. The cholesteric type is formed by optically active compounds that have the capability for molecular organizations of the nematic type.

Liquid crystals have some of the properties of liquids, such as fluidity, and some of the properties of crystals, such as optical anisotropy. A major use of liquid crystals is for digital displays, which consist of two sheets of glass separated by a sealed-in transparent liquid crystal material. The outer surface of the glass sheet is coated with a transparent conductive coating, with the viewing side coating etched into character-forming segments. A voltage applied between the two glass sheets disrupts the orderly arrangement of the molecules, thus darkening the liquid to form visible characters. Other typical applications of cholesteric liquid crystals are in skin thermography for tumor detection, in electronics for temperature mapping of circuits, and in nondestructive testing of laminates.

LIQUID CRYSTAL POLYMERS

Liquid-crystal polymers (LCP) are a unique class of wholly aromatic polyester polymers that provide previously unavailable high-performance properties. Particularly outstanding is their heat-deflection temperature at 1.79 MPa of 238 to 318°C. Structure of the LCPs consists of densely packed fibrous polymer "chains" that provide self-reinforcement almost to the melting point.

Before the commercial introduction of LCP resins, LCPs could not be injection-molded. Today's resin can be melt-processed on conventional equipment into thin-wall as well as heavy-wall components at fast speeds with excellent replication of mold details and efficient use of regrind.

Commercial melt-processible resins now available are Xydar formulations (biphenol based) and Vectra resins (naphthaline based).

Like most thermoplastics, molding these high-temperature resins requires heated tools and equipment capable of producing melt temperatures of 230 to 338°C for Vectra resins and 371 to 454°C for Xydar materials.

Properties

LCP resins are characterized by outstanding strength at extreme temperatures; excellent mechanical-property retention after exposure to weathering and radiation; good dielectric strength, arc resistance, and dimensional stability; low coefficient of thermal expansion; excellent flame retardance; and easy processibility. Underwriters' Laboratories continuous-use rating for electrical properties is as high as 240°C and, for mechanical properties, 220°C. The high heat-deflection value of biphenol-based resins permits molded parts to be exposed to intermittent temperatures as high as 315°C without affecting properties. Resistance to high-temperature flexural creep is excellent, as are fracture-toughness characteristics.

LCPs are exceptionally inert. They resist stress cracking in the presence of most chemicals at elevated temperatures, including aromatic or halogenated hydrocarbons, strong acids, bases, ketones, and other aggressive industrial substances. Hydrolytic stability in boiling water is excellent. Environments that deteriorate the polymers are high-temperature steam, concentrated sulfuric acid, and boiling caustic materials.

The oxygen index of LCP resins ranges from 35 to 50%. When exposed to open flame, the material forms an intumescent char that prevents dripping and results in extremely low generation of smoke containing no toxic by-products.

Easy processibility of the resins is attributed to its liquid-crystal molecular structure, which provides high melt flow and fast setup in molded parts. However, molded parts are highly anisotropic, and knit lines are much weaker than other areas. Properties are not affected by minor variations in processing conditions, and no postcuring is required.

LITHIA

Lithia is the oxide of lithium, (LiO_2), usually added to ceramic batches by means of chemically prepared lithium compounds.

Lithia is a very powerful flux, especially when used in conjunction with potash and soda feldspars. It is a valuable component in glasses having a low thermal expansion where its use permits the total alkali content to be kept to a minimum. The low thermal expansion properties also are exploited in flameproof ceramic bodies and glass ceramics where the formation of beta spodumene is the basis for oven-to-tableware production. It also enables the production of certain glasses having high electrical resistance and desirable working properties. A relatively high content of lithia allows the production of glasses that transmit ultraviolet light.

Glasses containing lithia are much more fluid in the molten state than those containing proportional amounts of sodium or potassium, and the successful use of lithia in glassmaking lies in the fact that much smaller amounts are required to produce a glass of the necessary fluidity for working without sacrificing the desired physical and chemical properties. In addition, lithia is being utilized to increase furnace capacity, decrease melting temperatures, and increase production capacities.

Uses

Although lithium batteries boast the highest energy density of any rechargeable, cobalt in the cathode keeps cost high — a lithium battery for an electric vehicle costs about $20,000. Computer modeling predicts a less expensive replacement material. Follow-on tests verify that a cathode made from a mixture of lithium aluminum oxide and lithium cobalt oxide could not only decrease battery cost by a significant margin, but also increase cell voltage.

Aluminum–lithium alloys are basically 2XXX and 7XXX aluminum alloys containing up to about 3% lithium. Because of the extremely light weight of lithium, they provide higher stiffness-to-density ratios than traditional structural aluminum alloys and, thus, have potential for aircraft applications. Because of the low weight, lithium compounds give the highest content of hydrogen, oxygen, or chlorine. Lithium hydride, LiH, a white or gray powder, is used for the production of hydrogen for signal balloons and floats. Lithium aluminum hydride, or lithium alanate, $LiAlH_4$, is used in the chemical industry for one-step reduction of esters without heat. Lithium metal is very sensitive to light, and is also used in light-sensitive cells.

Lithium is soluble in most commercial metals only to a slight extent: it is a powerful deoxidizer and desulfurizer of steel, but no lithium is left in the lithium-treated steel. In stainless steels it increases fluidity to produce dense castings. Cast iron treated with lithium has a fine grain structure and increased density with high impact value. Not more than 0.01% remains in the casting when treated with lithium–copper. In magnesium alloys the tensile strength is increased greatly by the addition of 0.05% lithium.

Lithium copper is a high conductivity, high-density copper containing a minute quantity of residual lithium, 0.005 to 0.008%, made by treating copper with a lithium–calcium–master alloy.

Lithia has been widely used in the production of pottery glazes of high quality. The addition of 1% lithium carbonate in the frit or the fluoride or silicate in the mill to dinnerware, electrical porcelain, and sanitaryware glazes has been found to increase the resulting gloss to a marked degree.

LITHIUM

This lightest of all metals, symbol Li, has a specific gravity of 0.534. It is found in more than 40 minerals, but is obtained chiefly from lepidolite, spodumene, and salt brines.

Lithium melts at 186°C and boils at 1342°C. It is unstable chemically and burns in the air with a dazzling white flame when heated to just above its melting point. The metal is silvery white but tarnishes quickly in the air. The metal is kept submerged in kerosene. Lithium resembles sodium, barium, and potassium, but has a wider reactive power than the other alkali metals. It combines easily with oxygen, nitrogen, and sulfur to form low melting-point compounds that pass off as gases, and is thus useful as a deoxidizer and degasifier of metals. In glass the small ionic radius of lithium permits a lithium ion coupled with an aluminum ion to displace two magnesium ions in the spinel structure. Lithium cobaltite, $LiCoO_2$, and lithium zirconate, Li_2ZrO_3, are also used in ceramics. Lithium carbonate, Li_2CO_3, is a powerful fluxing agent for ceramics, and is used in low-melting ceramic enamels for coating aluminum. It is used in medicine to treat mental depression.

Uses

Lithium metal, 99.4% pure, is produced by the reduction of lithium chloride, LiCl. The salts of lithium burn with a crimson flame, and lithium chloride is used in pyrotechnics. It is also used for dehumidifying air for industrial drying and for air conditioning, as it absorbs water rapidly. It is also employed in welding fluxes for aluminum and in storage batteries. The anode is lithium, the cathode is a lithium–tellurium alloy, and the electrolyte is a molten bath of lithium salts at 427°C. Lithium ribbon, for high-energy battery use, is 99.96% pure metal in continuous strip form, 0.05 cm thick. It comes on spools packed dry under argon. An anhydrous form of lithium hexafluoroarsenate powder is used as the anode in dry batteries.

LOW-ALLOY CARBON STEELS

Also known by other terms, including alloy constructional steels, these are generally limited to a maximum alloy content of 5%. One or more of the following elements may be present: manganese, nickel, chromium, molybdenum, vanadium, and silicon. Of these, nickel, chromium, and molybdenum are the most common. The steels are designated by a numerical code prefixed by AISI (American Iron and Steel Institute) or SAE (formerly Society of Automotive Engineers). The last two digits show the nominal carbon content. The first two digits identify the major alloying element(s) or group. For example, 2317 is a nickel-alloy steel with a nominal carbon content of 0.17%.

Whereas surface hardness attainable by quenching is largely a function of carbon content, the depth of hardness depends in addition on alloy content. Therefore, a principal feature of low-alloy steels is their enhanced hardenability compared to plain carbon steels. Like plain carbon steels, however, the mechanical properties of low-alloy steels are closely related to carbon content. In heat-treated, low-alloy steels, the alloying elements contribute to the mechanical properties through a secondary hardening process that involves the formation of finely divided alloy carbides. Therefore, for a given carbon content, tensile strengths of low-alloy steels can often be double those of comparable plain carbon steels.

Low-alloy steels may be surface-hardening (carburizing) or through-hardening grades. The former are comparable in carbon content to low-carbon steels. Grades such as 4023, 4118, and 5015 are used for parts requiring better core properties than are obtainable with the surface-hardening grades of plain carbon steel. The higher-alloy grades, such as 3120, 4320, 4620, 5120, and 8620, are used for still better strength and core toughness.

Most through-hardening grades are medium in carbon content and are quenched and tempered to specific strength and hardness levels. These steels also can be produced to meet specific hardenability limits as determined by end quench tests. Identified as H steels, they afford steel producers more latitude in chemical composition limits. The boron steels, which contain very small amounts of boron, are also H steels. They are identified by the letter B after the first two digits.

A few low-alloy steels are available with high carbon content. These are mainly spring-steel grades 9260, 6150, 5160, 4160, and 8655, and bearing steels 52100 and 51100. The principal advantages of low-alloy spring steels are their high degree of hardenability and toughness. The bearing steels, because of their combination of high hardness, wear resistance, and strength, are used for a number of other parts, in addition to bearings.

LOW-ALLOY, HIGH-STRENGTH STEELS

High-strength, low-alloy (HSLA) steels are low- to medium-carbon (0.10 to 0.30%)/manganese (0.6 to 1.70) steels containing small amounts of alloying elements, such as aluminum, boron, chromium, columbium, copper, molybdenum, nickel, nitrogen, phosphorus, rare earth metals, titanium, vanadium, and zirconium. Because of the small amount of some of these elements, these steels have been referred to as microalloyed steels. The chemical compositions and minimum mechanical properties of the steels are commonly designated by minimum tensile yield strength, which ranges from about 241 MPa to more than 552 MPa. They are available in most mill forms with hot-rolled sheet and plate probably the most common, and they are typically used in the as-supplied condition. Thus, they provide high strength without heat treatment by users, and that is the principal reason for their use, which includes structural applications in cars and cargo vessels, rail cars, and agricultural, earth-moving, and materials-handling equipment as well as office buildings and highway bridges.

HSLA steels are tougher than plain carbon steels; they are not quite as formable, although sheet grades having yield strengths to 345 MPa can be formed at room temperature to 1T (one times thickness) to 2T bends, depending on thickness. The most formable are those produced with inclusion-shape control. That is, with the use of special alloying ingredients, such as rare earth metals, titanium, and zirconium, and controlled-cooling practice, resulting inclusions are small dispersed globules rather than stringer-like in shape. They are also readily welded by all common methods, and can be brazed and soldered. Most of the steels are two to eight times more resistant to atmospheric corrosion than plain carbon steels, and those commonly called weathering steels naturally acquire a deep purple-brown corrosion-inhibiting surface that precludes painting for corrosion protection. The color is considered attractive, especially in rural areas, and, thus, the steels have found considerable use for exposed building members and highway applications. Dual-phase HSLA steel has a deformable martensite phase in the ferrite matrix and exhibits a high rate of strain hardening during cold working. In the as-rolled condition in which it is supplied, it has a tensile yield strength of about 345 MPa and, thus, the ductility (about 30% tensile elongation) and formability of conventional

HSLA steels of this strength level. But strains of 2 to 3% during forming operations will increase yield strength in the strained regions to 552 MPa or greater. Thus, the steel provides the formability of medium-strength HSLA steels and the opportunity to achieve strength levels in selected regions equivalent to those of stronger, but less formable, as-supplied grades.

As contrasted to the HSLA steels, quenched-and-tempered steels are usually treated at the steel mill to develop optimum properties. Generally low in carbon, with an upper limit of 0.2%, they have minimum yield strengths from 551 to 861 MPa. Some two dozen types of proprietary steels of this type are produced. Many are available in three or four different strength or hardness levels. In addition, there are several special abrasion-resistant grades. Mechanical properties are significantly influenced by section size. Hardenability is chiefly controlled by the alloying elements. Roughly, an increase in alloy content counteracts the decline of strength and toughness as section size increases. Thus, specifications for these steels take section size into account. In general, the higher-strength grades have endurance limits of about 60% of their tensile strength. Although their toughness is acceptable, they do not have the ductility of HSLA steels. Their atmospheric-corrosion resistance in general is comparable, and in some grades, it is better. Most quenched-and-tempered steels are readily welded by conventional methods.

Applications

High-strength steels can be used advantageously in any structural application where their greater strength can be utilized either to decrease the weight or increase the durability of the structure.

Although high-strength steels find application in all recognized market classifications, the largest single field of application has been in the manufacture of construction machinery and transportation equipment. One of the leading grades of high-strength steel has been used in the construction of railroad freight cars and railroad passenger cars.

In bridges, designers are increasingly recognizing the importance of reducing deadweight by using high-strength steels, particularly for bridges involving long spans in which a reduction of weight at the center permits additional savings in the weight of supporting members. High-strength steels also lend themselves to economical tower construction where the properties permit the use of sections smaller than would be required in structural carbon steel. This advantage is important in tall television towers where dynamic loading due to wind resistance is lessened by use of smaller sections, and in transmission towers where lighter weight is a substantial advantage in reducing freight and handling costs.

Another use of high-strength steels has been for columns in high-rise buildings. Judicious use of high-strength steels in place of, and in combination with, structural carbon steel can result in substantial cost savings and an increase in usable floor area. High-strength steels are also being used to advantage in framing members of industrial and farm buildings.

The weight of containers for liquefied petroleum gas has been reduced appreciably by the use of high-strength steel, making them easier and less costly to handle and ship. Almost all such containers are now made of high-strength steel.

Other applications of high-strength steels include the inner bottoms, floors, tanks, and hatch covers of ore boats; hulls and other structural members of small tankers, barges, tugs, launches, and river boats; coal bunkers; street-lighting poles; portable oil-drilling rigs; jet-blast fences; cable reels; automobile bumpers; pole-line hardware; air-conditioning equipment; stokers; agricultural-machinery parts; earthmoving equipment; military and domestic shipping containers; and air-preheater units.

LOW-CARBON FERRITIC STEELS

Low-carbon ferritic steels, which were developed by Inco, Ltd., are low-alloy steels containing nickel, copper, and columbium. They are precipitation-hardened and have yield strengths from 482 to 689 MPa in sections up to 1.9 cm.

They possess excellent welding and cold-forming characteristics. A major use of these steels has been for vehicle frame members. Atmospheric corrosion resistance is roughly three or four times that of carbon steels.

LOW-EXPANSION ALLOYS

These are alloys, mainly of iron and nickel, having low coefficients of thermal expansion, usually within a specific temperature range. Uses include precision-instrument parts requiring dimensional stability at various temperatures and glass-to-metal sealing applications, in which the thermal expansivity of the metal must closely match that of the glass. The best-known alloy is Invar, also known as Nilvar, an iron–36% nickel composition also containing (as impurities) minute amounts of carbon, manganese, and silicon. It has the lowest coefficient of thermal expansion of all metals in the –273 to 177°C range. In the annealed condition, the alloy has a coefficient of thermal expansion ranging from about 1.44 m/m/K $\times 10^{-6}$ at –17.8 to 25°C. At 149°C, the value is still only 1.8 m/m/K $\times 10^{-6}$. Expansivity is affected by heat treatment and cold work. Quenching from about 830°C, for example, reduces the coefficient of thermal expansion below that of annealed material, as does cold forming. A combination of quenching and cold work can even result in zero or negative coefficients. Invar has a thermal conductivity of 11 W/m · K from room temperature to 100°C and is quite soft, with a hardness of about Brinell 160. Tensile properties are about 517 MPa ultimate strength, 345 MPa yield strength, and 35 to 40% elongation. The alloy is ferromagnetic at room temperature but becomes paramagnetic with increasing temperature. Because the thermal expansivity of the alloy is rather constant within a specific temperature range, Invar is also known as a controlled-expansion alloy.

There are many other such alloys, each suited for specific coefficients of thermal expansion within certain temperature ranges. They include iron with 39% nickel, or Fe–39Ni, Fe–42Ni (Dumet and Alloy 42), Fe–48 Ni (Platinite), Fe–48.5Ni, Fe–50.5Ni, Fe–42Ni–6Cr, Fe–45Ni–6Cr, Fe–36Ni–12Cr (Elinvar), Fe–22Ni–3Cr, and Fe–42Ni–5.5Cr–2.5 Ti–0.40Al (NiSpan C and Elinvar Extra). Besides its low coefficient of thermal expansion, Elinvar is noted for its constant modulus of elasticity over a wide temperature range.

Cobalt in iron–nickel alloys increases the coefficient of thermal expansion at room temperature but enhances thermal stability over a wider temperature range. Kovar and Fernico, Fe–28Ni–18Co alloys, and Fernichrome (Fe–30Ni–25Co–8Cr) are used for applications requiring vacuum sealing to glass. A Co54–Fe37–Cr9 alloy is noted for its near-zero and sometimes negative coefficient of thermal expansion in the 0 to 100°C range. Elgiloy (40%Co–20Cr–15.5Ni–15.3Fe–7Mo–2Mn–0.15C–0.04Be), originally a watch-spring alloy, has found many other spring applications. Besides dimensional stability, the alloy is noted for its good fatigue strength, corrosion and heat resistance, and nonmagnetic characteristics. Incoloy 903 (42%Fe–38Ni–15Co–3Cb–1.4Ti–0.7Al), which is also heat-treatable, is noted for a near constant coefficient of thermal expansion, about 7.2 m/m/K $\times 10^{-6}$ from 100 to 427°C and a near-constant modulus of elasticity from –196 to 649°C.

Other low-expansion or controlled-expansion alloys that have been developed include Nivar, which contains 54% cobalt; the Swiss alloys Nivarox (Fe–37Ni–8Cr with small amounts of manganese, beryllium, silicon, and carbon) and Contracid (60%Ni–15Cr–15Fe–7Mo– 2Mn and small amounts of beryllium and silicon); Nicol (40Co–20Cr–16Fe–15Ni–7Mo–2Mn and small amounts of beryllium and carbon); the French iron–nickel alloys Dilvar and Adr; Super-Invar from Japan, a 5% cobalt; iron–nickel alloy; Sylvania 4 (Fe–42Ni–5.7Cr with small amounts of manganese, silicon, carbon, and aluminum) and the similar Sealmet HC-4; Niron 52 (52Ni–48Fe); Rodar (Fe–29Ni–l7Co–0.3Mn); and Nicromet (54Fe–46Ni).

LUBRICANTS

Lubricants are substances that facilitate the flow of nonplastic, or poorly plastic, materials in the formation of dense compacts under pressure. Lubricants, which may be liquids or solids, either organic or inorganic, are particularly

LUBRICANTS

useful in dry pressing. Pieces formed under high pressure are apt to stick to the die, and even more so if the die is of intricate design.

It has been shown that proper lubrication of the die and of the powders to be pressed does much to equalize pressure in the piece. Lubricants reduce the friction between particles, and particles and die surfaces. This results in denser compacts, possible use of lower forming pressures, and easier ejection.

According to some investigators, the major cause of pressure variation is due to die surface friction. It has been shown by them that a stearic acid lubricant applied to the die walls completely eliminated pressure variations. Since it is impractical, in many cases, to lubricate the die after each operation, lubricants must be added to the powders.

Lubricants can be added directly to the powder batch, or, in other cases, special techniques must be used, such as hot mixing to disperse low-melting-point solids. The total amount of lubricants, with other special additives, ranges from ~0.1 to ~10% of batch weight. The lubricants themselves usually are used in less than 5% amounts. Actual concentrations depend on the basic nature of the body and the complexity of the shape.

Some lubricants also serve as binders and plasticizers. These are beneficial secondary functions and might serve as a basis for selecting one lubricant over another. Other factors to be considered are the compatibility of the lubricant with the body and the manufacturing processes, possible discoloration in the fired state, undesirable residues, and the effect on glaze application.

Types

Descriptions of typical lubricants that have been used in industry follow:

- *Alginates*. Colloidal carbohydrate compounds, water soluble, that thicken ceramic bodies and facilitate pressing and extrusion operations.
- *Camphor*. A whitish, water insoluble material with a melting point of 174 to 197°C. It is soluble in several organic liquids.
- *Cetyl Alcohol*. A water insoluble, white crystalline powder with a melting point of 49.3°C. It also has a lubricating value as a result of its fatty nature.
- *Graphite, Talc, Clay, and Mica*. These are useful lubricants, particularly when finely pulverized, because of their platy nature. The plates tend to slide over one another and also deter sharp, hard particles from being embedded in the die surfaces.
- *Kerosene–Lard Oil*. These mixtures, sometimes known as die oil, can be added directly to dry powders, or else applied to die surfaces. It is the lard oil that provides the lubricating properties. This is a relatively inexpensive lubricant but it probably should not be used where a glaze is to be sprayed on the unfired piece.
- *Lignosulfonates*. An organic material derived from wood pulping that can provide improved plasticity and reduced forming friction. Some lignosulfonate products contain additives to enhance lubrication further and also to function as binders.
- *Methyl Cellulose*. Synthetic gum increases the viscosity of water phases and gives a body increased workability.
- *Mineral Oils*. Petroleum products that have viscosities similar to other oily liquids.
- *No. 4 Fuel Oil*. A petroleum product of moderate viscosity, which might be used as are kerosene–lard oil mixtures.
- *Polyvinyl Acetate*. Available in powder or emulsion form and can be made to be stable with water.
- *Polyvinyl Alcohol*. Colorless plastic available in several degrees of hydrolysis. These variants lend different properties in ceramic usage.
- *Starches*. Specially prepared for the ceramic industry; reportedly serve as lubricants due to a retained superficial water film.

LUBRICATING GREASE

Usually a compound of a mineral oil with a soap, lubricating grease is employed for lubricating machinery where the speed is slow or where it would be difficult to retain a free-flowing oil. The soap is one that is made from animal or vegetable oils high in stearic, oleic, and palmitic acids. The lime soaps give water resistance, or a mineral soap may be added for this purpose. Aluminum stearate gives high film strength to the grease. All of these greases are more properly designated as mineral lubricating grease. Originally, grease for lubricate purposes was hog fat or the inedible grades of lard, varying in color from white to brown. Some of these greases were stiffened with fillers of rosin, wax, or talc, which were not good lubricants. The stiffness of such a grease should be obtained with a mineral soap. ASTM specifications for heavy journal bearing grease require 45% soap content. About 2% calcium benzoate increases the melting point. Mineral lubricating grease may contain from 80 to 90% mineral oil and the remainder a lime soap.

Uses

Oronite GA-10, for example, is a sodium salt of terephthalic acid used as a gelling agent in high-temperature greases. It adds water resistance and stabilizes against emulsion. Ortholeum 300 is a mixture of complex amines, and small amounts added to a grease will give high heat stability. Braycote 617 is a synthetic grease for rockets subject to both heat and cold.

The lubricating grease known as trough grease, used in food plants for greasing trays, tables, and conveyors, contains no mineral oil and is edible.

Lime greases do not emulsify as readily as those made with a soda base, and are thus more suitable for use where water may be present.

Graphite grease contains 2 to 10% amorphous graphite, and is used for bearings, especially in damp places. For large ball and roller bearings a low-lime grease is used, sometimes mixed with a small percentage of graphite. Cylinder grease is made of about 85% mineral oil or mineral grease and 15% tallow. Compounded greases are also marketed containing animal and vegetable oils, or are made with blown oils and compounded with mineral oils. The fatty acids in vegetable and animal oils, however, are likely to corrode metals. Tannin holds graphite in solution; in gear grease Metaline is a compound of powdered antifriction metal, oxide, and gums, which is packed in holes in the bearings to form self-lubricating bearings. Lead-Lube grease has finely powdered lead metal suspended in the grease for heavy-duty lubrication.

Sett greases are mixtures of the calcium soaps of rosin acids with various grades of mineral oils. They are low-cost semisolid greases used for lubricating heavy gears or for greasing skidways. Clay fillers may be added to improve the film strength, or copper or lead powders may be incorporated for heavy load conditions. Solidified oil is also a name given to grease made from lubricating oil with a soda soap and tallow, used for heavy bearings.

M

MACHINABILITY OF METALS

The ease and economy with which a metal may be cut under average conditions is its machinability. Frequently, no truly quantitative assessment is made, but rather a rating or an index is established vis-à-vis a reference material. More quantitative comparisons are based on tool life. For example, maximum cutting speeds for a given tool life may be used as a rating of machinability. Alternatively, tool wear rate may be the basis for a machinability rating. Surface finish is sometimes used for assessing machinability.

MACHINING PROCESS

The wide range of metal-cutting processes may be represented, with some oversimplification, by the orthogonal cutting process.

Cutting Speed

The machining response of ductile metals is very sensitive to cutting speed. Below about 0.02 m/s, chips form discontinuously, metal chunks are lifted out of the surface, and the surface is scalloped or pockmarked. When the speed is in the 0.1 m/s range, chips are formed continuously, the shear zone is narrow, and the chip slides on the tool face. Under these conditions a cutting fluid can lubricate both the rake and flank faces of the tool.

Cutting Fluid

The interaction of the tool and the workpiece is considerably affected by the presence of cutting fluids. Cutting fluid has two primary functions. First, as long as the cutting speed is slow, the cutting fluid can act as a lubricant between the chip and the tool face. Even at higher speeds, some lubricating effect at the flank face may be present. Second, and perhaps more importantly, the cutting fluid serves as a coolant. In most instances the cutting fluid will be an emulsion of a lubricating phase (oil, graphite, and so on) in water, since water is the best heat-transfer medium readily available. Beyond lubrication and cooling, the cutting fluid can be used to flush out the cutting zone.

Surface Quality and Tool Wear

Cutting speed thus affects the all-important machinability considerations of surface quality and tool wear. Surface finish is best with a well-lubricated, moderately low-speed operation or high-speed cutting with no built-up edge. The high pressures and temperatures of operation, abetted by shock loading and vibrations, can lead to rapid tool wear. Tool wear is often sufficiently rapid to make tool replacement a major factor in machining economics. Tools must be replaced when they break or when they have worn to the point of producing an unacceptable surface finish or an unacceptable degree of surface heating.

METAL PROPERTIES

A machining operation can be optimized to effect metal removal with the least energy, or the best surface, or the longest tool life, or a reasonable compromise among these factors. Even so, it remains that the optimum ease and economy of machining some alloys is vastly different from that of others, and some basic attributes of easily machined alloys can be set forth.

Toughness

To ensure that chip separation occurs after minimum sliding, low ductility is required. To minimize cutting force, low strength and low ductility combine to mean low toughness.

Toughness is generally defined as energy per unit volume consumed en route to fracture. Ironically, materials of maximized toughness are desirable for most engineering applications and, thus, some of the most attractive alloys, such as austenitic stainless steels, are difficult to machine.

Adhesion

The degree to which the metal adheres to the tool material is important to its machinability.

Actually this attribute can work to advantage or disadvantage. If diffusion results, the tool can be weakened, and rapid wear occurs. Otherwise, high adhesion will stabilize the secondary shear zone.

Workpiece Second Phases

Small panicles or inclusions in the metal can have a marked effect on machinability. Hard sharp oxides, carbides, and certain intermetallic compounds abrade the tool and accelerate tool wear. On the other hand, soft second phases are beneficial because they promote localized shear and chip breakage.

Thermal Conductivity

In some cases, workpiece thermal conductivity can be important to machinahility. A low thermal conductivity generally results in high shear-zone temperature. This can be advantageous in reducing the strength of the metal or in softening second-phase particles. Of course, if adhesion and diffusional depletion of tool alloy content result, the high temperature is a problem. The workpiece temperature can be managed by cutting-speed and cutting-fluid manipulation.

Alloy Systems

Commercial alloys can be grouped into two categories, namely, those designed for ease of machining (so-called free machining grades), and the vast majority, which are of widely varying but generally less than optimum machinability. Considering these latter, ordinary alloys, it can be shown that their machinability may be considerably improved by metallurgical operations that limit strength or ductility or both. Of course, it is not often possible to reduce both strength and ductility simultaneously. Even so, machinability often can be improved by grossly reducing one or the other property.

MACHINING

Any one of a group of operations that change the shape, surface finish, or mechanical properties of a material by the application of special tools and equipment. Machining almost always is a process where a cutting tool removes material to effect the desired change in the workpiece. Typically, powered machinery is required to operate the cutting tools.

Although various machining operations may appear to be very different, most are very similar: they make chips. These chips vary in size from the long continuous ribbons produced on a lathe to the microfine sludge produced by lapping or grinding. These chips are formed by shearing away the workpiece material by the action of a cutting tool. Cylindrical holes can be produced in a workpiece by drilling, milling, reaming, turning, and electric discharge machining (EDM). Rectangular (or nonround) holes and slots may be produced by broaching, EDM, milling, grinding, and nibbling; and cylinders may be produced on lathes and grinders. Special geometries, such as threads and gears, are produced with special tooling and equipment utilizing the same turning and grinding mentioned above. Polishing, lapping, and buffing are variants of grinding where a very small amount of stock is removed from the workpiece to produce a high-quality surface.

In almost every case, machining accuracy, economics, and production rates are controlled by the careful evaluation and selection of tooling and equipment. Speed of cut, depth of cut, cutting-tool material selection, and machine-tool selection have a tremendous impact on machining. In general, the more rigid and vibration-free a machining tool is, the better it will perform. Jigs and fixtures are often used to support the workpiece. Since it relies on the plastic deformation and shearing of the workpiece by the cutting tool, machining generates heat that must be dissipated before it damages the workpiece or tooling. Coolants, which also act as lubricants, are often used.

Boring

This machining operation increases the size of an existing hole in a workpiece. The usual purpose of boring is to produce a hole with an accurate diameter and good surface finish. Boring can be performed on a special machine or a lathe, with either the workpiece or the boring tool being on a movable table. A rotating spindle, holding either a single-point cutting tool or the workpiece, is fed into the work. As the spindle rotates, the cutting tool engages the interior of the existing hole, and chips are formed as the tool cuts into the workpiece. The actual cutting action of a boring tool is very similar to a lathe turning tool.

Broaching

This is the removal of material to produce a slot (or other formed shape) in a workpiece by moving a multiple-tooth, barlike tool across the workpiece. The cutting action results from the configuration of each tooth being progressively higher than the preceding one. Each tooth of the broach removes a small, predetermined amount of stock, the chip. Broaching is a very economical machining operation, although tool costs can be high; accordingly, it is applied most often to high-volume production.

Drilling

One of the most common machining operations, drilling is a method of producing a cylindrical hole in a workpiece. A typical twist drill consists of a helically grooved steel rod with two cutting edges on the end. The helical flutes or grooves in the drill allow the chips to be removed from the cutting edge, conduct coolant to the cutting lips, and form part of the cutting edge geometry. Although drills may appear to be simple, their geometries are carefully controlled. Drilling is a very fast and economical process, but it usually does not produce a very accurate hole diameter or a fine surface finish.

Turning

This type of machining is performed on a lathe. The process involves the removal of material from a workpiece by rotating the workpiece under power against a cutting tool. The cutting tool is held in a tool-post that is supported on a cross slide and carriage. The tool may be moved radially or longitudinally in relation to the turning axis of the workpiece. Forms such as cones, spheres, and related workpieces of concentric shape as well as true cylinders can be turned on a lathe.

The most common lathe is an engine lathe, where the workpiece may be rotated and held between tapered centers or by means of a collet or chuck. A turret lathe is an engine lathe that has a multisided indexing tool holder or turret instead of a tailstock center. This adds to the versatility of the machine by allowing a greater variety of cutting tools to be applied to the rotating workpiece.

Automatic Screw Machines

Automatic screw machines are sophisticated lathes that have been designed to perform several turning operations automatically in rapid succession without removing the workpiece from the machine. They are used when the volume and complexity of the required workpieces justifies the expense of setting up and operating these very versatile but complex machines. A screw machine (named for the screw manufacturing role for which it was created) is cam-controlled or driven by computer numerical control (CNC).

Milling

This process removes material by feeding a workpiece through the periphery of a rotating circular cutter. Each tooth of the rotating multitoothed milling cutter removes a portion of material from the passing workpiece.

Milling cutters are designed for particular operations and are classified as either shell type or end type. Shell mills are disk shaped and usually produce continuous slots. When well supported in the machine tool to minimize vibration, they perform well and economically. The more versatile end mill is held by its shank only. End mills can be used for slotting just like a shell mill, but they also can cut pockets, contours, and even cylindrical holes. End mills with four or six flutes are stronger and more rigid than two-flute mills, and so are better able to machine tougher materials.

REAMING

This machining operation enlarges an existing hole by a few thousandths of an inch and produces a hole whose diameter is very accurately controlled. The cutting edges of a reamer may be ground on the apexes between longitudinal flutes or grooves, or cutting may take place on chamfered edges at the end of the reamer. Reaming is performed either manually or by machine, often as a finishing operation after drilling. Reamed holes have good surface finish.

SAWING

This is the parting of material by using blades, bands, or abrasive disks as the cutting tools. In the most common type of saw, a toothed blade is passed across the workpiece in either a reciprocating or continuous motion. The teeth can be mounted on continuous bands, short steel blades, or the periphery of a disk. Friction sawing is a rapid process used to cut steel as well as certain plastics. A very high speed blade softens the workpiece material with frictional heat. The material is then wiped away from the workpiece by the cutting blade. Since many teeth engage the workpiece, no single tooth overheats. Abrasive sawing looks similar to friction sawing, except that a thin rubber or bakelite bonded abrasive disk grinds the material away instead of simply softening and wiping it.

NIBBLING

Nibbling is the operation that cuts away small pieces of material by the action of a reciprocating punch. A nibbler takes repeated small bites from the workpiece (which is usually a thin sheet of material) utilizing a quickly reciprocating punch system. As the work is passed beneath the punch, a small nibble of material is removed by the punch during each punch cycle. After each quick nibble, the workpiece is advanced under the punch to allow another small bite to be taken away. In this manner, a great deal of material can be removed from the workpiece after several tens or hundreds of individual punching cycles.

SHAPING

This process cuts flat or contoured surfaces by reciprocating a single-point tool across the workpiece. The tool is mounted on a hinged unit known as a clapper box, which lifts up to disengage the tool from the surface on the return stroke. The cutting action of a shaper is actually similar to turning, except that the single-point cutting tool moves straight across a workpiece, instead of having the workpiece rotate against the cutting tool.

GRINDING

This process removes material by the cutting action of a solid rotating, grinding wheel. The abrasive grains of the wheel perform a multitude of minute machining cuts on the workpiece. Although grinding is sometimes used as the sole machining operation on a surface, it is generally considered a finishing process used to obtain a fine surface and extremely accurate dimensions.

Grinding is used to machine a wide range of metals, carbide materials, stone, and ceramics. The grinding process may be used on metals too hard to machine otherwise because commercial grinding abrasives are many times harder than the metals to be machined.

Grinding wheels are composed of abrasive grains plus a bonding material. The wheels are often very porous, with homogeneous open areas between the grains. The abrasives most commonly used are silicon carbide (SiC) and aluminum oxide (Al_2O_3). Coarse-grained wheels are used for rapid removal of stock; wheels with fine grains cut more slowly but give smoother finishes. Coolants are applied to the grinding point to dissipate the heat generated and to flush away the fine chips.

HONING

Honing is a grinding process that removes a small amount of material from a workpiece by means of abrasive stones. It is able to produce extremely close dimensional tolerances and very fine surface finishes. The abrading action of the fine grit stones occurs on a wide surface area rather than on a line of contact as in grinding.

Lapping

This is a precision abrading process used to finish a surface to a desired state of refinement or dimensional accuracy by removing an extremely small amount of material. Lapping is accomplished by abrading a surface with a fine abrasive grit rubbed about it in a random manner. A loose unbonded grit is used. It is traversed about on a lap, made of a somewhat softer material than the workpiece. The unbonded grit is mixed with a vehicle such as oil, grease, or soap and water.

Polishing

This is the smoothing of a surface by the cutting action of an abrasive grit that is either glued to or impregnated in a flexible wheel or belt. Polishing is not a precision process; it removes stock until the desired surface condition is obtained.

Superfinishing

For a mirrorlike surface, a precision abrading process known as superfinishing is used. Superfinishing removes minute flaws or inequalities because it is performed with an extremely fine-grit abrasive stone, shaped to match and cover a large portion of work surface.

Buffing

This is the smoothing and brightening of a surface by rubbing it with a fine abrasive compound carried in a soft wheel or belt. The abrasive is a fine powder or flour mixed with tallow or wax to form a smooth paste. Buffing is performed on the same types of machines as polishing; frequently both buffing and polishing wheels are included on the same machine. Buffing differs from polishing in that a finer grit is used and less material is removed from the workpiece.

Ultrasonic Machining

In this process, material is removed by abrasive bombardment and crushing in which a vibrating tool drives an abrasive grit against a workpiece. In ultrasonic machining, the tool never directly contacts the workpiece; rather, the vibrations (typically 20,000 Hz) drive the abrasive, which is suspended in a liquid, against the workpiece. Impressions may be economically sunk in glass, ceramics, carbides, and hard brittle metals by this method. The shape of the tool basically determines the shape of the impression.

Electric-Discharge Machining

This is a machining process in which electrically conductive materials can be removed by repeated electric sparks. Electric-discharge machining (EDM) is used to form holes of varied shape in materials of poor rnachinability. Unlike other machining operations, it does not rely on a cutting tool to shear away the workpiece. Instead, it uses electrical energy to melt or vaporize small areas on the workpiece. The sparks created by an EDM unit (at a rate greater than 20,000 per second) are discharged through the space between the tool (cathode) and the workpiece (anode). The small gap between the tool and workpiece is filled with a circulating dielectric hydrocarbon oil, which serves as a cooling medium and flushes away metal particles. Although not as fast as other machining processes, EDM has the unique ability to remove hard materials that otherwise would not be machinable.

MACHINING, HIGH SPEED

Machining operations such as turning, milling, and drilling involve the modification of a workpiece through the removal of material. In machining of metals, this material removal occurs via a concentrated shear flow initiated at the point of contact between the workpiece and a wedge-shaped tool (e.g., turning insert, milling flute, etc.) to produce a chip and a machined surface. The cutting action is generated by the rotation of the machine spindle. Increasing the power and speed of the spindle and axes has several advantages, which include

(1) shorter machining time, (2) improved surface finish, (3) reduced thermal and mechanical stresses on the workpiece and tool, and (4) increased dynamic stability. These potential advantages have driven a recent, rapid increase in the industrial adoption of high-speed machining processes and technology.

The definition of high-speed machining is not static but instead changes with time depending on available tool materials and machine technology. Reliable machining spindles capable of running continuously at speeds of up to 40,000 revolutions per minute (rpm) with a power output of 30 kW are now available on production equipment. In addition, axis speeds of production machining centers now exceed 60 m/min with accelerations of greater than 1 g (g = gravitational acceleration 9.8 m/s^2). The marked increase in material removal rates and the reduction in production time and costs afforded by this new technology have the potential of revolutionizing the manufacturing process. However, because of rapid tool wear, not all materials can be machined at high speeds. Figure M.1 illustrates this fact showing estimated machining speed ranges for a variety of different materials. In general, machining speeds on difficult-to-machine materials such as titanium and hardened steel are limited by the availability of suitable tool materials, whereas machining speeds in more machinable materials such as aluminum and plastic are limited by the available machine technology.

Tooling Materials

The increased use of high-speed machining has been enhanced by the development of suitable tool materials. In general, the higher the yield strength and melting temperature of the material being machined, the greater the stresses and temperatures that will be generated at the tool–chip interface. These increased stresses and temperatures cause increased rates of chemical and mechanical tool wear but with the introduction of Al_2O_3-TiC, Si_3N_4 (silicon nitride), and SiC whisker-reinforced Al_2O_3 ceramic cutting tools the high-speed machining of certain materials becomes practical. Speeds as high as 5000 sfm have been reported; and speeds 1000 to 3000 sfm are routine with essentially pure ceramics (i.e., low metallic impurities). However, the older cermets (ceramic-coated cemented carbides) have been limited to speeds below 1400 sfm. This is because plastic deformation of the metallic phase in these carbides occurs at temperatures as low as 600°C, which limits the cutting speed.

High toughness, strength, and thermal conductivity result in tools that have improved resistance to chipping, which makes it possible to do interrupted cutting when machining metal. High hardness generally correlates with improved wear resistance, although chemical compatibility is essential for good performance. Since temperatures can be extremely high at the tool/workpiece interface, it is important to retain high hardness and toughness as the temperature increases. The greater efficiency of many ceramics, compared with cermets, is that they maintain wear resistance (i.e., hardness) and strength at these elevated temperatures.

Controllers — Data Processing

The increased complexity of machined parts leads to a similar increase in the size and complexity of the programs required to generate necessary machine motion. The flow rate of the data required to control tool motions over complex surfaces has grown dramatically as spindle speeds have increased. This is especially true as high-speed machining becomes practical in the

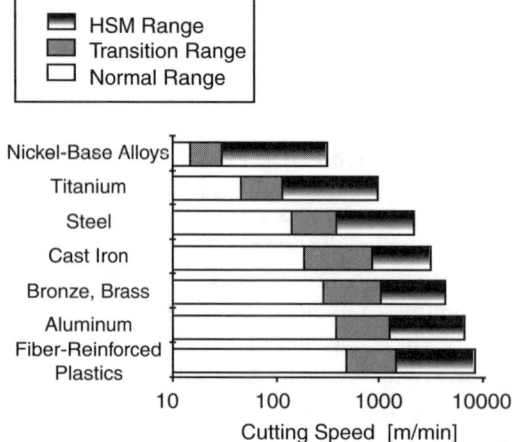

FIGURE M.1 Machining speeds as a function of material. (From Davies, M.A., in *McGraw-Hill Yearbook of Science and Technology*, McGraw-Hill, New York, 1988. With permission.)

MAGNESIUM AND ALLOYS

A silvery-white metal, symbol Mg, magnesium is the lightest metal that is stable under ordinary conditions and produced in quantity. It is the sixth most abundant element, specific gravity is 1.74, melting point is 650°C, boiling point about 1110°C, and electrical conductivity about 40% that of copper. Ultimate tensile strengths are about 90 MPa as cast, at least 158 MPa for annealed sheet, and 179 MPa for hard-rolled sheet, with corresponding elongations of about 4, 10, and 15%, respectively. The strength is somewhat higher in the forged metal. Magnesium has a close-packed hexagonal structure that makes it difficult to roll cold, and its narrow plastic range requires close control in forging. Repeated reheating causes grain growth. Sheet is usually formed at 150 to 200°C. It is the easiest to machine of the metals.

CHARACTERISTICS

The excellent machinability of magnesium makes its use economical in parts where weight saving may not be of primary importance, but where much costly machining is required. Such parts, when made of magnesium, can be machined at higher speeds and with greater economy than would be possible with most other commonly used metals. Chemical milling can be used on magnesium.

Magnesium can be cast and fabricated by practically every method known to the metal worker. The metal is cast in sand or permanent molds to obtain lightweight castings with good strength, stiffness, and resistance to impact or shock loading. Magnesium sand and permanent-mold castings are heat-treatable to further improve mechanical properties.

The die-casting process is similarly applicable to magnesium and this method of casting should always be considered when the quantities desired are in the range that indicates its use. Both hot and cold chamber processes are usable.

The metal can also be cast by some of the less common methods, including plaster mold, centrifugal, shell molding, and investment processes.

Magnesium is rolled into sheet and plate and can be extruded into rods, bars, tubing, and an almost endless variety of structural and special shapes (Table M.1).

Sheet and extrusions are very easily formed using techniques that have been developed especially for magnesium. Stamping, deep and shallow drawing, blanking, coining, spinning, and impact extrusion are just a few of the production forming operations regularly used on magnesium and these indicate the adaptability of the metal to a large variety of metalworking procedures.

The forging of magnesium is accomplished by methods much the same as those used for forging other metals. Both press and hammer equipment are used, but the former is most commonly employed because the physical structure of magnesium makes the metal better adapted to the squeezing action of the forging press. Magnesium forgings are chosen when a high strength-to-weight ratio, rigidity, or pressure tightness is required. The selection of forging, however, is governed by the fact that, like permanent mold and die castings, a sufficient number of parts must be needed to justify the cost of die equipment.

Magnesium parts can be joined by any of the common methods. Arc and electric resistance welding, adhesive bonding, and mechanical fastening are in daily production use. Brazing and gas welding, although not as frequently used as the other methods, are also suitable ways of joining magnesium.

Magnesium possesses relatively high thermal and electrical conductivities, very high damping capacities, and is nonferromagnetic. It has good stability to atmospheric exposure and good resistance to attack by alkalies, chromic and hydrofluoric acids, and many organic chemicals, including hydrocarbons, aldehydes, alcohols (except methyl), phenols, amines, esters, and most oils. Bare magnesium

TABLE M.1
Magnesium Shapes

	Typical Alloy			
	AZ91D-F	AZ31B-F ZK60A-T5	AZ31B-F HM21A-T5 AZ80A-T5 ZK60A-T6	AXZ31B-H24 HK31A-H24 HM21A-T8
	Form			
	Die Casting	Extrusion	Forging	Sheet, Plate
Density (lb/in.^{3})	0.065	0.064–0.066	0.065–0.065	0.064–0.065
Melting temperature (°F)	875–1105	970–1175	970–1202	1050–1202
Coefficient of thermal expansion (10^{-6}in./in.°F)	14.5	14.3–14.5	14.3–14.5	14.3–14.5
Thermal conductivity (Btu-ft/h-ft^2-°F)	31	44–70	—	44–79
Electrical resistivity ($\mu\Omega$-cm)	12.9	5.7–9.2	5.0–9.2	5.0–9.2
Yield strength (10^3 psi)	23	28–44	22–39	21–32
Tensile strength	34	38–53	34–50	33–42
Impact strength, Charpy (ft-lb)	2.0	3.2	3.2	—
Fatigue endurance limit (10^3 psi)	14	16–23	15–18	16–24
Creep strength, 0.1% in 100 h (10^3 psi)	3–4 (250°F)	Up to 5 (250°F)	5 (500°F)	Up to 8 (500°F)
Elongation in 2 in. (%)	3	11–15	6–11	9–21
Modulus of elasticity (10^6 psi)	6.5	6.5	6.5	6.5

Source: Mach. Design Basics Eng. Design, June, p. 760, 1993. With permission.

surfaces are nonsticking (snow, ice, sand, etc.) and nonmarking. Magnesium also has a low sparking tendency.

Magnesium develops a corrosion-inhibiting film upon exposure to clean atmospheres and fresh water, but that film breaks down in the presence of chlorides, sulfates, and other media, necessitating corrosion protection in many applications. Many protective treatments have been developed for this purpose. It is also rapidly attacked by mineral acids, except chromic and hydrofluoric acids, but is resistant to dilute alkalies; aliphatic and aromatic hydrocarbons; certain alcohols; and dry bromine, chlorine, and fluorine gases. Anodized magnesium is produced by immersing in a solution of ammonium fluoride and applying a current of 120 V. The fluoride film has a thickness of only 0.0003 cm, but it removes cathodic impurities from the surface of the magnesium, giving greater corrosion resistance and also better paint adhesion.

Magnesium is valued chiefly for parts where light weight is needed. It is a major constituent in many aluminum alloys, and very light alloys have been made by alloying magnesium with lithium.

The pure metal ignites easily, and even when alloyed with other metals the fine chips must be guarded against fire. In alloying, it cannot be mixed directly into molten metals because of flashing, but is used in the form of master alloys. The metal is not very fluid just above its melting point, and casting is done at temperatures considerably above the melting point so that there is danger of burning and formation of oxides. A small amount of beryllium added to magnesium alloys reduces the tendency of the molten metal to oxidize and burn. The solubility of beryllium in magnesium is only about 0.05%. As little as 0.001% lithium also reduces fire risk in melting and working the metal. Molten magnesium decomposes water so that green-sand molds cannot be used, because explosive hydrogen gas is liberated. For the same reason water sprays cannot be used to extinguish magnesium fires.

MAGNESIUM AND ALLOYS

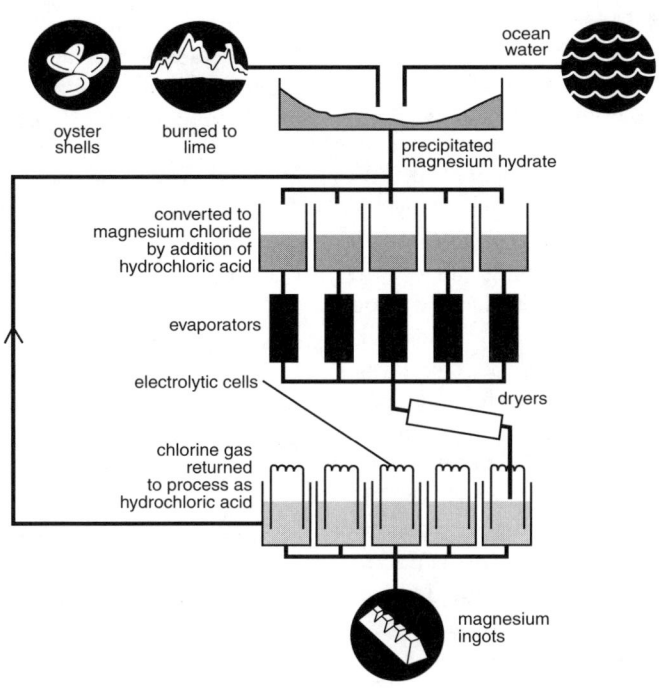

FIGURE M.2 The electrolytic extraction of magnesium from seawater. (From *McGraw-Hill Encyclopedia of Science and Technology*, Vol. 10, McGraw-Hill, New York, 295. With permission.)

The affinity of magnesium for oxygen, however, makes the metal a good deoxidizer in the casting of other metals.

PRODUCTION

Magnesium is produced commercially by the electrolysis of a fused chloride, or fluoride obtained either from brine or from a mineral ore, or it can be vaporized from some ores.

These two major methods of producing magnesium are used throughout the world. These can be seen in Figures M.2 and M.3 (electrolytic and the silicothermic processes). In the electrolytic process the electrolysis of magnesium chloride to yield chlorine and metallic magnesium is the basis of this process. Although magnesite, dolomite, and natural brines have been used as raw materials, the principal source is seawater, which contains about 0.13% magnesium.

The ferrosilicon process (silicothermic) was developed commercially during World War II in Canada.

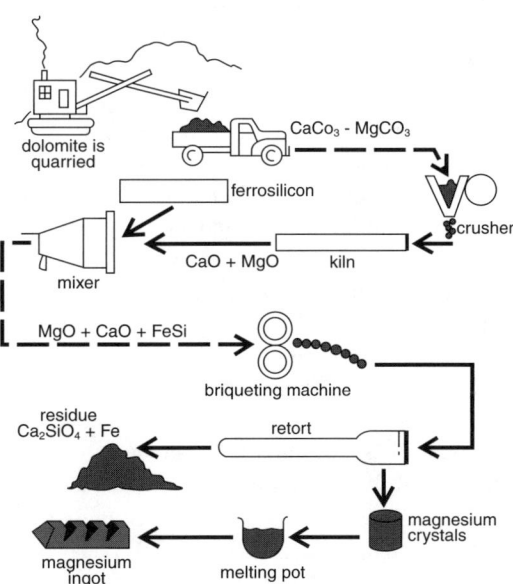

FIGURE M.3 Extraction of magnesium metal by silicothermic process. (From *McGraw-Hill Encyclopedia of Science and Technology*, Vol. 10, McGraw-Hill, New York, 296. With permission.)

Alloys and Alloy Designation

Although magnesium alloys are moderate in strength and rigidity, they have high specific strength and rigidity because of their low density, which, as it is in the range of 1771 to 1827 kg/m^3, is the lowest of common metals. Modulus of elasticity in tension is typically 44,800 MPa and ultimate tensile strengths range from 152 to 379 MPa, depending on the alloy and form. Both wrought and cast alloys are available, the former in sheet, plate, rod, bar, extrusions, and forgings, and the latter for sand, permanent-mold, investment, and die castings. Alloys are designated by a series of letters and numbers followed by a temper designation. The first part of the alloy designation indicates by letters the two principal alloying elements (or one if the alloy contains only one alloying element): A for aluminum, E (rare earth elements), H (thorium), K (zirconium), M (manganese), Q (silver), S (silicon), T (tin), and Z (zinc). The two (or one) numbers that follow indicate the amount (in percent rounded off to whole numbers) of these elements, respectively. These numbers are followed by a letter to distinguish among alloys having the same amount of these alloying elements. The temper designations that follow are similar to those for aluminum alloys: F (as fabricated); O (annealed); H10 and H11 (slightly strain-hardened); H23, H24, and H26 (strain-hardened and partially annealed); T4 (solution heat-treated); T5 (artificially aged); T6 (solution heat-treated and artificially aged); and T8 (solution heat-treated, cold-worked, and artificially aged). Thus, AZ91C-T6 is the designation for an alloy containing 8.7% aluminum and 0.7% zinc as the major alloying elements. The letter C indicates that it is the third such alloy to be standardized, and, in this case, it is in the solution heat-treated and artificially aged temper. These designations, however, do not distinguish between wrought and cast alloys.

In addition to AZ91C, other magnesium sheet and plate alloys include AZ31B, HK31A, HM2IA, of which AZ31B is the strongest at room temperature and the most commonly used. In the H24 temper, it has an ultimate tensile strength of 290 MPa, a tensile yield strength of 220 MPa, 15% elongation, and can be used at service temperatures to about 93°C. The others, however, especially HM2IA, are more heat resistant and can sustain temperatures to about 315°C. For HM21A, the 100-h creep strength for 0.1% deformation is 86 MPa at 214°C and 52 MPa at 316°C. None of the alloys is especially formable, with the minimum bend radii for AZ31B-O, the most formable, ranging from about five times thickness (5T) at room temperature to 2T at 260°C. Thus, heat is often required in forming operations, especially deep-drawing.

AZ31B is also widely used in the form of bar and other extruded shapes. The alloys for bar and extruded shapes are generally of two kinds: those alloyed principally with aluminum and zinc, and those alloyed with zinc and a bit of zirconium. In the former, strength increases with increasing aluminum content, and in the latter with increasing zinc content. Some of each kind respond to artificial aging, providing in the T5 temper ultimate tensile strengths in the 345 to 379 MPa range. AZ31B and many of the alloys for bar and extrusions are also suitable for forging. The hot-working range may be as low as 232 to 371°C or 293 to 538°C.

There are several magnesium die-casting alloys but more than a dozen magnesium sand-casting alloys and magnesium permanent-mold casting alloys. One composition, AZ91, is available in four grades: AZ91A, B, C, and D. AZ91C, for sand- and permanent-mold castings, contains 8.7% aluminum, as opposed to 9% in the die-casting alloys (A and B). Each also contains 0.7% zinc and 0.13% manganese. As die-cast, AZ91A and AZ91B provide an ultimate tensile strength of 228 MPa, a tensile yield strength of 152 MPa, and 24% elongation. For AZ91C-T6, these values are 276 MPa, 145 MPa, and 6%, respectively. AZ91D is a high-purity version, containing extremely low residual contents of iron, copper, and nickel, which markedly improves corrosion resistance, precluding the need for protective treatments in certain applications, such as auto underbody parts. The benefits of high purity are apparently applicable to other alloys as well, such as AS41, a more heat-resistant die-casting alloy (Table M.2).

The sand- and permanent-mold casting alloys generally use either aluminum, zinc, and manganese or zinc, thorium, zirconium, and, in

TABLE M.2
Properties of Magnesium Die Casting Alloys

	Alloy					
	AZ91B	AZ88	ZA124	ZA124	AZ55	AZ55
Composition, %						
Al	9	8	4	4	5	5
Zn	0.7	8	12	12	5	5
Mn	0.2	0.2	0.4	0.4	0.3	0.3
Temper	F	F	F[c]	T5[c]	F	T5[c]
Cast Test Bars						
Elongation, %	7	2	4	2	9	7
Yield Strength, MPa (ksi)	165 (24)	200 (29)	180 (26)	227 (33)	165 (24)	207 (30)
Tensile Strength, MPa (ksi)	269 (39)	255 (37)	248 (36)	276 (40)	276 (40)	290 (42)
Cut from Castings[a]						
Elongation, %	1.6	1.6	—	—	—	—
Yield Strength, MPa (ksi)	159 (23.1)	170 (24.7)	—	—	—	—
Tensile Strength, MPa (ksi)	190 (27.8)	210 (30.6)	—	—	—	—
Corrosion[b]						
Base	15	15	10	—	—	—
+2% Cu	92	12	7	—	—	—

[a] Average of 30 test bars cut from a chain saw handle.
[b] Percent weight loss of 1/8 in. diameter cast rod immersed five days in 3% NaCl.
[c] Aged 24 h at 325°F.

Source: Adv. Mater. Proc., 111, 1999. With permission.

some cases, silver and rare earth elements, as alloying elements. Almost all these alloys respond to artificial aging or solution heat treating and artificial aging. The strongest, ZE63A, in the T6 temper, and ZK61A, in the T5 or T6 temper, have an ultimate tensile strength of about 310 MPa, a tensile yield strength of about 193 MPa, and about 10% elongation. The alloys containing zirconium and rare earth elements — ZE33A, ZE41A, and ZE63A — are more creep resistant at higher temperatures than the aluminum-, zinc-, and manganese-bearing alloys, but are more difficult to cast. The magnesium–silver-bearing alloys, QE22A and QH21A, also are superior in elevated-temperature performance, have good castability and weldability, but are quite costly.

Like their base metal, magnesium alloys have outstanding machinability, providing faster cutting speeds and greater depths of cut at less power than all commonly machined metals. However, dust, chips, and turnings can pose a fire hazard, necessitating special precautions.

Although lithium is no longer used in magnesium alloys, a series of magnesium–lithium alloys, developed in the past for aerospace applications and ammunition containers, were noted for their extremely light weight and moderate ultimate tensile strengths — 100 to 248 MPa. One such alloy, LA141A, had a density of only 1245 kg/m^3, or three quarters that of magnesium.

Magnesium–nickel is a master alloy of magnesium and nickel used for adding nickel to magnesium alloys and for deoxidizing nickel and nickel alloys. One such alloy contains about 50% of each metal, is silvery white in color, and is furnished in round bar form. Magnesium–Monel contains 50% magnesium and 50% Monel metal. Alloys of magnesium with nickel, Monel, zinc, copper, or aluminum, used for deoxidizing nonferrous metals, are called stabilizer alloys.

Properties

Magnesium–aluminum–zinc alloys often have sufficient strength to give satisfactory service at temperatures as high as 177°C. For still higher temperature services (and sometimes lower temperature service) the alloys especially developed for such service are usually required. These are the magnesium–rare earth metal and the magnesium–thorium alloys. They have good short-time mechanical strength, good strength under long-time loading (creep strength), and good modulus of elasticity, at these higher temperatures. In addition, they retain these good properties throughout long exposures to elevated temperatures.

Magnesium alloys are similar to other nonferrous alloys in that they do not exhibit a definite endurance limit when subjected to fatigue loading. Instead, their fatigue strength continues to drop as the required life of the part increases. But the "signal–noise curves" for magnesium are much more horizontal at long lives than the curves of aluminum alloys, for example. Therefore, magnesium alloys are especially suited to applications requiring a high number of cycles of fatigue loading.

All magnesium alloys have excellent damping capacity compared to the same form in other metals. As with other metals, sand castings have the highest, die castings have lower, and the wrought forms have the lowest damping capacity.

K1A

A magnesium–zinc alloy, K1A was developed to give the greatest damping capacity in an alloy that has improved castability and better strength.

K1A magnesium casting alloy gives the greatest damping capacity in both sand and die castings. Where greater strength is required, especially creep strength, the elevated-temperature casting alloys can be used, as they have about the next highest damping capacity.

EZ33A-T5 sand castings where rare earth metals and zirconium with or without zinc are added to magnesium, results in casting alloys for use at temperatures falling roughly between 177 to 260°C. This alloy is often used for its high damping capacity combined with good strength.

It has damping capacity comparable to that of gray cast iron. The magnesium–zinc–zirconium and the magnesium–aluminum–zinc casting alloys have lower, but still good, damping capacity. No special high-damping wrought alloys have been developed, as the common alloys, AZ31B and ZE10A sheet and AZ31B extrusions, already have as high damping capacity as appears possible for wrought magnesium alloys.

Applications

Reviewing the nonstructural uses of magnesium provides an accurate perspective of the entire industry. Nonstructural uses are those in which magnesium is used for its particular chemical, electrochemical, and metallurgical properties. The principal nonstructural uses are as an alloying constituent in other metals (principally aluminum); nodularizing agent for graphite in cast iron; reducing agent in the production of other metals, such as uranium, titanium, zirconium, beryllium, and hafnium; desulfurizing agent in iron and steel production; sacrificial anode in the protection of other metals against corrosion; and anode material in reserve batteries and dry cells.

As structural materials, magnesium alloys are best known for their light weight and high strength-to-weight ratio. Accordingly, they are used generally in applications where weight is a critical factor and where high mechanical integrity is needed.

Some automobiles contain magnesium parts such as the distributor diaphragm, steering column brackets, and a lever cover plate. Magnesium-alloy die castings are also used on chain saws, portable power tools, cameras and projectors, office and business machines, tape reels, sporting goods, luggage frames, and many other products.

Use of sand and permanent mold castings are confined largely to aircraft engine and airframe components. Engine parts include gearboxes, compressor housings, diffusers, fan thrust reversers, and miscellaneous brackets. On the airframe, magnesium sand castings are used for leading edge flaps, control pulleys and brackets, entry door gates, and various cockpit components.

Magnesium is used extensively in wheels, auxiliary equipment, flooring, seating, electronic and instrument cases, etc. Other military uses are in ground handling equipment for aircraft and missiles, ordnance equipment such as vehicles and weapons, and portable shelters and hand-carried communications equipment.

In ground transportation vehicles, both military and commercial, magnesium is used in the engines, transmissions, differentials, pumps, and other parts of the power plant. It has also been used in the floors and body panels of trucks and trailers, while for many years magnesium wheels have been used on all the Indianapolis "500" racing cars.

Additionally, the use of magnesium wrought products has become rather limited. Magnesium sheet, extrusions, and forgings continue to be used in a few limited airframe applications. The high-temperature alloys are used in at least 20 different missiles (including ICBMs) and on various spacecraft.

In consumer goods magnesium is used in a wide variety of equipment. Among the household goods in which magnesium is used are portable appliances, furniture, luggage, griddles, ladders, and lawn mowers. Among types of office equipment that use magnesium are typewriters, dictating machines, adding machines, calculators, and furniture. In sporting goods, magnesium is used in such items as sleds, high-jump and pole vault cross bars, and baseball masks. Magnesium is also used in instruments and for such items as binoculars and camera bodies.

Because of its rapid, yet controlled, etching characteristics as well as its lightness, strength, and wear characteristics, magnesium finds usage in photoengraved printing plates.

In addition to the many structural uses, magnesium has several nonstructural uses. A large nonstructural use is in the cathodic protection of other metals from corrosion. It also functions well in dry-cell battery construction. Magnesium also has chemical and metallurgical uses. Among these are the Grignard reaction, pyrotechnics, high-energy fuels, alloying in aluminum, zinc, and lead alloys, and as an additive in the manufacture of nodular cast iron, lead, nickel alloys, and copper alloys. It is also used in the production of titanium, zirconium, beryllium, uranium, and hafnium.

MAGNESIUM OXIDE

Magnesium oxide (MgO) is a synthetic mineral produced in electric arc furnaces or by sintering of amorphous powder (periclase). Refractory applications consume a large quantity of MgO. Both brick and shapes are fabricated at least partially of sintered grain for use primarily in the metal-processing industries. Heating unit insulation is another major application for periclase. Principal advantages of periclase are its thermal conductivity and electrical resistivity at elevated temperatures.

Specialty crucibles and shapes also are fabricated from MgO. These are used in pyrometallurgical and other purifying processes for specialty metals. Both slip-casting and pressing techniques are employed to manufacture shapes.

Thermocouple insulation comprises still another outlet for periclase. Since most of these go into nuclear applications, a high-purity product is required. MgO is also an important glaze constituent.

Single crystals of MgO have received attention because of their use in ductile ceramic studies. Extreme purity is required in this area. Periclase windows are also of potential interest in infrared applications because of their transmission characteristics.

MAGNETIC FERROELECTRICS

These are materials that display both magnetic order and spontaneous electric polarization. Research on these materials has enabled considerable advances to be made in understanding the interplay between magnetism and ferroelectricity. The existence of both linear and higher-order coupling terms has been confirmed, and their consequences studied. They have given rise, in particular, to a number of magnetically induced polar anomalies and have even provided an example of a ferromagnet whose magnetic moment per unit volume is totally induced by its coupling via linear terms to a spontaneous electric dipole moment.

Most known ferromagnetic materials are metals or alloys. Ferroelectric materials, on the other hand, are nonmetals by definition because they are materials that can maintain a spontaneous electric moment per unit volume (called the polarization), which can be reversed by the application of an external electric field. It therefore comes as no surprise to find that there are no known room-temperature ferromagnetic ferroelectrics. In fact, there are no well-characterized materials that are known to be both strongly ferromagnetic and ferroelectric at any temperature. This is unfortunate because not only would a study of the interactions between ferromagnetism and ferroelectricity be valuable as basic research but such interplay could well give rise to important device applications.

Most antiferromagnetic materials, however, are nonmetals, so there is no apparent reason ferroelectricity and antiferromagnetism should not coexist. A study of antiferromagnetic ferroelectrics would also provide much information concerning the interplay of magnetic and ferroelectric characteristics even if the device potential were very much reduced. Somewhat unaccountably, antiferromagnetic ferroelectrics are also comparative rarities in nature.

The antiferromagnetic ferroelectrics $BaMnF_4$, $BaFeF_4$, $BaNiF_4$, and $BaCoF_4$ and their nonmagnetic magnesium and zinc counterparts are orthorhombic and all spontaneously polar (that is, pyroelectric) at room temperature. For all except the iron and manganese materials, which have a higher electrical conductivity than the others, the polarization has been reversed by the application of an electric field, so that they are correctly classified as ferroelectric, although their ferroelectric transition (or Curie) temperatures are in general higher than their melting points.

The importance of these magnetic ferroelectrics is the opportunity they provide to study and to separate the effects of a variety of magnetic and nonmagnetic excitations upon the ferroelectric properties and particularly upon the spontaneous polarization.

MAGNETIC HEATING

Magnetic heating is bidding for recognition as a mainstream thermal technology with the introduction of the Coreflux process. The equipment generates an intense low-frequency magnetic field throughout the cross section of magnetic and paramagnetic materials.

The process is sometimes called UMH for "uniform magnetic heating," because it heats the entire cross section of a workpiece simultaneously. Temperature distribution is uniform; there is little or no thermal gradient from the surface to the center.

However, UMH should not be confused with low-frequency induction heating. Although both processes require magnetic fields, that is where the similarity ends. Induction heating is a surface process. The induction coil generates magnetic eddy currents that are stronger on the surface of the workpiece than at the center. This produces a skin heating effect, and conductivity carries heat into the interior of the part.

Low frequency induction heating systems (around 60 Hz) can penetrate components to a depth of no more than 25 mm for most materials. This type of system still uses eddy currents to generate the heat — so uniform throughheating of thicker parts is not possible.

Process Principles

UMH is based on the principle of hysteresis loss, and heats without relying on thermal conductivity to transmit heat from the surface to the center. With UMH an alternating magnetic field causes domains or crystals in the metal to align and strain in reaction to the field (in the direction of the field for ferrous metals; in the opposite direction for nonferrous metals). As field polarity is reversed, these domains realign; the lag in the process, called hysteresis, is the mechanism that creates heat. This heat is spread rapidly and uniformly through the metal component.

Process Differences

Both induction heating and UMH use coils, but in very different ways. A part to be heated by induction is placed within the loops of a single coil. The magnetic field for UMH is generated by two coils that induce magnetic fields in two cores laminated from high-permeability, low-reluctance directional steel. The cores direct magnetic flux into the workpiece, and pneumatic cylinders clamp the workpiece between the laminated cores to prevent vibration when the field is energized.

For large masses that require uniform heating to produce optimum results, UMH is the logical alternative to induction.

Generating the field for uniform magnetic heating requires less energy and equipment than induction heating and can produce a field that can be varied from 40 to 400 Hz using a standard, variable-frequency motor drive inverter. It is less costly than the power supply for induction heating.

Another advantage is that UMH has no plumbing so the installation is simplified, and one of the other biggest advantages of this power supply is that water cooling is eliminated; only 480 VAC three-phase electric service and 90 psi shop air is needed. Also, a huge amount of induction heat is lost in water-cooling the coil. With the air-cooled coils, most of the energy flows to the workpiece.

UMH also has a tooling advantage. The same laminated cores can be used for a variety of parts. Induction heating, however, requires a separate coil customized for each component.

Temperatures can be maintained to an accuracy of ±2°C, which is generally a tighter tolerance than most users can measure. A proportional integral derivative (PID) controller tapers the percentage output of the power supply as the workpiece reaches the required temperature, to avoid overshooting the set point. The programmable logic controller includes a touch-screen interface and multiple thermocouple inputs.

Processing

UMH is capable of heating to 1100°C, well beyond the Curie point, at which magnetic materials lose their magnetic properties. But its benefits compared to induction heating are said to be more tangible at sub-Curie temperatures, at which induction tends to overheat part surfaces.

UMH can through-heat to 260 to 316°C very efficiently, while induction really struggles at these low temperatures because localized surface heating is very pronounced and there is little penetration. Above the Curie point, the eddy currents of induction heating penetrate deeper into the workpiece.

Applications

UMH machines can be applied to tempering, stress-relieving, and hardening of parts as well as bonding and shrink-fitting and are all sound applications. UMH can also preheat billets for extrusion and semisolid forming, and dies for extrusion and forging. Sintering of powder metals and brazing are among its other capabilities.

MAGNETIC IRONS AND STEELS

All magnetic materials fall into two general classes: they are either permanently or nonpermanently magnetized. The permanently magnetized materials retain their magnetization after being placed in a magnetic field and can be used as a constant source of magnetic field. On the other hand, the nonpermanently magnetized, or soft magnetic materials, retain their magnetization only while a magnetic field is applied to the material.

The basic magnetization process and hysteresis cycle is best understood in terms of the static domain theory, which is applicable under DC excitation. However, most applications of magnetic materials are for AC use. The predominant frequency commercially used in the United States is 60 cycles. For many special applications, particularly in the aircraft and military field, 400 cycles is a common frequency. The AC losses are not simple multiples of the DC hysteresis loss, for in addition to the basic magnetic losses.

Applications

The magnetic alloys used for AC applications are used in sheet form and are most often alloyed so that maximum resistivity is obtained commensurate with the induction required, ability to hot- and cold-work the material, minimum hysteresis loss, and cost.

The number of applications of magnetic materials is so large that a complete listing is

TABLE M.3
DC Magnetic Properties

Alloy	Sheet Thick, 0.001 in.	H_c	μ_1	μ_{10}	μ_{100}	B_r
Iron and Iron Silicon						
Comm pure Fe	62.5	0.83[a]	3600	1530	180	8362
Fe–0.35% Si	25.0	2.0[a]	1000	1340	176	8500
Fe–0.70% Si	25.0	0.92[a]	4300[c]	1400	175	8250
Fe–1.60% Si	25.0	1.00[a]	3900[c]	1380	175	7800
Fe–2.80% Si	18.5	0.75[a]	5000	1380	171	7500
Fe–3.25% Si	18.5	0.50[a]	6400	1380	171	6600
Fe–3.25% Si (oriented)	12.0	0.095[b]	16,000	1800	198	12,700
Fe–3.25% Si (oriented)	14.0	0.095[b]	15,700	1760	196	12,200
Fe–3.25% Si	5.0	0.70[a]	5200	1400	—	7600
Fe–3.25% Si (oriented)	4.0	0.28[b]	16,000	1800	—	14,200
Fe–3.25% Si (oriented)	2.0	0.40[b]	13,200	1630	—	13,800
Nickel–Iron						
50% Ni–50% Fe	14	0.04	13,000 (0.01)[d]	85,000 (0.1)[d]	12,000 (1.0)[d]	9000
50% Ni–50% Fe (oriented)	2–6	0.07	13,000 (0.1)[d]	15,000 (1.0)[d]	1500 (10.0)[d]	14,000
4% Mo–79% Ni, Balance Fe	20	0.015	40,000 (0.001)[d]	200,000 (0.01)[d]	60,000 (0.1)[d]	5000
Cobalt–Iron						
27% Co–Fe	14	2.0	2000 (3.0)[d]	1500 (10)[d]	208 (100)[d]	10,500
35% Co–Fe	17	1.8	2000 (3.0)[d]	1500 (10)[d]	230 (100)[d]	9000
50% Co–Fe	4	1.0	4400 (1.0)[d]	2000 (10)[d]	226 (100)[d]	10,000
50% Co–2% V	4	0.3	47,000 (0.17)[d]	19,500 (1.0)[d]	216 (10)[d]	19,000

[a] Hysteresis loop measured from $B_{max.}$ = 10,000 G.
[b] Hysteresis loop measured from $B_{max.}$ = 15,000 G.
[c] Permeability at H = 2 oe.
[d] Field at which permeability is measured.

not feasible. Furthermore, even for the same general application, such as a motor, many acceptable designs could be made, each requiring different-quality levels of magnetic material. Therefore, no attempt will be made to evaluate all the uses for a given alloy. However, several typical uses will be given. In some instances very special properties of a magnetic material may be required. The most common magnetic materials with their characteristic AC and DC properties are listed in Tables M.3 and M.4.

Tables M.3 and M.4 are designed to give the reader a ready reference source to identify the kinds of materials available and a brief resume of their most important properties. After preliminary selection, reference should be made to the detailed curves available from reputable suppliers before any design is anticipated. Many of the materials listed require special handling and annealing techniques to achieve optimum properties.

MAGNETIC IRONS AND STEELS

TABLE M.4
AC Magnetic Properties

Alloy	Sheet Thick, 0.001 in.	60 Cycles			40 Cycles			Applications
		W_a	W_b	W_c	W_a	W_b	W_c	
Comm pure Fe	62.5	—	—	—	—	—	—	DC devices, i.e., electromagnets, relays, pole pieces
Fe–0.35% Si	25.0	0.80 (5)[b]	2.8 (10)[b]	7.0 (15)[b]	—	—	—	Small motors, intermittently used electrical apparatus
Fe–0.70% Si	25.0	0.54 (5)[b]	1.9 (10)[b]	4.3 (15)[b]	—	—	—	Small motors, fractional horsepower motors
Fe–1.6% Si	25.0	0.49 (5)[b]	1.6 (10)[b]	3.5 (15)[b]	—	—	—	High-quality motors, medium efficiency motors, and generators
Fe–2.8% Si	18.5	0.33 (5)[b]	1.1 (10)[b]	2.4 (15)[b]	—	—	—	Motors and generators, reactors, small transformers
Fe–3.25% Si	18.5	0.26 (5)[b]	0.90 (10)[b]	1.95 (15)[b]	—	—	—	High-efficiency motors and generators, reactors, motors
Fe–3.25% Si (oriented)	12.0	0.15 (7)[b]	0.25 (10)[b]	0.57 (15)[b]	1.0 (4)[b]	5.0 (10)[b]	12.5 (15)[b]	High-quality, high-power, continuous duty transformers
Fe–3.25% Si	5.0	—	—	—	0.11 (1)[b]	1.8 (5)[b]	6.0 (10)[b]	Aircraft motors and transformers, television transformers
Fe–3.25% Si (oriented)	4.0	—	—	—	0.035 (1)[b]	0.70 (5)[b]	2.5 (10)[b]	Magnetic amplifiers, television transformers, power aircraft transformers
Fe–3.25% Si (oriented)	2.0	—	—	—	0.035 (1)[b]	0.84 (5)[b]	3.0 (10)[b]	Magnetic amplifiers, pulse transformers
50% Ni–50% Fe	14.0	0.0035 (1)[b]	0.054 (5)[b]	0.21 (10)[b]	0.64 (1)[b]	1.1 (5)[b]	5.1 (10)[b]	Instrument transformers, magnetic shields, sensitive low-current relays
50% Ni–50% Fe	2–6	Specialized use at 0 and 400 cycles requires *Constant Current Flux React Test*						Magnetic amplifiers, current transformers, pulse transformers
27% Co–Fe	0.014	3.8 (15)[b]	5.1 (18)[b]	6.6 (21)[b]	50 (15)[b]	62 (17)[b]	76 (19)[b]	
27% Co–Fe	0.004	2.9 (15)[b]	3.9 (18)[b]	5.0 (21)[b]	24 (15)[b]	29 (17)[b]	34 (19)[b]	High-temperature generators, high-temperature transformer transducers, aircraft equipment, pole pieces
35% Co–Fe	0.017	3.0 (15)[b]	3.9 (18)[b]	5.4 (21)[b]	52 (15)[b]	69 (18)[b]	110 (21)[b]	
50% Co–2% V, Balance Fe	0.004[c]	0.67 (18)[b]	0.80 (20)[b]	1.00 (22)[b]	8.5 (18)[b]	12.0 (20)[b]	17.0 (22)[b]	

[a] Dash indicates alloy is not usually used at this frequency.
[b] Induction in kilogauss at which loss is measured.
[c] Field annealed.

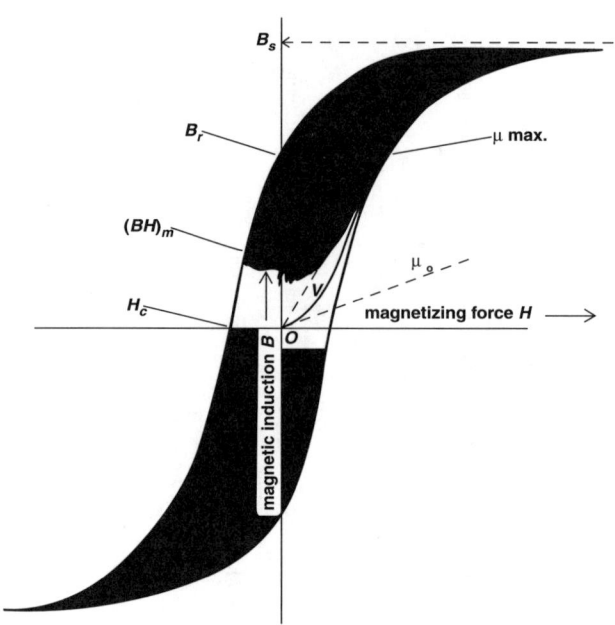

FIGURE M.4 Identification of properties related to the magnetic hysteresis loop. Colored area is the magnetization energy loss in every cycle (hysteresis loss). (*Source: McGraw-Hill Encyclopedia of Science and Technology*, 8th ed., Vol. 10, McGraw-Hill, New York, 521. With permission.)

MAGNETIC MATERIALS (HARD AND SOFT)

These are materials exhibiting ferromagnetism. The magnetic properties of all materials make them respond in some way to a magnetic field, but most materials are diamagnetic or paramagnetic and show almost no response.

The materials that are most important to magnetic technology are ferromagnetic and ferrimagnetic materials. Their response to a field H is to create an internal contribution to the magnetic induction B proportional to H, expressed as $B = \mu H$, where μ, the permeability, varies with H for ferromagnetic materials. Ferromagnetic materials are the elements iron, cobalt, nickel, and their alloys, some manganese compounds, and some rare earths. Ferrimagnetic materials are spinels of the general composition MFe_2O_4, and garnets, $M_3Fe_5O_{12}$, where M represents a metal.

Ferromagnetic materials are characterized by a Curie temperature above which thermal agitation destroys the magnetic coupling giving rise to the alignment of the elementary magnets (electron spins) of adjacent atoms in a crystal lattice. Below the Curie temperature, ferromagnetism appears spontaneously in small volumes called domains. In the absence of a magnetic field, the domain arrangement minimizes the external energy and the bulk material appears unmagnetized.

Magnetic materials are further classified as soft or hard according to the ease of magnetization. Soft materials are used in devices in which change in the magnetization during operation is desirable, sometimes rapidly, as in AC generators and transformers. Hard materials are used to supply a fixed field either to act alone, as in a magnetic separator, or to interact with others, as in loudspeakers and instruments. Both soft and hard materials are characterized by their magnetic hysteresis curve (Figure M.4).

These two broad groups — soft magnetic materials, sometimes called electromagnets, do not retain their magnetism when removed from a magnetic field; hard magnetic materials, sometimes referred to as permanent magnets, retain their magnetism when removed from a magnetic field. Cobalt is the major element used for obtaining magnetic properties in hard magnetic alloys.

MAGNETIC MATERIALS (HARD AND SOFT)

Soft Materials

These materials are characterized by their low loss and high permeability. There are a variety of alloys used with various combinations of magnetic properties, mechanical properties, and cost (Table M.5). There are seven major groups of commercially important materials: iron and low-carbon steels, iron–silicon alloys, iron–aluminum–silicon alloys, nickel–iron alloys, iron–cobalt alloys, ferrites, and amorphous alloys.

The behavior of soft materials is controlled by the pinning of domain walls at heterogeneities such as grain boundaries and inclusions. Thus, the common goal in their production is to minimize such heterogeneities. In addition, eddy-current loss is minimized through alloying additions that increase the electrical resistivity. Initial permeability, important in electronic transformers and inductors, is improved by minimizing all sources of magnetic anisotropy, for example, by using amorphous metallic alloys and by using zero magnetostrictive alloys. A high maximum permeability, which is necessary for motors and power transformers, is increased by the alignment of the anisotropy, for example, through development of crystal texture or magnetically induced anisotropy.

The class of alloys used in largest volume is by far iron and 1 to 3.5% silicon–iron for applications in motors and large transformers. In these applications the cost of the material is often the dominant factor, with losses and excitation power secondary but still important. Thus, the improvment over the years in these alloys has been in developing lower losses without increases in cost.

Common soft magnetic materials are iron, iron–silicon alloys, and nickel–iron alloys. Irons are widely used for their magnetic properties because of their relatively low cost. Common iron–silicon magnetic alloys contain 1, 2, 4, and 5% silicon. There are about six types of nickel–irons, sometimes called permeability alloys, used in magnetic applications. For maximum magnetostriction, the two preferred nickel contents are 42 and 79%. Additions of molybdenum give higher resistivities, and additions of copper result in higher initial permeability and resistivity.

Soft magnetic ceramics, also referred to as ceramic magnets, ferromagnetic ceramics, and ferrites (soft), were originally made of an iron oxide, Fe_2O_3, with one or more divalent oxides such as NiO, MgO, or ZnO. The mixture is calcined, ground to a fine powder, pressed to shape, and sintered. Ceramic and intermetal types of magnets have a square hysteresis loop and high resistance to demagnetization, and are valued for magnets for computing machines where a high remanence is desired. A ferrite with a square loop for switching in high-speed computers contains 40% Fe_2O_3, 40% MnO, and 20% CdO. Some intermetallic compounds, such as zirconium–zinc, $ZrZn_2$, which are not magnetic at ordinary temperatures, become ferromagnetic with properties similar to ferrites at very low temperatures, and are useful in computers in connection with subzero superconductors. Some compounds, however, are the reverse of this, being magnetic at ordinary temperatures and nonmagnetic below their transition temperature point. This transition temperature, or Curie point, can be arranged by the compounding to vary from subzero temperatures to above 100°C. Chromium–manganese–antimonide, $Cr_xMn_{2x}Sb$, is such a material. Chromium manganese alone is ferromagnetic, but the antimonide has a transition point varying with the value of x.

Vectolite is a lightweight magnet made by molding and sintering ferric and ferrous oxides and cobalt oxide. The weight is 3.2 g/cm³. It has high coercive force, and has such high electrical resistance that it may be considered as a nonconductor. Magnadur was made from barium carbonate and ferric oxide, and has the formula $BaO(Fe_2O_3)_6$. Indox and Ferroxdure are similar. This type of magnet has a coercive force to 127,200 A/m, with initial force to 206,700 A/m, high electrical resistivity, high resistance to demagnetization, and light weight, with specific gravity from 4.5 to 4.9. Ferrimag and Cromag are ceramic magnets. Strontium carbonate is superior to barium carbonate for magnets but is more costly. Lodex magnets are extremely fine particles of iron–cobalt in lead powder made into any desired shape by powder metallurgy.

TABLE M.5
Some Properties of Selected Soft Magnetic Materials

Material or Trade Name	Composition % by Weight, Remainder Fe	Relative Permeability		Coercive Force, H_c, A/m (oersteds)	Saturation Induction B_s, teslas (gauss)	Curie Temperature, T_c, °F (°C)	Electrical Resistivity ρ, $\mu\Omega$-cm	Core Loss at 60 Hz, 1.5 teslas (15 kilogauss), W/lb (W/kg)	Sample Thickness, in. (mm)
		Initial μ_0	Max. μ_{max}						
Iron, high-purity	0.05 impurity	10,000	200,000	88 (1.1)	2.15 (21,500)	1420 (770)	10	5.9 (13)	0.025 (0.64)
Iron, commercial-purity	0.2 impurity	250	9,000	72 (0.9)	2.14 (21,400)	1420 (770)	12	3.7 (8.1)	0.019 (0.47)
Carbonyl iron powder	—	60	150	—	—	—	10×10^6		
Armature M-43	0.95Si	—	4,100	75 (0.94)	2.13 (21,300)	1400 (760)	24	2.3 (5.1)	0.019 (0.47)
Electric M-36	2Si	—	7,500	40 (0.5)	2.11 (21,100)	1390 (755)	41	1.9 (4.2)	0.014 (0.36)
Dynamo M-22	2.8Si	—	9,400	32 (0.4)	2.03 (20,300)	1365 (740)	49	1.7 (3.7)	0.014 (0.36)
Transformer, M-15	2.2Si	1,500	7,000	28 (0.35)	1.95 (19,500)	1345 (730)	53	1.4 (3.0)	0.014 (0.36)
Oriented curbe-on-edge texture, M-4	3.2Si	7,500	55,000	8 (0.1)	2.01 (20,100)	1365 (740)	48	0.55 (1.2)	0.011 (0.27)
High-permeability G.O.	3.2Si	—	—	—	2.01 (20,100)	1365 (740)	45	0.35 (0.77)	0.008 (0.20)
Low-aluminum iron (3.5%)	3.5Al	500	19,000	24 (0.3)	1.90 (19,000)	1380 (750)	47		
16% Al-Fe	16.0Al	3,900	110,000	1.2 (0.015)	0.50 (5,000)	840 (450)	153		

Material	Composition	Column3	Column4	Column5	Column6	Column7	Column8
Sendust	5.0Al, 10.0Si	30,000	120,000	4 (0.05)	1.00 (10,000)	930 (500)	45
Thermoperm	30Ni	—	—	—	0.203 (2,030)	120 (50)	—
45 Permalloy	45Ni	2,500	25,000	8 (0.10)	1.60 (16,000)	750 (400)	45
50–50 Ni–Fe	50Ni	4,000	100,000	4 (0.05)	1.60 (16,000)	930 (500)	45
Mumetal	77Ni,5Cu,2Cr	20,000	100,000	4 (0.05)	0.65 (6,500)	—	62
78 Permalloy	78.5Ni	8,000	100,000	4 (0.05)	1.08 (10,800)	1110 (600)	16
Supermalloy	79Ni,5Mo	100,000	1,000,000	0.32 (0.004)	0.78 (7,800)	750 (400)	60
2–81 Moly permalloy powder	81Ni,2Mo	125	130	—	—	—	16×10^6
27% Co–Fe	27Co,1Cr	650	10,000	56 (0.70)	2.42 (24,200)	—	28
50% Co–Fe	49Co,2V, or 2Cr	800	5,000	160 (2.0)	2.45 (24,500)	1800 (980)	7
Supermendur	49Co,2V	800	70,000	24 (0.3)	2.40 (24,000)	1800 (980)	40
45–25 Perminvar	45Ni,25Co	400	2,000	95 (1.2)	1.55 (15,500)	1320 (715)	19
Mn–Zn Ferrite	Mn–δZn–δFe$_2$O$_4$	1,500	2,500	16 (0.2)	0.34 (3,400)	265 (130)	20×10^6
Ni–Zn Ferrite	—	2,500	5,000	8 (0.1)	0.32 (3,200)	285 (140)	10^{11}
Amorphous Fe–B–Si, METGLAS 2605S-2	4B,3Si	15,000	300,000	1.6 (0.02)	1.56 (15,600)	780 (415)	130	0.1 (0.2) at 1.4T (14kG)	0.001 (0.025)

Source: McGraw-Hill Encyclopedia of Science and Technology, 8th ed., Vol. 10, McGraw-Hill, New York, 322. With permission.

Hard Materials

Permanent magnets, or hard magnetic materials, strongly resist demagnetization once magnetized (Table M.6). They are used, for example, in motors, loudspeakers, meters, and holding devices, and have coercivities H_c from several hundred to many thousands of oersteds (10 to over 100 kA/m). The bulk of commercial permanent magnets are of the ceramic type, followed by the Alnicos and the cobalt–samarium, iron–neodymium, iron–chromium–cobalt, and elongated single-domain (ESD) types in decreasing sequence of usage. The overall quality of a permanent magnet is represented by the highest-energy product $(BH)_m$; but depending on the design considerations, high H_c, high residual induction B_r (the magnetic induction when H is reduced to zero), and reversibility of permeability may also be controlling factors.

To understand the relation between the resistance to demagnetization, that is, the coercivity, and the metallurgical microstructure, it is necessary to understand the mechanisms of magnetization reversal. The two major mechanisms are reversal against a shape anisotropy and reversal through nucleation and growth of reverse magnetic domains against crystal anisotropy. The Alnicos, the iron–chromium–cobalt alloys, and the ESD Lodex alloys are examples of materials of the shape anisotropy structure, whereas barium ferrites, the cobalt–samarium alloys, and the iron–neodymium–boron alloys are examples of the crystal anisotropy-controlled materials.

Neodymium–iron–boron is a powerful magnetic material that works best at room temperature. It begins losing its properties at higher temperatures, and at 312°C it loses its magnetism completely.

An investigation into the use of magnets in a Stirling engine to power deep space probes is under way. Stirling engines have a sealed cylinder in which hot gases move two pistons back and forth. By placing magnets on the ends of the pistons, and surrounding the cylinder with wire coils, the magnets would induce a current flow.

Research is being conducted whereby a magnetic material that would work at elevated temperatures is being produced in particles that each have two different compositions — one at the outer edge to resist demagnetization, and another at the core to retain magnetic power at higher temperatures. The "functionally graded" material would essentially have the outside material protecting the inside material.

The new magnets would be attractive for a variety of applications on Earth as well — cars, electronics, computers, and power tools.

Magnet steels, now largely obsolete, included plain high-carbon (0.65 or 1%) steels or high-carbon (0.7 to 1%) compositions containing 3.5% chromium–chromium magnet steels; 0.5% chromium and 6% tungsten–tungsten magnet steel; or chromium, tungsten, and substantial cobalt (17 or 36%) – cobalt magnet steels. They were largely replaced by ternary alloys of iron, cobalt, and molybdenum, or tungsten. Comol has 17% molybdenum, 12% cobalt, and 71% iron. Indalloy and Remalloy have similar compositions: about 20% molybdenum, 12% cobalt, and 68% iron. Chromindur has 28% chromium, 15% cobalt, and the remainder iron, with small amounts of other elements that give it improved strength and magnetic properties. In contrast to Indalloy and Remalloy, which must be processed at temperatures as high as 1250°C, Chromindur can be cold-formed.

Some cobalt magnet steels contain 1.5 to 3% chromium, 3 to 5% tungsten, and 0.50 to 0.80% carbon, with high cobalt. Alfer magnet alloys, first developed in Japan to save cobalt, were iron–aluminum alloys. MK alloy had 25% nickel, 12% aluminum, and the balance iron, close to the formula Fe_2NiAl.

Cunife is a nickel–cobalt–copper alloy that can be cast, rolled, and machined. It is not magnetically directional like the tungsten magnets, and thus gives flexibility in design. The electric conductivity is 7.1% that of copper, and it has good coercive force. Cunife 1 contains 50% copper, 21% nickel, and 29% cobalt. Cunife 2, with 60% copper, 20% nickel, and 20% iron, is more malleable. This alloy, heat-treated at 593°C, is used in wire form for permanent magnets for miniature apparatus. It has a coercive force of 39,750 A/m. Hipernom is a high-permeability nickel–molybdenum magnet alloy containing 79% nickel, 4% molybdenum, and the balance iron. It has a Curie temperature of

TABLE M.6
Representative Permanent-Magnet Properties

Material	Composition % by Weight	Curie Temperature T_c °F	Curie Temperature T_c °C	Coercive Force H_c kA/m	Coercive Force H_c Oe	Residual Induction B_c T	Residual Induction B_c G	Max Energy Product $(BH)_m$ kJ/m³	Max Energy Product $(BH)_m$ MGOe	Preparation	Mechanical Properties
Ba ferrite	BaO·6Fe$_2$O$_3$	840	450	170	2,100	0.43	4,300	36	4.5	Press, sinter	Brittle
Sr ferrite	SrO·6Fe$_2$O$_3$	860	460	250	3,100	0.42	4,200	36	4.5	Press, sinter	
Alnico 5	50Fe, 24Co, 15Ni, 8Al, 3Cu	1650	900	58	620	1.25	12,500	42	5.3	Cast, anneal	Hard, brittle
Alnico 8	34Fe, 35Co, 15Ni, 7Al, 5Ti, 4Cu	1580	860	130	1,600	0.83	8,300	40	5.0		
Alnico 9	34Fe, 35Co, 15Ni, 7Al, 5Ti, 4Cu			120	1,450	1.05	10,500	68	8.5		
Fe–Cr–Co	63Fe, 22Cr, 15Co	1165	630	51	640	1.56	15,600	66	8.3	Cast, anneal	Hard
Fe–Cr–Co–Cu	42Fe, 33Cr, 23Co, 2Cu			86	1080	1.30	13,000	78	9.8	Roll, anneal	Hard
Co$_5$Sm	66Co, 34Sm	1290	700	665	8,300	0.91	9,050	160	20	Press, sinter	Brittle
Co$_{17}$Sm$_2$	77Co, 23Sm	1470	800	670	8,400	1.08	10,800	223	28	Press, sinter	Brittle
Elongated single domain (ESD) Fe–Co (Lodex)	9.9Fe, 5.5Co, 77Pb, 8.6Sn	1795	980	70	870	0.8	8,000	25	3.2	Electroplate, distill, press	Soft
Mn–Al–C	70Mn, 29Al, 0.5Ni, 0.5C	570	300	220	2,700	0.61	6,100	56	7.0	Cast, extrude, anneal	
Co–Pt	77Pt, 23Co	895	480	360	4,500	0.65	6,500	73	9.2	Cast, anneal	Hard, strong
Fe–Nd–B	66Fe, 33Nd, 1B	570	300	905	11,300	1.21	12,100	280	35	Press, sinter, anneal	Brittle

Note: Oe = oersteds, T = teslas, G = gauss, MGOe = megagauss–oersteds.

Source: McGraw-Hill Encyclopedia of Science and Technology, 8th ed., Vol. 10, McGraw-Hill, New York, 323. With permission.

460°C and is used for relays, amplifiers, and transformers.

In the Alnico alloys, a precipitation hardening occurs with AlNi crystals dissolved in the metal and aligned in the direction of magnetization to give greater coercive force. This type of magnet is usually magnetized after setting in place. Alnico 1 contains 21% nickel, 12% aluminum, 5% cobalt, 3% copper, and the balance iron. The alloy is cast to shape, is hard and brittle, and cannot be machined. The coercive force is 31,800 A/m. Alnico 2, a cast alloy with 19% nickel, 12.5% cobalt, 10% aluminum, 3% copper, and the balance iron, has a coercive force of 44,520 A/m. The cast alloys have higher magnetic properties, but the sintered alloys are fine-grained and stronger. Alnico 4 contains 12% aluminum, 27% nickel, 5% cobalt, and the balance iron. It has a coercive force of 55,650 A/m, or ten times that of a plain tungsten magnet steel. Alnico 8 has 35% cobalt, 34% iron, 15% nickel, 7% aluminum, 5% titanium, and 4% copper. The coercive force is 115,275 A/m. It It has a hardness of Rockwell C59. The magnets are cast to shape and finished by grinding. Hyflux AlNico 9, of the same coercive force, has an energy product of 75,620 T · A/m. The magnets of this material, made by Indiana General Corp., are cylinders, rectangles, and prisms, usually magnetized and oriented in place. The Alnicus magnets are Alnico-type alloys with the grain structure oriented by directional solidification in the casting, which increases the maximum energy output. Ticonal, Alcomax, and Hycomax are Alnico-type magnet alloys produced in Europe.

Cobalt–platinum, as an intermetallic rather than an alloy, has a coercive force above 341,850 A/m, and a residual induction of 0.645 T. It contains 76.8% by weight of platinum and is expensive, but is used for tiny magnets for electric wristwatches and instruments. Placovar is a similar alloy that retains 90% of its magnetization flux up to 343°C. It is used for miniature relays and focusing magnets. Ultramag is a platinum–cobalt magnet material with a coercive force of 381,600 A/m. The Curie temperature is about 500 °C, and it has only slight loss of magnetism at 350°C, whereas cobalt–chromium magnets lose their magnetism above 150°C. The material is easily machined. Alloy 1751 is a cobalt–platinum intermetallic with a coercive force of 341,850 A/m, or of 540,600 A/m in single-crystal form. The metal is not brittle and can be worked easily. It is used for the motor and index magnets of electric watches.

Ceramic permanent magnets are compounds of iron oxide with oxides of other elements. The most used are barium ferrite, oriented barium ferrite, and strontium ferrite. Yttrium–iron garnets (YIG) and yttrium–aluminum garnets (YAG) are used for microwave applications.

Flexible magnets are made with magnetic powder bonded to tape or impregnated in plastic or rubber in sheets, strip, or forms. Magnetic tape for recorders may be made by coating a strong, durable plastic tape, such as a polyester, with a magnetic ferrite powder. For high-duty service, such as for spacecraft, the tape may be of stainless steel. For recording heads the ferrite crystals must be hard and wear resisistant.

Ferrocube is manganese zinc. The tiny crystals are compacted with a ceramic bond for pole pieces for recorders. Plastiform is a barium ferrite bonded with rubber in sheets and strips. Magnyl is vinyl resin tape with the fine magnetic powder only on one side. It is used for door seals and display devices.

Rare earth magnetic materials, used for permanent magnets in computers and signaling devices, have coercive forces up to ten times that ordinary magnets. They are of several types. Rare earth–cobalt magnets are made by compacting and extruding the powders with a binder of plastic or soft metal into small precision shapes. They have high permanency. Samarium–cobalt and cesium–cobalt magnets are cast from vacuum melts and are chemical compounds, $SmCo_5$ and $CeCo_5$. These magnets have intrinsic coercive forces up to 2.2 million A/m. The magnetooptic magnets for memory systems in computers are made in thin wafers, often no more than a spot in size. These are ferromagnetic ceramics of europium–chalcogenides. Spot-size magnets of europium oxide only 4 μm in diameter perform reading and writing operations efficiently. Films of this ceramic less than a wavelength in thickness are used as memory storage mediums.

Magnetic fluids consist of solid magnetic particles in a carrier fluid. When a magnetic field is applied, the ultramicroscopic iron oxide particles become instantly oriented. When the field is removed, the particles demagnetize within microseconds. Typical carrier fluids are water, hydrocarbons, fluorocarbons, diesters, organometallics, and polyphenylene ethers. Magnetic fluids can be specially formulated for specific applications such as damping, sealing, and lubrication.

MAGNETIC-SHIELDING METALS

The materials that fall under this category are used in airport environments, which deal with what is known in the industry as ELF (extremely low frequency) radiation interference to ticketing computer terminals caused by the close proximity of building utility electrical vaults as well as to main switchgear rooms.

Another application is in the production of handheld electronic equipment with and without radio cards.

Material Types

MuShield works primarily with an 80% nickel alloy. The alloy, with such trade names as HyMu80, Permalloy, and MuMetal, shields from low-frequency electromagnetic interference. For high-frequency shielding, refer to fabricators that work with copper, aluminum, and other conductive materials.

The term "B-40" is an arbitrary starting point for high-permeability alloys because it is a good average value of the flux density induced within a piece of high-permeability material.

It has also been found that time-varying (AC) magnetic fields need to be attenuated below 7 milligauss for most monitor applications. Also, larger monitors are more sensitive to EMI.

MAGNETOSTRICTION

The principle of magnetostriction is illustrated in Figure M.5. In the position sensor, a pulse is induced in a magnetostrictive waveguide by the momentary interaction of two magnetic fields: one from a magnet passing along the outside of the sensor tube; the other field from a current pulse launched along a waveguide within the tube. The interaction produces a strain pulse (twisting the waveguide) that travels at sonic speeds down the waveguide until detected at the sensor head. Measuring the elapsed time between the launching of the electronic pulse and the arrival of the strain pulse, or pulses, precisely determines the position of one, or more, magnets.

Such noncontact position sensing produces no wear in the sensing elements, cutting maintenance and extending sensor life. The encapsulated waveguide and electronics also provide durability in severe environments. And modularity gives mounting flexibility and easy integration.

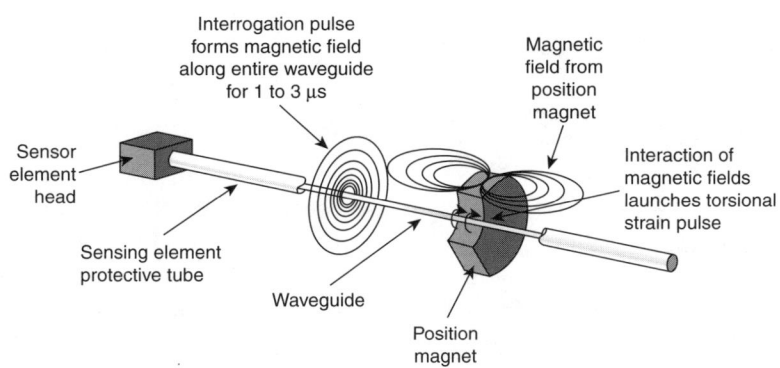

FIGURE 12.5 Principle of magnetostriction. (From *Design News*, June 22, p. 420, 1998. With permission.)

MALLEABLE IRONS

The malleable irons are a family of cast alloys — consisting primarily of iron, carbon, and silicon — which are cast as hard, brittle white iron and then rendered tough and ductile through a controlled heat conversion process. Because of their unique metallurgical structure, they possess a wide range of desirable engineering properties including strength, toughness, ductility, resistance to corrosion, machinability, and castability.

Three principal types of malleable iron are in wide use in this country: ferritic, pearlitic, and alloy malleable iron. A fourth type, called cupola malleable because of the method of manufacture, is also produced but only in small tonnages. Most important from the standpoint of production volume and use is standard malleable iron, which has a ferrite matrix. Pearlitic malleable, which, as the name implies, has a pearlitic matrix, is being produced in ever-increasing quantities. Alloy malleable iron is basically a specialty type iron with higher strength and corrosion resistance, finding primary use in railroad parts.

By far the largest tonnage of malleable castings normally is consumed by the automotive industry. The railroad, agricultural implement, electrical line hardware, pipe fittings, and detachable chain industries, and most other basic industries use standard and pearlitic malleable castings.

Advantages

Important attributes of malleable iron can be summarized as follows:

1. Malleable iron can be produced with a high yield strength, which is the static mechanical property upon which most mechanical design is based.
2. Ferritic and pearlitic malleable irons have a high ratio of yield strength to tensile strength. This means that the engineer can design to high applied strength values in service for materials of construction, concomitant with good machinability and low production cost for the final part.
3. Pearlitic malleable irons can be produced to a wide range of mechanical properties through carefully controlled heat treatments.
4. Malleable irons have a high modulus of elasticity and a low coefficient of thermal expansion, compared with the nonferrous metals.
5. Malleable and pearlitic malleable irons exhibit a low nil ductility transition temperature for brittle fracture.
6. Compared with steel, malleable and pearlitic malleable irons have considerably better damping capacity, which makes operation of moving components less prone to noise because of resonant vibration.
7. Pearlitic malleable irons exhibit good wear resistance and can be selectively hardened by flame, induction, or the carbonitriding process.
8. Pearlitic malleable irons will take a high-quality finish. Honed surfaces with 2 to 3 μin. finish at hardness values of 197 to 207 Bhn have been reported.
9. The uniformity of properties from surface to center is excellent, particularly in oil-quenched and tempered pearlitic malleable iron.
10. All malleable irons are substantially free from residual stresses as a result of long heat treatments at high temperatures.
11. Pearlitic malleable iron provides the properties of medium to high carbon steel coupled with a machinability rating unequaled for a material of similar hardness.

With respect to mechanical properties, minimum specification values are generally exceeded by a comfortable margin in the better-controlled malleable foundries.

Brinell hardness of ferritic malleable irons varies from about 110 to 145. Pearlitic malleable and alloyed malleable iron grades have higher values, ranging usually between 160 and 280 Bhn. Both hardness and tensile strength increase with combined carbon content.

Since the final properties of malleable iron castings are the result of thermal treatments, section thickness has no appreciable effect on strength. Therefore, mechanical properties will be essentially the same throughout the entire cross section.

Manufacture

The manufacture of malleable iron castings is fundamentally a two-phase operation. Phase one consists of producing the white iron castings, and the second phase involves the controlled heat treatment of these castings to obtain the desired finished product.

Structurally, malleable iron castings consist essentially of carbon-free iron (ferrite) and uniformly dispersed nodules of temper carbon. This combination of soft, ductile ferrite and nodular temper carbon accounts for the desirable mechanical properties of malleable iron. In pearlitic malleable iron the matrix is essentially pearlitic, resembling that of a medium-carbon steel.

Properties

Effect of Temperature

Studies of the behavior of ferritic malleable iron at both high and lower temperatures demonstrate, in general, that this material is well suited to applications in a temperature range from –51 to 649°C.

Low-temperature investigations have been concerned primarily with impact resistance and notch sensitivity; high-temperature studies have focused principally on tensile strength, yield point, elongation, stress rupture, and creep behavior.

Results of research have indicated a high level of performance at elevated temperatures, equal or superior to other ferritic materials for which data are available, particularly at 427°C. Strength at 538°C is adequate for many applications and strength is retained even at 649°C. No evidence was found in any of the investigations of changes in structure or performance during the test periods, which extended from 1 to over 2000 hr.

Surface Hardness

Many structural parts require high surface hardness backed up by a strong, tough core. In steel components this can be accomplished by carburizing or nitriding after machining, followed by a suitable heat treatment. In pearlitic malleable iron the combined carbon content is adequate for production of high surface hardnesses through quenching after either induction or flame heating. Many parts are preferentially hardened on wearing surfaces.

MANGANESE AND ALLOYS

A metallic element, symbol Mn, manganese is found in the minerals manganite and pyrolusite, and with most iron ores and traces in most rocks. Manganese has a silvery-white color with purplish shades.

It is brittle but hard enough to scratch glass. The specific gravity is 7.42, melting point 1245°C, and weight is 7418 kg/m^3. It decomposes in water slowly. It is not used alone as a construction metal. The electrical resistivity is 100 times that of copper or three times that of 18-8 stainless steel. It also has a damping capacity 25 times that of steel, and can be used to reduce the resonance of other metals.

It is the twelfthmost abundant element in the Earth's crust (approximately 0.1%) and occurs naturally in several forms, primarily as the silicate ($MnSiO_3$) but also as the carbonate ($MnCO_3$) and a variety of oxides, including pyrolusite (MnO_2) and hausmannite (Mn_3O_4). Pyrolusite is the most common and has been used in glassmaking since the time of the pharaohs in Egypt. Weathering of land deposits has led to large amounts of the oxide being washed out to sea, where they have aggregated into the so-called manganese nodules containing 15 to 30% manganese. Vast deposits, estimated at over 10^{12} metric tons, have been detected on the seabed, and a further 10^7 metric tons is deposited every year. The nodules also contain smaller amounts of the oxides of other metals such as iron, cobalt, nickel, and copper. The economic importance of the nodules as a source of these important metals is enormous.

Uses

All steels contain some manganese, the major advantage being an increase in hardness, although it also serves as a scavenger of oxygen and sulfur impurities that would induce induce defects and consequent brittleness in the steel. The manganese ore is first converted to a metallic alloy with iron known as ferromanganese.

Ferromanganese has a significantly lower melting point than pure manganese and is therefore more readily dissolved by molten iron for steel production. The demand for manganese by the steel industry is so great that about 95% of mined manganese ores is converted to ferromanganese.

In addition to its importance to the steel industry, manganese has a number of other industrial uses. Manganese dioxide (MnO_2) is employed in the manufacture of dry-cell batteries, most commonly of the carbon–zinc Leclanché type. It acts to suppress undesirable formation of hydrogen gas at the carbon (positive) electrode. Only very high quality manganese dioxide ore can be used directly in these batteries; consequently, increasing use is being made of synthetic manganese dioxide obtained by the electrolysis of manganese sulfate ($MnSO_4$) solutions. Manganese dioxide is also used in the brick industry to provide a range of red-to-brown and gray tints. Its venerable use in glassmaking has been mentioned; its role is to neutralize the effects of iron impurities that would impart a greenish tinge to the glass. The good oxidizing properties of manganese dioxide also make it useful for the oxidation of aniline to quinone, important as a photographic developer and also in the production of paints and dyes. Manganese even has some use in the electronics industry, where manganese dioxide either natural or synthetic, is employed to produce manganese compounds possessing high electrical resistivity; among other applications, these are utilized as components in every television set.

Manganese metal has very high sound-absorbing properties, and copper–manganese alloys with high percentages of manganese are used as sound-damping alloys for thrust collars for jackhammers and other power tools.

Alloys

A manganese alloy with 72% manganese, 18% copper, 10% nickel is noted for its high coefficient of thermal expansion, electrical resistivity, strength, and vibration damping. It is used for rheostat resistors and electrically heated expansion elements.

Manganese–aluminum is a hardener alloy employed for making additions of manganese to aluminum alloys. Manganese lowers the thermal conductivity of aluminum but increases strength. Manganese up to 1.2% is used in aluminum alloys when strength and stiffness are required. One manganese–aluminum contains 25% manganese and 75% aluminum. Manganese–boron is used for deoxidizing and hardening bronzes. It contains 20 to 25% boron, with small amounts of iron, silicon, and aluminum. For deoxidizing and hardening brasses, nickel bronze, and copper–nickel alloys, manganese copper, or copper manganese, may be used. The alloys used contain 25 to 30% manganese and the balance copper. The best grades of manganese copper are made from metallic manganese and are free from iron. For nickel bronzes and nickel alloys, the manganese copper must be free of both iron and carbon, but grades containing up to 5% iron can be used for manganese bronze. Grades made from ferromanganese contain iron.

Manganese bronze is wrought as well as cast alloys of copper and zinc mainly, with lesser amounts of iron, aluminum, silicon, tin, and lead. The two standard alloys, C67000 (2.5 to 5.0% manganese) and C67500 (0.05 to 0.50% manganese), were formerly designated manganese bronze B and A, respectively. Manganese bronze cast alloys constitute the C86100 to C86800 series, some of which also may contain as much as 5% manganese. Some manganese bronzes were formerly designated high-strength yellow-brasses and leaded high-strength yellow-brasses.

C67500, which contains 58.5% copper, 39% zinc, 1.4% iron, 1% tin, and 0.1% manganese, has an ultimate tensile strength of 448 MPa, a tensile yield strength of 207 MPa, and a tensile elongation of 33% in the annealed condition. The alloy is weldable, has good brazing and soldering characteristics, and good resistance to corrosion in rural, industrial, and marine atmospheres. Available in rod and shapes, it is used in pumps, clutches, and valves. Most of the cast alloys are castable by various methods, with C86200 the most versatile in this respect. As sand-cast, the alloys provide typical ultimate tensile strengths ranging from 448 to 793 MPa, and they are rather ductile as indicated by tensile elongations of 15 to 30%. The alloys are not hardenable by heat treatment; weldability, including brazing and soldering, is generally poor or fair; and their machinability is 8 to 65% that of free-cutting brass, with C86400 and C86700 the best in this respect.

MANGANESE DIOXIDE

Manganese dioxide (MnO_2) is soluble in water and HNO_3 and soluble in HCl. It occurs in nature as the blue-black mineral pyrolusite.

In glass, manganese dioxide is used as a colorant and decolorizer. As a coloring oxide in lead potash glasses, manganese produces an amethyst color, while in soda glass a reddish-violet is produced. Manganese suitable for such purposes should contain at least 85% MnO_2 and not more than 1% iron oxide.

The major use of manganese oxides is an ore of manganese for the manufacturing of steel; manganese serves to increase the hardness and decrease the brittleness of steel. Another important use of manganese oxides is as the cathode material of common zinc/carbon and alkaline batteries (such as flashlight batteries).

MANGANESE NODULES

These are concentrations of manganese and iron oxides found on the floors of many oceans. The complex growth histories of manganese nodules are revealed by the texture of nodule interiors.

Nodules from certain regions are significantly enriched in nickel, copper, cobalt, zinc, molybdenum, and other elements so as to make them important reserves for these strategic metals. Modern oceanographic surveys have delineated areas of the world's seafloors where nodule abundances and metal concentrations are highest.

Although manganiferous nodules and crusts have been sampled or observed on most seafloors, attention has focused on the nickel–copper-rich nodules (2 to 3 wt%) from the northern equatorial Pacific belt stretching from southeast Hawaii to Baja California, as well as the high-cobalt nodules and crusts from seamounts in the Central Pacific. Manganese nodules from the Atlantic Ocean and from higher latitudes in the Pacific Ocean have significantly lower concentrations of the minor strategic metals. Surveys of the Indian Ocean have revealed metal-enrichment trends comparable to those found in the Pacific Ocean nodules; high nickel–cobalt–copper-bearing nodules are found near the Equator. The ferromanganese nodules and crusts associated with submarine hydrothermal deposits have extremely low concentrations of nickel–cobalt–copper.

CHEMISTRY

Marine manganese nodules are usually classified by mode of formation into hydrogenetic, diagenetic, and hydrothermal types. A fourth, mixed type also exists. Hydrogenetic nodules form by direct precipitation of manganese–iron oxide phases onto an existing nucleus at the sediment–water interface; diagenetic nodules are believed to be biologically driven; and hydrothermal nodules form by submarine hydrothermal activity. Hydrothermal nodules are rarely reported; in submarine hydrothermal environments, manganese–iron oxides occur mainly as crusts. Cobalt is the only strategic metal reported in hydrogenetic nodules that have relatively low manganese/iron ratios. Nodules with high nickel, cobalt, and copper contents are diagenetic and have relatively high manganese/iron ratios.

MANGANESE STEELS

All commercial steels contain some manganese which has been introduced in the process of deoxidizing and desulfurizing with ferromanganese, but the name was originally applied to steels containing from 10 to 15% manganese. Steels with from 1.0 to 1.5% manganese are known as carbon–manganese steel, pearlitic manganese steel, or intermediate manganese steel. Medium manganese steels, with manganese from 2 to 9%, are brittle and are not ordinarily used, but steels with 1 to 2% manganese and with or without small amounts of chromium or molybdenum are used for air-hardening and oil-hardening cold-work tool steels. The original Hadfield manganese steel made in 1883 contained 10 to 14.5% manganese and 1% carbon.

Manganese increases the hardness and tensile strength of steel. In the absence of carbon, manganese up to 1.5% has only slight influence on iron; as the carbon content increases, the effect intensifies.

High-manganese steels are not commercially machinable with ordinary tools, but can be cut and drilled with tungsten carbide and high-speed-steel tools. The austenitic steels, with about 12% manganese, are exceedingly abrasion-resistant and harden under the action of tools. They are nonmagnetic.

High-manganese steels are brittle when cast and must be heat-treated. For castings of thin sections or irregular shapes where the drastic water quenching might cause distortion, nickel up to 5% may be added. The manganese–nickel steels have approximately the same characteristics as the straight manganese steels.

The manganese steels used for dipper teeth, tractor shoes, and wear-resistant castings contain 10 to 14% manganese, 1 to 1.4% carbon, and 0.30 to 1% silicon. The tensile strength is up to 861 MPa, elongation 45 to 55%, weight is 7916 kg/m^3, and Brinell hardness, when heat-treated, of 185 to 200.

A manganese–aluminum steel, has 30% manganese, 9% aluminum, 1% silicon, and 1% carbon. Its tensile strength is 840 MPa with elongation of 18%, but it work-hardens rapidly, and when cold-rolled and heat-aged the tensile strength is 2068 MPa with a yield strength of 1999 MPa. This alloy forms a special type of stainless steel, with high resistance to oxidation and sulfur gases to 760°C.

Tank car steel M-128 is a manganese–vanadium steel with 0.25% carbon, up to 1.5% manganese, and 0.02% or more vanadium. It has a minimum tensile strength of 558 MPa with elongation of 18%. This type of steel with up to 1.75% manganese is used for forgings.

MARAGING STEELS

The maraging steels develop unique combinations of properties that have not been obtained in conventional low-alloy steels. Some of these properties are (1) useful yield strengths to and above 2040 MPa; (2) high toughness and impact energy even at the 2040 MPa yield strength level; (3) low nil ductility temperature (NDT); (4) exceptional stress-corrosion resistance; (5) through hardening without quenching; (6) simple heat treatment; (7) good formability without prolonged softening treatments; (8) good machinability; (9) low distortion during maraging after forming or machining; (10) good weldability; and (11) freedom from decarburization problems.

The maraging steels are a family of low-carbon, high-alloy steels typically containing 12 to 18% nickel, 3 to 5% molybdenum, 0 to 12% cobalt, 0.2 to 1.6% titanium, and 0.1 to 0.3% aluminum (one cobalt-free grade also contains 5% chromium). They are noted for their high strength and toughness, simple heat treatment, dimensional stability during heat treating, good machinability, and excellent weldability. The term *maraging* refers to the martensitic structure that forms during heat treatment, which is a precipitation-hardening, or aging, treatment usually at 482°C. The 18% nickel steels, the most well known, are produced in four grades to provide tensile yield strengths of 1379, 1724, 2069, or 2413 MPa. Although the 18% nickel steels were originally developed for aerospace applications primarily, they also are now used for die-casting dies, cold-forming dies, and molds for forming plastics.

The 9% nickel, 4% cobalt alloys were designed to provide high strength and toughness at room temperature as well as at moderately elevated temperatures — to about 427°C. Weldability and fracture toughness are good, but the alloys are susceptible to hydrogen embrittlement. These steels are used in airframes, gears, and large aircraft parts (Table M.7).

MARBLE

Marble is a term applied commercially to any limestone or dolomite taking polish. Marble is

TABLE M.7
Ultrahigh-Strength Steels[a] — Mechanical Properties

	Typical Designations			
	Medium/Carbon Low-Alloys 4140M, 4130M, 4330V, D6AC, 300M, 98BV40, 4340	Mod 5 Cr–Mo–V Tool Steels H-11 (mod), H-13 (mod)	Maraging Steels (high nickel) 18Ni (250, Almar 30, Marvac 736, NiMark	9Ni–4Co Alloys HP-9–4–30, HP-9–4–20, HP-9–4–45
Yield strength (10^3 psi)	To 250	To 247	245	To 180
Tensile strength (10^3 psi)	To 300	To 311	255	To 280
Impact strength, RT, Charpy (ft-lb)	17	15–22	23	50
Elongation in 2 in. (%)	10	6.6–12	8	10–19

Note: RT = room temperature.

[a] Heat treated for maximum strength. Design data for specific alloys should be obtained from producers.

Source: Mach. Design Basics Eng. Design, June, p. 794, 1993. With permission.

a compact crystalline limestone used for ornamental building, for large slabs for electric-power panels, and for ornaments and statuary. In the broad sense, marble includes any limestone that can be polished, including breccia, onyx, and others. Pure limestone would naturally be white, but marble is usually streaked and variegated in many colors. Carrara marble, from Italy, is a famous white marble, being of delicate texture, very white, and hard. In the United States the marbles of Vermont are noted and occur in white, gray, light green, dark green, red, black, and mottled .

MATERIALS HANDLING

Materials handling involves the loading, moving, and unloading of materials. The loading, moving, and unloading of ore from a mine to a mill and of garments within a factory are examples of materials handling. There are hundreds of different ways of handling materials. These are generally classified according to the type of equipment used. For example, the International Materials Management Society has classified equipment as (1) conveyor; (2) cranes, elevators, and hoists; (3) positioning, weighing, and control equipment; (4) industrial vehicles; (5) motor vehicles; (6) railroad cars; (7) marine carriers; (8) aircraft; and (9) containers and supports.

Every materials-handling problem starts with the material — its dimensions, its nature, and its characteristics. Engineers who fail to start here usually end up trying to justify equipment rather than achieving safe and economical movement of the material. The quantity to be moved — both in total and in rate of moving desired — is next in selecting the appropriate handling method. Then comes the sequence of operations or the routing. Basically, this what, when (how much and how often), and where is the minimum information needed to evaluate or determine any handling system or equipment.

Materials handling is both a planning and an operating activity. These two activities are generally separated in industry; an analytical group designs or selects the system or equipment and the operating group puts it to use.

Equipment

This refers to devices used for handling materials in an industrial distribution activity. The equipment moves products as discrete articles, in suitable containers, or as solid bulk materials that are relatively free-flowing. Such equipment does not include the means employed to control the flow of fluids.

Many different types of machines result from combinations and permutations of the following factors:

1. The route over which the product is moved may be fixed or variable.
2. The path of travel may be horizontal, inclined, declined, or vertical.
3. Motion may be imparted to the product manually, by the force of gravity, by air pressure, by vacuum, by vibration, or by power-actuated components of the machine.
4. The motion may be continuous or intermittent (reciprocating).
5. The product may be supported or carried suspended during the handling operation.

Based on their most common characteristics, materials-handling equipment is classified into broad categories: bulk-handling machines, elevating machines, hoisting machines, industrial trucks, and monorail.

Improvements in handling techniques stem from the wide adoption by industry of palleting and of the forklift truck. These innovations have produced far-reaching effects. Among these are radical changes in plant layout, elevator design for multistory operations, and the increasing trend to single-story facilities.

Automation, in the sense of feedback control and advanced mechanization in the fabrication and transfer of products from one operation to the next, brings together two major industrial technologies.

Computer-controlled electronic data-processing devices facilitate the compiling of inventory records and the handling of orders. Use of photoelectric devices for counting and controlling the action of doors, conveyors, and other materials-handling equipment is another example of how electronic techniques are applied. Television and two-way radio improve the communication for materials handling in plants and yards.

MATERIALS PROCESSING

One of the most promising, fast, low-cost methods for processing materials is known as self-propagating high-temperature synthesis (SHS). Table M.8 presents a comparison of this combustion synthesis process and the other synthesis processes.

COMBUSTION SYNTHESIS

Nonoxide materials (such as carbides, nitrides, borides, chalcogenides, hydrides, intermetallic compounds, and sulfides) have been produced by combustion synthesis. This self-propagating high-temperature synthesis, originally developed in the late 1960s, usually involves exothermic reactions above 2500°C. The combustion process itself can be stable or unstable and does not require a furnace. The high temperatures remove any volatile contaminants by vaporization.

Combustion synthesis can involve several different types of reactions. An oxidation-reduction, or thermite, can produce multiphase compositions such as cermets. For example, thermite occurs when a mixture of aluminum powder and iron oxide powder reacts, causing strong heating and yielding aluminum oxide plus a white-hot molten mass of metallic iron.

The thermite reactions have been used to weld metals, as in the welding of cracks in railroad rails.

SYNTHESIS OF REFRACTORY MATERIALS

An example of this process is the reaction between transition metals and carbon powders. Upon ignition of the cold-pressed compact, a combustion wave rapidly (several seconds) propagates through the mass, converting the reactants to metal carbides (Figure M.6). Besides carbides, β-sialon, and microcomposites, silicon carbide + aluminum oxide and titanium carbide + aluminum oxide have been formed.

As a class of refractory materials, the nitrides, such as aluminum nitride, titanium nitride, zirconium nitride, hafnium nitride, boron nitride, and silicon nitride, have all been formed by self-propagating high-temperature synthesis. The process for synthesizing nitrides

MATERIALS PROCESSING

TABLE M.8
Comparison of Various Synthesis Processes

Process	Advantages	Disadvantages	Compositions
Carbothermal reduction	Possible automation; some control of chemistry	Can be somewhat energy and capital cost-intensive; usually requires milling, which can produce impurities; large scale-up may be difficult; expensive raw materials	Titanium boride, silicon carbide, other nonoxides
Solid–solid, solid–gas, combustion synthesis	Usually self-propagating (requiring no external heat source) and fast (within seconds)	Exothermic, volatile reactions; sometimes low density or low yields; densification may require high pressures; addition of dopants may be required	Titanium boride, titanium carbide, silicon nitride, other nonoxides, composites
Vapor-phase synthesis	High purity; no aggregation; ease of preparation; narrow size distribution; versatility; homogeneity	Limited chemistry, ternary compounds difficult; low yields; reactant gases expensive	Oxides, nonoxides (nitrides, carbides), metals, binary compounds
Laser synthesis	Wide range of composition; short reaction times; uniform heating rates; improved process control; uniform size distribution; minimum agglomeration	Volatile reactants; expensive equipment; powder yields can be low; contamination a problem for certain reactions	Refractory materials (nitrides, borides, silicides, carbides), transition metal compounds, oxides with other emission lives
Plasma synthesis	Highly efficient, simple, continuous; homogeneous mixtures; high surface areas; oxides with wide range of available starting materials: very fast quench rates	Requires high power (10^3 kW); large capital and operating costs; higher surface areas cause greater pyrophoricity; health hazards due to inhaling; carbides, nitrides are sensitive; low powder yields; some agglomeration; nonreproducibility	Oxides, carbides, nitrides, mixtures (silicon oxide + aluminum nitride, silicon carbide + silicon carbide, aluminum + aluminum nitride)

Source: Adv. Mater. Proc., 131(4), 53–58, 1987. With permission.

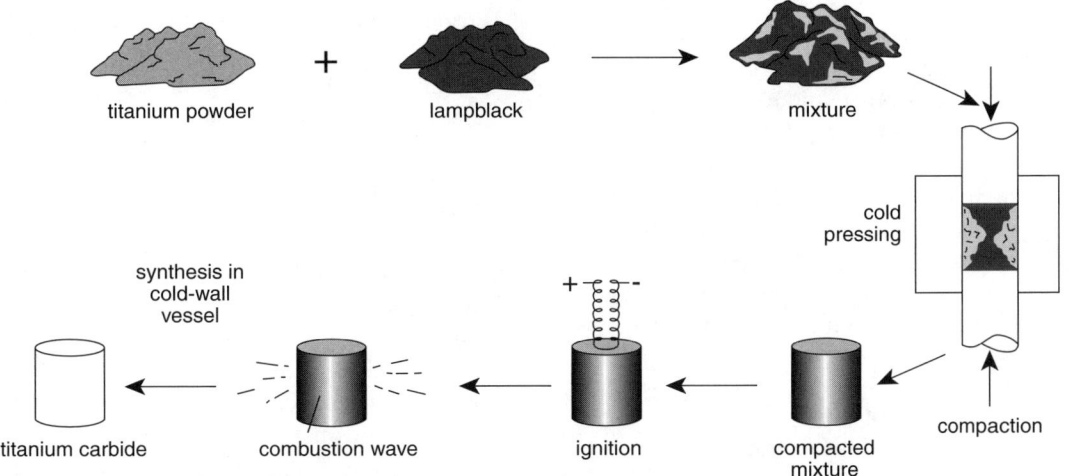

FIGURE M.6 Illustration of self-propagating, high-temperature synthesis. (From *McGraw-Hill Handbook of Structural Ceramics*, McGraw-Hill, New York, 1992, 423. With permission.)

proceeds in a fairly predictable manner. The metal powders are cold-pressed into porous compacts. Subsequently, these compacts are ignited in nitrogen gas or liquid nitrogen by a resistance-heated tungsten wire. The self-propagating combustion wave quickly transforms the metal powder into porous solid nitrides.

Control of Synthesis

Of four innovations to control the synthesis reactions, one is used to slow down (kinetic braking) and three to intensify the reactions (thermal explosion, chemical furnace, and chemical activators).

OTHER SELF-PROPAGATING HIGH-TEMPERATURE SYNTHESIS PROCESSES

- Reaction hot pressing, which utilizes self-propagating high-temperature synthesis reactions in place in a uniaxial hot press to form densified ceramic products. Reaction hot pressing takes advantage of the favorable thermodynamics of self-propagating high-temperature synthesis reactions to rapidly form and densify product phases.
- Reactive sintering, which occurs when pressure is applied during sintering. The process is also known as reactive hot isostatic pressing.
- Shock compression, where chemical reactions can also be initiated by shock compression of powder mixtures. This type of dynamic processing technique, referred to as shock-induced reaction synthesis, utilizes the simultaneous application of very high pressure and temperature generated during the passage of shock waves through a powder mixture.

MATERIALS SCIENCE AND ENGINEERING

This is a multidisciplinary field concerned with the generation and application of knowledge relating to the composition, structure, and processing of materials and their properties and uses. The field encompasses the complete knowledge spectrum for materials ranging from the basic end (materials science) to the applied end (materials engineering). It forms a bridge of knowledge from the basic sciences (and mathematics) to various engineering disciplines. New materials with special properties are constantly being discovered and developed, and thus materials science and engineering is a continually expanding field.

METALLIC MATERIALS

The study of metallic materials constitutes a major division of the materials science and engineering field. In general, metallic materials are inorganic substances composed of one or more metallic elements, but they may also contain nonmetallic elements. Most metals have a crystalline structure of closely packed atoms arranged in an orderly manner. Metals in general are good electrical and thermal conductors. Many are relatively strong at room temperature and retain good strength at elevated temperatures. Metals are commonly alloyed together in the liquid state so that, upon solidification, new solid metallic structures with different properties can be produced. Metals and alloys are often cast into the nearly final shape in which they will be used, and these products are called castings. However, most metals and alloys are first cast into shapes such as sheet ingots or extrusion billets, which are subsequently worked by processes such as rolling and extrusion into wrought products, for example, sheet, plate, and extrusions.

CERAMIC MATERIALS

The study of ceramic materials forms a second major division of the field of materials science and engineering. Ceramics are inorganic materials consisting of metallic and nonmetallic elements chemically bonded together. They can be crystalline, noncrystalline, or mixtures of both. Most ceramic materials have high hardness, high-temperature strength, and good chemical resistance; however, they tend to be brittle.

Ceramics in general have low electrical and thermal conductivities, which makes them useful for electrical and thermal insulative applications. Most ceramic materials can be classified into three groups: traditional ceramics, technical ceramics, and glasses.

Traditional Ceramics

These consist of three basic components: clay, silica, and feldspar. The clay provides the workability of the ceramic before it is hardened by the firing process. Clay makes up the major body material: it consists mainly of hydrated aluminum silicates ($Al_2O_3 \cdot SiO_2 \cdot H_2O$) with smaller amounts of other oxide impurities. The silica (Si_2O) has a high melting temperature and provides the refractory component of traditional ceramics. The third component, feldspar ($K_2O \cdot Al_2O_3 \cdot 6H_2O$), has a low melting temperature and produces a glass when the ceramic mix is fired; it bonds the refractory components together. Traditional ceramic products fabricated from whiteware such as electrical porcelain and sanitaryware are made from components of clay, silica, and feldspar for which the composition is controlled.

Technical Ceramics

Technical ceramics are based on pure or nearly pure ceramic components alone or in combination. The raw materials for technical ceramics must be processed carefully so that a controlled product can be produced, and they are made by using various composition mixes and processing procedures. Examples of technical ceramics are aluminum oxide (Al_2O_3), zirconia (ZrO_2), silicon carbide (SiC), silicon nitride (Si_3N_4), and barium titanate ($BaTiO_3$). Applications for technical ceramics include aluminas for auto spark-plug insulators and substrates for electronic circuitry, dielectric materials for capacitors, ceramic tool bits for machining, and high-performance ball bearings.

Glasses

Glasses differ from the other ceramic materials in that their constituents are heated to fusion and then cooled to a rigid state without crystallization. A characteristic of a glass is that it has a noncrystalline structure with no long-range order. Most inorganic glasses are based on the glass-forming silicon oxide, silica (SiO_2). About 90% of the glass produced is soda-lime glass, which has the basic composition of 1 to 3% SiO_2, 12 to 14% sodium oxide (Na_2O), and 10 to 12% calcium oxide (CaO). The sodium oxide and calcium oxide are added to lower the viscosity of the glass so that it becomes easier to work. Soda-lime glass is used, for example, for flat glass, containers, and light products, where high chemical durability and heat resistance are not required. Many other types of glasses with different compositions are produced for special applications.

POLYMERIC MATERIALS

The study of polymer materials forms a third major division of materials science and engineering. Most of these materials consist of carbon-containing long molecular chains or networks. Structurally, most of them are noncrystalline, but some are partly crystalline. The strength and ductility of polymeric materials vary greatly. Most polymers have low densities and relatively low softening or decomposition temperatures. Many are good thermal and electrical insulators. Polymeric materials have replaced metals and glasses for many applications.

Most polymeric materials can be classified as thermoplastics, thermosets, or elastomers.

Thermoplastics

These are polymeric materials that have a structure usually consisting of very long chains of carbon atoms strongly (covalently) bonded together, sometimes with other atoms, such as nitrogen, oxygen, and sulfur, also covalently bonded in the molecular chains. Weaker secondary bonds bind the chains together into a solid mass. Because of their weak intermolecular bonds, thermoplastics can be heated to a soft, viscous condition for forming into a desired shape and then cooled to a rigid state to retain that shape. Typical thermoplastics are polyethylenes, polyvinyl chlorides, and polyamides (nylons). Some examples of applications for thermoplastics are containers, electrical insulation, automotive internal parts, and appliance housings.

Thermosets

These are polymeric materials that have a network of mainly carbon atoms covalently bonded together to form a rigid solid. Sometimes nitrogen, oxygen, or other atoms are also covalently bonded into the network. Thermosets are formed into a permanent shape and are cured (set) by a chemical reaction that may require heat and pressure. Thermoset plastics cannot be remelted or reformed into another shape after curing. Thermosets such as phenolics (for example, Bakelite) and epoxies are used for handles, knobs, electrical connectors, and matrix materials for fiber-reinforced plastics.

Elastomers

This class of materials, also known as rubbers, can be deformed elastically by a large amount when a force is applied to them, and then they can return to approximately the same shape when the force is removed. Most elastomers consist of long, carbon-containing molecular chains with periodic strong bond links between the chains. Elastomers include both natural and synthetic rubbers, which are used for auto tires, electrical insulation, and industrial hoses and belts.

Composite Materials

A fourth major division of materials science and engineering comprises the study of composite materials. A composite material is a mixture of two or more materials that differ in form and chemical composition and are essentially insoluble in each other, and most are produced synthetically by combining various types of fibers with different matrices to increase strength, toughness, and other properties. Three important types of composite materials have polymeric, metallic, and ceramic matrices.

Polymeric–matrix composites are the most common type and find most applications where the temperature does not exceed about 100 to 200°C. For example, glass-fiber materials produced with short glass fibers embedded in a polyester plastic matrix are used in appliances, boats, and car bodies because of their light weight, ease of fabrication into complex shapes, corrosion resistance, and moderate cost. Other, more expensive advanced composites made with stronger carbon or aramid fibers, usually embedded in heat-resistant thermoset polymeric matrices, are used for aircraft surface material and structural members. Advanced polymeric–matrix composites are also finding use in sports equipment and other products, but their high cost has limited their use.

Metal–matrix composites have also been developed by embedding fibers such as silicon carbide and aluminum oxide into aluminum, magnesium, and other metal alloy matrices. The fibers strengthen the metal alloys and increase high-temperature stability. Metal–matrix composites are used, for example, for automotive pistons and missile guidance systems. Ceramic–matrix composites have also been considered to develop new and tougher ceramic materials. For example, the reinforcement of alumina with silicon carbide whiskers significantly improves its fracture toughness and strength.

Other Materials

In addition to metallic, ceramic, polymeric, and composite materials, materials science and engineering is also concerned with the research and development of other special classes of materials that are based on applications. Some major types of these materials are electronic materials, optical materials, magnetic materials, superconducting materials, dielectric materials, nuclear materials, biomedical materials, and building materials.

MECHANICAL ALLOYING

The application of powder metallurgy (P/M) techniques has greatly increased in recent years. In many cases, P/M allows for metal parts to be produced at lower costs with improved properties. Another advantage that P/M offers is that it can produce alloys and microstructures that are not possible to produce by standard metallurgical practices such as casting and forging. An important area of P/M is the production of the metal powders used. The physical and chemical nature of the initial powders will determine many of the final properties of the P/M parts produced.

MECHANICAL ALLOYING

Mechanical alloying is a materials-processing method that involves the repeated welding, fracturing, and rewelding of a mixture of powder particles, generally in a high-energy ball mill, to produce a controlled, extremely fine microstructure. The mechanical alloying technique allows alloying of elements that are difficult or impossible to combine by conventional melting methods. In general, the process can be viewed as a means of assembling metal constituents with a controlled microstructure. If two metals will form a solid solution, mechanical alloying can be used to achieve this state without the need for a high-temperature excursion. Conversely, if the two metals are insoluble in the liquid or solid state, an extremely fine dispersion of one of the metals in the other can be accomplished. The process of mechanical alloying was originally developed as a means of overcoming the disadvantages associated with using P/M to alloy elements that are difficult to combine. By using P/M, homogeneity is dictated by the size of the particle, but contamination and fire hazards become a concern when particle size is very small (Figure M.7).

Process

The process of mechanical alloying consists of repeated flattening, fracture, and rewelding of the powder particles in a high-energy ball charge. Every time two steel balls collide, they trap powder particles between them. This deforms the particles and creates minimal foreign species on the surface so that welding together of adjacent surfaces can occur easily. The alloying is effected by the ball-powder-ball collisions in several stages. In the first stage, intense cold welding predominates, and layered composite particulates of the starting constituents form (Figure M.8a). This is followed by a hardening of the particles and fracturing and cold welding leading to a finer composite particle size (Figure M.8b). Solid-solution formation begins at this stage. Next, in a moderate cold-welding period, the microstructure of the particles gets finer, with a typical spacing between adjacent regions of 1 μm (Figure M.8c). The compositions of the individual particles converge to the original blend composition, with the rate of refinement of the internal structure

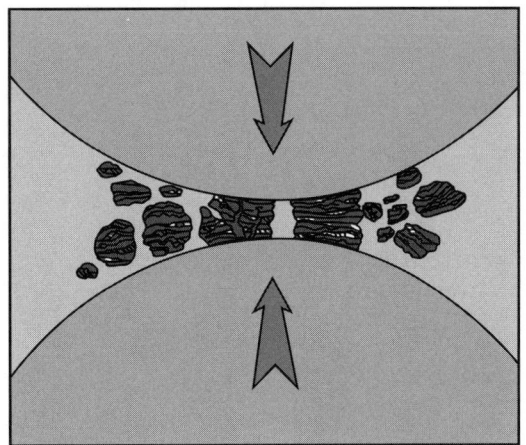

FIGURE M.7 Process of fusing and fracturing. (From *Adv. Mater. Proc.*, 131(4), 53–58, 1987. With permission.)

(reduction in scale of the microstructure) of the particles approximately logarithmic with processing time. The final stage is a steady-state period (Figure M.8d). By the time processing has been completed, the particles have an extremely deformed metastable structure, which can contain dispersoids. Generally, at this point the microstructural scale is significantly below the micrometer level.

Applications

Some oxides are insoluble in molten metals. Mechanical alloying was originally developed to provide a means of dispersing these oxides in the metals. Examples are nickel-based superalloys strengthened with dispersed thorium oxide or yttrium oxide (Y_2O_3), the latter being preferred. These superalloys have excellent strength and corrosion resistance at elevated temperatures, making them attractive candidate materials for use in applications such as jet-engine turbine blades, vanes, and combustors. This class of alloys is known as oxide-dispersion-strengthened (ODS) nickel-base superalloys. Iron- and aluminum-base dispersion-strengthened alloys have also been developed. Additionally, a number of other potential applications for mechanical alloying material are being explored, including powders for coating applications, alloys of immiscible systems, amorphous alloys, intermetallics, cermets, and

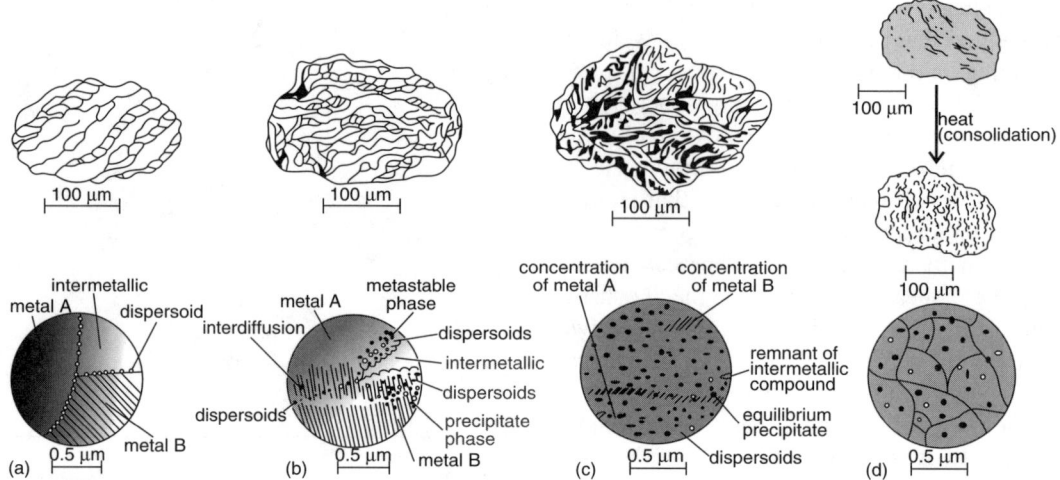

FIGURE M.8 Mechanical alloying. (From *McGraw-Hill Encyclopedia of Science and Technology*, 8th ed., Vol. 10, McGraw-Hill, New York, 607. With permission.)

organic–ceramic–metallic material systems in general.

Alloys like MA754 (Ni–20% Cr–0.3% Al–0.5% Ti–0.6% Y_2O_3) and MA6000 (Ni–15% Cr–4% W–2% Mo–2% Ta–4.5% Al–2.5% Ti–1.1% Y_2O_3) are commercially available as bar and plate for use in gas turbine vanes and turbine blades and for other applications in oxidizing or corrosive atmospheres. The vane alloy MA754 is capable of a service temperature of about 1100°C and is used in military engines. The iron-base alloy MA956 (Fe–20% Cr–4.5% Al–0.5% Ti–0.5% Y_2O_3), available as plate sheet, bar, and wire, is capable of withstanding temperatures up to 1300°C or even higher in corrosive environments; it has applications in aircraft and industrial gas turbine combustors, in swirlers and heat exchangers of power generation equipment, and in heat-treatment equipment.

Conventional coatings in advanced gas turbine engines fail by loss of aluminum in the coating or by interdiffusion between the coating and substrate. Coatings possessing a large amount of Y_2O_3, produced by mechanical alloying, could solve such problems.

Liquid or solid immiscible systems are difficult to process by conventional pyrometallurgy, for example, the copper–lead or copper–iron systems. In these cases, mechanical alloying provides a route to obtain a homogeneous distribution in the solid phase. Mechanical alloying has also been used to fabricate the superconducting intermetallic niobium–tin compound Nb_3Sn, which is difficult to produce conventionally because of the large melting-point difference between niobium and tin. Mechanical alloying can also be used to produce supercorroding magnesium-base alloys. These alloys are designed to corrode at a controlled rate for release of deep-sea equipment at specific depths. The requirement is to have the anode and cathode in proximity by using mechanical alloying. In another interesting application of mechanical alloying, titanium and magnesium have been combined by using mechanical alloying; usually it is very difficult to produce such an alloy because the boiling point of magnesium is lower than the melting point of titanium. This achievement could result in a lower-density titanium alloy for aerospace applications.

Amorphous alloys can also be produced by using the mechanical alloying technique. These include Nb_3Sn starting from elemental powders, various rare earth–cobalt combinations, and titanium–copper and titanium–copper–palladium systems.

MECHANICAL FINISHES

Finishing with Coated Abrasives

Coated abrasives or "sandpaper" can be used in many ways on many materials to produce a wide variety of finished surfaces. The method of application as well as the type of abrasive used greatly influences the results obtained. Following is a description of the three principal methods of using endless abrasive belts and the applications for each method.

Contact Wheel Application

The contact wheel method produces the best abrasive efficiency, including the fastest rate of cut and best abrasive life. The finish produced by this method is characterized by short individual scratches that are uniform across the width of contact and in line with the running direction of the belt. This method can be used on all types of parts having flat or lightly contoured surfaces such as hand tools, small household appliances, cutlery, and stainless–steel sheet for decorative trim in kitchens, hospitals, etc.

Platen Method

This method is used for finishing work with a flat surface that is either continuous or interrupted. Because individual abrasive grain remains in contact with the work from top to bottom, the method produces a continuous scratch running the length of the part in contact with the belt. The advantage of the platen method is that the whole surface is contacted simultaneously and it easily produces flatness with a high degree of accuracy.

Slack Belt

This method is used on work that is highly contoured such as metal furniture, light fixtures, and plumbing fixtures. The flexibility of the belt is useful in polishing hard-to-reach surfaces where contours and shapes are involved. The length of scratch pattern produced is between the short scratch pattern of the contact wheel method and the long scratch pattern of the platen.

Types of Abrasives and Finishes

Two principal types of abrasives are used for coated-abrasive types of finishing operations — aluminum oxide and silicon carbide. The former has a blocky shape and is very tough. It is used for grinding and finishing materials having a high tensile strength. Silicon carbide is harder and sharper but is more friable or brittle; consequently, it is used for grinding and finishing materials having low tensile strength.

Coated abrasives are used extensively for a vast multitude of grinding and finishing operations because of their inherent versatility and economy. The finishes produced with coated abrasives are primarily decorative but can also be used for functional purposes in cases where geometry of the finished parts is important.

Finishing with Loose Abrasives

Finishes produced by loose abrasives, such as aluminum oxide and silicon carbide, are available in many variations of surface roughness and appearance, depending on the methods used. The methods used to finish or polish a surface may include polishing wheels (referred to as "setup" wheels), lapping, pressure blasting, and barrel finishing (often referred to as tumbling).

Setup wheels can be likened to bonded grinding wheels. The basic difference is that setup wheels are resilient because they are made of cloth laminations that are stitched so as to give a certain rigidity or resiliency. Abrasive grain is cemented on the periphery of setup wheels with cold cement or hot glue. Because of its "cushioning" effect, the resiliency of setup wheels simplifies the blending of contoured surfaces.

Buffing

Buffing with loose abrasives usually follows the polishing operation, with the latter done with either setup wheels or coated belts. Buffing generates smooth surfaces that are essentially free of scratches and have high reflectivity. It is accomplished by bringing the surface into contact with the periphery of a buffing wheel (usually cloth laminations) to which abrasive materials have been applied in composition form.

Buffing generates smooth and reflective surfaces by removing or displacing relatively small amounts of metal. It is not designed to establish or retain dimensional accuracy.

Pressure Blasting

Pressure blasting is the process in which solid abrasive particles are propelled against a surface by means of expanding compressed air. The method always produces a matte finish. The finer the abrasive, the finer the finish; the coarser the abrasive, the faster the rate of stock removal and the rougher the finish.

Pressure blasting can be used for many finishing purposes. These include removal of heat-treat scale and oxides from metallic surfaces; preparation of surfaces for protective coatings; removal of contamination from nuclear devices; generation of decorative finishes; preparation of surfaces for visual inspection for surface defects; and removal of small burrs.

Lapping

Lapping can be used to produce very smooth finishes and geometrically true surfaces. Like grinding, the degree of finish obtained by lapping can be controlled by the particle size of the abrasive. Coarse grits plow deep furrows and therefore remove stock faster than fine particles. Fine particles, conversely, remove stock at a lower rate but produce finishes with a lower root mean square (rms). To reduce the surface roughness from, for example, 30 to 5 rms, usually requires using two to three progressively finer abrasives.

The finishes generated by lapping are functional rather than decorative. The surface smoothness dictates a low rms where wearing parts are involved. Flatness and scratch-free parts are essential for components such as high-pressure valves, and germanium and silicon transistor wafers.

Barrel Finishing

Barrel finishing is a surface-conditioning operation in which a mixture of metallic or nonmetallic parts, abrasive media, and various compounds is placed in a receptacle and rotated for the purpose of rounding corners, deburring, improving surface finish, cleaning, derusting, burnishing for high luster, and producing low microinch finishes. Many of these finishing operations are accomplished in simultaneous operations.

Unlike grinding wheel, setup wheel, or abrasive belt operations, tumbling produces a random scratch pattern. This is because the media and parts have a random motion while the barrel is rotating. But like grinding wheels and other types of abrasive finishing methods, the types and size of abrasive media (sometimes called nuggets, pellets, chips, etc.) influence the roughness and rate of stock removal.

FINISHING WITH BONDED ABRASIVES

Finishes produced by grinding wheels vary over a wide range both in surface roughness and in surface appearance. These finishes can be divided into two major classifications, those produced by rough grinding and those produced by finish grinding.

Rough-grinding operation is primarily intended for stock removal without any particular regard for appearance. The finish produced can be described as having the appearance of a plowed field with the direction of the furrows in the same direction as the path of the grinding wheel. The depth of the furrows is in direct proportion to the roughness of the grind. The rougher the grind, the deeper the furrow; the less severe the grind, the shallower the furrow. The surface roughness produced by rough grinding varies over a wide range, beginning in the 30 µin. area and becoming progressively rougher.

Finish grinding, as its name implies, is used for the production of finish and, in addition, the generation of geometry. This finish may be one that has a low surface roughness or a high luster and, in some cases, both a low roughness and a high luster.

The appearance of a finely ground surface can either be a dull matte finish or it can be a highly reflective surface. A distinction is made between the roughness of the dull matte ground finish and that of a highly reflective ground surface. A dull matte finish is generally as smooth and many times, smoother than a highly reflective surface. The matte finish is a truly ground surface and can be described as a plowed field where the furrows are very shallow in depth and all the ridges have been removed.

A highly reflective surface, on the other hand, can be described as a plowed field where some of the ridges have fallen into the adjacent furrow and are pushed into that furrow by the grinding wheel as it passes over it. This produces a luster but not necessarily a low surface roughness.

MELAMINE FORMALDEHYDE PLASTICS

The outstanding features of melamines, one of the amino resins, include excellent colorability, high hardness, and good electrical characteristics, including exceptional arc resistance.

Primary uses for melamine plastics as engineering materials are in decorative molded dinnerware, decorative thermosetting laminates, molded housings and wiring devices, closures, and buttons.

The resins are produced by reaction of melamine and formaldehyde. Crystals of melamine, an amino chemical, are reacted with formaldehyde and mixed with highly purified filler, dried, and ground. Color pigments, plasticizers, lubricant, and accelerators are added to produce the molding compound. Melamine is similar in many respects to urea-formaldehyde. Melamine resists water absorption and chemical attack to a greater extent than urea and its cost is higher.

APPLICATIONS

Molding Compounds

For molding applications, the hardness and colorability of melamine are outstanding advantages. Melamine is the hardest of plastics (Rockwell M118 to 124) and has unlimited colorability. Parts can be produced in any color in the spectrum (both translucent and opaque). Color quality ranges from pastels to bright, jewel-like tones, as well as two tones in the same object. Products such as dishes and tableware can be decorated by foil inlays.

Dinnerware (dishes, plates, cups, and saucers) make up 90% of the uses for melamine molding compounds. Hardness and colorability as well as resistance to attack by foods, detergents, and hot water account for its popularity in the home.

Melamine molding compounds are widely used as handles for flatware and kitchen utensils because of hardness, color, and resistance to heat and water. Handles can now be decorated with patterns applied integrally by foil.

Agitators for washing machines are being molded of melamines because of their resistance to hot water and detergents and excellent appearance. A colored agitator is more attractive than a black one.

Nondecorative applications for molding compounds include wiring fixtures and other electrical uses, where the arc resistance and nontracking characteristics of melamine, coupled with its nonflammability, make it useful. For such uses, strength and impact resistance is improved by incorporating fabrics and glass-fiber reinforcements.

High-Pressure Laminates

The majority of top-quality decorative thermosetting laminates, or high-pressure laminates (e.g., "Formica," "Micarta," and "Textolite"), use melamine resins impregnated in alpha-cellulose paper or other clear reinforcements to provide their decorative surfaces. With the proper resin selection, such laminates can be burn and stain resistant. Further, the high hardness of melamines makes them scuff and mar resistant.

MELAMINE RESIN

This is a synthetic resin of the alkyd type made by reacting melamine with formaldehyde. The resin is thermosetting, colorless, odorless, and resistant to organic solvents. It is more resistant to alkalies and acids than urea resins, has better heat and color stability, and is harder. The melamine resins have the general uses of molding plastics, and also are valued for dishes for hot foods or acid juices and they will not soften or warp when washed in hot water. Melamine, a trimer of cyanamide, has the composition $(N \equiv C \cdot NH_2)_3$.

The melamine resins have good adhesiveness but are too hard for use alone in coatings and varnishes. They are combined with alcohol-modified urea-formaldehyde resins to give coating materials of good color, gloss, flexibility, and chemical resistance. Urea-modified melamine-formaldehyde resins are used for coatings and varnishes. Melamine-formaldehyde molding resin, with cellulose filler, has a tensile strength of 51 MPa and dielectric strength per mil of 12.8×10^6 V/m.

Melamine-urea-formaldehyde resin with a lignin extender is used as an adhesive for water-resistant plywood. Phenol-modified melamine-formaldehyde resin solution is used for laminating fibrous materials. Highly translucent melamine-formaldehyde resin is used for molding high-gloss buttons. Methylol-melamine, made by alkylating a melamine-formaldehyde resin with methyl alcohol, is used for shrink-proofing woolen fabrics.

MERCURY

Mercury (symbol Hg), also called quicksilver, is a metallic element. It is the only metal that is a liquid at room temperature. Mercury has a silvery-white color and a high luster. Its specific gravity is 13.596. The solidifying point is –40°C, and its boiling point at 1 atm is 357°C. It does not oxidize at ordinary temperatures, but when heated to near its boiling point it absorbs oxygen and is converted into mercuric oxide, HgO, used as pigment in marine paints.

Metallic mercury is used as a liquid contact material for electrical switches, in vacuum technology as the working fluid of diffusion pumps, for the manufacture of mercury-vapor rectifiers, thermometers, barometers, tachometers, and thermostats, and for the manufacture of mercury-vapor lamps. Mercury-vapor lamps serve as sources for ultraviolet light.

For micro gas analyses, mercury is most often used as a sealing liquid for the evolved gases. Very large amounts of mercury are used as the electrode material for the electrolysis of aqueous solutions of alkali halides for the manufacture of chlorine and sodium hydroxide. Also, it finds application for the manufacture of silver amalgams for tooth fillings in dentistry.

Some mercury salts serve as catalysts for organic chemical reactions. Fulminate of mercury is used as a primer for explosives. Some complex salts find application as temperature indicators.

A large number of mercury compounds have been used for hundreds of years as medicines. The compounds of mercury with some organic substances are powerful diuretic substances and have applications as a disinfectant, and ammoniated mercury is used in the treatment of various diseases.

Of importance in electrochemistry are the standard calomel electrode, used as the reference electrode for the measurement of potentials and for potentiometric titrations, and the Weston standard cell, with which the accurate measurement of potential sources is possible. In the calomel electode and in the Weston standard cell, metallic mercury is used in contact with solutions that contain mercury salts as the solid phase.

Mercury is used for separating gold and silver from their ores, for coating mirrors, in tanning, in batteries, for the frozen-mercury molding process, mercury-vapor motors, as a circulating medium in atomic reactors, and in its compounds for paint pigments and explosives.

METAL

A metal is a solid or liquid (molten), opaque material with a lustrous surface and good electrical and thermal conductivities. Solid metal is usually crystalline and ductile and can be permanently deformed by shear on crystal planes; permanent deformation is accompanied by an increase in strength (work hardening). Metallic properties are related to the arrangements of positively charged ions bonded through a surrounding field (sea) of free electrons that draw the ions into a close-packed crystalline structure with planes appropriate for slip. Liquids are nearly close packed noncrystalline, with a thermal energy great enough to activate random, free movements of atoms.

The rapid developments of the metals industry since the late 19th century required an in-depth understanding of the fundamentals of metallurgy to improve metal properties and expand their usefulness.

Characteristics of Solids

The unique characteristics of the different classes of solids are directly related to the electron pattern and the types of bonds between the atoms of the solid. Atomic bonding is classified in two major categories: strong attraction (covalent, ionic, or metallic) and weaker, secondary bonding (van der Waals or hydrogen bonding).

Directional Bonds

A covalent bond between two or more atoms is a negative charge directed in space and equivalent in magnitude to one or more electrons. As a consequence, at the lowest energy state the corresponding bond is directional and usually strong, and considerable energy must be supplied to the material to free an electron from this bond or shift an ion from one fixed position to another. For covalent materials, bonding valence electrons require large energies to transport a current composed of many electrons through the aggregate atoms. Thus, electrical conductivity and thermal conductivity by electrons are low.

Metallic Bonding

In contrast to directional bonds, metallic bonds result from linking ions each having centrosymmetrical attraction to a surrounding atmosphere of the distributed charges of nearly free electrons. These electrons, usually valence electrons, have the same likelihood of being associated with any ion in the aggregate.

Crystal Structure

To visualize groups of atoms in an aggregate, the atoms are considered as spheres of the same radius.

This representation of metal atoms is reasonable, since bonds originating from metal atom cores with the superimposition of the electron charges in the intervening space are manifested by a nearly spherically symmetrical attraction between the ions.

Among metals, the face-centered-cubic crystalline structure is found in aluminum, nickel, copper, rhodium, palladium, silver, iridium, platinum, gold, and lead.

Because some metals display characteristics of other bonding types, they occur in other crystal structures primarily because of the interference of electrons at energy levels that are not centrally symmetric. All close-packed-hexagonal metals belong to two categories that can be illustrated by packing elongated or compressed spheres.

An ideal hexagonal-close-packed structure illustrates the hexagonal stacking of spheres, whereas spheres compressed in a direction perpendicular to a close-packed plane have an axial ratio less than the ideal of 1.633. Examples are cobalt, rhenium, titanium, gadolinium, hafnium, and beryllium, as well as lanthanum, cadmium, and zinc.

Other metals such as iron, titanium, tungsten, molybdenum, niobium, tantalum, potassium, sodium, vanadium, chromium, and zirconium (alkali and transition metals) are slightly less closely packed, with a body-centered cubic crystal structure that is 8.1% lower in packing density than face-centered cubic crystals. Furthermore, the equilibrium crystal structure for many metals is dependent on temperature and pressure. Consequently, a large number of elements are allotropic; that is, the element will appear as different crystal structures at various ranges of pressure–temperature conditions. Included in this group are calcium, titanium, manganese, iron, cobalt, yttrium, zirconium, tin, hafnium, thallium, polonium, thorium, and uranium.

Polycrystalline Metals

Common metallurgical production methods inevitably lead to solids made up of many small crystals separated by higher-energy boundaries containing non-lattice-positioned arrangement of atoms. The crystals in these polycrystalline metals are called grains, and the boundaries are grain boundaries. Since the grain boundaries interrupt periodicity of the lattice, they have a large effect on many of the physical and mechanical properties. They are origins (sources) and terminators (sinks) to all other defects, especially vacancies, solute atoms, and dislocations.

In an ideal polycrystalline arrangement composed of many randomly oriented grains, the anisotropic features of each grain are averaged, and the resulting metal solid would have isotropic properties. Actually, metal manufacturing processes such as casting and deformation forming (forging, rolling, and so forth) tend to develop grains with specific crystallographic directions distributed about certain directions that are distinctive of the forming process and the geometry of the bulk product. This nonrandom arrangement of grains, called preferred orientation, contributes to anisotropy of properties in the final polycrystalline product.

METAL ALLOYS

An alloy has metallic features and is composed of two or more chemical elements. Macroscopically, it appears homogeneous, but microscopically various thermodynamically stable arrangements of the chemical elements are possible, depending on the characteristics of the components.

METAL CASTING

Metal casting involves the introduction of molten metal into a cavity or mold where, upon solidification, it becomes an object whose shape is determined by mold configuration. Casting offers several advantages over other methods of metal forming: it is adaptable to intricate shapes, to extremely large pieces, and to mass production; it can provide parts with uniform physical and mechanical properties throughout; and, depending on the particular material being cast, the design of the part, and the quantity being produced, its economic advantages can surpass other processes.

CATEGORIES

Two broad categories of metal-casting processes exist: ingot casting (which includes continuous casting) and casting to shape. Ingot castings are produced by pouring molten metal into permanent or reusable molds. Following solidification, these ingots (or bars, slabs, or billets, as the case may be) are then further processed mechanically into many new shapes. Casting to shape involves pouring molten metal into molds in which the cavity provides the final useful shape, followed only by machining or welding for the specific application. Another mode of metal casting, zero gravity casting, has been developed for processing in a space station.

Ingot Casting

Ingot castings make up the majority of all metal castings and are separated into three categories: static cast ingots, semicontinuous or direct-chill cast ingots, and continuous cast ingots.

Static cast ingots. Static ingot casting simply involves pouring molten metal into a permanent mold. After solidification, the ingot is withdrawn from the mold and the mold can be reused. This method is used to produce millions of tons of steel annually.

Semicontinuous cast ingots. A semicontinuous casting process is employed in the aluminum industry to produce most of the cast alloys from which rod, sheet, strip, and plate configurations are made. In this process molten aluminum is transferred to a water-cooled permanent mold that has a movable base mounted on a long piston. After solidification has progressed from the mold surface so that a solid "skin" is formed, the piston is moved down, and more metal continues to fill the reservoir. However, technological advances have allowed major aluminum alloy producers to replace the metal mold (at least in part) by an electromagnetic field so that molten metal touches the metal mold only briefly, thereby making a product with a much smoother finish than that produced conventionally.

Continuous cast ingots. Continuous casting provides a major source of cast material in the steel and copper industry and is growing rapidly in the aluminum industry. In this process molten metal is delivered to a permanent mold, and the casting begins much in the same way as in semi-continuous casting. However, instead of the process ceasing after a certain length of time, the solidified ingot is continually sheared or cut into lengths and removed during casting. Thus, the process is continuous, with the solidified bar or strip being removed as rapidly as it is cast. This method has many economic advantages over the more conventional casting techniques; as a result, all modern steel mills produce continuous cast products.

Cast to Shape

Casting to shape is generally classified according to the molding process, molding material, or method of feeding the mold. There are four basic types of these casting processes: sand, permanent-mold, die, and centrifugal.

Sand casting. This is the traditional method that still produces the largest volume of cast-to-shape pieces. It utilizes a mixture of sand grains, water, clay, and other materials to make high-quality molds for use with molten metal. Other casting processes that utilize sands as a basic component are the shell, carbon dioxide, investment casting, ceramic molding, and plaster molding processes. In addition, there are a large number of chemically bonded sands that are becoming increasingly important.

Permanent-mold casting. Many high-quality castings are obtained by pouring molten metal into a mold made of cast iron, steel, or bronze. Semipermanent-mold materials such as aluminum, silicon carbide, and graphite may also be used. The mold cavity and the gating system are machined to the desired dimensions after the mold is cast; the smooth surface from machining thus gives a good surface finish and dimensional accuracy to the casting.

Die casting. A further development of the permanent molding process is die casting. Molten metal is forced into a die cavity under pressures of 0.7 to 700 MPa. Two basic types of die-casting machines are hot chamber and cold chamber. In the hot-chamber machine, a portion of the molten metal is forced into the cavity at pressures up to about 14 MPa. The process is used for casting low-melting-point alloys such as lead, zinc, and tin. In the cold-chamber process the molten metal is ladled into the injection cylinder and forced into the cavity under pressures that are about ten times those in the hot-chamber process. High-melting-point alloys such as aluminum-, magnesium-, and copper-base alloys are used in this process. Die casting has the advantages of high production rates, high quality and strength, surface finish on the order of 1.0 to 2.5-μm root mean square, and close tolerances with thin sections. Rheocasting is the casting of a mixture of solid and liquid. In this process the alloy to be cast is melted and then allowed to cool until it is about 50% solid and 50% liquid. Vigorous stirring promotes liquidlike properties of this mixture so that it can be injected in a die-casting operation. A major advantage of this type of casting is expected to be much reduced die erosion due to the lower casting temperatures.

Centrifugal casting. Inertial forces of rotation distribute molten metal into the mold cavities during centrifugal casting, of which there are three categories: true centrifugal casting, semicentrifugal casting, and centrifuging. The first two processes produce hollow cylindrical shapes and parts with rotational symmetry, respectively. In the third process, the mold cavities are spun at a certain radius from the axis of rotation; the centrifugal force thus increases the pressure in the mold cavity.

Evaporative Cooling

This is another casting process that utilizes sand as a basic component. However, it is unique in that the casting shape is not defined by molding the sand. Rather, a polystyrene foam pattern or replica is formed into the shape of the component to be cast and determines the size and shape of the final casting.

The evaporative casting process is compatible with bronze, steel, aluminum, and iron castings. In general, microstructural characteristics of castings made by the evaporative casting process tend to be the same as those of castings made by conventional sand casting processes.

The major advantages of the evaporative casting process are reduced costs and greater overall flexibility. Cost savings occur both in lower investment costs and in lower costs to make a casting. The lower investment costs come about because the equipment to make polystyrene foam replicas is less expensive than the equipment to make conventional sand molds and cores. Also, the polystyrene molding equipment operates at higher speeds. Finally, the tooling to make polystyrene foam replicas is also less expensive and has a longer life than tooling for conventional sand molds and cores. The costs for making a single casting by the evaporative casting process are lower because this process does not require cores and allows for easy reuse of the casting sand (no resins are added to the sand). Also, the casting made by this process is easier to clean than a conventional sand casting because there are no fins and parting lines that need to be ground down.

Zero Gravity Casting

Another mode of metal casting, zero gravity casting, has come of age with the advent of space stations containing laboratory equipment with metal-processing capabilities. In the absence of gravity the concept of melting and casting is drastically changed. The disadvantages of Earth-based processing that appear to be diminished or eliminated by processing under conditions of weightlessness are thermal and solutal convection and sedimentation; contamination of and contact with the container during processing; and hydrostatic pressure effects and deformation at high temperatures. Thus, there appears to be potential for processing liquid metals at zero gravity. However, commercial production is expected to be limited initially to small quantities of high-cost materials whose solidification characteristics are greatly improved by the lack of gravity. Semiconductor materials are thought to be excellent candidates for initial development efforts.

PRINCIPLES AND PRACTICE

Successful operation of any metal-casting process requires careful consideration of mold design and metallurgical factors. These include design consideration, gating systems, riser design, ease of pattern extraction, machining considerations, and, finally, metallurgical considerations, where the relationship between the processing history and the solidification event is illustrated in Figure M.9 for a part that is cast to shape.

APPLICATIONS

Ferrous alloys, steels, and cast irons constitute the largest volume of metals cast. Aluminum-, copper-, zinc-, titanium-, cobalt-, and nickel-base alloys are also cast into many forms, but in much smaller quantity than iron and steel.

Aluminum alloy castings have advantages such as resistance to corrosion, high electrical conductivity, ease of machining, and architectural and decorative uses. Magnesium alloy castings have the lowest density of all commercial casting allots. Copper alloy castings, although costly, have advantages such as corrosion resistance, high thermal and electrical conductivity, and wear properties suitable for antifriction bearing materials. Steel castings have more uniform (isotropic) properties than the same component obtained by mechanical working. On the other hand, because of high temperatures required, casting of steel is relatively expensive and requires considerable knowledge and experience. Cast irons, which constitute the largest quantity of all metals cast to shape, have properties such as hardness, wear resistance, machinability, and corrosion resistance. Gray iron castings are commonly produced for their low cost, machinability, good damping capacity, and uniformity. Nodular cast irons have significantly higher strengths than gray irons, but are more costly and not as machinable. There is an intermediate grade of cast iron called CG (compacted graphite) iron that shares the best properties of both gray and nodular irons.

METAL COATINGS

These are thin films of material bonded to metals to add specific surface properties, such as corrosion or oxidation resistance, color, attractive appearance, wear resistance, optical properties, electrical resistance, or thermal protection. In all cases proper surface preparation is essential to effective bonding between coating and basis metal, so that coated metals can function as duplex materials. The various methods of applying either metallic coatings (see Table M.9 showing methods for applying metallic coatings) or nonmetallic coatings, such as vitreous enamel and ceramics, and the conversion of surfaces to suitable reaction-product coatings.

METAL COATINGS

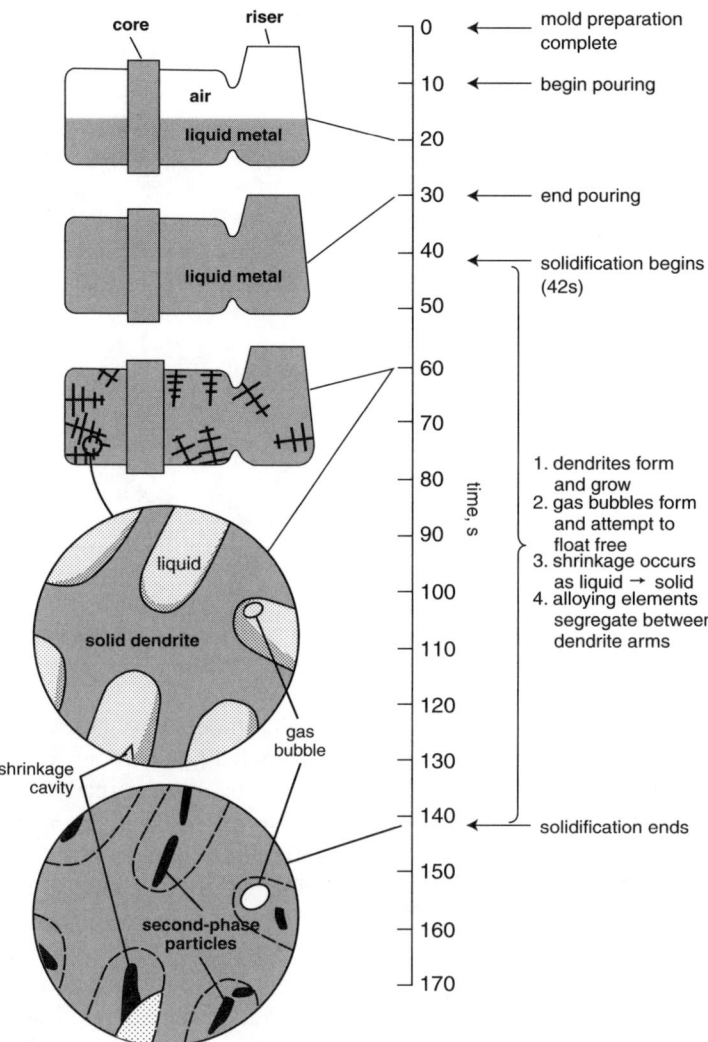

FIGURE M.9 Solidification and processing for a cast-to-shape part. (From *McGraw-Hill Encyclopedia of Science and Technology*, 8th ed., Vol. 11, McGraw-Hill, New York, 35. With permission.)

Hot-Dip

Low-melting metals provide inexpensive protection to the surfaces of a variety of steel articles. To form hot-dipped coatings, thoroughly cleaned work is immersed in a molten bath of the coating metal. The coating consists of a thin alloy layer together with relatively pure coating metal that adheres to the work as it is withdrawn from the bath. Sheet, strip, and wire are processed on a continuous basis at speeds of several hundred feet per minute. On the other hand, hardware and hollowware are handled individually or in batches.

Galvanized steel, that is, steel hot-dipped in zinc, is used for roofing, structural shapes, hardware, sheet, strip, and wire products. Hot-dipped tinplate is now largely supplanted by electroplated tinplate for tin cans. Terne plate, with coatings up to 20 μm, is used for roofing, chemical cabinets, and gasoline tanks. Hot-dipped aluminum-coated (aluminized) steel with coating up to 10 μm thick are used for oil refinery equipment and furnace and appliance parts where protection at temperatures up to 538°C is required.

TABLE M.9
Methods for Applying Metallic Coatings

Coating Metal	Hot Dip	Electroplate	Spray	Cementation	Vapor Deposition	Cladding	Immersion
Zinc	X	X	X	X			
Aluminum	X		X	X	X	X	
Tin	X	X					X
Nickel		X	X		X	X	X
Chromium		X		X	X		
Stainless Steel			X			X	
Cadmium		X			X		
Copper		X				X	X
Lead	X	X				X	
Gold		X			X	X	X
Silver		X			X	X	X
Platinum metals		X			X	X	X
Refractory metals		X	X			X	

Source: *McGraw-Hill Encyclopedia of Science and Technology*, 8th ed., Vol. 11, McGraw-Hill, New York, 40. With permission.

SPRAYED

A particular advantage of sprayed coatings is that they can be applied with portable equipment. The technique permits the coating of assembled steel structures to obtain corrosion resistance, the building up of worn machine parts for rejuvenation, and the application of highly refractory coatings with melting points in excess of 1650°C.

Nearly any metal or refractory compound can be applied by spraying. Coating material in the form of wire or powder is fed through a specially designed gun, where it is melted and subjected to a high-velocity gas blast that propels the atomized particles against the surface to be coated.

Coatings such as zinc and aluminum are applied with a gun that provides heat by burning acetylene, propane, or hydrogen in oxygen. Highly refractory coating materials, such as oxides, carbides, and nitrides, can be applied by plasma-arc spraying. In this process temperatures of 12,000°C or more may be produced by partially ionizing a gas (nitrogen or argon) in an electric arc and passing the gas through a small orifice to produce a jet of hot gas moving at high velocity. Another variation for applying refractory coatings is detonation-flame plating. In this process a mixture of oxygen and gas-suspended fine particles is fired four times per second by a timed spark.

Sprayed zinc or aluminum coatings up to 250 μm are used to protect towers, tanks, and bridges. Such coatings are normally sealed with an organic resin to enhance protection.

Sprayed refractory coatings have been developed for the high temperatures experienced in aerospace applications. They are also used for wear resistance, heat resistance, and electrical insulation.

CEMENTATION

These are surface alloys formed by diffusion of the coating metal into the base metal, producing little dimensional change. Parts are heated in contact with powdered coating material that diffuses into the surface to form an alloy coating, whose thickness depends on the time and the temperature of treatment. A zinc alloy coating of 25 μm is formed on steel in 2 to 3 h at 375°C. A chromium alloy (chromized) coating of 100 μm is formed in 1 h at 1000°C.

Chromized coatings on steel protect aircraft parts and combustion equipment. Sherardized (zinc–iron alloy) coatings are used in threaded parts and castings. Calorized (aluminum–iron alloy) coatings protect chemical equipment and furnace parts. Diffusion coatings are used to provide oxidation resistance to refractory metals, such as molybdenum and tungsten, in aerospace applications where reentry temperatures may exceed 1650°C. In addition to the pack process described above, such coatings may be applied in a fluidized bed. In forming disilicide coatings on molybdenum, the bed consists of silicon particles suspended in a stream of heated argon flowing at 0.15 m/s, to which a small amount of iodine is added. The hot gases react with the silicon to form SiI_4, which in turn reacts with the molybdenum to form $MoSi_2$.

Vapor Deposition

A thin specular coating is formed on metals, plastics, paper, glass, and even fabrics. Coatings form by condensation of metal vapor originating from molten metal, from high-voltage (500 to 2000 V) discharge between electrodes (cathode sputtering), or from chemical means such as hydrogen reduction or thermal decomposition (gas plating) of metal halides. Vacuums up to 10^{-4} Pa often are required.

Aluminum coatings of 0.125 µm are formed on zinc, steel, costume jewelry, plastics, and optical reflectors. Chemical methods can form relatively thick coatings, up to 250 µm.

Immersion

Either by direct chemical displacement or, for thicker coatings, by chemical reduction (electroless coating), metal ions plate out of solution onto the workpiece.

Tin coatings are displaced onto brass and steel notions and on aluminum-alloy pistons as an aid during the breaking-in period. Displacement nickel coatings of 1.25 µm are formed on steel articles. Electroless nickel, involving the reduction of a nickel salt to metallic nickel (actually a nickel–phosphorus alloy), permits the formation of relatively thick uniform coatings up to 250 µm on parts with recessed or hidden surfaces difficult to reach by electroplating.

Vitreous Enamel

Glassy but noncrystalline coatings for attractive durable service in chemical, atmospheric, or moderately high-temperature environments are provided by enamel or porcelain coating.

Dry enameling is used for castings, such as bathtubs. The casting is heated to a high temperature, and then dry enamel powder is sprinkled over the surface, where it fuses.

Firing temperatures for conventional enameling of iron or steel ranges up to 870°C. Low-temperature enamels have been developed, permitting the enameling of aluminum and magnesium.

Coatings of 75 to 500 µm are used for kitchenware, bathroom fixtures, highway signs, and water heaters. Vitreous coatings with crystalline refractory additives can protect stainless steel equipment at temperatures up to 950°C.

Ceramic

Essentially crystalline, ceramic coatings are used for high-temperature protection above 1100°C. The coatings may be formed by spraying refractory materials such as aluminum oxide or zirconium oxide, or by the cementation processes for coatings of intermetallic compounds such as molybdenum disilicide. Cermets are intimate mixtures of ceramic and metal, such as zirconium boride particles, dispersed throughout an electroplated coating of chromium.

Surface-Conversion

An insulating barrier of low solubility is formed on steel, zinc, aluminum, or magnesium without electric current. The article to be coated is either immersed in or sprayed with an aqueous solution, which converts the surface into a phosphate, an oxide, or a chromate. Modern solutions react so rapidly that sheet and strip materials can be treated on continuous lines.

Phosphate coatings, equivalent to 1 to 4 g/m^2, are applied to bare or galvanized steel and to zinc-base die castings as preparation for painting. The coating enhances paint adhesion and prevents underfilm corrosion. Phosphate coatings, containing up to 40 g/m^2 (lubricated), serve as an aid in deep-drawing steel and in other friction-producing processes or applications. Iridescent chromate coatings on zinc-coated steel improve appearance and reduce zinc corrosion. Chromate, phosphate, and oxide coatings on aluminum or magnesium are used to prepare the surface before painting.

Anodic

Coatings of protective oxide may be formed on aluminum or magnesium by making them the anode in an electrolytic cell. Anodized coatings on aluminum up to 75 μm thick are formed in sulfuric acid to form a porous oxide that may be sealed in boiling water or steam to provide a clear, abrasion-resistant, protective coating.

Such coatings are used widely on aluminum furniture, automobile trim, and architectural shapes. Thin, nonporous, electrically resistant coatings are formed on aluminum in a boric acid bath in the production of electrolytic capacitors. Anodized coatings on magnesium are thicker (up to 75 μm) and harder than those formed by chemical conversion. Anodic coatings 7.5 μm thick are often used as a paint base.

Powder

The term *powder coatings* refers to a process whereby organic polymers such as acrylic, polyester, and epoxies are applied to substrates for protection and beautification. It is essentially an industrial painting process that uses a powdered version (25- to 50-μm particle size) of the resin rather than the solvent solution represented by industrial baking enamels.

In the widest type of commercial usage, these powders are applied to electrically grounded substrates by means of an electrostatic spray gun. In these guns, high-voltage (60 to 100 kV) low-amperage charges are applied to the powders. The powder particles are attracted to and adhere to the substrate until it can be transported to an oven, where the powder particles melt, coalesce, flow, and form a smooth coating. Typical baking temperatures are 90°C for 20 min.

Coatings can be applied at thicknesses ranging from 25 to 250 μm, depending upon end-use requirements.

Although electrostatic spray is the principal method of commercial application, other techniques exist, including fluidized-bed coating, where a preheated part is immersed into an aerated bed of powder particles, and electrostatic fluidized bed, where a substrate is passed through an "electrified" cloud of the powder.

Powder coating uses no organic solvent, and the oversprayed material can be recaptured and reused. Thus, there are several notable advantages over conventional industrial solvent-based painting:

1. There are no hydrocarbon baking byproducts, and the process is environmentally acceptable.
2. Material conservation and economics are optimized because of the recycling of oversprayed material.
3. Powders are easy to apply and readily lend themselves to automation.

A wide variety of parts are finished with powder coatings. Outdoor lawn and patio furniture coated in this process display good weathering and abuse resistance. Powder-coated electrical transformers are insulated electrically and provided with corrosion protection. Powder coatings have also been developed for finishing major appliances and for automotive coatings. Also, laundry tops and lids have displayed excellent detergent resistance where organic powder coatings have replaced porcelain finishes.

METAL FOAMS

Metal foams are metallic cellular materials that have a high porosity fraction, typically ranging from 40 to 90 vol%. Because of their high stiffness and low specific weight, cellular materials are applied in construction: packaging, insulation, noise and vibration damping, and filtering. They are considered by many to be a new class of engineering material.

Typical foaming processes include casting, powder-pressing, metallic deposition, and sputter deposition. Metal foams can be fabricated in a variety of different ways, and many attempts have been made in the past to develop good foam structure. However, the choice frequently seems to be between high cost and poor quality.

METAL FOAMS

A new powder method allows for direct net-shape fabrication of foamed parts with a relatively homogeneous pore structure. Metallic foams fabricated by this approach exhibit a closed-cell microstructure with higher mechanical strength than open-cell foams. This type of microstructure is particularly appropriate for applications requiring reduced weight and energy-absorption capabilities.

The powder-metallurgy production method makes it possible to build metallic foam parts that have complex geometry. Sandwich structures composed of a porous metallic foam core and metallic face sheets can also be produced, with options exploiting combined materials and shapes. These foams enlarge the application range of cellular materials because of their excellent physical and mechanical properties, as well as their relative recyclability.

Processing

The process consists of mixing metal powders (either prealloyed metal powders or powder blends) with a small amount of foaming agent. When the agent is a metal hydride, a content of less than 1% is generally sufficient. After the foaming agent is uniformly distributed within the matrix powders, the mixture is compacted into a dense, semifinished product with no residual open porosity (Figure M.10 shows the production of metal foam). Typical compaction methods include uniaxial pressing, extrusion, and powder rolling. The foamable material may be further shaped through subsequent metalworking processes such as rolling, swaging, or extrusion.

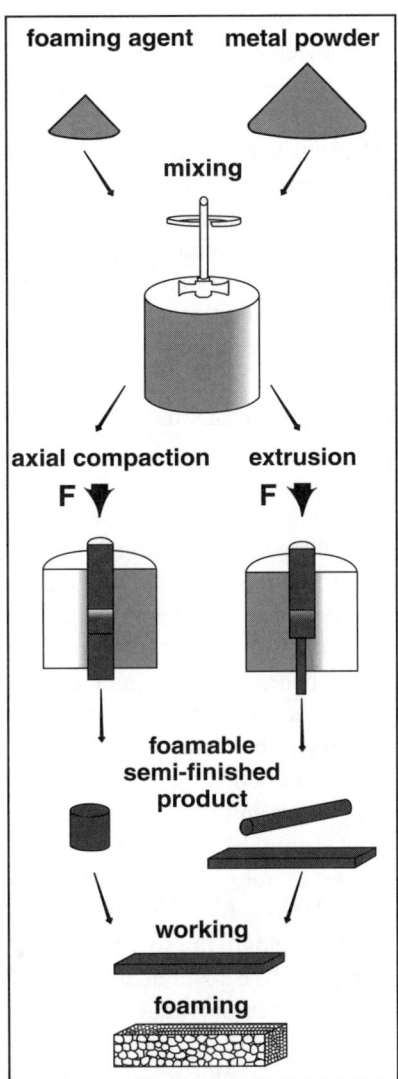

FIGURE M.10 Production of metal foam. (From *Adv. Mater. Proc.*, 154(5), 45, 1998. With permission.)

Potential Applications

The following is a list of applications currently being investigated:

- *Automotive industry*: Light, stiff structures made of aluminum foam and foam sandwich panels could help to reduce weight and increase stiffness. Examples are hoods, trunk lids, and sliding roofs. With regard to energy absorption, it is possible to engineer controlled deformation into the crash zone of cars and trains with maximum impact-energy dissipation. Possible applications include elements for side and front impact protection. In general, foam filling leads to higher deformation forces when profiles are bent and to higher energy absorption when profiles are axially crushed. Potential applications include bumpers, underside protection of trucks, A- and B-pillars, and other elements subjected to large deformation.

- *Aerospace industry*: Metal foam sandwich panels offer good potential to replace expensive honeycomb structures in the aerospace industry. The attributes of metal foams include the isotropy of the foam material and fire retardation to maintain the integrity of the structure. Roll cladding for direct metallic bonding can eliminate the need for adhesive bonding in the sandwich panels.
- *Building industry*: Many construction applications require light, stiff, and fire-resistant elements, or supports for such elements. Foamed sandwich panels could help to reduce the energy consumption for elevators by trimming the deck weight. Combining energy absorption and high specific stiffness, the foamed sandwich panels may be a good candidate for these applications. Another application of the metal foams is to fasten plugs into concrete walls. To fill the gap between the plugs and the wall, the foamable materials can be inserted into the gap and locally heated. The foamable material will expand and fill the space between the plugs and the concrete wall, provided the resultant foam density is high enough.
- *Further applications*: Additional applications can capitalize on the properties of foams made from other metals. For example, lead and nickel foams may be suitable for batteries, and gold or silver foams may be applicable in art and jewelry. Furthermore, open-cell foams can be fabricated by modifying the process scheme. This material would be excellent for several applications, such as heat exchangers, filters, and catalyst carriers. Foaming of high-temperature alloys such as nickel and titanium will also help expand the application range, especially for aerospace and biomedical components.

METAL FORMING

Metal forming is any manufacturing process by which parts or components are fabricated from metal stock. In the specific technical sense, metal forming involves changing the shape of a piece of metal. In general terms, however, it may be classified roughly into five categories: (1) mechanical working, such as forging, extrusion, rolling, drawing, and various sheet-forming processes; (2) casting; (3) powder and fiber metal forming; (4) electroforming; and (5) joining processes. The selection of a process, or combination of processes, requires a knowledge of all possible methods of producing the part if a serviceable part is to be produced at the lowest overall cost.

METAL HYDRIDES

A metal hydride is a compound in which hydrogen is bonded chemically to a metal or metalloid element. The compounds are classified generally as ionic, transition metal, and covalent hydrides.

IONIC HYDRIDES

The most reactive metals (alkali and alkaline earth metals, with the exception of magnesium and beryllium) combine directly on heating with hydrogen gas at a pressure of 10^2 kPa. Magnesium reacts at elevated hydrogen pressure. Examples of hydrides are shown in Table M.10.

TRANSITION METAL HYDRIDES

This group of compounds is less well understood than the ionic and covalent hydrides. When lanthanum is heated in hydrogen, the gas is readily taken up, forming first a material whose composition is approxosmtely $LaH_{1.86}$ (the "dihydride" phase) and finally LaH_3.

The ionic hydrides, such as NaH, exhibit much more limited nonstoichiometry. When sodium is converted to pure NaH, all electrons in the conduction band are consumed and the product is white and insulating.

Copper hydride, CuH, is unique in that it is the only hydride that can be precipitated from aqueous solution.

TABLE M.10
Physical Properties of Representative Hydrides

Compound	Formula	Form	Melting Point °C (°F)	Density, g/cm³	Boiling Point, °C (°F)	Temperature at Which Dissociation Pressure is 1 atm or 10^2 kPa, °C (°F)
Lithium hydride	LiH	White crystals	691 (1276)	0.77	—	~790 (1454)
Sodium hydride	NaH	White crystals	Decomposes	1.396	—	425 (797)
Calcium hydride	CaH$_2$	White crystals	>1000 (1830)	1.902	—	960 (1760)
Titanium hydride	TiH$_2$	Gray powder	Decomposes	3.78	—	630 (1166)
Cerium hydride	CeH$_2$	Dark gray powder or greenish metallic single crystals	1088 (1990) for CeH$_{1.73}$	5.45	—	750 (1382) for CeH$_{2.2}$
Uranium hydride	UH$_2$	Black powder	~1050 (1922) at 580 atm (58.8 MPa)	10.95	—	440 (824)
Lanthanum nickel hydride	LaNi$_5$H$_6$	Dark gray powder	Decomposes	—	—	~0 (32) [2.5 atm at 25°C or 0.25 MPa at 77°F]
Palladium hydride	PdH$_{0.66}$	Metallic	Decomposes	~10.8	—	25 (77) for PdH$_{0.56}$, −78 (−108) for PdH$_{0.83}$
Diborane	B$_2$H$_6$	Colorless gas	−165.5 (−265.9)	0.438 at bp	−92.5 (−135)	—
Silane	SiH$_4$	Colorless gas	−185 (−301)	0.68 at mp	−111.8 (−169.2)	—
Stannane	SnH$_4$	Colorless gas	−150 (−238)	—	−52 (−62)	—
Arsine	AsH$_3$	Colorless gas	−113.5 (−172.3)	—	−55 (−67)	—
Stibine	SbH$_3$	Colorless gas	−88.5 (−127.3)	2.2 at bp	−17 (1.4)	—
Tellurium hydride	TeH$_2$	Colorless gas	−51 (−59.8)	2.7 at −18°C (−0.4°F)	−4 (25)	—
Aluminum hydride	AlH$_3$	White solid	Decomposes	—	—	—
Copper hydride	CuH	Dark brown solid	Decomposes slowly even at 25°C (77°F)	6.39	—	—

Source: McGraw-Hill Encyclopedia of Science and Technology, 8th ed., Vol. 11, McGraw-Hill, New York, 47. With permission.

Titanium, zirconium, vanadium, and niobium all react with hydrogen, forming dihydrides at the limiting compositions. The materials are conducting at all compositions.

Common metals such as steel and copper dissolve small quantities of hydrogen at elevated temperatures. On cooling, the gas comes out of solution and results in severe degradation of the mechanical properties of the metals. This is called hydrogen embrittlement and can be prevented by degassing the metal while it is still molten.

A number of alloys or intermetallic compounds are known that react with hydrogen and form ternary hydrides. Examples are LaNi$_5$H$_6$, Mg$_2$NiH$_4$, AlTh$_2$H$_2$, and CaAg$_2$H.

Covalent Hydrides

Most evidence indicates that in ionic and metallic hydrides an electronic pair is associated primarily with the hydrogen as H⁻, while in covalent hydrides the electron pair is shared between the hydrogen atom and an atom of another element. In these compounds hydrogen is considerably smaller (radius 0.03 nm) than in the ionic hydrides. Covalent hydrides usually consist of small molecules, in which case they are gases (SiH_4 and SbH_3), but some form high polymers, in which case they are nonvolatile solids (AlH_3 and ZnH_2). All tend to decompose irreversibly rather easily on heating. Covalent hydrides are generally synthesized indirectly, not from direct combination of the elements. Gaseous hydrides such as SiH_4, PH_3, and AsH_3 can be generated by heating a solid mixture of the corresponding oxide and $LiAlH_4$.

METAL INJECTION MOLDING

Injection molding is one of the most productive techniques for shaping materials. Until recently, injection molding was restricted to thermoplastic polymers (polymers that melt on heating). However, metals have property advantages over polymers. They are stronger, stiffer, are electrically and thermally conductive, can be magnetic, and are more wear and heat resistant. The concept of metal injection molding (MIM) combines metal powders and a thermoplastic binder to allow shaping of complex objects in the high-productivity manufacturing setting associated with injection molding. After shaping, the metal powders are sintered to densities close to those listed in handbooks; that is, they have very few pores. Accordingly, the process delivers materials with metallic properties in an efficient manner.

Metal injection molding, and the related process of ceramic injection molding, is a derivative of powder metallurgy. Metal powders can be shaped in a semifluid state (powders that are poured into containers take on the shape of the container), but after heating to high temperatures the particles bond into a strong, coherent mass. In ceramics this is analogous to the manufacture of clay pottery, where shaping occurs with a water–clay system at the potter's wheel, but after kiln firing, the structure is strong and rigid.

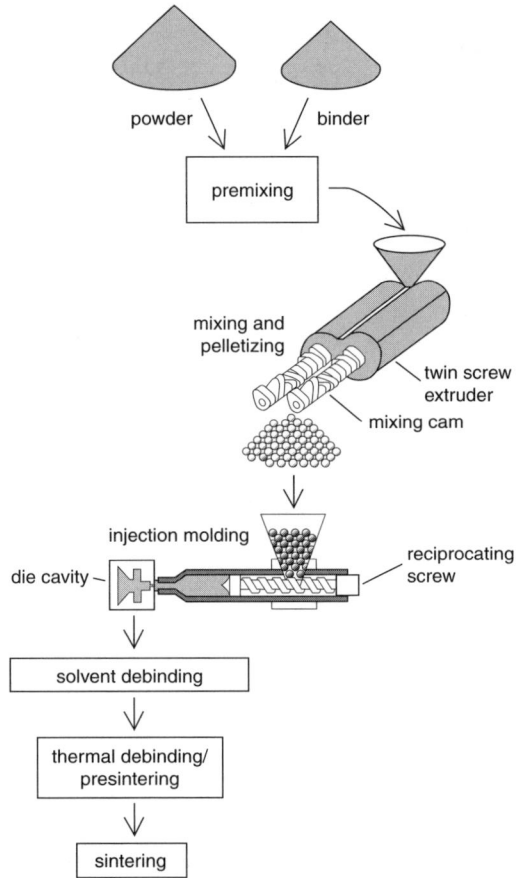

FIGURE M.11 Metal injection molding process. (From *Design News*, January 5, p. 66, 1998. With permission.)

Metal injection molding is a process in which metal in a powdered form is combined with a polymer binder to produce a homogeneous mixture known as feedstock. The feedstock is processed in a manner similar to plastic injection molding, except that parts are designed slightly larger to account for shrinkage during the final sintering step. The molded part is called the "green" part.

Once molded, three methods are available to remove the binder from the green parts: solvent, thermal, and catalytic debinding. Differences in these methods affect the outcome of the product. The debinding phase can be the most time-consuming aspect of the process. After debinding, parts (known as "brown" parts) are sintered — a process in which the components are exposed to heat over an extended period (Figure M.11).

Applications

Metal injection molding is applied in the production of high-performance components that have complex shapes. The most successful applications involve components that have complex shapes and also must be inexpensive to produce while providing high relative performance. Generally, metal injection molding can be used for all shapes that can be formed by plastic injection molding, especially for very small complex geometries.

Many applications have been developed for metal injection molding. They include components for use in diverse fields ranging from surgical tools to microelectronic packaging. Recently, automotive components such as sensor parts in the air-bag actuator mechanism have moved into production. The technology has been extended to utilize a wide range of materials, including steels, stainless steels, nickel, copper, cobalt alloys, tungsten alloys, niobium alloys, nickel-base superalloys, and intermetallics. Additionally, most ceramics can be processed in the same manner; these include silica, alumina, zirconia, silicon nitride, silicon carbide, aluminum nitride, cemented carbides, and various electronic ceramics.

Although metal injection molding is generally viable for all shapes that can be formed by plastic injection molding, it is not cost competitive for relatively simple or asymmetric geometries. Large components require larger molding and sintering devices that are more difficult to control. Accordingly, metal injection molding is largely applied to smaller shapes with masses below 100 g.

As the engineering community better appreciates its favorable attributes, metal injection molding applications will grow. Although the current industry is relatively small, the anticipated growth is impressive. One area of keen interest is in the processing of titanium alloys for biomedical applications. Another is in the co-molding of different materials, that is, forming part of a component from one material and then another portion from a second material. This option has merit for forming corrosion barriers, wear surfaces, and electrical interconnections in ceramics. Other major growth areas include high-performance magnets, technical ceramics, materials associated with heat dissipation in electronic circuits, hard materials, ultrahigh-strength metals, and biocompatible materials.

METAL POWDERS

By definition, a metal powder is an aggregate of discrete metal particles that are usually in the size range of 1 to 1000 µm.

The metal powders in common use are lead–tin alloys for solders; iron-, nickel-, and cobalt-base alloys for hard facing; copper-, silver-, and nickel-base alloys for brazing; aluminum, bronze, and stainless steel for paint pigments; magnesium for pyrotechnics; iron for welding rods, torch cutting and scarfing, and metal powder parts; copper for metal-powder parts; and carbide for tools. Used in smaller quantities are iron powders for radio and television tuning cores; nickel–iron and silicon–iron alloy powders for other soft magnetic parts; aluminum, iron, nickel, and cobalt powders for small permanent magnets in the "Alnico" series; nickel and cobalt powders as binders in the production of carbide tools and alloy and stainless steel powders for high-strength and special property parts.

Of these uses, the fabrication of structural parts or machinery components accounts for the largest single use of metal powders and competes with such other metal-forming methods as machining, casting, stamping, and forging. The powder-metallurgy technique may be selected by the designer as the best way to make a particular part for one of several reasons:

1. The process is ideal for mass-producing machine components at low unit cost.
2. Residual porosity can be controlled to provide long-wearing qualities through the self-lubricating feature of the "oil-less" sleeve bearing.
3. Metal combinations are possible through powder metallurgy that cannot be melted.
4. Powder-metallurgy techniques provide the only practical method of forming high-melting-point metals.

The good surface finish and close dimensional tolerances possible are additional reasons why the designer may specify metal powder parts.

Although most metal powder parts are made from iron, copper, or mixtures of these primary powders with or without graphite additions, many other powders are used to develop special properties like high strength, magnetic or electrical properties, corrosion resistance, or oxidation resistance. These special powders include brass, bronze, alloy steels, stainless steel, and various nickel-base alloys. Many of these special powders are also employed in the production of metal filters.

PRODUCTION METHODS

Electrolytic

These methods are used to make most of the copper powder used for metal powder parts and a special, high-purity iron powder often used for making high-density iron powder parts.

Atomization

This is a molten metal production method applied to a wide range of powders including aluminum, magnesium, brass, bronze, lead, tin, nickel–silver, and stainless and low-alloy steels. Other more complex iron-, nickel-, and cobalt-base alloys are also made by atomization.

METALLIC GLASSES

These are alloys with an amorphous or glassy structure. A glass is a solid material obtained from a liquid, which does not crystallize during cooling. It is therefore an amorphous solid, which means that the atoms are packed in a more or less random fashion similar to that in the liquid state. The word *glass* is generally associated with the familiar transparent silicate glasses containing mostly silica and other oxides of aluminum, magnesium, sodium, and so on. These glasses are not metallic; they are electrical insulators and do not exhibit ferromagnetism. One obvious answer to obtaining a glass having metallic properties is to start from a melt containing metallic elements instead of oxides.

ELECTRICAL AND SUPERCONDUCTIVITY PROPERTIES

The electrical resistivity of metallic glasses is high, for example 100 $\mu\Omega$-cm and higher, which is in the same range as the familiar nichrome alloys widely used as resistance elements in electric circuits. Another interesting characteristic of the electrical resistivity of metallic glasses is that it does not vary much with temperature. The temperature coefficient is of the order of $10^{-4} K^{-1}$ and can even be zero or negative, in which case the resistivity decreases with increasing temperature. Because of their insensitivity temperature variations, metallic glasses are suitable for applications in electronic circuits for which this property is an essential requirement.

Superconducting metallic glasses are much more stable, and some of them do not crystallize at temperatures as high as 500°C. Some superconducting metallic glasses contain only two metals, such as $Zr_{75}Rh_{25}$, and some are more complex alloys in which there is approximately 20% of metalloid elements, mostly boron, silicon, or phosphorus.

One of the main reasons for continuing research on new superconducting glasses is their projected usefulness in high-field electromagnets, which will be required to contain the high-temperature plasma in fusion reactors. The present requirement for these magnets is a field of at least 100,000 gauss, which is not attainable with conventional copper-wound electromagnets. Observations and results have shown that the electrical resistivity of metallic glasses is not affected by radiation.

MAGNETIC PROPERTIES

The ferromagnetic properties of metallic glasses have received a great deal of attention, probably because of the possibility that these materials can be used as transformer cores. One class of ferromagnetic metallic glasses is based on transition metals with zirconium or hafnium, for example, $(Co,Ni,Fe)_{90}Zr_{10}$.

Because metallic glasses do not have crystalline anisotropy, the core maker is able to anneal in a desired domain pattern. For example, metallic glasses can provide extremely high magnetic efficiency in motors, because their anisotropy can be built in during a final annealing to develop a low-reluctance path in the exact configuration required by the motor. The use of metallic glasses in motors can reduce core loss by as much as 90% as compared with conventional crystalline magnets. Alternatively, for transducer applications, one can just as well anneal in a domain structure that is transverse to the ribbon length by applying a transverse magnetic field during the annealing cycle. This will maximize the rotation of domains and hence the magnetostrain during any subsequent longitudinal excitation.

Mechanical and Other Properties

Although a metallic glass under tension will fracture with very small permanent deformation, scanning electron micrography reveals that a large local shear deformation occurs before rupture.

Metallic glasses thus are intrinsically ductile. The interest in the mechanical properties of metallic glasses is motivated by their high rupture strength and toughness. The fracture strength of metallic glasses approaches a theoretical strength that is about $1/50$ of Young's modulus.

Future and Applications

Considering the unusual physical and chemical properties of metallic glasses, there is no doubt that they will play an important role as an engineering material in the future. The melt-spinning process of producing metallic glasses should also result in substantial savings in labor and energy when compared with the present technology, because it can produce a thin sheet of material by direct casting from the liquid state.

Possible applications of metallic glasses have already been demonstrated on audio and video magnetic tape recording heads, sensitive and quick-response magnetic sensors or transducers, security systems, motors, and power transformer cores. The combination of excellent strength, resistance to corrosion and wear, and magnetic properties may lead to interesting applications, for example, the use of such glasses as inductors in magnetic separation equipment.

Besides the straightforward applications, there remain many less obvious potential applications of metallic glasses: the freedom from constraints imposed by the equilibrium phase diagram may well allow metallurgists to devise chemically interesting glassy alloys that could not be made otherwise in single-phase form. Glasses based on palladium–phosphorus alloys show a considerably higher catalytic activity for oxidation of methanol, and they are stabler than the fine-grained platinum electrodes. Another potential application of metallic glasses is their use when intentionally crystallized. Since glasses are highly supercooled liquids, crystallization can be effected via copious nucleation, which results in the formation of a fine-grained crystalline product that is microstructurally very homogeneous. An example is the use of a glass sheet of the alloy $Ni_{69}Cr_6Fe_3B_{14}Si_8$ as a brazing filler metal; the glass sheet can be stamped to shape and inserted between the parts to be joined, yielding excellent control of the joint properties after fusion of the products.

Another development is an application of powder-metallurgy methods to the fabrication of amorphous alloys. In the process of "atomizing" liquid alloys into a very fine powder, the rate of cooling may be high enough to produce an amorphous powder.

By using explosive loading, it is indeed possible to subject the powder to very high pressure, and probably some rather high temperature, for a short enough time to avoid crystallization, and to achieve near-theoretical densities. Massive ingots of metallic glasses, susceptible of being forged and rolled into various shapes, may become available within the not-too-distant future. Finally, since metallic glasses may be regarded as liquids whose structure has been frozen, they constitute ideal materials for low-temperature transport and critical behavior studies, and they are most suited for the study of electrons in noncrystalline metals.

METALLIC MATERIALS

About three quarters of the elements available can be classified as metals, and about half of these are of at least some industrial or commercial importance. Although the word *metal*, by strict definition, is limited to the pure metal elements, common usage gives it wider scope to include metal alloys. Although pure metallic elements have a broad range of properties, they are quite limited in commercial use. Metal alloys, which are combinations of two or more elements, are far more versatile and for this reason are the form in which most metals are used by industry.

Metallic materials are crystalline solids. Individual crystals are composed of unit cells repeated in a regular pattern to form a three-dimensional crystal-lattice structure. A piece of metal is an aggregate of many thousands of interlocking crystals (grains) immersed in a cloud of negative-valence electrons detached from the atoms of the crystals. These loose electrons serve to hold the crystal structures together because of their electrostatic attraction to the positively charged metal atoms (ions). The bonding forces, which are large because of the close-packed nature of metallic crystal structures, account for the generally good mechanical properties of metals. Also, the electron cloud makes most metals good conductors of heat and electricity.

Metals are often identified as to the method used to produce the forms in which they are used. When a metal has been formed or shaped in the solid, plastic state, it is referred to as a wrought metal. Metal shapes that have been produced by pouring liquid metal into a mold are referred to as cast metals.

There are two families of metallic materials — ferrous and nonferrous. The basic ingredient of all ferrous metals is the element iron. These metals range from cast irons and carbon steels, with over 90% iron, to specialty iron alloys, containing a variety of other elements that add up to nearly half the total composition.

Except for commercially pure iron, all ferrous materials, both irons and steels, are considered to be primarily iron–carbon alloy systems. Although the carbon content is small (less than 1% in steel and not more 4% in cast irons) and often less than other alloying elements, it nevertheless is the predominant factor in the development and control of most mechanical properties.

By definition, metallic materials that do not have iron as their major ingredient are considered to be nonferrous metals. There are roughly a dozen nonferrous metals in relatively wide industrial use. At the top of the list is aluminum, which next to steel is the most widely used structural metal today. It and magnesium, titanium, and beryllium are often characterized as light metals because their density is considerably below that of steel.

Copper alloys are the second nonferrous material in terms of consumption. There are two major groups of copper alloys: brass, which is basically a binary alloy system of copper and zinc, and bronze, which was originally a copper–tin alloy system. Today, the bronzes include other copper–alloy systems.

Zinc, tin, and lead, with melting points below 427°C, are often classified as low-melting alloys. Zinc, whose major structural use is in die castings, ranks third to aluminum and copper in total consumption.

Lead and tin are rather limited to applications where their low melting points and other special properties are required. Other low-melting alloys are bismuth, antimony, cadmium, and indium.

Another broad group of nonferrous alloys is referred to as refractory metals. Such metals as tungsten, molybdenum, and chromium, with melting points above 1649°C, are used in products that must resist unusually high temperatures. Although nickel and cobalt have melting points below 1649°C, they serve as the base metal or as alloying elements of many heat-resistant alloys.

Finally, the precious metals, or noble metals, have the common characteristic of high cost. In addition, they generally have high corrosion resistance, many useful physical properties, and generally high density.

METALLIC SOAP

This is a term used to designate compounds of the fatty acids of vegetable and animal oils with metals other than sodium or potassium. They are not definite chemical compounds like the alkali soaps, but may contain complex mixtures of free fatty acid, combined fatty acid, and free metallic oxides or hydroxides. The name distinguishes the water-insoluble soaps from the soluble soaps made with potash or soda. Metallic soaps are made by heating a fatty acid in the presence of a metallic oxide or carbonate, and are used in lacquers, leather and textiles, paints, inks, ceramics, and grease. They have the properties of being driers, thickening agents, and flattening agents. They are characterized by ability to gel in solvents and oils, and by their catalytic action in speeding the oxidation of vegetable oils.

When made with fatty acids having high iodine values, the metallic soaps are liquid, such as the oleates and linoleates, but the resinates and tungates are unstable powders. The stearates are fine, very stable powders. They are found as barium, strontium, chromium, manganese, cerium, nickel, lead, and lithium stearates. The fatty acid determines the physical properties, but the metal determines the chemical properties. Aluminum stearate is the most widely used metallic soap for colloid products. Aluminum soaps are used in polishing compounds, in printing inks and paints, for waterproofing textiles, and for thickening lubricating oils. The resinates, linoleates, and naphthanates are used as driers; the lead, cobalt, and manganese are the most common.

METALLIZING

Metallizing is the application of a metallic coating to a ceramic to permit subsequent brazing to a mating part. Various techniques are employed but the basic steps follow the common practices of ceramic decorating except that materials must be carefully chosen.

Metallizing Systems

- *Metals.* Reactivity being desirable, most metal powders are purchased as −325 mesh to ensure a high surface-to-volume ratio.
- *Copper.* May be mixed, as flake, with coarse glass particles that melt and seal to the ceramic preserving the integrity of the flake. This continuous electrical path through the glass seal is useful in the manufacture of spark plugs. Copper, as the oxide, will bond to a ceramic under precise firing conditions and, when mixed with silver, may be used as a metallizing preparation.
- *Gold.* Frequently combined with reactive glass for uses in the air-fired paste and solder category.
- *Iron.* Reactive material originally combined with tungsten. Very common additive for sintered powder process.
- *Manganese.* Reactive material combined with molybdenum and is a very common additive.
- *Molybdenum.* Most used as basic metal for sintered powder metallizing. Oxidation potential allows control of oxidation state in controlled atmosphere furnace. Coefficient of expansion of the metal and its reaction products are favorable.
- *Palladium.* Similar to platinum and sometimes added to platinum-frit mixtures for air firing.
- *Platinum.* Used with reactive glasses in air-fired metallizing.
- *Silver.* Basic metal for many air-fired pastes. Mixed as granular or flake material with a reactive glass; e.g., borosilicate.
- *Tin.* Basis for direct chemical bonding to numerous ceramics and high-temperature metals.
- *Titanium.* Basis of the active metals process. May be used as powder or foil prior to braze alloying. Titanium-bearing brazes will wet and flow over ceramic; in vacuum, almost as well as solder over copper. Frequently applied as the hydride, which dissociates at <800°C, providing nascent hydrogen, which tends to scour the surface to be wetted. May be added to sintered powder compositions to promote reaction.
- *Tungsten.* Similar in metallizing properties to molybdenum.
- *Zirconium.* Performs similarly to titanium but with less activity. Has lower coefficient of expansion.
- *Miscellaneous Reactive Powders.* Many companies have evolved proprietary

sintered powder metallizing systems that may be very complicated. Some of the likely additives to these compositions are aluminum, barium, boron, cadmium, magnesium, rare earths, and silicon. Most of these would be added as an oxide.

Glass Compositions. Most metallizing suppliers purchase glass-bearing metallizing compositions from commercial vendors. The glass compositions are regarded as proprietary and are not generally known.

Binders. Binders are fugitive materials that are required to increase viscosity and density of the vehicles developed for suspending metal powders during application of the ceramic. They must leave no deleterious residue following the firing operation and must hold the metal *in situ* until the particles become somewhat adherent. Typical binders are represented by acrylic polymers, commercial binders, nitrocellulose, and pyroxyline resins.

Solvents. Solvents are required to dissolve the binder material and usually have a high vapor pressure so they are effectively lost before firing of the metallizing begins. These materials, with the binders, constitute the vehicle. The best vehicles allow a smooth application with controlled thickness, and dry to a dense abrasion-resistant layer prior to firing. Solvents are selected typically from the following families: acetates, alcohols, commercial solvents, ethers, and ketones.

APPLICATION METHODS

Pastes, when used, are prepared by selecting the metal powders and blending with additives in a ball mill. Acetone is frequently used as the carrier during the milling operation. Attention is given to the desired degree of particle size reduction. The milled material is removed, dried, pulverized, and combined with the chosen binder and solvent. This preparation may be milled further to ensure complete blending.

The paste is then ready to be applied to the ceramic.

Many methods of application are in use. They vary depending on size of the order, configuration of the part, the precision required, and operating economy.

Brushing. A paintable slurry is prepared and applied by brush to the desired area of the ceramic. Useful for small lots or unusual configurations.

Decalcomania. Commercial sources offer patterns prepared for transfer to ceramic.

Dipping. Applicable where complete coverage is permissible or where subsequent methods of recovering a pattern are feasible.

Evaporation. Vacuum evaporation of metals offers a readily controllable metal thickness for thin-film and other applications not requiring high strength.

Printing. Banding or printing equipment may be successfully used with certain pastes. Limitations result from ink thickness requirements.

Screening. A slurry is forced through a fine-mesh screen by a squeegee. This technique yields excellent patterns on cylindrical or flat surfaces.

Solution Metallizing. A liquid solution of the desired metal is applied to the ceramic, subsequently dissociated, and then reduced to provide a metallic coating.

Spraying. Paint-spray techniques are adapted to metallizing. Masking is usually required on the part. High-volume production may be maintained.

Tapes. Metallizing compositions manufactured as a tape of controlled thickness are being offered commercially. The tape may be applied to the ceramic, adhered by solvent action, and fired.

Vapor Deposition. Reduction of chemical vapors, e.g., nickel carbonyl, produces a uniform metallic layer on the ceramic. Deposition can be conventional or electrostatic firing.

- *Air Furnaces.* Useful for glass-fritted pastes containing noble metal powders. Firing temperatures vary up to 1000°C.
- *Controlled Atmosphere Furnaces.* Hydrogen or cracked ammonia atmospheres, with dew point and temperature controls, are typically used to react sintered-powder metallized coatings with ceramic. These furnaces operate at 1300 to 1700°C.
- *Vacuum Furnaces.* Vacuum of 10^{-4} torr, or less, is required for metallizing and sealing by the active metals method. Vacuum evaporators operating in the 10^{-3} to 10^{-10} torr range are at present in use, experimentally, for studies of ultrapure metals deposited for two-dimensional circuitry requirements.
- *Others.* While not yet of full commercial significance, work in plasma-jet technology and electron beam machining have significant potential.
- *Electroplating.* Following firing, the active metal assembly hardware may be electroplated for ease of installation by user. Air-fired silver may be electroplated to improve solderability. Sintered-powder metallized ceramics must ordinarily be electroplated to allow wetting by solders or brazes. Many methods of electroplating are in use. Barrel plating is more economical than rack plating and is the preferred method. The thickness, purity, and type of electroplated metal has a significant effect on the effective strength of sintered-powder metallizing.
- *Etching.* A chemical process used in conjunction with a masking system to create fine lines.

METAL-MATRIX COMPOSITES

A metal-matrix composite (MMC) is a material in which a continuous metallic phase (the matrix) is combined with another phase (the reinforcement) that constitutes a few percent to around 50% of the total volume of the material. In the strictest sense, metal-matrix composite materials are not produced by conventional alloying. This feature differentiates most metal-matrix composites from many other multiphase metallic materials, such as pearlitic steels or hypereutectic aluminum–silicon alloys.

The particular benefits exhibited by metal-matrix composites, such as lower density, increased specific strength and stiffness, increased high-temperature performance limits, and improved wear-abrasion resistance, are dependent on the properties of the matrix alloy and of the reinforcing phase. The selection of the matrix is empirically based, using readily available alloys, and the major consideration is the nature of the reinforcing phase.

MATRICES AND REINFORCEMENTS

A large variety of metal-matrix composite materials exist. The reinforcing phase can be fibrous, platelike, or equiaxed (having equal dimensions in all directions), and its size can also vary widely, from about 0.1 to more than 100 μm. Matrices based on most engineering metals have been explored, including aluminum, magnesium, zinc, copper, titanium, nickel, cobalt, iron, and various aluminides. This wide variety of systems has led to an equally wide spectrum of properties for these materials and of processing methods used for their fabrication.

Reinforcements used in metal-matrix composites fall in five categories: continuous fibers, short fibers, whiskers, equiaxed particles, and interconnected networks.

Continuous Fibers

Several continuous fibers or filaments are used in metal-matrix composites. Their elastic moduli vary significantly, depending on the nature of the fiber and its fabrication process. For example, silica-alumina spinels and microcrystalline or amorphous polycarbosilane-derived fibers possess significantly lower elastic moduli than do pure alumina or crystalline β-silicon carbide produced by chemical vapor deposition. Carbon fiber strength and modulus also vary significantly with processing, depending on the level of graphitization of the microstructure.

Short Fibers

Short fibers are less expensive, especially when they are mass-produced for other applications such as high-temperature thermal insulation. Their physical properties can be similar to those of continuous fibers; however, their reinforcing efficiency in the matrix is also far lower. Short fibers used in engineering practice include chopped carbon fibers and alumina-silica fibers.

Whiskers

Whiskers are single-crystal short fibers, produced to feature highly desirable mechanical properties due to lack of microstructural defects. Whiskers have typically been made of silicon carbide, and they are often priced far higher than short fibers. The high price and toxicity of most whiskers have prevented their application in engineering practice.

Single-crystal whiskers, because of the absence of grain boundary defects, offer much higher tensile strength than other types of discontinuous reinforcements, and thus they are preferred for certain applications of discontinuously reinforced metal-matrix composites. The whiskers can be aligned to a preferred orientation by conventional metallurgical processes; higher directional strengths can be achieved in finished components where fabrication is by extrusion, rolling, forging, or superplastic forming. Whiskers tend to produce anisotropic properties due to their alignment during processing, whereas particulate materials usually produce essentially isotropic properties.

Equiaxed Particles

Equiaxed particles of several ceramics, including those containing silicon carbide, aluminum oxide, boron carbide, and tungsten carbide, do not provide the possibility for preferential strengthening of the matrix along selected directions; however, their price is low and their combination with the metal is relatively easier. These reinforcements are therefore used in many metal-matrix composite systems, including mass-produced aluminum-matrix composites.

Interconnected Cellular Networks

These can be produced by several methods, such as by chemical vapor deposition of ceramic onto a pyrolizable polymer foam or by conversion of a preceramic polymer foam prior to infiltration with the molten matrix. Alternatively, some processing techniques for in-place metal-matrix composites, including directional oxidation of aluminum melts, produce interconnected reinforcing networks.

MICROSTRUCTURES

The microstructure of a metal-matrix composite comprises the structure of matrix and reinforcement, that is, the interface and the distribution of the reinforcement within the matrix.

COMPOSITE PROPERTIES

Composite properties depend first and foremost on the nature of the composite; however, certain detailed microstructural features of the composite can exert a significant influence on its behavior.

Physical properties of the metal, which can be significantly altered by addition of a reinforcement, are chiefly dependent on the reinforcement distribution. A good example is aluminum–silicon carbide composites, for which the presence of the ceramic increases substantially the elastic modulus of the metal without greatly affecting its density. Elastic moduli for 6061 aluminum-matrix composites reinforced with discrete silicon carbide particles or whiskers have been calculated by using the rule of mixtures for the same matrix reinforced with two types of commercial continuous silicon carbide fibers. As a result, several general facts become apparent. First, modulus improvements are significant, even with equiaxed silicon carbide particles, which are far less expensive than fibers or whiskers. However, the level of improvement depends on the shape and alignment of the silicon carbide. Also, it depends on the processing of the reinforcement: for the same reinforcement shape (continuous fibers), microcrystalline polycarbosilane-derived silicon carbide fibers yield much lower improvements than do crystalline β-silicon carbide fibers. These features, which influence

METAL-MATRIX COMPOSITES

reinforcement shape, orientation, and processing of modules, are quite general; they are also observed, for example, in metal-matrix composites reinforced with aluminum oxide or carbon.

Other properties, such as the strength of metal-matrix composites, depend in a much more complex manner on composite microstructure. The strength of a fiber-reinforced composite, for example, is determined by fracture processes, themselves governed by a combination of microstructural phenomena and features. These include plastic deformation of the matrix, the presence of brittle phases in the matrix, the strength of the interface, the distribution of flaws in the reinforcement, and the distribution of the reinforcement within the composite. Consequently, predicting the strength of the composite from that of its constituent phases is generally difficult.

PRODUCTION

A variety of techniques are available for the production of continuous or discontinuous metal-matrix composites. These may be broadly classified as diffusion processes, deposition processes, and liquid processes. See Figure M.12 illustrating the methods used to make metal-matrix composites.

FABRICATION

Composite processing methods combine the reinforcement with the matrix. This is accomplished while the matrix is either solid or liquid.

Typical liquid-state processes include the dispersion processes, which are casting techniques. A second set of processes involves liquid-metal impregnation; these include squeeze

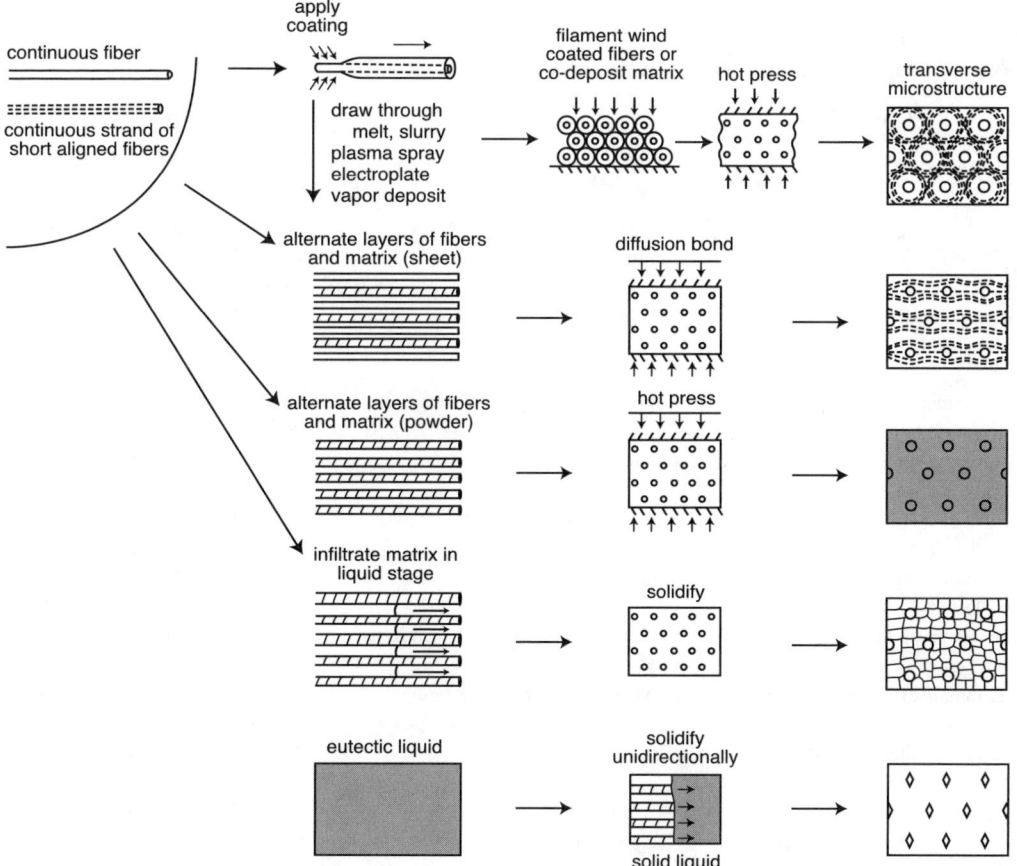

FIGURE M.12 Methods used to make MMCs. (From *McGraw-Hill Encyclopedia of Science and Technology*, 8th ed., Vol. 11, McGraw-Hill, New York, 57. With permission.)

casting, where a preform or a bed of dispersoids is impregnated by molten alloy under hydraulic pressure. A third set comprises spray processes. In one of these, a molten metal stream is fragmented by means of a high-speed cold inert-gas jet passing through a spray gun, and dispersoid powders are simultaneously injected. A stream of molten droplets and dispersoid powders is directed toward a collector substrate where droplets recombine and solidify to form a high-density deposit.

Depending on the process, the desired microstructure, and the desired part, metal-matrix composites can be produced to net or near-net shape; or alternatively they can be produced as billet or ingot material for secondary shaping and processing.

Applications

The combined attributes of metal-matrix composites, together with the costs of fabrication, vary widely with the nature of the material, the processing methods, and the quality of the product. In engineering, the type of composite used and its application vary significantly, as do the attributes that drive the choice of metal-matrix composites in design (Table M.11). For example, high specific modulus, low cost, and high weldability of extruded aluminum oxide particle-reinforced aluminum are the properties desirable for bicycle frames. High wear resistance, low weight, low cost, improved high-temperature properties, and the possibility for incorporation in a larger part of unreinforced aluminum are the considerations for design of diesel engine pistons.

TABLE M.11
Some Composite Components with Proven Potential

Composite	Components	Advantages
Aluminum–silicon carbide (particle)	Piston	Reduced weight, high strength and wear resistance
	Brake rotor, caliper, liner	High wear resistance, reduced weight
	Propeller shaft	Reduced weight, high specific stiffness
Aluminum–silicon carbide (whiskers)	Connecting rod	Reduced reciprocating mass, high specific strength and stiffness, low coefficient of thermal expansion
Magnesium–silicon carbide (particle)	Sprockets, pulleys, and covers	Reduced weight, high strength and stiffness
Aluminum–aluminum oxide (short fibers)	Piston ring	Wear resistance, high running temperature
	Piston crown (combustion bowl)	Reduced reciprocating mass, high creep and fatigue resistance
Aluminum–aluminum oxide (long fibers)	Connecting rod	Reduced reciprocating mass, improved strength and stiffness
Copper–graphite	Electrical contact strips, electronics packaging, bearings	Low friction and wear, low coefficient of thermal expansion
Aluminum–graphite	Cylinder, liner platon, bearings	Call resistance, reduced friction, wear and weight
Aluminum–titanium carbide (particle)	Piston, connecting rod	Reduced weight and wear
Aluminum–fiber flax	Piston	Reduced weight and wear
Aluminum–aluminum oxide fibers–carbon fibers	Engine block	Reduced weight, improved strength and wear resistance

Source: McGraw-Hill Encyclopedia of Science and Technology, 8th ed., Vol. 11, McGraw-Hill, New York, 53. With permission.

METAL, MECHANICAL PROPERTIES OF

Commonly measured properties of metals (such as tensile strength, hardness, fracture toughness, creep, and fatigue strength) associated with the way that metals behave when subjected to various states of stress. The properties are discussed independently of theories of elasticity and plasticity, which refer to the distribution of stress and strain throughout a body subjected to external forces.

STRESS STATES

Stress is defined as the internal resistance, per unit area, of a body subjected to external forces. The forces may be distributed over the surface of a body (surface forces) or may be distributed over the volume (body forces); examples of body forces are gravity, magnetic forces, and centrifugal forces.

Tension and Torsion

In simple tension, two of the three principal stresses are reduced to zero, so that there is only one principal stress, and the maximum shear stress is numerically half the maximum normal stress. Because of the symmetry in simple tension, every plane at 45° to the tensile axis is subjected to the maximum shear stress. For other kinds of loading, the relationship between the maximum shear stress and the principal stresses are obtained using the same method, with the results depending on the loading condition.

Stress–Strain Curve

If a metal is strained at a fixed rate, the resistance to deformation increases with straining at a diminishing rate. A plot of the resistance to deformation against the amount of strain is called a stress–strain curve, and the slope of this curve at any point is called the rate of strain hardening. Such a curve is normally obtained by making a tension test and plotting the maximum normal stress against the maximum normal strain. These are identical with the ordinary tensile stress (the tensile force divided by the area of the specimen) and the ordinary tensile strain (the extension per unit of length). The curve produced is known as an engineering stress–strain curve and is widely used to provide design strength information and as an acceptance test for materials.

Yield Strength

The elastic limit is rarely determined. Metals are seldom if ever ideally elastic, and the value obtained for the elastic limit depends on the sensitivity of strain measurement. The proportional limit, describing the limit of applicability of Hooke's law of linear dependence of stress on strain, is similarly difficult to determine. Modern practice is to determine the stress required to produce a prescribed inelastic strain, which is called the yield strength. The amount of strain used to define the yield strength varies with the application, but is most commonly taken as 0.2% (a unit strain of 0.002 in./in; 1 in. = 25 mm). Because upon unloading the behavior is almost linearly elastic, it is possible to use the offset method of determining the yield strength from a plotted stress–strain curve; a line with a slope equal to the elastic slope is drawn, displaced from the stress–strain curve at low stress levels by the amount of strain used in the definition of the yield strength, and the stress at the intersection of this line with the stress–strain curve is taken as the yield strength.

Tensile Strength

Tensile strength, usually called the ultimate tensile strength, is calculated by dividing the maximum load by the original cross-sectional area of the specimen. It is, therefore, not the maximum value of the true tensile stress, which increases continuously to fracture and which is always higher than the nominal tensile stress because the area continuously diminishes. For ductile materials, the maximum load, on which the tensile strength is based, is the load at which necking-down begins. Beyond this point, the true tensile stress continues to increase, but the force on the specimen diminishes.

Elongation

The tensile test provides a measure of ductility, by which is meant the capacity to deform by extension. The elongation to the point of necking-down is called the uniform strain or elongation because, until that point on the stress–strain curve, the elongation is uniformly distributed along the gauge length. The strain to fracture or total elongation includes the extension accompanying local necking.

Hardness

In all hardness tests, a standardized load is applied to a standardized indenter, and the dimensions of the indent are measured. To compare the resistance to deformation of a particular sample or lot with a standard material, indentation hardness tests are used. The most popular hardness test is the Rockwell test, which is carried out quite rapidly in a convenient machine by sacrificing any attempt to express the hardness as a resistance to deformation in units of force per area.

Fatigue

Fatigue is a process involving cumulative damage to a material from repeated stress (or strain) applications (cycles), none of which exceeds the ultimate tensile strength. The number of cycles required to produce failure decreases as the stress or strain level per cycle is increased. When cyclic stresses are applied, the results are expressed in the form of a signal–noise curve in which stress of logarithm of stress is plotted against the number of cycles to cause failure. The fatigue strength or fatigue limit is defined as the stress amplitude that will cause failure in a specified number of cycles, usually 10^7 cycles.

Fracture Toughness

Many modifications of conventional test procedures have arisen from an effort to obtain laboratory information that would be useful to designers of engineering structures. Since such structures often contain macroscopic defects, in particular weld flaws, large inclusion particles, or other manufacturing flaws, most effort has been directed toward studying the properties of notched bars. Unfortunately, neither the notched tensile test nor the several impact tests provide sufficiently quantitative relations between defect size, stress state, and the likelihood of brittle or otherwise unstable fracture, particularly in cases where crack propagation can occur below the macroscopic yield stress as measured in a tensile test. In the case of materials that exhibit little or no plastic deformation at fracture, the most successful approach has been that of A. A. Griffith, who deduced both thermodynamic and stress considerations.

Creep and Stress Rupture

Time-dependent deformation under constant load or stress is measured in a creep test. Creep tests are those in which the deformation is recorded with time, while stress rupture tests involve the measurement of time for fracture to occur. The two types of tests provide complementary information, and indeed both creep rate and time may be recorded in a single test.

METAL AND MINERAL PROCESSING

This is defined as the obtaining of useful metals from ores. Table M.12 summarizes one or more routes (for occurrence, extraction, and refining of metals) but not necessarily the only commercial route, used to obtain each of the useful metals by extractive metallurgy.

METHANE

Also known as marsh gas, in coal mines as firedamp, and chemically as methyl hydride, methane is a colorless, odorless gas, CH_4, employed for carbonizing steel, in the manufacture of formaldehyde, and as a starting point for many chemical compounds. The molecule has no free electrons and is the only stable carbon hydride, although it reacts easily on the No. 1 and No. 2 electrons of the carbon to form the hexagonal molecule called the benzene ring. It may thus be considered as the simplest of the vast group of hydrocarbons derived from petroleum, coal, and natural gas. Methane occurs naturally from the decomposition plant and animal life, and is also one of the chief constituents of illuminating gas. It is made synthetically by the direct union of carbon or carbon monoxide with hydrogen. It is also produced by the action of water on aluminum carbide.

Methane is also the chief constituent of natural gas from oil fields. Natural gas contains usually at least 75% methane. A new source of methane is a result of environmental cleanups. Methane is generated as a by-product during the biodegradation of polluted industrial waters in waste-treatment plants. It is also obtained by tapping the gases produced as biomass matter degrades in municipal landfills.

MICA

Mica has been known as an electrical material of excellent insulating and fabricating properties for many years. Almost everyone has seen or used mica in one form or another, yet it is surprising how little even the average technically trained person knows about its composition, origin, and properties.

It is, of course, a naturally occurring mineral having a wide range of possible chemical compositions and properties. All true mica, however, belongs to one mineral class of silicates having a sheet type of structure, and found in certain areas of the world growing from pegmatite deposits in "book" form. In recent years the increasing demands for high-quality sheet mica have led to the development and commercial production of synthetic mica as an electric furnace product.

Synthetic mica has recently been put into commercial production and is available in various powdered forms in organically bonded paper sheets; in a three-dimensional, hot-pressed, machinable synthetic mica ceramic; and in sheet form of limited area. Synthetic mica has unusually high thermal stability.

Of considerable interest to the refractories field is vermiculite, a form of mica that, on rapid heating, expands 16 times accordion-fashion to form a light and inert material suitable for use as a grog in refractories or as an aggregate for the production of lightweight casting refractory concretes.

Types

Mica comprises a group of hydrous aluminum silicate minerals with platy morphology and perfect basal (micaceous) cleavage. This group of silicate minerals has monoclinic crystals that break off easily into thin, tough scales, varying from colorless to black. Muscovite is the common variety of mica, and is called potash mica, or potash silicate. It has superior dielectric properties, and is valued for radio capacitors.

Properties

The most singular outstanding property of mica is its physical structure. As a sheet mineral it can be split into strong, flexible films with good high-temperature resistance and electrical insulating properties.

Uses

Commercial mica is of two main types: sheet, and scrap or flake. Sheet muscovite is used as a dielectric in capacitors and vacuum tubes in electronic equipment. Lower-quality muscovite is used as an insulator in home electrical products such as hot plates, toasters, and irons. Scrap and flake mica is ground for use in coatings on roofing materials and waterproof fabrics, and in paint, wallpaper, joint cement, plastics, cosmetics, well-drilling products, and a variety of agricultural products.

For many years, glass-bonded mica has been used in every type of electrical and electronic system where the insulation requirements are preferably low-dissipation factor at high frequencies, a high-insulation resistance and dielectric-breakdown strength, along with extreme dimensional stability. Glass-bonded micas are made in both machinable grades and precision-moldable grades. Basically, the material consists of natural mica flake bonded with a low-loss electrical glass.

The availability of synthetic mica resulted in the development of so-called ceramoplastics, consisting of high-temperature electrical glass filled with synthetic mica. Ceramoplastics provide an increase in the electrical characteristics over those of natural mica, and, in addition, are more easily molded and have greater thermal stability.

Glass-bonded mica and ceramoplastics have found use in many advanced components such as telemetering commutation plates, molded printed circuitry, high-reliability relay spacers and bobbins, coil forms, transducer

TABLE M.12
Occurrence, Extraction, and Refining of Metals

Metals	Occurrence	Beneficiation Operations	Extraction Processes			Refining (reduction) Operations
			Operation	Intermediate Product	Product Used in Refining	
Aluminum, Al	Bauxite, Al_2O_3, 55–65% Al	Dehydration roast	Caustic (Bayer process)	Sodium aluminate	Pure alumina	Fused salt electrolysis
	Alunite, $KAl_3(OH)_6(SO_4)_2$	Dehydration roast	Reduction roast, caustic digestion	Alumina hydrate	Pure alumina	
Antimony, Sb	Stibnite, Sb_2S_3, 50–60% Sb	Liquidation		Pure Sb_2S_3		Charcoal reduction at 1200–1300°F (650–700°C)
Arsenic, As	Enargite, $3Cu_2S \cdot As_2S_5$		Roasting	As_2O_3	Iron precipitation	
	Arsenopyrite, FeAsS					
Barium, Ba	Barite, $BaSO_4$	Flotation		Pure $BaSO_4$	$BaCl_2$	Fused salt electrolysis
Beryllium, Be	Beryl, $3BeO \cdot Al_2O_3 \cdot 6SiO_2$, 10–12% Be	Hand cobbing	Acid sinter plus leaching	BeO	BeF_2	Magnesium at 1652–2372°F (900–1300°C)
	Bertrandite, $Be_4Si_2O_7(OH)_2$					
Bismuth, Bi	Lead ores	Flotation, gravity	Smelting	Lead bullion	Anode residue from electrolyzed bullion	Chlorine at 932°F (500°C) to remove Pb and Zn
						Reduction with charcoal and flux
Boron, B	Borax, $Na_2B_4O_7 \cdot 10H_2O$	Evaporation	Crystallization from purified brine, calcining	B_2O_3	Potassium fluoroborate	Fused salt electrolysis
	Kernite, $Na_2B_4O_7 \cdot 4H_2O$	Brine purification				Reduction with hydrogen
	Colemanite, $Ca_2B_6O_{11} \cdot 5H_2O$					Reduction with sodium or magnesium
Cadmium, Cd	Sphalerite, ZnS, 1% Cd	Roasting, sintering	Leaching	Cd–Zn solution	Cd sponge by Zn dust addition	Distillation

Element	Source	Refining	Leaching	Intermediate	Final product
Calcium, Ca	Limestone, CaCO₃		Calcination	CaO	Electrolysis. Aluminum reduction in vacuum
Cerium, Ce	Monazite	Gravity, flotation	Caustic or acid leaching	$Ce_2(SO_4)_3$	Tetravalent Ce salts. Electrolysis of fused chloride at 1292°F (700°), or of fused fluoride baths
	Bastnaesite			Cerium oxides Cerium fluorides	Calcium, aluminum, or magnesium reduction
Cesium, Cs	Lepidolite, 1% Cs	Hand cobbing	Acid leaching	Cesium alum	CsCl
	Pollucite, $Cs_4Al_4Si_9O_{26} \cdot H_2O$	Selective mining of pegmatites			Cs hydroxide. Electrolysis
Chromium, Cr	Chromite, $FeO \cdot Cr_2O_3$, 35–50% Cr	Gravity and magnetic separation	Electric furnace treatment	Ferrochrome	Chromium alum solution. Chromium oxide. Aluminum reduction
Cobalt, Co	Cobalt or nickel ores, 0.8–10.0% Co	Flotation, leaching	Acid roast or smelting	$CoSO_4$ solution	$Co(OH)_2$ by milk of lime. Electrolysis of aqueous solution
	Lateritic ores				
Columbium, Cb (niobium, Nb)	Columbite, $Fe(Cb \cdot TaO_3)_2$	Gravity	Caustic fusion, then HCl, then KF	$CbKOF_2$	Columbium oxides. Columbium carbide in vacuum
		High-intensity magnetic or electrostatic separation	Chlorination		Columbium metal
			Solvent extraction		
			Hydrometallurgy		
Copper, Cu	Copper ores, 0.4–6.0% Cu	Flotation, fluid-bed roasting, leaching, pressure leaching, solvent extraction	Smelting, electrolysis, liquid ion exchange, cementation	Cu anodes	Cu cathodes. Electrolysis
				Ferric chloride leach solution	Stripped solvent. Other electrowinning methods to produce powdered metal
Cysprosium, Dy	Monazite	Gravity	Sulfuric acid treatment	$Dy_2(SO_4)_3$	Pure salts by fractional crystallization and ion exchange. Electrolysis of fused chloride

TABLE M.12 (CONTINUED)
Occurrence, Extraction, and Refining of Metals

Metals	Occurrence	Beneficiation Operations	Extraction Processes			Refining (reduction) Operations
			Operation	Intermediate Product	Product Used in Refining	
Erbium, Er	Monazite	Gravity	Sulfuric acid treatment	$Er_2(SO_4)_3$	Pure salts by fractional crystallization and ion exchange	Electrolysis of fused chloride
Europium, Eu	Monazite	Gravity	Sulfuric acid treatment	$Eu_2(SO_4)_3$	Pure salts by fractional crystallization and ion exchange	Electrolysis of fused chloride
Gadolinium, Gd	Monazite	Gravity	Sulfuric acid treatment	$Gd_2(SO_4)_3$	Pure salts by fractional crystallization and ion exchange	Electrolysis of fused chloride
Gallium, Ga	Al and Zn ores	Caustic	CO_2 treatment in acid Roasting	Ga_2O_3 and $GaCl_3$ Germanium oxide	Germanium chloride, then hydrolysis to pure oxide	Electrolysis
Germanium, Ge	Sphalerite, ZnS, 0.01–0.015% Ge	Flotation				Hydrogen reduction
Gold, Au	Elemental form, 0.001% Au	Gravity, flotation, leaching	Cyanide or smelting	AuCN or Au	Pure gold	Zinc reduction
Hafnium, Hf	Zircon, $ZrO_2 \cdot SiO_2$	Gravity	Acid or alkali leaching	Hf–Zr solutions	Hf solutions by ion exchange to $HfCl_4$; solvent extraction	Magnesium reduction
Holmium, Ho	Monazite	Gravity	Sulfuric acid treatment	$Ho_2(SO_4)_3$	Pure salts by fractional crystallization and ion exchange	Electrolysis of fused chloride
Indium, In	Zinc ores	Flotation	Acid roasting	In sulfate	Water solutions	Displacement by zinc metal
Iridium, Ir	Cu–Ni sulfide ores	Flotation, roasting, extraction	Electrolytic refining	Anode slimes	$(NH_4)_2IrCl_6$	Hydrogen reduction
Iron, Fe	Iron ores, 30–60% Fe	Gravity, flotation, magnetic and electrostatic separation, pelletizing	Blast furnace	Pig iron		Coke reduction in blast furnace Electric furnace reduction

MICA 433

Element	Source	Beneficiation	Prereduction or metallizing furnaces	Prereduced or metallized pellets		
Lanthanum, La	Monazite	Gravity	Sulfuric acid treatment	$La_2(SO_4)_3$	Pure salts by fractional crystallization and ion exchange	Electrolysis of fused chloride
Lead, Pb	Galena, PbS, 6–10% Pb	Gravity or flotation	Blast furnace Imperial furnace	Lead bullion	Lead free of Cu, Sn, As, and Sb by treatment with sulfur	Desilvering with Zn, Sb removal Basic oxygen furnace Direct reduction with Co, H_2, and CH_4
Lithium, Li	Spodumene, $Li_2O \cdot Al_2O_3 \cdot SiO_2$ Brines, 0.03–0.24% LiCl	Flotation Evaporation, leaching, flotation	Acid roasting	Li_2SO_4	Li_2, Co_3, then LiCl LiCl	Electrolysis of fused chloride
Lutetium, Lu	Monazite	Gravity	Sulfuric acid treatment	$Lu_2(SO_4)_3$	Pure salts by fractional crystallization and ion exchange	Electrolysis of fused chloride
Magnesium, Mg	Seawater, 0.13% Mg Brines Dolomite, $CaMg(CO_3)_2$		Lime slurry Evaporation	$Mg(OH)_2$	$MgCl$ with HCl $MgCl$	Electrolysis of fused chloride Thermal reduction of dolomite with ferrosilicon
Manganese, Mn	Manganese ores, 45–55% Mn	Flotation, gravity	Roasting in reducing atmosphere	Mn in solution	Pure Mn solutions by filtration	Electrolysis of aqueous solution
Mercury, Hg	Cinnabar, HgS, 1–3% Hg	Sorting, screening	Roasting	Hg		Retorting at 1290°F (700°C)
Molybdenum, Mo	Molybdenite, MoS_2, 1–3% Mo	Flotation	Roasting	MoO_3	Mo by hydrogen reduction	Power metallurgy or ore melting
Neodymium, Nd	Monazite	Gravity	Sulfuric acid treatment	$Nd_2(SO_4)_3$	Pure salts by fractional crystallization and ion exchange	Electrolysis of fused chloride
Nickel, Ni	Nickel ores, 1–35% Ni	Smelting, flotation of Cu–Ni–Fe product	Roasting, leaching	(NiCuFe)S or Ni	Pure NiS by molten Na_2S exchange	Carbon reduction
	Lateritic ores, lower nickel content	Leaching	Pressure leach with NH_3			Electrolytic refining

TABLE M.12 (CONTINUED)
Occurrence, Extraction, and Refining of Metals

Metals	Occurrence	Beneficiation Operations	Extraction Processes			Refining (reduction) Operations
			Operation	Intermediate Product	Product Used in Refining	
Osmium, Os	Cu–Ni sulfide ores	Flotation, roasting, extraction	Electrolytic refining	Anode slimes	$OsO_2 \cdot (NH_3)_4Cl_2$	Hydrogen reduction
Palladium, Pd	Cu–Ni sulfide ores	Flotation, roasting, extraction	Electrolytic refining	Anode slimes	$Pd(NH_3)_2Cl_2$	Hydrogen reduction
Platinum, Pt	Cu–Ni sulfide ores, 0.001% Pt	Flotation, roasting, extraction	Electrolytic refining	Anode slimes	$(NH_4)_2PtCl_2$	Hydrogen reduction
Potassium, K	Sylvite, KCl	Flotation, gravity				Sodium reduction
Praseodymium, Pr	Monazite	Gravity	Sulfuric acid treatment	$Pr_2(SO_4)_3$	Pure salts by fractional crystallization and ion exchange	Electrolysis of fused chloride
Radium, Ra	Pitchblende	Gravity, flotation	Acid digestion	Ra sulfate	Radium bromide	Repeated fractional crystallization
Rhenium, Re	Molybdenite, MoS_2	Flotation	Roasting	Re_2O_7	$KReO_4$	Hydrogen reduction
Rhodium, Rh	Cu–Ni sulfide ores	Flotation, roasting, extraction	Electrolytic refining	Anode slimes	$(NH_4)_3Rh(NO_2)_6$	Hydrogen reduction
Rubidium, Rb	Lepidolite	Hand cobbing	Chlorostannate process followed by pyrolysis, electrolysis, or reduction	Rubidium perchlorate	Chloride	High-temperature vacuum reduction with Na or Mg
	Pollucite				Carbonate or hydroxide	
Ruthenium, Ru	Cu–Ni sulfide ores	Flotation, roasting, extraction	Electrolytic refining	Anode slimes	RuO_2	Hydrogen reduction
Samarium, Sm	Monazite	Gravity	Sulfuric acid treatment	$Sm_2(SO_4)_3$	Pure salts by fractional crystallization and ion exchange	Electrolysis of fused chloride
						Precipitation of Ni as powder from leach solutions

MICA

Element	Source	Concentration	Processing	Intermediate	Final step
Scandium, Sc	Thortveitite, $(ScY)_2Si_2O_7$	Leaching with H_2SO_4, then solvent extraction	Roasting with C	Scandium carbide	Scandium oxide, then ScF_3	Electrolysis of fused salt
	Uranium ores		Stripping solvent with Hf	Sc_2O_3	ScF_3	Direct reduction of fluoride and distillation of Sc metal in vacuum induction furnace
Selenium, Se	Chalcopyrite, $CuFeS_2$	Flotation, roasting	Electrolytic refining	Anode muds	Sodium selenate	SO_2 precipitation from aqueous solution
Silicon, Si	Silica, SiO_2	Flotation, gravity	Smelting			Carbon reduction in electric furnace
Silver, Ag	Nonferrous ores, 0.07% Ag	Gravity or flotation	Blast furnace	Lead bullion	Ag–Zn mixture by Zn treatment	Zn distillation, electrolysis of Ag
Sodium, Na	Halite, NaCl	Flotation, leaching				Electrolysis of fused chloride
Strontium, Sr	Celestite, $SrSO_4$	Hand cobbing, flotation	Digestion with soda ash	$SrCO_3$		Electrolysis of fused chloride
			Calcining with coal; sulfide then leached with lime	$SrCO_3$	$SrCl_2$	
Tantalum, Ta	Tantalite, $Fe(Cb_2TaO_3)_2$	Gravity, electrostatic, magnetic separation	Digestion with HF or caustic, then KF	K_2TaF_7	Ta by sodium	Vacuum sintering at 1290°F (700°C)
	Tin slags					
Tellurium, Te	Chalcopyrite, $CuFeS_2$	Flotation, roasting	Electrolytic refining	Anode muds	Sodium telluride-tellurate	SO_2 from aqueous solution
Terbium, Tb	Monazite	Gravity	Sulfuric acid treatment	$Tb_2(SO_4)_2$	Pure salts by fractional crystallization and ion exchange	Electrolysis of fused chloride
Thallium, Tl	Spalerite, ZnS	Roasting, sintering	Leaching of flue dust	Thallium chloride	TlS by H_2S treatment, then converted to sulfate	Electrolysis of aqueous solution
Thorium, Th	Monazite	Gravity	Sulfuric acid treatment	$Th(SO_4)_3$	Pure salts by fractional crystallization and ion exchange	Reduction of ThF_4 by calcium
Thullium, Tm	Monazite	Gravity	Sulfuric acid treatment	$Tm_2(SO_4)_3$	Pure salts by fractional crystallization and ion exchange	Electrolysis of fused chloride
Tin, Sn	Cassiterite, SnO_2, 1.5–5.0% Sn	Gravity, flotation	Reverberatory furnace	Impure tin		Electrolytic refining

TABLE M.12 (CONTINUED)
Occurrence, Extraction, and Refining of Metals

Metals	Occurrence	Beneficiation Operations	Extraction Processes			Refining (reduction) Operations
			Operation	Intermediate Product	Product Used in Refining	
Titanium, Ti	Rutile, TiO_2, ilmenite, 1–8% Ti	Gravity, electrostatic, magnetic separation	Chlorination	$TiCl_4$		Magnesium or sodium reduction; Electrolysis of $TiCl_4$ in fused salt bath
Tungsten, W	Tungsten ores, 60–70% W	Gravity, flotation	Caustic fusion	Sodium tungstate	WO_3	Hydrogen reduction
Uranium, U	Uranium ores	Leaching with acid or caustic	Ion exchange, solvent extraction		UF_4	Ca, Al, Mg, or Na reduction
Vanadium, V	Carnotite, $K_2O \cdot 2UO_3 \cdot V_2O$	Salt roasting	Leaching, solvent extraction, ion exchange	Sodium vanadate	V_2O_5	Thermic reduction followed by electron-beam purification
Ytterbium, Yb	Monazite	Gravity	Sulfuric acid treatment	$Yb_2(SO_4)_3$	Pure salts by fractional crystallization and ion exchange	Electrolysis of fused chloride
Yttrium, Y	Monazite	Gravity	Sulfuric acid treatment	$Y_2(SO_4)_3$	Pure salts by fractional crystallization and ion exchange	Electrolysis of fused chloride
Zinc, Zn	Sphalerite, ZnS, 10–30% Zn	Flotation, gravity, pelletizing, roasting	Sulfuric acid leach; Fluxing for pyrolytic processes	Purified solution		Electrolysis to produce high-purity cathodes
Zirconium, Zr	Zircon, $ZrO_2 \cdot SiO_2$	Electrostatic, magnetic, or gravity concentration	Chlorination, or plasma fusion, then caustic leach		$ZrCl_4$	Magnesium reduction

Source: *McGraw-Hill Encyclopedia of Science and Technology*, 8th ed., Vol. 11, McGraw-Hill, New York, 27. With permission.

housing-miniature-switch cases, and innumerable other component applications.

MICROELECTROMECHANICAL SYSTEMS

A new discipline in miniaturization has emerged and is referred to as microelectromechanical systems (MEMS).

Three regions of "smallness" can now be identified. The "micro" region spans the dimensional range from micrometers to millimeters and is the most traditional. Here MEMS/MST systems are dominant and contain such elements as micropressure sensors, accelerometers, microactuators, biochips, microswitches, and traditional MEMS and optical MEMS (MOEMS) components.

Another region that is beginning to attract considerable interest is the "nano" regime. It spans the nominal range from 100 nm down to 0.5 nm, a region in which the properties of many materials transition from those of macroscopic systems to become dominated by the discrete electronic states of molecular and atomic domains.

The nano regime has rich potential for offering useful new materials characteristics including enhanced strength and hardness, reduced resistivity, superplasticity, self-assembly, enhanced catalytic intensity, and increased magnetic and dielectric performance, to name a few.

A very short list of potential applications for nanostructure devices includes field-emitter tips, carbon-fullerene nanotubular structures with electronic and superconducting potential, nanophase components for high-strength composite materials, more effective giant magnetoresistive structures for high capacity magnetic data storage applications, a variety of proximal probe methods for high density data reading, plus many more in the fields of chemistry, biology, and medicine.

Still a third regime of smallness, the mesoscale region, is beginning to achieve recognition. With dimensions in the submillimeter to centimeter range, mesoscale structures encompass complex composite systems of microcomponents. Examples, including some now being developed, include a mesoscale heat exchanger and a mesoscale vacuum pump. Both of these consist of arrays of individual microcomponents acting in concert. In addition, many microinstrumentation packages such as those now under development for medical applications and space exploration may qualify for recategorization as mesosystems.

MICROENGINEERING

Microengineering is the design and production of small, three-dimensional objects, usually for manufacture in high volumes at low cost. Such designs have a wide range of applications, including automobiles, medicine, aircraft, printing, security, and insurance. The techniques employed come from wide a range of disciplines, including biochemical, chemical, electrical, electronic, fluidic, mechanical, and optical engineering. Present designs occupy a volume about 1 mm^3 (40 mils^3), and are made to tolerances of a few micrometers (about 50 μin.).

The applications of the above-described sensors cover air-bag sensors, which detects crashes and sets off an air-bag.

Microengineering is also used to provide automatic braking systems and intelligent suspension for smoother rides, and to optimize the flame profile in an engine to reduce fuel consumption.

The process uses synchrotron radiation to produce sharp cuts in thick-film photoresists, which may then be plated with plastics, metals, or other materials. The use of nickel produces strong metal parts in substantial quantities. In one design these are assembled to produce a very small motor, which is suitable for use in microsurgery, watches, and camcorders.

Fiber clamps, gyroscopes, fuses, water-quality monitoring, patient monitoring, eye surgery, microsurgery, printing, aviation, and food production are just some of the applications and processing methods being utilized and considered under microengineering. In the United States, MEMS covers the electronic and mechanical parts of microengineering. In Japan, the term *micromachines* is used to cover a small specific subset that includes sensors and robots. In Europe, the terms *microsystems* and *microsystems technology* cover systems and the technologies used to make them, but usually only the use of semiconductor-like processing; they also exclude the many small devices that operate independently.

The term *nanoengineering* is usually used to describe even smaller devices, nominally 1000 times smaller, which are usually much more expensive and are not normally mass-produced. The term *mechatronics* covers the use of mechanical and electronic engineering irrespective of size.

MICROGRINDING

This technology offers benefits for optical materials manufacturers where lapping and finishing alternatives can provide reduced cycle time and increased quality in the manufacture of optical components.

In deterministic microgrinding, the infeed rate, rather than the nominal pressure, is kept constant. This manufacturing operation uses rigid, computer-controlled machining centers and high-speed tool spindles. The system eliminates the specialized tooling, special skills, and long cycle times required with conventional equipment — cycle time reductions of 50% have been routinely achieved.

The equipment is capable of surfacing, edging, centering, and beveling in one setup sequence. The grinding geometry uses a bound abrasive cup wheel that is rotated at high speed and tilted at a very precise angle. Specular surfaces, resulting after less than 5 min of deterministic microgrinding, have a typical root mean square (rms) microroughness of less than 20 nm.

Because the equipment uses sequential cuts with coarse, medium, and fine grinding wheels, it takes only about 10 min to grind each surface. The fine wheel is run at 15,000 rpm and is fed into the lens at 6 μm/min. In less than 3 min, a total of 15 μm of material is removed, which is equal to the depth of the subsurface cracks produced by the medium tool. When used at slow infeed rates and for appropriate removal tasks, the 2- to 4-μm tools can cut hundreds of parts without needing to be redressed.

Finishing can be defined as the production of the surface to within 0.25 μm (peak to valley) of the specified shape and under 20-Å rms finish. Conventional finishing processes are time-consuming and labor-intensive, and can require much rework. A new finishing technology (MR) has been developed that addresses these problems.

Called magnetorheological (MR) finishing, the method is based on an MR fluid, a suspension of noncolloidal magnetic particles (usually carbonyl iron in water with small concentrations of stabilizers) and finishing abrasives (cerium oxide). This viscous liquid becomes a solid when a DC magnetic field is applied to it; stiffness is directly proportional to field strength. The MR fluid can be thought of as a compliant replacement for the conventional rigid lap in the loose abrasive grinding and finishing process. A magnetic field applied to the fluid creates a temporary finishing surface, which can be controlled in real time by varying the strength and direction of the field.

Surface smoothing, removal of subsurface damage, and figure correction are accomplished by rotating the part on a spindle at a constant speed while sweeping the lens about its radius of curvature through the stiffened finishing zone.

With the standard MR fluid, MRF reduces the surface microroughness of fused silica and other optical materials to less than 10 Å rms. The time required to do this varies from 5 to 60 min, depending on the material and its initial roughness. If the initial rms surface microroughness is less than 30 nm, smoothing occurs in 5 to 10 min.

MICROINFILTRATED MACROLAMINATED COMPOSITES

Ceramics offer attractive properties, including good high-temperature strength and resistance to wear and oxidation. However, the major limitation to their use in structural applications is their inherent low fracture toughness, that is, the tendency to break (fracture) and produce low values. Ceramics have low fracture toughness, whereas steels and superalloys have better fracture toughness. Methods currently being used to improve the toughness of ceramics involve

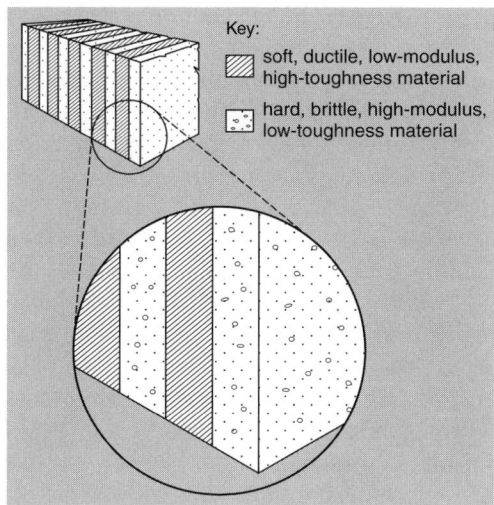

FIGURE M.13 Conceptual architecture of a microinfiltrated macrolaminated composite.

incorporation of reinforcing whiskers and fibers; inclusion of a phase that undergoes transformation within the stress field associated with a crack; and cermet technology (ceramic/metal composites), in which the tough metallic component absorbs energy. Improved toughness is attributed to various mechanisms, including crack branching, transformation-induced residual stresses, crack bridging, and energy absorption by plastic flow.

Another possible approach to improving the toughness of ceramics is via laminated construction. Sophisticated coating techniques have been used to produce laminated microstructures of ceramics and metals, but the cost of producing a bulk-laminated composite of useful size is prohibitive. Fabrication of microinfiltrated macrolaminated composites offers an economically feasible approach to produce a variety of cermet/metallic and ceramic/metallic bulk-laminated composites (Figure M.13). The basic architecture of a microinfiltrated macrolaminated composite is a double-layer structure. One layer consists of a soft, ductile material having a low modulus of elasticity, low strength, and high toughness; the second layer consists of a hard, brittle material having a high modulus of elasticity and low toughness. This double layer is repeated as many times as necessary to form the bulk composite; in addition, the brittle material is infiltrated with the ductile constituent.

ADVANTAGES

The architecture of the microinfiltrated macrolaminated composite offers a compromise to the conventional composite microstructure by providing repeated alternating layers of bulk tough metallic and brittle cermet and ceramic materials. Any crack introduced into the brittle constituent will, upon entering the metallic interlayer, be subjected to a potential crack, stopping higher toughness.

COMPOSITION AND FABRICATION

Fabrication of microinfiltrated macrolaminated composite materials offers a processing route to economical manufacture of large, bulk-laminated composites in which an interpenetrating microstructure is achievable. Use of both ceramic and metal powders in the process opens the door to a potentially wide range of material combinations. The same fundamental processing steps can be used to produce a wide variety of microinfiltrated macrolaminated composites, including ceramic/metal, metal/intermetallic, and ceramic/intermetallic systems.

Various combinations of composite properties, including hardness, strength, ductility, and fracture toughness, are possible by varying the laminate layer thicknesses. Microinfiltrated macrolaminated composites, when used in bulk form, are expected to have properties far superior to those of the individual monolithic constituents of the composite.

Alternative processing routes for microinfiltrated macrolaminated composites are available if the composite constituents have some solubility for one another, such as in the tungsten carbide–cobalt/cobalt (WC–Co/Co) and tungsten–nickel–iron heavy alloy/nickel (W–Ni–Fe/Ni) systems. In general, to obtain a large composite of this type, the best approach is to use the tape-casting process (well known in the ceramic industry) to produce large thin tapes (sheets) of the material.

Consolidation (compaction) helps the material achieve its full density by removing gas voids, porosity, and so forth. An attractive alternative for producing sheets consists of rolling the powder material followed by sintering to the required percent of theoretical density, retaining a certain level of interconnected porosity.

The fabrication steps used to make a W–Ni–Fe/Ni microinfiltrated macrolaminated composite illustrate the process. Tungsten powder is tape cast, and the sheets are sintered to 60 to 70% of theoretical density, which produces totally interconnected porosity. Small amounts of nickel can be used together with tungsten to promote activated sintering and to produce a porous tungsten sheet that can be handled safely. Sheet thickness ranges from 1 to 10 mm.

An 80:20 ratio of nickel and iron powders also is tape-cast and sintered to full density. Sheets of porous tungsten and fully dense nickel–iron alloy material are laid up in alternate layers and heated to about 1475°C, which is about 40°C above the melting point of the nickel–iron alloy. The molten Ni-Fe alloy infiltrates the porous tungsten sheets and takes into solution a fraction of the tungsten. A thin layer of the liquid phase can be retained between the tungsten sheets after the infiltration process is complete. This retention is possible if the tungsten sheet contains porosity levels that are lower than the volume of liquid formed by melting the nickel–iron sheets.

APPLICATIONS

The design of microinfiltrated macrolaminated composites could prove advantageous in applications involving impact or ballistic penetration. The metallic interlayer would function to hold together damaged portions of the brittle constituent. Thus, the development of ceramic armor having multihit capability becomes possible. Current metal/ceramic composite armor is layered on an extremely macroscopic scale. An approach involving microinfiltrated macrolaminated composites permits optimization of layer frequency and thickness to resist specific threats, as in antipersonnel weapons or armored vehicles.

In addition, high-temperature composites, consisting of a ceramic and intermetallic compound, such as aluminum oxide/nickel aluminide (Ni_3Al) plus boron, could be fabricated in the form of microinfiltrated macrolaminated composites to yield combinations with new properties. Ultrahigh-temperature composites incorporating high-temperature ceramics and ductile niobium is another potential application to obtain various combinations of high wear resistance and toughness. The hard ceramic outer layer would provide a wear-resistant surface, with toughness provided by the ductile material. A similar concept could lead to a new generation of cutting tools, in which the outermost cutting layer could be made of ultrahard materials suitably layered with a soft, ductile constituent for toughness.

Other potential applications are in heat- and oxidation-resistant, low-density structural components and aerospace parts having low density, high strength and modulus, and good fracture toughness.

Finally, the reinforcement of concrete with carbon fibers is a recently developed technique for the construction of structures such as buildings, bridges, and tunnels. Advantages of carbon-fiber reinforcement include outstanding electromagnetic shielding, high resistance to corrosive environments, light weight, and high mechanical property strength. Two methods are used to achieve the reinforcement: mixing carbon fibers, either chopped or mat, directly into cement (carbon fiber-reinforced concrete); and using carbon fiber/plastic composites in rod or tape form for strengthening the concrete.

MICROMACHINING

Magnetron plasma etching has been found to be a promising technique for micromachining of single-crystal silicon carbide to fabricate microscopic structures comprising integrated mechanical, electronic, and optical devices. Silicon carbide offers several advantages over silicon for the development of future devices. In comparison with silicon, silicon carbide is harder and stiffer, is less chemically reactive, and has greater thermal conductivity; moreover, silicon carbide-based electronic devices can operate at temperatures higher than silicon-based electronic devices can withstand. Etching techniques for micromachining of silicon are well known, but the lesser chemical reactivity of silicon carbide makes it necessary to devise alternative etching techniques.

In the present magnetron-plasma-etching technique as in the ECR etching technique, a magnetic field is used to increase the density of the plasma. Unlike in ECR etching, the main body of the plasma is confined in proximity to the sample. Such confinement results in a high concentration of reactive chemical species together with a high flux of bombarding ions at the sample, making it possible to achieve rapid etching.

Magnetron plasma etching of SiC was demonstrated in preliminary experiments in a vacuum chamber, using a magnetron sputter gun as a cathode. The duration of each etch was 7 min. At a radio-frequency power of 250 W, a maximum etch depth of 4.5 μm was achieved with a gas mixture of 0.6 CHF_3 + 0.4 O_2 at a total pressure of about 2.7 Pa. This depth corresponds to an etch rate of about 640 nm/min, which is a relatively high rate, significantly higher than that which has been achieved using an ECR plasma. The etched surface was found to have a root-mean-square surface roughness of only 20 Å. The reaction of aluminum and fluorine yields a nonvolatile product, which makes aluminum suitable for use as a mask material to provide selective etching in fluorine-based plasmas. At the radio-frequency power of 250 W, aluminum films were etched at a high rate. At a radio-frequency power of 50 W, silicon carbide was etched at a rate of 170 nm/min, while aluminum was etched at $1/12$ that rate.

MICROSPHERES

Microspheres are spherical particles used in plastics and other materials as fillers and reinforcing agents. They are made of glass or ceramics or resins. There are two different kinds of glass microspheres — solid and hollow. Solid spheres, made of soda-lime glass, range in size from 4 to 5000 μm in diameter and have a specific gravity of about 2.5. Hollow glass microspheres have densities ranging from 80 to 801 kg/m^3 and diameters from 20 to 200 μm.

The spheres in plastics improve tensile, flexural, and compressive strength and lower elongation in water absorption. They also serve as thermal and sound insulators. Plastic microspheres are used mostly in the production of syntactic foams. Polyvinylidene chloride microspheres are excellent resin extenders and are used in sandwich construction of boat hulls. Epoxy microspheres are used as low-density bulk fillers for plastics and ceramics, and were developed for use in submerged deep-water floats. They can withstand hydrostatic pressures 68 MPa. Phenolic microspheres filled with nitrogen are used for production of polyester foams and syntactic epoxy foams. Polystyrene microspheres are also used to produce syntactic foams.

MICROWAVE HEATING

A microwave-heating technique provides for batch processing of multiple, identically sized and shaped samples of the same material. The technique involves (1) excitation of a symmetrical electromagnetic mode or modes in a symmetrical microwave cavity and (2) positioning the samples symmetrically in the cavity so that all samples are exposed to the same electromagnetic-field conditions and thus the same heating conditions. Typically, the electromagnetic modes and the pattern for mounting the samples are chosen to maximize the heating effect and make it as nearly spatially uniform as possible.

The same principle can be applied to microwave heating of multiple spherical or disk-shaped samples.

MOLD STEEL

This is a term that generally refers to the steels used for making molds for molding plastics. Mold steel should have a uniform texture that will machine readily with die-sinking tools. It must have no microscopic porosity, and must be capable of polishing to a mirrorlike surface. When annealed, it should be soft enough to take the deep imprint of a hob, and when hardened it must be able to withstand high pressure without sinking and have sufficient tensile strength to prevent breakage of thin mold sections. And it should be dimensionally stable during heat treatment and corrosion-resistant to the plastics being formed.

The principal mold steels are the P-type mold steels, specifically P6, which contains 0.10% carbon, 3.50% nickel, and 1.50% chromium; P20 (0.35% carbon, 1.70% chromium, and 0.40% molybdenum); and P21 (0.20% carbon, 4.00% nickel, and 1.20% aluminum). P6 is a carburizing steel having moderate hardenability, P20 also has only moderate hardenability, but P21 is a deep-hardening steel. P20 and P21 are usually supplied hardened to Rc 30 to 36, in which condition they can be machined to complex shapes. P21, a precipitation-hardening steel, can be supplied moderately hard, then machined and posthardened by a moderate-temperature age.

MOLECULAR MATERIALS

Matter of nanoscale dimensions often displays properties unlike those of isolated gas-phase molecules or the bulk liquids or solids that they make up. These unusual properties are often attributable to the unique bonding, structure, and morphology that small systems of finite dimensions adopt, whereas in other cases they result because the small sizes bring about new phenomena arising from what is termed *quantum confinement*, the restriction of space available to electrons. Because of the special properties that these cluster systems (aggregates of atoms or molecules) display, they have become an active subject of basic investigation in recent years. Other impetus for such research has arisen from the promise that new advanced materials might be produced by assembling nanoscale systems.

Investigations into the reactions of small hydrocarbons with transition metal ions, atoms, and clusters led to the discovery of an unexpected dehydrogenation reaction leading to the assembly of a new molecular cluster family termed metallocarbohedrenes (met-cars).

Met-Car Composition and Structure

The surprising finding that a heretofore unknown compound composed of transition metal and carbon atoms was very stable, and displayed unique bonding far different than that of the well-known cubic structure of bulk titanium carbide, prompted a search for other clusters with similar compositions.

The first successful alternative method involved the laser vaporization of well-mixed powders of selected metals and graphite, followed by the entrainment of the formed product species in pulses of helium gas. Met-cars were readily produced with appropriate conditions of laser power and metal-to-carbon ratio mixtures. Thereafter, it was found that met-cars could be produced from a mixture of metal carbides along with carbon and metal powders by vaporizing them in a small crucible suspended in vacuum, without any need for a gas to transport the plasma or assist in its cooling.

An important advance toward exploring the potential use of these molecular clusters as new materials came from successful attempts to synthesize bulk quantities of met-cars. Synthesizing was first accomplished with an arc-discharge technique similar to ones which have been used to produce so-called buckyballs (fullerenes).

Future Status

Met-cars have been established as a new class of molecular cluster materials that are likely to have unusual properties and many potential applications. It has been observed that these molecular clusters readily ionize even at very low intensities of light of widely varying wavelengths, suggesting that they have a substantially delocalized electronic character and that they may display unique optical and perhaps electrical properties.

All the evidence points to their structure as one in which the metal and carbon atoms arc bound in a well-defined lattice with one type of metal site. The formation of these interesting materials appears to be dominated by kinetic effects involving MC_2 building units. The clusters are found to grow by a unique mechanism leading to the development of a multicaged extended network structure.

Work is actively being pursued to isolate large amounts of bulk material and study the exact geometry and properties of met-cars using conventional chemical and physical techniques. Of particular interest is elucidating their optical, electronic, and reactive properties, through the use of molecular beams, lasers, and flow reactors.

MOLECULAR NANOTECHNOLOGY

Molecular nanotechnology is an emerging interdisciplinary field combining principles of molecular chemistry and physics with the engineering principles of mechanical design, structural analysis, computer science, electrical engineering, and systems engineering. Molecular manufacturing is a method conceived for the processing and rearrangement of atoms to fabricate custom products. It would rely on the use of a large number of small manufacturing subsystems working in parallel and using commonly available chemicals. Built to atomic specifications, the products would exhibit order-of-magnitude improvements in strength, toughness, speed, and efficiency, and would be of high quality and low cost.

DISTINCTION FROM RELATED DISCIPLINES

It is useful to illustrate some of the key points of the technology by drawing distinctions between it and some related fields. Molecular nanotechnology is distinguished from solution chemistry by the manner in which the chemical reactions occur: instead of the statistical process of molecules bumping together in random orientations and directions in solution until a reaction occurs, discrete molecules are brought together in individually controlled orientations and trajectories to cause a reaction at a specific site. Furthermore, this process is performed under programmable control.

Molecular manufacturing systems will have attributes different from those of biological systems. They will be able to transport raw materials and intermediate products more rapidly and accurately by conveyor belts and robotic arms.

PROGRESS

There have been major advances toward developing molecular nanotechnology. With scanning probe microscopes, researchers can now manipulate individual atoms to nanometer precision and catalyze reactions at specific points on surfaces; molecular biologists can design and synthesize proteins from scratch; synthetic chemists can design and create uncomplicated molecules to trap ions and molecules; researchers on molecular electronic devices have developed a working prototype and spurred commercial development.

MOLYBDENUM AND ALLOYS

Molybdenum (Mo) is a silvery-white metal, occurring chiefly in the mineral molybdenite but also obtained as a by-product from copper ores. The metal has a specific gravity of 10.2 and a melting point of 2610°C. It is ductile, softer than tungsten, and is readily worked or drawn into very fine wire. It cannot be hardened by heat treatment, but only by working.

Its major use is in alloy steels, for example, as tool steels (≤10% molybdenum), stainless steel, and armor plate. Up to 3% molybdenum is added to cast iron to increase strength. Up to 30% molybdenum may be added to iron-, cobalt-, and nickel-base alloys designed for severe heat- and corrosion-resistant applications. It may be used in filaments for lightbulbs, and it has many applications in electronic circuitry.

Molybdenum forms mirrors and films on glass when it is produced by gas-phase reduction or decomposition of volatile molybdenum compounds in glass tubes. Also, because the molybdenum metal surface binds strongly to oxygen without weakening the metal structure, and because the metal has such a low coefficient of thermal expansion — virtually equal to hard glass — molybdenum metal is an extraordinarily good material for use in glass–metal seals. Molybdenum trioxide (MoO_3) dissolves in glass, allowing strong binding of molten glass with preoxidized metal surfaces. Annealing is very effective, with little or no difference in thermal expansion at the metal–glass interface. Molybdenum found early use in filaments for electric lightbulbs and later in the construction of electronic devices (for example, in vacuum tubes, contacts, electrodes, and transistors).

Protection from Oxidation

At temperatures over about 538°C, unprotected molybdenum oxidizes so rapidly in air or oxidizing atmospheres that its continued use under these conditions is impractical. Uncoated molybdenum is, however, used satisfactorily where very short lives are involved (as in some missile parts) or where the surrounding atmosphere is nonoxidizing (as in hydrogen and vacuum furnaces). Protective coatings seem to be the answer where oxidation is a problem. Various coatings differing in maximum time–temperature capabilities and in physical and mechanical characteristics are available. Selection of the proper coating for a specific application involves consideration of a number of factors, foremost among which is the service temperature. For temperatures up to 1204°C, nickel-base alloys applied as cladding or sprayed coatings, and chromium–nickel electroplates appear most generally suitable. For temperatures up to 1538°C, or short periods at higher temperatures, modified chromized coatings and sprayed aluminum–chromium–silicon are predominant. For longer periods at higher temperatures, the choice would probably rest between siliconizing and ceramic coatings. Component tests have been found most reliable for final selection of suitable coatings.

Applications

The applications of molybdenum are generally those based on its high melting point, high modulus of elasticity, high strength at elevated temperatures, high thermal conductivity, high resistance to corrosion, low specific heat, and low coefficient of expansion. Long-established uses are found in the electric and electronic industries for applications such as mandrels and supports in lamp fabrication, anodes and grids in electronic tubes, resistance elements and radiation shields in high-temperature furnaces, and electrical contacts and electrodes. In recent years molybdenum has become increasingly important in the missile field for nozzles, nozzle inserts, leading edges of control surfaces, and support vanes. Uses in the metalworking industry include boring bars, grinding quills, resistance-welding electrodes, and thermocouples. The nuclear-energy industry has been active in developing molybdenum heat exchangers, piping, heat shields, and structural parts. The glass industry is a major user of large molybdenum parts, especially for melting electrodes and stirrers. The chemical and petrochemical industry is beginning to use molybdenum for corrosion resistance. The usefulness of molybdenum as a metallized coating, either to improve bonding between the base metal and a sprayed top coating, or alone for improved wear resistance, is no longer confined to maintenance but is extending to original equipment.

Alloys

Four main classes of commercial molybdenum-base alloys exist. One class relies on the formation of fine metal carbides that strengthen the material by dispersion hardening, and extend the resistance of the microstructure to recrystallization above that of pure molybdenum.

The most common of the carbide-strengthened alloys is known as TZM, containing about 0.5% titanium, 0.08% zirconium, and 0.03% carbon. Other alloys in this class include TZC (1.2% titanium, 0.3% zirconium, 0.1% carbon), MHC (1.2% hafnium, 0.05% carbon), and ZHM (1.2% hafnium, 0.4% zirconium, 0.12% carbon). The high-temperature strength imparted by these alloys is their main reason for existence.

Both TZM and MHC have found application as metalworking tool materials. Their high-temperature strength and high thermal conductivity make them quite resistant to the collapse and thermal cracking that are common failure mechanisms for tooling materials. A particularly demanding application is the isothermal forging process used to manufacture nickel-base superalloy gas-turbine engine components. In this process the dies and workpiece are both heated to the hot working temperature, and the forging is performed in a vacuum by using large hydraulic presses.

A second class relies on solid-solution hardening to strengthen molybdenum. These two classes of materials are typically produced in both vacuum-arc-casting and powder metallurgy grades. In the solid solution class tungsten and rhenium are the two primary alloy additions. The most common compositions are 30% tungsten (Mo–30% W), 5% rhenium (Mo–5% Re), 41% rhenium (Mo–41% Re), and 47.5% rhenium (Mo–50% Re). With the exception of the Mo–30% W alloy, which is available as a vacuum-arc-cast product, these alloys are normally produced by powder metallurgy. The tungsten-containing alloys find application as components in systems handling molten zinc, because of their resistance to this medium. They were developed as a lower-cost, lighter-weight alternative to pure tungsten and have served these applications well over the years. The 5% rhenium alloy is used primarily as thermocouple wire, whereas the 41% and 47.5% alloys are used in structural aerospace applications.

The third class uses combinations of carbide formers and solution hardeners to provide improved high-temperature strength. This class of alloys is normally produced by powder metallurgy techniques, but some of the alloys are also amenable to vacuum-arc-casting processing.

The beneficial effects of solid-solution hardening and dispersion hardening found in the carbide-strengthened alloys have been combined in the HWM-25 alloy (25% tungsten, 1% hafnium, 0.07% carbon). This alloy offers high-temperature strength greater than that of carbide-strengthened molybdenum, but it has not found wide commercial application because of the added cost of tungsten and the expense of processing the material.

The final class of alloys, known as dispersion-strengthened alloys, relies on second-phase particles (usually an oxide of a ceramic material) introduced or produced during powder processing to provide resistance to recrystallization and to stabilize the recrystallized grain structure, enhancing high-temperature strength and improving low-temperature ductility. These latter materials by their very nature must be produced by powder metallurgy techniques.

The dispersion-strengthened alloys rely exclusively on powder metallurgy manufacturing techniques. This allows the production of fine stable dispersions of second phases that stabilize the wrought structure against recrystallization, resulting in a material having improved high-temperature creep strength as compared to pure molybdenum. Once recrystallization occurs, the dispersoids also stabilize the interlocked recrystallized grain structure. This latter effect produces significant improvements in the ductility of the recrystallized material.

The potassium- and silicon-doped alloys such as MH (150 ppm potassium, 300 ppm silicon) and KW (200 ppm potassium, 300 ppm silicon, 100 ppm aluminum) are the oldest of this category; they are analogues to the doped tungsten alloys in common use for tungsten lamp filament. They were first developed to satisfy the requirements of the lighting industry for creep-resistant molybdenum components. They do not possess particularly high strength at low temperatures, but they are quite resistant to recrystallization and they possess excellent creep resistance due to their stable, interlocked recrystallized grain structure. They are used in applications requiring high-temperature creep resistance such as nuclear fuel sintering boats.

Composites

A family of copper–molybdenum–copper (CMC) laminates has been developed to serve the needs of the high-performance electronics industry. Laminating molybdenum with copper raises the coefficient of thermal expansion and improves the heat-transfer characteristics of the composite relative to pure molybdenum. This allows better matching between the composite that serves as a packaging or mounting material and the solid-state devices being employed.

MOLYBDENUM DISILICIDE

The commercial product contains approximately 60% molybdenum, 31% silicon, 8% iron, with small amounts of carbon, sulfur, vanadium, phosphorus, and copper. Its principal use is for alloying iron and steel, but it has ceramic uses as well. Its outstanding resistance to oxidation in air at temperatures up to 1700°C makes it potentially useful for high-temperature applications.

The excellent resistance to oxidation at high temperatures combined with fairly good elevated temperature strength makes $MoSi_2$ promising for elevated temperature structural applications where impact resistance is not a major consideration. Possible uses are gas turbine nozzle blades and furnace heating elements. Also, a metal–ceramic combination of $MoSi_2$ and Al_2O_3 has been considered for use as kiln furniture, saggers, sand blast nozzles, hot draw or hot press dies, and induction brazing fixtures.

MOLYBDENUM STEEL

Next to carbon, molybdenum is the most effective hardening element for steel. It also has the property, like tungsten, of giving steel the quality of red-hardness, requiring a smaller amount for the same effect. It is also used in hot-work steels, and to replace part of the tungsten in high-speed steels. It is added to heat-resistant irons and steels to make them resistant to deformation at high temperatures and to creep at moderate temperatures. It increases the corrosion resistance of stainless steels at high temperatures. Molybdenum in small amounts also increases the elastic limit of steel, reduces grain size, strengthens crystalline structure, and gives deep-hardening. It goes into solid solution, but when other elements are present it may form carbides and harden the steel, giving greater wear resistance. It also widens the heat-treating range of tool steels. As it decreases the temper brittleness of aluminum steels, small amounts are added to nitriding steels.

Plain carbon–molybdenum steels are easier to machine than other steels of equal hardness. Molybdenum structural steels usually have from 0.20 to 0.75% molybdenum. SAE (formerly Society of Automotive Engineers) steels 4130 and 4140 contain about 1.0% chromium and 0.20% molybdenum, and are high-strength forging steels for such uses as connecting rods. Hollow-head screws are made of SAE 4150 steel. SAE steels 4615 and 4650 have no chromium but contain about 1.75% nickel and 0.25% molybdenum. SAE 4615 is used for molds and dies to be hobbed. It is easily worked and has a tensile strength of 1482 MPa. Finally, the old Damascus steel and Toledo steel were molybdenum steels; the molybdenum was in the original ore.

MULLITE

Artificial mullite, or synthetic mullite, a ceramic material made by a prolonged fusing in the electric furnace of a mixture of silica sand or diasphoric clay and bauxite, has the composition $3Al_2O_3 \cdot 2SiO_2$, has a melting point of 1810°C, and softens at 1650°C.

Mullite occurs in nearly all ceramic products containing alumina and silica but, with the exception of refractories, is seldom introduced as such except as calcined kyanite.

The bricks are resistant to flame and to molten ash, and have a low, uniform coefficient of thermal expansion and a heat conductivity only slightly above that of fireclays. Normally, mullite has very fine crystals that change form and become enlarged after prolonged heating, making the product porous and permeable. For stable high-temperature refractories the mullite is prefused to produce larger crystals. At very high temperatures mullite tends to decompose to form corundum and alkali-silicate minerals of lower heat resistance. Mullite is also used for making spark plugs, chemical crucibles, and extruding dies, and a foamed mullite is used as a uniformly latticed honeycomb structure for lightweight, heat-resistant structural parts.

MUNTZ METAL

Muntz metal is a yellow brass containing 60% copper and 40% zinc; invented in 1832, it is also called yellow metal and malleable brass. Muntz metal is also modified with small amounts of iron and manganese. Iron above 0.35% forms a separate iron-rich constituent that is stable and gives hardness and high strength to the alloys. The addition of manganese helps to absorb the iron and also hardens the alloy. These alloys were also called high-strength brass, and were employed for such uses as hydraulic cylinders and marine forgings, but are now largely replaced by manganese bronze or aluminum bronze.

Current standard wrought alloys are C28000 (59 to 63% copper, as much as 0.30% lead, 0.07% iron, and the balance zinc); four leaded alloys — C36500, C36600, C36700, and C36800 — each containing 58 to 61% copper, 0.25 to 0.70% lead, and as much as 0.15% iron and 0.25% tin, with the balance zinc except for 0.02 to 0.06% arsenic in C36600, 0.02 to 0.10% antimony in C36700, and 0.02 to 0.10% phosphorus in C36800.

The alloys are available in most wrought forms, are hardenable only by cold-working, and are used for forged parts, heat-exchanger tubing and tube sheets, baffles, valve stems, and fasteners. The corrosion resistance of C28000 is said to be generally similar to that of copper and better than that of higher-copper alloys to sulfur-bearing compounds.

NANOMATERIALS

Nanomaterials or nanocrystalline materials are commonly defined as crystalline materials that have an average particle or grain size of less than 100 nm (0.1 µm) and 10,000 times smaller than grains in conventional materials.

A deliberate distinction is made between nanocrystalline materials and crystalline materials that are submicron. Submicron materials have an average particle or grain size of less than 1 µm.

The resultant properties of nanocrystalline materials thus have a much greater dependence on the contributions of interfacial atoms than submicron materials. Interfacial atoms are those on the surface of a particle, or in the grain boundaries of a consolidated material. Some unconventional mechanical, chemical, electrical, optical, and magnetic properties that nanocrystalline materials exhibit are attributed to this greater dependence on the contributions of interfacial atoms.

The phase *nanophase materials* encompasses the new tools that are letting scientists and engineers characterize and manipulate materials at the nanoscale level.

APPLICATIONS

This whole new class of "nanophase" composites has the potential to create superior protective coatings, mechanical parts, electrical insulation, and even human bone replacements, bone implants, and stronger parts for cars, airplanes, and spacecraft, and to improve the performance of many conventional materials.

Aluminum components formed from ultrafine powders will be far stronger than conventional large-grained aluminum parts, and steel-cutting tools with ultrafine grains will last several times longer.

A new technique/process that stresses and breaks down microscopic "nanopowders" is called "severe plastic deformation," which puts materials and alloys under severe stresses that break down, or refine, grains under loads in excess of 400 tons/in.[2]. These high loads are combined with torsion straining or other pressing methods to produce materials of exceptional strength and toughness. Brittle ceramics can become pore-free and malleable under certain conditions when formed from these nanopowders. Parts made from nanopowder metals such as copper, iron, silver, aluminum, and titanium are up to ten times stronger than conventionally structured counterparts.

In addition to manufactured components, potential applications include new types of filter membranes, cutting blades, and drill bits, materials for joining ceramic parts, high-performance bearings, nonsilicon materials for active and passive elements in integrated circuits, and supermagnets to increase the efficiency of electric motors.

One study showed that nanotubes of pure carbon functioned as a two-terminal electronic device known as a diode. This appears to be the world's smallest room-temperature rectifier, one that is only a handful of atoms in size. When nanotubes are grown, electronic devices naturally form on them. Carbon nanotubes are created by heating ordinary carbon until it vaporizes, then allowing it to condense in a vacuum or an inert gas. Depending on its diameter, a pure carbon nanotube can conduct an electrical current as if it were a metal, or it can act as a semiconductor.

NANOSTRUCTURE

This is a material structure assembled from a layer or cluster of atoms with size on the order of nanometers. Interest in the physics of condensed matter at size scales larger than that of atoms and smaller than that of bulk solids (mesoscopic physics) has grown rapidly since the 1970s, owing to the increasing realization that the properties of these mesoscopic atomic ensembles are different from those of conventional solids. As a consequence, interest in artificially assembling materials from nanometer-sized building blocks, whether layers or clusters of atoms, arose from discoveries that by controlling the sizes in the range of 1 to 100 nm and the assembly of such constituents it was possible to begin to alter and prescribe the properties of the assembled nanostructures. (Many examples of naturally formed nanostructures can be found in biological systems, from seashells to the human body.)

Nanostructured materials are modulated over nanometer length scales in zero to three dimensions. They can be assembled with modulation dimensionalities of zero (atom clusters or filaments), one (multilayers), two (ultrafine-grained overlayers or coatings or buried layers), and three (nanophase materials), or with intermediate dimensionalities. Thus, nanocomposite materials containing multiple phases can range from the most conventional case in which a nanoscale phase is embedded in a phase of conventional sizes, to the case in which all the constituent phases are of nanoscale dimensions. All nanostructured materials share three features: atomic domains (grains, layers, or phases) spatially confined to less than 100 nm in at least one dimension, significant atom fractions associated with interfacial environments, and interaction between their constituent domains.

MULTILAYERS AND CLUSTERS

Multilayered materials have had the longest history among the various artificially synthesized nanostructures, with applications to semiconductor devices, strained-layer superlattices, and magnetic multilayers. Recognizing the technological potential of multilayered quantum heterostructure semiconductor devices helped to drive the rapid advances in the electronics and computer industries. A variety of electronic and photonic devices could be engineered utilizing the low-dimensional quantum states in these multilayers for applications in high-speed field-effect transistors and high-efficiency lasers, for example. Subsequently, a variety of nonlinear optoelectronic devices, such as lasers and light-emitting diodes, have been created by nanostructuring multilayers.

Examples

Examples of nanostructured materials that have been characterized include multilayers, individual and assembled atom clusters, and cluster-consolidated nanophase materials.

Magnetic Multilayers

Magnetic multilayers, such as those formed by alternating layers of ferromagnetic iron and chromium, can be nanostructured so that the electrical resistance is significantly decreased (by up to a factor of 2 depending on the chromium layer thickness) by the application of a magnetic field. Such an effect, called giant magnetoresistance, occurs when the magnetic moments of the neighboring alternating layers are arranged in an antiparallel fashion, so that application of the magnetic field overcomes the antiferromagnetic coupling and aligns the layers into a condition of parallel ferromagnetic ordering, strongly reducing the electron scattering in the system. Magnetoresistive materials have been introduced in the magnetic recording industry as read heads because of their lower noise and improved signal-handling capabilities. Other nanostructured materials besides multilayers also exhibit giant magnetoresistance, such as magnetic cobalt clusters embedded in a nonmagnetic matrix of copper or silver, or magnetic platelike nickel–iron deposits embedded in silver.

Optical Properties of Cluster Assemblages

Noninteracting assemblages of small semiconductor clusters have optical properties of both scientific and technological importance. The optical absorption behavior of cadmium sulfide clusters with diameters in the nanometer size regime made by any of a variety of methods, including chemical precipitation in solutions or in zeolite supports, is different from that for bulk cadmium sulfide.

Chemical Reactivity

Chemical reactivity of nanostructured materials, with their potentially high surface areas compared to conventional materials, can also be significantly altered and enhanced. Since clusters can be assembled by means of a variety of methods, there can be an excellent degree of control over the total available surface area in the resulting self-supported ensembles. Thus, it is possible to maximize porosity to obtain very high surface areas, remove most of it via consolidation but retain some to facilitate low-temperature doping or other processing, or fully densify the nanophase material.

Nanophase Materials

The assembly of larger atom clusters into bulk nanophase materials can also have dramatic effects upon properties. In this case, the clusters interact fully with one another, yet the effects of cluster size are still very important. Clusters of metals or ceramics in the size range 5 to 25 nm have been consolidated to form ultrafine-grained polycrystals that have mechanical properties remarkably different and improved relative to their conventional coarse-grained counterparts.

NANOTECHNOLOGY

This involves systems for transforming matter, energy, and information, based on nanometer-scale components with precisely defined molecular features. The term *nanotechnology* has also been used more broadly to refer to techniques that produce or measure features less than 100 nm in size; this meaning embraces advanced microfabrication and metrology. Although complex systems with precise molecular features cannot be made with existing techniques, they can be designed and analyzed. Studies of nanotechnology in this sense remain theoretical, but are intended to guide the development of practical technological systems.

BASIC PRINCIPLES

Nanotechnology based on molecular manufacturing requires a combination of familiar chemical and mechanical principles in unfamiliar applications. In conventional chemistry, molecules move by diffusion and encounter each other in all possible positions and orientations. The resulting chemical reactions are accordingly hard to direct. Molecular manufacturing, in contrast, can exploit mechanosynthesis, that is, using mechanical devices to guide the motions of reactive molecules. By applying the conventional mechanical principle of grasping and positioning to conventional chemical reactions, mechanosynthesis can provide an unconventional ability to cause molecular changes to occur at precise locations in a precise sequence. Reliable positioning is required in order for mechanosynthetic processes to construct objects with millions to billions of precisely arranged atoms.

NATURAL FIBER

A natural fiber is one obtained from a plant, animal, or mineral. The commercially important natural fibers are those cellulosic fibers obtained from the seed hairs, stems, and leaves of plants; protein fibers obtained from the hair, fur, or cocoons of animals; and the crystalline mineral asbestos.

The natural fibers may be classified by their origin as cellulosic (from plants), protein (from animals), and mineral. The plant fibers may be further ordered as seed hairs, such as cotton; bast (stem) fibers, such as linen from the flax plant; hard (leaf) fibers, such as sisal; and husk fibers, such as coconut. The animal fibers are grouped under the categories of hair, such as wood; fur, such as angora; or secretions, such as silk. The only important mineral fiber is asbestos, which because of its carcinogenic nature has been banned from consumer textiles.

NATURAL RUBBER

Rubber is characterized as being a highly elastic or resilient material, and the natural product is obtained mainly as a latex from cuts in the trunks of the *Hevea brasiliensis* tree. The latex consists of small particles (averaging about 2500 Å units in diameter) of rubber suspended in an aqueous medium (at about 35% solids content). The system also contains about 6 to 8% nonrubber constituents, some of which are emulsifiers, naturally occurring antioxidants, and proteins.

The rubber part of the composition, which is obtained from the latex by coagulation, washing, and drying, consists of long linear polymeric molecules of high average molecular weight. One report states that natural rubber contains molecules of molecular weights from 50,000 to 3,000,000 with 60% of it over 1,300,000. The rubber is built up of C_5H_8 units, each containing a double bond, and the overall structure is called *cis*-polyisoprene. Although a product almost identical to natural rubber has been made by polymerizing isoprene, there is no indication that a tree makes rubber from isoprene.

VULCANIZATION

Raw rubber as obtained from the latex is not useful directly in many commercial applications because of stiffening at low temperatures and softening at high temperatures. It must be vulcanized to give a wider temperature range for application and good physical properties. The most usual vulcanization is by the use of sulfur and organic accelerators of vulcanization in the presence of zinc oxide at temperatures in the neighborhood of 100 to 204°C depending on the curing system.

Also, for many applications the high rubber vulcanizates are not tough enough for good service. To improve the situation, reinforcing fillers are used for compounding the rubber, such as finely divided carbon black, silica, and silicates including clays. In addition, many other nonreinforcing fillers, such as whiting (calcium carbonate) and barytes, are used to lower cost or to lend special properties to finished articles.

ENVIRONMENTAL EFFECTS

Under service conditions, natural rubber articles are subject to some deteriorating effects, and the rubber must be compounded to minimize these effects. For example, it is subject to deterioration by the action of oxygen, ozone, heat, and light and to fatigue cracking under dynamic conditions. Antioxidants are used in compounds to retard air and heat deterioration, and special chemicals are added to protect against light deterioration, as well.

PROCESSING

Natural rubber is, for most applications, masticated and compounded on mills or in Banbury mixers; the masticated material is then passed through a tubing machine to form various articles or parts of articles that are then vulcanized at an elevated temperature in a mold or under other suitable conditions.

Natural rubber is furnished to fabricators in the form of large bales or as a latex. The bale rubber comes in a number of different grades, which are priced according to quality. The latex is sold at different solids contents; the higher solids contents are preferred partly because of shipping costs from the plantations.

APPLICATIONS

Natural rubber is used for making many types of articles. Because of its abrasion-resistant quality and low hysteresis in reinforced compounds, it is used in truck-tire tread stocks and in conveyor belts that which are employed in conveying abrasive material such as coal, crushed rock, ore, and cinders. In large tires, it has found application in carcass compounds because of the tack and building qualities of the raw polymer. It has also been used in carcass compounds because of the low heat buildup (low hysteresis) of the carcass compound vulcanizate during severe service conditions in tire usage.

Some of the many other applications for natural rubber include waterproof clothing and footwear, and wire insulation, as well as mechanical goods and sundries, products such as hot water bottles, surgical goods, rubber bands and thread, engine mounts, tank tread blocks, gaskets, sound and vibration damping equipment, rug underlay, etc. In fact, it can be used almost anywhere a highly resilient, elastic, durable material is required. A few applications in the latex field are (1) in making latex thread by coagulating a fine stream of latex compound coming from a nozzle and then vulcanizing it; (2) in making foam sponge such as mattresses by gelling and vulcanizing a foamed latex; (3) in tire cord dipping; (4) in making such articles as surgeon's gloves by the latex dipping method; and (5) in carpets where it is used as a backing material to provide anchorage of the yarns and to give dimensional stability.

NEOPRENE RUBBER

Except for polybutadiene and polyisoprene, neoprene is perhaps the most natural rubberlike of all, particularly with regard to dynamic response. Neoprenes are a large family of rubbers that have a property profile approaching that of natural rubber, and with better resistance to oils, ozone, oxidation, and flame. They age better and do not soften on heat exposure, although high-temperature tensile strength may be lower than that of natural rubber.

These materials, like natural rubber, can be used to make soft, high-strength compounds. A significant difference is that, in addition to neoprene being more costly than natural rubber by the pound, its density is about 25% greater than that of natural rubber. Neoprenes do not have the low-temperature flexibility of natural rubber, which detracts from their use in low-temperature shock or impact applications.

General-purpose neoprenes are used in hose, belting, wire and cable, footwear, coated fabrics, tires, mountings, bearing pads, pump impellers, adhesives, seals for windows and curtain-wall panels, and flashing and roofing. Neoprene latex is used for adhesives, dip-coated goods, and cellular cushioning jackets.

NICKEL ALUMINIDE

Nickel aluminide (NiAl) is available commercially in powder form in various mesh sizes from granular material to an average particle size of a few micrometers. Nickel aluminide is fairly strong at room temperature with modulus of transverse rupture values ranging from 204 to 952 MPa, depending on fabrication methods, while at 1000°C the corresponding values are approximately half the room temperature strength. Creep resistance at these temperatures is extremely poor. Thermal shock resistance is quite good.

Nickel aluminide can be formed by hot pressing or by cold pressing and sintering, but the former method produces by far the best results.

Excellent oxidation resistance and fairly good strength make this material of interest for turbine blading or other combustion chamber applications. It has impact resistance that is better than most ceramics, intermetallic compounds, and some cermets. NiAl is also resistant to attack by molten glass and red and white fuming nitric acid, which suggests possible uses in the glass-processing industry or in some high-temperature chemical processes.

NICKEL AND ALLOYS

A silvery-white metal, nickel (Ni), has been used in alloy form with copper since ancient times. Nickel has a specific gravity of 8.902, or a density of 8.902 g/cm^3, and is magnetic up to 360°C. The metal is highly resistant to atmospheric corrosion and resists most acids, although it is attacked by oxidizing acids, such as nitric. Its principal use is as an alloying element to stainless steels, alloy steels, and nonferrous metals. It is also used for electroplating.

There are several commercial high-purity nickels, containing at least 99% of the metal plus trace amounts of combined cobalt, and small amounts of other elements, such as carbon, copper, iron, manganese, silicon, and sulfur. They find use in electronic and aerospace applications, chemical- and food-processing equipment, and for anodes and cathodes, caustic evaporators, and heat shields. Nickel 200, also known as commercially pure nickel, may contain as much as 0.15% carbon, 0.25% copper, 0.40% iron, 0.35% manganese, 0.35% silicon, and 0.01% sulfur. It is especially resistant to caustics, high-temperature halogens and hydrogen halides, and salts, but not to oxidizing halides or ammonium hydroxide. Because it is susceptible to high-temperature embrittlement by carbon precipitation, however, it is restricted to a maximum service temperature of 316°C. Nickel 201, its low-carbon (0.02% maximum) counterpart, can be used at higher temperatures. The cast grade, designated CZ-100, is recommended for use at temperatures above 316°C.

Nickel 270 is the highest-purity grade, being at least 99.97% nickel and containing no more than 0.02% carbon, 0.005% iron, and 0.001% each of other ingredients. Although similar to the other nickels in general corrosion resistance, it may be more prone to sulfur embrittlement under some conditions.

At cryogenic temperatures, nickel alloys are strong and ductile. Several nickel-base superalloys are specified for high-strength applications at temperatures to 1093°C. High-carbon nickel-base casting alloys are commonly used at moderate stresses above 1204°C.

Dispersion-Strengthened Nickel

Dispersion-strengthened nickel, or DS nickel, contains about 2.2% thoria as the dispersion-strengthening phase and, thus, has also been called TD nickel. The thoria markedly increases high-temperature strength. Tensile strengths, about 483 MPa ultimate and 331 MPa yield at room temperature, are on the order of 117 MPa at 1093°C. The material also retains about 60% of its room-temperature tensile modulus, about 138×10^3 MPa, at 1093°C. DS nickel has poor oxidation resistance and, thus, requires coating for sustained use at high temperatures, such as for aircraft-turbine components and furnace equipment. It has been alloyed with about 20% chromium to improve oxidation resistance, but still may require coating for prolonged use at high temperatures.

Ferrous Alloys

The largest commercial application for nickel is as an alloying element in both ferrous and nonferrous alloys. In unhardened nonaustenitic steels, nickel has the effect of slightly strengthening ferrite and mildly improving corrosion resistance. In balanced combinations with small amounts of chromium and manganese; or chromium, molybdenum, and manganese; or molybdenum and manganese (AISI 3100, 3300, 4300, 4600, 4800, 8600, 8700, 9300, 9800 series) hardenabilities of steels are greatly increased. In carburizing steels (AISI 2300 and 2500 series) nickel is used to strengthen and toughen the core.

Nickel is a strong austenite stabilizer and with chromium is used to form the important AISI 300 series of nonmagnetic austenitic stainless steels. In these steels, chromium is the principal alloying element and is the element responsible for their excellent corrosion resistance. Nickel contributes to this corrosion resistance, but its primary function is to promote and retain an austenitic structure at all recommended temperatures of use and under most conditions of plastic deformation. The mechanical properties of an austenitic steel are generally much less adversely affected by moderately elevated temperatures and very low temperatures than are ferritic or martensitic steels. Wrought stainless steels (AISI listing) with chromium and nickel contents up to 22 and 26%, respectively, and in some cases with the addition of molybdenum, have been designed for specific heat- and corrosion-resisting service over and above the standard 18-8 type. Some cast stainless steels with extremely high chromium and nickel contents (up to 28% chromium, 10% nickel and 17% chromium, 66% nickel) are used for severe corrosion or heat-resisting applications. Nickel is also widely used as a property-improving addition to cast iron and as one of the many alloying elements in some ferrous high-temperature, high-strength superalloys.

Magnetic Alloys

Nickel and iron form a series of alloys with thermal-expansion and magnetic characteristics of commerical importance. The alloy of 36% nickel ("Invar") has a thermal coefficient of expansion of almost zero over a small temperature range. The addition of 12% chromium, in lieu of some of the iron, produces an alloy ("Elinvar") with an invariable modulus of elasticity over a considerable temperature range as well as a fairly low coefficient of expansion. Variations in content of nickel and other metals have produced a large number of alloys with particular coefficients for specific applications such as geodetic tapes, balance wheels in watches, tuning forks, and glass-to-metal seals. The alloy with 78% nickel (permalloy) and many modifications of this composition have very high magnetic permeabilities at low field strengths and find application in the electronics field. Nickel also plays an important role in permanent magnets of the Alnico type, which contain 14 to 18% nickel plus aluminum, cobalt, and sometimes copper.

Nonferrous Alloys

The principal groups of nonferrous alloys are those of slightly alloyed nickel, nickel and copper, nickel and chromium, and the superalloys. The slightly alloyed nickels ("Duranickel," "Permanickel," E nickel, electronic nickel, etc.) in most cases contain more than 94% nickel.

Duranickel 301, a precipitation-hardened, 94% nickel alloy, has excellent spring properties to 316°C. During thermal treatment, Ni_3AlTi particles precipitate throughout the matrix. This action enhances alloy strength. Corrosion resistance is similar to that of commercially pure wrought nickel.

Permanickel 300 is a high-nickel alloy, containing at least 97% nickel and small amounts of titanium and magnesium and most of the other elements found in high-purity nickels. It is used for high-strength parts in electrical and electronics applications. It has a tensile modulus of 207,000 MPa and, depending on form and condition, tensile yield strengths of 241 to 1034 MPa.

The above alloys retain much of the corrosion resistance and physical properties of nickel, but they are alloyed to make them heat-treatable by age hardening, more resistant to a specific form of corrosion (manganese to reduce susceptibility to sulfur embrittlement) or to produce desirable electronic characteristics.

Wrought beryllium nickel containing 97.5% nickel, 2% beryllium, and 0.5% titanium is used for springs, switches, bellows, diaphragms, and small valves. The age-hardenable alloy provides tensile yield strengths of 310 to 1586 MPa, 896 MPa at 538°C, and has good corrosion resistance in general atmospheres and to reducing media. Casting alloys, containing 2 to 3% beryllium and also age-hardenable, are about equally strong, and are used for molds to form glasses and plastics, and for metal-forming tools, aircraft fuel-pump impellers, and seal plates in aircraft engines.

Binary Nickel Alloys

The primary wrought alloys in this category are the Ni–Cu grades known as Monel alloy 400 (Ni–31.% Cu) and K-500 (Ni–29.% Cu), which also contain small amounts of aluminum, iron, and titanium. The Ni–Cu alloys differ from nickel 200 and 201 because their strength and hardness can be increased by age hardening. Although the Ni–Cu alloys share many of the corrosion characteristics of commercially pure nickel, their resistance to sulfuric and hydrofluoric acids and brine is better. Handling of waters, including seawater and brackish water, is a major application. Monel alloys 400 and K-500 are immune to chloride-ion stress-corrosion cracking, which is often considered in their selection

Other commercially important binary nickel compositions are Ni–Mo and Ni–Si. One binary type, Hastelloy alloy B-2 (Ni–28% Mo), offers superior resistance to hydrochloric acid, aluminum chloride catalysts, and other strongly reducing chemicals. It also has excellent high-temperature strength in inert atmospheres and vacuum.

Cast nickel–copper alloys comprise a low and high silicon grade. M-35-1 and QQ-N-288, Grades A and E (1.5% silicon), are commonly used in conjunction with wrought nickel-copper in pumps, valves, and fittings. A higher silicon grade, QQ-N-288, Grade B (3.5% silicon), is used for rotating parts and wear rings because it combines corrosion resistance with high strength and wear resistance. Grade D (4.0% silicon) offers exceptional galling resistance.

Two other binary cast alloys are ACI N-12M-1 and N-12M-2. These Ni–Mo alloys are commonly used for handling hydrochloric acid in all concentrations at temperatures up to the boiling point. These alloys are produced commercially under the tradenames Hastelloy alloy B and Chlorimet 2; see Table N.1.

Many chemical-resistant nickel alloys, both wrought and cast, are noted mainly for their

TABLE N.1
Nominal Composition of Some Nickel-Base Alloys, wt%

Trademark	Ni	Cu	Cr	Co	Mo	Ti	Al	Cb	Fe	Mn	Si	C	Other
Nickel 211	95	—	—	—	—	—	—	—	—	4.75	—	0.08	—
Duranickel alloy 301	93.7	0.05	—	—	—	0.4	4.4	—	0.35	0.3	0.5	0.17	—
Monel alloy 400	66	31.5	—	—	—	—	—	—	1.35	0.9	0.15	0.18	—
Monel alloy K-500	66	29	—	—	—	0.5	2.75	—	0.9	0.75	0.5	0.15	—
Chromel P	90	—	10	—	—	—	—	—	—	—	—	—	—
Nichrome V	80	—	19.5	—	—	—	—	—	—	2.5[a]	1	0.25	—
Alumel	94	—	—	—	—	—	2	—	—	3	1	—	—
Nimonic 75	Bal	0.5[a]	19.5	—	—	0.4	—	—	5[a]	1[a]	1[a]	0.12	—
Nimonic 80A	Bal	—	19.5	2[a]	—	2.2	1.1	—	5[a]	1[a]	1[a]	0.1[a]	—
Inconel alloy 600	Bal	0.5[a]	15.5	—	—	—	—	—	8	1[a]	0.5[a]	0.15[a]	—
Inconel alloy X-750	Bal	0.5[a]	15	—	—	2.5	0.9	0.9	7	0.7	0.4	0.04	—
Inconel alloy 718	53	0.3[a]	19	1.0[a]	3	0.9	0.5	5	Bal	0.35[a]	0.35[a]	0.08[a]	—
Alloy 713C	Bal	—	12	—	4	0.5	6	2	5[a]	1[a]	1[a]	0.2[a]	0.012 B, 0.10 Zr
Udimet 500	Bal	—	17.5	16.5	4	2.9	2.9	—	4[a]	0.75[a]	0.75[a]	0.15[a]	—
Waspaloy	Bal	—	19	14	3	2.5	1.2	—	2	0.7	0.4	0.05	—
M252	55	—	19	10	10	2.5	0.75	—	2	1	0.7	0.1	—
GMR 235	Bal	—	15.5	—	5	2.5	3	—	10	0.25[a]	0.6[a]	0.15	0.06B
Hastelloy B	61	—	1[a]	2.5[a]	27.5	2	—	—	5.5	1[a]	1	0.05[a]	0.4V
Hastelloy C	54	—	15.5	2.5[a]	15.5	—	—	—	5.5	1[a]	1[a]	0.08[a]	0.35 V[a], 4 W
Hastelloy D	82	3	1[a]	1.5[a]	—	—	—	—	2[a]	1	9	0.12[a]	—

[a] Maximum.

Source: McGraw-Hill *Encyclopedia of Science and Technology*, 8th ed., Vol. 11, McGraw-Hill, New York., 802. With permission.

resistance to chemicals that are aggressive toward other metals, and many of these also possess substantial strength at elevated temperature. Although trade names abound, the Hastelloys and the Incoloys and Inconels are probably the most well known. Many of these alloys are resistant to normally aggressive acids, such as hydrofluoric and sulfuric, as well as to acetic and phorphoric acids, mixed acids, chlorides, solvents, and high-temperature oxidation and, thus, are widely used for chemical-processing equipment.

Among the more common alloys and their principal ingredients are Hastelloy B, which contains 26 to 30% molybdenum, 4 to 6% iron, 2.5% cobalt, and 1% chromium; Hastelloy B-2, which contains the same amount of molybdenum and chromium but only 1% iron and 1% cobalt; Hastelloy C, having 15 to 18% molybdenum and about as much chromium plus 4 to 7% iron, 3 to 5% tungsten, and 2.5% cobalt; Hastelloy C-4, with molybdenum and chromium contents similar to those of Hastelloy C but no more than 3% iron and 2% cobalt; Hastelloy D, 8.5 to 10% silicon, 2 to 4% copper, and as much as 2% iron, 1.5% cobalt, and 1% chromium; Hastelloy G, 21 to 23% chromium, 18 to 21% iron, 5.5 to 7.5% molybdenum, 1.75 to 2.5% columbium plus tantalum, 1.5 to 2.5% copper, and as much as 2.5% cobalt; Hastelloy S, 14.5 to 17% chromium, 14 to 16.5% molybdenum, and as much as 3% iron and 2% cobalt; Hastelloy C-276, 15 to 17% molybdenum, 14.5 to 16.5% chromium, 4 to 7% iron, 3 to 4.5% tungsten, and as much as 2.5% cobalt. Incoloy 800, although iron base, is often grouped with these alloys. It contains 46% iron, 32.5% nickel, and 21% chromium. Nickel-base (42%) Incoloy 825 contains 30% iron, 21.5% chromium, 3% molybdenum, and 2.3% copper.

Hastelloy X, which provides substantial strength and oxidation resistance at temperatures to about 1204°C, contains 20 to 23% chromium, 17 to 20% iron, 8 to 10% molybdenum, 0.5 to 2.5% cobalt, and 0.2 to 1% tungsten. Solution-treated rapidly cooled sheet has room-temperature tensile properties of 786 MPa ultimate strength, 359 MPa yield strength, 43% elongation, and 197×10^3 MPa modulus. At 982°C, these properties are 155 MPa, 110 MPa, 45%, and 126×10^3 MPa, respectively. The alloy is widely used for gas-turbine parts and other applications requiring heat and oxidation resistance. Although mainly a wrought alloy, it also can be investment cast.

TERNARY NICKEL ALLOYS

Two primary wrought and cast compositions are Ni–Cr–Fe and Ni–Cr–Mo. Ni–Cr–Fe is known commercially as Haynes alloys 214 and 556, Inconel alloy 600, and Incoloy alloy 800. Haynes' new alloy No. 214 (Ni–16% Cr–2.5% Fe–4.5% A1–Y) has excellent resistance to oxidation to 1204°C, and resists carburizing and chlorine-contaminated atmospheres. Haynes alloy 556 (Fe–20% Ni–22% Cr–18% Co) combines effective resistance to sulfidizing, carburizing, and chlorine-bearing environments with good oxidation resistance, fabricability, and high-temperature strength. Inconel alloy 600 (Ni–15.5% Cr–8% Fe) has good resistance to oxidizing and reducing environments. Intended for severely corrosive conditions at elevated temperatures, Incoloy 800 (Ni–46% Fe–21% Cr) has good resistance to oxidation and carburization at elevated temperatures, and it resists sulfur attack, internal oxidation, scaling, and corrosion in many atmospheres.

A cast Ni–Cr–Fe alloy CY-40, known as Inconel, has higher carbon, manganese, and silicon contents than the corresponding wrought grade. In the as-cast condition, the alloy is insensitive to the type of intergranular attack encountered in as-cast or sensitized stainless steels.

Significant additions of molybdenum make Ni–Cr–Mo alloys highly resistant to pitting. They retain high strength and oxidation resistance at elevated temperatures, but they are used in the chemical industry primarily for their resistance to a wide variety of aqueous corrosives. In many applications, these alloys are considered the only materials capable of withstanding the severe corrosion conditions encountered.

In this group, the primary commercial materials are C-276, Hastelloy alloy C-22, and Inconel alloy 625. Hastelloy alloy C-22 (Ni–22% Cr–13% Mo–3% W–3% Fe) has better overall corrosion resistance and versatility than any other Ni–Cr–Mo alloy. Alloy C-276

(57% Ni–15.5% Cr–16% Mo) has excellent resistance to strong oxidizing and reducing corrosives, acids, and chlorine-contaminated hydrocarbons. Alloy C-276 is also one of the few materials that withstands the corrosive effects of wet chlorine gas, hypochlorite, and chlorine dioxide. Hastelloy alloy C-22, the newest alloy in this group, has outstanding resistance to pitting, crevice corrosion, and stress-corrosion cracking. Present applications include the pulp and paper industry, various pickling acid processes, and production of pesticides and various agrichemicals.

Superalloys

There are a great variety of high-temperature, high-strength nickel alloys; these are called superalloys because of their outstanding strength, creep resistance, stress-rupture strength, and oxidation resistance at high temperatures. They are widely used for gas turbines, especially aircraft engines. Most of these alloys contain substantial chromium for oxidation resistance; refractory metals for solid-solution strengthening; small amounts of grain-boundary-strengthening elements, such as carbon, boron, hafnium, and/or zirconium; and aluminum and titanium for strengthening by precipitation of an Ni(Al,Ti) compound known as "gamma prime" during age hardening. Among the well-known wrought alloys are D-979; GMR-235-D; IN 102; Inconel 625, 700, 706, 718, 722, X750, and 751; MAR-M 200 and 421; René 41, 95, and 100; Udimet 500 and 700; and Waspaloy. Cast alloys include B-1900; GMR-235-D; IN 100, 162, 738, and 792; M252; MAR-M 200, 246, and 421; Nicrotung; René 41, 77, 80, and 100; and Udimet 500 and 700. Some wrought alloys are also suitable for casting, primarily investment casting.

A third class of superalloys includes oxide-dispersion-strengthened (ODS) alloys such as IN MA-754 (Ni–20% Cr–0.6% yttria) and IN MA-6000 (Ni–15% Cr–2% Mo–4% W–2.5% Ti–4.5% Al), which are strengthened by dispersions such as yttria coupled (in some cases) with gamma prime precipitation (MA-6000).

An additional dimension of nickel-base superalloys has been the introduction of grain-aspect ratio and orientation as a means of controlling properties. In some instances, grain boundaries have been removed. Wrought powder-metallurgy alloys of the ODS class and cast alloys such as MAR M-247 have demonstrated property improvements due to grain morphology control by directional crystallization or solidification. Virtually all uses of the cast and wrought nickel-base superalloys are for gas-turbine components.

Specialty Nickel Alloys

In addition to the above families, there are specialty nickel alloys for glass sealing and other applications. A series of paramagnetic alloys, called Nitinol, are intermetallic compounds of nickel and titanium rather than nickel–titanium alloys. The compound TiNi contains theoretically 54.5% nickel, but the alloys may contain Ti_2Ni and $TiNi_3$ with about 50 to 60% nickel. The TiNi and nickel-rich alloys are paramagnetic, with a permeability value of 1.002, compared with the unity value of a vacuum. A 54.5% nickel alloy has a tensile strength of 758 MPa with elongation of about 15%, and hardness of Rockwell C35. The alloys close to the TiNi composition are ductile and can be cold-rolled.

The Nitinols, with nickel content ranging from 53 to 57%, are known as memory alloys because of their ability to "remember," or return to a previous shape upon being heated. This unusual behavior stems from a diffusionless transformation of the alloy. These shape-memory alloys have excellent fatigue strength, and damping around room temperature is reported to be one of the highest ever measured in a metal.

NICKEL-BONDED TITANIUM CARBIDE

This is a hard, strong material that retains its strength at temperatures up to 1100°C. It is the most commonly used of the so-called cermet materials for high-temperature structural applications.

Nickel-bonded titanium carbide is available in powder form. Nickel-bonded titanium carbides have hardnesses ranging from HRA 80 to 89, depending on binder content. The melting point is that of the binder material, and the density is 5.0 g/cm³ min, but can be higher depending on the amount and nature of the binder material.

Nickel-bonded titanium carbide is usually made by cold-pressing and sintering methods, but it can be hot-pressed and, if the binder content is very high, it can even be cast.

The material may be used for turbine blading, in such high-temperature structural applications as tool bits, and as a high-temperature bearing and seal material, where high compressive strength, low coefficient of friction, and high wear resistance are needed at elevated temperatures. Oxidation resistance of these materials is fair and can be improved somewhat by adding niobium or tantalum carbide to the hard phase or by adding cobalt or chromium to the nickel binder.

NICKEL BRASS

A number of alloys of copper, nickel, and zinc are termed nickel brass. Nickel–silicon brass contains very small percentages of silicon, usually about 0.60%, which forms a nickel silicide, Ni_2Si, increasing the strength and giving heat-treating properties. Rolled nickel–silicon brass, containing 30% zinc, 2.5% nickel, and 0.65% silicon, has a tensile strength of 785 MPa. Imitation silver, for hardware and fittings, was a nickel brass containing 57% copper, 25% zinc, 15% nickel, and 3% cobalt. The bluish color of the cobalt neutralizes the yellow cast of the nickel and produces a silver-white alloy. Silvel was a nickel brass containing 67.5% copper, 26% zinc, and 6.5% nickel, with sometimes a little cobalt. Nickel brass is an alloy used where white color and corrosion-resistance are desired.

NICKEL BRONZE

This is a name given to bronzes containing nickel, which usually replaces part of the tin, producing a tough, fine-grained, and corrosion-resistant metal. A common nickel bronze containing 88% copper, 5% tin, 5% nickel, and 2% zinc has a tensile strength of 330 MPa, elongation 42%, and Brinell hardness 86 as cast. When heat-treated or age-hardened, the tensile strength is 599 MPa, elongation 10%, and Brinell hardness 196. Small amounts of lead take away the age-hardening quality of the alloy, and also lower the ductility. But small amounts of nickel added to bearing bronzes increase the resistance to compression and shock without impairing the plasticity. A bearing bronze of this nature contains 73 to 80% copper, 15 to 20% lead, 5 to 10% tin, and 1% nickel.

NICKEL–CHROMIUM STEEL

This is defined as steel containing both nickel and chromium, usually in a ratio of 2 to 3 parts nickel to 1 part chromium. The 2:1 ratio gives great toughness, and the nickel and chromium are intended to balance each other in physical effects. The steels are especially suited for large sections that require heat treatment because of their deep and uniform hardening. Hardness and toughness are the characteristic properties of these steels. Nickel–chromium steel containing 1 to 1.5% nickel, 0.45 to 0.75% chromium, and 0.38 to 0.80% manganese is used throughout the carbon ranges for case-hardened parts and for forgings where high tensile strength and great hardness are required. Low nickel–chromium steels, having more carbon (0.60 to 0.80%) are used for drop-forging dies and other tools.

Nickel–chromium steels may have temper brittleness, or low impact resistance, when improperly cooled after heat treatment. A small amount of molybdenum is sometimes added to prevent this brittleness. A nickel–chromium coin steel, used by the Italian government for coins, was a stainless-steel type containing 22% chromium, 12% nickel, and some molybdenum.

Low-carbon nickel–chromium steels are water hardening, but those with appreciable amounts of alloying elements require oil quenching. Air-hardening steels contain up to 4.5% nickel and 1.6% chromium, but are brittle unless tempered in oil to strengths below 1378 MPa. The alloy known as Krupp analysis steel contains 4% nickel and 1.5% chromium.

NICKEL–MOLYBDENUM STEEL

This alloy steel is mostly used in compositions of 1.5% nickel and 0.15 to 0.25% molybdenum, with varying percentages of carbon up to 0.50%. These steels are characterized by uniform properties and are readily forged and heat-treated. Molybdenum toughens the steels, and in the case-hardening steels gives a tough core. Roller bearings are made of this class of steel. Superalloy steel is 3160 steel. A 5% nickel steel with 0.30% carbon and 0.60% molybdenum has a tensile strength of 1206 to 1585 MPa with elongation 12 to 22%, depending on the heat treatment. Molybdenum is more frequently added to the steels containing also chromium, the molybdenum giving air-hardening properties, reducing distortion and making the steels more resistant to oxidation.

NICKEL OXIDE

Two of the nickel oxides are useful as colorants in ceramics: (1) nickelous oxide or green nickel oxide, NiO, and (2) black nickel oxide, Ni_2O_3. At 400°C it oxidizes to Ni_2O_3, and at 600°C it is reduced back to NiO.

Nickel produces a bluish-violet in potash glasses and a violet tending toward brown in soda glasses. Nickel rates as one of the more powerful colorants, since 1 part in 50,000 produces a recognizable tint.

Nickel oxide is sometimes used to decolorize potash glass. Nickel oxide and nickel silicate have an advantage over manganese dioxide for decolorizing purposes in that they are not as sensitive in changing oxidizing and reducing environments.

NICKEL SILVER

Nickel silver is a name applied to an alloy of copper, nickel, and zinc, which is practically identical with alloys known in the silverware trade as German silver. Packfong, meaning white copper, is an old name for these alloys. The very early nickel silvers contained some silver and were used for silverware.

Some three dozen standard wrought alloys (C73150 to C79900) and four standard cast alloys (C97300 to C97800) are designated nickel silvers. Depending on the alloy, copper content of wrought alloys ranges from 48 to 80% and nickel content from about 7 to 25%, with zinc the balance except for smaller quantities of other elements, mainly manganese, iron, and lead. The cast alloys range from about 55 to 65% copper, 12 to 25% nickel, 2.5 to 21% zinc, 2 to 10% lead, with lesser amounts of other elements.

The most common alloy, nickel silver C75200, nominally contains 65% copper and 18% nickel and, thus, is often referred to as nickel silver 65-18. The electrical conductivity of the alloy is about 6% that of copper and its thermal conductivity is 33 W/m·K. Tensile properties for thin flat products in the annealed condition are about 414 MPa ultimate strength, 207 MPa yield strength, and 30% elongation. Cold working to the hard temper triples yield strength and markedly reduces ductility. Wire, which has similar tensile properties annealed, can be cold-worked to still greater tensile strength. Modulus of elasticity in tension is 124,110 MPa. All the cast alloys are suitable for sand and investment casting and some also for centrifugal and permanent-mold casting. The strongest of these alloys, nickel silver C97800, has typical tensile properties of 379 MPa ultimate strength, 207 MPa yield strength, 15% elongation, and 131,000 MPa modulus. Applications for wrought alloys include holloware and tableware, watch and camera parts, hardware, dairy equipment, costume jewelry, nameplates, keys, fasteners, and springs. The cast alloys are used for fittings, valves, ornaments, pump parts, and marine equipment.

Over the years, nickel silvers have been known by a variety of names. Benedict metal originally had 12.5% nickel, with 2 parts copper to 1 part zinc, but the alloy used for hardware and plumbing fixtures contains about 57% copper, 2% tin, 9% lead, 20% zinc, and 12% nickel. The cast metal has a strength of 241 MPa with elongation of 15%.

NICKEL STEEL

This is steel containing nickel as the predominant alloying element. The first nickel-steel armor plate, with 3.5% nickel, was known as Harveyized steel.

Nickel added to carbon steel increases the strength, elastic limit hardness, and toughness. It narrows the hardening range but lowers the critical range of steel, reducing danger of warpage and cracking, and balances the intensive deep-hardening effect of chromium. The nickel steels are also of finer structure than ordinary steels, and the nickel retards grain growth. When the percentage of nickel is high, the steel is very resistant to corrosion. At high nickel contents, the metals are referred to as iron–nickel alloys or nickel–iron alloys. The steel is nonmagnetic above 29% nickel, and the maximum permeability is at about 78% nickel. The lowest thermal expansion is at 36% nickel. The percentage of nickel in nickel steels usually varies from 1.5 to 5%, with up to 0.80% manganese. The bulk of nickel steels contains 2 and 3.5% nickel. They are used for armor plate, structural shapes, rails, heavy-duty machine parts, gears, automobile parts, and ordnance.

The standard ASTM structural nickel steel used for building construction contains 3.25% nickel, 0.45% carbon, and 0.70% manganese. This steel has tensile strength from 586 to 689 MPa and a minimum elongation 18%. An automobile steel contains 0.10 to 0.20% carbon, 3.25 to 3.75% nickel, 0.30 to 0.60% manganese, and 0.15 to 0.30% silicon. When heat-treated, it has a tensile strength up to 551 MPa and an elongation 25 to 35%. Forgings for locomotive crankpins, containing 2.5% nickel, 0.27% carbon, and 0.88% manganese, have a tensile strength 572 MPa, elongation 30%, and reduction of area 62%. A nickel–vanadium steel, used for high-strength cast parts, contains 1.5% nickel, 1% manganese, 0.28% carbon, and 0.10% vanadium. The tensile strength is 620 MPa and elongation 25%. Univan steel for high-strength locomotive castings is a nickel–vanadium steel of this type. Unionaloy steel is an abrasion-resistant steel.

The federal specifications for 3.5% nickel carbon steel call for 3.25 to 3.75% nickel, and 0.25 to 0.30% carbon. This steel has a tensile strength of 586 MPa and elongation 18%. When oil-quenched, a hot-rolled 3.5% nickel medium-carbon steel, SAE steel 2330, develops a tensile strength up to 1516 MPa, and Brinell hardness of 223 to 424, depending on the drawing temperature. Standard 3.5 and 5% nickel steels are regular products of the steel mills, although they are often sold under trade names. Steels with more than 3.5% nickel are too expensive for ordinary structural use. Steels with more than 5% nickel are difficult to forge, but the very high nickel steels are used when corrosion-resistant properties are required. Nicloy, used fork tubing to resist the corrosive action of paper-mill liquors and oil-well brines, contains 9% nickel, 0.10% chromium, 0.05% molybdenum, 0.35% copper, 0.45% manganese, 0.20% silicon, and 0.09% max carbon. The heat-treated steel has a tensile strength of 758 MPa, with elongation 35%. The cryogenic steels, or low-temperature steels, for such uses as liquid-oxygen vessels, are usually high-nickel steels. ASTM steel A-353, for liquid-oxygen tanks at temperatures to −196°C, contains 9% nickel, 0.85% manganese, 0.25% silicon, and 0.13% carbon. It has a tensile strength of 654 MPa with elongation of 20%. A 9% nickel steel, for temperatures down to −196°C, contains 9% nickel, 0.80% manganese, 0.30% silicon, and not over 0.13% carbon. It has a minimum tensile strength of 620 MPa and elongation of 22%.

NICKEL SULFATE

Nickel sulfate is the most widely used salt for nickel-plating baths, and is known in the plating industry as single nickel salt. It is easily produced by the reaction of sulfuric acid on nickel, and comes in pea-green water-soluble crystalline pellets of the composition $NiSO_4 \cdot 7H_2O$, of a specific gravity of 1.98, melting at about 100°C. Double nickel salt is nickel ammonium sulfate, $NiSO_4 \cdot (NH_4)_2 SO_4 \cdot 6H_2O$, used especially for plating on zinc. To produce a harder and whiter finish in nickel plating, cobaltous sulfamate, a water-soluble powder of the composition $Co(NH_2SO_3)_2 \cdot 3H_2O$, is used with the nickel sulfate. Nickel plate has a normal hardness of Brinell 90 to 140, but by controlled processes file-hard plates can be obtained from sulfate baths. In electroless plating, nickel sulfate, a reducing agent, a pH adjuster, and complexing and stabilizing agents are combined to deposit metallic nickel on an immersed object.

The electroless nickel coating is comparable to electrolytic chrome.

NIOBIUM AND ALLOYS

Note should be made of the fact that in the United States this element was originally called columbium (symbol Cb). The Nomenclature Committee of the International Union of Pure and Applied Chemistry in 1951 adopted a recommendation to name this element niobium (symbol Nb). American chemists use this name, but the metallurgists and metals industry still use the name columbium. Most niobium is used in special stainless steels, high-temperature alloys, and superconducting alloys such as Nb_3Sn. The low cross-section capture of niobium for thermal neutrons of only 1.1 barn makes it suitable for use in nuclear piles.

Niobium is a tough, shiny, silver-gray, soft, ductile metal that somewhat resembles stainless steel in appearance. Niobium is relatively low in density, yet can maintain its strength at high temperatures. It has excellent corrosion resistance to liquid metals, and can be easily fabricated into wrought products.

Over 95% of all niobium is used as additions to steel and nickel alloys for increasing strength. Only 1 to 2% of niobium is in the form of niobium-base alloys or pure niobium metal. Superconducting niobium–titanium alloys account for over half of that, and high-temperature and corrosion applications account for the remainder.

The density of niobium at 8.57 g/cm^3 is moderate compared with most other high melting point metals. It is less than molybdenum at 10.2 g/cm^3 and half that of tantalum at 16.6 g/cm^3.

Alloys

Commercial niobium alloys are relatively low in strength and extremely ductile, and can be cold-worked over 70% before annealing becomes necessary. The resulting ease of fabrication into complex parts combined with relatively low density frequently favors the selection of niobium alloys over other refractory metals such as molybdenum, tantalum, or tungsten.

High-temperature niobium alloys were developed in the 1960s for nuclear and aerospace applications and today serve in communications satellites, human body imaging equipment, and a variety of high-temperature components. Although niobium alloys have useful strength at temperatures hundreds of degrees above nickel-base superalloys, applications have been limited by their susceptibility to oxidation and to long-term creep.

Production

Typical production electron-beam furnaces generate 500 to 5000 kW of power, and are capable of purifying ingots with diameters of 305 to 508 mm and lengths over 2 m. "Drip-melting" is the current standard electron-beam method, but hearth melting may eventually become practical as higher-power furnaces are developed.

Alloys of niobium are made by vacuum-arc remelting with the appropriate elemental additions. The most common alloy additions are zirconium, titanium, and hafnium, which readily go into solution during arc melting.

Properties

The most common high-temperature niobium alloys are listed in Table N.2. All are hardened primarily by solid solution-strengthening; however, small amounts of second-phase particles are present. The composition of these particles varies, but they are generally associated with interstitial impurities that form oxides, nitrides, and carbides. These particles are important because the size and distribution of second phases can often have a strong influence on mechanical properties and recrystallization behavior. For example, a variation of the Nb–1% Zr alloy, commonly known as PWC-11, contains an intentional addition of 0.1 wt% carbon, specifically to form carbide precipitates that significantly improve high-temperature creep properties.

Another alloy listed, WC-3009, normally contains ~0.10 wt% oxygen, which is approximately five times more oxygen than other niobium alloys. This high level of oxygen, which is introduced during powder processing, is not

TABLE N.2
Commercially Available Niobium Alloys for High-Temperature Service

Alloy	Composition, wt%	Thermal Conductivity at 800°C (1470°F), W/m·°C	Thermal Conductivity at 1200°C (2190°F), W/m·°C	Total Emissivity at 800°C (1470°F)	Total Emissivity at 1200°C (2190°F)	Density, g/cm³ (lb/in.³)	Melting Point, °C (°F)	Coefficient of Thermal Expansion at 20°C × 10⁻⁶/°C
Pure niobium	Nb	—	—	—	—	8.57	2468	7.1
C-103	Nb–10Hf–1Ti	37.4	42.4	0.28	0.40 0.70–0.82 (silicide coated)	8.85 (0.320)	2350 ±50 (4260 ±90)	8.73
Nb–1Zr	Nb–1Zr	59.0	63.1	0.14	0.18	8.57 (0.310)	2410 ±10 (4365 ±15)	6.8
PWC-11	Nb–1Zr–0.1C	—	—	—	—	8.57 (0.310)	—	6.8
WC-3009	Nb–30Hf–9W	—	—	—	—	10.1 (0.365)	—	7.5
FS-85	Nb–28Ta–10W–1Zr	52.8	56.7	—	—	10.6 (0.383)	—	7.1

Source: Adv. Mater. Proc., 154(6), 27–31, 1998. With permission.

TABLE N.3
Typical Room-Temperature Tensile Properties of Niobium Alloys

Alloy	Yield Strength, MPa (ksi)	Ultimate Tensile Strength, MPa (ksi)	Elongation, %	Elastic Modulus at 20°C, GPa (Msi)	Elastic Modulus at 1200°C, GPa (Msi)
C-103	296 (42.93)	420 (60.91)	26	90 (13.1)	64 (9.3)
Nb–1Zr	150 (21.75)	275 (39.88)	40	80 (11.7)	28 (4.1)
PWC-11	175 (25.38)	320 (46.41)	26	80 (11.7)	28 (4.1)
WC-3009	752 (109.06)	862 (125.02)	24	123 (17.9)	NA
FS-85	462 (67.00)	570 (82.67)	23	140 (20.4)	110 (16.0)

Source: Adv. Mater. Proc., 154(6), 27–31, 1998. With permission.

deleterious to mechanical properties because the oxygen combines with hafnium in the alloy to form stable hafnium oxide precipitates. The WC-3009 alloy is unique in that it exhibits an oxidation rate less than one tenth that of most other niobium alloys. When WC-3009 was developed, it was speculated that such an alloy could survive a short supersonic mission even if its protective coating failed.

In general, niobium alloys are much less tolerant of impurity pickup than other reactive metals such as titanium and zirconium. Alloys containing second-phase particles that form a continuous boundary between grains can exhibit drastically reduced tensile elongation. This condition is usually caused by contamination or improper heat treatment. Copper, which can accidentally be introduced during welding, is particularly disastrous to mechanical properties. In fact, the total permissible interstitial oxygen, hydrogen, carbon, and nitrogen content of niobium alloys is typically one fifth to one tenth that of titanium or zirconium alloys.

All the commercial alloys are quite ductile at room temperature. Tensile properties of the common alloys at 20°C are shown in Table N.3. The highest tensile strength at room temperature and elevated temperature is exhibited by the WC-3009 alloy. Even though WC-3009 clearly exhibits the highest tensile strength, FS-85 has superior creep strength. Its high creep strength is due to its higher melting point, which is elevated by its high concentration of tantalum and tungsten.

Elastic modulus, thermal conductivity, coefficient of thermal expansion, and total hemispherical emissivity are also listed in Tables N.2 and N.3. The emissivity data is for smooth and nonoxidized surfaces, which exhibit much lower emissivity values than oxidized material. Also shown is an emissivity value of 0.7 to 0.82 for silicide-coated C-103. This value is for a common Si–20% Fe–20% Cr coating applied by the "slurry coat and fusion" method.

APPLICATIONS

The most common application for niobium alloys is in sodium vapor lamps. The Nb–1% Zr alloy demonstrates excellent formability, weldability, and long life in a sodium vapor environment.

These bulbs are used throughout the world for highway lighting, because of their high electrical efficiency and long life, which is typically in excess of 25,000 h. The high-carbon version of this alloy, PWC-11, has a creep rate approximately five times lower than that of Nb–1% Zr at 1100°C, because of the effects of carbide precipitates.

For aerospace applications at 1100 to 1500°C, alloy C-103 has been the workhorse of the niobium industry because of its higher strength. Excellent cold-forming and welding characteristics enable fabricators to construct very complex shapes, such as thrust cones and high-temperature valves. Closed die forgings are also easily produced. Most of these components in propulsion systems are exposed for relatively short times to temperatures between 1200 and 1400°C.

The service environment for propulsion systems often is less oxidizing than the normal atmosphere. Because C-103 has virtually no oxidation resistance, components are extensively coated with silicides.

Another very successful application for coated C-103 is thrust augmenter flaps in a turbine engine. These flaps, placed at the tail end of the engine to form a high-temperature liner in the afterburner section, typically reach 1200 to 1300°C and last for ~100 h of afterburner time.

Niobium alloys have also been evaluated for various high-temperature components of the National Aerospace Plane. Hypersonic leading edges and nose cones were fabricated to function as heat-pipe thermal management systems. The heat-pipe concept was designed to transport extreme heat away from hot spots, such as hypersonic leading edges, to cooler areas where heat could be expelled by radiation. A typical 500 g niobium heat pipe can dissipate over 10 kW of heat and operate isothermally at 1250 to 1350°C. These devices were successfully tested in combustion torches, high-velocity jet fuel burners, tungsten-quartz lamps, and even electric welding arcs at heat fluxes well over 1000 W/cm^2.

NITRIC ACID

Also called aqua fortis and azotic acid, nitric acid is a colorless to reddish fuming liquid of the composition HNO_3, having a wide variety of uses for pickling metals, etching, and in the manufacture of nitrocellulose, plastics, dyestuffs, and explosives. It has a specific gravity of 1.502 (95% acid) and a boiling point of 86°C, and is soluble in water. Its fumes have a suffocating action, and it is highly corrosive and caustic. Fuming nitric acid is any water solution containing more than 86% acid and having a specific gravity above 1.480. Nitric acid is made by the action of sulfuric acid on sodium nitrate and condensation of the fumes. It is also made from ammonia by catalytic oxidation, or from the nitric oxide produced from air.

NITRIDES

Nitrides are less stable than the oxides, carbides and sulfides, and their use in air at elevated temperature is limited because of their tendency to oxidize. However, in several instances, the oxide film is protective and deterioration is slow. Despite their limitations, nitrides have interesting properties and are sure to find many specialized uses as technology becomes more complex.

STABLE NITRIDES

Aluminum Nitride

Aluminum nitride is conveniently prepared by an electric arc between aluminum electrodes in a nitrogen atmosphere. Crucibles of the pressed powder, sintered at 1985°C, are resistant to liquid aluminum at 1985°C, to liquid gallium at 1316°C, and to liquid boron oxide at 1093°C. Aluminum nitride has good thermal shock resistance and is only slowly oxidized in air (1.3% converted to Al_2O_3 in 30 h at 1427°C). It is inert to hydrogen at 1705°C but is attacked by chlorine at 593°C.

Aluminum nitride is an excellent substrate for creating wide-band-gap semiconductors for wireless communications and power-industry applications. Since aluminum nitride withstands very high temperatures, this substrate material can be used for microelectronic devices on jet engines. Such substrates also would improve the production of blue and ultraviolet lasers that could be used to squeeze a full-length movie onto a CD. Aluminum nitride crystals have also been grown in a tungsten crucible at 2300°C.

Boron Nitride

The crystal structure of boron nitride is similar to that of graphite, giving the powder the same greasy feel. The platy habit of the particles and the fact that boron nitride is not wet by glass favors use of the powder as a mold wash, e.g., in the fabrication of high-tension insulators. It is also useful as thermal insulation in induction heating.

Boron nitride can be hot-pressed to strong ivory-white bodies that are easily machined. Hot-pressed boron nitride is stable in air to about 704°C. From 704 to 982°C the rate of oxidation increases moderately. It is also stable

TABLE N.4
Properties of Boron and Silicon Nitrides Compared with Graphite and Alumina

Property	Boron Nitride	Graphite	Silicon Nitride	Alumina
Melting point, °F	>4400 (subl.)	>6500 (subl.)	>3450 (subl.)	3722
Specific gravity				
Crystal	2.25	2.25	3.18	3.96
Body	2.10	1.6–2.0	1.5–2.7	2.6–3.9
Hardness, DPH[a]	30	20	1100	2800
Modulus of rupture (room temp.), 1000 psi[b]	16∥	2.2∥[c]		
	7.3∣	1.8∣	1–20	38
Coefficient of thermal expansion (avg at 70–1800°F)/°F × 10^{6b}	4.17∥	4.5–12∥		
	0.43∣	1.5–4∣	1.37	4.3
Thermal conductivity (room temp.), Btu-in./ft² h/°F[b]	105∥[d]			
	199∣[d]	120–360	130	20–30
Electrical resistance, Ω-cm				
At room temp.	1.7×10^{13}∥[b]	10^{-3c}	10^{13}	10^{16}
At 900°F	2.3×10^{10}∥		10^{13}	10^{12}
Dielectric constant	4.1–4.8	—	9.4	12.3

[a] Diamond pyramid hardness.
[b] ∥ = Parallel to molding pressure; ∣ = perpendicular.
[c] Varies widely with type of graphite.
[d] At 570°F.

in chlorine up to 704°C. However, at 982°C it is attacked rapidly.

Like commercial graphite, hot-pressed boron nitride is anisotropic. Thermal expansion parallel to the direction of pressing is ten times that in the perpendicular direction. The ratio for modulus of rupture is 2:1.

Although boron nitride resembles graphite in many respects, it differs uniquely in electrical characteristics, having high resistivity and high dielectric strength even at elevated temperatures (see Table N.4). This feature, combined with easy machinability has led to extensive use in high-temperature electronics.

Boron nitride is available as −325-mesh powder.

A cubic form of boron nitride (Borazon) similar to diamond in hardness and structure has been synthesized by the high-temperature, high-pressure process for making synthetic diamonds. Any uses it may find as a substitute for diamonds will depend on its greatly superior oxidation resistance.

Cutting tool materials like mixed ceramics or CBN cutting tools are already available for hard machining. To identify the proper cutting tool material, one must analyze the application.

In case of cutting interruptions, CBN cutting tools will be the appropriate choice. Continuous cuts allow the use of mixed ceramics or coated mixed ceramics for better efficiency. When producing a gear wheel, for example, turning with a CBN-tipped insert reduces the cost per wheel by more than 60%, compared to grinding. At the same time, disposal costs for grinding sludge vanish, because hard turning does not require coolant.

Silicon Nitride

Silicon nitride is most easily prepared by direct reaction of nitrogen at about 1316°C with finely divided elemental silicon (≤150 mesh), either as loose powder or as a slip-cast or otherwise preformed part. Conversion of silicon particles to the nitride Si_3N_4 is accompanied by the growth of a felt of interlocking needles in the void space between particles. Despite an overall porosity of 15 to 25%, silicon nitride bodies are effectively impervious in many applications because of the microscopic size of the pores.

Although silicon nitride is not machinable in its final form except by grinding, the partially converted body can be machined by conventional methods after which conversion can be completed without dimensional change.

Silicon nitride is indefinitely resistant to air oxidation up to 1649°C, but begins to sublime at about 1925°C. It is not attacked by chlorine at 899°C or hydrogen sulfide at 982°C nor by the common acids. Because of a low coefficient of thermal expansion, resistance to thermal shock is relatively good.

Uncoated and coated silicon nitride cutting tools dominate the high-performance end of gray cast iron machining. They typically offer metal removal rates at least three times higher than coated carbide grades.

The newly developed cutting tool combines a 6% cobalt substrate with a 10-μm-thick, medium-temperature $TiCN/Al_2O_3/TiN$ coating. Medium-temperature chemical vapor desposition TiCN coatings show a reduced tendency for the forming of eta-phase at the interface between coating and carbide substrate.

Recently developed silicon nitride cutting tools have a substantially improved fracture resistance. Because of their insufficient chemical wear resistance, however, they have a limited use in machining nodular cast irons, mainly in areas of severe cutting interruptions at higher speeds (>400 m/min).

Titanium and Zirconium Nitride

Titanium and zirconium nitrides for use in refractory bodies are most conveniently prepared by treating the corresponding metal hydrides with ammonia at 1000°C.

Sintered TiN can be heated to a bright red heat with only superficial oxidation, and then plunged into water without cracking; ZrN is less resistant to oxidation.

Combination coatings involving both chemical vapor deposition and physical vapor deposition technologies provide the wear-resistance advantages of chemical vapor deposition TiCN coatings with the compressive residual stress advantages of physical vapor deposition TiN coatings. The net result is improved wear and chipping resistance. These coatings increase the speed capabilities of carbide cutting tools in titanium turning by a factor of 2.

Eliminating coolants can turn an easy-to-machine material into a difficult drilling problem, when using standard cutting tools. The introduction of TiAlN coatings represents a significant step toward dry drilling.

The goal in all dry machining is to develop cutting tools with higher resistance to thermal load and fatigue. Cermet tools may be one of the most suitable materials for these applications.

NITRIDING STEELS

Nitriding steels are alloy steels (low- and medium-carbon steels with combinations of chromium and aluminum or nickel, chromium, and aluminum) designed particularly for optimum results when they are subjected to the nitriding operation. The composition is such that the required microstructure for optimum nitriding is produced after heat treatment. Nitrided parts made from nitriding steels have extremely high surface hardnesses of about 92 to 95 Rockwell N scale, wear resistance, and resistance to certain types of corrosion.

PROCESSING AND APPLICATIONS

Nitriding consists of exposing steel parts to gaseous ammonia at about 538°C to form metallic nitrides at the surface. The hardest coatings are obtained with aluminum-bearing steels. Nitriding of stainless steel is known as Malcomizing. After nitriding, these steels have extremely high surface hardnesses of about 92 to 95 Rockwell N. The nitride layer also has considerable resistance to corrosion from alkalies, the atmosphere, crude oil, natural gas, combustion products, tap water, and still salt water. Nitrided parts usually grow about 0.003 to 0.005 cm during nitriding. The growth can be removed by grinding or lapping, which also removes the brittle surface layer. Most uses of nitrided steels are based on resistance to wear. The steels can also be used at temperatures as high as 538°C for long periods without softening. The slick, hard, and tough nitrided surface also resists seizing, galling, and spalling. Typical applications are cylinder liners and barrels for aircraft engines, bushings, shafts, spindles and thread guides, cams, rolls, piston pins, rubber and paper-mill product rolls, special oil tool equipment, bearings, rollers, etc.

Fabrication

The fabrication characteristics of nitriding steels are basically the same as those of other steels of similar alloy content. They can be drilled, broached, tapped, milled, sawed, or ground. Light feeds and depth of cuts are recommended. Welding is done with rod or wire of similar composition. Flash welding is permissible.

If very heavy cuts are involved, they usually are made prior to heat treatment. Normal machining is done on heat-treated material, and is followed by a stress-relieving treatment of not less than 37.8°C above the nitriding temperature, before finish machining or grinding. It is essential that in machining, sufficient removal be allowed to remove all decarburization from the surface prior to nitriding. The surface also must be clean and free of any surface contamination.

Because nitriding is a low-temperature treatment, little or no warpage is encountered. If it is necessary to straighten because of residual stress, the part should be heated to 538 to 593°C to prevent surface cracking.

Nitrided parts normally grow about 0.03 to 0.05 mm during nitriding. This may be removed by grinding or lapping. This also has the advantage of removing a brittle layer on the surface and exposes a slightly harder layer immediately beneath it. This operation, however, will reduce corrosion resistance to a large degree.

Nitriding steels are available in all standard steel forms. They can be purchased heat-treated or annealed to desired physical properties.

Most uses of nitriding steels are based on resistance to wear. An outstanding property is that these steels can be heated to as high as 538°C for long periods without softening.

The slick, hard surface produced also makes it ideal to prevent seizing, galling, and spalling, and it is not readily attacked by combustion products.

NITRILE RUBBER

The nitriles are copolymers of butadiene and acrylonitrile, used primarily for applications requiring resistance to petroleum oils and gasoline. Resistance to aromatic hydrocarbons is better than that of neoprene but not as good as that of polysulfide. Nitrile butyl rubber (NBR) has excellent resistance to mineral and vegetable oils, but relatively poor resistance to the swelling action of oxygenated solvents such as acetone, methyl ethyl ketone, and other ketones. It has good resistance to acids and bases except those having strong oxidizing effects. Resistance to heat aging is good, often a key advantage over natural rubber.

With higher acrylonitrile content, the solvent resistance of an NBR compound is increased but low-temperature flexibility is decreased. Low-temperature resistance is inferior to that of natural rubber, and although NBR can be compounded to give improved performance in this area, the gain is usually at the expense of oil and solvent resistance. As with SBR, this material does not crystallize on stretching, and reinforcing materials are required to obtain high strength. With compounding, nitrile rubbers can provide a good balance of low creep, good resilience, low permanent set, and good abrasion resistance.

Tear resistance is inferior to that of natural rubber, and electrical insulation is lower. NBR is used instead of natural rubber where increased resistance to petroleum oils, gasoline, or aromatic hydrocarbons is required. Uses of NBR include carburetor and fuel-pump diaphragms and aircraft hoses and gaskets. In many of these applications, the nitriles compete with polysulfides and neoprenes.

Essentially the same techniques of emulsion polymerization employed in manufacture of general-purpose synthetic rubber may be used for production of nitrile polymers. Nitrile rubbers are supplied in various physical forms including sheet, crumb, powder, and liquid. The sheet is the most widely used type, with the other varieties offered for specialty applications.

BLENDS

An outstanding feature of nitrile rubber is its compatibility with many different types of resins permitting it to be easily blended with them. In combination with phenolic resins it provides adhesives with especially high strengths. Other resins used include resorcinol formaldehyde, urea formaldehyde, alkyd, epoxy, and polyvinyl chloride (to produce Type 2 rigid PVC). Both slab- and crumb-type nitrile rubber are used in this type of application, with the crumb type directly soluble and of special interest to adhesive manufacturers who do not have rubber-mixing equipment. Nitrile rubber–phenolic resin solvent solutions are used in shoe sole-attaching adhesives, for structural bonding in aircraft, adhering automotive brake lining to brake shoes, and many other industrial applications.

The powder-type rubbers were also developed for blending with phenolic resins, primarily for the manufacture of improved impact phenolic molding powders.

The liquid nitrile polymer finds use as a tackifier and nonextractable plasticizer in molded rubber parts, cements, friction, and calendered stocks.

Both the liquid and powder are of interest as curing-type plasticizers in vinyl plastisols. Nitrile rubber–PVC blends of various types are used in many other fields including cable jacket, retractable cord, abrasion-resistant shoe soles, industrial face masks, boat bumpers, and fuel lines.

COMPOSITION

The ratio of butadiene to acrylonitrile in the commercially available rubbers ranges from a low of about 20% to as high as 50% acrylonitrile. The various grades are usually referred to as high, medium-high, medium-low, and low acrylonitrile content.

The high acrylonitrile polymers are used in applications requiring maximum resistance to aromatic fuels, oils, and solvents. This would include oil well parts, fuel-cell liners, fuel hose, and other similar applications. The low acrylonitrile grade finds use in those areas requiring good flexibility at very low temperatures where oil resistance is of secondary importance. The medium types are most widely used and are satisfactory for all oil-resistant applications between these two extremes. Typical applications include conveyor belts, flexible couplings, soles, heels, floor mats, printing blankets, rubber rollers, sealing strips, aerosol bomb gaskets, milking inflations, seals, diaphragms, O-rings, packings, hose, washing machine parts, valves, and grinding wheels. These established uses give only a slight idea of products that are made of nitrile rubbers.

Physical properties of cured nitrile rubber parts are directly related to the ratio of butadiene and acrylonitrile in the polymer, as indicated below:

As acrylonitrile content increases:

1. Oil and solvent resistance improve.
2. Tensile strength increases.
3. Hardness increases.
4. Abrasion resistance improves.
5. Gas impermeability improves.
6. Heat resistance improves.

As acrylonitrile content decreases:

1. Low-temperature resistance improves.
2. Resilience increases.
3. Plasticizer compatibility increases.

PROPERTIES

The polymer with the highest acrylonitrile content produces the highest tensile strength and hardness; it also exhibits the best resistance to fuels and oils. As the percentage of acrylonitrile decreases, there is a corresponding decrease in resistance to fuels and oils; at the same time low-temperature flexibility characteristics are improved. Resiliency also increases. The lowest acrylonitrile polymer exhibits only moderate resistance to swelling in aromatic fluids but remains flexible at very low temperatures in the range of –57 to –62°C.

Thus, properly compounded nitrile polymers will provide high tensile strength, excellent resistance to abrasion, low compression set, very good aging under severe operating conditions, and excellent resistance to a wide range of fuels, oils, and solvents. They are practically unaffected by alkaline solutions, saturated salt solutions, and aliphatic hydrocarbons, both saturated and unsaturated. They are affected little by fatty acids found in vegetable fats and oils or by aliphatic alcohols, glycols, or glycerols.

Nitrile rubber is not recommended, generally, for use in the presence of strong oxidizing agents, ketones, acetates, and a few other chemicals.

NITROCARBURIZING

Salt bath nitrocarburizing is a thermochemical process for improving the properties of ferrous metals. However, some tools and other high-alloy steels are susceptible to reductions in core hardness after standard nitrocarburizing. To prevent such losses, a low-temperature salt bath nitrocarburizing process has been developed. With treatment temperatures as low as 480°C, this process not only maintains core hardness, but also can sometimes increase core hardness.

Processing

During salt bath nitrocarburizmg, the part is immersed in a vessel of molten salt. Nitrogen and carbon in the salt react with the iron on the surface, forming a compound layer with an underlying diffusion zone. The compound layer consists of iron nitrides, chromium nitrides, or other such compounds, depending on the alloying elements in the steel, and small amounts of carbides.

Ranging in depth from 2.5 to 20 μm, the compound layer provides improvements in wear and corrosion resistance, as well as in service behavior and hot strength. Hardness of the compound layer, measured on a cross section, ranges from 700 HV on unalloyed steels, up to 1600 HV on high chromium steels. Note that this layer is formed from the base metal and is an integral part of it, and is therefore not a coating. The diffusion zone can extend as deep as 1 mm, depending on the steel. This diffusion zone causes an increase in rotating-bending strength and rolling fatigue strength as well as pressure loadability.

Salt bath nitrocarburizing may be applied to a wide range of ferrous metals, from low-carbon to tool steels, cast iron to stainless steels. Specifically, the process:

- Improves wear and corrosion resistance
- Reduces or eliminates galling and seizing
- Increases fatigue strength
- Raises surface hardness
- Provides highly predictable, repeatable results
- Performs consistently, even with varying contours and thicknesses within the same part or load
- Maintains dimensional integrity
- Shortens cycle times
- Offers flexibility and ease of operation

Conventional treatment temperatures are in the range of 580°C, but for highly alloyed steels as well as stainless and tool steels, this temperature can cause a reduction in core hardness. The above benefits, derived both from the nitrogen and carbon diffused into the metal surface, as well as the processing in a liquid bath, are often necessary for applications in which a reduction in core hardness is not acceptable. For this reason, a new low-temperature process was developed.

The low temperature process normally takes place at 480°C, although it can operate at 480 to 520°C. This process has specific advantages:

- Core hardness and tensile strength are maintained in the tempered condition.
- Very thin compound layers can be formed.
- Distortion is extremely low.

- Formation of a compound layer on high-speed steels can be suppressed.
- Hardness of surface and diffusion layers can be customized.

This low-temperature process is beneficial for high-alloy steels such as stainless, tool, die, and high-speed steels (see Table N.5).

TABLE N.5
Suitable Steels for Low-Temperature Nitrocarburizing

Steel	Application
D2	Cold-work tool steel
D3	Cold-work tool steel
AISI 420	Cold-work tool steel
H11	Die-cast tools, machine pistons
H13	Die-cast tools, pistons, extrusion dies
HNV 3	Valves (martensitic)
17 4 PH	Planetary gears, press tools
HSS 36	High-speed steel drill bits
HSS M2	High-speed steel drill bits

Source: Adv. Mater. Proc., 154(3), 40–43, 1998. With permission.

NITROGEN

Nitrogen is an element (symbol N) that at ordinary temperatures is an odorless and colorless gas. The atmosphere contains 78% nitrogen in the free state. It is nonpoisonous and does not support combustion. Nitrogen is often called an inert gas, and is used for some inert atmospheres for metal treating and in lightbulbs to prevent arcing, but it is not chemically inert. It is a necessary element in animal and plant life, and is a constituent of many useful compounds. Lightning forms small amounts of nitric oxide from the air, which is converted into nitric acid and nitrates, and bacteria continuously convert atmospheric nitrogen into nitrates. Nitrogen combines with many metals to form hard nitrides useful as wear-resistant metals. Small amounts of nitrogen in steels inhibit grain growth at high temperatures, and also increase the strength of some steels. It is also used to produce a hard surface on steels.

APPLICATIONS

Because of the importance of nitrogen compounds in agriculture and chemical industry, much of the industrial interest in elementary nitrogen has been in processes for converting elemental nitrogen into nitrogen compounds. The principal methods for doing this are the direct synthesis of ammonia from nitrogen and hydrogen, the electric arc process, which involves the direct combination of N_2 and O_2 to nitric oxide, and the cyanamide process.

NODULAR CAST IRON

Nodular cast irons, such as GGG 40, GGG 50, and GGG 60, have become popular for parts such as housings, wheel parts, crankshafts, and camshafts. These metals offer higher strength and toughness than other cast irons, a result of spherical inclusions of carbon in the metal matrix. Generally easy to machine, GGG 40 irons with higher ferrite content tend to produce built-up edges on the cutting tool. For materials such as GGG 60 and higher, abrasiveness increases as the pearlite content increases, which can result in rapid insert wear. These nodular iron grades present unique machining characteristics.

NONMAGNETIC STEELS

These are steel and iron alloys used where magnetic effects cannot be tolerated. Manganese steel containing 14% manganese is nonmagnetic and casts readily but is not machinable. Nickel steels and iron–nickel alloys containing high nickel content are nonmagnetic. Many mills regularly produce nonmagnetic steels containing from 20 to 30% nickel. Manganese–nickel steels and manganese–nickel–chromium steels are nonmagnetic and may be formulated to combine desirable features of the nickel and manganese steels. One nonmagnetic steel with a composition of 10.5 to 12.5% manganese, 7 to 8% nickel, and 0.25 to 0.40% carbon has low magnetic permeability and low eddy-current loss, can be machined readily, and work-hardens only slightly. The tensile strength is 551 to 758 MPa, elongation 25 to 50%, and specific gravity 8.02. It is austenitic and cannot be hardened. The 18-8 austenitic chromium–nickel steels are also nonmagnetic. A nonmagnetic alloy used for watch gears and escapement wheels is not a steel but is a copper–nickel–manganese alloy containing 60% copper, 20% nickel, and 20% manganese. It is very hard, but can be machined with diamond tools.

NONWOVEN FABRICS

In the most general sense, nonwoven fabrics are fibrous-sheet materials consisting of fibers mechanically bonded together by interlocking or entanglement, by fusion, or by an adhesive. They are characterized by the absence of any patterned interlooping or interlacing of the yarns. In the textile trade, the terms *nonwovens* and *bonded fabrics* are applied to fabrics composed of a fibrous web held together by a bonding agent, as distinguished from felts, in which the fibers are interlocked mechanically without the use of a bonding agent. There are three major kinds of nonwovens based on the method of manufacture. Dry-laid nonwovens are produced by textile machines. The web of fibers is formed by mechanical or air-laying techniques, and bonding is accomplished by fusion-bonding the fibers or by the use of adhesives or needle punching. Either natural or synthetic fibers, usually 2.5 to 7.6 cm in length, are used. Wet-laid nonwovens are made on modified papermaking equipment. Either synthetic fibers or combinations of synthetic fibers and wood pulp can be used. The fibers are often much shorter than those used in dry-laid fabrics, ranging from 0.64 to 1.27 cm in length. Bonding is usually accomplished by a fibrous binder or an adhesive. Wet-laid nonwovens can also be produced as composites, for example, tissue-paper laminates bonded to a reinforcing substrate of scrim. Spin-bonded nonwovens are produced by allowing the filaments emerging from the fiber-producing extruder to form into a random web, which is then usually thermally bonded. These nonwovens are limited commercially to thermoplastic synthetics such as nylons, polyesters, and polyolefins. They have exceptional strength because the filaments are continuous and bonded to each other without an auxiliary bonding agent. Fibers in nonwovens can be arranged in a great variety of configurations that are basically variations of three patterns: parallel or unidirectional, crossed, and random. The parallel pattern provides maximum strength in the direction of fiber alignment, but relatively low strength in other directions. Cross-laid patterns (like wovens) have maximum strength in the directions of the fiber alignments and less strength in other directions. Random nonwovens have relatively uniform strength in all directions.

NYLON PLASTICS

Although nylon polymers are most familiar in fiber form, their combination of excellent chemical and mechanical properties, plus their ability to be molded and extruded into precise forms, have permitted their use in a wide variety of nontextile applications. These range from hammerheads, gears, and rifle stocks to miniature coil forms and delicately colored personal products.

CHARACTERISTICS

Nylons are a group of polyamide resins that are long-chain polymeric amides in which the amide groups form an integral part of the main polymer chain, and that have the characteristic that when formed into a filament the structural elements are oriented in the direction of the axis. Nylon was originally developed as a textile fiber, and high tensile strengths, above 344 MPa, are obtainable in the fibers and films. But this high strength is not obtained in the molded or extruded resins because of the lack of oriented stretching. When nylon powder that has been precipitated from solution is pressed and sintered, the parts have high crystallinity and very high compressive strength, but they are not as tough as molded nylon.

PRODUCTION

Nylons are produced from the polymerization of a dibasic acid and a diamine. The most common one of the group is that obtained by the reaction of adipic acid with hexamethylenediamine.

More specifically, nylons may be made by the amidation of diamines with dibasic acids, for example, hexamethylenediamine plus adipic acid (Type 6/6 nylon), or hexamethylenediamine and sebasic acid (Type 6/10 nylon). They can also be made by the polymerization of amino acids or their derivatives, for example, polycaprolactam (Type 6 nylon) and polymerized 11-aminoun-decanoic acid (Type 11 nylon).

Types

Nylon 6/6 is the most widely used of the nylon plastics because of its overall balance of properties. The second most widely used of the nylon family is nylon 6. Type 6/6 nylon resins have higher heat resistance, abrasion resistance, strength, stiffness, and hardness than type 6 nylons. The type 6 nylon resins are tougher and more flexible than type 6/6 nylons, and they have a wider processing window.

Nylon 6/12 absorbs less moisture and, therefore, maintains both mechanical and electrical properties better in high-humidity environments. But the reduced moisture sensitivity is accompanied by lower strength, lower stiffness, lower use temperatures, and higher cost.

Nylons 11 and 12 have lower moisture absorption combined with superior resistance to fuels, hydraulic oils, and most automotive fluids. The melting points of nylon 11 and 12 (180 to 185°C) are the lowest of the commercial polyamides. These two polyamides are often combined with plasticizers to generate a flexible, tough material suitable for tubing extrusion. Recently, nylon 12/12 was introduced with a slightly higher use temperature while maintaining good fuel resistance. Nylon 6/6T resins have low moisture absorption and they are much stronger, stiffer, tougher, fatigue resistant, and more heat resistant than type 6/6 nylons. The Ultramid type 6/6T resins also have better resistance to hot oils and fats than type 6/6 nylons. Reinforced grades of type 6/6T nylon resins also are available.

Nylon 4/6 is the latest version of the short-repeat-unit polyamides. Its melting point of 295°C is 12.2°C above that for nylon 6/6 and is the highest in the polyamide family. The inherent molecular symmetry of nylon 4/6 results in self-nucleation, rapid crystal growth, and, thus, a higher level of crystallinity. This higher level of crystallinity leads to faster setup and, hence, faster injection-molding cycles, up to 30% faster than for 6/6. Nylon 4/6 absorbs more water than nylon; however, its dimensional stability is similar to nylon 6/6 due to its high crystallinity.

Higher crystallinity has a major effect on nearly all properties leading to higher strength, higher stiffness, high heat-deflection temperature (HDT), high fatigue resistance, high wear resistance, and high creep resistance. Semicrystalline polymers maintain useful properties above the glass transition in contrast to amorphous polymers, which transform into a viscous mass. Nylon 4/6, with its unusually high crystallinity, maintains a higher level of performance at elevated temperatures. The HDT for reinforced nylon 4/6 is 545°C.

Nylon resin is available in a wide range of reinforcement levels, filler types, toughening agents, stabilizers, and flame-retardant additives. Newer flame retardants can provide good flammability ratings (UL 94V-0) while maintaining acceptable electrical properties.

Toughening technology has reduced notch sensitivity providing notched Izod values over 15 ft-lb/in.; see Table N.6.

Properties

Property comparisons among commercial grades of nylon vary widely because so many formulations are available. In general, however, nylons have excellent fatigue resistance, low coefficient of friction, good toughness (depending on degree of crystallinity), and they resist a wide spectrum of fuels, oils, and chemicals. They are inert to biological attack, and have adequate electrical properties for most voltages and frequencies.

The crystalline structure of nylons, which can be controlled to some degree in processing, affects stiffness, strength, and heat resistance. Low crystallinity imparts greater toughness, elongation, and impact resistance, but at the sacrifice of tensile strength and stiffness.

Nylons 6/6, 6/6T, and 4/6 have the lowest permeability of the nylons by gasoline, mineral oil, and fluorocarbon refrigerants. Nylon 6/12

TABLE N.6
Properties of Nylons (dry as molded)

ASTM or UL Test	Property	Type 4/6	6/6	6	6/12	11	Cast 6
	Physical						
D79	Specific gravity	1.13	1.14	1.14	1.07	1.04	1.15
	Specific volume (in.^{3}/lb)	23.5	24.3	24.3	25.9	26.6	24.1
D570	Water absorption, 24 h, 1/8 in. thick (%)	2.3	1.5	1.6	0.4	0.4	1.6
	Mechanical						
D638	Tensile strength (psi)	14000	12000	11500	8800	8600	11000
D638	Elongation (%)	30	60	100	150	300	15–50
D790	Flexural modulus (Kpsi)	460	440	420	150	150	400
	Thermal						
D2117	Melting point (crystalline) (°F)	663	509	428	419	374	419
D696	Coefficient of thermal expansion (1E-05 in./in./F)	4.2	4.4	4.5	5.0	5.1	5.0
D648	Deflection temperature (°F), at 264 psi	240	190	152	150	180	140
UL94	Flammability rating	V-2	V-2	V-2	-2	—	HB
	Electrical						
D150	Dielectric constant, 75°F, at 1 kHz	4.0	3.9	3.8	4.0	3.7	4.0
D257	Volume resistivity (Ω-cm) at 73°F, DAM	1E + 15	1E + 16	1E + 15	1E + 12	1E + 13	—

Source: Mach. Design Basics Eng. Design, June, p. 708, 1993. With permission.

and 6/6T are used where lower moisture absorption (and better dimensional stability) is needed.

All nylons absorb moisture from the environment; however, type 6/6T nylon has much lower moisture absorption than any other type of nylon resin. Moisture absorption leads to dimensional and property changes dependent upon the equilibrium level absorbed. At elevated temperatures, the moisture equilibrium level decreased above 88°C, nylon begins to dry out by a combination of internal diffusion and surface volatile emission. Extended exposure to temperatures above 121°C will reduce moisture content to about 0.1% (similar to dry-as-molded content).

Nylons are sensitive to ultraviolet radiation. Weatherability will be reduced unless ultraviolet stabilizers are incorporated into the formulation. Carbon black is the most commonly used ultraviolet stabilizer. Carbon black lowers the ductility and toughness as trade-off for ultraviolet stability.

Nylons have good resistance to creep and cold flow compared with many less rigid thermoplastics. Creep resistance is better at higher levels of crystallinity as demonstrated in nylon 4/6. Creep can be calculated from long-term apparent modulus under load data.

Processing

Nylons are generally fabricated by injection molding or by extrusion. Precise, intricate shapes of a variety of colors can be molded with little or no finishing required. These can often replace an assembly of several metal parts. Thus, even where nylons cost more on a per-volume basis than the common die-casting metals, economies in finishing and assembly often result in lower ultimate costs.

Tubing and rod stock manufacture, plus the coating of wire and cable, are the major forms of nylon extrusion. Film and relatively complex cross sections are also made, but in less volume. In general, tubing, film, and other unsupported shapes require higher melt viscosity than is desirable for injection molding. Most manufacturers of nylon supply these high-viscosity grades.

APPLICATIONS

Nylons are usually specified because of their combination of properties. Gears, bearings, cams, clutch facings, and similar mechanical parts require their strength, stiffness, low coefficient of friction, and resistance to fatigue and abrasion. In cases where oiling or greasing is apt to be neglected, as in home appliances, or is undesirable from a contamination standpoint, as in textile and food-handling machinery, nylon parts usually perform satisfactorily without any lubrication whatsoever.

Designers often utilize the mechanical properties of nylons, plus one or more characteristics of particular value. For example, the nylon housing for an electric drill must be tough, stiff, dimensionally stable, and resistant to commonly encountered lubricants and solvents. However, its electrical nonconductivity and safety are the critical advantages.

Similarly, one manufacturer makes an entire rifle stock, plus many of the moving parts of the rifle, out of nylon. It is far lighter and tougher than wood and provides moving surfaces that do not need lubrication. The ability of nylon to be molded into precise sections thus permits custom-quality guns to be mass-produced.

By utilizing different properties, marine electrical stuffing tubes of nylon capitalize on the durability, lightness, resistance to corrosion, and cost advantage of the resin over machined brass. Washing machine mixing valves and valves for the dispensing of hot beverages require the mechanical properties of nylon and its excellent resistance to the effects of hot water.

The coating of wire and cable construction is the most important extrusion application of nylon. Although it is most commonly used as a jacket over a primary insulator such as polyvinyl chloride or polyethylene to impart resistance to lubricants, abrasion, and to the effects of high temperature, its electrical properties are adequate for low-voltage uses.

Exercise bikes used in health clubs undergo about 10 h of use per day. Further, to be competitive in today's hot exercise-equipment marketplace one must design bikes to last from 5 to 7 years. The jump on the competition is a newly designed thermoplastic composite pillow block, the structure that supports the drive shaft of the bike. The material selected was a long-glass-fiber-reinforced nylon 6/6 structural composite. When compared to conventional short-fiber-reinforced thermoplastics, the nylon 6/6 structural composites have enhanced mechanical properties, while remaining lightweight. As an added bonus, this combination of properties reduces the weight of the pillow block from 3.68 to 0.69 g.

A nylon foot brace for kayakers has become the standard of excellence. The brace consists of a rail screwed into the bottom of the boat, and a foot pad that slides up and down the rail to fit the kayaker's leg. Today, most kayak makers feature the brace as standard equipment. It consists of glass-reinforced, nylon 6, injection-molded resins.

The nylon combines high-strength, stiffness, and heat-deflection characteristics, while extending the retention of these properties at high temperatures, and a lower cost than competitive materials.

A new sports car features a lightweight, high-performance nylon air-intake manifold with a molded-in fuel rail. The automaker claims the integrated air-fuel module with two fuel passages into the body of the intake manifold breaks the mold when it comes to under-the-hood components. It uses a nylon manifold/fuel rail using the fusible-core, injection-molding (FCIM) technique. The manifold, with its molded-in fuel rail, weighs about 3.68 g. That is nearly 50% less than a similar design based on pressure-cast aluminum would weigh. The smooth inner walls of the injection-molded manifold increase airflow. This, combined with the low thermal conductivity of the nylon manifold, helps improve engine performance up to 5% more than that of an aluminum counterpart. And the nylon manifold insulates the air inside from engine heat, allowing high-density air to flow into the engine. The specially formulated 6/6 nylon resists hot engine temperatures and attacks from oil, gasoline, and battery vapors. It also produces a manifold that better withstands engine vibration stresses, while reducing engine noise.

As a wire insulation, nylon is valued for its toughness and solvent resistance. Nylon fibers are strong, tough, and elastic, and have high gloss. The finer fibers are easily spun into yarns for weaving or knitting either alone or in blends with other fibers, and they can be crimped and heat-set. For making carpets, nylon staple fiber, lofted or wrinkled, is used to give the carpet a bulky texture resembling wool. Tire cord, made from nylon 6 of high molecular weight, has the yarn drawn to four or five times its original length to orient the polymer and give one-half twist per inch. Nylon film is made in thicknesses down to 0.005 cm for heat-sealed wrapping, especially for food products where tight impermeable enclosures are needed. Nylon sheet, for gaskets and laminated facings, comes transparent or in colors in thicknesses from 0.013 to 0.152 cm. Nylon monofilament is used for brushes, surgical sutures, tennis strings, and fishing lines. Filament and fiber, when stretched, have a low specific gravity down to 1.068, and the tensile strength may be well above 344 MPa. Nylon fibers made by condensation with oxalic acid esters have high resistance to fatigue when wet.

OLEFIN COPOLYMERS

The principal olefin copolymers are the polyallomers, ionomers, and ethylene copolymers. The polyallomers, which are highly crystalline, can be formulated to provide high stiffness and medium impact strength; moderately high stiffness and high impact strength; or extra-high impact strength. Polyallomers, with their unusually high resistance to flexing fatigue, have "hinge" properties better than those of polypropylenes. They have the characteristic milky color of polyolefins, are softer than polypropylene, but have greater abrasion resistance. Commonly injection-molded, extruded, and thermoformed, polyallomers are used for such items as typewriter cases, snap clasps, threaded container closures, embossed luggage shells, and food containers.

Ionomers are nonrigid plastics characterized by low density, transparency, and toughness. Unlike polyethylenes, density and properties are not crystalline dependent. Their flexibility, resilience, and high molecular weight combine to provide high abrasion resistance. They have outstanding low-temperature flexural properties but upper temperature use is limited to 71°C. Resistance to attack from organic solvents and stress cracking chemicals is high. Ionomers have high melt strength for thermoforming and extrusion-coating, and a broad temperature range for blow molding and injection molding. Representative ionomer parts include injection-molded containers, housewares, tool handles, and closures; extruded film, sheet, electrical insulation, and tubing; blow-molded containers and packaging.

There are four commercial ethylene copolymers, of which ethylene vinyl acetate (EVA) and ethylene ethyl acrylate (EEA) are the most common; Table O.1 shows typical properties of some olefin copolymers.

EVA copolymers approach elastomers in flexibility and softness, although they are processed like other thermoplastics. Many of their properties are density dependent, but in a different way from that of polyethylenes. Softening temperature and modulus of elasticity decrease as density increases, which is contrary to the behavior of polyethylene. Similarly, the transparency of EVA increases with density to a maximum that is higher than that of polyethylenes, which become opaque when density increases above around 0.935 g/cc. Although the electrical properties of EVA are not as good as those of low-density polyethylene, they are competitive with vinyl and elastomers normally used for electrical products. The major limitation of EVA plastics is their relatively low resistance to heat and solvents; the Vicat softening point is 64°C. EVA copolymers can be injection, blow, compression, transfer, and rotationally molded; they can also be extruded. Molded parts include appliance bumpers and a variety of seals, gaskets, and bushings. Extruded tubing is used in beverage vending machines and for hoses for air-operated tools and paint spray equipment.

EEA is similar to EVA in its density–property relationships. It is also generally similar to EVA in high-temperature resistance, and like EVA it is not resistant to aliphatic and aromatic hydrocarbons or their chlorinated versions. However, EEA is superior to EVA in environmental stress cracking and resistance to ultraviolet radiation. Similar to EVA, most of the applications of EEA are related to the outstanding flexibility and toughness of the plastic. Typical uses are household products such as trash cans, dishwasher trays, flexible hose and water pipe, and film packaging.

Two other ethylene copolymers are ethylene hexene (EH) and ethylene butene (EB). Compared with the other two, these copolymers

TABLE O.1
Typical Properties of Some Olefin Copolymers

	Polyallomer	Ionomer	EVA
Specific gravity	0.898–0.905	0.94	0.94
Tensile strength, 1000 psi	3–4.5	3–5	0.5–1.0
Elongation, %	350	450	650
Hardness, Phase D	—	60	35
Imp. strength, ft-lb/in. notch	1.5	9–14	—
Softening point, Vicat, F	250–275	162	147
Dielectric strength, V/mil	500–650	1000	525

have greater high-temperature resistance; their useful service range being between 66 and 88°C. They are also stronger and stiffer and, therefore, less flexible than EVA and EEA. In general, EH and EB are more resistant to chemicals and solvents than the other two, but their resistance to environmental stress cracking is not as good.

OPTICAL FIBERS

These are flexible transparent fiber devices, sometimes called light guides, used for either image or information transmission, in which light is propagated by total internal reflection. In its simplest form, the optical fiber or lightguide consists of a core of material with a refractive index higher than the surrounding cladding. The optical fiber properties and requirements for image transfer, in which information is continuously transmitted over relatively short distances, are quite different from those for information transmission, where typically digital encoding of information into on–off pulses of light (on = 1; off = 0) is used to transmit audio, video, or data over much longer distances at high bit rates. Another application for optical fibers is in sensors, where a change in light transmission properties is used to sense or detect a change in some property, such as temperature, pressure, or magnetic field.

Fiber Designs

There are three basic types of optical fibers (Figure O.1). Propagation in these lightguides is most easily understood by ray optics, although the wave or modal description must be used for an exact description. In a multimode, stepped-refractive-index-profile fiber (part a of the figure), the number of rays or modes of light that are guided, and thus the amount of light power coupled into the lightguide, is determined by the core size and the core-cladding refractive index difference. Such fibers, used for conventional image transfer, are limited to short distances for information transmission due to pulse broadening. An initially sharp pulse made up of many modes broadens as it travels long distances in the fiber, because high-angle modes have a longer distance to travel relative to the low-angle modes. This limits the bit rate and distance because it determines how closely input pulses can be spaced without overlap at the output end. At the detector, the presence or absence of a pulse of light in a given time slot determines whether this bit of information is a zero or one.

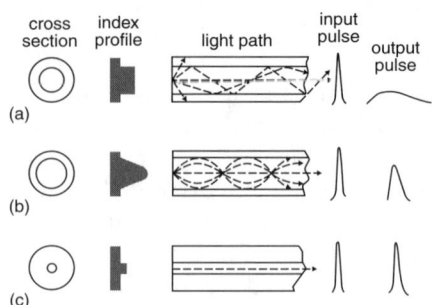

FIGURE O.1 Types of optical fiber designs. (a) Multimode, stepped-refractive-index profile. (b) Multimode, graded-index profile. (c) Single-mode, stepped-index. Graded index is possible. (From *McGraw-Hill Encyclopedia of Science and Technology*, 8th ed., Vol. 12, 431. With permission.)

A graded-index multimode fiber (part b), where the core refractive index varies across the core diameter, is used to minimize pulse broadening due to intermodal dispersion. Since light travels more slowly in the high-index region of the fiber relative to the low-index region, significant equalization of the transit time for the various modes can be achieved to reduce pulse broadening. This type of fiber is suitable for intermediate-distance, intermediate-bit-rate transmission systems. For both fiber types, light from a laser or light-emitting diode can be effectively coupled into the fiber.

A single-mode fiber (part c) is designed with a core diameter and refractive index distribution such that only one fundamental mode is guided, thus eliminating intermodal pulse-broadening effects. Material and waveguide dispersion effects cause some pulse broadening, which increases with the spectral width of the light source. These fibers are best suited for use with a laser source to couple light efficiently into the small core of the lightguide and to enable information transmission over long distances at very high bit rates. The specific fiber design and the ability to manufacture it with controlled refractive index and dimensions determine its ultimate bandwidth or information-carrying capacity.

Attenuation

The attenuation or loss of light intensity is an important property of the lightguide because it limits the achievable transmission distance, and is caused by light absorption and scattering. Every material has some fundamental absorption due to the atoms or molecules composing it. In addition, the presence of other elements as impurities can cause strong absorption of light at specific wavelengths. Fluctuations in a material on a molecular scale cause intrinsic Rayleigh scattering of light. In actual fiber devices, fiber-core-diameter variations or the presence of defects such as bubbles can cause additional scattering light loss.

Optical fibers based on silica glass have an intrinsic transmission window at near-infrared wavelengths with extremely low losses. Such fibers are used with solid-state lasers and light-emitting diodes for information transmission, especially for long distance (greater than 1 km). Plastic fibers exhibit much higher intrinsic as well as total losses, and are more commonly used for image transmission, illumination, or very short-distance data links.

Many other fiber properties are also important and their specification and control are dictated by the particular application. Good mechanical properties are essential for handling: plastic fibers are ductile, whereas glass fibers, intrinsically brittle, are coated with a protective plastic to preserve their strength. Glass fibers have much better chemical durability and can operate at higher temperatures than plastics.

ORGANIC COATINGS

Organic coatings are chiefly additive type finishes that find use on almost all types of materials. They can be monolithic consisting simply of one layer, or coat, or they can be composed of two or more layers. The total thickness of coating systems varies widely. Some run less than 1 mil thick. Others go as high as 10 and 15 mils thick. Generally, by definition, coatings that are more than 10 mils thick are referred to as linings, films, or mastics.

To function as a protective barrier against corrosion and oxidation, organic coatings depend principally on their chemical inertness and impermeability. In addition, however, some coatings provide protection with the use of inhibiting pigments that have a passivating action, particularly on metal surfaces. Also, some coatings contain metallic pigments that give electrochemical protection to metals.

Coating Application and Drying

Organic coatings are commonly applied by the following methods: brush, spray, dip, roller, flow coat, knife, tumbling, silk screen, and electrostatic means. Of these, application by brushing is the slowest; all the others are production methods.

Organic coatings dry, or cure, by one or more of the following mechanisms: (1) evaporation or loss of solvent, (2) oxidation, and (3) polymerization. After application of the coating, the volatile ingredients, which are almost always present in at least a small amount, evaporate. Some finishes, such as lacquers, dry completely by evaporation of solvents. After evaporation, coatings that do dry solely by evaporation are still in a semifluid state, and depend on oxidation or polymerization, or a combination of both, to convert to their final form.

Drying by oxidation, which is usually done at room temperature, is the slowest of the three methods. Polymerization, which involves polymer chain forming mechanism, can be done at normal or at elevated temperatures. Polymerization is speeded by use of heat. Radiation curing involving the use of an electron beam to polymerize the coating in a few seconds has found production usage.

COATING TYPES AND SYSTEMS

Coating Composition

An organic coating is made up of two principal components: a vehicle and a pigment. The vehicle is always there. It contains the film-forming ingredients that enable the coating to convert from a mobile liquid to a solid film. It also acts as a carrier and suspending agent for the pigment. Pigments, which may or may not be present, are the coloring agents and, in addition, contribute a number of other important properties.

Organic coatings are commonly divided into about a half dozen broad categories based on the types and combinations of vehicle and pigment used in their formulation. They are paints, enamels, varnishes, lacquers, dispersion coatings, emulsion coatings, and latex coatings. However, with the complexity in modern formulations, distinctions between these various types are often difficult to make.

As mentioned earlier, organic finish systems are frequently composed of more than one layer or coat. These various layers are commonly classified as primers, intermediate coats, and finish coats.

Primers

These are the first coatings placed on the surface (except for fillers, in some cases). Where chemical pretreatments are used, primer coats may often be unnecessary. Primers for industrial or production finishing are of two types: air-dry types and baking-types. The air-dry types have drying oil vehicle bases and are usually referred to as paints. They may or may not be modified with resins. They are not used as extensively as the baking-type primers, which have resin or varnish vehicle bases and dry chiefly by polymerization. Some primers, known as flash primers, are applied by spraying, and dry by solvent evaporation within 10 min. In practically all primers, the pigments impart most of the anticorrosion properties to the primer and, along with the vehicle, determine its compatibility and adherence with the base metal.

Intermediate Coats

These are fillers, surfacers, and sealers. They can be applied either before or after the primer, but more often after the primer and sometimes after the surfacer coat. Their function is to fill in large irregularities in the surface or local imperfections. They are usually puttylike substances, and a variety of materials are used. Their chief characteristics are (1) must harden with a minimum of shrinkage; (2) must have good adhesion; (3) must have good sanding properties; and (4) must work smoothly and easily.

Surfacers are often similar to primers; they usually have the same composition as the priming coat, except that more pigment is present. Surfacers are applied over the priming coat to cover all minor irregularities in the surface. Sealers as a rule are used either over the fillers or surfacers. The chief function of sealers is to fill up the pores of the undercoat to avoid "striking in" of the finish coats. This filling-in of the porous surfacer or sealer also tends to strengthen the entire coating system. The sealers when used over surfacers are usually formulated with the same type pigment and vehicle as used in the final coat.

Finish Coats

Finish or top coats are usually the decorative or functional part of a paint system. However, they often also have a protective function. The primer coats may require protection against the service conditions, because although the pigments used in primers are satisfactory for corrosion protection of the metal, they are frequently not satisfactory as top coats. Their color retention upon weathering or their physical durability may be poor. There are also one-coat applications where the finish coats are applied directly to the base material surface, and, therefore, provide the sole protective medium.

Vehicles

Vehicles are composed of film-forming materials and various other ingredients, including thinners (volatile solvents), which control viscosity, flow, and film thickness, and driers, which facilitate application and improve drying qualities.

The main concern is chiefly the film-forming part of the vehicle, because it is that part of the vehicle that to a large extent determines the quality and character of an organic finish. It determines the possible ways in which the finish can be applied and how the "wet" finish will dry to a hard film; it provides for adhesion to the metal surface; and it usually influences the durability of the finish.

Vehicles can be divided into three main types: (1) oil, (2) resin, and (3) varnish. The simplest and among the oldest vehicles are the straight drying oil types. Resins, as a class, can serve as vehicles in their own right, or can be used with drying oils to make varnish-type vehicles. Varnish vehicles are composed of resins and either drying or nondrying oils, together with required amounts of thinners and driers. They are often used alone as a full-fledged organic finish.

Drying Oils

Vehicles consisting of oil only are used to a limited extent in industrial finishes. Linseed oil is probably the most widely used oil. There are a number of different kinds that differ in rate of drying, and in such properties as water resistance, color, and hardness.

Tung oil or China wood oil, when property treated, excels all other drying oils in speed of drying, hardening, and water resistance. Oiticica oil is similar to tung oil in many of its properties. Dehydrated castor oil dries better than linseed oil, but slower than tung oil. Some of its advantages are good color and color retention, and flexibility. The oils from some fish are also used as drying oils. If processed properly, they dry reasonably well and have little odor. They are often used in combination with other oils. Perilla oil is quite similar in properties to fast-drying linseed oil. Its use is largely dependent upon its price and availability. Soybean oil is the slowest drying in the drying oil classes, and is usually used in combination with some faster-drying oil such as linseed oil.

Resins

Although both natural and synthetic resins can serve as organic coating vehicles, today the natural types, such as rosin, have been largely replaced by plastic resins. Nearly all the plastic resins — both thermosets and thermoplastics as well as many elastomets — can be used as film formers, and frequently two or more kinds are combined to give the set of properties desired. Typical thermoplastics used in vehicles are acrylics, acetates, butyrates, and vinyls. Commonly used thermosets for vehicles include phenolics, alkyds, melamines, ureas, and epoxies.

Pigments

Pigments are the second of the two principal components that make up most organic finishes. They contribute a number of important characteristics to a coating. They, first of all, serve a decorative function. The choice of color and shades of color by use of one or combinations of pigments is practically unlimited. Closely associated with color is the hiding power function or their ability to obscure the surface of the material being finished. In many primers the principal function of pigments is to prevent corrosion of the base metal. In other cases they may be added to counteract the destructive action of ultraviolet light rays. Pigments also help give body and good flow characteristics to the finish. And, finally, some pigments may give to organic coatings what is termed package stability — that is, they keep the coating material in usable condition in the container until ready for use.

Pigments can be conveniently divided into three classes as follows: (1) white hiding pigments, (2) colored pigments, and (3) extender or inert pigments. White pigments are used not only in white paints and enamels, but also in making white bases for the tinted and light shades. Colored pigments furnish the finish with both opacity and color. They may be used by themselves to form solid colors, or in combination with whites to produce tints, and often provide rust-inhibitive properties. For example, red lead, certain lead chromates, zinc chromates, and blue lead are used in iron and steel primers as rust inhibitors. There are two general classes of colored pigments — earth colors, which are very stable and are not readily affected by acids and alkalies, heat, light, and moisture, and chemical colors, which are produced under controlled conditions by chemical reaction. The metallic pigments can also be included within this class. Aluminum powder is perhaps the best known.

The chief functions of extender pigments are to help control consistency, gloss, smoothness, and filling qualities, and leveling and check resistance. Thus, particle size and shape, oil absorption, and flatting power are important selection considerations. Extender pigments are for the most part chemically inactive. They usually have little or no hiding power.

Enamels

By definition, enamels are an intimate dispersion of pigments in a varnish or a resin vehicle, or in a combination of both. Enamels may dry by oxidation at room temperatures and/or by polymerization at room or elevated temperatures. They vary widely in composition, in color and appearance, and in properties, and are available in all colors and shades. Although they generally give a high-gloss finish, there are some that give a semigloss or eggshell finish and still others that give a flat finish. Enamels as a class are hard and tough and offer good mar and abrasion resistance. They can be formulated to resist attack of most commonly encountered chemical agents and corrosive atmospheres.

Because of their wide range of useful properties, enamels are probably the most widely used organic coating in industry. One of their largest fields of use is for coating household appliances — washing machines, stoves, kitchen cabinets, and the like. A large portion of refrigerators, for example, are finished with synthetic baking enamels. These appliance enamels are usually white, and therefore must have a high degree of color and gloss retention when subjected to light and heat. Other products finished with enamels include automotive products, railway equipment, office equipment, toys and sport supplies, industrial equipment, and novelties.

Lacquers

The word *lacquer* comes from lac resin, which is the base of common shellac. Lac resin dissolved in alcohol was one of the first lacquers and has been in use for many centuries. Nowadays, shellac is called spirit lacquer. It is only one of several different kinds of lacquers; these, except for spirit lacquer, are named after the chief film-forming ingredient. The most common ones are cellulose acetate, cellulose acetate butyrate, ethyl cellulose, vinyl, and nitrocellulose.

A distinguishing characteristic of lacquers is that they dry by evaporation of the solvents or thinners in which the vehicle is dissolved. This is in contrast to oils, varnishes, or resin base finishes, which are converted to a hard film chiefly through oxidation or polymerization.

Because many modern lacquers have high resin content, the gap between lacquer and synthetic-type varnishes diminishes until finally one has modified synthetic air-drying varnishes. They may dry chiefly by oxidation or polymerization.

Lacquers normally dry hard and dust-free in a very few minutes at room temperature. In production line work, forced drying is often used. It is possible, therefore, to do a multicoat job without having to lose time between coats. Because of the speed of drying and the fact that they are permanently soluble in the solvents used for application, lacquers usually are not applied by brush. Spray application or dipping are the usual procedures.

Lacquers can be either clear and transparent or pigmented, and their color range is practically unlimited. Lacquers in themselves have good color retention, but sometimes the added pigments, modifying resins, and plasticizers may adversely affect this property. They are hard and mar resistant. Inherently, they lack good adhesion to metal, but modern lacquer formulations have greatly improved their adhesion properties. Lacquers can be made to be resistant to a large variety of chemicals, including water and moisture, alcohol, gasoline, vegetable, animal and mineral oils, and mild acids and alkalies. Because of the volatile solvents, lacquers are inflammable in storage and during application, and this sometimes limits their application.

Because of their fast drying speeds, lacquers find wide application in the protection and decoration of products that can be dipped, sprayed, roller-coated, or flow-coated. They are especially advantageous for coating metal hardware and fixtures, toys, and other articles that, because of volume production, must dry hard enough to handle and pack in a short period of time. Lacquers are widely used in automobile finishing and especially for refinishing autos and commercial vehicles where fast drying without baking equipment is a requirement. Lacquers also compete with enamels for coating metal stampings and castings, including die castings.

Varnishes

Varnishes consist of thermosetting resins and either drying or nondrying oils. They are clear and unpigmented and can be used alone as a coating. However, their major use in industrial finishing is as a vehicle to which pigments are added, thus formng other types of organic coatings.

The drying mechanisms of varnishes all follow the same general pattern. First, any volatile solvents that are present evaporate; then, drying by oxidation and/or polymerization takes place, depending on the nature of the resin and oil. At high temperatures, of course, there is more tendency to polymerize. So varnishes can be formulated for either air or bake drying. Varnishes may be applied by brushing or by any of the production methods.

It is evident that with the large variety of raw materials to choose from and the unlimited number of combinations possible that varnishes have an extensive range of properties and characteristics. They range from almost clear white to a deep gold; they are transparent, lacking any appreciable amount of opacity. Japan, a hard-baked black-looking varnish, is an exception. It is opaque, due to carbon and carbonaceous material being present.

There are some distinctions in properties between oil-modified alkyd varnishes and the other types. In general, oil-modified alkyds have better gloss and color retention and better resistance to weathering. They form a harder, tougher, more durable film and dry faster. On the other hand, they have less alkali resistance than the other varnishes. In such things as adhesion and rust inhibitiveness there is no distinctive difference.

The major use of varnishes, as coatings in their own right, is for food containers, closures such as bottle caps, and bandings of various kinds. Another large application is as a clear finish coat over lithographic coatings.

Paints

The word *paints* is sometimes used broadly to refer to all types of organic coatings. However, by definition a paint is a dispersion of a pigment or pigments in drying oil vehicle. They find little use these days as industrial finishes. Their principal use is for primers. Paints dry by oxidation at room temperature. Compared to enamels and lacquers their drying rate is slow; they are relatively soft and tend to chalk with age.

Other Organic Coatings

Dispersion and Emulsion Coatings

In recent years these coatings have become known as water-base paints or coatings because many of them consist essentially of finely divided ingredients, including plastic resins, fillers, and pigments, suspended in water. An organic media may also be involved. There are three types of water-base coatings. Emulsions, or latexes, are aqueous dispersions of high-molecular-weight resins. Strictly speaking, latex coatings are dispersions of resins in water, whereas emulsion coatings are suspensions of an oil phase in water.

Emulsion and latex coatings are clear to milky in appearance, have low gloss, excellent resistance to weathering, and good impact resistance. Chemical and stain resistance varies with composition. Dispersion coatings consist of ultrafine fine, insoluble resin particles present as a colloidal dispersion in an aqueous medium. They are clear or nearly clear. Weathering properties, toughness, and gloss are roughly equal to those of conventional solvent paints.

Water-soluble types, which contain low-molecular-weight resins, are clear finishes and they can be formulated to have high gloss, fair to good chemical and weathering resistance, and high toughness. Of the three types, they handle and flow most like conventional solvent coatings.

Plastic Powder Coatings

Several different methods have been developed to apply plastic powder coatings. In the most popular process — fluidized bed — parts are preheated and then immersed in a tank of finely divided plastic powders, which are held in a suspended state by a rising current of air. When the powder particles contact the heated part, they fuse and adhere to the surface, forming a continuous, uniform coating.

Another process, electrostatic spraying, works on the principle that oppositely charged materials attract each other. Powder is fed through a gun, where an electrostatic charge is applied opposite that applied to the part to be coated. When the charged particles leave the gun, they are attracted to the part where they cling until fused together as a plastic coating. Other powder application methods include flock coating, flow coating, flame and plasma spraying, and a cloud chamber technique.

Although many different plastic powders can be applied by the above techniques, vinyl, epoxy, and nylon are most often used. Vinyl and epoxy provide good corrosion and weather resistance as well as good electrical insulation. Nylon is used chiefly for its outstanding wear and abrasion resistance. Other plastics frequently used in powder coating include chlorinated polyether, polycarbonate, acetal, cellulosics, acrylic, and fluorocarbons.

Hot Melt Coatings

These consist of thermoplastic materials that solidify on the metal surface from the molten state. The plastic either is applied in solid form and then melted and flowed over the surface or is applied molten by spraying or flow coating. Since no solvent is involved, thick single coats are possible. Bituminous coatings are also commonly applied by the hot melt process.

Lining and Sheeting

Sheet, film, and tapes of various plastics and elastomers cemented to material shapes and parts are used to provide corrosion and abrasion resistance. Thicknesses usually range from 3.2 to 12.7 mm. The most widely used materials are polyvinyl chloride, polyethylene butadiene–styrene rubber, and neoprene.

Specialty Finishes

An almost infinite number of specialty or novelty finishes are available. Most of them are really lacquers or enamels to which special ingredients have been added or which are processed in some unique way to give the effects desired.

One of the most common types are those giving a roughened or wrinkle appearance, which is obtained by use of high percentages of driers causing wrinkling when the finish is baked. Another group of specialty finishes gives a crystalline effect. They are enamels in which impurities are purposely introduced during the baking process by retaining the products of combustion in the oven while the coating dries. The wrinkle and crystalline finishes are widely used on instrument panels, office equipment, and a variety of other industrial and consumer products.

Other unusual finishes are obtained by adding special ingredients to lacquers to give them a stringy or "veiled" appearance when applied by spraying. The application of the silk-screen process to organic finishing of metals has also resulted in unique finishes with multicolored effects.

ORGANOMETALLIC COMPOUNDS

These are members of a broad class of compounds whose structures contain both carbon (C) and a metal (M). Although not a required characteristic of organometallic compounds, the nature of the formal carbon–metal bond can be of the covalent type. Figure O.2 depicts carbon–metal formal bonds.

The term *organometallic* chemistry is essentially synonymous with organotransition metal chemistry; it is associated with a specific portion of the periodic table ranging from groups 3 through 11, and also includes the lanthanides.

This particular area has experienced exceptional growth since the mid-1970s largely because of the continuous discovery of novel structures as elucidated mainly by x-ray crystallographic analyses; the importance of catalytic processes in the chemical industry; and the development of synthetic methods based on transition metals that focus on carbon–carbon bond constructions. From the perspective of inorganic chemistry, organometallics afford seemingly endless opportunities for structural variations due to changes in the metal coordination number, alterations in ligand–metal attachments, mixed-metal cluster formation, and so forth. From the viewpoint of organic chemistry, organometallics allow for manipulations in the functional groups that in unique ways often result in rapid and efficient elaborations of carbon frameworks for which no comparable direct pathway using nontransition organometallic compounds exists.

As one moves across the periodic table, the metals that have found application include:

- Titanium
- Zirconium
- Chromium
- Molybdenum
- Tungsten
- Ruthenium
- Osmium
- Cobalt
- Rhodium
- Palladium
- Copper

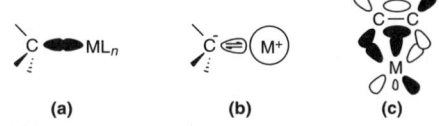

FIGURE O.2 Carbon–metal formal bonds. (From *McGraw-Hill Encyclopedia of Science and Technology*, 8th ed., Vol. 12, McGraw-Hill, New York, 574. With permission.)

Of interest to the ceramic industry are those organometallics, the hydroxide-free alkoxides, that can be used for the vapor-phase synthesis of hard ceramic oxide coatings, films, or freestanding bodies. Very fine particulate oxides also can be formed from these chemicals.

Compounds now available include aluminum isoproproxide, aluminum hexafluoroisoproproxide, lithium hexafluoroisoproproxide, sodium hexafluoroisoproproxide, zirconium hexafluoroisoproproxide, and zirconium tertiary amyloxide.

ORGANOSOL COATINGS

Organosol coatings are coatings in which the resin (usually polyvinyl chloride) is suspended rather than dissolved in an organic fluid. The dispersion technique permits the use of high-molecular-weight, relatively insoluble resins without the use of expensive solvents. In organosols, the fluid, or dispersant, consists of plasticizers together with a blend of inexpensive volatile diluents selected to give the desired fluidity, speed of fusion, and physical properties. The dispersant provides little or no solvating action on the resin particles until a critical temperature is reached at which point the resin is dissolved in the dispersant to form a single-phase solid solution. Since a portion of the liquid is made up of volatile diluents, the fusion process results in a proportional shrinkage. (Nonvolatile dispersions using only plasticizer dispersants are termed *plastisols*.)

Organosols are available with a wide range of flow characteristics and consequently may be formulated for application by any of the conventional techniques. Because they contain substantial quantities of volatile diluents, their thickness is limited to 10 to 12 mils per coat. Single-coat applications of greater thickness blister during bake as a result of trapped solvents.

Baking is generally accomplished in two stages. The volatiles are removed in the first stage at temperatures of from 93 to 121°C, but fusion does not occur until a temperature of 149 to 191°C is reached. The fusion stage accomplishes the union of the discrete vinyl particles into a single-phase solid. In addition to polyvinyl chloride, the organosol technique can be used with acrylonitrile-vinyl and polychlorotrifluoroethylene resins. A balance must be achieved in the baking operation between the removal of the volatiles and the solvation of the resins. Rapid heating results in solvent blistering, whereas the reverse causes a mud-cracking effect.

Application Methods

Organosol coatings can be applied by several methods.

Spread Coating

The bulk of the organosols is applied by spread-coating methods to fabrics and paper. Several plants are using spread coaters for the application of organosols and plastisols to strip steel to provide materials competitive with the light metals and plastics. Two basic processes are available: the knife coater and the roll coater. Knife coating is simple and fast but the product lacks the uniformity afforded by roller coating. Fusion is accomplished in a tunnel oven and embossing rolls may be employed at the oven exit to impart a texture or pattern to the hot gelled coating.

Strand Coating

Wires or filaments may be coated with organosols by first passing the strand through a dip tank, then through a wiping die to set the thickness. Fusion is accomplished in a drying tower. As many as nine to ten passes may be required to build a thickness of 20 mils.

Dip Coating

One of the major problems encountered in dip coating with organosols is the tendency of dried organosol to fall back into the dip tank and cause coating rejects. For this reason much of the dip coating is done with modified plastisols rather than organosols. A dipping formulation must have low viscosity with high yield value to prevent sags and drains. The rate at which the article is withdrawn from the dip tank is a determining factor in the thickness and quality of deposit. Withdrawal rates generally range between 1.7 to 5.5 mm/s. Special techniques such as inversion of the dipped article just prior to fusion may be employed to alleviate the drip problem when it is particularly troublesome.

Spray Coating

Organosols are readily handled in either suction or pressure spray equipment. For production-line spraying, pressure systems afford rapid delivery and are generally preferred. There is an increasing tendency to use electronic spray processes for handling organosols. A number of electrostatic processes are available including several "hand guns."

Properties

Organosols have the characteristic vinyl properties of toughness and moisture resistance. However, although the coatings possess good electrical resistance and are frequently used as secondary insulation to reduce shock hazards, they seldom meet the needs of primary wire insulation. Adhesive primers are required to bond the materials to metals and other dense substrates. Prolonged exposure to temperatures greater than 93°C causes thermal degradation, which is generally evidenced by a gradual darkening.

Uses

Vinyl organosols have found their widest usage in the coating of paper and fabric stock where their ease of handling has permitted the use of simplified, low-cost application equipment. Where fabric coating strike-through is to be avoided, as in open weave, thixotropic plastisols are employed rather than organosols. Closer weaves require some penetration.

Textured finishes for metals are a growing use and offer competition to the vinyl laminates and other decorative finishes.

Typical applications and related properties of organosols are shown in Table O.2.

TABLE O.2
Applications of Organosol Coatings

Application	Related Properties
Automotive interiors: station-wagon flooring, roof liners, dashboards, cowling, kick plates	Ease of application; uniformity of color and texture; resistance to gasoline, grease, and polishes; resistance to impact damage
Commercial vehicles: seat backs, trim, interior paneling, luggage racks, sill plates	Durability: resistance to abrasion and scuffing; cleanability; resistance to moisture and detergents
Appliance finishes: television and radio cabinets, slide projectors, refrigerator panels	Aesthetic qualities: novelty of appearance; durability: resistance to abrasion and scratching; resistance to moisture and staining
Business machines: typewriters, calculators, electronic computers, laboratory instruments	Durability: resistance to abrasion, scuffing, and chipping; resistance to chemicals and perspiration; sound-deadening qualities
Architectural applications: paneling, partitions, shower stalls, elevator doors, bathroom wall sections	Cleanability: resistance to moisture and detergents; aesthetic qualities: durability: resistance to abrasion and scuffing; sound-deadening qualities
Office furniture: desks, file cabinets, showcases, counters, waste baskets, chair finishes	Durability: resistance to abrasion, impact, and scuffing; resistance to moisture and staining
Luggage	Durability: toughness, resistance to abrasion and scuffing; aesthetic qualities
Paper and fabrics: floor and wall covering, place mats, bottle-cap liners, containers for food packaging, bandage dressings, upholstery fabrics, safety clothing, glove coatings	Ease of application: economy and simplicity of application equipment; resistance to abrasion and tearing; resistance to moisture and staining
Glass coatings: perfume bottles, bleach and chemical reagent bottles, photo flash bulbs	Resiliency, feel, cohesion strength; resistance to alcohol, moisture, and chemicals; ease of application

OSMIUM

A platinum-group metal, symbol Os, osmium is noted for its high hardness, about 400 Brinell. The heaviest known metal, it has a high specific gravity, 22.65, and a high melting point, 2698°C. The boiling point is about 5468°C. Osmium has a close-packed hexagonal crystal structure, and forms solid-solution alloys with platinum, having more than double the hardening power of iridium in platinum. However, it is seldom used to replace iridium as a hardener except for fountain-pen tips where the alloy is called osmiridium.

Osmium is not affected by the common acids, and is not dissolved by aqua regia. It is practically unworkable, and its chief use is as a catalyst.

Uses

Osmium tetraoxide, a commercially available yellow solid (melting point 40°C) is used commercially as a stain for tissue in microscopy. It is poisonous and attacks the eyes. Osmium metal is catalytically active, but it is not commonly used for this purpose because of its high price. Osmium and its alloys are hard and resistant to corrosion and wear (particularly to rubbing wear). Alloyed with other platinum metals, osmium has been used in needles for record players, fountain-pen tips, and mechanical parts.

OSPREY SPRAYFORMING

Copper alloy semifinished products made by the Osprey sprayforming process feature the ability to combine properties such as high strength, good corrosion resistance, and excellent machinability.

In the Osprey process, powder particles are liquefied and sprayed by an inert gas onto a substrate, where they are shaped into a billet. Because the droplets solidify very rapidly, the microstructure is much finer than that of conventionally cast material.

For example, the high yield strength and low elastic modulus of Cu–14% Sn alloy (with a small lead addition) make this alloy (BO5) ideal for spring applications. Its elastic modulus is 80 to 90 GPa, yield strength is 800 to 950 MPa, and ultimate tensile strength is 900 to 1000 MPa, with hardness of 240 to 280 HV.

For the fabrication of contact elements by stamping and bending, alloy Cu–15% Ni–8% Sn (CN8) would be a good choice. It has elastic modulus of 110 to 120 GPa, yield strength of 800 to 1300 MPa, ultimate tensile strength of 900 to 1400 MPa, and hardness of 240 to 400 HV.

OXIDE CERAMICS

Oxide ceramics can be divided into two groups — single oxides that contain one metallic element, and mixed or complex oxides that contain two or more elements. Examples include alumina, beryllia, magnesia, zirconia, and thoria. As a class they are low in cost compared to other technical ceramics, except for thoria and beryllia. Each of them can be produced in a variety of compositions, porosity, and microstructure, to meet specific property requirements.

Oxide ceramic parts are produced by slip casting or pressing or extrusion and then fired at about 1800°C. They are more difficult to fabricate than other types of ceramics, because of the usual requirement to obtain a high-density body with minimum distortion and dimensional error, except in the case of porous bodies for use as thermal insulation. Powder pressing produces bodies with the lowest porosity and highest strength, because of the high pressures and the small amount of binder required.

SINGLE OXIDES

Aluminum Oxide (Alumina)

Alumina is the most widely used oxide, chiefly because it is plentiful, relatively low in cost, and equal to or better than most oxides in mechanical properties. Density can be varied over a wide range, as can purity — down to about 90% alumina — to meet specific application requirements. Alumina ceramics are the hardest, strongest, and stiffest of the oxides. They are also outstanding in electrical resistivity, dielectric strength, are resistant to a wide variety of chemicals, and are unaffected by air, water vapor, and sulfurous atmospheres. However, with a melting point of only 2039°C, they are relatively low in refractoriness, and at 1371°C retain only about 10% of room-temperature strength. In addition to its wide use as electrical insulators and its chemical and aerospace applications, the high hardness and close dimensional tolerance capability of alumina make this ceramic suitable for such abrasion-resistant parts as textile guides, pump plungers, chute linings, discharge orifices, dies, and bearings.

Beryllium Oxide (Beryllia)

Beryllia is noted for its high thermal conductivity, which is about ten times that of a dense alumina (at 499°C), three times that of steel, and second only to that of the high-conductivity metals (silver, gold, and copper). It also has high strength and good dielectric properties. However, beryllia is costly and is difficult to work with. Above 1649°C it reacts with water to form a volatile hydroxide. Also, because beryllia dust and particles are toxic, special handling precautions are required. The combination of strength, rigidity, and dimensional stability make beryllia suitable for use in gyroscopes; and because of high thermal conductivity, it is widely used for transistors, resistors, and substrate cooling in electronic equipment.

Magnesium Oxide (Magnesia)

Magnesia is not as widely useful as alumina and beryllia. It is not as strong, and because of high thermal expansion, it is susceptible to thermal shock. Although it has better high-temperature oxidation resistance than alumina, it is less stable in contact with most metals at temperatures above 1705°C in reducing atmospheres or in a vacuum.

Zirconium Oxide (Zirconia)

There are several types of zirconia: a pure (monoclinic) oxide and a stabilized (cubic) form, and a number of variations such as yttria- and magnesia-stabilized zirconia and nuclear grades. Stabilized zirconia has a high melting point, about 2760°C, low thermal conductivity, and is generally unaffected by oxidizing and reducing atmospheres and most chemicals. Yttria- and magnesia-stabilized zirconias are widely used for equipment and vessels in contact with liquid metals. Monoclinic nuclear zirconia is used for nuclear fuel elements, reactor hardware, and related applications where high purity (99.7%) is needed. Zirconia has the distinction of being an electrical insulator at low temperatures, gradually becoming a conductor as temperatures increase.

Thorium Oxide (Thoria)

Thoria, the most chemically stable oxide ceramic, is only attacked by some earth alkali metals under some conditions. It has the highest melting point (3315°C) of the oxide ceramics. Like beryllia, it is costly. Also, it has high thermal expansion and poor thermal shock resistance.

MIXED OXIDES

Except for zircon the principal mixed oxides are composed of various combinations of magnesia, alumina, and silica.

Cordierite ($2MgO \cdot 2Al_2O_3 \cdot 5SiO_2$)

Cordierite is most widely used in extruded form for insulators in such parts as heating elements and thermocouples. It has low thermal expansion, excellent resistance to thermal shock, and good dielectric strength. There are three traditional groups of cordierite ceramics:

1. Porous bodies that have relatively little mechanical strength due to limited crystalline intergrowth and absence of ceramic bond. With long thermal endurance and low thermal expansion, they are used for radiant elements in furnaces, resistor tubes, and rheostat parts.

2. Low porosity bodies, developed principally for use as furnace refractory brick.
3. Vitrified bodies used for exposed electrical devices that are subjected to thermal variations.

Forsterite ($2MgO \cdot SiO_2$)

This mixed oxide has high thermal shock resistance, but good electrical properties and good mechanical strength. It is somewhat difficult to form and requires grinding to meet close tolerances.

Steatite

Steatites are noted for their excellent electrical properties and low cost. They are easily formed and fired at relatively low temperatures. However, compositions containing little or no clay or plastic material present fabricating problems because of a narrow firing range. Steatite parts are vacuum-tight, can be readily bonded to other materials, and can be glazed or ground to high-quality surfaces.

Zircon ($ZrO_2 \cdot SiO_2$)

This mixed oxide provides ceramics with strength, low thermal expansion, and relatively high thermal conductivity and thermal endurance. Its high thermal endurance is used to advantage in various porous-type ceramics.

OXIDE COATINGS

The black oxide finish on steel is one of the most widely used black or blue-black finishes. Some of the advantages of this type of finish are (1) attractive black color; (2) no dimensional changes; (3) corrosion resistance, depending on the final finish dip used; (4) nongalling surface; (5) no flaking, chipping, or peeling because the finish becomes an integral part of the metal surface; (6) lubricating qualities due to its ability to absorb and adsorb the final oil or wax dips; (7) ease and economy of application; and (8) nonelectrolytic solutions and a minimum of plain steel tank equipment required.

The black oxide finish produced on steel is composed essentially of the black oxide of iron (Fe_3O_4), and is considered by many to be a combination of FeO and Fe_2O_3. It can be produced by several methods: the browning process, carbonia process, heat treatment, and the aqueous alkali-nitrate process. Each of these processes produces a black oxide of iron finish, although the finish produced by each particular process differs in some characteristics. The chemical dip aqueous alkali-nitrate process is the most widely used to apply a black oxide finish on steel.

Aqueous Alkali-Nitrate Process

In this process a blackening solution is used that is highly alkaline and that also contains strong oxidizing chemicals. Refinements such as penetrants and rectifiers are also used to promote ease of operation, faster blackening, and trouble-free processing. At specific concentrations and boiling temperatures, these solutions will react with the iron in the steel to form the black oxide of iron (Fe_3O_4).

Because the reaction is directly with the iron in the steel, the finish becomes an integral part of the metal itself and, therefore, cannot flake, chip, or peel. For all practical purposes, there are also no dimensional changes. In transforming iron to black iron oxide, it is correct to assume that there is a change in volume. However, because of the highly alkaline nature of the blackening solution and because of the operating temperature, the blackening solution will dissolve a small amount of iron. Therefore, the amount of iron lost in this manner is compensated for by the buildup in volume from the change of iron to iron oxide — resulting in, for all intents and purposes, no dimensional changes. Extremely close measurements have shown that the actual change amounts to a buildup of only about 5 millionths of an inch.

There are types and conditions of steel and many different products that may require special procedures or additional steps.

It is important that only steel or steel alloys be immersed in the solution because other metals such as copper, zinc, cadmium, and aluminum will contaminate it. Many improvements have been incorporated into some proprietary black oxide salts and the latest one will rectify approximately 50 times more contaminants than heretofore. This particular product causes the contaminants to boil to the top, from which they can be skimmed, or dragged out and rinsed away during processing.

After a black oxide blackening solution has been mixed, there is a "breaking-in" period that can run from 24 to 48 h, depending on the volume of blackening solution and the amount of work being processed. During this period, if it occurs, erratic blackening may be encountered, resulting in some work being blackened and some remaining partially or totally unblackened.

Chemical black oxide finishes today are being used on a wide variety of consumer and military parts. Some of the most important applications are guns, firearms and components, metal stampings, toys, screws, spark plugs, machine parts, screw machine products, typewriter and calculating machine parts, auto accessories and parts, tools, gauges, and textile machinery parts. A black oxide finish can normally be used for indoor or semioutdoor applications on metal parts or fabrications that require an economical attractive finish, nominal corrosion resistance, and, in many cases, where "dimensional changes" cannot be tolerated.

Heat-Treatment Methods

These methods can be divided into three classes: (1) oven or furnace heating; (2) molten salt bath immersion; and (3) steam heat procss.

Oven or Furnace Method

The parts are heated to a temperature of 316 to 371°C at which temperature the metal surface is oxidized to a bluish-black color. The shade of color depends on the temperature and the analyses of the steel.

Molten Salt Bath Method

The oxide finish can be obtained in several ways, depending on the manufacturing or processing requirements:

1. In a molten salt bath composed essentially of nitrate salts maintained

OXIDE COATINGS

at 316 to 371°C, the pieces are first cleaned of any oils, greases, or objectionable oxides and then immersed in the molten bath. They will take on a blue-black finish, after which they are quenched in clean water and given a final oil dip.

2. In a molten nitrate bath in an austempering operation, the pieces are heated to the hardening temperature in a neutral salt hardening bath, after which they are quenched in the molten nitrate bath at 316 to 371°C. They are then removed, cleaned, and given a final oil dip.

Steam Process

In this method, the steel is placed in a retort and heated to a minimum temperature of 316°C. The retort is then purged with steam. Under these conditions a black oxide is formed on the metal surface.

BROWNING PROCESS

The browning process is commonly known as a rusting process. The pieces are first thoroughly cleaned and then swabbed with an acidic solution. After drying in a dry atmosphere at approximately 77°C, the pieces are then placed in an oven with an atmosphere of 100% humidity and temperature at around 77°C for about 1.5 h. They then become quite rusty and the surface is rubbed down to remove the loose rust.

This procedure is carried out three or four times, after which the surface will have taken on a bluish-black finish. The pieces are then given a final oil dip.

CARBONIA PROCESS

To apply a black oxide finish by this method, the pieces are placed in a rotary furnace heated to around 316 to 371°C. Charred bone or other carbonaceous material is placed in the furnace along with a thick oil known as carbonia oil. The floor or cover of the furnace is occasionally opened and closed to allow circulation of air. After the parts have been treated for approximately 4 h, they are removed and immersed in oil. This finish is essentially a black oxide of iron, but because the metal surface is in contact with carbonaceous material and oil, some black carbon penetrates into the metal surface.

PROCESSES FOR NONFERROUS METALS

Following are typical methods for applying black oxide coatings to nonferrous metals:

1. Stainless steel: aqueous alkali-nitrate and molten dichromate methods. In the aqueous alkali-nitrate method a solution is made up by using approximately 2.07 to 2.3 kg of blackening salt mixture to make up a gallon of blackening solution, which is operated at a boiling point of 124 to 127°C. The parts, after cleaning and acid pickling, are immersed in the boiling solution and will take on a black color. They are then given a rust-preventive oil dip. In the molten dichromate method a molten bath of sodium dichromate or a mixture of sodium and potassium dichromates is used at a molten temperature of 316 to 399°C. The parts are immersed in the molten bath until they take on a blue or blue-black color, after which they are removed and cooled in oil or water. They are then cleaned to remove the salt and oil (if cooled in oil), after which they are given a dip in a clean, rust-preventive oil.
2. Zinc, zinc plate, zinc-base die castings, and cadmium plate: hot molybdate blackening method, chromate and black dye method, and the black nickel plate method.
3. Copper and copper alloys (brasses and bronzes): alkali-chlorite aqueous solution method, cuprammonium carbonate method, anodic oxidation method, and aryl-sulfone monochloramide-sodium hydroxide method.
4. Aluminum and aluminum alloys: anodize and dye method.

OXYGEN

An abundant element, oxygen constitutes about 89% of all water, 33% of the Earth's crust, and 21% of the atmosphere. It combines readily with most of the other elements, forming their oxides. It is a colorless and odorless gas and can be produced easily by the electrolysis of water, which produces both oxygen and hydrogen, or by chilling air below −184°C, which produces both oxygen and nitrogen. The specific gravity of oxygen is 1.1056. It liquefies at −113°C at 59 atm. Liquid oxygen is a pale-blue, transparent, mobile liquid. As a gas, oxygen occupies 862 times as much space as the liquid. Oxygen is one of the most useful of the elements.

Oxygen is separated from air by liquefaction and fractional distillation. The chief uses of oxygen in order of their importance are (1) smelting, refining, and fabrication of steel and other metals; (2) manufacture of chemical products by controlled oxidation; (3) rocket propulsion; (4) biological life support and medicine; and (5) mining, production, and fabrication of stone and glass products.

Uncombined gaseous oxygen usually exists in the form of diatomic molecules, O_2, but oxygen also exists in a unique triatomic form, O_3, called ozone.

PRODUCTION AND DISTRIBUTION

Oxygen is produced on a large scale by liquefaction and fractional distillation of air. Minor quantities are produced by electrolysis of water with simultaneous production of hydrogen, which is usually the primary objective. Lower-purity oxygen can be produced by the pressure swing adsorption (PSA) process: nitrogen is preferentially adsorbed, increasing oxygen content of the remaining stream to over 90%. Plants with output up to 20 tons (18 metric tons) per day have been built, but the process is mainly attractive for very small units. Tens of thousands of these produce oxygen-enriched breathing air for home treatment of chronic pulmonary deficiencies.

Traditional methods of preparing oxygen often demonstrated in school chemistry courses are heating potassium chlorate and heating mercuric oxide (Priestley's original method). When oxygen is needed in laboratories, however, it is usually obtained from compressed-gas cylinders.

Oxygen is commonly distributed in three ways: (1) most oxygen is piped directly to users, (2) about 10% is liquefied for transportation and storage in insulated tanks, and (3) about 1% is compressed to high pressure more than 200 MPa for transport in steel cylinders or tube bundles.

Oxygen pipelines are usually short because the raw material for air separation is readily available. In industrial areas a single large plant may supply a dozen consumers through a network of pipelines.

For smaller or intermittent uses or for rocket engines, oxygen is produced and distributed as a liquid. In liquid form oxygen is about one third heavier than water. So long as it is kept at low temperature, the liquid can be stored, transported, pumped, or handled much as any other liquid. To keep heat away from this very cold liquid, the storage and transport tanks use the best possible insulating techniques.

USES

Oxygen is widely used in a variety of applications. While the fraction of oxygen present in the atmosphere is sufficient for many purposes, higher concentrations are necessary to improve some processes.

Metallurgical Uses

Oxygen is a component used in the metallurgical processes of smelting, refining, welding, cutting, and surface conditioning.

Smelting

Smelting of ore in the blast furnace involves the combustion of about 1 ton (0.9 metric ton) of oxygen for each ton of metal produced. When air is used, 3.5 tons (3.2 metric tons) of nitrogen accompany each ton (0.9 metric ton) of oxygen and must be compressed, heated, and blown into the furnace. A large amount of heat is lost with the exhaust gases, which also carry powdered ore and coke away as dust and limit the capacity of the furnace. By removing some or all of the nitrogen, the furnace capacity can be increased, less expensive fuels can replace some coke, and fuels can be used more efficiently.

Metal Refining

In refining copper and in making steel from pig iron various impurities such as carbon, sulfur, and phosphorus must be removed from the metal by oxidation. If air is blown through the molten metal, as in the Bessemer converter, nitrogen is picked up, limiting the product quality. Nitrogen also carries away a great deal of the heat produced by the oxidation process. Better-quality steel and copper can be produced injecting pure oxygen into the molten metal until the impurities are completely removed. Oxygen injection can be utilized in the open hearth or electric furnaces. However, steelmaking equipment has been developed that depends entirely on high-purity oxygen. All the heat for the furnace operation is supplied by oxidation of carbon and other impurities. The technique is called the basic oxygen process.

Reheat furnaces. With specially designed burners, oxygen can be utilized without raising flame temperature. With substantial fuel saving and less pollution, furnace temperature is more easily achieved and controlled.

Welding, Cutting, and Surface Conditioning

The high-temperature flame of the oxyacetylene torch can be used in welding steel, although most welding is done by the electric arc process.

In cutting, the point of the steel at which the cutting is to start is first heated by an oxygen-acetylene flame. A powerful jet of oxygen is then turned on. The oxygen burns some of the iron in the steel to iron oxide, and the heat of this combustion melts more iron; the molten iron is blown out of the kerf by the force of the jet. By feeding powdered iron into the oxygen stream, this cutting process can be extended to alloys such as stainless steel, which are not readily cut by oxygen alone, and to completely noncombustible materials such as concrete.

Steel ingots normally have oxide inclusions and other defects at the outer surface. After preliminary rolling, the steel in slab or billet form has the surface skin removed to eliminate these defects. This can be most easily accomplished by scarfing. Streams of oxygen from many nozzles are played on all sides of the billet at once. The oxygen burns off the surface defects and some of the steel in a spectacular shower of sparks. The billet is then ready for further rolling. Oxygen scarfing, also known as skinning, became a standard practice in most steel mills.

Hydrometallurgy

The mechanism by which metal values are leached from ores may involve oxidation. Oxygen gas is dissolved in the leach fluid to extract uranium from deeply buried ore deposits. Similarly copper is obtained from previously discarded mine waste.

Chemical Syntheses

Several syntheses in the chemical industry involve oxygen.

Partial Oxidation of Hydrocarbons

When natural gas or fuel oil is burned, the heat of combustion first cracks the hydrocarbon molecules into fragments. These fragments usually encounter oxygen molecules within a few hundredths of a second and are oxidized to water and carbon dioxide. However, if the supply of oxygen is carefully controlled and the passage of material through the combustion zone is very rapid, it is possible to freeze the reaction at various stages of completion.

In this manner natural gas (mostly methane, CH_4) can be converted to acetylene (C_2H_2), ethylene (C_2H_4), or propylene (C_3H_6). Ethylene (C_2H_4), in turn, can be partially oxidized to ethylene oxide (CH_2CH_2O).

Syngas Production

Reaction of carbon or hydrocarbons with oxygen and steam yields a mixture of carbon monoxide (CO) and hydrogen (H_2), that is, syngas. By use of suitable catalysts, syngas can be recombined to form various organic compounds such as methanol (CH_3OH), octane (C_8H_{16}), and many others. In the presence of other catalysts, carbon monoxide can combine with steam to form more hydrogen and carbon dioxide. After removal of the carbon dioxide, the hydrogen can be used for chemical reactions, such as the manufacture of ammonia (NH$_3$), hydrogenation of fats, and hydrocracking of petroleum.

Fuel Synthesis

Conversion of solid fuels to liquid or gaseous hydrocarbons requires addition of hydrogen. This normally involves combustion of some carbon to create high temperatures. Steam reacts with hot carbon to produce hydrogen. Use of oxygen for combustion yields higher-purity fuel gas or higher yields of liquids. In similar manner, use of oxygen for in-place combustion of coal or heavy oil may improve yields and product quality.

Manufacture of Pigments

Both titanium dioxide white and carbon black are useful primarily because of the characteristics of their small particles. The size, shape, and surface activity of these particles govern the ability of the material to perform properly as a pigment, bulking agent, or stiffener when blended into other materials. Formation of titanium dioxide or carbon in a flame process produces very fine, useful particles. Carefully controlled addition of oxygen to such burner operations can improve yield and quality of the product.

Liquid Fuel Rockets

In rocket engines, liquid oxygen is used as an oxidizer either with kerosine or liquid hydrogen fuels. While fluorine could theoretically provide somewhat improved performance in terms of specific impulse, oxygen is very nearly as good, is much cheaper, and is easier to handle.

Solid-fueled rockets, based on hydrocarbon polymers that contain sufficient oxidizer to effect self-combustion, dominate the short-range military uses. Liquid-fueled rockets are expected to remain dominant in space work until the full development of nuclear propulsion. The Saturn-Apollo launch vehicle has a fully loaded weight of about 3000 tons (2700 metric tons) of which more than 2000 tons (1800 metric tons) is liquid oxygen. Most of the liquid oxygen consumed by the aerospace industry has been used in the development and proof-testing of rocket engines mounted in static test stands. The usage of oxygen in this testing has been in excess of 1000 tons (900 metric tons) per day.

Biological Applications

Oxygen is a fundamental part of many biological processes. A few are described below.

Aerospace and Diving

Oxygen is necessary for life support of animals of this planet. Whenever humans desire to live or work in environments low or deficient in oxygen, it is necessary to carry oxygen along to supplement or substitute for the available atmosphere. High-altitude military aircraft normally provide oxygen for the aviators. Commercial transports carry oxygen for emergency use in case of failure of the cabin pressurizing system. Astronauts must of course carry their entire breathing gas requirements with them, which becomes one of the larger load requirements for any extended mission. Divers in shallow water are able to have air transmitted to them from the surface. However, for deeper diving the special breathing gases frequently are carried to the ocean bottom in special diving bells.

Medicine

In medical applications, patients breathe air enriched with oxygen. This is usually done to reduce the work of heart and lungs during the course of infectious disease, during or after major surgery, or in recovery from heart attack. Oxygen enrichment may be required to permit even moderate activity if lung function is chronically impaired, as in emphysema. Breathing oxygen at pressure up to 3 atm (300 kilopascals) absolute can increase dissolved oxygen content of body fluids, improving supply to tissues if circulation is impaired.

Treatment of Biological Pollutants

Experiments have demonstrated that eutrophication of ponds and lakes can be stopped, even reversed, by injecting oxygen into deeper water. The value of lakes for fishing and other water recreations can be preserved. Addition of oxygen to biological treatment equipment increases capacity and reduces power requirements of sewage treatment plants. Oxygen is sometimes pumped directly into sewer lines, rivers, or streams overloaded with biochemical contamination. With extra oxygen, naturally occurring bacteria may decompose wastes without further treatment.

Stone, Clay, and Glass Industries

Oxygen has a place in these industries as described below.

Glass Manufacture and Fabrication

The glass industry uses large quantities of oxygen in the manufacture and shaping of glass. Oxygen additions raise the combustion temperature in the furnace, speeding up and improving control over the melting of glass and its raw materials. Oxygen is used in the burners that heat glass for blowing, shaping, and flame-polishing rough edges.

Mining and Quarrying

An oxygen-kerosine burner can be used to heat and shape some types of stone. Granite and similar rocks expand when heated rapidly by such a burner so that the surface cracks loose, or spalls. The hot combustion gases blow the fine chips of rocks away, presenting a fresh surface, which is rapidly heated, continuing the process.

In this manner the extremely hard taconite iron ore can be pierced for blast holes more effectively than by conventional drilling methods. Granite for construction and decorative purpose can be quarried by special burners equipped to cut channels through the rock. Slabs of granite can be cut to desired dimension and given an even and pleasing surface using still other burner designs. A rock surface fouled with paint or tarry materials can easily be cleaned by this technique. Artists have used flame shaping to produce statuary.

Cement and Kiln Operations

In most kiln-type operations, such as manufacture of cement, roasting or sintering ore, and production of refractories, the essential reactions take place at rather high temperatures. When enough heat is provided to carry out the desired reaction, there is more than enough heat to raise the temperature of the fresh feed. Much heat is wasted at lower temperatures where it is not useful to the process. By using oxygen instead of air, the flame temperature is raised and much more heat is available for the high-temperature reaction from a given amount of fuel. Extensive tests have shown that large increases in capacity and reductions in fuel consumption are possible. However, certain changes in equipment are needed to achieve all the potential benefits.

P

PAINT

Paint is a term used since the dawn of history to designate cosmetics, marking chalks and pastes, tempera plaster, and colored fluids applied to surfaces for artistic, decorative, or weatherproofing purposes. The term survives today as a general marketing designation for decorative and protective formulations used in architectural, commercial, and industrial applications, with such diverse materials as lacquer, varnish, baking finishes, and specialty coating systems covered in a single category.

There is a general definition that suits most products: paint is a fluid, with viscosity, drying time, and flowing properties dictated by formulation, normally consisting of a vehicle or binder, a pigment, a solvent or thinner, and a drier, which may be applied in relatively thin layers and which changes to a solid in time. The change to a solid may or may not be reversible, and may occur by evaporation of the solvent, by chemical reaction, or by a combination of the two.

Paint is a general name sometimes used broadly to refer to all types of organic coatings. However, by definition, paint refers to a solution of a pigment in water, oil, or organic solvent, used to cover wood or metal articles either for protection or for appearance. Solutions of gums or resins, known as varnishes, are not paints, although their application is usually termed painting. Enamels and lacquers, in the general sense, are under the classification of paints, but specifically the true paints do not contain gums or resins. Stain is a varnish containing enough pigment or dye to alter the appearance or tone of wood in imitation of another wood, or to equalize the color in wood. It is usually a dye rather than a paint.

In modern technology, paint is classified in three major categories because of differing performance requirements: architectural paints, commercial finishes, and industrial coatings. A fourth category, artistic media, now admits inks, cements, pastes, dyes, plastics, semisolids, and conventional pigmented oils as acceptable materials.

Architectural Paints

These are air-drying materials applied by brush or spray to architectural and structural surfaces and forms for decorative and protective purposes. Materials are classified by formulation type as solvent thinned and water thinned.

Solvent-Thinned Paints

The drying mechanism of solvent-thinned paints predominantly may be by solvent evaporation, oxidation, or a combination of the two, and paints in this classification are subdivided accordingly.

Solvent-thinned paints that dry essentially by solvent evaporation rely on a fairly hard resin as the vehicle. Resins include shellac, cellulose derivatives, acrylic resins, vinyl resins, and bitumens. Shellac is usually dissolved in alcohol and is commonly used as shellac varnish. Paints based on nitrocellulose or other cellulose derivatives are usually called lacquers. Paints derived from acrylic and vinyl resins usually require a solvent such as ketone, and their architectural applications are limited. However, addition to the formulation of agents that result in emulsion polymerization has produced products with extensive architectural use. Bitumens or asphalts of petroleum or coal tar derivation are most often used in roofing and waterproofing applications where heavy layers are required and opportunity for renewal may be limited.

In paints that dry by oxidation, the vehicle is usually an oil or an oil-based varnish. These usually contain driers to accelerate drying of the oil. Paints based essentially on linseed oil with suitable pigments such as titanium dioxide and zinc oxide extenders once were the conventional exterior house paints. However, the successful development of polyvinyl acetate and acrylic emulsion types of paint reached the point where these materials became dominant in the exterior house paint market.

Water-Thinned Paints

This group of paints may be subdivided into those in which the vehicle is dissolved in water and those in which it is dispersed in emulsion form.

Paints with water-soluble vehicles include the calcimines, in which the vehicle is glue, and casein paints, in which the vehicle is casein or soybean protein. These paints are water-sensitive and have only limited use. Synthetic resins soluble to water and treatments rendering drying oils soluble in water are relatively recent developments. Water evaporates from paints having these materials in their formulation, and further chemical change, either oxidation or heat polymerization, converts the vehicle so that the film no longer is water-sensitive. These paints have a limited market because of the widespread utilization of latex emulsions.

Nearly any solvent-thinned paint may be emulsified by the addition of a suitable emulsifier and adequate agitation. The use of fugitive emulsifiers and of vehicles especially processed for use in emulsion paints has produced materials with excellent water resistance, color retention, and durability when cured. Materials formed by emulsion polymerization are described as a latex, and products are called latex paints. The most common latexes are made from a copolymer of butadiene and styrene, from polyvinyl acetate, and from acrylic resin. Properties and performance differ slightly among these paints, but as a group they dominate the architectural market.

Enamel Paints

Enamel paint is an intimate dispersion of pigments in either a varnish or a resin vehicle, or in a combination of both. Enamels may dry by oxidation at room temperature and/or by polymerization at room or elevated temperatures. They vary widely in composition, in color and appearance, and in properties. Although they generally give a high-gloss finish, some give a semigloss or eggshell finish and still others give a flat finish. Enamels as a class are hard and tough and offer good mar and abrasion resistance. They can be formulated to resist attack by the most commonly encountered chemical agents and corrosive atmospheres, and have good weathering characteristics.

Because of their wide range of useful properties, enamels are probably the most widely used organic coating in industry. One of their largest areas of use is as coatings for household appliances — washing machines, stoves, kitchen cabinets, and the like. A large proportion of refrigerators, for example, are finished with synthetic baked enamels. These appliance enamels are usually white, and therefore must have a high degree of color and gloss retention when subjected to light and heat. Other products finished with enamels include automotive products; railway, office, sports, and industrial equipment; toys; and novelties.

House Paints

House paint for outside work consists of high-grade pigment and linseed oil, with a small percentage of a thinner and drier. The volatile thinner in paints is for ease of application, the drying oil determines the character of the film, the drier is to speed the drying rate, and the pigment gives color and hiding power. Part or all of the oil may be replaced by a synthetic resin. Many of the newer house paints are water-based paints.

Paints are marketed in many grades, some containing pigments extended with silica, talc, barytes, gypsum, or other material; fish oil or inferior semidrying oils in place of linseed oil; and mineral oils in place of turpentine. Metal paints contain basic pigments such as red lead, ground in linseed oil, and should not contain sulfur compounds. Red lead is a rust inhibitor, and is a good primer paint for iron and steel, although it is now largely replaced by chromate primers. White lead has a plasticizing effect that increases adhesion. It is stable and not subject to flaking. Between some pigments and the vehicle there is a reaction that results in progressive hardening of the film with consequent flaking or chalking, or there may be a development of water-soluble compounds. Linseed oil reacts with some basic pigments, giving chalking and flaking. Fading of a paint is usually from chalking. The composition of paints is based on relative volumes because the weights of pigments vary greatly, although the custom is to specify pounds of dry pigment per gallon of oil.

BITUMINOUS PAINTS

Bituminous paints are usually coal tar or asphalt in mineral spirits, used for the protection of piping and tanks, and for waterproofing concrete. For line pipe heavy pitch coatings are applied hot, but a bitumen primer is first applied cold. The bituminous paints also have poor solvent resistance.

COMMERCIAL FINISHES

These include air-drying or baking-cured materials applied by brush, spray, or magnetic agglomeration to kitchen and laundry appliances, automobile, machinery, and furniture and used as highway marking materials. Paints in this group are subdivided by their drying mechanism.

Air-Drying Finishes

These materials once were the conventional factory-applied finishes. In the finishing of furniture, lacquers, varnishes, and shellac are still extensively used, but epoxy and unsaturated polyester materials have been adopted in recent years. Automobile manufacturers used lacquers and air-drying alkyd enamels until the adoption of baked acrylic and urea finishes. Epoxy, urethane, and polyester resins, converted at room temperature with a suitable catalyst, are beginning to replace conventional solvent-thinned paints as machinery finishes. The marking of center lines on highways and other painted areas for the control of traffic requires a finish that dries rapidly, adheres well to both asphalt and concrete, and resists abrasion and staining. Solvent-thinned materials especially formulated for this service from alkyds, modified rubbers, and other resins now are extensively used, but some latex formulations have been tried with success.

Baking Finishes

Urea and melamine resins polymerize by heat and are used in baking finishes where extreme hardness, chemical resistance, and color retention are required as on kitchen and laundry appliances. Baked acrylic resin formulations now dominate the automobile finish market. Certain phenolic resins are converted by heat to produce finishes with excellent water and chemical resistance.

INDUSTRIAL COATINGS

Industrial coatings are subdivided by their intended service: corrosion-resistant coatings, high temperature coatings, and coatings for immersion service. Materials in each subdivision are applied in a system that usually requires a base coat or primer, an intermediate coat or coats, and a top or finish coat.

Corrosion-Resistant Coatings

These are materials generally inert when cured to deterioration by acidic, alkaline, or other corrosive substances and applied in a system as a protective layer over steel or other substrates susceptible to corrosion attack.

The base, or prime, coat in the corrosion-resistance system is applied to dry surfaces prepared by abrasive blasting or other methods to a specified degree of cleanliness and toughness. The prime coat provides adhesion to the substrate for the entire coating system: adherence is by mechanical anchorage, chemical reaction, or a combination of the two. Prime coat materials once were predominately a red lead pigment dispersed in linseed oil and cured by oxidation. Today, the prime coat material commonly specified for application to steel surfaces is a dispersion of zinc in a suitable inorganic or organic vehicle. Although these zinc-rich primers now dominate the market where protection of steel is concerned, prime coat materials generically similar to subsequent coats in the system are specified when protection of nonferrous or nonmetallic substrates from corrosion attack is required.

An intermediate coat in the corrosion-resistant system is not always required. When used, intermediate coat materials usually are high-build layers of the same generic type specified for the top or surface coat, and are applied only to increase the dry-film thickness of the protective area in places where airborne corrosive fumes, particulates, or droplets are heavily concentrated or where splash and spillage of corrosive fluids are relatively frequent.

A top or surface coat in the corrosion-resistant system may be selected from a variety of formulations with vehicles that include phenolic resins, chlorinated rubber, coal tar and epoxy combination, epoxy resin cured from a solvent solution with polyfunctional amines, polyamide resins, vinyl resin in solvent solution, elastomers, polyesters, and polyurethanes.

Top coat materials are required to have good scaling properties; high resistance to corrosive deterioration, oxidation, erosion, and ultraviolet degradation; and relative freedom, when cured, from pinholes, blisters, and crevices. Color retention is a desirable but not a mandatory requirement.

Selection of materials able to withstand corrosive attack in a given environment and to maintain cohesion and integrity among the various coats in system has become a technological specialty.

High-Temperature Coatings

These are materials used to alleviate or prevent corrosion, thermal shock, fatigue, oxide sublimation, and embrittlement of metals at high temperature.

Formulation of high-temperature coatings is a relatively recent development in the art. Materials problems encountered since World War II in high-temperature chemical reactions, production of steel by the oxygen lance process, aerodynamic heating to complex high-speed aircraft, and aerospace vehicle launching and reentry have been alleviated to an extensive degree by the results of research toward finding high-strength resistant metals and cements and toward producing coatings serviceable at high temperatures. Research continues as higher and higher temperatures are required by technological developments and radiation effects in outer space are better understood.

Available high-temperature coatings include inorganic zinc dispersed in a suitable vehicle, serviceable to 400°C; a phosphate bonding system in which ceramic fillers are mixed with an aqueous solution of monoaluminum phosphate, applied by spraying, brushing, or dipping, and serviceable to 1538°C after curing at 204°C; and a "ceramic gold" coating used on jet engine shrouds and having a temperature limit near 538°C.

Ablative formulations that absorb heat through melting, sublimation decomposition, and vaporization, or that expand upon heating to form a foamlike insulation that replenishes itself until the coating is depleted, employ silicone rubber or silicone resins, or polyamide and tetrafluoroethylene polymers to provide short-term heat protection from 147 to 538°C.

Coatings for Immersion Service

Included here are materials used to coat or line interior surfaces of vessels of containers holding or storing corrosive fluids, pipelines in which corrosive fluids are the flowing medium, and hoppers or bins conveying or holding abrasive or corrosive pellets or particulates.

These coatings are usually applied in relatively high-build systems to carefully cleaned and prepared surfaces. Air-drying formulations are used when extensive coverage is required and where polymerization by automatic heating apparatus is impractical. Small vessels, pipe spools, and assembly components may be protected by coatings cured by oven baking.

Coatings for immersion service may be selected from a variety of formulations that include asphalt, chemically cured coal tar, thermoplastic coal tar, epoxy-furans, amine-cured epoxies, fluorocarbons, furfuryl alcohol resins, neoprene, baked unmodified phenolics, unsaturated polyesters, polyether resins, low-density polyethylene, chlorosulfonated polyethylene, polyvinyl chloride plastisols, resinous cements, rubber, and urethanes, among other vehicles.

Selection and specification of coatings for immersion service has become a technological specialty.

Paint Removers

Paint removers, for removing old paint from surfaces before refinishing, are either strong chemical solvents or strong caustic solutions. In general, the more effective they are in removing the paint quickly, the more damaging they are likely to be to the wood or other organic material base. The hiding power of a paint is measured by the quantity that must be applied to a given area of a black and white background to obtain nearly uniform complete hiding. The hiding power is largely in the pigment, but when some fillers of practically no hiding power alone, such as silica, are ground to microfine particle size, they may increase the hiding power greatly. Paint making is a highly developed art, and the variables are so many and the possibilities for altering the characteristics by slight changes in the combinations are so great that the procurement specifications for paints are usually by usage requirements rather than by composition.

PALLADIUM

A rare metal, palladium (symbol Pd) is found in the ores of platinum. It resembles platinum, but is slightly harder and lighter in weight and has a more beautiful silvery luster. It is only half as plentiful but is less costly. The specific gravity is 12.10 and the melting point is 1552°C. Annealed, the metal has a hardness of Brinell 40 and a tensile strength of 186 MPa.

It is highly resistant to corrosion and to attack by acids, but, like gold, it is dissolved in aqua regia. It alloys readily with gold and is used in some white golds. It alloys in all proportions with platinum and the alloys are harder than either constituent.

Physical and Chemical Properties

Palladium is soft and ductile and can be fabricated into wire and sheet. The metal forms ductile alloys with a broad range of elements. Palladium is not tarnished by dry or moist air at ordinary temperatures. At temperatures from 350 to 790°C a thin protective oxide forms in air, but at temperatures from 790°C this film decomposes by oxygen loss, leaving the bright metal. In the presence of industrial sulfur-containing gases a slight brownish tarnish develops; however, alloying palladium with small amounts of iridium or rhodium prevents this action.

At room temperature, palladium is resistant to nonoxidizing acids such as sulfuric acid, hydrochloric acid, hydrofluoric acid, and acetic acid but the metal is attacked by nitric acid. Palladium is also attacked by moist chlorine and bromine.

Uses

The major applications of palladium are in the electronics industry, where it is used as an alloy with silver for electrical contacts or in pastes in miniature solid-state devices and in integrated circuits. Palladium is widely used in dentistry as a substitute for gold. Other consumer applications are in automobile exhaust catalysts and jewelry.

The palladium–silver–gold alloys offer a series of noble brazing materials covering a wide range of melting temperatures. A palladium–silver alloy can be used as a diffusion septum for the separation of hydrogen from gas mixtures.

Palladium supported on carbon or alumina is used as a catalyst for hydrogenation and dehydrogenation in both liquid- and gas-phase reactions. Palladium finds widespread use in catalysis because it is frequently very active under ambient conditions and it can yield very high selectivities. Palladium catalyzes the reaction of hydrogen with oxygen to give water. Palladium also catalyzes isomerization and fragmentation reactions.

Palladium alloys are also used for instrument parts and wires, dental plates, and fountain-pen nibs. Palladium is valued for electroplating because it has a fine white color that is resistant to tarnishing even in sulfur atmospheres.

Although palladium has low electric conductivity, 16% that of copper, it is valued for its resistance to oxidation and corrosion. Palladium-rich alloys are widely used for low-voltage electric contacts. Palladium–silver alloys, with 30 to 50% silver, for relay contacts, have 3 to 5% the conductivity of copper. A palladium–silver alloy with 25% silver is used as a catalyst in powder or wire-mesh form. A palladium–copper alloy for sliding contacts has 40% copper with a conductivity 5% that of copper. Many of the palladium salts, such as sodium palladium chloride, Na_2PdCl_4, are easily reduced to the metal by hydrogen or carbon monoxide, and are used in coatings and electroplating.

PAPER

Paper is a flexible web or mat of fibers isolated from wood or other plants materials by the operation of pulping. Nonwovens are webs or mats made from synthetic polymers, such as high-strength polyethylene fibers, that substitute for paper in large envelopes and tote bags.

Paper is made with additives to control the process and modify the properties of the final product. The fibers may be whitened by bleaching, and the fibers are prepared for papermaking by the process of refining. Stock preparation involves removal of dirt from the fiber slurry and mixing of various additives to the pulp prior to papermaking. Papermaking is accomplished by applying a dilute slurry of fibers in water to a continuous wire or screen; the rest of the machine removes water from the fiber mat. The steps can be demonstrated by laboratory handsheet making, which is used for process control.

Although there is no distinct line to be drawn between papers and paperboard, paper is usually considered to be less than 0.15 mm thick. Most all fibrous sheets over 0.30 mm thick are considered to be board. In the borderline range of 0.15 to 0.30 mm, most are considered to be papers, although some are classified as board.

Although paper has numerous specialized uses in products as diverse as cigarettes, capacitors, and countertops (resin-impregnated laminates), it is principally used in packaging (~50%), printing (~40%), and sanitary (~7%) applications. Paper was manufactured entirely by hand before the development of the continuous paper machine; this development allowed the U.S. production of paper to increase by a factor of 10 during the 19th century and by another factor of 50 during the 20th century. In 1960, the global production of paper was 70 million tons (50% by the United States); in 1990 it was 210 million tons (30% by the United States). The annual per capita paper use in the United States is 300 kg, and about 40% is recovered for reuse.

Material of basis weight greater than 200 g/m^2 is classified as paperboard; lighter material is called paper. Production by weight is about equal for these two classes. Paperboard is used in corrugated boxes; corrugated material consists of top and bottom layers of paperboard called linerboard, separated by fluted corrugating paper. Paperboard also includes chipboard (a solid material used in many cold-cereal boxes, shoe boxes, and the backs of paper tablets) and food containers.

Mechanical pulp is used in newsprint, catalog, and other short-lived papers; they are only moderately white, and yellow quickly with age because the lignin is not removed. A mild bleaching treatment (called brightening) with hydrogen peroxide or sodium dithionite (or both) masks some of the color of the lignin without lignin removal. Paper made with mechanical pulp and coated with clay to improve brightness and gloss is used in 70% of magazines and catalogs, and in some enamel grades. Bleached chemical pulps are used in higher grades of printing papers used for xerography, typing paper, tablets, and envelopes; these papers are termed uncoated wood-free (meaning free of mechanical pulp). Coated wood-free papers are of high to very high grade and are used in applications such as high-quality magazines and annual reports; they are coated with calcium carbonate, clay, or titanium dioxide.

Like wood, paper is a hygroscopic material; that is, it absorbs water from, and also releases water into, the air. It has an equilibrium moisture content of about 7 to 9% at room temperature and 50% relative humidity. In low humidities, paper is brittle; in high humidities, it has poor strength properties.

General Features and Uses

The major attribute of paper is its extreme versatility. A wide range of end properties can be obtained by control of the variables in (1) original selection of the type and size of fiber, (2) the various pulp processing methods, (3) the actual web-forming operation, and (4) the treatments that can be applied after the paper has been produced.

Papers have been specifically developed for a number of engineered applications. These include gasketing; electrical, thermal, acoustical, and vibration insulation; liquid and air filtration; composite structural assemblies; simulated leathers and backing materials; cord or twine; and as yarns for paper textiles.

Wood

Wood, a diverse, variable material, is the source of about 90% of the plant fiber used globally to make paper. Straw, grasses, canes, bast, seed hairs, and reeds are used to make pulp, and in many regards their pulp is similar to wood pulp. Fibers are tubular elements of plants and contain cellulose as the principal constituent. Softwoods (gymnosperms) have fibers that are about 3 to 5 mm long, while in hardwoods (angiosperms) they are about 0.8 to 1.6 mm long. In both cases, the length is typically about 100 times the width. Softwoods are used in papers such as linerboard where strength is the principal intent. Hardwoods are used in papers such as tissue and printing to contribute to smoothness. Many papers also include some softwood pulp for strength and some hardwood pulp for smoothness.

Wood consists of three major components: cellulose, hemicellulose, and lignin. The first two are white polysaccharides of high molecular weight that are desirable in paper. Cotton is over 98% cellulose, while wood is about 45%. Hemicellulose, although not water soluble, is similar to starch. The hydroxyl groups of these materials allow fibers to be held together in paper by hydrogen bonding. Adhesives are not required to form paper, but some starch is usually used and has a similar effect to the hemicelluloses in helping the fibers bond together. Lignin makes up about 25 to 35% of softwoods and 18 to 25% of hardwoods. It is concentrated between fibers.

Paper Production

Most papers are made from crude fibrous wood pulps.

Types of Pulp

The type of pulp to a large extent determines the type of paper produced. Pulps are generally classified as mechanical or chemical wood pulps.

Mechanical wood pulps produced by mechanical processes include:

1. Ground wood, which is used in a number of papers where absorbency, bulk, opacity, and compressibility are primary requirements, and permanence and strength are secondary.
2. Defibrated pulps, which are used for insulating board, hardboard, or roofing felts where good felting properties are required.
3. Exploded pulps, used for building and insulation hardboards, or so-called "wood composition materials."

Chemical wood pulps are produced by "cooking" the fibrous material in various chemicals to provide certain characteristics. They include:

1. Sulfite pulps, used in the bleached or unbleached state for papers ranging from very soft or weak to strong grades. There are about 12 grades of sulfite pulps.
2. Neutral sulfite or monosulfite pulps, used for strong papers for bags, wrappings, and envelopes.
3. Sulfate or kraft pulps, providing high strength, fair cleanliness and, in some instances, high absorbency. Such pulps are used for strong grades of unbleached, semibleached, or bleached paper (called kraft paper) and board.

4. Soda pulps, used principally in combination with bleached sulfite or bleached sulfate pulp for book-printing papers.
5. Semichemical pulps, used for specialty boards, corrugating papers, glassine and greaseproof papers, test liners, and insulating boards and wallboards.
6. Screenings, used principally for coarse grades of paper and board, such as millwrapper, and as a substitute for chipboard, corrugating papers, and insulation board.

Papermaking

The two basic types of papermaking machines are the Fourdrinier and the cylinder machines; the Fourdrinier machine is the most commonly used.

In Fourdrinier papermaking, the pulp, mixed to a consistency of 97.5 to 99.5% water, is fed continuously to Fourdrinier, which consists of an endless belt of fine mesh screen called the "wire."

As the pulp web travels along the wire, water drains from it into suction boxes. As it leaves the wire (at about 83% water) it passes through presses, and usually a variety of other types of equipment, such as dryers and calender rolls, depending on the type of paper produced.

Cylinder machine papermaking differs from Fourdrinier in that the web of pulp is formed on a cylindrical mold surface instead of a continuous wire covered with fine wire cloth, which revolves in a vat of paper stock or pulp. The felt conveyor carries the resulting web to the press and dryers.

The cylinder machine is used to produce a greater variety of paper thicknesses, ranging from the thinnest tissue to the thickest building board.

Bleaching

Chemical pulp for printing paper has 3 to 6% lignin, which gives the pulp a brown color. This lignin is removed with bleaching chemicals in four to eight stages. Each stage consists of a pump to mix the bleaching agent with the pulp, a retention tower to allow the chemical to react with the pulp for 30 min to several hours, and a washing unit to remove the solubilized lignin and residual chemicals from the pulp. Usually an oxidizing material is followed by a stage of alkali extraction, because lignin becomes more soluble at high pH. A common bleaching sequence involves elemental chlorine, alkali extraction, and, finally, chlorine dioxide. Sodium hypochlorite, also found in household liquid bleach, is sometimes used. There is some pressure for the bleaching process to be elemental chlorine free. This is possible by using oxygen as the first bleaching chemical, chlorine dioxide in place of chlorine, and hydrogen peroxide as a bleaching agent.

TYPES OF PAPER

In the broadest classification, there are three basic types of papers: cellulose fiber, inorganic fiber, and synthetic organic fiber papers.

Cellulose Fiber Papers

These papers, made from wood pulp, constitute by far the largest number of papers produced. A great many of the engineering papers are produced from kraft or sulfate pulps. The term *kraft* is used broadly today for all types of sulfate papers, although it is primarily descriptive of the basic grades of unbleached sulfate papers, where strength is the chief factor, and cleanliness and color are secondary. By various treatments, kraft can be altered to produce various grades of condenser, insulating, and sheathing papers.

Other types of vegetable fibers used to produce papers include:

1. Rope, used for strong, pliable papers, such as those required in cable insulation, gasketing, bags, abrasive papers, and pattern papers.
2. Jute, used for papers possessing excellent strength and durability.
3. Bagasse, used for paper for wallboard and insulation, usually where strength is not a primary requirement.
4. Esparto, used for high-grade book or printing papers. A number of other types of pulps are also used for these papers.

Inorganic Fiber Papers

There are three major types of papers made from inorganic fibers:

1. Asbestos is the most widely used inorganic fiber for papers. Asbestos papers are nonflammable, resistant to elevated temperatures, and have good thermal insulating characteristics. They are available with or without binders and can be used for electrical insulation or for high-temperature reinforced plastics.
2. Fibrous glass can be used to produce porous and nonhydrating papers. Such papers are used for filtration and thermal and electrical insulation, and are available with or without binders. High-purity silica glass papers are also available for high-temperature applications.
3. Ceramic fiber (aluminum silicate) papers provide good resistance to high temperatures, low thermal conductivity, good dielectric properties, and can be produced with good filtering characteristics.

Synthetic Organic Fiber Papers

A great deal of research has been carried out on the use of such synthetic textile fibers as nylon, polyester, and acrylic fibers in papers. Some of the earliest appear highly promising for electrical insulating uses. Others appear promising for chemical or mechanical applications. They are most commonly combined with other fibers in a paper, primarily to add strength.

PAPER TREATMENTS

Papers can be impregnated or saturated, coated, laminated, or mechanically treated. The major treatments used are covered and indicate the extent of treatments available.

Impregnation or Saturation

Impregnation or saturation can be carried out either at the beater stage in the processing of the pulp, or after the paper web bas been formed. Beater saturation permits saturation of nonporous or adsorbent papers, whereas papers saturated after manufacture must be of the absorbent type to permit complete impregnation by the saturant.

Papers can be saturated or impregnated with almost any known resin or binder. Probably the most commonly used are asphalt for moisture resistance; waxes for moisture vapor and water resistance; phenolic resins for strength and rigidity; melamine and certain ureas for wet strength (not to be confused with moisture resistance); rubber latexes, both natural and synthetic, for resilience, flexibility, strength, and moisture resistance; epoxy or silicone resins for dielectric characteristics or dielectric characteristics at elevated temperatures; and ammonium salts, or other materials for flameproofing.

A number of proprietary beater saturated papers are currently available. They are used primarily for gasketing, filtration, simulated leathers, and backing materials. Most of these consist of cellulose or asbestos fibers blended with natural or synthetic rubbers. In some cases cork is added to the blend for increased compressibility. Another type of proprietary beater saturated paper consists of leather fibers blended with rubber latexes.

Coatings

Papers may be coated either by the paper manufacturer or by converters. Coating materials, which also impregnate the paper to a greater or lesser degree, include practically every known resin or binder and pigment used in the paint industry. Coatings can be applied in solvent or water solutions, water emulsions, hot melts, and extrusion coatings, or in the form of plastisols or organisols.

The most important properties provided by coatings are (1) gas and water vapor resistance, (2) water, liquid, and grease resistance, (3) flexibility, (4) heat sealability, (5) chemical resistance, (6) scuff resistance, (7) dielectric properties, (8) structural strength, (9) mold resistance, (10) avoidance of fiber contamination, and (11) protection of printing.

Coating materials range from the older asphalts, waxes, starches, casein, shellac, and natural gums, to the newer polyethylenes, vinyl copolymers, acrylics, polystyrenes, alkyds, polyamides, cellulosics, and natural or synthetic rubbers.

Laminations

Paper can be laminated to other papers or to other films to provide a variety of composite structures.

Probably the most common types of paper laminates are those composed of layers of paper laminated with asphalt to provide moisture resistance and strength. Simple laminations of paper can be so oriented that overall characteristics of the composite are isotropic.

Laminating paper with plastics or other types of films or with metal foils will, in many cases, combine the desirable properties of the film or foil with those of the paper.

Scrim is a mat of fibers, usually laminated as a "core" material between two faces of paper. It is usually used to provide strength but can also provide bulk for cushioning, or a degree of "hand" to the composite material.

Mechanical Treatments

Several mechanical treatments can be applied to papers to provide particular special properties.

Crimping, which can be done either on the paper web or on the individual fibers of the paper, essentially adds stretch or extensibility. Crimping the paper web results in crepe paper with improved strength, stretch, bulk, and conformability and with texture similar to that of cloth. The creping process usually consists of "crowding" the paper into small pleats or folds with a "doctor." Typical range of elongation or stretch obtainable is 20 to 300%. Cross-creping can provide controllable stretch in directions perpendicular to each other, further improving drapability.

A high degree of stretch, conformability, and flexibility is produced by a patented process that differs from creping in that the individual fibers in the paper web are crimped, rather than the web itself. The amount of stretch is variable, but about 10% stretch in the machine direction seems to be optimum for most industrial applications. The major advantages of this type of paper are reported to be a high degree of toughness, combined with a smooth surface, and a high resistance to tearing or punching.

Twisting

Twisting is used to convert paper to twine or yarns. Such yarns have substantially higher strength than the paper from which they are made.

Twisting papers are usually sulfate papers, either bleached or unbleached. High tensile strength is required in the machine direction, and the heavier-weight papers should usually be soft and pliable. Treatments to impart such characteristics as wear and moisture resistance can be applied during or after the spinning operation.

Embossing and Other Techniques

Decorative papers can be produced by embossing in a variety of patterns. Embossing does not usually improve strength significantly. Embossing or "dimpling" in certain patterns, followed by lamination, can produce composites with added strength as well as bulk and thermal insulation.

Other mechanical methods include (1) shredding for bulk or padding, (2) pleating, used as a forming aid and for strength in paper cups and plates, (3) die cutting and punching, and (4) molding, which consists of compressing the wet pulp web in a mold to form a finished shape, such as an egg crate.

APPLICATIONS

As an engineering material, paper has several important applications.

As filtration material, paper can be used either as a labyrinth barrier material to guide the fluid or gas to be filtered or, more commonly, as the filtering medium itself. Paper is used for filtering automotive air and oil, and air in room air conditioners, as well as for industrial plant filtration, filtering liquids in tea bags, in addition to machine-cooling oil and domestic hot water filters.

Papers are used for both light- and heavy-duty gaskets, for such applications as high- and low-pressure steam and water, high- and low-temperature oil, aromatic and nonaromatic fuel systems, and for sealing both rough and machined surfaces.

Electrical insulation represents one of the largest engineering uses of papers. For electrical uses, special types include coil papers or layer insulation, cable paper or turn insulation, capacitor papers, condenser papers, and high-temperature, inorganic insulating papers.

A large and growing use for paper is in structural sandwich materials, where papers are impregnated with a resin such as phenolic and formed in the shape of a honeycomb. The honeycomb is used as a core between facing sheets of a variety of materials including paperboard, reinforced plastics, and aluminum.

Another large-volume application of papers is as backing material for decorative films or other surfacing materials. These provide bulk and depth to the product in simulating leather or fabrics.

PARTICULATES

Particulates are solids or liquids in a subdivided state. Because of this subdivision, particulates exhibit special characteristics that are negligible in the bulk material. Normally, particulates will exist only in the presence of another continuous phase, which may influence the properties of the particulates. A particulate may comprise several phases. They can be categorized into particulate systems that relate them to commonly recognized designations. Fine-particle technology deals with particulate systems in which the particulate phase is subject to change or motion. Particulate dispersions in solids have limited and specialized properties and are conventionally treated in disciplines other than fine-particle technology.

The universe is made up of particles, ranging in size from the huge masses in outer space — such as galaxies, stars, and planets — to the known minute building blocks of matter — molecules, atoms, protons, neutrons, electrons, neutrinos, and so on. Fine-particle technology is concerned with those particles that are tangible to human senses, yet small compared to the human environment — particles that are larger than molecules but smaller than gravel. Fine particles are in abundance in nature (as in rain, soil, sand, minerals, dust, pollen, bacteria, and viruses) and in industry (as in paint pigments, insecticides, powdered milk, soap, powder, cosmetics, and inks). Particulates are involved in such undesirable forms as fumes, fly ash, dust, and smog and in military strategy in the form of signal flares, biological and chemical warfare, explosives, and rocket fuels.

PROCESSING AND USES

There are many cases where a particulate is either a necessary or an inadvertent intermediary in an operation. Areas that may be involved in processing particulates may be classified as follows:

Size Reduction
  Mechanical (starting with bulk material)
    Grinding
    Atomization
    Emulsification
  Physicochemical (conversion to molecular dispersion)
    Phase change (spray drying, condensation)
    Chemical reaction
Size Enlargement (agglomeration, compaction)
  Pelletizing
  Briqueting
  Nodulizing
  Sintering
Separation or Classification
  Ore beneficiation
  Protein shift
Deposition (collection, removal)
Coating or Encapsulation
Handling
  Powders
  Gas suspensions (pneumatic conveying)
  Liquid suspensions (non-newtonian fluids)

End products in which particulate properties themselves are utilized include the following:

Mass or Heat Transfer Agents (used in fluidized beds)
Recording (memory) Agents
    Electrostatic printing powders and toners (for optical images)
    Magnetic recording media (for electronic images)
Coating Agents (paints)
Nucleating Agents
Control Agents (for servomechanisms)
    Electric fluids
    Magnetic fluids
Charge Carriers (used in propulsion, magnetohydrodynamics)
Chemical Reagents
    Pesticides, fertilizers
    Fuels (coal, oil)
    Soap powders
    Drugs
    Explosives
Food Products

CHARACTERIZATION

The processing or use of particulates will usually involve one or more of the following characteristics:

Physical
    Size (and size distribution)
    Shape
    Density
    Packing or concentration
Chemical
    Composition
    Surface character
Physiochemical (including adhesive, cohesive)
Mechanical or Dynamic
    Inertial
    Diffusional
    Fluid drag
    Dilute suspensions (Stokes' law, Cunningham factor, drag coefficient)
    Concentrated suspensions (hindered settling, rheology, fluidization)
Optical (scattering, transmission, absorption)
    Refractive index (including absorption)
    Reflectivity
Electrical
    Conductivity
    Charge
Magnetic
Thermal
    Insulation (conductivity, absorptivity)
    Thermophoresis

Many of the characteristics of particulates are influenced to a major extent by the particle size. For this reason, particle size has been accepted as a primary basis for characterizing particulates. However, with anything but homogeneous spherical particles, the measured "particle size" is not necessarily a unique property of the particulate but may be influenced by the technique used. Consequently, it is important that the techniques used for size analysis be closely allied to the utilization phenomenon for which the analysis is desired.

PARTICLE SIZE

Size is generally expressed in terms of some representative, average, or effective dimension of the particle. The most widely used unit of particle size is the micrometer (μm), equal to 0.001 mm. Another common method is to designate the screen mesh that has an aperture corresponding to the particle size. The screen mesh normally refers to the number of screen openings per unit length or area; several screen standards are in general use; the two most common in the United States are the U.S. Standard and the Tyler Standard Screen Scales.

Particulate systems are often complex. Primary particulates may exist as loosely adhering (as by van der Waals forces) particles called flocs or as strongly adhering (as by chemical bonds) particulates called agglomerates. Primary particles are those whose size can only be reduced by the forceful shearing of crystalline or molecular bonds. Figure P.1 shows states of dispersion of particulates. The double arrows imply reversibility with application of light shearing forces.

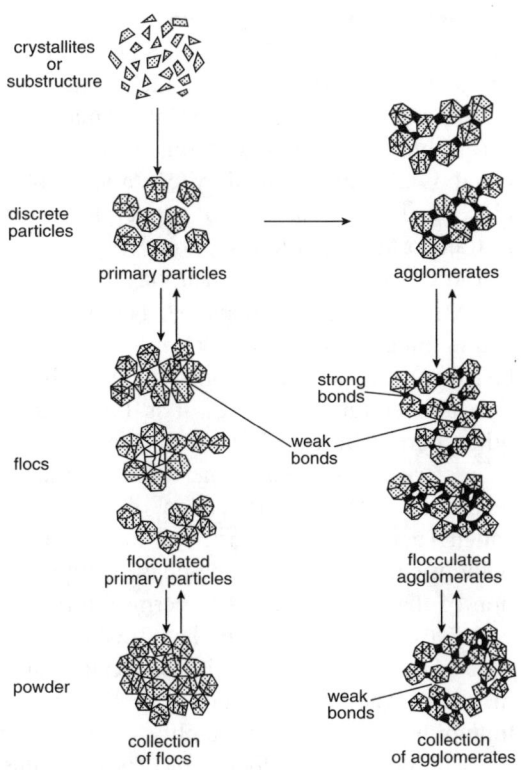

FIGURE P.1 States of dispersion of particulates. The double arrows imply reversibility with application of light shearing forces. (From *McGraw-Hill Encyclopedia of Science and Technology*, 8th ed., Vol. 13, McGraw-Hill, 168. With permission.)

PASTES

Conductor, resistor, dielectric, seal glass, polymer, and soldering compositions are available in paste or ink form. They are used to produce hybrid circuits, networks, and ceramic capacitors. The materials are often called thick-film compositions.

TYPES

Conductor pastes consist of metallic elements and binders suspended in an organic vehicle. Primarily, precious metals such as gold, platinum, palladium, silver, copper, and nickel are used singularly or in combination as the conductive element. The adhesion mechanism to the substrate is provided by either a frit bond, reactive bond, or mixed bond.

Important properties of conductor pastes include wire bondability, conductivity, solderability, solder leach resistance, and line definition.

Thick-film resistor pastes are composed of a combination of glass frit, metal, and oxides. These pastes are used in microcircuits, voltage dividers, resistor networks, chip resistors, and potentiometers.

Dielectric compounds are used as insulators for the fabrication of multilayer circuits, crossovers, or as protective coverings.

Solder pastes are one of the more common component attach products. They consist of finely divided solder powders of all common alloys of tin, lead, silver, gold, etc., suspended in a vehicle–flux system. The fluxes may be nonactivated or completely activated. The most popular is RMA (rosin, mildly activated).

PERMANENT MOLD CASTINGS

Permanent mold casting is performed in a mold, generally made of metal, that is not destroyed by removing the casting. Several types of casting can be included in the description: pressure die casting, centrifugal casting, and gravity die casting.

ADVANTAGES AND DISADVANTAGES

Gravity die castings are dense and fine grained and can be made with better surfaces and to closer tolerances than sand castings. Tolerances are wider than for pressure die castings and plaster mold castings, but narrower than for sand castings. Production rates are lower than those obtainable by die casting.

The process stands somewhere between sand and die casting with respect to possible complexity, dimensional accuracy, mold or die casts, etc. For many parts it provides an attractive compromise when the ultimate — whether in complexity of one-piece construction, narrow tolerances, or ultrahigh production rates — need not be met.

CASTING ALLOYS

Lead-, zinc-, aluminum-, magnesium-, and copper-base alloys, as well as gray cast iron, can be cast by permanent molding. Less than 1% of total gray iron production is permanent mold cast, 5 to 7% of copper and magnesium alloy, and up to 40% of total aluminum casting production. Here is a breakdown of problems to expect:

- *Gray iron.* Mechanical properties depend on thickness.
- *Aluminum.* Gates must be made larger to take the low specific gravity of aluminum into account.
- *Magnesium.* Extreme caution is needed when removing metal from the furnace. Ladles must be kept at red heat to exclude moisture and prevent an explosion.
- *Copper.* Aluminum bronze is the most popular permanent-mold-cast copper-base alloy. Turbulence must be prevented in the die cavity to reduce oxidation of the molten metal during solidification.

PERMEABILITY ALLOYS

This is a general name for a group of nickel–iron alloys with special magnetic properties. These soft magnetic materials possess a magnetic susceptibility much greater than iron. An early alloy was composed of 78.5% nickel and 21.5% iron. It also contained about 0.37% cobalt, 0.1% copper, 0.04% carbon, 0.03% silicon, and 0.22% manganese. It is produced sometimes with chromium or molybdenum, under the name of Permalloy, and is used in magnetic cores for apparatus that operates on feeble electric currents, and in the loading of submarine cables. It has very little magnetic hysteresis.

TYPES AND USES

Supermalloy for transformers contains 79% nickel, 15% iron, 5% molybdenum, and 0.5% manganese, with total carbon, silicon, and sulfur kept below 0.5%. It is melted in vacuum, and poured in an inert atmosphere. It can be rolled as thin as 0.00064 cm. The alloy has an initial permeability 500 times that of iron.

Supermendur contains 49% iron, 49% cobalt, and 2% vanadium. It is highly malleable, and has very high permeability with low hysteresis loss at high flux density. Duraperm is a high-flux magnetic alloy containing 84.5% iron, 9.5% silicon, and 6% aluminum. Perminvar is an alloy containing 45% nickel and 25% cobalt, intended to give a constant magnetic permeability for variable magnetic fields. A-metal is a nickel–iron alloy containing 44% nickel and a small amount of copper. It is used in transformers and loudspeakers to give nondistortion characteristics when magnetized.

Alfenol contains no nickel, but has 16% aluminum and 84% iron. It is brittle and cannot be rolled cold, but can be rolled into thin sheets at a temperature of 575°C. It is lighter than other permeability alloys, and has superior characteristics for transformer cores and tape-recorder heads. A modification of this alloy, called Thermenol, contains 3.3% molybdenum without change in the single-phase solid solution of the binary alloy. The permeability and coercive force are varied by heat treatment. At 18% aluminum, the alloy is practically paramagnetic. The annealed alloy with 17.2% aluminum has constant permeability.

Aluminum–iron alloys with 13 to 17% aluminum are produced in sheet form for transformers and relays. They have magnetic properties equal to the 50–50 nickel–iron alloys and to the silicon–iron alloys, and they maintain their magnetic characteristics under changes in ambient temperature.

Iron–nickel permeability alloys are used as loading in cable by wrapping layers around the full length of the cable. When nickel–copper alloys are used, they are employed as a core for the cable. Magnetostrictive alloys are iron–nickel alloys that will resonate when the frequency of the applied current corresponds to the natural frequency of the alloy. They are used in radios to control the frequency of the oscillating circuit. Magnetostriction is the stress that occurs in a magnetic material when the induction changes. In transducers it transforms electromagnetic energy into mechanical energy. Temperature-compensator alloys are iron–nickel alloys with about 30% nickel. They are fully magnetic at −29°C but lose their magnetic permeability in proportion to rise in temperature, until, at about 54°C, they are nonmagnetic. Upon cooling they regain permeability at the same rate. They are used in shunts in

PEWTER

Pewter is a very old name for tin–lead alloys used for dishes and ornamental articles; the term now refers to the use rather than to the composition of the alloy. Tin was the original base metal of the alloy, the ancient Roman pewter having about 70% tin and 30% lead, although iron and other elements were present as impurities. Pewter, or latten ware, of the 16th-century contained as much as 90% tin, and a strong and hard English pewter contained 91% tin and 9% antimony. This alloy is easily cold-rolled and spun, and can be hardened by long annealing at 225°C and quenching in cold water and tempering at 110°C. Pewter is now likely to contain lead and antimony, and very much less tin; when the proportion of tin is less than about 65%, the alloys are unsuited for vessels to contain food products, because of the separation of the poisonous lead.

Early pewter, with high lead content, darkened with age. With less than 35% lead, pewter was used for decanters, mugs, tankards, bowls, dishes, candlesticks, and canisters. The lead remained in solid solution with the tin so that the alloy was resistant to the weak acids in foods.

Addition of copper increases ductility; addition of antimony increases hardness. Pewter high in tin (91% tin and 9% antimony or antimony and copper, for example) has been used for ceremonial objects, such as religious communion plates and chalices, and for cruets, civic symbolic cups, and flagons.

PHENOL

Phenol is the simplest member of a class of organic compounds possessing a hydroxyl group attached to a benzene ring or to a more complex aromatic ring system.

Also known as carbolic acid or monohydroxybenzene, phenol is a colorless to white crystalline material of sweet odor, having the composition C_6H_5OH, obtained from the distillation of coal tar and as a by-product of coke ovens.

Phenol has broad biocidal properties, and dilute aqueous solutions have long been used as an antiseptic. At higher concentrations it causes severe skin burns; it is a violent systemic poison. It is a valuable chemical raw material for the production of plastics, dyes, pharmaceuticals, syntans, and other products.

PROPERTIES

Phenol melts at about 43°C and boils at 183°C. The pure grades have melting points of 39, 39.5, and 40°C. The technical grades contain 82 to 84% and 90 to 92% phenol. The crystallization point is given as 40.41°C. The specific gravity is 1.066. It dissolves in most organic solvents. By melting the crystals and adding water, liquid phenol is produced, which remains liquid at ordinary temperatures. Phenol has the unusual property of penetrating living tissues and forming a valuable antiseptic. It is also used industrially in cutting oils and compounds and in tanneries. The value of other disinfectants and antiseptics is usually measured by comparison with phenol.

USES AND DERIVATIVES

Phenol is one of the most versatile industrial organic chemicals. It is the starting point for many diverse products used in the home and industry. A partial list includes nylon, epoxy resins, surface active agents, synthetic detergents, plasticizers, antioxidants, lube oil additives, phenolic resins (with formaldehyde, furfural, and so on), cyclohexanol, adipic acid, polyurethanes, aspirin, dyes, wood preservatives, herbicides, drugs, fungicides, gasoline additives, inhibitors, explosives, and pesticides.

PHENOL–FORMALDEHYDE RESIN

This is a synthetic resin, commonly known as phenolic, made by the reaction of phenol and formaldehyde, and employed as a molding material for the making of mechanical and electrical parts. It was the earliest type of hard, thermoset synthetic resins, and its favorable combination of strength, chemical resistance, electrical properties, glossy finish, and nonstrategic abundance of low-cost raw materials has maintained the resin, with its many modifications and variations, as one of the most widely employed groups of plastics for a variety of products. The resins are also used for laminating, coatings, and casting resins.

Phenolic resins are used most extensively as thermosetting plastic materials, as there are only a few uses as thermoplastics. The polymer is composed of carbon, hydrogen, oxygen, and sometimes nitrogen. Its molecular weight varies from a very low value during its early state of formation to almost infinity in its final state of cure. The chemical configuration, in the thermoset state, is usually represented by a three-dimensional network in which the phenolic nuclei are linked by methylene groups. The completely cross-linked network requires three methylene groups to two phenolic groups. A lesser degree of cross-linking is attainable either by varying the proportions of the ingredients or by blocking some of the reactive positions of the phenolic nucleus by other groups, such as methyl, butyl, etc. Reactivity can be enhanced by increasing the hydroxyl groups on the phenolic nuclei, for example, by the use of resorcinol.

Characteristics

The outstanding characteristics of phenolics are good electrical properties, very rigid set, good tensile strength, excellent heat resistance, good rigidity at elevated temperature, good aging properties; also, good resistance to water, organic solvents, weak bases, and weak acids. All these characteristics are coupled with relatively low cost.

Phenolics are used in applications that differ widely in nature. For example, wood is impregnated to make "impreg" and "compreg"; paper is treated to make battery separators and oil and air filters; specific chemical radicals can be added to the molecule to make an ion-exchange material. Phenolics are also widely used in protective coating.

The hundreds of different phenolic molding compounds can be divided into six groups on the basis of major performance characteristics. General-purpose phenolics are low-cost compounds with fillers such as wood flour and flock, and are formulated for noncritical functional requirements. They provide a balance of moderately good mechanical and electrical properties, and are generally suitable in temperatures up to 149°C. Impact-resistant grades are higher in cost. They are designed for use in electrical and structural components subject to impact loads. The fillers are usually paper, chopped fabric, or glass fibers. Electrical grades, with mineral fillers, have high electrical resistivity plus good arc resistance, and they retain their resistivity under high-temperature and high-humidity conditions. Heat-resistant grades are usually mineral- or glass-filled compounds that retain their mechanical properties in the 190 to 260°C temperature range.

Special-purpose grades are formulated for service applications requiring exceptional resistance to chemicals or water, or combinations of conditions such as impact loading and a chemical environment. The chemical-resistant grades, for example, are inert to most common solvents and weak acids, and their alkali resistance is good. Nonbleeding grades are compounded specially for use in container closures and for cosmetic cases.

Fillers

Proper balance of fillers is important, because too large a quantity may produce brittleness. Organic fillers absorb the resin and tend to brittleness and reduced flexural strength, although organic fibers and fabrics generally give high impact strength. Wood flour is the most usual filler for general-service products, but prepared compounds may have mineral powders, mica, asbestos, organic fibers, or macerated fabrics, or mixtures of organic and mineral materials. Bakelite was the original name for phenol plastics, but trade names now usually cover a range of different plastics, and the types and grades are designated by numbers.

The specific gravity of filled phenol plastics may be as high as 1.70. The natural color is amber, and, as the resin tends to discolor, it is usually pigmented with dark colors. Normal phenol resin cures to single-carbon methylene groups between the phenolic groups, and the molded part tends to be brittle. Thus, many of the innumerable variations of phenol are now used to produce the resins, and modern phenol resins may also be blended or cross-linked with other resins to give higher mechanical and electrical characteristics. Furfural is frequently blended with the formaldehyde to give better flow, lower specific gravity, and reduced cost. The alkylated phenols give higher physical properties.

Phenol resins may also be cast and then hardened by heating. The cast resins usually have a higher percentage of formaldehyde and do not have fillers. They are poured in syrupy state in lead molds and hardened in a slow oven.

Some of the uses for phenolic resins are for making precisely molded articles, such as telephone parts, for manufacturing strong and durable laminated boards, or for impregnating fabrics, wood, or paper. Phenolic resins are also widely used as adhesives, as the binder for grinding wheels, as thermal insulation panels, as ion-exchange resins, and in paints and varnishes.

Molding Compounds

The largest single use for phenolic resins is in molding compounds. To make these products, either one- or two-stage resins are compounded with fillers, lubricants, dyes, plasticizers, etc. Wood flour is used as an inexpensive reinforcing agent in the general-purpose type of compounds. Cotton flock, chopped fabric, and sisal and glass fibers are used to improve strength characteristics; mineral fillers such as asbestos and mica are used where improvements in dimensional stability, heat resistance, or electrical properties are desired. The compounds are usually produced in granular, macerated, or nodular forms, depending on type of filler used. Since the color of the base resin is not stable to light, molding compounds are commonly produced only in dark colors such as black and brown.

Molding compounds are usually processed in hardened steel molds and molds can be designed to operate using the compression, transfer, or plunger-molding techniques, depending on the design of the article to be fabricated. Molded parts can be drilled, tapped, or machined.

About one third of all phenolic resins produced is processed into parts by molding. Compression and transfer molding are the principal processes used but they can also be extruded and injection molded.

Molded phenolic parts are used in bottle caps, automotive ignition and engine components, electrical wiring devices, washing machine agitators, pump impellers, electronic tubes and components, utensil handles, and a multitude of other products.

Adhesives

The thermosetting nature and good water- and fungus-resistant qualities of phenolic resins make them ideal for adhesive applications. Almost all exterior-grade plywood is phenolic resin bonded. This constitutes the second largest market for phenolics. The essential ingredient in many metal-to-metal, and metal-to-plastic adhesives is a phenolic resin. One-step phenol–formaldehyde resins are used predominantly for hot-pressed plywood. Special resorcin–formaldehyde resins curing at room temperature are employed for fabricating laminated timber.

Laminates

The third largest use for phenolic resins is in the manufacture of laminated materials. Many variations of paper, from cheap kraft to high-quality alpha cellulose, in addition to asbestos, cotton, linen, nylon, and glass fabrics are the most commonly used reinforcing filler sheets. The laminate is formed by combining under heat 177°C and pressure 3.4 to 14 MPa multiple layers of the various reinforcing sheet; after saturation with phenolic resin, generally of the one-step type dissolved in alcohol.

Paper laminates are used most extensively in the electrical and decorative fields. A large number of the laminates for the electrical industry are of the punching grade, making it possible to fabricate all kinds of small parts in a punch press. The laminate used for decorative purposes usually contains a surface sheet of melamine resin-treated paper for providing unlimited color or design configurations. Other fillers are used for special applications where superior dimensional stability, or water, fire, or chemical resistance, or extra strength is required.

Casting Resins

Plastic parts can also be manufactured by pouring resin into molds and heat-curing without pressure. Two basic grades of casting resins are manufactured commercially. Both are of the one-stage type and are cured under neutral to strongly acidic conditions depending on the application. The first grade is manufactured for its variegated color and artistic possibilities and is used primarily in the cutlery and decorative field. Since this type of cast material is noted for ease of machining, it is well adapted for small production runs or where machined prototypes are desired.

The second grade of casting resins includes all those modified by fillers and reinforcing agents. Designed primarily to have low-shrinkage characteristics during cure, they are usually set with a strong acid catalyst to obtain a low temperature set. Uses include containers, jigs, fixtures, and metal-forming dies.

Bonding Agents

Phenolic resins are noted for their excellent bond strength characteristics under elevated temperature conditions, and thus are used in such applications as thermal and acoustical insulation, grinding wheels, coated abrasives, brake linings, and clutch facings. Glass wool insulation is manufactured by spraying water-soluble one-stage resins on the glass fibers as they are formed. Heat given off by the fibers as they cool is sufficient to set the resin. Where organic fibers are used, finely pulverized, single-stage resins are distributed between the fibers, either by a mixing or a dusting operation, followed by an oven treatment.

Grinding wheels bonded with phenolic resin are commonly known as resinoid-bonded wheels. By combining a liquid single-stage resin and a powdered two-stage resin, the material can be evenly distributed with abrasive grit and fillers so that the mixture can be pressed into wheels and baked in ovens. Some wheels are also hot-pressed and cured directly in a press. Resinoid-bonded wheels are used primarily where the application requires a bond of exceptional strength such as in cut-off and snagging wheels.

Coated abrasives are manufactured by bonding abrasive grit to paper, fabric, or fiberboard by means of phenolic resins of the liquid one-stage type. Sander disks and belts are common applications.

Brake linings and clutch facings are made by bonding asbestos, fillers, metal shavings, and friction modifiers with phenolic resin, usually performed by a mixing, forming, and baking operation. Most resins for these applications are specially formulated, but simple two-stage resins are sometimes used. Formulation for each friction lining or clutch face must be determined carefully. Wood composition products of all descriptions are manufactured by hot-pressing sawdust, wood chips, or wood flour containing 8 to 20% resin. Two-step resins are most frequently used in composition boards. One-step resins are employed for special applications where the yellow color or the slight ammonia odor of two-step resins is undesirable.

Foundry Use

A relatively new application holding great promise for phenolic resins is in the shell mold process for the foundry industry. The basic principle involves binding sand grains with resin. The process is equally well adapted to making cores and is satisfactory for practically all metals, including magnesium and high-chrome alloys.

The process has many intriguing possibilities. Molds thus made can be reproduced in exact composition, detail, and size; they are rigid and, having no affinity for water, can be racked and stored indefinitely, with or without cores. Castings from these molds have excellent surface finish and detail and can be held to close dimensional tolerances. The process can be completely mechanized, yields more castings per ton of melt than sand casting, simplifies cleaning of castings, minimizes problems with sand control and handling, and is a relatively clean operation. It is revolutionizing the foundry industry.

Consumption of resin in the foundry industry has grown rapidly, and conceivably this could become the largest single outlet for phenolics. However, markets for phenolic resins in plywood adhesive, insulation, and wood composition board applications have also expanded.

PHOSGENE

The common name for carbonyl chloride, $COCl_2$, a colorless, poisonous gas made by the action of chlorine on carbon monoxide. It was used as a poison war gas. But it is now used in the manufacture of metal chlorides and anhydrides, pharmaceuticals, perfumes, isocyanate resins, and for blending in synthetic rubbers. It liquefies at 7.6°C, and solidifies at −118°C. It is decomposed by water. When chloroform is exposed to light and air, it decomposes into phosgene. One part in 10,000 parts of air is a toxic poison, causing pulmonary edema. For chemical warfare it is compressed into a liquid in shells.

Because of its toxicity, most phosgene is produced and employed immediately in captive applications. The biggest use of the material is for toluene diisocyanate (TDI), which is then reacted into polyurethane resins for foams, elastomers, and coatings. About 0.9 metric ton of phosgene is consumed to make a metric ton of polymethylene polyphenylisocyanate, also used for making polyurethane resins for rigid foams. Polycarbonate manufacturers require 0.42 metric ton phosgene per ton of product resin. Polycarbonate is used for making break-resistant housings, signs, glazings, and electrical tools. Phosgene also is a reactant for the isocyanates that are used in pesticides, and the di- and polyisocyanates are adhesives, coatings, and elastomers.

PHOSPHATE CONVERSION COATINGS

Phosphate coatings have been used commercially for approximately 50 years. They are used on iron, steel, zinc, and aluminum surfaces to increase corrosion protection, provide a base for paint, reduce wear on bearing parts, and aid in the cold forming and extrusion of metals.

Phosphate coatings are formed by chemically reacting a clean metal surface with an aqueous solution of a soluble metal phosphate of zinc, iron, or manganese, accelerating agents, and free phosphoric acid. For example, when steel is treated, the surface is converted into a crystalline coating consisting of secondary and tertiary phosphates, adherent to and integral with the base metal.

Properties of the coating such as size, weight, and uniformity of the crystals are influenced by many factors, for example, composition and concentration of the phosphating solution, temperature and processing time, type of metal, and the condition of its surface due to previous treatment.

The performance of a phosphate coating depends largely on the unique properties of the coating, which is integrally bound to the base metal and acts as a nonmetallic, adsorptive layer to hold a subsequent finish of oil or wax, paint, or lubricant. Heavy phosphate coatings are normally used in conjunction with an oil or wax for corrosion resistance. The combination of the coating with the oil film gives a synergistic effect, which affords much greater protection than that obtained by the sum of the two taken separately. The stable, nonmetallic, nonreactive phosphate coating provides an excellent base for paint. It is chemically combined with the metal surface, which results in increased adsorption of paint and materially reduces electrochemical corrosion normally occurring between the paint film and the metal.

The oil-absorptive phosphate coatings are useful in holding and maintaining a continuous oil film between metal-to-metal moving parts. They also permit rapid break-in of new bearing surfaces. Even after the coatings have been worn away, the controlled etched condition of the metal surface continues to hold the oil film between the moving parts. The ability of a well-anchored coating to hold a soap or oil-type lubricant is used in the cold forming and drawing of metals.

PROCESSES

Most phosphate coatings are formed from heated solutions following a hot cleaning cycle. Effective coatings, both the zinc phosphate and the iron phosphate types, are now produced by the relatively cold system with up to 70% savings in heating costs.

All steel must be thoroughly cleaned prior to phosphate coating to remove grease, oil, rust, and undesirable soils on the steel surface, which prevent or alter the formation of a satisfactory phosphate coating and interfere with paint adhesion. The cleaner used to remove oily soil prior to the formation of the zinc phosphate coating is a light-duty, mildly alkaline material especially formulated to function effectively at low temperature. The cleaner that is best suited for use in connection with the low-temperature iron phosphate coating process may be either a light-duty, low-foaming, mildly acidic mixture, or the above mild alkaline cleaner, depending upon the conditions of operation.

Proper formulation of the phosphating solutions allows the coatings to form rapidly at low temperature on the cleaned steel surface. Simple, on-the-job, chemical controls enable the operator to adjust the addition of coating chemicals to the requirements of the steel being treated.

Following the coating operation, a cold water rinse is used to remove excess coating chemicals. The flow of water through the rinse is regulated with the rate of production so that contamination of the main body of the rinse is minimized. An acidified rinse containing hexavalent chromium compounds follows the water rinse. This lapse has the specific effect of enhancing the corrosion resistance of the coating. An oven dry-off to remove surface moisture completes the process.

Typical products being treated by the cold phosphate system are automotive body and sheet metal parts, refrigerator cabinets, office furniture, lighting fixtures, commercial air conditioners, home heating equipment, home laundries, kitchen cabinets, desk and filing cabinets, steel drums, and window sash.

There are advantages of the cold phosphate system over conventional methods that require higher temperature operation:

1. Direct heat savings — up to 70%
2. Less heat-up time
3. Less maintenance due to decreased load on heating coils, steam traps, etc.
4. Reduced downtime; maintenance personnel can enter the units immediately after shutdown to make adjustments or repairs
5. Increased worker comfort near the installation
6. Reduced use of water through decreased evaporation
7. Elimination of exhaust fans

Applications

Phosphate conversion coatings have a wide range of application. Several examples follow.

Base for Plastisol Coatings

The domestic appliance industry has pointed toward the use of plastisol (polyvinyl chloride) coatings as a replacement for more costly porcelain enamel. During this period several major manufacturers of domestic dishwashers have standardized on plastisol films for coating tubs, lids, dish racks, etc.

The metal preparation method meeting this requirement was a zinc phosphate coating system. The fabricated parts are cleaned in a conditioned cleaner, phosphated, and rinsed thoroughly. The rinse procedure includes an acidified chromic-phosphoric acid rinse followed by a closely controlled deionized water rinse and thorough drying to remove surface moisture. This treatment produces a continuous, uniform, fine-grained coating of approximately 150 to 250 mg of zinc phosphate coating per square foot of surface area treated.

After the steel has been zinc phosphate coated, an especially formulated primer is applied to a very thin and closely controlled film build and followed by the subsequent application of 12 to 15 mils of plastisol with an intermediate and final baking operation.

Bond for Vinyl Coatings

Vinyl films of 5 to 15 mils thickness are applied to both steel and aluminum sheets by lamination of calendered and decorated films or by spray or roller coating. For the vinyl laminate as with the plastisol application, the base metal must be chemically treated to provide the necessary bond for laminate to metal surface. A controlled, accelerated iron phosphate treatment followed by an acidified chromate rinse has proved to be best for the preparation of steel for vinyl lamination to sheet or coil on a

PHOSPHATE CONVERSION COATINGS

continuous line. This method is now being used by several fabricators and rollers of steel.

Aluminum can also be finished with vinyl laminates and plastisols. Chromate conversion coatings of the "gold" oxide type provide excellent adhesion of the vinyl film as well as excellent corrosion resistance on the unprotected surface.

Metal preparation for both steel and aluminum is usually done by spray application in five-stage equipment in the following sequence:

1. Alkali clean — 30 s (smutty steel may require brush scrubbing)
2. Water rinse — 10 s
3. Phosphate coating — 10 s
4. Water rinse — 5 s
5. Acidified chromate rinse — 5 s

The same sequence and time cycles are used for aluminum except that an appropriate chromate conversion coating solution is in stage 3. Consequently, installations have been engineered to prepare both steel and aluminum for subsequent application of vinyl laminates by providing interchangeable coating solution tanks at that stage, thereby providing efficient and versatile "in-line" operation for this new decorative treatment.

Bolt Making

Phosphate coatings are also finding wide acceptance in the fastener industry as applied to rod to facilitate the forming of bolts. The coating is produced from a dilute, especially accelerated, zinc acid phosphate solution that reacts chemically with the surface of the rod to form an insoluble nonmetallic phosphate coating integral with the surface of the rod. The coating forms a porous bond to carry the extruding and heading lubricant and prevent metal-to-metal contact in subsequent heading operations. The result is longer tool life, increased percentage of reduction, and improved surface appearance of the finished product.

The methods of application vary from the conventional pickle-house immersion method to the in-line strand processes. In the immersion method, the coils of scaled rods are dipped in the treating solutions in the following sequence of operations:

1. Acid pickle — 10 min to 2.5 h
2. Water rinse-cold, overflowing or spray
3. Hot water rinse
4. Phosphate coating — 5 to 15 min, 71 to 93°C
5. Cold water rinse
6. Neutralizing rinse — lime, etc.
7. Bake dry

It has been found that a dip in a high-strength soap solution improves die life and reduces knockout pressures. The soap solution follows the neutralizing rinse, or in some instances, it is combined with the neutralizing rinse.

The in-line strand process eliminates intermediate handling from scaled rod to drawn wire and the necessity of acid disposal. The coils are handled continuously. Line speeds up to ~300 million m/s are required with this process so that the phosphate solution must deposit the desired heavy coating in about 15 s.

Even at this extremely short processing time, a line speed up to ~300 million m/s requires a special coating tank. Compact pre-rinse and neutralizing rinse stages are incorporated in this unit from which it is possible to obtain better than 750 mg/ft^2 of coating in 15 s by use of specially formulated phosphate solutions adapted to strand processing.

Wire produced by the strand method of processing produces wire superior in quality to that produced by conventional cleaning and drawing methods and does so at lower cost. As a consequence of the better quality, phosphate-base lubricated material will produce close tolerance bolts with full heads and sharp shoulders.

Wire Drawing

Phosphate coatings are also gaining acceptance as a lubricant carrier in the wire industry to permit increased drawing speeds and prolonged die life. This is particularly true in connection with both the dry and wet drawing of high carbon wire. Other advantages that the zinc phosphate coatings afford to the wire industry are increased corrosion resistance after drawing and closer dimensional tolerances. Improved dimensional tolerance is a major advantage in forming springs from spring wire.

The zinc phosphate coating may be applied by immersion in the conventional pickle house installation, or by the newly developed fast coating, continuous strand method. The continuous strand method fits in well with fast, in-line travel of the wire. This adaptability and the lower labor costs involved warrant the recommendation of strand phosphate lines in wire processing.

Extrusion

Another important development is the use of zinc phosphate coatings as an aid in the rapidly developing field of cold extrusion of steel and aluminum. In this application the phosphate coating solution is formulated to deposit a considerably heavier zinc phosphate coating than heretofore mentioned. The coating is then chemically reacted with a soap-base lubricant to form a water-insoluble lubricating film.

The use of the zinc phosphate coating is considered a basic requirement in the cold extrusion field where the pressure exerted by the tools on the steel being formed may be in excess of 2040 MPa. The phosphate and lubricant coating withstands the high unit pressure and temperatures developed in this type of cold forming. At the same time, it maintains the required separating film between the tools and workpiece being extruded to prevent scoring, galling, and tool breakage. A specially formulated phosphate coating solution produces a zinc phosphate coating on aluminum and is gaining wide acceptance as an aid in the cold extrusion of the heat-treatable alloys of this metal.

PHOSPHOR BRONZE

This, a copper-base alloys with low phosphorus content, originally called steel bronze, was 92-8 bronze deoxidized with phosphorus and cast in an iron mold. It is now any bronze deoxidized by the addition of phosphorus to the molten metal. It may or may not contain residual phosphorus in the final state. Ordinary bronze frequently contains cuprous oxide formed by the oxidation of the copper during fusion. By the addition of phosphorus, a powerful reducing agent, a complete reduction of the oxide takes place. Phosphor bronzes have excellent mechanical and cold-working properties and low coefficient of friction, making them suitable for springs, diaphragms, bearing plates, and fasteners. In some environments, such as salt water, they are superior to copper.

At present, there are 18 standard wrought phosphor bronzes, designated C50100 to C54800. Tin, which ranges from as much as 0.8 to 1% depending on the alloy, is the principal alloying element, although leaded alloys may contain as much lead (4 to 6%, for example) as tin. Phosphorus content is typically on the order of 0.1 to 0.35%, zinc 0 to 0.3% (1.5 to 4.5% in C54400), iron 0 to 0.1%, and lead 0 to 0.05% (0.8 to 6% in leaded alloys). The principal alloys were formerly known by letter designations representing nominal tin content: phosphor bronze A, 5% tin (C51000); phosphor bronze B, 4.75% tin (C53200); phosphor bronze C, 8% tin (C52100); phosphor bronze D, 10% tin (C52400); and phosphor bronze E, 1.25% tin (C50500). Phosphor bronze E, almost 99% copper, is one of the leanest of these bronzes in the way of alloying ingredients and is used for electrical contacts, pole-line hardware, and flexible tubing. Its electrical conductivity is about half that of copper and it is readily formed, soldered, brazed, and flash-welded. Thin flat products have tensile yield strengths ranging from about 83 MPa in the annealed condition to 517 MPa in the extra spring temper. More highly alloyed C54400 (4% tin, 4% lead, and 3% zinc, nominally) is about one fifth as electrically conductive as copper, has good forming characteristics, and 80% the machinability of C36000, a free-machining brass. Its ultimate tensile strength ranges from about 331 MPa in the annealed condition to 690 MPa in the extra-spring temper. Uses include bearings, bushings, gear shafts, valve components, and screw-machine products.

PHOSPHOR COPPER

An alloy of phosphorus and copper, phosphor copper was used instead of pure phosphorus for deoxidizing brass and bronze, and for adding phosphorus in making phosphor bronze. It comes in 5, 10, and 15% grades and is added directly to the molten metal. It serves as a powerful deoxidizer, and the phosphorus also hardens the bronze. Even slight additions of phosphorus to copper or bronze increase fatigue strength. Phosphor copper is made by forcing cakes of phosphorus into molten copper and holding until the reaction ceases. Phosphorus is soluble in copper up to 8.27%, forming Cu_3P, which has a melting point of about 707°C. A 10% phosphor copper melts at 850°C and a 15% at about 1022°C. Alloys richer than 15% are unstable. Phosphor copper is marketed in notched slabs or in shot.

Phosphor tin is a master alloy of tin and phosphorus used for adding to molten bronze in the making of phosphor bronze. It usually contains up to 5% phosphorus and should not contain lead. It has an appearance like antimony, with large glittering crystals, and is marketed in slabs.

PHOSPHORIC ACID

Also known as orthophosphoric acid, phosphoric acid is colorless, syrupy liquid of the composition H_3PO_4 used for pickling and rustproofing metals, for the manufacture of phosphates, pyrotechnics, and fertilizers, as a latex coagulant, as a textile mordant, as an acidulating agent in jellies and beverages, and as a clarifying agent in sugar syrup. The specific gravity is 1.65, melting point 73.6°C, and it is soluble in water. The usual grades are 90, 85, 75%, technical 50%, and dilute 10%. As a cleanser for metals, phosphoric acid produces a light etch on steel, aluminum, or zinc, which aids paint adhesion. Deoxidine is a phosphoric acid cleanser for metals. Nielite D is phosphoric acid with a rust inhibitor, used as a nonfuming pickling acid for steel. Albrite is available in 75, 80, and 85% concentrations in food and electronic grades, both high-purity specifications. DAB and Phosbrite are called Bright Dip grades, for cleaning applications. Phosphoric anhydride, or phosphorus pentoxide, P_2O_5, is a white, water-soluble powder used as a dehydrating agent and also as an opalizer for glass. It is also used as a catalyst in asphalt coatings to prevent softening at elevated temperatures and brittleness at low temperatures.

PHOSPHORUS

A chemical element, symbol P, phosphorus forms the basis of a very large number of compounds; the most important class is phosphates. For every form of life, phosphates play an essential role in all energy-transfer processes such as metabolism, photosynthesis, nerve function, and muscle action. The nucleic acids which, among other things, make up the hereditary material (the chromosomes) are phosphates, as are a number of coenzymes. Animal skeletons consist of a calcium phosphate.

Uses

About 90% of the total phosphorus (in all of its chemical forms) used in the United States goes into fertilizers.

Commercial phosphorus is obtained from phosphate rock by reduction in the electric furnace with carbon, or from bones by burning and treating with sulfuric acid. Phosphate rock occurs in the form of land pebbles and as hard rock.

The superphosphate used for fertilizers is made by treating phosphate minerals with concentrated sulfuric acid. It is not a simple compound, but may be a mixture of calcium acid phosphate, $CaHPO_4$, and calcium sulfate. Nitrophosphate for fertilizer is made by acidulating phosphate rock with a mixture of nitric and phosphoric acids, or with nitric acid and then ammoniation and addition of potassium or ammonium sulfate.

Other important uses are as binders for detergents, nutrient supplements for animal feeds, water softeners, additives for foods and pharmaceuticals, coating agents for metal-surface treatment, additives in metallurgy, plasticizers, insecticides, and additives for petroleum products. Except for the last four items, these uses involve phosphates.

Sodium tripolyphosphate is the major compound used in building synthetic detergents to achieve improved cleaning, primarily by dispersing inorganic soil and softening the water. The average phosphate-containing household detergent produced in the United States for washing clothes consists of about 40% by weight of sodium tripolyphosphate, $Na_5P_3O_{10}$. This compound is used extensively in water softening, as are other members of the homologous series of chain phosphates. The large-volume usage of phosphates in detergent building has led to unwanted growth of algae in inland waters (lakes and rivers) into which the dirty dishwashers are discharged. As a result of this fertilizing action, phosphates are considered water pollutants in those areas where such discharges occur, and in some areas phosphates have been eliminated from detergents by law. For reasonably fast-flowing rivers that discharge directly into the ocean, phosphates are not a problem.

An interesting water-softening application is found in "threshold treatment" in which tiny traces of a chain phosphate (much less than would be used in sequestering) are used to prevent the formation of pipe scale from hard waters. The application is related to the dispersing action of the phosphates, because traces of phosphate absorb on the growing surface of the pipe scale as it begins to form, and this inhibits its further growth.

A major pharmaceutical use of phosphates is in toothpastes, in which dicalcium phosphate is the most popular polishing agent. Monocalcium phosphate and sodium acid pyrophosphate, $Na_2H_2P_2O$ (the pyrophosphate is the second member of the phosphate family), are employed as leavening agents in cake mixes, refrigerated biscuits, self-rising flour, and baking powder.

Automobile bodies, for example, are generally phosphatized before they are painted to prevent rusting in use. Orthophosphate esters find wide use as plasticizers that have flameproofing properties and as gasoline and oil additives.

The phosphorus compound of major biological importance is adenosinetriphosphate (ATP), which is an ester of the salt sodium tripolyphosphate, widely employed in detergents and water-softening compounds. Practically every reaction in metabolism and photosynthesis involves the hydrolysis of this tripolyphosphate to its pyrophosphate derivative, called adenosinediphosphate (ADP).

Phosphorus is an essential element in the human body; a normal person has more than a pound of it in the system, but it can be taken into the system only in certain compounds. Nerve gases used in chemical warfare contain phosphorus, which combines with and inactivates the choline sterase enzyme of the brain. This enzyme controls the supply of the hormone that transmits nerve impulses, and when it is inactivated the excess hormone causes paralysis of the nerves and cuts off breathing. Organic phosphates are widely used in the food, textile, and chemical industries. Other phosphates are used as a plasticizer in plastics and as an antifoaming agent in paper coatings and textile sizings. They are also employed for scale and corrosion control, ore flotation, pigment dispersion, and detergents.

Flour and other foodstuffs are fortified with ferric phosphate, $FePO_4 \cdot 2H_2O$. Iron phosphate is used as an extender in paints. Tricalcium phosphate, $Ca_3(PO_4)_2$, is used as an anticaking agent in salt, sugar, and other food products and to provide a source of phosphorus. The tricalcium phosphate, used in toothpastes as a polishing agent and to reduce the staining of chlorophyll, is a fine white powder. Dicalcium phosphate, used in animal feeds, is precipitated from the bones used for making gelatin, but is also made by treating lime with phosphoric acid made from phosphate rock. Diammonium phosphate, $(NH_4)_2HPO_4$, is a mildly alkaline, white crystalline powder used in ammoniated dentifrices, for pH control in bakery products, in making phosphors, to prevent afterglow in matches, and for flameproofing paper.

Forms

There are two common forms of phosphorus, yellow and red. The former, also called white phosphorus, P_4, is a light-yellow waxlike solid, phosphorescent in the dark and exceedingly poisonous. Its specific gravity is 1.83 and it melts at 44°C. It is used for smoke screens in warfare and for rat poisons and matches. Yellow phosphorus is produced directly from phosphate rock in the electric furnace. It is cast in cakes of 0.45 to 1.36 kg each. Red phosphorus is a reddish-brown amorphous powder, with a specific gravity of 2.20 and a melting point of

725°C. Red phosphorus is made by holding white phosphorus at its boiling point for several hours in a reaction vessel. Both forms ignite easily. Amorphous phosphorus, or crystalline black phosphorus, is made by heating white phosphorus for extended periods. It resembles graphite, and is less reactive than the red or white forms, which can ignite spontaneously in air. Black phosphorus is made by this process. Phosphorus sulfide, P_4S_3, may be used instead of white phosphorus in making matches. Phosphorus pentasulfide, P_2S_5, is a canary-yellow powder of specific gravity 1.30, or solid of specific gravity 2.0, containing 27.8% phosphorus, used in making oil additives and insecticides. It is decomposed by water.

PHOTOGRAPHIC MATERIALS

These are the light-sensitive recording materials of photography, that is, photographic films, plates, and papers. They consist primarily of a support of plastic sheeting, glass, or paper, respectively, and a thin, light-sensitive layer, commonly called the emulsion, in which the image will be formed and stored. The material will usually embody additional layers to enhance its photographic or physical properties.

Supports

Film support, for many years made mostly of flammable cellulose nitrate, is now exclusively made of slow-burning "safety" materials, usually cellulose triacetate or polyester terephthalate, which are manufactured to provide thin, flexible, transparent, colorless, optically uniform, tear-resistant sheeting. Polyester supports, which offer added advantages of toughness and dimensional stability, are widely used for films intended for technical applications. Film supports usually range in thickness from 0.06 to 0.23 mm and are made in rolls up to 1.5 m wide and 1800 m long.

Glass is the predominant substrate for photographic plates, although methacrylate sheet, fused quartz, and other rigid materials are sometimes used. Plate supports are selected for optical clarity and flatness. Thickness, ranging usually from 1 to 6 mm, is increased with plate size as needed to resist breakage and retain flatness. The edges of some plates are specially ground to facilitate precise registration.

Photographic paper is made from bleached wood pulp of high α-cellulose content, free from ground wood and chemical impurities. It is often coated with a suspension of barium sulfate in gelatin for improved reflectance and may be calendered for high smoothness. Fluorescent brighteners may be added to increase the appearance of whiteness. Many paper supports are coated on both sides with water-repellent synthetic polymers to preclude wetting of the paper fibers during processing. This treatment hastens drying after processing and provides improved dimensional stability and flatness.

Emulsions

Most emulsions are basically a suspension of silver halide crystals in gelatin. The crystals, ranging in size from 2.0 to less than 0.05 μm, are formed by precipitation by mixing a solution of silver nitrate with a solution containing one or more soluble halides in the presence of a protective colloid. The salts used in these emulsions are chlorides, bromides, and iodides. During manufacture, the emulsion is ripened to control crystal size and structure. Chemicals are added in small but significant amounts to control speed, image tone, contrast, spectral sensitivity, keeping qualities, fog, and hardness; to facilitate uniform coating; and, in the case of color films and papers, to participate in the eventual formation of dye instead of metallic silver images upon development. The gelatin, sometimes modified by the addition of synthetic polymers, is more than a simple vehicle for the silver halide crystals. It interacts with the silver halide crystals during manufacture, exposure, and processing and contributes to the stability of the latent image.

After being coated on a support, the emulsion is chilled so that it will set, then dried to a specific moisture content. Many films receive more than one high-sensitive coating, with individual layers as thin as 1.0 μm. Overall thickness of the coatings may range from 5 to 25 μm, depending on the product. Most x-ray films are sensitized on both sides, and some black-and-white films are double coated on one side.

Color films and papers are coated with at least three emulsion layers and sometimes six or more plus filter and barrier layers. A thin, non-sensitized gelatin layer is commonly placed over film emulsions to protect against abrasion during handling. A thicker gelatin layer is coated on the back of most sheet films and some roll films to counteract the tendency to curl, which is caused by the effect of changes in relative humidity on the gelatin emulsion. Certain films are treated to reduce electrification by friction because static discharges can expose the emulsion. The emulsion coatings on photographic papers are generally thinner and more highly hardened than those on film products.

Another class of silver-based emulsions relies on silver-behenate compounds. These materials require roughly ten times more exposure than silver halide emulsions having comparable image-structure properties (resolving power, granularity); are less versatile in terms of contrast, maximum density, and spectral sensitivity; and are less stable both before exposure and after development. However, they have the distinct advantage of being processed through the application of heat (typically at 116 to 127°C) rather than a sequence of wet chemicals. Hence, products of this type are called Dry Silver films and papers.

PIEZOELECTRICITY

Piezoelectricity is electricity, or electric polarity, resulting from the application of mechanical pressure on a dielectric crystal. The application of a mechanical stress produces in certain dielectric (electrically nonconducting) crystals an electric polarization (electric dipole moment per cubic meter) that is proportional to this stress. If the crystal is isolated, this polarization manifests itself as a voltage across the crystal, and if the crystal is short-circuited, a flow of charge can be observed during loading. Conversely, application of a voltage between certain faces of the crystal produces a mechanical distortion of the material. This reciprocal relationship is referred to as the piezoelectric effect. The phenomenon of generation of a voltage under mechanical stress is referred to as the direct piezoelectric effect, and the mechanical strain produced in the crystal under electric stress is called the converse piezoelectric effect.

Piezoelectric materials are used extensively in transducers for converting a mechanical strain into an electrical signal. Such devices include microphones, phonograph pickups, vibration-sensing elements, and the like. The converse effect, in which a mechanical output is derived from an electrical signal input, is also widely used in such devices as sonic and ultrasonic transducers, headphones, loudspeakers, and cutting heads for disk recording. Both the direct and converse effects are employed in devices in which the mechanical resonance frequency of the crystal is of importance. Such devices include electric wave filters and frequency-control elements in electronic oscillator circuits.

PIG IRON

Pig iron is the iron produced from the first smelting of the ore. The melt of the blast furnace is run off into rectangular molds, forming, when cold, ingots called pigs. Pig iron contains small percentages of silicon, sulfur, manganese, and phosphorus, besides carbon. It is useful only for resmelting to make cast iron or wrought iron. Pig iron is either sand-cast or machine-cast. When it is sand-cast, it has sand adhering and fused into the surface, giving more slag in the melting. Machine-cast pig iron is cast in steel forms and has a fine-grained chilled structure, with lower melting point. Pig irons are classified as Bessemer or nonBessemer, according to whether the phosphorus content is below or above 0.10%. There are six general grades of pig iron: low-phosphorus pig iron, with less than 0.03%, used for making steel for steel castings and for crucible steelmaking; Bessemer pig iron, with less than, 0.10% phosphorus, used for Bessemer steel and for acid open-hearth steel; malleable pig iron, with less than 0.20%, used for making malleable iron; foundry pig iron, with from 0.5 to 1%, for cast iron; basic pig iron, with less than 1%, and low-silicon, less than 1%, for basic open-hearth steel; and basic Bessemer, with from 2 to 3%, used for making steel by the basic Bessemer process employed in England.

Because silicon is likely to dissolve the basic furnace lining, it is kept as low as possible, 0.70 to 0.90%, with sulfur not usually over 0.095%. Pig irons are also specified on the basis of other elements, especially sulfur. The sulfur may be from 0.04 to 0.10%, but high-sulfur pig iron cannot be used for the best castings. The manganese content is usually from 0.60 to 1%. Most of the iron for steelmaking is now not cast but is carried directly to the steel mill in car ladles. It is called direct metal.

PIGMENT (MATERIAL)

A finely divided material, pigment contributes to optical and other properties of paint, finishes, and coatings. Pigments are insoluble in the coating material, whereas dyes dissolve in and color the coating. Pigments are mechanically mixed with the coating and are deposited when the coating dries. Their physical properties generally are not changed by incorporation in and deposition from the vehicle. Pigments may be classified according to composition (inorganic or organic) or by source (natural or synthetic). However, the most useful classification is by color (white, transparent, or colored) and by function.

WHITE PIGMENTS

These pigments are essentially transparent to visible light. Because of the difference in refractive index between the pigment particles and the vehicles, white pigments refract the light from a multitude of surfaces and return a substantial portion in the direction of illumination without significant change in the spectral composition of the light.

The common white pigments are titanium dioxide, derived from titanium ores; white lead, from corrosion of metallic lead; zinc oxide, from burning of zinc metal; and lithopone, a mixture of zinc sulfide and barium sulfate. Pure zinc sulfide and antimony oxide are less commonly used.

Titanium dioxide may be crystallized in the rutile or anatase form, depending on the method of production. It may be further modified by surface treatment to control the rate of chalking and other properties. Rutile titanium dioxide has a higher refractive index than anatase and therefore higher hiding power, but it has a somewhat yellow color. Anatase titanium dioxide provides a purer white.

White lead pigments are the oldest of white pigments and were used extensively to provide excellent hiding power, flexibility, and durability to interior and exterior paints and enamels. Consumer protection rulings have all but removed white lead paints from the market, because leaded paint particles were ingested by children, with toxic effects.

Zinc oxide and lithopone pigments were extensively used in paint formulation, but have been superseded by titanium dioxide. Pure zinc oxide pigment is rarely used. Antimony oxide pigment is used chiefly in certain fire-retardant paints.

TRANSPARENT PIGMENTS

The refractive indexes of these pigments are very close to the index of the paint vehicle (about 1.54). They are used to provide bulk, control setting, and contribute to the hardness, durability, and abrasion resistance of the paint film. Because they are commonly used to add bulk to other pigments, they are called extenders. Most transparent pigments are natural minerals reduced to pigment particle size. Among the most commonly used transparent pigments are calcium carbonate (ground limestone, whiting, or chalk), magnesium silicate, bentonite clay, silica, or barites (barium sulfate). Transparent pigments often constitute a substantial portion of a protective coating.

COLORED PIGMENTS

These pigments are available in a wide variety of colors and properties, depending upon the end use. Several hundred have been used; the following are the most common.

> *Red*. Iron oxides, often classified by color, include Indian red, Spanish red, Persian Gulf red, and Venetian red, a mixture of iron oxide and calcium red sulfate. Other red pigments include cadmium red (cadmium selenide) and organic reds, which are usually coal tar derivatives either precipitated in pigment form (toners)

or deposited on a transparent pigment (lakes). Organic reds include toluidines and lithols.

Orange. Chrome orange (basic lead chromate), molybdatee orange (lead chromate-molydate), and various organic toners and lakes are the most common orange pigments.

Brown. Browns are nearly always iron oxides, although certain lakes and toners are used for special purposes.

Yellow. These pigments include natural iron oxides such as ocher or sienna, or synthetic iron oxides, which are stronger and brighter, such as chrome yellow (normal lead chromate) and cadmium yellow (cadmium sulfide), and organic toners and lakes such as Hansa yellow and benzidene yellow.

Green. The most important green pigments are chrome green, a mixture of chrome yellow and Prussian blue; chromium oxide, duller but more permanent; phthalocyanine green, an organic pigment containing copper; and various other organic toners or lakes, often precipitated with phosphotungstic or phosphomolybdic acid.

Blue. The blue pigments include Prussian blue (ferric ferrocyanide, sometimes called milori or Chinese blue, depending on the shade); ultramarine, an inorganic pigment made by fusing soda sulfur and other materials under controlled conditions; phthalocyanine blue, an organic pigment containing copper; and numerous organic toners and lakes.

Purple and Violet. These are nearly all organic toners or lakes. Manganese phosphate is a very weak, inorganic purple pigment.

Black. The vast majority of black pigments consist of finely divided carbon—carbon black, lampblack, and bone black — usually obtained by allowing a smoky flame to impinge on a cold surface. Black iron oxide and certain organic pigments are used where special properties are required.

Special Pigments

Anticorrosive pigments are used to prevent the formation or spread of rust on iron when the metal is exposed by a break in the coating. The most common are red lead, an oxide of lead, and zinc yellow or zinc chromate, a basic chromate of zinc. Other colored chromates are sometimes used. The color of red leads fades rapidly, and the anticorrosive chromates are usually very weak in tinting strength. Metallic lead is sometimes used for anticorrosive paint.

Metallic pigments are small, usually flat particles of metal, prepared for dispersal in coatings. Aluminum is most commonly used because it leafs and forms a smooth, metallic film. The flakes are sometimes colored. Bronze, copper, lead, nickel, stainless steel, and silver appear occasionally. Zinc dust, or powdered zinc, is used more often because of its excellent adhesion to galvanized iron than because of its appearance.

Luminous pigments radiate visible light when exposed to ultraviolet light. Phosphorescent pigments continue to glow for a period after the exciting light has been removed; these are usually sulfides of zinc and other materials, with small amounts of additives that control the phosphorescent properties. Fluorescent pigments lose luminosity as soon as the exciting light is removed; these pigments may be sulfides, although many organic pigments have this property.

Other specialized pigments include pigments that change color at some predetermined temperature, used to indicate hot areas on motors; pigments that give a pearly appearance; and pigments that conduct electricity for printed circuits.

Coarse materials such as pumice are often added when a nonslippery coating is required. Glass beads give a very high degree of refractivity in the direction of illumination and are often used in center-line paints or for signs where night visibility is required. Intumescent pigments puff up under heat, giving a fire-resistant coating.

PLASMA-ARC COATINGS

In this process, a flow of gas, such as argon, is directed through the nozzle of a device called the plasma arc torch. When a high-current electric arc is struck within the torch between a negative tungsten electrode and the positive water-cooled copper nozzle, electrical and aerodynamical effects force the arc through the nozzle, which concentrates and stabilizes it. A substantial portion of the gas flows through the arc and is heated to temperatures as high as 16,649°C and accelerated to supersonic speeds to form an ionized gas jet called plasma. A cool layer of gas next to the nozzle wall effectively insulates the torch from the tremendous heating effect of the arc column.

Particles of refractory coating material, introduced into the plasma in either powder or wire form, are melted and accelerated to high velocity. When these molten particles strike the workpiece, they impact to form a dense, high-purity coating. Sprays of cold carbon dioxide gas, played on the workpiece, keep it from overheating during the process and protect the purity of the coating from air oxidation.

CHARACTERISTICS

The primary advantage of the process is its ability to combine the bulk properties of a base material with the surface properties of a refractory material. Furthermore, the application of the thin, tenacious coatings can be limited to the specific areas of the base material where a coating is needed, and warpage or distortion of precision parts is eliminated because of the low base material temperature maintained during coating.

Whether as-coated or finished, the refractory coatings have extremely good resistance to wear, abrasion, and corrosion and erosion, even under the adverse conditions of high temperature, high load, and lack of lubrication and cooling. When ground and lapped, the coatings give superior performance under conditions of fretting corrosion. Finished coating, when mated with proper materials, have generally lower coefficients of friction than most metal-to-metal combinations. This ratio is also true at elevated temperatures. The coatings have a porosity of less than 1% and an as-coated surface finish of approximately 150 μin. rms (root mean square), which can be finished down to better than 1 μin. rms.

FABRICATION

Coatings can be applied in practically any desired thickness. But only areas that allow the particles free access will be coated evenly. This limitation excludes narrow holes, blind cavities, and deep V-shaped grooves. All corners and edges should be rounded by a minimum 0.38 mm rad or have a minimum chamfer of 0.38 mm by 45° to prevent weak spots.

Several types of parts that can be coated:

1. Long external cylindrical parts
2. Short external cylindrical parts
3. Internal diameters
4. Rectangular flat surfaces
5. Circular flat surfaces

Since plasma-arc coatings can be deposited in practically any desired thickness, it is also possible to fabricate parts by this method. The required thickness is built up on a mandrel formed to the desired internal shape of the finished part, and the mandrel is then removed chemically from the part with acid or caustic.

This method allows intricate shapes to be made of materials that are normally difficult to fabricate. But, as with flame spraying, only areas that allow the particles of coating material sufficient access will be plated evenly.

MATERIALS

Almost any base material can be coated, even certain reinforced plastics, and any known inorganic solid which will melt without decomposition can be used as a coating material. Many basic coatings have already been established including tantalum, palladium, platinum, molybdenum, tungsten, alumina, zirconium diboride, and oxide, and three combinations of tungsten with additives to improve its properties. These additives are zirconia, chromium and alumina.

Other coatings include the refractory metals such as columbium; some of the refractory metal compounds such as the borides of tungsten, columbium, tantalum, titanium, and chromium; the refractory carbides of columbium, hafnium, tantalum, zirconium, titanium, tungsten, and vanadium; the refractory oxides of thorium, hafnium, magnesium, cerium, and aluminum; and other pure metals such as aluminum, copper, nickel, chromium, and boron.

Properties

The properties of the coatings or parts made from all of these materials are equivalent to those of the pure materials themselves.

PLASTER MOLD CASTINGS

Plaster mold casting is primarily used for producing parts in quantities that are too small to justify the use of permanent molds, yet large enough to outweigh the machining costs of sand castings. The process is noted for its ability to produce parts with high dimensional accuracy, smooth and intricate surfaces, and low porosity. On the other hand, it is limited to nonferrous metals (aluminum and copper alloys) and relatively small parts. Also production times are relatively high because the molds take relatively long to make and are not reusable.

The Process

The plaster used for molding generally consists of water mixtures of gypsum or plaster of paris (calcium sulfate) and strengthening binders such as asbestos, magnesium silicate, silicate flour, and others. Impurities such as salts and section thickness are about 40 to 60 mils and bosses and undercuts can be incorporated into the design.

Applications

Plaster mold castings are usually used for medium-production applications, and their cost falls between sand castings and permanent mold castings. Typical parts where the process has been used include gears, ratchet teeth, cams, handles, small housings, pistons, wing nuts, locks, valves, hand tools, and radar parts for aircraft, railroad, household, and electrical uses.

PLASTER OF PARIS

The material (calcined gypsum), $CaSO_4 \cdot 0.5H_2O$, is a white, gray, or pinkish-colored powder prepared by heating gypsum ($CaSO_4 2H_2O$) to remove 75% of its water of crystallization.

When mixed with water and allowed to rehydrate to the dihydrate ($CaSO_4 2H_2O$), there is no apparent action at first, but soon a slight stiffening takes place and shortly after that it "sets" to a solid mass. As set progresses, the mass begins to heat and expand, and final set is not reached until the evolution of heat has ceased and expansion is complete.

Through changes in the manufacturing process, the time of set can be varied widely (from a few minutes to many hours), and linear setting expansion also is controllable from 0.05 to 2.0%. The normal linear setting expansion of pottery plasters is –0.20% in all directions if the cast is unconfined, but under conditions of confinement, all the setting expansion may take place in one direction only.

Applications

Plasters are used in a variety of ceramic industry applications:

1. In a limited way, as chemical additives to glazes, supplying neutral, slightly soluble calcium and sulfate sulfur.
2. As a glass batching material to replace part or all of the salt cake when combined with soda ash in proper proportions. Here, use of plaster eliminates saltwater scumming, retaining the desirable fluxing property of salt cake.
3. As a bedding and leveling agent in grinding and polishing plate glass, plaster cements the glass to the grinding bed during the operation while also being easy to remove from the glass surface.
4. Optical glass mounting. Used to retain optical glass, lenses, prisms, and oculars in position while surfaces are formed to the desired curves by grinding and polishing.

5. Model making. Used in the ceramic industry generally for preparing original models.
6. Metal mold making. When suitably compounded with refractory substances, molds for the casting of nonferrous alloys such as white metal, brass, aluminum alloys, etc. are made with plaster.
7. Low-density insulation. Used to provide green strength to mixtures of clays, nonplastic refractories, and organics.
8. Potter mold and die making. This use constitutes the principal ceramic application of plasters.

PLASTIC ALLOYS AND BLENDS

Plastics, like metals, can be alloyed. And like metal alloys, the resulting materials have different, and often better, properties than those of the base materials making up the alloys.

These alloys consist of two thermoplastics compounded into a single resin. The two polymers must be melt-compatible. Some polymers are naturally compatible; others require the use of compatibilizing agents. The purpose of alloying polymers is to achieve a combination of properties not available in any single resin. There are a great many alloys available, and the list continues to grow. At present, some of the more widely used alloys are ABS/polycarbonate, ABS/polyurethane, polyvinyl chloride (PVC)/acrylic, PVC/CPE (chlorinated polyethylene), polyphenylene oxide (PPO)/polystyrene, nylon/ABS, PPO/PBT thermoplastic polyester, polycarbonate/PBT thermoplastic polyester, polycarbonate/ASA, and polysulfone/ABS.

The plastics most widely used in alloys today are polyvinyl chloride (PVC), ABS, and polycarbonate. These three plastics can be combined with each other or with other types of polymers.

ABS

ABS, in addition to its use with polycarbonate, can also be alloyed with polyurethane. ABS–polycarbonate alloys extend the exceptionally high impact strength of carbonate plastics to section thicknesses over 0.16 cm. ABS–polyurethane alloys combine the excellent abrasion resistance and toughness of the urethanes with the lower cost and rigidity of ABS.

The materials can be injection molded into large parts but cannot be extruded. Typical applications for which they are suitable include such parts as wheel treads, pulleys, low load gears, gaskets, automotive grilles, and bumper assemblies.

ABS is also being successfully combined with PVC and is available commercially in several grades. One of the established grades provides self-extinguishing properties, thus eliminating the need for intumescent (nonburning) coatings in present ABS applications, such as power tool housings, where self-extinguishing materials are required. A second grade possesses an impact strength about 30% higher than general-purpose ABS. This improvement, plus its ability to be readily molded, has resulted in its use for automobile grilles.

PVC

ABS–PVC alloys are available commercially in several grades. Two of the established grades are described above. ABS–PVC alloys also can be produced in sheet form. The sheet materials have improved hot strength, which allows deeper draws than are possible with standard rubber-modified PVC base sheet. They also are nonfogging when exposed to the heat of sunlight. Some properties of ABS–PVC alloys are lower than those of the base resins. Rigidity, in general, is somewhat lower, and tensile strength is more or less dependent on the type and amount of ABS in the alloy.

Another sheet material, an alloy of about 80% PVC and the rest acrylic plastic, combines the nonburning properties, chemical resistance, and toughness of vinyl plastics with the rigidity and deep drawing merits of the acrylics. The PVC–acrylic alloy approaches some metals in its ability to withstand repeated blows. Because of its unusually high rigidity, sheets ranging in thickness from 1.5 to 0.5 cm can be formed into thin-walled, deeply drawn parts.

PVC is also alloyed with CPE to gain materials with improved outdoor weathering or to obtain better low-temperature flexibility. The PVC–CPE alloy applications include wire and cable jacketing, extruded and molded shapes, and film sheeting. Acrylic-base alloys with a polybutadiene additive have also been developed, chiefly for blow-molded products. The acrylic content can range from 50 to 95%, depending on the application. Besides blow-molded bottles, the alloys are suitable for thermoformed products such as tubs, trays, and blister pods. The material is rigid and tough and has good heat-distortion resistance up to 82°C.

PPO

Another group of plastics, PPO, can be blended with polystyrene to produce a PPO–polystyrene alloy with improved processing traits and lower cost than nonalloyed PPO. The addition of polystyrene reduces tensile strength and heat deflection temperature somewhat and increases thermal expansion.

PLASTIC LAMINATES

These are resin-impregnated paper or fabric, produced under heat and high pressure; they are also referred to as high-pressure plastic laminates. Two major categories are decorative thermosetting laminates and industrial thermosetting laminates. Most of the decorative thermosetting laminates are a paper base, and are known generically as papreg. Decorative laminates are usually composed of a combination of phenolic- and melamine-impregnated sheets of paper. The final properties of the laminate are related directly to the properties of the paper from which the laminate is made.

Early laminates were designated by trade names, such as Bakelite, Textolite, Micarta, Condensite, Dilecto, Phenolite, Haveg, Spauldite, Synthane, and Formica. These are designated as various types of laminates with a decorative facing layer for such uses as tabletops. Trade names now usually include a number or symbol to describe the type and grade. Textolite, for example, embraces more than 70 categories of laminates subdivided into use-specification grades, all produced in many sizes and thicknesses. Textolite 11711 is an electronic laminate for such uses as multilayer circuit boards. It is made with polyphenolene oxide resin, and may have a copper or aluminum cladding.

Forms

Industrial thermosetting laminates are available in the form of sheet, rod, and rolled or molded tubing. Impregnating resins commonly used are phenolic, polyester, melamine, epoxy, and silicone. The base material, or reinforcement, is usually one of the following: paper, woven cotton or linen, asbestos, glass cloth, or glass mat.

Laminating resins may be marketed under one trade name by the resin producer and other names by the molders of the laminate. Paraplex P resins, for example, comprise a series of polyester solutions in monomeric styrene that can be blended with other resins to give varied qualities. But Panelyte refers to the laminates that are made with phenolic, melamine, silicone, or other resin, for a variety of applications.

PLASTIC POWDER COATINGS

Although many different plastic powders can be applied as coatings, vinyl, epoxy, and nylon are most often used. Vinyl and epoxy provide good corrosion and weather resistance as well as good electrical insulation. Nylon is used chiefly for its outstanding wear and abrasion resistance. Other plastics frequently used in powder coating include chlorinated polyethers, polycarbonates, acetals, cellulosics, acrylics, and fluorocarbons.

Several different methods have been developed to apply these coatings. In the most popular process, fluidized bed, parts are preheated and then immersed in a tank of finely divided plastic powders, which are held in a suspended state by a rising current of air. When the powder particles contact the heated part, they fuse and adhere to the surface, forming a continuous, uniform coating. Another process, electrostatic spraying, works on the principle that oppositely charged materials attract each other. Powder is fed through a gun, which applies an electrostatic charge opposite to that applied to the part to be coated. When the charged particles leave the gun, they are attracted to the part where they cling until fused together as a plastic coating. Other powder application methods include flock and flow coating, flame and plasma spraying, and a cloud-chamber technique.

PLASTICS

Plastics are a major group of materials that are primarily noncrystalline hydrocarbon substances composed of large molecular chains whose major element is carbon. The three terms — *plastics*, *polymers*, and *resins* — are sometimes used interchangeably to identify these materials. However, the term *plastics* has now come to be the commonly used designation.

The first commercial plastic, Celluloid, was developed in 1868 to replace ivory for billiard balls. Phenolic plastics, developed by Baekeland and named Bakelite after him, were introduced around the turn of the century. A plastic material, as defined by the Society of the Plastics Industry, is "any one of a large group of materials consisting wholly or in part of combinations of carbon with oxygen, hydrogen, nitrogen, and other organic and inorganic elements which, while solid in the finished state, at some stage in its manufacture is made liquid, and thus capable of being formed into various shapes, most usually through the application, either singly or together, of heat and pressure."

There are two basic types of plastics based on intermolecular bonding. Thermoplastics, because of little or no cross-bonding between molecules, soften when heated and harden when cooled, no matter how often the process is repeated. Thermosets, on the other hand, have strong, intermolecular bonding. Therefore, once the plastic is set into permanent shape under heat and pressure, reheating will not soften it.

Within these major classes, plastics are commonly classified on the basis of base monomers. There are over two dozen such monomer families or groups. Plastics are also sometimes classified roughly into three stiffness categories: rigid, flexible, and elastic. Another method of classification is by the "level" of performance or the general area of application, using such categories as engineering, general-purpose, and specialty plastics, or the two broad categories of engineering and commodity plastics.

Some major characteristics of plastics that distinguish them from other materials, particularly metals:

1. They are essentially noncrystalline in structure.
2. They are nonconductors of electricity and are relatively low in heat conductance.
3. They are, with some important exceptions, resistant to chemical and corrosive environments.
4. They have relatively low softening temperatures.
5. They are readily formed into complex shapes.
6. They exhibit viscoelastic behavior — that is, after an applied load is removed, plastics tend to continue to exhibit strain or deformation with time.

Polymers can be built of one, two, or even three different monomers, and are termed homopolymers, copolymers, and terpolymers, respectively. Their geometrical form can be linear or branched. Linear or unbranched polymers are composed of monomers linked end-to-end to form a molecular chain that is like a simple string of beads or a piece of spaghetti. Branched polymers have side chains of molecules attached to the main linear polymer. These branches can be composed either of the basic linear monomer or of a different one. If the side molecules are arranged randomly, the polymer is atactic; if they branch out on one side of the linear chain in the same plane, the polymer is isotactic; and if they alternate from one side to the other, the polymer is syndiotactic.

Plastics are produced in a variety of different forms. Most common are plastic moldings, which range in size from 2 cm to several meters. Thermoplastics, such as polyvinyl chloride (PVC) and polyethylene, are widely used in the form of plastic film and plastic sheeting. The term *film* is used for thicknesses up to and including 0.25 cm, while sheeting refers to thicknesses over that.

Both thermosetting and thermoplastic materials are used as plastic coatings on metal, wood, paper, fabric, leather, glass, concrete, ceramics, or other plastics. There are many coating processes, including knife or spread coating, spraying, roller coating, dipping, brushing, calendering, and the fluidized-bed process. Thermosetting plastics are used in high-pressure laminates to hold together the reinforcing materials that comprise the body of the finished product. The reinforcing materials may be cloth, paper, wood, or glass fibers. The end product may be plain flat sheets, or decorative sheets as in countertops, rods, tubes, or formed shapes.

PLASTICS ADDITIVES

Almost all plastics contain one or more additive materials to improve their physical properties, processing characteristics, or to reduce costs. There is a wide range of additives for use with plastics, including antimicrobials, antistatic agents, clarifiers, colorants, fillers, flame retardants, foaming agents, heat stabilizers, impact modifiers, light stabilizers, lubricants, mold-release agents, odorants, plasticizers, reinforcements, and smoke retardants.

Fillers

Fillers are probably the most common of the additives. They are usually used to either provide bulk or modify certain properties. Generally, they are inert and thus do not react chemically with the resin during processing. The fillers are often cheap and serve to reduce costs by increasing bulk. For example, wood flour, a common low-cost filler, sometimes makes up 50% of a plastic compound. Other typical fillers are chopped fabrics, asbestos, talc, gypsum, and milled glass. Besides lowering costs, fillers can improve properties. For example, asbestos increases heat resistance, and cotton fibers improve toughness.

Plasticizers

Plasticizers are added to plastics compounds either to improve flow during processing by reducing the glass transition temperature or to improve properties such as flexibility. Plasticizers are usually liquids that have high boiling points, such as certain phthalates. Substances that are themselves polymers of low molecular weight, such as polyesters, are also used as plasticizers.

Stabilizers

Stabilizers are added to plastics to help prevent breakdown or deterioration during molding or when the polymer is exposed to sunlight, heat, oxygen, ozone, or combinations of these. Thus there is a wide range of compounds, each designated for a specific function. Stabilizers can be metal compounds, based on tin, lead, cadmium, barium, and others. And phenols and amines are added antioxidants that protect the plastic by diverting the oxidation reactions to themselves.

Catalysts

Catalysts, by controlling the rate and extent of the polymerization process in the resin, allow the curing cycle to be tailored to the processing requirements of the application. Catalysts also affect the shelf life of the plastics. Both metallic and organic chemical compounds are used as catalysts.

Colorants

Colorants, added to plastics for decorative purposes, come in a wide variety of pigments and dyestuffs. The traditional colorants are metal-base pigments such as cadmium, lead, and selenium. More recently, liquid colorants, composed of dispersions of pigments in a liquid, have been developed.

Flame Retardants

Flame retardants are added to plastic products that must meet fire-retardant requirements, because polymer resins are generally flammable, except for such notable exceptions as PVC. In general, the function of fire retardants is limited to the spread of fire. They do not normally increase heat resistance or prevent the plastic from charring or melting. Some fire-retardant additives include compounds containing chlorine or bromine, phosphate-ester compounds, antimony thrioxide, alumina trihydrate, and zinc borate.

REINFORCED MATERIALS

Reinforcement materials in plastics are not normally considered additives. Usually in fiber or mat form, they are used primarily to improve mechanical properties, particularly strength. Although asbestos and some other materials are used, glass fibers are the predominant reinforcement for plastics.

PLASTICS PROCESSING

Plastics processing includes those methods and techniques used to convert plastics materials in the form of pellets, granules, powders, sheets, fluids, or preforms into formed shapes or parts. The plastic materials may contain a variety of additives that influence the properties as well as the processibility of the plastics. After forming, the part may be subjected to a variety of ancillary operations such as welding, adhesive bonding, and surface decorating (painting, metallizing).

As with other materials of construction, processing of plastics is but one step in the normal design-to-finished-part sequence. The choice of process is influenced by economic considerations, number and size of finished parts, and complexity of postfinishing operations, as well as the adaptability of the plastics to the process.

INJECTION MOLDING

This process consists of heating and homogenizing plastics granules in a cylinder until they are sufficiently fluid to allow for pressure injection into a relatively cold mold where they solidify and take the shape of the mold cavity. For thermoplastics, no chemical changes occur within the plastic, and consequently the process is repeatable. Injection molding of thermosetting resins differs primarily in that the cylinder heating is designed to homogenize and preheat the reactive materials, and the mold is heated to complete the chemical cross-linking reaction to form an intractable solid. Solid particles, in the form of pellets or granules, constitute the main feed for injection moldable plastics. The major advantages of the injection molding process are the speed of production, minimal requirements for postmolding operations, and simultaneous multipart molding.

The development of reaction injection molding (RIM) allowed the rapid molding of liquid materials. In these processes, cold or warm, two highly reactive, low-molecular weight, low-viscosity resin systems are first injected into a mixing head and from there into a heated mold, where the reaction to a solid is completed.

Polymerization and cross-linking occur in the mold. This process has proved particularly effective for high-speed molding of such materials as polyurethanes, epoxies, polyesters, and nylons.

EXTRUSION

In this process, plastic pellets or granules are fluidized, homogenized, and continuously formed. Products made this way include tubing, pipe, sheet, wire and substrate coatings, and profile shapes. The process is used to form very long shapes or a large number of small shapes that can be cut from the long shapes. The homogenizing capability of extruders is used for plastics blending and compounding. Pellets used for other processing methods, such as injection molding, are made by chopping long filaments of extruded plastic.

BLOW MOLDING

This process consists of forming a tube (called a parison) and introducing air or other gas to cause the tube to expand into a free-blown hollow object or against a mold for forming into a hollow object with a definite size and shape. The parison is traditionally made by extrusion, although injection molded tubes have gained prominence because they do not require postfinishing, have better dimensional tolerances and wall thicknesses, and can be made unsymmetrical and in higher volume production.

Thermoforming

Thermoforming is the forming of plastic sheets into parts through the application of heat and pressure. The pressure can be obtained through use of pneumatics (air) or compression (tooling) or vacuum. Tooling for this process is the most inexpensive compared to other plastic processes, accounting for the popularity of the method. It can also accommodate very large parts as well as small parts, which are useful in low-cost prototype fabrication.

Rotational Molding

In this process, finely ground powders are heated in a rotating mold until melting or fusion occurs. If liquid materials, such as vinyl plastisols, are used, the process is often called slush molding. The melted or fused resin uniformly coats the inner surface of the mold. When cooled, a hollow finished part is removed. The processes require relatively inexpensive tooling, are scrap-free, and are adaptable to large, double-walled, hollow parts that are strain-free and of uniform thickness. The processes can be performed by relatively unskilled labor. On the other hand, the finely ground plastics powders are more expensive than pellets or sheet, thin-walled parts cannot be easily made, and the process is not suited for large production runs of small parts.

Compression and Transfer Molding

Compression molding is one of the oldest molding techniques and consists of charging a plastics powder or preformed plug into a mold cavity, closing a mating mold half, and applying pressure to compress, heat, and cause flow of the plastic to conform to the cavity shape. The process is primarily used for thermosets, and consequently the mold is heated to accelerate the chemical cross-linking.

Transfer molding is an adaptation of compression molding in that the molding powder or preform is charged to a separate preheating chamber and, when appropriately fluidized, injected into a closed mold. The process predates, yet closely parallels, the early techniques of ram injection molding of thermoplastics. It is most used for thermosets, and is somewhat faster than compression molding. In addition, parts are more uniform and more dimensionally accurate than those made by compression molding. See Figures P.2A and P.2B showing compression molding and transfer molding.

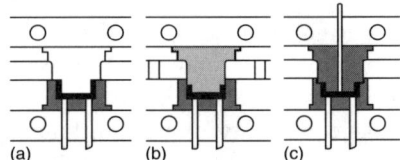

FIGURE P.2A Three types of compression molds. (a) Flash-type, (b) positive, (c) semipositive. (From *McGraw-Hill Encyclopedia of Science and Technology*, 8th ed., Vol. 14, McGraw-Hill, New York, 43. With permission.)

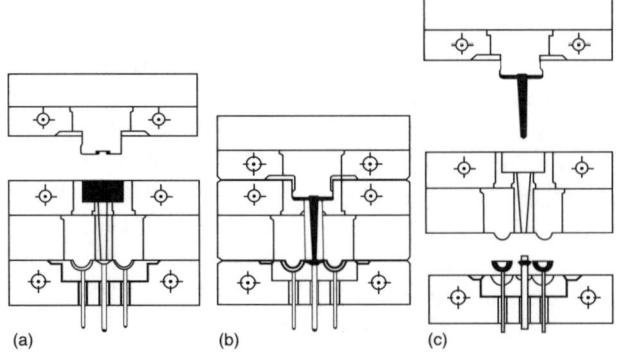

FIGURE P.2B Transfer molding. (a) In the molding cycle, material is first placed in the transfer pot. (b) It is then forced through an orifice into the closed mold. (c) When the mold opens, the cull and sprue are removed as a unit, and the part is lifted out of the cavity by ejector pins. (From *McGraw-Hill Encyclopedia of Science and Technology*, 8th ed., Vol. 14, McGraw-Hill, New York, 43. With permission.)

Foam Processes

Foamed plastics materials have achieved a high degree of importance in the plastics industry. Foams can be made in a range from soft and flexible to hard and rigid. There are three types of cellular plastics: blown (expanded matrix, such as a natural sponge), syntactic (the encapsulation of hollow organic or inorganic microspheres in the matrix), and structural (dense outer skin surrounding a foamed core).

There are seven basic processes used to generate plastics foams. They include the incorporation of a chemical blowing agent that generates gas (through thermal decomposition) in the polymer liquid or melt; gas injection into the melt, which expands during pressure relief; generation of gas as a by-product of a chemical condensation reaction during cross-linking; volatization of a low-boiling liquid (for example, Freon) through the exothermic heat of reaction; mechanical dispersion of air by mechanical means (whipped cream); incorporation of nonchemical gas-liberating agents (adsorbed gas on finely divided carbon) into the resin mix, which is released by heating; and expansion of small beads of thermoplastic resin containing a blowing agent through the external application of heat.

Structural foam differs from other foams in that the part is produced with a hard integral skin on the outer surfaces and a cellular core in the interior. They are made by injection-molding liquefied resins containing chemical blowing agents. The initial high injection pressure causes the skin to solidify against the mold surface without undergoing expansion. The subsequent reduction in pressure allows the remaining material to expand and fill the mold. Coinjection (sandwich) molding permits injection molding of parts containing a thermoplastic core within an integral skin of another thermoplastic material. When the core is foam, an advanced form of structural foam is produced.

Reinforced Plastics/Composites

These are plastics whose mechanical properties are significantly improved because of the inclusion of fibrous reinforcements. The wide variety of resins and reinforcements that constitute this group of materials led to the more generalized description "composites."

Composites consist of two main components, the fibrous material in various physical forms and the fluidized resin, which will convert to a solid. There are fiber-reinforced thermoplastic materials, and these are typically processed in standard thermoplastic processing equipment.

The first step in any composite fabrication procedure is the impregnation of the reinforcement with the resin. The simplest method is to pass the reinforcement through a resin bath and use the wet impregnate directly. For easier handling and storage, the impregnated reinforcement can be subjected to heat to remove impregnating solvents or advance the resin cure to a slightly tacky or dry state. The composite in this form is called a prepreg. This B-stage condition allows the composite to be handled, yet the cross-linking reaction has not proceeded so far as to preclude final flow and conversion to a homogeneous part, when further heat or pressure is applied.

Premixes, often called bulk molding compounds, are mixtures of resin, inert fillers, reinforcements, and other formulation additives that form a puttylike rope, sheet, or preformed shape.

Converting these various forms of composite precursors to final part shape is achieved in a number of ways. Hand layup techniques entail an open mold onto which the impregnated reinforcement or prepreg is applied layer by layer until the desired thicknesses and contours are achieved; see part a of Figure P.3 depicting techniques for producing reinforced plastics and composites. The thermoset resin is then allowed to harden (cure). Often the entire configuration will be enclosed in a transparent sealed bag (vacuum bag) so that a vacuum can be applied to remove unwanted volatile ingredients and entrained air for improved densification of the composite (part b of Figure P.3). External heat may be applied to accelerate the process. Often a bagged laminate will be inserted into an autoclave so that the synergistic effects of heat, vacuum, and pressure can be obtained. At times, a specially designed spray apparatus is used that simultaneously mixes and applies a coating of resin and chopped reinforcement to a mold surface (part c). This

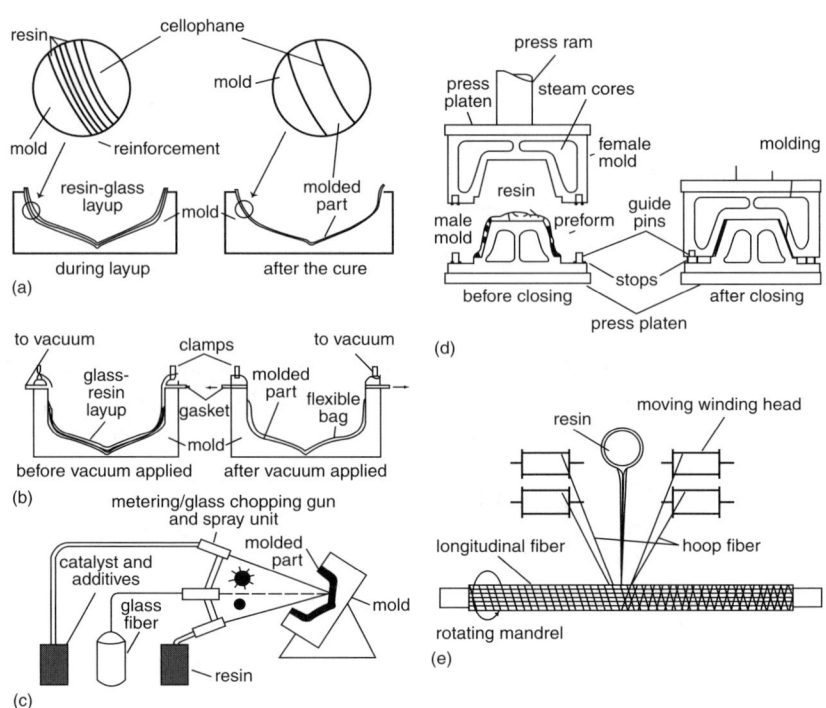

FIGURE P.3 Techniques for producing reinforced plastics and composites. (a) Hand lay-up technique for reinforced thermosets; (b) vacuum bag molding method; (c) spray-up method; (d) matched metal die molding; (e) filament winding. (From *McGraw-Hill Encyclopedia of Science and Technology*, 8th ed., Vol. 14, McGraw-Hill, New York, 45. With permission.)

technique is particularly useful for large structures such as boat hulls and truck cabs, covering complex shapes as readily as simple configurations.

Matched die compression molding resembles normal compression molding, although the pressures are considerably lower (part d). Premix molding is essentially the same process, except that premix compounds are used. *Pultrusion* is a term coined to describe the process for the continuous extrusion of reinforced plastics profiles. Strands of reinforcement are drawn (pulled) through an impregnating tank, the forming die, and finally a curing area (radio-frequency exposure). Filament winding is a process in which the continuous strands of reinforcement are drawn through an impregnating bath and then wound around a mandrel to form the part (part e). This technique is most used for the formation of hollow objects such as chemical storage tanks or chemically resistant pipe. Advanced automated processes, such as ply cutting, tape laying and contouring, and ply lamination are providing improved parts and reduced costs particularly in the aerospace industry.

Casting and Encapsulation

Casting is a low-pressure process requiring nothing more than a container in the shape of the desired part. For thermoplastics, liquid monomer is poured into the mold and, with heat, allowed to polymerize in place to a solid mass. For vinyl plastisols, the liquid is fused with heat. Thermosets, usually composed of liquid resins with appropriate curatives and property-modifying additives, are poured into a heated mold in which the cross-linking reaction completes the conversion to a solid. Often a vacuum is applied to gasify the resultant part for improved homogeneity.

Encapsulation and potting are terms for casting processes in which a unit or assembly is encased or unpregnated, respectively, with a liquid plastic, which is subsequently hardened by fusion or chemical reaction; Figure P.4 depicts low-pressure plastics processes. These

PLASTISOL COATINGS

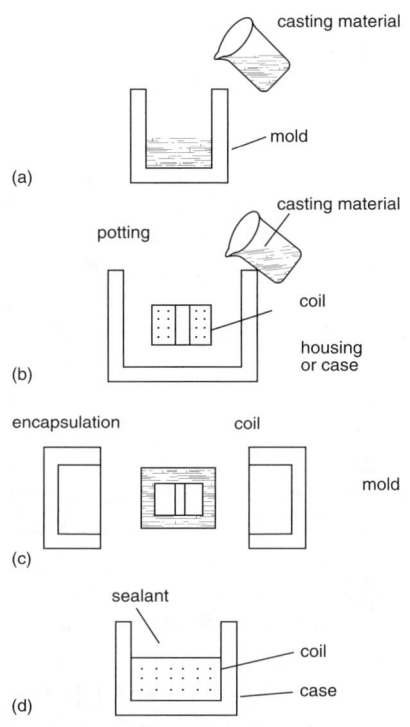

FIGURE P.4 Low-pressure plastics processes: (a) casting, (b) potting, (c) encapsulation, and (d) sealing. (From *McGraw-Hill Encyclopedia of Science and Technology*, 8th ed., Vol. 14, McGraw-Hill, New York, 44. With permission.)

processes are predominant in the electrical and electronic industries for the insulation and protection of components.

CALENDERING

In the calendering process, a plastic is masticated between two rolls that squeeze it out into a film that then passes around one or more additional rolls before being stripped off as a continuous film. Fabric or paper may be fed through the latter rolls, so that they become impregnated with the plastic.

PLASTISOL COATINGS

Vinyl plastisols, or pastes, as they are described in Europe, are suspensions of vinyl resin in nonvolatile oily liquids known as plasticizers. They vary in viscosity from a motor oil consistency to a puttylike dough. In the more viscous state, the plastisol is termed *plastigel*, while the more fluid materials, to which volatile diluents have been added, are known as *modified plastisols*. Modified plastisols differ from organosols in the function of the volatile components. In organosols, the volatiles are used as resin dispersants, whereas in modified plastisols they serve as diluents to adjust fluidity and are generally present in small quantities.

The polyvinyl chloride resin, resembling confectioner's sugar, is blended into a mixture of one or more plasticizers to form a suspension. This fluid remains essentially unchanged until heat is applied. During the heating process, the dispersion first sets or gels; this is followed by solution or fusion of the resin in the hot plasticizer to form a single-phase solid solution. Upon cooling, the coating assumes the properties of a tough, rubbery plastic.

These plastisols have no adhesion to metals or dense, nonporous substrates and consequently require the use of adhesive primers for bonding. To a large extent, the nature of these primers determines the suitability of plastisol coatings for specific applications.

APPLICATION METHODS

The fluidity or absence of fluidity in the liquid plastisol is sometimes deceiving. These materials are supplied at very high solids content and consequently exhibit non-Newtonian flow (viscosity varies with applied shear). While the viscosity cup is satisfactory for many paints and lacquers, and may even suffice for organosols, it can only serve to mislead the plastisol user. Viscosity of plastisols should be specified and measured using a viscosimeter capable of operating over a range of shear rates preferably within the area of use.

Spread Coating

Roller and knife coaters are the two major types of spread-coating equipment used for handling plastisols. Fabric, paper, and even strip steel are all being coated with this type of process. Compound viscosity characteristics, speed of coating, clearance between the web and the knife or roll, and type and angle of the knife are all factors in determining the quality of the coating. Heavy paper and fabric coatings, which will withstand folding and forming, may be applied to porous stock without danger of penetration or strike through. The momentary application of heat to the coated side of the stock will fuse the plastisol with a minimum of thermal action on the paper.

Plastisol-coated strip steel is currently being produced by roller-coating processes for use by appliance and other manufacturers.

Dip Coating

Two dip-coating processes are available. In hot dipping, the object for coating is prebaked, prior to immersion in the plastisol. The heat content in the article serves to gel a deposit on the surface of the object. This gelled coating must then be fused by baking. Plastisol formulation and temperature, dipping rate, mass, shape, and heat content of the article to be coated all serve to determine the thickness and nature of deposit.

Cold-dip processes permit the application of from 1 to 60 mils per coat without the necessity of a prebake operation. Cold dips lend themselves to conveyor line coating of products. These coatings have a high yield value and permit controlled film thicknesses without the presence of sags or drips to mar the appearance.

Spray Coating

Plastisols may be spray-applied through either pressure or suction guns, but generally the pressure equipment is preferred because it permits faster delivery with a minimal use of volatile diluents. Airless spray equipment is currently available that operates at fluid pressures in excess of 13.6 MPa and requires no atomizing air. This airless-spray process is reported to give extreme smoothness to highly thixotropic formulations.

Properties

Although plastisols may be modified with slight additions of volatile diluents, their fluidity is mainly due to the presence of large quantities of plasticizer. Unlike the fluid phase of the organosol, which is largely volatile, the plasticizers remain behind after baking, as a portion of the fused film. Thus, the plastisol tends to be softer and more resilient than the organosol. Its low volatile content (or its absence altogether) permits wide baking latitude by eliminating the problem of mud cracking and reducing the solvent entrapment tendency found in organosols. Film thicknesses may range from 2 to 250 mils per coat.

Uses

Plastisols, because of their ability to be applied readily in heavy thicknesses, found early success in the electroplating field as rack coatings. The plastisol serves as an insulation, confining current to the work being plated, and is resistant to chemical attack by plating solutions. The use of plastisols as linings for tanks, chemical equipment, and steel drums followed.

In fabric coating, they have replaced solution coatings by eliminating the need for expensive solvents lost in the baking operation. Rubber has been replaced by vinyl plastisols and organosols as coatings for wire baskets because of their superior resistance to moisture and detergents.

One of the most dramatic applications of plastisols is as a lining for kitchen dishwashers. The use of a plastisol lining permitted a lightweight tub design not possible with porcelain enameling in which firing resulted in buckling and warping of light-gauge steel. Plastisols also served to reduce scrap units since defects may be readily patched and repaired. The resistance to impact damage, etching, and enamel erosion are other factors that prompted manufacturers to select plastisols for this application.

A list of applications and the properties related to the specific application follow:

Application	Related Property
Industrial: Tool handles, stair treads, conveyor hooks, conveyor rollers, railings	Resiliency, thermal and electrical insulating qualities, resistance to abrasion
Electrical: Bus bars, conduit boxes, battery clamps and cases, toggle switches, electroplating racks, and plating barrels	Dielectric strength, electrical resistivity, resistance to moisture and chemicals
Linings: Tanks, duct-work, pumps, filter presses, centrifugal cleaners, dishwasher tubs, piping, drums, and shipping containers	Resistance to abrasion and impact, resistance to moisture and chemicals
Wire Goods: Dish-drain baskets, egg baskets, deep-freeze baskets, refrigerator shelves, record racks, clothes hangers	Resistance to moisture, detergents, and staining, resiliency
Miscellaneous: Bottles and glassware, glove coating, bobby-pin coatings	Resiliency and aesthetic qualities, abrasion resistance, softness

PLASTISOLS

Plastisols are dispersions of high-molecular-weight vinyl chloride polymer or copolymer resins in nonaqueous liquid plasticizers, which do not dissolve the resin at room temperature. Plastisols are converted from liquids to solids by fusing under heat, which causes the resin to dissolve in the plasticizers.

There are many advantages in molding with plastisols each varying in importance according to the particular type of molding application. Vinyl plastisol is supplied as a liquid and consequently is easy to handle. The material requires no catalysts or curing agents to convert it to a solid, only moderate heat in the range of 149 to 204°C. Vinyl plastisol does not require a long baking cycle nor high pressures to fuse and shape it. Consequently, lightweight inexpensive molds are suitable for molding. It can be formulated to have a virtually indefinite shelf life. Plastisols are usually 100% solids and shrinkage from the mold is at an absolute minimum, thus assuring that the molded object is exact and consistent.

PROPERTIES

Chemical and physical properties of plastisols can be varied throughout a wide range. This versatility makes plastisols adaptable to a multitude of end uses.

The following general ranges indicate the properties that may be compounded into plastisols:

- Specific Gravity: 1.05 to 1.35
- Tensile Strength: As required to 27.2 MPa
- Elongation: As required to 600%
- Flexibility: Good to a temperature as low as −55°C
- Hardness: From 10 to 100 on the Shore A Durometer Scale; up to 80 on the Shore D Durometer Scale
- Chemical Resistance: Outstanding to most acids, alkalies, detergents, oils, and solvents
- Heat Resistance: Can resist 107°C for as long as 2000 h and 232°C for over 2 h
- Electrical Properties: Dielectric strength at a minimum of 400 v/mil in thicknesses of 3 mils and over
- Flammability: Slow burning to self-extinguishing
- Colors: All colors available including phosphorescent and fluorescent shades

MOLDING METHODS

There are several different methods by which plastisols may be molded.

Pour and Injection

Two of the simplest methods are pour molding and low-pressure injection molding. The first method entails merely pouring plastisol into a cavity until it is filled and subsequently fusing the compound. This system is used in manufacturing products such as plastic doilies, sink stoppers, and display plaques.

If the mold is closed, a low-pressure injection system such as a grease gun can be used to inject the liquid plastisol into the cavity. A low-pressure injection mold should be designed with bleeders at the extremities of the cavity to ensure complete filling of the mold, as well as to relieve the minor pressure on the mold surface caused by expansion during the heating. Laboratory models, novelties, and electrical harnesses are products that are commonly low-pressure injection-molded with vinyl plastisol.

Heating sources used for molding plastisols vary according to the particular product under consideration. For shallow, open molds, such as those for plastic doilies, radiant heat would be satisfactory. When this type of heat is used, the material thickness should not be so great that the open surface exposed to the heat overfuses in the time taken for the temperature to reach the mold surface of the part. Conductive heat is another source for fusing plastisol, in particular when closed molds are employed. Immersing the mold in a hot bath or using cartridge heaters are two conductive heating methods. A more commonly used method of fusion is convection heat. The advantage of a convection oven, in particular a forced air type, is that the entire inner area of the oven, and consequently the entire surface of the mold, is maintained at a constant temperature, ensuring a more even heat transfer through the mold.

Pour and low-pressure injection molds usually are made of aluminum, electroformed copper, brass, or steel. The thickness of the metal should be kept to a minimum for good heat transfer yet should be thick enough to withstand expansion pressure during fusing.

In-Place Molding

Another molding process, which is somewhat similar to the foregoing methods, is in-place molding. This method permanently attaches plastisol to another component during the fusing process. The combination serves an important functional purpose and usually eliminates the need for several assembly steps. In-place molding is most commonly used in forming seals and gaskets of various sorts.

Gaskets are applied to vitrified clay pipe by pouring plastisol into special molds on the bell and spigot ends. Such gaskets compensate for inherent out-of-roundness of the pipe. Flowed in gaskets also are applied to bottle caps and jar lids.

Dip Molding

When a hollow object is to be molded and the internal dimensions are of importance, many times a dip molding process is employed. The metal molds are shaped according to the interior design of the molded object. They are usually solid and are made of cast or machined aluminum, machined brass, steel, or ceramic. These molds are preheated to a temperature in the range of 149 to 204°C, dipped into the plastisol, and allowed to dwell until the proper thickness has gelled on the mold. To eliminate drips or sags, the mold is withdrawn at a rate that does not exceed the rate at which the liquid residue drains from the gelled coating. The thickness of this coating can be varied by altering the preheating time and temperature, as well as the dwell time. A dip molding system can easily be conveyorized. In such a system mandrels holding the molds would be conveyed through a preheat oven, dipping station, fusing oven, cooling station, and stripping station.

Slush Molding

Another method for molding hollow pieces is slush molding. In this process an open-end metal mold is heated to a temperature in the range of 149 to 204°C and then filled with plastisol. The plastisol is allowed to dwell until the desired thickness has gelled on the inner surface of the mold and then the remaining liquid in the mold is poured back into a reservoir for use again. The mold with the gelled inner coating is placed in an oven where the plastisol is fused. Upon cooling, the plastisol part is stripped from the mold, retaining the design of its inner surface on the exterior of the piece. Molds of electroformed copper or fine sand-cast aluminum are usually used for slush molding.

The above process is generally known as the single-pour system of slush molding. For a mold of intricate detail, a two-pour method often is used. In this process, the mold is filled when it is cold, vibrated to remove bubbles, and then emptied, leaving a thin film of plastisol on the inner surface. In this way, the plastisol does not have a chance to gel before flowing into the mold extremities.

Rotational Molding

Completely enclosed hollow parts can be produced by rotational molding. A measured amount of plastisol is poured into one half of a two-piece mold. The mold is closed and rotated in two or more planes while being heated. During this rotation, the plastisol flows, gels, and fuses evenly over the interior walls of the mold. Molds for this operation are either electroformed copper or cast or machined aluminum, and the molds are arranged in clusters or "gangs" so that the maximum number of molds can be operated per spindle.

In rotational molding it is possible to vary the thickness of the walls of the molded piece. One way this can be accomplished is by rotating the mold more in one plane than in another.

A few familiar products manufactured by this process are toys and novelties such as dolls and beach balls, swimming pool floats, and artificial fruit.

Combinations

In many cases several of the above molding methods are combined to produce a product made from plastisol. For example, vinyl foam products such as armrests, toys, or electrical harnesses are manufactured by first forming a tough vinyl skin by spraying, slush molding, or rotational molding. The interior then is formed by casting, low-pressure injection, or rotational molding a vinyl plastisol foam within the pregelled skin.

PLATINUM

A whitish-gray metal, symbol Pt, platinum is more ductile than silver, gold, or copper, and is heavier than gold. The melting point is 1769°C, and the specific gravity is 21.45. The hardness of the annealed metal is 45 Brinell, and its tensile strength is 117 MPa; when hard-rolled, the Brinell hardness is 97 and tensile strength 234 MPa. Electrical conductivity is about 16% that of copper. The metal has a face-centered cubic lattice structure and it is very ductile and malleable. It is resistant to acids and alkalies, but dissolves in aqua regia. Platinum is widely used in jewelry, but because of its heat resistance and chemical resistance it is also valued for electric contacts and resistance wire, thermocouples, standard weights, and laboratory dishes. Generally too soft for use alone, it is almost always alloyed with harder metals of the same group, such as osmium, rhodium, and iridium. An important use of the metal, in the form of gauze, is as a catalyst. Platinum gauze is of high purity in standard meshes of 18 to 31/cm, with wire from 0.020 to 0.008 cm in diameter. Dental foil is 99.99% pure and of maximum softness. Platinum foil for other uses is made in thicknesses as thin as 0.0005 cm. Platinum powder comes in fine submesh particle size. It is made by chemical reduction and is at least 99.9% pure, with amorphous particles 0.3 to 3.5 μm in diameter. Platinum flake has the powder particles in the form of tiny laminar platelets that overlap in the coating film.

Because of the high resistance of the metal to atmospheric corrosion even in sulfur environments, platinum coatings and electroplating are used on springs and other functioning parts of instruments and electronic devices where precise operation is essential. Coatings are also produced by vapor deposition of platinum compounds; thin coatings, 0.0005 cm or less, are made by painting the surface with a solution of platinum powder in an organic vehicle and then firing to drive off the organic material, leaving an adherent coating of platinum metal.

Platinum is sometimes used in glazes to obtain luster and metallic effects. Liquid bright platinum and liquid bright palladium (an element of the platinum group) are preparations used in metallic decorations. As platinum produces a better silver effect than silver itself and is less likely to tarnish, platinum is preferred to that metal. A luster produced from a strong solution of platinum chloride and spirits or oil of lavender upon firing gives a steely appearance which is nearly opaque. Another method consists of precipitating the metal from its solution in water by heating it with a solution of caustic soda and glucose. The metal is mixed with 5% bismuth subnitrate, applied to the ware by painting, and fired in a reducing atmosphere.

PLATINUM ALLOYS

Platinum is alloyed to obtain greater hardness, strength, and electrical resistivity. Because most applications require freedom from corrosion, the other platinum metals are usually employed as alloying agents.

Platinum–Iridium

Iridium is the addition to platinum most often used to provide improved mechanical properties. It increases resistance to corrosion while the alloy retains its workability. Up to 20% iridium, the alloys are quite ductile. With higher iridium content fabrication becomes difficult.

Platinum–iridium alloys are employed for instruments, magneto contacts, and jewelry. The alloys are hard, tough, and noncorrosive. An alloy of 95% platinum and 5% iridium, when hard-worked, has a Brinell hardness of 170; an alloy with 30% iridium has a hardness of 400. The 5 and 10% alloys are used for jewelry manufacture; the 25 and 30% alloys are employed for making surgical instruments. An alloy of 80% platinum and 20% iridium is used for magneto contact points, and the 90–10 alloy is widely used for electric contacts in industrial control devices. The addition of iridium does not alter the color of the platinum. The 5% alloy dissolves readily in aqua regia; the 30% alloy dissolves slowly.

Platinum–Rhodium

The addition of rhodium to platinum also provides improved mechanical properties to platinum and increases its resistance to corrosion. For applications at high temperatures, the platinum–rhodium alloys are preferred because of retention of good mechanical properties including good hot strength and very little tendency toward volatilization or oxidation.

Platinum–rhodium alloys are used for thermocouples for temperatures above 1100°C. The standard thermocouple is platinum vs. platinum–10% rhodium. Other thermocouples for higher operating temperatures use platinum–rhodium alloys in both elements. The alloys of platinum–rhodium are widely used in the glass industry, particularly as glass-fiber extrusion bushings. Rhodium increases the high-temperature strength of platinum without reducing its resistance to oxidation. Platinum–rhodium gauze for use as a catalyst in producing nitric acid from ammonium contains 90% platinum and 10% rhodium.

Platinum–Ruthenium

Alloying platinum with ruthenium has the most marked effect upon both hardness and resistivity. However, the limit of workability is reached at 15% ruthenium. The lower cost and the lower specific gravity of ruthenium offer an appreciable economic benefit as an alternate to other platinum alloys.

Platinum–ruthenium alloy, with 10% ruthenium, has a melting point of 1800°C, and an electrical conductivity 4% that of copper.

Platinum–Gold

Platinum–gold alloys cover a wide range of compositions and provide distinct chemical and physical characteristics.

Platinum–Cobalt and Platinum–Nickel

These alloys, with about 23% cobalt, are used for permanent magnets. Platinum–nickel alloys, with as much as 20% nickel, are noted for high strength. With 5% nickel, for example, tensile strength of the annealed alloy is about 621 MPa, and with 15% it increases to 896 MPa. Strength almost doubles with appreciable cold work.

Platinum–Rhenium

These alloys are efficient catalysts for reforming operations on aromatic compounds. The platinum alloys have lower electric conductivity than pure platinum, but are generally harder and more wear resistant, and have high melting points. A platinum–rhenium alloy with 10% rhenium has an electrical conductivity of only 5.5% that of copper compared with 16% for pure platinum. Its melting point is 1850°C, and the Rockwell T hardness of the cold-rolled metal is 91 compared with 78 for cold-rolled platinum.

Platinum–Tungsten

These alloys, with 2 to 8% tungsten, have been used for aircraft-engine spark plug electrodes, radar-tube grids, strain gauges, glow wires, switches, and heating elements. The tungsten markedly increases electrical resistivity while decreasing the temperature coefficient of resistivity. It also substantially increases tensile strength — to 896 MPa for platinum (8% tungsten alloy in the annealed condition) — and tensile strength more than doubles with appreciable cold work.

PLATINUM GROUP METALS

The platinum group metals — ruthenium, rhodium, palladium, osmium, iridium, and platinum — are found in the second and third long period in Group VIII of the periodic table. Platinum and palladium are the most abundant of the group although all are generally found together.

The outstanding characteristics of platinum, the most important member of the group, are its remarkable resistance to corrosion and chemical attack, high melting point, retention of mechanical strength, and resistance to oxidation in air, even at very high temperatures. These qualities, together with the ability of the metal to greatly influence the rates of reaction in a large number of chemical processes, are the basis of nearly all its technical applications. The other five metals of the platinum group are also characterized by high melting points, good stability, and resistance to corrosion. Addition of these metals to platinum forms a series of alloys that provide a wide range of useful physical properties combined with the high resistance to corrosion that is characteristic of the parent metals.

PLATINUM (PT)

When heated to redness, platinum softens and is easily worked. It is virtually nonoxidizable and is soluble only in liquids generating free chlorine, such as aqua regia. At red heat platinum is attacked by cyanides, hydroxides, sulfides, and phosphides. When heated in an atmosphere of chlorine, platinum volatizes and condenses as the crystalline chloride. Reduction of platinum chloride with zinc gives platinum black, which has a high adsorptive capacity for hydrogen. Platinum sponge is finely divided platinum.

PALLADIUM (PD)

Palladium is silvery white, very ductile, and slightly harder than platinum. It is readily soluble in aqua regia and is attacked by boiling nitric and sulfuric acids. Palladium has the remarkable ability to occlude large quantities of hydrogen. When properly alloyed it can be used for the commercial separation and purification of hydrogen. Palladium and platinum can both be worked by normal metalworking processes.

IRIDIUM (IR)

This is the most corrosion-resistant element known. It is a very hard, brittle, tin-colored metal with a melting point higher than that of platinum. It is soluble in aqua regia only when alloyed with sufficient platinum. Iridium has its greatest value in platinum alloys where it acts as a hardening agent. By itself it can be worked only with difficulty.

RHODIUM (RH)

Rhodium serves an important role in high-temperature applications up to 1649°C. Platinum–rhodium thermocouple wire makes possible high-temperature measurement with great accuracy. Rhodium and rhodium alloys are used in furnace windings and in crucibles at temperatures too high for platinum. It is a very hard, white metal and is workable only under certain conditions, and then with difficulty. Applied to a base metal by electroplating, it forms a hard, wear-resistant, permanently brilliant surface. Solubility is slight even in aqua regia.

RUTHENIUM (RU)

Ruthenium is hard and brittle with a silver-gray luster. Its tetraoxide is very volatile and poisonous. When alloyed with platinum, its effect on hardness and resistivity is the greatest of all the metals in the group. It is unworkable in the pure state.

OSMIUM (OS)

This element has the highest specific gravity and melting point of the platinum metals. It oxidizes readily when heated in air to form a very volatile and poisonous tetraoxide. Application has been predominantly in the field of catalysis. As a metal it is also practically unworkable.

PLYWOOD

Plywood is a term generally used to designate glued wood panels made up of layers, or plies, with the grain of one or more layers at an angle, usually 90°, with the grain of the other. The outside plies are called faces or face and back, the center plies are called the core, and the plies immediately below the face and back, laid at right angles to them, are called the crossbands.

The core may be veneer, lumber, or various combinations of veneer and lumber; the total thickness may be less than 1.5 mm or more than 76 mm; the different plies may vary as to number, thickness, wood species. Also, the shape of the members may vary. The crossbands and their arrangement generally govern both the properties (particularly warping characteristics) and uses of all such constructions.

Plywood is an outgrowth of the laminated wood known as veneer, which consists of an outside sheet of hardwood glued to a base of lower-cost wood. The term *veneer* actually refers only to the facing layer of selected wood, used for artistic effect or for economy in the use of expensive woods. Veneers are generally marketed in strip form in thicknesses of less than 0.32 cm in mahogany, oak, cedar, and other woods. The usual purpose of plywood now is not aesthetic but to obtain high strength with low weight. The term *laminated wood* generally means heavier laminates for special purposes, and such laminates usually contain a heavy impregnation of bonding resin that gives them more of the characteristics of the resin than of the wood.

Composition

The composition of a plywood panel is generally dependent on the end use for which it is intended. The number of plywood constructions is almost endless when one considers the number of wood species available, the many thicknesses of wood veneers used in the outer plies or cores, the placement of the adjacent plies, the types of adhesives and their qualities, various manufacturing processes, and more technical variations.

Conventional plywood generally consists of an odd number of plies with the grains of the alternate layers perpendicular to each other. The use of an odd number permits an arrangement that gives a substantially balanced effect; that is, when three plies are glued together with the grain of the outer two plies at right angles to that of the center ply, the stresses are balanced and the panel tends to remain flat with changes in moisture content. These forces may be similarly balanced with five, seven, or some other uneven number of plies. If only two plies are glued together with the grain of one ply at right angles to the other, each ply tends to distort the other when changes in moisture content occur; cupping will result.

Low-cost plywoods may be bonded with starch pastes, animal glues, or casein, and are not water-resistant, but are useful for boxes and for interior work. Waterproof plywood for paneling and general construction is now bonded with synthetic resins, but when the plies are heavily impregnated with the resin and the whole cured into a solid sheet, the material is known as a hardboard or as a laminated plastic rather than a plywood.

Grades and Types

Broadly speaking, two classes of plywood are available — hardwood and softwood. Most softwood plywood is composed of Douglas fir, but western hemlock, white fir, ponderosa pine, redwood, and other wood species are also used. Hardwood plywood is made of many wood species.

Various grades and types of plywood are manufactured. "Grade" is determined by the quality of the veneer and "Type" by the moisture resistance of the glue line. For example, there are two types of Douglas fir plywood — interior and exterior. The interior type is expected to retain its form and strength properties when occasionally subjected to wetting and drying periods. It is commonly bonded with urea–formaldehyde resin adhesives. On the other hand, the exterior type is expected to retain its form and strength properties when subjected to cyclic wetting and drying and to be suitable for permanent exterior use. It is commonly bonded with hot-pressed phenolic-resin glues.

For construction purposes, where plywood is employed because of its unit strength and nonwarping characteristics, the plies may be of a single type of wood and without a hardwood face. The Douglas Fir Plywood Association sets up four classes of construction plywood under general trade names. Plywall is plywood in wallboard grade; Plypanel is plywood in three standard grades for general uses; Plyscord is unsanded plywood with defects plugged and patched on one side; and Plyform is plywood in a grade for use in concrete forms.

The bulk of commercial plywood comes within these classes; the variations are in the type of wood used, the type of bonding adhesive, or the finish. Etch wood, for example, is a paneling plywood with the face wire brushed to remove the soft fibers and leave the hard grain for two-tone finish. Paneling plywoods with faces of mahogany, walnut, or other expensive wood have cores of lower-cost woods, but the woods of good physical qualities are usually chosen.

ENGINEERING PROPERTIES

The mechanical and physical properties of plywood are dependent upon the particular construction employed. Plywood may be designed for beauty, durability, rigidity, strength, cost, or many other properties. With practically an unlimited variety of constructions to choose from, there is a wide range of differing characteristics in any given plywood panel. Most important among properties are the following.

High Strength–Weight Ratio

Perhaps the most notable feature of plywood is its high strength–weight ratio. Plywood is given special consideration whenever lightness and strength are desired. Plywood is widely used for concrete form work, floor underlayment, roof decks, siding, and many other applications because of its high strength–weight ratio. A comparison between birch plywood and other structural materials shows that its strength–weight ratio is 1.52 times that of 100,000-lb test heat-treated steel and 1.36 times that of l0,000-lb test aluminum.

Bending Properties

A most desirable characteristic of plywood is its flatness but it can and will support substantial curvatures without appreciable loss of strength. Standard construction plywood can be bent or shaped to nominal radii and held in place with adhesives, nails, screws, or other fixing methods.

The radius of curvature to which a panel can be formed varies with panel thickness and the species of wood employed in the panel construction. The arc of curvature is limited by the tension force in the outer plies of the convex perimeter and by the compression forces in the outer plies of the concave perimeter.

Waterproofed plywood, soaked or steamed before bending, exhibits approximately 50% greater flexibility than panels bent when dry.

Resistance to Splitting

Because plywood has no line of cleavage, it cannot split. This is an exceptional property when one considers its effect on fastening. The crisscross arrangement of wood plies in plywood construction develops extraordinary resistance to pull-through of nail or screw heads.

Resistance to Impact

The absence of a cleavage line has a pronounced effect on the impact resistance of plywood. Plywood will fracture only when the impact force is greater than the tensile strength of the wood fibers in the panel composition. Under an impact force, the side of the panel opposite the impact point will rupture along the long grain fiber followed by successive shattering of the various plies. Splintering usually does not take place because pressure is dispersed throughout the panel at the point of impact. Under similar conditions, solid lumber will show complete rupture.

Beauty

Plywood has certain intrinsic qualities that add much to any structure or construction in which it is used. Because of improved modern methods of manufacture, there is practically no limit to the decorative potential of plywood.

The entire range of fine woods is at the designer's disposal; they vary in shade from golden yellow to ebony, from pastels to reds and browns. The use of bleaches, toners, and stains in manufacturing procedures gives the designer even greater latitude with respect to design freedom.

The fine woods employed in the manufacture of plywood offer warmth and charm to any decorative scheme because the surface of the wood variously absorbs, reflects, and refracts light rays, giving the wood pattern depth and making it restful to the eye. This phenomenon accounts for the play of color and pattern when a plywood panel is viewed from different angles.

Dimensional Stability

The absorption of water causes wood to swell and this movement is much greater across the grain than along the grain. The alternating layers of veneers in standard plywood construction inhibits this cross-grain movement because the cross-grain weakness is reinforced by the long-grain stability. Therefore, the dimensional stability of a plywood panel can be controlled by controlling its moisture content. In the field, this control can be achieved by applying coatings such as paints, lacquers, and sealers of various types.

Thermal Insulating Qualities

The thermal insulating qualities of plywood are the same as those of the wood of which it is composed.

The use of plywood as an insulating material can be attributed to two factors: (1) The use of large sheets reduces the numbers of cracks and joints and thereby inhibits wind leakage; and (2) the resistance of plywood to moisture vapor transmission stabilizes the moisture content of the trapped air and maintains its insulating qualities.

Fire Resistance

Fire-resistant plywood is manufactured by impregnating the core stock with a salt solution, which, upon evaporation, leaves a salt deposit in the wood. Plywood or wood treated in this manner will not support combustion but will char when heated beyond the normal charring point of the wood.

Nonimpregnated plywood can be made fire resistant by applying surface coatings such as intumescent paint and chemicals such as borax. An intumesent paint has a silicate of soda base that bubbles or intumesces in the presence of heat, thus forming a protective coating. At high temperatures, borax releases a gas such as carbon dioxide, which blankets the fire. It must be noted that fire-resistant coatings are effective only in direct ratio to their thickness. Highly resistant plywood can be manufactured by using an incombustible core such as asbestos.

Resistance to Borers

Plywood panels are subject to attack by borers to the same extent as the wood species of which they are composed, but phenol–formaldehyde resin glue lines are fairly effective barriers against further penetration. Panels may also be treated with pentachlorophenol for increased resistance to these pests.

Fatigue Resistance

Plywood has the same resistance to fatigue as the wood of which it is composed.

FABRICATION

The fact that almost anyone can use plywood has contributed greatly to its wide acceptance in many varied applications. The utilization of plywood does not require special tools, special skills, or safeguards, and practically anyone capable of handling a saw and hammer can make use of its inherent engineering properties.

Since plywood does not exhibit the typical cross-grain weakness of lumber, it can very often be used in place of lumber for various applications. For example, 0.312-mm-thick plywood replaces conventional 0.750-mm sheathing; 0.250-mm plywood can be used for interior wall paneling without sheathing and 0.375-mm plywood can be used for shipping containers, furniture, and case goods instead of 0.750-mm lumber. The use of the thinner plywood reduces weight, bulkiness, and is less fatiguing for tradespeople to handle. The use of large plywood sheets instead of narrow boards also reduces the amount of cutting, fitting, and fastening involved in a particular job.

Plywood is especially adaptable to the portable power-driven saws, drills, and automatic hammers normally used on production or construction jobs. Multiple cutting with band saws may be done with assurance because even the thinnest plywood has strength in all directions and the danger of splitting or chipping is reduced to a minimum. This quality is particularly important when fine fitting is required.

Whenever plywood is employed in a structure, fewer and smaller fastenings can be specified because consideration need be given only to the holding power of the fastener and the tensile strength of the fastening itself.

AVAILABLE FORMS, SIZES, SHAPES

Plywood is available in practically any size, but the 1.2 × 2.4 m panel has become the standard production unit of the industry. Larger panels usually demand a price premium and are available on special order. Panels with continuous cores and faces can be produced in one piece up to 4 m long.

Oversize panels up to 2.4 m in width and of unlimited length can be manufactured by scarf jointing. (A scarf is an angling joint, made either in veneers or plywood, where pieces are spliced or lapped together. The length of the scarf is usually 12 to 20 times the thickness. When properly made, scarf joints are as strong as the adjacent unspliced material.)

Decorative plywood is now commercially available with a plant-applied finish. Prefinished wall paneling is supplied with the finish varying from offset printing to polyester film.

GENERAL FIELDS OF APPLICATION

There are many uses and applications for plywood in industry today. Owing to the wide diversity of plywood applications, only the more prominent ones are mentioned here:

Architectural	Marine construction
Aviation	Mock-ups, models
Boatbuilding	Paddles
Building construction	Panel boards
Cabinet work	Patterns
Concrete forms	Prefabrication
Containers, cases	Remodeling
Die boards	Sheathing
Display	Signs
Fixtures	Sporting goods
Floor underlayment	Table tops
Furniture	Toys
Hampers	Trays
Luggage	Truck floors, bodies
Machine bases	Wall paneling

POLONIUM

Polonium (symbol Po) is a rare metallic element belonging to the group of radioactive metals, but emitting only alpha rays. The melting point of the metal is about 254°C. It is used in meteorological stations for measuring the electrical potential of the air. Polonium-plated metal in strip and rod forms has been employed as a static dissipator in textile-coating machines. The alpha rays ionize the air near the strip, making it a conductor and drawing off static electric charges. Polonium-210 is obtained by irradiating bismuth; 45 kg yields 1 g of polonium-210. It is used as a heat source for emergency auxiliary power such as in spacecraft. The metal is expensive, but can be produced in quantity from bismuth.

POLYACRYLATE RESIN

Useful polymers can be obtained from a variety of acrylic monomers, such as acrylic and methacrylic acids, their salts, esters, and amides, and the corresponding nitriles. Polymethyl methacrylate, polyethyl acrylate, and a few other derivatives are the most widely used.

Polymethyl methacrylate is a hard, transparent polymer with high optical clarity, high refractive index, and good resistance to the effects of light and aging. It and its copolymers are useful for lenses, signs, indirect lighting fixtures, transparent domes and skylights, dentures, and protective coatings.

Solutions of polymethyl methacrylate and its copolymers are useful as lacquers. Aqueous latexes formed by the emulsion polymerization of methyl methacrylate with other monomers are useful as water-based paints and in the treating of textiles and leather.

Polyethyl acrylate is a tough, somewhat rubbery product. The monomer is used mainly as a plasticizing or softening component of copolymers. Ethyl acrylate is usually produced by the dehydration and ethanolysis of ethylene cyanohydrin.

Modified acrylic resins with high impact strengths can be prepared. Blends or "alloys" with polyvinyl chloride are used for thermoforming impact-resistant sheets.

Methyl methacrylate is of interest as a polymerizable binder for sand or other aggregates, and as a polymerizable impregnant for concrete: usually a cross-linking acrylic monomer is also incorporated.

Polymers of methyl acrylate or acrylamide are water-soluble and useful for sizes and finishes. Addition of polylauryl methacrylate to petroleum lubricating oil improves the flowing properties of the oil at low temperatures and the resistance to thinning at high temperatures.

POLYACRYLIC RUBBER

The first types of polyacrylic rubbers were proposed as oxidation-resistant elastomeric materials. Chemically, they were polyethyl acrylate, and a copolymer of ethyl acrylate and 2-chloro ethyl vinyl ether.

The development of polyacrylic rubbers was accelerated by the interest expressed throughout the automotive industry in the potential applications of this type of polymer in special types of seals. An effective seal for today's modern lubricants must be resistant not only to the action of the lubricant but to increasingly severe temperature conditions. It must also resist attack of highly active chemical additives that are incorporated in the lubricant to protect it from deterioration at extreme temperature.

Polyacrylic rubber compounds were developed to provide a rubber part that would function in applications where oils and/or temperatures as high as 204°C would be encountered. These were also very resistant to attack by sulfur-bearing chemical additives in the oil. These properties have resulted in general use of polyacrylic rubber compounds for automotive rubber parts as seals for automatic transmission fluids and extreme pressure lubricants.

Polyacrylic rubber will prove most useful in fields where these special properties are used to the maximum. It is recommended for products such as automatic transmission seals, extreme pressure lubricant seals, searchlight gaskets, belting, rolls, tank linings, hose, O-rings and seals, white or pastel-colored rubber parts, solution coatings, and pigment binders on paper, textiles, and fibrous glass.

CURING

A typical polyacrylic rubber, such as the copolymer of ethyl acrylate and chloroethyl vinyl ether, is supplied as a crude rubber in the form of white sheets having a specific gravity of approximately 1.1. It may be mixed and processed according to conventional rubber practice.

However, polyacrylic rubber is chemically saturated and cannot be cured in the same manner as conventional rubbers. Sulfur and sulfur-bearing materials act as retarders of cure and function as a form of age resistor in most formulations. Polyacrylic rubber is cured with amines; "Trimene Base" and triethylene tetramine are most widely used. Aging properties may be altered by balancing the effect of the amine and the sulfur.

Like other rubber polymers, reinforcing agents such as carbon black or certain white pigments are necessary to develop optimum physical properties in a polyacrylic rubber vulcanizate. Selection of pigments is more critical in that acidic materials, which would react with the basic amine curing systems, must be avoided. The SAF or FEF carbon blacks are most widely used, while hydrated silica or precipitated calcium silicate are recommended for light-colored stocks.

Typical curing temperatures are from 143 to 166°C at cure times of 10 to 45 min depending on the thickness of the part. Polished, chromium-plated molds are recommended. For maximum overall physical properties, the cured parts should be tempered in an air oven for 24 h at 149°C.

FORMING

To obtain smooth extrusions, more loading and lubrication are necessary than for molded goods, because of the inherent nerve of the polymer. Temperatures of 43°C in the barrel and 77°C on the die are recommended.

Generally, those compounds that extrude well are also good calendering stocks. Suggested temperatures for calendering are in the range of 37.8 to 54°C. Higher temperatures will result in sticking of the stock to the rolls. Under optimum conditions, 15-mil films may be obtained.

Polyacrylic rubber may be coated on nylon either by calendering or from solvent solution. It also has excellent adhesion to cotton and is often used as a solvent solution applied to cotton duck to be used as belting. Solvents generally used include methylethyl ketone, toluene, xylene, or benzene.

Polyacrylic rubber is most widely used in many types of seals because of its excellent resistance to sulfur-bearing oils and lubricants.

In general, polyacrylic rubber vulcanizates are resistant to petroleum products and animal and vegetable fats and oils. They will swell in aromatic hydrocarbons, alcohols, and ketones. Polyacrylic rubber is not recommended for use in water, steam, ethylene glycol, or in alkaline media.

Laboratory tests indicate that polyacrylic vulcanizates become stiff and brittle at a temperature of –23°C. But in actual service, these same polyacrylic rubbers have been found to provide satisfactory performance at engine start-up and operation in oil at temperatures as low as –40°C.

For those applications requiring improvement in low-temperature brittleness by as much as –4.0°C and that can tolerate considerable sacrifice in overall chemical oil and heat resistance, a copolymer of butyl acrylate and acrylonitrile may be used.

POLYACRYLONITRILE RESINS

The polyacrylonitrile resins are hard, horny, relatively insoluble, and high-melting materials. Polyacrylonitrile (polyvinyl cyanide) is used almost entirely in copolymers. The copolymers fall into three groups: fibers, plastics, and rubbers. The presence of acrylonitrile in a polymeric composition tends to increase its resistance to temperature, chemicals, impact, and flexing.

Acrylonitrile is generally prepared by several methods, including the catalyzed addition of hydrogen cyanide to acetylene. The polymerization of acrylonitrile can be readily initiated by means of the conventional free-radical catalysts such as peroxides, by irradiation, or by the use of alkali metal catalysts. Although polymerization in bulk proceeds too rapidly to be commercially feasible, satisfactory control of a polymerization or copolymerization may be achieved in suspension and in emulsion, and in aqueous solutions from which the polymer precipitates. Copolymers containing acrylonitrile may be fabricated in the manner of thermoplastic resins.

The major use of acrylonitrile is in the form of fibers. By definition an acrylic fiber must contain at least 85% acrylonitrile: a modacrylic fiber may contain less than 35 to 85% acrylonitrile. The high strength; high softening temperature; resistance to aging, chemicals, water, and cleaning solvents; and the soft wool-like feel of fabrics have made the product popular for many uses such as sails, cordage, blankets, and various types of clothing. Commercial forms of the fiber probably are copolymers containing minor amounts of other vinyl derivatives, such as vinyl pyrrolidone, vinyl acetate, maleic anhydride, or acrylamide. The comonomers are included to produce specific effects, such as improvement of dyeing qualities.

Copolymers of vinylidene chloride with small proportions of acrylonitrile are useful as tough, impermeable, and heat-sealable packaging films.

Extensive use is made of copolymers of acrylonitrile with butadiene, often called NBR (formerly Buna N) rubbers, which contain 15% acrylonitrile. Minor amounts of other unsaturated esters, such as ethyl acrylate, which yield carboxyl groups on hydrolysis, may be incorporated to improve the curing properties. The NBR rubbers resist hydrocarbon solvents such as gasoline and abrasion, and in some cases show high flexibility at low temperatures.

In the 1960s development of blends and interpolymers of acrylonitrile-containing resins and rubbers represented a significant advance in polymer technology. The products, usually called ABS resins, typically are made by blending acrylonitrile–styrene copolymers with a butadiene–acrylonitrile rubber, or by interpolymerizing polybutadiene with styrene and acrylonitrile. Specific properties depend on the proportions of the comonomer, on the degree of grattings, and on molecular weight. In general, the ABS resins combine the advantages of hardness and strength of the vinyl resin component with toughness and impact resistance of the rubbery component. Certain grades of the ABS resin are used for blending with brittle thermoplastic resins such as polyvinyl chloride to improve impact strength.

The combination of low cost, good mechanical properties, and ease of fabrication by a variety of methods, including typical metalworking methods such as cold stamping, led to the rapid development of new uses for ABS resins. Applications include products requiring high impact strength, such as pipe, and sheets for structural uses, such as industrial duct work and components of automobile bodies. ABS resins are also used for housewares and appliances, because of their ability to be electroplated for decorative items in general.

POLYAMIDES

These are horny, whitish, translucent, high-melting polymers. Polyamide resins can be essentially transparent and amorphous when their melts are quenched. On cold drawing and annealing, most become quite crystalline and translucent. However, some polyamides based on bulky repeating units are inherently amorphous. The polymers are used for fibers, bristles, bearings, gears, molded objects, coatings, and adhesives. The term *nylon* formerly referred specifically to synthetic polyamides as a class. Because of many applications in mechanical engineering, nylons are considered engineering plastics.

Nylon-6.6 and nylon-6.10 are products of the condensation reaction of hexamethylenediamine (6 carbon atoms) with adipic acid (6 carbon atoms), and with sebacic acid (10 carbon atoms), respectively.

Nylon-6.6, nylon-6.10, nylon-6.12, and nylon-6 are the most commonly used polyamides for general applications as molded or extruded parts; nylon-6.6 and nylon-6 find general application as fibers.

As a group, nylons are strong and tough. Mechanical properties depend in detail on the degree and distribution of crystallinity, and may be varied by appropriate thermal treatment or by nucleation techniques. Because of their generally good mechanical properties and adaptability to both molding and extrusion, the nylons described above are often used for gears, bearings, and electrical mountings. Nylon bearings and gears perform quietly and need little or no lubrication. Sintering (powder metallurgy) processes are used to make articles such as bearings and gears that have controlled porosity, thus permitting retention of oils or inks. Nylon resins are used extensively as filaments, bristles, wire insulation, appliance parts, and film. Properties can be modified by copolymerization.

Reinforcement of nylons with glass fibers results in increased stiffness, lower creep, and improved resistance to elevated temperatures. Such formulations, which can be readily injection-molded, can often replace metals in certain applications. The use of molybdenum sulfide and polytetrafluoroethylene as fillers increases wear resistance considerably. For uses requiring impact resistance, nylons can be blended with a second, toughening phase.

Other types of nylon are useful for specialty applications. Solubility may be increased by interference with the regularity and hence intermolecular packing. This may be accomplished by copolymerization, or by the introduction of branches on the amide nitrogen, for example, by treatment with formaldehyde. The latter type of resin may be subsequently cross-linked. Nylons incorporating aromatic structures, for example, based on isophthalic acid, are becoming more common for applications requiring resistance to very high temperatures.

POLYAMIDE-IMIDES

Polyamide-imides are engineering thermoplastics characterized by excellent dimensional stability, high strength at high temperature, and good impact resistance. Molded parts can maintain structural integrity in continuous use at temperatures to 260°C.

Polyamide-imide, produced and called Torlon, is available in several grades including a general-purpose, injection-molding grade; three polytetrafluoroethylene (PTFE)/graphite wear-resistant compounds; a 30% graphite-fiber-reinforced grade; and a 30% glass-fiber-reinforced grade. Additional grades are developed to meet special requirements.

Torlon resins are moldable on screw-injection-molding machines. Molds must be heated to 218°C, and the barrel and nozzle should be capable of being heated to about 371°C. High injection speed and pressure (136 MPa or greater) are desirable. Developing optimum physical properties in injection-molded or extruded parts requires postcuring through an extended, closely controlled temperature program gradually reaching 260°C. The specific time/temperature program depends on part configuration and thickness.

PROPERTIES

Room-temperature tensile strength of unfilled polyamide-imide is about 190 MPa and compressive strength is 213 MPa. At 232°C, tensile strength is about 61.2 MPa — as strong as many engineering plastics at room temperature — and continued exposure at 260°C for up to 8000 h produces no significant decline in tensile properties.

Flexural modulus of 36.5 MPa of the unfilled grade is increased, with graphite-fiber reinforcement, to 19,913 MPa. Retention of modulus at temperatures to 260°C is on the order of 80% for the reinforced grade. Creep resistance, even at high temperature and under load, is among the best of the thermoplastics; dimensional stability is extremely good.

Polyamide-imide is extremely resistant to flame and has very low smoke generation. Reinforced grades have surpassed Federal Aviation Administration requirements for flammability, smoke density, and toxic gas emission.

Radiation resistance of polyamide-imide is good; tensile strength drops only about 5% after exposure to 10^9 rad of gamma radiation. Chemical resistance is good; the resin is virtually unaffected by aliphatic and aromatic hydrocarbons, halogenated solvents, and most acid and base solutions. It is attacked, however, by some acids at high temperature, steam at high pressure and high temperature, and strong bases.

Torlon moldings absorb moisture in humid environments or when immersed in water, but the rate is low, and the process is reversible. Parts can be restored to original dimensions by drying.

POLYARYLATES

These high-heat-resistant thermoplastics are derived from aromatic dicarboxylic acids and diphenols. When molded, they become amorphous, providing a combination of toughness, dimensional stability, high dielectric properties, and ultraviolet stability. Polyarylates have heat deflection temperatures up to 175°C at 264 lb/in.[2]. The resins can be injection-molded, extruded, and blow-molded, and sheet can be thermoformed. These resins also are blended with other engineering thermoplastics and reinforcements.

POLYARYLETHERKETONES (PAEK)

A glass-fiber-reinforced (polyaryletherketone) semicrystalline polymer has been designed into the sensor housing for a new conductivity measuring cell.

Conductivity measuring cells are used to determine the electrolytic conductivity of media in the food and pharmaceutical industries. The primary requirements for the cell are resistance to corrosive media and biocompatible surface quality to remain in compliance with U.S. and European hygienic requirements. The PAEK polymer is insoluble in all common solvents, and can be immersed for thousands of hours at temperatures in excess of 250°C in steam or high-pressure water environments without significant degradation. The addition of glass fibers increases the chemical resistance of the base resin as well as mechanical strength at elevated temperatures.

POLYBENZIMIDAZOLE

Polybenzimidazole (PBI) was used in a variety of applications for the U.S. space program, including flight suits and other protective clothing, webbings, straps, and tethers. Research in raw materials and process development continued, along with applications development, during the 1960s and 1970s.

In 1983, commercial production of PBI fiber commenced. Today, PBI fiber has been used successfully in firefighters' gear, industrial protective clothing, fire-blocking layers for aircraft seats, braided pump packings, and other high-performance products. Research has continued to develop other forms of PBI. In addition to Celazole molded parts, these forms include polymer additives, films, fibrids, papers, microporous resin, sizing, and coatings; see Table P.1 (Properties of Polybenzimidazole).

Scientists say the material has no known melting point and can withstand pressures of up to 394 MPa. It also has demonstrated resistance to steam at 343°C.

POLYBUTADIENE RUBBER

Polybutadiene may be prepared in several ways to yield different products. The method of polymerization can have a marked effect on polymer structure, which in turn controls the properties of the polymer and thus its ultimate end use.

Composition

When polybutadiene is made, the butadiene molecule may enter the polymer chain by either 1,4-addition or 1,2-addition. In 1,4-addition, the unsaturated bonds may be either of *cis* or of *trans* configuration. Polybutadienes containing a more or less random mixture of these polymer units can be prepared with alkali metal catalysts or with emulsion polymerization systems but these have not achieved commercial significance as general-purpose rubbers in the United States.

TABLE P.1
Properties of Polybenzimidazole

ASTM Test	Property	Celazole U-60, Unified
	Mechanical	
D638	Tensile strength	23,000 psi
D638	Elongation	3%
D638	Tensile modulus	850 kpsi
D790	Flexural strength	32,000 psi
D790	Flexural modulus	950 kpsi
D695	Compressive strength	58,000 psi
D695	Compressive modulus	900 kpsi
D256	Izod impact strength	
	Notched	0.5-ft-lb/in.
	Unnotched	11 ft-lb/in.
	Poisson's ratio	0.34
	Electrical	
D149	Dielectric strength	550 V/mil
D257	Volume resistivity	8×10^{14}
D150	Dissipation factor (10 kHz)	0.003
D495	Arc resistance	186 s
	Thermal	
D648	Heat deflection temperature	815°F
DMA	Glass transition temperature	800°F
	Thermal conductivity (77°F)	2.8 BTU/in.h-ft^2-°F
TMA	Coefficient of linear thermal expansion	13×10^{-6} in./in.-°F
	Physical	
	Specific gravity	1.3
D785	Hardness	115 Rockwell K
D570	Water absorption	0.4%

Source: Mach. Design Basics Eng. Design, June, p. 710, 1993. With permission.

In recent years, new catalysts have been developed that allow the structure of the polymer chain to be controlled. Polybutadienes containing in excess of 85% of either *cis, trans,* or vinyl unsaturation have been studied and, in the case of *cis* and *trans* polymers, produced commercially. It is also possible to prepare a number of other polymers with various combinations of these three component structures. Of current importance are those types commercially available, which are (1) polymers of high *cis*-1,4 content (more than 85%) with low *trans* and vinyl content, (2) polymers of more than 80% *trans*, low *cis,* and low vinyl content, and (3) polymers of intermediate *cis* content (approximately 40% *cis*, 50% *trans*) with low vinyl content.

HIGH-*CIS* POLYBUTADIENE

Outstanding properties of high-*cis* polybutadiene are high resilience and high resistance to abrasion. The high resilience is indicative of low hysteresis loss under dynamic conditions, i.e., low heat rise in the polymer under rapid, repeated deformations. In this respect *cis*-polybutadiene is similar to natural rubber. The combination of good hysteresis properties and resistance to abrasion makes this polymer attractive for use in tires, especially in heavy-duty tires where heat generation is a problem.

Another favorable factor, particularly for use in tire bodies, is that high-*cis* polybutadiene imparts good resistance to heat degradation under heavy loads in dynamic applications. Tensile strength (in reinforced stock) is lower than for natural rubber or styrene–butadiene rubber (SBR) but is adequate for many uses. Modulus depends on the degree of cross-linking but *cis*-polybutadiene generally requires relatively low stress to reach a given elongation (at slow deformation). Hardness may be varied depending on the compound formulation and is similar to that of SBR and natural rubbers. Ozone resistance is typical of unsaturated polymers and is inferior to that of saturated rubbers. Oil resistance is comparable to that of SBR and natural rubber. Permeability to gases is higher than that of most other rubbers, which may be either an advantage or a disadvantage, depending on the application. Freeze point is quite low; *cis*-polybutadiene tread compounds do not become brittle until very low temperatures are reached, lower than −101°C.

Abrasion resistance of the high-*cis* polybutadienes varies with the type of service and in nearly all cases is superior to that of natural rubber or SBR. The advantage for these polybutadienes in relation to natural rubber and SBR increases as the severity of the service increases. From 30% to as high as 100% improvement in wear resistance has been reported through the substitution of these polybutadienes for natural rubber to tire treads.

Processing

Polybutadiene rubbers in general are more difficult to process in conventional equipment than natural rubber, particularly with regard to milling and extrusion operations. The processing problems have been overcome in many cases by treatment of the polymer, changes in compounding formulations, or by blending with other rubbers. Blends of natural rubber and high-*cis* polybutadiene are particularly attractive. The presence of natural rubber alleviates the processing difficulties and improves tensile and tear properties of the polybutadiene while the latter improves abrasion resistance and complements the already good hysteresis properties of natural rubber.

Processibility can also be improved by the use of higher levels of reinforcing fillers and oils than normally employed. Consideration must also be given to changes in quality, but extension with 35 to 70 phr oil is possible with retention of properties suitable for tires and many rubber goods. Even in blends with natural rubber, where processing is not a problem, increases in carbon black and oil content have proved practical and often desirable.

Applications

For most uses, vulcanization is necessary to develop the desired strength and elastic qualities. The polymer chain is unsaturated and can be readily vulcanized with sulfur (in conjunction with the usual activators) or with other cross-linking agents such as peroxides. With some exceptions, admixture of the rubber with a reinforcing pigment is required to obtain high strength. Antioxidants or antioxidants should be used for most applications. Lower than normal sulfur levels provide a better balance of properties in many instances, particularly with blends of high-*cis* polybutadiene and natural rubber.

In large tires, the use of blends of high-*cis* polybutadiene and natural rubber in treads improves both abrasion resistance and resistance to tread groove cracking compared to natural rubber alone. Substitution of high-*cis* polybutadiene for a portion of the natural rubber in the tire body has improved resistance to blowout or other heat failures (in some cases to a remarkable degree).

The high-*cis* polybutadienes are also used in increasing quantity in blends with SBR for the production of tires for passenger cars and small trucks. In this use, advantages include improved resistance to abrasion and cracking as well as adaptability to extension with large amounts of oil and carbon black.

As indicated, properties of high-*cis* polybutadiene are well suited for tire use, and this is expected to be the major application. Its use should be considered in other areas where high resilience, resistance to abrasion, or low-temperature resistance is required. Sponge stocks, footwear, gaskets and seals, and conveyor belts are possible applications. In shoe heels high-*cis* polybutadiene may be used to provide good resilience and better abrasion resistance than realized with other commonly used rubbers. In shoe soles designed to be soft and resilient, high-*cis* polybutadiene has been used as a partial or total replacement for natural rubber to maintain high resilience, improve abrasion resistance, and to provide better resistance to crack growth.

An important use for high-*cis* polybutadiene is in blends with other polymers to improve low-temperature properties. Replacement of 35% of an acrylonitrile (nitrile) rubber in a compound with *cis*-polybutadiene can reduce the brittleness failure temperature from –40°C to –55°C. Similarly, it is possible to reduce the brittle point of neoprene compounds by some –6.7°C with the substitution of *cis*-polybutadiene for one fourth of the neoprene rubber. Such substitutions usually reduce resistance to swelling in hydrocarbons but this effect can be lessened by compounding with oil-resistant resins or by using *cis*-polybutadiene as a replacement for plasticizer in the compound rather than as a replacement for the polymer. In many cases, of course, oil resistance is not required and *cis*-polybutadiene may be blended with various polymers to give low-temperature properties approaching those of special arctic rubbers such as butadiene–styrene copolymers with low styrene content.

Partial substitution of *cis*-polybutadiene for other rubbers used in light-colored or black-reinforced mechanical goods can provide a product that displays better snap. Also, resilience, abrasion resistance, and low-temperature properties are usually improved; tensile strength and tear strength may be reduced but show less decline after aging.

High-*cis* polybutadienes have been used successfully as base components of caulking compounds and sealants. Another use is as a base polymer for graft polymerization of styrene to produce high-impact polystyrene.

Commercially available *cis*-polybutadiene rubbers are supplied in talc form. The rubber contains a small amount of antioxidant to provide stability during storage.

High-*trans* Polybutadiene

In contrast to the soft, rubbery nature of *cis*-polybutadiene, polybutadiene of high *trans* content is a hard, horny material at room temperature. It is thermoplastic and thus can be molded without addition of vulcanization agents. The softening point, hardness, and tensile strength increase with increasing *trans* content. A polymer of approximately 90% *trans* content, for example, displays a softening point near 93°C, Shore A hardness of about 98, and tensile strength in excess of 6.8 MPa without addition of other ingredients.

High-*trans* polybutadiene can be readily vulcanized with sulfur or peroxides. It remains a hard, partially thermoplastic material when lightly vulcanized but can be made rubbery by increasing the curative level. Tensile strength is increased by the addition of reinforcing pigments (carbon black, silica, or clay). Compounded in such a manner, high-*trans* polybutadienes are characterized by high modulus, tensile strength, elongation, and hardness, by moderate resilience, and by excellent abrasion resistance.

In vulcanized stocks, properties such as hardness, resilience, and heat buildup can be modified.

Uses

Applications include those where balata has been used such as golf ball covers and wire and cable coverings. Properties have also been demonstrated to be suitable for shoe soles (high hardness, good abrasion resistance), floor tile (high hardness, low compression set), gasket stocks, blown sponge compounds, and other molded or extruded items. High-*trans* polybutadiene requires processing temperatures above its softening point but is relatively easy to mix, mill, or extrude at these temperatures. Care should be taken to avoid scorch or precure when handling stocks containing curatives at elevated temperatures.

POLYBUTADIENE OF INTERMEDIATE *CIS* CONTENT

This type of product has a structure of approximately 40% *cis*, 50% *trans*, and 10% (or less) vinyl content.

The raw polymer is soft and somewhat waxlike in character. This polybutadiene displays high resilience, high dynamic modulus, and a low brittle point.

Processing difficulties in milling and extrusion operations may be encountered and it is usually recommended that polybutadienes of intermediate *cis* content be used in blends with other rubbers. In tire treads substitution of 40% *cis*-polybutadiene for a portion of the natural rubber or SBR 1500 improves resilience and abrasion resistance and gives some reduction in operating temperature. Other applications include products where similar changes in properties to those above are desired or in products where improvements in low-temperature properties are desired.

LIQUID POLYMERS

Liquid polybutadienes can be prepared in most of the systems used to make solid rubbers. Such polymers may be cross-linked with chemicals or solidified with heat. One type of liquid polybutadiene has been manufactured on a small scale utilizing sodium catalyst. Uses for this type polymer include coatings, binders, adhesives, potting agents, casting and laminating resin, or vulcanizable plasticizer for rubber.

POLYCARBONATES

Polycarbonate resins offer a combination of properties that extends the usefulness and fields of application for thermoplastic materials. This relatively new plastic material is characterized by very high impact strength, superior heat resistance, and good electrical properties.

In addition, the low water absorption, high heat distortion point, low and uniform mold shrinkage, and excellent creep resistance of the material result in especially good dimensional stability. Of value for many applications is the its transparency, shear strength, stain resistance, colorability and gloss, oil resistance, machinability, and maintenance of good properties over a broad temperature range from less than $-73°C$ to 132 to 138°C. The fact that polycarbonate resin is self-extinguishing is important in many applications.

Polycarbonates are amorphous engineering thermoplastics that offer exceptional toughness over a wide temperature range. The natural resins are water-clear and transparent.

Polycarbonate resins are available in general-purpose molding and extrusion grades and in special grades that provide specific properties or processing characteristics. These include flame-retardant formulations as well as grades that meet Food and Drug Administration regulations for parts used in food-contact and medical applications. Other special grades are used for blow-molding, weather and UV-resistance, glass-reinforcement, EMI, RFI, ESD-shielding, and structural-foam applications. Polycarbonate is also available in extruded sheet and film.

Composition

The polycarbonate name is taken from the carbonate linkage that joins the organic units in the polymer. This is the first commercially useful thermoplastic material that incorporates the carbonate radical as an integral part of the main polymer chain.

There are several ways polycarbonates can be made. One method involves a bifunctional phenol, bisphenol A, which combines with carbonyl chloride by splitting out hydrochloric acid to give a linear polymer consisting of bisphenol groups joined together by carbonate linkages. Bisphenol A, which is the condensation product of phenol and acetone, is the basic building block used also in the preparation of epoxy resins.

Other members of the polycarbonate family may be made by using other phenols and other ketones to modify the isopropylidene group, or to replace this bridge entirely 1)'N' other radicals Subatttutiotr on the benzene ring offer further possibilities for variations.

Properties

Polycarbonate is a linear, low-crystalline, transparent, high-molecular-weight plastic. It is generally considered to be the toughest of all plastics. In thin sections, up to about 0.478 cm, its impact strength is as high as 24 kg·m. In addition, polycarbonate is one of the hardest plastics. It also has good strength and rigidity, and, because of its high modulus of elasticity, is resistant to creep. These properties, along with its excellent electrical resistivity, are maintained over a temperature range of about −170 to 121°C (see Table P.2). It has negligible moisture absorption, but it also has poor solvent resistance and, in a stressed condition, will craze or crack when exposed to some chemicals. It is generally unaffected by greases, oils, and acids. Polycarbonate plastics are easily processed by extrusion, by injection, blow, and rotational molding, and by vacuum forming. They have very low and uniform mold shrinkage. With a white light transmission of almost 90% and high impact resistance, they are good glazing materials. They have more than 30 times the impact resistance of safety glass.

Other typical applications are safety shields and lenses. Besides glazing, the high impact strength of polycarbonate makes it useful for air-conditioner housings, filter bowls, portable tool housings, marine propellers, and housings for small appliances and food-dispensing machines.

Humidity changes have little effect on dimensions or properties of molded parts. Even boiling water exposure does not change dimensions more than 0.30 mm/mm after parts are returned to room temperature. Creep resistance is excellent throughout a broad temperature range and is improved by a factor of 2 to 3 in glass-reinforced compounds.

The insulating and other electrical characteristics of polycarbonate are excellent and almost unchanged by temperature and humidity conditions. One exception is arc resistance, which is lower than that of many other plastics.

Polycarbonates are generally unaffected by greases, oils, and acids. Nevertheless, compatibility with specific substances in a service environment should be checked with the resin supplier. Water at room temperature has no effect, but continuous exposure in hot (65°C) water causes gradual embrittlement. The resins are soluble in chlorinated hydrocarbons and are attacked by most aromatic solvents, esters, and ketones, which cause crazing and cracking in stressed parts. Grades with improved chemical resistance are available, and special coating systems can be applied to provide additional chemical protection.

Fabrication

Polycarbonate resin has been molded in standard injection equipment using existing molds designed for nylon, polystyrene, acrylic, or other thermoplastic materials. Differences in mold shrinkage must be considered. And, in fabrication, the polycarbonate does have its own unique processing characteristics. Most important among these are the broad plastic range and high melt viscosity of the resin. Production runs in molds designed for nylon or acetal resin are not recommended.

Like other amorphous polymers, polycarbonate resin has no precise melting point. It softens and begins to melt over a range from

TABLE P.2
Properties of Polycarbonates

ASTM or UL Test	Property	General Purpose	High Flexural Modulus	20% Glass Reinforced
	Physical			
D792	Specific gravity	1.2	1.25	1.35
D792	Specific volume (in.^{3}/lb)	23	22.2	20.5
D570	Water absorption, 24 h, 1/8-in. thk (%)	0.15	0.12	0.16
	Mechanical			
D638	Tensile strength (psi)	9,000–10,500	8,000–9,600	16,000
D638	Elongation (%)	110–125	10–20	4–6
D790	Flexural strength (psi)	11,000–15,000	15,000	19,000
D790	Flexural modulus (10^5 psi)	3.0–3.4	5.0	8.0
D256	Impact strength, Izod (ft-lb/in. of notch)	12–16	2	2
D671	Fatigue endurance limit, 10^7 cycles (psi)	1,000	2,000	5,000
D785	Hardness, Rockwell M	62–70	85	91
	Thermal			
C177	Thermal conductivity (Btu-in./h-ft^2-°F)	1.35	1.41	1.47
D696	Coefficient of thermal expansion (10^{-5} in./in.-°C)	6.6–7.0	3.2	2.7
D648	Deflection temperature (°F)			
	At 264 psi	260–270	288	295
	At 66 psi	280	295	300
UL94	Flammability rating	HB, V-0	V-2, V-0	V-2, V-0
	Electrical			
D149	Dielectric strength (V/mil) Short time, 1/8-in. thk	380–400	450	490
D150	Dielectric constant			
	At 1 kHz	3.02	—	—
D150	Dissipation factor			
	At 1 kHz	0.0021	—	—
D257	Volume resistivity (Ω-cm)			
	At 73°F, 50%RH	>10^{16}	>10^{16}	>10^{16}
D495	Arc resistance	10–120	5–120	5–120
	Optical			
D542	Refractive index	1.586	—	—
D1003	Transmittance (%)	85–89	—	—
	Frictional			
—	Coefficient of friction			
	Self	0.52	—	—
	Against steel	0.39	—	—

[a] A + 1 MHz.

Source: Mach. Design Basics Eng. Design, June, p. 711, 1993. With permission.

216 to 227°C. Optimum molding temperatures lie above 271°C. The most desirable range of cylinder temperatures for molding the resin is in the area of 275 to 316°C.

Mold design must take into consideration the high melt viscosity of the material. Large sprues, large full-round runners, and generous gates with short lands usually give best results.

Tab gating is good for filling large thin sections. The gate to the tab should be large.

In injection molding polycarbonate, the following conditions are desirable:

1. *Heated molds.* Generally, hot water heat is adequate for heating molds, with typical mold temperatures ranging from 77 to 93°C. Molds for large areas, thin sections, or complex shapes or multiple cavity molds with long runners may require higher temperatures. Some molds have run best at 110 to 121°C.
2. *Cylinder temperatures.* The most usual cylinder temperature for molding polycarbonates is in the range of 275 to 316°C. Few parts require cylinder temperatures above 316°C. A heated nozzle and adequate mold temperature are helpful in keeping cylinder temperatures below 316°C. In most cases rear cylinder temperatures higher than front cylinder temperatures give best results.
3. *Heated nozzle.* In general, nozzle temperature equal to front cylinder temperature gives good results.
4. *Adequate injection pressure.* Injection pressures used in molding polycarbonate resin range from 80 to 204 MPa. Most usual range is in the 103 to 136 MPa range. Typical pressure setting is 3/4 to full pressure capacity of the press.
5. *Fast fill time.* For most parts molded, a fast ram travel time has been found desirable for thick as well as thin sections. For very thick sections, it is better to utilize somewhat slower ram speed.

Polycarbonate must be well dried to obtain optimum properties in the molded part. For this reason the resin is packaged in sealed containers.

Preheating pellets in the can to 121°C for 4 to 8 h and using of hopper heaters at 121°C are recommended for production operations to prevent moisture pickup.

Although polycarbonates are fabricated primarily by injection molding, other fabricating techniques may be used. Rod, tubing, shapes, film, and sheet may be extruded by conventional techniques. Films and coatings may be cast from solution. Parts can readily be machined from rod or standard shapes. Cementing, painting, metalizing, heat sealing, welding, machining operations, and other standard finishing operations may be employed. Film and sheet can be vacuum-formed or cold-formed.

Applications

The properties of polycarbonate resin make this new plastic suitable for a wide variety of applications. It is now being used in business machine parts, electrical and electronic parts, military components, and aircraft parts, and is finding increasing use in automotive, instrument, pump, appliance, communication equipment, and many other varied industrial and consumer applications.

One of the applications is in molded coil forms, which take advantage of the electrical properties, heat and oxidation resistance, dimensional stability, and resistance to deformation under stress of the resin.

A transparent plastic with the heat resistance, the dimensional stability, and the impact resistance of polycarbonate resin has created considerable interest for optical parts, such as outdoor lenses, instrument covers, lenses, and lighting devices.

Housings make use of the impact resistance of the material and its attractive appearance and colorability. In many cases, also, heat resistance and dimensional stability are important.

An interesting application area for plastic materials is the use of polycarbonate resin for fabricating fasteners of various types. Such uses as grommets, rivets, nails, staples, and nuts are in production or under evaluation. The ability of polycarbonate parts to be cold-headed has developed considerable interest in rivet applications.

Terminal blocks, connectors, switch housings, and other electrical parts may advantageously be molded of polycarbonate resin to take advantage both of the electrical properties of the material and the unusual physical properties, which give strength and toughness to the parts over a range of temperatures. Because of the heat resistance of polycarbonate, and the fact that it is self-extinguishing, molded parts can be used in current-carrying support applications.

Another application is in bushings, cams, and gears. Here, dimensional stability is important as is the high impact strength of the resin. Good physical properties over a broad temperature range, low water absorption, resistance to deformation under load, and resistance to creep suggest its use for many applications of this type. However, the resin has a higher coefficient of friction and a lower fatigue endurance limit than do some other plastics used in these types of applications. For this reason, it should not be considered a general-purpose gear and bearing material, but might be considered for applications subject to light loading, or to heavier but intermittent loading.

In most heart-bypass operations, the saphenous vein from a patient's leg replaces blocked blood vessels in the heart. The separate surgical procedure, performed during the heart-bypass operation, involves removing the vein through a long incision. Following the surgery, patients frequently complain of ongoing leg pain, potentially leading to reduced mobility and delayed rehabilitation while the large incision heals.

A new generation of surgical instruments not only makes the procedure less invasive, it helps speed the patient's return to normal activities. The system uses endoscopic techniques to harvest the vein. This, in turn, requires smaller incisions. The potential benefits are less postoperative pain, fewer wound-healing complications, minimal scarring, and quicker recovery.

The materials chosen for these applications have provided several significant benefits and, from the medical side, the materials were biocompatible. The materials resist chemicals and withstand gamma sterilization. The balloon mount is made with polycarbonate. On the orbital dissection cannula, the end piece embodies a polycarbonate resin.

POLYESTER FILM

Polyester film is a transparent, flexible film, ranging from 0.15 to 14 mils in thickness, used as a product component, in industrial processes, and for packaging. The most widely used type is produced from polyethylene terephthalate (i.e., Mylar). Polyester films based on other polymers or copolymers or manufactured by other methods are not identical, although they are similar in nature.

It is the strongest of all plastic films and strength is probably the outstanding property. However, it is useful as an engineering material because of its combination of desirable physical, chemical, electrical, and thermal properties. For example, strength combined with heat resistance and electrical properties makes it a good material for motor slot liners.

Polyester film is made by the condensation of terephthalic acid and ethylene glycol. The extremely thin film, 0.00063 to 0.0013 cm, used for capacitors and for insulation of motors and transformers, has a high dielectric strength, up to 236×10^6 V/m. It has a tensile strength of 137 MPa with elongation of 70%. It is highly resistant to chemicals, and has low water absorption. The material is thermoplastic, with a melting point at about 254°C. Polyester fibers are widely used in clothing fabrics. The textile fiber produced from dimethyl terephthalate is known as Dacron.

For magnetic sound-recording tape, polyester tape has the molecules oriented by stretching to give high strength. The 0.013-cm tape has a breaking strength of 3.4 kg/0.64 cm of width. Electronic tape may also have a magnetic-powder coating on the polyester. But where high temperatures may be encountered, as in spacecraft, the magnetic coating is applied to metal tapes.

PROPERTIES

Polyester film has excellent resistance to attack and penetration by solvents, greases, oils, and many of the commonly used electrical varnishes. At room temperature, permeability to such solvents as ethanol, ethyl acetate, carbon tetrachloride, hexane, benzene, acetone, and acetic acid is very low. It is degraded by some strong alkali compounds and embrittles under severe hydrolysis conditions.

Moisture absorption is less than 0.8% after immersion for a week at 25°C. Water-vapor permeability is similar to that of polyethylene film and permeability to gases is very low. The film is not subject to fungus attack and copper corrosion is negligible.

An outstanding feature of the film is the fact that good physical and mechanical properties are retained over a wide temperature range. Service temperature range is –60 to 150°C. The effect of temperature is relatively small between –20 and 80°C. No embrittlement occurs at temperatures as low as –60°C, and useful properties are retained up to 150 to 175°C. Tensile modulus drops off sharply at 80 to 90°C.

Melting point is 250 to 255°C; thermal coefficient of expansion is 15×10^{-6} in./in./°F; and shrinkage at 150°C is 2 to 3%.

Fabrication and Forms

Polyester film can be printed, laminated, metallized, coated, embossed, and dyed. It can be slit into extremely narrow tapes (0.038 mm and narrower), and light gauges can be wound into spiral tubing. Heavy gauges can be formed by stamping or vacuum (thermo-)forming. Matte finishes can be applied, and adhesives for bonding the film to itself and practically any other material are available.

Because of its desirable thermal characteristics, polyester film is not inherently heat sealable. However, some coated forms of the film can be heat-sealed, and satisfactory seals can be obtained on the standard film by the use of benzyl alcohol, heat, and pressure.

Polyethylene terephthalate film is available in several different types:

- A. General-purpose and electrical film for wide variety of uses
- C. Special electrical applications requiring high insulation resistance
- D. Highly transparent film, minimum surface defects
- K. Coated with a polymer for heat sealability and outstanding gas and moisture impermeability
- HS. Shrinks uniformly about 30% when heated to approximately 100°C; after shrinking, it has substantially the same characteristics as the standard film
- T. A film with high tensile strength (available in some thin gauges) with superior strength characteristics in the machine direction; designed for use in tapes requiring high-strength properties
- W. For outdoor applications; resistant to degradation by ultraviolet light

Applications

The range of properties outlined above has made polyester film functional in many totally different industrial applications and suggests its use in numerous other ways.

The largest current user of the film is the electrical and electronic market, which uses it as slot liners in motors and as the dielectric for capacitors, replacing other materials that are less effective, bulkier, and more expensive. It is found in hundreds of wire and cable types, sometimes used primarily as an insulating material, sometimes for its mechanical and physical contributions to wire and cable construction. Reduced cost of materials and processing and improved cable performance result.

Magnetic recording tapes for both audio and instrumentation uses are based on polyester film. In audio applications they contribute toughness, durability, and long play; for instrumentation tapes, the film ensures maximum reliability. The film has proved to be a highly successful new material for the textile industry since it can be used to produce metallic yarns that are nontarnishing and unusually strong. They can be run unsupported, knit, dyed at the boil, and either laundered or dry-cleaned. Yarns are made by laminating the film to both sides of aluminum foil or by laminating metallized and transparent film. The structure is then slit into the required yarn widths.

As a surfacing material, polyester film is used on both flexible and rigid substrates for both protective and decorative purposes. Metallized, laminated to vinyl, and embossed, the film becomes interior trim for automobiles, for example.

With a special coating, the film becomes a drafting material that is tougher and longer-lasting than drafting cloth. It is used for mapmaking, templates, and other applications in which its dimensional stability becomes a significant factor.

Strength in thin sections makes the film advantageous for sheet protectors, card holders, sheet reinforcers, and similar stationery products. Pressure-sensitive adhesives make the properties of the film available for uses ranging from decorative trim to movie splicing.

The weatherable form (Type W) has a life of 4 to 7 years. It is principally used in greenhouses, where it cuts construction costs by as much as two thirds because a simple, inexpensive structure suffices and maintenance costs are at a minimum.

In the packaging field, polyester film serves in areas where other materials fail or have functional disadvantages. In window cartons it lasts longer and does not break as other materials do; its toughness permits transparent packaging of heavy items; and, coated with polyethylene, it has made possible the "heat-in-the-bag" method of frozen-food preparation.

An optically clear form of polyester film in thicknesses of 4 to 7 mils is used as a base for the coating of light-sensitive emulsions in the manufacture of photographic film. The outstanding qualities of toughness and dimensional stability make this film especially well suited as a base for graphic arts films, motion picture film, engineering reproduction films, and microfilm. Other advantages include excellent storage and aging characteristics.

POLYESTER PLASTICS

The materials are commonly called polyester resins but this simple name does not distinguish between at least two major classes of commercial materials. Also, the same name is used within the unsaturated class to designate both the cured and uncured state. These plastics may be defined by identifying the materials as "unsaturated polyester resins which, when cured, yield thermoset products as opposed to thermoplastic products." The latter, as exemplified by Dacron and Videne, are saturated polyesters.

Composition

Unsaturated polyester resins of commerce are composed of two major components, a linear, unsaturated polyester and a polymerizable monomer. The former is a condensation-type of polymer prepared by esterification of an unsaturated dibasic acid with a glycol. Actually, most polyesters are made from two or more dibasic acids.

The most commonly used unsaturated acids are maleic or fumaric with limited quantities of others used to provide special properties. The second acid does not contain reactive unsaturation. Phthalic anhydride is most commonly used but adipic acid and, more recently, isophthalic acid are employed. The properties of the final product can be varied widely, from flexible to rigid, by changing the ratios and components in the polyester portion of the resin.

The polyesters vary from viscous liquids to hard, brittle solids but, with a few exceptions, are never sold in this form. Instead, the polyester is dissolved in the other major component, a polymerizable monomer. This is usually styrene; diallyl phthalate and vinyl toluene are used to a lesser extent. Other monomers are used for special applications.

Polyester resins may contain numerous minor components such as light stabilizers and accelerators, but must contain inhibitory components so that storage stability is achieved. Otherwise, polymerization will take place at room temperature.

The most widely used polyester resins are supplied with viscosities in the range of 300 to 5000 centipoise. They are clear liquids varying in color from nearly water-white to amber. They can be colored with certain common pigments, which are available ground in a vehicle for ease of dispersion. Inert, inorganic fillers such as clays, talcs, calcium carbonates, etc. are often added, usually by the fabricator, to reduce shrinkage and lower costs. Resins with thixotropic properties are also available.

Curing

The liquid resins are cured by the use of peroxides with or without heat to form solid materials. During the cure the monomer copolymerizes with the double bonds of the unsaturated polyester. The resulting copolymer is thermoset and does not flow easily again under heat and pressure. Heat is evolved during the cure. This must be considered when thick sections are made. Also, a volume shrinkage occurs and the density increases to 1.20 to 1.30. The amount of shrinkage varies between 5 to 10% depending on monomer content and degree of unsaturation in the polyester. The most popular catalyst is benzoyl peroxide but methylethyl ketone peroxide, cumene hydroperoxide, and others find application.

REINFORCEMENT

The development of this field of commercial resins owes a great deal to the commercial production of two other products: The first was the production of low-cost styrene for the synthetic rubber program. Other cheap monomers will work but not nearly as well. Styrene and its homologues have relatively high boiling points and fast polymerization rates. Both properties are important in this field.

The second development was the commercial production of fibrous glass. Polyester resins are not widely used as cast materials nor are the physical properties of such castings particularly outstanding.

The unique property of polyesters is their ability to change from a liquid to a hard solid in a very short time under the influence of a catalyst and heat. This property was not available in any of the earlier plastic materials. Polyesters flow easily in a mold with little or no pressure so that expensive, high-pressure molds are not required. Alternatively, very large parts can be made because the total pressure required to form the material is low.

Glass fibers of fine diameter have high tensile strengths, good electrical resistance properties, and low specific gravity when compared to metals. When such fibers are used as reinforcement for polyester resins (like steel in concrete) the resulting product possesses greatly enhanced properties. Specific physical properties of the polyester resins may be increased by a factor varying from 2 to 10. Naturally the increase in physical strengths obtained will depend upon both the amount of glass fibers used and their form. Strengths approaching those of metals on an equal weight basis are obtained with some constructions.

FABRICATION TECHNIQUES

Hand Layup

This method involves the use of either male or female molds. Products requiring highest physical properties are made with glass cloth. The cloth may be precoated with resin but usually one ply is laid on or in the mold, coated with resin by brushing or spraying, and then a second ply of cloth laid on top of the first. The process is continued until the desired thickness is built up. Aircraft parts, such as radomes, usually require close tolerances on dimensions and resin-to-glass weight ratio. After the layup has been completed, pressure is applied either by covering the assembly with flexible, extensible blanket and drawing it down by vacuum or the mold is so made that a rubber bag can be contained above the part. This is then blown up to apply pressure on the laminate. Thereafter, the cure is accomplished by heating in an oven, by infrared lamps, or by heating means built into the mold.

Corrugated Sheet Molding

Glass-reinforced polyester sheets are sold in large volume and are made by a relatively simple intermittent or continuous method. The process consists of placing a resin-impregnated mat between cellophane sheets, rolling or squeegeeing out the air, placing the assembly between steel or aluminum molds of the desired corrugation, and curing the assembly in an oven.

Matched Die Molding

This method produces parts rapidly and generally of uniform quality. The molds used are somewhat similar to those employed in compression molding, usually of two-piece, mating construction. The process consists of two steps. A "preform" of glass fibers is prepared by collecting fibers on a screen, which has the shape of the finished article. Suction is used to hold the fibers on the screen; fibers are either blown at it or fall on it from a cutter. Commercial equipment is available for this operation. When the desired weight of glass fibers has been collected, a resinous binder is applied. The preform and screen are then baked to cure the binder, after which the preform is ready for use. The second step is the actual molding. The preform is placed in the mold, the catalyzed resin poured on the preform in correct amount, the mold closed, and the cure effected by heat. Design of the mold is very important and some features are different from those in other molding fields.

Molding cycles vary from 2 to 5 min at temperatures from 110 to 149°C. Trimming, sanding, and buffing are usually required at the flash line.

Premix Molding

This branch of the field provides parts for automotive and similar end uses. The parts are strictly functional and are usually pigmented black. The strength properties required are relatively low except that impact strength must be good. As the name infers, the unsaturated resin is first mixed with fillers, fibers, and catalyst to provide a nontacky compound. The mixers used are of the heavy-duty Day or Baker-Perkins type. The "premix" is usually extruded to provide a "rope" or strip of material easily handled at the press. The fibers used most extensively are cut sisal, but glass and asbestos are also used, and frequently all three are present in a compound. The fillers are clays, carbonates, and similar cheap inorganic materials. A typical premix will contain 38% catalyzed unsaturated resin, 12% total fiber, and 50% filler. However, rather wide variations in composition are practiced to obtain specific end use properties.

The premix is molded at pressures of 103 to 350 MPa and at temperatures ranging from 121 to 154°C. Cure cycles are short, usually from 30 to 90 s. Again, the fact that the resin starts as a liquid makes possible the molding of intricate parts because of ease of flow in the mold. Heater housings for autos are the largest use, but housings of many types and electrical parts are produced in volume. This process provides a cheap but very serviceable molded material.

Properties

The strengths obtainable in the finished product are of prime interest to the engineer. However, the numerous forms of reinforcement materials and the variations possible in the polyester constituent present a whole spectrum of obtainable properties. As an example, products are available that will resist heat for long periods of time at temperatures varying from 66 to 177°C; but the higher the temperature, the more expensive the resin.

In general, the commercial resins have good electrical properties and are resistant to dilute chemicals. Alkali resistance is poor, as is resistance to strong acids. The strength-to-weight ratio of polyester parts and their impact resistance are outstanding physical properties. The data in Table P.3 (General Properties of Polyesters) are intended to be illustrative of the properties obtainable by different fabrication techniques.

Available Forms

Probably over 95% of the unsaturated polyester resins sold are liquids in the uncured state. However, certain types are available in solid or paste form for special uses. The cured resins are available in laminate form as corrugated sheeting, which is sold widely for partitions, windows, patio roofs, etc. Rod stock may be purchased for fishing rods and electrical applications. Paper and glass cloth laminates are sold to fabricators. The boat end use is making the material more familiar to the general public. However, a large part of the industrial production is concerned with custom-molded parts. These are perhaps best classified by industries rather than specific products. The aircraft industry uses substantial quantities but automotive end uses are larger. The chemical industry is using increasing amounts in fume ducts and corrosion-resistant containers. The electrical industry uses the material in laminate and molded forms and as an encapsulation medium. Furniture applications are growing. The machinery industry uses moldings as housings and guards; a recent volume use is motor boat shrouds. There are few fields that have not found the material useful in some application.

TABLE P.3
General Properties of Polyesters

Type of Reinforcement	None	Glass Cloth	Mat or Preform	Mat[a]	Parallel Yarn or Roving	Premix
Glass content, % by wt	—	60–70	35–45	20–30	60–80	10–40[b]
Specific gravity range	1.20–1.30	1.7–1.9	1.5–1.6	1.4–1.5	1.7–1.95	1.6–1.9
Rockwell Hardness	M100–110	M100–110	M90–100	M85–95	M90–110	M55–75
Flexural strength, 1000 psi	13–17	40–85	25–35	15–25	80–115	5–20
Tensile strength, 1000 psi	8–12	30–55	15–25	10–15	70–100	3–6
Compressive strength, 1000 psi	18–23	20–45	17–28	20–25	50–75	10–16
Flexural modulus, 10^6 psi	0.50–0.60	2.0–3.0	1.0–1.8	0.8–1.5	3.0–6.0	0.8–1.2
Tensile modulus, 10^6 psi	0.45–0.55	1.8–3.0	0.8–1.6	0.7–1.4	3.0–6.0	0.8–1.2
Impact, notched, ft-lb/in. notch	0.17–0.25	15–30	10–20	6–10	—	1.5–3.0
Shear strength, 1000 psi	—	15–25	12–18	8–12	—	—
Water absorption (24 h), %	0.15–0.25	0.10–0.20	0.2–0.5	0.2–0.4	0.15–0.30	0.3–1.0

[a] Corrugated sheet-type laminate. [b] Total fiber content.

POLYESTER THERMOPLASTIC RESINS

There are several types of melt-processible thermoplastics, including polybutylene terephthalate, polyethylene terephthalate, and aromatic copolyesters.

POLYBUTYLENE TEREPHTHALATE (PBT)

This plastic material is made by the transesterification of dimethyl terephthalate with butanediol through a catalyzed melt polycondensation. These molding and extrusion resins have good resistance to chemicals, low moisture absorption, relatively high continuous-use temperature, and good electrical properties (track resistance and dielectric strength). PBT resin is sensitive to alkalies, oxidizing acids, aromatics, and strong bases. Various additives, fillers, and fiber reinforcements are used with PBT resins, in particular flame retardants, mineral fillers, and glass fibers. PBT resins and compounds are used extensively in automotive, electrical-electronic, appliance, military, communications, and consumer product applications.

POLYETHYLENE TEREPHTHALATE (PET)

This is a widely used thermoplastic packaging material. Beverage bottles and food trays for microwave and convection oven use are the most prominent applications. PET resins are made from ethylene glycol and either terephthalic acid or the dimethyl ester of terephthalic acid. Most uses for PET require the molecular structure of the material to be oriented. Orientation of PET significantly increases tensile strength and reduces gas permeability and water vapor transmission. For packaging uses, PET is processed by blow molding and sheet extrusion. Typical trade names are Cleartuf, Traytuf, Tenite, and Kodapak. Injection-molding grades of PET, with fillers and/or reinforcements, also are available for making industrial products.

Aromatic polyesters (liquid crystal polymers, or LCP) have high mechanical properties and heat resistance. Commercial grades are aromatic polyesters, which have a highly ordered or liquid crystalline structure in solution and molten states. A high degree of molecular orientation develops during processing and, hence, anistropy in properties. Typically, these melt-processible resins can be molded or extruded to form products capable of use at temperatures over 260°C. Tensile strengths up to 240.3 MPa and flexural moduli up to 3586.3 MPa are reported for LCPs. Chemical resistance also is excellent. Trade names for LCPs include Vectra and Xydur Granlar. Applications are in chemical processing, electronic, medical, and automotive components.

POLYESTER THERMOSETTING RESINS

These are a large group of synthetic resins produced by condensation of acids such as maleic, phthalic, or itaconic with an alcohol or glycol such as allyl alcohol or ethylene glycol to form an unsaturated polyester which, when polymerized, will give a cross-linked, three-dimensional molecular structure, which in turn will copolymerize with an unsaturated hydrocarbon, such as styrene or cyclopentadiene, to form a copolymer of complex structure of several monomers linked and cross-linked. At least one of the acids or alcohols of the first reaction must be unsaturated. The polyesters made with saturated acids and saturated hydroxy compounds are called alkyd resins, and these are largely limited to the production of protective coatings and are not copolymerized.

The resins undergo polymerization during cure without liberation of water, and do not require high pressure for curing. Through the secondary stage of modification with hydrocarbons a very wide range of characteristics can be obtained. The most important use of the polyesters is as laminating and molding materials, especially for glass-fiber-reinforced plastic products. The resins have high strength, good chemical resistance, high adhesion, and capacity to take bright colors. They are also used, without fillers, as casting resins, for filling and strengthening porous materials such as ceramics and plaster of paris articles, and for sealing the pores in metal castings. Some of the resins have great toughness, and are used to produce textile fibers and thin plastic sheet and film. Others of the resins are used with fillers to produce molding powders that cure at low pressure of 3 to 6 MPa with fast operating cycles.

POLYESTER LAMINATES

These are usually made with a high proportion of glass-fiber mat or glass fabric, and high-strength reinforced moldings may also contain a high proportion of filler. A resin slurry may contain as high as 70% calcium carbonate or calcium sulfate, with only about 11% of glass fiber added, giving an impact strength of 165 MPa in the cured material. Bars and structural shapes of glass-fiber-reinforced polyester resins of high tensile and flexural strengths are made by having the glass fibers parallel in the direction of the extrusion. Rods and tubes are made by having the glass-fiber rovings carded under tension, then passing through an impregnating tank, an extruding die, and a heat-curing die. The rods contain 65% glass fiber and 35% resin. They have a flexural strength of 441 MPa and a Rockwell M hardness of 65.

Physical properties of polyester moldings vary with the type of raw materials used and the type of reinforcing agents. A standard glass-fiber-filled molding may have a specific gravity from 1.7 to 2.0, a tensile strength of 27 to 68 MPa with elongation of 16 to 20%, a flexural strength to 206 MPa, a dielectric strength to about 16×10^6 V/m, and a heat distortion temperature of 177 to 204°C. The moldings have good acid and alkali resistance. But, because an almost unlimited number of fatty-type acids are available from natural fatty oils or by synthesis from petroleum, and the possibilities of variation by combination with alcohols, glycols, and other materials are also unlimited, the polyesters form an ever-expanding group of plastics.

Some of the polyester-type resins have rubberlike properties, with higher tensile strengths than the rubbers and superior resistance to oxidation. These resins have higher wear resistance and chemical resistance than GRS rubber. They are made by reacting adipic acid with ethylene glycol and propylene glycol and then adding diisocyanate to control the solidifying action. They can be processed like rubber, but solidify more rapidly.

POLYETHERETHERKETONES

The latest developments in polyetheretherketone (PEEK) resin formulations let manufacturers take advantage of its wear and strength qualities for pump applications.

Uses

Pumps with mating components made of cast iron, stainless steel, and bronze have long been a source of problems for manufacturers and end users. The abrasive quality of metals leads to significant wear in pump parts, which frequently results in failures from galling and seizing. As anyone working in a manufacturing environment knows, one's worst nightmare can come true when a pump runs dry or a bearing fails, shutting down an entire assembly line or plant operation. And the costs are hardly trivial for repairing all of the pump components destroyed or carrying out periodic maintenance on the equipment to head off such a nightmare.

Engineers are starting to replace metals with advanced plastics for pump components such as bushings, line shaft bearings, and case and impeller wear rings. The resins they use are based on PEEK chemistry and were initially developed for aerospace and defense applications. Using these engineered resins, pump manufacturers improve performance, boost output, and cut costs by taking advantage of better wear and friction qualities, mechanical strength, and the ability to resist chemicals.

The chemical makeup of PEEK resins helps ease engineers' concerns about galling and seizing and are safer to work with. Taking advantage of these qualities, along with an understanding of some basic design guidelines, opens new doors in pump performance and reliability.

Types, Fabrication, and Properties

Engineers often choose PEEK resin because of its overall balance of qualities, including hydrolysis resistance, dimensional stability, wear resistance, temperature stability, strength, and chemical resistance. However, switching from metal to plastic in pump parts requires engineers to change their mind-sets.

PEEK can be injection-molded, compression-molded, or extruded into stock shapes, which are typically machined to final-part dimensions. Most pump components can be injection-molded as long as they do not require tight tolerances. The strict tolerances required for bushings and wear rings require parts to be machined from molded tube stock.

Although manufacturers may be used to machining metals, speeds and feeds are much different for plastics, and the process generally calls for high-end carbide and diamond tooling. Machining operations vary depending on which fillers and reinforcements are used in each resin grade.

Even though plastics can be machined more quickly than metals, it is not as easy to hold tolerance because plastics tend to spring when parted off. Tolerances in the range of ±0.05 mm can be held on parts with diameters up to 254 mm.

The most common blends of PEEK resin used in pump applications are carbon filled and carbon-fiber reinforced. Carbon-filled PEEK compounds can be molded in solids or tubes up to 101.6 cm in diameter. Carbon-fiber reinforced resins, in contrast, are robotically wound and formed around a mandrel. This limits fiber-reinforced resins to hollow shapes no smaller than 9.5 mm in diameter. Splitting hollow shapes is not recommended because the strength of the composite relies on the carbon fibers being molded and kept in tension.

Another difference between carbon-filled and carbon-fiber-reinforced PEEK resins is how they react to temperature changes. Carbon-filled compounds have a radial thermal expansion rate of 16×10^{-6} in./in./°F, while that of fiber-reinforced resins is less than 3×10^{-6} in./in./°F. These two values fall on either side of the thermal expansion rate of carbon steel,

TABLE P.4
Properties of PEEK Composites

	WR 300 Carbon-Filled PEEK	WR 525 Carbon-Fiber Reinforced PEEK
Specific gravity	1.47	1.60
Temperature range, °F	–100–300	–100–525
Tensile strength, psi	17,875	300,000
Tensile modulus, psi	1,560,000	20,000,000
Elongation, %	1.41	1.0
Compressive strength, psi	15,000	197,000
Flexural strength, psi	38,000	290,000
Coefficient of friction	0.20–0.28	0.10–0.15
Coefficient of thermal expansion from 73 ot 290°F, in./in.-°F	16×10^{-6}	$<3 \times 10^{-6}$
Color	Black	Black

Source: Mach. Design, July 23, pp. 69–71, 1998. With permission.

which is 6×10^{-6} in./in./°F, making the choice between the two resins critical when combining PEEK resins with metal pump housings and components; see Table P.4 (Properties of PEEK Composites).

Carbon-filled resins are recommended for press-in applications such as bushings and case wear rings. As temperatures rise, PEEK components will become tighter because they are expanding at a faster rate than the steels surrounding the bushings or rings. In contrast, carbon-fiber-reinforced resins work better for press-on applications such as impeller wear rings. The inside diameter of the wear ring fits on the impeller skirt, which has a slightly larger outside diameter than the ring inside diameter. As operating temperatures rise, the steel impeller skirt expands more quickly than the PEEK wear rings, creating a tighter fit.

In summary, when compared to carbon-fiber-reinforced resin, carbon-filled PEEK is less expensive, easier to machine, and can be split without compromising physical properties. However, fiber-reinforced resins have the advantage when it comes to higher operating temperatures. Reinforced resins are also stronger than carbon-filled blends. Fiber-reinforced PEEK is a true metal replacement because it combines the physical properties of steel with the processing ease and flexibility associated with thermoplastics.

APPLICATIONS

One area where metal-to-plastic conversions are reaping benefits is in the petroleum industry where multistage horizontal pumps often experience high vibration levels that frequently cause seals to fail. Vibrations wear down bronze bushings and case wear rings, which, in turn, reduces overall pump efficiency and increases maintenance costs.

Replacing bronze and steel wear rings with carbon-filled or carbon-reinforced PEEK components, allows manufacturers to increase pump efficiency by achieving tighter clearances. PEEK resins also reduce vibrations caused by fluid flow across wear rings and excess rotor runout.

The ductility of plastic pump components helps them outperform metal versions when it comes to reducing vibration, as well as resisting impact. For example, a petroleum refinery that once considered vibrations of 6.67 mm/s acceptable for two multistage pipeline diesel-fuel feed pumps was able to reduce vibration by more than 80% using carbon-filled PEEK case wear rings and bushings.

An added benefit when designing pump components from plastic is safety, particularly for industries working with flammable fluids. Plastics eliminate the risk of sparks when pumps run dry because of momentary suction losses during system upsets, which is a concern with metal-on-metal contact in conventional designs.

Another area is boiler-feed water pumps. Galling and even seizure are common concerns when using pumps with steel components and wear parts. This is especially true in water pumps where, because of the corrosion resistance of the material, stainless steels are used for pump parts as well as wear components. Although they protect pumps from corrosion, stainless steels gall more easily than other metals.

Engineers often design boiler-feed water pumps with tight diametrical running clearances to maximize discharge pressures. However, the tight clearances on stainless steel wear rings and bushings severely damage pumps during brief periods of cavitation. Cavitation takes place when suction pressures drop below the net positive suction head required by a pump. Momentary cavitations are common in pumps running close to the vapor pressure of a fluid, which is often the case in boiler-feed water pumps.

Wear, or hard rubs, resulting from cavitation often show up initially in the first-stage wear rings and the center-stage bushings. As a result, rotors deflect more, which causes pumps to seize.

Impeller wear rings and sleeves made of fiber-reinforced PEEK eliminate galling because the material can run for short periods without lubrication. As a result, PEEK wear parts and mating steel components remain undamaged during momentary system upsets.

POLYETHERIMIDES

Polyetherimide (PEI) is an amorphous engineering thermoplastic characterized by high heat resistance, high strength and modulus, excellent electrical properties that remain stable over a wide range of temperatures and frequencies, and excellent processability. Unmodified PEI resin is transparent and has inherent flame resistance and low-smoke evolution. The resin is produced as Ultem.

Polyetherimide resin is available in an unreinforced grade for general-purpose injection molding, blow molding, foam molding, and extrusion, in four glass-fiber-reinforced grades (10, 20, 30, and 40% glass), in bearing grades, and in several high-temperature grades. The unreinforced grade is available as a transparent resin and in standard and custom colors.

The resin can be processed on conventional thermoplastic molding equipment. Melt temperatures of 349 to 427°C are typical for injection-molding applications. Mold temperatures of 66 to 177°C are required.

Polyetherimide is extruded to produce profiles, coated wire, sheet, and film. Film thicknesses as low as 0.25 mil are obtained by solvent-casting techniques. Molded and extruded parts can be machined using either conventional or laser techniques and can be bonded together or to dissimilar materials using ultrasonic, adhesive, or solvent methods.

PROPERTIES

The Underwriter's Laboratories continuous-use listing of PEI is 170°C; glass-transition temperature is 215°C, and heat-deflection temperature is 200°C at 1.78 MPa, contributing to its high strength and modulus retention for service under load at elevated temperatures; see Table P.5 (Properties of Polyetherimids).

A key feature of polyetherimide is maintenance of properties at elevated temperatures. For example, at 179°C, tensile strength and flexural modulus are 48 and 2040 MPa. Moduli and strengths of the glass-reinforced grades are still higher. For example, flexural modulus is 8840 MPa with 30% glass reinforcement, and more than 80% of this is retained at 179°C.

Polyetherimide has good creep resistance as indicated by its apparent modulus of 2403 MPa after 1000 h at 82°C under an initial applied load of 34.3 MPa.

The resin resists a broad range of chemicals under varied conditions of stress and temperatures. Compatibility has been demonstrated with aliphatic hydrocarbons and alcohols including gasoline and gasohol, mineral–salt solutions, dilute bases, and fully halogenated hydrocarbons. Resistance to mineral acids is outstanding. The polymer is attacked by partially halogenated solvents such as methylene chloride and trichloroethane and by strong bases.

Resistance to ultraviolet radiation is good; change in tensile strength after 1000 h of xenon arc exposure is negligible. Resistance to gamma radiation is also good; strength loss is less than 6% after 500 Mrad exposure to cobalt-60 at the rate of one Mrad/h.

Hydrolytic-stability tests show that more than 85% of tensile strength is retained after 10,000 h of boiling-water immersion. The material is also suitable for applications requiring short-term or repeated steam exposure.

POLYETHER RESINS

These are thermoplastic or thermosetting materials that contain ether–oxygen linkages,

TABLE P.5
Properties of Polyetherimides

ASTM or UL Test	Property	Unmodified Resin	Glass-Reinforced 10%	Glass-Reinforced 30%
	Physical			
D792	Specific gravity	1.27	1.34	1.51
D570	Water absorption			
	At 24h, 73°F	0.25	0.21	0.16
	At equilibrium, 73°F	1.25	1.20	0.90
—	Mold shrinkage (in./in.)	0.007	0.005–0.006	0.002–0.004
	Mechanical			
D638	Tensile strength (psi)	15,200	16,600	24,500
D638	Elongation, ultimate (%)	60	6	3
D638	Tensile modulus (10^5 psi)	4.3	6.5	13.0
D790	Flexural strength (psi)	22,000	28,000	33,000
D790	Flexural modulus (10^5 psi)	4.8	6.5	13.0
D256	Impact strength, Izod (ft-lb/in. of notch)	1.0	1.1	2.0
D785	Hardness, Rockwell M	109	114	114
	Thermal			
C177	Thermal conductivity (Btu-in./h-ft^2-°F)	0.85	1.22	1.56
D696	Coefficient of thermal expansion (10^{-5}in./in.-°F)	3.1	1.8	1.1
D248	Deflection temperature (°F)			
	At 264 psi	392	405	410
	At 66 psi	410	410	414
UL94	Flammability rating	V-0 (0.016 in.)	V-0 (0.016 in.)	V-0 (0.010 in.)
D2863	Oxygen index, 0.060 in.	47	47	50
	Electrical			
D149	Dielectric strength (V/mil) 1/16 in.	830	—	770
D150	Dielectric constant			
	At 1 kHz, 50% RH	3.15	3.5	3.7
D150	Dissipation factor			
	At 1 kHz, 50% RH	0.0013	0.0014	0.0015
	At 2450 MHz, 50% RH	0.0025	0.0046	0.0053
D257	Volume resistivity (Ω-cm)	6.7×10^{17}	1.0×10^{17}	3.0×10^{16}
D495	Arc resistance(s)	128	85	85

Source: Mach. Design Basics Eng. Design, June, p. 714, 1993. With permission.

–C–O–C–, in the polymer chain. Depending upon the nature of the reactants and reaction conditions, a large number of polyethers with a wide range of properties may be prepared.

The main groups of polyethers in use are epoxy resins, prepared by the polymerization and cross-linking of aromatic diepoxy compounds; phenoxy resins, high-molecular-weight epoxy resins; polyethylene oxide and polypropylene oxide resins; polyoxymethylene, a high polymer of formaldehyde; and polyphenylene oxides, polymers of xylenols.

Epoxy Resins

The epoxy resins form an important and versatile class of cross-linked polyethers characterized by excellent chemical resistance, adhesion to glass and metals, electrical insulating properties, and ease and precision of fabrication.

The type of curing agent employed has a marked effect on the optimum temperature of curing and has some influence on the final physical properties of the product. By judicious selection of the curing system, the curing operation can be carried out at almost any temperature from 0 to 200°C.

Various fillers such as calcium carbonate, metal fibers and powders, and glass fibers are commonly used in epoxy formulations to improve such properties as the strength and resistance to abrasion and high temperatures.

Because of their good adhesion to substrates and good physical properties, epoxies are commonly used in protective coatings. Because of the small density change on curing and because of their excellent electrical properties, the epoxy resins are used as potting or encapsulating compositions for the protection of delicate electronic assemblies from the thermal and mechanical shock of rocket flight. Because of their dimensional stability and toughness, the epoxies are used extensively as dies for stamping metal forms, such as automobile gasoline tanks, from metal sheeting. Foams are also made.

By combining epoxies, especially the higher-performance types based on polyfunctional monomers, with fibers such as glass or carbon, exceedingly high moduli and strengths may be obtained. Thus, a typical carbon-fiber-reinforced epoxy may have a tensile modulus and strength of 500 and 2 GPa, respectively. Such composites are of importance in aerospace applications where high ratios of a property to density are desired, as well as in such domestic applications as sports equipment.

The adhesive properties of the resins for metals and other substrates and their relatively high resistance to heat and to chemicals have made the epoxy resins useful for protective coatings and for metal-to-metal bonding.

Polyolefin Oxide Resins

Polyethylene oxide and polypropylene oxide are thermoplastic products whose properties are greatly influenced by molecular weight. Oxides, such as tetrahydrofuran, can also be polymerized to give polyethers.

Low to moderate-molecular-weight polyethylene oxides vary in form from oils to waxlike solids. They are relatively nonvolatile, are soluble in a variety of solvents, and have found many uses as thickening agents, plasticizers, lubricants for textile fibers, and components of various sizing, coating, and cosmetic preparations. The polypropylene oxides of similar molecular weight have somewhat similar properties, but tend to be more oil soluble (hydrophobic) and less water soluble (hydrophilic).

Phenoxy Resins

Phenoxy resins differ from the structurally similar epoxy resins based on the reaction of epichlorohydrin with bisphenol A mainly by possessing a much higher molecular weight, about 25,000. The polymers are transparent, strong, ductile, and resistant to creep, and, in general, resemble polycarbonates in their behavior. Cross-linking may be effected by the use of curing agents that can react with the OH groups. Molding and extrusion may be used for fabrication. The major application is as a component in protective coatings, especially in metal primers.

Polyphenylene Oxide Resins

Polyphenylene oxide (PPO) is the basis for an engineering plastic characterized by chemical, thermal, and dimensional stability. The PPO resin is normally blended with another compatible but cheaper resin such as high-impact polystyrene. The blends are cheaper and more processible than PPO and still retain many of the advantages of PPO by itself.

PPO is outstanding in its resistance to water, and in its maximum useful temperature range (about 170 to 300°C). In spite of the high softening point, the resins, including glass-reinforced compositions, can be molded and extruded in conventional equipment. Uses include medical instruments, pump parts, and insulation. Structurally modified resins are also of interest and the general family of resins may be expected to replace other materials in many applications.

Commercial products can be processed by injection or compression molding, as well as by slurry or electrostatic coating; reinforcement with glass fibers is also commonly practiced. The combination of properties is useful in such applications as electronic pumps and automotive components and coatings, especially where environmental stability is required.

Polyoxymethylene

Polyoxymethylene, or polyacetal, resins are polymers of formaldehyde. With high molecular weights and high degrees of crystallinity, they are strong and tough and are established in the general class of engineering thermoplastics.

Although somewhat similar to polyethylene in general molecular structure, polyacetal molecules pack more closely, and attract each other to a much greater extent, so that the polymer is harder and higher-melting than polyethylene. Polyacetals are typically strong and tough, resistant to fatigue, creep, organic chemicals (but not strong acids or bases), and have low coefficients of friction. Electrical properties are also good. Improved properties for particular application may be attained by reinforcement with fibers of glass or polytetrafluoroethylene.

The combination of properties has led to many uses such as plumbing fittings, pump and valve components, bearings and gears, computer hardware, automobile body parts, and appliance housings. Other aldehydes may be polymerized in a similar way.

POLYETHYLENE

Polyethylene thermoplastic resins include low-density polyethylenes (LDPE), linear low-density polyethylenes (LLDPE), high-density polyethylenes (HDPE), and ethylene copolymers, such as ethylene-vinyl acetate (EVA) and ethylene-ethyl acrylate (EEA), and ultrahigh-molecular-weight polyethylenes (UHMWPE). In general, the advantages gained with polyethylenes are light weight, outstanding chemical resistance, good toughness, excellent dielectric properties, and relatively low cost compared to other plastics. The basic properties of polyethylenes can be modified with a broad range of fillers, reinforcements, and chemical modifiers, such as thermal stabilizers, colorants, flame retardants, and blowing agents. Further, polyethylenes are considered to be very easy to process by such means as injection molding, sheet extrusion, film extrusion, extrusion coating, wire and cable extrusion coating, blow molding, rotational molding, fiber extrusion, pipe and tubing extrusion, and powder coating. Major application areas for polyethylenes are packaging, industrial containers, automotive, materials handling, consumer products, medical products, wire and cable insulation, furniture, housewares, toys, and novelties.

Polyethylene is a water-repellent, white, tough, leathery, thermoplastic resin very similar in appearance to paraffin wax. Properties vary from a viscous liquid at low molecular weights to a hard waxlike substance at high molecular weights. It is used as a coating for glass bottles and fiberglass fabrics (special treatments for glass are required to obtain good adhesion between polyethylene and glass) and is also used as an injection-molding material for ceramics.

The basic building blocks for polyethylenes are hydrogen and carbon atoms. These atoms are combined to form the ethylene monomer, C_2H_4, i.e., two carbon atoms and four hydrogen atoms. In the polymerization process, the double bond connecting the carbon atoms is broken. Under the right conditions, these bonds reform with other ethylene molecules to form long molecular chains. Ethylene copolymers, EVA, and EEA are made by the polymerization of ethylene units with randomly distributed comonomer groups, such as vinyl acetate (VA) and ethyl acrylate (EA).

Properties

Three basic molecular properties affect most polyethylene properties: crystallinity (density), average molecular weight, and molecular weight distribution; see Table P.6 (Properties of Polyethylenes).

Molecular chains in polyethylenes in crystalline areas are arranged somewhat parallel to

TABLE P.6
Properties of Polyethylenes

ASTM Test	Property	Low Density	Medium Density	High Density	Ultrahigh Molecular Weight
	Physical				
D792	Specific gravity	0.910–0.925	0.926–0.940	0.941–0.965	0.928–0.941
D792	Specific volume (in.³/lb)	30.4–29.9	29.9–29.4	29.4–28.7	29.4
D570	Water absorption, 24 h, 1/8-in. thk (%)	0.01	0.01	0.01	0.01
	Mechanical				
D638	Tensile strength (psi)	600–2,300	1,200–3,500	3,100–5,500	4,000–6,000
D638	Elongation (%)	90–800	50–600	20–1,000	200–500
D638	Tensile modulus (10^5 psi)	0.14–0.38	0.25–0.55	0.6–1.8	0.20–1.10
D790	Flexural modulus (10^5 psi)	0.08–0.60	0.60–1.15	1.0–2.0	1.0–1.7
D256	Impact strength, Izod (ft-lb/in. of notch)	No break	0.5–16	0.5–20	No break
D785	Hardness, Rockwell R	10	15	65	67
	Thermal				
C177	Thermal conductivity (10^{-4} cal-cm/s-cm²-°C)	8.0	8.0–10.0	11.0–12.4	11.0
D696	Coefficient of thermal expansion (10^{-5}in./in.–°F)	5.6–12.2	7.8–8.9	6.1–7.2	7.8
D648	Deflection temperature (°F)				
	At 264 psi	90–105	105–120	110–130	118
	At 66 psi	100–121	120–165	140–190	170
	Electrical				
D149	Dielectric strength (V/mil) Short time, 1/8-in. thk	460–700	460–650	450–500	900[a]
D150	Dielectric constant At 1 kHz	0.0002	0.0002	0.0003	0.0002
D257	Volume resistivity (Ω-cm) At 73°F, 50% RH	10^{15}	10^{15}	10^{15}	10^{18}
D495	Arc resistance(s)	135–160	2,000–235	—	
	Optical				
	Refractive index	1.51	1.52	1.54	—
	Transmittance (%)	4–50	4–50	10–50	—

[a] kV/cm.

Source: *Mach. Design Basics Eng. Design*, June, p. 717, 1993. With permission.

each other. In amorphous areas, they are randomly arranged. High-density polyethylene resins have molecular chains with comparatively few side chain branches. Therefore, the chains are packed more closely together. The result is crystallinity up to 95%. LDPE resins have crystallinity from 60 to 75%. LLDPE resins have crystallinity from 60 to 85%. For polyethylenes, the higher the degree of crystallinity, the higher the resin density. Higher density, in turn, influences numerous properties. With increasing density, heat-softening point, resistance to gas and moisture vapor permeation, and stiffness increase. However, increased density results in a reduction of stress cracking resistance and low-temperature toughness. LLDPE resins have densities ranging from 0.915 to 0.940 g/cms³; LDPE resins range from 0.910 to 0.930 g/cms³; and HDPE from 0.941 to 0.965 g/cm³.

High-molecular-weight LDPE resins are used to extrude high-clarity, tough film used in shrink packaging and for making heavy-duty bags. High-molecular-weight HDPE resins (ave. MW 200,000 to 500,000) have excellent environmental stress cracking resistance, toughness, high moisture barrier properties, and high strength and stiffness. The high-molecular-weight-HDPE resins are used in film, pressure pipe, large blow moldings, and sheet for thermoforming. UHMWPEs generally are considered to be those resins with molecular weights greater than 2 million (materials with MW up to 6 million are available). These resins have excellent abrasion resistance, stress cracking resistance, and toughness. They are, however, much more difficult to process than standard polyethylenes and require special forming techniques. UHMWPE resins are used in applications requiring high wear resistance, chemical resistance, and low coefficient of friction.

Polyethylene rubbers are rubberlike materials made by cross-linking with chlorine and sulfur, or they are ethylene copolymers. Chlorosulfonated polyethylene is white spongy material, which has chlorine atoms and sulfonyl chloride groups spaced along the molecule. It is used to blend with rubber to add stiffness, abrasion resistance, and resistance to ozone, and also for wire covering. Ethylene-propylene rubber, produced by various companies, is a chemically resistant rubber of high tear strength. The ethylene butadiene resin can be vulcanized with sulfur to give high hardness and wide temperature range. For greater elongation a terpolymer with butene can be made.

Polyethylene of low molecular weight is used for extending and modifying waxes, and also in coating compounds especially to add toughness, gloss, and heat-sealing properties. Such materials are called polyethylene wax, but they are not chemical waxes. They can be made emulsifiable by oxidation, and they can be given additional properties by copolymerization with other plastics. The polymethylene waxes are microcrystalline and have sharper melting points than the ethylene waxes. They are more costly, but have high luster and durability. Polybutylene plastics are rubberlike polyolefins with superior resistance to creep and stress cracking. Films of this resin have high tear resistance, toughness, and flexibility, and are used widely for industrial refuse bags.

Chemical and electrical properties are similar to those of polyethylene and polypropylene plastics. Polymethyl pentene is a moderately crystalline polyolefin plastic resin that is transparent even in thick sections. Almost optically clear, it has a light transmission value of 90%. Parts molded of this plastic are hard and shiny with good impact strength down to −29°C. Its specific gravity (0.83) is the lowest of any commercial solid plastic. A major use is for molded food containers for quick frozen foods that are later heated by the consumer.

Low-Density Polyethlyene

LDPE, the first of the polyethylenes to be developed, has good toughness, flexibility, low-temperature impact resistance, clarity in film form, and relatively low heat resistance. Like the higher-density grades, LDPE has good resistance to chemical attack. At room temperature, it is insoluble in most organic solvents but is attacked by strong oxidizing acids. At higher temperatures, it becomes increasingly more susceptible to attack by aromatic, chlorinated, and aliphatic hydrocarbons.

Polyethylene is susceptible to environmental and some chemical stress cracking. Wetting agents such as detergents accelerate stress cracking. Some copolymers of LDPE are available with improved stress-crack resistance.

About half of LDPE production goes into packaging applications such as industrial bags, shrink bundling, soft goods, and produce and garment bags. Other applications include blow-molded containers and toys, hot-melt adhesives, injection-molded housewares, paperboard coatings, and wire insulation. LDPE resins are rotationally molded into large agricultural tanks, chemical shipping containers, tote boxes, and battery jars.

One of the fastest-growing plastics is linear LLDPE, used mainly in film applications but also suitable for injection, rotational, and blow molding. Properties of LLDPE are different from those of conventional LDPE and HDPE in that impact, tear, and heat-seal strengths and environmental stress-crack resistance of LLDPE are significantly higher. Major uses at present are grocery bags, industrial trash bags, liners, and heavy-duty shipping bags for such products as plastic resin pellets.

HIGH-DENSITY POLYETHYLENE

Rigidity and tensile strength of the HDPE resins are considerably higher than those properties in the low- and medium-density materials. Impact strength is slightly lower, as is to be expected in a stiffer material, but values are high, especially at low temperatures, compared with those of many other thermoplastics.

HDPE resins are available with broad, intermediate, and narrow molecular-weight distribution, which provides a selection to meet specific performance requirements. As with the other polyethylene grades, very-high-molecular-weight copolymers of HDPE resins are available with improved resistance to stress cracking.

Applications of HDPE range from film products to large, blow-molded industrial containers. The largest market area is in blow-molded containers for packaging milk, fruit juices, water, detergents, and household and industrial liquid products. Other major uses include high-quality, injection-molded housewares, industrial pails, food containers, and tote boxes; extruded water and gas-distribution pipe, and wire insulation; and structural-foam housings.

HDPE resins are also used to rotationally mold large, complex-shaped products such as fuel tanks, trash containers, dump carts, pallets, agricultural tanks, highway barriers, and water and waste tanks for recreational vehicles.

A special category of HDPE known as high-molecular-weight HDPE (HMW-HDPE) offers outstanding toughness and durability, particularly at low temperatures. These characteristics result from a unique combination of high average molecular weight (250,000 to 500,000), and a bimodal molecular-weight distribution.

In blow-molding applications, HMW-HDPE allows drum manufacturers to meet Department of Transportation and Occupational Health and Safety Administration specifications. In pipe production, HMW-HDPE meets PE-3408, currently the highest strength rating for polyethylene pipe. Extruded sheet applications include pond liners, truck-bed liners, and outdoor leisure products. The primary market, however, for HMW-HDPE is in film applications, where its toughness allows down-gauging in merchandise bags and trash bags. The material is also suited for use in T-shirt-type grocery sacks requiring high handle strength.

UHMWPE

UHMWPE was originally defined as a polyethylene whose average molecular weight, as measured by the solution-viscosity method, is greater than 2,000,000. (Molecular weight of HDPE ranges from 100,000 to 500,000.) Over the years, producers and processors of UHMWPE materials have tried to reach agreement on just how high is "ultrahigh." Values proposed in the past have ranged from as low as 1,000,000 to over 3,500,000. Also in dispute was the question of the relationship between molecular weight and properties of UHMWPE in finished parts.

Several years ago, a value of 3,100,000 was agreed upon for the molecular weight as the dividing line, above which the UHMW description should apply. Resin properties increase with increasing molecular weight and start to level off at the 3,100,000 value. Also, processibility is more difficult above that dividing line, and material cost rises more rapidly.

As with most high-performance polymers, processing of UHMWPE is not easy. Because of its high melt viscosity (it does not register a melt-flow index), conventional molding and extrusion processes would break the long molecular chains that give the material its excellent properties. Methods used currently are compression molding, ram extrusion, and warm forging of extruded slugs. Developmental work is being done on injection molding of UHMWPE resins, but the process forces the polymer to behave in a manner that is not conducive to maintaining its molecular structure. Compression-molded sheets as large as 1.43×4 m are available.

UHMWPE has outstanding abrasion resistance and a low coefficient of friction. Impact strength is high, and chemical resistance is excellent. The material does not break in impact strength tests using standard notched specimens; double-notched specimens break at 20 ft-lb/in. Crystalline melting point of the material is 130°C. Recommended maximum service temperature is about 93°C, however, because of a high coefficient of thermal expansion.

Applications of UHMWPE include conveyor wear strips and guide rails, paper-machine suction-box covers, chute linings, snowmobile-track sprockets, bearings, parts for textile looms, pipe for distribution of slurry materials, and other components requiring abrasion resistance, impact strength, and a low-friction coefficient. These plastics have outperformed other materials, including metals, in these applications.

POLYETHYLENE GLYCOLS

Polyethylene glycols are water-soluble, nonvolatile liquids and solids. Polyethylene glycols dissolve in water to form transparent solutions and are also soluble in many organic solvents. They do not hydrolyze or deteriorate.

CHARACTERISTICS

The intermediate members of the series with average molecular weights of 200 to 20,000 are produced as residue products by the sodium or potassium hydroxide-catalyzed batch polymerization of ethylene oxide onto water or mono- or diethylene glycol. These polymers are formed by stepwise anionic addition polymerization and, therefore, possess a distribution of molecular weights. Examples of commercial uses for products in this range are in ceramic, metal forming, and rubber-processing operations; as druf suppository bases and in cosmetic creams, lotions, and deodorants; as lubricants; as dispersants for casein, gelatins, and inks; and as antistatic agents. These polyethylene glycols generally have low human toxicity.

The highest members of the series have molecular weights from 100,000 to 10,000,000. They are produced by special anionic polymerization catalysts that incorporate metals such as aluminum, calcium, zinc, and iron, and coordinated ligands such as amides, nitriles, or ethers. These members of the polyethylene glycol series are of interest because of their ability at very low concentrations to reduce friction of flowing water.

APPLICATIONS

Polyethylene glycol has a range of properties making it suitable for medical and biotechnical applications. These polymers of ethylene oxide are frequently described as amphiphilic, meaning that they are soluble both in water and in most organic solvents. The medical and biotechnical applications derive from this amphiphilicity, from a lack of toxicity and immunogenicity, and from a tendency to avoid other polymers and particles also present in aqueous solution.

Attaching polyethylene glycol to another molecule provides the latter with enhanced solubility in solvents in which polyethylene glycol is soluble, and thus polyethylene glycol is attached to drugs to enhance water and blood solubility. Similarly, polyethylene glycol is attached to insoluble enzymes to impart solubility in organic solvents. These polyethylene glycol enzymes are used as catalysts for industrial reactions in organic solvents.

The tendency to avoid other polymers results in formation of two immiscible aqueous layers when a solution of a polyethylene glycol is mixed with certain other polymer solutions. These aqueous two-phase systems are used to purify biological materials such as proteins, nucleic acids, and cells by partitioning of desired and undesired materials between the two phases.

The tendency of polyethylene glycols to avoid interaction with cellular and molecular components of the immune system results in the material being nonimmunogenic.

Ceramic applications are varied and interesting. For example, they may be used to bind glass tubes into bundles prior to cutting into lengths. They also have been used as components of ceramic slips and glazes for the manufacture of porcelain signs and other vitreous coatings.

As binders for colors, excellent adhesion is obtained when the coatings are sprayed on the ceramic surface. A crayon for applying identification marks or decorations to ceramic articles, prior to firing, may be made from pigmented polyethylene glycol. As an extrusion binder, it aids in slurrying the clay, promotes die life, and gives good binding properties. The automatic extrusion of tile, pipe, and electrical insulators has been improved through its use. Polyethylene glycol is an effective additive for the electrophoretic coating of certain components for vacuum tubes.

POLYFLUOROLEFIN RESINS

These resins are distinguished by their resistance to heat and chemicals and by the ability to crystallize to a high degree. Several main products are based on tetrafluoroethylene (TFE), hexafluoropropylene (HFP), and monochlorotrifluoroethylene.

Copolymers of TFE and HFP with each other and of TFE and HEP with ethylene are available commercially.

Polytetrafluoroethylene

The polymer is insoluble, resistant to heat (up to 275°C) and chemical attack, and, in addition, has the lowest coefficient of friction of any solid. Because of its resistance to heat, the fabrication of polytetrafluoroethylene requires modification of conventional methods. After molding the powdered polymer using a cold press, the moldings are sintered at 360 to 400°C by procedures similar to those used in powder metallurgy. The sintered product can be machined or punched. Extrusion is possible if the powder is compounded with a lubricating material. Aqueous suspensions of the polymer can also be used for coating various articles. However, special surface treatments are required to ensure adhesion because polytetrafluoroethylene does not adhere well to anything.

Polytetrafluoroethylenc (TFE resin) is useful for applications under extreme conditions of heat and chemical activity. Polytetrafluoroethylene bearings, valve seats, packings, gaskets, coatings, and tubing can withstand relatively severe conditions. Fillers such as carbon, inorganic fibers, and metal powders may be incorporated to modify the mechanical and thermal properties.

Because of its excellent electrical properties, polytetrafluoroethylene is useful when a dielectric material is required for service at a high temperature. The nonadhesive quality is often turned to advantage in the use of polytetrafluoroethylene to coat articles such as rolls and cookware to which materials might otherwise adhere.

Polymonochlorotrifluoroethylene

The properties of polymonochlorotrifluoroethylene (CTFE resin) are generally similar to those for polytetrafluoroethylene; however, the presence of the chlorine atoms in the former causes the polymer to be a little less resistant to heat and to chemicals. The polymonochlorotrifluoroethylene can be shaped by use of conventional molding and extrusion equipment, and it is obtained in a transparent, noncrystalline condition by quenching. Dispersions of the polymer in organic media may be used for coating.

The applications of polychlorotrifluoroethylene are in general similar to those for polytetrafluoroethylene. Because of its stability and inertness, the polymer is useful in the manufacture of gaskets, linings, and valve seats that must withstand hot and corrosive conditions. It is also used as a dielectric material, as a vapor and liquid barrier, and for microporous filters.

Copolymers

Copolymers of TFE and HFP propylene (fluorinated ethylenepropylene, or FEP resins) and copolymers of TFE with ethylene (ETFE) are often used in cases where ease of fabrication is desirable. The copolymers can be processed by conventional thermoplastic techniques, and except for some diminution in the level of some properties, properties generally resemble those of the TFE homopolymer.

Copolymers of ethylene with CTFE can also be processed by conventional methods, and have better mechanical properties than TFE, FEP, and PFA resins.

Polyvinylidene Fluoride

The properties are generally similar to those of the other fluorinated resins: relative inertness, low dielectric constant, and thermal stability (up to about 150°C). The resins (PVF_2 resins) are, however, stronger and less susceptible to creep and abrasion than TFE and CTFE resins.

Applications of polyvinylidene fluoride are mainly as electrical insulation, piping, process equipment, and as a protective coating in the form of a liquid dispersion.

Fluorinated Elastomers

Several types of fluorinated, noncrystallizing elastomers were developed to meet needs (usually military) for rubbers that possess good low-temperature behavior with a high degree of resistance to oils and to heat, radiation, and weathering.

Copolymers of hexafluoropropylene with vinylidene fluoride make up an important class with such applications as gaskets and seals. Copolymers of nitrosomethane with tetrafluoroethylene are showing considerable promise for similar applications.

POLYIMIDES

Available both as thermoplastic and thermoset resins, polyimides (PIs) are a family of some of the most heat- and fire-resistant polymers known. Moldings and laminates are generally based on thermoset resins, although some are made from thermoplastic grades. Unlike most plastics, polyimides are available as laminates and shapes, molded parts, and stock shapes from some materials producers. Thin-film products — enamel, adhesives, and coatings — are usually derived from thermoplastic polyimide resins.

Laminates are based on continuous reinforcements including woven glass and quartz fabrics, or fibers of graphite, boron, quartz, or organic materials. Molding compounds, on the other hand, contain discrete fibers such as chopped glass or asbestos, or particulate fillers such as graphite powders, MoS_2, or polytetrafluoroethylene.

Polyimide films and wire enamels are generally unfilled. Coatings may be pigmented or filled with particles such as polytetrafluoroethylene for lubricity. Most adhesives contain aluminum powder to provide a closer match to the thermal-expansion characteristics of metal substrates and to improve heat dissipation.

Polyimide parts are fabricated by techniques that range from powder-metallurgy methods to conventional injection, transfer, and compression molding, and extrusion methods. Porous polyimide parts are also available. Generally, those compounds that are the most difficult to fabricate have the highest heat resistance.

Properties

Polyimide parts and laminates can serve continuously in air at 260°C; service temperature for intermittent exposure can range from cryogenic to as high as 482°C. Glass-fiber-reinforced versions retain over 70% of their flexural strength and modulus at 249°C. Creep is almost nonexistent, even at high temperatures, and deformation under load (27.2 MPa) is less than 0.05% at room temperature for 24 h.

These materials have good wear resistance and low coefficients of friction, both of which are further improved by polytetrafluoroethylene fillers. Self-lubricating parts containing graphite powders have flexural strengths above 68 MPa, which is considerably higher than those of typical thermoplastic bearing compounds.

Electrical properties of polymide moldings are outstanding over a wide range of temperature and humidity conditions.

Polyimide parts are unaffected by exposure to dilute acids, aromatic and aliphatic hydrocarbons, esters, ethers, alcohols, Freons, hydraulic fluids, JP-4 fuel, and kerosene. They are attacked, however, by dilute alkalies and concentrated inorganic acids.

Polyimide adhesives maintain useful properties for over 12,000 h at 260°C, 9,000 h at 302°C, 500 h at 343°C, and 100 h at 371°C. Resistance of these adhesives to combined heat (to 302°C) and saltwater exposure is excellent; see Table P.7 (Properties of Polyimides).

POLYKETONES

Polyketones are partially crystalline engineering thermoplastics that can be used at high temperatures. They also have excellent chemical resistance, high strength, and excellent resistance to burning. Although they require high melt temperatures, polyketones can be extruded and injection-molded with standard processing equipment.

TABLE P.7
Properties of Polyimides

| | | Moldings | | | |
| | | Filler | | | |
ASTM Test	Property	Glass Fiber[a]	25% Graphite Powder[b]	15% Graphite Powder[c]	Laminates
		Physical			
D792	Specific gravity	1.9	1.45	1.51	1.95
D792	Specific volume (in.³/lb)	14.5	19.0	18.3	14.1
D570	Water absorption, 24 h, 1/8-in. thk (%)	0.2	0.6	0.2	0.3
		Mechanical			
D638	Tensile strength (psi)				
	At 70°F	27,000	5,700	9,500	50,000
	At 480°F	23,000	3,600	5,500	35,000
D638	Elongation (%)	1	1	4.5	1
D790	Flexural strength (psi)				
	At 70°F	50,000	12,800	16,000	70,000
	At 480°F	36,000	9,200	9,000	50,000
D790	Flexural modulus (10^5 psi)				
	At 70°F	32.5	9	5.5	40
	At 480°F	24.2	7.5	3.7	32
D256	Impact strength, Izod (ft-lb/in. of notch)	17	0.25	0.8	13
D785	Hardness, Rockwell M	120	110	88	—
		Thermal			
C177	Thermal conductivity (Btu-in./h-ft²-°F)	3.48	11.3	6.0	2.2
D696	Coefficient of thermal expansion (10^{-5}in./in.-°C)	1.4	3.4	4.7	1
D648	Deflection temperature (°F)				
	At 264 psi	660	550	680	660
		Electrical			
D149	Dielectric strength (V/mil)				
	Short time, 1/8-in. thk	500	—	250	625
D150	Dielectric constant				
	At 1 kHz	4.70	—	13.3	4.5
D150	Dissipation factor				
	At 1 kHz	0.003	—	0.006	0.015
D257	Volume resistivity (Ω-cm)				
	At 73°F, 50% RH	5×10^{15}	—	10^{14}	6×10^{14}
D495	Arc resistance(s)	180	—	—	180
		Frictional			
	Coefficient of friction				
	Against steel (PV–10,000)	—	0.24	0.24	—

[a] Rhone-Poulenc Kinel 5504.
[b] Kinel 5505.
[c] DuPont SP-21 (Vespel parts).
[d] Rhone-Poulenc Kerimid 601.

Source: Mach. Design Basics Eng. Design, June, p. 715, 1993. With permission.

Several polyketones are commercially available:

- Polyaryletherketones (PAEK or PEK), repeating ether and ketone groups combined by phenyl rings
- Polyetheretherketones (PEEK), repeating monomers of two ether groups and a ketone group
- Polyetherketoneketones (PEKK), repeating monomers of one ether group and two ketone groups

In addition, other materials with various combinations of the ether and ketone groups are designed to provide a balance of high heat resistance and good processibility, e.g. poly[ether][ketone][ether][ketone][ketone].

Polyketone resins are available as natural resins and glass- or carbon-fiber-reinforced and mineral-filled grades.

Properties

Glass transition temperatures (T_g) and melting temperatures (T_m) for polyketones depend on the ratio of ketone to ether groups. With increasing ether groups, both temperatures decrease. With increasing ketone groups, the temperatures go up. The PEEK T_g is 148°C and its T_m is 335°C; the PAEK T_g is 190°C and its T_m 380°C; and PEKEKK has a T_g of 177°C and a T_m of 375°C.

Polyketones are stronger and more rigid than most other engineering plastics. They are tough and impact resistant over a wide range of temperatures. Polyketones have very high fatigue strength. Both coefficients of friction and wear rates for polyketones are very low.

The thermal oxidative stability of polyketones is excellent. Typically, continuous-use temperatures are above 249°C (i.e., 50% of strength and rigidity are retained up to this temperature). Moduli of polyketones remain almost constant until the temperature is close to the T_g. Polyketones have comparatively low thermal coefficients of linear expansion. Polyketones have excellent resistance to burning and very low flame spread. These materials have good dielectric properties, with high volume and surface resistivities, and high dielectric strength.

Polyketones are extremely resistant to numerous inorganic and organic chemicals. They are dissolved or decomposed by concentrated, anhydrous, or strong oxidizing acids. Common solvents do not attack polyketones even at elevated temperatures. They have very good resistance to hydrolysis, even in hot water. Like most plastics composed of aromatic building blocks, polyketones are affected by ultraviolet radiation. Polyketones have extremely high resistance to beta-rays, gamma-rays, and x-rays over a wide range of temperatures.

Processing

Applicable processing methods are injection molding, extrusion, rotational molding, and powder coating. Conventional injection-molding machines can be used with polyketones. Melt temperatures for polyketones vary depending on the resin type. PEK melt temperatures range from 390 to 430°C; for PEEK, the range is 380 to 399°C. Mold temperatures from 180 to 216°C are recommended for molding PEK, and 180 to 198°C for PEEK. Parts for high-temperature applications should be molded at the high end of the temperature range.

Polyketones can be extruded to form sheet, cast-film, stock for machining, and wire coatings. Typical melt temperatures for extrusion are 399 to 430°C.

Polyketone parts can be assembled using various adhesives and welding techniques. The adhesives can be epoxies, cyanoacrylates, polyurethanes, or silicones. Welding techniques include heated-tool welding (399 to 538°C for 10 to 90 s); spin welding; hot-air welding (449 to 499°C); and ultrasonic welding.

POLYMER

The terms *polymer, high polymer, macromolecule,* and *giant molecule* are used to designate high-molecular-weight materials of either synthetic or natural origin. Plastics are relatively stiff at room temperature, rubbers or elastomers are flexible and retract quickly after stretching, and fibers are especially strong filamentary materials. Coatings are generally either plastics or rubbers that have been applied as a thin layer on a substrate. In practice, plastics, rubbers, fibers, and coatings are used as formulations of the polymers with other ingredients such as fillers, pigments, plasticizers, flow improvers, and stabilizers against aging and degradation.

Historical Development

The first modified natural polymers, cellulose nitrate and casein-formaldehyde, were commercially produced about 1860, and the first fully synthetic polymer, phenol-formaldehyde, was made about 1910. The major development of present polymer science and technology has taken place since about 1920.

Interest in the synthesis of products similar to natural products but possessing more useful properties has been continually stimulated by the successful synthesis of polyamide fibers and rubbers equivalent to natural rubber, and by increasing understanding of the nature of proteins, nucleoproteins, carbohydrates, and enzymes in living tissues.

Several striking advances have been the development of cheap raw materials for plastics and of new polymerization processes, and remarkable advances in understanding the relationships among molecular structure, morphology, and physical and chemical behavior. As properties have been improved, plastics have been developed that can be readily and economically fabricated, and that can be used for hitherto inappropriate engineering purposes such as gears, bearings, and structural members. Such engineering plastics may frequently be used advantageously to replace metals or other materials.

There has been considerable interest in polymers that can withstand even more extreme environments, or possess other specialized properties, for use under the sea, in space, and in biological systems. These interests have emphasized the continuing need for still deeper understanding of bonding (in organic, inorganic, and organic–inorganic systems) and of the implications of bonding and structure to morphology, and, in turn, to the overall response to applied stresses in given environments.

Properties

The properties of polymeric materials are determined by the molecular properties of the macromolecules, the morphology, and the type of formulation involving plasticizers or fillers. The morphology in turn depends on the conditions of fabrication and which molecular orientation or crystallization may be induced. Properties also depend on the temperature and elapsed time of the measurement.

Environmental Stability

The stability of a polymer depends on the chemical nature, on morphology, and on molecular properties. Thus, polyamides are susceptible to hydrolysis on long exposures to dilute acids at high temperatures. Resistance to high temperatures under oxidizing conditions is enhanced by minimizing the hydrogen content, by increasing the molecular softness, and by increasing the content of aromatic structures or heterocyclic rings.

Compounding

Plastic masses, rubber formulations, coatings, and other polymeric compositions may contain age inhibitors, strengthening and coloring pigments, flow improvers, and plasticizing or softening agents. Roll-mill, sigma-blade, and dough mixers are generally employed to mix the resin with plasticizers and pigments at temperatures usually between about 50 and 250°C.

Additives

The addition of plasticizers to a polymeric material causes it to be softer and more rubbery in character. Plasticizers are held in association with the polymer chains by secondary valence forces. They separate the molecules, thus reducing the effective intermolecular attractive forces.

Age inhibitors are almost always incorporated in polymeric compositions. Oxygen, ozone, light, and electric discharge produce free radicals that cause degradation of the polymer chains. Free-radical inhibitors and light-masking agents are therefore commonly used.

Polyblends

These are produced by the addition of small amounts of a rubbery polymer to a polymeric glass. The impact strength of the glass thus is substantially increased. The rubbery polymer is not truly compatible and exists as a finely dispersed separate phase. Interpenetrating polymer networks (IPNs), in which two different networks are intertwined, constitute an interesting special case of polyblends. The phenomenon has somewhat similar counterparts in inorganic glass technology and in physical metallurgy. The study of polyblends combines the rigor of physics, for example, fracture mechanics, with organic and physical chemistry and the engineering behavior of the systems of interest.

FABRICATION

Polymer formulations can be fabricated into useful forms or articles by a variety of methods.

In the use of molded thermosetting compositions in which heat and pressure are required for the production of sound articles, some form of compression molding is usually employed. The physically compacted composition containing the resin in the fusible stage is forced into a mold cavity of the desired shape and is held under heat and pressure until the curing or vulcanization is complete. When high pressure is not required, as in the preparation of epoxy compositions and polyester–styrene–glass fiber compounds, it is desirable to use moderate pressure to obtain a uniform molding.

Various forms of injection molding are used for the shaping of thermoplastic (permanently fusible) compositions. The composition is first softened temporarily by forcing the compounded resin granules through a heated chamber, after which it is driven into a relatively cold mold. Under proper conditions, the resin remains soft long enough to fill the mold completely and then rapidly hardens as it cools. The injection molding of engineering plastics often makes it possible to replace an assembly of metal parts by one plastic piece. Variations of the injection molding process are used in the extrusion of films, rods, and pipe and in the spinning of fibers.

Blow molding, a process in which air is blown into a hot tube held in a mold, is used for the manufacture of bottles and other shapes. Certain articles are molded conveniently by rotating a hot mold filled with resin.

In extrusion, a hot melt is forced through a die with an opening shaped to produce the cross section desired. Extrusion and molding techniques may be combined to form foams into sheets and other shapes.

Films are also produced by extrusion of a tube into which air is forced. The process is called bubble extrusion. The pipe expands because of the air pressure, to a wall thickness equivalent to the film thickness desired. The walls of the expanded bubble are pressed together in nip rolls, and later the large, thin-walled, collapsed pipe is slit to yield flat film. Films and sheets are also produced by calendering; in this process the hot resin is forced between tightly fitted rolls.

A technique useful for shaping sheets into various shapes is thermoforming, in which a heated sheet is forced against the contours of a mold by a positive pressure or vacuum.

In the casting process, fluid compositions are poured into molds of the desired shape and then allowed to cool or cure. This process is used for the production of foams and in encapsulation, such as in the protection of electronic components or the mounting of biological specimens. In broad terms, coating, slush molding, and painting may be considered to be casting operations.

Some polymers can be formed by methods similar to those used in metallurgical forming. Forging and cold-forming are adaptable to the mass production of an increasing number of plastics. Sintering techniques and variations are also used for coating metals and forming polymers that cannot be formed by conventional processes.

In the shaping of thermoplastics particularly, the conditions of molding (temperature, time, and pressure) have a marked effect on the properties of the final product. Uneven cooling produces strains, and the flow of the plasticized resin can cause some orientation of the molecules. In bubble molding of film, biaxial orientation is produced. In the drawing of films and fibers, either uniaxial or biaxial orientation may be obtained with equipment similar to the tentors used in the textile industry. Orientation results in a substantial increase in the strength of the product in the direction of stretching; in many polymers, crystallization is induced during cold-drawing or stretching. On occasion, discontinuities or areas of strain are produced by partially orienting the molecules, and the product becomes more subject to stress cracking in the presence of solvents or other agents. Thus, the optimum condition of a product results from a judicious combination of mechanical treatment and thermal annealing.

POLYMERIC COMPOSITE

A polymeric composite is any of the combinations or compositions that comprise two or more materials as separate phases, at least one of which is a polymer. By combining a polymer with another material, such as glass, carbon, or another polymer, it is often possible to obtain unique combinations or levels of properties. Typical examples of synthetic polymeric composites include glass-, carbon-, or polymer-fiber-reinforced thermoplastic or thermosetting resins, carbon-reinforced rubber, polymer blends, silica- or mica-reinforced resins, and polymer-bonded or polymer-impregnated concrete or wood. It is also often useful to consider as composites such materials as coatings (pigment–binder combinations) and crystalline polymers (crystallites in a polymer matrix). Typical naturally occurring composites include wood (cellulosic fibers bonded with lignin) and bone (minerals bonded with collagen). On the other hand, polymeric compositions compounded with a plasticizer or very low proportions of pigments or processing aids are not ordinarily considered composites.

Typically, the goal is to improve strength, stiffness, or toughness, or dimensional stability by embedding particles or fibers in a matrix or binding phase. A second goal is to use inexpensive, readily available fillers to extend a more expensive or scarce resin; this goal is increasingly important as petroleum supplies become costlier and less reliable. Still other applications include the use of some fillers such as glass spheres to improve processibility, the incorporation of dry-lubricant particles such as molybdenum sulfide to make a self-lubricating bearing, and the use of fillers to reduce permeability.

Emphasis on the development of polymeric composites has been stimulated by the need for greatly improved mechanical and environmental behavior, especially on a strength- or stiffness-to-weight basis. Such composites are also often more efficient in their energy requirements for production than traditional materials. The high absolute and specific (per unit of weight) values of properties such as strength and stiffness have made composites ideal candidates for new applications in aircraft and boats, in passenger vehicles and farm equipment, and in machinery tools, and appliances. Composites based on chemically resistant matrixes are used in chemical process equipment.

Mechanical Properties

The behavior of composites depends on the volume fractions of the phases, their shape, and on the nature of the constituents and their interfaces. With anisotropic phases, the orientation with respect to the direction of stressing or exposure to permeants is also important. In general, given an appropriate preferred direction, the greater the anisotropy, the greater the effect on a given property, at least up to some point. Thus, all high-modulus reinforcements will stiffen a lower-modulus matrix, but fibers and platelets are more effective than spheres; a similar role of shape holds for the ability to reduce permeability at right angles to the anisotropic particles. Anisotropic high-modulus inclusions invariably increase strength if adhesion is good, but the effect is more complex with particulate fillers such as spheres. Rubbery inclusions lower the stiffness of a high-modulus matrix, but may enhance toughness by stimulating a combination of localized crazing and shear deformation.

Many fibers, for example, glass and carbon, are very stiff and strong. However, the maximum strength of these brittle materials cannot be realized in practical objects because of a high sensitivity to the inevitable small cracks and flaws that are ordinarily present. Most polymers are much less sensitive to such flaws, even though they are inherently less strong. More energy is needed to fracture a polymer than a ceramic or glass, and a crack tends to grow much less readily in a polymer.

If, for example, the mass of a strong but flaw-sensitive ceramic or glass is divided into many parts, typically into a fiber, and embedded in a polymer matrix, a growing crack may break one fiber but its progress may be hindered by the matrix or diverted along the interfaces. Thus, even though the matrix contributes only insignificantly to the total strength, it permits a closer approach to the theoretical maximum strength of the glass. Similar considerations apply to short fibers and, with some qualifications, to other forms of reinforcement.

The matrix has several other functions besides the dissipation of energy, which would otherwise cause a catastrophic failure. It protects the fiber against damage by mechanical action, such as rubbing, or by environmental agents, such as water. It must also transfer an applied stress or force to the filament so that they, being much stronger, can bear most of the load. To transfer the stress the matrix must adhere well to the fiber, although the optimum strength of the interfacial bond desired may vary depending on the application.

The strength of such a composite depends on the orientation of the fibers with respect to the applied force and on the nature of the stress (tensile or compressive). In the selection of materials and design for a composite, the ultimate application must be known. For example, strength of a composite based on longitudinal aligned fibers will be greatest in the direction of the fibers. Rupture under tension will require the pulling out of many fibers; this implies a fiber matrix bond that can yield fairly readily, and in yielding increase the energy required for rupture in compression. On the other hand, a stronger interfacial bond is needed to prevent buckling. Compromises in design are often necessary. Thus, at the expense of some strength, fibers are often crisscrossed to minimize the directionality of strength.

Reinforced Thermosetting Resins

The most common fiber-reinforced polymer composites are based on glass fibers, cloth, mat, or roving embedded in a matrix of an epoxy or polyester resin. Usually, the glass surface is treated with a coupling agent to promote adhesion to the matrix. Fabrication may be effected in several ways. Frequently resin-dipped continuous filaments are wound on a mandrel in one of several patterns. Tapes or bundles of fiber coated with resin can also be positioned by hand or machine in layers, or pulled through a die. After the desired shape is obtained, curing is effected under heat and pressure.

Boron, polyaramids, and especially carbon fibers confer especially high levels of strength and stiffness. Hybrid fiber combinations, for example, glass with carbon, are often used to achieve a desired balance between cost and performance.

Some examples of properties for several experimental composites (unidirectional fibers) in comparison with metals are striking; see Table P.8 (Epoxy Composites in Comparison to Metals). In fact, the relative properties of the polymer composites can exceed those of other materials. Although relative stiffness is less favorable for glass-fiber composites, carbon-fiber composites have a relative stiffness five times that of steel. Because of these excellent properties, many applications are uniquely suited for epoxy and polyester composites, such as components in new jet aircraft, parts for automobiles, boat hulls, rocket motor cases, and chemical reaction vessels.

Reinforced Thermoplastic Resins

Although the most dramatic properties are found with reinforced thermosetting resins such as epoxy and polyester resins, significant improvements can be obtained with many thermoplastics. Usually short fibers of glass are used — about 3 to 50 mm in length.

Nylon resins are particularly acceptable. Reinforcement of nylon-6, -10 with 20% fiber glass raises the tensile strength and heat-deflection temperature from 59 to 138 MPa and from 150 to 210°C.

Polycarbonates, acetals, polyethylene, and polyesters are among the resins available as glass-reinforced composition.

The combination of inexpensive, one-step fabrication, by injection molding, with improved

TABLE P.8
Epoxy Composites in Comparison to Metals

Material	Tensile Strength, lb/in.² (GPa)		Young's Modulus, lb/in.² (GPa)	
	Actual	Relative to Density	Actual	Relative to Density
Fiber in epoxy laminate				
High-strength glass	47,000 (0.3)	0.8	5.6×10^6 (38)	80
Boron	83,000 (0.6)	1.2	30×10^6 (210)	430
Carbon	160,000 (1.1)	2.5	33×10^6 (230)	520
Metal				
Steel	320,000 (2.2)	1.2	30×10^6 (210)	109
Titanium	280,000 (1.9)	1.6	17.5×10^6 (120)	109

Source: McGraw-Hill Encyclopedia of Science and Technology, 8th ed., Vol. 14, McGraw-Hill, New York, 182. With permission.

properties has made it possible for reinforced thermoplastics to replace metals in many applications in appliances, instruments, automobiles, and tools.

OTHER COMPOSITES

In the development of other composite systems, various matrices are possible; for example, polyimide resins are excellent matrices for glass fibers, and give a high-performance composite. Different fibers are of potential interest, including polymers (such as polyvinyl alcohol), single-crystal ceramic whiskers (such as sapphire), and various metallic fibers.

POLYMERIZATION

Polymerization is the linking of small molecules (monomers) to make larger molecules. Polymerization requires that each small molecule have at least two reaction points or functional groups. There are two distinct major types of polymerization processes: condensation polymerization, in which the chain growth is accompanied by elimination of small molecules such as water (H_2O) or methanol (CH_3OH); and addition polymerization, in which the polymer is formed without the loss of other materials.

POLYOLEFIN RESINS

These resins are polymers derived from hydrocarbon molecules that possess one or more alkenyl (or olefinic) groups. The term *polyolefin* typically is applied to polymers derived from ethylene, propylene, and other alpha-olefins, isobutylene, cyclic olefins, and butadiene and other diolefins. Polymers produced from other olefinic monomers, such as styrene, vinyl chloride, and tetrafluoroethylene, generally are considered separately.

Polyolefin homopolymers are made from ethylene, propylene, butylene, and methyl pentene. Other olefin monomers such as pentene and hexene are used to make copolymers.

Because the chemical and electrical properties of all olefins are similar, they often compete for the same applications. They differ from each other primarily in their crystalline structure. However, since strength properties vary with the type and degree of crystallinity, the tensile, flexural, and impact strength of each polyolefin may be quite different. Stress-crack resistance and useful temperature range also vary with crystalline structure.

In addition to the solid polyolefin resins, these materials are also available as beads from which very low density (1.25 to 5.0 lb/ft³) foam shapes and blocks are produced. Resilience and energy-absorption properties of these products are exceptional compared with those of conventional polystyrene foams. Polymers available as moldable beads include polyethylene, polypropylene, and a polyethylene/polystyrene copolymer alloy.

The bead forms can be processed by the same methods used for expandable polystyrene (EPS). After the beads are expanded (20 to 40 times that of the solid resin) and conditioned, they are poured into a mold and heated, usually by direct injection of steam. This softens, expands further, and fuses the particles together, forming a uniform, void-free, closed-cell shape. After molding, the shapes are usually annealed by being stored at 49 to 71°C to stabilize shape and dimensions.

Because they contain 80 to 95% air by volume, the foamed shapes are not nearly as strong as solid moldings. They are used primarily to cushion impact, insulate thermally, and provide high stiffness-to-weight core materials in composite components. An important application for polypropylene foam is in bumper cores for automobiles. A 76- to 102-mm-thick section of foam at 2 to 4 lb/ft^3 can absorb the energy of a 5-mph impact. Package cushioning for fragile and valuable products such as electronic or audio components is another application for either polyethylene or polypropylene foam. The toughness of polyethylene/polystyrene alloy foams qualifies them for material-handling applications where repeated use is required.

The principal resins of the polyolefin family are polyethylene and polypropylene. Other polyolefin polymers and copolymers are ethylene-vinyl acetate, ionomer, polybutylene, and polymethyl pentene.

POLYPHENYLENE ETHER

Alloys, or blends of polyphenylene ether (PPE) and polystyrene in various proportions, are marketed under the trade names of Noryl and Prevex. The resins are processed by conventional injection-molding, extruding, and thermoforming methods. Structural-foam parts are processed in standard foam-molding systems, using either direct induction of nitrogen gas or conventional chemical-blowing agents.

PPE is produced by a process based on oxidative coupling of phenolic monomers. The result, a resin that has good mechanical properties and thermal stability, is then blended with polystyrene to improve processability. Available grades of molding resins include glass-reinforced and platable compounds and heat-resistant grades (containing nylon), in addition to extrusion and foamable grades. All resins can be furnished in a wide range of colors.

PROPERTIES

PPE blends are characterized by outstanding dimensional stability, the lowest water absorption of the engineering thermoplastics, broad temperature ranges, excellent mechanical and thermal properties, and excellent dielectric properties over a wide range of frequencies and temperatures; see Table P.9 (Properties of PPE Resins).

Because of their excellent hydrolytic stability, both at room and elevated temperatures, PPE blend parts can be repeatedly steam-sterilized with no significant change in properties. In exposure to aqueous environments, dimensional changes are low and predictable. Resistance to acids, bases, and detergents is excellent. The material is attacked, however, by many halogenated or aromatic hydrocarbons. Prototype testing of components requiring exposure to such environments is recommended.

POLYPHENYLENE OXIDE

Polyphenylene oxide is a plastic that is notable for its high strength and broad temperature resistance. There are two major types: phenylene oxide (PPO) and modified phenylene oxide (Noryl). These materials have a deflection temperature ranging from 100 to 174°C at 2 MPa. Their coefficients of linear thermal expansion are among the lowest for engineering thermoplastics. Room-temperature strength and modulus of elasticity are high and creep is low. In addition, they have good electrical resistivity. Their ability to withstand steam sterilization and their hydrolytic stability make them suitable for medical instruments, electric dishwashers, and food dispensers. They are also used in the electrical and electronic fields and for business-machine housings.

Tensile strength and modulus of phenylene oxides rank high among engineering thermoplastics. They are processed by injection-molding, extrusion, and thermoforming techniques. The foam grades, with their high rigidity, are suitable for large structural parts. Because of good dimensional stability at high temperatures and under moisture conditions, these plastics are readily plated without blistering.

TABLE P.9
Properties of PPE Resins

ASTM or UL Test	Property	Standard	Glass Reinforced	Extrusion	Platable
		Physical			
D792	Specific gravity	1.06–1.10	1.21–1.36	1.06–1.10	1.05
D792	Specific volume (in.^{3}/lb)	25.2–26.2	20.4–22.9	25.2–26.2	—
D570	Water absorption, 24 h, 1/8-in. thk (%)	0.07	0.07	0.07	0.07
		Mechanical			
D638	Tensile strength (psi)	7,000–9,600	14,500–17,800	7,800–11,000	7,000
D638	Elongation (%)	50–60	4.6	15–60	60
D638	Tensile modulus (10^5 psi)	3.55–3.80	9.25–1.2	3.55–3.80	—
D790	Flexural strength (psi)	8,200–15,000	8.500–20,000	12,800–16,000	9,800
D790	Flexural modulus (10^5 psi)	3.0–3.6	7.5–11	3.6	3.0
D256	Impact strength, Izod (ft-lb/in. of notch)	5.0–10.0	2.3	3.10	5.0
D671	Fatigue endurance limit, 2×10^6 cycles (psi)	1,850–2,500	4,000–5,000	1,850–2,500	—
D785	Hardness, Rockwell	R115–119	L106–108	R115–119	—
		Thermal			
C177	Thermal conductivity (Btu-in./h-ft^2-°F)	1.50	1.10	1.50	—
D696	Coefficient of thermal expansion (10^{-5}in./in.-°C)	3.3–3.8	1.4–2.0	3.3–3.8	3.5
D648	Deflection temperature (°F)				
	At 264 psi	190–300	270–300	185–300	235
	At 66 psi	220–315	280–317	230–315	—
UL94	Flammability class	HB, V-1, V-0	HB, V-1	HB-V-1, V-0	HB
		Electrical			
D149	Dielectric strength (V/mil)				
	Short time, 1/8-in. thk	400–630	420–600	400–550	—
D150	Dielectric constant				
	At 1 MHz	2.64–2.68	2.85–3.11	2.64–2.68	—
D150	Dissipation factor				
	At 1 MHz	0.0009–0.0024	0.0014–0.0021	0.0009–0.0047	—
D257	Volume resistivity (Ω-cm)				
	At 73°F, 50% RH	10^{17}	10^{17}	10^{17}	—
D495	Arc resistance(s)	75	70–120	75	—

Source: Mach. Design Basics Eng. Design, June, p. 720, 1993. With permission.

POLYPHENYLENE SULFIDE

Polyphenylene sulfide (PPS) is a crystalline, high-performance engineering thermoplastic characterized by outstanding high-temperature stability, inherent flame resistance, and broad chemical resistance. PPS resins and compounds manufactured in the United States are tradenamed Ryton.

A wide range of injection-molding grades of PPS is available. The series with various glass-fiber levels (designated R-3, R-4, and R-5) is recommended for mechanical electronic applications requiring high mechanical strength, impact resistance, and insulating characteristics. All other compounds contain various mineral fillers in addition to glass-fiber reinforcement. Ryton R-7 and R-8 are suitable for electrical applications requiring high arc resistance and low arc tracking. The R-10 series of pigmented compounds includes several grades suitable for support of current-carrying parts in electrical components. PPS is essentially transparent to microwave radiation, so the R-11 series is suited specifically for microwave ovenware and appliance components; see Table P.10 (Properties of PPS Compounds).

Unreinforced PPS resins are also available as powders for slurry coating and electrostatic spraying. The resin coatings are suitable for food-contact applications as well as for chemical-processing equipment.

The injection-moldable PPS compounds require processing temperature of 316 to 343°C. Mold temperatures can range from 37.8 to 135°C to control the crystallinity. Cold-molding parts deliver optimum mechanical strength, and hot-molded highly crystalline parts provide optimum dimensional stability at high temperatures.

PPS is also available in long-fiber-reinforced forms. One type, called stampable sheet, contains fiber-mat reinforcement for processing by compression molding. The other form contains reinforcement, and is designed for laminating and thermoforming. Reinforcement in both forms can be glass or carbon fiber.

PROPERTIES

Most PPS compounds are used for their combination of high-temperature stability, chemical resistance, dimensional reliability, and flame retardance. The compounds all have excellent stability at very high temperatures. Underwriter's Laboratories (UL) temperature index is 200 to 240°C, depending on compound, thickness, and end use. In short-term excursions, the compounds have heat-deflection temperatures of 260°C or higher, depending on the crystallinity of the molding. Mechanical strength of the compounds remains high at high temperatures. For example, the flexural modulus of Ryton R-4 is as high at 260°C as that of ABS (acrylonitrile, butadiene, styrene) at room temperature.

PPS has excellent resistance to a broad variety of chemicals, even at high temperatures. In fact, the resin has no known solvent below 204°C. The resins are flame-retardant without additives (UL 94V-0/5). The oxygen index of the resin is 44, and indexes of the compounds range from 47 to 53. Because flame retardance is inherent, regrind is as flame resistant as virgin material.

Mechanical properties of the various PPS compounds are tailored for target applications. The balance of properties can be controlled by the degree of crystallinity of a molded part. Amorphous moldings have optimum mechanical strength at room temperature, and crystalline moldings deliver optimum dimensional stability at high temperatures.

APPLICATIONS

A metal plate spins at 7200 rpm inside a computer's hard disk drive. Less than three virus-widths above that surface resides a rock-hard ceramic block on the end of an aluminum arm. The temperature inside is 260 to 316°C. Under these conditions, a new grade of PPS, Ryton, R-4-230 NA, provided connectors with the proper electrical insulation, dimensional stability, flame retardancy, rigidity, and creep resistance for the drives. It is also a material that would not out-gas inside the drives.

The Ryton PPS stiffness and creep resistance also make the material ideal for load-

TABLE P.10
Properties of PPS Compounds

ASTM or UL Test	Property	Glass-Reinforced					Glass and Mineral Filled		
		R-3	R-4	R-5	R-7	R-8	R-9	R-10
		Physical							
D792	Specific gravity	1.57	1.67	1.72	1.9	1.8	1.9	2.0
D570	Water absorption, 24 h, in. thk (%)	0.05	0.05	0.05	0.03	0.03	—	0.07
		Mechanical							
D638	Tensile strength (psi)	16,000	17,500	19,000	14,000	10.75	11,000	10,000–13,500
D638	Elongation (%)	1.0	0.9	0.8	0.7	0.5	0.5	0.7
D790	Flexural modulus (10^5 psi)	14	17	21	24	22	21	20
D256	Impact strength, Izod (ft-lb/in. of notch)							
	Notched	1.0	1.3	1.3	1.0	0.6	0.7	0.7–1.0
	Unnotched	3.3	4.5	5.0	3.5	1.9	2.1	1.7–2.8
D785	Hardness, Rockwell R	120	123	123	121	121	—	120
		Thermal							
C177	Thermal conductivity (Btu-in./in.-ft^2-°F)	—	2.0	—	4.0	—	—	3.9
D696	Coefficient of thermal expansion (10^{-5}in./in.-°F)	1.7	1.6	1.5	1.1	1.6	1.1	1.2

POLYPHENYLENE SULFIDE

D648	Deflection temperature							
	At 264 psi (°F)	500	500	500	500	500	500	500
D2863	Oxygen index (%)	46	47	48	53	50	45	53
UL946B	Temperature index (°C)	200/220	200/220	200/220	200/220	200/220	200/240	200/240
UL94	Flammability rating	V-0	V-0/5-V	V-0	V-0	V-0/5-V	V-0	V-0/5-V
	Electrical							
D149	Dielectric strength (V/mil)	450	450	420	340	340	400	320–400
D150	Dielectric constant							
	At 1 kHz	3.9	3.9	4.0	5.1	4.6	4.6	4.7
	At 1 MHz	3.8	3.8	3.9	4.6	4.3	4.5	4.5
D150	Dissipation factor							
	At 1 kHz	0.002	0.002	0.002	0.058	0.017	0.014	0.008
	At 1 MHz	0.0014	0.0014	0.0014	0.0088	0.016	0.0072	0.007
D257	Volume resistivity (Ω-cm)	1×10^{16}	1×10^{16}	1×10^{16}	5×10^{15}	2×10^{15}	6×10^{15}	6×10^{15}
D495	Arc resistance(s)	10	34	128	167	182	180	116–182
UL746A	Arc-tracking rate (in./min)	11.3	7.1	6.3	0	0	0	0–0.6

Note: Test specimens prepared using hot (275°F) mold. R-10 values listed cover a series of compounds.

Source: Mach. Design Basics Eng. Design, June, p. 721, 1993. With permission.

bearing applications, such as the actuator latch and stop.

POLYPROPYLENE PLASTICS

Polypropylene (PP) is made by polymerization of propylene using catalysts and similar to the low-pressure polymerization of ethylene. It is a linear polymer, more than 95% of which has a spatially ordered structure. The commercial polymer contains over 99% C_3H_6, with the remainder stabilizing additives. Molded pieces or molding pellets of unpigmented material may be identified by density (0.905) and by maximum crystalline melting point (168 to 170°C).

Produced from propylene gas, polypropylene resins are semitranslucent and milky white in color and have excellent colorability. Most polypropylene parts are produced by injection molding, blow molding, or extrusion of either unmodified or reinforced compounds. Other applicable processes are structural-foam molding and solid-phase and hot-flow stamping of glass-reinforced sheet stock (a product of Azdel).

Polypropylene plastics are an important group of synthetic plastics employed for molding resins, film, and texture fibers. Propylene is a methyl ethylene, $CH_3CH:CH_2$, produced in the cracking of petroleum. It belongs to the class of unsaturated hydrocarbons known as olefins, which are designated by the word ending -ene. Thus, propylene is known as propene as distinct from propane, the corresponding saturated compound of the group of alkanes from petroleum and natural gas. These unsaturated hydrocarbons tend to polymerize and form gums, and are thus not used in fuels although they have antiknock properties.

Properties

Polypropylene is a low-density resin that offers a good balance of thermal, chemical, and electrical properties, along with moderate strength and moderate cost. Strength properties are increased significantly with glass-fiber reinforcement. Increased toughness is provided in special, high-molecular-weight, rubber-modified grades.

Electrical properties of polypropylene moldings are affected to varying degrees by service temperature. Dielectric constant is essentially unchanged, but dielectric strength increases and volume resistivity decreases with increased temperature.

Polypropylene has limited heat resistance, but heat-stabilized grades are available for applications requiring prolonged use at elevated temperatures. Useful life of parts molded from such grades may be as long as 5 years at 121°C, 10 years at 110°C, and 20 years at 99°C. Specially stabilized grades are Underwriter's Laboratories-rated at 120°C for continuous service.

Polypropylene resins are unstable in the presence of oxidative conditions and ultraviolet radiation. Although all grades are stabilized to some extent, specific stabilization systems are often used to suit a formulation for a particular environment. Polypropylenes resist chemical attack and staining and are unaffected by aqueous solutions of inorganic salts or mineral acids and bases, even at high temperatures. They are not attacked by most organic chemicals, and there is no solvent for the resin at room temperature. The resins are attacked, however, by halogens, fuming nitric acid, and other active oxidizing agents, and by aromatic and chlorinated hydrocarbons at high temperatures; see Table P.11 (Properties of Polypropylene).

Polypropylene is low in weight. The molded plastic has a density of 0.910, a tensile strength of 34 MPa, with elongation of 150%, and hardness of Rockwell R95. The dielectric strength is 59×10^6 V/m, dielectric constant 2.3, and softening point 150°C. Blow-molded bottles of polypropylene have good clarity and are nontoxic. The melt flow is superior to that of ethylene. A unique property is their ability in thin sections to withstand prolonged flexing. This characteristic has made polypropylenes popular for "living hinge" applications. In tests, they have been flexed over 70 million times without failure.

Molecular Types

In polypropylene plastics each carbon atom linked in the molecular chain between the CH_2 units has a CH_3 and an H attached as side links,

POLYPROPYLENE PLASTICS

TABLE P.11
Properties of Polypropylenes

ASTM or UL Test	Property	Unmodified Resin	Glass Reinforced	Impact Grade
	Physical			
D792	Specific gravity	0.905	1.05–1.25	0.89–0.91
D792	Specific volume (in.^{3}/lb)	30.8–30.4	24.5	30.8–30.5
D570	Water absorption, 24 h, 1/8-in. thk (%)	0.010–0.03	0.01–0.05	0.01–0.03
	Mechanical			
D638	Tensile strength (psi)	5,000	6,000–14,500	2,800–4,400
D638	Elongation (%)	10–20	2.0–3.6	350–500
D638	Tensile modulus (10^5 psi)	1.6	4.5–9.0	1.0–1.7
D790	Flexural modulus (10^5 psi)	1.7–2.5	3.8–8.5	1.2–1.8
D256	Impact strength, Izod (ft-lb/in. of notch)	0.5–2.2	1.0–5.0	1.0–15
D785	Hardness, Rockwell	80–110	110	50–85
	Thermal			
C177	Thermal conductivity (Btu-in./h-ft^2-°F)	2.8	—	3.0–4.0
D696	Coefficient of thermal expansion (10^{-5}in./in.-°C)	3.2–5.7	1.6–2.9	3.3–4.7
D648	Deflection temperature (°F)			
	At 264 psi	125–140	230–300	120–135
	At 66 psi	200–250	310	160–210
UL94	Flammability rating[a]	HB	HB	HB
	Electrical			
D149	Dielectric strength (V/mil)			
	Short time, 1/8-in. thk	500–660	475	500–650
D150	Dielectric constant			
	At 1 MHz	2.2–2.6	2.36	2.3
D150	Dissipation factor			
	At 1 MHz	0.0005–0.0018	0.0017	0.0003
D257	Volume resistivity (Ω-cm)			
	At 73°F, 50% RH	10^{17}	2×10^{16}	10^{15}
D495	Arc resistance(s)	160	100	—

[a] V-2, V-1, and V-0 grades are also available.

Source: Mach. Design Basics Eng. Design, June, p. 718, 1993. With permission.

with the bulky side groups spiraled regularly around the closely packed chain. The resulting plastic has a crystalline structure with increased hardness and toughness and a higher melting point. This type of stereosymmetric plastic has been called isostatic plastic. It can also be produced with butylene or styrene, and the general term for the plastics is polyolefins.

GRADES

The many different grades of polypropylenes fall into three basic groups: homopolymers, copolymers, and reinforced and polymer blends. Properties of the homopolymers vary with molecular-weight distribution and the degree of crystallinity. Commonly, copolymers are produced by adding other types of olefin monomers to the propylene monomers to improve properties such as low-temperature toughness. Polypropylenes are frequently reinforced with glass fibers and fillers to improve mechanical properties and increase resistance to deformation at elevated temperatures. Biaxially oriented polypropylene (BOPP) film (film stretched in two ways) has greatly improved moisture resistance, clarity, and stiffness. It is used for packaging tobacco products, snack foods, baked goods, and pharmaceuticals. Metallized grades also are available for packaging designed for extended shelf life.

Foamed polypropylenes include expandable polypropylene (EPP) bead and injected-molded structural foam. EPP provides greater energy absorption and flexibility than expandable polystyrene. Structural polypropylene foam moldings consist of a solid outer skin and foam core. They are used to achieve greater stiffness in larger, lightweight parts (strength-to-weight ratios are three to four times greater than solids parts).

Polypropylene fiber unless modified is more brittle at low temperatures and has less light stability than polyethylene, but it has about twice the strength of high-density linear polyethylene. Monofilament fibers are used for filter fabrics, and have high abrasion resistance and a melting point at 154°C. Multifilament yarns are used for textiles and rope. Polypropylene rope is used for marine hawsers, will float on water, and does not absorb water like Manila rope. It has a permanent elongation, or set, of 20%, compared with 19% for nylon and 11% for Manila rope, but the working elasticity is 16%, compared with 25% for nylon and 8% for Manila. The tensile strength of the rope is 406 MPa. Fine-denier multifilament polypropylene yarn for weaving and knitting dyes easily and comes in many colors. Chlorinated polypropylene is used in coatings, paper sizing, and adhesives. It has good heat and light stability, high abrasion resistance, and high chemical resistance.

Fabrication

Compression Molding

This is seldom used with polypropylene except for making heavy slab in multiple daylight presses.

Injection Molding

Standard techniques for molding apply to polypropylene. Cylinder temperatures of 288°C or less, and fast rams operating in the neighborhood of half the available machine pressure, generally give good molding at fast cycles. No special metals are needed for cylinders or molds. Since polypropylene shows a definite change in melt viscosity between 232 and 274°C, the indicated cylinder temperature should be balanced with the machine heater capacity, the size of each shot, and the cycle time to maintain such melt temperatures.

Molds for polypropylene should embody the best techniques used with thermoplastics: uniform wall thicknesses; avoidance of heavy ribs, bosses, and fillets; use of channels or curved walls instead of ribs to increase rigidity.

Extrusion

Heavy sheet, shapes, thin film, and monofilament are all being produced commercially by extrusion. Equipment required can be made of the usual steel alloys with no danger of corrosive degradation products. For even melting, screws of length-to-diameter ratio of 20:1 are best.

Uses

Polypropylene homopolymer, random copolymer, and impact copolymer resins are tailored for specific polymer applications and fabrication methods and also to achieve desired end-product performance. Low-molecular-weight resins, used for melt-spun and melt-blown fibers and for injection-molding applications, are produced by oxidative degradation of higher-molecular-weight polymers at elevated temperatures. These materials, often called controlled rheology resins, have narrower molecular-weight distributions and lower viscoelasticity. The brittleness of polypropylene homopolymer, particularly at temperatures below 0°C, is greatly reduced by blending it with ethylene-propylene rubber. Compounding with mineral fillers and glass fibers improves product stiffness and other properties. Higher-stiffness resins also are produced by increasing the stereoregularity of the polymer or by the addition of nucleating agents. Polypropylene resins are used in extrusion and blow-molding processes and to make cast, slit, and oriented films. Stabilizers are added to polypropylene to protect it from attack by oxygen, ultraviolet light, and thermal degradation; other additives improve resin clarity, flame retardancy, or radiation resistance.

New applications for polypropylene include in-line skates. The low-cost skate consists of a boot and frame injection molded as a single unit. Using ACC-TUF polypropylene impact copolymer to make the one-piece skate has produced a significant cost savings, while retaining the performance attributes of the more expensive material.

The copolymer solved the problem of balance between impact resistance and stiffness in thermoplastic materials. Typically, the greater the impact resistance of a material, the less its stiffness. The balanced properties of ACC-TUF result in a high degree of toughness, while resisting low-temperature impacts and thermal deformation.

In another application a polypropylene composite was used to keep boats ice-free; it used a submersible circulation unit that is able to keep water around boats and docks from freezing. With a powerful flow of water around a boat or dock the formation of ice is prevented. The submersible, motor-driven unit features a one-piece plastic shroud made with a long-glass-fiber-reinforced polypropylene composite material.

Because the shroud protects the propeller blade from damage by wood and other debris, a material was needed with excellent strength, stiffness, and impact resistance.

Suspended by two lines over the side of the boat, the D-Icer unit works by continuously propelling warmer subsurface water up to the surface. Because it lies under the water, a material that was available in bright colors for easy visibility was needed. It also had to be ultraviolet-stabilized so that any ultraviolet light that penetrates the water would not destroy it.

In addition, the structural composites have good dimensional stability to help maintain mechanical performance in wet environments. This is important because the shroud forms the upper half of the motor housing, which contains a gasket. The stable material helps prevent leaks. And, most important, it resists subzero temperatures.

Finally, recently introduced TSL snowshoes, made of lightweight, injection-molded polypropylene, combine strength and reliability with a convex shape that prevents snow buildup on either side. Each model offers several other features that ensure trouble-free hiking, even in challenging snow conditions and over difficult terrain. They include steel crampons attached to the shoe base that resist temperatures as low as $-20°C$ for traction on icy or hard-packed snow; radial snow fins that aid walking in deep powder; and a binding system that exhibits excellent grip in traverses or when scrabbling up steep slopes.

An assortment of models make the shoes an easy fit for various users and conditions. For example, the TSL 510 fits children, the TSL 710 adapts to lighter-weight adults or firmer-packed snow, the TSL 810 accommodates heavier weights or powder-snow use, and the TSL 225 multipurpose shoe is designed for steep, irregular terrain.

Different bindings also adapt to differing needs. The Trappeur model, a high-strength, expandable-rubber shoe, conforms to any kind of boot. The Rando binding, designed for use with hiking, pack, and after-ski boots, features a climbing arc for steep ascents and a heel lock for descent and jumping. And the Aventure binding, built for extreme terrains, adapts to mountaineering boots notched in the heel and toe; see Figure P.5.

POLYSTYRENES

Polystyrenes comprise one of the largest and most widely used families of plastics. Often called the "workhorse" of the thermoplastics, polystyrenes consist of the basic general-purpose materials, plus a wide variety of modified grades of polymers, copolymers, and blends.

All polystyrenes generally have several features in common: (1) low cost, (2) unexcelled electrical insulating properties, (3) virtually unlimited colorability, (4) ability to be made crystal clear (in general-purpose grades), and (5) high hardness and gloss. Table P.12 lists typical property ranges of several types of polystyrenes.

Polystyrenes have excellent molding and extrusion characteristics, and are formed readily and inexpensively by any of the thermoplastic forming methods. Types available

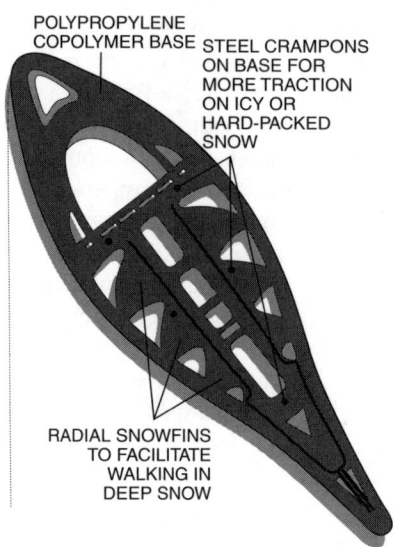

FIGURE P.5 TSL snowshoe. (From *Design News*, March 23, p. 45, 1998. With permission.)

include molding and extrusion grades, foams, sheet and film, and fiber.

CHARACTERISTICS AND PROPERTIES

A thermoplastic resin used for molding, in lacquers, and for coatings, polystyrene is formed by the polymerization of monomeric styrene, which is a colorless liquid of the composition $C_6H_5CH:CH_2$, specific gravity 0.906, and boiling point 145°C. It is made from ethylene, and is ethylene with one of the hydrogen atoms replaced by a phenyl group. It is also called phenyl ethylene and vinyl benzene. As it can be made by heating cinnamic acid, $C_6H_5CH:CHCO_2H$, an acid found in natural balsams and resins, it is also called cinnamene. In the form of vinyl toluene, which consists of mixed isomers of methyl styrene, the material is reacted with drying oils to form alkyd resins for paints and coatings.

The polymerized resin is a transparent solid very light in weight with a specific gravity of 1.054 to 1.070. The tensile strength is 27 to 68 MPa, compressive strength 82 to 117 MPa, and dielectric strength 18 to 24×10^6 V/m. Polystyrene is notable for water resistance and high dimensional stability. It is also tougher and stronger at low temperatures than most other plastics. It is valued as an electrical insulating material, and the films are used for cable wrapping.

TYPES

General-Purpose Types

So-called general-purpose polystyrenes are characterized by clarity, luster, colorability, rigidity, unexcelled dielectric properties, and moldability. They are used where rigidity and appearance are important, but where toughness is not required. Typical uses include wall tile, container lids, and brush backs.

Grades are available with a wide range of processing characteristics. Higher heat-resistance types, which also have improved toughness, are available in crystal and a full range of transparent, translucent, and opaque colors.

Impact Grades

To overcome the relatively low impact strength of general-purpose polystyrene, combinations of polystyrene and rubber provide grades whose impact strength depends on the proportion of rubber added. Grades are generally characterized as medium-, high-, and extra-high-impact types. As impact strength increases, rigidity or modulus decreases. Such materials are available in virtually unlimited colors, but cannot be produced crystal clear.

Medium-impact polystyrenes are used where moderate toughness plus good translucency is required. They are used for such products as containers, closures, and table-model radio cabinets. Types of medium-impact polystyrenes are available with improved heat resistance, surface gloss, and moldability.

High-impact polystyrenes include special grades with improved heat resistance, moldability, and surface gloss. They are used for refrigerator inner-door liners, crisper trays, containers, appliance housings, and toys. Higher heat-resistant, high-impact grades (generally suitable for sections greater than 2.2 mm) are used for television masks and housings, portable radio cabinets, auto heater ducts, and automatic washer soap dispensers.

Extra-high-impact strength grades have relatively low moduli about 1360 to 1700 MPa

Chemical-Resistant Grades

Copolymers of styrene and acrylonitrile provide resistance to chemicals such as carbon tetrachloride, aliphatic hydrocarbons, and food stains, and provide much better stress-crack resistance than polystyrene. The copolymers are transparent and haze-free, but are slightly yellow; they are available also in a wide variety of colors.

Primary uses for the copolymers are in drinking tumblers and cups that have high resistance to crazing by butter fat and staining by coffee oils. Industrial uses include water filter parts, oil filter bowls, storage battery containers, and washing machine parts.

Other Special Grades

Light-stabilized types of styrene-methyl methacrylate copolymers and glass-reinforced molding materials are other special grades of polystyrene.

Light-stabilized grades were developed specifically for applications involving exposure to intense fluorescent-light radiation, e.g., "egg crate" light diffusers. Such formulations prevent the yellowing that occurs in unstabilized polystyrene when it is exposed to fluorescent light.

Styrene-methyl methacrylate copolymers were developed to provide a material with weatherability approaching that of acrylics, but at a lower cost. They have been used primarily for escutcheons, instrument panels, decorative medallions, brush blocks, auto tail light lenses, and advertising signs.

Glass-reinforced polystyrenes (incorporating chopped glass fibers) are available for injection molding or extrusion, and provide higher strength, greater durability, and higher strength at elevated temperatures than do unreinforced polystyrenes.

Styrene can be polymerized with butadiene, acrylonitrile, and other resins. The terpolymer, acrylonitrile–butadiene–styrene (abbreviated ABS), is one of the common combinations. Styrene–acrylonitrile (SAN) has excellent resistance to acids, bases, salts, and some solvents. It also is among the stiffest of the thermoplastics with a tensile modulus of 2757 to 3791 MPa. Acrylate–styrene–acrylonitrile (ASA) has very good weathering resistance (nonyellowing), good toughness, and stress cracking resistance. Styrene–butylene resins are copolymers that mold easily and produce thermoplastic products of low water absorption and good electrical properties. They have strength equal to the vinyls with greater elongation.

Foamed Polystyrene

It is used in many forms, including extruded sheet (which is then thermoformed, e.g., egg cartons, trays); expandable polystyrene (EPS) beads, which contain a blow agent (usually pentane), and which are processed into low-density foamed products, such as hot and cold drink cups and protective packaging; and block and heavy sheet, used for thermal insulation.

POLYSULFIDE RESINS

These resins vary in properties from viscous liquids to rubberlike solids. Organic polysulfide resins are prepared by the condensation of organic dihalides with a polysulfide.

The linear, high-molecular-weight polymers can be cross-linked or cured by reaction with zinc oxide. Compounding and fabrication of the rubbery polymers can be handled on conventional rubber machinery. The polysulfide rubbers are distinguished by their resistance to solvents such as gasoline, and to oxygen and ozone. The polymers are relatively impermeable to gases. The products are used to form coatings that are chemically resistant and special rubber articles, such as gasoline bags.

APPLICATIONS

The polysulfide rubbers were among the very first commercial synthetic rubbers. Although the products are not as strong as other rubbers, their chemical resistance makes them useful in various applications. The polysulfide rubbers were also among the first polymers to be used in solid-fuel compositions for rockets.

TABLE P.12
Properties of Polystyrenes

ASTM or UL Test	Property	Polymers		Copolymers		
		General Purpose	Impact Modified	Crystal Clear	Impact Modified	Glass Reinforced[a]
Physical						
D792	Specific gravity	1.04–10.9	1.03–1.10	1.08–1.10	1.05–10.8	1.13–1.22
D792	Specific volume (in.³/lb)	26.0–25.6	28.1–25.2	—	—	—
D570	Water absorption, 24 h, 1/8-in. thk (%)	0.03–0.10	0.05–0.6	0.1	0.1	0.08
Mechanical						
D638	Tensile strength (psi)	5,000–12,000	1,500–7,000	7,000–7,600	4,800–7,200	10,500–12,500
D638	Elongation (%)	0.5–2.0	2–60	1.4–1.7	2.0–20.0	1.3–2.0
D638	Tensile modulus (10^5 psi)	4.0–6.0	1.4–5.0	4.4–4.7	2.8–4.2	6.3–10.0
D790	Flexural strength (psi)	8,000–17,000	3,000–12,000	12,000–12,600	8,500–12,200	12,200–19,700
D790	Flexural modulus (10^5 psi)	4.0–4.7	1.5–4.6	4.6–4.7	3.2–4.5	5.5–9.8
D256	Impact strength, Izod (ft-lb/in. of notch)	0.2–0.45	0.5–4.0	0.3–0.5	0.5–4.4	1.8–2.6
D785	Hardness, Rockwell M	65–80	10–90	108	80	101
Thermal						
C177	Thermal conductivity (Btu-in./h-ft°F)	2.4–3.3	1.0–3.0	2.4–3.3	1.0–3.0	—
D696	Coefficient of thermal expansion (10^{-5} in./in.-°F)	3.3–4.4	1.9	3.5–3.7	3.5–3.7	2.0–2.2

ASTM	Property				
D648	Deflection temperature (°F)				
	At 264 psi	190–220	160–200	235–249	235–249	235–260
	At 66 psi	180–230	180–220	—	—	—
UL94	Flammability rating[b]	HB	HB	HB	HB	HB
Electrical						
D149	Dielectric strength (V/mil)					
	Short time, 1/8-in. thk	500–700	300–600	500–700	300–600	—
D150	Dielectric constant					
	At 1 kHz	2.40–2.64	2.4–4.5	—	—	—
D150	Dissipation factor					
	At 1 kHz	0.0001–0.0003	0.0004–0.0020	—	—	—
D257	Volume resistivity (Ω-cm)					
	At 73°F, 50% RH	10^{17}–10^{19}	10^{16}	—	—	—
D495	Arc resistance(s)	60–135	20–100	95	95	—
Optical						
D542	Refractive index	1.60	—	1.59	—	—
D1003	Transmittance (%)	87–92	35–57	92	—	—

[a] 10–20%.
[b] V-2, V-1, and V-0 grades are also available.

Source: Mach. Design Basics Eng. Design, June, p. 722, 1993. With permission.

POLYSULFIDE RUBBER

Basically, polysulfide rubbers are chemically saturated polymers with reactive terminals through which conversion to a thermoset, elastomeric state can be effected by means of suitable catalysts or curing agents. More specifically, these rubbers are the products of a condensation polymerization in which one or more organic dihalides are reacted with an aqueous solution of sodium polysulfide.

PROPERTIES

As with a number of other synthetic elastomers, polysulfide rubbers require reinforcing fillers to achieve optimum physical properties. However, the high tensile values possible with unsaturated rubbers are not obtainable with polysulfides. For practical purposes, valves of 10.3 to 12.4 MPa are the upper maximum limit.

The primary assets of these rubbers are outstanding oil, gasoline, and solvent resistance, as well as very low permeability to gases and solvent vapors. Also because of their saturated structure, polysulfides possess excellent resistance to oxidation, ozone, and weathering. Performance as regards temperature is somewhat dependent on polymer structure, but with few exceptions the serviceability range is from −51 to 121°C and intermittently up to 149°C. The principal limitation of polysulfide rubber compounds is relatively low resistance to compression set.

AVAILABLE FORMS

Polysulfide rubbers are novel in that they are available not only as solids but also as liquids. The liquids are 100% polymer, unadulterated with solvents or diluents, which also can be converted to highly elastic rubbers with properties closely approaching those of the cured, solid polysulfides. Viscosities range from a very fluid 5 poise to a heavy molasses-like 700 poise.

FABRICATION AND APPLICATIONS

Design with both polysulfide crudes and liquids is quite similar, but because of the difference in physical form, the products made from these materials are processed and fabricated somewhat differently.

Crudes or Solid Polymers

Polysulfide crude rubbers are processed in the same way as other synthetic rubbers, on conventional mixing and fabricating equipment. The incorporation of reinforcing fillers, curing agents, and other additives is accomplished on two-roll mills or Banbury-type internal mixers. Subsequent operations such as extruding, calendering, molding, or steam vulcanization can be carried out in the normal manner, except that somewhat closer factory control must be exercised than might be necessary with the larger volume, general-purpose rubbers.

Because polysulfide crudes may be classified as specialty rubber, their use is usually limited to those applications that demand exceptional solvent resistance. Such products include gaskets, washers, diaphragms, various types of oil and gasoline hose, and other mechanical rubber goods items. However, the solvent combinations encountered in various paints, coatings, and inks are responsible for the major consumption of these rubbers. A great number of the rollers employed for can lacquering, wood and metal coating, and the application of quick-drying inks are fabricated with a polysulfide rubber covering. Much of the hose used with hot lacquer and paint spraying equipment is made with a polysulfide tube or inner liner. One other unique commercial use of crudes is in the form of non-hardening putties, which make effective solvent-resistant seals for static type joints.

Liquid Polymers

Liquid polymer compounds are mixed by a three-roll paint mill, colloid mill, ball mill, or internal mixer. The resulting products may be applied by brushing, spraying, casting, or caulking gun depending on the characteristics of the specific compound involved and the type of application for which it was designed. Curing or conversion of these materials to highly elastic rubbers can be accomplished over a relatively wide temperature range, but their ability to cure at room temperature has been a primary factor in the employment of these polymers for a diversity of industrial applications. Despite their similarity in performance properties, there has been little overlapping use of the polysulfide crudes and liquid polymers in the same application areas; however, because of versatility in end-use product fabrication, liquids are threatening to intrude into solid-rubber fields of application.

One of the major uses of liquid polysulfide polymers is in the manufacture of sealants for the aircraft, building, and marine industries. Such products may be compounded to bond to most building materials, and provide flexible, elastic seals for joints that are subject to a high degree of movement and vibration. Curing or conversion to the rubbery state can be regulated to occur in minutes or hours at normal atmospheric temperatures depending on the demands of the application techniques involved.

Other applications include cold-setting casting, potting, and molding compounds, which exhibit flexibility, very low shrinkage, and excellent dimensional stability. The impregnation of leather with polysulfide liquid polymers imparts water and solvent resistance without loss of pliability. The liquid polysulfides may also be employed as coatings and adhesives, but more commonly they are used as modifiers of epoxy resins for these and many other industrial applications. Modification is chemical rather than physical and results from an addition reaction between the polysulfide thiol terminals and epoxide groups. Greatly improved flexibility and impact resistance, lower shrinkage, less internal strain, better wetting properties, and lower moisture vapor transmission are the advantages gained.

POLYSULFONE RESINS

These are plastics whose molecules contain sulfone groups ($-SO_2-$) to the main chain, as well as a variety of aromatic or aliphatic constituents such as ether or isopropylidene groups. Polysulfones based on aromatic backbones constitute a useful class of engineering plastics, owing to their high strength, stiffness, and toughness together with high thermal and oxidative stability, low creep, transparency, and the ability to be processed by standard techniques for thermoplastics. Aliphatic polysulfones are less stable, for example, to hydrolysis, but are of interest for some biomedical applications. Four major types of polysulfones have been of commercial interest: polysulfone (1), polyarylsulfone (2), polyethersulfone (3), and polyphenylsulfone (4).

The aromatic structural elements and the presence of sulfone groups are responsible for the resistance to heat and oxidation; ether and isopropylidene groups contribute some chain flexibility. Aromatic polysulfones can be used over wide temperature ranges, to ~150°C for polysulfone and to 200°C for polyethersulfone. In fact, the high-temperature performance of polyethersulfones is surpassed by few other polymers. Resistance to hydrolysis at high temperatures and to most acids, alkalies, and nonpolar organic solvents is excellent; however, the resins may be attacked or dissolved by polar solvents, especially if the component is under stress. Although resistance to ionizing radiation is high, protection against ultraviolet light is recommended for outdoor applications. Dimensional stability and electrical properties such as dielectric loss and strength are retained well during service, and flammability is low.

Because of the combination of properties discussed, polysulfone resins find many applications in electronic and automotive parts, medical instrumentation subject to sterilization, chemical and food processing equipment, and various plumbing and home appliance items. Coating formulations are also available, as well as grades reinforced with glass beads or fibers.

POLYTETRAFLUOROETHYLENE

Polytetrafluoroethylene (PTFE) lubricants dispersed into a thermoplastic base resin greatly improve surface-wear characteristics. Molecular weight and particle size of the PTFE lubricant are designed to provide optimum improvements in wear, friction, and PV values for selected resin systems. PTFE has the lowest coefficient of friction (0.02) of any known internal lubricant. Its static coefficient of friction is lower than its dynamic coefficient, which accounts for the slip/stick properties associated with PTFE/metal sliding action. During the initial break-in period, the PTFE particles embedded in the thermoplastic matrix shear to form a high-lubricity film over the mating surface. The PTFE cushions asperities from shock and minimizes fatigue failure.

Forms

Glass fibers are frequently used in combination with silicone and PTFE lubricants, which offset the negative wear effects that the glass fibers have on surface characteristics. The use of silicone only, in conjunction with glass fibers, is not recommended, however. PTFE provides far more protection to the mating surface and should be used (with or without silicone) if the wear rate of the mating surface is important (Table P.13).

Uses

The first thermoplastics that were recognized for their inherent lubricity were nylon, acetal, and PTFE. These materials perform well, but for the more critical uses, their coefficients of friction may be too high, or wear may be too rapid.

The next generation of self-lubricated thermoplastics was formulated of various base resins that contained molybdenum disulfide, graphite, or PTFE particles to improve both lubrication and wear characteristics. Although wear resistance was indeed upgraded considerably, mechanical strength and dimensional stability of these compounds are often insufficient.

To minimize these deficiencies, reinforcing fibers of glass or carbon are added. The resulting composites are several times stronger than the unreinforced materials, and they are extremely stable in a wide range of service environments. But these materials too have a shortcoming: In service, a period of time is required for the internal lubricant to become exposed and to be burnished over the wear surfaces. During this run-in period, as a bearing or wear member is put into service, the unlubricated surfaces are in contact, and damage may occur.

Two approaches to eliminating these problems use silicone fluids to provide the lubrication function.

Thermoplastic composites can also be internally lubricated with a variety of systems to improve wear resistance. PTFE and silicone, separately or in combination, provide the best improvements in wear characteristics. Graphite powder and molybdenum disulfide are also used, primarily in nylons. The PTFE lubricants are specially modified to enhance their lubricious nature in the compound. The optimum level of lubricating filler varies depending on filler type and resin, but typical ranges follow:

PTFE	15 to 20%
Silicone	1 to 5%
PTFE/silicone	15 to 20%
Graphite	10%
MoS_2	2 to 5%

Addition of these lubricants further improves wear characteristics of good bearing materials such as nylon and acetal. The lubricants also allow the use of poor-wearing, but close-tolerance materials, such as polycarbonate, in gear or bearing applications. Lubricants can be used by themselves or in conjunction with glass or carbon-fiber reinforcements.

PTFE and silicone fluids; glass, aramid, and carbon fibers; and graphite powder are the primary reinforcements and lubricants used in internally lubricated composites. The composites are based on engineering resins for injection-molded wear and structural parts.

POLYURETHANES

Extremely wide variations in forms and in physical and mechanical properties are available in polyurethanes. Grades can range in density from 13.88 g/cm^3 in cellular form, to over 1937.6 g/cm^3 in solid form, and in hardness from rigid solids at 85 Shore D to soft, elastomeric compounds.

Polyurethane polymers, produced by the reaction of polyisocyanates with polyester or polyether-based resins can be either thermoplastic or thermosetting. They have outstanding flex life, cut resistance, and abrasion resistance. Some formulations are as much as 20 times more resistant to abrasion than metals.

The noncellular grades — millable gums and viscous, castable, liquid urethanes — are elastomeric thermoset types, processed by conventional rubber methods.

Processing

There are three major types of polyurethane elastomers. One type is based on ether- or ester-type prepolymers that are chain-extended and cross-linked using polyhydroxyl compounds or amines; alternatively, unsaturated groups may be introduced to permit vulcanization with common curing agents such as peroxides. All of these can be processed by methods commonly used for rubber. A second type is obtained by first casting a mixture of prepolymer with chain-extending and cross-linking agents, and then cross-linking further by heating. The third type is prepared by reacting a dihydroxy ester- or ether-type prepolymer, or a diacid, with a diisocyanate such as diphenylmethane diisocyanate and a diol; these thermoplastic elastomers can be processed on conventional plastics equipment. In general, urethane elastomers are characterized by outstanding mechanical properties and resistance to ozone, although they may be degraded by acids, alkalies, and steam.

A wide variety of tough and abrasion-resistant urethane coatings are available. Many are based on the reaction of castor oil, a triol, with an excess of diisocyanate; the resulting triisocyanate undergoes cross-linking by reaction with atmospheric moisture. Urethane alkyds can also be made by reacting an unsaturated drying oil with glycerol, and then reacting the product with a diisocyanate; curing is effected by atmospheric oxidation of the double bonds. Still other coatings are based on the use of prepolymers. Polyurethanes are also used as adhesives, for example, in the bonding of rubber and of nylon.

Foams

Polyurethane foams are thermoset materials that can be made soft and flexible or firm and rigid at equivalent densities. These foams, made from either polyester or polyether-type compounds, are strong, even at low density, and have good chemical resistance. Polyether-based foams have greater hydrolysis resistance, are easier to process, and cost less. Polyester-based foams have higher mechanical properties, better oil resistance, and more uniform cell structure. Both types can be sprayed, molded, foamed in place, or furnished as sheets cut from slab stock.

Flexible Foams

Glass-transition temperatures (the temperature at which an elastomeric material becomes stiff and brittle) of flexible foams are well below room temperature. The foams can be pigmented to any color, but, regardless of pigmentation, they yellow when exposed to air and light. Some types of flexible foams are excellent liquid-absorbing media, and can hold up to 40 times their weight of water.

Polyether-type foams are not affected by high-temperature aging, either wet or dry, but UV exposure produces brittleness and reduces properties. In use, these foams are always covered with a fabric or other material.

Most solvents and corrosive solutions decrease tear resistance and tensile strength and cause swelling of flexible foams. Swelling is not permanent, however, if the solvent is removed and the foam dried. However, the foams can be destroyed by strong oxidizing agents and hydrolyzed in strong acids or bases. Generally, the polyether foams are more resistant to hydrolytic degradation; the polyester foams are more resistant to oxidative attack.

Applications for polyester flexible urethane foam include gasketing, air filters, sound-absorbing elements, and clothing interliners (laminated to a textile material). The polyether types are used in automobile and recreational-vehicle seats, carpet underlay, furniture upholstering, bedding, and packaging.

Rigid Foams

Bases for rigid foams are polymers having glass-transition temperatures higher than room temperature. The cells of rigid foam are about the same size and uniformity as those of flexible foam, but rigid foams usually consist of 90% closed cells. For this reason, water absorption is low. Compressing the foam beyond its elastic limit damages the cellular structure.

Rigid foams are blown with either carbon dioxide or fluorocarbons. Gas generated by vaporization of fluorocarbons, entrapped in the closed cells, gives the foam a very low thermal

TABLE P.13
Properties of Glass Fillers

Base Resin	Specific Gravity D792	Mold Shrinkage (in./in.) D955	Water absorption, 24-h (%) D570	Tensile Strength (10³ psi) D638	Flexural Modulus (10⁶ psi) D790	Impact Strength, Izod (ft-lb/in.) D256		Thermal Expansion (10⁻⁵ in./in.-°F) D696	Deflection Temperature, 264 psi (°F) D648
						Notched	Unnotched		
ABS	1.28 (1.05)	0.001 (0.006)	0.14 (0.30)	14.5 (5.0)	1.10 (0.32)	1.4 (4.4)	6–7 (—)	1.6 (5.3)	220 (195)
Acetal	1.63 (1.42)	0.003 (0.020)	0.30 (0.22)	19.5 (8.8)	1.40 (0.40)	1.8 (1.3)	8–10 (20)	2.2 (4.5)	325 (230)
ETFE	1.89 (1.70)	0.003 (0.018)	0.02 (0.02)	14.0 (6.5)	1.10 (0.20)	7.5 (>40)	17–18 (—)	1.6 (4.0)	460 (160)
Nylon 6	1.37 (1.14)	0.004 (0.016)	1.1 (1.8)	23.0 (11.8)	1.20 (0.40)	2.3 (1.0)	20 (—)	1.7 (4.6)	420 (167)
Nylon 6 alloy	1.37	0.004	1.1	22.0	1.00	3.2	22–24	1.7	415
Nylon 6/6	1.37 (1.14)	0.004 (0.016)	0.90 (1.50)	26.0 (11.8)	1.30 (0.41)	2.0 (0.9)	17 (—)	1.3 (4.5)	480 (170)
Nylon 6/6 alloy	1.37	0.004	0.90	27.0	1.10	3.01	9–21	1.8	485
Nylon 6/6 copolymer	1.3 (1.09)	0.004 (0.013–0.018)	0.6 (1.50)	11.6 (8.0)	0.9 (0.275)	4.52 (14)	0–22 (—)	1.9 (4.5)	490 (160)
Nylon 6/12	1.30 (1.06)	0.004 (0.011)	0.21 (0.25)	22.0 (8.8)	1.20 (0.295)	2.4 (1.0)	20 (—)	1.5 (5.0)	415 (194)
Modified PPO	1.27 (1.06)	0.002 (0.005)	0.06 (0.07)	21.0 (9.5)	1.30 (0.36)	1.7 (5.0)	9–10 (—)	1.4 (3.3)	310 (265)

Polycarbonate	1.43	0.001	0.07	18.5	1.20	3.7	17	1.3	300
	(1.20)	(0.006)	(0.15)	(9.0)	(0.33)	(2.7)	(60)	(3.7)	(265)
Polyester (PBT)	1.52	0.003	0.06	19.5	1.40	2.5	16–18	1.2	430
	(1.31)	(0.020)	(0.08)	(8.5)	(0.34)	(1.2)	(—)	(5.3)	(130)
Polyetherimide	1.51	0.002	0.18	28.5	1.25	1.7	16	1.1	420
	(1.27)	(0.006)	(0.25)	(15.2)	(0.48)	(1.0)	(25)	(3.1)	(392)
Polyetheretherketone	1.49	0.003	0.10	25.0	1.10	2.1	17	—	600
	(1.32)	(0.011)	(0.15)	(14.5)	(0.55)	(—)	(—)	(—)	(360)
Polyether sulfone		1.60	0.003	0.20	19.0	1.20	1.5–10	18	415
	(1.37)	(0.007)	(0.20)	(12.0)	(0.37)	(1.6)	(—)	(3.1)	(400)
Polyethylene (HD)		1.17	0.003	0.02	10.0	0.90	1.18–9	2.7	260
	(0.95)	(0.020)	(0.02)	(2.6)	(0.20)	(0.4)	(—)	(6.0)	(120)
PPS	1.56	0.002	0.04	20.0	1.60	0.3	1.4	1.1	500
	(1.34)	(0.010)	(0.05)	(10.8)	(0.60)	(3–4)	(8–9)	(3.0)	(280)
Polypropylene	1.13	0.004	0.03	9.8	0.80	1.6	5–6	2.0	295
Polypropylene (chemically coupled)	1.12	0.004	0.03	9.7	0.55	3.01	1–12	2.0	295
Polystyrene	1.28	0.001	0.05	13.5	1.30	1.0	2–3	1.9	215
	(1.07)	(0.004)	(0.10)	(7.0)	(0.45)	(0.45)	(1.2)	(3.6)	(180)
Polysulfone	1.45	0.003	0.20	18.0	1.20	1.8	14	1.4	365
	(1.24)	(0.007)	(0.20)	(10.0)	(0.40)	(1.2)	(60)	(3.1)	(340)
SAN	1.31	0.001	0.10	17.4	1.50	1.0	3–4	1.8	215
	(1.08)	(0.005)	(0.25)	(9.8)	(0.50)	(0.4)	(—)	(3.4)	(200)

Note: Property values for the unreinforced resins are shown, for comparison, in parentheses. Values given are representative; both higher and lower values may be obtained in commercially available resins and compounds. Values for the reinforced compounds are typical of 30% glass-reinforced formulations. Although there is no unnotched Izod test under ASTM D256, values from this test are often useful for material comparisons and selection.

Source: *Mach. Design Basics Eng. Design*, June, p. 729, 1993. With permission.

conductivity of 0.11 to 0.14 Btu-in./h-ft²-°F. Conductivity increases with age, however, to a constant value of about 0.16.

Rigid urethane foams are used for thermal insulation of refrigerators, refrigerated trucks and railroad cars, cold-storage warehouses, and process tanks because of their low conductivity and high strength-to-weight ratio. Other applications include flotation devices, encapsulation, structural and decorative furniture components, and sheathing and roof insulation for buildings.

Integral-Skin Foam

Urethane foams that are formed with integral skins range from soft and flexible types to impact-absorbing grades and rigid foams used in structural parts. Color can be added, but since the foams yellow on aging, black is most practical for the surface color. If other colors are required, coatings are recommended. The tough, high-density, integral skin is formed against the mold surface and the low-density core is produced by a blowing agent — usually a fluorocarbon.

Elastomeric foams of this type are used in automotive bumper and fascia systems and, most recently (reinforced with milled-glass fibers), in fenders and other exterior body panels. The semirigid types are used in athletic protective gear, in automotive crash-protection areas, horn buttons, sun visors, and arm rests. Applications for the rigid structural foams include housings for computer systems, chair shells, furniture drawers, and sports equipment.

Part Fabrication

A low-pressure molding process — reaction-injection molding (RIM) — is used almost exclusively to produce urethane parts, some weighing as much as 45.4 kg. In the process, two or more highly reactive liquid systems are injected with high-pressure impingement mixing into a closed mold at low pressure, where they react to form a finished polymer. Depending on formulation, the polymer can be a rigid, integral-skin, microcellular urethane foam with a flexural modulus of over 680 MPa, or a soft, flexible elastomer with a flexural modulus as low as 49.06 MPa, or a rigid structural foam having a density of 830.4 g/cm³. Cycle time is short; parts can be demolded in less than a minute.

Reinforcement in the form of milled glass fiber, glass flake, or mineral filler increases the stiffness, thermal properties, and dimensional stability of RIM parts. Maximum glass content in reinforced reaction-injection molding (RRIM) is about 25% — a limit determined by the increased viscosity with increasing glass. Natural color of unpigmented RIM urethane parts is tan.

Applications

Reticulated polyester foams are used for explosion suppression. The material protects aircraft fuel tanks from small arms gunfire. Over the years, several technical advances were made to the foam, lowering weight, and increasing service life and conductivity (to prevent static buildup).

The material is an open-pore, reticulated polyurethane foam that contains a network of skeletal strands with 98% void space at any pore size. The material functions essentially as a three-dimensional fire screen similar to a safety fire screen over a lighted Bunsen burner.

In a fuel tank, the empty space above the ullage may readily contain an explosive mixture of fuel vapor and air. Since the liquid fuel itself does not explode, a completely filled tank is far less likely to explode than one that is not. The lower the fuel level in the tank, the greater amount of explosive vapor present. When an ignition source is present, the vapor adjacent to the spark ignites rapidly. This ignition, in turn, ignites the vapor around it, creating a chain reaction as the ignition, or flame front, gets larger and moves faster as it propagates through the vapor. The rapid ignition and propagation of the flame results in an ever-growing compression wave in front of it, compressing the unignited vapor, thus adding even greater force to an explosion. This sequence occurs in milliseconds.

This chain reaction is prevented from occurring; instead, vapor ignition is confined to the area immediately around the ignition source. Flame and wave propagation are mitigated by the foam, thus preventing an explosion.

The foam may be easily fabricated into any configuration by conforming to the inside of complex, rigid or flexible, bladder-type fuel tanks. Once installed, the foam can remain in the fuel tank for many years without degradation or loss of physical properties and performance characteristics. It can be readily removed for internal fuel tank maintenance and reinstalled repeatedly.

In the Volkswagen Beetle polyurethane raw materials protect the finishes of plastic and metal components, both exterior and interior. Polyurethane foam is used inside the doors and instrument panel, and polyurethane raw materials help protect the instrument panel.

POLYVINYL CHLORIDE

Among the vinyl polymers and copolymers, the polyvinyl chloride (PVC) thermoplastics are the most commercially significant. With various plasticizers, fillers, stabilizers, lubricants, and impact modifiers, PVC is compounded to be flexible or rigid, opaque or transparent, to have high or low modulus, or to have any of a wide spectrum of properties or processing characteristics.

PVC resin can also be chlorinated (CPVC) and it can be alloyed with other polymers such as ABS (acrylonitrile–butadiene–styrene), acrylic, polyurethane, and nitrile rubber to improve impact resistance, tear strength, resilience, heat-deflection temperature, or processibility.

Forms

PVC is a hard, flame-resistant, and chemical-resistant thermoplastic resin. The resin is available in the powder form, as a latex, or in the form of plastisol. PVC resin, pigments, and stabilizers are milled into plasticizers to form a viscous coating material (plastisol) that polymerizes into a tough elastic film when heated. Plastisols are used extensively for coating glass bottles and glass fabrics. The dispersion types of resins are used in flexible molding compounds. Such formulations consist of a vinyl paste resin, a suitable plasticizer such as dioctyl phthalate, and a stabilizer (usually a compound of lead). Flexible molds are widely applied to plaster casting and encapsulation of electronic circuits with epoxy resins.

Processing

PVC compounds are processed by extrusion, injection molding, calendering, compression molding, and blow molding. PVC coatings are applied by fluidized-bed and electrostatic powder-coating methods. The resins are also used for dip molding and coating, in the form of plastisols and organosol dispersions or water dispersions (latexes). Cellular PVC products are made by introducing gas into the resin during molding or extrusion. Foams can be open or closed cell, and can be elastomeric or rigid, depending on plasticizer content.

PVC compounds can be made water-white in flexible compounds, very clear in rigid compounds, and they can be pigmented to almost any color.

Properties

With so many property variations attainable by compounding methods, no single compound can be considered typical of PVC. For example, creep rate of rigid compounds is so low and predictable that they can be used to make pressure pipe for water distribution; flexible compounds can be soft enough, yet impermeable, so that they are used for baby pants and for an excellent imitation suede, or they can be transparent, nontoxic, and tough enough to be used for mineral-water bottles.

Rigid PVC, sometimes called the "poor man's engineering plastic," is a hard, tough material that can be compounded to a wide range of properties. Noteworthy among its properties is low combustibility; it has high resistance to ignition and is self-extinguishing. It also provides good corrosion and stain resistance, thermal and electrical insulation, and weatherability. However, PVC is attacked by aromatic solvents, ketones, aldehydes, naphthalenes, and some chloride, acetate, and acrylate esters. Some impact modifiers used in rigid PVC reduce chemical resistance. In general, normal-impact grades have better chemical resistance than the high-impact grades.

Most PVC compounds are not recommended for continuous use above 60°C. Chlorination increases heat-deflection temperature, flame retardancy, and density and extends the continuous-use temperature to 80 to 100°C, depending on the amount of chlorination.

POLYVINYL FLUORIDE FILM

Polyvinyl fluoride film is a flexible, transparent film, which, unlike most films, is inherently weatherable as well as tough, inert, and easy to fabricate. All its properties derive from its chemical structure and are not dependent on additives or plasticizers.

ENGINEERING PROPERTIES

Two types of the film are available, which vary slightly in properties. The principal difference lies in the percent elongation at break and the dimensional stability at higher fabricating temperatures. The manufacturer is able to recommend specific types for specific uses. Properties listed below are for one particular type, but those of the other type are of the same order.

Polyvinyl fluoride film is strong, flexible, and fatigue-resistant. Tensile strength is 91.3 MPa; tensile modulus, 1900 MPa; break elongation, 185% at 20°C. Flex life is 200,000 cycles (1 to 2 mils, at 20°C). Toughness and flexibility are retained over a wide range of temperatures. The surface is stain resistant, easily cleaned, and highly resistant to abrasion.

The film is impermeable to greases and oils, and retains its film form and strength even when boiled in strong acids and bases. At ordinary temperatures it is not affected by many classes of common solvents including hydrocarbons and chlorinated solvents. It is partially soluble in a few highly polar solvents at temperatures above 149°C. Its resistance to hydrolysis is excellent.

Impermeability to gases, water vapor, and organic vapors (with the exception of ketones and esters) is good. Resistance to thermal embrittlement is excellent. The film remains flexible at 204°C and resists flexural fatigue at –17.8°C. Service temperature range is –73 to 107°C.

Resistance to degradation by sunlight is an outstanding characteristic — the result of chemical inertness and the fact that the material is essentially transparent to and unaffected by the near-ultraviolet, visible, and near-infrared regions of the spectrum. Unsupported film is not discolored and is still flexible and strong with 50% residual tensile strength after 10 years of Florida exposure. Outdoor life of two or three times that period is predicted under less severe exposure conditions and for film used as the surface in laminates.

The film can be colored as desired by the addition of pigments and will be supplied in standard colors.

The film can be sealed by electronic or impulse methods and some types by the hot-bar process as well. The film can be metallized, embossed, vacuum-formed, delustered, printed, postformed as a laminate, and fabricated in simple processes using existing equipment.

APPLICATIONS

The film is used for metal prefinishing, building board prefinishing, and for roofing — applications in which its weatherability permits it to make a unique contribution as an exterior finish. It can be applied to galvanized steel and to aluminum as well as to plywood, hardboard, cement-asbestos, fibrous glass, and similar building materials to provide both a decorative and a long-lived protective surface. Laminated to flexible substrates the material offers promise as an easily installed, highly reflective, maintenance-free roofing material.

Chemical, physical, and thermal properties combine to make the film a unique material for bag molding and as a parting sheet. Shapes can be fabricated by pressure or vacuum forming.

Polyvinyl fluoride film in either transparent or pigmented form provides a weatherable, soil-resistant surface layer for flexible and rigid plastics. It can be combined with glass-reinforced polyester panels during manufacture and remains as a permanent skin on the panel. A special polymer containing permanent UV absorber is used to produce this film.

The material appears suitable for some difficult packaging applications and offers useful properties in tape form. Used as a jacketing material over insulation on tanks and pipelines, it gives superior performance at lower cost than other materials in use.

Outdoor glazing is another application for the film because of its strength, transparency, and weatherability, combined with ease of fabrication through heat sealing. Solar energy structures, transparent tarpaulins, air-supported structures, storm windows, greenhouses, poultry sheds, and crop covers can be made from it.

In the electrical industry, capacitors, transformers, motors, wire, and cable are all uses in which the film's high dielectric constant, high dielectric strength, outstanding resistance to thermal degradation and the effects of hydrolysis, and low moisture absorption of the film offer substantial benefits.

POLYVINYL RESINS

These resins are polymeric materials generally considered to include polymers derived from monomers.

Many of the monomers can be prepared by addition of the appropriate compound to acetylene. For example, vinyl chloride, vinyl fluoride, vinyl acetate, and vinyl methyl ether may be formed by the reactions of acetylene with HCl, HF, CH_3OOH, and CH_3OH, respectively. Processes based on ethylene as a raw material have also become common for the preparation of vinyl chloride and vinyl acetate.

The polyvinyl resins may be characterized as a group of thermoplastics, which, in many cases, are inexpensive and capable of being handled by solution, dispersion, injection molding, and extrusion techniques. The properties vary with chemical structure, crystallinity, and molecular weight.

Polyvinyl Acetals

These are relatively soft, water-insoluble thermoplastic products obtained by the reaction of polyvinyl alcohol with aldehydes. Properties depend on the extent to which alcohol groups are reacted. Polyvinyl butyral is rubber and tough and is used primarily in plasticized form as the inner layer and binder for safety glass. Polyvinyl formal is the hardest of the group. It is used mainly in adhesive, primer, and wire-coating formulations, especially when blended with a phenolic resin.

Polyvinyl butyral is usually obtained by the reaction of butyraldehyde with polyvinyl alcohol. The formal can be produced by the same process, but is more conveniently obtained by the reaction of formaldehyde with polyvinyl acetate in acetic acid solution.

Polyvinyl Acetate

Polyvinyl acetate is a leathery, colorless thermoplastic material that softens at relatively low temperatures and that is relatively stable to light and oxygen. The polymers are clear and noncrystalline. The chief applications of polyvinyl acetate are as adhesives and binders for water-based or emulsion paints.

Vinyl acetate is conveniently prepared by the reaction of acetylene with acetic acid.

Polyvinyl Alcohol

Polyvinyl alcohol is a tough, whitish polymer that can be formed into strong films, tubes, and fibers that are highly resistant to hydrocarbon solvents. Although polyvinyl alcohol is one of the few water-soluble polymers, it can be rendered insoluble in water by drawing or by the use of cross-linking agents.

Polyvinyl Chloride

Polyvinyl chloride is a tough, strong thermoplastic material that has an excellent combination of physical and electrical properties. The products are usually characterized as plasticized or rigid types. Polyvinyl chloride (and copolymers) is the second most commonly used polyvinyl resin and one of the most versatile plastics.

Polyvinylidene Chloride

Polyvinylidene chloride is a tough, horny thermoplastic with properties generally similar to those of polyvinyl chloride. In comparison with the latter, polyvinylidene chloride is softer and less soluble: it softens and decomposes at lower temperatures, crystallizes more readily, and is more resistant to burning.

Because of its relatively low solubility and decomposition temperature, the material is most widely used in the form of copolymers with other vinyl monomers, such as vinyl chloride. The copolymers are employed as packaging film, rigid pipe, and as filaments for upholstery and window screens.

Films of polyvinylidene chloride, and especially the copolymer containing about 15% of vinyl chloride, are resistant to moisture and gases. Also, they can be heat-sealed and, when oriented, have the properties of shrinking on heating. By warming a food product wrapped loosely with a film of the polymer, a skintight, tough, resistant coating is produced.

By cold-drawing, the degree of crystallinity, strength, and chemical resistance of sheets, filaments, and even piping can be greatly increased.

POLYVINYL ETHERS

Polyvinyl ethers exist in several forms varying from soft, balsamlike semisolids to tough, rubbery masses, all of which are readily soluble in organic solvents. Polymers of the alkyl vinyl ethers are used in adhesive formulations and as softening or flexibilizing agents for other polymers.

POLYVINYL FLUORIDE

Polyvinyl fluoride is a tough, partially crystalline thermoplastic material that has a higher softening temperature than polyvinyl chloride. Films and sheets are characterized by high resistance to weathering.

Films are used in industrial and architectural applications. Coatings, for example, on pipe, are resistant to highly corrosive media.

PORCELAIN

Porcelains and stoneware are highly vitrified ceramics that are widely used in chemical and electrical products. Electrical porcelains, which are basically classical clay-type ceramics, are conventionally divided into low- and high-voltage types. The high-voltage grades are suitable for voltages of 500 and higher, and are capable of withstanding extremes of climatic conditions. Chemical porcelains and stoneware are produced from blends of clay, quartz, feldspar, kaolin, and certain other materials. Porcelain is more vitrified than stoneware and is white in color. A hard glaze is generally applied. Stonewares can be classified into two types: a dense, vitrified body for use with corrosive liquids, and a less dense body for use in contact with corrosive fumes. Chemical stoneware may range from 30 to 70% clay, 5 to 25% feldspar, and 30 to 60% silica. The vitrified and glazed product will have a tensile strength up to 16 MPa and a compressive strength up to 551 MPa. Industrial stoneware is made from specially selected or blended clays to give desired properties.

Both chemical porcelains and stoneware resist all acids except hydrofluoric. Strong, hot, caustic alkalies mildly attack the surface. These ceramics generally show low thermal-shock resistance and tensile strength. Their universal chemical resistance explains their wide use in the chemical and processing industries for tanks, reactor chambers, condensers, pipes, cooling coils, fittings, pumps, ducts, blenders, filters, and so on.

Ceratherm is an acid-resistant and heat-shock-resistant ceramic with a base of high-alumina clay. It is strong and nonporous, and is used for pump and chemical-equipment linings.

PROPERTIES

Porcelain is distinguished from other fine ceramic ware, such as china, by the fact that the firing of the unglazed ware (the bisque firing) is done at a lower temperature (1000 to 1200°C) than the final or glost firing, which may be as high as 1500°C. In other words, the ware reaches its final state of maturity at the maturing temperature of the glaze.

The white color is obtained by using very pure white-firing kaolin or china clay and other pure materials, the low absorption results from the high firing temperature, and the translucency results from the glass phase.

PORCELAIN ENAMELS AND CERAMIC COATINGS

Earlier definitions of ceramic materials usually stressed their mineral origin and the need for heat to convert them into useful form. As a consequence, only porcelain enamels and glazes were recognized as ceramic coatings until recently, when the principles of phase relations, bonding mechanisms, and crystal structure were applied to ceramic materials and to coatings made from them. In consequence, ceramic materials can now be most safely defined as solid substances that are neither metallic nor organic in nature, a definition that is somewhat more inclusive than older ones, but more accurately reflects modern scientific usage.

Most ceramics are metal oxides, or mixtures and solutions of such oxides. Certain ceramic materials, however, contain little or no oxygen. As a whole, ceramic materials are harder, more inert, and more brittle than organic or metallic substances. Most ceramic coatings are employed to exploit the first two properties while minimizing the third.

Low-Temperature Coatings

The outstanding resistance to corrosion of certain metals, notably aluminum and chromium, is attributable to the remarkable adherence of their oxide films. Aluminum does not corrode because its oxidation product, unlike that of iron, is a highly protective coating. It was once believed that some mysterious kinship between a metal and its own oxide was needed for this protection, but recent knowledge relating to the structure of metals and metal oxides has enabled metallurgists to develop alloys that form even more stable and adherent films. Methods for thickening or stabilizing these oxide coatings by heat treatment, electrolysis, or chemical reaction are widely accepted.

The presence of such coatings on metals and alloys strongly influences such properties as their emission and absorption of radiation, frictional and wetting characteristics, and electrical and electrochemical properties.

Usually, ceramic coatings contain compounds of metals other than the substrate. Most of these are oxides, which are amorphous or crypto-crystalline in nature. They can be divided into two groups: chemical reactants (in which a new compound or complex is formed) and inorganic colloids.

Chemical Reactants

These coatings usually involve the chemical modification of the natural metal oxide into a coating that is stabler or denser. This classification includes the treatment of aluminum, tin, and zinc, with chromates, electrolytic "anodizing," the treatment of iron with soluble phosphates, etc. The films that are formed are more inert to most chemicals than normal oxide films, and the process is termed "passivation."

Inorganic Colloids

Colloidal particles are sufficiently small so that their surface energy is sufficient for bonding. These particles can be made into colloidal suspensions called "slurries" or "slips." They contain natural colloids such as hydrophilic clays, or finely ground materials of a fibrous or platelike nature such as asbestos, mica, graphite, potassium titanate, alumina monohydrate, zirconia hydrate, or molybdenum sulfide.

Some inorganic colloids are so finely dispersed as to be true sols. "Water glass" is a familiar example of the colloidal sol; depending on its sodium content, alkalinity, and dilution it may be a thin fluid, a sticky, viscous liquid, or a translucent semisolid. Newer aqueous sols include other alkali silicates, aluminum acid hydrates and phosphates, alkyl titanates and silicates, and lime hydrate.

These coatings are usually processed by drying or flocculation. The dried film and substrate are usually heated sufficiently to drive off all traces of moisture and irreversibly "set" the coating for its final use, as for lubricants, thermionic emitters, and fluorescent lamp coatings. The relatively weak bonding power of these dried films can be used to hold them in place for further heating; this is the basis of "wet process" porcelain enameling.

Moderate Temperature Processes

In the temperature range between 538 and 1093°C, most silicate glasses melt. In this range the fusion of a vitreous powder (usually called a frit) or the formation of a glass from its component materials is then used as the basis for porcelain enameling and glazing processes.

Glazes differ from enamels principally in the substrates on which they are applied. When the substrate is metallic, the vitreous coating is called an enamel; when applied to ceramic bodies such as porcelain, china, terra cotta, pottery, and electrical ceramics, the coating is termed a glaze. Since these coatings are vitreous they may be transparent, but most enamels and many glazes contain finely divided crystalline materials that color them and make them more or less opaque.

Enamels are used primarily to provide resistance to corrosion, heat, and/or abrasion and wear; they are frequently employed for their attractive appearance as well.

Vitreous enamels cannot easily be classified; a typical enameling slip may contain ten or more ingredients, including those that are glassy or form a vitreous network (feldspars, frits, borax, and other mineral sources of silicates, phosphates, and borates). The modifiers usually consist of alkali and alkaline earth compounds or lead. They also contain fluxing agents, opacifiers, suspending agents, and clays, as well as refractory oxides intended to dissolve in the melting glass and increase its viscosity. The oxides or other compounds of cobalt, nickel, manganese, arsenic, or antimony may be included to promote adhesion between glass and metal.

Most porcelain enamels consist of two or more layers of glass, separately applied and fused. The first layer is called the ground coat or base coat; its purpose is to attach firmly to the metal substrate and prevent undesirable interactions between substrate and enamel or the evolution of gas from the metal. It is in the ground coat that the adherence-promoting additives are used, and it is these additives that produce the blue, brown, or black color of most ground coats. Where light colors are not required (range parts, heat-resisting coatings, and certain abrasion-resistant applications), a ground coat may suffice over the base metal.

Although "cover coat" may be required to resist chemical attack, abrasion, impact, heat, or weathering, enamels are most commonly used to provide a durable and attractive finish for steel and ferrous alloys. Most of these are white, hence the name "porcelain" enamels. Earlier enamels contained antimony or zircon to provide opacity and whiteness, but zirconium oxide and titanium oxide are now most frequently used. The latter not only provides superior opacity but usually improves acid resistance as well. As a consequence, titania-bearing enamels may be applied in a single coating, not only over a suitable base coat but directly upon special steels. Such enamels can now be fused at less than 760°C to titanium-bearing steels, to steels precoated with a thin nickel film, or to steels pretreated with iron phosphate.

On cast iron, enamels are used for range parts and high-grade sanitaryware. Certain chemicalware (tanks, pumps, etc.) are also made with cast iron. The rigidity and good acoustical damping of cast iron, together with its resistance to distortion by heat, permits heavier layers of the protective glass to be used than are possible on steel or enameling iron.

Special enamels have been developed for the chemical industry and hot water tanks. Such equipment is frequently called "glass lined," and can contain all acids except hydrofluoric and hot phosphoric, moderately alkaline solutions up to their boiling point, and water under pressures up to 3500 MPa.

Enamels are not limited to use on ferrous substrates; the earliest decorative enamels were used on precious metals and copper artware and jewelry. Enamels have recently been developed suitable for aluminum. Such enamels can be fired at temperatures well below 538°C. Although not so hard or corrosion resistant as sheet-steel enamels, these aluminum enamels provide attractive and durable finishes for sheet, extrusions and castings.

Although most enamels are applied to the substrate by a wet-process technique, some coatings may be applied by dusting or sieving the powdered composition directly upon the heated surface. This "dry process" is principally used with chemicalware and cast iron sanitaryware.

The mechanical properties of enamels are strongly influenced by the composition, thickness, and geometry of the substrate as well as by the kind, thickness, method of application, and firing conditions of the enamel layer or layers. In general, thin enamel coatings are best able to resist thermal and mechanical shock and stress. Very thin (1 to 4 mils thick) coatings on steel or aluminum may even be bent, punched, sheared, or drilled without damage. Thicker coatings are usually required for ordinary applications (appliances, curtain walls or structural panels, and signs); the thickness of the glassy layers is usually between 3 and 20 mils. For cast iron sanitaryware the vitreous coating may be 40 mils thick or more, and some chemical tanks employ 6.4 mm of protective glass.

For maximum resistance to chipping, the enamel must be supported by a hard, relatively thick substrate. Where the enameled article must resist bending or twisting, thinner and more ductile substrates are indicated.

The selection of metal for porcelain enamel and its preparation strongly influences the properties of the composite. "Enameling irons" are essentially very low carbon, basic open-hearth, rimmed steels. Regular SAE (Society of Automobile Engineers) and AISI (American Iron and Steel Institute) low-carbon steels can seldom be perfectly enameled especially when hot-rolled. Premium enameling stock may contain titanium sufficient to further lower the available carbon content and improve resistance to warping or sagging.

For thicker substrates a basic open-hearth plate steel of flange quality may be used. Higher impurity levels, however, may require that the enamel be fired in an inert atmosphere to eliminate "boiling" defects. In some cases aluminized steel can be used in moderately oxidizing furnace atmospheres.

Cast steel and cast iron can be enameled acceptably; carbon is oxidized from the surface during the relatively long firing period needed to fuse and consolidate the ground coat. Purity of the metal is not so stringent a requirement for cast iron as for steel.

Surface treatment of the metal usually requires the removal of all scale and dirt (this may require sandblasting for heavy-gauge stock) followed by a light pickling for ultimate control of the oxide layer thickness. For some metals and cast irons, sandblasting or grit blasting alone may be sufficient pretreatment. Special enamels may require phosphate bath treatment or the deposition of a nickel-, copper-, or aluminum-base coating over the metal.

Metals to be enameled should be reasonably strain-free. Burrs, sharp edges, or small external radii and large variations in substrate thickness should be avoided. Welds must be sound and metallurgically similar to the parent area.

Refractory Enamels and Glazes

Because the fused porcelain enamel cools with its substrate, it is important that the total contraction that occurs during cooling be approximately equal in both metal and glass. If the metal contracts more than the enamel, the latter will be forced into compression and may shatter or chip easily; if the enamel contracts more than the metal, it will crack or craze.

Among the silicate glasses, the most refractory generally have a low thermal expansion coefficient, and they must therefore be used on metals that have low expansion. Most of the refractory enamels contain finely powdered silica and chromium oxide, which dissolve in the glass on fusion, further lowering expansion. In consequence, such enamels are largely restricted to refractory substrate metals such as certain stainless steels and nickel and chromium alloys. The service temperature of enamel-coated stainless steel and nickel alloys is about 954°C in aircraft engine exhaust manifolds, turbosupercharger linings, jet engine combustors, and commercial burners.

Thin, electrically conductive ceramic coatings can be used as resistance heaters on aircraft windshields and the like. Most of these consist of a mixture of the oxides and suboxides of tin with traces of bismuth, antimony, cadmium, or arsenic. They are applied by heating the glass to a dull red heat and spraying it with a fusible tin halide under slightly reducing conditions. As it cools in air, the tin halide decomposes into the conductive complex. When sufficiently thin, these coatings are quite transparent.

High-Temperature Coatings

To attach a ceramic coating to any substrate by enameling, both substrate and glass must be heated to the fusion temperature of the glass. However, fusion methods have not been successful for the more refractory materials. Because most refractory ceramic coatings are amorphous or crystalline in nature, they have to be applied by relatively novel techniques.

Although most ceramic materials are refractory, some of them can be vaporized in an electric arc or hot vacuum. Thin coatings of amorphous silica can be applied readily to relatively cool substrates by vaporizing metallic silicon or silicon halides in the presence of small quantities of oxygen. Apparently the transfer is accomplished largely as silicon monoxide, which recombines with oxygen on cooling. The process is used to obtain thin, protective, optically transparent films on lenses, certain electrical components, and metal reflectors.

Other ceramic coatings may be produced by vaporizing one or more components of the coating. In this way coatings of the respective carbides of silicon, boron, aluminum, and chromium can be deposited on graphite, silicon nitride can be formed on metallic silicon, and silicide coatings can be deposited on metals such as tungsten and molybdenum. These processes are necessarily expensive and are poorly adaptable to large specimens or complex shapes.

Flame- or Arc-Spraying

Many metallic oxides and interstitial compounds can be heated to or above their melting temperatures with a chemical flame or electric arc. The coatings obtained by directing a spray of nearly molten ceramic particles toward an otherwise unheated substrate are interesting and useful for a variety of purposes. Most of the processes are proprietary and differ chiefly in the form in which material is fed into the heat source.

The coatings obtained in this way may be quite porous or they may approach the theoretical density of the material being sprayed. Since the substrate need not be heated, the need for a close match of thermal expansion is less important than with vitreous coatings. The porous coatings are surprisingly immune to thermal and mechanical shock, but confer little chemical protection.

Substrates must usually be roughened before application of these coatings, but heat-resistant glasses, glazes, and some porcelain enamels make excellent substrates. Adherence seems to be largely mechanical, and the adherence tests used for porcelain enamels are not applicable to these coatings. No standards for testing or performance have yet been established.

Coatings obtained by flame spraying may consist of pure ceramic materials, metals, and some organic polymers; a modification of flame spraying can be used to produce pyrolytic graphitic coatings of high density and resistance to hot gas erosion. Mixtures of ceramic and metal powders may be used to produce cermet coatings or even graded coatings, and unusual electrical, magnetic, and dielectric properties can be obtained with such mixtures or with multilayer application.

Because each particle is suddenly and individually chilled, the structures of flame-sprayed ceramics may be unlike those of the bulk materials. Nonstoichiometry is common, and the stresses in the coating and between the coating and substrate are complex. Nevertheless, certain flame-sprayed and arc plasma-sprayed coatings have already found acceptance in missile technology, metalworking and foundry applications, and for heat- or wear-resistant coatings on metallic, ceramic, and polymer substrates.

POROUS METALS

These are metals with uniformly distributed controlled pore sizes, in the form of sheets, tubes, and shapes, used for filtering liquids and gases. They are commonly made by powder metallurgy, and the pore size and density are controlled by the particle size and the pressure used. Stainless steel, nickel, bronze, silver, and other metal powders are used, depending on the corrosion resistance required of the filter. Pore sizes can be as small as 0.2 µm, but the most generally used filters have pores of 4, 8, 12, and 25 µm. Pore sizes have a uniformity within 10%. The density range is from 40 to 50% of the theoretical density of the metal. Standard filter sheet is 0.76 to 1.52 cm, but thinner sheets are available. Sheet as thin as 0.010 cm, and with void fractions as high as 90%, have been made for fuel cells and catalytic reactors. Porous steel is made from 18-8 stainless steel, with pore openings from 20 µm to 65 µm. The fine-pore sheet has a minimum tensile strength of 69 MPa, and the coarse sheet has a strength of 48 MPa. Felted metal is porous sheet made by felting metal fibers, pressing, and sintering. It gives a high strength-to-porosity ratio, and the porosity can be controlled over a wide range. In this type of porous metal the pores may be from 0.003 to 0.038 cm in diameter, and of any metal to suit the filtering conditions. A felted fiber filter of 430 stainless steel with 25% porosity has a tensile strength of 172 MPa.

POTASSIUM ALLOYS AND COMPOUNDS

An elementary metal, symbol K, and atomic weight 39.1, potassium is also known as kalium. It is silvery white in color, but oxidizes rapidly in the air and must be kept submerged in ether or kerosene.

It stands in the middle of the alkali metal family, below sodium and above rubidium, in group II of the periodic table of the elements. It is a lightweight, soft, low-melting, reactive metal. It is very similar to sodium in its behavior in metallic form, and its uses are limited by the availability of low-cost sodium in large volume.

Physical Properties and Alloys

Potassium has a low melting point, 63°C, and a boiling point of 756°C. The specific gravity is 0.855 at 20°C. It is soluble in alcohol and in acids. It decomposes water with great violence. Potassium is obtained by the electrolysis of potassium chloride. Potassium metal is used in combination with sodium as a heat-exchange fluid in atomic reactors and high-temperature processing equipment. A potassium–sodium alloy contains 78% potassium and 22% sodium. It has a melting point of −11°C and a boiling point of 756°C, and is a silvery mobile liquid. Cesium–potassium–sodium alloys are called BZ Alloys. Potassium hydride is used for the photosensitive deposit on the cathode of some photoelectric cells. It is extremely sensitive and will emit electrons under a flash so weak and so rapid as to be imperceptible to the eye. Potassium diphosphate, KH_2PO_4, a colorless, crystalline, or white powder soluble in water, is used as a lubricant for wool fibers to replace olive oil in spinning wool. It has the advantages that it does not become rancid like oil and can be removed without scouring. Potassium, like sodium, has a broad range of use in its compounds, giving strong bonds. Metallurgically it is listed as having a body-centered cubic structure, but the atoms arrange themselves in pairs in the metal as K_2, and the structure is cryptocrystalline.

Chemical Properties

Potassium is even more reactive than sodium. It reacts vigorously with the oxygen in the air to form the monoxide, K_2O, and the peroxide, K_2O_2. In the presence of excess oxygen, it readily forms the superoxide, KO_2 (formerly believed to be K_2O_4).

Potassium does not react with nitrogen to form a nitride, even at elevated temperatures. With hydrogen, potassium reacts slowly at 200°C and rapidly at 350 to 400°C. It forms the least stable hydride of all the alkali metals.

Principal Compounds

- Potassium chloride, KCl, is the most important potassium compound. It is not the only form in which potassium is often found in nature, but it is the form in which potash is used as a fertilizer.
- Potassium hydroxide, KOH, is also known as caustic potash. It is usually made by the electrolysis of aqueous solutions of potassium chloride.
- Potassium carbonate, K_2CO_3, is made from potassium hydroxide and carbon dioxide. It cannot be made by the Solvay process used for sodium carbonate, because potassium bicarbonate is too soluble in ammonium chloride solution.
- Potassium nitrate, KNO_3, is made from fractional crystallization of an aqueous solution containing sodium nitrate and potassium chloride.

Handling

Handling of potassium metal is much the same as that of sodium metal, with two major exceptions. First, the formation of the superoxide, KO_2, causes difficulties because it can react vigorously with hydrocarbons and other organic matter. Second, potassium is generally more reactive than sodium. Potassium forms an explosive carbonyl with carbon monoxide, and the metal deteriorates in contact with bromine. Usually sodium, potassium, and the sodium–potassium (NaK) alloys are considered to be in the same general class of reactivity, allowing for the chemical differences outlined above and for the liquid (and hence more reactive) nature of the NaK alloys over a wide range of composition.

Uses

- Potassium chloride finds its main use in fertilizer mixtures. It also serves as the raw material for the manufacture of other potassium compounds.
- Potassium hydroxide is used in the manufacture of liquid soaps, and potassium carbonate in making soft soaps.
- Potassium carbonate is also an important raw material for the glass industry.
- Potassium nitrate is used in matches, in pyrotechnics, and in similar items that require an oxidizing agent.

Biological Activity

The recognition of potassium ions by biological molecules is optimal when there is a good match between the cavity size provided by the host molecule and the ionic diameter of the potassium ion; other factors, such as the water structure surrounding the potassium ion and the number and nature of donor atoms in the host molecule, may, however, be important. Novel physical methods, such as nuclear magnetic resonance spectroscopy based on the ^{39}K isotope, are being used to help achieve an improved understanding of potassium ion transport in biological systems.

Potassium deficiency may occur in several conditions including malnutrition and excessive vomiting or diarrhea, and in patients undergoing dialysis; supplementation with potassium salts is sometimes required. Although toxicity caused by therapeutic doses of potassium is rare, it may lead to cardiac arrest if left untreated; potassium supplementation should be administered with caution to patients suffering from cardiovascular diseases or those with impaired renal function.

Potassium Compounds

Potash

The compound K_2O is very soluble in water and other solvents. The most important original source of commercial potash is natural potassium salts. These are prepared as potassium nitrate and potassium carbonate for use in ceramics, but most of the potash is automatically introduced into batches in feld-soda glass; this is especially true with the use of manganese, nickel oxide, and selenium. In the potash glasses much less cobalt oxide is required in connection with manganese to secure a good neutral tint for crystal glass. Similarly, nickel is a suitable decolorizer for glasses high in potash, whereas its effect in the soda glasses is decidedly ugly.

The alkali content of commercial glasses runs about 15% in window glass, 15 to 17% in container glass, and 20% in thin blown glass. Most of the alkali is soda, and while a higher potash content is often desirable, its greater cost limits its wider application. The growth of the American potash industry may allow a price reduction that will make this material more available to glass manufacturers, who now limit its general use to the more expensive glass products.

In optical glass, a ratio of 7 parts potash to 3 parts soda gives good durability and color to a number of commercial compositions, in which the total potash content of the glass may vary from 7 to 16% for some crown types. It probably is not possible to derive a potash–soda ratio suitable for all optical glasses. Some high-lead glasses, for example, contain no soda at all, yet show high durability. The discoloring effect of ferrous iron is much less noticeable in a potash–soda optical glass than in a high-soda glass.

It has been found that glasses containing both Na_2O and K_2O give lower thermal conduction than either alone; the minimum conductivity is obtained with a potash:soda ratio of 4:1. This factor is becoming increasingly important in view of developments in fiberglass.

The behavior of colorants in colored glass is often superior in potash glass to that in spar.

In enamels the alkali content averages 10% in sheet ground coats, 20% in cover coats, 15% in cast iron enamels, and as much as 36% in jewelry enamels. In the last type, all the alkali is potash, which is believed to increase brilliance and luster, but in other enamels all or most of the potash is merely accessory to alumina in the addition of feldspar. The same may be said for the potash content of glazes. As a flux in glazes, potash is only about 85% as active as soda. If present in excess, K_2O may cause peeling and crazing if the other constituents are not in suitable proportions. Potash is reported not as conducive as soda to the formation of crystals in crystalline glazes.

Potash in the hydroxide or carbonate form is an important deflocculating agent. It is used at ordinary temperatures to prepare casting slips, glaze slips, and engobes, to purify clay, to reduce the plasticity of excessively plastic clays, and to neutralize any acid present.

Potassium Carbonate

Also called pearl ash, this is a white alkaline granular powder, which is a potassium carbonate, K_2CO_3 or $K_2CO_3:H_2O$. It is used in soft soaps, for wool washing, and in glass manufacture.

Kalium is a high-purity grade that results from an evaporation–crystallization step; it has about 62.4% K_2O and less than 1% salt. The material is a free-flowing white powder of 91 to 94% K_2CO_3, or is the hydrate at 84%, or calcined at 99% purity. The specific gravity of potash is 2.33 and its melting point is 909°C.

In glass manufacture, potassium carbonate is supplied in both calcined and hydrated form. The product sold to the glass industry is easy to handle as it is of granular particle form, and has entirely eliminated the dusty material, formerly supplied from abroad, with its irritating handling problems.

The present domestic materials are supplied with very low chloride and sulfate content and are entirely suitable for all types of glass production. Although the viscosity of the potash glasses is high, thus making them somewhat difficult to work, the viscosity is easily remedied by introducing lead oxide. Hence, the combination of potash and lead oxide leads to the production of a glass that lends itself well to handworking.

This combination possesses a long working range. All of the potash glasses, which from their nature must be melted in closed pots, exert a different sort of corrosive action on the clay wall from that exhibited by soda glasses.

The corrosion by soda glass proceeds quite smoothly, but the potash glasses produce a honeycombing or pitting effect, and the thin partitions between these pits, finally reduced to small pinnacles, float out into the glass, forming stones. It seems to be an almost unavoidable characteristic of the potash–lead glasses to produce a great deal of stony ware.

In its influence on the physical properties of glass, potash does not differ greatly from soda. Compared with soda in equal weight percentages, potash seems to confer a little more density, less hardness, and less tenacity. The two are by far the most expansible oxides in glass.

In glazes potassium carbonate appears as an ingredient when it is desirable to modify the effect of a colorant such as copper oxide, which may thus be brought through tints of green toward yellow.

When potassium carbonate is used in glazes in combination with sodium oxide, lead oxide, or calcium oxide, the potassium oxide derivative cannot exceed 0.15 equivalent without affecting the color. If the foregoing glaze composition were modified by decreasing CaO to 0.30 equivalent and increasing K_2O to 0.30 equivalent, a brilliant robin's egg blue is achieved at cone 2-3. When potassium carbonate is used in colored glazes, it is advisable to frit about 90% of the clay, but none of the color.

In enamels potassium carbonate tends to produce high luster, but it decreases strength and elasticity, making the enamel soft.

In general, enamels containing potassium are more readily fusible than those with sodium.

Potassium carbonate has largely been replaced in enamels, however, by sodium carbonate, due to the difference in price, except in occasional cases where it is used to alter colors.

Potassium Chlorate

Also known as chlorate of potash and potassium oxymuriate, this is a white crystalline powder, or lustrous crystalline substance, of the composition $KClO_3$, employed in explosives, chiefly as a source of oxygen. It is also used as an oxidizing agent in the chemical industry, as a cardiac stimulant in medicine, and in toothpaste. It melts at 357°C and decomposes at 400°C with the rapid evolution of oxygen. It is odorless but has a slightly bitter saline taste. The specific gravity is 2.337. It is not hygroscopic, but is soluble in water. It imparts a violet color to the flame in pyrotechnic compositions.

Potassium Chloride

Potassium chloride is a colorless or white crystalline compound of the composition KCl, used for molten salt baths for the heat treatment of steels. The specific gravity is 1.987. A bath composed of three parts potassium chloride and two parts barium chloride is used for hardening carbon-steel drills and other tools. Steel tools heated in this bath and quenched in a 3% sulfuric acid solution have a very bright surface. A common bath is made up of potassium chloride and common salt and can be used for temperatures up to 900°C.

Potassium chloride is used in the porcelain enamel industry as a setting-up agent in titanium cover coats. In general, the quantities of potassium chloride, when used as an electrolyte, will be approximately the same as sodium nitrite, which it replaces. However, KCl does not aid tearing resistance as does nitrite. The main advantage in using potassium chloride is the freedom from yellowing or creaming when used in a blue-white enamel. Potassium chloride may exert an adverse effect on the gloss and may cause a slight decrease in the acid-resisting properties of the enamel, although the latter effect is somewhat debatable.

Potassium Cyanide

A white amorphous or crystalline solid of the composition KCN, potassium cyanide is employed for carbonizing steel for case hardening and for electroplating. The specific gravity is 1.52, and it melts at about 843°C. It is soluble in water and is extremely poisonous, giving off the deadly hydrocyanic acid gas. For cyaniding steel the latter is immersed in a bath of molten cyanide and then quenched in water, or the cyanide is rubbed on the red-hot steel.

Commercial potassium cyanide is likely to contain a proportion of sodium cyanide. Potassium ferrocyanide, or yellow prussiate of potash, can also be used for case-hardening steel. It has the composition $K_4Fe(CN)_6$ and comes in yellow crystals or powder. The nitrogen as well as the carbon enters the steel to form the hard case. Potassium ferricyanide, or red prussiate of potash, is a bright-red granular powder of the composition $K_3Fe(CN)_6$, used in photographic reducing solutions, in etching solutions, in blueprint paper, and in silvering mirrors. Redsol crystals is the name of this chemical for use as a reducer and mild oxidizing agent, or toner, for photography. Potassium cyanate, KCNO, is a white crystalline solid used for the production of organic chemicals and drugs. It melts at 310°C. The potassium silver cyanide used for silver plating comes in white, water-soluble crystals of the composition $KAg(CN)_2$. Sel-Rex is this material. Potassium gold cyanide has a similar function in gold plating. Platina comes as colorless tablets that are soluble in both water and alcohol.

Potassium Dichromate

This material, $K_2Cr_2O_7$, decomposes at 500°C. Bright yellowish-red crystals are soluble and poisonous. Sometimes potassium chromate, $K_2Cr_2O_4$, and the dichromate are utilized in ceramics as coloring agents.

Potassium dichromate is used in glass for aventurine effects. It is said that 20 or 21 parts to 100 parts sand will give a chrome aventurine. This glass is characterized by glittering metallic scales of chromium oxide. Potassium dichromate is also used in glass to give a green color. However, it has been shown that it may cause considerable trouble by formation of black, chrome corundum crystals in the glass. Air-floated chromite is suggested to avoid this problem.

Potassium dichromate is used in glazes to produce chrome-tin pinks, low-fire reds, greens, and purplish-red colors.

Potassium Nitrate

Potassium nitrate is also called niter and saltpeter, although these usually refer to the native mineral. A substance of the composition KNO_3, it is used in explosives, for bluing steel, and in fertilizers. A mixture of potassium nitrate and sodium nitrate is used for steel-tempering baths. The mixture melts at 250°C. Potassium nitrate is made by the action of potassium chloride on sodium nitrate. It occurs in colorless prismatic crystals, or as a crystalline white powder. It has a sharp saline taste and is soluble in water. The specific gravity is 2.1 and the melting point is 337°C.

Potassium nitrate contains a large percentage of oxygen, which is readily given up and is well adapted for pyrotechnic compounds. It gives a beautiful violet flame in burning. It is used in flares and in signal rockets.

Most enamels contain some oxidizing agent in the form of potassium or sodium nitrate. Only a small amount of nitrate is necessary; 2 to 4% is sufficient to maintain oxidizing conditions in most smelting operations.

In glazes it is sometimes used as a flux in place of potassium oxide, but, owing to its cost and solubility, very little of it is contained in glaze. Where conditions prevent the use of sufficient potash feldspar, potassium oxide is introduced into the mix, usually in the form of the nitrate in a frit.

Potassium nitrite is a solid of the composition KNO_2 used as a rust inhibitor, for the regeneration of heat-transfer salts, and for the manufacture of dyes.

POWDER-METALLURGY PARTS

Powder-metallurgy parts, commonly referred to as P/M parts, are produced by the powder-metallurgy process, which involves blending of powders, pressing the mixture in a die, and then sintering or heating the compact in a controlled atmosphere to bond the contacting surfaces of the particles. Where desirable, parts can be sized, coined, or re-pressed to closer tolerances; they can be impregnated with oil or plastic or infiltrated with a lower-melting metal; and they can be heat-treated, plated, and machined. Production rates range from several hundred to several thousand per hour.

Shapes that can be fabricated in conventional P/M equipment range up to about 16.1 kg. Parts of over 460 kg can be produced with special techniques such as isostatic compacting and extrusion. However, most P/M parts weigh less than 2.3 kg. While most of the early P/M parts were simple shapes, such as bearings and washers, developments over the years in equipment and materials now make economical the production of more intricate and stronger parts. And shapes with flanges, hubs, cores, counterbores, and combinations of these are fairly commonplace.

P/M parts are made from a wide range of materials, including combinations not available in wrought or cast form. And these materials can be processed by P/M techniques to provide tailored densities in parts ranging from porous components to high-density structural and mechanical parts. In addition, almost any conceivable alloy system under equilibrium or nonequilibrium conditions can be achieved, and segregation effects (nonhomogeneities) are avoided or minimized.

Most metal powders are produced by atomization, reduction of oxides, electrolysis, or chemical reduction. Metals available include iron, nickel, copper, and aluminum, as well as refractory and reactive metals. These metals can be blended together to form different alloy compositions during sintering. Also, prealloys such as low-alloy steels, bronze, brass, nickel–silver, and stainless steel are produced in which each particle is itself an alloy, thus ensuring a homogeneous metallurgical structure in the part. And it is possible to combine metal and nonmetal powders to provide composite materials with the desirable properties of both in the finished part.

The method selected is dictated primarily by composition, intended application, and cost. Typically, metal powders for commercial usage

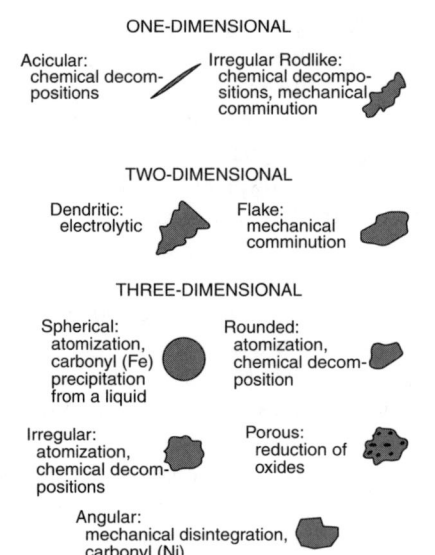

FIGURE P.6 Powder particle shape as a function of the method of production. (From *McGraw-Hill Encyclopedia of Science and Technology*, 8th ed., Vol. 14, McGraw-Hill, New York, 273. With permission.)

range from 1 to 1200 μm. Depending on the method of production, metal powders exhibit a diversity of shapes (see Figure P.6 depicting powder particle shape as a function of the method of production) from spherical to acicular. Particle shape is an important property, because it influences the surface area of the powder, its permeability and flow, and its density after compaction.

Chemical composition and purity also affect the compaction behavior of powders. For most applications, powder purity is higher than 99.5%.

The P/M process is being used to produce many thousands of different parts in most product and equipment manufacturing industries, including automotive, business machines, aircraft, consumer products, electrical and electronic, agricultural equipment, machinery, ordnance, and atomic energy.

The applications of P/M parts in these industries fall into two main groups. The first are those applications in which the part is impossible to make by any other method. For example, parts made of refractory metals like tungsten and molybdenum or of materials such as tungsten carbide cannot be made efficiently by any other means. Porous bearings and many types of magnetic cores are exclusively products of the P/M process. The second group of uses consists of mechanical and structural parts that compete with other types of metal forms such as machined parts, castings, and forgings.

Processes

P/M processes include pressing and sintering, powder injection molding, and full-density processing.

Pressing and Sintering

The basic P/M process requires two steps involving powder compaction followed by sintering. First, the rigid steel die (mold) is filled with powder. Pressure is then applied uniaxially at room temperature via steel punches located above and below the powder, after which the compact is ejected from the mold. Commercial compaction pressures are normally in the range from 140 to 900 MPa. The powder compact is porous. Its density depends on compaction pressure and the resistance of the powder particles to deformation; compact densities about 90% of the theoretical level are common.

In the sintering process, the powder compact is heated below its melting point to promote bonding of the solid powder particles. The major purpose of sintering is to develop strength in the compact. Normally an increase in the density of the compact also occurs, but the compact still contains pores. The internal architecture (microstructure) of the material is developed during sintering. As a rule of thumb, the sintering temperature must be higher than one half the melting temperature (Kelvin scale) of the powder.

Normally, parts made by pressing and sintering require no further treatment. However, properties, tolerances, and surface finish can be enhanced by secondary operations. Examples of secondary or finishing operations are re-pressing, resintering, machining, heat treatment, and various surface treatments such as deburring, plating, and sealing.

Powder Injection Molding

This is a process that builds on established injection-molding technology used to fabricate plastics into complex shapes at low cost. The metal powder is first mixed with a binder consisting of waxes, polymers, oils, lubricants, and surfactants. After granulation, the powder–binder mix is injection-molded into the shape desired. Then, the binder is removed and the remaining metal powder skeleton sintered. Secondary operations may be performed on the sintered part, similar to those for conventional press-plus-sinter parts. The viscosity of the powder–binder mix should be below 100 Pa · s (1000 poise) with about 40% by volume of binder, and spherical powders less than 20 μm in diameter are preferred. Powder injection molding produces parts that have the shape and precision of injection-molded plastics but that exhibit superior mechanical properties such as strength, toughness, and ductility.

Full-Density Powder Processing

Parts fabricated by pressing and sintering are used in many applications. However, their performance is limited because of the presence of porosity. To increase properties and performance and to better compete with products manufactured by other metalworking methods (such as casting and forging), several powder metallurgy techniques have been developed that result in fully dense materials; that is, all porosity is eliminated. Three examples of full-density processing are hot isostatic pressing, powder forging, and spray forming.

In the hot isostatic pressing process the metal (or ceramic) powder is sealed in a metallic or glass container, which is then subjected to isostatic pressure at elevated temperature. Temperatures up to 2200°C and pressures up to 200 MPa are possible in modern presses. After complete densification of the powder, the compact is removed and the container stripped from the densified compact. Hot isostatic pressing is applied to the consolidation of nickel and titanium alloys, tool steels, and composites.

Powder forging is an adaptation of conventional forging in which a compact is prepared that is then sintered to retain about 20% by volume of porosity. This preform is transferred to a forging press and formed in a closed die cavity to its final shape with one stroke of the press. All porosity is removed in the forging operation. Powder forging is used primarily to fabricate components from low-alloy steels.

In the spray-forming process a stream of liquid metal is gas-atomized into a spray of droplets that impinge on a substrate. The spraying conditions are such that the droplets arrive at the substrate in the semisolid state. By control of the geometry and motion of the substrate, it is possible to fabricate sheet or plate, tubes, and circular billets. It is a rapid process with a deposition rate up to about 2 kg/s. Applications include aluminum, copper, iron, and nickel alloys.

CHARACTERISTICS AND APPLICATIONS

P/M competes with several more conventional metalworking methods in the fabrication of parts, including casting, machining, and stamping. Characteristic advantages of P/M are close tolerances, low cost, net shaping, high production rates, and controlled properties. Other attractive features include compositional flexibility, low tooling costs, available shape complexity, and a relatively small number of steps in most P/M production operations. Control of the level of porosity is the intrinsic feature of P/M that enables the parts producer to predict and specify physical and mechanical properties during fabrication. Broad areas of usage for P/M parts include structural (load-bearing) components, and controlled-porosity, electrical, magnetic, thermal, friction, corrosion-resistant, and wear-resistant applications. Industries that make extensive use of P/M parts are aerospace, agriculture, automotive, biomedical, chemical processing, and electrical. In addition, the fabricators of domestic appliances and office equipment are dependent on the availability of a wide range of sizes and geometries of P/M parts exhibiting unique combinations of physical and mechanical properties.

Ferrous parts fabricated by pressing and sintering make up the largest segment of the P/M industry, and a majority of the applications are for the automotive market. Representative P/M parts include gears, bearings, rod guides, pistons, fuel filters, and valve plates. In the nonautomotive sector, examples of parts include gears, levers, and cams in lawn tractors and garden appliances; and gears, bearings, and sprockets in office equipment.

Full-density processing has been used in the powder forging of steel automobile connecting rods. The iron–copper–carbon rod is forged to a minimum density of 7.84 g/cm^3.

Important characteristics of powder forging include straightness, elimination of surface defects inherent in conventional forging, uniform microstructure, dimensional control, and superior machinability. Secondary operations include deflashing, shot peening, and machining.

An example of a part manufactured by powder injection molding is the miniature read/write latch arm made from stainless steel powder and used in a small, high-capacity hard-disk drive. The geometric form of the entire outside profile of the part is very critical relative to the pivot hole. Coining is the only secondary operation.

SAFETY AND HEALTH CONSIDERATIONS

Because metal powders possess a high surface area per unit mass, they can be thermally unstable in the presence of oxygen. Very fine metal powders can burn in air (pyrophoricity) and are potentially explosive. Therefore the clean handling of powder is essential; methods include venting, controlled oxidation to passivate particle surfaces, surface coating, and minimization of sparks or heat sources. Some respirable fine powders pose a health concern and can cause disease or lung dysfunction: the smaller the particle size, the greater the potential health hazard. Control is exercised by the use of protective equipment and safe handling systems such as glove boxes. There is no recognized hazard associated with the normal handling of the common grades of metal and alloy powders such as copper and iron.

POWDER PLASTICS MOLDINGS

The term *powder molding* broadly describes a technique of sintering or fusing finely divided thermoplastic materials to conform to the surface of a mold.

PROCESSING

There are a number of processes and techniques loosely termed powder molding. The Engel or Thermofusion process was developed in Europe and sublicensed in the United States in 1960. This patented technique employs inexpensive sheet-metal molds and a hot air oven heated by either gas or electricity. In essence, the process consists of filling the mold with powdered thermoplastic material, fusing the layer of material next to the mold walls, removing the excess powder, and then smoothing the inside surface with heat.

Another older technique, known as the Heisler process, involves rotating a heated mold. Both of these patented techniques employ an *excess* of powdered material over that required for the object being fabricated; the extra material is removed as one of the processing steps.

More recently, it was discovered that powdered polyethylene behaves enough like a liquid to permit use of rotational casting equipment designed for use with vinyl plastisols. The technique involves multiaxial rotation of a closed mold filled with a *measured* charge of material, all of which is fused during the heating step. This permits fabrication of totally enclosed articles. A hobby horse for children is the most striking example of the versatility of the rotational casting process. There is at present a variety of rotational casting apparatus available. As soon as it is engineered specifically for use with powdered thermoplastic materials, it is anticipated that this technique will find widespread acceptance.

The selection of a particular powder-molding technique depends on the application under consideration. The selection depends on factors such as size, wall thickness, geometric configuration, and quantity of parts desired. Rotational casting with a measured charge is obviously indicated where automation for a large volume of production is desired, due to the reduction of material-handling requirements.

ADVANTAGES

Because of low tooling costs, all these various powder-molding processes offer significant economies over conventional fabricating techniques such as injection or blow molding in small or moderately large quantities (up to 250,000 pieces, depending on the application concerned).

Typically, inexpensive sheet metal or cast aluminum molds may be used, as compared with expensive matched metal molds made of tool steels. The heat source is usually an electric or gas-fired air oven, a small capital cost in comparison to an injection-molding machine or extruder. Moreover, the large size of some objects that are fabricated by powder molding defy production by most other techniques. The size of an object to be fabricated by this particular process is limited only by the size of the oven. Large tanks, refuse containers, and even boats are included in the wide variety of products being fabricated by this method. These parts range in wall thickness from 1.5 to 6.4 mm with a tolerance of ±10% considered reasonable.

Materials Used

Although powder-molding processes are theoretically applicable to any thermoplastic material, current practice has been restricted primarily to polyethylenes of low and intermediate density. As soon as effective stabilizing systems are found for other materials, particularly the broader range of polyolefin materials, it is anticipated that such materials will also find use where their particular properties offer advantages in specific applications. The particles of powder range in size from 30 to 100 mesh (U.S. Standard), with the peak of the bell-shaped distribution at approximately 60 mesh. The powder is made by grinding the material in pellet form as an additional manufacturing step.

Production cycles can be cited only generally, as they depend on the application concerned. Cycles range from roughly three minutes for small, thin-walled parts made by rotational casting, to as much as 30 minutes for very large, heavy-walled parts made by the Engel process.

PRECIOUS METALS

These are the metals gold, silver, and platinum, which are used for coinage, jewelry, and ornaments, and also for industrial applications. Expense or rarity alone is not the determining factor; rather, a value is set by law, with the coinage having an intrinsic metal value as distinct from a copper coin, which is merely a token with little metal value. The term noble metal is not synonymous, although a metal may be both precious and noble, as platinum. Although platinum was once used in Russia for coinage, only gold and silver fulfill the three requisites for coinage metals. Platinum does not have the necessary wide distribution of source. The noble metals are gold, platinum, iridium, rhodium, osmium, and ruthenium. Unalloyed, they are highly resistant to acids and corrosion. Radium and certain other metals are more expensive than platinum but are not classed as precious metals. Because of the expense of the platinum noble metals, they may be alloyed with gold for use in chemical crucibles.

PREFINISHED METALS

Prefinished metals are sheet metals that are precoated or treated at the mill so as to eliminate or minimize final finishing by the user. The metals are made in a ready-to-use form with a decorative and/or functional finish already applied. Prefinished metals provide many advantages including: (1) better product appearance, (2) lower product cost, (3) greater product uniformity, and (4) improved product function. (See Table P.14, which reflects types of prefinished metals.) It is possible to obtain base metals preplated with almost any decorative or functional metal, from bright, shiny chromium to dull, rich-looking brass. Similarly, it is possible to obtain sheet with prepainted surfaces in almost every color and in a wide variety of special-purpose plastic resins. Also, where extra durability or a special decorative effect is needed some of these resins, notably polyvinyl chloride, are available in sheet or film, laminated to several different metals.

Furthermore, almost every metal is available in a limitless number of textured, patterned, and

TABLE P.14
Types of Prefinished Metals

Material	Surface Composition, Appearance	Base Metals
Preplated metals	**Nickel, Chromium** Excellent appearance; available in dull, satin, and bright (sometimes provided with a clear lacquer coating for added protection) finishes, many of which can be embossed with wide range of patterns)	Steel, zinc, brass, copper, aluminum
	Brass, Copper Excellent appearance; available in dull, satin, and highly polished finishes, many of which can be embossed with wide range of patterns	Steel, zinc
	Zinc Natural grayish finish that can be used to improve product appearance by providing with semilustrous finish and coating with clear lacquer	Steel
Prepainted metals	Almost every organic coating is available or can be ordered in prepainted form. Selection of coating resin depends on end use requirements. Five most popular prepainted metals now in use are alkyds, acrylics, vinyls, epoxies, and epoxy-phenolics. Many coatings can be pigmented to provide metal-like appearance.	Most common ferrous and nonferrous metals. For added protection reverse side of ferrous metals is usually given a rust-preventive treatment or provided with an organic or metallic coating. Selection of base metal depends on cost, appearance, and product life requirements. Most popular base metals are cold-rolled steel, tinplate, tin mill backplate, hot-dipped and electrogalvanized steel, and standard aluminum alloys.
Plastic–metal laminates	**Vinyl (PVC)–Metal** Polyvinyl chloride sheet laminated to base metal with thermosetting adhesives under heat and pressure (25 to 60 psi)	Can be applied to most metals; popular are: Steel — Provides strength at low cost; reverse side can be painted or provided with corrosion-resistant coating; Aluminum — Light weight and/or corrosion resistance; Magnesium — Light weight
	Vinyl (PVF)–Metal Polyvinyl fluoride film laminated to base metal with thermosetting adhesive under heat and pressure	Cold-rolled steel — Provides strength at low cost; Galvanized, aluminized, and tin-plated steel — Corrosion resistance; Aluminum — Light weight and corrosion resistance
	Polyester–Metal Polyester film laminated to base metal	Usually steel
Textured and embossed metals	Surfaces available in hundreds of different textures and patterns. Available with texture on one surface only or with pattern that extends completely through cross section of metal. Also available in perforated form. Surfaces can be provided with dull satin or highly polished finish or combinations thereof (e.g., dull background with polished highlights). Can also be painted, porcelain enameled, or oxidized; these finishes can be buffed off high spots to provide two-tone effect.	All common sheet metals, including: Carbon steel — Strength at low cost; Stainless steel — Corrosion resistance plus strength; Aluminum — Light weight plus corrosion resistance; Copper — Pleasing appearance plus strength and corrosion resistance.
Hot-dipped other plated metals	**Galvanized (Zinc-coated)**	

PREFINISHED METALS

TABLE P.14 (CONTINUED)
Types of Prefinished Metals

Material	Surface Composition, Appearance	Base Metals
	Zinc surface with intermediate zinc–iron alloy layer	Steel, ingot iron
	Aluminized	
	Aluminum surface; intermediate aluminum–iron layer forms above 900°F	Steel
	Tin-coated	
	Hot-dipped or electroplated tin	Mild carbon steel
	Terne or Lead-coated	
	Lead–tin alloys, pure lead	Steel
Specially finished aluminum	**Color Anodized**	
	Anodized aluminum available in clear, yellow gold (70:30 brass), red gold, rich low brass (85:15 brass), copper, blue, green, red and black. Colors are obtainable over standard mill, satin, and bright finishes, as well as over embossed and perforated textures.	All commercial aluminum alloys and tempers
	Spangled	
	Uncoated surface containing large grains that stand out in relief and facets that break up and reflect light. Available in wide variety of colors and mill finishes.	Wrought aluminum alloys

embossed finishes right from the mill. These textured metals can be used as is, but even they can be supplied preplated, prepainted, or even with a colored preanodized finish, as in the case of aluminum.

Many sheet metals are also available with galvanized, aluminized, tin, and terne coatings. These materials, especially galvanized, were among the first prefinished metals to be developed and remain today as basic prefinished metals.

FUNCTIONAL ADVANTAGES

Many prefinished metals are used for functional applications. Practically all zinc-plated sheet, for example, is used in functional applications where good corrosion resistance, rather than a bright, decorative finish is wanted (e.g., condenser cans and hidden parts in door locks). The zinc coating also provides a good paint base, provided the surface is first given a chemical conversion treatment.

Copper-plated steel is another good example of a functional finish. It provides a good lubricating surface for deep-drawing operations and also makes a good base for further electroplating. The material is also used for its electrical conductivity, for its usefulness in low-temperature tinning and high-temperature brazing operations, and as a stop-off coating in carburizing operations.

Prepainted metals are also popular for functional applications; this is borne out by the wide number of functional resins that are now available. In addition to providing decorative appearance, these coatings prevent the base metal from corroding and can provide good resistance to chemicals and foods (epoxies), toughness and resistance to forming damage (vinyls), and good resistance to outdoor exposure (acrylics).

TYPES AND APPLICATIONS

Prepainted metals are produced using various organic coatings on many common ferrous and nonferrous metals. Extra durability or special decorative effects are provided by plastic–metal laminates. Polyvinyl chloride, polyvinyl fluoride, and polyester are the plastics commonly used.

Black-coated steel is used to give a high thermal emittance in electronic equipment. The base metal is aluminum-deoxidized steel containing 0.13% carbon, 0.45% manganese, 0.04% max phosphorus, and 0.05% max sulfur. The steel is coated with a 5% by weight layer of nickel oxide, which is reduced in a hydrogen furnace to form a spongy layer of nickel. This sponge is impregnated with a carbon slurry to form a black carbonized surface.

Preplated metals consist of a thin electrodeposited plate of one metal or alloy on a base, or substrate, usually sheet, of another metal. Steel is the most common base metal and it is commonly plated with brass, chromium, copper, nickel, nickel–zinc, or zinc. Other common preplated metals are chromium-plated brass, copper, nickel, or zinc, and nickel-plated brass, copper, or zinc. The surface of the plate may be mirror bright, satin, bright satin, embossed, antique, or black, or have some other finish. The plate, usually 0.025 to 0.127 mm thick, is sufficiently ductile to withstand shearing, bending, drawing, and stamping operations. Common joining methods include lock-seaming, stud welding, adhesive bonding, and spot welding. One of the earliest groups of preplated metals included Brassoid, Nickeloid, and Chromaloid, which were brass-, nickel-, and chromium-plated zinc sheet.

Prefinished metals are now available with almost any metal plated or bonded to almost any other metal, or single metals may be had prefinished in colors and patterns. They come in bright or matte finishes, and usually have a thin paper coating on the polished side, which is easily stripped off before or after forming. The metals are sold under a variety of trade names and are used for decorative articles, appliances, advertising displays, panels, and mechanical parts.

PREIMPREGNATED DECORATIVE FOIL

Special papers preimpregnated with melamine resin make possible a broad range of molded-in decoration for thermoset plastic products. By far the leading application to date is the use of melamine-impregnated rayon paper. It is used to decorate about 70% of all melamine dinnerware. Development of new materials and techniques to shorten the production cycle is extending use of prepreg paper in closures, cutlery handles, and other markets.

Fabrication

The rayon paper is impregnated with resin under closely controlled conditions at the impregnator's plant to provide a foil with precise resin content, even weight and resin distribution, specified penetration, and absorption. The melamine resin is partially cured, then held at a stable, flexible B-stage, cut into sheets convenient for printing, and shipped to printers for decoration.

Designs are printed either by lithography or silk screen with special inks and techniques developed by specialty printers. Virtually any design can be reproduced, including photographs in black and white or full color. Even moldable silver and gold inks have been perfected.

The printed foil is cut to final size and shape and delivered to the molder ready for the mold. Normally the product to be decorated is formed and partially cured; then the mold is opened, the foil inserted, and the mold closed to fuse the foil to the product and complete the cure. Except when metallic inks are used, the paper is inserted so the designs face the product. The foil becomes transparent but forms a protective melamine surface.

Stretch of the foil is carefully limited to avoid distortion of the design, but the effect of "deep draw" can be achieved by shaping the printed foil. Thus, "doughnut" shapes are used effectively to decorate the inside of plates and bowls. Wrap-around shapes put molded designs on the outside of cups and pitchers. Mold design must allow for the decoration of side surfaces.

Decorating Closures

A relatively new development is the use of prepreg foils to decorate closures such as cosmetic jar covers. Previously, decoration was limited to the one or two colors possible with hot stamping and direct screening or to lithography on metal, which is subject to rust. Molding the foil into a thermosetting plastic closure gives the package designer full color range and combination, and later permits change of design without altering the molds.

The demand of the closure producer for high-volume, high-speed molding encouraged a research program that has yielded a new one-shot decorative foil. Instead of preforming the product and opening the mold for insertion of the foil, the new one-shot foil is put into the mold first, the molding powder poured on top of it, and the mold closed for a complete, uninterrupted cycle. As yet, the new technique has been used only for closures, but the same principles are believed to be technically feasible for larger products such as dinnerware. More immediate markets for the one-shot foil are control knobs and drawer pulls.

Materials

Prepreg foils can be used to apply decoration to a wide variety of thermosetting resins, including all melamines, all ureas, and most phenolics.

Although the self-effacing rayon foil is most popular for dinnerware and most melamine and urea products, opaque prepreg paper is favored for such products as cutlery handles, clock faces, switch plates, and trays. The opaque paper shields the darker resins and even permits use of varicolored odds and ends in a molder's inventory for such applications as a clock face. Often, of course, the opaque paper is used to provide a decorative color contrast.

The opaque print-base paper is made with a high-purity alpha pulp, fillers, pigments, and other additives. Like the rayon foils, this paper is preimpregnated with resin and advanced to a stable B-stage for shipment and printing. Among other products for which the opaque prepreg papers are suitable are handles for pots and pans, organ keys, and wall tile.

PREIMPREGNATED MATERIALS FOR REINFORCED PLASTICS

So-called *prepregs* are ready-to-mold reinforced plastics with the resin and reinforcement precombined into one easy-to-handle material. They are fabricated by what is called dry layup techniques; the alternative is the wet layup method of combining a liquid resin and reinforcement at the mold.

Fabrication

Prepregs are produced by impregnating continuous webs of fabric or fiber with synthetic resins under close control. The resins are then partially cured to the B-stage, or partly polymerized. At this stage the preimpregnated material remains stable, and may be shipped or stored, ready for forming into the final shape by heat and pressure.

Most of the controllable variables in a reinforced plastics structure are taken into account in the production of a prepreg, e.g., resin type and content, reinforcement, and finish, which are largely determined by the requirements of the final laminate. Prepregs are preengineered to meet performance and processing requirements.

Major Advantages

The major advantages of using preimpregnated materials for molding reinforced plastics products are (1) high and more uniform strength, (2) uniform quality, (3) simplified production, and (4) design freedom.

Key to the majority of the advantages of prepreg is the close control maintained through all stages of the preimpregnation process. For example, the amount of resin solution picked up by the reinforcing web as it passes through one or more resin baths is influenced not alone by resin viscosity and resin solids, but by web tension and speed (which regulate the time the reinforcement is in the bath) and by the temperature of the resin. Because of the high degree of control afforded by preimpregnating equipment over tension, speed, and temperature, resin content can be controlled ±2%.

Such controls also permit the production of prepregs with a reinforcement content from as high as 85% to as low as 20%. It is the high reinforcement content possible with prepregs that accounts for the high strength of these materials. Further, since the reinforcement is thoroughly saturated during preimpregnation, little air is entrapped in the reinforced plastic material, assuring better quality in the molded part.

Handling characteristics of prepregs can also be closely controlled by controlling temperature and speed of the material as it passes through the second stage of the process where the resin is partially cured. Temperature of the drying oven, speed of travel, and resin type and content determine the degree of polymerization or B-stage, which, in turn, determines such handling and molding characteristics as resin flow, gel time, tack, and drape. In prepregs these factors can be tailored to suit a variety of production molding requirements.

Prepregs also facilitate use of such resins as phenolics, melamines, and silicones, which when available in liquid form, are usually solvent solutions (except for newer solventless silicones). Wet layup of solvent solutions of resins is usually unsatisfactory, at best, because solvents are released during cure. As B-staged prepregs, such resin systems are virtually solvent-free.

Prepregs can also be made using resins whose viscosities are so low that they would be difficult, or impossible, to handle by wet layup techniques.

Prepregs can be particularly advantageous to the smaller reinforced plastics fabricator, eliminating the need for resin fomulation. Scrap loss can be minimized by chopping up or macerating leftover prepreg to form molding compound.

Because the resin is already properly distributed throughout the reinforcement, prepregs offer distinct advantages in molding odd-shaped parts, with varying thicknesses, undercuts, flanges, etc. They preclude the problem of excessive flow of resin, eliminating resin-rich or resin-starved areas of such parts. Cut to the proper pattern, the prepreg can be accurately prepositioned in the mold to provide optimum finished parts.

Possibly the greatest advantage of prepregs is that they lend themselves to automated, high-production molding. The continuous prepreg web can be prepared for molding by cutting, slitting, or blanking operations. Often parts can be molded directly from such blanks. Die-cut parts can also be used to preassemble a complete layup in advance of molding, thus minimizing press time.

Materials Used

A wide range of reinforcing materials and resins are available in prepreg form. Principal resins are polyesters, epoxies, phenolics, melamines, silicones, and several elastomers. The most commonly used reinforcements are glass cloth, asbestos, paper, and cotton. Specialty fibrous reinforcements include high-silica glass, nylon, rayon, and graphite.

Reinforcing Materials

Reinforcing materials available in prepregs today include the following:

1. Glass fiber is by far the largest volume reinforcing material in use today. Prepregs incorporating glass fiber are available in the form of pre-impregnated roving (for filament winding), cloth, and mat. It provides a good balance of properties: outstanding strength-to-weight characteristics, high tensile strength, high modulus of elasticity compared to other fibers, resilience, and excellent dimensional stability. It is used for such products as airplane and missile parts, ducts, trays, electrical components, truck body panels, and construction materials.
2. Asbestos is used in prepregs in the form of paper, felt, and fabrics. All provide good thermal insulation. Papers provide cost advantage where structural characteristics are not critical. Felts provide maximum tensile and flexural strength, and good ablation resistance. Asbestos fabrics are used principally in flat laminates.

Prepregs employing asbestos webs have won an important place in rocket and missile components in good part because of their remarkable short-term resistance to extremes of temperature and flame exposure.
3. Paper, a leading reinforcement for high-pressure prepreg laminates, is inexpensive, prints well, and has adequate strength for its principal uses as counter surfacing, furniture and desktops, and wallboards.
4. The specialty reinforcements, like high-silica glass, quartz, graphite, nylon, have their main use in missile and rocket parts. Some exhibit unusual high-temperature resistance; some are good ablative materials.

Resins

The major plastic resins used in prepregs are the polyesters, epoxies, phenolics, melamines, and silicones, although any thermosetting resin or elastomer may be used.

1. Polyester resin prepregs are comparatively low in cost and easy to mold at low temperatures. They have good mechanical, chemical, and electrical properties, and some types are flame resistant.
2. Epoxy prepregs provide high mechanical strength, excellent dimensional stability, corrosion resistance and interlaminar bond strength, good electrical properties, and very low water absorption.
3. Phenolic prepregs have high mechanical strength, excellent resistance to high temperature, good thermal insulation and electrical properties, and high chemical resistance.
4. Melamine prepregs have excellent color range and color retention, high abrasion resistance, good electrical properties, and are resistant to alkalies, and flame.
5. Silicone prepregs provide the highest electrical properties available in reinforced plastic and are the most heat stable. They retain strength and electrical characteristics under long-term exposure to 260 to 316°C.

Actually any thermosetting resin or elastomer may be used and any continuous-length reinforcement. Even sheet material not capable of supporting its own impregnated weight in a drying oven can be used. The web may be a woven fabric, nonwoven sheet, or continuous strand or roving. After impregnation, the prepreg may be slit into tape, chopped, or macerated, or cut to specified pattern, depending on the requirements of the part and the desire of the molders.

Molding

The one requirement for using prepregs is that some heat and pressure must be used to mold them to final shape. They cannot be room-temperature cured. Prepregs can be formulated to fit all other methods of processing; the four major methods are vacuum-bag molding, pressure-bag molding, matched metal-die molding, and filament and tape winding.

The choice of method of forming prepregs is most often governed by production volume. The exception is the critical part, like a missile component, where the higher performance qualities achieved by matched metal-die molding justifies use of this process even for a few hundred parts.

The vacuum-bag molding method uses the simplest and most economical molds and equipment and is used where the number of parts needed is low or where frequent design changes occur.

The pressure-bag method of molding uses pressures up to 0.70 MPa to achieve greater strength than is possible with the vacuum-bag method. Mold and equipment costs remain comparatively low. In both vacuum-bag and pressure-bag molding, additional pressure with resulting higher strengths can be achieved through use of an autoclave.

Matched metal-die molding is used for long runs where it gives the lowest cost and the highest production rates. This method permits accurate control of dimensions, density, rate of cure, and surface smoothness.

Tape and filament winding are used for cylindrical, spherical, and conical shapes and can produce high physical properties and high strength-to-weight characteristics. Techniques range from simple winding to highly automated operations. In fact, it comes closest to automation of current fabrication methods and achieves homogeneous, void-free products. It does not involve the expense of matched metal dies.

PREMIX MOLDINGS

Premix molding materials are physical mixtures of a reactive thermosetting resin (usually polyester), chopped fibrous reinforcement (usually fibrous glass, asbestos, or sisal), and powdered fillers (usually carbonates or clays). Such mixtures, when properly formulated, can exhibit a wide range of performance properties at variable costs. In general, the resin type will determine the corrosion-resistance properties of the premix; increasing amounts of glass reinforcement will increase strength and increasing amounts of filler will reduce cost and corrosion resistance.

COMPARATIVE BENEFITS

In comparison with such corrosion-resistant metals as brass, bronze, and stainless steels, premixes offer, primarily, cost reductions due to ease of manufacture of complex shapes, and reductions in actual material costs. In some cases, design latitude, colorability, resistance to abrasion, and reduction in weight are also important factors in favor of the premix parts.

In comparison with aluminum, premixes can offer superior corrosion resistance, especially to alkaline detergents, in addition to the advantages noted above.

In comparison with thermoplastic resins, the premix materials offer considerably greater hardness, rigidity, and heat resistance. The premix materials are also stronger. The thermoplastics usually have superior colorability, surface smoothness, and gloss, and a broader range of corrosion resistance. Thermoplastics are also somewhat lower in cost for the molded item when the number of parts is large. In general, tooling costs for injection-molded thermoplastics are higher, but molding, material-handling, and finishing costs are lower for the thermoplastics.

Phenolics, ureas, and melamines are considerably less strong than the glass-fiber-reinforced premixes and, in general, they do not have as good a range of resistances to aqueous solutions as do properly formulated premixes. The phenolics, of course, also have limited color possibilities.

In comparison with most other plastics, premixes are usually more difficult to handle because they are more difficult to meter or preweigh automatically. Generally, they must be preweighed by hand to an exact mold charge. Recently, however, there has been a trend toward extrusion and chopping of premix to logs of predetermined charge weight.

MOLDING

Because premixes are essentially heterogeneous materials, certain well-known molding phenomena such as weld lines and orientation during flow take on added significance. Fiber orientation tends to give somewhat greater effects of anisotropy and more care must be taken with premix molded parts to prevent excessive orientation. Similarly, weld lines in premix parts tend to be proportionately weaker than weld lines in thermoplastics. Much of the "art" of premix molding is concerned with reducing or eliminating orientation and weld-line effects by proper mold design, molding conditions (closing speed, pressure, temperature), and by selection of charge shape and location in the mold. Again, recent advances by glass-fiber manufacturers have resulted in superior glass fibers, which reduce orientation and weld-line effects.

Premixes are generally molded in conventional compression molds and presses using pressures in the order of 3.5 to 13.6 MPa and temperatures in the range of 121 to 177°C.

Premixes are also occasionally molded in transfer presses, usually at somewhat higher pressures. Transfer molding tends to reduce strength properties by degrading the glass fibers in the compounds.

Premixes are suitable for molding very complex shapes over a wide range of sizes. Filter plates and frames weighing over 45.4 kg have been molded successfully.

A large volume of premix is consumed in automotive air-conditioner and heater housings and ducts. These parts are low-cost, low-strength items requiring high rigidity and good heat resistance. Other large-volume applications include electrical insulators and housings of various types. Most premixes are manufactured captively; i.e., molders prepare their own premix material. The advantages of captive premix are lower initial raw-material costs, reduced packaging costs, ability to formulate faster curing compounds, and ability to tailor-make special compounds for each part. The major disadvantage of captive premix very often is lack of good quality control. Another disadvantage of the captive operation is that not enough compound development effort is spent in working out the "bugs" in a new formulation before such a formulation becomes commercial.

Compounding

Premixes are prepared by mixing glass, filler, and resin mix in a 180° spiral double-arm dough mixer. Blade clearance should be about 6.4 to 12.7 mm. The resin mix is usually prepared separately and includes resin, lubricant, catalyst, inhibitor, and other minor additives. The resin mix is charged to the mixer, followed by the filler. After the filler is thoroughly blended (usually 10 min is required) the fibrous material can be added. If the fibrous material is glass, care must be taken to disperse the fiber to prevent clumping; close control of mixing time must be maintained to prevent fiber breakdown. Recently, a high strand integrity fiber has been introduced, which reduces fiber-breakdown tendency.

After mixing, the premix must be stored in airtight containers to prevent styrene monomer loss. Cellophane bags may be used to store small units of premix (6.7 to 9 kg), and larger bins or hoppers may be used for larger quantities.

Premix of high filler loadings may be extruded through screw-type or ram-type extruders and automatically chopped to logs of predetermined weight.

The storage life of premix that is sealed against styrene loss may vary from 1 or 2 days to 1 year depending upon formulation. The premix with short storage life can be molded at lower temperatures and shorter cycles than the more stable premix.

PRESSURE-SENSITIVE ADHESIVES

Advanced pressure-sensitive adhesives (PSAs) are finding their way into every aspect of modern automobile manufacturing. Automakers and their suppliers are turning to new assembly methods to speed manufacturing, improve quality, lower cost, and meet more stringent environmental regulations. Automotive applications for today's adhesives include mirrors, brakes, headliners, flexible circuit bonding, sound deadening, carpets, door panels, instrument panels, seats, and windows. In a remarkable example of the potential of automotive applications, advanced structural adhesives have even replaced certain chassis-to-body panel welds in new automobiles.

Automotive air bags are yet another component where PSAs play a critical role. In fact, today's air bags are operating more reliably and safely thanks, in part, to the use of the latest adhesive formulations.

The millions of air bags in service today must resist years of shock, vibration, temperature fluctuations, and humidity extremes, and then work flawlessly when needed.

Various adhesives are used to assemble and seal the components of air bags, initiators, ignitors, gaskets, O-rings, high pressure seals, and vibration dampers.

In general, the adhesives serve to:

- Position and set the air bag in place
- Hermetically seal the propellant so that it remains protected from moisture
- Attach the propellant cover, which limits the chance of ignition during a vehicle fire or other nonimpact event

Assembly Benefits

Although acrylic and heat-activated adhesives offer excellent benefits in other applications, the advantages of silicone PSAs increasingly make them the adhesive system of choice for air bag inflators. Manufacturer's testing and field operation indicate that for air bag inflators, silicone PSAs offer numerous advantages over acrylic and heat-activated adhesives. Silicone PSAs simply provide the best seal for air bag inflators. In this application, silicone PSA benefits also include:

- Thermal stability not found with acrylic or heat-activated adhesives
- Uniform wet-out and adhesion
- Noncombustibility
- Lower outgassing
- Fewer seal leaks

PSA technology lends itself to manual, semiautomatic, and fully automated manufacturing. Environmental releases of volatile organic compounds are also lower with the new adhesives, good for both worker safety and air quality.

Many manufacturers find utilizing PSAs in place of heat-activated adhesives saves them money. That is because when using silicone, the reject rate due to leaks, adhesive voids, high outgassing, and other manufacturing problems are also dramatically lower. Postassembly testing shows a 15-fold decline in rejects after postassembly testing.

Added together, the cost factor of utilizing silicone PSAs (not to mention the performance benefits) makes them an attractive option for this application.

Other PSAs

Polymers used in the preparation of PSAs can be divided into two general classes of raw materials. In the first class are polymers that undergo chemical cross-linking following coating to achieve high cohesive strength. Examples include solvent-based acrylic PSAs and solvent-based natural rubber PSAs. Polymers in the second general class do not undergo chemical cross-linking, but instead have high molecular weight (e.g., water-based acrylic, SBR) or physical cross-links (e.g., hot melt styrenic block copolymers) to achieve adequate cohesive strengths.

PYROLYTIC GRAPHITE

Pyrolytic graphite is a high-purity form of carbon produced by thermal decomposition of carbonaceous gases.

The commercial manufacture of pyrolytic graphite products is a relatively new division of the graphite industry. Although this material has been known for some 50 years (glance coal, deposited carbon in gas retorts, etc.), it is only within the last 10 to 15 years that the necessary production techniques have been developed.

The manufacturing process essentially consists of bringing a relatively cold, carbonaceous gas into contact with a heated surface (mandrel) and thus extracting the carbon, in the form of graphite, directly from the gas. This process is not to be confused with the pyrolysis of resins and pitches, which is involved in the manufacture of the more common bulk or polycrystalline graphites.

PYROLYTIC MATERIALS

Essentially, pyrolytic deposition (literally, deposition by thermal decomposition) is a form of so-called gas or vapor plating. Gas or vapor plating can be accomplished by (1) hydrogen reduction, (2) displacement, or (3) thermal decomposition. Pyrolytic deposition is accomplished by the last mechanism.

The process involves passing the vapors of a compound over a surface maintained at a temperature above the decomposition temperature of the compound, in a vacuum furnace. The surface provides a source for nucleation of the desired material, which is built up to the desired section thickness.

Elemental materials are deposited from single-compound vapors, e.g., carbon, from a hydrocarbon; metal, from its halide; compound materials, from mixtures of compounds, e.g., boron nitride from boron halide and ammonia.

In producing coatings, the substrate serves as the surface on which the coating is deposited.

In producing self-supporting parts, the substrate serves as a mandrel or mold from which the pyrolytic material is removed after deposition.

The pyrolytic deposition process is used to produce pyrolytic graphite, as well as coatings or self-supporting structures of an extremely broad range of materials. Theoretically, the only limitation on the type of material that can be produced is that (1) it must be available in the form of a compound whose vaporization temperature is below its decomposition temperature, and (2) the desired material must separate cleanly from the vapor of the compound.

GENERAL PROPERTIES

The major benefits of pyrolytic materials are the following:

1. Highly directional properties are obtained in some materials by the substantial degree of orientation of crystals or grains. (*Note:* Highly directional properties are only obtained in materials such as pyrolytic graphite and boron nitride, which possess the unique and anisotropic graphite crystal structure.)
2. High densities, equivalent to theoretical densities, are obtainable.
3. High purity of material and close control of ingredients in "alloys" are obtainable by control of the reactant gases.

All initial work has been aimed at producing high-temperature materials, primarily for aerospace use. The unique properties of these materials make them attractive for a number of commercial applications.

MATERIALS AND FORMS

To date, pyrolytic graphite has been the largest-volume material produced, but newer materials include the following:

1. Graphite–boron compounds: pyrolytic graphite to which less than 2% boron has been added for increased strength and oxidation resistance as well as lower electrical resistivity.
2. Other graphite compounds: pyrolytic graphite to which varying percentages of columbium, molybdenum, or tungsten have been added.
3. Boron nitride: pyrolytic boron nitride containing 50 atm% of boron and nitrogen.
4. Carbides: pyrolytic carbides of tantalum, columbium, hafnium, and zirconium.
5. Tungsten: pyrolytic tungsten has been produced in the form of coatings and parts such as crucibles and tubes.

Coatings can be deposited on complex surfaces, providing gas impermeability in extremely thin sections. On the other hand, only those surfaces of the shape that can be exposed to the flow of gases will be coated. Also, differences in coefficients of thermal expansion between substrate and coating materials must be carefully considered.

Self-supporting shapes are limited by the fact that the material must be produced by deposition on a mandrel that must be removed after the part is formed.

Directionality Depends on Crystal Structure

The directional properties of the produced part or coating depend on the inherent crystal structure of the material. In general, materials are either highly anisotropic or nearly isotropic.

Pyrolytic deposition of material with hexagonal graphite crystal structures (i.e., pyrolytic graphite and boron nitride) results in preferred crystal orientation producing a high degree of directionality of properties. The hexagonal or layer plane alignment of the grains is essentially parallel to the substrate surface. Directionality results from strong atomic bonds within layer planes and weak bonds between layer planes, and also from the mode of heat transfer through the material, which is predominantly by lattice vibration.

Pyrolytic deposition of materials with face-centered or body-centered cubic structures (i.e., carbides or tungsten) results in relatively isotropic properties. Although such materials can have a high degree of crystal orientation, orientation does not necessarily produce directional properties.

Properties of Anisotropic Materials

Pyrolytic graphite and its compounds and boron nitride all have substantial directionality of properties. The ratio of the number of crystallites with layer planes parallel to the deposition surface (i.e., a axis) to the number normal to the surface (i.e., c axis) can be varied by process control. For example, in graphite, ratios may range from 100 to 1000 to 1, compared with ratios of 3 or 4 for some commercial graphites. Orientation obtained in boron nitride has been as high as 1900 to 1.

Conductivity

One of the most useful properties of pyrolytic graphite is its insulating ability. In the direction normal to the deposition surface, pyrolytic graphite is a better insulator than the most refractory ceramic materials. In addition, the thermal conductivity parallel to the deposition surface is comparable to the more *conductive* metals, tungsten and copper. This high conductivity evens out hot spots over the total surface.

The high destruction temperature of pyrolytic graphite combined with low conductivity normal to the surface allows the surface temperature to become very high. This cuts down heat absorbed by the component by reradiating heat back to the atmosphere, thus acting as a "hyperinsulator."

Tensile Strength Improved

Tensile strengths in the a axis are orders of magnitude higher than in the c axis. For example, at room temperature, pyrolytic graphite has an a-axis average tensile strength of about 95.2 MPa, compared with about 3.5 MPa in the c direction. The graphite–boron compound has room temperature tensile strengths of 112.2 and 4.9 MPa in the a and c axes; boron nitride has a- and c-axis tensile strengths of 84 and 4.5 MPa, respectively.

Pyrolytic graphite (like conventional graphite) offers the singular advantage of increasing in strength with increasing temperature. Preliminary data indicate that the c-axis strength decreases with temperature.

Addition of alloying elements has been found to improve c-axis strength. Additions of low concentration of less than 1% of tungsten and molybdenum have increased c-axis tensile strength by 50 to 90%.

Oxidation Resistance Improved

Pyrolytic graphite has somewhat greater oxidation resistance than normal graphite due largely to its imperviousness. Addition of boron improves oxidation resistance of pyrolytic graphite by a factor of 1. Oxidation resistance of boron nitride is superior to other pyrolytic materials produced to date specifically at temperatures below 2000°C.

Other Properties

Owing to the atom-by-atom deposition process, pyrolytic materials are all near theoretical density, are impervious, and have extremely high purity levels. The materials do exhibit substantial directionality of thermal expansion, which must be carefully considered in designing components.

Properties of Isotropic Materials

Performance data on the isotropic pyrolytic materials, tungsten, and the carbides of tantalum, hafnium, and columbium are much more limited than those for the anisotropic pyrolytic materials.

As mentioned before, although such materials are considered to be isotropic (in comparison with the anisotropic materials), crystal growth does tend to provide a preferred orientation in a plane normal to the deposition surface. Thus, generally speaking, the strength of such materials is greater in the plane normal to the surface than in the plane parallel to the surface.

Carbides

The pyrolytic deposition process can be controlled to produce carbides of varying metal-to-carbon ratio, resulting in carbides of differing microhardness. Following are Knoop hardness number (K_{100}) for three carbides, hardness increasing with increasing carbon-to-metal ratio:

TaC	1400 to 3500
HfC	2000 to 3600
NbC	1700 to 4000

The very hard grades of carbide are extremely brittle and difficult to handle. Their strength is low but they promise to be useful as thin, well-bonded coatings.

As hardness decreases, the ductility and strength of the carbides increase. Bend strengths, except for the high-hardness grade, were found to fall in the range of 68 to 272 MPa; a value as high as 1030 MPa was observed for tantalum carbide low in carbon.

Tungsten

Pyrolytic tungsten, which can now be produced in thicknesses over 6.4 mm as coatings on components to 457 mm in diameter, is being evaluated for missile application. The increased strength and lower impurity level of pyrolytic tungsten it believed to be a major factor permitting production of sound coatings on large rocket nozzles.

RADIANT HEATING

Traditionally, radiant heating tubes have been fabricated from Ni–Cr–Fe alloys and are welded, cast, or extruded. These materials have the benefit of possessing good heat/strength properties and are very easy to weld and handle.

The protective oxide layer in these alloys consists mainly of Cr_2O_3. This layer is easily displaced revealing a fresh surface of metal which, in turn, oxidizes. However, this results in a thinning of the tube walls and contamination of the tube interior by loose oxides.

The oxide scaling from the tube affects the heat transfer in a negative way, which could affect the function of the recuperator and burner. It might also be exhausted into the furnace atmosphere. The useful life of the tube is also limited through the carburization of the wall or by stress cracking due to reduced strength in the remaining material.

Another common material is mullite which is used for ceramic tubes. These are especially popular in the straight-through design as the material has a higher temperature capability than Ni–Cr–Fe. However, these ceramic tubes suffer from thermal shock sensitivity. Also, fabrication is more complicated than metallic materials since mullite cannot be welded to fit flanges and end closures.

SiC tubes are increasing in popularity, both in straight-through applications and in single-ended recuperative (SER) systems. The advantage of this material is the possibility of increasing load. However, the limitations are fragility and difficulties in fabrication and handling.

Combinations of metallic and SiC materials in SER systems exploit the benefits of both materials and offer a range of interesting possibilities.

Alloy Characterization

A metallic Fe–Cr–Al-based alloy (APM) with a higher maximum temperature, better loading capabilities and longer life than Ni–Cr alloys is being utilized for the production of radiant tubes for gas-fired heating systems. The material offers better ductility and thermal shock resistance than ceramic materials.

APM is a powder metal, dispersion strengthened ferrite alloy (Fe–21% Cr–5.7% Al) able to withstand a maximum temperature of 1250°C in tube form. The radiant tubes are seamless and extruded.

This alloy forms a thin adherent Al_2O_3 surface oxide, which provides a good protective layer in most corrosive atmospheres, especially those that have a high carbon potential or that contain sulfur, which is common in heat-treatment processes.

A particular advantage of the Al_2O_3 layer is that it does not scale and therefore there is no contamination of the burner or the product.

The high-temperature creep strength is also improved compared with conventional Fe–Cr–Al alloys. Therefore, the tubes can be used both vertically and horizontally, in straight-through as well as SER applications.

The majority of the tubes in use today are of the SER design, where the burner and exhaust system are placed on same side of the tube. With this solution, the exhaust gases are used to preheat the gas and air, which is required for combustion resulting in major improvements in system efficiency.

Other Alloys

There is a wide array of resistance material that easily covers the wide range of industrial requirements.

Nikrothal® and other nickel-based alloys: Ni–Cr–Fe formulations in wire, ribbon, strip, and foil form for temperatures up to 1200°C.

Cuprothal® alloys: Copper-based alloys, in wire and ribbon form for temperatures up to 600°C.

Kanthal SUPER®: Molybdenum disilicide elements covering a range of temperatures up to 1900°C. SUPER is ideal for the high heating power applications and increased furnace capacities, or atmospheres that preclude many metallic elements.

Superthal® modules: SUPER element mounted in vacuum-formed, ceramic fiber modules provide compact, high density heating power to 1550°C.

RAPID PROTOTYPING

The terms *agility* or *agile manufacturing* have been applied to the operation of forward-looking companies. These terms imply a number of characteristics about the organization and operation of a company that provide for quick response to changes in customer needs, and especially the ability to minimize the time between conception and marketing of new models or new products.

One of the most important tools of an agile manufacturing organization is rapid prototyping (RP). Rapid prototyping refers to the production of a part directly from a computer file generated with a computer-aided design program. While rapid prototyping is sometimes considered to include computer numerical control machining, in most cases it refers to building up a part from a stack of thin layers, each of which is patterned from the computer-generated three-dimensional model of the part — a process that is also called "solid free-form fabrication."

Rapid Tooling

Rapid prototyping is evolving from just a means of making prototypes to a technique for making production tooling, chiefly dies for plastic parts.

Initially, the prototypes were quite fragile and the various techniques produced products that could only show form. As materials improved and prototypes became stronger, the products could be tried for fit and measured. Today some systems can produce parts strong enough to run briefly in a machine and do low-volume production of parts. The good news is that there are a wide variety of rapid tooling systems; the bad news is no black-and-white answers exist on how to match a project with a system.

There are three levels of rapid tooling. Some users need only a few prototypes. Called soft tooling, these parts can usually be made from room-temperature-vulcanizing materials. The next, "bridge tooling," includes those production situations that require up to several hundred items to cover the time between early prototype and full production. The third type, hard tooling, employs tools for actual production.

Future

Recent developments aim at reducing the amount of postprocessing required for rapidly produced molds. Researchers are also experimenting with new materials such as thermoplastics, composites, and perhaps even cement. Bolstering this work are improvements in underlying rapid prototyping hardware, particularly lasers.

Other developments in rapid tooling focus on processing new materials in existing machines. For example, one experimental rapid-production method showing good results so far marries powder injection-metal feed stocks, from injection-metal molding methods, with a rapid prototyping machine. Powder-injection metals contain a polymer that lubricates and suspends the powder as it is deposited in the rapid prototyping machine.

Other efforts in rapid tooling concern the making of production molds out of ordinary SLA resins. Although SLA epoxies would melt at typical injection-molding temperatures, judicious cooling can let molds last long enough to make a few dozen pads (Table R.1).

RARE EARTH ELEMENTS/METALS

The rare earths are a closely related group of highly reactive metals comprising about one sixth of the known elements. The rare earths form a transition series including the elements of atomic number 57 through 71, all having three outer electrons and differing only in the inner electronic structure. Because chemical properties are determined by the outer electronic structure, it is evident why these metals are chemically alike. Although not truly members of this series, scandium and yttrium (atomic numbers 21 and 39, respectively) are frequently included with this grouping (shown in Table R.2, which shows the rare earth elements and the density, melting point, boiling point, and heat of vaporization).

Element	Atomic No.	Element	Atomic No.
Scandium	21	Gadolinium	64
Yttrium	39	Terbium	65
Lanthanum	57	Dysprosium	66
Cerium	58	Holmium	67
Praseodymium	59	Erbium	68
Neodymium	60	Thulium	69
Promethium	61	Ytterbium	70
Samarium	62	Lutetium	71
Europium	63		

TABLE R.1
Materials for SLA Machines

Developer and Build Style	Material Name	Characteristics	Applications
3D Systems SLA	SLA 5210 vinyl ether	Cures fast, resists humidity, water and elevated temperatures to 225°F	Under hood and wind tunnels
	SLA 5220 epoxy	Accurate, fast, humidity resistant	Multipurpose and master patterns
	SLA 5520 durable and flexible	Prototyping snap fits; good for room-temperature testing	
Stratasys Inc. Fused-deposition modeling	E20 elastomer	Mechanical strength and durability in flexible components; flex modulus of 20,000 psi	Seals, bushings, protective boots, impact-absorbing devices
DTM Corp. Selective laser sintering	RapidSteel 2.0	Long-run tooling	Durable metal mold inserts capable of more than 100,000 parts
	Copper polyamide	Short-run tooling	Metal-based mold inserts capable of about 200 parts from plastics such as polyethylene, polypropylene, and ABS
	DuraForm GF	Combines fine feature detail, smooth surface finish, and easy processing with mechanical integrity	Functional prototypes
DuPont Stereo-lithography	Somos 8120 epoxy	Fast photo speeds and broad processing latitudes; humidity and water resistant	Snap fit and functional tasks
	Somos 7110	Fast photo speeds; high or low humidity and almost no bubbles	For HeCd lasers

Source: Mach. Design, July 23, p. 62, 1998. With permission.

TABLE R.2
Some Rare Earth Metal Properties

Symbol	Melting Point, °C (°F)	Boiling Point, °C (°F)	Heat of Vaporization ($\Delta H_{v,0}$), kcal/g-atm	Density 25°C, g/cm³
Sc	1541	2831 (5128)	89.9	2.9890
Y	1522	3338 (6040)	101.3	4.4689
La	921 (1690)	3457 (6255)	103.1	6.1453
Ce	799 (1470)	3426 (6199)	101.1	6.672
Pr	931 (1708)	3512 (6353)	85.3	6.773
Nd	1021 (1870)	3068 (5554)	78.5	7.007
Pm	1168 (2134)	2700 (est.) (4892)	—	—
Sm	1077 (1971)	1791 (3256)	49.2	7.520
Eu	822 (1531)	1597 (2907)	(41.9) $\Delta H° = 29$	5.2434
Gd	1313 (2395)	3266 (5911)	95.3	7.9004
Tb	1356 (2473)	3123 (5653)	93.4	8.2294
Dy	1412 (2574)	2562 (4646)	70.0	8.5500
Ho	1474 (2685)	2695 (4883)	72.3	8.7947
Er	1529 (2874)	2863 (5145)	76.1	9.066
Tm	1545 (2813)	1947 (3537)	55.8	9.3208
Yb	819 (1506)	1194 (2181)	36.5	6.9654
Lu	1663 (3025)	3395 (6143)	102.2	9.8404

Source: McGraw-Hill Encyclopedia of Science and Technology, 8th ed., Vol. 15, McGraw-Hill, New York,, 206–207. With permission.

Scandium and yttrium occur together in nature with the rare earths and are similar in properties.

The rare earths are neither rare (even the scarcest are more abundant than cadmium or silver) nor earths. The term *earth* stems from the oxide mineral in which these elements were first discovered. More suitable names such as lanthanons (after lanthanum, the first member of the group) have been proposed, but the term *rare earths* persists.

Two Groups

This series of metals is frequently divided into two groups based upon atomic weight and chemical properties. The "light" rare earths consist of elements with atomic number 57 to 63 and may be called the cerium group. The "heavy" rare earths consist of elements 64 through 71 as well as scandium and yttrium because of similar chemical behavior. The metals of the heavy or yttrium group are more difficult to separate from each other and, as a result, have found commercial interest only recently. Other historical terms will be found in commercial usage and can lead to confusion. For example, *didymium* is not actually another metal in this series, but refers to the neodymium-rich rare earth mixture left after lanthanum and cerium are removed.

Properties

The rare earth elements are metals possessing distinct individual properties that make them potentially valuable as alloying agents. They are usually reduced thermally by treating the anhydrous hydride with calcium, lithium, or other alkali metals and then remelting under vacuum to volatilize the last traces of the reductant. They can also be reduced electrolytically from fused-salt baths.

Mechanical

For the various metals of 99.5% purity, the room-temperature ultimate tensile strength ranges from 103 to 272 MPa with elongations of 5 to 25%. Strength at 427°C is about one half that at room temperature. Therefore, the rare earths do not appear to offer any outstanding mechanical properties that would indicate their use as a base for structural alloys. Scandium has a density similar to aluminum, but the potential for high strength-to-density materials is unknown. Yttrium provides an interesting combination of properties: density similar to titanium, melting point of 1549°C, transparency to neutrons, and formation of one of the most stable hydrides.

Fabricability

The rare earths may be hot-worked, and some of them can be fabricated cold. Small arc-cast ingots of yttrium have been reduced 95% at room temperature. The metals are poor conductors of electricity. All the metals are paramagnetic, and some are strongly ferromagnetic below room temperature.

Oxidation Resistance

Yttrium and the higher atomic number rare earths maintain a typical metallic appearance at room temperature. Lanthanum, cerium, and europium oxidize rapidly under ordinary atmospheric conditions; the other metals of the light group form a thin oxide film. Some of the rare earths, notably samarium, form a stable protective oxide in air at temperatures up to at least 593°C.

Applications

Applications of the rare earths may be divided into two general categories: the long-established uses and the newer developments that frequently require the higher-purity separated elements. Three of the older applications that still account for three fourths of total output are rare earth–cored carbons for arc lighting; lighter flints, which are misch metal–iron alloys; and cerium oxide for polishing of glass and also salts for coloring or decolorizing of glass.

Yttrium oxide and cerium sulfide are high-melting refractories. Yttria-stabilized zirconia ceramic materials with excellent properties are commercially available.

Misch metal is used in making aluminum alloys, and in some steels and irons. In cast iron it opposes graphitization and produces a malleablized iron. It removes the sulfur and the oxides and completely degasifies steel. In stainless steel it is used as a precipitation-hardening agent. An important use of misch metal is in magnesium alloys for castings. From 3 to 4% of misch metal is used with 0.2 to 0.6% zirconium, both of which refine the grain and give sound castings of complex shapes. The cerium metals also add heat resistance to magnesium castings.

Ceria, cerium oxide, or ceric oxide, CeO_2, is used in coloring ceramics and glass fiber, producing distortion-free optical glass. It is used also for decolorizing crystal glass, but, when the glass contains titania, it produces a canary-yellow color.

Neodymium is used in magnesium alloys to increase strength at elevated temperatures, and is used in some glasses to reduce glare. Neodymium glass, containing small amounts of neodymium oxide, is used for color television filter plates because it transmits 90% of the blue, green, and red light rays and no more than 10% of the yellow. It thus produces truer colors and sharper contrasts in the pictures and decreases the tendency toward gray tones. Neodymium is also a dopant for yttrium–aluminum–garnet, or YAG, lasers as well as for glass lasers.

Lanthanum oxide, La_2O_3, is a white powder used for absorbing gases in vacuum tubes. Lanthanum boride, LaB_6, is a crystalline powder used as an electron emitter for maintaining a constant, active cathode surface. It has high electric conductivity.

Dysprosium has a corrosion resistance that is higher than that of other cerium metals. It also has good neutron-absorption ability, with a neutron cross section of 1100 barn. The metal is paramagnetic. It is used in nuclear-reactor control rods, in magnetic alloys, and in ferrites for microwave use. It is also used in mercury vapor lamps. With argon gas in the arc area, it balances the color spectrum and gives a higher light output. Samarium has a higher neutron cross section, 5500 barn, and is used for neutron absorption in reactors.

Ytterbium metal is produced in lumps and ingots. Yttrium is more abundant in nature than lead, but is difficult to extract. It is found associated with the elements 57 to 71, although its atomic number is 39. It is the lightest of the cerium metals except scandium. The metal is corrosion-resistant to 400°C. It has a hexagonal close-packed crystal structure. Ytterbium oxide, Yb_2O_3, and yttrium oxide, Y_2O_3, are the usual commercial forms of these metals.

Rare earths are also used in the petroleum industry as catalysts.

Another very important use of individual rare earths is in the manufacture of solid-state microwave devices widely used in radar and communications systems. Yttrium–iron garnets are especially good, because they transmit shortwave energy with low energy losses. The devices, however, are very small, so the total use of rare earths is not large.

Still another important use of individual rare earths is in the construction of lasers. A high percentage of the patent applications for new lasers involves the use of a rare earth as the active constituent of the laser.

The rare earths show interesting magnetic properties. Alloys of cobalt with the rare earths, such as cobalt–samarium, produce permanent magnets that are far superior to most of the varieties on the market, and many uses for these magnets are developing.

Many phosphors contain rare earths, and barium–phosphate–europium phosphor finds applications in x-ray films that form satisfactory images with only half the exposure time of conventional x-rays.

Yttrium–aluminum–garnets (YAG) are used in the jewelry trade as artificial diamonds. Single crystals of YAG have a very high refractive index, similar to diamond, and when these crystals are cut in the form of diamonds, they sparkle in the same manner as a diamond. Also, they are very hard, so that they scratch glass, and only an expert can tell the difference between a YAG diamond and a real one.

RARE GASES

Also known as inert gases in the metallurgical industry, and as noble gases, rare gases is a general name applied to the five elements helium, neon, argon, krypton, and xenon. They are rare in that they are highly rarified gases at ordinary temperatures and are found dissipated in minute quantities in the atmosphere and in some substances. All have zero valence and normally form no chemical combinations. The rare gases are colorless, odorless, and tasteless at ambient temperatures. However, they exhibit very different properties when cooled to extreme low temperatures. When saturated helium, or helium-4, is cooled to 2.17 K it becomes a superfluid. One unique property of superfluids is the ability to pass undetected through very small openings. Helium-3, the unsaturated counterpart, differs in that it is magnetic.

Types

Neon

Neon is also used in voltage-regulating tubes for radio apparatus, and will respond to low voltages. In television the neon lamp will give fluctuations from full brilliancy to total darkness as many as 100,000 times a second. Colored electric advertising signs are often referred to as neon signs, but the colors other than orange are produced by different gases.

Argon

Argon is obtained by passing atmospheric nitrogen over red-hot magnesium, forming magnesium nitride and free argon. It is also obtained by separation from industrial gases. Argon is employed in incandescent lamps to give increased light and to prevent vaporization of the filament, and is used instead of helium for shielding electrodes in arc welding and as an inert blanket for nuclear fuels.

Krypton

Krypton, which occurs in the air to the extent of 1 part in 1 million, is a heavy gas used as a filler for fluorescent lamps to decrease filament evaporation and heat loss and to permit higher temperatures in the lamp. Krypton-85, obtained from atomic reactions, is a beta-ray emitter used in luminous paints for activating phosphors and also as a source of radiation.

Xenon

Xenon, another gas occurring in the air to the extent of 1 part in 11 million, is the heaviest of the rare gases. When atomic reactors are operated at high power, xenon tends to build up as a reaction product, poisoning the fuel and reducing the reactivity. Xenon lamps for military use give a clear white light known as sunlight plus north-sky light. This color does not change with the voltage, and thus the lamps require no voltage regulators. Xenon is a mild anesthetic; the accumulation from air helps to induce natural sleep, but it cannot be used in surgery since the quantity needed produces asphyxiation.

RARE METALS

These are metals that are rare in the sense that they are difficult to extract and are rare and expensive commercially. They include the elements astatine, technetium, and francium. The silvery metal technetium, element 43, has been produced by bombardment of molybdenum with neutrons. Although radium is a widely distributed metal, it is classed as a rare metal. All of the ultraheavy metallic elements, such as plutonium, which are produced synthetically, are classed as rare metals. They are called transuranic metals because they are above the heavy metal uranium in weight. They are all radioactive.

Element 99, called einsteinium, was originally named ekaholmium because it appears to have chemical properties similar to holmium. It is produced by bombarding uranium-238 with stripped nitrogen atoms. It decays rapidly to form the lighter berkelium, or element 97. Neptunium (element 93), californium (element 98), and illinium (element 61), are also made atomically. The latter also has the names florentium and promethium.

Plutonium is made from uranium-238 by absorption of neutrons from recycled fuel. The metal, 99.8% pure, is obtained by reduction of plutonium fluoride, PuF_4, or plutonium chloride, $PuCl_3$. Plutonium-238 has a low radiation level and is used as a heat source for small water-circulating heat exchangers for naval undersea diving suits.

Plutonium-241 emits beta and gamma rays. Because all the allotropic forms are radioactive, it is a pure nuclear fuel in contrast to uranium, which is only 0.7% directly useful for fission. It is thus necessary to dilute plutonium for control. For fuel elements it may be dispersed in stainless steel and pressed into pellets at about 871°C, or pellets may be made of plutonium carbide. Plutonium–iron alloy, with 9.5% iron, melts at 410°C. It is encased in a tantalum tube for use as a reactor fuel. Plutonium–aluminum alloy is also used.

Element 102, called nobelium, has a half-life of only 12 min. Other transuranic metals produced synthetically are americium (element 95), and curium (element 96). Curium is used as a heat source in remote applications. Curium 244 is obtained as curium nitrate in the reprocessing of spent reactor fuel. It is converted to curium oxide. The byproduct americium is used as a component in neutron sources. Other transuranic metals that have been produced by nuclear reactions and synthesis include fermium (element 100), mendelevium (element 101), lawrencium (element 103), rutherfordium or kurchatovium (element 104), and hahnium or nielsbohrium (element 105).

RAYON

Rayon is a general name for artificial-silk textile fibers or yarns made from cellulose nitrate, cellulose acetate, or cellulose derivatives. In general, the name *rayon* is limited to the viscose, cuprammonium, and acetate fibers, or to fibers having a cellulose base. Other synthetic-fiber groups have their own group names, such as azlon for the protein fibers and nylon for the polymeric amine fibers, in addition to individual trade names.

Viscose rayon is made by treating the cellulose with caustic soda and then with carbon disulfide to form cellulose xanthate, which is dissolved in a weak caustic solution to form the viscose.

Rayons manufactured by the different processes vary both chemically and physically. They are resistant to caustic solutions that would destroy natural silk. They are also mildewproof, durable, and easily cleaned. But they do not have the permeability and soft feel of silk. The acetate rayons are more resistant than the viscose or cuprammonium. The lack of permeability of the fibers is partly overcome by having superfine fibers so that the yarns are permeable.

REACTION-INJECTION MOLDING

Reaction-injection molding (RIM) is a process in which two or more liquid-chemical components from separate tanks are metered through high-pressure supply lines to a chamber. They mix in the chamber via high-velocity impingement. The resulting mixture is then injected into a mold at low pressure, about 0.48 MPa, and at low temperature, about 66°C. This leads to a lower tooling cost and less lead time than for most other molding processes.

The process is similar in concept to an ordinary two-part epoxy mix. RIM reactants become a low-viscosity, exothermic, expanding material that readily flows into a mold. The mold-clamping machine typically is equipped with electronics that turn the mold, so that material more easily flows into all cavities. When a successful series of movements is perfected, they are stored as a program to ensure uniform molding of the part.

As the mold fills, the reactants are still releasing heat while tiny bubbles swell the plastic into the finest details of the mold. Cooling coils built into the mold transfer heat to maintain an ideal working temperature. The reactants quickly harden and parts can be removed in a few minutes.

RIM polyurethane has an integral, low-density, cellular core and a solid, high-density skin. RIM structural foam has a high strength-to-weight ratio and excellent chemical resistance. A variety of shapes can be molded this way without molded-in stresses.

The process accommodates rapid thickness variations, without sink marks. Substrates of wood, steel, or other materials can be encapsulated for fillers or for increased strength. Available RIM material systems are varied and range from flexible to rigid to structural composite. In general, RIM best serves in larger, thick-walled (6.4 mm or more) structural parts.

RED BRASS

A series of copper alloys including one wrought alloy (C23000) and several cast alloys (C83300 to C84800). Red brass C23000 contains 84 to 86% copper and the balance zinc except for small amounts (0.05% maximum) of lead and iron. The cast alloys, which include leaded red brasses, have a zinc content that can range from 1.0 to 2.5% (C83500) to 13 to 17% (C84800), and all but C83400, which is limited to 0.20% tin, contain substantial amounts of this element: 1.0 to 6.5%, depending on the alloy. Lead content of the cast alloys can be as little as 0.50% maximum (C83400) or as much as 8.0% (C84400). Some of these alloys also contain about 1% nickel, and most contain smaller amounts of other metals.

C23000 is available in most wrought mill forms and is used for condenser and heat-exchanger tubing, plumbing lines, electrical conduit, fasteners, and architectural and ornamental applications. It is quite ductile in the annealed condition, and can be appreciably hardened by cold work.

Standard gilding metal (C21000), used for making cheap jewelry and small-arms ammunition, contains 94 to 96% copper. It has a golden-red color, is stronger and harder than copper, but has only about half the electrical conductivity.

REFRACTORIES, SPECIALTY

These are ceramic materials used in high-temperature structures or equipment. The term *high temperatures* is somewhat indefinite but usually means above about 1000°C, or temperatures at which, because of melting or oxidation, the common metals cannot be used. In some special high-temperature applications, the so-called refractory metals such as tungsten, molybdenum, niobium, and tantalum are used.

Materials, usually ceramics, are employed where resistance to very high temperature is required, as for furnace linings and metal-melting pots. Materials with a melting point above 1580°C are called refractory, and those with melting points above 1790°C are called highly refractory. However, in addition to the ability to resist softening and deformation at the operating temperatures, other factors are considered in the choice of a refractory, especially load-bearing capacity and resistance to slag attack and spalling. Heat transfer and electrical resistivity are sometimes also important. Many of the refractories are derived directly from natural minerals, but synthetic materials are much used. To manufacture refractory products, powders of the raw materials are mixed and usually dry-pressed to form the desired shape.

The greatest use of refractories is in the steel industry, where they are used for construction of linings of equipment such as blast furnaces, hot stoves, and open-hearth furnaces. Other important uses of refractories are for cement kilns, glass tanks, nonferrous metallurgical furnaces, ceramic kilns, steam boilers, and paper plants. Special types of refractories are used in rockets, jets, and nuclear power plants. Many refractory materials, such as aluminum oxide and silicon carbide, are also very hard and are used as abrasives; some applications, for example, aircraft brake linings, use both characteristics.

Refractory materials are commonly grouped into (1) those containing mainly aluminosilicates; (2) those made predominately of silica; (3) those made of magnesite, dolomite, or chrome ore, termed basic refractories (because of their chemical behavior); and (4) a miscellaneous category usually referred to as special refractories.

Aluminosilicate Refractories

Fireclay is the raw material from which the bulk (about 70%) of refractories is manufactured. Different grades are distinguished according to the softening temperature or the pyrometric cone equivalent (PCE), the number of the standard pyrometric cone that deforms under heat treatment in the same manner as the fireclay. Thus, the minimum PCEs for low, intermediate, high, and superduty fireclays are 19, 29, 31/32, and 33, respectively. Fireclays are also classified by their working properties into two other classes: plastic (those that form a moldable mass when mixed with water) and flint (a hard, rocklike clay that does not become plastic when mixed with water). In general, flint clays have higher PCEs than plastic clays and are mixed with them to form higher-grade fireclay brick.

High-alumina refractories are made from clays that contain, in addition to the alumina (Al_2O_3) in the clay minerals, hydrates of aluminum oxide. These raise the total Al_2O_3 content and make the material more refractory.

Different grades are distinguished on the basis of the total Al_2O_3 content (50, 60, 70% alumina refractories).

Sillimanite and kyanite are anhydrous aluminosilicate minerals used to make special refractory objects, such as crucibles, tubes, and muffles, or as an addition to fireclay to increase refractoriness and to control its shrinkage during firing.

Silica Refractories

These account for about 15% of total production. They are made from crushed and ground quartzite (ganister) to which about 2% lime has been added to assist in bonding, both before and after firing. The quality of silica refractories is to a great extent determined by the amount of Al_2O_3 impurity; even small amounts have a deleterious effect on refractoriness. This is just opposite to the case of alumina in fireclays, where a higher alumina content means greater refractoriness. High-grade silica brick contains less than 0.6% Al_2O_3, and even the standard grade contains less than 1%. During firing, the mineral quartz transforms to cristobalite and tridymite, the high-temperature forms of silica.

The outstanding characteristic of silica is its ability to withstand high loads at elevated temperatures, for example, as a sprung-arch roof 9 to 12 m wide over an open hearth. The hearth may be operated within (50°C) of the melting point of silica.

Basic Refractories

Magnesite refractories are so named because magnesium carbonate mineral was for many years the sole raw material. Since World War II seawater has become a significant source of magnesium oxide refractory, and such material is often called seawater magnesite. In any case, the raw material is calcined to form a material largely magnesium oxide, MgO; about 5% iron oxide is usually added before calcining.

Chrome refractories are made from chrome ore, a complex mineral containing oxides of chromium, iron, magnesium, aluminum, and other oxides crystallized in the spinel structure. These crystals are usually embedded in a less refractory matrix called gangue.

In an attempt to combine the best properties of each, magnesite and chrome are often mixed to form chrome-magnesite or magnesite-chrome refractories (the first-named is the dominant constituent).

Dolomite is a mixed calcium–magnesium carbonate, $CaMg(CO_3)_2$, which, when calcined to a mixture of MgO and CaO, is used in granular form to patch the bottoms of open hearths and also to make bricks.

Miscellaneous Materials

Special refractories are made of a great many materials, and it is possible to mention here only a few of the more important.

Silicon carbide, SiC, is used for many refractory shapes; its outstanding properties are good thermal and electrical conductivity (it is used to make electric heating elements for furnaces), good heat-shock resistance, strength at high temperatures, and abrasion resistance. The first silicon carbide refractories were bonded with clay, so that the refractory properties of the bond placed the ultimate limit on the material. A method of making self-bonded silicon carbide has been developed to remove this limitation. Although silicon carbide tends to oxidize to form SiO_2 and either CO or CO_2, the silica-oxidation product forms a glassy coating on the remaining material and to a certain extent protects it from further oxidation.

Insulating firebrick is made from refractory clays to which a combustible material (sawdust, cork, coal) has been added; when this burns during the firing operation, it leaves a brick of high porosity.

The low thermal conductivity of insulating brick reduces heat losses from furnaces, and the low bulk density and consequent low heat capacity reduce the amount of heat needed to bring the furnace itself up to temperature. The main disadvantage of such bricks is their low strength, but even this is useful in that they can be cut or ground to shape quite readily.

Pure oxides, of which alumina, Al_2O_3, is the prime example, are used for many special refractories. Zircon, $ZrSiO_2$, and zirconia, ZrO_2, are finding increased and significant uses as refractory materials. Some, such as beryllia, BeO, thoria, ThO_2, and uranium oxide, UO_2, are of particular interest for nuclear applications.

Carbides, nitrides, borides, silicides, and sulfides of various sorts have been considered refractory materials, and some study has been made of them; aside from a few carbides and nitrides, however, none has found much use.

Cermets are an intimate mixture of a metal and a nonmetal, for example, Al_2O_3 and chromium.

Although the nonmetal may be an oxide, it is more commonly a carbide or nitride (as in cemented tungsten carbide).

Manufacture

Standard ceramic techniques are used. Hand molding, once widely used, is used only for special shapes and small orders. The extrusion or stiff mud process is used for plastic fireclays; very often the extruded blanks are repressed or hydraulically rammed to form special shapes, for example, T-sections of refractory pipe. Power pressing of simple shapes is the most widely used forming method. Hot pressing and hydrostatic pressing are used for some special refractories. Slip casting is used for special refractory shapes. Fusion casting is commonly used for glass tank blocks; these are mainly either Al_2O_3 or Al_2O_5, with significant amounts of SiO_2, ZrO_2, or both.

Refractories are generally fired in tunnel kilns, but some periodic kilns are still used, particularly for special shapes.

Some types of basic refractories, known as chemically bonded, are pressed with a chemical binder, such as magnesium oxychloride, and installed without firing. Some of these, the steel-clad refractories, are encased in a metal sheath at the time of pressing. When the refractory is heated after installation, the iron oxidizes and reacts with the refractory, forming a tight bond between the individual bricks.

In all refractory products and in unfired brick in particular, the maximum possible formed density is desired. To this end, careful crushing and sizing of raw materials are carried out so that, as far as possible, the gaps between large pieces are filled with smaller particles, and the space between these with still smaller, and so on. In the case of clay refractories, it is customary to use prefired (calcined) clay or crushed, fired rejects (both are known as grog) to increase the density and to reduce the firing shrinkage.

Properties

A high melting point is, of course, necessary in a refractory, but many other properties must be considered in choosing a refractory for a specific application.

A definite melting point is characteristic of pure materials; however, actual minerals from which refractories are made, for example, clay, are far from pure and hence do not melt at a specific temperature. Rather, they form increasing amounts of liquid as the temperature is increased above a certain minimum temperature at which liquid first appears. This characteristic of gradual softening is indicated by the PCE of the material and the underload test.

High-temperature strength is important for refractories, but most materials become plastic and flow at elevated temperatures. Therefore, the rate of flow (creep rate) at a given temperature under a given load is a more important design criterion.

Thermal conductivity determines the amount of heat that will flow through a furnace wall under given conditions, and a knowledge of this property is essential to furnace design.

Thermal-shock resistance is the ability of a specimen to withstand, without cracking, a difference in temperature between one part and another. For example, if a red-hot brick is dropped into cold water, it is likely to shatter since the outside cools and contracts while the center is still hot. This cracking is often referred to as thermal spalling, the term *spalling* meaning any cracking off of large pieces of brick. Other causes of spalling are mechanical (hitting the brick and knocking off a piece) and structural (a reaction in the brick that changes the mineral structure and causes cracking). Thermal-shock resistance is enhanced by high strength, low Young's modulus, low thermal expansion, and sometimes, depending on conditions, high thermal conductivity. Whether or not a given specimen cracks under heat shock depends not only on the material of which it is made, but also on its size and shape and on the test conditions, for example, whether it is dropped into water or into still air at the same temperature.

Various chemical properties are important in refractories. For example, the tendency of the magnesium oxide in basic brick to hydrate, that is, to react with water to form $Mg(OH)_2$, should be as low as possible. Turning to high-temperature chemistry, the rate of corrosion of refractories by molten slags and iron oxide fumes is vital to the length of service.

Carbon deposition is another chemical reaction that affects the life of refractories. The reaction is not with the refractory, but is catalyzed by substances in it. When carbon monoxide, perhaps in the top of a blast furnace, comes in contact with certain iron compounds, which can occur in fireclays, its reduction to carbon is catalyzed. This carbon deposits at the site of the catalyst in the brick, and causes the brick to shatter.

REFRACTORY CEMENT

A large proportion of the commercial refractory cements used for furnace and oven linings and for fillers are fireclay–silica–ganister mixtures with a refractory range of 1427 to 1538°C. Cheaper varieties may be mixtures of fireclay and crushed brick, fireclay and sodium silicate, or fireclay and silica sand. An important class of refractory cements is made of silicon carbide grains or silicon carbide-fire sand with clay bonds or synthetic mineral bonds. The temperature range of these cements is 1482 to 1871°C. Silicon carbide cements are acid resistant and have high thermal and electric conductivity. For crucible furnaces the silicon carbide cements are widely used except for molten iron. Alumina and alumina–silica cements are very refractory and have high thermal conductivity. Calcined kaolin, diaspore clay, mullite, sillimanite, and combinations of these make cements that are neutral to most slags and to metal attacks. They are electrical insulators. Chrome-ore cements are difficult to bond unless mixed with magnesite. A chrome–magnesite cement is made of treated magnesite and high-grade chrome ore. It sets quickly and forms a hard, dense structure. The melting point is above 1982°C. It is used particularly for hot repairs in open-hearth furnaces.

Zircon–magnesite cement is made with 25% refined zircon sand, 10% milled zircon, 15% fused magnesia, and 50% low-iron dead-burned magnesite bonded with sodium silicate. A wide range of refractory cements of varying compositions and characteristics is sold under trade names, and these are usually selected by their rated temperature resistance.

Carbofrax cement is silicon carbide with a small amount of binder in various grades for temperatures from 871 to 1760°C, depending on the fineness and the bond. Firefrax cement is an aluminum silicate, sometimes used in mixtures with ganister for lining furnaces. It is for temperatures to 1649°C. Alfrax cement is fused silica also in various grades for temperatures from 899 to 1816°C.

REFRACTORY HARD METALS

Refractory hard metals (RHMs) are a ceramic-like class of materials made from metal–carbide particles bonded together by a metal matrix. Often classified as ceramics and sometimes called cemented or sintered carbides these metals were developed for extreme hardness and wear resistance.

These are true chemical compounds of two or more metals in the form of crystals of very high melting point and high hardness. Because of their ceramic-like nature, they are often classified as ceramics, and they do not include the hard metallic carbides, some of which, with metal binders, have similar uses; nor do they include the hard cermets. The refractory hard metals may be single large crystals, or crystalline powder bonded to itself by recrystallization under heat and pressure. In general, parts made from them do not have binders, or contain only a small percentage of stabilizing binder. The intermetallic compounds, or intermetals, are marketed regularly as powders of particle size from 150 to 325 mesh for pressing into mechanical parts or for plasma-arc deposition as refractory coatings, and the powders are referred to chemically, such as borides, beryllides, and silicides. The oxides and carbides of the metals are also used for sintering and for coatings, and the oxides are called cermets.

The RHMs are more ductile and have better thermal shock resistance and impact resistance than ceramics, but they have lower compressive strength at high temperatures and lower operating temperatures than most ceramics. Generally, properties of RHMs are between those of conventional metals and ceramics. Parts are made by conventional powder-metallurgy compacting and sintering methods.

Many metal carbides such as SiC and BC are not RHMs but are true ceramics. The fine distinction is in particle bonding: RHMs are always bonded together by a metal matrix, whereas ceramic particles are self-bonded. Some ceramics have a second metal phase, but the metal is not used primarily for bonding.

RHM System

Tungsten Carbide

Tungsten carbide with a 3 to 20% matrix of cobalt is the most common structural RHM. The low-cobalt grades are used for applications requiring wear resistance; the high-cobalt grades serve where impact resistance is required.

Tantalum Carbide/Tungsten Carbide

Tantalum carbide and tungsten carbide combined in a matrix of nickel, cobalt, and/or chromium provide an RHM formulation especially suited for a combination of corrosion and wear resistance. Some grades are almost as corrosion resistant as platinum. Nozzles, orifice plates, and valve components are typical uses.

Titanium Carbide

Titanium carbide in a molybdenum and nickel matrix is formulated for high-temperature service. Tensile and compressive strengths, hardness, and oxidation resistance are high at 1093°C. Critical parts for welding and thermal metalworking tools, valves, seals, and high-temperature gauging equipment are made from grades of this RHM.

Tungsten–Titanium Carbide

Tungsten–titanium carbide ($WTiC_2$) in cobalt is used primarily for metal-forming applications such as draw dies, tube-sizing mandrels, burnishing rolls, and flaring tools. The $WTiC_2$ is a gall-resistant phase in the RHM containing tungsten carbide as well as cobalt.

Borides

Zirconium boride is a microcrystalline gray powder of the composition ZrB_2. When compressed and sintered to a specific gravity of about 5.3, it has a Rockwell A hardness of 90, a melting point of 2980°C, and a tensile strength of 241 to 276 MPa. It is resistant to nitric and hydrochloric acids, to molten aluminum and silicon, and to oxidation. At 1204°C it has a transverse rupture strength of 379 MPa. It is used for crucibles and for rocket nozzles.

Chromium boride occurs as very hard crystalline powder in several phases: the CrB orthorhombic crystal, the hexagonal crystal Cr_2B, and the tetragonal crystal Cr_3B_2. Chromium boride parts produced by powder metallurgy have a specific gravity of 6.20 to 7.31, with a Rockwell A hardness of 77 to 88. They have good resistance to oxidation at high temperatures, are stable to strong acids, and have high heat-shock resistance up to 1316°C. The transverse rupture strength is from 552 to 931 MPa. CrB is used for oil-well drilling. A sintered material, used for gas-turbine blades, contains 85% CrB with 15% nickel binder. It has a Rockwell A hardness of 87 and a transverse rupture strength of 848 MPa.

Molybdenum boride, Mo_2B, has a specific gravity of 9.3, a Knoop hardness of 1660, and a melting point of about 1660°C. Tungsten boride, W_2B, has a specific gravity of 16.7 and a melting point of 2770°C. Titanium boride, TiB_2, is light in weight with a specific gravity of 4.5. It has a melting point at about 2593°C. Molded parts made from the powder have a Knoop hardness of 3300 and a flexural strength of 241 MPa, and they are resistant to oxidation to 982°C with a very low oxidation rate above that point to about 1371°C. They are inert to molten aluminum. Intermetal powders of beryllium–tantalum, beryllium–zirconium, and beryllium–columbium are also marketed, and they are lightweight and have high strength. Sintered parts resist oxidation to 1649°C.

Disilicides

Molybdenum disilicide, $MoSi_2$, has a crystalline structure in tetragonal prisms, and has a Knoop hardness of 1240. The decomposition point is above 1870°C. It can be produced by sintering molybdenum and silicon powders or by growing single crystals from an arc melt. The specific gravity of the single crystal is 6.24. The tensile strength of sintered parts is 276 MPa and compressive strength is 2296 MPa. The resistivity is 29 $\mu\Omega \cdot cm$. It is used in rod form for heating elements in furnaces. The material is brittle, but can be bent to shape at temperatures above 1093°C.

Kanthal Super is an $MoSi_2$ rod. In an inert atmosphere the operating temperature is 1600°C. Furnace gases containing active oxygen raise the operating temperature to about 1700°C, while gases containing active hydrogen lower it to about 1350°C.

Tungsten disilicide, WSi_2, is not as hard and not as resistant to oxidation at high temperatures, but has a higher melting point, 2050°C.

Others (Nitrides, Aluminides)

Titanium nitride, TiN, is a light-brown powder with a cubic lattice crystal structure. Sintered bars are extremely hard and brittle, with a hardness above Mohs 9 and a melting point of 2950°C. It is not attacked by nitric, sulfuric, or hydrochloric acid, and is resistant to oxidation at high temperatures. In recent years, titanium-nitride coatings have been used to markedly extend the life of tool-steel cutters and forming tools. The coatings, golden in color, are deposited by chemical or physical vapor deposition. Aluminum nitride, AlN, when molded into shapes and sintered, forms a dense, nonporous structure with a hardness of Mohs 6. It resists the action of molten iron or silicon to 1704°C, and molten aluminum to 1427°C, but is attacked by oxygen and carbon dioxide at 760°C. At least 1% oxygen causes AlN properties to deteriorate rapidly. However, if oxygen is removed, AlN is highly thermally conductive, lightweight, and a good insulator. A newly developed process has removed oxygen concentration from AlN down to 0.5%, and subsequent nitriding and a self-purification reaction has produced commercial-grade AlN.

Nickel aluminide is a chemical compound of the two metals, and when molded and sintered into shapes, has good oxidation resistance and heat resistance at high temperatures. It is a sintered nickel–aluminide material with a specific gravity of 5.9, and a transverse rupture strength of 1034 MPa at 1093°C, twice that of cobalt-bonded titanium carbide at the same temperature. The melting point is 1649°C. It resists oxidation at 1093°C. It is used for highly stressed parts in high-temperature equipment. This compound in wire form is used for welding, flame coating, and hard surfacing. An aluminum powder coated with nickel may be mixed with zirconia or alumina, which will increase the hardness and heat resistance of the nickel–aluminide coating. Columbium aluminide is used as a refractory coating as it is highly resistant for long periods at 1427°C. Tin aluminide is oxidation resistant to 1093°C, but a liquid phase forms at about this point.

Tribaloy intermetallic materials are composed of various combinations of nickel, cobalt, molybdenum, chromium, and silicon. For example, one composition is 50% nickel, 32% molybdenum, 15% chromium, and 3% silicon. Another is 52% cobalt, 28% molybdenum, 17% chromium, and 3% silicon. Supplied as alloy-metal powder, welding rod, or casting stock, they can be cast, deposited as a hard-facing surface, plasma-sprayed, or consolidated by powder metallurgy. The materials have exceptional wear and corrosion resistance properties in corrosive media and in air up to 1093°C. Typical applications are in pumps, valves, bearings, seals, and other parts for chemical process equipment. Also, the materials are suited for marine and saltwater applications and for parts subject to wear in atomic energy plants.

REFRACTORY METALS

Various new techniques are being utilized to produce refractory metals and their alloys. One of these processes is powder injection molding (PIM). Over 250 PIM operations serve a diverse range of applications worldwide: cemented carbides for wear and cutting applications, tungsten or molybdenum heat sinks, niobium rocket nozzles, heavy alloy projectiles and radiation shields, and titanium golf clubs. It is interesting to note the frequent PIM successes involving components with about 100 dimensions, such as a wristwatch case.

PIM. Currently, complex-shaped parts of tungsten–iron–nickel with 93% tungsten are also being produced by PIM. Special efforts are focused on the sintering process to develop high density and keep tight tolerances in the final components.

PAS. Plasma-activated sintering (PAS) as a means to consolidate tungsten powders is another effective process to produce dynamic compression properties.

VPS. Vacuum plasma spray (VPS) forming/sintering as a viable fabrication alternative to near-net shape manufacturing of fully dense components of W–3.5% Ni–1% Fe, Ta–10% W, Mo–40% Re, and Nb–1% Zr alloys. Because the process allows direct shape-making in vacuum, it results in much lower oxygen and moisture pickup than conventional powder–metallurgy consolidation methods.

LPPS. Low-pressure plasma spray (LPPS) techniques for depositing tungsten coatings on metallic, ceramic, and composite substrates is another novel technique.

SANS. Tungsten and molybdenum and their ongoing critical role have dominated lighting technology since the development of potassium-doped tungsten as a special dispersion-strengthened system, in which hardening centers are soft particles in the form of potassium-filled bubbles. Such bubbles act as a barrier against boundary migration, and are essential for the mechanical stability of elongated grains in the doped (or nonsag) tungsten wire. A new method, based on small angle neutron scattering (SANS), has been developed to determine shape distribution and the state of deformation of second-phase dispersoids. Correlation between the results of the SANS measurements and metallographical examination has proved SANS to be suitable for characterizing the morphology of the bubbles in doped tungsten.

REINFORCED CONCRETE

This is portland cement concrete containing higher-strength, solid materials to improve its structural properties. Generally, steel wires or bars are used for such reinforcement, but for some purposes glass fibers or chopped wires have provided desired results.

Unreinforced concrete cracks under relatively small loads or temperature changes because of low tensile strength. The cracks are unsightly and can cause structural failures. To prevent cracking or to control the size of crack openings, reinforcement is incorporated in the concrete. Reinforcement can also be used to help resist compressive forces or to improve dynamic properties.

Steel usually is used in concrete. It is elastic, yet has considerable reserve strength beyond its elastic limit. Under a specific axial load, it changes in length only about one tenth as much as concrete. In compression, steel is more than ten times stronger than concrete, and in tension, more than 100 times stronger.

As reinforcement for concrete, steel may be used in the form of bars or rods, wire, fibers, pipe, or structural shapes such as wide-flange beams.

Bars are the most commonly used form. The size of a bar usually is specified by a number that is about eight times the nominal diameter; sizes range from No. 2, a nominal 0.6 cm bar, to No.18, a nominal 5.7 cm bar.

During construction, the bars are placed in a form and then concrete from a mixer is cast to embed them. After the concrete has hardened, deformation is resisted and stresses are transferred from concrete to reinforcement by friction and adhesion along the surface of the reinforcement. The bonding may be improved mechanically by giving reinforcing bars raised surfaces; such bars are known as deformed bars. They are generally manufactured to standard specifications, such as those of the American Society for Testing and Materials (ASTM).

Similarly, deformed wires used as reinforcement provide greater bond than smooth wires. Deformed wires are designated by D followed by a number equal to 100 times the nominal area in square inches.

Individual wires or bars resist stretching and tensile stress in the concrete only in the direction in which such reinforcement extends. Tensile stresses and deformations, however, may occur simultaneously in other directions. Therefore, reinforcement must usually be placed in more than one direction. For this purpose, reinforcement sometimes is assembled as a rectangular grid, with clips or welded joints at the intersections of the wires or bars. For example, prefabricated wire grids, called welded-wire fabric, often are used for slab reinforcement in highway pavement and in buildings.

Under some conditions, fiber-reinforced concrete is an alternative to such arrangements. The fiber, made of fine steel wires, glass fibers, or plastic threads, is embedded in the concrete in short lengths, often only about 2.5 cm. They are added to the concrete mixer along with the other ingredients. To achieve crack control in all directions, the fibers should be uniformly spaced, at close intervals, and randomly oriented throughout the hardened concrete. Such reinforcement also improves concrete tensile strength, ductility, and dynamic properties.

RESIN

Originally a category of vegetable substances soluble in ethanol but insoluble in water, resin in modern technology is generally an organic polymer of indeterminate molecular weight. The class of flammable, amorphous secretions of conifers or legumes are considered true resins, and include kauri, copal, dammar, mastic, guaiacum, jalap, colophony, shellac, and numerous less well-known substances. Water-swellable secretions of various plants, especially the Burseraceae, are called gum resins and include myrrh and olibanum The official resins are benzoin, guaiac, mastic, and resin. Other natural resins are copal, dammar, dragon's blood, elaterium, lac, and sandarac. The natural vegetable resins are largely polyterpenes and their acid derivatives, which find application in the manufacture of lacquers, adhesives, varnishes, and inks.

The synthetic resins, originally viewed as substitutes for copal, dammar, and elemi natural resins, have a large place of their own in industry and commerce. Phenol-formaldehyde, phenolurea, and phenol-melamine resins have been important commercially for a long time. Any unplasticized organic polymer is considered a resin, thus nearly any of the common plastics may be viewed as a synthetic resin. Water-soluble resins are marketed chiefly as substitutes for vegetable gums and in their own right for highly specialized applications. Carboxymethylcellulose, hydroxyalkylated cellulose derivatives, modified starches, polyvinyl alcohol, polyvinylpyrrolindone, and polyacrylamides are used as thickening agents for foods, paints, and drilling muds, as fiber sizings, in various kinds of protective coatings, and as encapsulating substances.

Oleoresins are natural resins containing essential oils of the plants. Gum resins are natural mixtures of true gums and resins and are not as soluble in alcohol. They include rubber, gutta percha, gamboge, myrrh, and olibanum. Some of the more common natural resins are rosin, dammar, mastic, sandarac, lac, and animi. Fossil resins, such as amber and copal, are natural resins from ancient trees, which have been chemically altered by long exposure. The synthetic resins differ chemically from natural resins, and few of the natural resins have physical properties that make them suitable for mechanical parts.

RESISTANCE HEATING

Resistance heating is the generation of heat by electric conductors carrying current. The degree of heating for a given current is proportional to the electrical resistance of the conductor. If the resistance is high, a large amount of heat is generated, and the material is used as a resistor rather than as a conductor.

Resistor Materials

In addition to having high resistivity, heating elements must be able to withstand high temperatures without deteriorating or sagging. Other desirable characteristics are low-temperature coefficient of resistance, low cost, formability, and availability of materials. Most commercial resistance alloys contain chromium or aluminum or both, because a protective coating of chrome oxide or aluminum oxide forms on the surface upon heating and inhibits or retards further oxidation. Some commercial resistor materials are listed in Table R.3.

Heating Element Forms

Because heat is transmitted by radiation, convection, conduction, or combinations of these, the form of element is designed for the major mode of transmission. The simplest form is the helix, using a round wire resistor, with the pitch of the helix approximately three wire diameters. This form is adapted to radiation and convection and is generally used for room or air heating. It is also used in industrial furnaces, utilizing forced convection up to about 650°C. Such helices are stretched over grooved high-aluminum refractory insulators and are otherwise open and unrestricted. These helices are suitable for mounting in air ducts or enclosed chambers, where there is no danger of human contact.

TABLE R.3
Electric Furnace Resistor Materials

35% nickel, 20% chromium[a]
60% nickel, 16% chromium[a]
68% nickel, 20% chromium, 1% cobalt[a]
78% nickel, 20% chromium[a]
15% chromium, 4.6% aluminum[a]
22.5% chromium, 4.6% aluminum[a]
22.5% chromium, 5.5% aluminum[a]
Silicon carbide
Platinum
Molybdenum[b]
Tungsten[b]
Graphite[b]

[a] Balance is largely iron, with 0.5–1.5% silicon.
[b] Usable only in pure hydrogen, nitrogen, helium, or oxygen or in vacuum because of inability to form protective oxide.

Source: McGraw-Hill Encyclopedia of Science and Technology, 8th ed., Vol. 15, McGraw-Hill, New York, 426. With permission.

For such applications as water heating, electric range units, and die heating, where complete electrical isolation is necessary, the helix is embedded in magnesium oxide inside a metal tube, after which the tube is swaged to a small diameter to compact the oxide and increase thermal conductivity. Such units can then be formed and flattened to desired shapes. The metal tubing is usually copper for water heaters and stainless steel for radiant elements, such as range units. In some cases the tubes may be cast into finned aluminum housings, or fins may be brazed directly to the tubing to increase surface area for convection heating.

Modification of the helix for high-temperature furnaces involves supporting each turn in a grooved refractory insulator, with the insulators strung on stainless alloy rods. Wire sizes for such elements are 5 mm in diameter or larger, or they may be edge-wound strap. Such elements may be used up to 980°C furnace temperature.

Another form of furnace heating element is the sinuous grid element, made of heavy wire or strap or casting and suspended from refractory or stainless supports built into the furnace walls, floor, and roof.

Silicon carbide elements are in rod form, with low-resistance integral terminals extending through the furnace walls.

Direct Heating

When heating metal strip or wire continuously, the supporting rolls can be used as electrodes, and the strip or wire can be used as the resistor.

Molten Salts

The electrical resistance of molten salts between immersed electrodes can be used to generate heat. Limiting temperatures are dependent on decomposition or evaporization temperatures of the salt. Parts to be heated are immersed in the salt. Heating is rapid and, since there is no exposure to air, oxidation is largely prevented. Disadvantages are the personnel hazards and discomfort of working close to molten salts.

Major Applications

A major application of resistance heating is in electric home appliances, including electric ranges, clothes dryers, water heaters, coffee percolators, portable radiant heaters, and hair dryers. Resistance heating has also found application in home or space heating; some homes are designed with suitable thermal insulation to make electric heating practicable.

RHENIUM AND ALLOYS

Rhenium (symbol Re) is a rare, silvery-white element with a metallic luster. It is one of the hardest metals at 580 HV (30% cold work), is one of the heaviest with a density of 21 g/cm^3, and has one of the highest melting points at 5760°C. It also has no known ductile-to brittle transition temperature, which enables rhenium parts to be thermally cycled without being degraded. It remains strong at high temperatures, and does not become brittle after welding. Although it is a very ductile material, it has the third highest modulus of elasticity of any metal. The addition of rhenium to molybdenum and tungsten has enhanced their strength, ductility, and formability.

The crystal structure is closely packed hexagonal, making it more difficult to work than the cubic-structured tungsten, but the crystal grains are tiny, and small amounts of rhenium added to tungsten improve ductility and increase high-temperature strength of tungsten used in lamp filaments and wire.

During the past decade, interest has increased dramatically in rhenium as a construction material for parts with complex shapes in high-temperature service environments. A typical example of such an application is rocket thrusters, which have operating temperatures up to 2230°C. Other applications include welding rods, thermocouples, and cryogenic magnets. When rhenium is added to superalloys in amounts of 3 to 6%, the turbine-inlet temperature in gas turbine engines is increased by 125°C.

Processing

The primary source of rhenium is the by-product molybdenite obtained from porphyry copper deposits.

Virtually all of the wrought products of pure rhenium are produced from bars made by powder metallurgy. Although rhenium can be arc-melted in an inert atmosphere, the resultant metal is not well suited for fabrication due to the coarse as-melted grain size and the possible segregation of small amounts of rhenium oxide at the grain boundaries. Even with vacuum melting, the workability has not equaled that of bars produced by powder metallurgy. The latter bars have a small grain size and are much more amenable to fabrication.

Rhenium powder produced by hydrogen reduction of ammonium perrhenate is consolidated by pressing in split rectangular dies at pressures of 25 to 30 tsi using nonlubricated powder. These pressed bars are vacuum-presintered at 1200°C and a pressure of 0.5 to 1 μ. They are subsequently resistance-sintered in dry hydrogen at a maximum temperature of approximately 90% of the melting point in a manner almost identical to that used for tungsten. The sintered bars are normally 90 to 96% of the theoretical density, having undergone 15 to 20% shrinkage in each dimension. These bars are then ready for fabrication.

Pure rhenium is fabricated by cold working with frequent intermediate recrystallizing anneals. Hot working of rhenium results in "hot short" failures, probably because of the presence of rhenium heptoxide at the grain boundaries. This volatile oxide has a melting point of 297°C and a boiling point of 363°C.

The fabrication of rhenium wire requires processing by swaging and drawing. Work hardening is so great in cold swaging that reductions are limited to one 10% reduction between anneals to avoid serious damage to the swaging dies. The reductions are somewhat greater in cold drawing beginning with 10% but increasing to about 40% between anneals. Diamond drawing dies are used for all pure rhenium wire drawing.

The production of rhenium strip is likewise accomplished by cold rolling. After several light initial reductions on the sintered bars, reductions up to 40% can be effected between anneals. Strip thinner than 0.13 mm is rolled using small-diameter tungsten carbide work rolls in a "4 high" mill. Strip can be processed in this manner to thicknesses of 0.03 mm.

Because of the rapid rate of work hardening that accompanies the cold working of rhenium, annealing constitutes a very important part of the fabrication processes. Frequent anneals are necessary at temperatures of 1550 to 1700°C for times varying from 10 to 30 min in an atmosphere of dry hydrogen or a hydrogen–nitrogen mixture. Furnaces utilizing molybdenum heating elements and aluminum oxide (Al_2O_3) refractories are used most extensively. Anneals performed as indicated result in complete recrystallization with the grain size usually between 0.010 and 0.040 mm and a hardness of about 250 to 275 VHN.

It is virtually impossible to apply conventional ingot casting and hot/warm forming techniques to rhenium because of its high melting point, poor oxidation resistance, and the difficulty of working an as-cast structure.

Rhenium and the alloys shown in Table R.4 are generally manufactured in the United States and Europe through the powder-metallurgy route, with a tolerance on the rhenium addition in the range of ±0.5% or better. Various methods have been chosen to suppress second phase formation. In Russia, vacuum melting is the preferred production process for alloys such as Mo–47% Re and W–27% Re, which have rhenium content tolerances of ±3%.

In alloys W–3 to 5% Re for wire products, the tungsten-base material is intentionally doped with potassium, aluminum, and silicon to control recrystallization behavior. Some European companies and Russians also manufacture wire of these alloys in the undoped condition. In recent years, W–3 to 5% Re, and up to 10% Re undoped compositions have been found to be ideal for fabricating x-ray targets, heat sinks, and other complex shapes.

Better performance can be achieved by additions of 11 to 14% rhenium to molybdenum, and additions of 10, 15, and 20% rhenium to tungsten, as so-called "dilute or moderate alloys." Mo–41 to 44% Re is the basis of another group of alloys for consideration.

TABLE R.4
Mechanical Properties of Rhenium and Its Alloys

Parameters	Rhenium	Mo–47.5Re	W-3Re	W-5Re	W-25Re
Young's modulus, GPa (ksi × 10³), at 20°C/68°F	461 (66.9); 464 (67.3) 469ᴿ (68.1); 460 (66.7)–510 (74); 406 (58.9); 431 (62.5)	357ⱽ (51.8); 346ⱽ (50.2) 314ᶜ (45.6)/299 (43.4)–314ᶜⱽ (45.6); 361 (52.4)/367ᴿ (53.3)	403 (58.5)	411 (59.6) 400 (58.0)	410ᶜⱽ (59.4) 430 (62.4); 370ˢ (53.6)/431 (62.5)
Shear modulus, GPa (ksi × 10³), at 20°C / 68°F	176; 179 (25.5; 25.9)	137ⱽ, 132ⱽ	—	—	159ᶜⱽ (23.1)
Poisson's ratio	0.49; 0.296	0.285ᶜ	(19.9; 19.1)	—	0.3 0.29ᶜ
Tensile strength, MPa (ksi), at 20°C / 68°F					
Deformed	1800–2400 (261–348)	2450–3900ⱽ (355.5–566)	—	—	2750–4900ᶜⱽ (399–711)
Annealed	700–1300 (101.5–188.5)	980–1180ⱽ (142–171)	—	—	1370ᶜⱽ (198.8)
Elongation, % at 20°C (68°F)					
Deformed	1–3	1.5–3ⱽ	—	—	1–2ᶜⱽ
Annealed	10–25	20–25ⱽ	—	—	20ᶜⱽ
Tensile strength, MPa (ksi)					
At 1500°C / 2730°F	360 (52.2)	147ⱽ (21.3)	—	—	330ᶜⱽ (47.8)
At 1800°C / 3270°F	200 (29.0)	59ⱽ (8.5)	—	—	196ᶜⱽ (28.4)
Impact strength, MPa (ksi)	—	2.45ⱽ (0.35)	—	—	1.76ᶜⱽ (0.25)

Note; R = recrystallized (otherwise, material in as wrought condition), S = sintered P/M (otherwise, worked P/M), V = vacuum melted, C = material with composition Mo–50Re and W–27Re. Values are based on results of an extensive search of the literature. For details of these values, please contact the author.

Source: Adv. Mater. Proc., 156(6), 127, 1999. With permission.

Properties

Rhenium is a refractory metal with properties similar in many ways to tungsten and molybdenum. Pure rhenium has significantly greater room-temperature ductility combined with good high-temperature strength. When used as an alloying addition with tungsten or molybdenum forming body-centered cubic solid solutions, rhenium produces significant improvements in the ductility of those metals.

Mechanical properties of rhenium and rhenium alloys are given in Table R.4.

It is significant to note that the melting point of rhenium is second highest of the metals, exceeded only by tungsten. Its density is surpassed only by osmium, iridium, and platinum and its modulus of elasticity only by osmium and iridium. Moreover, the hardness and strength of rhenium are very high and its rate of work hardening is greater than that for any other metal.

The combination of high strength and ductility exhibited by rhenium and rhenium alloys is of great interest. Furthermore, these materials can be inert-arc-welded with a resultant weld that is ductile and possesses corrosion resistance as good as the parent metal. No ductile-to-brittle transition is observed in pure rhenium or the rhenium–molybdenum alloys containing 30 to 50 wt% Re as occurs in molybdenum at about room temperature and in tungsten at approximately 300°C. Although rhenium additions lower the transition temperature of tungsten, it still occurs above room temperature. However, even below the transition temperature for the recrystallized material, the ductility of the rhenium–tungsten alloys is higher than that of pure tungsten.

Fabrication

Rhenium cannot be machined by conventional methods because it work-hardens rapidly. However, electrical discharge machining is an effective manufacturing method, and parts can be brought to final finish through ceramic or diamond grinding. Rhenium can be plastically cold-formed, although frequent annealings are required to restore formability after relatively small deformation.

Chemical vapor deposition is also a practical method for fabricating rhenium components with complex shapes. This method is particularly advantageous for products with thin metal structures, and for applications that require an oxidation-resistant protective layer on the rhenium surface. In addition, recently developed multiform powder-metallurgy products have the advantage of being isotropic, and they can be produced with a high relative density and consistent mechanical properties from lot to lot by well-established technologies.

Thermal spray is another alternative to the conventional powder-metallurgy route. Specifically, the cold gas dynamic spray method (CGSM) compacts rhenium powder as a coating onto a suitable substrate by ballistic impingement. Unfortunately, little success has been achieved because of the morphology and small particle size of rhenium metal powder.

Low-pressure plasma spray (LPPS) experiments have produced tantalum forms with thickness of 6 mm and a relative density exceeding 95%. These experiments indicate that LPPS may be a realistic way to produce near-net shape components of refractory metals, including rhenium. A preliminary attempt to make rhenium tubing resulted in a relative density of 83.5%, which was increased to over 95% after sintering. The prospects for manufacturing high-density isotropic products by plasma spraying of rhenium appear to be very good, although it is difficult to produce rhenium powder with good flow properties.

Among the newest technologies, the directed light fabrication (DLF) process is a multidimensional, laser-controlled deposition technique. Good results in casting complex near-net shape products have been achieved for metals with moderate melting points. The unidirectional solidification of a small molten pool makes this a single-step, almost waste-free process without risk of oxidation. Further studies of the DLF process are encouraged, to develop the processing technique for refractory metals, and to study the final properties of its products.

A long-standing effort has been made to apply the favorable combination of mechanical properties (high strength, outstanding plasticity, and formability) exhibited by rhenium at ambient temperature to pressure-shape the required profile from semifinished products such as sheet. Rhenium tube-spinning technology has recently emerged as a developmental effort.

Another example is the trend toward adopting explosive metalworking techniques for joining and fabricating rhenium parts. However, because rhenium work-hardens extraordinarily fast at low levels of cold deformation, frequent intermediate annealing constitutes a major and important part of such fabrication processes. As an interesting alternative, the appropriate shape design has been achieved by winding rhenium wire in layers, and filling the interwire void space with fine-grain particles by chemical vapor deposition.

Direct hot isostatic pressing (HIP) of near-net shape rhenium products is an intense plastic-deforming technique applied to powder that has not been explored in much detail. Little data have been published on the effect of HIP on material properties of rhenium powder-metallurgy ingots and mill products. It is known that a combination of sintering and HIP procedures has demonstrated the possibility of developing both high density and good formability in powder-metallurgy rhenium parts obtained after further machining.

One of the challenges with direct HIP consolidation of near-net geometry is to reach near-full density of rhenium at temperatures far below typical sintering temperatures. During the development of the technique, careful evaluation of process parameters such as pressure, temperature, time, and heating/cooling rates may make it possible to control grain size, material properties, and dimensional variations. Because of the need to control so many parameters, success may depend on the nascent technology of intelligent HIP, in which the response of the material to processing parameters dictates the selection of optimal processing conditions.

APPLICATIONS

The outstanding properties of rhenium and rhenium alloys suggest their use in many specialized applications.

Thermocouples

Rhenium vs. tungsten thermocouples can be used for temperature measurement and control to approximately 2200°C, whereas previous thermocouple use was limited to temperatures below 1750°C. Indeed, 74% W–26% Re vs. W thermocouples can be used to temperatures of at least 2750°C with a high emf output ensuring accurate precise temperature measurements with excellent reproducibility and reliability.

Electronic

Rhenium is now widely used for filaments for mass spectrographs and for ion gauges for measuring high vacuum. Its ductility, chemical properties, including the fact that it does not react with carbon to form a carbide, and its emission characteristics make it superior to tungsten for these applications.

The elevated temperature properties of rhenium and rhenium alloys, their weldability, and electrical resistivity suggest their use for various components in electronic tubes.

Electrical

The superconducting properties of Re–Mo alloys has prompted their use for coils in compact-size electromagnets with high field strength. Although this application is still in the development stages it looks most promising.

Rhenium has received considerable acclaim as an electrical contact material. It possesses excellent resistance to wear as well as arc erosion. Furthermore, the contact resistance of rhenium is extremely stable because of its good corrosion resistance in addition to the fact that possible formation of an oxide film on the contacts would not cause any appreciable change in the contact resistance since the resistivity of the oxide is almost the same as that of the metal. Extensive tests for some types of make-and-break switching contacts has shown rhenium to have 20 times the life of platinum–palladium contacts currently in use.

Heating elements for resistance heated vacuum or inert atmosphere furnaces seems to be a potential application for rhenium and especially rhenium–tungsten alloys. These materials would not undergo the embrittlement that occurs in tungsten upon heating, and they would allow for usage in vacuum, hydrogen, or inert atmospheres.

Welding Filler Rod

Rhenium and Re–Mo alloys can be readily welded, permitting fabrication of the welds. In fact, Re–Mo alloy can be used as a filler material for obtaining ductile welds in molybdenum. Although not all of the problems associated with the heat-affected zone are readily overcome, the properties of welds made with Re–Mo alloy filler wire are very much better.

RHODIUM

Metallic rhodium is the whitest of the platinum metals and does not tarnish under atmospheric conditions. This rare metal, (symbol Rh), found in platinum ores, is very hard and is one of the most infusible of the metals. The melting point is 1963°C. It is insoluble in most acids, including aqua regia, but is attacked by chlorine at elevated temperatures and by hot fuming sulfuric acid. Liquid rhodium dissolves oxygen, and ingots are made by argon-arc melting. At temperatures above 1200°C, rhodium reacts with oxygen to form rhodium oxide, Rh_2O_3. The specific gravity is 12.44. Rhodium is used to make the nibs of writing pens, to make resistance windings in high-temperature furnaces, for high-temperature thermocouples, as a catalyst, and for laboratory dishes. It is the hardest of the platinum-group metals; the annealed metal has a Brinell hardness of 135. Rhodium also has considerable strength and rigidity, ultimate tensile strengths ranging from 952 to 2068 MPa, and tensile modulus from 28.9×10^4 to 37.9×10^4 MPa, depending on condition or hardness. Rhodium is also valued for electroplating jewelry, electric contacts, hospital and surgical instruments, and especially reflectors.

The most important alloys of rhodium are rhodium–platinum. They form solid solutions in any proportion, but alloys of more than 40% rhodium are rare. Rhodium is not a potent hardener of platinum but increases its high-temperature strength. It is easily workable and does not tarnish or oxidize at high temperatures. These alloys are used for thermocouples and in the glass industry.

RIVET

A rivet is a short rod with a head formed on one end. A rivet is inserted through aligned holes in two or more parts to be joined; then by pressing the protruding end, a second head is formed to hold the parts together permanently. The first head is called the manufactured head and the second the point. In forming the point, a hold-on or dolly bar is used to back up the manufactured head and the rivet is driven, preferably by a machine riveter. For high-grade work such as boiler-joint riveting, the rivet holes are drilled and reamed to size, and the rivet is driven to fill the hole completely. Structural riveting uses punched holes.

Small rivets (11 mm and under) are for general-purpose work with head forms as follows: flat, countersunk, button, pan, and truss (Figure R.1). These rivets are commonly made of rivet steel, although aluminum and copper are used for some applications. The fillet under the head may be up to 0.8 mm in radius.

Large rivets (13 mm and over) are used for structural work and in boiler and ship construction with heads as follows: roundtop countersunk, button (most common), high button or acorn, pan, cone (truncated), and flattop countersunk.

Boiler rivets have heads similar to large rivets with steeple (conical) added but have different proportions from large rivet heads in some cases.

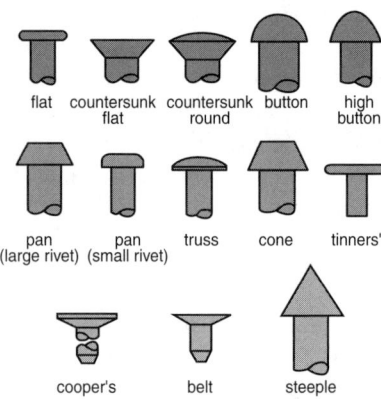

FIGURE R.1 Standard rivet heads. (From *McGraw-Hill Encyclopedia of Science and Technology*, 8th ed., Vol. 15, 589. With permission.)

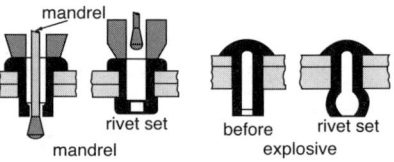

FIGURE R.2 Two types of blind rivet. (From *McGraw-Hill Encyclopedia of Science and Technology*, 8th ed., Vol. 15, 589. With permission.)

Special-purpose rivets are tinners' rivets, which have flat heads for use in sheet-metal work; cooper's rivets, which are used for riveting hoops for barrels, casks, and kegs; and belt rivets, used for joining belt ends.

Blind rivets are special rivets that can be set without access to the point. They are available in many designs but are of three general types: screw, mandrel, and explosive (Figure R.2). In the mandrel type the rivet is set as the mandrel is pulled through. In the explosive type an explosive charge in the point is set off by a special hot iron; the explosion expands the point and sets the rivet.

Standard material for rivets is open-hearth steel (containing manganese, phosphorus, and sulfur) with tensile strength of 310 to 380 MPa. Standards include acceptance tests for cold and hot ductility and hardness. Materials for some special-purpose rivets are aluminum and copper. In the past four decades aluminum and titanium rivets have been used in the aerospace industry very effectively.

ROBOTICS

Robotics is the field of engineering concerned with the development and application of robots, as well as computer systems for their control, sensory feedback, and information processing. There are many types of robotic devices, including robotic manipulators, robot hands, mobile robots, walking robots, aids for disabled persons, telerobots, and microelectromechanical systems.

The term *robotics* has been very broadly interpreted. It includes research and engineering activities involving the design and development of robotic systems. Planning for the use of industrial robots in manufacturing or evaluation of the economic impact of robotic automation can also be viewed as robotics. This breadth of usage arises from the interdisciplinary nature of robotics, a field involving mechanisms, computers, control systems, actuators, and software.

ROBOTIC MECHANISMS

Robots produce mechanical motion that, in most cases, results in manipulation or locomotion. For example, industrial robots manipulate parts or tools to perform manufacturing tasks such as material handling, welding, spray painting, or assembly; automated guided vehicles are used for transport of materials in factories and warehouses. Telerobotic mechanisms provide astronauts with a large manipulator for space applications. Walking machines have applications in hazardous environments. Mechanical characteristics for robotic mechanisms include degrees of freedom of movement, size and shape of the operating space, stiffness and strength of the structure, lifting capacity, velocity, and acceleration under load. Performance measures include repeatability and accuracy of positioning, speed, and freedom from vibration.

MANIPULATOR GEOMETRIES

An important design for robotic manipulators is the articulated arm, a configuration with rotary joints that resembles a human arm. An extension of this idea is the selective compliance assembly robot arm (SCARA) configuration, which has been extensively applied to assembly in manufacturing. Manipulators with this configuration have high stiffness during vertical motions. Cylindrical configurations with a rotary joint at the base and a prismatic joint at the shoulder are useful for simple material transfer and assembly. Overhead gantry (Cartesian) robots are used for high-payload applications and those requiring rapid positioning over large work spaces.

DEGREES OF FREEDOM

Some manipulators have simple mechanical designs involving only two or three degrees of freedom of movement. Most robotic manipulators, however, have at least six degrees of freedom so that the device can position a part or approach a part with any desired position or orientation. The wrist is positioned at any x, y, z position in the work space. Then, the end effector is rotated to a desired orientation (roll, pitch, yaw). In effect, the wrist represents the origin of a three-axis coordinate fixed to the gripper. Moving the first three points of the arm translates this origin to any point in a three axis coordinate system fixed to the work space; motion of the final three joints (in the wrist) orients the gripper coordinate system in rotation about an origin at the wrist point. Some robots have wrist-joint axes that do not intersect at a single point. However, the mathematics of mechanical motion (kinematics) is considerably simplified if these axes do intersect at a point.

TYPES OF JOINTS

Robotic mechanisms can have joints that are prismatic (linear motion), articulated (rotary motion), or a combination of both types. Although many robots use only articulated joints to imitate the human arm, limited actions can be produced by using prismatic joints alone. Robot locomotion can also be obtained by the use of wheels and treads. Walking robots have been developed that use articulated legs.

JOINT ACTUATORS

Actuators for moving joints of a robotic mechanisrn are typically electric or hydraulic motor. Although actuators can be placed directly at the joints, the weight and bulk of these motors and associated gear transmissions limit the performance of the robot, particularly at the wrist of industrial robots. Another design involves placement of the actuators in the base of the robot and transmission of motion to the joints through mechanical linkages such as shafts, belts, cables, or gears. This approach overcomes many of the problems associated with locating actuators at the joints, but requires the design of backlash-free mechanical linkages that can transmit power effectively through the arm in all of its positions and orientations. Other devices include end effectors, robotic hands, and mobile robots.

RUBBER

This term originally referred to a natural or tree rubber, which is a hydrocarbon polymer of isoprene units. With the development of synthetic rubbers having some rubbery characteristics but differing in chemical structure as well as properties, a more general designation was needed to cover both natural and synthetic rubbers. The term *elastomer*, a contraction of the words elastic and polymer, was introduced, and defined as a substance that can be stretched at room temperature to at least twice its original length and, after having been stretched and the stress removed, returns with force to approximately its original length in a short time.

Three requirements must be met for rubbery properties to be present in both natural and synthetic rubbers; long threadlike molecules, flexibility in the molecular chain to allow flexing and coiling, and some mechanical or chemical bonds between molecules.

A useful way of visualizing rubber structure is to consider as a model a bundle of wiggling snakes in constant motion. When the bundle is stretched and released, it tends to return to its original condition. If there were no entanglements, stretching the bundle would pull it apart. The more entanglements, the greater the tendency to recover, corresponding to cross-links in rubber. In rubbers the characteristic property of reversible extensibility results from the randomly coiled arrangement of the long polymer chains. Upon extension, the chains are elongated in a more or less orderly array. The tendency to revert to the original coiled disarray upon removal of the stress accounts for the elastic behavior. Vulcanizing rubber increases the number of cross-links and improves the properties.

Natural rubber and most synthetic rubbers are also commercially available in the form of latex, a colloidal suspension of polymers in an aqueous medium. Natural rubber comes from trees in this form and many synthetic rubbers are polymerized in this form; some other solid polymers can be dispersed in water.

Latexes are the basis for a technology and production methods completely different from the conventional methods used with solid rubbers.

Processing

In the crude state, natural and synthetic rubbers possess certain physical properties that must be modified to obtain useful end products. The raw or unmodified forms are weak and adhesive. They lose their elasticity with use, change markedly in physical properties with temperature, and are degraded by air and sunlight. Consequently, it is necessary to transform the crude rubbers by compounding and vulcanization procedures into products that can better fulfill a specific function.

Curatives and Vulcanization

After the addition of curing or vulcanizing agents to rubber, the application of heat causes cross-linking, yielding a durable product by binding the long chains together. Sulfur, the first successful curing ingredient, is the basis of vulcanization. However, various chemicals or combinations of chemicals are also capable of vulcanizing rubber. These include oxidizing agents, such as selenium, tellurium, organic peroxides, and nitro compounds, and also generators of free radicals, such as organic peroxides, azo compounds, and certain organic sulfur compounds such as the alkyl thiuram disulfides.

Sulfur alone results in very slow cure rates and develops less than optimum physical properties in the rubber. Several classes of compounds have been found that accelerate the rate of vulcanization, improve the efficiency of the vulcanization reaction, and reduce the sulfur requirements, in general enhancing the physical properties of the vulcanized rubber.

Accelerators are substances that act as catalysts in the vulcanization process, initiating free radicals. The first accelerators to be used in the rubber industry were inorganic chemicals, such as the basic carbonates and oxides of lead supplemented by magnesium or lime.

Most accelerators can be grouped into the following general classifications: aldehyde-amines, guanidines, thiuram sulfides, thiazoles, thiazolines, dithiocarbamates, and mercaptoimidazolines.

Some rubbers can be vulcanized by gamma radiation.

Pigments

Vulcanization improves the elasticity and aging properties of rubber, but in most cases it is necessary to enhance further such properties as tensile strength, abrasion resistance, and tear resistance by incorporation of fillers. Those that improve specific physical properties are known as reinforcing fillers; those that serve primarily as diluents are classed as inert fillers. The physical properties of the resulting vulcanizates are affected by both the type and amount of filler.

The most universally used reinforcing filler in the rubber industry is carbon black. The types used commercially in the greatest bulk for this purpose include furnace and thermal blacks.

In addition to the carbon blacks, inorganic reinforcing agents, such as zinc oxide and the silicas, are used for reinforcement of light-colored end products.

SPECIFIC TYPES OF RUBBER

One rubber is obtained from natural sources in commercial quantities. In addition, a number of synthetic compounds are classified as rubbers.

Smoked sheet and pale crepe represent the forms in which the major portion of natural rubber is commercially available.

Natural Rubber (NR)

Although natural rubber may be obtained from hundreds of different plant species, the most important source is the rubber tree. Natural rubber is *cis*-1,4-polyisoprene, containing approximately 5000 isoprene units in the average polymer chain.

The economic competition from synthetic rubbers has stimulated research and development in natural rubber by increasing productivity in the field, improving uniformity and quality of the product and packaging, and developing modified natural rubbers with specific properties.

Increased productivity has been achieved by increasing the yield of the trees by cross-pollination of high-rubber-yielding clones, use of chemical stimulants, and better tapping and collection methods. These methods have resulted in large increases in yields on Malaysian estates, with much greater increases anticipated.

With improved productivity and better processing methods and controls, much better quality and uniformity have been achieved. This has made possible the development of standard Malaysian rubbers (SMR) to meet specifications on a number of properties, including dirt and ash content, viscosity, and copper and manganese content.

In addition, experimental work has been done on improved collection of latex by micro-tapping and collecting in polyethylene bags, and trees having three parts: a high-yielding trunk system grafted onto a strong root system and an improved, more prolific leaf system grafted onto the trunk.

Despite competition from synthetic rubbers and a great reduction in percentage use worldwide of natural rubber, the tonnage of natural rubber continues a steady growth in parallel with the growth of the industry.

Styrene–Butadiene Rubbers (SBR)

The extensive development of the synthetic rubber industry originated with the World War II emergency, but continued expansion has been the result of the superiority of the various synthetic rubbers in certain properties and applications. The most important synthetic rubbers and the most widely used rubbers in the entire world are the styrene–butadiene rubbers (SBR).

Formerly designated GR-S, SBRs are obtained by the emulsion polymerization of butadiene and styrene in varying ratios.

When SBR is used in light-duty tires such as passenger car tires, the cold-rubber compounds have proved equal or superior to natural rubber treads. However, they are inferior to natural rubber for truck tires because of the greater heat buildup during flexing. The cold, oil-extended type, prepared by the replacement of a portion of the polymer by a heavy-traction oil, accounts for more than 50% of the cold-rubber production. Its advantage is primarily economic. Master batches, containing both oil and carbon black, have the advantage of process simplification.

In general, the compounding and processing methods for all the SBR types are similar to those of natural rubber. Although natural rubber is superior with respect to lower heat buildup, resilience, and hot tear strength, the SBR types are more resistant to abrasion and weathering. However, carbon black or some other reinforcing fillers must be added to the SBR to develop the best physical properties. Unlike natural rubber, SBR does not crystallize on stretching and thus has low tensile strength unless reinforced. The major use for SBR is in tires and tire products. Other uses include belting, hose, wire and cable coatings, flooring, shoe products, sponge, insulation, and molded goods.

cis-1,4-Polybutadiene (BR)

Work on the duplication of natural rubber stimulated interest in stereoregulated polymerization of butadiene, particularly of the high cis-1,4 structure.

Black-loaded vulcanizates of cis-1,4-polybutadiene rubbers exhibit good physical properties, such as lower heat generation, higher resilience, improved low-temperature properties, and greatly improved abrasion resistance. Processing properties are rather poor, but can be greatly improved by employing blends with natural rubber or SBR. Tests on retreaded passenger tires gave outstanding abrasion resistance and increased resistance to cracking as compared with natural rubber. In passenger tires, cis-1,4-polybutadiene improves treadwear by about 1% for each percent of polybutadiene in the tread compound. In truck tires, a blend with natural rubber gives 14% more wear than natural rubber alone.

Butyl Rubber

Isobutylene and isoprene or butadiene obtained from cracked refinery gases are the primary raw materials required for the manufacture of butyl rubber.

Neither neoprene nor butyl rubber requires carbon black to increase its tensile strength, but the reinforcement of butyl rubber by carbon black or other fillers does improve the modulus and increases the resistance to tear and abrasion.

The excellent resistance of butyl rubbers to oxygen, ozone, and weathering can be attributed to the smaller amount of unsaturation present in the polymer molecule. In addition, these rubbers exhibit good electrical properties and high impermeability to gases. The high impermeability to gases results in use of butyl as an inner liner in tubeless tires. Other widespread uses are for wire and cable products, injection-molded and extruded products, hose, gaskets, and sealants, and where good damping characteristics are needed.

Ethylene-Propylene Polymers (EPM, EPDM)

Stereospecific catalysts are employed to make synthetic rubbers by the copolymerization of ethylene and propylene. Either monomer alone polymerizes to a hard, crystallizable plastic, but copolymers containing 35 to 65% of either monomer are amorphous, rubbery solids. Special catalysts must be employed because ethylene polymerizes many times faster than propylene. Best results seem to be obtained with complex catalysts derived from an aluminum alkyl and a vanadium chloride or oxychloride.

Processing techniques and factory equipment used with other rubbers can also be applied to these copolymers. The mechanical properties of their vulcanizates are generally approximately equivalent to those of SBR.

Terpolymers containing ethylene, propylene, and a third monomer, such as dicyclopentadiene, have become more popular because they contain unsaturation and thus may be sulfur-cured by using more or less conventional curing systems.

Neoprene (CR)

One of the first synthetic rubbers used commercially to the rubber industry, neoprene is a polymer of chloroprene, 2-chlorobutadiene-1,3. In the manufacturing process, acetylene, the basic raw material, is dimerized to vinylacetylene and then hydrochlorinated to the chloroprene monomer.

Sulfur is used to vulcanize some types of neoprene, but most of the neoprenes are vulcanized by the addition of basic oxides such as magnesium oxide and zinc oxide. The cure proceeds through reaction of the metal oxide with the tertiary allylic chlorine that arises from the small amount of 1,2-polymerization that occurs. Other compounding and processing techniques follow similar procedures and use the same equipment as for natural rubber. One of the outstanding characteristics of neoprene is the good tensile strength without the addition of carbon black filler. However, carbon black and other fillers can be used when reinforcement is required for specific end-use applications that require increased tear and abrasion resistance.

The neoprenes have exceptional resistance to weather, sun, ozone, and abrasion. They are good in resilience, gas impermeability, and resistance to heat, oil, and flame. They are fairly good in low temperature and electrical properties. This versatility makes them useful in many applications requiring oil, weather, abrasion, or electrical resistance or combinations of these properties, such as wire and cable, hose, belts, molded and extruded goods, soles and heels, and adhesives.

Nitrile Rubber (NBR)

Much of the basic pioneering research on emulsion polymerization systems was with nitrile-type rubbers. These rubbers, first commercialized as the German Buna N types in 1930, are copolymers of acrylonitrile and a diene, usually butadiene.

Both sulfur and nonsulfur vulcanizing agents may be used to cure these rubbers. Carbon black or other reinforcing agents are necessary to obtain the optimum properties. If proper processing methods are followed, the nitrile rubbers can be blended with natural rubber, polysulfide rubbers, and various resins to provide characteristics such as increased tensile strength, better solvent resistance, and improved weathering resistance.

The nitrile rubbers have outstanding oil, grease, and solvent resistance. Consequently, the commercial usage of these rubbers is largely for items in which these properties are essential. Another major usage is the utilization of the latex form for adhesives and for the finishing of leather, impregnation of paper, and the manufacture of nonwoven fabrics.

cis-1,4-Polyisoprene (IR)

In 1954, synthetic *cis*-1,4-polyisoprene was made from isoprene with two different classes of catalysts. The first class includes lithium and the lithium alkyls. The second class uses a mixture of an aluminum alkyl and titanium tetrachloride, the system first used for the low-pressure polymerization of ethylene. Both catalyst system polymerizations are carried out in hydrocarbon solution and require highly purified monomer and solvent. Traces of air, moisture, and most polar compounds adversely affect reaction rates, polymer properties, and structure.

The *cis*-1,4 polymer structure obtained with these catalysts is also characteristic of natural *Hevea* rubber. The presence of high *cis* content appears necessary for the desirable physical properties with *Hevea* in contrast to the inferior properties of emulsion-type polyisoprene, which contains mixed *cis*-, *trans*-, 1,2-, and 3,4-isoprene units.

This polymer and the corresponding butadiene polymer discussed above are called stereorubbers because of their preparation with stereospecific catalysts. The emulsion polymers are formed by a free-radical mechanism that does not permit control of the molecular structure. Stereorubbers are formed by anionic mechanisms that permit nearly complete control of the structure of the growing polymer chain in stereoregular fashion.

The type of catalyst employed influences the structure of the polymer. The aluminum–titanium–catalyzed rubbers exhibit, on stretching, a gradual crystallization somewhat slower than natural, and they are readily processed because the molecular weight range tends to be less than that of natural rubber. The lithium-catalyzed rubbers contain about 93% cis-1,4 structure; they exhibit very little tendency to crystallize and cannot be processed satisfactorily until they have undergone substantial mastication to reduce their very high molecular weight values to a level more nearly resembling that of well masticated natural rubber.

The differences in structure influence the properties of these rubbers. The aluminum–titanium-catalyzed rubber, because it crystallizes more readily, exhibits better hot tensile strength. The lithium-catalyzed polymer, as it is of higher molecular weight, exhibits higher resilience and less heat generation. Both types, when compounded and cured, produce physical properties that closely approach, or are equivalent to, natural rubber.

Tire tests with heavy-duty tires for trucks, buses, and airplanes have shown that with respect to wear and heat buildup the isoprene rubbers are comparable to natural rubber during tire operation. Polyisoprene rubber has passed qualification tests in high-speed jet aircraft tires to withstand landing speeds as high as 250 mi/h (112 m/s).

Other Rubbers

There are other specialty rubbers available that are important because of specific properties, but in the aggregate they make up only approximately 2% of the world production of synthetic rubbers.

Silicone Rubbers

Silicone rubber is a linear condensation polymer based on dimethyl siloxane. In the preparation, dimethyl dichlorosilane is hydrolyzed to form dimethyl silanol, which is then condensed to dimethyl siloxane, and this, upon further condensation, yields dimethyl polysiloxane, the standard silicone rubber.

Various types of silicone rubbers are produced by substituting some of the methyl groups in the polymer with other groups such as phenyl or vinyl. Advantages of this type of substitution are evidenced by improvements in specific properties. For example, the presence of phenyl groups in the polymer chain gives further improvement in low-temperature properties. Fluorine-containing side groups improve chemical resistance. Many types are commercially available, ranging from fluid liquids to tough solids.

Because sulfur is not effective for the vulcanization of most silicone rubbers, a strong oxidizing agent, such as benzoyl peroxide, is used; the cross-linking produced is random.

Although the standard silicone rubbers are not reinforced by carbon black, the physical properties can be improved by the incorporation of various inorganic fillers such as titania, zinc oxide, iron oxide, and silica, which act as reinforcing and modifying agents. The physical, chemical, and electrical properties can be altered by varying the type and amount of these fillers. Carbon black can be used as a filler with vinyl-containing polymers.

In general, the silicone rubbers have relatively poor physical properties and are difficult to process. However, they are the most stable of rubbers and are capable of remaining flexible over a temperature range of –90 to 316°C. They are unaffected by ozone, are resistant to hot oils, and have excellent electrical properties. Their most extensive uses are for wire and cable insulation, tubing, packings, and gaskets in aerospace and aircraft applications. In the form of dispersions and pastes, they are used for dip-coating, spraying, brushing, and spreading. Silicone rubbers are important in medical and surgical devices because of their property, unique among elastomers, of being compatible with body tissues. Fast, automatic, economical injection molding of liquid silicones has been developed.

Hypalon

Hypalon is the Du Pont trade name for a family of chlorosulfonated polyethylenes prepared by treating polyethylene with a mixture of chlorine and sulfur dioxide, whereby a few scattered chlorine and sulfonyl chloride groups are introduced into the polyethylene chain. By this treatment, polyethylene is converted to a rubberlike material in which the undesirable degree of crystallinity is destroyed but other desirable properties of polyethylene are retained. The outstanding chemical stability of Hypalon results from the complete absence of unsaturation in the polymer chain.

Vulcanization is accomplished by means of metallic oxides, such as litharge, magnesia, or red lead, in the presence of an accelerator as a stock in itself. Hypalon can be blended with other types of rubbers to provide a wide range of properties.

Hypalon has extreme resistance to ozone. Its chemical resistance to strong chemicals, such as nitric acid, hydrogen peroxide, and strong bleaching agents, is superior to any of the commonly used rubbers. These vulcanizates also have good heat resistance, mechanical properties, and unlimited colorability. Typical applications include white sidewall tires and a variety of sealing, waterproofing, insulating, and molded items.

Epichlorohydrin Elastomers

These elastomers are polymers of epichlorohydrin. The two main types of epichlorohydrin elastomers are the homopolymer (CO) and the copolymer (ECO) of epichlorohydrin and ethylene oxide. The copolymer has a lower brittle point, better resilience, and a lower specific gravity than the homopolymer, but is not as good in high-temperature properties.

Since the backbone of the molecule is saturated, these elastomers cannot be vulcanized with sulfur. Cross-linking is achieved by reaction of the chloromethyl side group with diamines or thioureas. A metal oxide is also required.

They exhibit outstanding ease of processing; extreme impermeability to gases; moderate tensile strength and elongation; high modulus; good abrasion resistance; good heat aging; excellent resistance to hot oil, water, perchloroethylene, acids, bases, and ozone; good low-temperature properties; and electrical properties ranging between those of a poor insulator and a good conductor, depending on compounding.

These properties make the epichlorohydrin elastomers useful in gaskets for oil field specialties, diaphragms, and pump and valve parts; hose for low-temperature flexibility, oil and fuel resistance, or gas impermeability applications, and mechanical goods such as belting, wire, and cable.

Thermoplastic Elastomers

Proper choice of catalyst and order of procedure in polymerization have led to development of thermoplastic elastomers. The leading commercial types are styrene block copolymers with a structure that, unlike the random distribution of monomer units in conventional polymers, consists of polystyrene segments or blocks connected by rubbery polymers such as polybutadiene, polyisoprene, or ethylene-butylene polymer. These types can be designated SBS, SIS, and SEBS, respectively. Other types of thermoplastic elastomers, such as the polyester type, have also been developed.

SBS, SIS, and SEBS polymers, when heated to a temperature above the softening point of the styrene blocks, can be processed and shaped like thermoplastics but, when cooled, act like vulcanized rubber, with the styrene blocks serving as cross-links. Thus, it is unnecessary to vulcanize. Also scrap or excess material can be reused.

Thermoplastic elastomers are very useful in providing a fast and economical method of producing a variety of products, including molded goods, toys, sporting goods, and footwear. One of the disadvantages for many applications is the low softening point of the thermoplastic elastomers. Attempts to modify the composition and structure to raise the softening temperature have generally resulted in inferior rubbery properties.

Fluoroelastomers

These are basically copolymers of vinylidene fluoride and hexafluoropropylene.

Because of their fluorine content, they are the most chemically resistant of the elastomers and also have good properties under extremes of temperature conditions. They are useful in the aircraft, automotive, and industrial areas.

Polyurethane Elastomers

Polyurethane elastomers are of interest because of their versatility and variety of properties and uses. They can be used as liquids or solids in a number of manufacturing methods. The largest use has been for making foam for upholstery and bedding, but difficulties have been encountered in public transportation due to flammability.

Polysulfide Rubbers

These rubbers have a large amount of sulfur in the main polymer chain and are therefore very chemically resistant, particularly to oils and solvents. They are used in such applications as putties, caulks, and hose for paint spray, gasoline, and fuel.

Polyacrylate Rubbers

Polyacrylate rubbers are useful because of their resistance to oils at high temperatures, including sulfur-bearing extreme-pressure lubricants.

PRODUCTS

A wide variety of products have a rubber, or elastomer, as an essential component. Most rubber products contain a significant amount of nonrubber materials, used to impart processing, performance, or cost advantages.

The automotive industry is the biggest consumer of rubber products. About 60% of all rubber used goes into the production of passenger, bus, truck, and off-the-road tires. In addition, a typical automobile contains about 68 kg of rubber products such as belts, hose, and cushions. Outside the automotive area, a wide variety of rubber products are produced, including such familiar items as rubber bands, gloves, and shoe soles.

The formulation and manufacture of rubber products is as diverse as the wide variety of applications for which they are intended This discussion is limited to the more common methods of manufacture and those that apply to the widest variety of products. In general, the manufacture of rubber products involves compounding, mixing, processing, building or assembly, and vulcanization.

Compounding

The technical process of determining what the formulation components of a rubber product should be is called compounding. This term is also applied to the actual process of weighing out the individual components in preparing for mixing.

Rubber is the backbone of any rubber product. Natural rubber, obtained from the latex of rubber trees, accounts for about 23% of all rubber consumed in the United States. The balance is synthetic rubber. Of the synthetic rubbers, SBR (styrene–butadiene rubber) and polybutadene rubber are most important, accounting for 71% of synthetic rubber production. A variety of specialty synthetic rubbers, such as butyl, EPDM, polychloroprene, nitrile, and silicone, account for the balance of synthetic rubber production.

Choice of the rubber to be used depends on cost and performance requirements. The specialty rubbers often give superior performance properties but do so at higher product cost.

Many rubber products contain less than 50% by weight of rubber. The balance is a selection of fillers, extenders, and processing or protective coatings.

In products other than tires, clays may be added as extenders, silicas as reinforcing agents, and plasticizers for flex or fire-retardant properties; colors or brightening agents may also be used.

Mixing

This step accomplishes an intimate and homogeneous mix of the formulation components. Most large-volume stocks are mixed in internal mixers; these operate with two winged rotors with the compound ingredients forced into the rotors by an external ram.

There are thousands of different recipes, each designed for different purposes and each requiring a mixing procedure of its own. In addition, various manufacturers differ in their ideas as to precisely how the mixing operation should be conducted, and the equipment in various plants differs widely. The trend is toward more automation in this process, both in weighing and changing the ingredients into the mixer and in handling the stock after discharge.

Forming

Usually, forming operations involve either extrusion into the desired shape or calendering to sheet the material to some specified gauge or to apply a sheet of the material to a fabric.

Building

Building operations, varying from simple to complex, are required for products such as tires, shoes, fuel cells, press rolls, conveyor belts, and life rafts. These products may be built by combining stocks of different compositions or by combining rubber stocks with other construction materials such as textile cords, woven fabric, or metal. For a few products no building operations are required. For example, many molded products are made by extruding the rubber, cutting it into lengths, and placing them in the mold cavities. Also, extruded products that are given their final shape by the die used in the extruder are ready for the final step of vulcanization, with no intermediate building operations.

The building operations for tires vary widely, depending on the kind of tire to be built and the equipment available in the particular plant. For example, the cord reinforcement may be of polyester, nylon, or steel wire; the number of plies required increases as the size and service requirements increase; the tread proper may be of a different composition than the tread base and sidewalls; one of the sidewalls may be made of a white stock; the construction may be of the tubeless variety, in which case an air-barrier ply of rubber is applied to the inside surface of the first ply; a puncture-sealing layer may be applied to the inside surface; or there may be several bead assemblies instead of only one.

Increasing popularity of radial tires has caused some basic changes in tire manufacturing operations. In a radial tire, the reinforcement cords in the body, or carcass, of the tire are placed in a radial direction from bead to bead. Radial tires require special building drums and result in a green tire of a different appearance from the barrel-shaped green bias-ply tire.

Vulcanization

The final process, vulcanization, follows the building operation or, if no building operations are involved, the forming operation.

Vulcanization is the process that converts the essentially plastic, raw rubber mixture to an elastic state. It is normally accomplished by applying heat for a specified time at the desired level. The most common methods for vulcanization are carried out in molds held closed by hydraulic presses and heated by contact with steam-heated platens, which are a part of the press in open steam in an autoclave; under water maintained at a pressure higher than that of saturated steam at the desired temperature; in air chambers in which hot air is circulated over the product; or by various combinations of these methods.

The vast majority of products are sulfur-cured; that is, sulfur cross-links join the rubber chairs together. For special applications, vulcanization may take place without the use of sulfur, for example, in resin- or peroxide-initiated cross-links or metal-oxide-cured polychloroprene.

The time and temperature required for vulcanization of a particular product may be varied over a wide range by proper selection of the vulcanizing system. The usual practice is to use as fast a system as can be tolerated by the processing steps through which the material will pass without "scorching," that is, without premature vulcanization caused by heat during these processing steps. Rapid vulcanization affects economies by producing the largest volume of goods possible from the available equipment. This is particularly the case for products made in molds, because molds are costly, and their output is determined by the number of heats that can be made per day.

The rate of vulcanization increases exponentially with an increase in temperature; hence, the tendency is to vulcanize at the highest temperature possible. In practice, this is limited by many factors, and the practical curing temperature range is 127 to 171°C. There are numerous exceptions both below and above this range, but it probably covers 95% of the products made.

Finishing operations following vulcanization include removal of mold flash, sometimes cutting or punching to size, cleaning, inspection for defects, addition of fittings such as valves or couplings, painting or varnishing, and packing.

Latex Technology

In addition to the technology of solid rubber, there is a completely different, relatively small but important part of the rubber industry involving the manufacture of products directly from natural rubber latex from the tree, synthetic latex from emulsion polymerization, or aqueous dispersions made from solid rubbers.

In latex technology, materials to be added to the rubber are colloidally dispersed in water and mixed into the latex, a process involving the use of lighter equipment and less power than the mixing of solid rubber compounds. The latex compound can then be used in a variety of processes such as coating or impregnating of cords, fabrics, or paper; in paints or adhesives; molding (such as in toys); dipping (for thin articles like balloons, or household and surgeon's gloves); rubber thread (for garments); and production of foam. Latex technology is particularly important in producing articles for medical and surgical uses.

RUST PREVENTIVES

Petroleum rust preventives have been used for the temporary preservation of corrodible metal surfaces ever since Colonel Drake discovered the first oil well over 100 years ago. From this first oil production waxes were separated, refined, and made into "petroleum jelly," bringing into being the first petroleum rust preventives. The use of petroleum products as protective coatings for oil-well equipment spread into all areas requiring low-cost, temporary rust protection.

The raw materials used today are produced in modern petroleum refineries in large volume at low cost. Petroleum has provided the wide range of physical properties required to meet the exacting and varied engineering specifications of the complex industrial and diversified military rust-preventive requirements. These physical properties are viscosity, melting point, pour point, consistency, adhesion, volatility, and density. Chemical additives are usually incorporated with petroleum rust preventives to improve lubricity, to modify viscosity–temperature relationships, to reduce pour point, to displace water from metal surfaces, to dissolve salts, to prevent foaming, to prevent oxidation of the compound, to impart antiwear properties, to improve detergency and dispersancy, and to prevent corrosion and rusting.

Atmospheric corrosion of ferrous metals accounts for a substantial part of why rust prevention is important and critical. The rust prevention of machined parts presents an overall cost saving because the cost of protection is much less than the cost of cleaning to remove rust, of reprocessing, or of scrapping parts that cannot be salvaged.

Petroleum rust preventives provide temporary but complete protection against rusting, are easy to apply, easy to remove, and are low in cost. The results of rusting may result in dimensional change, structural weakening, and loss of decorative finish.

Rust-Preventive Types

Petroleum rust preventives may be conveniently separated into four major types: (1) solid or semisolid waxes; (2) oils; (3) solvent cutbacks; and (4) emulsions.

Solid or Semisolid Waxes

These compounds are generally prepared from combinations of petrolatums and microwaxes with heavy lube oils added to obtain the desired consistency. Three consistencies are specified for military use: hard, medium, and soft.

1. The hard-film grade is applied in the molten state and is intended for long-term, temporary outdoor storage of heavy arms, jigs, and forms, for protecting small metal parts, either packaged or unpackaged, and for long-term indoor storage protection of brightly finished surfaces.
2. The medium-film grade is applied in the molten state or at about ambient temperature by dipping or brushing. It is intended for unshielded outdoor storage in relatively moderate climates at temperatures not exceeding the flow point temperature of the compound.
3. The soft-film grade is applied either by brushing or dipping at room temperature or from the molten state. It is intended primarily for the preservation of friction bearings and for use on machined surfaces indoors or for packaged parts.

Additives may or may not be present, depending on the type of service for which the compound is intended. Several advantages are claimed for the solid type compounds:

1. They form coatings of controllable thickness, they do not drain off during storage, and they provide a substantial physical barrier to both moisture and corrosive gases.
2. They form coatings resistant to mild abrasion, airborne contaminants, and ultraviolet light (sunlight) for extended periods.
3. The fast-setting wax coatings eliminate prolonged draining or drying periods during processing.

There are also several disadvantages:

1. Special thermostatically controlled dip tanks must be used to control thickness.
2. The coating, once removed accidentally during handling, is not self-healing and must be reapplied.
3. The compounds are difficult to remove by wiping or rinsing with solvents. Special facilities must be available for the efficient removal of these hard grades. They are not to be applied to intricate parts or inaccessible surfaces.

Preservative Oils

These are available in as many types as there are special uses for oil-type products. Examples:

1. Preservative lubricants may contain detergents, pour depressants, bearing corrosion inhibitors, and antioxidants, as well antirust additives for gasoline or diesel engine use.
2. Special preservative oils for both reciprocating and turbojet aviation engines are available containing nonmetallic (nonashing), antirust additives.
3. Light lubricants for automatic rifles, machine guns, and aircraft, suitable for operation at temperatures as low as $-55°C$, are available in several viscosity grades.
4. General-purpose preservatives are produced specifically for marine use where salt spray as well as high humidity may rust machinery and shipboard cargo.
5. Preservative hydraulic fluids are compounded to preserve hydraulic systems during storage, shipment, and manufacture. These oils contain additives that will permit temporary operational use of the hydraulic systems. In addition to antirust additives, oxidation inhibitors, viscosity index improvers, antiwear agents, and pour depressants are present in hydraulic fluids.
6. Slushing oils are used extensively for the preservation of rolled steel and aluminum sheet rods and bar stock. These oils require careful selection of additives to prevent "black staining" of stacked or rolled sheet. This form of corrosion is a common problem in steel mills.

Metal-conditioning oils are used to loosen and soften scale and rust from heavily corroded surfaces such as marine bulkheads, decks, and structural supports of ballast tanks. These oils appreciably reduce the pitting and general corrosion resulting from the combination of humid atmospheres and frequent immersion in salt water.

Solvent Cutbacks

These are generally of three types and can be applied by spraying, dipping, or slushing. The volatile solvent used in these compounds normally has a flash point above $37.8°C$ and will evaporate completely in several hours leaving the residual protective coating. They include:

1. Asphalt cutbacks formulated to dry "to handle" in several hours leaving a smooth black coating approximately 2 mils thick. These coatings are recommended for severe outdoor

exposure. They are resistant for several years to salt spray, sunlight, rain, and high humidity.

2. Petrolatum and waxlike coatings that can be either hard and "dry to touch" or soft and easily removed. The dry-to-touch or tack-free coating is usually transparent and provides an excellent preservative coating for spare parts that are stored for long periods indoors and are handled frequently. The soft coatings provide excellent protection against corrosive atmospheres, high humidity, and airborne contamination. These coatings are easily removed by wiping or solvent degreasing but must be protected by protective packaging or storage arrangements. These coatings are not fully self-healing if damaged.

3. Oil coatings are used largely for short-term, in-shop use where parts are still being processed. Humidity protection is provided, with additional emphasis placed on suppression of fingerprint rusting and on displacement of the water remaining from processing.

Emulsifiable Rust Preventives

The emulsifiable rust preventive is an oil or wax concentrate that on proper dilution and on mixing with water will provide a ready-to-use product in the form of an oil-in-water emulsion system. These emulsion systems are fire resistant, provide excellent protection, and are economical. They are intended for use whenever a low-cost, fire-resistant, soft-film, corrosion preventive can be utilized. The compound is formulated to be emulsified in distilled or deionized water, normally at a 1:4 ratio, reducing the volatile content of the corrosion preventive to a minimum. Its fire-resistant properties make it especially adaptable as a substitute for volatile solvent cutbacks where fire hazards increase the expense of application. It has proved efficient for use on small hardware or on parts of uniform contour. Caution should be exercised in applying it to parts of radical design with cracks, crevices, or depressions, which would prevent adequate drainage and cause subsequent corrosion by trapping excess amounts of water.

Selection of Rust Preventives

The selection of petroleum rust preventives is based upon their temporary nature, ease of application and removal, the economy of providing protection against corrosion, and the type of protection desired (e.g., indoor protection or outdoor exposure).

Indoor Protection

Preservatives of this group must provide a high degree of protection against high humidity, corrosive atmospheres, and moderate amounts of airborne contaminants. Resistance to abrasion is occasionally required where spare parts are stored in bins or on uncovered racks. Normally, however, the more easily removed products are used, such as oils, soft petrolatums, or soft-film solvent cutbacks.

Outdoor Exposure

Preservatives of this group must withstand the effects of rain, snow, and sunlight as well as provide good resistance to abrasion. The products used for this service are the hard-film, asphalt solvent cutbacks and the hard grades of hot-dip waxes.

Short-Term Storage

These preservatives are commonly used to protect parts between machining operations or prior to final preservation with a heavy-duty product. These light oils or solvent cutbacks usually have fingerprint-suppressing properties to protect the parts during production handling.

Special-Purpose Lubricants

Lubricants for many purposes are formulated to provide primary or secondary rust-prevention properties. These special lubricants include nonashing lubricants for aircraft engine use, high-detergency oils for motor vehicles, low-temperature machine gun or rifle oils, and special salt-spray-resistant lubricating oils for shipboard use.

Functional Uses

Hydraulic oils possessing rust-preventive properties are used during the shipment or storage of equipment containing hydraulic power-transmission systems. This equipment is found in aerospace systems, submarines, aircraft carriers, and industrial machine tools. Various viscosity grades are available.

RUTHENIUM

A hard silvery-white metal (symbol Ru), ruthenium has a specific gravity of 12.4, a melting point of about 2310°C, and a Brinell hardness of 220 in the annealed state. The metal is obtained from the residue of platinum ores by heat reduction of ruthenium oxide, RuO_2, in hydrogen. Ruthenium is the most chemically resistant of the platinum metals, and is not dissolved by aqua regia. It is used as a catalyst to combine nitrogen in chemicals. As ruthenium tetroxide, RuO_4, it is a powerful catalyst for organic synthesis, oxidizing alcohols to acids, ethers to esters, and amides to imides. Ruthenium has a close-packed hexagonal crystal structure. It has a hardening effect on platinum; a 5% addition of ruthenium raises Brinell hardness from 30 to 130 and the electrical resistivity to double that of pure platinum.

Ruthenium is used commercially to harden alloys of palladium and platinum. The alloys are used as electrical contacts (12% ruthenium in platinum) and in jewelry and fountain-pen tips. Application of ruthenium to industrial catalysis (hydrogenation of alkenes and ketones) and to automobile emission control (catalytic reduction of nitric oxide) and detection have been active areas of research.

RUTILE

Rutile is the most frequent of the three polymorphs of titania, TiO_2, crystallizing in the tetragonal system is the rutile structure type.

The mineral occurs as striated tetragonal prisms and needles, commonly repeatedly twinned. The color is deep blood red, reddish brown, to black, rarely violet or yellow. Specific gravity is 4.2, and hardness 6.5 on Mohs scale. Melting point is 3317°C. It is synthesized from $TiCl_4$ at red heat, and colorless crystals are grown by the flame-fusion method, which because of their adamantine luster are used as gem material.

In ceramic applications where titanium oxide is desirable (but where a pure white or certain shades are not required) the more economical rutile is frequently substituted for the pure chemical. Rutile is used to stain pottery bodies and glazes in colors ranging from ivory through yellows to dark tan, according to the amount introduced. Artificial teeth are among the ceramics so tinted.

Rutile is also used in the glass and porcelain enamel industries as a colorant, and to introduce TiO_2. The largest use of rutile is as a constituent in welding rod coatings.

S

SALT

A salt is a compound formed when one or more of the hydrogen atoms of an acid are replaced by one or more cations of the base. The common example is sodium chloride in which the hydrogen ions of hydrochloric acid are replaced by the sodium ions (cations) of sodium hydroxide. There is a great variety of salts because of the large number of acids and bases that has become known.

CLASSIFICATION

Salts are classified in several ways. One method — normal, acid, and basic salts — depends upon whether all the hydrogen ions of the acid or all the hydroxide ions of the base have been replaced:

Class	Examples
Normal salts	$NaCl$, NH_4Cl, Na_2SO_4, Na_2CO_4, Na_3PO_4
Acid salts	$NaHCO_4$, NaH_2PO_4, Na_2HPO_4, $NaHSO_4$
Basic salts	$Pb(OH)Cl$, $Sn(OH)Cl$

The other method — simple salts, double salts (including alums), and complex salts — depends upon the character of completeness of the ionization:

Class	Examples
Simple salts	$NaCl$, $NaHCO_3$, $Pb(OH)Cl$
Double salts	$KCl \cdot MgCl_2$
Alums	$KAl(SO_4)_2$, $NaFe(SO_4)_2$, $NH_4Cr(SO_4)_2$
Complex salts	$K_3Fe(CN)_6$, $Cu(NH_3)_4Cl_2$, $K_2Cr_2O_7$

In general, all salts in solution will give ions of each of the metal ions: an exception is the complex type of salt such as $K_3Fe(CN)_6$ and $K_2Cr_2O_7$.

Any aggregate of molecules, atoms, or ions joined together with a coordinate covalent bond can be correctly called salt; however, by common parlance, the term *salt* usually refers to an electrovalent compound, the classical example of which is sodium chloride.

SAND

Sand is an unconsolidated granular material consisting of mineral or rock fragments between 6.3 μm and 2 mm in diameter. Finer material is referred to as silt and clay, coarser material as gravel. Sand is usually produced by the chemical or mechanical breakdown of older source rocks, but may also be formed by the direct chemical precipitation of grains or by biological processes. Accumulations of sand result from hydrodynamic sorting of sediment during transport and deposition.

ORIGIN AND CHARACTERISTICS

Most sand originates from the chemical and mechanical breakdown, or weathering, of bedrock. Chemical weathering is most efficient to soils, and most sand grains originate within soils. Rocks may also be broken into sand-size fragments by mechanical processes, including diurnal temperature changes, freeze–thaw cycles, wedging by salt crystals or plant roots, and ice gouging beneath glaciers.

TRANSPORT

Sand can be transported by any medium with sufficient kinetic energy to keep the grains in movement. In nature, sand is commonly transported by rivers, ocean currents, wind, or ice. Rivers are responsible for transporting the greatest volume of sand over the greatest distances. When rivers deposit sand at the ocean edge, it is commonly remobilized by high-energy breaking waves or transported along the shore by powerful longshore currents. Sand, like coarser and finer sediments, may also be entrained by glacial ice and transported to the sea by glacial flow. Where sand is widely exposed to wind currents, such as in river bars, on beaches, and in glacial outwash plains, it may be picked up and transported by the wind. As it moves, whether subaqueously or subaerially, sand commonly is organized into complex moving structures (bedforms) such as ripples and dunes.

Economic Importance

Sand and gravel production is second only to crushed-stone production among nonfuel minerals in the United States. Although sand and gravel have one of the lowest average per ton values of all mineral commodities, the vast demand makes it among the most economically important of all mineral resources. Sand and gravel are used primarily for construction purposes, mostly as concrete aggregate. Pure quartz sand is used in the production of glass, and some sand is enriched in rare commodities such as ilmenite (a source of titanium) and gold.

SAND CASTINGS

Sand casting can be broken down into three general processes: green-sand molding, dry-sand molding, and pit molding. Green-sand molding, in all probability, produces the largest tonnage of castings but is limited in that long, thin projections are very difficult to cast. Dry-sand molds, on the other hand, can produce castings of any desired intricacy (within the normal dimensional limits of the process) because the sand is baked with a binder, increasing the strength of the thin mold sections. Pit molding is the sand process used to make very large, heavy castings.

The Process

Green-Sand Molding

The essential parts of a green-sand mold are shown in Figure S.1. The pattern is placed in the flask (consisting of the cope and drag) and molding sand is tightly rammed around it. Usu-

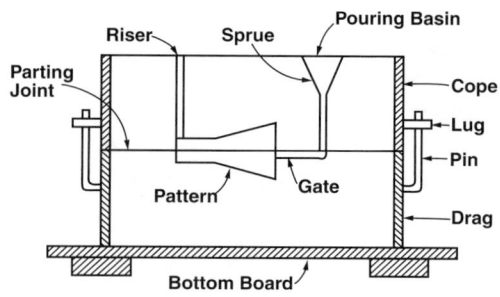

FIGURE S.1 Cross-section of a sand casting mold.

ally, the mold is made in two halves, with the pattern lying at the parting line. Removing the pattern from the rammed sand leaves the desired cavity from which the casting is produced. The molder then produces a sprue for pouring metal into the cavity and an opening for a riser to permit air in the cavity to be expelled. Both the sprue and riser also act as a source of hot metal during solidification and help eliminate shrinkage cavities in the casting. Cores can be added to the mold cavity to shape internal casting surfaces.

Among the advantages of green-sand molding are the following: one pattern can be used to produce any number of castings; most nonferrous alloys, gray irons, ductile irons, malleable irons, and steel can be cast; and finally, the fragility of the sand cores permits them to collapse after metal is cast around them, thereby eliminating or reducing casting stresses and tendency for hot tearing.

The limitation of the process is its inability to support long, thin projections in the sand.

Dry-Sand Molding

Core boxes, not patterns, are used to make the various parts of the mold. A mixture of sand and a binder is formed in a core box acid baked at 204 to 260°C to harden the sand. The various pieces are then assembled into a mold.

Dry-sand molds, because of the way in which they are produced, can be used to make intricate castings. The baking operation strengthens the sand and permits thin projections to be cast without danger of collapsing the mold walls. Cores used in dry-sand molding are collapsible and help reduce hot tearing tendencies.

Pit Molding

If large, intricate parts must be produced, pit molding is considered the most economical production method. (Castings weighing up to 230,000 kg have been produced in pit molds.) Pit molding is a highly specialized operation and the equipment used — sand slingers, molding machines, etc. — precludes the use of much hand labor. The mold usually is dried, increasing sand strength and, consequently, the ability to resist mold erosion during pouring as well as the weight of the casting being poured.

SANDWICH MATERIALS

These are a type of laminar composite composed of a relatively thick, low-density core between faces of comparatively higher density. Structural sandwiches can be compared to I beams. The facings correspond to the flanges; the objective is to place a high-density, high-strength material as far from the neutral axis as possible, thus increasing the section modulus. The bulk of a sandwich is the core. Therefore, it is usually lightweight for high strength-to-weight and stiffness-to-weight ratios. However, it must also be strong enough to withstand normal shear and compressive loadings, and it must be rigid enough to resist bending or flexure.

Core Materials

Core materials can be divided into three broad groups: cellular, solid, and foam. Paper, reinforced plastics, impregnated cotton fabrics, and metals are used in cellular form. Balsa wood, plywood, fiberboard, gypsum, cement-asbestos board, and calcium silicate are used as solid cores. Plastic foam cores — especially polystyrene, urethane, cellulose acetate, phenolic, epoxy, and silicone — are used for thermal-insulating and architectural applications. Foamed inorganics such as glass, ceramics, and concrete also find some use. Foam cores are particularly useful where the special properties of foams are desired, such as insulation. And the ability to foam in place is an added advantage in some applications, particularly in areas that are difficult to reach.

Cellular Cores

Of all the core types, however, the best for structural applications are the rigid cellular cores. The primary advantages of the cellular core are that (1) it provides the highest possible strength-to-weight ratio, and (2) nearly any material can be used, thereby satisfying virtually any service condition.

There are, essentially, three types of cellular cores: honeycomb, corrugated, and waffle. Other variations include small tubes or cones and mushroom shapes. All these configurations have certain advantages and limitations. Honeycomb sandwich materials, for example, can be isotropic, and they have a high strength-to-weight ratio, good thermal and acoustical properties, and excellent fatigue resistance. Corrugated-core sandwich is anisotropic and does not have as wide a range of application as honeycomb, but it is often more practical than honeycomb for high production and fabrication into panels.

Construction

Theoretically, any metal that can be made into a foil and then welded, brazed, or adhesive-bonded can be made into a cellular core. A number of materials are used, including aluminum, glass-reinforced plastics, and paper. In addition, stainless steel, titanium, ceramic, and some superalloy cores have been developed for special environments.

One of the advantages of sandwich construction is the wide choice of facings, as well as the opportunity to use thin sheet materials. The facings carry the major applied loads and therefore determine the stiffness, stability, and, to a large extent, the strength of the sandwich. Theoretically, any thin, bondable material with a high tensile- or compressive-strength-to-weight ratio is a potential facing material. The materials most commonly used are aluminum, stainless steel, glass-reinforced plastics, wood, paper, and vinyl and acrylic plastics, although magnesium, titanium, beryllium, molybdenum, and ceramics have also been used.

Theory

The theory of sandwich materials and functions of the individual components may best be described by making an analogy to an I-beam. The high-density facings of a sandwich correspond to the flanges of the I-beam; the objective is to place a high-density, high-strength material as far from the neutral axis as possible to increase the section modulus without adding much weight. Honeycomb in a sandwich is comparable to the I-beam web that supports the flanges and allows them to act as a unit. The web of the I-beam and honeycomb of the sandwich carry the beam shear stress. Honeycomb in a sandwich differs from the web of an I-beam in that it maintains a continuous area support for the facings, allowing them to carry stresses up to or above the yield strength without crippling or buckling. The adhesive that bonds honeycomb to its facings must be capable of transmitting shear loads between these two components, thus making the entire structure an integral unit.

When a sandwich panel is loaded as a beam, the honeycomb and the bond resist the shear loads while the facings resist the moments due to bending forces, and hence carry the beam bending as tensile and compressive load. When loaded as a column, the facings alone resist the column forces while the core stabilizes the thin facings to prevent buckling, wrinkling, or crippling.

Advantages

The largest single reason for the use of sandwich construction and its rapid growth to one of the standard structural approaches during the past 40 years is its high strength-to-weight or stiffness-to-weight ratio. As an example consider a 0.6-m-span beam with a width of 0.3 m and supporting a load of 1634.4 kg at the midspan. This beam, if constructed of solid steel, would have a deflection of 14.2 mm and weigh 31.15 kg. A honeycomb-sandwich beam using aluminum skins and aluminum cores and carrying the same total load at the same total deflection would weigh less than 3.6 kg. As an interesting further comparison, a magnesium plate to the same specifications would weigh 11.9 kg and an aluminum plate 15.7 kg. Although such clear and simple cases of comparative strength, weight, and stiffness are not normally found in actual designs, it has generally been found that equivalent structures of sandwich construction will weigh from 5% to 80% less than other minimum weight structures and frequently possess other significant advantages.

Other advantages of sandwich construction include extremely high resistance to vibration and sonic fatigue, relatively low noise transmission, either high or low heat transmission depending on the selection of core materials, electrical transparency (varying from almost completely transparent in the case of radom structures to completely opaque in the case of metal sandwich structures), relatively low-cost tooling when producing complex aircraft parts, ability to mass-produce complicated shapes, ability to absorb damage and absorb energy while retaining significant structural strength, and flexibility of design available.

SAPPHIRE

Sapphire is any gem variety of the mineral corundum (Al_2O_3) except those called ruby because of their medium-to dark-red color. Sapphire has a hardness of 9 (Mohs scale), a specific gravity near 4.00, and refractive indices of 1.76 to 1.77.

Although blue sapphires are most familiar to the public, transparent sapphires occur in many other colors, including yellow, brown, green, pink, orange, and purple, as well as colorless, and black.

General Nature

Synthetic sapphire, chemically and physically, is the same as the natural mineral, which is now uneconomical to mine because of the low cost of the synthetic ($0.05/g). It is a single crystal of very high purity aluminum oxide and crystallizes in the hexagonal-rhombohedral crystal system. It has a hardness that is second only to diamond among naturally occurring minerals. It has unusual chemical stability, high thermal conductivity, excellent high-temperature strength, broad band optical transmission, and low electrical loss.

CHEMICAL PROPERTIES

Sapphire is inert to most chemical agents. For example, it is not attacked by any acids, including hydrofluoric acid. It is resistant to alkalies and is not attacked by uranium hexafluoride up to 600°C. Sapphire will dissolve in water at the rate of 1 mg/day at 240 MPa and 700°C. It is also attacked by anhydrous sodium borate–sodium carbonate at 1000°C, by silicon at 1500°C, by sodium hydroxide at 700°C, and by boiling phosphoric acid.

FORMS AVAILABLE AND USES

Synthetic sapphire is available in the form of rods, boule, and disks as well as optical lens shapes. Sapphire is fabricated with diamond grinding techniques into a variety of surface roughnesses, can be optically polished, and can be chemically or flame polished.

Most applications are for watch and instrument bearings, gemstones, phonograph needles, ball pointpen balls, infrared and ultraviolet optics, as well as electronic insulators and masers. Others include orifices, arc lamp tubes, radiation pyrometry, crucibles, film and textile guides, solar cell covers, optical flats, liquid nitrogen heat sinks, klystron windows, and optical filter substrates.

By deliberately adding impurities to the starting material, other modifications of sapphire crystals can be obtained. The most important today is ruby, which is alumina with 0.05% chromium oxide added and is the basic solid-state crystal in microwave and optical maser systems.

The submicron size high-purity alumina powder, which is used as a starting material for producing crystals, is also used as a polishing agent in dentifrice formulation, metallurgical polishing, as supports for high-purity chemical catalyst carriers, and as the raw material in specialty ceramic manufacture.

SBR RUBBER

SBR synthetic rubbers are versatile, general-purpose elastomers, developed as a substitute for natural rubber during World War II. Known as Buna-S (BU-tadiene, NA-trium, or sodium catalyzed, S-tyrene) in Germany, they were designated GR-S (G-overnment R-ubber S-tyrene) by the United States. The materials have been designated SBR (S-tyrene B-utadiene R-ubber) by the ASTM (American Society for Testing and Materials). They constitute about 80% of domestic synthetic-rubber consumption.

SBR TYPES

SBR polymers are generally available as dry rubbers (bale or crumb form) and its latices. So-called "hot" polymers (polymerized at 50°C) have been steadily replaced in recent years by the 20 to 30% higher tensile, longer-wearing "cold" rubbers (polymerized at 5°C). Cold rubbers generally accept higher pigment loadings than the hot types. The remaining 12% usage of hot polymers is maintained by certain demands for the inherently easier processing characteristics of these materials.

Styrene content varies from as low as 9% in so-called "Arctic" or low-temperature-resistant rubbers up to 44% in a rubber designed for its excellent flow characteristics. Materials above an arbitrary 50% level are termed plastics, butadiene-styrene, or high-styrene resins. These resins are used as stiffening agents for SBR rubbers in applications such as shoe soles, and latices in this composition range are used in the paint industry and paper coating.

Latices vary in styrene content from 0 to 46%; the higher levels are used where greater strength is needed. Solids vary from 28 to 60%. The higher solids content, large particle size, cold types provide the largest single outlet for SBR latices as they are rapidly taking over the major share of the natural rubber foam market. Other applications for SBR latices include adhesives, coatings, impregnating or spreading compounds for rug backing, textiles, and paper saturation.

PHYSICAL PROPERTIES

With natural and SBR compounds of the same hardness, it is found that the synthetic formulations not only extrude faster, more smoothly, and with a greater degree of dimensional stability, but also with less danger of sticking if the article is to be coiled during curing.

In molded items SBR compounds are more nearly free of taste and odor than those made with natural rubber.

The inherently superior physical properties of SBR over natural rubber (or those superior characteristics obtainable with special polymers or through compounding of SBR) include the following:

1. Enhanced resistance to attack by organisms (e.g., soil bacteria)
2. Lower compression set
3. Less subject to scorch in curing operations
4. Superior flex resistance
5. Better resistance to animal and vegetable oils
6. Higher abrasion resistance
7. Superior water resistance
8. Lower freeze temperatures
9. Better dampening effect
10. Increased filler tolerance
11. Less high-temperature discoloration
12. Superior aging, including:
    a. Heat resistance (steam or dry heat)
    b. Sunlight checking
    c. No reversion or tackiness
    d. Resistance to oxidation
    e. Ozone cracking
13. Enhanced electrical insulation properties

Natural rubber excels SBR in building tack, tensile strength, tear resistance, elongation, hysteresis (heating on flexing), and resilience properties.

Although SBR has a higher thermal conductivity than natural rubber, it is not sufficient to overcome its higher hysteresis value. Whereas the low conductivity of natural rubber is an engineering advantage in gaskets or seals used in low-temperature equipment, it becomes a distinct disadvantage in an application such as electrical insulation where overheating at the surface of the conductor is experienced during an overload condition.

Another drawback of the low thermal conductivity of natural rubber is encountered when thick slabs or sections are molded, as the time required to bring the stock up to temperature and reduce it again is excessive. This is particularly significant when the compound used is a hard rubber, containing large amounts of sulfur, which liberate internal heat as well. Although the internal heating factor is not important with soft compounds containing small amounts of sulfur, the polymer content of hard rubber may actually be destroyed by excessive internal temperatures.

COMPOUNDING

After selection of the proper polymer from the wide diversity of types available, the compounder is afforded a further latitude in building in desired properties through selection of vulcanizing agents, fillers, deodorants, age resistors, blends of other rubbers and resins, etc.

One of the outstanding differences between natural rubber (*Hevea*) and SBR is the need for reinforcement in the latter. Whereas the *Hevea* tensile strength might be increased 25% with the addition of carbon black from its uncompounded value of about 24.03 to 27.2 MPa, the tensile strength of SBR is increased to above 20.6 MPa from an initial value of less than 6.9 MPa.

Although not as effective as the carbon blacks, other inorganic fillers including clays, whiting, coated calcium carbonate and precipitated silicas, barytes, zinc oxide, magnesia, lithopone, and mica are used. Titanium dioxide pigment is used where covering power is desired in white or very brightly colored stocks.

Various degrees of oil resistance may be imparted to SBR compounds (or, put conversely, lower-cost, oil-resistant rubbers can be produced) through blending SBR with the higher-cost neoprene or butadiene-acrylonitrile copolymer rubbers.

SCANDIUM

Scandium (symbol Sc, element number 21) is as plentiful in the Earth's crust as tungsten and more than cadmium, and it is considered a rare element because it is so widely distributed that commercial quantities are difficult to obtain.

Chemically, scandium is somewhat similar to aluminum but is more nearly like yttrium and the rare earth metals. It is quite reactive, forming very stable compound. The halides, however, can be reduced by calcium, sodium, and potassium, and it is in this manner that scandium metal is produced. For example, scandium fluoride is generally reduced with calcium metal in a tantalum or tungsten crucible in an inert atmosphere. The scandium is purified by distillation at 1650 to 1700°C in a vacuum of 10^{-5} mmHg. This produces a material with a metallic luster that has a slight yellow tinge but the general appearance of aluminum.

CHEMICAL PROPERTIES

Scandium reacts very rapidly with dilute acids but reacts slowly in a mixture of concentrated nitric and hydrofluoric acids owing to the formation of an insoluble layer of scandium fluoride on the surface of the metal.

Scandium metal is quite stable in air at room temperature but oxidizes fairly rapidly at higher temperatures. Scandium oxidizes at a rate of 0.0187, 0.304, and 0.421 mg/cm^2/h at 400, 600, and 800°C, respectively. The oxides of 400 and 600°C are very dense, adherent films, while the coatings at 800°C or above are loose and nonprotective.

FABRICATION

Scandium metal can be arc-melted in an inert atmosphere such as argon at pressures of 50 mmHg without loss by evaporation. It is extremely active, however, and will "getter" the furnace atmosphere, producing a dark film on the surface of the metal.

Pure scandium can be cold-rolled or swaged into thin pieces without annealing if the oxygen content is kept below 200 ppm (parts per million). Excessive oxygen occurs as dispersed oxides in the grain boundaries and makes fabrication difficult by causing intergranular cracking.

METAL PREPARATION AND PURIFICATION

Scandium is normally prepared by reduction of ScF_3 by heating with calcium or lithium in an inert atmosphere using tantalum crucibles. Unfortunately, molten scandium dissolves up to 5 wt% tantalum, and the metal must be further processed to remove this tantalum. Tungsten is more inert, but tungsten crucibles are seldom used because they are more difficult to fabricate and are more brittle than tantalum.

The as-reduced metal is purified by vacuum-induction melting to remove the more volatile impurities. Then the scandium is distilled away from the residual tantalum and tungsten crucible contamination and other less volatile impurities. The scandium vapor is condensed in a closed-end tantalum tube inverted over the crucible that holds the metal to be purified, and as long as the condensed scandium is not melted it will not pick up any tantalum. Scandium can be purified by zone refining. But the purest metal is achieved by electrotransport purification of distilled scandium metal.

USES

The amount of scandium produced and consequently used annually is less than a few hundred kilograms. This is due to a lack of a good source of a raw material containing scandium. If scandium were less costly, much more would be used, but in most applications other, less-expensive elements are substituted. The demand for scandium has begun to increase and may spur further development of supply.

Metallurgy

Scandium modifies the grain size and increases the strength of aluminum. If added to magnesium together with silver, cadmium, or yttrium, it also strengthens magnesium alloys. Scandium inhibits the oxidation of the light rare earths and, if added along with molybdenum, inhibits the corrosion of zirconium alloys in high-pressure steam.

Ceramics

The addition of 1 mol of ScC to 4 mol TiC has been reported to form the second-hardest material known at the time. Sc_2O_3 can be used in many other oxides to improve electrical conductivity, resistance to thermal shock, stability, and density. In most of these applications other rare earth oxides are used because of cost. Laboratory balance knife-edges made from Sc_2O_3 are reported to be better than those made from sapphire.

Electronics

Scandium is used in the preparation of the laser material $Gd_3ScGa_4O_{12}$, gadolinium–scandium–gallium–garnet (GSGG). This garnet when doped with both Cr^{3+} and Nd^{3+} ions is said to be $3\frac{1}{2}$ times as efficient as the widely used Nd^{3+}-doped yttrium–aluminum–garnet (YAG:Nd^{3+}) laser. Ferrites and garnets containing scandium are used in switches in computers; in magnetically controlled switches that modulate light passing through the garnet; and in microwave equipment.

Lighting and Phosphors

The largest use of scandium outside of the research laboratory is probably in high-intensity lights. Scandium iodide is added because of its broad emission spectrum. Bulbs with mercury, NaI, and ScI_3 produce a highly efficient light output of a color close to sunlight. This is especially important when televising presentations indoors or at night. When used with night displays, the bulbs give a natural daylight appearance.

Scandium compounds may be used as host for phosphors or as the activator ion. Sc_2O_3, $ScVO_4$, and Sc_2O_2S are typical host materials, while $ZnCdS_2$ activated with a mixture of silver and scandium is a red, luminescent phosphor suitable for use in television. In most cases, other materials are used because of economics.

COMPOUNDS

Scandium Carbide

This is a gray powder of density 3.59 g/cm³ with a hexagonal structure, that is soluble in mineral acids and has potential as a high-temperature semiconductor.

Scandium Nitride

This gray to red powder with density of 3.6 g/cm³ has a face-centered (NaCl) structure. The material is of interest in space technology by virtue of its light weight and high melting point (2700°C). Also, scandium nitride crucibles are used for preparing single-crystal gallium arsenide or phosphide.

Scandium Oxide

This is a lightweight refractory oxide with a melting point of ~2300°C. The single-crystal density is 3.91 g/cm³ with a cubic structure.

Scandium oxide is now prepared in quantity from a variety of sources. It is now routinely separated from certain uranium tailings. Some recovery from beryllium ores can be expected plus potential recovery from some phosphate ores with up to 1 to 2% scandium values.

It appears to provide better service than high-alumina compositions. The oxide can be flame-sprayed onto a variety of surfaces where it shows heat and thermal shock resistance superior to zirconia, alumina, and magnesia. Single crystals of the oxide are superior to sapphire for balance knife edges or laser crystal host matrices.

SCREW

A screw is a cylindrical body with a helical groove cut into its surface. For practical purposes, a screw may be considered to be a wedge wound in the form of a helix so that the input motion is a rotation while the output remains translation. The screw is to the wedge much the same as the wheel and axle is to the lever in that it permits the exertion of force through a greatly increased distance.

The screw is by far the most useful form of inclined plane or wedge and finds application in the bolts and nuts used to fasten parts together; in lead and feed screws used to advance cutting tools or parts in machine tools; in screw jacks used to lift such objects as automobiles, houses, and heavy machinery; in screw-type conveyors used to move bulk materials; and in propellers for airplanes and ships.

SCREW-MACHINE PARTS

Among the many screw-machine parts are such things as bushings, bearings, shafts, instrument parts, aircraft fittings, watch and clock parts, pins, bolts, studs, and nuts. Screw-machine parts can be produced at rates up to 4000 parts per hour, and tolerances of ±0.03 mm are common. In addition, these parts can be made of practically any material that is machinable and that can be obtained in rod form. In comparison, die casting is restricted to a limited number of alloys of relatively low melting point, and cold heading requires materials that are ductile at room temperature.

However, screw-machine parts do have certain limitations: Because the material must be in the form of bar or rod, the cross section of the part is generally limited to a circle, hexagon, square, or other readily available extruded cross section, although a special extruded cross section can sometimes be ordered. The stock used must be as large as the greatest cross section of the part; depending on the shape, this may or may not result in large scrap losses. In general, irregular and nonsymmetrical parts are not good screw-machine parts, although some parts of this type are produced.

Parts can become quite expensive unless certain restrictions are observed in selection of material and in design of parts.

MATERIALS

Selection of the material is very important. Cost of producing a part is influenced greatly by ease of machining; poor machinability reduces tool life and causes frequent shutdown to replace tools. If service requirements permit, readily machinable materials, such as resulfurized steels or leaded brasses, should be given preference.

Some of the many metals used in screw-machine parts are carbon and low-alloy steels; stainless and heat-resisting steels; cast and malleable irons; copper alloys, particularly brass, bronze, and nickel–silver; aluminum and magnesium alloys; nickel and cobalt superalloys; and gold and silver. Among the nonmetallic materials used are vulcanized fiber, hard rubber, nylon, Teflon, methyl methacrylate, polyethylene, polystyrene, and phenolics.

Lot Size

Lot size should be considered in selecting the material, as it influences not only the cost of the part but the material that can be used.

DESIGN

Screw-machine parts should be designed to be made from standard compositions, standard stock sizes, and standard shapes. Specifying nonstandard bar not only increases costs but may result in delays in obtaining the parts.

Closer dimensions can be held but should be specified only when necessary, because they increase the cost of producing and inspecting the part.

Holes

Standard drill sizes should be specified if possible. When holes require reaming, sizes that can be finished with standard reamers should be specified.

Threads

A major item in the design of a screw-machine product is the selection of a proper standard thread, with the aim of providing interchangeability. In new designs, use the Unified Screw Thread System, adopted by the United States, Canada, and Great Britain. For existing designs, however, continue to use the American National System. Indicate threads by size, pitch, series, and class, only; using a special pitch and specifying major and minor diameters is often unnecessary and always expensive.

Concentricity

When concentricity is required, specify it in terms of total indicator reading (TIR) rather than as a dimension. It is good engineering practice to indicate the diameters that must run true by an arrow leading to a note reading "diameters marked A must be concentric to within __ TIR measured at points __." Specifying concentricity when it is not required increases cost of the part unnecessarily.

Burrs

Do not specify removal of burrs unless this operation is essential. Since there is some doubt about the definition of a burr, the screw-machine industry has established certain criteria:

1. Sharp corners are not considered burrs unless they have ragged edges and interfere with operation of the part.

2. Slight tears or roughness on the first two threads of a tapped hole or the male thread are not burrs unless they interfere with assembly.
3. A projection is not a burr if it must be found with a magnifying glass.

Burrs can be eliminated in most cases without additional cost by specifying chamfered or rounded corners on the parts.

Finish

In general, cost of finishing will depend on the degree of finish required; the finer the finish, the higher the cost. A finish for appearance only need not be as fine as a finish required on mating or bearing surfaces. In any case, finish should be specified in terms of microinch root-mean-square (rms) units. Microinch finish depends on material and operation and without secondary operations can range from the usual 125 rms to as low as 16 rms. The latter finish requires special care and is a very high cost operation on a production basis.

Dimensions

If a part requires heat treatment or plating, dimensions supplied to the part producer will depend on where these operations are to be done. If the supplier is to finish the part, give dimensions to apply after heat treatment or plating; if the buyer is to finish the part, give dimensions to apply before heat treating or plating.

SEALS (CERAMIC-TO-METAL SEALS)

The resultant fabrication after brazing depends on the integrity of the seal. Application in electronics demands that the seal withstand high vacuum and that the materials used in its fabrication do not outgas to poison the vacuum. There are several types of ceramic-to-metal seals.

Butt Seals

A flat metal washer brazed to a flat metallized ceramic surface is typical of this class. The principal stress is a shear stress that may break the ceramic in brittle fracture. Ceramic backup rings are frequently used to establish a uniform stress distribution, which materially improves the ruggedness of the joint. Ductility of the braze and of the metal hardware is also significant.

External Seals

A flange, or collar, surrounding a cylindrical ceramic part is typical of this class. The metal is selected for its slightly higher coefficient of expansion and produces the resultant seal.

Glass-to-Metal Seals

The coefficient of linear expansion of the glass is matched as closely as possible to that of the metal it is to join. Both materials are heated prior to forming the seal: the glass to flowing and the metal to a dull red heat. A thin oxide coating on the metal often is needed for good adhesion.

Internal Seals

These seals can be very reliable if a low-expansivity metal (tantalum, for example) can be used for the lead. Low-expansion metals result in compressive stresses but their poor oxidation resistance frequently precludes their use. Kovar and nickel–iron alloys, plus the high expansivity braze metals, ordinarily result in tensile stresses at the inner wall of the ceramic. As the ceramic is weakest in tension, this design may be troublesome unless properly engineered.

Ram Seal (Crunch Seal)

The metal sleeve with a soft plating such as copper, silver, or nickel is distorted while moving over the sharp ceramic edge. Ceramic–metal contact is limited to a thin circumferential line.

Taper Seal (Telescoping Seal)

A thin metal sleeve is forced onto a thick ceramic cylinder whose outside diameter is tapered. Negligible stresses are introduced into the ceramic but the metal may be stressed beyond its yield point. During thermal cycling, the metal may fail through fatigue.

SELENIUM

Selenium (symbol Se) is the the third member of group VIB of the periodic arrangement of the elements; it is more metallic than sulfur but less metallic than tellurium, the two adjacent elements of this group.

It is rarely found in its native state but is usually associated with lead, copper, and nickel from which it is recovered as a by-product.

CHEMICAL PROPERTIES

The chemical properties of selenium are very similar to other Group VIB elements such as sulfur and tellurium. It reacts readily with oxygen and halogens to form their respective compounds, i.e., oxides and halides. Selenous and selenic acids are easily prepared and in many respects resemble sulfurous and sulfuric acids. Selenium is not very soluble in aqueous alkali solutions, but selenites and selenates are readily prepared by fusion with oxidizing salts of the alkalis.

Selenium metal is odorless and tasteless, but the vapor has a putrid odor. The material is highly poisonous, and is used in insecticides and in ship-hull paints. Foods grown on soils containing selenium may have toxic effects, and some weeds growing in the Western states have high concentrations of selenium and are poisonous to animals eating them. Selenium burns in air with a bright flame to form selenium dioxide, SeO_2, which is in white, four-sided, crystalline needles. The oxide dissolves in water to form selenous acid, H_2SeO_3, resembling sulfurous acid but very weak.

FORMS

Commercial grades (99.5%) of selenium are usually sold as a powder of various mesh sizes packed in steel drums. High-purity selenium (99.99%), used primarily in the electronics industry, is available as small pellets.

Ferroselenium containing 50 to 58% selenium and nickel–selenium containing 50 to 60% selenium are available in the form of metallic chunks packed in wooden barrels or boxes.

APPLICATIONS

The most important uses for selenium are in the electronics industry, in such components as current rectifiers, photoelectric cells, xerography plates, and as a component in intermetallic compounds for thermoelectric applications.

A selenium dry-plate rectifier consists of a base plate of aluminum or iron, either nickel plated or coated with a thin layer of bismuth, a layer of 0.05 to 0.08 mm of halogenated selenium, an artificial barrier layer, and a counter-electrode of cadmium or a low-melting alloy. The purpose of the controlled amount of halogen in the selenium is to accelerate the transformation from amorphous to the hexagonal form, which is done at temperatures slightly above 210°C during the manufacturing process.

The photoelectric properties of selenium make it useful for light-measuring instruments and for electric eyes. Amorphous or vitreous selenium is a poor conductor of electricity, but when heated, it takes the crystalline form, its electrical resistance is reduced, and it changes electrical resistance when exposed to light. The change of electric conductivity is instantaneous; even the light of small lamps has a marked effect since the resistance varies directly as the square of the illumination. The pure amorphous powder is also used for coating nickel-plated steel or aluminum plates in rectifiers for changing alternating current to pulsating direct current. The coated plates are subjected to heat and pressure to change the selenium to the metallic form, and the selenium coating is covered with a layer of cadmium bismuth alloy. Selenium rectifiers are smaller and more efficient than copper oxide rectifiers, but they require more space than silicon rectifiers and are limited to an ambient temperature of 85°C.

Selenium is also used in steels to make them free-machining, with up to 0.35% used. Up to 0.05% of selenium may also be used in forging steels. From 0.017 to 0.024 kg of selenium per 0.9 metric ton of glass may be used in glass to neutralize the green tint of iron compounds. Large amounts produce pink and ruby glass. Selenium gives the only pure red color for signal lenses. Pigment for glass may be in the form of the black powder, barium selenite, $BaSeO_3$, or as sodium selenite, Na_2SeO_3, and may be used with cadmium sulfide. Selenium is also used as an accelerator in rubber and to increase abrasion resistance.

In copper alloys selenium improves machinability without hot-shortness. Selenium copper is a free-cutting copper containing about 0.50% selenium. It machines easily, and the electric conductivity is nearly equal to that of pure copper. The tensile strength of annealed selenium copper is about 207 MPa. Small amounts of selenium salts are added to lubricating oils to prevent oxidation and gummimg.

Xerography is a dry printing method for the production of images by light or other rays. A thin layer of selenium is given a strong positive electrical charge and the design is projected on it, which discharges the selenium in proportion to the light intensity, producing a latent image. This image is developed by dusting with negatively charged particles that adhere to the plate where it was not struck by the projected light. Some further simple steps allow the image to be used as a master for many additional prints.

One of the oldest uses of selenium is in the glass and ceramic industries. The familiar ruby red that is commonly seen in glass is produced by cadmium-selenide. These same cadmium-selenide reds find broad application in the field of ceramics and enamels.

SEMICONDUCTING MATERIALS

Semiconductors may be defined as materials that conduct electricity better than insulators, but not as well as metals. An enormous range of conductivities can meet this requirement. At room temperature, the conductivities characteristic of metals are on the order of 10^4 to 10^6 Ω/cm, while those of insulators range from 10^{-25} to 10^{-9} Ω/cm. The materials classed as semiconductors have conductivities that range from 10^{-9} to 10^4 Ω/cm. The conductivity of metals normally decreases with an increase of temperature, but semiconductors have the distinctive feature that in some range of temperature their conductivity increases rather than decreases with an increase of temperature.

Another criterion for a semiconductor is that the conduction process be primarily by electrons and not ionic, since the latter involves the transfer of appreciable mass as well as charge.

Materials used are ones that are capable of being partly conductors of electricity and partly insulators, and are used in rectifiers for changing alternating current to pulsating direct current, and in transistors for amplifying currents. They can also be used for the conversion of heat energy to electric energy, as in the solar battery. In an electric conductor the outer rings of electrons of the atoms are free to move, and provide a means of conduction. In a semiconductor the outer electrons, or valence electrons, are normally stable, but, when a doping element that serves to raise or lower energy is incorporated, the application of a weak electric current will cause displacement of valence electrons in the material. Silicon and germanium, each with a single stable valence of four outer electrons, are the most commonly used semiconductors. Elements such as boron, with a lower energy level but with electrons available for bonding and thus accepting electrons into the valence ring, are called hypoelectronic elements. Elements such as arsenic, which have more valence electrons than are needed for bonding and may give up an electron, are called hyperelectronic elements. Another class of elements, like cobalt, can either accept or donate an electron, and these are called buffer atoms. All of these types of elements constitute the doping elements for semiconductors.

In a nonconducting material, used as an electrical insulator, the energy required to break the valence bond is very high, but there is always a limit at which an insulator will break the bond and become a conductor with high current energy. The resistivity of a conductor rises with increasing temperature, but in a semiconductor the resistivity decreases with temperature rise, and the semiconductor becomes useless beyond its temperature limit. Germanium can be used as a semiconductor to about 93°C, silicon can be used to about 204°C, and silicon carbide can be used to about 343°C.

Extrinsic and Intrinsic Semiconductors

Metals for use as semiconductors must be of great purity, since even minute quantities of impurities would cause erratic action. The highly purified material is called an intrinsic metal, and the desired electron movement must come only from the doping element, or extrinsic conductor, that is introduced. The semiconductors are usually made in single crystals, and the positive and negative elements need be applied only to the surfaces of the crystal, but methods are also used to incorporate the doping element uniformly throughout the crystal.

The process of electron movement, although varying for different uses and in different intrinsic materials, can be stated in general terms. In the silicon semiconductor, the atoms of silicon with four outer valence electrons bind themselves together in pairs surrounded by eight electrons. When a doping element with three outer electrons, such as boron or indium, is added to the crystal, it tends to take an electron from one of the pairs, leaving a hole and setting up an imbalance. This forms the *p*-type semiconductor. When an element with five outer electrons, such as antimony or bismuth, is added to the crystal it gives off electrons, setting up a conductive band which is the *n*-type semiconductor. Fusing together the two types forms a *p-n* junction, and a negative voltage applied to the *p* side attracts the electrons of the three-valence atoms away from the junction so that the crystal resists electronic flow. If the voltage is applied to the *n* side it pushes electrons across the junction and the electrons flow. This is a diode, or rectifier, for rectifying alternating current into pulsating direct current. When the crystal wafers are assembled in three layers, *p-n-p* or *n-p-n*, a weak voltage applied to the middle wafer increases the flow of electrons across the whole unit. This is a transistor. Germanium and silicon are bipolar, but silicon carbide is unipolar and does not need a third voltage to accelerate the electrons.

Semiconductors can be used for rectifying or amplifying, or they can be used to modulate or limit the current. By the application of heat to ionize the atoms and cause movement they can also be used to generate electric current; or in reverse, by the application of a current they can be used to generate heat or remove heat for heating or cooling purposes in air conditioning, heating, and refrigeration. But for uses other than rectifying or altering electric current the materials are usually designated by other names and not called semiconductors. Varistors are materials, such as silicon carbide, whose resistance is a function of the applied voltage. They are used for such applications as frequency multiplication and voltage stabilization. Thermistors are thermally sensitive materials. Their resistance decreases as the temperature increases, which can be measured as close as 0.001°C, and they are used for controlling temperature or to control liquid level, flow, and other functions affected by rate of heat transfer. They are also used for the production or the removal of heat in air conditioning, and may then be called thermoelectric metals.

HALL EFFECT

Whether a given sample of semiconductor material is *n*- or *p*-type can be determined by observing the Hall effect. If an electric current is caused to flow through a sample of semiconductor material and a magnetic field is applied in a direction perpendicular to the current, the charge carriers are crowded to one side of the sample, giving rise to an electric field perpendicular to both the current and the magnetic field. This development of a transverse electric field is known as the Hall effect. The field is directed to one or the opposite direction depending on the sign of the charge of the carrier.

The magnitude of the Hall effect gives an estimate of the carrier concentration. The ratio of the transverse electric field strength to the product of the current and the magnetic field strength is called the Hall coefficient, and its magnitude is inversely proportional to the carrier concentration. The coefficient of proportionality involves a factor that depends on the energy distribution of the carriers and the way in which the carriers are scattered in their motion. However, the value of this factor normally does not differ from unity by more than a factor of 2. The situation is more complicated when more than one type of carrier is important for the conduction. The Hall coefficient then depends on the concentrations of the various of carriers and their relative mobilities.

The product of the Hall coefficient and the conductivity is proportional to the mobility of the carriers when one type of carrier is dominant. The proportionality involves the same factor that is contained in the relationship between the Hall coefficient and the carrier concentration. The value obtained by taking this factor to be unity is referred to as the Hall mobilty.

TYPES

Although some of the chemical elements are semiconductors, most semiconductors are chemical compounds. Table S.1 lists the elemental semiconductors and some of their characteristic properties. Compound semiconductors can be simple compounds, such as compounds formed from elements of the III-V, V-VI, II-IV, II-VI, I-VI, II-V, II-VI groups of the periodic table, as well as alloys of these compounds, ternary compounds, complicated oxides, and even organic complexes.

Characteristics of III-V compounds are shown in Table S.2 and examples of other compound semiconductors are given in Table S.3.

For organic semiconductors it is generally true that the larger the molecule, the higher the conductivity. Inclusion of atoms other than carbon, hydrogen, and oxygen in the molecule often appears to increase the conductivity. The phthalocyanine molecule, which is much like the active center of several biologically active compounds, e.g., chlorophyll, can accommodate a metallic atom in its center.

An organic semiconductor may be either n-type or p-type. For example, among the organic dyes, the cationic dyes appear to be n-type semiconductors and the anionic dyes p-type.

Magnetic ferrites are usually polycrystalline ceramics, composed of mixed semiconducting oxides, and possess useful magnetic properties combined with high electrical resistivities. The magnetically soft ferrites have the general form, MFe_2O_4, where M represents one or more of the following divalent metals, copper, magnesium, manganese, nickel, iron, and zinc. The crystallites have the cubic structure of the spinel system.

FABRICATION

Although some electronic devices require extremely small semiconductor elements, others require large masses of material. For example, some diodes and thermistors have volumes, exclusive of the lead wires, of less than 1 ten-millionth of a cubic inch, while some silicon domes require as much as a 1000 mm^3 of material, and xerographic and luminescent panels cover many square centimeters. Semiconductors are prepared in the form of single crystals (as large as several millimeters in diameter and several centimeters in length), polycrystalline ingots, thin films, or sintered ceramic shapes. Semiconductors are generally brittle materials and some make excellent abrasives. However, at higher temperatures (near their melting points) they can be deformed plastically in a manner analogous to engineering metals at room temperature.

Single crystals and polycrystalline ingots can be cut into the desired shapes by using diamond or other appropriate abrasive wheels or by using ultrasonic cutting tools and a boron carbide abrasive. A continuous wire and an abrasive slurry are sometimes used for cutting semiconductors. Lapping and etching are commonly used for preparing small semiconductor elements of the desired shape and surface condition. Large-area selenium xerographic plates are prepared by vacuum evaporation, whereas ceramic molding and sintering techniques are used in the preparation of ferrites and some luminescent materials.

Single crystals are grown by the horizontal Bridgman, the vertical Bridgman, Czochralski, floating zone, and vapor-phase deposition techniques. The impurities may be added to the starting materials or to the melt during growth. Controlled regions of impurities are created by diffusion, alloying, or vapor-phase decomposition to form epitaxial layers on a single-crystal substrate.

Some semiconductor elements are formed by use of a semiconductor powder and an appropriate binder. In the manufacture of magnetic ferrites the oxides are mixed in required proportions and milled to obtain a fine particle size. The powder is then partially sintered in a prefiring kiln, and subsequently crushed,

TABLE S.1
Properties of Elemental Semiconductors

Element	Crystal Structure	Density, g/cm³	Melting Point, °C	Linear Coefficient of Expansion, 10⁻⁶/°C	Energy Band Gap at 300 K, ev	Electron Mobility, cm²/v-s	Hole Mobility Light Mass, cm²/v-s	Hole Mobility Heavy Mass, cm²/v-s
B	—	2.34	2075	—	1.4	1	—	2
C (diamond)	Cub. (f.c.), O_h^7	3.51	3800	1.18	5.3	1800	—	1600
Si	Cub. (f.c.), O_h^7	2.33	1417	4.2	1.09	1500	1500	480
Ge	Cub. (f.c.), O_h^7	5.32	937	6.1	0.66	3900	14,000	1860
α-Sn	Cub. (f.c.), O_h^7	5.75	231.9	—	0.08	144,000[a]	—	1600[a]
As	Hex. (rhomb.), D_{3d}^5	5.73	814	3.86	1.2	—	—	—
Sb	Hex. (rhomb.), D_{3d}^5	6.68	630.5	10.88	0.11	—	—	—
α-S	Rhomb. (f.c.), V_h^{24}	2.07	112.8	64.1	2.6	—	—	—
Se	Hex., D_3^4	4.79	217	36.8	1.8	—	—	1
Se	Amorphous	4.82	—	—	2.3	0.005	—	0.15
Te	Hex., D_3^4	6.25	452	16.8	0.38	1100	10,000	700

[a] Values for carrier mobilities are experimental values obtained at 300 K, except in the case of gray tin where the 77 K values are given.

TABLE S.2
III–V Compounds and Their Properties

Compound	Crystal Structure	Density, g/cm³	Melting Point, °C	Linear Coefficient of Expansion, 10^{-6}/°C	Energy Band Gap at 300 K, ev	Hole Mobility[a]		Electron Mobility[a]	
						Light Mass, cm²/v-s	Heavy Mass, cm²/v-s	Light Mass, cm²/v-s	Heavy Mass, cm²/v-s
BN	Hex. (graphite), D_{6h}^4	2.2	3000	—	—	—	—	—	—
BN	Cub. (ZnS), T_d^2	—	—	—	4.6	—	—	—	500–1000
BP	Cub. (ZnS), T_d^2	—	>3000	—	—	—	—	—	—
BAs	Cub. (ZnS), T_d^2	—	—	—	—	—	—	—	—
AlN	Hex., C_{6v}^4	3.26	>2700	—	—	—	—	—	—
AlP	Cub. (ZnS), T_d^2	—	>2100	—	2.42	—	—	—	—
AlAs	Cub. (ZnS), T_d^2	—	1600	—	2.16	—	—	—	—
AlSb	Cub. (ZnS), T_d^2	4.28	1065	—	1.6	—	180–230	—	420–500
GaN	Hex., C_{6v}^4	—	1500	—	3.25	—	—	—	150–250
GaP	Cub. (ZnS), T_d^2	4.13	1450	5.3	2.25	—	120–300	—	70–150
GaAs	Cub. (ZnS), T_d^2	5.32	1238	5.7	1.43	8600–11,000	1000	3000	426–500
GaSb	Cub. (ZnS), T_d^2	5.62	706	6.9	0.70	5000–40,000	1000	7000	700–1200
InN	Hex., C_{6v}^4	—	1200	—	—	—	—	—	—
InP	Cub. (ZnS), T_d^2	4.79	1062	4.5	1.27	4800–6800	—	—	150–200
InAs	Cub. (ZnS), T_d^2	5.67	942	5.3	0.33	33,000–40,000	—	8000	450–500
InSb	Cub. (ZnS), T_d^2	5.78	530	5.5	0.17	78,000	—	12,000	750

[a] Since InSb is the only III–V compound that has been prepared with an impurity concentration low enough that the characteristic lattice carrier mobility can be determined, the best experimental value is given followed by the theoretical estimate for higher-purity material. All mobility values are for 300 K.

TABLE S.3
Compound Semiconductors

Formula	Typical Compounds
I-V	KSb, K$_3$Sb, CsSb, Cs$_3$Sb, Cs$_3$Bi
I-VI	Ag$_2$S, Ag$_2$Se, Cu$_2$S, Cu$_2$Te
II-IV	Mg$_2$Si, Mg$_2$Ge, Mg$_2$Sn, Da$_2$Si, Ca$_2$Sn, Ca$_2$Pb, MnSi$_2$, CrSi$_2$
II-V	ZnSb, CdSb, Mg$_3$Sb$_2$, Zn$_3$As$_2$, Cd$_3$P$_2$, Cd$_3$As$_2$
II-VI	CdS, CdSe, CdTe, ZnS, ZnSe, ZnTe, HgS, HgSe, HgTe, MoTe$_2$, RuTe$_2$, MnTe$_2$, BeS, MgS, CaS
III-VI	Al$_2$S$_3$, Ga$_2$S$_3$, Ga$_2$Se$_3$, Ga$_2$Te$_3$, In$_2$Se$_3$, In$_2$Se$_3$, In$_2$Te$_3$, GaS, GaSe, GaTe, InS, InSe, InTe
IV-IV	SiC
IV-VI	PbS, PbSe, PbTe, TiS$_2$, GeTe
V-VI	Sb$_2$S$_3$, Sb$_2$Se$_3$, Sb$_2$Te$_3$, As$_2$Se$_3$, As$_2$Te$_3$, Bi$_2$S$_3$, Bi$_2$Se$_3$, Bi$_2$Te$_3$, Ce$_2$S$_3$, Gd$_2$Se$_3$
A^{I}B^{II}C$_2^{VI}$	CuFeS$_2$
A^{I}B^{III}C$_2^{VI}$	C$_4$AlS$_2$, CuInS$_2$, CuInSe$_2$, CuInTe$_2$, AgInSe$_2$, AgInTe$_2$, CuGaTe$_2$
A^{I}B^{V}C$_2^{VI}$	AgSbSe$_2$, AgSbTe$_2$, AgBiS$_2$, AgBiSe$_2$, AgBiTe$_2$, AuSbTe$_2$, Au(Sb,Bi)Te$_2$
A$_3^I$B^{V}C$_3^{VI}$	Cu$_3$SbS$_3$, Cu$_3$AsS$_3$
A$_3^I$B^{V}C$_4^{VI}$	Cu$_3$AsSe$_4$
A^{II}B^{IV}C$_2^V$	ZnSnAs$_2$
Oxides	SrO, BaO, MnO, NiO, Fe$_2$O$_3$, BaFe$_{12}$O$_{19}$, Al$_2$O$_3$, In$_2$O$_3$, TiO$_2$, BeO, MgO, CaO, CdO, ZnO, SiO$_2$, GeO$_2$, ZnSiO$_3$, MgWO$_3$, CuO

milled, and granulated before being pressed or extruded into the required shape. The main firing cycle takes place at temperatures around 1250°C in a controlled atmosphere. As a result the parts are sintered into a dense, homogeneous ceramic. During sintering, the parts shrink some 20% in linear dimensions.

Applications

The technology and practical importance of semiconductor devices have been growing steadily in the past decade. The major applications of semiconductors can be divided into ten categories: diodes and transistors, luminescent devices, ferrites, special resistors, photovoltaic cells, infrared lenses and domes, thermoelectric devices, piezoelectric devices, xerography, and electron emission. The materials most commonly used for these applications are listed in Table S.4.

Indium antimonide, InSb, has a cubic crystal structure, and it is used for infrared detectors and for amplifiers in galvanomagnetic devices. Indium arsenide, InAs, also has a very high electron mobility, and is used in thermistors for heat-current conversion. It can be used to 816°C. Some materials can be used only for relatively low temperatures. Copper oxide and pure selenium have been much used in current rectifiers, but they are useful only at moderate temperatures, and they have the disadvantage of requiring much space. Indium phosphide, InP, has a mobility higher than that of germanium, and can be used in transistors above 316°C. Aluminum antimonide, AlSb, can be used at temperatures to 538°C. In lead selenide, PbSe, the mobility of the charge-carrying electrons decreases with rise in temperature, increasing resistivity. It is used in thermistors.

Bismuth telluride, Bi$_2$Te$_3$, maintains its operating properties between −46 and 204°C, which is the most useful range for both heating and refrigeration. When doped as a p-type conductor it has a temperature difference of 601°C and an efficiency of 5.8%. When doped as an n-type conductor the temperature difference is lower, 232°C, but the efficiency within this range is more than doubled. Lead telluride, PbTe, has a higher efficiency, 13.5%, and temperature difference of 582°C, but it is not usable below 177°C, and is employed for conversion of the waste heat atomic reactors at about 371°C.

Gallium arsenide has high electron mobility, and can be used as a semiconductor. Cadmium sulfide, CdS, is thus deposited as a semiconductor film for photovoltaic cells, or solar batteries, with film thickness of about 2 μm. When radioactive isotopes, instead of solar rays, are added to provide the activating agent, the unit is called an atomic battery, and the large area of transparent backing for the semiconductor is not needed.

Manganese telluride, MnTe, with a temperature difference of 982°C, has also been used as a semiconductor. Many other materials can be used, and semiconductors with temperature differences at different gradients can be joined in series electrically to obtain a wider gradient, but the materials must have no diffusion at the junction.

Cesium sulfide, CeS, has good stability and thermoelectric properties at temperatures to 1093°C, and has a high temperature difference, 1110°C. It can thus be used as a high-stage unit in conversion devices. High conversion efficiency is necessary for transducers, while a high dielectric constant is desirable for capacitors. Low thermal conductivity makes it easier to maintain the temperature gradient, but for some uses high thermal conductivity is desirable. Silver–antimony–telluride, $AgSbTe_2$, has a high energy-conversion efficiency for converting heat to electric current, and it has a very low thermal conductivity, about 1% that of germanium.

TABLE S.4
Semiconductor Applications and Materials Used (including some insulators)

1. Transistors, diodes, rectifiers, and related devices
   - Transistors — Ge, Si, GaAs
   - Diodes
     - Switching diodes — Ge, Si, GaAs
     - Varactor diodes — Ge, Si, GaAs
     - Tunnel diodes — Ge, Si, GaAs, GaSb
     - Photodiodes — Ge, Si, GaAs
     - Zenner diodes — Ge, Si, GaAs
     - Microwave diodes — Ge, Si, GaAs
     - Magnetodiodes — InSb
   - Power rectifiers — Cu_2O, Se, Si, Ge
   - Varistors — SiC, Cu_2O
2. Luminescent devices
   - Electroluminescence — ZnS, ZnO
   - Phosphers — ZnS, CdS, ZnO, $ZnSiO_3$, $MgWO_3$, $SrWO_4$
   - Lasers — Al_2O_3, CaF_2, $CaWO_4$, BaF_2, $SrMoO_4$, SrF_2, LaF_2, LaF_3, As_2S_3, $CaMoO_4$, $KMgF_3$, $(Ba,Mg)_2P_2O_7$
3. Ferrites
   - Soft — ZnO, MnO, NiO, Fe_2O_3
   - Permanent — $BaFe_{12}O_{19}$
4. Special resistors
   - Thermistors — B, U_3O_8, Si, $(NiMn)O_2$
   - Photoconductors — Ge, Se, CdS, CdSe, GaAs, InSb, PbSe, PbTe
   - Particle detectors — CdS, diamond, Si, GaAs
   - Magnetoresistors — InSb, InAs
   - Piezoresistors — Si, PbTe, BaSb
   - Cryosars — Ge
   - Bokotrons — Ge
   - Helicons — Fe_2S
   - Oscillators — Ge, Si, InSb
   - Chargistors — Ge, Si
5. Photovoltaic and hall effect
   - Photovoltaic cells — Se, Si, GaAs
   - Photoelectromagnetic cells — InSb, InAs
   - Hall effect devices — InSb, InAs, GaAs
6. Optical materials
   - Infrared lenses and domes — Ge, Si, Se, Al_2O_3, As_2S_3, MgO, TiO_2, $SrTiO_3$
7. Thermoelectric devices
   - Generators — PbTe, Bi_2Te_3, ZnSb, GeTe, MnTe, CeS, Bi_2Te_3
   - Refrigerators — Bi_2Te_3
8. Piezoelectric devices — $BaTiO_3$, $PbTiO_3$, $(PbZr)TiO_3$, $PbNbO_2$, CdS, GaAs
9. Xerography — Se, ZnO
10. Electron emission — BaO, SrO

The semiconductor-type intermetals are also used in magnetic devices, since the ferroelectric phenomenon of heat conversion is the electrical analog of ferromagnetism. Chromium–manganese–antimonide is nonmagnetic below about 250°C and magnetic above the temperature. Various compounds have different critical temperatures. Below the critical temperature the distance between the atoms is less than that which determines the line-up of magnetic forces, but with increased temperature the atomic distance becomes greater and the forces swing into a magnetic pattern.

Organic semiconductors fall into two major classes: well-defined substances, such as molecular crystals and crystalline complexes, isotatic and syndiotactic polymers; and disordered materials, such as atactic polymers and pyrolitic materials. Few of these materials have yet found commercial application.

Amorphous silicon containing hydrogen is promising for use in solar cells because of its low cost and suitable electrical and optical properties.

SENSITIZING COMPOUNDS

Supersensitizing compounds are metal salts in aqueous or organic solutions, which form an invisible film on the surface of glass and other ceramic surfaces. This film is not completely understood but is believed to be an ionizing or electronic effect that serves to initiate and hasten surface treatments such as silvering and plating. Supersensitizing refers to a second step involving the use of noble metal compounds, which further enhances the reduction properties of the metals about to be formed on the glass or ceramic surface.

Sensitizing compounds include aluminum compounds (basic aluminum acetate, aluminum chloride, aluminum formoacetate, aluminum nitrate), barium salts, boron trichloride, cadmium compounds, iron sulfate, tin chloride, titanium sulfate, and triethanolamine titanate.

Supersensitizing compounds include gold chloride, iridium salts, osmium compounds, palladium chloride, silver nitrate, and silver oxide.

SHAPE MEMORY ALLOYS

These are a group of metallic materials that can return to some previously defined shape or size when subjected to the appropriate thermal procedure. That is, shape memory alloys (SMA) can be plastically deformed at some relatively low temperature and, upon exposure to some higher temperature, will return to their original shape. Materials that exhibit shape memory only upon heating are said to have one-way shape memory, whereas those that also undergo a change in shape upon recooling have a two-way memory. Typical materials that exhibit the shape memory effect include a number of copper alloy systems and the alloys of gold–cadmium, nickel–aluminum, and iron–platinum.

A shape memory alloy may be further defined as one that yields a thermoelastic martensite, that is, a martensite phase that is crystallographically reversible. In this case, the alloy undergoes a martensitic transformation of a type that allows the alloy to be deformed by a twinning mechanism below the transformation temperature. A twinning mechanism is a herringbone structure exhibited by martensite during transformation. The deformation is then reversed when the twinned structure reverts upon heating to the parent phase.

Transformation Characteristics

The martensite transformation that occurs in shape memory alloys yields a thermoelastic martensite and develops from a high-temperature austenite phase with long-range order. The martensite typically occurs as alternately sheared platelets, which are seen as a herringbone structure when viewed metallographically. The transformation, although a first-order phase change, does not occur at a single temperature but over a range of temperatures that is characteristic for each alloy system.

There is a standard method of characterizing the transformation and naming each point in the cycle. Most of the transformation occurs over a relatively narrow temperature range, although the beginning and end of the transformation during heating and cooling actually extends over a much larger temperature range. The transformation also exhibits hysteresis in

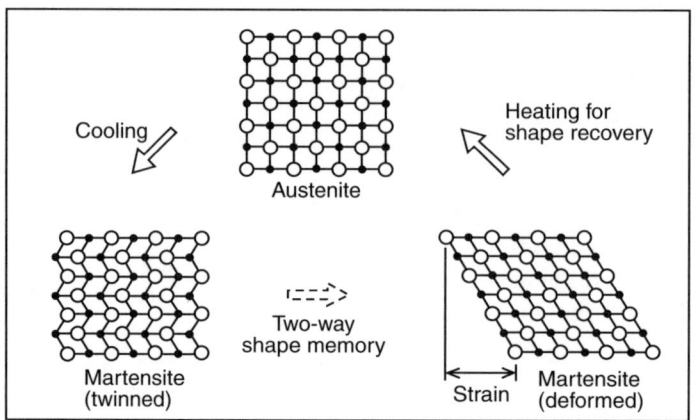

FIGURE S.2 Microscopic view of shape memory process. (*Aerospace Eng.*, April 2000, p. 26. With permission.)

that the transformation on heating and on cooling does not overlap. This transformation hysteresis varies with the alloy system.

THERMOMECHANICAL BEHAVIOR AND ALLOY PROPERTIES

The mechanical properties of shape memory alloys vary greatly over the temperature range spanning their transformation. The only two alloy systems that have achieved any level of commercial exploitation are the nickel–titanium alloys and the copper-base alloys. Properties of the two systems are quite different. The nickel–titanium alloys have greater shape memory strain, tend to be much more thermally stable, have excellent corrosion resistance compared to the copper-base alloys, and have higher ductility. The copper-base alloys are much less expensive, can be melted and extruded in air with ease, and have a wider range of potential transformation temperatures. The two alloy systems thus have advantages and disadvantages that must be considered in a particular application.

MATERIALS

The most common shape memory alloy material is Nitinol, an acronym for Ni (nickel)–Ti (titanium)–NOL (Naval Ordnance Laboratory). As suggested by the name, the material consists of approximately equal parts of nickel and titanium and was originally developed by the Naval Ordnance Laboratory.

Two different phases, martensite and austenite, are typically associated with the crystalline structure of shape memory alloys. The austenite phase is a highly ordered phase that occurs above a certain transition temperature. In this phase, the crystalline bonds must be at right angles with one another, as shown in the upper sketch in Figure S.2. The martensite phase occurs below the transition temperature, and right angle bonds are not required, as indicated by the twinned case in Figure S.2 (lower left). The lower right of the figure shows that strain is required to achieve a particular alignment, since the martensite phase is not in an ordered state.

The term *shape memory* is derived from the fact that a shape memory alloy can recover its original shape when heated above a certain transition temperature. Before a shape memory alloy such as Nitinol can present shape memory behavior, it must first be trained. The training process is illustrated in the upper drawing of Figure S.3, in which the shape memory alloy is annealed while it is constrained in the shape that it is to memorize. From a microscopic view, this corresponds to the austenite phase. For the Nitinol wire, the annealing process corresponds to a contraction in the length of the wire. After annealing, the material is said to have a one-way shape memory.

The Nitinol recovers its memorized shape when heated above the transition temperature and will remain in that shape upon cooling (lower left of Figure S.3). For the shape memory alloy to return to its initial shape, an external

SHAPE MEMORY ALLOYS

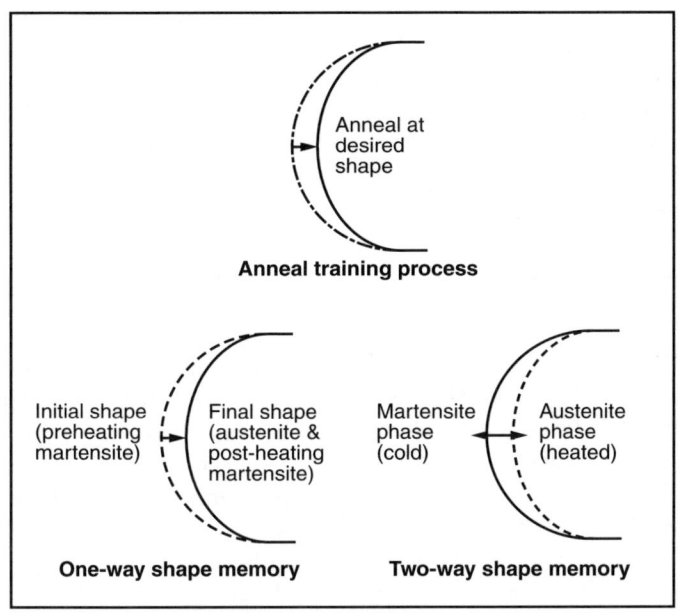

FIGURE S.3 Macroscopic view of shape memory process. (*Aerospace Eng.*, April 2000, p. 27. With permission.)

force must be applied to deform it. The two-way shape memory (lower right of Figure S.3) corresponds to a martensite phase to austenite phase by heating (shape recovery) followed by a mechanically produced deformation of the material upon cooling.

APPLICATIONS

The manufacturing and the technology associated with both of the two commercial classes of shape memory alloys are quite different as are the performance characteristics. Therefore, in applications where a highly reliable product with a long fatigue life is desired, the nickel–titanium alloys are the exclusive materials of choice. Typical applications of this kind include electric switches and actuators. However, if high performance is not mandated and cost considerations are important, then the use of copper–zinc–aluminum shape memory alloys can be recommended. Typical applications of this kind include safety devices such as temperature fuses and fire alarms.

Heating a shape memory alloy product to a temperature above some critical temperature is not recommended. The critical temperature for nickel–titanium is approximately 250°C, and for copper–zinc–aluminum approximately 90°C. Extended exposure to thermal environments above these critical temperatures results in an impaired memory function regardless of the magnitude of the load.

Another technical consideration in the practical application of shape memory alloys pertains to fastening or joining these materials to conventional materials. This is a significant issue because shape memory alloys undergo expansions and contractions not encountered in traditional materials. Therefore, if shape memory alloys are welded or soldered to other materials, they can easily fail at the joint when subjected to repeated loading. Alloys of nickel–titanium and copper–zinc–aluminum can also be brazed by using silver filler metals; however, the brazed region can fail because of cyclic loading. It is therefore desirable to devise some other mechanism for joining shape memory alloys to traditional materials. Shape memory alloys cannot be plated or painted for similar reasons.

Free Recovery

In this case, a component fabricated from a shape memory alloy is deformed while martensitic, and the only function required of the shape memory is that the component return to its previous shape upon heating. A prime application is the blood-clot filter in which a nickel–titanium wire is shaped to anchor itself in a vein and catch passing clots. The part is chilled so it can be collapsed and inserted into the vein; then body heat is sufficient to return the part to its functional shape.

Constrained Recovery

The most successful example of this type of product is undoubtedly a hydraulic coupling. These fittings are manufactured as cylindrical sleeves slightly smaller than the metal tubing that they are to join. Their diameters are then expanded while martensitic and, upon warming to austenite, they shrink in diameter and strongly hold the tube ends. The tubes prevent the coupling from fully recovering its manufactrured shape, and the stresses created as the coupling attempts to do so are great enough to create a joint that, in many ways, is superior to a weld.

Force Actuators

In some applications, the shape memory component is designed to exert force over a considerable range of motion, often for many cycles. Such an application is the circuit-board edge connector. In this electrical connector system, the shape memory alloy component is used to force open a spring when the connector is heated. This allows force-free insertion or withdrawal of a circuit board in the connector. Upon cooling, the nickel–titanium actuator becomes weaker, and the spring easily deforms the actuator while it closes tightly on the circuit board and forms the connections.

An example based on the same principle is a fire safety valve, which incorporates a copper–zinc–aluminum actuator designed to shut off toxic or flammable gas flow when fire occurs.

Other Applications

A number of applications are based on the pseudoelastic (or superelastic) property of shape memory alloys. Some eyeglass frames use superelastic nickel–titanium alloy to absorb large deformations without damage. Guide wires for steering catheters into vessels in the body have been developed using wire fashioned of nickel–titanium alloy, which resists permanent deformation if bent severely. Arch wires for orthodontic correction also use this alloy.

Shape memory alloys have found application in the field of robotics. The two main types of actuators for robots using these alloys are biased and differential. Biasing uses a coil spring to generate the bias force that opposes the unidirectional force of the shape memory alloy. In the differential type, the spring is replaced with another shape memory alloy, and the opposing forces control the actuation. A microrobot was developed with five degrees of freedom corresponding to the capabilities of the human fingers, wrist, elbow, and shoulder. The variety of robotic maneuvers and operations are coordinated by activating the nickel-titanium coils in the fingers and the wrist in addition to contraction and expansion of straight nickel–titanium wires in elbow and shoulders. Digital control techniques in which a current is modulated with pulse-width modulation are employed in all of the components to control their spatial positions and speeds of operation.

In medical applications, in addition to mechanical characteristics, highly reliable biological and chemical characteristics are very important. The material must not be vulnerable to degradation, decomposition, dissolution, or corrosion in the organism, and must be biocompatible.

Nickel–titanium shape memory alloys have also been employed in artificial joints such as in artificial hip joints. These alloys have also been used for bone plates, for marrow pins for healing bone fractures, and for connecting broken bones.

SHEET METAL FORMING

This is the process of shaping thin sheets of metal (usually less than 6 mm) by applying pressure through male or female dies or both. Parts formed of sheet metal have such diverse geometries that it is difficult to classify them. In all sheet-forming processes, excluding shearing, the metal is subjected to primarily tensile or compressive stresses or both. Sheet forming is accomplished basically by processes such as stretching, bending, deep drawing, embossing, bulging, flanging, roll forming, and spinning. In most of these operations there are no intentional major changes in the thickness of the sheet metal.

There are certain basic considerations that are common to all sheet forming. Grain size of the metal is important in that too large a grain produces a rough appearance when formed, a condition known as orange peel. For general forming, an American Society for Testing and Materials (ASTM) no. 7 grain size (average grain diameter 32 μm) is recommended. Another type of surface irregularity observed in materials such as low carbon steel is the phenomenon of yield-point elongation that results in stretcher strains or Lueder's bands, which are elongated depressions on the surface of the sheet. This is usually avoided by cold-rolling the original sheet with a reduction of only 1 to 2% (temper rolling). Since yield-point elongation reappears after some time, because of aging, the material should be formed within this time limit. Another defect is season cracking (stress cracking, stress corrosion cracking), which occurs when the formed part is in a corrosive environment for some time. The susceptibility of metals to season cracking depends on factors such as type of metal, degree of deformation, magnitude of residual stresses in the formed part, and environment.

Anisotropy or directionality of the sheet metal is also important because the behavior of the material depends on the direction of deformation. Anisotropy is of two kinds: one in the direction of the sheet plane, and the other in the thickness direction. These aspects are important, particularly in deep drawing.

Formability of sheet metals is of great interest, even though it is difficult to define this term because of the large number of variables involved. Failure in sheet forming usually occurs by localized necking or buckling or both, such as wrinkling or folding. For a simple tension-test specimen, the true (natural) necking strain is numerically equal to the strain-hardening exponent of the material: thus, for example, commercially pure annealed aluminum or common 304 stainless steel stretches more than cold-worked steel before it begins to neck. However, because of the complex stress systems in most forming operations, the maximum strain before necking is difficult to determine, although some theoretical solutions are available for rather simple geometries.

Considerable effort has been expended to simulate sheet-forming operations by simple tests. In addition to bend or tear tests, cupping tests have also been commonly used, such as the Swift, Olsen, and Erichsen tests. Although these tests are practical to perform and give some indication of the formability of the sheet metal, they generally cannot reproduce the exact conditions to be encountered in actual forming operations.

Stretch Forming

In this process the sheet metal is clamped between jaws and stretched over a form block. The process is used in the aerospace industry to form large panels with varying curvatures. Stretch forming has the advantages of low die cost, small residual stresses, and virtual elimination of wrinkles in the formed part.

Bending

This is one of the most common processes in sheet forming. The part may be bent not only along a straight line, but also along a curved path (stretching, flanging). The minimum bend radius, measured to the inside surface of the bend, is important and determines the limit at which the material cracks either on the outer surface of the bend or at the edges of the part. This radius, which is usually expressed in terms of multiples of the sheet thickness, depends on the ductility of the material, width of the part, and its edge conditions.

Springback in bending and other sheet-forming operations is due to the elastic recovery of the metal after it is deformed. Determination of springback is usually done in actual tests. Compensation for springback in practice is generally accomplished by overbending the part; adjustable tools are sometimes used for this purpose.

In addition to male and female dies used in most bending operations, the female die can be replaced by a rubber pad. In this way die cost is reduced and the bottom surface of the part is protected from scratches by a metal tool. The roll-forming process replaces the vertical motion of the dies by the rotary motion of rolls with various profiles. Each successive roll bends the strip a little further than the preceding roll. The process is economical for forming long sections in large quantities.

Rubber Forming

Although many sheet-forming processes are carried out in a press with male and female dies usually made of metal, there are four basic processes that utilize rubber to replace one of the dies. Rubber is a very effective material because of its flexibility and low compressibility. In addition, it is low in cost, is easy to fabricate into desired shapes, has a generally low wear rate, and also protects the workpiece surface from damage.

The simplest of these processes is the Guerin process. Auxiliary devices are also used in forming more complicated shapes. In the Verson–Wheelon process hydraulic pressure is confined in a rubber bag, the pressure being about five times greater than that in the Guerin process. For deeper draws the Marform process is used. This equipment is a packaged unit that can be installed easily into a hydraulic press. In deep drawing of critical parts the Hydroform process is quite suitable, where pressure in the dome is as high as 100 MPa. A particular advantage of this process is that the formed portions of the part travel with the punch, thus lowering tensile stresses, which can eventually cause failure.

Bulging of tubular components, such as coffee pots, is also carried out with the use of a rubber pad placed inside the workpiece; the part is then expanded into a split female die for easy removal.

Deep Drawing

A great variety of parts are formed by this process, the successful operation of which requires a careful control of factors such as blank-holder pressure, lubrication, clearance, material properties, and die geometry. Depending on many factors, the maximum ratio of blank diameter to punch diameter ranges from about 1.6 to 2.3.

This process has been extensively studied, and the results show that two important material properties for deep drawability are the strain-hardening exponent and the strain ratio (anisotropy ratio) of the metal.

Spinning

This process forms parts with rotational symmetry over a mandrel with the use of a tool or roller. There are two basic types of spinning: conventional or manual spinning, and shear spinning. The conventional spinning process forms the material over a rotating mandrel with little or no change in the thickness of the original blank. Parts can be as large as 6 m in diameter. The operation may be carried out at room temperature or higher for materials with low ductility or greater thickness. Success in manual spinning depends largely on the skill of the operator. The process can be economically competitive with drawing; if a part can be made by both processes, spinning may be more economical than drawing for small quantities.

In shear spinning (hydrospinning, floturning) the deformation is carried out with a roller in such a manner that the diameter of the original blank does not change but the thickness of the part decreases by an amount dependent on the mandrel angle. The spinnability of a metal is related to its tensile reduction of area. For metals with a reduction of area of 50% or greater, it is possible to spin a flat blank to a cone of an included angle of 3° in one operation. Shear spinning produces parts with various shapes (conical, curvilinear, and also tubular by tube spinning on a cylindrical mandrel) with good surface finish, close tolerances, and improved mechanical properties.

Miscellaneous Processes

Many parts require one or more additional processes: some of these are described briefly here. Embossing consists of forming a pattern on the sheet by shallow drawing. Coining consists of putting impressions on the surface by a process that is essentially forging; the best example is the two faces of a coin.

Coining pressures are quite high, and control of lubrication is essential to bring out all the fine detail in a design. Shearing is separation of the material by the cutting action of a pair of sharp tools, similar to a pair of scissors. The clearance in shearing is important to obtaining a clean cut. A variety of operations based on shearing are punching, blanking, perforating, slitting, notching, and trimming.

Die materials used are cast alloys, die steels, and cemented carbides for high-production work. Nonmetallic materials such as rubber, plastics, and hardwood are also used as die materials. The selection of the proper lubricant depends on many factors, such as die and workpiece materials, and severity of the operation. A great variety of lubricants are commercially available, such as drawing compounds, fatty acids, mineral oils, and soap solutions.

Pressures in sheet-metal forming generally range between 7 and 55 MPa (normal to the plane of the sheet); most parts require about 10 MPa.

SHEET METAL PARTS

Stamping and pressing make up a large family of metal-forming processes. Included in this group are blanking, pressing, stamping, and drawing, all of which are used to cut or form metal plate, sheet, and strip. The steps common to all stamping and pressing operations are the preparation of a flat blank and shearing or stretching the metal into a die to attain the desired shape.

In drawing, the flat stock is either formed in a single operation, or progressive drawing steps may be needed to reach the final form. In spinning, flat disks are dished by a tool as they revolve on a lathe.

Stamping involves placing the flat stock in a die and then striking it with a movable die or punch. Besides shaping the part, the dies can perform perforating, blanking, bending, and shearing operations. Almost all metals can be stamped. In general, stampings are limited to metal thicknesses of 9.5 mm or less. Pressing and drawing operations can be performed on cold metals up to 19 mm thick and up to about 89 mm on hot metals.

In recent years many new press-forming and drawing techniques have been developed. A number of them make use of rubber pads, bags, and diaphragms as part of the die or forming elements. Some involve stretch forming over dies. Others combine forming and heat-treating operations. And still other methods, known as high energy rate forming, employ explosive, electrical, or magnetic energy to produce shock waves that form the material into the desired shape.

STAMPING

Frequently, secondary operations, such as annealing in furnaces, trimming in lathes or rolls, brake bends, tapping, etc., make it difficult to define a part as a stamping. Such operations, as well as finishing operations, may cost more than comparatively fast and economical press operations.

Sizes and Materials

Presses and presslike machines are not necessarily limited to sheet-metal forming. They punch paper doilies and cut uppers for shoes. Multislide machines form round or flat wire; some presses impact-extrude aluminum, zinc, and steel into deep shells for toothpaste tubes or shell bodies, using slugs cut from bars, or cast for the purpose, or sometimes punched from sheets and plates, in bellmouth dies to burnish the edges. Presses forge from billets and they compress powdered metal and carbon into compacts.

Types of Work

Presses perform such operations as blanking (some blanks are made on shears) and cutting off; piercing (punching) holes, cutouts, or extruding holes; bending to almost any angle (this would include lance forming of tabs); embossing; or forming strengthening ribs or shallow pockets and hemming (bending edges up and then back flat on themselves). The edges of shells may be curled in or out. Coining changes thicknesses, for the raw material of stamping is almost always uniform.

Related to blanking are trimming operations to remove excess stock and shaving, a slight removal of stock. Examples are the first step of broaching to obtain close tolerances, small teeth, and straight edges where the breakout on stampings is objectionable, or to improve edge appearance to about 125 µin. on thinner materials, and about 150 μin. on stock 2.3 mm and over.

SHEET MOLDING COMPOUND

Sheet molding compound (SMC) combined with matched-metal-die compression molding has been the most successful glass fiber composite in automotive exterior body panel applications. In the medium and heavy truck market, SMC is a clear winner. In the more aesthetically demanding car market, however, SMC remains a niche player relegated to medium-volume speciality vehicles. More recent successes point to a shift in focus to more functional and structural applications.

The most successful applications address the total system needs and offer unmatched value in terms of high quality, dependability and cost effectiveness. The radiator support assembly is such a case. This illustrates that SMC can win and retain large volume applications resulting in profitable growth. SMC solutions are perhaps the best known and most developed and thus offer valuable lessons for all composites.

SMC Developments

Compression molded SMC is perhaps the most talked about composite system for automotive applications. SMC is a versatile composite capable of satisfying structural and aesthetic needs at fast molding cycles, and it is suitable for large production volumes.

SMC is an accepted and commercially proven material for automotive part manufacture, serving the role of composite ambassador to the automotive world. SMC offers capital efficiency vs. steel. Capital efficiency makes SMC a viable option for both styling differentiation and new capacity expansion.

Because of longer cycles and more defects, however, the cost of low-density SMC remains high, relegating it to but a few niche applications. This illustrates that the tension resulting from simultaneous needs of the original equipment manufacturers (OEMs) for low cost and weight reduction cannot be resolved with higher-cost technology.

SMC applications fall into three major categories: functional, structural, and appearance. Recent gains in functional applications, such as oil pans and heat shields, have been impressive. The main driver of this achievement has been resin development. Both elevated temperature and oil resistance needs are now being met with polyester resins, which are less expensive than the vinyl esters previously employed.

Structural applications offer deeper insights into the capabilities and limitations of the SMC. Two such applications, the radiator support assembly and cross car beam, both illustrate the capability of SMC for parts consolidation. Yet, the former leads to a low-cost product, whereas the latter remains a premium niche component. Some reasons behind this outcome can be explained from the total system perspective.

Appearance or Class A applications are of importance to both growth and survival of SMC. The opportunity for SMC in this arena is huge and significant in-roads have been made. Yet, SMC has penetrated only a small fraction of this potential and only, to a significant degree, in the United States. The battle with steel in this arena has been examined using cost models.

Structural Applications

Structural applications attempt to exploit the design freedom and parts consolidation potential of SMC leading to integration of multiple functional roles. Integrated front-end system design, for the Ford Taurus and Mercury Sable, incorporates an upper and lower radiator support molded in SMC. The lower radiator support consolidates 22 steel parts into two SMC parts, resulting in a 14% cost reduction vs. steel.

Another application for SMC is cross car beams, which are also known as supported instrument panels. Because of their use on light truck platforms, cross car beams consume significant amounts of SMC. The level of functional integration is as large as in the case of the radiator support but the end result is very different. The added complexity pushes the boundaries of SMC flow and moldability. To make matters worse, attempts have been made to use very low density SMC, resulting in even more molding difficulties. In the end, low cost has not been achieved, making both OEMs and molders unhappy. Perhaps a simpler design, incorporating a steel structural insert, might be more successful.

These two examples illustrate that parts consolidation is a useful tool for achieving low cost; however, there are limits to this approach and often the increased complexity leads to higher cost than in an equivalent system of many specialized components.

SHELL MOLD CASTINGS

Shell mold casting is a process that uses relatively thin-wall mold made by bonding silica or zircon sand with a thermosetting phenolic or urea resin. It has gained widespread use because it offers many advantages over conventional sand castings. Shell mold casting is a practical and economical way to meet the demand for weight reduction, thinner sections, and closer tolerances.

Advantages

There are five basic advantages:

1. Lower costs. High production rates and fewer finishing operations result in a lower unit cost for applicable parts.
2. Closer tolerances. Shell castings have closer tolerances than sand castings. Draft allowances are also reduced.
3. Smoother surface finish. Shell molding provides an improved surface compared with sand casting (250 to 1000 μin. rms for sand, 125 to 250 for shell).
4. Less machining. The precision of shell molding reduces, and in many cases eliminates, machining or grinding operations.
5. Uniformity. Insulating properties of the molds produce casting surfaces free of chill and with a more uniform grain structure.

Although a shell mold is more expensive than a green sand mold, the possibility of reducing weight, minimizing machining, and eliminating cores often results in savings sufficient to offset the mold cost.

The Process

The shell molding process can be broken down into five operations:

1. A match plate pattern is made from tool steel with dimensions calculated to allow for subsequent metal shrinkage.
2. The resin–sand mixture is applied to the metal pattern, which is then heated to 218 to 232°C. The hot pattern melts the resin, which flows between the grains of sand and binds them together. Thickness of the mold increases with time. After the desired thickness is reached, excess unbonded sand is poured from the pattern.
3. The pattern, with the soft shell adhering, is placed in an oven and heated to 566 to 649°C for 30 s to 1 min. This cures the shell and produces a hard, smooth mold that reproduces the pattern surface exactly. The shell is stripped from the pattern by ejector pins. The other half of the mold is produced in the same way.
4. Sprues and risers are opened and cores inserted to complete the cope and drag halves of the mold.
5. The two shell halves are glued together under pressure to form a tightly sealed mold that can be stacked and stored indefinitely.

Shell Molded Cores

Shell molded cores are an offshoot of the shell molding process. These cores have several advantages over the sand cores that they replace: the shell core usually costs less than the sand core; strength and rigidity permit handling without damage or distortion; sharp details, including threads, are accurately reproduced; core weight is reduced; cured cores are unaffected by moisture and can be stored for long periods of time; most shell cores are hollow and can function as vents during casting.

SILICA

Silica, as the dioxide of silicon is commonly termed, is the principal constituent of the solid crust of the Earth. Consequently, it is a major ingredient of most of the nonmetallic, inorganic materials used in industry. Silica occurs in a number of allotropic forms, which have different properties and different uses.

The principal modifications of silica are quartz, silica glass, cristobalite, and tridymite. The last two are sometimes combined under the term of inverted silica. The valuable physical properties of these four principal modifications lead to a wide variety of applications for silica in industry and technology.

QUARTZ

The common low-temperature form of silica, quartz, is strong and insoluble in water. Consequently, sandstone, which is composed of grains of quartz held together by a siliceous cement, is an excellent building stone. Huge quantities of silica stone are used as aggregate for concrete. One curious variety has a structure that renders the rock flexible and is known as itacolumite or flexible sandstone. Quartzite, in which the quartz grains of an original sandstone have recrystallized and grown to a compact mass, and vein quartz are too hard and difficult to shape for use as building stones.

Rock crystal, as the enhedral quartz found in nature is termed, is often carved into ornaments of great beauty because of its perfect transparency and because of the high luster given it by polishing. The "crystal ball" of story and romance is a polished quartz sphere. Colored varieties of quartz are much valued as semiprecious stones. The purple amethyst, blue sapphire quartz, yellow citrine or false topaz, red or pink rose quartz, smoky quartz, and the dark brown morion are examples.

Sandstone is also useful as an abrasive. It is used for grindstones and pulpstones. "Berea" grit from northern Ohio and novaculite from Arkansas are examples of coarse and fine-grained natural stones that are still preferred for some purposes to the artificial abrasives made from silicon carbide or aluminum oxide.

The strength and hardness of various natural forms of silica leads to use for crushing and for grinding. Flint pebbles are used in ball mills for grinding all sorts of materials. Agate mortars are invaluable in the chemical laboratory for pulverizing minerals before analysis. Sandstone and quartzite are used for millstones and buhr-stones to crush and grind grain, paint pigments, fertilizers, and many other products. Quartz sand, driven by compressed air, constitutes the useful sandblast for cleaning metal, decorating glass, and refacing stone buildings. Coated on paper, quartz grains provide the carpenter and cabinet maker with the familiar "sandpaper," although much modern sandpaper is really coated with crushed glass containing only about 70% silica.

The abrasive properties of quartz grains are also employed by locomotives to gain traction on steel rails, in billiard cue chalk, and in the chicken gizzard, where quartz grains are used as a kind of natural ball mill.

The thermal properties of the various forms of silica also lead to important uses. Its high melting point (1710°C) makes it a good refractory, while its low cost and low thermal expansion have brought about a wide use in industrial furnaces, particularly for melting steel and glass.

Properties

In general, the thermal expansion of all of the common forms of silica is low at high temperatures. This makes silica refractories capable of withstanding sudden temperature changes and large thermal gradients very well in these high-temperature ranges. At lower temperatures, large volume changes, due to rapid inversions from the high-temperature crystalline forms of quartz, tridymite, and cristobalite to corresponding low-temperature forms, make these materials rather sensitive to sudden temperature changes. The low-temperature forms of quartz, tridymite, and cristobalite also have higher thermal expansions than the high-temperature forms and the amorphous form of silica. This form of silica, variously known as silica glass, vitrosil, and fused quartz, is thus the only form of silica that may be heated rapidly from room temperature without fear of breakage. A further limitation on the thermal behavior of silica refractories is occasioned by the large volume increase that occurs as quartz changes slowly to tridymite or cristobalite at temperatures above 870°C. This may lead to swelling and warping of silica bricks if they are not first converted to the forms stable at high temperatures by prolonged firing. Because the volume changes that occur when tridymite changes from its high-temperature form to the forms stable at lower temperatures are less than the corresponding change for cristobalite, an effort is made to convert the silica refractory as completely to tridymite as is feasible before using it.

Uses

Silica bricks are made from crushed quartzite rock, known as ganister, which is bonded with 1.5 to 3.0% of lime. They are molded smaller than the dimensions desired in the finished brick to allow for an expansion of about $3/8$ in./ft as the quartz inverts to tridymite during firing. A firing schedule of about 20 days at 1450°C (about cone 16) is required. Study of the phase diagram of the silica-lime system shows why considerable quantities of lime may be used to bond the quartzite in silica bricks without loss of refractoriness. The lime is taken up in an immiscible lime-rich glass phase of which only a small amount is formed because of its high lime content.

In use, silica bricks are characterized by retention of rigidity and load-bearing capacity to temperatures above 1600°C, without the slow yield characteristic of fireclay brick. If a cold silica brick is heated suddenly, it spalls and disintegrates owing to the sudden volume changes taking place at the high-low inversions of tridymite and cristobalite. If heated cautiously through this sensitive temperature region, silica brick is very resistant to temperature shock.

The refractory properties of quartz sand are also employed in molds for cast iron. Huge quantities of sand, with varying amounts of clay impurity to bond the grains, are used for this purpose in the foundries of the world.

Crystal quartz in piezoelectric oscillators have become indispensable for the control of radio transmitters, radar equipment, and other electronic "timing" gear.

This use and other electrical and optical uses of crystal quartz depend upon the symmetry of the material. Some experts state that low quartz is the most important and most familiar example of the trigonal enantiomorphous hemihedral or trigonal trapezohedral class of symmetry. This class is characterized by one axis of threefold symmetry with 3 axes of twofold symmetry perpendicular thereto and separated by angles of 120°. This class has no plane of symmetry and no center of symmetry.

Optical Properties

The transparency of crystal quartz, particularly to the short waves of the ultraviolet, makes it very useful in optical instruments. Quartz prisms are employed in spectographs for analysis of light waves varying in length from almost 5 μm in the infrared to almost 0.2 μm in the ultraviolet. Lenses made of crystal quartz are also useful over this range for photography and microscopy. Crystal quartz is more transparent than fused quartz for these purposes, but its birefringence introduces complications in design.

Quartz also has the property of rotating the plane of vibration of polarized light traveling along its axis. This property is associated with its left- or right-handed character. Sugar solutions have the same optical rotating power and quartz plates or wedges are employed as standards in optical devices used for analyzing sugar solutions by means of this rotary power. Many other organic chemicals that have asymmetric right- or left-handed molecules can be studied and assayed by this optical means.

The variation of this rotary power of quartz with wavelength or color of light makes it possible to construct monochromators capable of selecting light of a desired color from a white source with very large optical apertures.

AMORPHOUS SILICA

The amorphous form of silica also has mechanical, optical, thermal, and electrical properties, which make it very useful in hundreds of technical applications. Vitreous silica is made by three different processes. The earliest process, which is still used, consists simply of melting quartz by application of high temperature. The oxyhydrogen flame was the first source of heat for this process and is still used to melt fragments of rock crystal that are combined to form rods and other shapes of silica glass. Electrical heat from graphite resistors is more commonly employed at present, however, since there is less trouble from volatilization of the silica. Large masses can be made by this technique and shaped into various useful forms although the very high temperature required for working silica glass makes this method of manufacture difficult and expensive.

To obtain a clear transparent product, crystal must be melted. If sand is used, the product is white and opaque because of the numerous air bubbles trapped in the very viscous glass. It is impossible to heat silica glass hot enough to drive out the bubbles or, in glass parlance, to "fine" it. This opaque "vitrosil" made by fusing sand, however, is very useful for chemical apparatus, when the low thermal expansion and insolubility of the silica glass play a role.

Recently, the Corning Glass Works developed a new process for making silica glass by hydrolysis of silicon tetrachloride in a flame. The resulting silica is deposited directly on a support in the form of transparent silica glass, which is more homogeneous and purer than the glass made by fusing quartz.

Large pieces can be made by this process and the unusual homogeneity of the product makes it especially useful for optical parts and for sonic delay lines where striations in the older fused quartz are detrimental. The sonic delay lines are polygons of the silica glass in which acoustic waves travel on a long path, being reflected at the numerous polygonal faces.

Finally, silica glass of 96% purity or better is made by an ingenious process invented by the Corning Glass Works. In this third process for making vitreous silica, an object is shaped first from a soft glass containing about 30% of borax and boric acid as fluxes. A suitable heat treatment causes a submicroscopic separation of two glass phases. One phase, which is composed chiefly of the fluxes, is then leached from the glass in a hot, dilute nitric acid bath. If the composition and heat treatment are exactly right, the high silica phase is left as a porous "sponge" with the shape and size of the original soft glass article. This is carefully washed and dried and fired by heating to about 1200°C.

In the firing step, the porous silica sponge shrinks to a dense transparent object of glass containing 96% SiO_2 or more, which has the desirable properties of silica glass made by either of the other processes. The 4% of impurity in the glass made by this process consists chiefly of boric oxide. Because of its presence this glass is appreciably softer than pure silica glass. For corresponding viscosities the "Vycor" 96% silica glass requires about 100°C lower temperature.

Silica glass made by any of these processes has many uses because of its unique combination of good properties.

Fibers of silica glass have very high tensile strength and almost perfect elasticity. This makes them useful in constructing microbalances, electrometers, and similar instruments. Silica fiber in diameters as small as 0.000076 cm comes in random matted form or in rovings.

Because of its small thermal expansion, fused silica is very resistant to sudden temperature changes. It is also a very hard glass so that it may be used in the laboratory for crucibles and combustion tubes to much better advantage than ordinary glass. Vitreous silica makes possible the construction of thermometers operating up to 1000°C. The small thermal expansion and durability of fused silica have led to its use in fabrication of standards of length. Fused silica plates, ground to optical flatness, are used in interferometers for measurement of thermal expansion.

The fact that fused silica is an excellent insulator with little or no tendency to condense surface films of moisture makes it valuable in the construction of electrical apparatus. It has a very high dielectric strength and low dielectric loss.

Silica glass also has very good transmission for visible and ultraviolet light. This makes it useful for the construction of mercury lamps and other optical equipment. In the mercury lamps the strength and heat resistance of silica glass make it possible to operate with high internal pressures, producing very high light intensities and great efficiency. The 96% silica glass made by the Vycor process may be fired in a reducing atmosphere or in vacuum to modify and improve its light-transmitting properties.

Silica glass is among the most chemically resistant of all glasses. This makes it particularly useful in the analytical laboratory where there is the added advantage that any contamination of contained solutions can only be by the one oxide, silica. Crucibles of fused silica glass may be used for pyrosulfate fusions. Condensers of fused silica are extremely useful for distilling acids (except hydrofluoric) and for preparation of extremely pure water.

Another useful form of silica is obtained by dehydrating silicic acid. In this way a porous "gel" is obtained that has an enormous surface area and is capable of adsorbing various gases and vapors, particularly water vapor. Silica gel is used in dehumidifiers to remove water vapor from the air. It is also used as a catalyst and as a support for other catalysts.

The porous 96% silica skeleton obtained in the Vycor process also has a very large surface and is useful as an adsorbent and drying agent. It has less capacity than silica gel, but better mechanical strength.

A very finely divided form of silica known as silica soot is obtained by hydrolizing silicon tetrachloride in a flame without heating the product hot enough to consolidate it as a glass. This material is valuable as a thermal insulator and as a white filler for rubber and plastics.

OTHER USES

The silica (actually ground silica) used in the pottery industry is called flint. The addition of flint affects warpage very little.

In ceramic bodies, potters' flint or pulverized quartz or sand is the constituent that reduces drying and burning shrinkage and assists promotion of refractoriness. Flint has an important bearing on the resistance of bodies to thermal and mechanical shock, because of the volume changes that accompany crystal transformation. In the unburned body, it lowers plasticity and workability, lowers shrinkage, and hastens drying. A coarse crystalline form of quartz, called macrocrystalline quartz, is more often used for potters' flint than the cryptocrystalline form.

Silica is used in all glazes as the chief, and often the only, acid radical (RO_2 group). It may be adjusted to regulate the melting temperature of the glaze. In common glazes, the ratio of silica to bases (RO group) is never less than 1:1 nor more than 3:1. By varying the relative proportions of the RO group and balancing the group against any desired silica content, the maturing temperature of a glaze may be quite closely controlled. In other words, the fusibility of the glazes used in the presence of equal proportions of fluxes depends on their relative silica contents.

In porcelain enamels, it may be taken as a general rule that, other things remaining constant, the higher the percentage of silica, the higher will be the melting point of the enamel and the greater its acid resistance. Silica has a low coefficient of expansion and increasing it in an enamel lowers the coefficient of expansion of that enamel. One method of regulating an enamel coating is to increase the silica content when the enamel is inclined to split off in cooling. Silica in the form of flint or quartz is used in both ground-coat and cover-coat enamels, and it has the same effect in either type.

The temperature required for melting an enamel is materially affected by the fineness of the silica. Cryolite, antimony, and tin oxide give their maximum value as opacifiers with minimum heat treatment in the smelter. The form and fineness of the silica should, therefore, be carefully watched and allowed for in compounding the batch. All forms of silica may be used with good results, but experience has shown that, in the same enamel, a smaller quantity of sand than of powdered quartz is necessary. Similarly, less sand than flint should be used, but the difference in this case is less than in the former. High SiO_2 tends to harden the enamel. The lower limit established in the usual run of enamels is 1.1 equivalents.

In the manufacture of semiconductors, monolithic circuits, and integrated circuits for the electronics industry, the use of fused quartz is widespread for plumbing and diffusion furnace muffles. The need to prevent product contamination makes this choice mandatory.

Irish Refrasil is 98% silica and has a green color. It is used for ablative protective coatings. It resists temperatures to 1588°C. Silica flour, made by grinding sand, is used in paints, as a facing for sand molds, and for making flooring blocks. Silver bond silica is water-floated silica flour of 98.5% SiO_2, ground to 325 mesh. In zinc and lead paints it gives a hard surface. Pulverized silica, made from crushed quartz, is used to replace tripoli as an abrasive. Ultrafine silica, a white powder having spherical particles of 4 to 25 μm, is made by burning silicon tetrachloride. It is used in rubber compounding, as a grease thickener, and as a flatting agent in paints. Aerosil is this material.

A polymer-impregnated silica, Polysil, has twice the dielectric strength of porcelain as well as better strength. It is also cheaper to make, and its composition can be tailored to meet specific environmental and operating conditions.

Silica aerogel is a fine, white, semitransparent silica powder; its grains have a honeycomb structure, giving extreme lightness. It weighs 40 kg/m^3 and is used as an insulating material in the walls of refrigerators, as a filler in molding plastics, as a flatting agent in paints, as a bodying agent in printing inks, and as a reinforcement for rubber. It is produced by treating sand with caustic soda to form sodium silicate, and then treating with sulfuric acid to form a jellylike material called silica gel, which is washed and ground to a fine dry powder. It is also called synthetic silica.

Silicon monoxide, SiO, does not occur naturally but is made by reducing silica with carbon in the electric furnace and condensing the vapor out of contact with the air. It is lighter than silica, having a specific gravity of 2.24, and is less soluble in acid. It is brown powder valued as a pigment for oil painting, as it takes up a higher percentage of oil than ochres or red lead. It combines chemically with the oil. Fumed silica is a fine translucent powder of the simple amorphous silica formula made by calcining ethyl silicate. It is used instead of carbon black in rubber compounding to make light-colored products, and to coagulate oil slicks on water so that they can be burned off. It is often called white carbon, but the "white carbon black" called Cab-O-Sil, used for rubber, is a silica powder made from silicon tetrachloride. Cab-O-Sil EH5, a fumed colloidal form, is used as a thickener in resin coatings. The thermal expansion of amorphous fused silica is only about one-eighth that of alumina. Refractory ceramic parts made from it can be heated to 1093°C and cooled rapidly to subzero temperatures without fracture.

SILICA MINERALS

Silica (SiO_2) occurs naturally in at least nine different varieties (polymorphs), which include tridymite (high-, middle-, and low-temperature forms), cristobalite (high- and low-temperature forms), coesite, and stishovite, in addition to high (β) and low (α) quartz. These forms have distinctive crystallography and optical characteristics.

The transformation between the various forms are of two types. Displacive transformations, such as inversions between high-temperature (β) and low-temperature (α) forms, result in a displacement or change in bond direction but involve no breakage of existing bonds between silicon and oxygen atoms. These transformations take place rapidly over a small temperature interval and are reversible. Reconstructive transformations, in contrast, involve disruption of existing bonds and subsequent formation of new ones. These changes are sluggish, thereby permitting a species to exist metastably outside its defined pressure-temperature stability field. Two examples of reconstructive transformations are tridymite $\rightleftharpoons$ quartz and quartz $\rightleftharpoons$ stishovite.

SILICIDES

Silicides are a group of substances, usually compounds, comprising silicon in combination with one or more metallic elements. These hard, crystalline materials are closely related to intermetallic compounds and have, therefore, many of the physical and chemical characteristics and some of the mechanical properties of metals.

Silicides are not natural products. They received but little attention prior to the development of the electrical furnace, which provided the first practical means of attaining and controlling the high temperatures generally required in their preparation.

Composition

Although a majority of the metals react with silicon, many of the resulting silicides do not have the properties usually required in engineering materials. The silicides that appear most promising for practical utilization in engineering and structural applications are, with few exceptions, limited to those of the refractory, or high melting, metals of groups IV, V, and VI of the periodic table. Included in this category are the silicides of titanium, zirconium, hafnium, vanadium, columbium, tantalum, chromium, molybdenum, and tungsten.

Silicides can be prepared by direct synthesis from the elements, by reduction of silica or silicon halogenides and the appropriate metal oxide or halogenide with silicon, carbon, aluminum, magnesium, hydrogen, etc., and by electrolysis of molten compounds. They are also obtained as by-products in many metallurgical processes. High-purity silicides of stoichiometric composition are difficult to prepare.

The chemical composition of silicides cannot, in general, be predicted from a consideration of the customary valences of the elements. The zirconium–silicon system, for example, is reported to include the compounds Zr_4Si, Zr_2Si, Zr_3Si_2, Zr_4Si_3, Zr_6Si_5, $ZrSi$, and $ZrSi_2$. The disilicide composition (MSi_2) occurs in all of the refractory metal–silicon systems and probably will prove the most important, particularly in those applications requiring high temperature stability.

General Properties

Silicides resemble silicon in their chemical properties with the degree of similarity roughly proportional to the silicon content. At normal temperatures the refractory metal-disilicides are inert to most ordinary chemical reagents. The compounds are not thermodynamically stable in the presence of oxygen and in the finely pulverized state they oxidize readily. However, in massive form they are oxidation resistant because of the formation of a protective surface layer of silica. Bodies of $MoSi_2$ and WSi_2 are highly resistant to oxidation even at temperatures approaching their melting points.

The physical and mechanical characteristics of silicides are, to a large extent, determined by the properties of the component metal. The refractory metal silicides have highly crystalline structures and moderate densities. Their melting points are intermediate to relatively high. They have low electrical resistance, high thermal conductivity, and fair thermal shock resistance. They have high hardness, high compressive strength, and moderate tensile strength at both room and elevated temperatures. Their elevated temperature stress-rupture and creep properties are good. Brittleness and low impact resistance are the most serious disadvantages of these materials.

Excellent oxidation resistance has prompted detailed studies of molybdenum disilicide. Quantitative information on the properties of most silicides has been developed within the past two decades.

Fabrication

The general methods used for consolidating powders can be applied to the silicides. High-density parts are obtained by cold pressing and sintering and also by hot pressing. Slip casting and extrusion are convenient methods for preparing certain shapes and sizes. Casting in the molten state is difficult due to the partial decomposition of silicides at their melting points. Silicide coatings can be prepared by vapor deposition techniques. Dense, fully sintered silicide parts are extremely difficult to work. They can be cut using silicon carbide or diamond wheels. Grinding has shown some promise but it is a slow process. "Green" or presintered compacts can, however, be shaped by conventional methods.

Availability

Although the availability of silicides is gradually increasing, the varieties, quantities, and shapes are limited. Molybdenum disilicide is commercially available in powder form and as furnace heating elements, and certain shapes and sizes have been produced on a custom basis. Some of the other compositions can be obtained on special order. The small market that has developed for $MoSi_2$ can be expected to stimulate general interest in the silicides and to foster their commercial development and production.

The practical utilization of silicides, with few exceptions, notably MoSi$_2$ furnace heating elements, has been implemented. Molybdenum disilicide is a structural material for gas turbine and missile components, which do not require high impact and thermal shock resistance. Igniter elements, thermocouple shields, gas probes, and nozzles are other potential applications. The high hardness of these materials has found usage in metalworking dies and tooling.

Silicide coatings, prepared by vapor-phase deposition, afford excellent oxidation protection to molybdenum and tungsten. This method of providing oxidation resistance is versatile because fabrication can be completed before the coating is applied. Similar coatings can be produced on other materials such as graphite by using silicide powders. Silicide coatings are commercially available and are a major item in the present market for silicide products.

SILICON

A metallic element (symbol Si), silicon is used chiefly in its combined state and is the most abundant solid element in the Earth's crust (28%).

The metal has been prepared by reducing the tetrachloride by hydrogen using a hot filament, or by aluminum, magnesium, or zinc. The fluoride or alkali fluorosilicates have been reduced with alkali metals or aluminum. Silica can be converted to metal by reduction in the electric furnace with carbon, silicon carbide, aluminum, or magnesium. The element has also been produced by the fusion electrolysis of silica in molten alkali oxide–sodium chloride–aluminum chloride baths. Extremely pure metal has been made by the treatment of silanes with hydrogen.

It is a gray-white, brittle, metallic-appearing element, not readily attacked by acids except by a mixture of HF and HNO$_3$. It is soluble in hot NaOH or KOH and is prepared in the pure crystalline form by reduction of fractionally distilled SiCl$_4$.

Silicon combines with many elements including boron, carbon, titanium, and zirconium in the electric furnace. It readily dissolves in molten magnesium, copper, iron, and nickel to form silicides. Most oxides are reduced by silicon at high temperatures.

Fabrication

Silicon can be cast by melting in a vacuum furnace and cooling in vacuum by withdrawing the element from the heated zone. Single-crystal ingots have been prepared by drawing from the melt and by the "floating zone" technique. It is claimed that silicon exhibits some workability above 1000°C, but at room temperature it is very brittle. Metals can be coated with a silicon-rich layer by reducing the tetrachloride with hydrogen on the hot metal surface.

Types

In very pure form silicon is an intrinsic semiconductor, although the extent of its semiconduction is greatly increased by the introduction of minute amounts of impurities.

Pure silicon metal is used in transistors, rectifiers, and electronic devices. It is a semiconductor, and is superior to germanium for transistors as it will withstand temperatures to 149°C and will carry more power. Rectifiers made with silicon instead of selenium can be smaller, and will withstand higher temperatures. Its melting point when pure is about 1434°C, but it readily dissolves in molten metals. It is never found free in nature but, combined with oxygen, it forms silica, SiO$_2$, one of the most common substances on Earth. Silicon can be obtained in three modifications.

Form

Amorphous silicon is a brown-colored powder with a specific gravity of 2.35. It is fusible and dissolves in molten metals. When heated in the air, it burns to form silica. Graphitoidal silicon consists of black glistening spangles, and is not easily oxidized and not attacked by the common acids, but is soluble in alkalies. Crystalline silicon is obtained in dark, steel-gray globules of crystals or six-sided pyramids of specific gravity 2.4. It is less reactive than the amorphous form, but is attacked by boiling water. All these forms are obtainable by chemical reduction. Silicon is an important constituent of commercial metals. Molding sands are largely silica, and silicon carbides are used as abrasives. Commercial silicon is sold in the graphitoidal flake form, or as ferrosilicon, and silicon–copper. The latter forms are employed for adding silicon to iron and steels. Commercial refined silicon contains 97% pure silicon and less than 1% iron. It is used for adding silicon to aluminum alloys and for fluxing copper alloys. High-purity silicon metal, 99.95% pure, made in an arc furnace, is too expensive for common uses, but is employed for electronic devices and in making silicones. For electronic use, silicon must have extremely high purity, and the pure metal is a nonconductor with a resistivity of 300,000 $\Omega \cdot$ cm. For semiconductor use it is "doped" with other atoms yielding electron activity for conducting current. Epitaxial silicon is higher purified silicon doped with exact amounts of impurities added to the crystal to give desired electronic properties.

Principal Compounds

Silicon is reported to form compounds with 64 of the 96 stable elements, and it probably forms silicides with 18 other elements. Besides the metal silicides, used in large quantities in metallurgy, silicon forms useful and important compounds with hydrogen, carbon, the halogen elements, nitrogen, oxygen, and sulfur. In addition, useful organosilicon derivatives have been prepared.

Hydrides

The hydrides of silicon are named silanes; the compound SiH_4 is called monosilane, Si_2H_6 disilane, Si_3H_8 trisilane, and so on. Compounds in which oxygen atoms alternate with silicon atoms in the principal part of the structure are called siloxanes, and those with nitrogen between silicon atoms are called silazanes. All other covalent compounds of silicon are considered for the purpose of nomenclature to be derived from these silanes, and modified silanes and are named according to substituent groups and their placement along the principal silicon-containing chain or ring.

Use

Because of its inherent brittleness, there are no engineering applications of silicon. The chief use of highly purified metal is as a semiconductor in transistors, rectifiers, and solar cells. It may be fired with ceramic materials to form heat-resistant articles. Silicon can serve as an autoxidation catalyst and as an element in photocells. Mirrors for dental use are formed with a reflecting surface of silicon. The metal is also employed to prepare silicides and alloys, and to coat various materials. In commercial quantities it is also used as a starting material for the synthesis of silicones.

By far the largest use of silicon is as compounds in the ceramic industry. It also is employed as an alloying element in ferrous metals, and is the basis of the family of chemicals known as silicones.

An important application of silicon is in the electronics industry where it has been widely employed in the manufacture of crystal rectifiers and integrated circuits. Sufficiently pure silicon has been produced by carefully controlled zone refining and crystal growth to make possible its use as transistors. Since the energy gap in silicon is 1.1 eV, compared with 0.75 eV for germanium, silicon transistors may be operated at higher temperatures and power levels than those made of germanium.

SILICON BRONZE

Silicon bronze is a family of wrought copper-base alloys (C64700 to C66100) and one cast copper alloy (C87200), the wrought alloys containing from 0.4 to 0.8% silicon (C64700) to 2.8 to 4.0% silicon (C65600), and the cast alloy 1.0 to 5.0%, along with other elements, usually lead, iron, and zinc. Other alloying elements may include manganese, aluminum, tin, nickel, chromium, and phosphorus. The most well-known alloys are probably silicon bronze C65100, or low-silicon bronze B, and silicon bronze C65500, or high-silicon bronze A, as they were formerly called. As these names imply, they differ mainly in silicon content: 0.8 to 2.0% and 2.8 to 3.8%, respectively, although the latter alloy also may contain as much as 0.6% nickel. C87200 contains at least 89% copper, 1.5%

silicon, and as much as 5% zinc, 2.5% iron, 1.5% aluminum, 1.5% manganese, 1% tin, and 0.5% lead. Regardless of alloying ingredients, copper content is typically 90% or greater.

Both of the common wrought alloys are quite ductile in the annealed condition, C65500 somewhat more ductile than C65100, and both can be appreciably strengthened by cold working. Annealed, tensile yield strengths are on the order of 103 to 172 MPa depending on mill form, with ultimate tensile strengths to about 414 MPa and elongations of 50 to 60%. Cold working can increase yield strength to as much as 483 MPa. Electrical conductivity is 12% for C65100 and 7% for C65500 relative to copper, and thermal conductivity is 57 and 36 W/m · K, respectively. The alloys are used for hydraulic-fluid lines in aircraft, heat-exchanger tubing, marine hardware, bearing plates, and various fasteners.

Silicon bronze C87200 is suitable for centrifugal, investment, and sand-, plaster-, and permanent-mold castings. As sand-cast, typical tensile properties are 379 MPa ultimate strength, 172 MPa yield strength, and 30% elongation. Hardness is Brinell 85, electrical conductivity 6%, and, relative to free-cutting brass, machinability is 40%. Uses include pump and valve parts, marine fittings, and bearings.

SILICON CARBIDE

The reduction of silica with excess carbon under appropriate conditions gives silicon carbide (SiC), which crystallizes in a number of forms but is best known in the cubic-diamond form with spacing a_o of 0.435 nm (compared with 0.356 nm for diamond). In the pure form, silicon carbide is green (α-hexagonal) or yellow (β-cubic), but the commercial product is black and has a bluish or greenish iridescence. The carbide is not easily oxidized by air except above 1000°C, and retains its physical strength up to this temperature. For these reasons it is a favorite structural refractory material for the ceramic arts. It also is extremely hard, with a Mohs hardness in excess of 9, and so has found wide application as an abrasive.

Silicon carbide does not melt without decomposition at atmospheric pressure, but does melt at 2830°C at 3.5 MPa.

FORMS

β-SiC (cubic) forms at 1400 to 1800°C and α-SiC (hexagonal) forms at temperatures >1800°C.

SiC is used as an abrasive as loose powder, coated abrasive cloth and paper, wheels, and hones. It will withstand temperatures to its decomposing point of 2301°C and is valued as a refractory. It retains its strength at high temperatures, has low thermal expansion, and its heat conductivity is ten times that of fireclay. Silicon carbide is made by fusing sand and coke at a temperature above 2204°C.

Unlike aluminum oxide, the crystals of silicon carbide are large, and they are crushed to make the small grains used as abrasives. They are harder than aluminum oxide, and because they fracture less easily, they are more suited for grinding hard cast irons and ceramics. The standard grain sizes are usually from 100 to 1000 mesh. The crystalline powder in grain sizes from 60 to 240 mesh is also used in lightning arrestors. Carborundum, Crystolon, and Carbolon are trade names for silicon carbide.

TYPES

Three main types are produced commercially. Green SiC is an entirely new batch composition made from a sand and coke mixture, and is the highest purity of the three. Green is typically used for heating elements. Black SiC contains some free silicon and carbon and is less pure. A common use is as bonded SiC refractories. The third grade is metallurgical SiC, and is not very pure. It typically is used as a steel additive.

FABRICATION AND PROPERTIES

Silicon carbide is manufactured in many complex bonded shapes, which are utilized for superrefractory purposes such as setter tile and kiln furniture, muffles, retorts and condensors, skid rails, hot cyclone liners, rocket nozzles and combustion chambers, and mechanical shaft seals. It is also used for erosion- and corrosion-resistant uses such as check valves, orifices, slag blocks, aluminum diecasting machine parts, and sludge burner orifices. Electrical uses of SiC include lightning arrestors, heating elements, and nonlinear resistors.

Silicon carbide refractories are classified on the basis of the bonds used. Associated-type bonds are oxide or silica, clay, silicon oxynitride, and silicon nitride, as well as self-bonded.

A process for joining high-temperature-resistant silicon carbide structural parts that have customized thermomechanical properties has been developed and the materials include SiC-based ceramics and composites reinforced by different fibers.

The method begins with the application of a carbonaceous mixture to the joints. The mixture is cured at a temperature between 90 and 110°C. The joints are then locally infiltrated with molten silicon or with alloys of silicon and refractory metals. The molten metal reacts with the carbon in the joint to form silicon carbide and quantities of silicon and refractory disilicide phases that can be tailored by choosing the appropriate reactants.

In mechanical tests, the joints were found to retain their strength at temperatures from ambient to 1370°C. The technique can also be used in the repair of such parts.

USES AND APPLICATIONS

SiC is used in the manufacture of grinding wheels and coated abrasives. Large tonnages are used in cutting granite with wire saws and as a metallurgical additive in the foundry and steel industries.

Other uses are in the refractory and structural ceramic industries. As an abrasive, silicon carbide is best used on either very hard materials such as cemented carbide, granite, and glass, or on soft materials such as wood, leather, plastics, rubber, etc.

Refrax Silicon Carbide and KT

The first material is bonded with silicon nitride. It is used for hot-spray nozzles, heat-resistant parts, and for lining electrolytic cells for smelting aluminum. Silicon carbide KT is molded without a binder. It has 96.5% SiC with about 2.5% silica. The specific gravity is about 3.1, and it is impermeable to gases. It is made in rods, tubes, and molded shapes.

Silicon Carbide Foam

This is a lightweight material made of self-bonded silicon carbide foamed into shapes. It is inert to hot chemicals and can be machined.

Silicon Carbide Crystals

These are used for semiconductors at temperatures above 343°C. As the cathode of electronic tubes instead of a hot-wire cathode, the crystals take less power and need no warm-up.

Silicon Carbide Fibers

SiC fiber is one of the most important fibers for high-temperature use. It has high strength and modulus and will withstand temperatures even under oxidizing conditions up to 1800°C, although the fibers show some deterioration in tensile strength and modulus properties at temperatures above 1200°C. It has advantages over carbon fibers for some uses, having greater resistance to oxidation at high temperatures, superior compressive strength, and greater electrical resistance.

There are two forms of SiC fibers, neither of which is available commercially. One consists of a pyrolytic deposit (chemical vapor deposition) of SiC on an electrically conductive, usually carbon, continuous filament. Fiber diameter is about 140 μm. This technology has been used to make filaments with both graded and layered structures, including surface layers of carbon, which provide a toughness-enhancing parting layer in composites with a brittle matrix (silicon nitride, for example).

The other form of filamentary SiC is still in the development stage. Fibers are extruded from sinterable SiC powder, and allowed to sinter during free fall from the extruder. Commercialization is not yet possible because the fibers produced to date are too large: 0.13 to 0.25 mm in diameter.

There are two commercial processes for making continuous silicon carbide fibers: (1) by coating silicon carbide on either a tungsten or carbon filament by vapor deposition to produce a large filament (100 to 150 μm in diameter), or (2) by melt spinning an organic polymer containing silicon atoms as a precursor fiber followed by heating at an elevated temperature to produce a small filament (10 to 30 µm in diameter). Fibers from the two processes differ considerably from each other but both are used commercially.

Improved composites of SiC fibers in Si/SiC matrices have been invented for use in applications in which there are requirements for materials that can resist oxidation at high temperatures in the presence of air and steam. Such applications are likely to include advanced aircraft engines and gas turbines.

The need for the improved composites arises as follows: Although both the matrix and fiber components of older SiC/(Si/SiC) composites generally exhibit acceptably high resistance to oxidation, these composites become increasingly vulnerable to oxidation and consequent embrittlement whenever mechanical or thermomechanical loads become large enough to crack the matrices. Even the narrowest cracks become pathways for the diffusion of oxygen.

Typically, to impart toughness to an SiC/(Si/SiC) composite, the SiC fibers are coated with a material that yields at high stress to allow some slippage between the fibers and matrix. If oxygen infiltrates through the cracks to the fiber coatings, then, at high temperature, the oxygen reacts with the coatings (and eventually with the fibers), causing undesired local bonding between fibers and the matrix and consequent loss of toughness.

The improved composites incorporate matrix additives and fiber coatings that retard the infiltration of oxygen by reacting with oxygen in such a way as to seal cracks and fiber/matrix interfaces at high temperatures. These matrix additives and coating materials contain glass-forming elements — for example, boron and germanium.

Boron is particularly suitable for use in fiber coatings because it can react with oxygen to form boron oxide, which can, in turn, interact with the silica formed by oxidation of the matrix and fiber materials to produce borosilicate glasses. Boron and SiB_6 may prove to be the coating materials of choice because they do not introduce any elements beyond those needed to form borosilicate glasses.

One way to fabricate an SiC/(Si/SiC) composite object is first to make a preform of silicon carbide fibers interspersed with a mixture of silicon carbide and carbon particles, then infiltrate the preform with molten silicon. Boron can be incorporated by chemical vapor deposition onto the fibers prior to making the preform.

Boron and germanium can be incorporated into the matrix by adding these elements to either the matrix or the molten silicon infiltrant. Inasmuch as the solubility of boron in silicon is limited, it may be necessary to add the boron via the preform in a typical case.

Silicon Carbide Platelets

Single crystals of α-phase hexagonal crystal structure and four size ranges currently are produced: –100, +200 mesh (100 to 300 µm in diameter, 5 to 15 µm thick); –200, +325 mesh (50 to 150 µm in diameter, 1 to 10 µm thick); –325 mesh (5 to 70 mm in diameter, 0.5 to 5 mm thick), and –400 mesh (3 to 30 µm in diameter, 0.5 to 3 µm thick). The finest size is a research product and additional development work is being conducted to produce an even smaller diameter platelet in the 0.5 µm range, which would be an ideal reinforcement material for ceramic-matrix composites.

In addition to reinforcing ceramics, silicon carbide platelets also are used to increase the strength, wear resistance, and thermal shock performance of aluminum matrices, and to enhance the properties of polymeric matrices. Because platelets are very free-flowing, they can be processed in the same manner as particulates.

Silicon Carbide Whiskers

Whiskers as small as 7 µm in diameter can be made by a number of different processes. Although these whiskers have the disadvantage in some applications of not being in continuous filament form, they can be made with higher tensile strength and modulus values than continuous silicon carbide filament.

Silicon carbide whiskers are single crystals of either α- or β-phase crystal structure. The SiC whiskers tend to exhibit a hexagonal, triangular, or rounded cross section and may contain stacking faults.

SiC whiskers can be fabricated by the reaction of silicon and carbon to form a gaseous species that can be transported and reacted in the vapor phase. This type of formation is referred to as a vapor-solid reaction.

The reactions occur at temperatures greater than 1400°C and in an inert or nonoxidizing atmosphere. In addition, a catalyst is added to assure the formation of whiskers rather than particulate during the reaction.

Although SiC whiskers can be coated with several different materials, such as carbon, to enhance their performance, the as-produced SiC whiskers generally contain a 5 to 30 SiO_2 coating, which forms during synthesis.

SiC whiskers are added to a variety of matrices to increase the toughness and high-temperature strength of these materials. The elastic modulus for SiC whiskers is 400 to 500 GPa and the tensile strength ranges from 1 to 5 GPa. A variety of ceramic matrices, such as Al_2O_3, Si_3N_4, $MoSi_2$, AlN, mullite, cordierite, and glass ceramics, are combined with SiC whiskers to increase the overall mechanical properties of the resulting composite. For example, the wear resistance, toughness, and thermal shock of Al_2O_3 is increased by the addition of SiC whiskers. The resulting composite has been used for such applications as high-performance cutting tool inserts. The addition of SiC whiskers to an alumina matrix can double the fracture toughness of the resulting composite, depending on whisker content and processing conditions.

SiC whiskers also can be combined with metals to increase the high-temperature strength of the material as well as provide a comparable substitution for heavier traditional materials, such as steel. Metal-matrix composites (MMCs) are being tested for such applications as piston ring grooves, cylinder block liners, brake calipers, and aerospace components. MMCs can be fabricated by infiltrating an SiC whisker preform with aluminum or by the addition of SiC whiskers to molten aluminum.

Polymer-matrix composites combine the strength and impact resistance of polymers with the thermal conductivity fatigue and wear resistance of the whiskers. Whisker-reinforced polymers have strong potential to replace traditional plastics in automotive, aerospace, and recreational applications.

SILICON CAST IRON

This is an acid-resistant cast iron containing a high percentage of silicon. When the amount of silicon in cast iron is above 10%, there is a notable increase in corrosion and acid resistance. The acid resistance is obtained from the compound Fe_3Si, which contains 14.5% silicon. The usual amount of silicon in acid-resistant castings is from 12 to 15%. The alloy casts well but is hard and cannot be machined. These castings usually contain 0.75 to 0.85% carbon.

A 14 to 14.5% silicon iron has a silvery-white structure, and is resistant to hot sulfuric acid, nitric acid, and organic acids. Silicon irons are also very wear resistant, and are valued for pump parts and for parts for chemical machinery.

SILICON COPPER

An alloy of silicon and copper used for adding silicon to copper, brass, or bronze, silicon copper is also employed as a deoxidizer of copper and for making hard copper. Silicon alloys in almost any proportion with copper, and is the best commercial hardener of copper. A 50–50 alloy of silicon and copper is hard and extremely brittle and black in color. A 10% silicon, 90% copper alloy is as brittle as glass; in this proportion silicon copper is used for making the addition to molten copper to produce hard, sound copper-alloy castings of high strength. The resulting alloy is easy to cast in the foundry and does not dross. Silicon-copper grades in 5, 10, 15, and 20% silicon are also marketed. A 10% silicon-copper melts at 816°C; a 20% alloy melts at 623°C.

SILICON HALIDES

Silicon tetrachloride, $SiCl_4$, is perhaps the best known monomeric covalent compound of silicon. It is readily available commercially. It can be prepared by chlorinating elementary silicon, or by the action of chlorine on a mixture of silica with finely divided carbon, or by the chlorination of silicon carbide. It is a volatile liquid that fumes in moist air and hydrolyzes rapidly to silica and hydrochloric acid.

Pure silica with very high surface area, produced by this method, is used as a reinforcing filler (white carbon black) in silicone rubber and as a thickening agent in organic solutions. Silicon tetrachloride reacts readily with alcohols and glycols, for example, to form the corresponding ethers, which may also be considered to be esters of silicic acid.

SILICON MANGANESE

An alloy employed for adding manganese to steel, and also as a deoxidizer and scavenger of steel, silicon manganese usually contains 65 to 70% manganese and 12 to 25% silicon. It is graded according to the amount of carbon, generally 1, 2, and 2.5%. For making steels low in carbon and high in manganese, silicomanganese is more suitable than ferromanganese. A reverse alloy, called manganese–silicon, contains 73 to 78% silicon and 20 to 25% manganese, with 1.5% max iron and 0.25% max carbon. It is used for adding manganese and silicon to metals without the addition of iron. Still another alloy is called ferromanganese–silicon, containing 20 to 25% manganese, about 50% silicon, and 25 to 30% iron, with only about 0.50% or less carbon. This alloy has a low melting point, giving ready solubility in the metal.

SILICON NITRIDE

Silicon nitride (Si_3N_4) dissociates in air at 1800°C and at 1850°C under 1 atm N_2. There are two crystal structures: α (1400°C) and β (1400 to 1800°C), both hexagonal. Its hardness is approximately 2200 on the Knoop K100 scale; and it exhibits excellent corrosion and oxidation resistance over a wide temperature range. Typical applications are molten-metal-contacting parts, wear surfaces, special electrical insulator components, and metal forming dies. It is under evaluation for gas turbine and heat engine components as well as antifriction bearing members.

Processing

Pure silicon nitride powders are produced by several processes, including direct nitridation of silicon, carbothermal reduction — $C + SiO_2 + N_2$ yields Si_3N_4 (gas atmosphere) — and chemical vapor deposition — $3 SiH_4 + 4NH_3$ yields $Si_3N_4 + 12H_2$. Reacting SiO_2 with ammonia or silanes with ammonia will also produce silicon nitride powders. It is found that the highest-purity powders come from gas-phase reactions.

Types

Sinterable/Hot Pressed/ Hot Isostatically Pressed Silicon Nitride

These types are SSN, HPSN, and HIPSN, respectively. They are used mainly in higher performance applications. Powdered additives, known as sintering aids, are blended with the pure Si_3N_4 powder and allow densification to proceed via the liquid state. Pore-free bodies can be so produced by sintering or hot pressing. Of course, the properties of the material and dense pieces are dependent on the chemical nature of the sintering aids employed.

Sinterable silicon nitrides are a more recent innovation, and allow more flexibility in shape fabrication than does HPSN. Highly complex shapes can be die-pressed or isostatically pressed. Densification can be performed by either sintering or hot isostatic pressing (HIP). Properties of the dense piece are dependent on the additives, but in general the strength below 1400°C, as well as oxidation resistance of HPSN and SSN, far exceed those properties for reactor bonded silicon nitride (RBSN).

Reaction Bonded Silicon Nitride

More common today is RBSN. Silicon powder is pressed, extruded, or cast into shape, then carefully nitrided in a N_2 atmosphere at 1100 to 1400°C, so as to prevent an exothermic reaction, which might melt the pure silicon.

The properties of RBSN are usually lower than those of HPSN or SSN, due mainly to the fact that bodies fabricated in this manner only reach 85% of the theoretical density of silicon nitride and no secondary phase between grains is present.

Silicon Nitride Fibers

Si_3N_4 fibers have been prepared by reaction between silicon oxide and nitrogen in the presence of a reducing agent in an electrical resistance furnace at 1400°C. Silicon nitride short fibers are used in composites for specialty electrical parts, aircraft parts, and radomes (microwave windows). Silicon nitride whiskers have also been grown as a result of the chemical reaction between nitrogen and a mixture of silicon and silica.

SILICON OXIDES

Silicon dioxide is perhaps best known as one of its crystalline modifications known as quartz, colorless crystals of which are also known as rhinestones and Glens Falls diamonds. Purple or lavender-colored quartz is called amethyst, the pink variery is rose quartz, and the yellow type, citrine.

Because rock crystal has been collected and admired for thousands of years, large and perfectly formed natural crystals of quartz are now very rare. With the growth of radio broadcasting and the electronics industry, piezoelectric crystals cut from perfect specimens of quartz have been used in increasing quantities, to the point of scarcity of natural crystals. As the supply diminished, considerable effort was devoted to the problem of growing crystals of quartz by artificial means. Some success has been achieved by growing the crystals hydrothermally from a solution of silica glass in water containing an alkali or a fluoride.

A number of natural noncrystalline varieties of silicon dioxide are also known, such as the hydrated silica known as opal and the dense unhydrated variety known as flint. Onyx and agate represent still other semiprecious forms.

SILICON STEEL

All grades of steel contain some silicon and most of them contain from 0.10 to 0.35% as a residual of the silicon used as a deoxidizer. But from 3 to 5% silicon is sometimes added to increase the magnetic permeability, and larger amounts are added to obtain wear-resisting or acid-resisting properties. Silicon deoxidizes steel, and up to 1.75% increases the elastic limit and impact resistance without loss of ductility. Silicon steels within this range are used for structural purposes and for springs, giving a tensile strength of about 517 MPa and 25% elongation.

The structural silicon steels are ordinarily silicon–manganese steel, with the manganese above 0.50%. Low-carbon steels used as structural steels are made by careful control of carbon, manganese, and silicon and with special mill heat treatment.

The value of silicon steel as a transformer steel occurs where silicon increases the electrical resistivity and also decreases the hysteresis loss, making silicon steel valuable for magnetic circuits where alternating current is used.

SILICONE RESINS

Silicone resins are synthetic materials capable of cross-linking or polymerizing to form films, coatings, or molded shapes with outstanding resistance to high temperatures.

They are a group of resinlike materials in which silicon takes the place of the carbon of the organic synthetic resins. Silicon is quadrivalent like carbon. But, while the carbon also has a valence of 2, silicon has only one valence of 4, and the angles of molecular formation are different. The two elements also differ in electronegativity, and silicon is an amphoteric element, with both acid and basic properties. The molecular formation of the silicones varies from that of the common plastics, and they are designated as inorganic plastics as distinct from the organic plastics made with carbon.

In the long-chain organic synthetic resins the carbon atoms repeat themselves, attaching on two sides to other carbon atoms, while in the silicones the silicon atom alternates with an oxygen atom so that the silicon atoms are not tied to each other. The simple silane formed by silicon and hydrogen corresponding to methane, CH_4, is also a gas, as is methane, and has the formula SiH_4. But, in general, the silicones do not have the SiH radicals, but contain CH radicals as in the organic plastics.

COMPOSITION

Silicones are made by first reducing quartz rock (SiO_2) to elemental silicon in an electric furnace, then preparing organochlorosilane monomers ($RSiCl_3$, R_2SiCl_2, R_3SiCl) from the silicon by one of several different methods. The monomers are then hydrolyzed into cross-linked polymers (resins) whose thermal stability is based on the same silicon–oxygen–silicon bonds found in quartz and glass. The properties of these resins will depend on the amount of crosslinking, and on the type of organic groups (R) included in the original monomer. Methyl, vinyl, and phenyl groups are among those used in making silicone resins.

PROPERTIES

Silicone resins have, in general, more heat resistance than organic resins, have higher dielectric strength, and are highly water resistant. Like organic plastics, they can be compounded with plasticizers, fillers, and pigments. They are usually cured by heat. Because of the quartzlike structure, molded parts have exceptional thermal stability. Their maximum continuous-use service temperature is about 260°C. Special grades exceed this and go as high as 371 to 482°C. Their heat-deflection temperature for 1.8 MPa is 482°C. Their moisture absorption is low, and resistance to petroleum products and acids is good. Nonreinforced silicones have only moderate tensile and impact strength, but fillers and reinforcements provide substantial improvement. Because silicones are high in cost, they are premium plastics and are generally limited to critical or high-performance products such as high-temperature components in the aircraft, aerospace, and electronics fields.

Common characteristics shared by most silicone resins are outstanding thermal stability, water repellency, general inertness, and electrical insulating properties. These properties, among others, have resulted in the use of silicone resins in the following fields:

1. Laminating (reinforced plastics), molding, foaming, and potting resins
2. Impregnating cloth coating, and wire varnishes for Class II (high-performance) electric motors and generators
3. Protective coating resins
4. Water repellents for textiles, leather, and masonry
5. Release agents for baking pans

Laminating Resins

Laminates made from silicone resin and glass cloth are lightweight, strong, heat-resistant materials used for both mechanical and dielectric applications. Silicone-glass laminates have low moisture absorption and low dielectric losses, and retain most of their physical and electrical properties for long periods at 260°C.

Laminates may be separated into three groups according to the method of manufacture: high pressure, low pressure, and wet layup.

High pressure. Silicone-glass laminates (industrial thermosetting laminates) have excellent electric strength and arc resistance, and are normally used as dielectric materials. To prepare high-pressure laminates, glass cloth is first impregnated by passing it through a solvent solution of silicone resin. The resin is dried of solvent and precured by passing the fabric through a curing tower. Laminates are prepared by laying up the proper number of plies of preimpregnated glass cloth and pressing them together at about 6.8 MPa and 177°C for about 1 h. They are then oven-cured at increasing temperatures, with the final cure at about 249°C. The resulting laminates can be drilled, sawed, punched, or ground into insulating components of almost any desired shape. Typical applications include transformer spacer bars and barrier sheets, slot sticks, panel boards, and coil bobbins.

Low pressure. Silicone-glass laminates made by low-pressure reinforced plastics molding methods usually provide optimum flexural strength, e.g., about 272 MPa even after heat aging. They are used for mechanical applications such as radomes, aircraft ductwork, thermal barriers, covers for high-frequency equipment, and high-temperature missile parts. In making low-pressure laminates, glass cloth is first impregnated and laid up as described above. Since the required laminating pressure can be as low as 0.068 MPa, matched-metal-molding and bag-molding techniques can be used in laminating, making possible greater variety in laminated shapes. Lamination should be after-cured as already described.

Wet layup. Silicone-glass laminates can now be produced by wet layup techniques because of the solventless silicone resins recently developed. Such laminates can be cured without any pressure except that needed to hold the laminate together. This technique should prove especially useful in making prototype laminates, and in short production runs where expensive dies are not justified. Laminates are prepared by wrapping glass cloth around a form and spreading on catalyzed resin, repeating this process until the desired thickness is obtained. The laminate surface is then wrapped with a transparent film, and air bubbles are worked out. Laminates are cured at 149°C and, after the transparent film is removed, postcured at 204°C.

Molding Compounds

Silicone molding compounds consist of silicone resin, inorganic filler, and catalyst, which, when molded under heat and pressure, form thermosetting plastic parts. Molded parts retain exceptional physical and electrical properties at high temperatures, resist water and chemicals, and do not support combustion. Specification MIL-M-14E recognizes two distinctly different types of silicone molding compounds: type MSI-30 (glass-fiber filled) and type MSG (mineral filled). Requirements for both are as follows.

Type MSI-30. Glass-filled molded parts have high strengths, which become greater the longer the fiber length of the glass filler. Where simple parts can be compression-molded from a continuous fiber length compound, strengths will be approximately twice as great. Properly cured glass-fiber-filled molded parts can be exposed continuously to temperatures as high as 371°C and intermittently as high as 538°C. The heat-distortion temperature after postcure is 482°C. Because of the flow characteristics of these fiber-filled compounds, their use is generally limited to compression molding.

Type MSG. Mineral-filled compounds are free-flowing granular materials. They are suitable for transfer molding, and can be used in automatic preforming and molding machines. They are used to make complex parts that retain their physical and electrical properties at temperatures above 260°C, but that do not require impact strength. Silicone molding compounds are excellent materials for making Class H electrical insulating components such as coil forms, slot wedges, and connector plugs. They have many potential applications in the aircraft, missile and electronic industries.

Foaming Powders

Silicone foaming powders are completely formulated, ready-to-use materials that produce heat-stable, nonflammable, low-density silicone foam structures when heated. Densities vary from 160 to 288 kg/m^3, compressive strengths from 0.68 to 2.23 MPa. Electrical properties are excellent, and water absorption after 24-h immersion is only 2.5%. The maximum continuous operating temperature of these foams is about 343°C.

Foams are prepared by heating the powders to between 149 and 177°C for about 2 h. The powder can be foamed in place, or foamed into blocks and shaped with woodworking tools. Foams are normally after-cured to develop strength, but can often be cured in service.

Silicone foams are being used in the aircraft and missile industries to provide lightweight thermal insulation and to protect delicate electronic equipment from thermal shock. They can also be bonded to silicone-glass laminates or metals to form heat- and moisture-resistant sandwich structures.

Potting Resins

Solventless silicone resins can be used for impregnating, encapsulating, and potting of electrical and electronic units. Properly catalyzed, filled, and cured, they form tough materials with good physical and electrical properties, and will withstand continuous temperatures of 204°C and intermittent temperatures above 260°C.

Typical physical properties of cured resins include flexural strength of 48.6 MPa, compressive strength of 117 MPa, and water absorption of 0.04%.

Before use, resins are catalyzed with dicumyl peroxide or ditertiary butyl peroxide.

Resins can be simply poured in place, although vacuum impregnation is suggested where fine voids must be filled. Fillers such as glass beads or silica flour are added to extend the resin; their use increases physical strength and thermal conductivity, but decreases electrical properties. The resin is polymerized by heating it to about 149°C, and postcured, first at 204°C, then at the intended operating temperature if higher.

Electrical Varnishes

Silicone varnishes (solvent solutions of silicone resins) have made possible the new high-temperature classes of insulation for electrical motors and generators. Silicone insulating varnishes will withstand continuous operating temperatures at 177°C or higher.

Electrical equipment that operates at higher temperatures makes possible motors, generators, and transformers that are much smaller and lighter, or equipment that delivers 25 to 50% more power from the same size and still has a much longer service life.

The resinous silicone materials used in Class H electrical insulating systems include the following:

1. Silicone bonding varnish for glass-fiber-covered magnet wire.
2. Silicone varnishes for impregnating and bonding glass cloth, mica, and asbestos paper. Sheet insulations made of these heat-resistant materials are used as slot liners for electric motors and as phase insulation.
3. Silicone dipping varnish that impregnates, bonds, and seals all insulating components into an integrated system.

Other silicone materials used in electrical equipment include silicone rubber lead wire, silicone-adhesive-backed glass tape, and temperature-resistant silicone bearing greases. Silicone insulated motors, generators, and transformers are now being produced by the leading electrical equipment manufacturers.

Uses

The wide range of structural variations of silicone resins makes it possible to tailor compositions for many kinds of applications. Low-molecular-weight silanes containing amino or other functional groups are used as treating or coupling agents for glass fiber and other reinforcements to cause unsaturated polyesters and other resins to adhere better.

The liquids, generally dimethyl silicones of relatively low molecular weight, have low surface tension, great wetting power and lubricity for metals, and very small change in viscosity with temperature. They are used as hydraulic fluids, as antifoaming agents, as treating and waterproofing agents for leather, textiles, and masonry, and in cosmetic preparations. The greases are particularly desired for applications requiring effective lubrication at very high and at very low temperatures.

Silicone resins are used for coating applications in which thermal stability in the range 300 to 500°C is required. The dielectric properties of the polymers make them suitable for many electrical applications, particularly in electrical insulation that is exposed to high temperatures and as encapsulating materials for electronic devices.

Silicone enamels and paints are more resistant to chemicals than most organic plastics, and when pigmented with mineral pigments will withstand temperatures up to 538°C. For lubricants the liquid silicones are compounded with graphite or metallic soaps and will operate between –46 and 260°C. The silicone liquids are stable at their boiling points, between 399 and 427°C, and have low vapor pressures, so that they are also used for hydraulic fluids and heat-transfer media. Silicone oils, used for lubrication and as insulating and hydraulic fluids, are methyl silicone polymers. They retain a stable viscosity at both high and low temperatures. As hydraulic fluids they permit smaller systems to operate at higher temperatures. In general, silicone oils are poor lubricants compared with petroleum oils, but they are used for high temperatures, 150 to 200°C, at low speeds and low loads.

Silicone resins are blended with alkyd resins for use in outside paints, usually modified with a drying oil. Silicone-alkyd resins are also used for baked finishes, combining the adhesiveness and flexibility of the alkyd with the heat resistance of the silicone. A phenyl ethyl silicone is used for impregnating glass-fiber cloth for electrical insulation and it has about double the insulating value of ordinary varnished cloth.

SILICONE RUBBER

Silicone rubbers are a group of synthetic elastomers noted for their (1) resilience over a very wide temperature range, (2) outstanding resistance to ozone and weathering, and (3) excellent electrical properties.

COMPOSITION

The basic silicone elastomer is a dimethyl polysiloxane. It consists of long chains of alternating silicon and oxygen atoms, with two methyl ($-CH_3$) side chains attached to each silicon atom. By replacing a part of these methyl groups with other side chains, polymers with various desirable properties can be obtained. For example, where flexibility at temperatures lower than $-57°C$ is desired, a polymer with about 10% of the methyl side chains replaced by phenyl groups ($-C_6H_5$) will provide compounds with brittle points below $-101°C$. Side-chain modification can also be used to produce elastomers with lower compression set, increased resistance to fuels, oils, or solvents, or to permit vulcanization at room temperature.

Curing, or vulcanization, is the process of introducing cross-links at intervals between the long chains of the polymer. Silicone rubbers are usually cross-linked by free radical-generating curing agents, such as benzoylperoxide, which are activated by heat, or the cross-linking can be accomplished by high-energy radiation beams. Room temperature vulcanized compounds are cross-linked by the condensation reaction resulting from the action of metal-organic salts, such as zinc or tin octoates. Pure polymers upon cross-linking change from viscous liquids into elastic gels with very low tensile strength. To attain satisfactory tensile strength, reinforcing agents are necessary. Synthetic and natural silicas and metallic oxides are commonly used for this purpose. In addition to the vulcanizing agents and reinforcing fillers described, other additives may be incorporated into silicone compounds to pigment the stock, to improve processing, or to reduce the compression set of certain types of silicone gum.

Ordinary silicone rubber has the molecular group ($H \cdot CH_2 \cdot Si \cdot CH_2 \cdot H$) in a repeating chain connected with oxygen linkages, but in the nitrile-silicone rubber one of the end hydrogens of every fourth group in the repeating chain is replaced by a C:N radical. These polar nitrile groups give a low affinity for oils, and the rubber does not swell with oils and solvents. It retains strength and flexibility at temperatures from $-73°C$ to above $260°C$, and is used for such products as gaskets and chemical hose. As lubricants, silicones retain a nearly constant viscosity at varying temperatures. Fluorosilicones have fluoroalkyd groups substituted for some of the methyl groups attached to the siloxane polymer of dimethyl silicone. They are fluids, greases, and rubbers, incompatible with petroleum oils and insoluble in most solvents. The greases are the fluids thickened with lithium soap, or with a mineral filler.

TYPES

Silicone-rubber compounds can be conveniently grouped into several major types according to characteristic properties. Typical properties of several types are shown in Table S.5. According to one such classification system, types are (1) general purpose, (2) extremely low temperature, (3) extremely high temperature, (4) low compression set, (5) high strength, (6) fluid resistant, (7) electrical, and (8) room temperature vulcanizing rubbers.

1. General-purpose compounds are available in Shore A hardnesses from 30 to 90, tensile strengths of 4.8 to 8.24 MPa, and ultimate elongations of 100 to 500%. Their service temperature range extends from -55 to $260°C$, and they have good resistance to heat and oils, along with good electrical properties. Many of these compounds contain semireinforcing

or extending fillers to lower their cost.
2. Extremely low temperature compounds have brittle points near −118°C and are quite flexible at −84 to −90°C. Their physical properties are usually about the same as those of the general-purpose stocks, with some reduction in oil resistance.
3. Extremely high temperature compounds are considered serviceable for over 70 h at 343°C and will withstand brief exposures at higher temperatures; for example, 4 to 5 hr at 371°C and 10 to 15 min at 399°C. In comparison, general-purpose compounds are limited to about 260°C for continuous service and 316°C for intermittent service.
4. Low compression set compounds provide typical values of 10 to 20% compression set after 22 h at 149°C. They have improved resistance to petroleum oils and various hydraulic fluids, and are particularly suitable for use in O-rings and gaskets.
5. High-strength compounds, in Shore A hardnesses of 25 to 70, provide tensile strengths from 0.82 to over 136 MPa and elongations from 400 to 700%, with tear strengths from 26,790 to 58,045 g/cm. Compounds of this class may operate over a service temperature range from −90 to 316°C.
6. Excellent resistance to a wide range of fuels, lubricants, and hydraulic fluids is offered by compounds based on a silicone polymer with 50% of its side methyl groups substituted by trifluoropropyl groups. Physical properties are similar to properties of other types of silicone compounds. However, service temperature range is somewhat limited. Its low-temperature properties are about the same as those of the dimethyl polymer, with a brittle point around −68°C and an upper service temperature of around 260°C.
7. In general, silicone-rubber compounds have excellent electrical properties, which, along with their resistance to high temperatures, make them suitable for many electrical applications. With proper compounding, dielectric constant can be easily varied from about 2.7 to 5.0 or higher, while the power factor can be varied from 0.0005 up to 0.1 or higher. The volume resistivity of a typical silicone compound will be in the range 10^{14} to 10^{16} Ω-cm and its dielectric strength will be about 450 to 550 V/mil thickness (measured on a slab 2.0 mm). Resistance to corona is excellent and water absorption is low. In most cases, excellent electrical properties are retained over a wide temperature and frequency range. Compounds can also be prepared with very low resistivity, as low as about 10 Ω-cm, for special applications. Insulated tapes for cable-wrapping applications can be prepared from electrical-grade compounds with a partial cure or from a completely cured self-adhering silicone compound.
8. Room-temperature vulcanizing silicone rubbers are available to provide most of the performance characteristics of silicone rubbers in compounds that cure at room temperature.

In addition to having excellent heat resistance, silicone rubbers retain their properties to a much greater extent at high temperatures than do most organic rubbers. For example, a silicone compound with a tensile strength at temperature of 1.36 MPa will have a tensile strength at 316°C of 0.48 MPa or over. Most organic rubbers, although their initial properties are much higher, will be virtually useless at 260°C (except for the fluoroelastomers). In applications where a silicone rubber part operates in low oxygen atmosphere, such as sealing on high-altitude aircraft, its heat resistance will be still further improved.

In using silicone rubber at high temperatures, care must be taken to prevent reversion

SILICONE RUBBER

TABLE S.5
Properties of Some Typical Silicone Rubbers

	General Purpose	Extreme Low Temp.	Extreme High Temp.	Low Compression Set	High Strength	Fluorosilicone (General Purpose)	RTV[a]
Hardness (Shore A)	50 ± 5	25 ± 5	50 ± 5	60 ± 5	50 ± 5	60 ± 5	65 ± 5
Tensile strength, psi	1000	1000	1200	900	2000	950	750
Elongation, %	400	600	300	130	600	225	110
Tear strength, lb/in.	80	120	150	60	300	75	40
Compression set (22 h at 300°F), %	20	20	15	10	30	15	13[b]
Max service temp., °F							
Continuous	500	500	550	500	500	500	500
Intermittent	600	600	700	600	600	550	600
Low temp. flex., °F	−65	−130	−130	−65	−130	−65	−65
Volume swell (ASTM No. 1 Oil; 70 h at 300°F), %	7	10	+9	+5	+10	+1	+3

[a] Room temperature vulcanizing silicone rubber.
[b] After additional postcure 24 h at 480°F.

or depolymerization, which may occur where a part is required to operate in an enclosed environment. Here again, where this problem cannot be eliminated by the design engineer, the silicone-rubber fabricator can produce compounds with relatively high resistance to reversion.

Fabrication and Uses

In general, silicone rubbers may be handled on standard rubber-processing equipment. Their fabrication differs from that of organic rubbers chiefly in that uncured silicone compounds are softer and tackier and have much lower green strength. Also, an oven postcure in a circulating air oven is often required after vulcanization to obtain optimum properties. Silicone rubbers can be extruded, molded, calendered, sponged, and foamed. Since compounding of the rubber stock determines to a great extent the processing characteristics of the material, the fabricator should be consulted before the material is specified, to determine whether compromises are necessary to obtain the best combination of physical properties and the most desirable shape. Silicone-rubber compounds can also be applied to fabrics by calender coating, knife spreading, or solvent dispersion techniques.

For certain applications, very soft or low durometer materials are required. Suitable for such applications are low durometer solid silicone rubbers, closed or open cell expanded silicone rubbers (designated here as sponge and foam, respectively), and fibrous silicone rubber.

Silicone-rubber sponge is available in molded sheets, extrusions, and simple molded shapes. As in the case of solid silicone rubber, improved resistance to fluids or abrasion can be obtained by bonding molded or extruded sponge to a fabric or plastic cover. Silicone foam rubber can be fabricated in heavy cross sections and complex shapes, and can be foamed and vulcanized either at ambient or elevated temperature. It is suitable for use where an extremely soft, low-density silicone material is required. Like sponge, it can be bonded to fabrics and plastics.

For some applications a solid, low durometer material has advantages over sponge or foam. For example, it should probably be specified for gaskets or seals where the low compression set of foam, the compress-ion-deflection characteristics of sponge, and the higher tensile and tear strength of solid silicone rubbers must be combined.

The last highly compressible material, fibrous silicone rubber, consists of hollow rubber fibers sprayed in a random manner and bonded into a low-density porous mat. Its properties include excellent compression set combined with good tear and tensile strength, and very high porosity. As manufactured at present, it is serviceable from –55°C to over 260°C, and is available in mats 3.2 mm thick and 228 mm wide.

Silicone rubbers are most widely used in the aircraft, electrical, and automotive industries, although their unique properties have created many other applications. Specific examples would include seals for aircraft canopies or access doors, insulation for wire and cable, dielectric encapsulation of electronic equipment, and gaskets or O-rings for use in aircraft or automobile engines. An example of another field in which they are useful is the manufacture of stoppers for pharmaceutical vials, since silicone rubbers are tasteless, odorless, and nontoxic.

SILK

Silk is the fibrous material in which the silkworm, or larva of the moth, envelops itself before passing into the chrysalis state. Silk is closely allied to cellulose and resembles wool in structure, but unlike wool it contains no sulfur. The natural silk is covered with a wax or silk glue, which is removed by scouring in manufacture, leaving the glossy fibroin, or raw-silk fiber. The fibroin consists largely of the amino acid alanine, $CH_3CH(NH_2)CO_2H$, which can be synthesized from pyruvic acid. Silk fabrics are used mostly for fine garments, but are also valued for military powder bags because they burn without a sooty residue.

The fiber is unwound from the cocoon and spun into threads. Each cocoon has from 1829 to 2743 m of thread. The chief silk-producing countries are China, Japan, India, Italy, and France. Floss silk is a soft silk yarn practically without twist, or is the loose waste silk produced by the worm when beginning to spin its cocoon.

Satin is a heavy silk fabric with a close twill weave in which the fine warp threads appear on the surface and the weft threads are covered by the peculiar twill. Common satin is of eight-leaf twill, the weft intersecting and binding down the warp at every eighth pick, but 16 to 20 twills are also made. In the best satins a fine quality of silk is used.

SILVER AND ALLOYS

A white metal (symbol Ag), silver is very malleable and ductile, and is classed with the precious metals. It occurs in the native state, and also combined with sulfur and chlorine. Copper, lead, and zinc ores frequently contain silver; about 70% of the production of silver is a by-product of the refining of these metals.

Silver is the whitest of all the metals and takes a high polish, but easily tarnishes in the air because of the formation of a silver sulfide. It has the highest electrical and heat conductivity: 108% IACS relative to 100% for the copper standard and about 422 W/m · K, respectively. Cold work reduces conductivity slightly. The specific gravity is 10.7, and the melting point is 962°C. When heated above the boiling point (2163°C), it passes off as a green vapor. It is soluble in nitric acid and in hot sulfuric acid. The tensile strength of cast silver is 282 MPa, with Brinell hardness 59. The metal is marketed on a troy-ounce value.

Pure silver has the highest thermal and electrical conductivity of any metal, as well as the highest optical reflectivity. Next to gold it is the most ductile and most malleable of any metal. Silver can be hammered into sheet 0.01 mm thick or drawn out in wire so fine 120 m would weigh only 1 g. Classified as one of the most corrosion-resistant metals, silver, under ordinary conditions, will not be affected by caustics or corrosive elements, unless hydrogen sulfide is present, causing silver sulfide to form. Silver will dissolve rapidly in nitric acid and more slowly in hot concentrated sulfuric acid. Unless oxidizing agents are present, the action of diluted or cold solutions of sulfuric acid is negligible. Organic acids generally do not attack the metal and caustic alkalies have but a slight effect on pure silver.

Although silver tarnishes quickly in the presence of sulfur and sulfur-bearing compounds, it oxidizes slowly in air and the oxide decomposes at a relatively low temperature.

CLASSIFICATION

Silver is classified by grades in parts per thousand based on the silver content (impurities are reported in parts per hundred). Commercial grades are Fine Silver and High Fine Silver. As ordinarily supplied, fine silver contains at least 999.0 parts silver per 1000, and may go as high as 999.3 parts per 1000. Any of the common base metals may be present, although copper is usually the major impurity. Any silver of higher purity than commercial fine contains its purity in its description, i.e., 999.7 High Fine Silver. The purest silver obtainable in quantity is 999.9 plus; the impurities are less than 0.01 part per 1000. Fine silver may also contain small percentages of oxygen or hydrogen; deoxidized silver is available for applications where these elements may be a detriment.

Fabricability

Silver can be cold-worked, extruded, rolled, swaged, and drawn. It can be cold-rolled or cold-drawn drastically between anneals, and can be annealed at relatively low temperatures. To prevent oxidation when casting by conventional methods, silver should be protected by a layer of charcoal or by melting under neutral or reducing gas. Deoxidation by adding lithium or phosphorus can be obtained leaving a residual content of 0.01% max. The excellent ductility of silver makes it readily workable hot or cold.

Molten silver will absorb approximately 20 times its own volume of oxygen. Most of this oxygen is given up when the silver solidifies in cooling, but care should be taken in melting and casting because any oxygen left in the cast bars will cause cracking when they are fabricated and the castings may have blow holes.

Galling, seizing of the tool, and surface tearing are problems encountered when machining fine silver. This can be somewhat alleviated by using material cold-worked as much as possible.

Joining

Fine silver can be soldered without difficulty using tin–lead solders. Boron–silver filler metal can be used in brazing, and welding can be done by resistance methods and by atomic hydrogen or inert-gas shielded arc processes. A range of 204 to 427°C is recommended for annealing, with best strength and ductility achieved between 371 and 427°C. Little additional softening occurs at higher temperatures, which may induce welding of adjacent surfaces. The lighter the gauge, the lower should be the annealing temperature.

Applications of Alloys and Compounds

Because silver is a very soft metal, it is not normally used industrially in a pure state, but is alloyed with a hardener, usually copper. Sterling silver is the name given to a standard high-grade alloy containing a minimum of 925 parts in 1000 of silver. It is used for the best tableware, jewelry, and electrical contacts. This alloy of 7.5% copper work-hardens and requires annealing between roll passes. Silver can also be hardened by alloying with other elements.

The standard types of commercial silver are fine silver, sterling silver, and coin silver. Fine silver is at least 99.9% pure, and is used for plating, making chemicals, and for parts produced by powder metallurgy. Coin silver is usually an alloy of 90% silver and 10% copper, but when actually used for coins the composition and weight of the coin are designated by law. Silver and gold are the only two metals that fulfill all the requirements for coinage. The so-called coins made from other metals are really official tokens, corresponding to paper money, and are not true coins. Coin silver has a Vickers hardness of 148 compared with a hardness of 76 for hard-rolled pure silver. It is also used for silverware, ornaments, plating, for alloying with gold, and for electric contacts. Silver is not an industrial metal in the ordinary sense. It derives its coinage value from its intrinsic aesthetic value for jewelry and plate, and in all civilized countries silver is a controlled metal.

Silver powder, 99.9% purity, for use in coatings, integrated circuits, and other electrical and electronic applications, is produced in several forms. Amorphous powder is made by chemical reduction and comes in particle sizes of 0.9 to 15 µm. Powder made electrolytically is in dendritic crystals with particle sizes from 10 to 200 µm. Atomized powder has spherical particles and may be as fine as 400 mesh. Silver-clad powder for electric contacts is a copper powder coated with silver to economize on silver. Silver flake is in the form of laminar platelets and is particularly useful for conductive and reflective coatings and circuitry. The tiny flat plates are deposited in overlapping layers permitting a metal weight saving of as much as 30% without reduction in electrical properties.

Nickel-coated silver powder, for contacts and other parts made by powder metallurgy, comes in grades with $1/4$, $1/2$, 1, and 2% nickel by weight.

The porous silver comes in sheets in standard porosity grades from 2 to 55 µm. It is used for chemical filtering.

Silver plating is sometimes done with a silver–tin alloy containing 20 to 40 parts silver and the remainder tin. It gives a plate having the appearance of silver but with better wear resistance. Silver plates have good reflectivity at high wavelengths, but reflectivity falls off at about 350 nm, and is zero at 3000, so that it is not used for heat reflectors.

Silver-clad sheet, made of a cheaper nonferrous sheet with a coating of silver rolled on, is used for food-processing equipment. It is resistant to organic acids but not to products containing sulfur. Silver-clad steel, used for machinery bearings, shims, and reflectors, is made with pure silver bonded to the billet of steel and then rolled. For bearings, the silver is 0.025 to 0.889 cm thick, but for reflectors the silver is only 0.003 to 0.008 cm thick. Silver-clad stainless steel is stainless-steel sheet with a thin layer of silver rolled on one side for electrical conductivity.

Silver iodide is a pale-yellow powder of the composition AgI, best known for its use as a nucleating agent and for seeding rain clouds. Silver nitrate, formerly known as lunar caustic, is a colorless, crystalline, poisonous, and corrosive material of the composition $AgNO_3$. It is used for silvering mirrors, for silver plating, in indelible inks, in medicine, and for making other silver chemicals. The high-purity material is made by dissolving silver in nitric acid, evaporating the solution, and crystallizing the nitrate, then re-dissolving the crystals in distilled water and recrystallizing. It is an active oxidizing agent. Silver chloride, AgCl, is a white granular powder used in silver-plating solutions. This salt of silver and other halogen compounds of silver, especially silver bromide, AgBr, are used for photographic plates and films. Silver chloride is used in the preparation of yellow glazes, purple of Cassius, and silver lusters. A yellowish-silver luster is obtained by mixing silver chloride with three times its weight of clay and ochre and sufficient water to form a paste.

Silver chloride crystals in sizes up to 4.5 kg are grown synthetically. The crystals are cubic, and can be heated and pressed into sheets. The specific gravity is 5.56, index of refraction 2.071, and melting point of 455°C. They are slightly soluble in water and soluble in alkalies. The crystals transmit more than 80% of the wavelengths from 50 to 200 µm.

Silver sulfide, Ag_2S, is a gray-black, heavy powder used for inlaying in metal work. It changes its crystal structure at about 179°C, with a drop in electrical resistivity, and is also used for self-resetting circuit breakers. Silver potassium cyanide, $KAg(CN)_2$, is a white, crystalline, poisonous solid used for silver-plating solutions. Silver tungstate, Ag_2WO_4, silver manganate, $AgMnO_4$, and other silver compounds are produced in high-purity grades for electronic and chemical uses.

Silver nitrate, $AgNO_3$, with a melting point of 212°C, decomposes at 444°C and is soluble, corrosive, and poisonous. It is prepared by the action of nitric acid on metallic silver. Silver nitrate is the most convenient method of introducing silver into a glass; a solution of the compound is poured over the batch.

The photosensitive halides used in photography, the cyanides used in electroplating, and most of the minor silver salts are prepared from silver nitrate.

In advanced ceramic applications, silver is unsurpassed as a conductor of heat and electricity. Silver is used in conductive coatings for capacitors, printed wiring, and printed circuits on titanites, glass-bonded mica, steatite, alumina, porcelain, glass, and other ceramic bodies. These coatings also are used to metallize ceramic parts to serve as hermetically sealed enclosures, becoming integral sections of coils, transformers, semiconductors, and monolithic and integrated circuits.

Two types of conductive coatings can be used on ceramic parts: those that are fired on and those that are baked on or air dried. The fired-on type contains, in addition to silver powder, a finely divided low-melting glass powder, temporary organic binder, and liquid solvents in formulations with direct soldering properties and others suitable for electroplating, both possessing excellent adhesion and electrical conductivity. The baked-on and air-dry types contain, in addition to silver powder, a permanent organic binder and liquid solvents. These preparations have somewhat less adhesion, electrical conductivity, and solderability than the fired-on type, but can be electroplated if desired. The air-dry type is used when it is not desirable to subject the base material to elevated firing temperatures.

Any of the above silver compositions are available in a variety of vehicles suitable for application by squeegee, brushing, dipping, spraying, bonding wheel, roller coating, etc.

Firing temperatures for direct-solder silver preparations range from 677 to 788°C. Silver compositions to be copper plated are fired at 1200 to 1250F. The firing cycle used with these temperatures will vary from 10 min to 6 h, depending on the time required to equalize the temperature of the furnace charge.

A 62% Sn–36% Pb–2% Ag solder is generally used with the direct-solder silver compositions. It is recommended that this solder be used at a temperature of 213 to 219°C. Soldering to the plated silver coating is less critical and 50% Sn–50% Pb or 40% Sn–60% Pb, as well as other soft solders, are being used with good results.

The air-dried silver compositions will, as the designation implies, air dry at room temperature in approximately 16 h. This drying time can be shortened by subjecting the coating to temperatures of 60 to 93°C for 10 to 30 min. The baked-on preparations must be cured at a minimum temperature of 149°C for 5 to 16 h. The time may be shortened to 1 h by raising the temperature to 301°C.

The same soft solders and techniques as recommended for the fired-on coatings may be used for the electroplated air-dried and baked-on preparations. It is extremely difficult to solder to air-dried or baked-on coatings without first electroplating.

The surface conductivity of the fired silver coating is far better than that of the air-dried or baked-on coating. Fired coatings have a surface electrical square resistance of approximately 0.01 Ω while the surface electrical square resistance of air-dried or baked-on is about 1 Ω.

Usually Alloyed

Because pure silver is so soft, it is usually alloyed with other metals for strength and durability. The most common alloying metal is copper, which imparts hardness and strength without appreciably changing the desirable characteristics of silver. Sterling silver has applications in manufacturing processes. For example, sterling silver plus lithium has been used in the aircraft industry for brazing honeycomb sections. Other silver–copper alloys are coin silver, 90.0% silver, 10.0% copper, and the silver–copper eutectic, 72% silver, 28% copper. This latter alloy has the highest combination of strength, hardness, and electrical properties of any of the silver alloys.

Silver braze filler metals are widely used for joining virtually all ferrous and nonferrous metals, with the exception of aluminum, magnesium, and some other lower-melting-point metals. Whereas pure silver melts at 960°C, silver alloys, with compositions of 10 to 85% silver (the alloying metals are copper, zinc, cadmium, and/or other base metals), have melting points of 618 to 960°C. These alloys have ductility and malleability and can be rolled into sheet or drawn into wire of very small diameter. They may be employed in all brazing processes and are generally free flowing when molten. Recommended joint clearances are 0.05 to 0.13 mm when used with flux. Whereas fluxes are usually required, zinc and cadmium-free alloys can be brazed in a vacuum or in reducing or inert atmospheres without flux. Joints made with silver brazing alloys are strong, ductile, and highly resistant to shock and vibration. With proper design there is no difficulty in obtaining joint strength equal to or greater than that of the metals joined. The strongest joints have but a few thousandths of an inch of the alloy as bonding material. Typical joints made with silver brazing alloys, giving the greatest degree of safety, are scarf, lap, and butt joints.

For electrical contacts, silver is combined with a number of other metals, which increase hardness and reduce the tendency to sulfide tarnishing. Silver–cadmium, for example, is extensively used for contacts, with the cadmium ranging from 10 to 15%. The advantages of these alloys are resistance to sticking or welding, more uniform wear, and a decreased tendency for metal transfer.

Alloys used for contact and spring purposes are the silver–magnesium–nickel series (99.5% silver), which are used where electrical contacts are to be joined by brazing without loss of hardness, in miniature electron tubes for spring clips where high thermal conductivity is essential, and for instruments and relay springs requiring good electrical conductivity at high temperatures. These are unique, oxidation-hardening alloys. Before being hardened, the silver–magnesium–nickel alloy can be worked by standard procedures. After hardening in an oxidizing atmosphere, the room-temperature tensile properties are similar to those of hard rolled sterling silver or coin silver.

Gold and palladium are also combined with silver for contact use because they reduce welding and tarnishing and, to some extent, increase hardness.

When certain base metals do not combine with silver by conventional methods, powder metal processes are employed. This is particularly true of silver–iron, silver–nickel, silver–graphite, silver–tungsten, etc. These alloys are used in electrical contacts because of the desirable conductivity of silver and the mechanical properties of the base metals. They can be pressed, sintered, and rolled into sheet and wire that is ductile and suitable for forming into contacts by heading or stamping operations.

Other silver products are those produced chemically — powder, flake, oxide, nitrate, and paint. Silver powder and flake are composed of large amounts of silver with 0.03 or 0.04% copper and traces of lead, iron, and other volatiles.

Silver Paints

These are used as conductive coatings that are pigmented with metallic silver flake or powder and bonding agents that are specially selected for the type of base material to which they are applied. These coatings are used to make conductive surfaces on such materials as ceramics, glass, quartz, mica, plastics, and paper, as well as on some metals. They are used for making printed circuits, resistor and capacitor terminals, and in miniature electrical instruments and equipment. Silver paints fall into two classifications: (1) fired-on types for base materials that can withstand temperatures in the 399 to 927°C range, and (2) air-dried or baked-on types for organic base materials that are dried at temperatures ranging from 21.1 to 427°C. The bonding agent in the fired-on type of coating is a powdered glass frit, whereas in the air-dried or baked-on type of coating organic resins are used. The viscosity and drying rate of each type varies, depending on the method of application, such as spraying, dipping, brushing, roller coating, or screen stenciling.

Silver-Type Batteries

Because they are six times lighter and five times smaller than other batteries of similar capacity, silver–zinc batteries have found wide use in guided missiles, telemetering equipment, and guidance control circuits and mechanisms. Where longer life and ruggedness are more important than the weight, silver–cadmium rechargeable batteries are specified. Where seawater activation is required, silver chloride–magnesium couples are used. Another silver type of battery is the solid electrolyte type made with silver, silver iodide, and vanadium pentoxide. This battery, designed for low-current applications, weighs less than 1 oz and has almost unlimited shelf life.

SINGLE CRYSTALS

In crystalline solids the atoms or molecules are stacked in a regular manner, forming a three-dimensional pattern, which may be obtained by a three-dimensional repetition of a certain pattern unit called a unit cell. When the periodicity of the pattern extends throughout a certain piece of material, one speaks of a single crystal. A single crystal is formed by the growth of a crystal nucleus without secondary nucleation or impingement on other crystals.

Growth Techniques

Among the most common methods of growing single crystals are those of P. Bridgman and J. Czochralski. In the Bridgman method the material is melted in a vertical cylindrical vessel that tapers conically to a point at the bottom. The vessel then is lowered slowly into a cold zone. Crystallization begins in the tip and continues usually by growth from the first formed nucleus. In the Czochralski method, a small single crystal (seed) is introduced into the surface of the melt and then drawn slowly upward into a cold zone. Single crystals of ultrahigh purity have been grown by zone melting. Single crystals are also often grown by bathing a seed with a supersaturated solution; the supersaturation is kept lower than is necessary for sensible nucleation.

When grown from a melt, single crystals usually take the form of their container. Crystals grown from solution (gas, liquid, or solid) often have a well-defined form that reflects the symmetry of the unit cell.

Physical Properties

Ideally, single crystals are free from internal boundaries. They give rise to a characteristic x-ray diffraction pattern. For example, the Laue pattern of a single crystal consists of a single characteristic set of sharp intensity maxima.

Many types of single crystal exhibit anisotropy, that is, a variation of some of their physical properties according to the direction along which they are measured. For example, the electrical resistivity of a randomly oriented aggregate of graphite crystallites is the same in all directions. The resistivity of a graphite single crystal is different, however, when measured along different crystal axes. This anisotropy exists for both structure-sensitive properties, which are not affected by imperfections (such a elastic coefficients).

Anisotropy of a structure-sensitive property is described by a characteristic set of coefficients that can be combined to give the macroscopic property along any particular direction in the crystal. The number of necessary coefficients can often be reduced substantially by consideration of the crystal symmetry; whether anisotropy, with respect to a given property, exists depends on crystal symmetry.

The structure-sensitive properties of crystals (for example, strength and diffusion coefficients) seem governed by internal defects, often on an atomic scale.

Industry and government are developing one of the first accurate computer-model predictions of molten metals and molding materials used in casting.

Currently, the computer information is being used to design and cast aircraft turbine blades. Similarly, another company is using the information from the computer models to improve the casting of automobile and light-truck engine blocks.

Cast-metal parts are used in 90% of all durable goods such as washing machines, refrigerators, stoves, lawn mowers, cars, boats, and aircraft. The goal of the partnership is to produce accurate models for all alloys used by the casting industry. The information can then be used by manufacturers to standardize metal-mixing recipes, allowing more effective competition in the marketplace.

SINTER HARDENING

Recognizing the great potential of sinter hardening, researchers continue to develp useful data that will help to attain benefits derived from this heat treatment. Improved properties can be achieved by effective control of material composition, density, section size, sintering temperature, and cooling rate. By controlling these variables a variety of microstructures and resultant properties are achievable, enabling particular powder metallurgy parts to favorably perform under specific service conditions. However, materials options, process flexibility, and application requirements demand a better understanding of process, microstructure, and mechanical property relationships to capitalize fully on the opportunity of sinter hardening.

Sinter hardening refers to a process where the cooling rate experienced in the cooling zone of the sintering furnace is fast enough that a significant portion of the material matrix transforms to martensite. Interest in sinter hardening has grown because it offers good manufacturing economy by providing a one-step process and a unique combination of strength, toughness, and hardness.

A variety of microstructures and properties can be obtained by varying both the alloy type and content as well as the postsintering cooling rate. By controlling the cooling rate, the microstructure can be manipulated to produce the required proportion of martensite, which will lead to desired mechanical properties. By understanding how the sintering conditions affect the microstructure, materials can be modeled to produce the final properties that are desired.

Alloying

Alloying elements are used in cast, wrought, and P/M materials to promote hardenability and increase the mechanical strength of the parts. A graphical way of examining the effects of alloying elements on the final microstructure of a steel is by using the characteristic isothermal transformation (I-T) diagram. This indicates the time necessary for the isothermal transformation of phases in the material from start to finish, as well as the cooling time and temperature combinations needed to produce the final microstructure. A similar diagram, known as a continuous cooling (CCT) curve, also is frequently used to determine the variation in microstructure as a function of cooling rate.

Materials and Processing

By evaluating the materials and properties shown in Tables S.6 and S.7, the apparent hardness, ultimate tensile strength, yield strength, total elongation, and martensite content of the various materials are seen. As expected, increasing the cooling rate resulted in increased apparent hardness and strength values. On the whole, hardness values were increased between 2 to 10 HRC for a given material.

As expected, in all materials, the percent of martensite present increased significantly with the increase in cooling rate. The effect of the increased martensite levels is apparent in the hardness values for each of the materials. The effect of the higher levels of martensite on tensile properties is less obvious. In several cases, materials with significantly lower percentages of martensite and lower hardness values demonstrated higher tensile strengths.

Some results show the following trends:

1. Materials with 2 wt% Ni and 0.5 wt% Graphite Admixed
   - Accelerated cooling resulted in increased strength and apparent hardness while decreasing elongation values only slightly. This result was the consequence of increased martensite content and finer pearlitic microstructures. In these materials, the martensite was the result of transformation of nickel-rich areas in the microstructure.
   - The increase in prealloyed alloy content from 0.85 wt% to 1.5 wt% Mo resulted in a larger increase in strength than the addition of 1.0 wt% admixed Cu.
   - Although the 0.85 wt% Mo materials exhibited higher percentages of martensite than identical chemistries based on the 1.5 wt% Mo prealloyed material, the higher molybdenum materials had higher apparent hardness and strength values. This surprising result was explained by the presence in the 1.5 wt% Mo-based material of significantly finer pearlite.
2. Materials with 2 wt% Cu and 0.9 wt% Graphite Admixed
   - As the cooling rate was increased for these materials, the apparent hardness increased. This was associated with higher martensite contents in the faster-cooled materials. Martensite contents of greater than 50% were found in all three base materials when accelerated cooling was utilized.
   - The materials with the highest apparent hardness values (0.5 wt% Ni, 1.5 wt% Mo prealloy) did not exhibit the highest tensile strength values. The highest UTS values were determined for the

fast-cooled version of the 0.85 wt% Mo prealloyed material. Retained austenite may be one potential cause for the fall off in strength for the Mo–Ni material.

SIZING AGENTS

These coatings are applied to glass textile fibers in the forming operation. The sizes used may contain one or more or any combination of binders, lubricants, and coupling agents.

Dextrin. A starch derivative commonly called starch in the fiberglass industry. It is of the low viscosity variety when used in fiberglass sizes. Practically all the fiberglass yarn that has been woven has been sized with this starch size. In industrial fiberglass fabrics, the size is removed in a process called heat cleaning, whereas in decorative woven material the size is burned off in the coronizing operation.

TABLE S.6
Premix Compositions

		Prealloyed Additions			Premix Additions	
Mix	Base	Nickel (wt%)	Molybdenum (wt%)	Copper (wt%)	Nickel (wt%)	Graphite (wt%)
1	Ancorsteel® 85 HP	—	0.85	—	2.00	0.50
2	Ancorsteel 150 HP	—	1.50	—	2.00	0.50
3	Ancorsteel 85 HP	—	0.85	1.00	2.00	0.50
4	Ancorsteel 150 HP	—	1.50	1.00	2.00	0.50
5	Ancorsteel 85 HP	—	0.85	2.00	—	0.90
6	Ancorsteel 150 HP	—	1.50	2.00	—	0.90
7	Ancorsteel 4600V	1.85	0.55	2.00	—	0.90

Source: Ind. Heating, May, p. 12, 1998. With permission.

TABLE S.7
Properties of Material Matrix

Mix	VARICOOL Setting (%)	Apparent Hardness (HRC)	0.2% Offset YS (psi × 10³/Mpa)	UTS (psi × 10³/Mpa)	Elg. (%)	Martensite Content (%)
1	50	6	66.5/459	90.5/624	2.4	7.3
1	100	9	70.2/484	97.1/669	2.3	20.8
2	50	12	80.5/555	103.5/714	1.6	10.1
2	100	16	87.0/600	110.3/760	1.5	11.8
3	50	7	71.6/494	98.1/676	2.0	23.0
3	100	11	78.5/541	107.9/744	1.9	38.8
4	50	14	89.7/618	114.3/688	1.5	15.7
4	100	19	98.1/676	122.4/844	1.4	20.2
5	50	21	95.2/656	109.5/755	1.1	22.5
5	100	30	112.6/776	135.9/937	1.2	66.3
6	50	25	102.7/708	132.0/910	1.5	29.8
6	100	35	114.7/791	127.1/876	1.0	60.1
7	50	35	102.4/706	118.9/820	1.1	71.9
7	100	37	106.3/732	117.9/813	0.9	95.5

Source: Ind. Heating, May, p. 12, 1998. With permission.

Gelatin. Used in small amounts as a constituent of a standard glass fiber textile size. Together with dextrin (starch), the gelatin acts as a binder for the other materials formulated into a size for the fiberglass textile yarns.

Polyvinyl Acetate. As a latex, it is used as a binder in a standard size formulation for continuous fiberglass yarn. Usually this continuous yarn is made into roving and a large package is usually made up of 60 ends of continuous yarn, although other end counts such as 30 ends, 20 ends, and 8 ends are also available. Roving is used primarily as reinforcement in reinforced plastics, and in preforming operations.

SODIUM

A metallic element (symbol Na and atomic weight 23), sodium occurs naturally only in the form of its salts. The most important mineral containing sodium is the chloride, NaCl, which is common salt. It also occurs as the nitrate, Chile saltpeter, as a borate in borax, and as a fluoride and a sulfate. When pure, sodium is silvery white and ductile, and it melts at 97.8°C and boils at 882°C. The specific gravity is 0.97. It can be obtained in metallic form by the electrolysis of salt. When exposed to the air, it oxidizes rapidly, and it must therefore be kept in airtight containers. It has a high affinity for oxygen, and it decomposes water violently. It also combines directly with the halogens, and is a good reducing agent for the metal chlorides. Sodium is one of the best conductors of electricity and heat.

The metal is a powerful desulfurizer of iron and steel even in combination. For this purpose it may be used in the form of soda ash pellets or in alloys. Desulfurizing alloys for brasses and bronzes are sodium–tin, with 95% tin and 5% sodium, or sodium–copper. Sodium–lead, used for adding sodium to alloys, contains 10% sodium, and is marketed as small spheroidal shot. It is also marketed as sodium marbles, which are spheres of pure sodium up to 2.54 cm in diameter coated with oil to reduce handling hazard. Sodium bricks contain 50% sodium metal powder dispersed in a paraffin binder. They can be handled in the air, and are a source of active sodium. Sodium in combination with potassium is used as a heat-exchange fluid in reactors and high-temperature processing equipment. A sodium–potassium alloy, containing 56% sodium and 44% potassium, has a melting point of 19°C and a boiling point of 825°C. It is a silvery mobile liquid. High-surface sodium is sodium metal absorbed on common salt, alumina, or activated carbon to give a large surface area for use in the reduction of metals or in hydrocarbon refining. Common salt will adsorb up to 10% of its weight of sodium in a thin film on its surface, and this sodium is 100% available for chemical reaction. It is used in reducing titanium tetrachloride to titanium metal. Sodium vapor is used in electric lamps. When the vapor is used with a fused alumina tube it gives a golden-white color. A 400-W lamp produces 42,000 lumens and retains 85% of its efficiency after 6000 h.

INORGANIC REACTIONS

Sodium reacts rapidly with water, and even with snow and ice, to give sodium hydroxide and hydrogen. The reaction liberates sufficient heat to melt the sodium and ignite the hydrogen.

When exposed to air, freshly cut sodium metal loses its silvery appearance and becomes dull gray because of the formation of a coating of sodium oxide. Sodium probably oxidizes to the peroxide, Na_2O_2, which reacts with excess sodium present to give the monoxide, Na_2O. When sodium reacts with oxygen at elevated temperatures, sodium superoxide, NaO_2, is formed; this reacts with more sodium to form the peroxide.

Sodium does not react with nitrogen, even at very high temperatures. Sodium and hydrogen react above about 200°C to form sodium hydroxide. This compound decomposes at about 400°C and cannot be melted. Sodium hydride can be formed by the direct reaction of hydrogen and molten sodium or by hydrogenating dispersions of sodium metal in hydrocarbons. Sodium reacts with carbon with difficulty, if at all, and this reaction may be said to have been adequately studied.

At room temperature fluorine and sodium ignite, dry chlorine and sodium react slightly, bromine and sodium do not react, and iodine and sodium do not react. However, in the presence of moisture or at elevated temperatures all reactions take place at very high rates.

Sodium reacts with ammonia, forming sodium amide and liberating hydrogen. The reaction may be carried out between molten sodium and gaseous ammonia (−30°C) in the presence of catalysts of finely divided metals. Sodium reacts with ammonia in the presence of coke to form sodium cyanide.

Carbon monoxide reacts with sodium, but the resulting carbonyl, NaCO, is stable only at liquid ammonia temperatures. At high temperatures sodium carbide and sodium carbonate are formed from carbon monoxide and sodium.

The reactions of sodium with various metal halides to give the metal plus sodium chloride are very important. Thus, titanium tetrachloride is reduced to titanium metal. Similarly, the halides of zirconium, beryllium, and thorium can be reduced to the corresponding metals by sodium. The interaction between sodium and potassium chloride is used in the commercial production of potassium metal.

Sodium hydroxide, NaOH, is also commonly known as caustic soda, and also as sodium hydrate. Lye is an old name used in some industries and in household uses. It readily absorbs water from the atmosphere and must be protected in storage and handling. It is corrosive to the skin and must be handled with extreme care to avoid caustic burns.

Most sodium hydroxide is produced by the electrolysis of sodium chloride solutions in one of several types of electrolytic cells. An older proces is the soda-lime process whereby soda ash is converted to caustic soda.

Organic Reactions

Sodium does not react with paraffin hydrocarbons but does form additional compounds with naphthalene and other polycyclic aromatic compounds and with arylated alkenes. It reacts with acetylene, replacing the acetylenic hydrogens to form sodium acetylides. Sodium adds to dienes, the reaction which forms the basis of the buna synthetic rubber process.

Principal Compounds

Sodium compounds are widely used in industry, particularly sodium chloride, sodium hydroxide, and soda ash. Sodium bichromate, $Na_2Cr_2O_7 \cdot 2H_2O$, a red crystalline powder, is used in leather tanning, textile dyeing, wood preservation, and in pigments. Sodium metavandate, $NaVO_3$, is used as a corrosion inhibitor to protect some chemical-processing piping. It dissolves in hot water, and a small amount in the water forms a tough impervious coating of magnetic iron oxide on the walls of the pipe. Sodium iodide crystals are used as scintillation probes for the detection and analysis of nuclear energies. Sodium oxalate is used as an antienzyme to retard tooth decay. In the drug industry sodium is used to compound with pharmaceuticals to make them water-soluble salts. Sodium is a plentiful element, easily available, and is one of the most widely used.

Sodium carbonate, Na_2CO_3, is best known under the name soda ash because sodium carbonate occurs in (and once was extracted from) plant ashes. Most sodium carbonate is produced by the Solvay or ammonia-soda process. In an initial reaction, salt is converted to sodium carbonate, which precipitates and is then separated.

Some soda ash is made synthetically by the Solvay process although an increasing amount is obtained from lake brines. Commercial grades of soda ash are available as 48% (Na_2O) light and dense and as 58% (Na_2O) light and dense; light and dense refers to apparent bulk density. Ordinary 48 to 58% grades are available in either light or dense but contain NaCl, which may affect certain ceramic uses. A 48% special grade is available in granular and extra light forms; it contains Na_2SO_4. The material derived from natural sources is almost NaCl-free.

About one half of the total American soda ash production is used as a fluxing ingredient by the glass industry. The quality of soda ash in glass batches varies with the type of glass being made.

Sodium sulfate, Na_2SO_4, is also known in the anhydrous form as salt cake. The decahydrate, $Na_2SO_4 \cdot 10H_2O$, is known as glauber salt.

Most sodium sulfate is produced synthetically as a by-product or coproduct in various industries.

Sodium aluminate, $Na_2OAl_2O_3$, whose melting point is 1650°C, is soluble in water and sodium carbonate. Sodium aluminate has found use as a settling-up agent for acid-resistant enamel. It is prepared by heating together bauxite and slips. When used in this capacity, it affords easier control of the slip than can be obtained by the use of alum or sulfuric acid, because of its tendency to stabilize the mobility and yield values. Sodium aluminate is also used as a substitute for sodium silicate and sodium carbonate in pottery slips.

Sodium antimonate (sodium meta-antimonate), $Na_2OSb_2O_5 0.5H_2O$, is a white powder insoluble in water and fruit acids. Sodium antimonate is extremely stable at high temperatures and does not decompose below 1427°C. It is usually made from antimony oxide, caustic soda, and sodium nitrate. Sodium antimonate is used as the principal opacifier in dry-process enamel frits for cast iron sanitaryware and in some of the acid-resistant enamel frits for sheet steel. It is used in cast iron enamels; sodium antimonate is generally recognized as being more desirable than antimony trioxide.

Sodium cyanide is a salt of hydrocyanic acid of the composition NaCN, used for carbonizing steel for case hardening, for heat-treating baths, for electroplating, and for the extraction of gold and silver from their ores. For carburizing steel it is preferred to potassium cyanide because of its lower cost and its higher content of available carbon. It contains 53% CN, as compared with 40% in potassium cyanide. The nitrogen also aids in forming the hard case on the steel. The 30% grade of sodium cyanide, melting at 679°C, is used for heat-treating baths instead of lead, but it forms a slight case on the steel. Sodium cyanide is very unstable, and on exposure to moist air liberates the highly poisonous hydrocyanic acid gas, HCN. For gold and silver extraction it easily combines with the metals, forming soluble double salts, $NaAu(CN)_2$. Sodium cyanide is made by passing a stream of nitrogen gas over a hot mixture of sodium carbonate and carbon in the presence of a catalyst. It is a white crystalline powder, soluble in water. The white copper cyanide used in electroplating has the composition $Cu_2(CN)_2$, containing 70% copper. It melts at 474.5°C and is insoluble in water, but is soluble in sodium cyanide solution. Sodium ferrocyanide, or yellow prussiate of soda, is a lemon-yellow crystalline solid of the composition $Na_4Fe(CN)_6 \cdot 10H_2O$, used for carbonizing steel for case hardening. It is also employed in paints, in printing inks, and for the purification of organic acids; in minute quantities, it is used in salt to make it free-flowing. It is soluble in water. Calcium cyanide in powder or granulated forms is used as an insecticide. It liberates 25% of hydrocyanic acid gas.

Sodium nitrate (soda niter), $NaNO_3$, with a melting point of 208°C, decomposes at 380°C and is soluble. Sodium nitrate is used in enamel frits in quantities of 2 to 8%. It is highly important that sufficient nitrate be present in enamels to prevent reduction of any easily reducible compounds in the batch, especially lead or antimony compounds. The function of sodium nitrate in glass is to oxidize organic matter that may contaminate batch materials, to prevent reduction of some of the batch constituents, to help maintain colors, and to speed the melt. It is the lowest melting of all glassmaking materials. Common applications of sodium nitrate are to ensure the pink color of manganese oxide and to prevent reductions of lead in potash lead glasses.

Sodium nitrite, $NaNO_2$, is soluble in water. It is prepared from sodium nitrate by reduction with lead. Sodium nitrite has been used for some years as a mill addition, or as an addition after milling, to enamel ground coats to prevent rust while drying, and also as a setting-up agent. More recently, sodium nitrite has been used rather generally in cover coats to correct for tearing.

Sodium phosphate, $Na_2HPO_4 12H_2O$, has a melting point of 346°C and is soluble in water. Sodium phosphate has been recently added to glass batches, producing an opal glass of unusual properties. Three other forms of the phosphate are available — monobasic, tribasic, and pyrophosphate. The last is most adaptable because it melts at 970°C in the anhydrous form. It is derived by the fusion of disodium phosphate.

Sodium silicate, Na_2OxSiO_2, is commonly made by melting sand and soda ash in a reverberatory furnace. Various proportions of the two ingredients are used and widely divergent characteristics result. The most alkaline liquid silicate made by this furnace process has a ratio of $1Na_2O:1.6SiO_2$ and the most siliceous liquid grade has a ratio of $1Na_2O:3.75SiO_2$.

Uses

The largest single use for sodium metal, accounting for about 60% of total production, is in the synthesis of tetraethyllead, an antiknock agent for automotive gasolines.

A second major use is in the reduction of animal and vegetable oils to long-chain fatty alcohols; these alcohols are raw materials for detergent manufacture. This use has been decreasing in favor of production of such alcohols by high-pressure catalytic hydrogenation.

Another major use is in the reduction of titanium and zirconium halides to the respective metals. Here the use of sodium is increasing at the expense of magnesium as the preferred reducing agent in such operations.

Sodium metal is also used in making sodium hydride, sodium amide, and sodium cyanide. It is also used in the synthesis of "isosebacic acid." The use of liquid sodium metal as a heat-transfer agent in nuclear reactors is also becoming increasingly important.

Sodium chloride is used in the manufacture of sodium hydroxide, sodium carbonate, sodium sulfate, and sodium metal. In sodium sulfate manufacture, hydrogen chloride is the coproduct; in metallic sodium manufacture, chlorine gas is the coproduct.

Rock salt is used in curing fish, in meat packing, in curing hides, and in making freezing mixtures. Food preparation, including canning and preserving, consumes much salt. Table salt accounts for only a small percentage of sodium chloride consumption, most of it going into the industrial uses outlined above.

Sodium hydroxide is perhaps the most important industrial alkali. Its major use is in the manufacture of chemicals, about 30% attributed to this category. The next major use is the manufacture of cellulose film and rayon, both of which proceed through soda cellulose (the reaction product of sodium hydroxide and cellulose); this accounts for about 25% of the total caustic soda production. Soap manufacture, petroleum refining, and pulp and paper manufacture each account for a little less than 10% of total sodium hydroxide use.

Sodium carbonate finds its major use in the glass industry, which takes about one third of total production. Approximately another third goes into the manufacture of soap, detergents, and various cleansers. The manufacture of paper and textiles, nonferrous metals, and petroleum products accounts for much of the balance.

The major consumer of sodium sulfate (salt cake) is the kraft pulp industry. Increasing quantities of sodium sulfate are used in the manufacture of flat glass. Other uses of salt cake are in detergents, ceramics, mineral stock feeds, and pharmaceuticals.

In the area of biological activity, the sodium ion (Na^+) is the main positive ion present in extracellular fluids and is essential for maintenance of the osmotic pressure and of the water and electrolytic balances of body fluids.

SOLDER ALLOYS

These are alloys of two or more metals used for joining other metals together by surface adhesion without melting the base metals as in welding and without requiring as high a temperature as in brazing. However, there is often no definite temperature line between soldering alloys and brazing filler metals. A requirement for a true solder is that it have a lower melting point than the metals being joined and an affinity for, or be capable of uniting with, the metals to be joined.

Types and Forms

The most common solder is called half-and-half, plumbers' solder, or ASTM (American Society for Testing and Materials) solder class 50A, and is composed of equal parts of lead and tin. It melts at 182°C. The density of this solder is 8802 kg/m^3, the tensile strength is 39 MPa, and the electrical conductivity is 11% that of copper. SAE (Society of Automotive Engineers) solder No. 1 has 49.5 to 50.0% tin, 50% lead, 0.12% max antimony, and 0.08% max copper. It melts at 181°C. Much commercial half-and-half, however, usually contains larger proportions of lead and some antimony, with less tin. These mixtures have higher melting points, and solders with less than 50% tin have a wide melting range and do not solidify quickly. Sometimes a wide melting range is desired, in which case a wiping solder with 38 to 45% of tin is used. A narrow-melting-range solder, melting at 183 to 185°C, ASTM solder class 60A, contains 60% tin and 40% lead. A 42% tin and 58% lead solder has a melting range of 183 to 231°C. Slicker solder is the best quality of plumbers' solder, containing 63 to 66% tin and the balance lead.

Solder alloys are available in a wide range of sizes and shapes, enabling users to select that one that best suits their application. Among these shapes are pig, slab, cake or ingot, bar, paste, ribbon or tape, segment or drop, powder, foil, sheet, solid wire, flux cored wire, and preforms. There are 11 major groups of solder alloys:

Tin–antimony. Useful at moderately elevated operating temperatures, around 149°C, these solders have higher electrical conductivity than the tin–lead solders. They are recommended for use where lead contamination must be avoided. A 95% Sn–5% Sb alloy has a solidus of 235°C, a liquidus of 240°C, and a resulting pasty range of –11.1°C.

Tin–lead. Constituting the largest group of all solders in use today, the tin–lead solders are used for joining a large variety of metals. Most are not satisfactory for use above 149°C under sustained load.

Tin–antimony–lead. These may normally be used for the same applications as tin–lead alloys with the following exceptions: aluminum, zinc, or galvanized iron. In the presence of zinc, these solders form a brittle intermetallic compound of zinc and antimony.

Tin–silver. These have advantages and limitations similar to those of tin–antimony solders. The tin silvers, however, are easier to apply with a rosin flux. Relatively high cost confines these solders to fine instrument work. Two standard compositions: 96.5% Sn–3.5% Ag, the eutectic; 95% Sn–5% Ag, with a solidus of 221°C and liquidus of 245°C.

Tin–zinc. These are principally for soldering aluminum since they tend to minimize galvanic corrosion.

Lead–silver. Tensile, creep, and shear strengths of these solders are usually satisfactory up to 177°C. Flow characteristics are rather poor and these solders are susceptible to humid atmospheric corrosion in storage. The use of a zinc chloride base flux is recommended to produce a good joint on metals uncoated with solder.

Cadmium–silver. The primary use of cadmium–silver solder is in applications where service temperature will be higher than permissible with lower melting solder. Improper use may lead to health hazards. The solder has a composition of 95% Cd–5% Ag. Solidus is 338°C and liquidus is 393°C.

Indium–lead alloys. These are alkali-resistant solders. A solder with 50% lead and 50% indium melts at 182°C, and is very resistant to alkalies, but lead–tin solders with as little as 25% indium are resistant to alkaline solutions, have better wetting characteristics, and are strong. Indium solders are expensive. Adding 0.85% silver to a 40% tin soft solder gives equivalent wetting on copper alloys to a 63% tin solder, but the addition is not effective on low-tin solders. A gold–copper solder used for making high-vacuum seals and for brazing difficult metals such as iron–cobalt alloys contains 37.5% gold and 62.5% copper.

Palladium–nickel. A palladium–nickel alloy with 40% nickel has a melting point about 1237°C. The brazing filler metals containing palladium are useful for a wide range of metals and metal to ceramic joints.

The remaining four groups of Zn–Al, Cd–Zn, and solders containing bismuth and indium were covered earlier.

Silver Solder

Silver solder is high-melting-point solder employed for soldering joints where more than ordinary strength and, sometimes, electric conductivity are required. Most silver solders are copper–zinc brazing filler metals with the addition of silver. They may contain from 9 to 80% silver, and the color varies from brass yellow to silver white. Cadmium may also be added to lower the melting point. Silver solders do not necessarily contain zinc, and may be braze filler metals of silver and copper in proportions arranged to obtain the desired melting point and strength. A silver braze filler metal with a relatively low melting point contains 65% silver, 20% copper, and 15% zinc. It melts at 693°C, has a tensile strength of 447 MPa, and elongation 34%. The electrical conductivity is 21%

that of pure copper. A solder melting at 760°C contains 20% silver, 45% copper, and 35% zinc. ASTM silver solder No. 3 is this solder with 5% cadmium replacing an equal amount of the zinc. It is general-purpose solder. ASTM silver solder No. 5 contains 50% silver, 34% copper, and 16% zinc. It melts at 693°C, and is used for soldering electrical work and refrigeration equipment.

Any tin present in silver solders makes them brittle; lead and iron make the solders difficult to work. Silver solders are malleable and ductile and have high strength. They are also corrosion resistant and are especially valuable for use in food machinery and apparatus where lead is objectionable. Small additions of lithium to silver solders increase fluidity and wetting properties, especially for brazing stainless steels or titanium. Sil-Fos is a phosphor–silver brazing solder with a melting point of 704°C. It contains 15% silver, 80% copper, and 5% phosphorus. Lap joints brazed with Sil-Fos have a tensile strength of 206 MPa. The phosphorus in the alloy acts as a deoxidizer, and the solder requires little or no flux. It is used for brazing brass, bronze, and nickel alloys.

Another grade, Easy solder, contains 65% silver, melts at 718°C, and is a color match for sterling silver. TL silver solder has only 9% silver and melts at 871°C. It is brass yellow in color, and is used for brazing nonferrous metals. Sterling silver solder, for brazing sterling silver, contains 92.8% silver, 7% copper, and 0.2% lithium. Flow temperature is 899°C.

A lead–silver solder to replace tin solder contains 96% lead, 3% silver, and 1% indium. It melts at 310°C, spreads better than ordinary lead–silver solders, and gives a joint strength of 34 MPa. Silver–palladium alloys for high-temperature brazing contain from 5 to 30% palladium. With 30%, the melting point is about 1232°C. These alloys have exceptional melting and flow qualities and are used in electronic and spacecraft applications.

Cold Solders

Cold solder, used for filling cracks in metals, may be a mixture of a metal powder in a pyroxylin cement with or without a mineral filler, but the strong cold solders are made with synthetic resins, usually epoxies, cured with catalysts, and with no solvents to cause shrinkage. The metal content may be as high as 80%. Devcon F, for repairing holes in castings, has 80% aluminum powder and 20% epoxy resin. It is heat-cured at 66°C, giving high adhesion. Epoxyn solder is aluminum powder in an epoxy resin in the form of a putty for filling cracks or holes in sheet metal. It cures with a catalyst. The metal–epoxy mixtures give a shrinkage of less than 0.2%, and they can be machined and polished smooth.

Lead-Free Solder Replacements

In spite of the sustained efforts of researchers and technology leaders in packaging, to date there is no lead-free solder alloy that is a drop-in replacement for tin–lead solder in assembly processes. Because tin–lead solder has been used for so long and is so much a part of the typical process engineer's thinking, the quest for an affordable lead-free alloy replacement is facing mounting pessimism that the effort will be successful.

On the other hand, optimism that adhesive-type solders will prove to be a serious alternative to metallurgical materials continues to grow. These polymer-based conductive adhesives are being used in various applications previously "reserved" for tin–lead solders. Regular production equipment and traditional assembly processes are producing high-quality assemblies with demonstrated long-term reliability using the new solders. And for some products, polymers are considered an enabling technology. One major market is the polyester-based flexible circuit market, particularly those built using polymer thick film.

Polymer Solders

Polymer solders are also known as "conductive adhesives," or materials that provide the dual functions of electrical connection and mechanical bond. The adhesive components of a typical material are some form of polymer, i.e., long-chain molecules widely used to produce structural products, which are also known for their excellent dielectric properties. Already used extensively as electrical insulators, as solders their necessary conductivity is accomplished by adding highly conductive fillers to the polymer binders.

The most common polymer solders are silver-filled thermosetting epoxies supplied as one-part thixotropic pastes. Silver is used not only because it is usually cost-effective, but also for its unique conductive oxide. A blend of silver powder and flakes achieves high conductivity while maintaining good printability. Because the mechanical strength of the joint is provided by the polymer, the challenge in a formulation is to use the maximum metal loading without sacrificing the required strength. (Some polymer solders contain more than 80% metal filler by weight.)

Polymer solders do not typically form metallurgical interfaces in the usual sense. Electrical integrity requires that the metal filler particles be in close contact to form a conductive path between the circuit trace and the component lead. Ideally, the silver flakes will overlap and smaller particles will fill in the gaps to form a conductive chain.

In the past, polymer solders were successful only on circuits using precious-metal conductors. This was because junction resistance was seen to increase to unacceptable levels when ordinary printed circuit boards and components were joined with these materials. To solve the instability problem it was necessary to create stable, nonmetallurgical junctions with oxidizable surfaces. For example, one polymer-solder formulation, used on flexible circuits and recently optimized for rigid boards, provides junction stability between solder-coated and bare-copper surfaces via polymer shrinkage during curing to force irregular particles through the interface oxides.

Advantages in using polymer solders include compatibility with a range of surfaces including some nonsolderable substrates; low-temperature processing, resulting in lower thermal stress; no pre- or postclean requirements, thereby reducing equipment needs and cycle times, and lowering or eliminating the release of volatile organic compounds.

SOLDER MATERIALS

FLUXES

Fluxes range from very mild substances to those of extreme chemical activity. For centuries rosin, a pine product, has been known as an effective and practically harmless flux. It is used widely for electrical connections in which utmost reliability, freedom from corrosion, and absence of electrical leakage are essential. When less stringent requirements exist and when less carefully prepared surfaces are to be soldered, rosin is mixed with chemically active agents that aid materially in soldering.

The rosin-type fluxes may be incorporated as the core of wire solders or dissolved in various solvents for direct application to joints prior to soldering.

Inorganic salts are widely used where stronger fluxes are needed. Zinc chloride and ammonium chloride, separately or in combination, are most common. They may also be obtained as so-called acid-core solder wire or in petroleum jelly as paste flux. All of the salt-type fluxes leave residues after soldering that may be a corrosion hazard. Washing with ample water accomplished by brushing is generally wise.

SOL-GEL PROCESS

This is a chemical synthesis technique for preparing gels, glasses, and ceramic powders. The synthesis of materials by the sol-gel process generally involves the use of metal alkoxides, which undergo hydrolysis and condensation polymerization reactions to yield gels.

The production of glasses by the sol-gel method is an area that has important scientific and technological implications. For example, the sol-gel approach permits preparation of glasses at far lower temperatures than is possible by using conventional melting. It also makes possible synthesis of compositions that are difficult to obtain by conventional means because of problems associated with volatization, high melting temperatures, or crystallization. In addition, the sol-gel approach is a high-purity process that leads to excellent homogeneity. Finally, the sol-gel approach is adaptable to producing films and fibers as well as bulk pieces, that is, monoliths (solid materials of macroscopic dimensions, at least a few millimeters on a side).

GLASS FORMATION

Formation of silica-based materials is the most widely studied system. However, an enormous range of multicomponent silicate glass compositions have also been prepared.

The sol-gel process can ordinarily be divided into the following steps: forming a solution, gelation, drying, and densification.

Hydrolysis and Condensation

In general, the processes of hydrolysis and condensation polymerization are difficult to separate. The hydrolysis of the alkoxide need not be complete before condensation starts; and in partially condensed silica, hydrolysis can still occur at unhydrolyzed sites. Several parameters have been shown to influence the hydrolysis and condensation polymerization reactions: these include the temperature, solution pH, the particular alkoxide precursor, the solvent, and the relative concentrations of each constituent. In addition, acids and bases catalyze the hydrolysis and condensation polymerization reactions; therefore, they are added to help control the rate and the extent of these reactions.

Microstructural Development

The conditions under which hydrolysis and condensation occur have a profound effect on gel growth and morphology. These structural conditions greatly influence the processing of sol-gel glasses into various forms. It is well established, for example, that acid-catalyzed solutions with low water content (that is, conditions that produce linear polymers) offer the best type of solution for producing fibers.

Gelation and Aging

As the hydrolysis and condensation polymerization reactions continue, viscosity increases until the solution ceases to flow. The time required for gelation to occur is an important characteristic that is sensitive to the chemistry of the solution and the nature of the polymeric species. This sol-to-gel transition is irreversible, and there is little if any change in volume.

Drying

The drying process involves the removal of the liquid phase; the gel transforms from an alcogel to a xerogel. Low-temperature evaporation is frequently employed, and there is considerable weight loss and shrinkage. The drying stage is a critical part of the sol-gel process. As evaporation occurs, drying stresses arise that can cause catastrophic cracking of bulk materials.

Densification

The final stage of the sol-gel process is densification. At this point the gel-to-glass conversion occurs and the gel achieves the properties of the glass. As the temperature increases, several processes occur, including elimination of residual water and organic substances, relaxation of the gel structure, and, ultimately, densification.

APPLICATIONS

The sol-gel process offers advantages for a broad spectrum of materials applications. The types of materials go well beyond silica and include inorganic compositions that possess specific properties such as ferroelectricity, electrochromism, or superconductivity. The most successful applications utilize the composition control, microstructure control, purity, and uniformity of the method combined with the ability to form various shapes at low temperatures. Films and coatings were the first commercial applications of the sol-gel process. The development of the sol-gel-based optical materials has also been quite successful, and applications include monoliths (lenses, prisms, lasers), fibers (waveguides), and a wide variety of optical films. Other important applications of sol-gel technology utilize controlled porosity and high surface area for catalyst supports, porous membranes, and thermal insulation.

SOYBEAN OIL

Biodiesel fuel, a combination of natural oil or fat with an alcohol such as methanol or ethanol, could help the U.S. reduce air pollution and its dependence on imported oil.

Soybean oil is the most commonly used feedstock in the U.S. Biodiesel works in all conventional diesel engines and can be distributed through the existing industry infrastructure.

Using 100% biodiesel fuel reduces carbon dioxide emissions by more than 75% compared to petroleum diesel. Using a blend of 20% biodiesel reduces them by 15%. The fuel also produces less particulate, carbon monoxide, and sulfur emissions, all targeted as public health risks. On the downside, biodiesel is more expensive than petroleum diesel and it produces slightly higher amounts of nitrogen oxide, a pollutant.

SPACE PROCESSING

The carrying out of various processes on materials aboard orbiting spacecraft is known as space processing. Until the space age, the Earth's gravity had always been considered a constant in the fluid-flow equations that govern heat and mass transport in materials processing. When the space shuttle became operational in the early 1980s, the potential benefits of suppressing the acceleration of gravity in certain processes began to be seriously considered. There have been numerous flight opportunities for microgravity experimentation.

Protein Crystal Growth

The importance of crystallography as a mechanism for determining three-dimensional structure of complex macromolecules has placed new demands on the ability to grow large (approximately 0.5 mm on a side), highly ordered crystals of a vast variety of biological macromolecules to obtain high-resolution x-ray diffraction data.

The growth of protein crystals in reduced gravity has the potential advantages of (1) the ability to suspend the growing crystals in the growth solution to provide a more uniform growth environment and (2) the ability to reduce the convective mass transport so that growth can take place to a diffusion-control led environment. The effect of convection on the growth of crystals is not well understood, but it is generally accepted that unsteady growth conditions that can result from convective flows are harmful to crystal growth.

Attempts to grow protein crystals in space produce mixed results. Sometimes no crystals or crystals that are inferior to those grown on Earth are produced. However, occasionally, the space-grown crystals are larger and better ordered than the best ever grown on Earth. In fact, the improvement in internal order obtained in proteins grown in reduced gravity can be so dramatic as to allow structure to be solved or refined to higher resolution than is possible by using the diffraction data from the best available Earth-grown crystals.

There is still the question of why attempts to grow protein crystals in space produce superior results only part of the time. (It should be remembered that many unreported experiments on the ground are not successful either.) One possible explanation is that the growth process is developed and optimized on the ground before committing the experiment to flight. However, the conditions that are optimum under normal gravity may not take advantage of the microgravity environment. Therefore, it may be necessary to actually develop the optimal growth processes in space to improve the yield of protein crystal growth experiments there.

Electronic and Photonic Materials

Single-crystalline materials suitable for electronic and photonic applications have received much attention as candidates for microgravity processing for several reasons. These are critical, high-value materials whose applications demand extreme control of composition, purity, and defects. Despite the rapid advances made in electronic materials, progress on many fronts is still limited by available materials. Gravity-driven flows certainly influence mass transport in growth processes. Even though the primary purpose of most of the space processing with these materials has been to gain insight and understanding that can be used in Earth-based processing, limited production of certain specialty materials in space is a possibility if it turns out that there is no Earth-based alternative.

Vapor Growth

Several vapor crystal growth experiments on the shuttle produced some interesting results that are not at all understood. An example is the growth of unseeded germanium–selenium (GeSe) crystals by physical vapor transport using an inert noble gas as a buffer to a closed tube. When this is done on the ground, many small crystallites form a crust inside the growth ampoule at the cold end. Growth in space produces dramatically different results: The crystals apparently nucleate away from the walls and grow as thin platelets, which eventually become entwined with one another, forming a web that is loosely contained by the tube. Even more striking is the appearance of the surfaces of the space-grown crystals. The surfaces are mirrorlike and almost featureless, exhibiting only a few widely spaced growth terraces. By contrast, crystallites grown on the ground under identical thermal conditions have many pits and irregular, closely spaced growth terraces.

Another example is the growth of $Hg_{0.4}Cd_{0.6}Te$ by closed-tube chemical vapor deposition on mercury–cadmium–tellurium (HgCdTe) substrates using mercuric iodide (HgI_2) as the transport agent. Again, considerable improvements in the space-grown samples are observed, relative to those grown on the ground, in terms of surface morphology, chemical microhomogeneity, and crystalline perfection.

In the growth of thin films of copper phthalocyanine on copper substrates by physical vapor deposition, a dramatic difference is found in the appearance and morphology of the space-grown film as compared with the films produced on the ground. Scanning electron microscopy reveals a close-packed columnar structure for the space-grown films, roughly resembling a thick pile carpet. The ground-grown samples have a lower-density, randomly oriented structure that resembles a shag carpet.

Mercuric iodide crystals grown during orbital flight by physical vapor transport exhibit sharp, well-formed facets indicating good internal order. This is confirmed by gamma-ray rocking curves, which are approximately one third the width of those taken on samples grown on the ground. Both electron and hole mobility are significantly enhanced in the flight crystals.

Solution Growth

Triglycine sulfate crystals can be grown from solution during orbital flight by using a novel cooled sting method. Supersaturation is maintained by extracting heat through the seed mounted on a small heat pipe, which in turn is attached to a thermoelectric device. Growth under diffusion-controlled transport conditions may avoid liquid and gas inclusions, the most common type of defect in solution-grown crystals, which are believed to be caused by unsteady growth conditions resulting from convective flows.

This growth technique has produced crystals of exceptional quality. The usual growth defects in the vicinity of the seed that form during the transition from dissolution to growth (the so-called ghost of the seed) are notably absent. High-resolution x-ray topographs taken with synchrotron radiation indicate a high degree of perfection. In pyroelectric detection for far-infrared radiation, the detectivity of the space-grown crystal is significantly higher than the seed crystal and the Q (ratio of energy stored to energy loss per cycle) is more than doubled.

METALLIC ALLOYS AND COMPOSITES

Processing of metallic alloys and composites in space has been carried out to study dendrite growth, monotectic alloys, liquid-phase sintering, and electrodeposition.

Dendrite Growth

The microgravity environment provides an excellent opportunity to carry out critical tests of fundamental theories of solidification without the complicating effects introduced by buoyancy-driven flows. For example, one investigator carried out a series of experiments to elucidate dendrite growth kinetics under well-characterized diffusion-controlled conditions in pure succinonitrile. This constituted a rigorous test of various nonlinear dynamical pattern formation theories, which provide the basis for the prediction of the microstructure and physical properties achieved in a solidification process.

Comparison of dendrite tip velocities, measured as a function of undercooling over a range from 0.05 to 1.5°C with ground-based measurements, shows that effects convection are more significant at the smaller undercoolings and are still important up to undercoolings as large as 1.3 K. Even in microgravity, there is a slight departure in the data at the smallest undercooling, which is attributed to the residual acceleration of the spacecraft. These data also allow the determination of the scaling constant important in the selection of the dynamic operating state, which the present theories have been unable to provide.

Monotectic Alloys

Some of the first microgravity experiments in metallurgy were attempts to form fine dispersions in monotectic alloys, that is, alloy systems that have liquid-phase immiscibilities. Attempts to solidify such alloys from the melt in normal gravity always result in macroscopic segregation because the densities of the two liquid phases are invariably different. It was thought that this phase separation could be avoided in microgravity and intimate mixtures of the two phases would result that might have interesting and unusual properties.

Liquid-Phase Sintering

Composites formed by liquid-phase sintering have many commercial applications, from cutting tools to electrical switch contacts. Fine particles of the more refractory phase are mixed with particles of a lower-melting material, which, when melted, forms a matrix to bind the nonmelting particles together. The system is stabilized during the sintering process by using a large volume fraction (80 to 85%) of nonmelting particles to support the structure while the molten phase interpenetrates the intergranular spaces. Fortunately, for many applications it is desirable to have a large volume fraction of the more refractory phase. However, there are some applications where there is a requirement to increase the volume fraction of the matrix material to amplify its properties. Space processing can be used to prepare composites (such as tungsten particles in a copper–nickel matrix, cobalt particles in a copper matrix, and iron particles in a copper matrix) with host-material volume fractions ranging from 30 to 50%, to increase the sintering time, and to provide valuable insight into evolution of pores and other defects that occur in sintered products produced on Earth.

Electrodeposition

Electrodeposition experiments in reduced gravity have produced some intriguing results. With higher current densities than can normally used on Earth, nickel with a nanocrystalline structure can be deposited on gold substrates. Attempts to duplicate this result in normal gravity by the use of convectively stable geometries and porous media have not been successful. It is speculated that the morphology of the hydrogen bubbles that form on the cathode in microgravity somehow promotes the formation of nickel hydride, which produces the nanocrystalline structure.

Attempts have been made to codeposit diamond dust with copper, and small particles of Co_2C_3 with cobalt, to form cermets that would be extremely hard and wear resistant. A ground-based technique for depositing a bonelike hydroxyapatite coating on prosthetic implants, which is based on this work, has significantly better adhesion than currently available coatings. Another derivative of this work is a plating process using Cr(III), which poses significantly fewer environmental problems than the common Cr(VI) process.

SPINEL

Spinel is any of a family of important AB_2O_4 oxide minerals, where A and B represent cations. Spinel minerals are widely distributed in the Earth, in meteorites, and in rocks from the moon. The ideal spinel formula is $MgOAl_2O_3$ or $MgAl_2O_4$. Spinel has a melting point of 2135°C and the mineral is found in small deposits. It is formed by solid-state reaction between MgO and Al_2O_3 and is an excellent refractory showing high resistance to attack by slags, glass, etc.

High-purity spinel is a chemically derived spinel powder made by the coprecipitation of magnesium and aluminum complex sulfates, with subsequent calcination to form the oxide compound. Purities range from 99.98 to

99.995%. The ceramic powders prepared by this process can be hot-pressed into transparent window materials with exceptional infrared transmission range.

APPLICATIONS

The major ceramic applications for spinels are the magnetic ferrospinels (ferrites), chromite brick, and spinel colors. Magnetic recording tape coated with $\alpha\text{-}Cr_2O_3$ is a relatively recent development. It is also used as a porous protective coating in oxygen sensors for automotive emission controls.

The material is available as fused spinel in special refractory applications and also in a special particle shape and distribution for flame and plasma-arc spraying. The magnetic spinels are of special importance because of the widespread interest and application of the ceramic ferrospinels (ferrites). Two classes of ferrospinels occur: magnetic and nonmagnetic.

The magnetic are related to the inverse structure and the nonmagnetic to the normal structure.

SPRAY METAL FORMING

Spray metal forming is a rapid solidification technology for producing semifinished tubes, billets, plates, and simple forms in a single integrated operation. In contrast to other powder metallurgy processes, spray metal forming offers the distinct advantage of skipping the intermediate steps of atomization and consolidation by atomizing and collecting the spray in the form of a billet in a single operation. Also, the elimination of powder handling reduces oxide content and enhances ductility.

THE PROCESS

Spray forming involves converting a molten metal stream into a spray of droplets by high-pressure gas atomization (Figure S.4). The droplets cool rapidly in flight and ideally arrive at a collector plate with just enough liquid content to spread and completely wet the surface. The metal then solidifies into an almost fully dense preform with a very fine, uniform microstructure. Steel, copper, nickel-based superalloys, and aluminum alloys have been

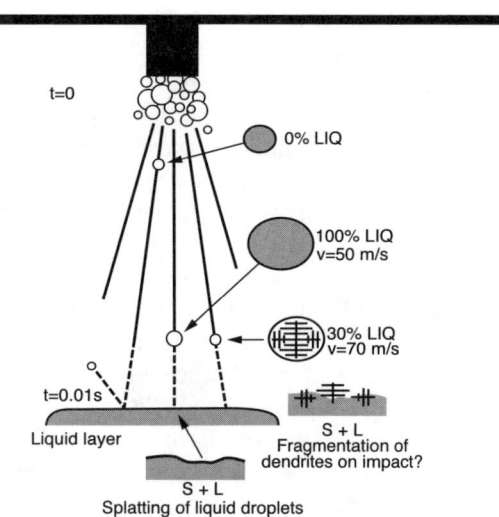

FIGURE S.4 Atomization and deposition process for spray metal forming. (From *NASA Tech. Briefs*, 21(5), 81, 1997. With permission.)

successfully spray-formed. These billets are combined, in the downstream manufacturing process, with extrusion or forging, and then coupled with high-speed machining to produce components in final form.

ADVANTAGES

Rapid solidification processes such as spray metal forming offer some distinct advantages over conventional ingot metallurgy processing. Superior properties due to fine grain sizes; a fine, homogeneous distribution of second-phase precipitates; and the absence of macrosegregation result from cooling rates on the order of 10^3 to 10^5 K/s (gas atomization processes approach 10^6 K/s). The high cooling rate in spray forming is obtained by higher gas-to-metal ratios. In certain alloy systems, a high volume fraction of fine (0.05 to 0.2 μm) intermetallic dispersoids may be obtained with high gas-to-metal ratios.

APPLICATIONS

Many different aluminum alloys and SiC-particulate-reinforced aluminum metal-matrix composites have been processed. They include conventional alloys in the 2XXX, 3XXX, 5XXX, 6XXX, and 7XXX series; high-temperature and high-strength alloys; and high-silicon-content alloys. The nonconventional alloys have been developed specifically for spray forming or adapted from alloys developed for rapid solidification rate (RSR) processes such as gas atomization or melt spinning. These alloys have shown superior properties such as wear resistance, room- and high-temperature strength, and creep resistance.

Alloys developed especially for spray forming, such as the ultrahigh zinc content alloys and 7050 aluminum with additional zinc, have been processed. Many of these alloys have been processed with the addition of SiC to improve the stiffness. These alloys have shown superior strength properties over conventionally processed material.

Four principal alloy systems originally developed for RSR processing have been extensively spray-formed. These are Al–Fe–V–Si (FVS), Al–Fe–Ce–W (FCW), Al–Ce–Cr–Co (CCC), and Al–Ni–Co (NYC) alloys. To optimize mechanical properties, these alloys were spray-formed over a range of processing parameters. For ultrahigh-temperature aluminum alloys, the melt superheat, or pour temperature, gas-to-metal ratio, and the injection of SiC particulate, have the greatest effect on microstructure (e.g., droplet and dispersoid size and volume fraction) and resultant properties.

Uses

Two government-sponsored programs have been completed using spray-formed material to produce components for use in Department of Defense vehicles. The first program produced track pins for advanced tracked ground combat vehicles using an ultrahigh-strength aluminum alloy. The second program developed the spray-forming processing parameters to produce an ultrahigh-temperature aluminum alloy for stator vanes in high-performance jet aircraft engines.

The goal of the track pin program was to develop processing that produces an ultrahigh-strength aluminum alloy that can replace the steel currently used in the manufactured pins for tracked ground combat vehicles. The objective was to replace the hollow steel pin with a solid aluminum pin that has similar properties. Several alloys were spray-formed, extruded, heat-treated, and tested to determine the tensile strength, ductility, and modulus. The two alloys with the best combination of properties were selected for additional processing. To increase the stiffness, SiC was added during spray forming. The spray-formed ultrahigh-strength alloys have yield strengths in excess of 690 MPa and show good ductility even with the addition of SiC. Track pins have been produced and are being tested. The corrosion resistance of these materials is improved by spray forming. The use of the ultrahigh-strength aluminum would result in a weight reduction of over 204.3 kg.

The NYC alloy shows promise for use in static parts of jet engines. The material properties at high temperatures provide an opportunity to replace heavier titanium parts with a lighter-weight aluminum alloy resulting in substantial saving in life-cycle costs.

The high-temperature aluminum alloy has been produced as extrusions and forgings. Machining contractors have developed the most efficient and economical processes to deliver a finished part. The alloy has been extruded and machined into the final component configuration and the alloy has good machining characteristics. Forging trials have shown that the material can be readily deformed offering increased material yield.

SPRING STEEL

This is a term applied to any steel used for springs. The majority of springs are made of steel, but brass, bronze, nickel silver, and phosphor bronze are used where their corrosion resistance or electric conductivity is desired. Carbon steels, with from 0.50 to 1.0% carbon, are much used, but vanadium and chromium–vanadium steels are also employed, especially for heavy car and locomotive springs. Special requirements for springs are that the steel be low in sulfur and phosphorus, and that the analysis be kept uniform. For flat or spiral springs that are not heat-treated after manufacture, hard-drawn or rolled steels are used. These may be tempered in the mill shape. Music wire is widely employed for making small spiral springs. A much-used straight-carbon spring steel has 1% carbon and 0.30 to 0.40% manganese, but becomes brittle when overstressed. ASTM (American Society for

Testing and Materials) carbon steel for flat springs has 0.70 to 0.80% carbon and 0.50 to 0.80% manganese, with 0.04% max each of sulfur and phosphorus. Motor springs are made of this steel rolled hard to a tensile strength of 1723 MPa. Watch spring steel, for mainsprings, has 1.15% carbon, 0.15 to 0.25% manganese, and in the hard-rolled condition, has an elastic limit above 2068 MPa.

Silicon Steels

These are used for springs and have high strength. These steels average about 0.40% carbon, 0.75% silicon, and 0.95% manganese, with or without copper, but the silicon may be as high as 2%. A steel, used for automobile leaf springs and recoil springs, contains 2% silicon, 0.75% manganese, and 0.60% carbon. The elastic limit is 689 to 2068 MPa, depending on drawing temperature, with hardness 250 to 600 Brinell.

Manganese Steels

These steels for automotive springs contain about 1.25% manganese and 0.40% carbon, or about 2% manganese and 0.45% carbon. When heat-treated, the latter has a tensile strength of 1378 MPa and 10% elongation. Part of the manganese may be replaced by silicon and the silicon–manganese steels have tensile strengths as high as 1861 MPa. The addition of chromium or other elements increases ductility and improves physical properties. Manganese steels are deep-hardening but are sensitive to overheating. The addition of chromium, vanadium, or molybdenum widens the hardening range.

Forms

Wire for coil springs ranges in carbon from 0.50 to 1.20%, and in sulfur from 0.028 to 0.029%. Bessemer wire contains too much sulfur for spring use. Cold working is the method for hardening the wire and for raising the tensile strength.

The highest grades of wire are referred to as music wire. The second grade is called hard-drawn spring wire. The latter is a less expensive basic open-hearth steel with manganese content of 0.80 to 1.10%, and an ultimate strength up to 2068 MPa.

Applications

For jet-engine springs and other applications where resistance to high temperatures is required, stainless steel and high-alloy steels are used. But, while these may have the names and approximate compositions of standard stainless steels, for spring-wire use their manufacture is usually closely controlled. For example, when the carbon content is raised in high-chromium steels to obtain the needed spring qualities, the carbide tends to collect in the grain boundaries and cause intergranular corrosion unless small quantities of titanium, columbium, or other elements are added to immobilize the carbon. Types include Type 302 stainless steel of highly controlled analysis for coil springs. Alloy NS-355 is a stainless steel having a typical analysis of 15.64% chromium, 4.38% nickel, 2.68% molybdenum, 1% manganese, 0.32% silicon, 0.12% copper, with the carbon at 0.14%. The modulus of elasticity is 205,100 MPa at 27°C and 168,000 MPa at 427°C. 17-7 PH stainless steel has 17% chromium, 7% nickel, 1% aluminum, and 0.07% carbon. Wire has a tensile strength up to 2378 MPa. Spring wire for high-temperature coil springs may contain little or no iron. Alloy NS-25, for springs operating at 760°C, contains about 50% cobalt, 20% chromium, 15% tungsten, and 10% nickel, with not more than 0.15% carbon.

SPUN PARTS

Metal spinning, essentially, involves forming flat sheet metal disks into seamless circular or cylindrical shapes. It is a useful processing technique when quantity does not warrant investment needed for draw dies.

The Process

The first step in the spinning process is to produce a form to the exact shape of the inside contours of the part to be made. The form can be of wood or metal. This form is secured to the headstock of a lathe and the metal blank is, in turn, secured to the form. Manual spinning techniques exist, and mechanical spinning lathes usually can be set up to force the blank against the form mechanically.

In addition to manual or power spinning, hot spinning is sometimes used either to anneal a spun part, eliminating the need to remove a partially formed blank from the lathe, or else to increase the plasticity of the metal being formed. In the latter category, some metals such as titanium or magnesium must be spun hot because their normal room-temperature crystal structure lacks ductility. Heavy parts (up to 122 mm in some cases) can also be spun with increased facility at elevated temperature.

SHAPES AND TOLERANCES

Basically, a component must be symmetrical about its axis to be adaptable to spinning. The three basic spinning shapes are the cone, hemisphere, and straight-sided cylinder. The shapes are listed in order of increasing difficulty to be formed by spinning.

Available spinning equipment is the limiting factor in determining the size of parts. Parts can be made ranging in diameter from 25.4 mm to almost 3.6 m. Thickness ranges from 0.010 to 122 mm. Most commonly, spun parts range in thickness from 0.059 to 4.7 mm.

SPUTTER TEXTURING

Texturing of a metal surface improves the bonding of surfacing material to the substrate or the attachment of parts to the textured piece. One of many applications is the texturing of the metal surface of medical hip implants. These devices require an irregular surface to stimulate bone attachment.

Texturing of complex shapes can be improved with a sputter-etching method that uses temporarily attached ceramic particles. By controlling the size and distribution of the ceramic particles, the width and depth of the texture can be regulated.

The first stage of the process (see Figure S.5) is the spraying or dripping of adhesive on the area to be textured. Microspheres of ceramic are forced into the adhesive and the area is heated. The part is then placed in a discharge chamber where it is bombarded with argon ions. This operation produces an etching on the surface of the part not covered by the ceramic spheres. The etch depth is controlled by voltage, current density, and sputtering duration. The adhesive, which is charred by the sputter-etch process, is removed with atomic oxygen in a plasma asher. The brushing away of the ceramic particles reveals a textured surface. See Figure S.5.

STAINLESS STEEL

Stainless steel comprises a large and widely used family of iron–chromium alloys known for their corrosion resistance — notably their "non-rusting" quality. This ability to resist corrosion

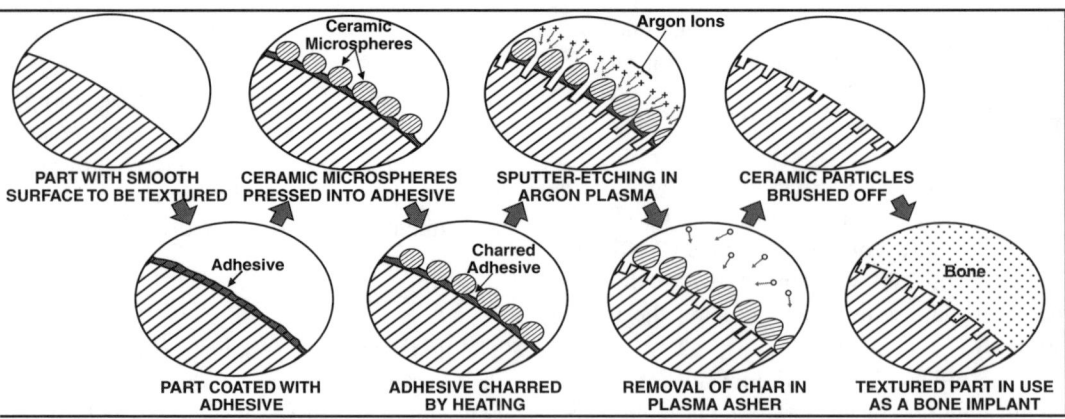

FIGURE S.5 Sputtering technique for improving surface for bonding. (From *Ind. Heating*, January, p. 45, 2000. With permission.)

is attributable to a chromium-oxide surface film that forms in the presence of oxygen. The film is essentially insoluble, self-healing, and nonporous. A minimum chromium content of 12% is required for the formation of the film, and 18% is sufficient to resist even severe atmospheric corrosion. Chromium content, however, may range to about 30% and several other alloying elements, such as manganese, silicon, nickel, or molybdenum, are usually present. Most stainless steels are also resistant to marine atmospheres, fresh water, oxidation at elevated temperatures, and mild and oxidizing chemicals. Some are also resistant to salt water and reducing media. They are also quite heat resistant, some retaining useful strength to 981°C. And some retain sufficient toughness at cryogenic temperatures. Thus, stainless steels are used in a wide range of applications requiring some degree of corrosion and/or heat resistance, including auto and truck trim, chemical- and food-processing equipment, petroleum-refining equipment, furnace parts and heat-treating hardware, marine components, architectural applications, cookware and housewares, pumps and valves, aircraft and aircraft-engine components, springs, instruments, and fasteners.

The 18-8 chromium–nickel steels were called super stainless steels in England to distinguish them from the plain chromium steels. Today, wrought stainless steels alone include some 70 standard compositions and many special compositions. They are categorized as austenitic, ferritic, martensitic, or precipitation-hardening (PH) stainless steels, depending on their microstructure or, in the case of the PH, their hardening and strengthening mechanism. There are also many cast stainless steels having these metallurgical structures. They are also known as cast corrosion-resistant steels, cast heat-resistant steels, and cast corrosion- and heat-resistant steels. Several compositions are also available in powder form for the manufacture of stainless steel powder-metal parts. See Tables S.8 and S.9.

Fabrication

As with steels in general, the so-called wrought stainless steels come from the melting furnaces

TABLE S.8
Standard Designations for Corrosion-Resistant Castings

Cast Alloy Designation	Wrought Alloy Type[a]
CA–15	410
Ca–40	420
CB–30	431
CB–7Cu	—
CC–50	446
CD–4MCu	—
CE–30	—
CF–3	304L
CF–8	304
CF–20	302
CF–3M	316L
CF–8M	316
CF–12M	316
CF–8C	347
CF–16F	303
CG–8M	317
CH–20	309
CK–20	310
CN–7M	—

[a] Wrought alloy type numbers are listed only for the convenience of those who want to determine corresponding wrought and cast grades. Because the cast alloy chemical composition ranges *are not the same* as the wrought composition ranges, buyers should use cast alloy designations for proper identification of castings.

in the form of ingot or continuously cast slabs. Ingots require a roughing or primary hot working, which the other form commonly bypasses. All then go through fabricating and finishing operations such as welding, hot and cold forming, rolling, machining, spinning, and polishing. No stainless steel is excluded from any of the common industrial processes because of its special properties; yet all stainless steels require attention to certain modifications of technique.

Hot Working

Hot working is influenced by the fact that many of the stainless steels are heat-resisting alloys. They are stronger at elevated temperatures than ordinary steel. Therefore, they require greater roll and forge pressure, and perhaps less reductions per pass or per blow. The austenitic steels are particularly heat resistant.

TABLE S.9
Physical Properties of Corrosion-Resistant Grades

Alloy Type	Density, lb/cu in.	Specific heat at 70°F, Btu/lb/°F	Thermal Conductivity at 212°F, Btu/h/ft²/°F	Thermal Expansion 70–1000°F, in./in./°F × 10⁶	Mag Perm	Electrical Resistance μΩ-cm at 70°F
CA–15	0.275	0.11	14.5	6.4	Ferromagnetic	78
CA–40	0.275	0.11	14.5	6.4	Ferromagnetic	76
CB–30	0.272	0.11	12.8	6.5	Ferromagnetic	76
CC–50	0.272	0.12	12.6	6.4	Ferromagnetic	77
CD–4Mcu	0.277	0.12	8.8	6.5	Ferromagnetic	75
CE–30	0.277	0.14	—	9.6	1.5	85
CF–3	0.280	0.12	9.2	10.0	1.0 to 2.0	76
CF–8	0.280	0.12	9.2	10.0	1.0 to 2.0	76
CF–20	0.280	0.12	9.2	10.4	1.01	78
CF–3M	0.280	0.12	9.4	9.7	1.5 to 2.5	82
CF–8M	0.280	0.12	9.4	9.7	1.5 to 2.5	82
CF–12M	0.280	0.12	9.4	9.7	1.5 to 2.5	82
CF–8C	0.280	0.12	9.3	10.3	1.2 to 1.8	71
CF–16F	0.280	0.12	9.4	9.9	1.0 to 2.0	72
CG–8M	0.281	0.12	9.4	9.7	1.5 to 2.5	82
CH–20	0.279	0.12	8.2	9.6	1.71	84
CK–20	0.280	0.12	8.2	9.2	1.02	90
CN–7M	0.289	0.11	12.1	9.7	1.01 to 1.10	90

Source: ACI Data Sheets.

Welding

Welding is influenced by another aspect of high-temperature resistance of these metals — the resistance to scaling. Oxidation during service at high temperatures does not become catastrophic with stainless steel because the steel immediately forms a hard and protective scale. But this, in turn, means that welding must be conducted under conditions that protect the metal from such reactions with the environment. This can be done with specially prepared coatings on electrodes, under cover of fluxes, or in vacuum; the first two techniques are particularly prominent. Inert-gas shielding also characterizes widely used processes among which at least a score are now numbered. As for weld cracking, care must be taken to prevent hydrogen absorption in the martensitic grades and martensite in the ferritic grades, whereas a small proportion of ferrite is almost a necessity in the austenitic grades. Metallurgical "phase balance" is an important aspect of welding the stainless steels because of these complications from a two-phase structure. Thus, a minor austenite fraction in ferritic stainless can cause martensitic cracking, while a minor ferrite fraction in austenite can prevent hot cracking. However, the most dangerous aspect of welding austenitic stainless steel is the potential "sensitization" affecting subsequent corrosion.

Machining and Forming

These processes adapt to all grades, with these major precautions: First, the stainless steels are generally stronger and tougher than carbon steel, such that more power and rigidity are needed in tooling. Second, the powerful work-hardening effect gives the austenitic grades the property of being instantaneously strengthened upon the first touch of the tool or pass of the roll. Machine tools must therefore bite surely and securely, with care taken not to "ride" the piece. Difficult forming operations warrant careful attention to variations in grade that are available, also in heat treatment, for

Finishing

These operations produce their best effects with stainless steels. No metal takes a more beautiful polish, and none holds it so long or so well. Stainlessness is not just skin-deep, but body through. And, of course, coatings are rendered entirely unnecessary.

STAINLESS STEEL (CAST)

Cast stainless steels are divided into two classes: those intended primarily for uses requiring corrosion resistance and those intended mainly for uses requiring heat resistance. Both types are commonly known by the designations of the Alloy Casting Institute of the Steel Founders Society of America, and these designations generally begin with the letter C for those used mainly for corrosion resistance and with the letter H for those used primarily for heat resistance. All are basically iron–chromium or iron–chromium–nickel alloys, although they may also contain several other alloying ingredients, notably molybdenum in the heat-resistant type, and molybdenum, copper, and/or other elements in the corrosion-resistant type. The corrosion-resistant cast stainless steel type follows the general metallurgical classifications of the wrought stainless steels, that is, austenitic, ferritic, austenitic-ferritic, martensitic, and precipitation hardening. Specific alloys within each of these classifications are austenitic (CH-20, CK-20, CN-7M), ferritic (CB-30 and CC-50), austenitic-ferritic (CE-30, CF-3, CF-3A, CF-8, CF-8A, CF-20, CF-3M, CF-3MA, CF-8M, CF-8C, CF-16F, and CG-8M), martensitic (CA-15, CA-40, CA-15M, and CA-6NM), and precipitation hardening (CB-7Cu and CD-4MCu). The chromium content of these alloys may be as little as 11% or as much as 30%, depending on the alloy. The heat-resistant cast stainless steel types may contain as little as 9% chromium (Alloy HA), although most contain much greater amounts, as much as 32% in HL. Although nickel content rarely exceeds chromium content in the corrosion-resistant type, it does in several heat-resistant types (HN, HP, HT, HU, HW, and HX). In fact, nickel is the major ingredient in HU, HW, and HX. Several of the heat-resistant types can be used at temperatures as high as 1149°C. C-series grades are used in valves, pumps, and fittings. H-series grades are used for furnace parts and turbine components.

APPLICATIONS

Iron–chromium alloys containing from 11.5 to 30% chromium and iron–chromium–nickel alloys containing up to 30% chromium and 31% nickel are widely used in the cast form for industrial process equipment at temperatures from –257 to 649°C. The largest area of use is in the temperature range from room temperature to the boiling points of the materials handled.

Typical stainless castings are pumps, valves, fittings, mixers, and similar equipment. Chemical industries employ them to resist nitric, sulfuric, phosphoric, and most organic acids, as well as many neutral and alkaline salt solutions. The pulp and paper industry is a large user of high alloy castings in digesters, filters, pumps, and other equipment for the manufacture of pulp. Fatty acids and other chemicals involved in soap-making processes are often handled by high alloy casting. Bleaching and dyeing operations in the textile industry require parts made from high alloys. These corrosion-resistant alloys are also widely used in making synthetic textile fibers. Pumps and valves cast of various high alloy compositions find wide application in petroleum refining. Other fields of application are food and beverage processing and handling, plastics manufacture, preparation of pharmaceuticals, atomic-energy processes, and explosives manufacture. Increasing use is being made of cast stainless alloys for handling liquid gases at cryogenic temperatures.

STAINLESS STEEL (WROUGHT)

Except for the precipitation-hardening (PH) stainless steels, wrought stainless steels are commonly designated by a three-digit numbering system of the American Iron and Steel Institute. Wrought austenitic stainless steels constitute the 2XX and 3XX series and the wrought ferritic stainless steels are part of the 4XX series. Wrought martensitic stainless steels belong either to the 4XX or 5XX series. Suffix letters, such as L for low carbon content or Se for selenium, are used to denote special compositional modifications. Cast stainless steels are commonly known by the designations of the Alloy Casting Institute of the Steel Founders Society of America, which begin with letters CA through CN and are followed by numbers or numbers and letters. Powder compositions are usually identified by the designations of the Metal Powder Industries Federation.

Of the austenitic, ferritic, and martensitic families of wrought stainless steels, each has a general-purpose alloy. All of the others in the family are derivatives of the basic alloy, with compositions tailored for special properties. The stainless steel 3XX series has the largest number of alloys and stainless steel 302, a stainless "18-8" alloy, is the general-purpose one. Besides its 17 to 19% chromium and 8 to 10% nickel, it contains a maximum of 0.15% carbon, 2% manganese, 1% silicon, 0.4% phosphorus, and 0.03% sulfur. 302B is similar except for greater silicon (2 to 3%) to increase resistance to scaling. Stainless steels 303 and 303Se are also similar except for greater sulfur (0.15% minimum) and, optionally, 0.6% molybdenum in 303, and 0.06 maximum sulfur and 0.15 minimum selenium in 303Se. Both are more readily machinable than 302. 304 and 304L stainless steels are low-carbon (0.08% and 0.03 maximum, respectively) alternatives, intended to restrict carbide precipitation during welding and, thus, are preferred to 302 for applications requiring welding. They may also contain slightly more chromium and nickel. 304N is similar to 304 except for 0.10 to 0.16% nitrogen. The nitrogen provides greater strength than 302 at just a small sacrifice in ductility and a minimal effect on corrosion resistance. 305 has 0.12% maximum carbon but greater nickel (10.5 to 13%) to reduce the rate of work hardening for applications requiring severe forming operations. S30430, as designated by the Unified Numbering System, contains 0.08 maximum carbon, 17 to 19% chromium, 8 to 10% nickel, and 3 to 4% copper. It features a still lower rate of work hardening and is used for severe cold-heading operations. 308 contains more chromium (19 to 21%) and nickel (10 to 12%) and, thus, is somewhat more corrosion and heat resistant. Although used for furnace parts and oil-refinery equipment, its principal use is for welding rods because its higher alloy content compensates for alloy content that may be reduced during welding. See Table S.10 (Wrought Stainless Steels).

Stainless steels 309, 309S, 310, 310S, and 314 have still greater chromium and nickel contents. 309S and 310S are low-carbon (0.08% maximum) versions of 309 and 310 for applications requiring welding. They are also noted for high creep strength. Stainless steel 314, which like 309 and 310 contains 0.25% maximum carbon, also has greater silicon (1.5 to 3%), thus providing greater oxidation resistance. Because of the high silicon content, however, it is prone to embrittlement during prolonged exposure at temperatures of 649 to 816°C. This embrittlement, however, is only evident at room temperature and is not considered harmful unless the alloy is subject to shock loads. These alloys are widely used for heaters and heat exchangers, radiant tubes, and chemical and oil-refinery equipment.

Stainless steels 316, 316L, 316F, 316N, 317, 317L, 321, and 329 are characterized by the addition of molybdenum, molybdenum and nitrogen (316N), or titanium (321). Stainless steel 316, with 16 to 18% chromium, 10 to 14% nickel, and 2 to 3% molybdenum, is more corrosion and creep resistant than 302- or 304-type alloys. Type 316L is the low-carbon version for welding applications; 316F, because of its greater phosphorus and sulfur, is the "free-machining" version; and 316N contains a small amount of nitrogen for greater strength. Stainless steels 317 and 317L are slightly richer in chromium, nickel, and molybdenum and, thus, somewhat more corrosion and heat resistant. Like 316, they are used for processing equipment in the oil, chemical, food, paper, and pharmaceutical industries. Type 321 is titanium-stabilized to inhibit carbide precipitation and provide greater resistance to intergranular corrosion in welds. Type 329, a high-chromium (25 to 30%) low-nickel (3 to 6%) alloy with 1 to 2% molybdenum, is similar to 316 in general corrosion resistance but more resistant to stress corrosion. Stainless steel 330, a high-nickel (34 to 37%), normal chromium (17 to 20%), 0.75 to 1.5% silicon, molybdenum-free alloy, combines good resistance to carburization, heat, and thermal shock.

Stainless steels 347 and 348 are similar to 321 except for the use of columbium and tantalum instead of titanium for stabilization. Type 348 also contains a small amount (0.2%) of copper. Both have greater creep strength than 321 and they are used for welded components, radiant tubes, aircraft-engine exhaust manifolds, pressure vessels, and oil-refinery equipment. Stainless steel 384, with nominally 16% chromium and 18% nickel, is another low-work-hardening alloy used for severe cold-heading applications.

The stainless steel 2XX series of austenitics comprises 201, 202, and 205. They are normal in chromium content (16 to 19%), but low in nickel (1 to 6%), high in manganese (5.5 to 15.5%), and with 0.12 to 0.25% carbon and some nitrogen. Types 201 and 202 have been called the low-nickel equivalents of 301 and 302, respectively. Type 202, with 17 to 19% chroinium, 7.5 to 10% manganese, 4 to 6% nickel, and a maximum of 1% silicon, 0.25% nitrogen, 0.15% carbon, 0.06% phosphorus, and 0.03% sulfur, is the general-purpose alloy. Type 201, which contains less nickel (3.5 to 5.5%) and manganese (5.5 to 7.5%), was prominent during the Korean war due to a nickel shortage. Type 205 has the least nickel (1 to 1.75%), and the most manganese (14 to 15.5%), carbon (0.12 to 0.25%), and nitrogen (0.32 to 0.40%) contents. It is said to be the low-nickel equivalent of 305 and has a low rate of work hardening that is useful for parts requiring severe forming operations.

Like stainless steels in general, austenitic stainless steels have a density of 7750 to 8027 kg/m^3. Unlike some other stainless steels, they are essentially nonmagnetic, although most alloys will become slightly magnetic with cold work. Their melting range is 1371 to 1454°C, specific heat at 0 to 100°C is about 502 J/kg · K, and electrical resistivity at room temperature ranges from 69×10^{-8} to 78×10^{-8} Ω · m. Types 309 and 310 have the highest resistivity, and 201 and 202 the lowest.

Most are available in many mill forms and are quite ductile in the annealed condition, tensile elongations ranging from 35 to 70%, depending on the alloy. Although most cannot be strengthened by heat treatment, they can be strengthened appreciably by cold work. In the annealed condition, the tensile yield strength of all the austenitics falls in the range of 207 to 552 MPa, with ultimate strengths in the range of 517 to 827 MPa. But cold-working 201 or 301 sheet just to the half-hard temper increases yield strength to 758 MPa and ultimate strength to at least 1034 MPa. Tensile modulus is typically 193×10^3 to 199×10^3 MPa and decreases slightly with severe cold work. As to high-temperature strength, even in the annealed condition most alloys have tensile yield strengths of at least 83 MPa at 815°C, and some (308, 310) about 138 MPa. Types 310 and 347 have the highest creep strength, or stress-rupture strength, at 538 to 649°C. Annealing temperatures range from 954 to 1149°C, initial forging temperatures range from 1093 to 1260°C, and their machinability index is typically 50 to 55, 65 for 303 and 303Se, relative to 100 for 1112 steel.

Among the many specialty wrought austenitic stainless steels are a number of nitrogen-strengthened stainless steels: Nitronic 20, 32, 33, 40, 50, and 60; 18-18 Plus and Marinaloy HN and 22; and SAF 2205 and 253MA. Nitrogen, unlike carbon, has the advantage of increasing strength without markedly reducing ductility. Some of these alloys are twice as strong as the standard austenitics and also provide better resistance to certain environments. All are normal or higher than normal in chromium content. Some are also normal or higher than normal in nickel content, whereas others are low in nickel and, in the case of 18-18 Plus, nickel-free. Nitronic 20, a 23% chromium, 8% nickel, 2.5% manganese alloy, combines high resistance to oxidation and sulfidation and was developed for engine exhaust valves. Unlike austenitics in general, it is hardenable by heat treatment. Solution treating at 1177°C, water quenching, and aging at 760°C provide tensile strengths of 579 MPa yield and 979 MPa ultimate. SAF 2205, an extra-low-carbon (0.03%), 22% chromium, 5.5% nickel, 3% molybdenum alloy, is a ferritic-austenitic alloy with high resistance to chloride- and hydrogen-sulfide-induced stress corrosion, pitting in chloride

TABLE S.10
Wrought Stainless Steels

Designation AISI or Co.	UNS	Yield Strength 70°F (10^3 psi)	Tensile Strength (10^3 psi) 70°F	Tensile Strength (10^3 psi) -320°F	Elongation 70°F (%)	Impact Strength, Izod (ft-lb) 70°F	Impact Strength, Izod (ft-lb) -320°F	Fatigue Endurance Limit (10^3 psi)	Creep Strength 0.001% at 1000°F (10^3 psi)	Thermal Conductivity 32–212°F (Btu-ft/h-ft²-°F)	Coefficient of Thermal Expansion 32–212°F (10^{-6}in./in.-°C)
Austenitic Grades											
202	S20200	38	90	220	40	115	120	—	—	9.4	9.7
301	S30100	30	75	275	60	100	110	35	—	9.4	9.4
302	S30200	30	75	220	60	80	110	34	17	9.4	9.6
303/303/SE	S30323	35	85	230	50	80	85(Se)	35	17	9.4	9.6
304	S30400	30	75	220	60	110	115	34	17	9.4	9.6
304L	S30403	25	70	220	60	160[a]	110	—	—	—	—
309/	S30900										
309S	S30908	30	75	—	45	110	—	—	16	9.0	8.3
310/	S31000										
310S	S31008	30	75	150	50	110	90	31.5	20	8.2	8.8
316	S31600	30	75	185	60	110	110	38	25	9.4	8.8
316L	S31603	25	70	—	60	—	—	—	—	—	—
317	S31700	35	85	—	50	—	—	—	—	9.4	8.8
317L	S31703	30	80	—	5	—	—	—	—	8.3	9.2
321	S32100	35	90	205	45	110	—	38	—	9.3	9.3
347/	S34700										
348	S34800	30	75	200	50	110	110	39	17	9.3	9.2
AL-GX	N08366	40	90	—	45	—	—	—	—	7.9	8.5
254S MO	S31254	44	95	—	35	—	—	—	—	—	9.4
18-9LW	—	30	78	—	65	—	—	—	—	—	—
Nitrogen-Strengthened Grades											
18Cr-2Ni-12Mn	—	60	115	—	55	230[a]	—	52	—	—	9.0
21Cr-6Ni-9Mn	S21904	57	100	220	53	240[a]	115[a]	49	—	8.0	6.3
18-18 Plus	S28200	65	120	228	60	240[a]	28[a]	50	—	—	—
22Cr-13Ni-5Mn	S20910	65	120	228	60	240[a]	28[a]	50	—	—	—
Nitronic 30	—	50	108	—	56	240[a]	18[a]	—	—	—	9.0
Nitronic 60	S21800	55	105	213	60	240[a]	160[a]	37	—	—	8.8

STAINLESS STEEL (WROUGHT)

Grade	UNS										
Precipitation-Hardened Grades											
15-5PH	S15500	85–185	120–200	240–260	15–22	16–100[a]	3–28	73–60	—	10.6	6.0
17-4PH	S17400	85–185	120–200	240–260	15–22	16–100[a]	3–23	73[b]–60[c]	—	10.6	6.0
AM362	S36200	108–182	125–188	—	16–21	6–80	—	95	—	—	—
Custom 450	S45000	117–184	194–196	130–250	13–25	18–105[a]	1–36	75–78	—	—	—
Custom 455	S45500	115–235	145–250	—	3–14	9–70[a]	3–5[a]	—	—	—	5.9
PH 13-8 Mo	S13800	82–215	130–235	—	12–22	24–120[a]	2–30[a]	100	—	—	6.0
17-7 PH	S17700	40–175	130–265	—	2–35	5–35[a]	3–5[a]	82–110	77[e]	9.5	9.2
A286	—	100	150	250	8–50	64[a]	57[a]	63[d]	80	8.2	
Ferritic Grades											
405	S40500	25	60	—	30	20	—	—	8	15.6	6.0
409	S40900	30	55	—	22	—	—	—	—	14.4	6.5
430	S43000	30	65	—	22	35	—	40	8.5	15.1	5.8
446/	S44600	40	75	—	20	2	—	47	6.4	12.1	5.8
18Cr-2Mo[g]	S18200	75–80	80–87	—	14–15	—	—	—	—	—	—
18SR	—	65	85	—	27	—	—	—	—	—	5.9
Sea-Cure	S44660	75	90	—	32	—	—	—	—	11.4	5.4
Monit	S44635	75	90	—	21	—	—	—	—	9.9	6.1
AL29-4C	S44735	75	90	—	25	—	—	—	—	—	5.2
AL29-4-2	S44800	85	95	—	22	—	—	—	—	—	5.2
Martensitic Grades											
403/	S40300										
410	S41000	40	75	—	35	100	—	40	9.2	14.4	5.5
416/	S41600										
416 Se	S41623	40	75	—	30	100	—	40	9.2	14.4	5.5
440A	S44002	60	105	—	20	2	—	40	—	14.0	5.7
440C	S44004	65	110	—	14	2	—	40	—	14.0	5.7

Note: Unless otherwise indicated, data are for annealed wrought bar. Properties of other mill forms may vary somewhat.

[a] Charpy V-notch.
[b] For 10⁸ cycles, condition H900, RT.
[c] At 600°F.
[d] For 10⁸ cycles aged.
[e] For 15 × 10⁸ cycles, condition RH 950, RT.
[f] Stabilized low interstitial (sheet, strip).
[g] Free-maching bar.

Source: Mach. Design Basics Eng. Design, June, p. 789, 1993. With permission.

environments, and intergranular corrosion in welded applications.

The wrought ferritic stainless steels are magnetic and less ductile than the austenitics. Although some can be hardened slightly by heat treatment, they are generally not hardenable by heat treatment. All contain at least 10.5% chromium and, although the standard alloys are nickel-free, small amounts of nickel are common in the nonstandard ones. Among the standard alloys, stainless steel 430 is the general-purpose alloy. It contains 16 to 18% chromium and a maximum of 0.12% carbon, 1% manganese, 1% silicon, 0.04% phosphorus, and 0.03% sulfur. Stainless steel 430F and 430FSe, the "free-machining" versions, contain more phosphorus (0.06% maximum) and sulfur (0.15% minimum in 430F, 0.06 maximum in 430FSe). 430FSe also contains 0.15% minimum selenium, and 0.6% molybdenum is an option for 430F. The other standard ferritics are stainless steels 405, 409, 429, 434, 436, 442, and 446. 405 and 409 are the lowest in carbon (0.08% maximum) and chromium (11.5 to 14.5% and 10.5 to 11.75%, respectively), the former containing 0.10 to 0.30% aluminum to prevent hardening on cooling from elevated temperatures, and the latter containing 0.75% maximum titanium. Type 429 is identical to 430 except for less chromium (14 to 16%) for better weldability. Types 434 and 436 are identical to 430 except for 0.75 to 1.25% molybdenum in the former and this amount of molybdenum plus 0.70% maximum columbium and tantalum in the latter; these additives improve corrosion resistance in specific environments. Types 442 to 446 are the highest in chromium (18 to 23% and 23 to 27%, respectively) for superior corrosion and oxidation resistance, and in carbon (0.20% maximum). Type 446 also contains more silicon (1.50% maximum).

These standard alloys melt in the range of 1427 to 1532°C, thermal conductivities of 21 to 27 W/m · K at 100°C, and electrical resistivities of 59 to 67 μΩ · cm at 21°C. In the annealed condition, tensile yield strengths range from 241 to 276 MPa for 405 to as high as 414 MPa for 434, with ultimate strengths of 448 to 586 MPa and elongations of 20 to 33%. For 1% creep in 10,000 h at 538°C, 430 has a stress-rupture strength of 59 MPa. Typical applications include automotive trim and exhaust components, chemical-processing equipment, furnace hardware and heat-treating fixtures, turbine blades, and molds for glass.

Wrought martensitic stainless steels are also magnetic and, as they are hardenable by heat treatment, provide high strength. Of those in the stainless steel 4XX series, 410, which contains 11.5 to 13.0% chromium, is the general-purpose alloy. The others, 403, 414, 416, 416Se, 420, 420F, 422, 431, 440A, and 440C, have similar (403, 414) or more chromium (16 to 18% in the 440s). Most are nickel-free or, as in the case of 414, 422, and 431, low in nickel. Most of the alloys also contain molybdenum, usually less than 1%, plus the usual 1% or so maximum of manganese and silicon. Carbon content ranges from 0.15% maximum in 403 through 416 and 416Se, to 0.60 to 0.75% in 440A, and as much as 1.20% in 440C. Type 403 is the low-silicon (0.50% maximum) version of 410; 414 is a nickel (1.25 to 2.50%)-modified version for better corrosion resistance. Types 416 and 416Se, which contain 12 to 14% chromium, also contain more than the usual sulfur or sulfur, phosphorus, and selenium to enhance machinability. Type 420 is richer in carbon for greater strength, and 420F has more sulfur and phosphorus for better machinability. Type 422, which contains the greatest variety of alloying elements, has 0.20 to 0.25% carbon, 11 to 13% chromium, low silicon (0.75% maximum), low phosphorus, and sulfur (0.025% maximum), 0.5 to 1.0% nickel, 0.75 to 1.25% of both molybdenum and tungsten, and 0.15 to 0.3% vanadium. This composition is intended to maximize toughness and strength at temperatures to 649°C. Type 431 is a higher-chromium (15 to 17%) nickel (1.25 to 2.50%) alloy for better corrosion resistance. The high-carbon, high-chromium 440 alloys combine considerable corrosion resistance with maximum hardness. The stainless steel 5XX series of wrought martensitic alloys — 501, 501A, 501B, 502, 503, and 504 — contain less chromium, ranging from 4 to 6% in 501 and 502, to 8 to 10% in 501 B and 504. All contain some molybdenum, usually less than 1%, and are nickel-free.

Most of the 4XX alloys can provide yield strengths greater than 1034 MPa and some, such as the 440s, more than 1724 MPa. The martensitic stainless steels, however, are less machinable than the austenitic and ferritic alloys and they are also less weldable. Forging temperatures range from 1038 to 1232°C. Most of the alloys are available in a wide range of mill forms and typical applications include turbine blades, springs, knife blades and cutlery, instruments, ball bearings, valves and pump parts, and heat exchangers.

The wrought PH stainless steels are also called age-hardenable stainless steels. Three basic types are now available: austenitic, semi-austenitic, and martensitic. Regardless of the type, the final hardening mechanism is precipitation hardening, brought about by small amounts of one or more alloying elements, such as aluminum, titanium, copper, and, sometimes, molybdenum. Their principal advantages are high strength, toughness, corrosion resistance, and relatively simple heat treatment.

Of the austenitic PH stainless steels, A-286 is the principal alloy. Also referred to as an iron-base superalloy, it contains about 15% chromium, 25% nickel, 2% titanium, 1.5% manganese, 1.3% molybdenum, 0.3% vanadium, 0.15% aluminum, 0.05% carbon, and 0.005% boron. It is widely used for aircraft turbine parts and high-strength fasteners. Heat treatment (solution treating at 981°C, water or oil quenching, aging at 718 to 732°C for 16 to 18 h and air cooling) provides an ultimate tensile strength of about 1035 MPa and a tensile yield strength of about 690 MPa, with 25% elongation and a Charpy impact strength of 87 J. The alloy retains considerable strength at high temperatures. At 649°C, for example, tensile yield strength is 607 MPa. The alloy also has good weldability and its corrosion resistance in most environments is similar to that of 3XX stainless steels.

The semiaustenitic PH stainless steels are austenitic in the annealed or solution-treated condition and can be transformed to a martensitic structure by relatively simple thermal or thermomechanical treatments. They are available in all mill forms, although sheet and strip are the most common. True semiaustenitic PH stainless steels include PH 14–8 Mo, PH 15–7 Mo, and 17-7PH. AM-350 and AM-355 are also so classified, although they are said not truly to have a precipitation-hardening reaction.

The above PH steels are lowest in carbon content (0.04% nominally in PH 14-8Mo, 0.07% in the others). PH 14-8Mo also nominally contains 15.1% chromium, 8.2% nickel, 2.2% molybdenum, 1.2% aluminum, 0.02% manganese, 0.02% silicon, and 0.005% nitrogen. PH 15-7Mo contains 15.2% chromium, 7.1% nickel, 2.2% molybdenum, 1.2% aluminum, 0.50% manganese, 0.30% silicon, and 0.04% nitrogen. 17-7PH is similar to PH 15-7Mo except for 17% chromium and being molybdenum-free. AM-350 contains 16.5% chromium, 4.25% nickel, 2.75% molybdenum, 0.75% manganese, 0.35% silicon, 0.10% nitrogen, and 0.10% carbon. AM-355 has 15.5% chromium, 4.25% nickel, 2.75% molybdenum, 0.85% manganese, 0.35% silicon, 0.12% nitrogen, and 0.13% carbon. In the solution-heat-treated condition in which these steels are supplied, they area readily formable. They then can be strengthened to various strength levels by conditioning the austenite, transformation to martensite, and precipitation hardening. One such procedure, for 17-7PH, involves heating at 760°C, air cooling to 16°C, then heating to 565°C and air-cooling to room temperature. In their heat-treated conditions, these steels encompass tensile yield strengths ranging from about 1241 MPa for AM-355 to 1793 MPa for PH 15-7Mo.

After solution treatment, the martensitic PH stainless steels always have a martensitic structure at room temperature. These steels include the progenitor of the PH stainless steels, Stainless W, PH 13-8Mo, 15-5PH, 17-4PH, and Custom 455. Of these, PH 13-8Mo and Custom 455, which contain 11 to 13% chromium and about 8% nickel plus small amounts of other alloying elements, are the higher-strength alloys, providing tensile yield strengths of 1448 MPa and 1620 MPa, respectively, in bar form after heat treatment. The other alloys range from 15 to 17% in chromium and 4 to 6% in nickel, and typically have tensile yield strengths of 1207 to 1276 MPa in heat-treated bar form. They are used mainly in bar form and forgings, and only to a small extent in sheet. Age hardening, following high-temperature solution treating, is performed at 427 to 677°C.

STAINLESS STEEL PRODUCTS

Metal fibers are used for weaving into fabrics for arctic heating clothing, heated draperies, chemical-resistant fabrics, and reinforcement in plastics and metals. Stainless-steel yarn made from the fibers is woven into stainless-steel fabric that has good crease resistance and retains its physical properties to 427°C. The fiber may be blended with cotton or wool for static control, particularly for carpeting.

STEEL

Steel is iron alloyed with small amounts of carbon, 2.5% maximum, but usually much less. The two broad categories are carbon steels and alloy steels, but they are further classified in terms of composition, deoxidation method, mill-finishing practice, product form, and principal characteristics. Carbon is the principal influencing element in carbon steels, although manganese, phosphorus, and sulfur are also present in small amounts, and these steels are further classified as low-carbon steels and sometimes referred to as mild steel (up to 0.30% carbon), medium-carbon steels (0.30 to 0.60%), and high-carbon steels (more than 0.60%). The greater the amount of carbon, the greater the strength and hardness, and the less ductility. Alloy steels are further classified as low-alloy steels, alloy steels, and high-alloy steels; those having as much as 5% alloy content are the most widely used. The most common designation systems for carbon and alloy steels are those of the American Iron and Steel Institute and the Society of Automotive Engineers, which follow a four- or five-digit numbering system based on the key element or elements, with the last two digits indicating carbon content in hundredths of a percent.

Plain carbon steels (with 1% maximum manganese) are designated 10XX; resulfurized carbon steels, 11XX; resulfurized and rephosphorized carbon steels, 12XX; and plain carbon steels with 1 to 1.65% manganese, 15XX. Alloy steels include manganese steels (13XX), nickel steels (23XX and 25XX), nickel–chromium steels (31XX to 34XX), molybdenum steels (40XX and 44XX), chromium–molybdenum steels (41XX), nickel–chromium–molybdenum steels (43XX, 47XX, and 81XX to 98XX), nickel–molybdenum steels (46XX and 48XX), chromium steels (50XX to 52XX), chromium–vanadium steels (61XX), tungsten–chromium steels (72XX), and silicon–manganese steels (92XX). The letter B following the first two digits designates boron steels and the letter L leaded steels. The suffix H is used to indicate steels produced to specific hardenability requirements. High-strength, low-alloy steels are commonly identified by a 9XX designation of the SAE, where the last two digits indicate minimum tensile yield strength in 1000 psi (6.8 MPa).

In contrast to rimmed steels, which are not deoxidized, killed steels are deoxidized by the addition of deoxidizing elements, such as aluminum or silicon, in the ladle prior to ingot casting, thus, such terms as aluminum-killed steel. Deoxidation markedly improves the uniformity of the chemical composition and resulting mechanical properties of mill products. Semikilled steels are only partially deoxidized, thus intermediate in uniformity to rimmed and killed steels. Capped steels have a low-carbon steel rim characteristic of rimmed-steel ingot and central uniformity more characteristic of killed-steel ingot, and are well suited for cold-forming operations.

Steels are also classified as air-melted, vacuum-melted, or vacuum-degassed. Air-melted steels are produced by conventional melting methods, such as open hearth, basic oxygen, and electric furnace. Vacuum-melted steels are produced by induction vacuum melting and consumable electrode vacuum melting. Vacuum-degassed steels are air-melted steels that are vacuum processed before solidification. Vacuum processing reduces gas content, nonmetallic inclusions, and center porosity, and segregation. Such steels are more costly, but have better ductility and impact and fatigue strengths.

Steel-mill products are reduced from ingot into such forms as blooms, billets, and slabs, which are then reduced to finished or semifinished shape by hot-working operations. If the final product is produced by hot working, the steel is known as hot-rolled steel. If the final product is shaped cold, the steel is known as cold-finished steel or, more specifically, cold-rolled steel, or cold-drawn steel. Hot-rolled mill products are usually limited to low- and medium-nonheat-treated carbon steels. They are the most economical steels, have good formability and weldability, and are widely used. Cold-finished steels, compared with hot-rolled products, have greater strength and hardness, better surface finish, and less ductility. Wrought steels are also classified in terms of mill-product form, such as bar steels, sheet steels, and plate steels. Cast steels refer to those used for castings, and P/M (powder metal) steels refer to powder compositions used for P/M parts. Steels are also known by their key characteristic from the standpoint of application, such as electrical steels, corrosion-resistant stainless steels, low-temperature steels, high-temperature steels, boiler steels, pressure-vessel steels, etc.

STEEL POWDER

Steel powder is used mainly for the production of steel powder metal parts made by consolidating the powder under pressure and then sintering, and, to a limited extent, for steel-mill products, principally tool-steel bar products. For powder metal parts, the powder may be admixed for the desired composition or prealloyed, that is, each powder particle is of the desired composition. For mill products, prealloyed powder is used primarily. Steel powder is widely used to make small- to moderate-size powder metal parts, with compositions closely matching those of wrought steels. Among the more common are carbon steels, copper steels, nickel steels, nickel–molybdenum steels, and stainless steels.

STEEL WOOL

Steel wool consists of long, fine fibers of steel used for abrading, chiefly for cleaning utensils and for polishing. It is made from low-carbon wire that has high tensile strength, usually having 0.10 to 0.20% carbon and 0.50 to 1% manganese. The wire is drawn over a track and shaved by a stationary knife bearing down on it, and may be made in a continuous piece as long as 30,480 m. Steel wool usually has three edges but may have four or five, and strands of various types are mixed. There are nine standard grades of steel wool, the finest of which has no fibers greater than 0.0027 cm thick; the most commonly used grade has fibers that vary between 0.006 and 0.010 cm. Steel wool comes in batts, or in flat ribbon form on spools usually 10 cm wide. Stainless steel wool is also made, and copper wool is marketed for some cleaning operations.

STEREOLITHOGRAPHY

Stereolithography uses a laser beam to convert a special photosensitive polymer from liquid to solid. The liquid polymer is held in a tank that also contains a platform that can be raised and lowered and onto which the part will be built.

The platform is first positioned just below the surface of the liquid polymer, and the laser beam is rastered across the surface of the polymer to solidify a two-dimensional image of the bottom layer of the part. The platform is then lowered a small distance to allow a thin layer of liquid polymer to cover the solidified layer, and the laser beam is again rastered to solidify the next layer on top of and bonded to the initial layer. The platform is lowered again to form the third layer, and so on.

When the final (top) layer of the part has been solidified, the platform is raised from the tank to drain away the liquid polymer from the finished part. Very complex shapes can be formed, including holes and even internal hollows (if a means is provided to drain out the unsolidified polymer).

Problems

Overhangs and undercuts are a challenge to form by stereolithography because the lower layers of such features will not be connected to the main body of the part, and temporary supporting structures (which can later be removed) must be fabricated as the layers are built up. By loading the polymer with ceramic or metal powder, stereolithography can be used to fabricate greenware that can be debinded and sintered to form a finished part.

Because this and the other rapid prototyping processes are rather slow, they would be appropriate production methods only for limited runs. By the same token, however, the avoidance by rapid prototyping of expensive tooling makes it all the more attractive for limited runs. Stereolithography is perhaps the most well-developed technique for rapid prototyping, and a number of vendors provide hardware, software, and special polymers to accomplish this technique. Some of the hardware is small enough that the term *desktop manufacturing* can legitimately be applied to it.

STRIPPABLE COATINGS

Strippable coatings are those that are applied for temporary protection and that can be readily removed. They are composed of such resins as cellulosics, vinyl, acrylic, and polyethylene; they can be water based, solvent based, or hot-melt. The choice of base depends on the surface to be protected. Water-base grades are neutral to plastic and painted surfaces, whereas solvent-base types affect those surfaces. Clear vinyl strippable coatings, perhaps the most widely used, are usually applied by spraying in thicknesses of 30 to 40 mils. Acrylic strippable coatings impart a clear, high-gloss, high-strength temporary film to metal parts. Polyethylene strippable coatings are relatively low cost and can be used on almost all surfaces except glass. Cellulosic strippable coatings are designed for hot-dip application. Film thicknesses range widely and can go as high as 200 mils. The mineral oil often present in these coatings exudes and coats the metal surface to protect it from corrosion over long periods.

TYPES

Vinyl coatings like those described in MIL-C-3254 specification were first developed for ships. These are called cocooning systems and are applied over chicken wire or a similar frame over the object to be protected. The interstices of the chicken wire are coated by a process known as webbing. This consists of spraying a specially designed vinyl coating in a web fashion so that it coats the interstices with a very thin, spiderweb-like, fragile covering. This in turn is coated with a material similar to an ordinary strip coat vinyl by spraying. A more rigid protective coat of asphalt is then applied, following which coats of vinyl or aluminum enamel are applied. The advantages of this coating system are long life, ability to cover irregular surfaces, and easy removal. Disadvantages include a cumbersome structure, which is expensive.

Strippable, Sprayable, Vinyl Coatings

These materials are generally applied by spraying in thicknesses of 30 to 40 mils. Their tensile strength runs 3.4 MPa minimum with an elongation of 200%, minimum. These materials are designed to be strippable after years of protective service. They are suitable for bright steel, aluminum, painted surfaces, wood, etc. They can also be used to protect spray booths, for aircraft protection, and on tanks, trucks, ships, and similar equipment.

Coatings can be produced in a translucent, colored effect or in a clear form so as to show any defects in the substrate. These coatings are generally sprayed from 1 to 2 mils thick and are designed for protection in covered storage or in the transportation or fabrication of tools. They can be readily removed even though film thickness is low.

Ethyl Cellulose, Type I, and Cellulose Acetobutyrate, Type II

These 100% coatings are designed for dip application in a hot-melt bath of 177°C. Thickness ranges from 100 to 200 mils, depending on the protection desired. The mineral oil generally present in these coatings exudes and coats the metal surface to keep it from corroding and in a strippable condition for long periods. Tensile strength of the coatings is about 2.06 MPa minimum, with an elongation of about 90% when originally made, and 70% when aged.

The coatings are generally used on tools, steel and aluminum parts, and many other parts that can withstand the temperature of dipping. Variations of specification types can be formulated that do not exude oil, which can be objectionable in handling, particularly with electrical equipment. Pourable variations are also commercially available.

Some special types of ethyl cellulose strippable materials can be used on painted surfaces and are formulated so that these surfaces are not affected. They can be applied by spraying. Other specially designed strippable materials are also practical; some of them can be used for packaging.

STRONTIUM AND ALLOYS

A chemical element (symbol Sr), strontium is the least abundant of the alkaline-earth metals. It has a melting point about 770°C, and it decomposes in water. The metal is obtained by electrolysis of the fused chloride, and small amounts are used for doping semiconductors. Its compounds have been used for deoxidizing nonferrous alloys and for desulfurizing steel. But the chief uses have been in signal flares to give a red light, and in hard, heat-resistant greases. Strontium-90, produced atomically, is used in ship-deck signs as it emits no dangerous gamma rays. It gives a bright sign, and the color can be varied with the content of zinc, but it is short-lived. Strontium is very reactive and used only in compounds.

COMPOUNDS

Strontium nitrate is a yellowish-white crystalline powder, $Sr(NO_3)_2$, produced by roasting and leaching celestite and treating with nitric acid. The specific gravity is 2.96, the melting point is 645°C, and it is soluble in water. It gives a bright crimson flame, and is used in railway-signal lights and in military flares. It is also a source of oxygen, pyrotechnics, as well as a precursor for ceramic powders.

The strontium sulfate used as a brightening agent in paints is powdered celestite. Strontium sulfide, SrS, used in luminous paint, gives a blue-green glow, but it deteriorates rapidly unless sealed. Strontium carbonate, $SrCO_3$, is used in pyrotechnics, ceramics, and ceramic permanent magnets for small motors.

The development of glazes for low-temperature vitreous bodies can be materially aided through the use of strontia. The added fluidity provided by strontia when replacing calcium and/or barium should promote interface reaction, improve glaze fit, while offsetting the slightly higher thermal expansion evidenced in some cases in the dinnerware glaze tested. Strontia additions to such glazes should materially increase glaze hardness and lower the solubility. Scratch resistance should be improved when replacements are made, especially at the expense of calcium and barium, which would be due, in part, to the earlier reaction of strontia enabling the glaze to clear with a minimum of pits.

Strontium hydrate, $Sr(OH)_2 \cdot 8 H_2O$, loses its water of crystallization at 100°C and melts at 375°C. It is used in making lubricating greases and as a stabilizer in plastics. Strontium fluoride is produced in single crystals for use as a laser material. When doped with samarium it gives an output wavelength around 650 nm.

Strontium hexaboride, which is generally known as SrB_6, is stable to temperatures up to 2760°C, above which decomposition initiates. Possible uses for the material are energy sources when using the radioisotope, high-temperature insulation, nuclear reactor control rods, and control additives.

Strontium titanate, $SrTiO_3$, has a melting point of 2080°C. Methods of compounding are (1) from mixed strontium carbonate and titanium dioxide, (2) from mixed strontium oxalate and titanium dioxide, and (3) from strontium titanyl oxalate. Strontium titanate is a high-dielectric-constant material (225 to 250), which at lower temperatures has a temperature coefficient of dielectric constant somewhat higher than that of calcium titanate.

Strontium titanate can be used by itself or in combination with barium titanate in applications for capacitors and other parts.

The power factor of strontium titanate is unusually high at low frequencies with a great improvement in power factor in the neighborhood of 1 MHz. The thermal expansion of strontium titanate is linear over a wide temperature range (100 to 700°C).

STRUCTURAL FOAM

Extending the size capabilities of molded parts beyond the limits of conventional injection molding is one of the main advantages of structural foam molding. Whereas injection moldings are usually referred to in terms of ounces and inches, foam moldings usually involve pounds and feet, despite the fact that foam density is much lower. Parts weighing 22.7 kg are not uncommon by low-pressure structural foam methods, and some molders can produce 45.4-kg parts in a single shot.

FOAM PROCESSING

Although structural foam parts are produced by several different methods, all systems disperse a gas into the polymer melt during processing, either by adding a chemical blowing agent to the compound or by inducing a gas directly into the melt. The gas creates the cellular core structure in the part. Regardless of the type or form of foaming agent used or when it is added to the melt, structural foam processes are classified as either low-pressure or high-pressure methods. Such a classification relates directly to size range, surface finish, economics, and properties of the molded part.

Low Pressure

Also known as short-shot, conventional structural foam processing methods are the most commonly used because they are the simplest and best suited for economical production of large, three-dimensional parts. In this process, a controlled mixture of resin and gas is injected into a mold creating a low cavity pressure — from 1.36 to 3.43 MPa. The mixture only partially fills the mold, and the bubbles of gas, having been at a higher pressure, expand immediately and fill the cavity. As the cells collapse against the mold surface, a solid skin of melt is formed over the rigid, foamed core.

Skin thickness is controlled by amount of melt injected, mold temperature, type and amount of blowing agent, and temperature and pressure of the melt. With the use of multiple injection nozzles, extremely large parts can be molded; alternatively, several parts of varying sizes can be molded simultaneously in multicavity molds. Standard nominal wall thickness is 3.2 mm.

Another process variation, coinjection, involves the separate injection of two compatible resins. First, a solid resin is injected to form the solid, smooth skin against the mold surfaces. Then, the second material, a measured short shot containing a blowing agent, is injected to form the foamed part interior. The core material is usually a lower-cost resin than the skin material.

High Pressure

This is an expandable-mold, structural foam molding that is closer to conventional injection molding. The heating melt (with a blowing agent) is injected into the mold, creating cavity pressures of between 34 and 136 MPa. The mold is entirely filled, and the pressure prevents any foaming from occurring while the skin portion solidifies against the mold surfaces.

At this point, the method departs from conventional injection molding. Mold pressure must be reduced and space provided to allow foaming to take place between the solid-skin surfaces. Depending on the type of equipment and size and configuration of the part, these two provisions are made either by withdrawing cores or by special press motions that partially open the mold halves.

ADVANTAGES

In addition to large-size capability and low-density structure, structural foam parts offer high stiffness-to-weight advantages. A 25% increase in wall thickness (over that of a solid section) can provide twice the rigidity — at equal weight — of a solid part. Strength-to-weight ratios of foamed sections can be two to five times those of structural metals. Foamed parts made by any of the various methods are relatively stress-free because the foaming is done at a low pressure. For the same reason, sink marks do not occur in foamed parts behind ribs or at wall intersections.

Tooling for low-pressure structural foam molding is generally less expensive than that for injection molding because the low pressure permits the use of lighter-weight mold materials. Tooling for high-pressure systems is more expensive than for conventional molding because of the special tooling motions involved to accommodate the foaming cycle.

Limitations

Surface finish is the most apparent difference in low-pressure structural foam parts, compared with conventional moldings. Part surfaces have a characteristic swirl pattern caused by the blowing agent, some of which becomes trapped between the mold surface and the skin of the part. The swirl pattern is both visual and tactile; surface roughness can be as much as 1,000 μin. Parts that require smooth, finished surfaces require secondary operations, usually sanding, filling, and painting.

The principal process variables that control the swirl pattern are mold temperature, melt temperature, injection rate, and the nature and concentration of the blowing agent. Control of these variables can produce foamed parts with surfaces that replicate any mold surface, smooth or patterned. But this improvement is not achieved without trade-offs: changes involve slower injection rates, heating and cooling of the mold, and other alterations that increase mold costs.

Parts produced by gas counterpressure molding have a significantly reduced swirl pattern because the foaming gases are kept in solution until the solid skin is formed against mold surfaces. Surfaces of coinjected parts are comparable to those of solid, injection-molded parts. Surfaces of parts made by the high-pressure processes are comparable to those of injection-molded parts because the surface of the melt in contact with the mold solidifies while under pressure. But such parts have a "witness line," which has the appearance of a wide parting line at the edges — where the mold opening made provision for foaming. Such areas may require touching up.

Another limitation of high-pressure foam molding is part size and shape. The high pressures and the cost of tools limit these systems to much smaller parts than can be molded economically by low-pressure systems. Most high-pressure parts are relatively flat.

STRUCTURAL MATERIALS

These are construction materials that, because of their ability to withstand external forces, are considered in the design of a structural framework. Materials used primarily for decoration, insulation, or other than structural purposes are not included in this group.

Clay Products

The principal products in this class are the solid masonry units such as brick and the hollow masonry units such as clay tile or terra-cotta.

Brick is the oldest of all artificial building materials. It is classified as face brick, common brick, and glazed brick. Face brick is used on the exterior of a wall and varies in color, texture, and mechanical perfection. Common brick consists of the kiln run of brick and is used principally as backup masonry behind whatever facing material is employed. It provides the necessary wall thickness and additional structural strength. Glazed brick is employed largely for interiors where beauty, ease of cleaning, and sanitation are primary considerations.

Structural clay tiles are burned-clay masonry units having interior hollow spaces termed *cells*. Such tile is widely used because of its strength, light weight, and insulating and fire protection qualities. Its size varies with the intended use.

Load-bearing tile is used in walls that support, in addition to their own weight, loads that frame into them, for example, floors and the roof. Tiles manufactured for use as partition walls, for furring, and for fireproofing steel beams and columns are classed as non-load-bearing tile. Special units are manufactured for floor construction: some are used with reinforced-concrete joists, and others with the steel beams in flat-arch and segmental-arch construction.

Architectural terra-cotta is a burned-clay material used for decorative purposes. The shapes are molded either by hand in plaster-of-paris molds or by machine, using the stiff-mud process.

Building Stones

Building stones generally used are limestone, sandstone, granite, and marble. Until the advent of steel and concrete, stone was the most important building material. Its principal use now is as a decorative material because of its beauty, dignity, and durability.

Concrete

Concrete is a mixture of cement, mineral aggregate, and water, which, if combined in proper proportions, form a plastic mixture capable of being placed in forms and of hardening through the hydration of the cement.

Wood

The cellular structure of wood is largely responsible for its basic characteristics, unique among the common structural materials. The strength of wood depends on the thickness of the cell walls. Its tensile strength is generally greater than its compressive strength. The ratio of its strength to its stiffness is much higher than that of steel or concrete; therefore, it is important that deflection be carefully considered in the design of a wooden floor system.

Laminated structural lumber is formed by gluing together two or more layers of wood with the grain of all layers parallel to the length of the member. Both laminated lumber and plywood make use of modern gluing techniques to produce a greatly improved product. The principal advantages derived from lamination are the ease with which large members are fabricated and the greater strength of built-up members. Laminated lumber is used for beams, columns, arch ribs, chord members, and other structural members.

Plywood, while also laminated, is formed from three or more thin layers of wood that are cemented or bonded together, with the grain of the several layers alternately perpendicular and parallel to each other. Plywood is generally used as a replacement for sheathing or as form lumber for reinforced concrete structures. Both laminated structural lumber and plywood have the advantage of minimizing the effects of knots, shakes, and other lumber defects by preventing them from occurring in more than one lamination at a given cross section.

Structural Metals

Of importance in this group are the structural steels, steel castings, aluminum alloys, magnesium alloys, and cast and wrought iron.

Steel castings are used for rocker bearings under the ends of large bridges. Shoes and bearing plates are usually cast in carbon steel, but rollers are often cast in stainless steel.

Aluminum alloys are strong, lightweight, and resistant to corrosion. The alloys most frequently used are comparable with the structural steels in strength. However, because aluminum alloys have a modulus of elasticity one third that of steel, the danger of local buckling is likely to determine the design of aluminum compression members.

Magnesium alloys are produced as extruded shapes, rolled plate, and forgings. The principal structural applications are in aircraft, truck bodies, and portable scaffolding.

Composite Materials

These are engineered materials synthesized with two distinct phases and comprising a load-bearing material housed in a relatively weak protective matrix. The combination of two or more constituent materials yields a composite material with engineering properties superior to those of the constituents. The associated materials are termed polymer-matrix composites (PMCs), ceramic-matrix composites (CMCs), and metal-matrix composites (MMCs).

The principal features of a fibrous composite material are the fibers, the matrix material, and the interface region between these two dissimilar materials. This class of structural material can be classified as metallic, ceramic, or polymeric, depending on the load-bearing or reinforcing material employed. The reinforcement may be particulates, whiskers, laminated fibers, or a woven fabric. These reinforcements are bonded together by the matrix, which distributes the loading between them. Generally, the reinforcement is a fibrous or particulate material, with the latter category permitting far superior structural properties to be achieved at the expense of more-challenging fabricating technologies and higher costs.

There are numerous examples in nature where this type of microstructure, comprising a load-bearing structural phase housed in a protective matrix, is present. An example is a tree, where the trunk and branches comprise flexible cellulose fibers in a rigid lignin matrix. Development of the class of synthetic materials with this type of structure, such as fiberglass composites and graphite-epoxy laminates, revolutionized the automotive, aerospace, and sporting goods industries. However, it should be noted that composite materials have been used for centuries; for example, bricks were manufactured in ancient Egypt that featured a composite material of clay and straw.

The latest generation of advanced composite materials includes some of the lightest, strongest, stiffest, and most corrosion-resistant materials available to the engineering community. For example, there is a striking contrast of the magnitudes of the stiffness-to-weight ratio or the strength-to-weight ratio of the commercial metals relative to those of the advance composite materials. Whereas the specific stiffness of aluminum can be increased threefold by the addition of silicon carbide fibers to create a metal-matrix composite, the specific stiffness of graphite-epoxy, fiber-reinforced, polymeric materials can be over four times greater than the specific strength of steel. The ramifications of this comparison are immense, because lightweight, high-strength, high-stiffness structures can be fabricated with these advanced polymeric composite materials with a weight savings of approximately 50%. This class of designs translates into superior performance for diverse products such as those in the aerospace, defense, automotive, biomedical, and sporting goods industries.

SULFONE POLYMERS

Sulfones are amorphous engineering thermoplastics noted for high heat-deflection temperatures and outstanding dimensional stability. These strong, rigid polymers are the only thermoplastics that remain transparent at service temperatures as high as 204°C.

Three commercially important sulfone-based resins are: polysulfone (PSU), including Udel and Ultrason S; polyarylsulfone (PAS), including Radel; and polyethersulfone (PES), including Ultrason E. These materials are claimed to offer the highest performance profiles of any thermoplastics processible on conventional screw-injection and extrusion machinery. Processing temperatures, however, are higher than those of other thermoplastics; the sulfones are processed on equipment that can generate and monitor stock temperatures in the range of 343 to 382°C.

Properties

Heat resistance is the outstanding performance characteristic of the sulfones. Service temperature is limited by heat-deflection temperature, which ranges from 174 to 204°C. A high percentage of physical, mechanical, and electrical properties is maintained at elevated temperatures, within limits defined by the heat-deflection temperatures. The strength and stiffness of PSU and PES are virtually unaffected up to their glass-transition temperature. For example, the flexural modulus of molded parts remains above 2040 MPa at service temperatures as high as 160°C. Even after prolonged exposure to such temperatures, the resins do not discolor or degrade. Thermal stability and oxidation resistance are excellent at service temperatures well above 149°C.

The continuous service temperature limit (CSTL) for PSU is 160°C, and 180°C for PES. With respect to flammability, PES is rated V-0 per UL 94, and PSU is rated at V-2 (Table S.11).

Electrical insulating properties are generally in the midrange among those of other thermoplastics, and they change little after heat aging at the recommended service temperatures. Dissipation factor and dielectric constant — and thus, loss factor — are not affected significantly by increased temperature or frequency.

Creep of the sulfones compared with that of other thermoplastics is exceptionally low at elevated temperatures and under continuous load. For example, creep at 99°C is less than that of acetal or heat-resistant ABS (acrylonitrile–butadiene–styrene) at room temperature. This excellent dimensional stability qualifies the sulfone resins for precision-molded parts.

The hydrolytic stability of these resins makes them resistant to water absorption in aqueous acidic and alkaline environments. The combination of hydrolytic stability and heat resistance results in exceptional resistance to boiling water and steam, even under autoclave pressures and cyclic exposure of hot-to-cold and wet-to-dry. PES resins have excellent resistance to hot lubricants, engine fuels, and radiator fluids, and they are resistant to gasoline.

The aromatic resins are also resistant to aqueous inorganic acids, organic acids, alkalies, aliphatic hydrocarbons, alcohols, and most cleaners and sterilizing agents.

The sulfones also share a common drawback: they absorb ultraviolet rays, giving them poor weather resistance. Thus, they are not recommended for outdoor service unless they are painted, plated, or UV-stabilized.

SULFUR

Sulfur (symbol S) is one of the most useful of the elements. Its occurrence in nature is little more than 1% that of aluminum, but it is easy to extract and is relatively plentiful. In economics, it belongs to the group of "S" materials — salt, sulfur, steel, sugars, starches — whose consumption is a measure of the industrialization and the rate of industrial growth of a nation.

Strict environmental laws are driving the production of sulfur recovered as a by-product of various industrial operations. It is also obtained by the distillation of iron pyrites, as a by-product of copper and other metal smelting, natural gas, and from gypsum. The sterri exported from Sicily for making sulfuric acid is broken rock rich in sulfur. Brimstone is an ancient name still in popular use for solid sulfur.

TABLE S.11
Properties of Sulfone Polymers

ASTM Test	Property	Polysulfone	Polyarylsulfone	Polyethersulfone
	Physical			
D792	Specific gravity	1.24	1.37	1.37
D570	Water absorption, 24 h, 1/8-in. thk (%)	0.3	0.40	0.43
	Mechanical			
D638	Tensile strength (psi)	10,200	12,000	12,200
D638	Elongation at break (%)	50–100	40	40–80
D638	Tensile modulus (10^5 psi)	3.6	3.9	3.9
D790	Flexural strength (psi)	15,400	16,100	18,650
D790	Flexural modulus (10^5 psi)	3.9	4.0	3.8
D256	Impact strength, Izod (ft-lb/in. of notch)	1.3	1.6	1.6
D785	Hardness, Rockwell M	69	85	88
	Thermal			
C177	Thermal conductivity (10^4 cal-cm/s-cm^2-°C)	6.2	—	3.2–4.4
D696	Coefficient of thermal expansion (10^{-5}in./in.-°F)	3.1	2.7	3.1
D648	Deflection temperature (°F)			
	At 264 psi	345	400	398
D2863	Oxygen index rating	30	33	34–38
	Electrical			
D149	Dielectric strength (V/mil)			
	Short time, (1/8-in. thk)	425	383	400
D150	Dielectric constant			
	At 60 Hz to 1 MHz	3.07–3.03	3.51–3.54	3.5
D150	Dissipation factor			
	At 60 Hz to 1 MHz	0.0008–0.0034	0.00171–0.00564	0.001–0.0035
D257	Volume resistivity (Ω-cm)	5×10^{16}	7.71×10^{16}	10^{17}–10^{18}

Source: Mach. Design Basics Eng. Design, June, p. 725, 1993. With permission.

Sulfur forms a crystalline mass of a pale-yellow color, with a melting point of 111°C. It forms a ruby vapor at about 416°C. When melted and cast, it forms amorphous sulfur with a specific gravity of 1.955. The tensile strength is 1 MPa, and compressive strength is 22 MPa. Since ancient times it has been used as a lute for setting metals into stone. Sulfur also condenses into light flakes known as flowers of sulfur, and the hydrogen sulfide gas, H_2S, separated from sour natural gas, yields a sulfur powder.

Elemental sulfur is widely used for the synthesis of sulfur compounds. It reacts directly with virtually all elements except the noble gases. In addition to its use as a chemical intermediate, sulfur is used increasingly as a construction material. For example, sulfur-impregnated concrete is much more resistant to acid corrosion than is conventional concrete. Highways have been paved with high-sulfur asphalts.

Properties

Sulfur has twice the atomic weight of oxygen but has many similar properties and has great affinity for most metals. The crystalline sulfur is orthorhombic, which converts to monoclinic crystals if cooled slowly from 120°C. This form remains stable below 120°C. When molten sulfur is cooled suddenly, it forms the amorphous sulfur, which has a ring molecular structure and is plastic, but converts gradually to the rhombic form. Sulfur has a wide variety of uses in all industries. The biggest outlet is for sulfuric acid, mainly for producing phosphate fertilizers.

Uses

Sulfur is used for making gunpowder and for vulcanizing rubber, but for most uses it is employed in compounds, especially as sulfuric acid or sulfur dioxide.

Sulfur is used in glass as a colorant to produce golden yellows and ambers, and also with cadmium sulfide in selenium ruby glass. In sulfur amber glasses, the element is introduced as flowers of sulfur, cadmium sulfide, or sodium sulfide. Its manufacture necessitates several precautions.

Compounds

Sulfur dioxide, or sulfurous acid anhydride, is a colorless gas of the composition SO_2, used as a refrigerant, as a preservative, in bleaching, and for making other chemicals. It liquefies at about −10°C. As a refrigerant it has a condensing pressure of 23.5 kg at 30°C. The gas is toxic and has a pungent, suffocating odor, so that leaks are detected easily. It is corrosive to organic materials but does not attack copper or brass. The gas is soluble in water, forming sulfurous acid, H_2SO_3, a colorless liquid with suffocating fumes. The acid form is the usual method of use of the gas for bleaching.

SULFURIC ACID

An oily, highly corrosive liquid of the composition H_2SO_4, sulfuric acid has a specific gravity of 1.841 and a boiling point of 330°C. It is miscible in water in all proportions, and the color is yellowish to brown according to the purity. It may be made by burning sulfur to the dioxide, oxidizing to the trioxide, and reacting with steam to form the acid. It is a strong acid, oxidizing organic materials and most metals. Sulfuric acid is used for pickling and cleaning metals, in electric batteries and plating baths, for making explosives and fertilizers, and for many other purposes. In the metal industries it is called dipping acid, and in the automotive trade it is called battery acid.

Uses

Sulfuric acid is used in the enameling industry for pickling purposes. The solutions vary in strength from 5 to 8%, although it is said that a 6% solution of sulfuric acid heated to 71 to 77°C will be the most effective in the pickling of sheet iron.

In making up H_2SO_4 solutions, always add the acid to the water, and never the water to the acid, as the latter method may cause a violent reaction. Sulfuric acid also has been used as a mill addition for acid-resisting enamels.

Compounds

Sulfur trioxide, or sulfuric anhydride, SO_3, is the acid minus water. It is a colorless liquid boiling at 46°C, and forms sulfuric acid when mixed with water.

Niter cake, which is sodium acid sulfate, $NaHSO_4$, or sodium bisulfate, contains 30 to 35% available sulfuric acid and is used in hot solutions for pickling and cleaning metals. It comes in colorless crystals or white lumps, with a specific gravity of 2.435 and melting point 300°C. Sodium sulfate, or Glauber's salt, is a white crystalline material of the composition $Na_2SO_4 \cdot 10H_2O$, used in making kraft paper, rayon, and glass. Salt cake, Na_2SO_4, is impure sodium sulfate used in the cooking liquor in making paper pulp from wood. It is also used in freezing mixtures. Sodium sulfite, Na_2SO_3 or $Na_2SO_3 \cdot 7H_2O$, is a white to tan crystalline powder very soluble in water but nonhygroscopic.

Sodium sulfide, Na_2S, is a pink flaky solid, used in tanneries for dehairing, and in the manufacture of dyes and pigments. The commercial product contains 60 to 62% Na_2S, 3.5% NaCl, and other salts, and the balance water of crystallization.

SUPERALLOYS

The term *superalloy* is broadly applied to iron-base, nickel-base, and cobalt-base alloys, often quite complex, which combine high-temperature mechanical properties and oxidation resistance to an unusual degree. Alloy requirements for turbosuperchargers and, later, the jet engine, largely provided the incentive for superalloy development.

Because of their excellent high-temperature performance, they are also known as high-temperature, high-strength alloys. Their strength at high temperatures is usually measured in terms of stress-rupture strength or creep resistance. For high-stress applications, the iron-base alloys are generally limited to maximum service temperature of about 649°C, whereas the nickel- and cobalt-base alloys are used at temperatures to about 1093°C and higher. In general, the nickel alloys are stronger than the cobalt alloys at temperatures below 1093°C and the reverse is true at temperatures above 1093°C. Superalloys are probably best known for aircraft-turbine applications, although they are also used in steam and industrial turbines, nuclear-power systems, and chemical- and petroleum-processing equipment. A great variety of cast and wrought alloys are available and, in recent years, considerable attention has been focused on the use of powder-metallurgy techniques as a means of attaining greater compositional uniformity and finer grain size.

STRENGTHENING MECHANISMS

Superalloys of the nickel–chromium and iron–nickel–chromium types usually contain sufficient chromium to provide the needed oxidation resistance and are further strengthened by the addition of other elements. The strengthening mechanisms include solid solution, precipitation, and carbide hardening. Most of the superalloys combine at least two and frequently all three of the above mechanisms.

Solid Solution

This is accomplished by introducing elements having different atomic sizes than those of the matrix elements, to increase the lattice strain. In addition to chromium, needed also for oxidation resistance, the elements molybdenum, columbium, vanadium, cobalt, and tungsten are effective in varying degree as solid-solution strengtheners when added in proper balance.

Superalloys of this type, such as 16-25-6, were used in gas turbines of older design for disks, employing "hot-cold work" to obtain the required yield strength in the hub area. However, hot-cold work is not effective as a means of getting high strength at temperatures much above about 538°C. Therefore, as gas turbine operating temperatures increased and turbine disk rim temperatures appreciably exceeded 538°C, other approaches were needed.

Cobalt-base alloys are strengthened principally by solid solution hardening, usually combined with a dispersion of stable carbides.

Precipitation Hardening

Precipitation hardening is the method now employed to impart high strength at high temperatures to most of the superalloys used for critical components of aircraft gas turbines. This includes alloys ranging from the 25% nickel A-286 wheel alloy to such recent and complex high-temperature nickel-base wrought and cast turbine blade alloys as "Udimet" 700, "Nimonic" alloy 115, and IN-100.

The presence of aluminum and titanium, usually jointly, in a nickel–chromium base, with or without iron, imparts unique age-hardening characteristics through the precipitation of gamma prime phase {$Ni_3(Al,Ti)$}.

The outstanding difference between the previously used age-hardening systems, typified by duralumin or beryllium–copper, and those utilizing gamma prime hardening lies in the fact that an alloy of the latter type can be heated in service appreciably above the optimum aging temperature without permanent loss of strength. In the conventional critical dispersion type age-hardening alloys, strength can only be restored by a complete cycle of heat treatment involving a high-temperature solution treatment and reaging.

Figure S.6 shows the way in which the two types of age-hardening systems differ, with the aluminum–titanium hardened alloy (curve B) regaining most of its initial hardness (and high-temperature properties) when the overheat is removed. However, an alloy typical of the other type of system (such as beryllium–copper or beryllium–nickel) overages and does not regain its hardness when the overheat is removed. It is this "reversible" aging behavior, along with high resistance to overaging (agglomeration of gamma-prime hardening phase), that has led to the widespread use of aluminum–titanium age-hardened nickel-base alloys for first-stage turbine blades in commercial and advanced design jet engines. The somewhat less complex iron–nickel–chromium alloys, such as A-286 and "Incoloy" alloy 901, used for turbine disks, are similarly age-hardened, with Ni_3Ti comprising most of the age-hardening component in these two alloys. Sufficient aluminum is also present in the gamma-prime precipitate to improve structural stability.

Carbide Hardening

Carbides provide a major source of dispersed-phase strengthening in cobalt-base alloys. These alloys do not respond to age hardening with aluminum and titanium since the gamma-prime phase does not form unless substantial

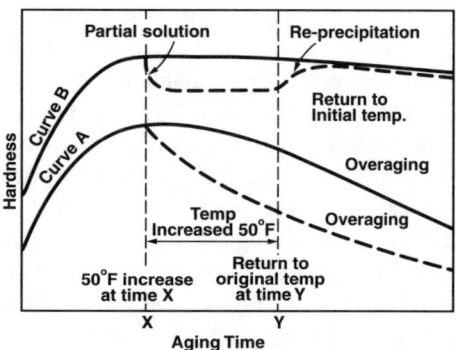

FIGURE S.6 Effect of aging time on hardness varies with type of age-hardening system.

amounts of nickel are present. Although gamma prime {$Ni_3(Al,Ti)$} is the principal strengthener in the age-hardenable nickel-base alloys, important auxiliary strengthening can be obtained by precipitation of quite complex carbides. The nature of the carbides formed and their mode of distribution can usually be controlled by alloy formulation and heat treatment.

Deoxidizers and Malleabilizers

In addition to the major alloying elements, small but effective amounts of malleabilizers such as boron and zirconium must be present to neutralize the effects of impurities that adversely affect hot ductility. As little as 0.005% boron is highly effective in nickel-base alloys, and zirconium, in a concentration about ten times that of boron, is also useful. In air melting of nickel-base alloys, magnesium is also added to "fix" any sulfur picked up during melting.

The high-temperature properties of the age-hardenable alloys are governed to a large degree by the hardener (aluminum plus titanium) content. However, as the hardener increases, the temperature of incipient fusion decreases and the lower temperature limit of forgeability rises, because of increased high-temperature strength, until forging is no longer practical by conventional procedures. The use of cast turbine blade alloys to meet very high temperature requirements is a natural consequence.

While other co-present elements and the amount of hot and cold workability required are also factors, it may be considered that alloys with up to 4% total hardener (aluminum plus titanium) are generally available in all common mill forms and are fabricable by conventional methods. As the hardener increases to about 8 or 9%, wrought alloys are still available, but in progressively fewer forms, ultimately being limited to small forgings such as turbine blades, at increasingly greater cost. The commercial cast turbine blade alloys contain from about 7 to 11% total hardener and are often chosen over forgings for reasons of cost or necessity or both.

COMPOSITIONS AND PROPERTIES

Although factors such as tensile properties, corrosion resistance, fatigue strength, expansion characteristics, etc. are important, the creep-rupture characteristics are usually the prime requisite in the selection of a superalloy.

Vacuum Melting

Vacuum induction-melted superalloys are generally more ductile than air-melted material, permitting the use of higher aluminum plus titanium levels while still retaining adequate ductility. Vacuum induction melting accomplishes refining and also permits closer control of composition than does air melting. The vacuum arc (consumable electrode) method is widely used for superalloys, especially those employed for turbine disks. Often the electrode for the vacuum arc-melting charge is obtained from a vacuum induction melt, the end product thus combining the benefits of refining occurring in vacuum induction melting with the controlled solidification and sound ingot structure associated with the vacuum arc process. Some refining is also accomplished in the vacuum arc process, but to a lesser degree than in vacuum induction melting.

Iron-Base Superalloys

The iron-base superalloys include solid-solution alloys and precipitation-hardening (PH), or precipitation-strengthened, alloys. Solid-solution types are alloyed primarily with nickel (20 to 36%) and chromium (16 to 21%), although other elements are also present in lesser amounts. Superalloy 16-25-6, for example, the alloy designation indicating its chromium, nickel, and molybdenum contents, respectively, also contains small amounts of manganese (1.35%), silicon (0.7%), nitrogen (0.15%), and carbon (0.06%). Superalloy 20Cb-3 contains 34% nickel, 20% chromium, 3.5% copper, 2.5% molybdenum, as much as 1% columbium, and 0.07% carbon. Incoloy 800, 801, and 802 contain slightly less nickel and slightly more chromium with small amounts of titanium, aluminum, and carbon. N-155, or Multimet, an early sheet alloy, contains about equal amounts of chromium, nickel, and cobalt (20% each), plus 3% molybdenum, 2.5% tungsten, 1% columbium, and small amounts of carbon, nitrogen, lanthanum, and zirconium. At 732°C, this alloy has a 1000-h stress-rupture strength of about 165 MPa.

PH iron-base superalloys provide greater strengthening by precipitation of a nickel–aluminum–titanium phase. One such alloy, which may be the most well known of all iron-base superalloys, is A-286. It contains 26% nickel, 15% chromium, 2% titanium, 1.25% molybdenum, 0.3% vanadium, 0.2% aluminum, 0.04% carbon, and 0.005% boron. At room temperature, it has a tensile yield strength of about 690 MPa and a tensile modulus of 145×10^3 MPa. At 649°C, tensile yield strength declines only slightly, to 607 MPa, and its modulus is about the same or slightly greater. It has a 1000-h stress-rupture strength of about 145 MPa at 732°C. Other PH iron-base superalloys are Discoloy, Haynes 556 (whose chromium, nickel, cobalt, molybdenum, and tungsten content is similar to that N-155); Incoloy 903 and Pyromet CTX-1, which are virtually chromium-free but high in nickel (37 to 38%) and cobalt (15 to 16%); and V-57 and W-545, which contain about 14% chromium, 26 to 27% nickel, about 3% titanium, 1 to 1.5% molybdenum, plus aluminum, carbon, and boron. V-57 has a 1000-h stress-rupture strength of about 172 MPa at 732°C and greater tensile strength, but similar ductility, than A-286 at room and elevated temperatures.

Nickel-Base Superalloys

Nickel-base superalloys are solid-solution, precipitation, or oxide-dispersion strengthened. All contain substantial amounts of chromium, 9 to 25%, which, combined with the nickel, accounts for their excellent high-temperature oxidation resistance. Other common alloying elements include molybdenum, tungsten, cobalt, iron, columbium, aluminum, and titanium. Typical solid-solution alloys include Hastelloy. X (22 to 23% chromium, 17 to 20% iron, 8 to 10% molybdenum, 0.5 to 2.5% cobalt, 2% aluminum, 0.2 to 1% tungsten, and 0.15% carbon); Inconel 600 (15.5% chromium, 8% iron, 0.25% copper maximum, 0.08% carbon); and Inconel 601, 604, 617, and 625, the latter containing 21.5% chromium, 9% molybdenum, 3.6% columbium, 2.5% iron, 0.2% titanium, 0.2% aluminum, and 0.05% carbon. At 732°C, wrought Hastelloy X (it is also available for castings) has a 1000-h stress-rupture strength of about 124 MPa, and has high oxidation resistance at temperatures to 1204°C.

The precipitation-strengthened alloys, which are the most numerous, contain aluminum and titanium for the precipitation of a second strengthening phase, the intermetallic $Ni_3(Al,Ti)$ known as gamma prime (γ') or the intermetallic Ni_3Cb known as gamma double prime (γ''), during heat treatment. One such alloy, Inconel X-750 (15.5% chromium, 7% iron, 2.5% titanium, 1% columbium, 0.7% aluminum, 0.25% copper maximum, and 0.04% carbon), has more than twice the tensile yield strength of Inconel 600 at room temperature and nearly three times as much at 760°C. Its 1000-h stress-rupture strength at 760°C is in the range of 138 to 207 MPa. Still great tensile yield strength at room and elevated temperatures and a 172-MPa stress-rupture strength at 760°C are provided by Inconel 718 (19% chromium, 18.5% iron, 5.1% columbium, 3% molybdenum, 0.9% titanium, 0.5% aluminum, 0.15% copper maximum, 0.08% carbon maximum), a wrought alloy originally that also has been used for castings. Among the strongest alloys in terms of stress-rupture strength is the wrought or cast IN-100 (10% chromium, 15% cobalt, 5.5% aluminum, 4.7% titanium, 3% molybdenum, 1% vanadium, less than 0.6% iron, 0.15% carbon, 0.06% zirconium, 0.015% boron). Investment cast, it provides a 1000-h stress-rupture strength of 517 MPa at 760°C, 255 MPa at 871°C, and 103 MPa at 982°C.

Other precipitation-strengthened wrought alloys include Astroloy; D-979; IN 102; Inconel 706 and 751; M252; Nimonic 80A, 90, 95, 100, 105, 115, and 263; René 41, 95, and 100; Udimet 500, 520, 630, 700, and 710; Unitemp AF2-1DA; and Waspaloy. Other cast alloys, mainly investment-cast, include B-1900; IN-738X; IN-792; Inconel 713C; M252; MAR-M 200, 246, 247, and 421; NX-188; René 77, 80, and 100; Udimet 500, 700, and 710; Waspaloy; and WAZ-20.

A few of the cast alloys, such as MAR-M 200, are used to produce directionally solidified castings, that is, investment castings in which the grain runs only unidirectionally, as along the length of turbine blades. Eliminating transverse grains improves stress-rupture properties and fatigue resistance. Grain-free alloys, or single-crystal alloys, also have been cast, further improving high-temperature creep resistance. Regarding powder-metallurgy techniques, emphasis has been the use of prealloyed powder made by rapid solidification techniques (RST) and mechanical alloying (MA), a high-energy milling process using attrition mills or special ball mills. Dispersion-strengthened nickel alloys are alloys strengthened by a dispersed oxide phase, such as thoria, which markedly increases strength at very high temperatures but only moderately so at intermediate elevated temperatures, thus limiting applications. TD-nickel, or thoria-dispersed nickel, was the first of such superalloys, and it was subsequently modified with about 20% chromium, TD-NiCr, for greater oxidation resistance. The recent MA 754 and MA 6000E alloys combine dispersion strengthening with yttria and gamma-prime strengthening.

Cobalt-Base Superalloys

Cobalt-base superalloys are for the most part solid-solution alloys, which, when aged, are strengthened by precipitation of carbide or intermetallic phases. Most contain 20 to 25% chromium, substantial nickel and tungsten and/or molybdenum, and other elements, such as iron, columbium, aluminum, or titanium. One of the most well known, L-605, or Haynes 25, is mainly a wrought alloy, although it is also used for castings. In wrought form, it contains

20% chromium, 15% tungsten, 10% nickel, 3% iron, 1.5% manganese, and 0.1% carbon. At room temperature, it has a tensile yield strength of about 462 MPa, and at 871°C about 241 MPa. Its 1000-h stress-rupture strength at 815°C is 124 MPa. The more recent Haynes 188 (22% chromium, 22% nickel, 14.5% tungsten, 3% iron, 1.5% manganese, 0.9% lanthanum, 0.35% silicon, and 0.1% carbon), which was developed for aircraft-turbine sheet components, provides roughly similar strength and high oxidation resistance to about 1093°C. MP35N (35% nickel, 20 chromium, 10 molybdenum) is a work-hardening alloy used mainly for high-temperature corrosion-resistant fasteners. Another alloy, S-816, contains equal amounts of chromium and nickel (20% each), equal amounts of molybdenum, tungsten, columbium, and iron (4% each), and 0.38% carbon. Primarily a wrought alloy, although also used for castings, it has a 1000-h stress-rupture strength of 145 MPa at 815°C. Other casting alloys include AiResist 13, 213, and 215; Haynes 21 and 31, the latter also known as X-40; Haynes 151; J-1650; MAR-M 302, 322, 509, and 918; V-36; and W1-52. Their chromium content ranges from 19% (AiResist 215) to 27% (Haynes 21) and some are nickel-free or low in nickel. Most contain substantial amounts of tungsten or tantalum, and various other alloying elements. Among the strongest in terms of 1000-h stress-rupture strength at 815°C are Haynes 21 and 31 — 290 MPa and 352 MPa, respectively.

FABRICATION

Forging and Hot Working

Many of the commercial nickel-base age-hardenable superalloys can be forged or hot-worked with varying degrees of ease. As has been indicated, the top side of the forging temperature range is limited by such considerations as incipient fusion temperature, grain size requirements, tendency for "bursts," etc., and the lower side by the stiffness and ductility of the alloy. The recommendations of the metal producer should be sought for optimum forging practice for a given alloy.

The hot extrusion process increases the gamut of superalloy compositions that can be hot-worked. By the use of a suitable sheath on the extrusion billet, otherwise unworkable alloys can be reduced to bar by hot extrusion. In some instances, mild steel has been used for a sheath while in others nickel–chromium alloys such as Nimonic alloy 75 and Inconel alloy 600 have been employed. Some of the very high-speed hot extrusion processes may be of value in hot-working the more refractory superalloys.

Heat Treatment

The conventional heat-treating equipment and fixtures generally suitable for nickel alloys and austenitic stainless steels are also applicable to the nickel-base high-temperature alloys. Nickel-base alloys are more susceptible to sulfur and lead embrittlement than iron-base alloys. It is therefore essential that all foreign material, such as grease, oil, cutting lubricants, marking paints, etc., be removed by suitable solvents, vapor degreasing, or other methods, before heat treatment.

When fabricated parts made from thin sheet or strip of age-hardening alloys such as Incond alloy X-750 must be annealed during and after fabrication, it is desirable, especially in light gauges, to provide a protective atmosphere such as argon or dry hydrogen to lessen the possibility of surface depletion of the age-hardening elements. This precaution may not be as necessary in heavier sections, since the surface oxidation involves a much smaller proportion of the effective cross section.

It is usually necessary after severe forming, or after welding, to apply a stress relief anneal (above 899°C) to assemblies fabricated from aluminum–titanium age-hardenable nickel-base alloys prior to aging. It is vitally important to heat the structure rapidly through the age-hardening temperature range of 649 to 760°C (which is also the low ductility range) so that stress relief can be achieved before any appreciable aging takes place. This is conveniently done by charging into a furnace at or above the desired annealing temperature. It has been found at times that the efficacy of this procedure has been vitiated in large welded structures by charging on to a cold car, resulting in a slower and nonuniform heating of the fabricated part when run into the hot furnace. Contrary to expectations, little difficulty has been encountered with distortion under the above rapid heating conditions. In fact, distortion of weldments of substantial size has been reported to be less than by conventional slow-heating heating methods.

Forming

All of the wrought nickel-base alloys available as sheet can be formed successfully into quite complex shapes involving much plastic flow. The lower-strength Inconel alloy 600 and Nimonic alloy 75 offer few problems. The high-strength age-hardening varieties, processed in the annealed condition, can be subjected to a surprising amount of cold work and deformation, provided sufficient power is available. Explosive forming has also been successfully employed on a number of nickel-base alloys.

Machining

All of the alloys discussed can be machined, the strongest and highest hardener content materials causing the most difficulty. The recommendations of the metal producer should be followed with respect to optimum condition of heat treatment, type of tool, speed and feed, cutting lubricant, etc. Wrought alloys of quite high hardener content, such as Inconel alloy 700 and Udimet 500, although difficult to handle, can be machined with reasonable facility using high-speed-steel tools of the tungsten–cobalt type, and cemented carbide tools of the tungsten–cobalt anal tungsten–tantalum–cobalt type.

Various electroerosion processes have been successfully used on a number of the age-hardened superalloys and, at high hardener levels, may be necessary for some operations such as drilling.

Welding

Inconel alloy 600, Nimonic alloy 75, and other nickel-base alloys of the predominantly solid solution strengthened type offer no serious problems in welding. All of the common resistance and fusion welding processes (except submerged arc) are regularly and successfully employed.

In handling the wrought superalloys age-hardened with gamma prime [$Ni_3(Al,Ti)$], it is necessary to observe certain precautions. Material should be welded in the annealed condition to minimize the hazard of cracking in weld or parent metal. If the components to be joined have been severely worked or deformed they should be stress-relief-annealed before welding by charging into a hot furnace to ensure rapid heating to the stress-relieving temperature. Similarly, weldments should be stress-relieved before attempting to apply the 704°C age-hardening treatment.

Where subassemblies must be joined in the age-hardened condition, the practice of "safe ending" with a compatible nonaging material prior to age hardening can be usefully employed. The final weldment joining the fully age-hardened components is then made on the "safe ends."

Such new welding processes as "short arc," electron, and laser beam have come into increasing use and have been helpful in joining some of the very high hardener content alloys.

Brazing

The solid solution type chromium-containing alloys, such as Inconel alloy 600, are quite readily brazed, using techniques and brazing filler metals applicable to the austenitic stainless steels. Generally speaking, it is desirable to braze annealed (stress-free) material to avoid embrittlement by the molten braze metal. Where brazing filler metals are employed that melt above the stress-relieving temperature, a prior anneal is usually not needed. As with the stainless steels, dry hydrogen, argon, and helium atmospheres are used successfully; and vacuum brazing is also very successfully employed.

The age-hardened nickel-base alloys containing titanium and aluminum are rather difficult to braze, unless some method of fluxing, solid or gaseous, is used. Alternatively, the common practice is to preplate the areas to be furnace brazed with 0.01 to 0.03 mm of nickel, which prevents the formation of aluminum or titanium oxide films and permits ready wetting by the brazing filler metal.

Silver brazing filler metals can be used for lower temperature applications. However, since the nickel-base superalloys are usually employed for high-temperature applications, the higher melting point and stronger and more oxidation-resistant brazing filler metals of the Ni–Cr–Si–B type are generally used. The silver–palladium–manganese and palladium–nickel filler metals also provide useful brazing materials for intermediate service temperatures.

SUPERCONDUCTIVITY

Superconductivity is a phenomenon occurring in many electrical conductors, in which the electrons responsible for conduction undergo a collective transition into an ordered state with many unique and remarkable properties. These include the vanishing of resistance to the flow of electric current, the appearance of a large diamagnetism and other unusual magnetic effects, substantial alteration of many thermal properties, and the occurrence of quantum effects otherwise observable only at the atomic and subatomic level.

The ability of certain materials, when cooled to extremely low or cryogenic temperatures, to conduct electricity with essentially zero resistance to DC current, and to AC current below certain critical high-frequency ranges, was found in the scientific community. This current is referred to as supercurrent, and is carried on the surface of the superconductor within a particular depth characteristic of the material.

Until recently, superconductivity was only seen in materials at fantastically cold temperatures not exceeding 23 K (–250°C) in the intermetallic compound Nb_3Ge. This meant that all superconductors had to be cooled with liquid helium, which is expensive and cumbersome to handle. Applications for superconductors were quite limited. The main use was for nuclear magnetic resonance scanners used by hospitals to examine soft tissue without surgery. Nuclear magnetic resonance scanners tap the intensely powerful magnetic fields that superconductors can be made to generate.

However, in August 1986, researchers discovered a compound of the metals lanthanum, barium, and copper, along with oxygen, that would superconduct at 35 K (–238°C).

Following this discovery, researchers brought about a dramatic jump in T_c to 93 K (with an onset temperature of 98 K) by substituting yttrium for lanthanum, while roughly switching the +2 and +3 ion ratios.

SUPERCONDUCTORS

Superconductors are solid crystalline materials whose electrical resistance drops significantly as temperature decreases. Until recently, temperatures approaching close to absolute zero (–272°C) were required for the resistivity to vanish. Some of the metals exhibiting superconductivity at near absolute zero include iridium, lead, mercury, columbium, tin, tantalum, vanadium, and many alloys and chemical compounds. Alloys considered among the best commercially available are lead–molybdenum–sulfur, columbium–tin, and columbium–titanium.

In recent years, alloys and compounds have been developed that are superconductive at temperatures substantially above absolute zero. These include a compound of lanthanum, strontium, copper, and oxygen, which is superconductive at –240°C, and a barium–yttrium–copper oxide, which is superconductive at –183°C. A two-phase ceramic superconductor has been developed in which one of the phases is superconductive at –33°C. It is basically copper oxide containing barium and yttrium. Composition of the second phase is yet to be determined.

PROPERTIES AND PROCESSING

Because J_c is higher in single crystals than in polycrystalline materials, it can be increased by minimizing the presence of grain boundaries. This is done by single-crystal growth or techniques such as melt texturing, where grain boundaries become highly directionalized to allow for significantly less random disruption of current flow. Here, the grain boundaries become predominantly parallel to the copper–oxygen chains (the direction of current flow) in all the crystals. Consequently, the microstructure consists of long, needlelike grains that are parallel and intermeshed.

Single crystals have been grown epitaxially by sputtering (as well as by pulsed excimer laser deposition and electron beam epitaxy). Good deposition can be achieved when the material is reactively sputtered in an oxygen-containing environment from three separate metallic targets (yttrium, barium, copper) simultaneously. This is known as triode sputtering, and allows for tight control of stoichiometry. Following this, as with bulk samples, the films typically must be annealed in pure oxygen at –500°C. Samples also must be slow cooled to 300°C, as gravimetric analysis shows that the capability of the compound to absorb more oxygen into its crystal structure increases with decreased temperature down to 300°C. Annealing can be successfully accomplished below 300°C by use of an oxygen-ion bombardment.

The greatest reproducibility of results in single-crystal formation has come when using single-crystal $SrTiO_3$ as the epitaxy substrate. Some success has been found in depositing single crystals on single-crystal MgO, which has a much lower dielectric constant than $SrTiO_3$

APPLICATIONS

There are a number of practical applications of superconductivity. Powerful superconducting electromagnets guide elementary particles in particle accelerators, and they also provide the magnetic field needed for magnetic resonance imaging (MRI). Ultrasensitive superconducting circuits are used in medical studies of the human heart and brain and for a wide variety of physical science experiments. A completely superconducting prototype computer has even been built.

Most superconductive applications that have been considered to have reasonable possibility of being achieved in the next few years involve thin-film deposition of these materials. Thin films clearly have the higher J_c advantage over bulk superconductors.

Photovoltaic substances, for example, if interfaced with a superconductor, can act as signal detectors (e.g., infrared devices) because they will be sensitive to the most minute electrical fields.

One of the potential uses of the newer superconductors is for making more powerful and efficient electromagnets that could be used in trains to levitate them above their tracks, and thus make train speeds of hundreds of miles per hour possible.

SUPERPLASTIC FORMING

This is a process for shaping superplastic materials, a unique class of crystalline materials that exhibit exceptionally high tensile ductility. Superplastic materials may be stretched in tension to elongations typically in excess of 200% and more commonly in the range of 400 to 2000%. There are rare reports of higher tensile elongations reaching as much as 8000%. The high ductility is obtained only for superplastic materials and requires both the temperature and rate of deformation (strain rate) to be within a limited range. The temperature and strain rate required depend on the specific material. A variety of forming processes can be used to shape these materials; most of the processes involve the use of gas pressure to induce the deformation under isothermal conditions at the suitable elevated temperature. The tools and dies used, as well as the superplastic material, are usually heated to the forming temperature. The forming capability and complexity of configurations producible by the processing methods of superplastic forming greatly exceed those possible with conventional sheet forming methods in which the materials typically exhibit 10 to 50% tensile elongation.

PROCESSES

Superplastic forming typically utilizes a gas pressure differential across the superplastic sheet to induce the superplastic deformation and cause forming. Two processes have been developed: blow forming and movable-tool forming.

Blow Forming

Where gas pressure alone is used, the process is termed *blow forming*. Blow forming utilizes tooling heated to the superplastic temperature, and the gas pressure differential is usually applied according to a time-dependent schedule designed to maintain the average strain rate within the superplastic range. The tools and the superplastic sheet are heated to the same temperature, and the gas pressure is applied to cause a creeplike plastic stretching of the sheet that eventually contacts and takes the shape of the configuration die.

Movable-Tool Forming

For relatively deep shapes, forming methods involving movable tools combined with gas-pressure forming may permit greater thinning control and reduced forming times as compared with the blow-forming method. One method is essentially the same as the blow-forming process, except that the die may be moved during the forming process. Another method uses a more complex sequence. The bubbleplate holds the superplastic sheet in place and prevents gas breakage.

The plug-assisted forming method involves a movable die that is pushed into and stretches the superplastic sheet material, followed by the application of gas pressure on the same side of the sheet as the movable die. In snap-back forming, the sheet is first billowed by tree forming with gas pressure imposed on the movable-tool side of the sheet; the tool is then moved into the billowed sheet, and finally the gas pressure is imposed on the opposite side of the sheet to form the superplastic material onto the tool.

Diffusion Bonding

Diffusion bonding, also known as diffusion welding, is sometimes used in conjunction with superplastic forming to produce parts of complexity not possible with a single-sheet forming process. Diffusion bonding is a solid-state joining process in which two or more materials are pressed together under sufficient pressure and at a sufficiently high temperature to result in joining. In diffusion bonding, there is usually little permanent deformation in the bulk of the parts being joined, although local deformation does occur at the interfaces on a microscopic scale. Because interfacial contamination, such as oxidation, will interfere with the bonding mechanisms, the process is usually conducted under an inert atmosphere, such as vacuum or inert gas.

Some superplastic materials are ideally suited for processing by diffusion bonding, because they deform easily at the superplastic temperature and this temperature is consistent with that required for diffusion bonding. The most suitable alloys tend to have a high solubility for oxygen and nitrogen, so that these contaminants can be removed from the surface by diffusion into the base metal. For example, titanium alloys fall into this class and are readily diffusion bonded. Aluminum alloys form a very thin but tenacious oxide film and are therefore quite difficult to diffusion bond. For certain materials, and under conditions of proper processing, the diffusion-bond interfacial strength can be equal to that of the parent base material. It has been found that metals processed to fine grain size, as required for superplastic deformation, are the most suitable for diffusion bonding because they require lower bonding pressure than coarse-grained metal of the same alloy composition.

Combined Methods

The processing conditions for superplastic forming and diffusion bonding are similar, both requiring an elevated temperature and benefiting from the fine grain size. Consequently, a combined process of superplastic forming with diffusion bonding has been developed that can produce parts of greater complexity than single-sheet forming alone. The combined process of superplastic forming and diffusion bonding can involve multiple-sheet forming after localized diffusion bonding, producing expanded structures and sandwich configurations of various types. It is also possible to form a sheet superplastically onto, and diffusion bonded to, a separate piece of material thereby producing structural configurations much more like forgings than sheet metal structures. There are other joining methods that have also been utilized as alternatives to diffusion bonding, such as spot welding, and have been combined with superplastic forming to produce complex structures.

APPLICATIONS

There are a number of commercial applications of superplastic forming and combined superplastic forming and diffusion bonding, including aerospace, architectural, ground transportation, and numerous miscellaneous uses. Examples are wing access panels in the Airbus A310 and A320, bathroom sinks in the Boeing 737, turbo-fan-engine-cooling duct components, external window frames in the space shuttle, front covers of slot machines, and architectural siding for buildings.

SUPERPOLYMERS

Many plastics developed in recent years can maintain their mechanical, electrical, and chemical resistance properties at temperatures over 213°C for extended periods of time. Among these materials are polyimide, polysulfone, polyphenylene sulfide, polyarylsulfone, novaloc epoxy, aromatic polyester, and polyamide-imide. In addition to high-temperature resistance, they have in common high strength and modulus of elasticity, and excellent resistance to solvents, oils, and corrosive environments. They are also high in cost. Their major disadvantage is processing difficulty. Molding temperatures and pressures are extremely high compared to conventional plastics. Some of them, including polyimide and aromatic polyester, are not molded conventionally. Because they do not melt, the molding process is more of a sintering operation. One indication of the high-temperature resistance of the superpolymers is their glass transition temperature of well over 260°C, as compared to less than 177°C for most conventional plastics. In the case of polyimides, the glass transition temperature is greater than 427°C and the material decomposes rather than softens when heated excessively. Aromatic polyester, a homopolymer also known as polyoxybenzoate, does not melt, but at 427°C can be made to flow in a nonviscous manner similar to metals. Thus, filled and unfilled forms and parts can be made by hot sintering, high-velocity forging, and plasma spraying. Notable properties are high thermal stability, good strength at 316°C, high thermal conductivity, good wear resistance, and extra-high compressive strength. Aromatic polyesters have also been developed for injection and compression molding. They have long-term thermal stability and a strength of 20 MPa at 288°C. At room temperature polyimide is the stiffest of the group with a top modulus of elasticity of 51,675 MPa, followed by polyphenylene sulfide with a modulus of 33,072 MPa. Polyarysulfone has the best impact resistance of the superpolymers with an impact strength of 0.27 kg · m/cm (notch).

Polyetherimide (PEI) is an amorphous thermoplastic that can be processed with conventional thermoplastic processing equipment. Its continuous-use temperature is 170°C and its deflection temperature is 200°C at 2 MPa. The polymer also has inherent flame resistance without use of additives. This feature, along with its resistance to food stains and cleaning agents, makes it suitable for aircraft panels and seat component parts. Tensile strength ranges from 103 to 165 MPa. Flexural modulus at room temperature is 3300 MPa.

Polyimide (PI) foam is a spongy, lightweight, flame-resistant material that resists ignition up to 427°C and then only chars and decomposes. Some formulations result in harder materials that can be used as lightweight wallboard or floor panels while retaining fire resistance.

Aromatic polyketones are high-performance thermoplastics, which include polyetheretherketone (PEEK), glass transition temperature of 143°C and melting point of 335°C; polyetherketone (PEK), glass transition temperature of 154°C; polyetherketoneketone (PEKK), glass transition temperature of 154°C and melting point of 335°C; polyaryletherketone (PAEK), glass transition temperature of 170°C and melting point of 380°C. Glass fiber reinforcement improves the strength, stiffness, and dimensional stability of these materials. In addition, there are various ketone-based copolymers.

SURFACE PIGMENTS (FOR BRICK)

Brick has a long history of durability. Environmental issues such as energy consumption and waste disposal are increasing in importance, producing opportunities for brick usage to increase.

Facing brick need no rendering, painting, or regular maintenance. They look good year after year and often century after century with no energy or product input. The brick can be recycled again and again, and at the end of their lives can be used as aggregate. Brick are the material of the future, and as demand for brick increases, so will demand for choice. The range of choices can be expanded by using surface pigments.

COMPOSITIONS

Surface pigments can be classified into several groups. Pigment mixtures, which on firing develop their intrinsic color, fall into the largest category of stains. The second class includes pigments that react with the brick surface sand or grog to produce a specific shade of color. Usually the iron in the surface layers of the clay is the critical factor. There are also colors that have been stabilized by prior treatment. These are normally used where it is necessary to have the maximum volume of brick one uniform color.

Ferrous and ferric oxides are the source of a large number of colors ranging from very bright yellows through reds and maroons to blacks. The tone of the red surface colors depends on the firing temperature, the compound from which the oxide was formed, and the kiln atmosphere.

The presence of zinc or titanium compounds modifies the iron color in pigments and produces a range of buffs and yellows.

The spectral range of brown pigments is extremely wide, ranging in shade from cream to dark farmhouse brown.

Black pigments are mixtures of compounds of iron and manganese. Occasionally, cobalt oxide is added to intensify the blue-black quality of some surface stains.

Normally, gray pigments are produced by diluting selected black stains with inert and reactive fillers. The dominant tone of any black pigment is apparent when it is diluted, and some satisfactory blacks show undesirable tones when used diluted as grays.

DRY AND WET APPLICATIONS

Dry

There are several methods of applying pigments in dry form. Surface pigments are trickle-fed through a sieve and then brushed or lightly rolled into the brick surface. Surface effects are applied by vibrating the granules onto the column and rolling them into the surface.

For surface pigments or frits added to aggregates, the aggregate (a sand, grog, or other suitable material) is mixed with the pigment or frit at a predetermined ratio.

Wet

Surface pigments can be added to water-based suspensions and applied as a spray, slurry, or engobe. For spraying, a ratio of one part by weight of pigment to four parts by weight of water is generally acceptable.

This thorough wetting of the finely powdered pigment ensures a smooth suspension.

SYNTHETIC NATURAL RUBBER (ISOPRENE)

Stereoregular polyisoprene and polybutadiene elastomers, high in *cis*-1,4 content, are of growing interest to the engineer, both because of engineering performance, and their competitive price. IR (*cis*-polyisoprene) has been called "synthetic natural rubber" because chemically and physically it is similar to *Hevea*.

General properties and examples of end-use performance show it to be a satisfactory supplement to natural rubber in a wide variety of products. Molecular weight can be controlled within quite wide limits and linearity can be maintained even with the longest chains. Higher-molecular-weight materials have been satisfactorily extended with oil to yield compositions with a desirable combination of low cost and attractive properties.

The development of IR latex is a noteworthy advance in latex technology. The low emulsifier level, stereoregularity of the polymer, large particle size, and low viscosity have not hitherto been available in a general-purpose synthetic latex. These properties, combined with high gum strength and elongation, offer advantages for many latex applications.

VULCANIZATE PROPERTIES

IR can be processed in a manner similar to that used for natural rubber. Vulcanization can be carried out by means of curatives commonly used with natural or SBR (styrene–butadiene rubber). Properties of gum vulcanizates are quite similar to those of natural rubber, although IR has somewhat lower modulus and higher extensibility.

There are excellent hysteresis properties, low heat buildup and high resilience of both the IR and the polybutadiene tread vulcanizates, and IR is second only to natural rubber in tear strength.

Processing and compounding procedures for oil-extended IR are similar to those for unextended polymer, except that lower curative levels are recommended for maximum tensile strength and flat curing characteristics. Properties of extended vulcanizates approach those of the non-extended materials.

APPLICATIONS

Similarity of performance between IR and natural rubber has permitted use of IR as a supplement for natural rubber in uses such as tire treads, carcasses, and white sidewalls.

In nontire uses the low ash content, light color, and good mold flow characteristics of IR are of particular advantage. Good electrical properties and low moisture absorption make it suitable for a number of electrical insulating uses. Parts molded in IR exhibit sharp definition and excellent color stability to light.

The low cost and good performance of oil-extended IR are promising for tire carcass compounds, molded mechanical goods, and footwear.

In latex form, IR is the first synthetic that possesses an average particle size as large and a particle size distribution as broad as that of natural rubber latex. It is highly promising for a number of coating and dipping applications, as well as foaming.

T

TALC

Talc is a hydrous magnesium silicate, with the composition 63.4% SiO_2, 31.9% MgO, and 4.7% H_2O when found in pure form. It is an extremely soft mineral with a Moh hardness of 1.

Talc is a soft friable mineral of fine colloidal particles with a soapy feel. It has a composition of $4SiO_2 \cdot 3MgO \cdot H_2O$ and a specific gravity of 2.8. It is white whelf pure, but may be colored gray, green, brown, or red with impurities.

Talc is now used for cosmetics, for paper coatings, as a filler for paints and plastics, and for molding into electrical insulators, heater parts, and chemical ware. The massive block material, called steatite talc, is cut into electrical insulators. It is also called lava talc. The more impure block talcs are used for firebox linings and will withstand temperatures to 927°C. Gritty varieties contain carbonate minerals and are in the class of soapstones. Varieties containing lime are used for making porcelain.

Talc used in ceramics is usually mined, sorted, crushed, and milled to 95 to 99% −200 mesh.

APPLICATIONS

The major applications for talc are tile and hobbyware bodies, cordierite catalyst supports, kiln furniture, and electrical porcelains. There are minor applications in electronic packaging, sanitaryware, dinnerware, and glazes.

Talc is used as a flux for high alumina ceramics, sanitaryware, and dinnerware. It is a low-cost source of magnesium in these applications and helps to produce less porous bodies at lower firing temperatures.

TANTALUM AND ALLOYS

Tantalum is a white lustrous metal (symbol Ta), resembling platinum. Tantalum is a high-density, ductile, refractory metal that exhibits exceptional corrosion resistance and good high-temperature strength over 1663°C. The annealed wrought metal in its pure form is easily worked and can be cold-worked in much the same manner as fully annealed mild steel.

It is one of the most acid-resistant metals and is classed as a noble metal. Its specific gravity is 16.6, or about twice that of steel and, because of its high melting temperature (2996°C), it is called a refractory metal. In sheet form, it has a tensile yield strength of 345 MPa and is quite ductile. At very high temperatures, however, it absorbs oxygen, hydrogen, and nitrogen and becomes brittle. Its principal use is for electrolytic capacitors, but because of its resistance to many acids, including hydrochloric, nitric, and sulfuric, it is also widely used for chemical-processing equipment. It is attacked, however, by hydrofluoric acid, halogen gases at elevated temperatures, fuming sulfuric, and strong alkalies. Because of its heat resistance, tantalum is also used for heat shields, heating elements, vacuum-furnace parts, and special aerospace and nuclear applications. It is a common alloying element in superalloys. The metal is also used for prosthetic applications.

Tantalum metal is used in the manufacture of capacitors for electronic equipment, including citizen band radios, smoke detectors, heart pacemakers, and automobiles. An extremely stable film of tantalum oxide acts as an insulator in the capacitor. It is also used for heat-transfer surfaces in chemical production equipment, especially where extraordinarily corrosive conditions exist. Its chemical inertness has led to dental and surgical applications.

Tantalum forms alloys with a large number of metals. Of special importance is ferrotantalum, which is added to austenitic steels to reduce intergranular corrosion. Tantalum is used extensively in the chemical industry where its excellent fabrication and joining properties permit the application to acid-resistant heat exchangers, condensers, ductwork, chemical lines, and other chemical process equipment. Tantalum also finds use in the medical profession. Because of its nontoxic properties and immunity to body chemicals, tantalum is used for sutures, gauze, pins, and plates.

FABRICABILITY

Hot Working

High-purity tantalum sintered bar, cast ingot, and annealed wrought forms can be worked at room temperature, although the working of large ingots and billets is sometimes performed at elevated temperature to permit working within equipment strength capacity. Cast ingots, protected by canning or coating materials, have been forged and extruded at temperatures up to 1316°C. Cold-worked tantalum can be stress-relieved or annealed at a variety of time–temperature schedules depending upon stress level and chemical purity of the material. A temperature of at least 1204°C is generally used for full annealing while temperatures between 816 and 927°C can be used for stress relieving. Annealing atmosphere must be high-purity argon, helium, or, preferably, a vacuum of one tenth of a micron or less.

Cold Working

The excellent room-temperature ductility of stress-relieved and fully annealed tantalum makes the forming of tantalum comparatively simple. But the grain size of the material must be carefully considered for requirements where surface finish is of importance. The combination of the higher tensile strength and fair uniform elongation of stress-relieved tantalum sometimes makes it more satisfactory for drawing and forming than fully annealed material. It is very important to consider the temper properties of the material in the design of forming tools.

Joining

Tantalum can be joined by electron beam, tungsten-inert gas (TIG), and spot and resistance welding, but must be carefully protected from the effects of oxidation during welding. Uncontaminated welds are ductile and usually can be worked at room temperature.

Electron-beam-melted 90% tantalum–10% tungsten alloy can be formed and joined in a similar manner as pure tantalum provided that the higher strength and more rapid work-hardening characteristics of the alloy are considered.

ALLOYS

Tantalum alloys, including tungsten and tungsten–hafnium compositions, such as Ta-10W, T-111 (8% tungsten, 2% hafnium), and T-222 (9.6% tungsten, 2.4% hafnium, and 0.01% carbon), are used for rocket-engine parts and special aerospace applications. The tensile yield strength of Ta-10W is about 1089 MPa at room temperature and 621 MPa at 871°C.

TANTALUM BERYLLIDES AND CARBIDES

TANTALUM BERYLLIDE

$TaBe_{12}$ and $TaBe_{17}$ are intermetallic compounds with good strengths at elevated temperatures. $TaBe_{12}$ is tetragonal; density is 4.18 g/cm^3; melting point is 1849°C; and coefficient of thermal expansion is 8.42×10^{-6}/°C. $TaBe_{17}$ is hexagonal; density is 5.05 g/cm^3; melting point is 2045°C; and coefficient of thermal expansion is 8.72×10^{-6}/°C. Both compounds can be formed by all of the known ceramic forming methods plus flame and plasma-arc spraying. The materials are subject to safety requirements for all beryllium compounds.

TANTALUM CARBIDE

TaC and Ta_2C are the two primary carbides. The ore tantalum carbide with a congruent melting point is TaC. It is dark-to-light brown in color with a metallic luster. Ta_2C melts incongruently and is gray with a metallic luster. Ta_2C melts at 3400°C; the melting point of TaC has been reported to be as high as 4820°C.

TaC burns in air with a bright flash and is only slightly soluble in acids. The tensile strength at room temperature is 13.6 to 27.2 MPa.

Tantalum carbide is used in cemented-carbide cutting tools.

TAPE CASTING

Tape casting is a familiar technique to most ceramic engineers. It has been widely used as a method for fabricating improved capacitors. Conceptually, the process is simple. First, a slurry is prepared that contains ceramic powder (or powders) suspended in a solution of polymers. The slurry is spread into a relatively thin liquid coating on a smooth surface. The solution is then removed, usually through evaporation, although absorption into a porous medium can also be used, and a dry film is formed that is a green ceramic powder compact, termed *green-sheet*.

The tape usually has three qualitatively distinct phases: an inorganic powder, a continuous polymer matrix, and a porosity phase. (The powder is dispersed within the polymer matrix.) Each phase is important. In considering each of these phases, it is useful to recognize that the process to produce green-sheet is usually best regarded as a primary process, and one or more secondary operations are necessary to complete the fabrication of a green part ready for firing.

APPLICATIONS

The "classic" applications of tape casting are centered in the electronics industry and include the production of flat, smooth substrates for either thick-film or integrated circuitry, capacitors, dielectrics, and piezoelectric elements.

LAYERED MANUFACTURING

Although the ability to mix materials and stack sheets has long been recognized in the electronics industry, a recent and important development has been the recognition that green-sheet can play an important role as a feedstock for rapid prototyping (RP), also known as solid freeform fabrication (SFF). Although much that is useful can be derived from the work done for electronic applications, there are a number of distinct requirements for rapid prototyping that produce additional constraints on green-sheet physical properties and, therefore, formulations.

The field of rapid prototyping was developed in response to the high cost of tooling that is often an obstacle to implementing design changes or material substitution. All RP processes start with a CAD (computer-aided design) file (A in Figure T.1), computationally

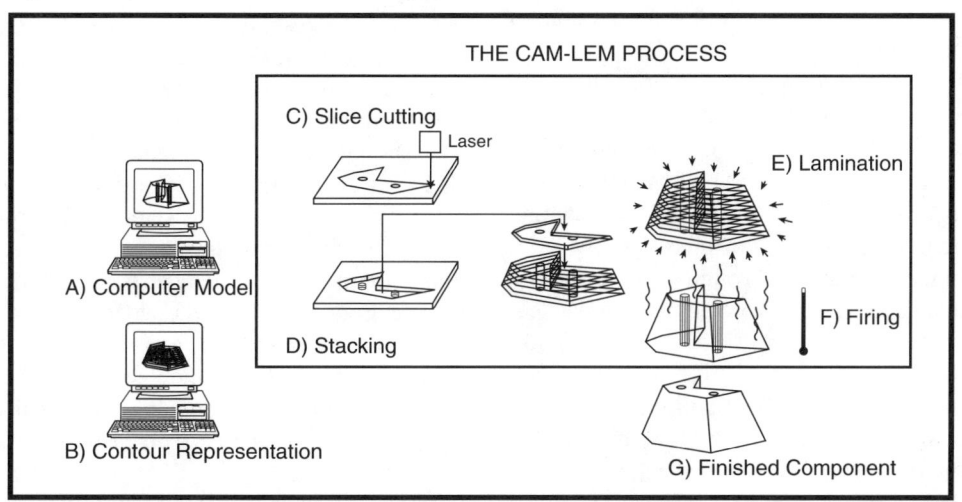

FIGURE T.1 Schematic of the CAM-LEM process.

approximate it by a series of closely spaced two-dimensional outlines (B in Figure T.1), then use a process to construct each outline sequentially, assemble them, and suitably post-process the assemblage to yield the final part (G in Figure T.1).

Frequently, the term *layered manufacturing* is also applied to these processes. This name is particularly appropriate since each process involves the fabrication of a three-dimensional object by automatic sequential stacking of appropriately contoured thin (pseudo-two-dimensional) sections. By controlling x–y motions in each layer, an (in principle) arbitrary component can be built up.

Each RP technology is distinguished by the process and machinery used to define each thin section and create the final stack.

RP AND CERAMICS

The application of solid freeform fabrication to engineering ceramics is motivated by a desire to take advantage of the striking advances in these materials over the last 20 years (such as transformation-toughened oxides, high-toughness silicon nitrides, and ceramic-matrix composites) as a result of government-funded research programs throughout the world.

CAM-LEM

The CAM-LEM process (computer-aided-manufacturing of laminated engineering materials) was developed specifically for production of ceramic parts.

Among the ceramic formulations that have been used with CAM-LEM are alumina, silicon nitride, zirconia-toughened oxides, and PZT.

The essence of the CAM-LEM is illustrated in steps C through F in Figure T.1: (C) feeding the green-sheet onto a movable platform; (D) creating a cut outline through relative motion of the table and laser; (E) automatic and selective extraction of the cut outline; and (F) addition to the build stack. Conversion to a dense ceramic article requires lamination of the green-sheets followed by binder burnout and firing.

The total time to produce a ceramic part is the sum of that required to produce the green part plus the firing time, which includes binder burnout and sintering. The CAM-LEM process produces green parts that are compatible with conventional firing cycles for laminated tape-cast parts.

TELLURIUM

An elementary metal (symbol Te), tellurium is obtained as a steel-gray powder of 99% purity by the reduction of tellurium oxide, or tellurite, TeO_2. The specific gravity is about 6.2 and the melting point is 450°C. The chief uses are in lead to harden and toughen the metal, and in rubber as an accelerator and toughener. Less than 0.1% tellurium in lead makes the metal more resistant to corrosion and acids, and gives a finer grain structure and higher endurance limit. Tellurium–lead pipe, with less than 0.1% tellurium, has a 75% greater resistance to hydraulic pressure than plain lead. Tellurium copper (C14500, C14510, and C14520) is a free-machining copper confining 0.3 to 0.7% tellurium. It machines 25% more easily than free-cutting brass. The tensile strength, annealed, is 206 MPa, and the electric conductivity is 98% that of copper. A tellurium bronze containing 1% tellurium and 1.5% tin has a tensile strength, annealed, of 275 MPa, and is free-machining. Tellurium is used in small amounts in some steels to make them free-machining without making the steel hot-short as do increased amounts of sulfur. But tellurium is objectionable for this purpose because inhalation of dust or fumes by workers causes garlic breath for days after exposure, although the material is not toxic. As a secondary vulcanizing agent with sulfur in rubber, tellurium in very small proportions, 0.5 to 1%, increases the tensile strength and aging qualities of the rubber. It is not as strong an accelerator as selenium, but gives greater heat resistance to the rubber.

Tellurium is an important component of many thermoelectric devices, and such devices can be used for both power generation and cooling. The requirements of a good thermoelectric element are high thermoelectric power, low thermal conductivity, and low electrical resistivity. Lead telluride (PbTe), bismuth telluride (Bi_2Te_3), and silver antimony telluride meet these requirements better than any other currently known materials. By the addition of

OXIDES

The oxides of tellurium are tellurium monoxide, TeO; tellurium dioxide, TeO_2; and tellurium trioxide, TeO_3. The monoxide is reported as a black, amorphous powder that is stable in dry air in the cold but that is oxidized in moist air to the dioxide. On being heated in vacuum, it apparently disproportionates into the dioxide and elemental tellurium. It can be formed by heating the mixed oxide $TeSO_3$. The dioxide is the most stable oxide and is formed when tellurium is burned in air or oxygen or by oxidation of tellurium with cold nitric acid. It has two crystalline forms.

TEXTILE FIBERS

Natural fibers constitute one of man's oldest sources of building materials. There is evidence to indicate that weaving and probably spinning were not unknown to our Stone Age ancestors. It is important to realize that there is no such thing as a natural textile fiber, although today there are human-made textile fibers. There are only natural fibers that have been diverted from their original function by mankind for use in textiles.

In man's search for fibers that can be used to further our own ends, literally dozens of naturally occurring fibers have been investigated. Only 23 are readily recognized by most textile authorities as being of commercial importance, and one fiber alone, cotton, accounts for approximately 70% of the total fibers consumed by the world population for textile purposes. If, however, this listing of natural fibers is carefully reviewed, it will be found that all fibers can be grouped into six different types of spinnable fibers, each differing fundamentally with respect to molecular and morphological structure. The distinctive characteristics possessed by the fibers in these six groups are such that the groups may be subjectively described as cottonlike, linenlike, sisal-like, wool-like, silk-like, and asbestos-like.

TEXTILE

A textile is a material made mainly of natural or synthetic fibers. Modern textile products may be prepared from a number of combinations of fibers, yarns, plies, sheets, foams, furs, or leather. They are found in apparel, and household and commercial furnishings, articles, and industrial products. Materials made directly from plastic sheet or film, leather, fur, or film are not usually considered to be textiles.

The term *fabric* may be defined as a thin, flexible material made of any combination of cloth, fiber, or polymer (film, sheet, or foams); *cloth* as a thin, flexible material made from yarns; *yarn* as a continuous strand of fibers; and *fiber* as a fine, rodlike object in which the length is greater than 100 times the diameter. The bulk of textile products is made from cloth.

The natural progression from raw material to finished product requires the cultivation or manufacture of fibers; the twisting of fibers into yarns (spinning); the interlacing (weaving) or interlooping (knitting) of yams into cloth; and the finishing of cloth prior to sale.

Spinning Processes

The ease with which a fiber can be spun into yarn is dependent upon its flexibility, strength, surface friction, and length. Exceedingly stiff fibers or weak fibers break during spinning. Fibers that are very smooth and slick or fibers that are very short do not hold together. To varying degrees, the common natural fibers (wool, cotton, and linen) have the proper combinations of the above properties. The synthetic fibers are textured prior to use to improve their spinning properties by simulating the convolutions of the natural fibers. Natural and synthetic filament fibers, because of their great length, need not be twisted to make useful yarns.

The properties of a yarn are influenced by the kind and quality of fiber, the amount of processing necessary to produce the required fineness, and the degree of twist. The purpose of the yarn determines the amount and kind of processing. The yarn number (yarn count) is an indication of the size of a yarn — the higher the number, the finer the yarn. The degree of twist is measured in turns per inch (tpi) and is varied from three to six times the square root of the yarn number for optimum performance.

The conversion of staple fiber into yarn requires the following steps: picking (sorting, clearing, and blending), carding and combing (separating and aligning), drawing (reblending), drafting (reblended fibers are drawn out into a long strand), and spinning (drafted fibers are further attenuated and twisted into yarn).

THALLIUM

A soft bluish-white metal (symbol Tl), thallium resembles lead but is not as malleable. The specific gravity is 11.85, and melting point 302°C. At about 316°C it ignites and burns with a green light. Electrical conductivity is low. It tarnishes in air, forming an oxide coating. It is attacked by nitric acid and by sulfuric acid. The metal has a tensile strength of 9 MPa and a Brinell hardness of 2. Thallium–mercury alloy, with 8.5% thallium, is liquid with a lower freezing point than mercury alone, –60°C, and is used in low-temperature switches. Thallium–lead alloys are corrosion resistant, and are used for plates on some chemical-equipment parts.

Applications

The major use for thallium is as a rodenticide and insecticide. The sulfate compound is most commonly employed for this application; therefore, the largest commercial sale of the element is in the form of the sulfate. Thallium sulfate is a heavy white crystalline powder, odorless, tasteless, and soluble in water. The advantage of this compound over many other rodenticides is that it is not detected by the rodent.

Other commercially available thallium chemicals are thallous nitrate and thallic oxide. Further uses of thallium compounds are as follows: (1) thallium oxisulfide, employed in a photosensitive cell that has high sensitivity to wavelengths in the infrared range; (2) Thallium bromide-iodide crystals, which have a good range of infrared transmission and are used in infrared optical instruments; and (3) alkaline earth phosphors, which are activated by the addition of thallium.

Other minor uses for thallium are in glasses with high indices of refraction, in the production of tungsten lamps as an oxygen getter, and in high-density liquids used for separating precious stones from ores by flotation.

Toxicity

Thallium and thallium compounds are toxic to humans as well as other forms of animal life. Therefore, special care must be taken that thallium is not touched by persons handling it. Rubber gloves should be used in handling both the metal and its compounds. Proper precautions should be taken for adequate ventilation of all working areas.

THERMAL SPRAYING

Thermal spray comprises a group of processes in which a heat source converts metallic or nonmetallic materials into a spray of molten or semimolten particles that are deposited onto a substrate. Any material that does not sublimate or decompose at temperatures close to its melting point can be applied by thermal spray, as long as it is available in wire or powder form.

Thermal spray coatings offer practical and economical solutions to a variety of industrial problems. They are most commonly applied to resist wear, heat, oxidation, and corrosion; provide electrical conductivity or resistance; and restore worn or undersized dimensions. Although the coating techniques have been around for some time, ongoing improvements are leading to lower application costs and a better understanding of how these coatings work. When properly selected and applied, thermally sprayed coatings can reduce downtime, lower production costs, and improve production yields.

Thermal spray is somewhat related to the welding process. In welding, the added material is actually fused to the base metal, forming a metallurgical bond, whereas a thermally sprayed coating generally adheres to the substrate through a mechanical bond. Nonetheless, some thermal spray processes are capable of achieving mechanical bond strengths that exceed 70 MPa.

The basic thermal spray technologies include plasma spray, wire arc spray, flame spray, detonation gun, and high-velocity oxygen fuel.

Plasma Spray

The plasma spray process requires a plasma gun or torch to generate an arc, which creates the plasma by ionizing a continuous flow of argon gas that is injected into the arc. The arc is struck between a water-cooled copper anode and a tungsten cathode. This type of process is also referred to as nontransferred arc spraying, because the arc is confined to the plasma gun. It is generally operated at energies in the neighborhood of 40 to 100 kW.

The plasma is a conductive gas with an extremely high internal working temperature (around 10,000°C). However, little heat is transferred by the plasma, so the part being sprayed remains relatively cool. For example, the process temperature of an 8-kg part will stay around 100°C. Because of the high internal operating temperature, this process is ideally suited for spraying high-melting-point materials such as ceramics and refractory metals.

The high heat of plasma causes a large increase in the volume of inert gas introduced, and this produces a high-speed gas jet that accelerates the molten particles and propels them toward the substrate at high velocities. High particle velocities result in dense coatings with high bond strengths.

The plasma transferred arc (PTA) process is somewhat of a hybrid between plasma spraying and welding. In this process, an arc is struck between the nonconsumable electrode of the plasma torch and the workpiece itself. The feedstock, in the form of wire or powder, is introduced into the resulting external plasma. The material is melted and puddled onto the substrate, producing a metallurgical bond similar to welding, but with a lot less dilution. This process is capable of producing dense and smooth coatings, but it is not capable of applying ceramics.

Wire Arc Spray

The wire arc spray process, like plasma spraying, requires an electrical heat source to melt materials. In this case, the feedstock consists of two conductive metal wires. These two wires act as electrodes that are continuously consumed as the tips are melted by heat from the electrical arc that is struck between them. An atomizing gas shears off the molten droplets and propels them toward the substrate.

The atomizing gas is usually compressed air, but it can also be an inert gas such as nitrogen or argon. Compressed air causes oxidation of metal particles, resulting in a large amount of metal oxide in the coating. Because of this, the coating is harder and more difficult to machine than the source material of the coating. This can be a disadvantage because some coatings have to be ground. However, the increased hardness can also enhance wear resistance.

In addition, the temperature of the arc far exceeds the melting point of the sprayed material, resulting in the formation of superheated particles. Consequently, localized metallurgical interactions or diffusion zones develop, which enable achievement of good cohesive and adhesive strengths.

The wire-arc process also operates at higher spray rates than the other thermal spray processes. The spray rate, which is dependent on the applied current, makes this process relatively economical.

Flame Spray

In the flame spray process, powder or wire materials are melted through the release of chemical energy triggered by a combustion process. A fuel gas (or liquid) is burned in the presence of oxygen or compressed air. Acetylene fuel gas is most frequently selected, due to its high combustion temperature of 3100°C and low cost. Propane, hydrogen, MAPP, and natural gas are also common choices. The flame melts the feedstock, and also accelerates and propels the molten particles. Compressed shop-air is also used to assist and boost the particle velocities. However, a compressed inert gas such as argon or nitrogen is preferred if oxidation is a concern.

The setup of a flame spray system is relatively inexpensive and mobile. A basic setup requires only a flame spray torch, a supply of oxygen, and a fuel gas. To increase safety, the setup might have to be augmented with an enclosed spray booth and exhaust.

Because its particle velocities are lower than those of the other thermal spray processes, flame spray coatings are usually of lower quality; they have higher porosity and lower cohesive and adhesive strengths. However, coating quality can be improved by a "spray-and-fuse" process. After the coating is applied by flame spray, the combustion process is repeated to raise the substrate temperature to the point at which the previously applied coating starts to melt. Fusing temperatures exceed 1040°C. The final coating is extremely dense and well-bonded by a metallurgical bond. A disadvantage of this technique is the high substrate temperature required and the possibility for deformation of the part.

Detonation Gun

The detonation gun (D-gun) process involves an intermittent series of explosions, which melt and propel the particles onto the substrate. Specifically, a spark plug ignites a mixture of powder and oxygen-acetylene gas in a barrel. After ignition, a detonation wave accelerates and heats the entrained powder particles. After each detonation, the barrel is purged with nitrogen gas, and the process is repeated several times per second.

Coatings produced by the detonation gun process are of excellent quality. The particle velocities are high, so the coatings are dense and exhibit high bond strengths. The drawback is that the process is relatively expensive to operate. It also produces noise levels that can exceed 140 decibels, and requires special sound- and explosion-proof chambers.

HVOF Spray

The high-velocity oxygen fuel (HVOF) thermal spray process is closely related to the flame spray process, except that combustion takes place in a small chamber rather than in ambient air. The HVOF combustion process generates a large volume of gas caused by the formation and thermal expansion of such exhaust gases as carbon dioxide and water vapor.

These gases must exit the chamber through a narrow barrel several inches long. Because of the extremely high pressure created in the combustion chamber, the gases exit the barrel at supersonic velocities, thereby accelerating the molten particles. Although the particles do not reach the speed at which the gases are traveling, they do reach very high velocities. Particle velocities of over 750 m/s have been measured. These high particle speeds, and subsequent high kinetic energy, translate into dense coatings with some of the highest bond strengths possible.

Coating Characteristics

The goal of all these thermal spray processes is to provide a functional coating that meets all of the necessary requirements; see Table T.1. The quality of a coating depends on the final function of the coating, and can be determined by evaluating a number of coating characteristics.

Characteristics that can be evaluated to determine coating quality include:

- Microstructure (porosity, unmelts, oxidation level)
- Macrohardness (Rockwell B or C) and microhardness (Vickers or Knoop)
- Bond strength (adhesive and cohesive)
- Corrosion resistance
- Wear resistance
- Thermal shock resistance
- Dielectric strength.

Coating Selection

Metal forming, paper and pulp, paper converting, printing (including offset and flexographic), chemical, petrochemical, textile, infrastructure, food processing, automotive, medical, power generation, and aerospace all take advantage of thermally sprayed coatings. For each application, the coating is selected to perform one or more functions. The five most encountered functions are wear resistance, heat and/or oxidation resistance, corrosion resistance, electrical conductivity or resistance, and the restoration of worn or undersized dimensions.

Basically, coatings fall into three categories: metals/alloys, ceramics, and cermets. Almost every metal and alloy available can be sprayed

TABLE T.1
The Functions and Applications of Thermal Spray Coatings

Function	Application	Coating
Wear resistance		
Adhesive wear	Bearings, piston rings, hydraulic press sleeves	Chrome oxide, babbitt, carbon steel
Abrasive wear	Guide bars, pump seals, concrete mixer screws	Tungsten carbide, alumina/titania, steel
Surface fatigue wear	Dead centers, cam followers, fan blades (jet engines), wear rings (land-based turbines)	Tungsten carbide, copper–nickel–indium alloy, chromium carbide
Erosion	Slurry pumps, exhaust fans, dust collectors	Tungsten carbide, Stellite (Deloro-Stellite Co.)
Heat resistance	Burner cans/baskets (gas turbines), exhaust ducts	Partially stabilized zirconia
Oxidation resistance	Exhaust mufflers, heat treating fixtures, exhaust valve stems	Aluminum, nickel–chromium alloy, Hastelloy (Haynes International Co.)
Corrosion resistance	Pump parts, storage tanks, food handling equipment	Stainless steel (316), aluminum, Inconel (Inco Alloys International), Hastelloy
Electrical conductivity	Electrical contacts, ground connectors	Copper
Electrical resistance	Insulation for heater tubes, soldering tips	Alumina
Restoration of dimensions	Printing rolls, undersize bearings	Carbon steel, stainless steel

Source: Adv. Mater. Proc., 154(6), p. 32, 1998. With permission.

in some form. Frequent choices include copper, tungsten, molybdenum, tin, aluminum, and zinc. Frequently sprayed alloys include steels (carbon and stainless), nickel/chromium, cobalt-base alloys, nickel-base alloys, bronzes, brass, and babbitts. Ceramic materials are usually metal-oxide ceramics such as chromium oxide, aluminum oxide (also called alumina), alumina–titania composites, and stabilized zirconias.

Cermets are coatings that combine a ceramic and a metal or alloy. Two examples include tungsten carbide (the ceramic constituent) in a cobalt matrix, and chromium carbide in a nickel–chromium matrix.

Overall, thousands of different products and components are coated with great success. In addition, new applications are developed daily. Although it is not possible to provide examples of every application and what coating is best for the specific function, the Table T.1 illustrates the broad range of applications and industries that are served.

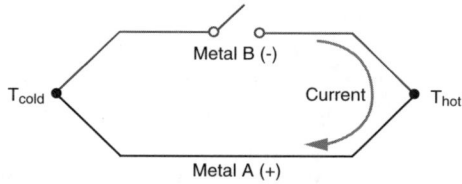

FIGURE T.2 Schematic drawing of a thermocouple.

THERMOCOUPLES

Thermocouples are the most common type of temperature sensor used and nearly 16% of all process instrumentation measures, indicates, or controls temperature. Thomas Johann Seebeck is credited with inventing the thermocouple in 1821. His experiment consisted of two dissimilar metal wires joined at the ends to form a loop with each end held at a different temperature; Figure T.2.

Seebeck detected the induced current by the displacement of a compass needle that was near one of the wires. Further study revealed that the

TABLE T.2
Standard Thermocouple Designations

Letter Code	Conductor Material	Color Code	Magnetic	Temperature Range	Environment
J	Iron	+white	Yes	0–760°C	Oxidizing or reducing
	Constantan	–red	No	32–1400°F	
K	Chromel	+yellow	No	0–1260°C	Oxidizing
	Alumel	–red	Yes	32–2300°F	
T	Copper	+blue	No	–150–370°C	Oxidizing or reducing
	Constantan	–red	No	–300–700°F	
E	Chromel	+purple	No	–150–870°C	Oxidizing
	Constantan	–red	No	–300–1600°F	
R	Platinum–rhodium	+green	No	0–1480°C	Oxidizing or inert
	Platinum	–red	No	32–2700°F	
S	Platinum–rhodium	+black	No	0–1480°C	Oxidizing or inert
	Platinum	–red	No	32–2700°F	
B	Platinum–rhodium	+gray	No	0–1700°C	Oxidizing
	Platinum–rhodium	–red	No	32–3100°F	Inert or vacuum
W	Tungsten	+white	No	0–2300°C	Vacuum or inert
	Tungsten–rhenium	–red	No	32–4200°F	

temperature gradient induced an electric current and when this circuit was broken at the center, an open circuit voltage was measured, i.e., the Seebeck electromotive force. The thermocouple is based on the concept that for small changes in temperature (T_{hot} to T_{cold}), the voltage is proportional to the temperature difference.

Operating environment and temperature are important considerations for picking the correct thermocouple. Table T.2 provides some practical guidelines for selection.

When using a thermocouple, it is very important to understand that the measured voltage is developed along the entire length of the thermocouple. Steep temperature gradients should be avoided because any defect in the wire within the gradient will contribute a large error. Steep gradients may also induce recrystallization and grain growth, thus changing the calibration. In this regard, feeding thermocouples through insulation is critically important because deformation of the wires may produce recrystallization during operation.

THERMOFORMED PLASTIC SHEET

Thermoplastic sheet forming consists of the following three steps; (1) a thermoplastic sheet or film is heated above its softening point; (2) the hot and pliable sheet is shaped along the contours of a mold, the necessary pressure being supplied by mechanical, hydraulic, or pneumatic force or by vacuum; and (3) the formed sheet is removed from the mold after being cooled below its softening point.

Sheet Materials

The following five groups of thermoplastic materials account for the major share of the thermoforming business:

1. *Polystyrenes:* High-impact polystyrene sheet, ABS (acrylonitrile–butadiene–styrene) sheet, biaxially oriented polystyrene film, and polystyrene foam
2. *Acrylics:* Cast and extruded acrylic sheet, and oriented acrylic film
3. *Vinyls:* Unplasticized rigid PVC, vinyl copolymers, and plasticized PVC sheeting
4. *Polyolefins:* Polyethylene, polypropylene, and their copolymer films
5. *Cellulosics:* Cellulose acetate, cellulose acetate butyrate, and ethyl cellulose sheet.

THERMOFORMED PLASTIC SHEET

A sixth group of increasing importance is the linear polycondensation products, such as polycarbonates, polycaprolactam (type 6 nylon), polyhexamethylene adipamide, oriented polyethylene terephthalate (polyester), and polyoxymethylene (acetal) films.

GENERAL PROCESS CONSIDERATIONS

Thermoplastic sheets soften between 121 and 232°C. It is important that the sheets be heated rapidly and uniformly to the optimum forming temperature. The fastest heating is brought about with infrared radiant heaters. Some thermoplastics cannot tolerate such intense heat and require convection heating in air-circulating ovens or conduction heating between platens. In a few instances, the sheets are formed "in line," making use of the heat of extrusion.

Four basic forming methods and more than 20 modifications are known:

Matched Mold Forming

This process, in which the hot sheet is formed between a registering male and female mold section, employs mechanical or hydraulic pressure (Figure T.3a).

It is used for corrugating flat rigid sheeting either "in line" or in a separate operation. For continuous longitudinal corrugation, the hot sheet is pulled through a matched mold with registering top and bottom teeth. Transverse corrugation is accomplished with matched top and bottom rolls or molds mounted on an endless conveyor belt or chains.

For stationary molding operations, a rubber blanket, backed with a liquid or inflated by air,

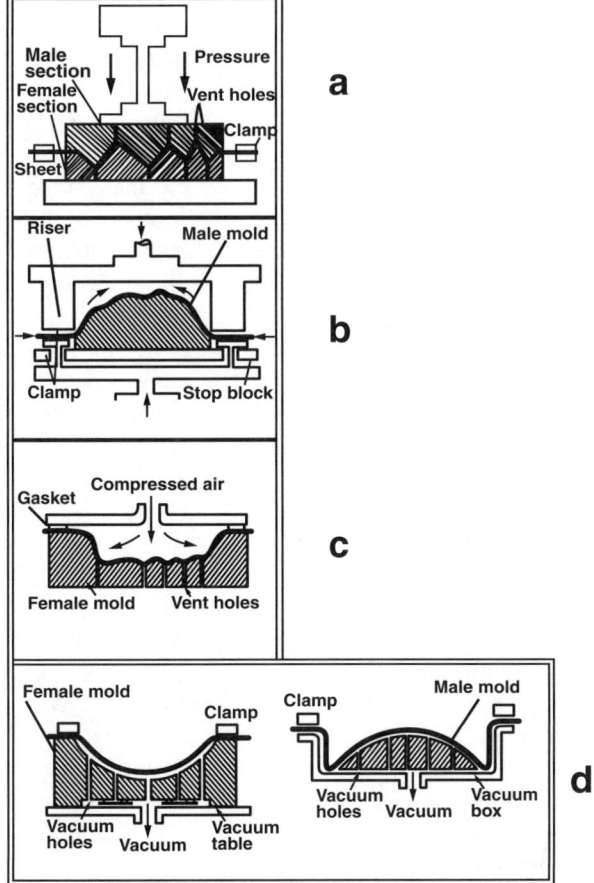

FIGURE T.3 The four basic methods of thermoplastic sheet forming. (a) Matched mold forming; (b) slip forming; (c) air blowing; (d) vacuum forming.

frequently replaces the male section. Another modification is the *plug and ring* technique in which the ring acts as a stationary clamping device and the moving plug resembles the top portion of the male mold.

Elastic and oriented thermoplastic sheets possess a plastic memory; i.e., they tend to draw tight against the force that stretches them. This property occasionally permits forming against a single mold only.

Slip Forming

A loosely clamped sheet is allowed to slip between the clamps and is "wiped" around a male mold (Figure T.3b). This process has been in use to avoid excessive thinning when forming articles with deep draws.

Air Blowing

A hot sheet is blown with preheated compressed air into a female mold (Figure T.3c).

Variations of this process are *free blowing* without a mold into a bubble, *plug-assist blowing* in which a cored plug pushes the hot sheet ahead before blowing, and *trapped sheet forming* in which a clamping ring slides over the mold before applying the compressed air. The last process is employed in automatic roll-fed packaging machines with biaxially oriented films.

Vacuum Forming

This is the most common sheet-forming process with many modifications (Figure T.3d). *Modifications of vacuum forming:*

- In *straight vacuum forming*, the hot thermoplastic sheet is clamped tight to the top of a female mold or of a vacuum box that contains a male mold. Drawing sheets into a female mold results in excellent replica of fine details on the outer surface of the drawn articles. Straight vacuum forming has proved excellent for shallow draws; however, in articles with small radii or deeper dimensions, the corners and bottom are excessively thinned out. Employing a female mold with multiple cavities is more economical than forming over a number of male molds, because it permits smaller spacing between cavities (without bridging), which in turn allows more pieces per sheet. Drawing sheets over a male mold produces articles with the thickest section on top. It is used for the production of three-dimensional geographical maps, because it gives a greater accuracy of registration due to restricted shrinkage.
- *Free vacuum forming* into a hemisphere, without a mold, is similar to free air blowing and is employed with acrylic sheets where perfect optical clarity has to be maintained.
- *Vacuum snap-back forming* makes use of the plastic memory of the sheet. The hot sheet is drawn by vacuum into an empty vacuum box, while a male mold on a plug is moved from the top into the box. The vacuum is released and the sheet, still hot and elastic, snaps back against the male mold and cools along the contours of the mold. This method is employed with ABS and plasticized vinyl sheets for the production of cases and luggage shells.
- *Drape forming* is a technique that allows deeper drawing. After clamping and heating, the sheet is mechanically stretched over a male mold, then formed by vacuum, which picks up the detailed contours of the mold. Acrylic and polystyrene sheets slide easily over the mold, whereas polyethylene sheets tend to freeze on contact with the mold, causing differences in thickness.
- To overcome the problem of thinning, *vacuum plug-assist forming* into a female mold has been developed. Its principle is to force a heated and clamped sheet into a female mold using a plug-assist before applying the vacuum. This technique may be considered a reverse of the snapback method.

- *Vacuum air-slip forming* represents a modification of drape forming and is designed to reduce thinning on deep-drawn articles. It consists of prestretching the sheet pneumatically prior to vacuum forming over a male mold. Prestretching is accomplished either with entrapped air by moving the male mold like a piston in the vacuum chamber or by compressed air.
- There are at least three variations of the *reverse-draw technique*. All employ the principle of blowing a bubble of the hot plastic sheet and pushing a plug in reverse direction into the outside of the hot bubble. This accomplishes a folding operation permitting deeper draws than any other common practice.
- The technique of *reverse-draw with plug-assist* consists of heating the clamped sheet, raising a female mold so that a sealed cavity is formed while a bubble is blown upward. The preheated plug-assist is lowered and pushes the sheet into the cavity. The final shaping is accomplished with vacuum.
- A variation of this method is *reverse-draw with air-cushion*. The plug-assist is furnished with holes through which hot air is blown downward and pushes the hot sheet ahead of the plug-assist, minimizing mechanical contact. This technique is used in forming materials with sharp softening ranges and limited hot strength, such as polyethylene and polypropylene.
- *Reverse-draw on a plug* uses a male mold on the plug to preserve the finish of the sheet.

Design and Construction of Vacuum Molds

Depth of draw is a prime factor controlling the wall thickness of the formed article. During straight vacuum forming into a female mold, the depth of draw should not exceed one half of the cavity width. For drape forming over a male mold, the height-to-width ratio should be 1:1 or less. With plug-assist, air slip, or one of the reverse-draw techniques, the ratio may exceed the 1:1 ratio.

Proper air evacuation assists material flow in the desired direction and in uniform wall thickness. In general, deep corners require intensified evacuation. The diameter of vacuum holes should be 0.25 to 0.6 mm for polyethylene sheets, 0.6 to 1 mm for other thin-gauge materials, and may increase to 1.5 mm for heavier rigid materials. Sharp bends and corners should be avoided, because they result in excessive stress concentration and in reduction of strength. The forming cycle can be accelerated and maintained by the use of mold temperature controls. Molds for permanent use are cast from aluminum or magnesium alloy.

Finishing

After the article has been formed, it must be cooled, removed from the forming machine, and separated from the remainder of the sheet. Trimming of thin-walled articles can be carried out hot or cold, in the forming machine or after removal. Heavy-gauge articles should be trimmed only after cooling. Clicker dies, high dies, and Walker dies are frequently used. Decorating formed articles is generally accomplished by printing the flat sheet before the forming operation. Formed articles may also be spray-coated.

THERMOPLASTIC

One definition of thermoplastic covers the type of decorating glass enamels or overglaze applied through a hot screen. The media is a wax composition that is heated and added to the enamel powder. On cooling, it forms a solid case that is broken into cubes. The cubes are heated and flow onto a resistance-heated metal screen. When applied on ware, the thermoplastic freezes immediately and other colors can be superimposed without a drying cycle. Thermoplastic permits wraparound decorations and reduces ware handling losses.

Other thermoplastic materials are used in conjunction with the injection molding of technical ceramics such as spark plugs.

THERMOPLASTIC CONTINUOUS-FIBER-REINFORCED MATERIALS

These materials are tougher and withstand impacts better. They mold readily and can be recycled. They also have unlimited shelf life and emit no hazardous solvents during processing.

Processing

This continuous-fiber-reinforced thermoplastic (CFRTP) process weaves together strands of powder-resin-coated fibers to produce TowFlex fabrics. This is in contrast to other fabrics where raw reinforcement fibers are first woven then coated. Weaving individual coated strands makes the fabric highly drapable. This is because each strand within the fabric moves freely relative to adjacent strands. In addition, the strands remain flexible because they are not fully wet out prior to molding. Complete wet out of the fabric comes during the compression-molding process (Figure T.4).

CFRTP Basics

Fiber-reinforced thermoplastics are widely used in injection molding. The vast majority of these products use short or chopped fibers. These fibers generally measure less than 6.4 mm and will randomly orient themselves during molding. Typical injection molded parts contain only 20 to 30% reinforcement fiber. Short, randomly oriented fibers in low percentages do not provide much reinforcement. And it is often difficult to mold such material into complex, large, or thick-walled parts without voids or knit lines.

In comparison, parts molded from new CFRTPs often contain more than 60% reinforcement. Reinforcement fibers run continuously throughout the entire part in specified directions to help optimize strength and stiffness. CFRTPs contain continuous-reinforcement fiber filaments that are powder-coated with melt-fusible thermoplastic particles. The uniformly coated filaments are woven into fabrics or braid, formed into semirigid unidirectional tapes or ribbons, or laminated into panels. The resin particles wet out and consolidate quickly when compression molded.

The first step in compression molding CFRTP fabric is to cut and assemble fabric plies. The plies create a preform that approximates the flat pattern shape of the molded part. Automated cutting equipment such as reciprocating knives or ultrasonic gear may be an option for complicated patterns manufactured in high volume. Steel-rule dies, electric rotary shears, or hand shears/scissors might work best for lower-volume applications. It is often useful to machine a simple preform fixture with a cavity or recess in the shape of the flat pattern. The assembly fixture helps keep the precut fabric plies in proper order, location, and orientation. This includes any partial plies needed to build up additional thickness in specific areas.

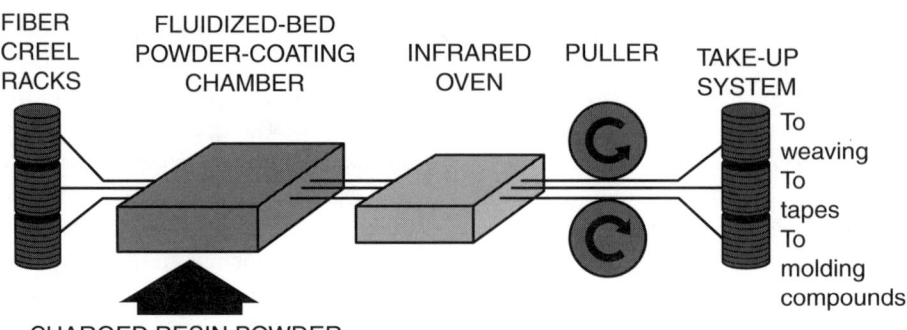

FIGURE T.4 Continuous-fiber construction.

Stacked plies are tack-welded ultrasonically or via a soldering iron before removal from the fixture. This helps keep them in the right orientation during handling and storage. The preforms have unlimited shelf life and can be produced in large quantities independent of the molding process or molds. They also need no additional layup labor before being compression-molded into finished parts. The fabric drapes easily and makes for a straightforward molding process that forms and shapes the flat preform. In contrast, thermoset preimpregnated materials generally require cutting and splicing to mold complex shapes.

Molds

Matched steel metal molds give the best surface finish and mold life when molding CFRTP parts. Nickel-plated aluminum molds are good for moderate volumes of material processing below 316°C. Steel molds are mandatory for higher-temperature molding of matrix resins such as polyphenylene sulfide or polyetherether ketone. Registration of the upper to lower mold halves takes place through either guide pins or the mold configuration itself.

Generally, molds are designed to fully "bottom out" on the CFRTP material rather than on thickness stops. This helps maintain pressure on the material throughout the molding process. Thickness stops, however, are used to maintain flatness and help ensure that the mold does not "rock" during the process. They also establish a minimum part thickness. Stops should typically be 0.25 to 0.50 mm lower than the desired nominal part thickness. For relatively thin parts or plates of less than 3.81 mm quantity of fabric plies loaded into the mold primarily controls thickness. Thicker parts or plates use a specified preform weight along with the number of plies to control part thickness.

Compression Molding

CFRTP fabrics generally use three variations of the compression-molding process. All employ conventional equipment and do not need rapid closing speeds or excessive press tonnage.

Single-press/heated-cooled platens — The platens in a single press are heated and cooled to reach the right processing conditions. This approach applies best in situations requiring a variety of different parts in relatively low volumes that do not need a quick molding cycle. The molding process begins with loading of the preform into the lower mold half. Operators next install the upper mold half and load the complete mold into the press. It is then heated with pressures on the order of 68.6 to 343 kPa. Low pressure during heating helps ensure good heat transfer and initiates the forming of the preform. Full pressure, 0.68 to 5.44 MPa, comes during final compaction as the mold reaches its required processing temperature. Next the press platen is cooled to bring the mold and CFRTP part to the removal temperature.

Hot/cold shuttle press — Hot/cold shuttle presses separate the basic segments of the compression-molding process — heating, cooling, and loading/unloading — for better efficiency. Separate heated and cooled platen presses apply pressure to the mold and CFRTP material. Cycle times are short because molds shuttle between preheated hot presses and precooled cooling presses. Cycle times are particularly fast for relatively thin parts and low-mass molds. Molds do not need individual heating and cooling systems, which helps reduce cost. Molds for deep-draw parts may need to use heated or cooled press platens or bolsters. These approximate the part shape and mount to the hot and cold presses. This approach helps keep heating and cooling sources close to the material to increase processing speed. The hot/cold shuttle-press approach is useful for combinations of different parts in moderate to high volumes. Here economics do not justify the expense of heating and cooling provisions in each mold. The hot press station is first preheated to 10 to 37.8°C higher than the desired mold temperature. A preform goes in the lower mold half, the upper mold half lowers into place, and the mold shuttles into the preheated press. Low pressure, 68.6 to 343 kPa, is applied as the mold heats for good heat transfer and to start preform shaping. Full pressure of 0.68 to 3.4 MPa or more forces final compaction as the mold reaches the processing temperature range. Next, the hot press opens, pressure releases, and the mold shuttles into the cold press station.

Full pressure again applies as the mold and material cool enough for part removal.

Single-press/heated-cooled molds — Here integrally heated and cooled molds mount directly onto press platens. Processing cycles for complex-shaped parts can be fast because mold heating and cooling systems can be close to the CFRTP. Individual molds are more expensive, because each mold must contain integral heating and cooling. Press platens must also have sufficient travel to open wide enough for easy preform loading and part removal. Otherwise molds would need to be removed from the press for part removal. The single-press/heated-cooled mold approach is especially appropriate for large production runs of a specific part or plate. Manufacturers can amortize molds over a large part count and speedy processing helps drive cost down. It is also appropriate for large or deep molds where the shuttle process is impractical. The integrally heated and cooled mold halves are usually attached to the press platens to minimize handling. The mold temperature cycles between the processing and part-removal temperatures. The CFRTP preform loads into the mold and the press closes with low pressure, 68.6 to 343 kPa, for good heat transfer and to initiate preform shaping. The heated mold goes under full pressure, 0.68 to 3.4 MPa or more, for final compaction and forming. The mold and material then cool and the part is demolded. Cycle times can be under 5 min depending on the mold mass, part thickness, and part geometry.

APPLICATIONS

There have been few commercial applications for CFRTP, however. One reason has been that early product forms of these materials were tougher to process and mold. The development of highly drapable, conformable CFRTP fabrics addresses such shortcomings. They are made from thermoplastics such as nylon, polypropylene, polyphenylene sulfide, polyetherimide, and polyetheretherketone with carbon, glass, or aramid reinforcement fibers. They can easily be molded into complex structural shapes and are ideally suited for production quantities of between 1000 and 50,000 parts annually.

A CFRTP fabric called RF6 is used to make an eight-dihedral-faced kayak paddle. The paddle surface is said not only to grab the water more effectively than conventional blades but also to release surface pressure at eight precise locations along the outside edge of the blade. This produces a blade that has zero flutter while providing more bite per square inch of surface area. The thin blade design was made possible by the high stiffness and impact resistance of the CFRTP material.

THERMOPLASTIC ELASTOMERS

Thermoplastic elastomers (TPEs) are a group of polymeric materials having some characteristics of both plastics and elastomers. They are also called elastoplastics. Requiring no vulcanization or curing, they can be processed on standard plastics processing equipment. They are lightweight, resilient materials that perform well over a wide temperature range.

There are a half-dozen different types of elastoplastics.

TYPES

TPEs have been traditionally categorized into two classes: block copolymers, which include styrenics, copolyesters, polyurethanes, and polyamides; and thermoplastic/elastomer blends and alloys, comprised of thermoplastic polyolefins and thermoplastic vulcanizates.

Conventional TPEs are considered two-phase materials composed of a hard thermoplastic phase that is mechanically or chemically mixed with a soft elastomer phase. The resulting material shares the characteristics of both.

ADVANTAGES AND DRAWBACKS

TPEs give engineers several advantages over thermoset rubbers, including lower fabrication costs, faster processing times, little or no compounding, recyclable scrap, and processing by conventional thermoplastic equipment. In addition, the processing equipment consumes less energy and maintains tighter tolerances than that of rubber processing.

There are several drawbacks, however. TPEs have relatively low melting temperatures, making them unusable in high-temperature applications. They usually require drying before molding because they are hygroscopic, or moisture absorbing. Manufacturers must use TPEs for high-volume applications for it to be economical. And molders accustomed to working with rubber have little experience using TPE materials and equipment.

PROPERTIES

As for material properties, TPEs come in durometers, or hardnesses, ranging from 3 Shore A to 70 Shore D, which roughly translates into going from very soft and flexible to semirigid. Although they have relatively low melt temperatures, TPEs resist continuous exposure to temperatures up to 135°C, with spikes up to 149°C. On the other end of the temperature spectrum, they remain flexible at temperatures down to –69°C. The polymers have outstanding dynamic fatigue resistance, good tear strength, and resist acids and alkalis, ultraviolet light, fuels and oils, and ozone, and maintain their grip in wet and dry conditions (Table T.3).

ADVANCEMENTS AND APPLICATIONS

The latest advancements in TPE formulations are resin blends that adhere directly to engineering thermoplastics without any special features and can be applied by insert molding (overmolding) and two-shot injection molding. The new formulations let designers mold grips onto a wider range of thermoplastics, particularly high-strength materials such as high-impact polystyrene, glass-filled nylon, and polycarbonate, which are commonly used in power tools, sporting goods, and electronics.

New materials are also easier to mold in thin-wall sections, which is crucial for consumer electronics where weight is a concern. Consumer-electronics manufacturers can use TPEs for grip strips, for example, because new resins flow easily across long, thin channels in part molds. Engineers prefer to mold "soft-touch" elastomers onto rigid substrates instead of using adhesives and mechanical locks; yet most elastomers do not provide the necessary adhesion, compatibility, and durability,

To answer these needs, a material system was developed consisting of a rigid and a flexible TPU formulation. The two materials — Estaloc thermoplastic and Estane elastomer — have similar chemical makeup. This chemical compatibility helps the two materials form a strong bond without adhesives, giving automotive manufacturers, for example, a "one-stop shop" for interior applications such as ignition bezels.

Copolyether-ester thermoplastic elastomer applications include tubing and hose, V belts, couplings, oil-field parts, and jacketing for wire and cable. Their chief characteristic is toughness and impact resistance over a broad temperature range.

THERMOPLASTIC OLEFINS

The olefinics, or TPOs, are produced in durometer hardnesses from 54A to 96A. Specialty flame-retardant and semiconductive grades are also available. The TPOs are used in autos for paintable body filler panels and air deflectors, and as sound-deadening materials in diesel-powered vehicles.

The TPOs have room-temperature hardnesses ranging from 60 Shore A to 60 Shore D. These materials, as they are based on polyolefins, have the lowest specific gravities of all thermoplastic elastomers. They are uncured or have low levels of cross-linking. Material cost is midrange among the elastoplastics.

These elastomers remain flexible down to –51°C and are not brittle at –68°C. They are autoclavable and can be used at service temperatures as high as 135°C in air. The TPOs have good resistance to some acids, most bases, many organic materials, butyl alcohol, ethyl acetate, formaldehyde, and nitrobenzene. They are attacked by chlorinated hydrocarbon solvents. Compounds rated V-0 by UL 94 methods are available.

THERMOPLASTIC POLYESTERS

Known chemically as polybutylene terephthalate (PBT) and polyethylene terephthalate

TABLE T.3
Engineering Properties of Elastomeric Materials

	TPV	Thermoset Rubbers					
		Polychloroprene	EPDM	Chlorosulfonated Polyethylene	Nitrile	Butyl
Specific gravity	0.97	1.4	1.2	1.4	1.2	1.1
Durometer (Shore)	45A to 50D	50A to 90A	40A to 90A	40A to 90A	40A to 90A	40A to 90A
Compression set	Excellent to good	Good to fair	Excellent to good	Fair	Good to fair	Fair
Continuous service temperature, high (°F)	275	250	275	250	240	240
Continuous service temperature, low (°F)	−81	−50	−65	−50	−40	−65
Cycle time	Excellent	Fair	Fair	Fair	Fair	Fair
Recyclability	Yes	No	No	No	No	No
Environmental Resistance						
Ozone	Excellent	Good	Excellent	Good	Good to fair	Excellent
Ultraviolet light	Excellent to good	Good	Excellent to good	Excellent to good	Fair	Excellent to good
Acids	Excellent	Good	Excellent	Good	Good to fair	Excellent
Alkalis	Excellent	Good	Excellent	Excellent	Good to fair	Excellent
Lubricating oils	Fair	Fair	Poor	Good	Excellent	Fair
Gas permeability	Fair	Good to fair	Fair	Fair	Fair	Excellent

	Thermoplastic Polyolefin Rubber (TPR)	Styrene–Butadiene Block Copolymer	Thermoplastics Copolyester	Flexible Vinyl	Polyurethane
Specific gravity	0.97	0.94	1.2	1.2	1.1
Durometer (Shore)	60A to 90A	3A to 55D	40D to 70D	60A to 90A	60A to 60D
Compression set	Fair	Good	Good to fair	Fair to poor	Good
Continuous service temperature, high (°F)	250	210	230	200	250
Continuous service temperature, low (°F)	−50	−50	−90	−30	−40
Cycle time	Excellent to good	Good	Excellent to good	Excellent to good	Fair to good
Recyclability	Yes	Yes	Yes	Yes	Yes
Environmental Resistance					
Ozone	Excellent	Excellent	Excellent	Good	Good
Ultraviolet light	Good	Good to fair	Excellent to good	Good	Good
Acids	Excellent	Excellent	Poor	Good	Fair
Alkalis	Excellent	Excellent	Good to fair	Good	Poor
Lubricating oils	Poor	Poor	Fair	Fair	Fair
Gas permeability	Fair	Fair	Fair	Good to fair	Good

Source: Mach. Design, March 10, pp. 174-175, 1998. With permission.

(PET), the thermoplastic-polyester molding compounds are crystalline, high-molecular-weight polymers. They have an excellent balance of properties and processing characteristics and, because they crystallize rapidly and flow readily, mold cycles are short.

In addition to several unreinforced molding resins, the polyesters are available in glass-reinforced grades. Unreinforced and glass-filled grades are available with UL flammability ratings of 94 HB and 5V.

Properties

Thermoplastic polymers have excellent resistance to a broad range of chemicals at room temperature including aliphatic hydrocarbons, gasoline, carbon tetrachloride, perchloroethylene, oils, fats, alcohols, glycols, esters, ethers, and dilute acids and bases. They are attacked by strong acids and bases.

High creep resistance and low-moisture absorption give the polyesters excellent dimensional stability. Equilibrium water absorption, after prolonged immersion at 22.75°C, ranges from 0.25 to 0.50% and, at 66°C, is 0.52 to 0.60%. Black-pigmented grades are recommended for maximum strength retention in outdoor uses.

THERMOPLASTIC POLYPROPYLENE

With long-fiber-reinforced materials, thermoplastic composite extrusions reportedly offer properties superior to those of PVC and polypropylene — as well as more costly engineered thermoplastics like polycarbonates. This development is referred to as Very High Modulus Extrusion (VHME).

Parts reinforced by long-fiber materials have been injection-molded for 15 years. But VHME thermoplastic is the first commercial application of long-fiber extrusions. The material consists of a long-fiber-based core for strength and impact resistance, sheathed by inner and outer layers of ABS, polypropylene, or weatherable PVC.

To date, the fibers extruded have been limited to glass and carbon and the glass fibers are about 1.27 cm long. Quite a number of thermoplastics can be used as the matrix in VHME work. They include polyurethane, polypropylene, ABS, and polycarbonate.

The nylons or polyacetals have not been done because current equipment is not capable of processing at that high a temperature.

Applications

Recent applications for VHME include a commercial refrigeration system and the subfloor of a refrigerated semi-truck trailer. In the commercial refrigeration application, VHME was used to make a corner support to hold glass in a glass assembly. Engineers chose the material because it offered low thermal conductivity and enough strength to retain the glass. In the trailer application, the material replaced a heavy wood that tended to rot. Aside from its resistance to rot, VHME produced other benefits. It improved the thermal performance of the unit, and it also reduced the weight about 1 lb/lineal foot. Thus, on the semi-trailer it saved about 184 kg.

THERMOPLASTIC POLYURETHANES

Thermoplastic polyurethane (TPU) is often the choice for critical tasks because it offers a broad range of high-performance properties. Moreover, it is known for reliability and has a long working life even in harsh end uses. TPUs routinely provide low-temperature flexibility, high abrasion and moisture resistance, and a long storage life.

TPU is a thermoplastic elastomer with many of the same physical and mechanical properties of vulcanized rubber, but with the wider range of processing options common to other thermoplastic polymers. TPUs, like vulcanized rubber, have low hysteresis, high elongation, and good tensile strength. Processors can also tailor their durometer, or hardness, by varying their internal structure without adding special chemicals or plasticizers.

Urethane chemistry is built around four of the most common elements: carbon, hydrogen, nitrogen, and oxygen. And it uses the molecular urethane linkage ($NHCO_2$) to connect a series of block copolymers with alternating hard and soft segments. The ratio and molecular structure

TABLE T.4
Flexible Material Comparison

	Natural Rubber	Synthetic Rubber	PVC	PE/Metallocene	TPU Ether	TPU Ester	Aliphatic
Flex resistance	VG	VG	G to VG	G	E	E	E
Low temperature	F	F to G	F	P	E	G to VG	E
Total strength	G	G to VG	F to G	F to G	VG	E	G
Lamination ability	P	G	G to VG	F	E	E	E
Manufacturability	P	G to VG	VG	F	E	VG	VG
Abrasion resistance	F	F	F to G	F	VG	E	G to VG
Chemical resistance	F to G	G	F to G	E	G to VG	G to VG	G to VG
Soft hand feel	VG	VG	F to G	F	E	VG	E
Recyclability	P	G to VG	VG	G	E	E	E
Biogradability	P	P	P	P	F	VG	F
Clean incineration	F	G to VG	P	G to VG	E	E	E
Migration	G to VG	G to VG	P	G to VG	E	E	E
Ultraviolet resistance	F to G	G	G to VG	G to VG	F	F	VG to E
Moisture resistance	G	G	G to VG	G to VG	G to VG	F to G	G to VG

Source: *Mach. Design*, January 13, pp. 111–112, 2000. With permission.

of these segments determine the specific properties of a TPU grade. The hard segments contain an isocyanate structure while the soft segments consist of different polyols. These polyol segments, either polyether or polyester, are used to distinguish different types of TPUs (Table T.4).

TYPES AND PROPERTIES

Polyether TPU provides a softer "feel" or drape than polyester, and is generally preferred where there is skin contact. Compared with polyesters, it offers better moisture vapor transmission rates (MVTR) and superior low-temperature properties. It is inherently stable when exposed to high humidity, and is naturally more resistant to fungus, mildew, and microbe attack.

However, polyester TPUs have better abrasion resistance with higher tensile and tear strengths for a given durometer. The polyester version also stands up better to fuels and oils, has superior barrier properties, and does not age as fast thanks to better oxidation resistance. However, polyester TPUs will eventually break down in high humidity.

TPUs are further identified by the chemical makeup of their isocyanate, or hard segment components. TPUs are classified into aliphatics (linear or branched chains) or aromatics (containing benzene rings). Aromatic TPUs are strong, general-purpose resins that resist attack by microbes, stand up well to chemicals, and are easily processed. An aesthetic drawback, however, is the tendency of aromatics to discolor, or yellow, when exposed to ultraviolet (UV) light or low-level gamma sterilization. The addition of UV stabilizers or absorbers can reduce this discoloration.

Aliphatic urethanes, on the other hand, are inherently light stable and resist discoloration from UV exposure or gamma sterilization. They are also optically clear, which makes them suitable laminates for encapsulating glass and security glazings.

From an environmental standpoint, use of TPU often makes sense.

Urethane burns more cleanly than polyvinyl chloride (PVC) and other films that compete in the disposable medical-supply market. Using urethane helps avoid the toxic by-products resulting from incinerating disposables fabricated from PVC. In addition, TPU is also readily recycled.

THERMOPLASTIC STYRENES

The styrenics are block copolymers, composed of polystyrene segments in a matrix of polybutadiene or polyisoprene. Lowest in cost of the elastoplastics, they are available in crumb grades and molding grades, and are produced in durometer hardnesses from 35A to 95A.

THERMOPLASTIC URETHANES

Thermoplastic urethanes are of three types: polyester-urethane, polyether-urethane, and caproester-urethane. All three are linear polymeric materials, and therefore do not have the heat resistance and compression set of the cross-linked urethanes. They are produced chiefly in three durometer hardness grades — 55A, 80A, and 90A. The soft 80A grade is used where high flexibility is required, and the hard grade, 70D, is used for low-deflection load-bearing applications.

THERMOPLASTIC VULCANIZATES

Thermoplastic vulcanizates (TPVs), a type of TPE, are more challenging to color because they consist of a blend of polypropylene and very fine EPDM rubber particles. The EPDM particles make it impossible to develop a clear resin. However, recent developments offer whiter and cleaner TPVs that can produce bright colors. This class of thermoplastic elastomers consists of mixtures of two or more polymers that have received a proprietary treatment to give them properties significantly superior to those of simple blends of the same constituents. The two types of commercial elastomeric alloys are melt-processible rubbers (MPRs) and TPVs. MPRs have a single phase; TPVs have two phases.

PROPERTIES

Thermoplastic vulcanizates are essentially a fine dispersion of highly vulcanized rubber in a continuous phase of a polyolefin. Critical to the properties of a TPV are the degree of vulcanization of the rubber and the fineness of its dispersion. The cross-linking and fine dispersion of the rubber phase gives a TPV high tensile strength (7.55 to 26.78 MPa), high elongation (375 to 600%), resistance to compression and tension set, oil resistance, and resistance to flex fatigue. TPVs have excellent resistance to attack by polar fluids and fair-to-good resistance to hydrocarbon fluids. Maximum service temperature is 135°C.

Elastomeric alloys are available in the 55A to 50D hardness range, with ultimate tensile strengths ranging from 5.44 to 27.2 MPa. Specific gravity of MPRs is 1.2 to 1.3; the TPV range is 0.9 to 1.0.

USES

In 1981, a line of TPVs, called Santoprene, was commercialized based on EPDM rubber and polypropylene, designed to compete with thermoset rubbers in the middle performance range. In 1985, a second TPV, Geolast, based on polypropylene and nitrile rubber was introduced. This TPV alloy was designed to provide greater oil resistance than that of the EPDM-based material. The nitrile-based TPV provides a thermoplastic replacement for thermoset nitrile and neoprene because oil resistance of the materials is comparable.

The MPR product line, called Alcryn, was introduced in 1985. It is a single-phase material, which gives it a stress–strain behavior similar to that of conventional thermoset rubbers. MPRs are plasticized alloys of partially cross-linked ethylene interpolymers and chlorinated polyolefins.

THERMOSET PLASTICS

THERMOSET COMPOSITES

Thermoset matrix systems dominate the composites industry because of their reactive nature and ease of impregnation. They begin in a monomeric or oligomeric state, characterized by very low viscosity. This allows ready impregnation of fibers, complex shapes, and a means of achieving cross-linked networks in the cured part. The early high-performance thermoset-matrix materials were called advanced composites, differentiating them from the glass/polyester composites that were emerging commercially in the 1950s. The "advanced" term has come to denote, to most engineers, a resin-matrix material reinforced with high-strength, high-modulus fibers of glass, carbon, aramid, or even boron, and usually laid up in layers to form an engineered component. More specifically, the term has come to apply principally to epoxy-resin-matrix materials reinforced with oriented, continuous fibers of carbon or of a combination of carbon and glass fibers, laid up in multilayer fashion to form extremely rigid, strong structures.

Resin Systems

More than 95% of thermoset composite parts are based on polyester and epoxy resins; of the two, polyester systems predominate in volume by far. Other thermoset resins used in reinforced form are phenolics, silicones, and polyimides.

Polyesters

They can be molded by any process used for thermosetting resins. They can be cured at room temperature and atmospheric pressure, or at temperatures to 177°C and under higher pressure. These resins offer a balance of low cost and ease of handling, along with good mechanical, electrical, and chemical properties, and dimensional stability.

Polyesters can be compounded to be flexible and resilient, or hard and brittle, and to resist chemicals and weather. Halogenated (chlorinated or brominated) compounds are available for increased fire retardance. Low-profile (smoother surface) polyester compounds are made by adding thermoplastic resins to the compound.

Polyesters are also available in ready-to-mold resin/reinforcement forms — bulk-molding compound (BMC), and sheet-molding compound (SMC). BMC is a premixed material containing resin, filler, glass fibers, and various additives. It is supplied in a doughlike, bulk form and as extruded rope.

SMC consists of resin, glass-fiber reinforcement, filler, and additives, processed in a continuous sheet form. Three types of SMC compounds are designated by Owens-Corning Fiberglas Corp. as random (SMC-R), directional (SMC-D), and continuous fiber (SMC-C). SMC-R, the oldest and most versatile form, incorporates short glass fibers (usually about 25.4 mm long) in a random fashion. Complex parts with bosses and ribs are easily molded from SMC-R because it flows readily in a mold. SMC-C contains continuous glass fibers oriented in one direction, and SMC-D, long fibers (203 to 305 mm long), also oriented in one direction.

Moldings using SMC-C and SMC-D have significantly higher unidirectional strength but are limited to relatively simple shapes because the long glass fiber cannot stretch to conform to a shape. These two types of SMC are usually, but not always, used in combination with SMC-R. Various combinations are available that contain a total of as much as 65% glass by weight. These materials are used for structural, load-bearing components.

High-glass-content SMCs are also produced by PPG Industries, designated as XMC. These compounds contain up to 80% glass (or glass/carbon mixtures) as continuous fibers in an X pattern.

Epoxies

These are low-molecular-weight, syruplike liquids that are cured with hardeners to cross-linked thermoset structures that are hard and tough. Because the hardeners or curing agents become part of the finished structure, they are chosen to provide desired properties in the molded part. (This is in contrast to polyester formulations wherein the function of the catalyst is primarily to initiate cure.) Epoxies can also be formulated for room-temperature curing, but heat-curing produces higher properties.

Epoxies have outstanding adhesive properties and are widely used in laminated structures. The cured resins have better resistance than polyesters to solvents and alkalies, but less resistance to acids. Electrical properties, thermal stability (to 288°C in some formulations), and wear resistance are excellent.

Phenolics

The oldest of the thermoset plastics have excellent insulating properties and resistance to moisture. Chemical resistance is good, except to strong acids and alkalies.

Reinforced phenolics are processed principally by high-pressure methods — compression molding and continuous laminating — because volatiles are condensed during the molding process. Recently developed injection-moldable grades, however, have made the processing of phenolics competitive with thermoplastic molding in some applications.

Silicones

These have outstanding thermal stability, even in the range of 260 to 371°C. Water absorption is low, and dielectric properties are excellent. Chemical resistance (except to strong alkalies) is very good.

Properties

Typical thermoset composites are brittle and have poor impact resistance. An impact may cause little visible surface damage but makes the part dramatically weaker.

Thermoset resins need a chemical reaction, usually brought on by heat, to harden, or cure. The chemical reaction is irreversible; once cured, a thermoset material cannot be reprocessed or reformed. And molding cycle time for thermoset materials is largely determined by the curing time.

THIXOMAG

Auto and aerospace parts producers, in particular, are constantly being challenged with finding new ways to provide state-of-the-art, lightweight parts.

Casting is a huge part of this process. When thinking of the casting process, most envision the standard high-pressure die process that forms components from liquid metal. A recently introduced technique called Thixomag transforms metal blanks into an intermediate semisolid before casting.

Aluminum and magnesium offer the attractive combination of high strength and low weight, but magnesium is particularly reactive at the melting point of 650°C in the presence of oxygen; it requires an inert atmosphere for stability. Casting this temperamental metal is difficult; however, this technology offers the possibility to those who want to diversify products, magnesium parts included, without investing much money to provide a fireproof working environment.

Speaking of money, there are cost concerns in switching to a different process. Fireproofing facilities is an expensive process; this is another reason companies are a little apprehensive when it comes to working with magnesium. With the traditional die casting machine, one cannot make parts with thixotropic properties. Because of this product, those who previously could not manufacture magnesium parts, now can. It is possible to reuse the cold-chamber casting machine and buy only the Thixomag module along with the induction heating power supply.

The Process

To work with magnesium, a low casting temperature must be achieved. Thixomag can cast at 550°C. Instead of casting from a liquid state, the material is put into a thixotropic or semisolid state. Because the material does not need to be melted, lower temperatures suffice. This semisolid also allows the finished product to be free of porosity defects. As the material flows, it is thick enough to resist gas pockets.

The thixotropic state is achieved by proper stirring during solidification. After the stirring process, the thixotropic material is then reheated by induction to the temperature interval between a solid and a liquid. Upon heating the rheocast ingot to the two-phase region (semisolid, semiliquid), the material flows under the application of shear.

Induction is the only type of heating process possible to obtain the required thixotropic state, viscous enough so the product does not flow before injection, but not so solid that injection would be impossible. The present developments will allow work rates of several parts per minute, depending on the material used.

THIXOTROPIC PROCESSING

Thixoforming

A method of producing aluminum castings with performance characteristics said to meet or exceed those of steel forgings with 65% less weight, and up to 25% lower cost, has been developed.

During the process, marketed under the name Thixoforming, aluminum ingots are heated to the precise temperature at which the material begins to be transformed from the solid to the liquid state. In this form, it can be cast to make strong, lightweight parts that are cheaper and perform better than many forged steel parts.

The critical technology enabling this process is the temperature control system, which is capable of maintaining a uniform temperature throughout the ingot to within ±1°C. It is critical to the process that the entire ingot be precisely in the thixotropic state (borderline liquid/solid). If the center is harder than the exterior, the part will have less beneficial metallurgical properties. The process has reportedly had very good success in the automotive industries.

Thixotropic Casting

Ranging from setter tiles and saggers for use in the electronic ceramics industry, to furnace forehearth shapes for the container glass industry, to kiln-car furniture and structural ceramics, precast shapes offer significant benefits.

A new generation of precast shapes has been developed using a thixotropic casting process. This process provides a smoother surface, more precise tolerances, greater strength, improved thermal shock resistance, higher heat tolerances, and more resistance to chemical attack than traditional refractory castable shapes.

The Process

Thixotropic casting uses a dispersion agent that allows the ceramic mix to flow when vibrated without requiring high water content. The mix, vibrated into a plaster of paris mold, deairs and consolidates without the use of a cement bonding agent. The finely crafted plaster of paris mold provides greater dimensional uniformity than common wooden or metal molds and results in a much smoother surface finish.

The crafting process for plaster molds generally involves four separate steps. The first step is to make an exact model of the end product. A master mold, which is a negative of the model, is then made. Then a case mold is made. The case mold is a replica of the original model and is made to protect the original from damage. From the case mold comes the durable working mold, which is a copy of the master mold and a negative of the case mold.

The four-step process is in place to ensure consistency from mold to mold. Depending on the various manufacturing processes, each working mold is used to produce roughly 10 to 50 pieces. The working mold is then replaced with another, again to ensure manufacturing uniformity.

All shapes produced by the thixotropic casting process are high-temperature fired between 1412 and 1524°C. This develops a ceramic bond that enables the products to withstand greater temperatures for a longer period of time.

The fact that these shapes are made without a cement bond — they use a high-fired ceramic bond, instead — accounts for their high heat tolerance. Above 1371°C, cement-bonded refractories begin to soften, the result of glass phase formation. But firing the shapes to up to 1524°C results in a completely formed ceramic bond that will not readily soften in temperatures above 1371°C.

Firing at high temperatures also has the advantage of adding stability to precast shapes. When you first fire a castable shape with a cement bond, a lot of mineralogical changes take place as the product heats up. It can change the overall size and shape of the product, which can lead to an inconsistent product for the customer. By firing the shapes to 1412°C and beyond, there will be no unexpected or unwanted mineralogical changes when first used. The changes that occur in firing are taken into consideration when the molds are designed. The result is that the product performs as it was designed.

The firing process for precast fired shapes is a highly controlled operation. All shapes are fired in a 7-day cycle that includes a precise drying process. The periodic kilns that are used to dry and fire the shapes are temperature-controlled to within 20°, which is close tolerance for this type of manufacturing. Even the kiln cars are loaded to facilitate a specific airflow pattern that enhances strength and consistency.

Applications

Typical applications for precast kiln furniture shapes include saggers, setter plates, and pusher tiles.

The most common use is with technical ceramic products that have to be fired at higher-temperature areas — above 1301°C.

Other applications include kiln furniture and structural ceramics. The process can handle very small pieces to the very large pieces like kiln furniture, kiln car shapes, posts, beams — even tracks.

Other applications for precast shapes include glass contact parts in the container glass industry — stirrers, cover blocks, spouts, plungers, and tubes. These glass tank forehearth shapes enable glass manufacturers to maintain consistency in manufacturing their products.

Shapes are also available with a variety of chemical compositions, depending on the requirements of the application. Compositions range from 50% alumina products to shapes with ultrahigh alumina and alumina/zirconia/silica content.

THORIUM

A soft, ductile, silvery-white metal (symbol Th), thorium occurs in nature to about the same extent as lead but so widely disseminated in minute quantities difficult to extract that it is considered a rare metal.

It was once valued for use in incandescent gas mantles in the form thorium nitrate, $Th(NO_3)_4$, but is now used chiefly for nuclear electronic applications.

For many years thorium oxide has been incorporated in tungsten metal, which is used for electric light filaments; small amounts of the oxide have also been found to be useful in other metals and alloys. The oxide is employed in catalysts for the promotion of certain organic chemical reactions. Thorium oxide has special uses as a high-temperature ceramic material.

Thoria-urania ceramics are used for reactor-fuel elements. They are reinforced with columbium or zirconium fibers to increase thermal conductivity and shock resistance. The metal or its oxide is employed in some electronic tubes, photocells, and special welding electrodes. The metal can serve as a getter in vacuum systems and in gas purification, and it is also used as a scavenger in some metals,

Because of its high density, chemical reactivity, mediocre mechanical properties, and relatively high cost, thorium metal has no market value as a structural material. However, many alloys containing thorium metal have been studied in some detail and thorium does have important applications as an alloying agent in some structural metals. Perhaps the principal use for thorium metal, beyond its use in the nuclear field, is in magnesium technology. Approximately 3% thorium, added as an alloying ingredient, imparts to magnesium metal high-strength properties and creep resistance at elevated temperatures. The magnesium alloys containing thorium, because of their light weight and desirable strength properties, are being used in aircraft engines and in airframe construction.

Thorium can be converted in a nuclear reactor to uranium-233, an atomic fuel. The system of thorium and uranium-233 gives promise of complete utilization of all thorium in the production of atomic power. The energy available from the world supply of thorium has been estimated as greater than the energy available from all of the world's uranium, coal, and oil combined.

THYRISTORS

Thyristors (semiconductor controlled rectifiers) made from silicon carbide have been fabricated and tested as prototypes of power-switching devices capable of operating at temperatures up to 350°C. The highest-voltage-rated of these thyristors are capable of blocking current at forward or reverse bias as large as 900 V, and can sustain forward current as large as 2 Å with a forward potential drop of −3.9 V. The highest-power-rated of these thyristors (which are also the highest-power-rated SiC thyristors reported thus far) can block current at a forward or reverse bias of 700 V and can sustain an "on" current of 6 Å at a forward potential drop of −3.67 V. The highest-current-rated of these thyristors can block

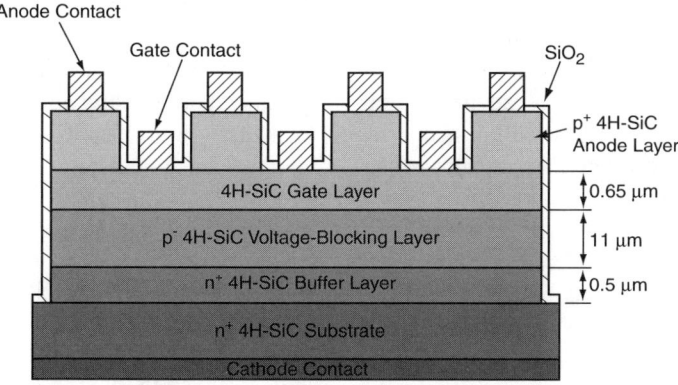

FIGURE T.5 This cross section (not to scale) shows the *npnp*-layer structure of a representative thyristor of the present type. The n^+-, p^+-, n-, and p^+-doped 4H-SiC layers are formed by epitaxy on the n^+ 4H-SiC substrate, which is cut at an angle of 8° off axis. *Note:* Layer thicknesses are not to scale.

current at a forward or reverse bias of 400 V and can sustain an "on" current of 10 Å.

These thyristors feature epitaxial *n*- and *p*-doped layers of 4H-SiC in the sequence *npnp* starting on the substrate; this structure (Figure T.5) stands in contrast to the *pnpn* structure of common silicon thyristors. The fabrication of the high-quality crystalline structures needed in these layers has been made possible by advances in growth of crystals, epitaxial growth of thin films, doping by both *in situ* and ion-implantation techniques, oxidation, formation of electrical contacts, and other techniques involved in the fabrication of electronic devices. The above *npnp* 4H-SiC thyristors have been found to exceed the speed of the fastest inverter-grade silicon thyristors.

Two other important parameters for a thyristor are (1) the maximum rate of increase of forward applied voltage that can be applied before the thyristor latches on and (2) the time taken to achieve a high forward current density. The 4H-SiC thyristors show no turn-on even when forward bias was ramped up at a rate of 900 V/μs. Measurements in pulsed operation showed that it took between 3 and 5 ns for these devices to start carrying currents at densities of 2800 A/cm².

TILE

As a structural material, tile is a burned clay product in which the coring exceeds 25% of the gross volume; as a facing material, any thin, usually flat, square product. Structural tile used for load bearing may or may not be glazed; it may be cored horizontally or vertically. Two principal grades are manufactured: one for exposed masonry construction, and the other for unexposed construction. Among the forms of exposure is frost; tile for unexposed construction where temperatures drop below freezing is placed within the vapor barrier or otherwise projected by a facing, in contrast to roof tile.

Structural tile with a ceramic glaze is used for facing. The same clay material that is molded and fired into structural tile is also made into pipe, glazed for sewer lines, or unglazed for drain tile.

As a facing, clay products are formed into thin flat, curved, or embossed pieces, which are then glazed and burned. Commonly used on surfaces that are subject to water splash or that require frequent cleaning, such vitreous glazed wall tile is fireproof. Unglazed tile is laid as bathroom floor. By extension, any material formed into a size comparable to clay tile is called tile. Among the materials formed into tile are asphalt, cork, linoleum, vinyl, and porcelain.

TIN AND ALLOYS

A silvery-white lustrous metal (symbol Sn), with a bluish tinge, tin is soft and malleable, and can be rolled into foil as thin as 0.0051 cm. Tin melts at 232°C. Its specific gravity is 7.298, close to that of steel. Its tensile strength is

27 MPa. Its hardness is slightly greater than that of lead, and its electric conductivity is about one seventh that of silver. It is resistant to atmospheric corrosion, but is dissolved in mineral acids. The cast metal has a crystalline structure, and the surface shows dendritic crystals when cast in a steel mold.

It alloys readily with nearly all metals.

Tin is a nontoxic, soft, and pliable metal adaptable to cold working such as rolling, extrusion, and spinning. It is highly fluid when molten and has a high boiling point, which facilitates its use as a coating for other metals. It can be electrodeposited readily on all common metals.

Properties

Tin reacts with strong acids and strong bases but is relatively inert to nearly neutral solutions. In indoor and outdoor exposure it retains its white silvery color because of its resistance to corrosion. A thin film of stannic oxide is formed in air which provides surface protection.

Two allotropic forms exist: white tin (β) and gray tin (α). Although the transformation temperature is 13.2°C, the change does not take place unless the metal is of high purity, and only when the exposure temperature is well below 0°C. Commercial grades of tin (99.8%) resist transformation because of the inhibiting effect of the small amounts of bismuth, antimony, lead, and silver present as impurities.

Alloying elements such as copper, antimony, bismuth, cadmium, and silver increase its hardness. Tin tends rather easily to form hard, brittle intermetallic phases, which are often undesirable. It does not form wide solid solution ranges in other metals in general, and there are few elements that have appreciable solid solubility in tin. Simple eutectic systems, however, occur with bismuth, gallium, lead, thallium, and zinc.

Tin is rarely used alone; rather, it is generally used as a coating for a baser metal or as a constituent of an alloy. The range of useful alloys is extensive and extremely important.

Forms

Tin can be obtained in a number of forms: granulated, mossy, fine powder, sheet, foil, and wire. Tin-base alloys are available in many forms: solder can be obtained in 0.46-kg bars, solid and cored wire, powder, sheet, and foil; babbitt type metal and casting alloys in bars and ingots; bronze in ingot; continuously cast bars and shapes, sheet, and foil.

Applications

Full use is made of the ductility, surface smoothness, corrosion resistance, and hygienic qualities of tin in the form of foil, pipe, wire, and collapsible tubes. Tin foil is devoid of springiness and is ideal for wrapping food products, as liners for bottle caps, and for electrical condensers. Heavy-walled tin pipe and tin-lined copper pipe are used by the food and beverage industries for conveying distilled water, beer, and soft drink syrups. Tin wire is used for electrical fuses and for packing glands in pumps of food machinery. Collapsible tubes, made by impact extrusion from disks of pure tin, are used for pharmaceutical and food products. Additionally, tin is used in brasses, bronzes, and babbitts, and in soft solders.

The most important use of tin is for tin-coated steel containers (tin cans) used for preserving foods and beverages. Other important uses are solder alloys, bearing metals, bronzes, pewter, and miscellaneous industrial alloys. Tin chemicals, both inorganic and organic, find extensive use in the electroplating, ceramic, plastic, and agricultural industries.

Tinplate manufacture is now largely a continuous electrolytic process with only a small percentage of production in hot-tinning machines. The coating thickness may be less than 0.01 mm. Heavily coated tinned steel sheet is used in making gas meters and automotive parts such as filters and air coolers.

The electrical industry is a large user of tin-coated steel and copper in the form of connectors, capacitor and condenser cans, and tinned copper wire. Many kinds of food-handling machinery, including holding tanks, mixers, separators, milk cans, pipes, and valves, are made of tinned steel, cast iron, copper, or brass. The tin coating may be applied by dipping in molten tin or by electrodeposition.

The plating industry utilizes tin as anodes for the electrodeposition of pure tin and tin alloy coatings. Plated tin, either as a matt or bright finish, provides easily solderable surfaces for steel, copper, or aluminum. Tin alloy coatings (tin–copper, tin–lead, tin–nickel, tin–zinc, tin–cadmium, tin–cobalt) have advantages over single metal plates. They are denser and harder, more corrosion-resistant, brighter or more easily buffed, and more protective to basis metals. Tin–copper (12% tin) has the appearance of 24-karat gold and, when lacquered, serves as an attractive finish for jewelry, trophies, wire goods, and hardwares. Tin–lead electroplates (40 to 65% tin) have excellent corrosion resistance and solderability, and are well adapted to the plating of printing circuits and electronic parts. Tin–zinc coatings (75% tin) are a good alternative coating for cadmium in particular applications, and they provide galvanic protection to steel in contact with aluminum. Tin–cadmium coatings (25% tin) are especially resistant to salt vapors, and have a number of applications in the aircraft industry and as a coating for fasteners. A tin–nickel coating (65% tin) finds use as an etchant resist in the manufacture of printed circuit boards, as well as an ornamental and highly corrosion-resistant finish for watch parts, scientific instruments, and power connectors. The tin–cobalt alloy (80% tin) has an appearance similar to a chromium deposit, and is used to plate fasteners, ancillary office equipment, hinges, kitchen utensils, hand tools, and tubular furniture.

STANNIC OXIDE (SnO_2)

SnO_2 has by far its largest commercial outlet in the ceramic industry, where it is used either as a white pigment (i.e., opacifier) or as a constituent of colored pigments in the glazes applied to, for example, crockery, lavatoryware, and decorative wall tile.

Tin oxide is an important constituent of ceramic stains for enamels, glazes, and bodies. Pink and maroon colors are obtained with tin oxide, chrotin oxide, and vanadium compounds. Tin oxide also is an important color stabilizer for some of the tin-bearing pink, gray, yellow, and blue coloring stains for glazes.

This oxide was formerly an important opacifier for enamels on cast iron and sheet steel, although it has been replaced by substitute materials such as antimony, zirconium, titanium, and other compounds. The substitution has been largely for economic reasons, as tin oxide is still recognized as the superior opacifier from the standpoint of quality for both glazes and enamels.

In glass, stannic oxide is an important addition to cadmium–selenium and gold colors, especially reds. Stannous oxide, SnO, is a necessary ingredient in the development of copper ruby glass and is also used to produce black glasses. Stannous oxide is a black powder and is also a component of ruby-red and black glasses. Tin oxide, because of its resistance to solution in most glasses, especially those high in lead oxide, is being used for refractories for special applications, such as glass feeders, and conducting electrodes for electrical resistance melting of glass.

ALLOYS

Soft Solders

Soft solders constitute one of the most widely used and indispensable series of tin-containing alloys. Common solder is an alloy of tin and lead, usually containing 20 to 70% tin. It is made easily by melting the two metals together. With 63% tin, a eutectic alloy melting sharply at 169°C is formed. This is much used in the electrical industry. A more general-purpose solder, containing equal parts of tin and lead, has a melting range of 31°C. With less tin, the melting range is increased further, and wiping joints such as plumbers make can be produced. Lead-free solders for special uses include tin containing up to 5% of either silver or antimony for use at temperatures somewhat higher than those for tin–lead solders and tin–zinc base solders often used in soldering aluminum.

Bronzes

Bronzes are among the most ancient of alloys and still form an important group of structural metals. Of the true copper–tin bronzes, up to 10% tin is used in wrought phosphor bronzes, and from 5 to 10% tin in the most common cast bronzes. Many brasses, which are basically copper–zinc alloys, contain 0.75 to 1.0% tin for additional corrosion resistance in such wrought alloys as Admiralty Metal and Naval brass, and up to 49% tin in cast leaded brasses. Among special cast bronzes are bell metal, historically 20 to 24% tin for best tonal quality, and speculum, a white bronze containing 33% tin that gained fame for high reflectivity before glass mirrors were invented. Although soft, conformable, and corrosion resistant, the low mechanical strength of bronze must be boosted by bonding to steel, cast iron, or bronze backing materials.

Pewters

Pewter is an easily formed tin-base alloy that originally contained considerable lead. Thus, because Colonial pewter darkened and because of potential toxicity effects, its use was discouraged. Modern pewter is lead-free. The most favorable composition, Britannia Metal, contains about 7% antimony and 2% copper. This has desired hardness and luster retention, yet it can be readily cast, spun, and hammered.

Alloys that contain from 1 to 8% antimony and 0.5 to 3% copper have excellent castability and workability. For spun pewter products, antimony content is usually below 7%, and pewter casting alloys contain 7.5% antimony and 0.5% copper. Because of the excellent drawing and spinning properties of tin, wrought parts are usually made from pewter that is first cast into slabs, then rolled into sheet.

Babbitt/Bearing Metal

Babbitt or bearing metal for forming or lining a sleeve bearing is one of the most useful tin alloys. It is tin containing 4 to 8% each of copper and antimony to give compressive strength and a structure desired for good bearing properties. An advantage of this alloy is the ease with which castings can be made or bearing shells relined with simple equipment and under emergency conditions. Aluminum–tin alloys are used in bearing applications that require higher loads than can be handled with conventional babbitt alloys.

Type Metals

Type metals are lead-base alloys containing 3 to 15% tin and a somewhat larger proportion of antimony. As with most tin-bearing alloys, these are used and remelted repeatedly with little loss of constituents. Tin adds fluidity, reduces brittleness, and gives a structure that reproduces fine detail.

Flake and nodular gray iron castings are improved by adding 0.1% tin to give a fully pearlitic matrix with attendant higher hardness, heat stability, and improved strength and machinability.

Tin is commonly an ingredient in costume jewelry, consisting of pewterlike alloys and bearing metal compositions often cast in rubber molds, in die castings hardened with antimony and copper for applications requiring close tolerances, thin walls, and bearing or nontoxic properties; and in low-melting alloys for safety appliances. The most common dental amalgam for filling teeth contains 12% tin.

Die-Cast Tin Base

Historically, these were the first materials to be die cast. Low melting point and extreme fluidity of these alloys produce sound, intricate castings inexpensively and with little wear on molds. Antimony, copper, and lead are the principal additions to tin in die-casting alloys. These alloys are mainly gravity or centrifugally die cast. Cast tin-alloy parts can be held to tolerances of 0.015 mm/mm, with wall thicknesses down to 0.79 mm. Shrinkage is negligible.

Fusible Tin

Melting temperatures for these alloys are usually below the solidus of eutectic-base tin–lead solders (184°C). Primary alloying elements include bismuth, lead, cadmium, and indium. Most of these alloys provide electrical or mechanical links in safety devices. Other applications include low-temperature solders, seals for glass and other heat-sensitive materials, foundry patterns, molds for low-volume production of plastic parts, internal support for tube bending, and localized thermal treatment of parts.

Tin Powders

Produced by atomization techniques, these powders are available in a number of mesh sizes.

They are used in the manufacture of powder-metallurgy parts, in tinning and solder pastes, and in the spray metallization of surfaces.

In small amounts, tin is also combined with titanium, zirconium, and other metals to provide special properties. It is used as an alloy in nodular and gray irons to provide greater strength, increased and uniform hardness, and improved machinability. Tin–nickel and tin–zinc coatings are used in the braking systems of automobiles. The tin–nickel alloy is coated on disk-brake pistons because of its good resistance to wear and corrosion. Tin–zinc is used to plate master cylinders in automotive braking systems.

Noncritical parts such as costume jewelry and small decorative items such as figurines can be made by casting pewter and other low-melting tin-base alloys in rubber molds. Tin die-casting alloys are suitable for low-strength precision parts and bearings for household appliances, engines, motors and generators, and gas turbines. These bearings perform well even at start-up and run-down periods of operation, at which times they carry a heavy, unidirectional load without the benefit of a fully formed hydrodynamic film. Other applications for tin-base die castings include parts for food-handling equipment, instruments, gas meters, and speedometers.

TINPLATE

Tinplate is soft-steel plate containing a thin coating of pure tin on both sides. A large proportion of the tinplate used goes into the manufacture of food containers because of its resistance to the action of vegetable acids and its nonpoisonous character. It solders easily, and also is easier to work in dies than terneplate, so that it also is preferred over terneplate for making toys and other cheap articles in spite of a higher cost. Tinplate is made by the hot-dip process using palm oil as a flux, or by a continuous electroplating process.

Processing and Uses

Tinplate manufacture in the United States is largely a continuous, high-speed electrolytic process, with less than 1% of production from hot-tinning machines. Electrolytic tinplate can be produced from either alkaline or acid electrolytes. Tinplates have either tin on each steel surface or a differential tin-coating thickness.

Hot-dipped tinplate is used for special corrosive packs, kitchen utensils, hardware items, and automotive parts. For other industrial applications, hot-dip tin coatings are applied to copper wire and sheet, as well as steel and cast iron parts. Examples are tinned copper and copper alloy strip for manufacture of electrical connectors and tinned food processing equipment. Hot-dip tin–lead (terne) coatings find service as coatings for gasoline tanks, roofing materials, electronic applications, radiator water tubes, and component leads.

TITANATES

Titanates are compounds made by heating a mixture of an oxide or carbonate of a metal and titanium dioxide. High dielectric constants, high refractive indices, and ferroelectric properties contribute primarily to their commercial importance.

Ferroelectricity may be described as the electric analogue of ferromagnetism. As a field is applied to a ferroelectric material, a nonlinear relationship between polarization and field (similar to the magnetization curve for iron) is observed. This increase in polarization is a function of the orientation of ferroelectric domains within the crystal.

As these domains become aligned, a saturation point is reached. If the field is now removed, the domains tend to remain aligned and a finite value of polarization (called remanent polarization) can be measured. Extrapolation of the polarization at high field strength to zero field gives a somewhat higher value (spontaneous polarization).

To eliminate the remanent polarization, the field must be applied in the opposite direction, and the field required to return to the original state is called the coercive field. On further increase in the electric field, polarization in the opposite direction is achieved. This behavior leads to a characteristic ferroelectric hysteresis loop as the field is alternated.

Type of Materials

Titanium Dioxide

Titanates, for the most part, are prepared by heating a mixture of the specific oxide or carbonate with titanium dioxide. Titanium dioxide has an exceptionally high refractive index for a white oxide (2.6 to 2.9 for the rutile form, 2.5 for anatase) and due to its high refractive index finds wide application as a white pigment of high reflectance for the opacification of paint, plastics, rubber, paper, and porcelain enamels.

For electronics, polycrystalline (ceramic) titanium dioxide with its moderately high dielectric constant (~95) has been used as a capacitor; it does not show ferroelectric behavior. Titanium dioxide is normally an insulator, but in the oxygen deficient state where some of the Ti^{4+} sites are occupied by Ti^{3+} ions, it becomes an "n-type" semiconductor with conductivities in the range 1 to 10/Ω-cm.

Barium Titanate

Barium titanate crystals, $BaTiO_3$, are made by die-pressing titanium dioxide and barium carbonate a sintering at high temperature. This crystal belongs to the class perovskite in which the closely packed lattice of barium and oxygen ions has a barium ion in each corner and an oxygen ion in the center of each face of a cube with the titanium ion in the center of the oxygen octahedron. For piezoelectric use the crystals are subjected to a high current, and they give a quick response to changes in pressure or electric current. They also store electric charges, and are used for capacitors. Glennite 103 is a piezoelectric ceramic molded from barium titanate modified with temperature stabilizers.

Ceramic barium titanate can be made piezoelectric by applying a polarizing field of about 30 kV/cm at room temperatures. The remanent polarization after removal of the polarizing voltage is permanent unless the material is overheated or subjected to high reverse voltages.

The advantages of ceramic barium titanate as a transducer lie in its mechanical strength, chemical durability, and ease of fabrication into virtually any shape desired. Barium titanate transducers, as ultrasonic generators, are used in various applications (emulsification, mixing, cleaning, drilling); other applications include such things as phonograph pickups and accelerometers.

Calcium Titanate

Calcium titanate, $CaTiO_3$, occurs in nature as the mineral perovskite. As a ceramic it has a room-temperature dielectric constant of about 160. It is frequently used as an addition to barium titanate or by itself as a temperature compensating capacitor.

Single crystals have been grown by the flame fusion technique; calcium titanate crystals show a strong tendency toward twinning and, although the material is not ferroelectric, the twinning in the crystal shows a marked resemblance to the domain structure observed in barium titanate crystals.

Strontium Titanate

Strontium titanate, $SrTiO_3$, has a cubic perovskite structure at room temperature. It has a dielectric constant of about 230 as a ceramic, and is commonly used as an additive to barium titanate to decrease the Curie temperature. By itself, it is used as temperature-compensating material because of its negative temperature characteristics. Strontium titanate has been used as a brilliant diamond-like gemstone and is a strontium mesotrititanate. Stones are made up to 4 karat. The refractive index is 2.412. It has a cubic crystal similar to the diamond but the crystal is opaque in the x-ray spectrum. Single crystals of strontium titanate have been grown by the flame fusion process and strontium titanate is essentially colorless.

Because of its cubic structure, it is somewhat more satisfactory as an optical material than rutile (tetragonal). Strontium titanate crystals show a room-temperature dielectric constant of about 300 with a loss tangent of 0.0003.

Magnesium Titanate

Magnesium titanate, $MgTiO_3$, crystallizes as an ilmenite rather than perovskite structure. It is not ferroelectric, and is used with titanium dioxide to form temperature-compensating capacitors. It has also been used as an addition agent to barium titanate.

Lead Titanate

Lead titanate, $PbTiO_3$, is used as a less costly substitute for titanium oxide. It is yellowish in color and has only 60% of the hiding power, but is very durable and protects steel from rust. Good ceramic specimens of lead titanate are somewhat difficult to prepare owing to the volatility of lead oxide at the firing temperature. Lead titanate is commonly used as a solid solution additive to increase the Curie temperature of barium titanate.

By substitution of zirconium for titanium in lead titanate a solid solution, lead zirconium titanate, may be produced. Lead zirconate–lead titanate is a piezoelectric ceramic that can be used at higher temperatures than barium titanate.

Miscellaneous Titanates

The metatitanates of cadmium, manganese, iron, nickel, and cobalt all have an ilmenite rather than perovskite structure. None of these, as far as is known, is ferroelectric and they are not particularly important electrically other than perhaps as addition agents to barium titanate. Nickel titanate has been grown as a single crystal, and its use as a rectifier has been suggested after the addition of suitable impurities during growth.

Bismuth stannate, $Bi_2(SnO_3) \cdot 5H_2O$, a crystalline powder that dehydrates at about 140°C, may be used with barium titanate in capacitors to increase stability at high temperatures.

Butyl titanate is a yellow viscous liquid used in anticorrosion varnishes and for flameproofing fabrics. It is a condensation product of the tetrabutyl ester of ortho-titanic acid, and contains about 36% titanium dioxide.

Titanate fibers can be used as reinforcement in thermoplastic moldings. The fibers, called Fybex, can also be used in plated plastics to reduce thermal expansion, warpage, and shrinkage. Titanate fibers in plastics also provide opacity.

TITANIUM AND ALLOYS

A metallic element (symbol Ti), titanium occurs in a great variety of minerals. The chief commercial ores of titanium are rutile and ilmenite. In rutile it occurs as an oxide. It is an abundant element but is difficult to reduce from the oxide. High-purity titanium (99.9%) has a melting point of about 1668°C, a density of 4.507 g/cm^3, and tensile properties at room temperature of about 234 MPa ultimate strength, 138 MPa yield strength, and 54% elongation. It is paramagnetic and has low electrical conductivity and thermal expansion.

The commercial metal is produced from sponge titanium, which is made by converting the oxide to titanium tetrachloride followed by reduction with molten magnesium. The metal can also be produced in dendritic crystals of 99.6% purity by electrolytic deposition from titanium carbide. Despite its high melting point, titanium reacts readily in copper and in other metals, and is much used for alloying and for deoxidizing. It is a more powerful deoxidizer of steel than silicon or manganese.

Melting

Because titanium has a great affinity for oxygen, nitrogen, and hydrogen at elevated temperatures, particularly when molten, all melting operations must be conducted in a vacuum and/or an inert-gas atmosphere. The last 2 years has seen a dramatic shift in titanium melting capacity. At least 181,818,181 kg of cold hearth melting capacity has been added to an industry dominated historically by vacuum arc remelting (VAR). Cold hearth melting is a more complex melting process that brings advantages, such as scrap recycling, shape casting (rectangular slabs), and unique alloying capability. Titanium presents special problems during melting because of its high reactivity with oxygen, nitrogen, and carbon. Exposure to these elements during melting causes severe embrittlement, even at low concentrations. This means that all titanium melting must occur in a vacuum or inert atmosphere (argon or helium). It also means that ceramic or graphite containers or liners are not permissible, limiting the design of melting containers to water-cooled copper vessels (Figure T.6a).

Vacuum arc remelting (VAR) in water-cooled copper crucibles quickly became the standard melting method (Figure T.6b). When discussing titanium cold hearth melting, one

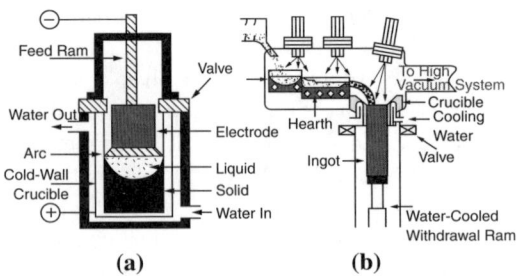

FIGURE T.6 (a) Schematic of cold hearth furnace. (b) Vacuum arc remelting furnace. (From *Ind. Heating*, January, p. 49, 2000. With permission.)

must divide the subject into two major divisions: (1) the production of commercially pure (CP) titanium and (2) the production of titanium alloys. Cold hearth production of CP titanium is done principally using electron beam (EB) melting and occurs at very high melting rates, in some cases greater than 2300 kg/h.

Cold hearth melting of CP titanium provides several advantages. First is the use of economical raw materials, such as machined turnings contaminated with tungsten carbide particles. The cold hearth allows the tungsten carbide particles to be removed efficiently by gravity separation. The second advantage, as mentioned earlier, is high production rates. A third advantage is the ability to produce rectangular slabs instead of round ingots.

Although titanium alloys are successfully produced in both plasma and EB cold hearth processes, advantages for cold hearth melting titanium alloys are similarly impressive. As with CP, cold hearth melting of titanium alloys allows the use of many low-cost raw material forms with excellent yield. If EB melting holds an advantage for producing CP, plasma is simpler to use for titanium alloy melting. Because no alloy changes occur during plasma melting, chemistry control is relatively easy. Electron beam melting of alloys containing high vapor pressure components is more problematic. For a common alloy such as Ti–6% Al–4%V, as much as 30% of the aluminum content may evaporate during EB melting. Alloy control in this case requires accurate prediction and compensation for elemental losses, and then very precise process control to meet the predicted losses uniformly throughout the ingot.

A very important technical justification for the use of cold hearth melting for titanium is in the production of premium quality titanium alloys for aircraft turbine engine rotating parts. Inclusion defects contained in the titanium used to produce these parts can be detrimental, leading to low cycle fatigue cracking, engine failure, and even aircraft loss. In the past 10 years, the aircraft engine producers have increasingly specified the use of cold hearth melting for the most critical titanium parts.

Cold hearth melting provides important mechanisms for the removal of inclusion-forming contaminants from titanium raw materials. High-density inclusion sources, such as tungsten carbide, tungsten, tantalum, and molybdenum, are easily removed by gravity separation (settling) in the cold hearth. By contrast, VAR provides virtually no removal of these defects.

The titanium melting industry has undergone a revolution in melting technology with plasma and electron beam cold hearth melting replacing traditional VAR melting. All major U.S. titanium producers now partly or wholly own large-scale cold hearth melting facilities.

Cold hearth melting provides both economic and technical advantages to the titanium melter. Inexpensive raw materials not usable in other melt methods are readily utilized with excellent yields at high production rates. Potentially damaging inclusion sources may be removed assuring high-quality titanium products for the most demanding and critical applications.

STRUCTURE AND PROPERTIES

Titanium is one of the few allotropic metals (steel is another); that is, it can exist in two different crystallographic forms. At room temperature, it has a close-packed hexagonal structure, designated as the alpha phase. At around 884°C, the alpha phase transforms to a body-centered cubic structure, known as the beta phase, which is stable up to the melting point of titanium of about 1677°C. Alloying elements promote formation of one or the other of the two phases. Aluminum, for example, stabilizes the alpha phase; that is, it raises the alpha to the beta transformation temperature. Other alpha stabilizers are carbon, oxygen, and nitrogen. Beta stabilizers, such as copper, chromium, iron, molybdenum, and vanadium, lower the transformation temperature, therefore allowing the beta phase to remain stable at lower temperatures, and even at room temperature. The mechanical properties of titanium are closely related to these allotropic phases. For example, the beta phase is much stronger, but more brittle than the alpha phase. Titanium alloys therefore can be usefully classified into three groups on the basis of allotropic phases: alpha, beta, and alpha-beta alloys.

Titanium and its alloys have attractive engineering properties. They are about 40% lighter than steel and 60% heavier than aluminum. The combination of moderate weight and high strengths, up to 1378 MPa, gives titanium alloys the highest strength-to-weight ratios of any structural metal. Furthermore, this exceptional strength-to-weight ratio is maintained from −216°C up to 538°C. A second outstanding property of titanium materials is corrosion resistance. The presence of a thin, tough oxide surface film provides excellent resistance to atmospheric and sea environments as well as a wide range of chemicals, including chlorine and organics containing chlorides. As it is near the cathodic end of the galvanic series, titanium performs the function of a noble metal. Titanium and its alloys, however, can react pyrophorically in certain media. Explosive reactions can occur with fuming nitric acid containing less than 2% water or more than 6% nitrogen dioxide and, on impact, with liquid oxygen. Pyrophoric reactions also can occur in anhydrous liquid or gaseous chlorine, liquid bromine, hot gaseous fluorine, and oxygen-enriched atmospheres.

FABRICATION

Fabrication is relatively difficult because of the susceptibility of titanium to hydrogen, oxygen, and nitrogen impurities, which cause embrittlement. Therefore, elevated-temperature processing, including welding, must be performed under special conditions that avoid diffusion of gases into the metal. Heat is usually required in postforming operations.

Generally, titanium is welded by gas-tungsten arc or plasma-arc techniques. Metal inert-gas processes can be used under special conditions. Thorough cleaning and shielding are essential because molten titanium reacts with nitrogen, oxygen, and hydrogen, and will dissolve large quantities of these gases, which embrittles the metal. In all other respects, gas-tungsten arc welding of titanium is similar to that of stainless steel. Normally, a sound weld appears bright silver with no discoloration on the surface or along the heat-affected zone.

Like stainless steel, titanium sheet and plate work harden significantly during forming. Minimum bend radius rules are nearly the same for both, although springback is greater for titanium. Commercially pure grades of heavy plate are cold-formed or, for more severe shapes, warm-formed at temperatures to about 427°C. Alloy grades can be formed at temperatures as high as 760°C in inert-gas atmospheres. Tube can be cold bent to radii three times the tube outside diameter, provided that both inside and outside surfaces of the bend are in tension at the point of bending. In some cases, tighter bends can be made.

Despite their high strength, some alloys of titanium have superplastic characteristics in the range of 816 to 927°C. The alloy used for most superplastically formed parts is the standard Ti-6Al-4V alloy. Several aircraft manufacturers are producing components formed by this method. Some applications involve assembly by diffusion bonding.

Titanium plates or sheets can be sheared, punched, or perforated on standard equipment. Titanium and Ti–Pd alloy plates can be sheared subject to equipment limitations similar to those for stainless steel. The harder alloys are more difficult to shear, so thickness limitations are generally about two-thirds those for stainless steel.

Titanium and its alloys can be machined and abrasive-ground; however, sharp tools and continuous feed are required to prevent work hardening. Tapping is difficult because of the metal galls. Coarse threads should be used where possible.

FORMS

Commercially pure titanium and many of the titanium alloys are now available in most common wrought mill forms, such as plate, sheet, tubing, wire, extrusions, and forgings. Castings can also be produced in titanium and some of the alloys; investment casting and graphite-mold (rammed graphite) casting are the principal methods. Because of the highly reactive nature of titanium in the presence of such gases as oxygen, the casting must be done in a vacuum furnace. Because of their high strength-to-weight ratio primarily, titanium and titanium alloys are widely used for aircraft structures requiring greater heat resistance than aluminum alloys. Because of their exceptional corrosion resistance, however, they (unalloyed titanium primarily) are also used for chemical-processing, desalination, and power-generation equipment, marine hardware, valve and pump parts, and prosthetic devices.

Unalloyed and Alloy Types

There are five grades of commercially pure titanium, also called unalloyed titanium: ASTM (American Society for Testing and Materials) Grade 1, Grade 2, Grade 3, Grade 4, and Grade 7. They are distinguished by their impurity content, that is, the maximum amount of carbon, nitrogen, hydrogen, iron, and oxygen permitted. Regardless of grade, carbon and hydrogen contents are 0.10 and 0.015% maximum, respectively. Maximum nitrogen is 0.03%, except for 0.05% in Grades 3 and 4. Iron content ranges from as much as 0.20% in Grade 1, the most pure (99.5%) grade, to as much as 0.05% in Grade 4, the least pure (98.9%). Maximum oxygen ranges from 0.18% in Grade 1 to 0.40% in Grade 4. Grade 7, 99.1% pure based on maximum impurity content, is actually a series of alloys containing 0.12 to 0.25% palladium for improved corrosion resistance in hydrochloric, phosphoric, and sulfuric acid solutions. Palladium content has little effect on tensile properties, but impurity content, especially oxygen and iron, has an appreciable effect. Minimum tensile yield strengths range from 172 MPa for Grade 1 to 483 MPa for Grade 4.

There are three principal types of titanium alloys: alpha or near-alpha alloys, alpha-beta alloys, and beta alloys. All are available in wrought form and some of each type for castings as well. In recent years, some also have become available in powder compositions for processing by hot isostatic pressing and other powder-metallurgy techniques. Titanium alpha alloys typically contain aluminum and usually tin. Other alloying elements may include zirconium, molybdenum, and, less commonly, nitrogen, vanadium, columbium, tantalum, or silicon. Although they are generally not capable of being strengthened by heat treatment (some will respond slightly), they are more creep resistant at elevated temperature than the other two types, are also preferred for cryogenic applications, and are more weldable but less forgeable. Ti–5% Al–2% Sn, which is available in regular and EL1 grades (extra-low inertial) in wrought and cast forms, is the most widely used. In wrought and cast form, minimum tensile yield strengths range from 621 to 793 MPa and tensile modulus is on the order of 107,000 to 110,000 MPa. It has useful strength to about 482°C and is used for aircraft parts and chemical-processing equipment. The EL1 grade is noted for its superior toughness and is preferred for containment of liquid gases at cryogenic temperatures. Other alpha or near-alpha alloys and their performance benefits include Ti–8% Al–1% Mo–1% V (high creep strength to 482°C), Ti–6% Al–2% Sn–4% Zr–2% Mo (creep resistance and stress stability to 593°C, Ti–6% Al–2% Cb–1% Ta–0.8% Mo (toughness, strength weldability), and Ti–2.25% Al–11% Sn–5% Zr–1% Mo (high tensile strength—931 MPa yield, superior resistance to stress corrosion in hot salt media at 482°C. Another alpha alloy, Ti–0.3% Mo–0.8% Ni, also known as TiCode 12, is noted for its greater strength than commercially pure grades and equivalent or superior corrosion resistance, especially to crevice corrosion in hot salt solutions.

Titanium alpha-beta alloys, which can be strengthened by solution heat treatment and aging, afford the opportunity of parts fabrication in the more ductile annealed condition and then can be heat-treated for maximum strength. Ti–6% Al–4% V, which is available in regular and EL1 grades, is the principal alloy, its production alone having accounted for about half of all titanium and titanium alloy production. In the annealed condition, tensile yield strength is about 896 MPa and 13% elongation. Solution treating and aging increase yield strength to about 1034 MPa. Yield strength decreases steadily with increasing temperature, to about 483 MPa at about 510°C for the aged alloy. At

454°C, aged bar has a 1000-h stress-rupture strength of about 345 MPa. Uses range from aircraft and aircraft-turbine parts to chemical-processing equipment, and marine hardware. The alloy is also the principal alloy used for superplastically formed, and superplastically formed and simultaneously diffusion-bonded parts. At 899 to 927°C and low strain rates, the alloy exhibits tensile elongations of 600 to 1000%, a temperature range also amenable to diffusion bonding the alloy.

Following are other alpha-beta alloys and their noteworthy characteristics:

- Ti–6% Al–6% V–2% Sn: high strength to about 315°C but low toughness and fatigue resistance
- Ti–8% Mn: limited use for flat mill products, not weldable
- Ti–7% Al–4% Mo: a forging alloy mainly, but limited use; 1034 MPa yield strength in the aged condition
- Ti–6% Al–2% Sn–4% Zr–6% Mo: high strength, 1172 MPa yield strength, decreasing to about 759 MPa at 427°C; for structural applications at 400 to 540°C
- Ti–5% Al–2% Sn–2% Zr–4% Mo–4% Cr and Ti–6% Al–2% Sn–2% Zr–2% Mo–2% Cr: superior hardenability for thick-section forgings; high modulus — about 117,000 to 124,000 MPa, respectively; tensile yield strength of about 1138 MPa
- Ti–10% V–2% Fe–3% Al: best of the alloys in toughness at a yield strength of 896 MPa; can also be aged to a yield strength of about 1186 MPa; intended for use at temperatures to about 315°C
- Ti–3Al–2.5V: a tubing and fastener alloy primarily, moderate strength and ductility, weldable.

Beta titanium alloys, fewest in number, are noted for their hardenability, good cold formability in the solution-treated condition, and high strength after aging. On the other hand, they are heavier than titanium and the other alloy types, their density ranging from about 4.84 g/cm^3 for Ti–13% V–11% Cr–3% Al, Ti–8% Mo–8% V–2% Fe–3% Al, and Ti–3% Al–8% V–6% Cr–4% Zr–4% Mo to 5.07 g/cm^3 for Ti–11.5% Mo–6% Zr–4.5% Sn, which is also known as Beta III. They are also the least creep-resistant of the alloys. Ti–13% V–11% Cr–3% Al, a weldable alloy, can be aged to tensile yield strengths as high as 1345 MPa and retains considerable strength at temperatures to 315°C, but has limited stability at prolonged exposure to higher temperatures.

APPLICATIONS

One of the chief uses of the metal has been in the form of titanium oxide as a white pigment. It is also valued as titanium carbide for hard facings and for cutting tools. Small percentages of titanium are added to steels and alloys to increase hardness and strength by the formation of carbides or oxides or, when nickel is present, by the formation of nickel titanide.

The major portion of the commercial applications of titanium has been in the chemical processing industries in the form of reactors, vessels, and heat exchangers. The pulp and paper industry has used it in bleaching equipment primarily as chlorine dioxide mixers, while the electrochemical industry has utilized heating and cooling (tubing) coils and anodizing and plating racks. Pumps, valve, thermowells, and other miscellaneous items are additional examples of commercial applications of titanium (Table T.5).

TITANIUM CARBIDE

A hard crystalline powder of the composition TiC, titanium carbide is made by reacting titanium dioxide and carbon black at temperatures above 1800°C. It is compacted with cobalt or nickel for use in cutting tools and for heat-resistant parts. It is lighter in weight and less costly than tungsten carbide, but in cutting tools it is more brittle. When combined with tungsten carbide in sintered carbide tool materials, however, it reduces the tendency to cratering in the tool.

PROPERTIES

TiC theoretically contains 20.05% carbon and is light metallic gray in color. It is chemically

TABLE T.5
Typical Titanium Applications for Various Markets

Market	Application
Gas turbine engines	Compressor blades, disks, ducts, cases
Airframes	Landing gear beams, wing structure, hydraulic tubing
General chemical	Heat exchangers, condensers, mixers, piping
Organic/petrochemical	Strippers, reboilers, condensers, reactors
Power plants	Surface condensers, turbine blades
Electrolysis	Anodes for chlorine, chlorate, manganese dioxide; cathodes for copper, manganese; cathodic protection of bridges
Pulp/paper	Bleach tanks, wet chlorine systems, drum washers
Water technology	Flash desalination, heat exchangers for desalination
Metal recovery	Plating of chromium, nickel, silver, gold, zinc, and galvanizing; hydrometallurgy of copper
Energy extraction	Logging tools, seals, springs, tubulars for sour gas and geothermal power
Medical	Prosthetic implants, instruments
Environmental	Flue gas desulfurization, wet air oxidation of waste, incinerator stacks, nuclear waste
Marine	Heat exchangers, piping systems, ball valves, sonsar masts
High-performance vehicles	Racing valves, springs, retainers, connecting rods

Source: *McGraw-Hill Encyclopedia of Science and Technology*, 8th ed., Vol. 18, McGraw-Hill, New York, . With permission.

stable, and is almost inert to hydrochloric and sulfuric acids. In oxidizing chemicals, such as aqua regia and nitric or hydrofluoric acids, TiC is readily soluble. It also dissolves in alkaline oxidizing melts. When heated in atmospheres containing nitrogen, nitride formation occurs above –1500°C. TiC is attacked by chlorine gas and is readily oxidized in air at elevated temperatures.

The density of TiC is 4.94 g/cm^3, Mohs hardness is 9+, microhardness is 3200 kg/mm^2, and modulus of elasticity is 309,706 MPa. The modulus of rupture at room temperature has been reported as 499.8 to 843.2 MPa for materials sintered at 2600 to 3000°C. Hot modulus of rupture values are given as 107.78 to 116.96 MPa at 982°C and 54.4 to 63.92 MPa at 2200°C.

The melting point of TiC is 3160°C, and electrical resistivity at room temperature is 180 to 250 μΩ-cm. It can be used as a conductor at high temperatures. Coefficient of thermal expansion between room temperature and 593°C is 4.12×10^{-6}/°F. Thermal conductivity is 0.041 cal/cm · s/°C.

Grades and Uses

A general-purpose cutting tool of this type contains about 82% tungsten carbide, 8% titanium carbide, and 10% cobalt binder. Kentanium is titanium carbide in various grades with up to 40% of either cobalt or nickel as the binder, used for high-temperature, erosion-resistant parts. For highest oxidation resistance only about 5% cobalt binder is used. Other grades with 20% cobalt are used for parts where higher strength and shock resistance are needed, and where temperatures are below about 982°C. This material has a tensile strength of 310 MPa, compressive strength of 3789 MPa, and Rockwell hardness A90. Another for resistance to molten glass or aluminum has a binder of 20% nickel. A titanium carbide alloy for tool bits has 80% titanium carbide dispersed in a binder of 10% nickel and 10% molybdenum. The material has a hardness of Rockwell A93, and a dense, fine-grained structure. Ferro-Tic has the titanium carbide bonded with stainless steel. It has a hardness of Rockwell C55. Machinable carbide is titanium carbide in a matrix of Ferro-Tic C tool steel. Titanium carbide tubing is produced in round or rectangular form 0.25 to 7.6 cm in diameter. It is made by vapor deposition of the carbide without a binder. The tubing has a hardness above 2000 Knoop and a melting point of 3249°C.

Grown single crystals of titanium carbide have the composition TiC$_{0.94}$, with 19% carbon.

The melting point is 3250°C, density 4.93, arid Vickers hardness 3230.

The material is finding new uses in cermet components such as jet engine blades and cemented carbide tool bits. Titanium carbide has a relatively low electrical resistivity (1×10^{-4}) and can be used as a conductor of electricity, especially at high temperatures.

Extreme hardness of titanium carbide makes it suitable for wear-resistant parts such as bearings, nozzles, cutting tools, etc. It also serves for special refractories under either neutral or reducing conditions.

TITANIUM DIBORIDE

The material has a melting point of 2980°C, is stable in HCl and HF acid, but decomposes readily in alkali hydroxides, carbonates, and bisulfates. It reacts with hot H_2SO_4.

Processing

Sintered parts of titanium diboride, TiB_2, are usually produced by either hot pressing, pressureless sintering, or hot isostatic pressing. Hot pressing of titanium diboride parts is conducted at temperatures >1800°C in vacuum or 1900°C in an inert atmosphere. Hot pressed parts generally have a final density of >99% of theoretical. Typical sintering aids used for hot pressed parts include iron, nickel, cobalt, carbon, tungsten, and tungsten carbide.

Pressureless sintering of TiB_2 is a less expensive method for producing net shape parts. Because of the high melting point of titanium diboride, sintering temperatures in excess of 2000°C often are necessary to promote sintering. Several different sintering aids have been developed to produce dense pressureless sintered parts by liquid-phase sintering. A combination of carbon and chromium, iron, or chromium carbide can be used as a sintering aid to produce pressureless sintered parts with a final density >95% of the theoretical density. Boron carbide also is added to inhibit grain growth during sintering. These sintering aids as well as atmospheric conditions can be used to lower the sintering temperature necessary for full densification.

Properties

Typical mechanical properties for hot pressed titanium diboride include a flexural strength of 350 to 575 MPa, a hardness of 1800 to 2700 kg/mm^2, and a fracture toughness of 5 to 7 MPa · m$^{-1/2}$. The mechanical property values are dependent on the type of fabrication method used (pressureless sintering vs. hot pressing), the purity of the synthesized powder and the amount of porosity remaining in the finished part.

The elastic modulus of titanium diboride can range from 510 to 575 GPa and the Poisson ratio is 0.18 to 0.20. Titanium diboride has a room-temperature electrical resistivity of 15×10^{-6} Ω-cm and a thermal conductivity of 25 Ω/mK.

Uses

Titanium diboride is used for a variety of structural applications including lightweight ceramic armor, nozzles, seals, wear parts, and cutting tool composites. Titanium diboride also has shown exceptional resistance to attack by molten metals, including molten aluminum, which makes it a useful material for such applications as metallizing boats, molten metal crucibles, and Hall–Heroult cell cathodes because of its intrinsic electrical conductivity. TiB_2 can be combined with a variety of other nonoxide ceramic materials, such as silicon carbide (SiC) and titanium carbide (TiC), and oxide materials, such as alumina (Al_2O_3), to increase the mean strength and fracture toughness of the matrix material.

TITANIUM NITRIDE

Titanium nitride whiskers (TiN_w) are single-crystal, acicular-shaped particles of titanium nitride that typically range in size from 0.3 to 1.0 μm in diameter with aspect ratios ranging from 5:1 to 50:1. TiN_w have been produced using several different approaches including carbothermic reduction, laser synthesis, plasma synthesis, solid/solid and solid/gas combustion synthesis, and vapor phase reactions. TiN_w are reported to form from both vapor–solid and vapor–liquid–solid mechanisms.

TiN_w purity can vary from near stoichiometric (nitrogen of 22.7 wt%) to solid solutions of TiCN and TiON where carbon or oxygen substitutes for nitrogen. TiN_w typically form with smooth surfaces and are relatively free from internal defects, which limit the strength of other whiskers such as SiC_w. Unlike SiC_w, TiN_w are electrically conductive and have a coefficient of thermal expansion that is closer to steel and intermetallic materials. TiN_w also have been found to have increased stability with iron-containing metals, alloys, and intermetallic compounds.

Potential applications include reinforcements in various materials including alumina and tungsten–carbide-base cutting tool inserts for machining ferrous alloys, iron, nickel, and titanium aluminides (intermetallics), iron, nickel, and titanium metals/alloys and polymers.

Improved chemical resistance of TiN_w with iron compounds makes it an excellent candidate for use as tool inserts for the machining of cast iron and tool steels. The reactivity of SiC_w with iron compounds has limited alumina/SiC_w cutting tool insert use primarily to superalloys. Alumina/TiN_w composites have been shown to be machinable using electric discharge machining.

Other areas of interest include stable reinforcements for iron-, nickel- and titanium-based intermetallic compounds and metals/alloys, and electrically conducting reinforcements for polymers to promote electrical conduction and charge dissipation. TiN_w provide chemical stability and current carrying characteristics desirable in many applications.

TITANIUM OXIDE

The white titanium dioxide, or titania, of the composition TiO_2, is an important paint pigment. The best quality is produced from ilmenite, and is higher in price than many white pigments but has great hiding power and durability. Off-color pigments, with a light buff tone, are made by grinding rutile ore. The pigments have fine physical qualities and may be used wherever the color is not important. Titania is also substituted for zinc oxide and lithopone in the manufacture of white rubber goods, and for paper filler. The specific gravity is about 4.

Titania crystals are produced in the form of pale-yellow, single-crystal boules for making optical prisms and lenses for applications where the high refractive index is needed. The crystals are also used as electric semiconductors, and for gemstones. They have a higher refractive index than the diamond, and the cut stones are more brilliant but are much softer. The hardness is about 925 Knoop, and the melting point is 1825°C. The refractive index of the rutile form is 2.7 and that of the anatase is 2.5; the synthetic crystals have a refractive index of 2.616 vertically and 2.903 horizontally.

Titanium oxide is a good refractory and electrical insulator. The finely ground material gives good plasticity without binders, and is molded to make resistors for electronic use. A micro sheet is titanium oxide in sheets as thin as 0.008 cm for use as a substitute for mica for electrical insulation where brittleness is not important. Titania-magnesia ceramics have been made in the form of extruded rods and plates and pressed parts.

Uses

Titanium dioxide is a most important ceramic finish coat for sheet metal products. The opacity of this enamel imparted by titanium dioxide has lowered film thickness of these finishes to the range of organic coatings while retaining the durability of porcelain. These enamels are self-opacified. That is, titanium dioxide is not dispersed as an insoluble suspension during smelting nor is it added at the mill. Rather, titanium dioxide is taken into solution during smelting of the batch and is held in supersaturated solution through fritting. Upon firing the enamel, titanium dioxide crystallizes or precipitates from the glassy matrix.

Trimmers or trimmer condensers employing TiB_2 bodies are used for minute adjustments of capacitance. Normally, the rotor consists of a TiO_2 body. Parts are made with extreme accuracy, and are usually supplied in one of three temperature coefficient types. The base is a low-loss ceramic composition.

TABLE T.6
General Characteristics of Tool Steels

AISI Type (quench)	Hardening Depth	Toughness	Wear Resistance	Decarb Resistance	Distortion in Heat Treatment
A2	Deep	Medium	Medium	Medium	Low
A6	Deep	High	Low	Medium	Lowest
A8	Deep	High	Low	Medium	Lowest
D2	Deep	Low	High	Medium	Lowest
D3	Medium	Low	High	Medium	Medium
H11	Deep	Highest	Low	Med, high	Very low
L2 (Water)	Medium	High	Low	High	High
L2 (Oil)	Medium	High	Low	High	Medium
L6	Medium	High	Low	High	Low
S1	Medium	High	Low	Low	Medium
S7 (Air)	Med, deep	High	Low	Medium	Low
S7 (Oil)	Med, deep	High	Low	Medium	Low
O2	Medium	Medium	Medium	Medium	Medium

Source: Mach. Design Basics Eng. Design, June, p. 792, 1993. With permission.

Mechanical and physical properties of TiO_2 include relatively low strength (MOR 123.5 to 150.9 MPa; tensile strength 40.8 to 54.4 MPa, low thermal conductivity (0.14 cal/cm/s/°C), and a coefficient of thermal expansion (for rutile) of 7 to 9×10^{-6}/°C.

TOOL STEEL

To develop their best properties, tool steels are always heat treated. Because the parts may distort during heat treatment, precision parts should be semifinished, heat-treated, then finished. Severe distortion is most likely to occur during liquid quenching, so an alloy should be selected that provides the needed mechanical properties with the least severe quench (Table T.6).

Steels are used primarily for cutters in machining, shearing, sawing, punching, and trimming operations, and for dies, punches, and molds in cold- and hot-forming operations. Some are also occasionally used for nontool applications. Tool steels are primarily ingot-cast wrought products, although some are now also powder-metal products. Regarding powder-metal products, there are two kinds: (1) mill products, mainly bar, produced by consolidating powder into "ingot" and reducing the ingot by conventional thermomechanical wrought techniques, and (2) end products tools, produced directly from powder by pressing and sintering techniques. There are seven major families of tool steels as classified by the American Iron and Steel Institute: (1) high-speed tool steels, (2) hot-work tool steels, (3) cold-work tool steels, (4) shock-resisting tool steels, (5) mold steels, (6) special-purpose tool steels, and (7) water-hardening tool steels.

HIGH-SPEED TOOL STEELS

These steels are subdivided into three principal groups or types: the molybdenum-type, designated M1 to M46; the tungsten-type (T1 to T15); and the intermediate molybdenum-type (M50 to M52). Virtually all M-types, which contain 3.75 to 9.5% molybdenum, also contain 1.5 to 6.75% tungsten, 3.75 to 4.25% chromium, 1 to 3.2% vanadium, and 0.85 to 1.3% carbon. M33 to M46 also contain 5 to 8.25% cobalt, and M6, 12% cobalt. The T-types, which are molybdenum-free, contain 12 to 18% tungsten, 4 to 4.5% chromium, 1 to 5% vanadium, and 0.75 to 1.5% carbon. Except for T1, which is cobalt-free, they also contain 5 to 12% cobalt.

Both M50 and M52 contain 4% molybdenum and 4% chromium; the former also contain 0.85% carbon and 1% vanadium, the latter 0.9% carbon, 1.25% tungsten, and 2% vanadium.

The molybdenum types are now by far the most widely used, and many of the T-types have M-type counterparts. All of the high-speed tool steels are similar in many respects. They all can be hardened to at least Rockwell C63, have fine grain size, and deep-hardening characteristics. Their most important feature is hot hardness: they all can retain a hardness of Rockwell C52 or more at 538°C. The M-types, as a group, are somewhat tougher than the T-type at equivalent hardness but otherwise mechanical properties of the two types are similar. Cobalt improves hot hardness, but at the expense of toughness. Wear resistance increases with increasing carbon and vanadium contents. The M-types have a greater tendency to decarburization and, thus, are more sensitive to heat treatment, especially austenitizing. Many of the T-types, however, are also sensitive in this respect, and they are hardened at somewhat higher temperatures. The single T-type that stands out today is T-15, which is rated as the best of all high-speed tool steels from the standpoint of hot hardness and wear resistance. Typical applications for both the M-type and T-type include lathe tools, end mills, broaches, chasers, hobs, milling cutters, planar tools, punches, drills, reamers, routers, taps, and saws. The intermediate M-types are used for what somewhat similar cutting tools but, because of their lower alloy content, are limited to less-severe operating conditions.

Hot-Work Tool Steels

These steels are subdivided into three principal groups: (1) the chromium type (H10 to H19), (2) the tungsten type (H21 to H26), and (3) the molybdenum type (H42). All are medium-carbon (0.35 to 0.60%) grades. The chromium types contain 3.25 to 5.00% chromium and other carbide-forming elements, some of which, such as tungsten and molybdenum, also impart hot strength, and vanadium, which increases high-temperature wear resistance. The tungsten types, with 9 to 18% tungsten, also contain chromium, usually 2 to 4%, although H23 contains 12% of each element. Tungsten hot-work tool steels with higher contents of alloying elements are more heat resistant at elevated temperatures than H11 and H13 chromium hot-work steels but the higher percentage also tends to make them more brittle in heat treating.

The one molybdenum type, H42, contains slightly more tungsten (6%) than molybdenum (5%), and 4% chromium and 2% vanadium. These alloying elements (chromium, molybdenum, tungsten, and vanadium) make the steel more resistant to heat checking than tungsten hot-work steels. Also, their lower carbon content in relation to high-speed tool steels gives them a higher degree of toughness.

Typical applications include dies for forging, die casting, extrusion, heading, trim, piercing and punching, and shear blades.

Cold-Work Tool Steels

There are also three major groups of cold-work tool steels: (1) high carbon (1.5 to 2.35%); high chromium (12), which are designated D2 to D7; (2) medium alloy air-hardening (A2 to A10), which may contain 0.5 to 2.25% carbon, 0 to 5.25% chromium, 1 to 1.5% molybdenum, 0 to 4.75% vanadium, 0 to 1.25% tungsten, and, in some cases, nickel, manganese or silicon, or nickel and manganese; and (3) oil-hardening types (O1 to O7). They are used mainly for cold-working operations, such as stamping dies, draw dies, and other forming tools as well as for shear blades, burnishing tools, and coining tools.

Shock-Resistant Tool Steels

These steels (S1 to S7) are, as a class, the toughest, although some chromium-type hot-work grades, such as H10 to H13, are somewhat better in this respect. The S-types are medium-carbon (0.45 to 0.55%) steels containing only 2.50% tungsten and 1.50% chromium (S1), only 3.25% chromium and 1.40% molybdenum (S7), or other combinations of elements, such as molybdenum and silicon, manganese and silicon, or molybdenum, manganese, and silicon. Typical uses include chisels, knockout pins, screwdriver blades, shear blades, punches, and riveting tools.

Mold Steels

There are three principal mold steels: (1) P6, containing 0.10% carbon, 3.5% nickel, and

1.5% chromium; P20, 0.35% carbon, 1.7% chromium, and 0.40% molybdenum; and P21, 0.20% carbon, 4% nickel, and 1.2% aluminum. P6 is basically a carburizing steel produced to tool-steel quality. It is intended for hubbing — producing die cavities by pressing with a male plug — then carburizing, hardening, and tempering. P20 and P21 are deep-hardening steels and may be supplied in hardened condition. P21 may be carburized and hardened after machining. These steels are tough but low in wear resistance and moderate in hot hardness; P21 is best in this respect. All three are oil-hardening steels and they are used mainly for injection and compression molds for forming plastics, but they also have been used for die-casting dies.

SPECIAL-PURPOSE TOOL STEELS

These steels include L2, containing 0.50 to 1.10% carbon, 1.00% chromium, and 0.20% vanadium; and L6, having 0.70% carbon, 1.5% nickel, 0.75% chromium, and, sometimes, 0.25% molybdenum. L2 is usually hardened by water quenching and L6, which is deeper hardening, by quenching in oil. They are relatively tough and easy to machine and are used for brake-forming dies, arbors, punches, taps, wrenches, and drills.

WATER-HARDENING TOOL STEELS

The water-hardening tool steels include W1, which contains 0.60 to 1.40% carbon and no alloying elements; W2, with the same carbon range and 0.25% vanadium; and W5, having 1.10% carbon and 0.50% chromium. All are shallow-hardening and the least qualified of tool steels in terms of hot hardness. However, they can be surface-hardened to high hardness and, thus can provide high resistance to surface wear. They are the most readily machined tool steels. Applications include blanking dies, cold-striking dies, files, drills, countersinks, taps, reamers, and jewelry dies.

Coatings

To prolong tool life, tool-steel end products, such as mills, hobs, drills, reamers, punches, and dies, can be nitrided or coated in several ways. Oxide coatings, imparted by heating to about 566°C in a steam atmosphere or by immersion in aqueous solutions of sodium hydroxide and sodium nitrite at 140°C, are not as effective as traditional nitriding, but do reduce friction and adhesion between the workpiece and tool. The thickness of the coating developed in the salt bath is typically less than 0.005 mm, and its nongalling tendency is especially useful for operations in which failure occurs this way. Hard-chromium plating to a thickness of 0.0025 to 0.0127 mm provides a hardness of DPH 950 to 1050 and is more effective than oxide coating, but the plate is brittle and, thus, not advisable for tools subject to shock loads. Its toughness may be improved somewhat without substantially reducing wear resistance by tempering at temperatures below 260°C, but higher tempering temperatures impair hardness, thus wear resistance, appreciably. An antiseize iron sulfide coating can be applied electrolytically at 191°C using a bath of sodium and potassium thiocyanate. Because of the low temperature, the tools can be coated in the fully hardened and tempered condition without affecting hardness. Tungsten carbide is another effective coating. One technique, called Rocklinizing, deposits 0.0025 to 0.0203 mm of the carbide using a vibrating arcing electrode of the material in a hand-held gun. Titanium carbide and titanium nitride are the latest coatings. The nitride, typically 0.008 mm thick, has stirred the greatest interest, although the carbide may have advantages for press tools subject to high pressure. In just the past few years, all sorts of tools, primarily cutters but also dies, have been titanium nitride-coated, which imparts a gold- or brasslike look. The coating can be applied by chemical vapor deposition (CVD) at 954 to 1066°C or by physical vapor deposition (PVD) at 482°C or less. Thus, the PVD process has an advantage in that the temperature involved may be within or below the tempering temperature of the tool steels so that the coating can be applied to fully hardened and tempered tools. Also, the risk of distortion during coating is less.

Another method being used to prolong tool life is to subject the tools to a temperature of −196°C for about 30 h. The cryogenic treatment, which has been called Perm-O-Bond and

Cryo-Tech, is said to rid the steel of any retained austenite — thus the improved tool life.

PROPERTIES

Toughness

Toughness in *tool steels* is best defined as the ability of a material to absorb energy without fracturing rather than the ability to deform plastically without breaking. Thus, a high elastic limit is required for best performance since large degrees of flow or deformation are rarely permissible in fine tools or dies. Hardness of a tool has considerable bearing on the toughness because the elastic limit increases with an increase in hardness. However, at very high hardness levels, increased notch sensitivity and brittleness are limiting factors.

In general, lower carbon tool steels are tougher than higher carbon tool steels. However, shallow hardening carbon (W-1) or carbon–vanadium (W-2) tool steels with a hard case and soft core will have good toughness regardless of carbon content. The higher alloy steels will range between good and poor toughness depending upon hardness and alloy content.

Abrasion Resistance

Some tool steels exhibit better resistance to abrasion than others. Attempts to measure absolute abrasion resistance are not always consistent, but in general, abrasion resistance increases as the carbon and alloy contents increase. Carbon is an influential factor. Additions of certain alloying elements (chromium, tungsten, molybdenum, and vanadium) balanced with carbon have a marked effect on increasing the abrasion resistance by forming extremely hard carbides.

Hardness

Maximum attainable hardness is primarily dependent upon the carbon content, except possibly in the more highly alloyed tool steels. Tool steels are generally used somewhat below maximum hardness except for deep-drawing dies, forming dies, cutting tools, etc. Battering or impact tools are put in service at moderate hardness levels for improved toughness.

Hot Hardness

The ability to retain hardness with increasing temperature is defined as hot hardness or red hardness. This characteristic is important in steels used for hot-working dies. Generally, as the alloy content of the steel is increased (particularly in chromium, tungsten, cobalt, molybdenum, and vanadium, which form stable carbides), the resistance to softening at elevated temperatures is improved. High-alloy tool steels with a properly balanced composition will retain high hardness up to 593°C. In the absence of other data, hardness after high-temperature tempering will indicate the hot hardness of a particular alloy.

HEAT TREATMENT

Hardenability

Carbon tool steels are classified as shallow hardening, i.e., when quenched in water from the hardening (austenitizing) temperature, they form a hardened case and a soft core. Increasing the alloy content increases the hardenability or depth of hardening of the case. A small increase in alloy content will result in a steel that will harden through the cross-sections when quenched in oil. If the increase in alloy content is great enough, the steels will harden throughout when quenched in still air. For large tool or die sections, a high-alloy tool steel should be selected if strength is to be developed throughout the section in the finished part.

For carbon tool steels that are very shallow in hardening characteristics, the P/F test, Disc test, and PV test are methods for rating this characteristic. Oil-hardening tool steels of medium-alloy content are generally rated for hardenability by the Jominy End Quench test.

Dimensional Changes during Heat Treatment

Carbon tool steels are apt to distort because of the severity of the water quench required. In general, water-hardening steels distort more than oil hardening, and oil hardening distort more than air-hardening steels. Thus, if a tool or die is to be machined very close to final size before heat treatment and little or no grinding is to be performed after treatment, an air-hardening tool steel would be the proper selection.

Resistance to Decarburization

During heat treatment, steels containing large amounts of silicon, molybdenum, and cobalt tend to lose carbon from the surface more rapidly than steels containing other alloying elements. Steels with extremely high carbon content are also susceptible to rapid decarburization. Extra precaution should be employed to provide a neutral atmosphere when heat treating these steels. Otherwise, danger of cracking during hardening will be present. Also, it would be necessary to allow a liberal grinding allowance for cleanup after heat treatment.

Machinability

Since most tool steels, even in the annealed state, contain wear-resistant carbides, they are generally more difficult to machine than the open-hearth grades or low-alloy steels. In general, the machinability tends to decrease with increasing alloying content. Microstructure also has a marked effect on machinability. For best machinability, a spheroidal microstructure is preferred over pearlitic.

The addition of small amounts of lead or sulfur to the steels to improve machinability has gained considerable acceptance in the tool steel industry. These free machining steels not only machine more easily but give a better surface finish than the regular grades. However, some caution is advised in applications involving transverse loading since lead or sulfur additions actually add longitudinal inclusions in the steel.

Available Forms

Tool steels are available in billets, bars, rods, sheets, and coil. Special shapes can be furnished upon request. Generally, the material is furnished in the soft (or annealed) condition to facilitate machining. However, certain applications require that the steel be cold-drawn or prehardened to a specified hardness.

A word of caution: Mill decarburization is generally present on all steel except that guaranteed by the producer to be decarburization-free. It is important that all decarburized areas be removed prior to heat treating or the tool or die may crack during hardening.

TUNGSTEN AND ALLOYS

In many respects, tungsten (symbol W) is similar to molybdenum. The two metals have about the same electrical conductivity and resistivity, coefficient of thermal expansion, and about the same resistance to corrosion by mineral acids. Both have high strength at temperatures above 1093°C, but because the melting point of tungsten is higher, it retains significant strength at higher temperatures than molybdenum does. The elastic modulus for tungsten is about 25% higher than that of molybdenum, and its density is almost twice that of molybdenum. All commercial unalloyed tungsten is produced by powder-metallurgy methods; it is available as rod, wire, plate, sheet, and some forged shapes. For some special applications, vacuum-arc-melted tungsten can be produced, but it is expensive and limited to relatively small sections.

Fabrication

Fabrication is a multistep process that converts tungsten metal from the original massive state (bars or ingots) to a more useful shape (sheet, tube, wire) and, at the same time, improves its physical properties. The exact details of fabrication depend on the method used for consolidating the metal and the type of product desired. Arc- or electron-beam-melted tungsten normally is extruded or forged to increase its ductility, whereas powder-processed material, because of its finer-grained structure and smaller tendency to crack, is less likely to require this initial step.

Tungsten is usually worked below its recrystallization temperature because the recrystallized metal tends to be brittle. Because increased working decreases the recrystallization temperature, successive lower temperatures are used in each fabrication step.

Full-density wrought tungsten can be hot-forged, swaged, extruded, rolled, and drawn as secondary fabrication steps used to produce the final shape. Working temperature is usually

1100°C or above depending on the grain size and type of deformation.

Sintered billets are forged, swaged, or rolled initially at temperatures in excess of 1400°C. Working temperature can be progressively lowered as the amount of work increases, but consideration must be given to equipment capacity because of the high strength of tungsten.

Several tungsten alloys are produced by liquid-phase sintering of compacts of tungsten powder with binders of nickel–copper, iron–nickel, iron–copper, or nickel–cobalt–molybdenum combinations; tungsten usually comprises 85 to 95% of the alloy by weight. These alloys are often identified as heavy metals or machinable tungsten alloys. In compact forms, the alloys can be machined by turning, drilling, boring, milling, and shaping; they are not available in mill product forms because they are unable to be wrought at any temperature.

Properties

Tungsten, element 74 on the periodic chart, has a melting point of approximately 3410°C, with values ranging between 3387 and 3422°C reported in the literature. This value easily makes it the highest-melting-point metal. It has the lowest coefficient of thermal expansion of all metals, and with a density of 19.25 g/cm^3 it is one of the heaviest. It has the lowest vapor pressure of all metals, and high thermal and electrical conductivity.

Single crystals of tungsten are elastically isotropic and have very high tensile and bulk moduli, but mechanical properties are strongly temperature dependent, with the yield strength and ultimate tensile strength decreasing significantly with increasing temperature. At elevated temperatures, tungsten reacts rapidly with oxygen, forming a series of oxides that have stoichiometries ranging between WO_2 and WO_3.

The unique properties of tungsten make it the element of choice for such applications as filaments for incandescent lamps and x-ray tubes, electron sources for scanning and transmission electron microscopes, and connectors for circuit boards. Although these characteristics might suggest an even wider range of applications, several actually limit its utility. For example, the high density of tungsten makes it unsatisfactory for any weight-conserving application, and its aggressive reaction with oxygen limits its service at high temperatures. Welding is difficult because of the reactivity of tungsten with oxygen, and the presence of oxygen and other interstitials in the metal can make it very brittle at room temperature. Nonetheless, the special properties of tungsten are so beneficial that in many cases it has been worth the cost and effort to engineer around the problems.

Tungsten retains a tensile strength of about 344 MPa at 1371°C, but because of its heavy weight is normally used in aircraft or missile parts only as coatings, usually sprayed on. It is also used for x-ray and gamma-ray shielding. Electroplates of tungsten or tungsten alloys give surface hardnesses to Vickers 700 or above.

Applications

Tungsten has a wide usage for alloy steels, magnets, heavy metals, electric contacts, rocket nozzles, and electronic applications. Tungsten resists oxidation at very high temperatures, and is not attacked by nitric, hydrofluoric, or sulfuric acid solutions. Flame-sprayed coatings are used for nozzles and other parts subject to heat erosion.

Tungsten is usually added to iron and steel in the form of ferrotungsten, made by electric-furnace reduction of the oxide with iron or by reducing tungsten ores with carbon and silicon. Standard grades with 75 to 85% tungsten have melting points from 1760 to 1899°C. Tungsten powder is usually in sizes from 200 to 325 mesh, and may be had in a purity of 99.9%. Parts, rods, and sheet are made by powder metallurgy, and rolling and forging are done at high temperature.

The tungsten powder is used for spray coatings for radiation shielding and for powder-metal parts. Tungsten wire is used for spark plugs and electronic devices. Tungsten wire as fine as 0.00046 cm is used in electronic hardware. Tungsten whiskers, which are extremely fine fibers, are used in copper alloys to add strength. Copper wire, which normally has a tensile strength of 206 MPa, will have a strength of 827 MPa when 35% of the wire is tungsten whiskers. Tungsten yarns are made up of fine fibers of the metal. The yarns are flexible and can be woven into fabrics. Continuous tungsten filaments, usually 10 to 15 µm in diameter, are used for reinforcement in metal, ceramic, and plastic composites. Finer filaments of tungsten are used as cores, or substrates, for boron filaments.

The metal is also produced as arc-fused grown crystals, usually no larger than 0.952 cm in diameter and 25.4 cm long, and worked into rod, sheet, strip, and wire. Tungsten crystals, 99.9975% pure, are ductile even at very low temperatures, and wire as fine as 0.008 cm and strip as thin as 0.013 cm can be cold-drawn and cold-rolled from the crystal. The crystal metal has nearly zero porosity and its electrical and heat conductivity are higher than ordinary tungsten.

One tungsten–aluminum alloy is a chemical compound made by reducing tungsten hexachloride with molten aluminum.

Tungsten wire is not used exclusively for lamp filaments. Because of its high melting temperature, tungsten can be heated to the point where it becomes a thermionic emitter of electrons, without losing its mechanical integrity. Consequently, tungsten filaments are often used as electron sources in scanning electron microscopes and transmission electron microscopes, and also as filaments in x-ray tubes.

In x-ray tubes, electrons produced from the tungsten filament are accelerated so that they strike a tungsten or tungsten–rhenium anode, which emits the x-rays. Again, this application takes advantage of the high melting point of tungsten, since the energy of the electron beam required to generate x-rays is very high, and the spot where the beam hits the surfaces becomes very hot. In most tubes, the anode is rotated to limit the peak temperature and to allow for cooling.

Finally, tungsten filaments of a much larger size are often selected as the heating elements in vacuum furnaces. Again, because of the high melting point of tungsten, these furnaces can achieve much higher temperatures than furnaces made with other heating elements. It is important to note that in vacuum furnaces, as well as all of the other applications, the tungsten is in a controlled environment that inhibits its oxidation.

For example, tungsten heavy alloys are materials in which tungsten powder is liquid-phase sintered, usually with nickel–iron powders, to produce a composite material in which tungsten occupies about 95% of the volume. As the sintering process proceeds, the nickel–iron powder melts. Although the solubility of liquid nickel–iron in solid tungsten is small, solid tungsten readily dissolves in liquid nickel–iron. As the liquid wets the tungsten particles and dissolves part of the tungsten powder, the particles change shape, and internal pores are eliminated as the liquid flows into them. As processing continues, the particles coalesce and grow, producing a final product that is approximately 100% dense and has an optimized microstructure.

One of the main products made by this method is kinetic energy penetrators of military armored vehicles. This application takes advantage of the high density of tungsten, and it has been found that the liquid sintered materials have better impact properties than pure tungsten made by traditional powder processing.

Cutting tools and parts that must resist severe abrasion are often made of tungsten carbide. Tungsten carbide chips or inserts, with the cutting edges ground, are attached to the bodies of steel tools by brazing or by screws. The higher cutting speeds and longer tool life made feasible by the use of tungsten carbide tools are such that the inserts are discarded after one use.

Tungsten compounds (5% of tungsten consumption) have a number of industrial applications. Calcium and magnesium tungstates are used as phosphors in fluorescent lights and television tubes. Sodium tungstate is employed in the fireproofing of fabrics and in the preparation of tungsten-containing dyes and pigments used in paints and printing inks. Compounds such as WO_3 and WS_2 are catalysts for various chemical processes in the petroleum industry. Both WS_2 and WSe_2 are dry, high-temperature lubricants. Other applications of tungsten compounds have been made in the glass, ceramics, and tanning industries.

A completely new and different approach to produce bulk tungsten products from the powder-metallurgy process is through chemical vapor deposition (CVD), which provides a tungsten coating on a substrate.

Tungsten hexafluoride is the most common tungsten source for CVD processing. This compound is a liquid at room temperature, but its vapor pressure is high enough that the vapor can be continuously extracted and passed across the part that is to be coated.

$$WF_6 + 3H_2 \rightarrow W + 6HF$$

The reaction requires temperatures above approximately 300°C and a surface that causes the dissociation of molecular hydrogen into atomic hydrogen. Therefore, sections of a part may be selectively coated by having surfaces that either catalyze or prevent this reaction.

One of the most important applications of this process has been in the electronics industry, in which tungsten vias are placed in integrated circuits. The vias are small metal plugs that connect one level of wiring to another in the circuit board. They are generally about 0.4 mm in diameter, with an aspect ratio of about 2.5. In future applications, the diameter may shrink to less than 0.1 mm, and have an aspect ratio greater than five. The metal for this application must have good electrical conductivity, must not react with the surrounding materials, must adhere to the wiring or silicon above or below the via, and must be deposited by a CVD reaction, as that is the only way to fill such small holes.

The most common method in the electronics industry is blanket CVD. In this technology, an adhesion layer is first put down to make certain that the CVD tungsten will stick to the surface. This adhesion layer is often titanium nitride, TiN. Tungsten is deposited on top of this layer, covering the surface and filling the vias. After the CVD is complete, the tungsten on the entire surface is removed by chemical-mechanical polishing. This procedure leaves the vias filled, but cleans the surface of the unnecessary tungsten.

Alloys

A large number of tungsten-based alloys have been developed. Binary and ternary alloys of molybdenum, niobium, and tantalum with tungsten are used as substitutes for the pure metal because of their superior mechanical properties. Adding small amounts of other elements such as titanium, zirconium, hafnium, and carbon to these alloys improves their ductility. Tungsten–rhenium alloys possess excellent high-temperature strength and improved resistance to oxidation, but are difficult to fabricate. This problem is ameliorated somewhat by the addition of molybdenum, a common composition being W (40 at%)–Re (30%)–Mo (30%). The strengths of tungsten or tungsten–rhenium systems can be increased by small amounts of a dispersed second phase such as an oxide (ThO_2, Ta_2O_5), carbide (HfC, TaC), or boride (HfB, ZrB). The so-called heavy alloys are three-component systems composed mainly of tungsten in combination with a nickel–copper or nickel–iron matrix. These materials are characterized by high density (17 to 19 g/cm^3), hardness, and good thermal conductivity.

Tungsten is used widely as a constituent in the alloys of other metals, since it generally enhances high-temperature strength. Several types of tool steels and some stainless steels contain tungsten. Heat-resistant alloys, also termed superalloys, are nickel-, cobalt-, or iron-base systems containing varying amounts (typically 1.5 to 25 wt%) of tungsten. Wear-resistant alloys having the trade name Stellites are composed mainly of cobalt, chromium, and tungsten.

Cobalt–tungsten alloy, with 50% tungsten, gives a plate that retains a high hardness at red heat. Tungsten RhC is a tungsten–rhenium carbide alloy containing 4% rhenium carbide. It is used for parts requiring high strength and hardness at high temperatures. The alloy retains a tensile strength of 517 MPa at 1927°C.

TUNGSTEN CARBIDE

Tungsten carbide is an iron-gray powder of minute cubical crystals with a Mohs hardness above 9.5 and a melting point of about 2982°C. It is produced by reacting a hydrocarbon vapor with tungsten at high temperature. The composition is WC, but at high heat it may decompose into W_2C and carbon, and the carbide may be a mixture of the two forms. Other forms may also be produced, W_3C and W_3C_4. Tungsten carbide is used chiefly for cutting tool bits and for heat- and erosion-resistant parts and coatings.

One of the earliest of the American bonded tungsten carbides was Carboloy, which was used for cutting tools, gauges, drawing dies, and wear parts. The carbides are now often mixed carbides. Carboloy 608 contains 83% chromium carbide, 2% tungsten carbide, and 15% nickel binder. It is lighter in weight than tungsten carbide, is nonmagnetic, and has a hardness to Rockwell A93. It is used for wear-resistant parts, and resists oxidation to 1092°C. Titanium carbide is more fragile, but may be mixed with tungsten carbide to add hardness for dies. Kennametal K601 is used for seal rings and wear parts, and is a mixture of tantalum and tungsten carbides without a binder. It has a compressive strength of 4650 MPa, rupture strength of 689 MPa, and Rockwell hardness A94. Kennametal K501 is tungsten carbide with a platinum binder for parts subject to severe heat erosion.

Tungsten carbide LW-1 is tungsten carbide with about 6% cobalt binder used for flame-coating metal parts to give high-temperature wear resistance. Deposited coatings have a Vickers hardness to 1450, and resist oxidation at 538°C. Tungsten carbide LW-1N, with 15% cobalt binder, has a much higher rupture strength, but the hardness is reduced to 1150.

TUNGSTEN STEEL

Tungsten steel is any steel containing tungsten as the alloying element imparting the chief characteristics to the steel. It is one of the oldest of the alloying elements in steel.

Tungsten increases the hardness of steel, and gives it the property of red hardness, stabilizing the hard carbides at high temperatures. It also widens the hardening range of steel, and gives deep hardening. Very small quantities serve to produce a fine grain and raise the yield point. The tungsten forms a very hard carbide and an iron tungstite, and the strength of the steel is also increased, but it is brittle when the tungsten content is high. When large percentages of tungsten are used in steel, they must be supplemented by other carbide-forming elements. Tungsten steels, except the low-tungsten chromium–tungsten steels, are not suitable for construction, but they are widely used for cutting tools, because the tungsten forms hard abrasion-resistant particles in high-carbon steels. Tungsten also increases the acid resistance and corrosion resistance of steels. The steels are difficult to forge, and cannot be readily welded when tungsten exceeds 2%. Standard tungsten–chromium alloy steels 72XX contain 1.5 to 2% tungsten and 0.50 to 1% chromium. Many tool steels rely on tungsten as an alloying element, and it may range from 0.50 to 2.50% in cold-work and shock-resisting types to 9 to 18% in the hot-work type, and 12 to 20% in high-speed steels.

ULTRAHIGH-STRENGTH STEELS

These are the highest-strength steels available. Arbitrarily, steels with tensile strengths of around 1378 MPa or higher are included in this category, and more than 100 alloy steels can be thus classified. They differ rather widely among themselves in composition or the way in which the ultrahigh strengths are achieved.

Medium-carbon low-alloy steels were the initial ultrahigh-strength steels, and within this group, a chromium–molybdenum steel (4130) grade and a chromium–nickel–molybdenum steel (4340) grade were the first developed. These steels have yield strengths as high as 1654 MPa and tensile strengths approaching 2068 MPa. They are particularly useful for thick sections because they are moderately priced and have deep hardenability. Several types of stainless steels are capable of strengths above 1378 MPa, including a number of martensitic, cold-rolled austenitic, and semiaustenitic grades. The typical martensitic grades are types 410, 420, and 431, as well as certain age-hardenable alloys. The cold-rolled austenitic stainless steels work-harden rapidly and can achieve 1241 MPa tensile yield strength and 1378 MPa ultimate strength. Semiaustenitic stainless steels can be heat-treated for use at yield strengths as high as 1516 MPa and ultimate strengths of 1620 MPa.

Maraging steels contain 18 to 25% nickel plus substantial amounts of cobalt and molybdenum. Some newer grades contain somewhat less than 10% nickel and between 10 and 14% chromium. Because of the low-carbon (0.03% max) and nickel content, maraging steels are martensitic in the annealed condition, but are still readily formed, machined, and welded. By a simple aging treatment at about 482°C, yield strengths of as high as 2068 and 2413 MPa are attainable, depending on specific composition. In this condition, although ductility is fairly low, the material is still far from being brittle.

Among the strongest of plain carbon sheet steels are the low- and medium-carbon sheet grades called MarTinsite. Made by rapid water quenching after cold rolling, they provide tensile yield strengths to 1517 MPa but are quite limited in ductility.

There are two types of ultrahigh-strength, low-carbon, hardenable steels. One, a chromium–nickel–molybdenum steel, named Astralloy, with 0.24% carbon is air-hardened to a yield strength of 1241 MPa in heavy sections when it is normalized and tempered at 260°C. The other type is an iron–chromium–molybdenum–cobalt steel and is strengthened by a precipitation hardening and aging process to levels of up to 1654 MPa in yield strength. High-alloy quenched-and-tempered steels are another group that have extra-high strengths. They contain 9% nickel, 4% cobalt, and from 0.20 to 0.30% carbon, and develop yield strengths close to 2068 MPa and ultimate strengths of 2413 MPa. Another group in this high-alloy category resembles high-speed tool steels, but are modified to eliminate excess carbide, thus considerably improving ductility. These so-called matrix steels contain tungsten, molybdenum, chromium, vanadium, cobalt, and about 0.5% carbon. They can be heat-treated to ultimate strengths of over 2757 MPa — the highest strength at present available in steels, except for heavily cold-worked high-carbon steel strips used for razor blades and drawn wire for musical instruments, both of which have tensile strengths as high as 4136 MPa.

ULTRAVIOLET-CURABLE HOT-MELT ADHESIVES

For years, ultraviolet (UV)-curable pressure-sensitive adhesives (PSAs) have been recognized as a fixture alternative to solvent-borne products. The idea of achieving the solvent and heat resistance of an acrylic without facing the various safety and environmental ramifications has always been enticing to both PSA formulators and users. The promise of this technology has led to the development of a variety of adhesive technology platforms.

Photoinitiators can now be purchased that offer much better thermal stability for improved pot life and coatability. Polymers have been developed that are much more chemically active, dramatically reducing the amount of photoinitiator required to achieve proper cure (and consequently the total cost of the adhesive). Thanks to the response of their suppliers, adhesive manufacturers are making UV-curable products that are more versatile than ever.

Converting a pressure-sensitive hot-melt coating line over to UV-curing no longer demands growing accustomed to radically different adhesives.

PROPERTIES AND APPLICATIONS

Conventional hot-melt PSAs are widely used for tape and label applications. Their room-temperature performance is difficult to match with alternative chemistries. They possess an outstanding combination of high tack, peel, and shear and adhere well to wet or low-energy surfaces.

In addition to the performance advantages, conventional hot melts possess some significant processing advantages because they require no solvent vehicle for application. The lack of a combustible solvent makes them safer and more environmentally friendly than any other adhesive.

Because they require no drying, they can be applied more easily at high depositions for use on slick or rough surfaces. Finally, since they require no dryers, hot-melt coaters are generally more compact and lower in cost than liquid coaters.

Unfortunately, users of conventional hot melts can only enjoy these benefits over a limited range of conditions. The products are hobbled by poor resistance to solvents, plasticizers, and heat. This precludes their use in some industrial applications where their high room-temperature peel and shear could make them otherwise well suited.

TYPES AND FORMS

Traditional pressure-sensitive hot melts are formulated primarily with block copolymers and various tackifying resins. The cohesive strength of the product is largely determined by the block copolymer used. Some of the most common block copolymers used are styrenic triblock copolymers. These are long polymer chains with polystyrene molecules grouped together to form two end blocks surrounding one mid-block made of an elastomeric material. Frequently used triblock copolymers are styrene–isoprene–styrene (SIS) or styrene–butadiene–styrene (SBS).

FUTURE

UV offers an outstanding combination of versatile performance and ease of use. The UV-curable hot melt is comparable to the solvent-borne acrylic. In addition, a significant performance advantage has been seen on low-energy surfaces. This makes the technology even more appealing because of the ever-increasing use of plastics.

With new tools at their disposal, the performance of UV-curables is now up to any adhesive task required. Formulators have crafted newer and better adhesives capable of a variety of tasks. These products capture the traditional advantages of hot melts while meeting many of the standards of acrylics.

UNIFORM MAGNETIC HEATING

Uniform magnetic heating (UMH) is a system by which electrical energy is converted to heat within metallic materials in a very efficient and flexible manner. Although this system and conventional induction heating both require

UNIFORM MAGNETIC HEATING

electrical coils to convert electricity to magnetic flux energy, the similarities stop there

UMH vs. Induction

The CoreFlux UMH system utilizes two coils that are permanently fixed around a C-shaped laminate core. Similar to induction systems, the coils convert electrical energy into magnetic flux energy. However, in conventional induction, the component to be heat treated is placed inside the coil. The energy is directly transferred to the part in the form of surface eddy currents, which are generated as the current flows around the component.

The CoreFlux system transfers energy in a different way. Energy is transferred into a laminate core (similar to a transformer core) and channeled directly to the part in a linear manner. In this way, the magnetic flux energy is distributed throughout the entire part. Key to the technology is that the flux direction in the core oscillates at a user-defined frequency from 20 to 400 Hz. As a result, the polarity of the flux field changes at this defined frequency. Each time the polarity changes, heat is released throughout the component via "hysteresis loss."

Simply put, hysteresis loss is the energy released throughout the material as the magnetic domains in the microstructure are forced to realign continually with the alternating magnetic field. The effect of this phenomenon is the uniform production of heat throughout the component.

Because of the inherent nature of this method, the core and surface temperatures show minimal thermal gradients throughout the entire heating process. With induction, localized overheating of components with holes and unusual characteristics is a common problem, but overheating is not typically a concern with UMH.

In addition to the benefits of uniform through-heating, UMH also provides several other key advantages:

- The coils are permanently fixed to the machine and require no maintenance or coil changeovers.
- The same coils can run a wide range of processes and bring the benefit of flexibility by running families of parts with no coil changes required.
- Metallurgical results are typically significantly better than conventional systems.
- The machine requires no water cooling, which results in lower facility costs, less maintenance, and no energy lost to heated water.
- Energy efficiency can be as much as twice that of conventional induction. The power supply is basically a standard AC variable motor drive. This eliminates costly custom power supplies and the problems associated with their maintenance. It also greatly reduces overall capital equipment cost (Figure U.1).

Applications

Heating Press Dies

One simple application of the technology that offers significant benefits is the heating of press dies. A variety of die shapes may be placed in

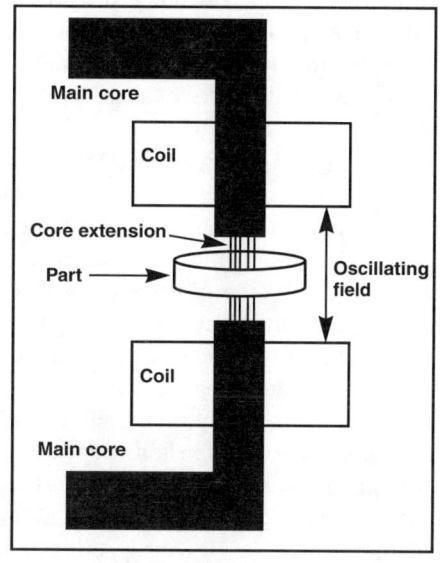

FIGURE U.1 The component to be treated may be placed around the core extension, which provides a secondary field that causes very rapid and uniform heating with no part contact. UMH generates a uniform field from the inner core extension outward through the entire part. (From *Adv. Mater. Proc.*, 154(5), 41–43, 1998. With permission.)

a common machine that can heat the dies to the required temperature in a matter of minutes. As an alternative to oven preheating, the UMH process offers much faster, cleaner, and more efficient heating, and allows customers to change dies more quickly. Machines may also be mobile, so that one machine can service multiple press locations. Current size and shape capabilities range from very small to approximately $180 \times 45 \times 45$ cm. Larger components may be accommodated with custom designs.

Tempering Gears and Bearings

One of the most-promising areas of application for this technology is the tempering of gears and bearings. Parts do not need to be rotated, and taller parts do not require scanning. For example, if a component has a round shape and an inner bore, the CoreFlux UMH technology can be applied by utilizing a core extension. The component is placed around the core extension, which provides a secondary field that causes very rapid and uniform heating with no part contact. UMH generates a uniform field throughout the component, from the inner core extension. On the other hand, if the part were induction tempered, the coil would be placed around the outside diameter, and the field would be generated from the outside inward to the core.

Fortunately, the CoreFlux UMH process overcomes time-at-temperature and the skin-effect phenomenon. Because of the uniformity of heating, the core of the gear comes to temperature at virtually the same rate as the teeth. Metallurgical results typically exceed expectations and return properties similar, and sometimes superior, to oven tempering. An additional benefit is the ability to run multiple parts around the same core extension. Depending on the actual geometry, parts can be stacked around the core extension with minimal effect on the total cycle time. Obviously, this can have a profound effect on production throughput and machine utilization.

Hardening Gears and Bearings

The same core extension approach described above may be utilized for higher-temperature applications. Although the research and documentation related to hardening is less mature than the lower-temperature applications, the technology has again demonstrated substantial benefits over today's alternatives. Targeted at through-hardening applications only, the technology offers the same flexibility as in tempering. In fact, UMH has potential for the design of an entire hardening and tempering line that could allow for the flexible running of components inside a large "family" grouping, with no setup changes.

Heating Aluminum Billets

The properties of aluminum make billets difficult to through-heat quickly and uniformly by conventional technology. With the CoreFlux UMH process, an aluminum billet may be placed directly on the insulated core cap, and the top core/coil assembly may then be lowered to make light contact with the billet. This clamping effect is utilized to create the most efficient transfer of energy into the billet, and to facilitate holding the billet in place during heating. Clamping pressure is adjustable to eliminate any marking or deformation.

Capabilities have been documented and proved, and application development continues in the aluminum field, with preheating applications ranging from 370°C to semisolid temperatures. Although steel forging offers similar promise, development is still under way in this area to assure that the machine cores will endure long exposures to extreme forging temperatures.

Shrink-Fit Applications

Another simple through-heat application is preheating components for shrink fitting. Again, when compared with any other alternative available for lower-temperature through-heating, the CoreFlux UMH process is an improvement.

Press Tempering

Although a relatively new development, press-temper applications have recently drawn considerable attention. In cases where thin parts must be stacked and held flat during tempering or stress relieving, the CoreFlux process is worth considering. The top coil and core assembly are typically lowered via pneumatic cylinders, and the clamping pressure is limited to as little as a pound. However, the pressure can be adjusted to provide more than a ton of damping force if necessary. If higher force is required, the machine can be customized. In addition, this system could allow for the pressure to be controlled as a function of temperature.

UNSATURATED POLYESTER RESIN

The use of unsaturated polyester resins in structural applications is well documented. There are, however, significant quantities of unsaturated polyester resins used in specialist compounded products, which are more likely to be unreinforced. The most well known of these technologies are formulated gel-coats, a technology that has been changing rapidly in recent years with improvements in gloss retention, color retention, and volatile organic compound emissions.

The introduction of granite effect coatings and solid surface material is a further example of the versatility of unsaturated polyester resins. Although these materials have been predominantly used for interior applications, their potential for exterior use on buildings provides exciting possibilities for a new and varied range of composite building materials providing stone effects at a fraction of the weight of conventional building materials.

Other compounded resins that are especially important to the building and construction market are those with fire resistant characteristics. In addition, the improvements in smoke reduction from unsaturated polyester resin systems make such materials attractive for cladding applications. Combining the advantages of these resins with decorative coatings and sandwich construction provides the basis for structural, insulating components.

Markets

The markets for reinforced plastics are frequently split into a number of generally accepted sectors, such as marine, land transport, building and construction, and chemical containment. There are, of course, subdivisions in each sector, for example, powered pleasure boats, powered work boats, sailboats, and offshore applications in the marine market, but most of the discussion in the literature is about the use of fiber-reinforced composites in these market sectors and market subgroups. In general, unfilled resins with good mechanical properties are preferred, but there are, very often, requirements for compounded products to provide special characteristics to meet specific performance requirements. Obviously, compounded fire-resistant materials fall into such a category and are used to impart resistance to ignition, resistance to surface spread of flame, and, increasingly, reduction in emissions of smoke and toxic fumes. Although such materials are often highly filled, they are used with fiber reinforcement for the manufacture of structural and semistructural components. The importance of these resins and their developments together with two other important compounded unsaturated polyester resin-based products has been disclosed. These latter materials are not used in conjunction with fiber reinforcement but are usually simply filled or pigmented; they are gel-coats and are mainly used as "in-mold" coatings and solid surface materials for the manufacture of synthetic granite-type products.

Resin concrete and repair putties are also large consumers of unsaturated polyester resin in non-fiber-reinforced compounds.

Resistance to Fire

The use of glass-fiber-reinforced plastics (GRP) in applications where fire resistance was particularly important was introduced into the building industry five decades ago. Generally, the structural performance of the material was not questioned for building applications because it had been well proven for the construction of boats. However, as with most plastic materials, its ability to perform under fire conditions was in question for use in buildings even though it had been documented that fires in buildings originate from the contents and in a vast majority of circumstances the structure does not contribute to loss of life.

One of the most successful means to improve the resistance of plastics to fire is by the incorporation of fillers, which break down with heat to produce heavy vapors to prevent oxygen reaching the surface of the material and hence reduce the possibility of burning. The major problem associated with the high levels of filler required to render resins fire retardant is the increase in their viscosity, which results in handling difficulties when manufacturing structural components.

The use of halogenated additives, which work synergistically with some fire-retardant fillers, help to overcome handling problems but result in the potential for toxic fume production under fire conditions.

The availability of improved viscosity modifiers is now enabling resins filled with high levels of nontoxic fillers, such as alumina trihydrate, to be used to manufacture laminates containing reasonable levels of reinforcement to produce, at least, semistructural components.

Such systems will meet the new International Maritime Organization (IMO) requirements for use on passenger ships. Under the test conditions, the material has to exhibit low surface spread of flame characteristics, low smoke emissions, and low emissions of carbon monoxide.

Gel-Coat Protection

In the early days of the GRP industry, the need for resin-rich surfaces was established to:

- Improve the durability of components
- Protect the laminate from the environment
- Reduce fiber pattern
- Provide a smooth aesthetic finish
- Eliminate the need for painting

As a result of these requirements, a market for ready-formulated coatings was established and gel-coat product ranges became established.

The availability of quality "in-mold" coatings, such as gel-coats, to fabricators saves labor and wastage in the workshop and improves the quality of molded components. Gel-coats are available in brush and spray versions with a variety of properties and performance characteristics to meet a range of needs. They must be applied carefully and correctly to avoid faults.

Gel-Coat Developments

Over the years the need for improved gloss and color retention in gel-coats has been recognized and developments in ultraviolet (UV) resistance and color fastness have resulted in a range of gel-coats that can be weathered under the severest tropical weather conditions without changes in appearance.

Solid Surfaces

Resins have often been used to bind together fillers and aggregates to produce materials such as resin concrete and synthetic cultured and onyx marble. For decorative surfaces a clear (translucent) gel-coat is used to improve the quality of the surface finish and remove the effects of surface porosity. Although the gel-coat used is usually based on good quality, water-resistant resins, the inferior quality of the backing systems often results in a material that is susceptible to crazing, cracking, poor water resistance, and poor thermal resistance. Because the gel-coat surface is too thin for repairs to be effectively carried out, the problems cannot be easily rectified.

The monopoly of the acrylic-based solid surface material has been gradually eroded by the introduction of unsaturated polyester-based solid surface, which offer a much wider range of colors to provide improved customer choice. Raw materials and manufacturing processes have been designed to eliminate voids in polyester-based solid surfaces.

Traditionally, solid surface materials have been used for the manufacture of kitchen surfaces, sinks, and bathroom units. However, there is increasing interest in more diverse applications such as furniture, table tops, tiles, paneling, cutlery, and pens. It is also possible to use the material as a 2- to 3-mm-thick coating for other materials and the granite effect finish is reviving interest in GRP for cladding for buildings.

The unsaturated polyester resin-based material comprises of three components:

1. *Chips.* The colored fillers or chips, used to provide the granite effect, can be based on thermoplastic or thermoset materials.
2. *Resins.* Resin must be clear and near "water white" to allow the depth of color of the chips to be appreciated. The resin must also be resistant to elevated temperature, water, staining, UV light, and cigarette burns. Hence, typical formulations giving an acceptable level of performance are based on isophthalic acid and neopentyl glycol (NPG).
3. *Fillers.* Only alumina trihydrate can be used in addition to the colored "chips" because it is translucent. It also offers fire-retardant characteristics.

Solid surface systems are nonreinforced and can be machined and cut with conventional woodworking equipment. Patterns can be routed in solid surface materials and cast resin "in laid" to provide a variety of customized finishes.

It is important to ensure when manufacturing solid surface that the resin is formulated to accept high filler loading without air entrapment and will develop hardness rapidly. The final product must be resistant to chipping, cracking, hot-cold water cycling, "blushing," and UV light. It must also be easy to machine for shaping and finishing.

Future

Unsaturated polyester resin-based compounded products provide a range of materials with tailored performance characteristics for a variety of markets.

Gel-coats are essential for most applications for GRP, providing aesthetic finishes in the marine, transport, building, and construction markets. They have well-proven durability but improvements in gloss and color retention will ensure their position as the major coating for fiber-reinforced composite materials in the future.

Fire-retardant resins with exceptionally low smoke production under fire conditions are becoming a reality with unsaturated polyester resin-based systems. New standards are providing new challenges, which are being met successfully to ensure materials meet new requirements for surface spread of flame for materials for use in construction applications.

URANIUM

An elementary metal (symbol U), uranium never occurs free in nature but is found chiefly as an oxide in the minerals pitchblende and carnotite where it is associated with radium. The metal has a specific gravity of 18.68 and atomic weight 238.2. The melting point is about 1133°C. It is hard but malleable, resembling nickel in color, but related to chromium, tungsten, and molybdenum. It is soluble in mineral acids.

Uranium has three forms. The alpha phase, or orthorhombic crystal, is stable to 660°C; the beta, or tetragonal, exists from 660 to 760°C; and the gamma, or body-centered cubic, is from 760°C to the melting point. The cast metal has a hardness of 80 to 100 Rockwell B, work-hardening easily. The metal is alloyed with iron to make ferrouranium, used to impart special properties to steel. It increases the elastic limit and the tensile strength of steels, and is also a more powerful deoxidizer than vanadium. It will denitrogenize steel and has also carbide-forming qualities. It has been used in high-speed steels in amounts of 0.05 to 5% to increase the strength and toughness, but because of its importance for atomic applications its use in steel is now limited to the by-product nonradioactive isotope uranium-238.

Uses

Metallic uranium is used as a cathode in photoelectric tubes responsive to ultraviolet radiation. Uranium compounds, especially the uranium oxides, were used for making glazes in the ceramic industry and also for paint pigments. It produces a yellowish-green fluorescent glass, and a beautiful red with yellowish tinge is produced on pottery glazes. Uranium dioxide, UO_2, is used in sintered forms as fuel for power reactors. It is chemically stable, and has a high melting point at about 2760°C, but a low thermal conductivity. For fuel use the particles may be coated with about 0.003 cm of aluminum oxide. This coating is impervious to xenon and other radioactive isotopes so that only the useful power-providing rays can escape. These are not dangerous at a distance of about 15 cm, and thus less shielding is needed. For temperatures above 1260°C, a coating of pyrolitic graphite is used.

Uranium has isotopes from 234 to 239, and uranium-235, with 92 protons and 143 neutrons, is the one valued for atomic work.

UREA

Also called carbamide, urea is a colorless to white crystalline powder, $NH_2 \cdot CO \cdot NH_2$, best known for its use in plastics and fertilizers. The chemistry of urea and the carbamates is very complex, and a great variety of related products are produced. Urea is produced by combining ammonia and carbon dioxide, or from cyanamide, $NH_2 \cdot C \cdot N$. It is a normal waste product of animal protein metabolism, and is the chief nitrogen constituent of urine. It was the first organic chemical ever synthesized commercially. It has a specific gravity of 1.323, and a melting point at 135°C.

TYPES

The formula for urea may be considered to be $O \cdot C(NH_2)_2$, and thus as an amide substitution in carbonic acid, $O \cdot C(OH)_2$, an acid that really exists only in its compounds. The urea-type plastics are called amino resins. The carbamates can also be considered as deriving from carbamic acid, NH_2COOH, an aminoformic acid that likewise appears only in its compounds. The carbamates have the same structural formula as the bicarbonates, so that sodium carbamate has an NH_2 group substituted for each OH group of the sodium bicarbonate. The urethanes, used for plastics and rubber, are alkyl carbamates made by reacting urea with an alcohol, or by reacting isocyanates with alcohols or carboxyl compounds. They are white powders of the composition $NH_2COOC_2H_5$, melting at 50°C.

Isocyanates are esters of isocyanic acid, $H \cdot N \cdot C \cdot O$, which does not appear independently. The dibasic diisocyanate is made from a 36-carbon fatty acid. It reacts with compounds containing active hydrogen. With modified polyamines it forms polyurea resins, and with other diisocyanates it forms a wide range of urethanes. Tosyl isocyanate for producing urethane resins without a catalyst is toluene sulfonyl isocyanate. The sulfonyl group increases the reactivity.

Methyl isocyanate, CH_3NCO, known as MIC, is a colorless liquid with a specific gravity of 0.9599. It reacts with water. With a flash point of less than –6.6°C, it is flammable and a fire risk. It is a strong irritant and is highly toxic. One of its principal uses is as an intermediate in the production of pesticides.

Urea is used with acid phosphates in fertilizers. It contains about 45% nitrogen and is one of the most efficient sources of nitrogen. Urea reacted with malonic esters produces malonyl urea, which is the barbituric acid that forms the basis for the many soporific compounds such as luminal, phenobarbital, and amytal. The malonic esters are made from acetic acid, and malonic acid derived from the esters is a solid of the composition $CH_2(COOH)_2$, which decomposes at about 160°C to yield acetic acid and carbon dioxide.

For plastics manufacture, substitution on the sulfur atom in thiourea is easier than on the oxygen in urea. Thiourea, $NH_2 \cdot CS \cdot NH_2$, also called thiocarbamide, sulfourea, and sulfocarbamide, is a white, crystalline, water-soluble material of bitter taste, with a specific gravity of 1.405. It is used for making plastics and chemicals. On prolonged heating below its melting point, 182°C, it changes to ammonium thiocyanate, or ammonium sulfocyanide, a white, crystalline, water-soluble powder of the composition NH_4SCN, melting at 150°C. This material is also used in making plastics, as a mordant in dyeing, to produce black nickel coatings, and as a weed killer. Permafresh, used to control shrinkage and give wash-and-wear properties to fabrics, is dimethylol urea, $CO(NHCH_2OH)_2$, which gives clear solutions in warm water.

Urea-formaldehyde resins are made by condensing urea or thiourea with formaldehyde. They belong to the group known as aminoaldehyde resins made by the interaction of an amine and an aldehyde. An initial condensation product is obtained that is soluble in water, and is used in coatings and adhesives. The final condensation product is insoluble in water and is highly chemical resistant. Molding is done with heat and pressure. The urea resins are noted for their transparency and ability to take translucent colors. Molded parts with cellulose filler have a specific gravity of about 1.50, tensile strength from 41 to 89 MPa, elongation 15%, compressive strength to 310 MPa, dielectric strength to 16×10^6 V/m, and heat distortion temperature to 138°C. Rockwell hardness is about M 118. Urea resins are marketed under a wide variety of trade names. The Uformite resins are water-soluble thermosetting resins for adhesives and sizing. The Urac resins, and the Casco resins and Cascamite, are urea-formaldehyde. They are used as adhesives for plasterboard, plywood, and in wet-strength paper.

URETHANES

Also termed polyurethanes, urethanes are a group of plastic materials based on polyether or polyester resin. The chemistry involved is the reaction of a diisocyanate with a hydroxyl-terminated polyester or polyether to form a higher-molecular-weight prepolymer, which in turn is chain-extended by adding difunctional compounds containing active hydrogens, such as water, glycols, diamines, or amino alcohols. The urethanes are block polymers capable of being formed by a literally indeterminate number of combinations of these compounds. The urethanes have excellent tensile strength and elongation, good ozone resistance, and good abrasion resistance. Combinations of hardness and elasticity unobtainable with other systems are possible in urethanes, ranging from Shore hardnesses of 15 to 30 on the "A" scale (printing rolls, potting compounds) through the 60 to 90 A scale for most industrial or mechanical goods applications, to the 70 to 85 Shore "D" scale. Urethanes are fairly resistant to many chemicals such as aliphatic solvents, alcohols, ether, certain fuels, and oils. They are attacked by hot water, polar solvents, and concentrated acids and bases.

Urethane Foams

Urethane foams are made by adding a compound that produces carbon dioxide or by reaction of a diisocyanate with a compound containing active hydrogen. Foams can be classified somewhat according to modulus as flexible, semiflexible or semirigid, and rigid. No sharp lines of demarcation have been set on these different classes as the gradation from the flexibles to the rigids is continuous. Densities of flexible foams range from about 16 kg/m³ at the lightest to 64 to 80 kg/m³ depending on the end use. Applications of flexible foams range from comfort cushioning of all types, e.g., mattresses, pillows, sofa seats, backs, and arms, automobile topper pads, and rug underlay, to clothing interliners for warmth at light weight.

Flexible Types

The techniques of manufacture of flexible urethane foam vary widely, from intermittent hand mixing to continuous machine operation, from prepolymer to one-shot techniques, from slab-forming to molding, from stuffing to foamed-in-place.

Future applications envision the flexible foam not as a substitute for latex rubber foam or cotton, but as a new material of construction allowing for design of furniture, for example, that is essentially all foam with a simple cloth cover and a very simple metal-supporting framework.

Rigid Types

Densities from about 24 to 800 kg/m³ on the semirigid side have been produced with corresponding compression strengths again for particular end uses ranging from insulation to fully supporting structural members. The usefulness of the urethane system has been in the foam-in-place principle using a host of containing wall materials.

Applications in the more rigid foam field have been thermal insulation of all types (low-temperature refrigeration ranging from liquid nitrogen temperatures up to the freezing point of water and high temperature insulation of steam pipes, oil lines, etc.); shock absorption such as packaging, crash pads, etc., where the higher hysteresis values produce either a better one-time high impact "crash" use or, more often, lower amplitude but higher frequency container end use; filtration (air, oil, etc., where a large surface-to-volume ratio is needed with a simple technique to produce a reusable filter to allow for its initially higher cost factor); structural (building applications of all kinds combining a good thermal as well as structural behavior, filling of building voids, and curtain walls are some basic applications); flotation (boats, buoys, and every other imaginable object afloat represents some possible application of urethane foams); and, finally, general-purpose applications that include all other uses such as decorative applications.

Rigid foams can be produced using a simple spray technique and a number of machines are sold on the market for this technique. Time-consuming layup of foam is eliminated using this method. Insulation of walls, tanks, etc. are applications in use today. With the use of low-vapor-pressure isocyanates such as MDI (4,4'-diphenylmethane diisocyanate), the potential irritant hazard during spraying is greatly lowered. Self-adhesion of the sprayed foam is a valuable asset of this type of system.

Urethane foams offer advantages over many of the better-known foams such as latex foam rubber, polystyrene, and polyethylene, with the combination of excellent properties and lower installed costs. Depending on the application, a lower foam density can be used with similar load-bearing properties, also one having an extremely low thermal conductivity can be fabricated. The oil resistance, high-temperature resistance, good high-tensile properties, good permanence properties, resistance to mildew, resistance to flammability, and so on are in general the types of properties that, combined with foamed-in-place technology, put urethane foam far ahead of competitive materials.

OTHER URETHANES

Thermoplastic polyurethanes (TPU) include two basic types: esters and ethers. Esters are tougher, but hydrolyze and degrade when soaked in water. There also are TPUs based on polycaprolactone, which while technically being esters, have better resistance to hydrolysis. TPUs are used when a combination of toughness, flex resistance, weatherability, and low-temperature properties are needed. These materials can be injection-molded, blow-molded, and extruded as profiles, sheet, and film. Further, TPUs are blended with other plastic resins, including polyvinyl chloride, ABS, acetal, SAN, and polycarbonate.

Urethane elastomers are made with various isocyanates, the principal ones being TDI (tolylene diisocyanate) and MDI (4,4'-diphenylmethane diisocyanate), reacting with linear polyols of the polyester and polyether families. Various chain extenders, such as glycols, water, diamines, or aminoalcohols, are used in either a prepolymer or a one-shot type of system to form the long-chain polymer.

Flexible urethane fibers, used for flexible garments, are more durable than ordinary rubber fibers or filaments, and are 30% lighter in weight. They are resistant to oils and to washing chemicals, and also have the advantage that they are white in color. Spandex fibers are stretchable fibers produced from a fiber-forming substance in which a long chain of synthetic molecules are composed of a segmented polyurethane. Stretch before break of these fibers is from 520 to 610%, compared to 760% for rubber. Recovery is not as good as in rubber. Spandex is white and dyeable. Resistance to chemicals is good but it is degraded by hypochlorides.

There are six basic types of polyurethane coatings, or urethane coatings, as defined by the American Society for Testing and Materials (ASTM), Specification D16. Types 1, 2, 3, and 6 have long storage life and are formulated to cure by oxidation, by reaction with atmospheric moisture, or by heat. Types 4 and 5 are catalyst-cured and are used as coatings on leather and rubber and as fast-curing industrial product finishes. Urethane coatings have good weathering characteristics as well as high resistance to stains, water, and abrasion.

FABRICATION

Urethane elastomers can be further characterized by the method of fabrication of the final article. Three principal types of fabrication are possible: (1) casting technique where a liquid prepolymer or a liquid mixture of all initial components (one-shot) is cast into the final mold, allowed to "set" and harden, and is then removed for final cure; (2) millable gum technique where conventional rubber methods and equipment are used to mill the gum, add fillers, color, etc., and/or banbury, extrude, calender, and compression mold the final shaped item; (3) thermoplastic processing techniques where the resin can be calendered, extruded, and injection- or blow-molded on conventional plastic machinery in final form (an important benefit here is that scrap can be reground and reused in fabricating other parts).

The choice of the proper method of fabrication largely depends on the economics of the process, because the properties of the final product may be about the same regardless of the method of fabrication. If a few large-volume items are needed, casting these into a single mold is usually more economical. However, if many thousands of small, intricate pieces are needed, usually injection molding is the preferred, more economical method of fabrication.

Uses

Applications of urethane elastomers have been developed where high abrasion resistance, good oil resistance, and good load-bearing capacity are of value, as in solid tires and wheels, especially of industrial trucks, the shoe industry, drive and belting applications, printing rolls, gasketing in oil, etc. Other applications include vibration dampening; for example, in hammer heads, air hammer handles, shock absorption underlays for heavy machinery, etc.; low coefficient of friction with the addition of molybdenum disulfide for self-lubricating uses as ball and socket joints, thrust bearings, leaf spring slide blocks, etc. In the electrical industry, cable jacketing and potting compounds are developing as important uses. Various systems of urethane elastomers with specific fillers have been developed into an important class of caulks and sealants, which is just beginning to take hold in applications such as concrete road-expansion joints, building caulking, and so on, in direct competition with such older materials as the polysulfides but at a much lower price and superior properties.

A host of other applications varies from adhesive bonding of fibers of all kinds to rocket fuel binders of the more exotic variety, which are becoming so important in the U.S. national defense picture. Therefore, it is imperative that design engineers understand fully the material they are using and how they intend to utilize it in the final piece of equipment. For example, one recommendation is to limit the use of urethanes to below 82°C in water for continuous exposures. Dry uses can go somewhat higher, e.g., to 107°C for certain systems. In oil, exposures can be up to 121°C. Disregard of such limitations can result in failures, but the design engineer can eliminate these by the proper choice of material. On the other hand, the design engineer should choose the urethanes for their virtues, such as hardness and elasticity, where other materials such as natural and other synthetic rubbers may fail.

Properties

The urethanes have excellent tensile strengths and elongation, good ozone resistance, and good abrasion resistance. Knowledge of these properties is mandatory for good engineering design.

The greater load-bearing capacity of urethanes as compared to other elastomers is noteworthy, for it leads to smaller, less costly, lower-weight parts in equivalent applications. Tear strength is extremely high, which may be important in particular applications along with the very high tensile strengths. The high abrasion resistance has made possible driving parts for which no other materials could compete. However, in every such dynamic application, the engineer must design the part to allow for the higher hysteresis losses in the urethane. Whereas in some applications such as dampening, the higher hysteresis works to advantage, in others hysteresis will lead to part failure if the upper temperature limit is thereby exceeded. Redesign of the part (thinner walls, etc.) to allow for greater dissipation of the heat generated will permit the part to operate successfully. This has proved to be the case many times.

Urethane elastomers generally have good low-temperature properties. The same hysteresis effect works in reverse here so that a part in dynamic use at temperatures as low as –51°C, while stiff in static exposure, immediately generates enough heat in dynamic use to pass through its second-order transition and does not show any brittleness but becomes elastic and usable. By proper choice of the polyester or polyether molecular backbone, lower use temperatures (as low as –62°C) have been formulated in urethane elastomers.

In addition to good mechanical properties, urethanes have good electrical properties, which suggest a number of applications. Oxygen, ozone, and corona resistances of this system are generally excellent.

V

VACUUM ASSIST MOLD PROCESSING

The use of atmospheric pressure to hold closed molds together during injection was the early process of vacuum assisted resin injection (VARI). Adding vacuum has enabled resin-transfer molding (RTM) to challenge compression molding and autoclaving systems capable of making the best high-performance composites. Vacuum is used in two ways. First, it mixes resin and hardener under a 91-Pa vacuum just prior to injection. Using an impeller, it agitates the mixture to drive air to the surface where the vacuum removes it. Degassing, which takes about 2 h per tank, also removes volatiles and low-molecular-weight by-products.

Some companies place their RTM tools in a vacuum chamber rather than using a vacuum tool. The chamber creates a vacuum that does not vary, even as resin fills the tool. The hard vacuum pulls any air and water vapor off the preform and sucks resin into the mold.

This combination of degassed resin and vacuum keeps voids under 3% and often better. This ensures consistently high structural integrity because voids concentrate stresses that initiate fractures and cause premature failure. It takes only a 2% increase in voids to drain interlaminar shear strength 20% and flexural modulus 10%. The process matches equivalent compression molding and autoclaving fiber volumes and voids.

Compared with compression molding and autoclaving, VARI does not need to apply pressure over the entire skin surface to vanquish voids, simplifying cocuring. For example, the process can fabricate cores and reinforced skins in a single step rather than bonding them after fabrication.

More importantly, the process makes composites in fewer, more controllable steps. These resin injection and molding processes — RTM, VARI, vacuum resin-transfer molding (VRTM), and vacuum-assisted resin-transfer molding (VARTM) — are much simpler processes to use.

Each process step is also independent and controllable. In VRTM, air evacuation depends on only the vacuum, and resin preparation depends on only the resin mixer. Preform production varies with automated fabric weaving and preform placement, whereas core manufacture depends on molding or a machining process. The cure depends on a programmable heat source.

VRTM composites also show excellent resistance to water, solvents, and chemicals. This is largely a function of resin type and surface finish. Rough surfaces pitted with micropores trap water and chemicals and act as tiny reaction chambers that set in motion their own destruction. VRTM yields parts with less than 20-μin. rms (root mean square) porosity. VRTM can achieve this fine finish repeatedly on all surfaces, depending on the finish of the tool. For high-quality finishes, compression and autoclaving processes depend on uniform resin flow under pressure, which they cannot always maintain.

The main attraction of VRTM, despite its competitive properties, remains cost, where it offers real advantages over compression molding and autoclaving.

Another production process is low-cost VARTM infusion technology.

VARTM is becoming a manufacturing method of choice because of its ability to produce fairly large structures out of the autoclave with the high quality usually associated with higher-priced processes.

VACUUM CARBURIZING

Heat treatment with gas quenching has already been an established heat-treatment process for two or three decades in the field of full hardening. At first, it was limited to the hardening of high-alloyed tool steels whose alloy structure enabled them to be hardened satisfactorily with a rather slow gas cooling rate.

The enhanced quenching action achieved with gas pressures above 10 bar has allowed successful extension of gas quenching to the field of low alloyed tool steels, steels for hardening and tempering, antifriction-bearing steels, and case-hardening steels. The capability to carburize and gas-quench in vacuum furnace installations has provided the industry with a new, environmentally friendly case-hardening process.

THE PROCESS

Like plasma carburizing, vacuum carburizing can also be performed in a vacuum furnace system. Vacuum carburizing can be succinctly described by the following key points:

- Carburizing gas is propane.
- Pressure ranges up to 20 mbar (absolute).
- Temperature range is usually 900 to 1050°C, but higher temperatures are also possible.

Once the charge has been heated to the carburizing temperature under a neutral atmosphere (vacuum or nitrogen), propane is admitted into the evacuated heating chamber. Propane very rapidly undergoes 100% dissociation into more stable hydrocarbons and hydrogen. Carbon is also released and diffuses through the surface of the steel or component.

Vacuum carburizing is characterized by a high carbon mass flow rate, which carburizes the surface layer to near the carbon saturation limit within a short treatment time. In the subsequent diffusion phase, no more carburizing gas is fed in — rather the existing carbon diffuses farther into the steel in accordance with the diffusion law until the desired carbon profile has been attained.

PROCESS COMPARISONS

In contrast to protective-gas carburizing, vacuum carburizing can be performed with substantially higher case carbon contents. The case carbon percentage is already over 1.3% after a short period of carburization and then is held at 1.4 to 1.5% at 930°C, which is about 0.2% higher than in protective-gas carburizing with a carbon level just below the sooting limit. The higher case carbon content in vacuum carburizing results in shortened treatment times, even at the same carburizing temperature. Raising the carburizing temperature results in a further considerable time savings.

Vacuum carburizing systems readily permit a carburizing temperature of over 1050°C, although the heat treatment racks made of heat-resistant cast steel (which are currently in use) are no longer usable at such high temperatures. With racks made of CFC (carbon-fiber-reinforced carbon), the limit is shifted to much higher temperatures. CFC material can only be used in an oxygen-free atmosphere such as that prevailing during vacuum carburizing. Of course, the carburizing action at the point of contact with the component must be taken into consideration.

The grain growth of case-hardening steels also does not permit such high temperatures over a long period of time. The vacuum furnace offers a pearlitizing treatment to refine the grain. Despite the time cost for pearlitizing, the result in comparison to the time required in a multipurpose protective-gas chamber furnace is a time savings of about 4 h for the case hardening of a 25% Cr Mo 4 steel to a case depth (550 HV) of 1.7 mm.

DISTORTION

Vacuum carburizing with gas quenching offers a potential for reduced parts distortion. A large number of experiments, conducted primarily on transmission parts, have shown that the scatter of the dimensional and shape changes after gas quenching is narrower than after oil quenching.

For example, a clutch body (O.D., 84 mm; I.D., 50 mm; height, 15 mm; mass, 0.2 kg each) made of 16% Mn Cr 5 with a case depth (550 HV) of 0.4 to 0.8 mm was tested. The study

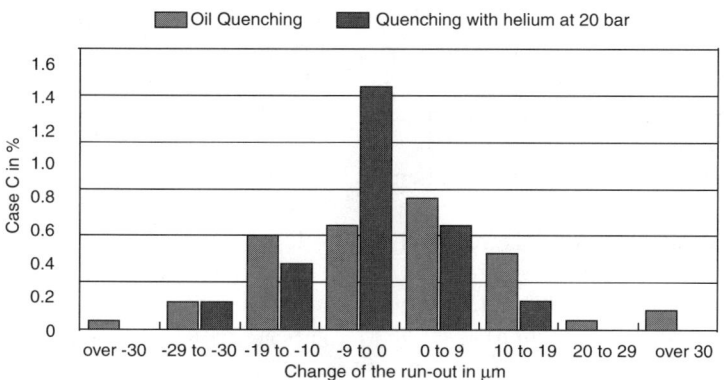

FIGURE V.1 Comparison of run-out between case hardening in the vacuum furnace with gas quenching and protective-gas carburizing with oil quenching. (From *Ind. Heating*, January, 54, 2000. With permission.)

evaluated so clutch bodies after case hardening in the vacuum furnace (quenching with helium at 20 bar) and, for comparison, 50 others were evaluated after case hardening in the protective-gas furnace (oil quenching). The radial run-out of the clutch bodies was measured in the soft and hard states. The difference is illustrated in Figure V.1.

Advantages

Case hardening in vacuum heat-treatment systems with gas quenching offers the user many advantages in comparison to conventional protective-gas carburizing with oil quenching. Parts are clean and dry after treatment requiring no washers or management or disposal of liquid waste. Leidenfrost phenomenon is avoided and with more uniform quenching, distortion is minimized. Vacuum carburizing also allows for carburizing at up to 1000°C.

As a protective atmosphere, vacuum can prevent case oxidation and eliminate toxic off-gases. The vacuum carburizing process also provides a high carbon mass flow rate with low consumption of carburizing gas. With regard to productivity, the vacuum process can be integrated into a production line without the burdening requirements for fire-protection and fire-extinguishing systems, excessive heat removal to the surroundings, or extensive exhaust gas handling.

In determining the carbon mass flow rate, it becomes clear that in the first few minutes of carburizing in this process (up to 30 min) there is a very high carbon mass flow rate of up to 100 g/m²h. The case-hardening steel can be carburized up to its limit of solubility in the surface layer without any sooting occurring. The system technology also makes it possible to carburize at temperatures above 1000°C. These two factors result in an enormously shortened process duration.

The carburization results are comparable with those of the protective-gas process with regard to case depth, case carbon content, and surface hardness. The advantages for component quality lie in reduced distortion. Investigations of various transmission parts have shown that the scatter of the dimensional and shape changes can be narrowed with gas quenching in comparison to oil quenching. The clean surface of the component and the absence of case oxidation after heat treatment are additional advantages of this technology.

VACUUM COATINGS

The process of vacuum coating is used to modify a surface by evaporating a coating material under vacuum and condensing it on the surface. It is normally carried out under high vacuum conditions (at approximately 1 millionth of an atmosphere pressure). The material to be evaporated is heated until its vapor pressure appreciably exceeds the residual pressure within the vacuum system.

Vacuum coating can be used for many applications. For example, optical lenses are coated with magnesium fluoride to a fraction of a wavelength to prevent glare and provide much better transmission of light and a more reliable optical system. The deposited film is extremely adherent and will withstand normal cleaning.

Silicon monoxide is frequently used as an abrasion-resistant coating material. As deposited, it is soft and requires postheat treatment in air to convert it to silicon dioxide, which is transparent and extremely hard. It is frequently used to protect front surface mirrors and increases abrasion resistance by a factor of 5000 to 10,000, while maintaining equal or higher reflectivity. Similarly, titanium is sometimes used for coating and is subsequently oxidized to yield a titanium dioxide abrasion-resistant surface.

By far the most common type of vacuum coating is the process of vacuum metallizing. In this process metal is evaporated and used as deposited without further treatment as opposed to the evaporation of compounds or materials that require posttreatment.

Vacuum metallizing has generally been used as a decorative process whereby costume jewelry, toys, etc. are given a metallic sheen and are made highly reflective. The base material may be either plastic or metal. In either case, the part is frequently lacquered before metallizing to prevent the evolution of gas from the base and to provide a smooth surface without mechanical buffing. Because the metal deposit is only about 2 or 3 millionths of an inch thick, the smooth surface is necessary to give a specular reflection.

When the metal is on the outside of the coated part, it is referred to as front surface. However, in applications where it is used on the back of a transparent plastic (e.g., dashboards and taillight assemblies on automobiles) it is referred to as a second surface coating. The advantage of second surface coating is provided by using the plastic as the exposed surface. Front surface coatings must generally be protected with a transparent lacquer overcoat (applied after metallizing) because the thin decorative coatings are not wear resistant in themselves.

Aluminum is the most popular vacuum-metallizing coating material for most applications. However, other metals may be used, such as zinc, cadmium, copper, silver, gold, or chromium. Of all these metals, aluminum has the best general combination of reflectivity, conductivity, and stability in air. By adding color to the topcoat lacquer, the aluminum deposit may be made to appear like copper or gold as well as metallic sheens of blues, reds, yellow, etc. By using separate sources for each constituent, it is possible to deposit alloys as well as pure metals.

The above applications are for parts produced by batch metallizing; i.e., the individual parts are mounted on racks inserted in the vacuum system and after the necessary vacuum and evaporation temperatures are obtained, the parts are rotated so that they are uniformly coated by the evaporating metal. In batch metallizing the aluminum is evaporated from tungsten filaments, which are heated by direct resistance. Because of this, the amount of aluminum that can be charged is limited and only thin coatings can be produced. Similarly, only small surfaces (a few square centimeters), such as can be exposed within a matter of seconds, can be metallized.

When it is desirable to coat larger surfaces, e.g., rolls of flexible material, a semicontinuous metallizing process must be employed. For semicontinuous metallizing a roll of material is mounted in the vacuum chamber and unrolled under vacuum to coat either or both sides of the web, which is subsequently rewound in vacuum. This process is currently in use for coating rolls of plastic sheeting and paper. To coat continuously over a period of hours, it is necessary to have larger volumes of aluminum available for evaporation than can be held on resistance-heated tungsten filaments. Therefore, the aluminum is generally heated by induction in crucibles.

Coating of rolls of materials provided one of the first functional applications for coatings that used the electrical conductivity of the metal deposited. This conductive layer was deposited on thin insulating layers of either paper or plastic and could be used for winding miniature condensers. The electrical conductivity is also used in the metallizing process itself as a means of measuring the amount of metal deposited. Since the conductivity is a function of the thickness of the metal, continuously measuring conductivity provides a control for the amount of metal deposited. Other functional uses of the coating are based on its reflectivity (e.g., reflective insulation).

Vacuum metallizing has recently been extended to include thick films, i.e., in the range of 1 to 3 mils. Such coatings serve as corrosion-resistant barriers, particularly on high-tensile-strength steel exposed to marine atmospheres. Where the temperature requirements of steel are less than 260°C, cadmium deposits can be used. For temperatures in excess of this, aluminum shows much better protection and does not react with the base steel as cadmium does.

Truly continuous operation is necessary for coating rolled steel. Here the rolls are unwound and rewound in air with the strip passing through seals into the vacuum chamber where it is coated. The metallizing of rolled stock allows separate control on each side of the web and the composition of the coating, as well as thickness, may be changed from one side to the other. There are several typical advantages of vacuum metallizing:

1. Close control of coating thickness and composition
2. Uniform deposits without buildup at sharp discontinuities
3. High coating rate
4. Low coating costs in volume production
5. Long life of equipment since few moving parts

There are also disadvantages of the process:

1. Part must be extremely clean.
2. Surfaces to be metallized must not evolve gas under vacuum.
3. Parts must not be temperature sensitive, i.e., must be stable to about 125°C.
4. Deposits form well only on a surface exposed to hot metal; reentrant angles are not well coated.

The cost for metallizing in production lots for corrosion-resistant coatings is comparable to electroplating. Decorative metallizing is generally much less expensive than electroplating.

VACUUM PROCESSING

Vacuum processing is used in many industrial applications. Some of these processes and their typical working pressure ranges are shown in Figure V.2. The application of vacuum technology is especially critical to the success of the various coating processes.

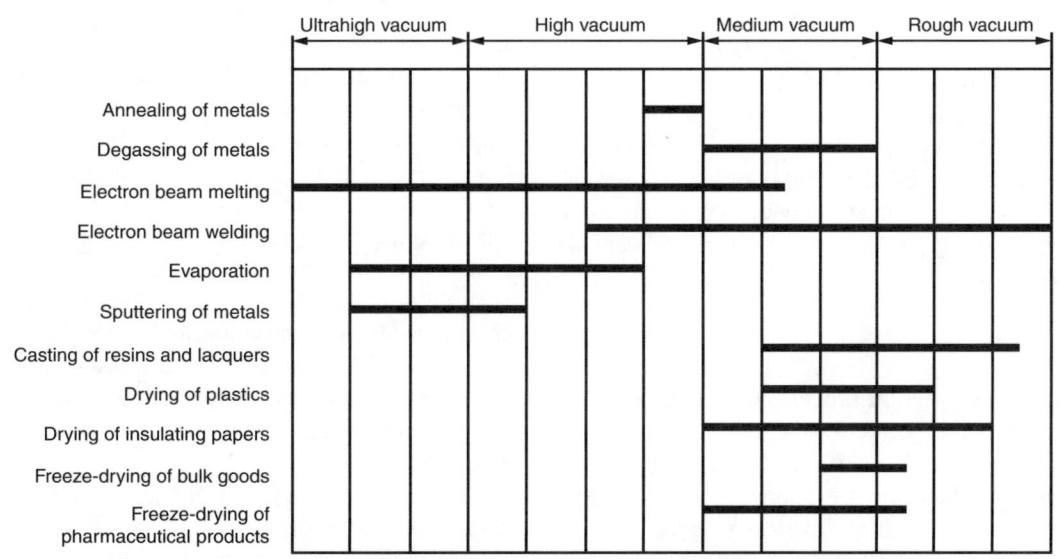

FIGURE V.2 Pressure ranges for various industrial processes. (From *Ind. Heating*, September, 113, 2000. With permission.)

Through the use of vacuum it is possible to create coatings with a high degree of uniform thickness ranging from several nanometers to more than 100 mm while still achieving very good reproducibility of the coating properties. Flat substrates, web and strip, as well as complex molded-plastic parts, can be coated with virtually no restrictions as to the substrate material.

The variety of coating materials is also very large. In addition to metal and alloy coatings, layers may be produced from various chemical compounds or layers of different materials applied in sandwich form. A significant advantage of vacuum coating over other methods is that many special coating properties desired, such as structure, hardness, electrical conductivity, or refractive index, are obtained merely by selecting a specific coating method and the process parameters for a certain coating material.

Deposition of thin films is used to change the surface properties of a base material or substrate. For example, optical properties such as transmission or reflection of lenses and other glass products can be adjusted by applying suitable coating layer systems. Metal coatings on plastic web produce conductive coatings for film capacitors. Polymer layers on metals enhance the corrosion resistance of the substrate.

COATING SOURCES

In all vacuum coating methods, layers are formed by deposition of material from the gas phase. The coating material may be formed by physical processes such as evaporation and sputtering, or by chemical reaction. Therefore, a distinction is made between physical vapor deposition (PVD) and chemical vapor deposition (CVD).

Thermal Evaporators

In the evaporation process, the material to be deposited is heated to a temperature high enough to reach a sufficiently high vapor pressure and the desired evaporation or condensation rate is set. The simplest sources used in evaporation consist of wire filaments, boats of sheet metal, or electrically conductive ceramics that are heated by passing an electrical current through them. However, there are restrictions regarding the type of material to be heated. In some cases, it is not possible to achieve the necessary evaporator temperatures without significantly evaporating the source holder and thus contaminating the coating. Furthermore, chemical reactions between the holder and the material to be evaporated can occur resulting in either a reduction of the lifetime of the evaporator or contamination of the coating.

Electron Beam Evaporators (Electron Guns)

To evaporate coating material using an electron beam gun, the material, which is kept in a water-cooled crucible, is bombarded by a focused electron beam and thereby heated. Since the crucible remains cold, in principle, contamination of the coating by crucible material is avoided and a high degree of coating purity is achieved. With the focused electron beam, very high temperatures of the material to be evaporated can be obtained and thus very high evaporation rates. Consequently, high-melting point compounds such as oxides can be evaporated in addition to metals and alloys. By changing the power of the electron beam, the evaporation rate is easily and rapidly controlled.

Cathode Sputtering

In the cathode sputtering process, the target, a solid, is bombarded with high energy ions in a gas discharge. The impinging ions transfer their momentum to the atoms in the target material, knocking the atoms off. These displaced atoms — the sputtered particles — condense on the substrate facing the target. Compared to evaporated particles, sputtered particles have considerably higher kinetic energy. Therefore, the conditions for condensation and layer growth are very different in the two processes. Sputtered layers usually have higher adhesive strength and a denser coating structure than evaporated ones.

Sputter cathodes are available in many different geometric shapes and sizes as well as electrical circuit configurations. What all sputter cathodes have in common is a large particle source area compared to evaporators, and the capability to coat large substrates with a high degree of uniformity. In this type of process, metals and alloys of any composition, as well as oxides, can be used as coating materials.

Chemical Vapor Deposition

In contrast to physical vapor deposition methods, where the substance to be deposited is either solid or liquid, in chemical vapor deposition, the substance is already in the vapor phase when admitted to the vacuum system. To deposit it, the substance must be thermally excited, i.e., by means of appropriate high temperatures or with plasma. Generally, in this type of process, a large number of chemical reactions take place, some of which are taken advantage of to control the desired composition and properties of the coating. For example, by using silicon–hydrogen monomers, soft silicon–hydrogen polymer coatings, hard silicon coatings, or — by the addition of oxygen — quartz coatings can be created by controlling process parameters.

Web Coating

Metal-coated plastic webs and papers play an important role in food packaging. Another important area of application of metal-coated web is the production of film capacitors for electrical and electronics applications.

Metal coating is carried out in vacuum web coating systems. The unit consists of two chambers, the winding chamber with the roll of web to be coated and the winding system, as well as the coating chamber, where the evaporators are located. The two chambers are sealed from each other, except for two slits through which the web runs. This makes it possible to pump high gas loads from the web roll using a relatively small pumping set. The pressure in the winding chamber may be more than a factor of 100 higher than the pressure simultaneously established in the coating chamber.

During the coating process, the web, at a speed of more than 10 m/s, passes a group of evaporators consisting of ceramic boats from which aluminum is evaporated. To achieve the necessary aluminum coating thickness at these high web speeds, very high evaporation rates are required. The evaporators must be run at temperatures in excess of 1400°C. Thermal radiation of the evaporators, together with the heat of condensation of the growing layer, yields a considerable thermal load for the web. With the help of cooled rollers, the foil is cooled during and after coating so that it is not damaged during coating and has cooled significantly prior to winding.

During the entire coating process, the coating thickness is continuously monitored with an optical measuring system or by means of electrical resistance measurement devices. The measured values are compared with the coating thickness set points in the system, and the evaporator power is thus automatically controlled.

Optical Coatings

Vacuum coatings have a broad range of applications in production of ophthalmic optics, lenses for cameras, and other optical instruments as well as a wide variety of optical filters and special mirrors. To obtain the desired transmission of reflection properties, at least 3, but sometimes up to 50, coatings are applied to the glass or plastic substrates. The coating properties, such as thickness and refractive index of the individual coatings, must be controlled very precisely and matched to each other.

Most of these coatings are produced using electron beam evaporators in single-chamber units. The evaporators are installed at the bottom of the chamber, usually with automatically operated crucibles, in which there are several different materials. The substrates are mounted on a rotating calotte above the evaporators. Application of suitable shielding, combined with relative movement between evaporators and substrates, results in a very high degree of coating uniformity. With the help of quartz coating thickness monitors and direct measurement of the attained optical properties of the coating system during coating, the coating process is fully controlled automatically.

One of the key requirements of coatings is that they retain their properties under usual ambient conditions over long periods of time. This requires the production of dense coatings, into which neither oxygen nor water can penetrate. Using glass lenses, this is achieved by

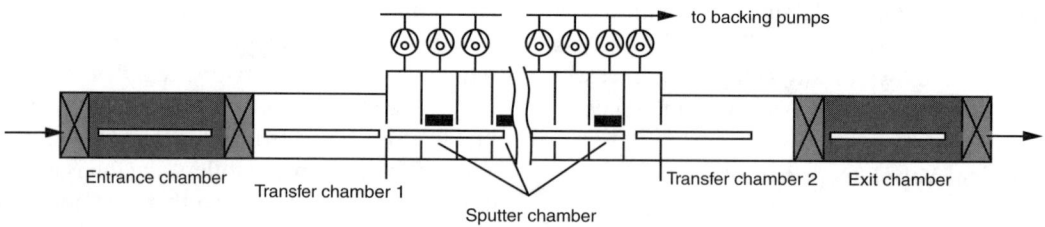

FIGURE V.3 Plant for coating glass panes — three-chamber in-line system with throughput up to 3,600,000 m^2/year. (From *Ind. Heating*, September, 118, 2000. With permission.)

keeping the substrates at temperatures up to 300°C during coating by means of radiation heaters. However, plastic lenses, as those used in eyeglass optics, are not allowed to be heated above 80°C.

To obtain dense, stable coatings these substrates are bombarded with argon ions from an ion source during coating. Through ion bombardment, the right amount of energy is applied to the growing layer so that the coated particles are arranged on the energetically most favorable lattice sites, without the substrate temperature reaching unacceptably high values. At the same time, oxygen can be added to the argon. The resulting oxygen ions are very reactive and ensure that the oxygen is included in the growing layer as desired.

Glass Coating

Coated glass plays a major role in a number of applications such as heat-reflecting coating systems on windowpanes to lower heating costs; solar protection coatings to reduce air-conditioning costs in countries with high-intensity solar radiation; coated car windows to reduce the heating-up of the interior; and mirrors used both in the furniture and the automobile industry. Most of these coatings are produced in large in-line vacuum systems such as that shown in Figure V.3.

The individual glass panes are transported into an entrance chamber at atmospheric pressure. After the entrance valve is closed, the chamber is evacuated with a forepump set. As soon as the pressure is low enough, the valve to the evacuated transfer chamber can be opened. The glass pane is moved into the transfer chamber and from there at constant speed to the process chambers, where coating is carried out by means of sputter cathodes. On the exit side, there is, in analogy to the entrance side, a transfer chamber in which the pane is held until it can be transferred out through the exit chamber.

Most of the coatings consist of a stack of alternative layers of metal and oxide. Because the metal layers may not be contaminated with oxygen, the individual process stations have to be vacuum-isolated from each other and from the transfer stations. To avoid frequent and undesirable starting and stopping of the glass panes, the process chambers are vacuum-separated through so-called "slit locks," i.e., constantly open slits combined with an intermediate chamber with its own vacuum pump. The gaps in the slits are kept as small as technically possible to minimize clearance and therefore conductance as the glass panes are transported through them. The pumping speed at the intermediate chamber is kept as high as possible to achieve a considerably lower pressure in the intermediate chamber than in the process chambers. This lower pressure greatly reduces the gas flow from a process chamber via the intermediate chamber to the adjacent process chamber. For very stringent separation requirements, it may be necessary to place several intermediate chambers between two process chambers.

The glass coating process requires high gas flows for the sputter processes as well as low hydrocarbon concentration. Turbomolecular pumps are used almost exclusively because of their high pumping speed stability over time.

While the transfer and process chambers are constantly evacuated, the entrance and exit chambers must be periodically vented and then evacuated again. Because of the large volumes of these chambers and the short cycle times, a combination of rotary vane pumps and Roots

Data Storage Disks

Coatings for magnetic- or magneto-optic data storage media usually consist of several functional coatings that are applied to mechanically finished disks. Most disks must be coated on both sides, and there are substantially greater low particle contamination requirements as compared to glass coating. The sputter cathodes in the process stations are mounted on both sides of the carrier so that the front and back of the disk can be coated simultaneously.

An entirely different concept is applied for coating of single disks. In this case, the different process stations are arranged in a circle in a vacuum chamber. The disks are transferred individually from a magazine to a star-shaped transport arm. The transport arm cycles one station farther after each process step and in this way transports to substrates from one process station to the next.

During cycling, all processes are switched off and the stations are vacuum-linked to each other. As soon as the arm has reached the process position, the individual stations are separated from each other by closing seals. Each station is pumped by means of its own turbomolecular pump and the individual processes are started. By sealing off the process stations, excellent vacuum separation of the individual processes can be achieved. However, since the slowest process step determines the cycle interval, two process stations may have to be dedicated for particularly time-consuming processes.

VANADIUM AND ALLOYS

An elementary metal (symbol V), vanadium is widely distributed, and is a pale-gray metal with a silvery luster. Its specific gravity is 6.02, and it melts at 1780°C. It does not oxidize in the air and is not attacked by hydrochloric or dilute sulfuric acid. It dissolves with a blue color in solutions of nitric acid. It is marketed as 99.5% pure, in cast ingots, machined ingots, and buttons. The as-cast metal has a tensile strength of 372 MPa, yield strength of 310 MPa, and elongation of 12%. Annealed sheet has a tensile strength of 537 MPa, yield strength of 455 MPa, and elongation of 20%, and the cold-rolled sheet has a tensile strength of 827 MPa with elongation of 2%. Vanadium metal is expensive, but is used for special purposes such as for springs of high flexural strength and corrosion resistance.

Commercially important as an oxidation catalyst, vanadium also is used in the production of ceramics and as a colorizing agent. Studies have demonstrated the biological occurrence of vanadium, especially in marine species; in mammals, vanadium has a pronounced effect on heart muscle contraction and renal function.

FABRICATION

Hot Working

Since vanadium oxidizes rapidly at hot-working temperatures, forming a molten oxide, it must be protected during heating. This is most easily accomplished by heating in an inert-gas atmosphere. Other common practices have been found less suitable.

Vanadium ingots up to 152 mm in size have been successfully hot-worked, but the degree of contamination is a modifying factor. Generally, the procedures used in working alloy steels apply.

In view of the difficulties involved in heating the metal, reheating is generally avoided and the starting temperature is a function of the amount of hot work to be accomplished and of the desired finishing temperature. Starting temperatures can range as high as 1260°C and the finishing temperatures is limited by the beginning of recrystallization. Straightening is performed between 371 and 427°C but not at room temperature.

Cold Working

Vanadium has excellent cold-working properties, provided its surfaces are uncontaminated. They are therefore machined clean by removing between 0.50 to 1 mm.

Strip can be readily made from hot-rolled sections 31 × 152 mm in cross section, and 0.25 mm material has been produced without and 0.03 mm with intermediate annealing. Where incipient cracking is observed, vacuum annealing at 899°C becomes necessary.

Extrusion is one of the most suitable fabricating methods for vanadium, since warm extrusion followed by cold rolling or drawing avoids hot working with the troublesome heating step. At temperatures below 538°C, tube blanks 50.8 mm outside diameter × 6.4 mm wall thickness have been produced from hot-rolled and turned bars as well as from ingots.

Wires can be drawn from 9.5-mm-diameter stock down to 0.025 mm, especially after copper plating. Reductions are usually 10% per pass.

In machining, vanadium resembles the more difficult stainless steels. Low speeds with light to moderate feed are used and very light finishing cuts at higher speeds are possible.

Welding is not difficult but contamination of the metal must be avoided by shielding from air by means of an inert gas, i.e., argon.

Uses and Applications

The greatest use of vanadium is for alloying. Ferrovanadium, for use in adding to steels, usually contains 30 to 40% vanadium, 3 to 6% carbon, and 8 to 15% silicon, with the balance iron, but may also be had with very low carbon and silicon. Vanadium–boron, for alloying steels, is marketed as a master alloy containing 40 to 45% vanadium, 8% boron, 5% titanium, 2.5% aluminum, and the balance iron, but the alloy may also be had with no titanium. Van-Ad alloy, for adding vanadium to titanium alloys, contains 75% vanadium and the balance titanium. It comes as fine crystals. The vanadium–columbium alloys containing 20 to 50% columbium, have a tensile strength above 689 MPa at 700°C, 482 MPa at 1000°C, and 275 MPa at 1200°C.

Vanadium salts are used to color pottery and glass and as mordants in dyeing. Red cake, or crystalline vanadium oxide, is a reddish-brown material, containing about 85% vanadium pentoxide, V_2O_5, and 9% Na_2O, used as a catalyst and for making vanadium compounds. Vanadium oxide is also used to produce yellow glass; the pigment known as vanadium–tin yellow is a mixture of vanadium pentoxide and tin oxide.

Vanadium is used in the cladding of fuel elements in nuclear reactors because it does not alloy with uranium and has good thermal conductivity as well as satisfactory thermal neutron cross section.

Because the metal alloys with both titanium and steel, it has found application in providing a bond in the titanium-cladding of steel. Also, the good corrosion resistance of vanadium offers interesting possibilities for the future; it has excellent resistance to hydrochloric and sulfuric acids and resists aerated salt water very well. But its stability in caustic solutions is only fair and, in nitric acid, inadequate.

Borides, Carbides, and Oxides

Vanadium boride, VB, has a melting point of 2100°C with oxidation at 1000 to 1100°C; density 5.1 g/cm³; Mohs hardness 8 to 9; electrical resistivity 16 Ω-cm. It is also formed as VB_2.

Vanadium carbide, VC, has a density 5.81 g/cm³ and is silver gray in color. It is chemically very stable; among the cold acids, it is attacked only by HNO_3. Below 499°C, Cl_2 reacts with VC. It burns in oxygen or air, but is stable to 2500°C in nitrogen. VC is harder than corundum.

Vanadium pentoxide, V_2O_5, has a melting point of 690°C and is slightly soluble in water. V_2O_5 is used by the ceramic industry as coloring agents producing various tints of yellow and greenish yellow. Vanadium pentoxide is an excellent flux and small amounts may be helpful in promoting vitrification of ceramic products. Vanadate glasses are relatively fusible when compared with other oxide types.

VANADIUM STEEL

Vanadium was originally used in steel as a cleanser, but is now employed in small amounts, 0.15 to 0.25%, especially with a small quantity of chromium, as an alloying element to make strong, tough, and hard low-alloy steels. It increases the tensile strength without lowering the ductility, reduces grain growth, and increases the fatigue-resisting qualities of steels. Larger amounts are used in high-speed steels and in special steels. Vanadium is a powerful deoxidizer in steels, but is too expensive for this purpose alone. Steels with 0.45 to 0.55% carbon and small amounts of vanadium are used for forgings, and cast steels for aircraft parts usually contain vanadium. In tool steels, vanadium widens the hardening range, and by the formation of double carbides with chromium makes hard and keen-edge die and cutter steels. All these steels are classed as chromium–vanadium steels. The carbon–vanadium steels for forgings and castings, without chromium, have slightly higher manganese.

Vanadium steels require higher quenching temperatures than ordinary steels or nickel steels. Society of the Automotive Engineers (SAE) 6145 steel, with 0.18% vanadium and 1% chromium, has a fine grain structure and is used for gears. It has a tensile strength of 799 to 2013 MPa when heat-treated, with a Brinell hardness 248 to 566, depending on the temperature of drawing, and an elongation of 7 to 26%. In cast vanadium steels it is usual to have from 0.18 to 0.25% vanadium with 0.35 to 0.45% carbon. Such castings have a tensile strength of about 551 MPa and an elongation of 22%. A nickel–vanadium cast steel has much higher strength, but high-alloy steels with only small amounts of vanadium are not usually classed as vanadium steels.

VAPOR-DEPOSITED COATINGS

These are thin single or multilayer coatings applied to base surfaces by deposition of the coating metal from its vapor phase. Most metals and even some nonmetals, such as siliconoxide, can be vapor-deposited. Vacuum-evaporated films or vacuum-metallized films were produced by vacuum evaporation. In addition to vacuum evaporation, vapor-deposited films can be produced by ion sputtering, chemical-vapor plating, and a glow-discharge process. The first two are discussed under vacuum processing.

In the glow-discharge process, applicable only to polymer films, a gas discharge deposits and polymerizes the plastic film on the base material.

Applications

Vapor plating is not considered to be competitive with electroplating. Its chief use (present and future) is to apply coating materials that cannot be electroplated, or cannot be applied in a nonporous condition by other techniques. Such materials include titanium, zirconium, columbium, tantalum, molybdenum, and tungsten, and refractory compounds such as the transition metal carbides, nitrides, borides, and silicides. Vapor plating will also continue to be useful in the preparation of ultrahigh-purity metals and compounds for use in electronics applications and in alloy development.

A few of the main commercial uses of vapor plating are as follows:

1. The application of high-chromium alloy coatings to iron and steel articles by the displacement-diffusion coating process (known as pack chromizing), for abrasion resistance and for protection from corrosion by food products, strong oxidizing acids, alkalies, salt solutions, and gaseous combustion products at temperatures up to about 800°C.
2. The application of molybdenum disilicide coatings to molybdenum by gas-phase siliconizing, for protection against air oxidation at temperatures between 800 and 1700°C.
3. The preparation of ultrahigh-purity titanium, zirconium, chromium, thorium, and silicon by iodide vapor decomposition processes.
4. The preparation of junction transistors by the controlled diffusion of boron from boron halide into the surface of silicon or germanium wafers.
5. The preparation of oriented graphite plates and shapes (pyrolytic graphite), by the high-temperature pyrolysis of hydrocarbon gases, for use in rocket and missile applications.

In addition, the following coatings have been developed:

1. Tantalum coatings on iron and steel for corrosion resistance
2. Vanadized, tungstenized, and molybdenized iron for wear resistance
3. Tungsten coatings on copper x-ray and cyclotron targets
4. Conductive metallic coatings on glass, porcelain, alundum, porous bodies, rubber, and plastics
5. Refractory metal coatings on copper wires
6. Oxidation-resistant carbide coatings on graphite tubes, nozzles, and vanes
7. Metallic coatings of all types on metallic and nonmetallic powders
8. Decorative, colored coatings on glass
9. High-purity boron, rhenium, vanadium, germanium, and aluminum

The displacement-diffusion plating processes, such as pack-chromizing, can plate uniformly somewhat larger pieces and more complex shapes with inaccessible areas. Sheets and rod up 0.6 to 0.9 m in dimension have been coated, and no technical obstacles are seen to scaling up the processes to coat even larger pieces. The pack coating processes have the advantage of minimizing the problems of specimen support and warpage during plating.

Plating uniformity varies somewhat with the particular process used, with the shape of the object being plated, and with the attention given to providing proper gas flow around the object. A variation in thickness of 10 to 25% is usually obtained. However, some coating processes can be made self-limiting so that the variation in coating thickness is much less than this range.

DISADVANTAGES

Vapor plating has the following disadvantages:

1. Relative instability and air and moisture sensitivity of most of the compounds used as plating agents
2. A tendency to produce nonuniform deposits due to unfavorable gas flow patterns around the work, or to uneven specimen temperature
3. Alteration of physical properties of the substrate due to the elevated processing temperatures
4. The possibility of poor coating quality arising from undesired side reactions in the plating process

In general, the materials used as plating compounds in vapor plating are relatively unstable and easily decomposed by air and moisture, thus rendering them more expensive and more difficult to store and handle than the compounds used in other plating techniques. Also, some of the metal carbonyls, hydrides, and organometallic compounds are highly toxic, and some of the hydrides and metal alkyls inflame spontaneously upon contact with air.

To develop optimum properties in deposits of many materials, the plating compounds, particularly the moisture-sensitive metal halides, must be purified and used without contamination from the atmosphere. This apparent disadvantage is sometimes put to good use, however, when intentional contamination of the coating atmosphere with nitrogen or moisture is used to produce harder deposits (e.g., of titanium or tantalum), or to reduce the codeposition of carbon (e.g., with molybdenum from the carbonyl).

Nonuniform plating may result in all vapor-plating processes, except the displacement-diffusion process, if consideration is not given to the gas-flow pattern around, or through, the article being coated. The shape factor may also have to be taken into account in selecting the method of heating the article, to avoid nonuniform deposition due to nonuniform heating of the part. These difficulties can be overcome in extreme cases by applying more than one coating and using a different direction of gas flow over the specimen for each application.

If necessary, the displacement-diffusion type of coating process can be carried out at very low gas flow rates (since solid-state diffusion is the rate-controlling factor), and still produce uniform coatings. For this reason, this type of coating process is ideally suited for coating large, or highly irregular objects, or large numbers of small objects.

The elevated processing temperatures required in vapor plating may produce undesired physical changes in the article, such as loss of temper, grain growth, warping, dimensional change, or precipitation or solution of alloying constituents. However, in many instances a vapor-plating procedure can be selected that will avoid marked undesirable change.

Undesirable side reactions in vapor-plating processes must be watched for and avoided. A particularly troublesome one in plating from metal halide vapors is the interaction of the base material and the halide vapor to form lower-valent halides, either of the base material or of the coating vapor. If the substrate temperature is too low, or the plating atmosphere too rich in plating vapor, these lower-valent halides will condense at the surface of the substrate, producing a plate underlaid or contaminated with halide salts. Such deposits are always poorly adherent, porous, and sensitive to moisture. Incomplete reduction or decomposition of the plating vapor alone can produce the same result. Contamination of this type is less likely to occur when plating inert base materials such as graphite, glass, and some ceramics.

After having been plated in a hydrogen atmosphere, some metals such as tantalum, columbium, and titanium, with a strong affinity for hydrogen, must be vacuum-annealed, or at least cooled in an inert-gas atmosphere to avoid excessive hydrogen absorption and embrittlement.

Also, carburization of the substrate may occur in processes employing the metal carbonyls to coat metals with a strong affinity for carbon, if the substrate temperature is too high. The metal of the deposit itself may be partially carburized in some cases, as when depositing molybdenum, chromium, and tungsten from their carbonyl vapors.

VINYL ACETATE ETHYLENE

Since their introduction, vinyl acetate ethylene (VAE) copolymer emulsions have been a staple base for adhesive manufacturers. As the performance requirements within the packaging and construction markets have increased and diversified, so too has the use of these emulsions.

First, VAE copolymer emulsions offer a tremendous balance between performance properties and ease of use. The internal plasticization of the vinyl acetate with ethylene gives these emulsions adhesion to many difficult-to-adhere substrates while the polyvinyl alcohol (PVOH) stabilization system provides for high wet tack, good setting speeds, and excellent machinability.

Second, manufacturers of VAE emulsions have continued to advance the performance capabilities of these materials. Available today are functionalized VAE systems for adhesion to metalized surfaces, a range of glass transition temperatures for specific film properties, low volatile organic compound emulsions for sensitive food packaging applications, and higher solids technologies as an alternative to nonwater-based systems. With these new VAE copolymer emulsions, adhesive compounders are better able to address the ever-changing needs of the adhesive industry.

Last, these types of VAE emulsions are made even more versatile by their ability to be compounded with other raw materials and polymer systems. The additional formulations that can result from their compatibility with plasticizers, resins, fillers, humectants, surfactants, polyvinyl alcohol, etc. can offer various improvements in adhesion, tack, heat/cold resistances, flame retardancy, and range.

PROCESSING AND APPLICATIONS

The most recent advance in VAE emulsion technology has been the introduction of a PVOH stabilized, ultrahigh-solids copolymer emulsion that is polymerized at 72% solids and a 2000 cps viscosity. Its composition, structure, and colloidal properties provide faster setting speeds, higher wet tack, and improved adhesion to difficult-to-adhere substrates than was thought possible for VAE emulsions a few years ago. These performance features are allowing adhesive compounders to broaden greatly the applications utilizing waterborne technologies.

Polyurethanes have been available as adhesives for quite some time and are commonly found in vacuum-forming and plastics-bonding operations within the automotive and footwear industries. During the last 5 years waterborne urethane chemistry has undergone a significant transformation from solvent-borne or high cosolvent-containing polymer systems to 100% waterborne systems.

These aqueous polyurethane dispersions, like their solvent-borne counterparts, have some unique performance characteristics. They offer low heat reactivation temperatures, good adhesion to difficult-to-bond substrates, rapid green strength development, and high temperature heat resistance. However, they also have some significant drawbacks. They are low in solids, low in wet tack, slow drying, and relatively high in cost.

The blending of polymers to improve adhesive properties is already widely done in the industry. With the commercially available aqueous polyurethane dispersions on the market today, the opportunity exists to enhance the performance of ultrahigh-solids VAE emulsions through blending because these technologies are so complementary. They are both 100% waterborne, and the characteristics of the ultrahigh-solids VAE emulsion can compensate for the disadvantages of urethane with its speed of set, wet tack, and minimal water content.

Future

Through the blending of an ultrahigh-solids VAE emulsion with many of the commercially available aqueous polyurethane dispersions on the market today, adhesive compounders can create a new class of stable high-performance waterborne adhesives.

Depending on the urethane grade selected and the level incorporated in the blend, the performance properties of the ultrahigh-solids VAE emulsion can be dramatically enhanced in several areas:

- Cohesive strength
- Adhesion (vinyl)
- Heat sealability
- Cross-linker performance

VINYL RESINS AND PLASTICS

These are a group of products varying from liquids to hard solids, made by the polymerization of ethylene derivatives, employed for finishes, coatings, and molding resins, or it can be made directly by reacting acetic acid with ethylene and oxygen. In general, the term *vinyl* designates plastics made by polymerizing vinyl chloride, vinyl acetate, or vinylidene chloride, but may include plastics made from styrene and other chemicals. The term is generic for compounds of the basic formula $RCH:CR'CR''$. The simplest are the polyesters of vinyl alcohol, such as vinyl acetate. This resin is lightweight, with a specific gravity of 1.18, and is transparent, but it has poor molding qualities and its strength is no more than 34 MPa. But the vinyl halides, $CH_2:CHX$, also polymerize readily to form vinylite resins, which mold well, have tensile strengths to 62 MPa, high dielectric strength, and high chemical resistance, and a widely useful range of resins is produced by copolymers of vinyl acetate and vinyl chloride.

The possibility of variation in the vinyl resins by change of the monomer, copolymerization, and difference in compounding is so great that the term *vinyl resin* is almost meaningless when used alone. The resins are marketed under a continuously increasing number of trade names. In general, each resin is designed for specific uses, but is not limited to those uses.

Vinyl Alcohol

Vinyl alcohol, $CH_2:CHOH$, is a liquid boiling at 35.5°C. Polyvinyl alcohol is a white, odorless, tasteless powder which on drying from solutions forms a colorless and tough film. The material is used as a thickener for latex, in chewing gum, and for sizes and adhesives. It can be compounded with plasticizers and molded or extruded into tough and elastic products. Hydrolyzed polyvinyl alcohol has greater water resistance, higher adhesion, and its lower residual acetate gives lower foaming. Soluble film, for packaging detergents and other water-dispersible materials to eliminate the need of opening the package, is a clear polyvinyl alcohol film. Textile fibers are also made from polyvinyl alcohol, either water soluble or insolubilized with formaldehyde or another agent. Polyvinyl alcohol textile fiber is hot-drawn by a semimelt process and insolubilized after drawing. The fiber has a high degree of orientation and crystallinity, giving good strength and hot-water resistance.

Vinyl alcohol reacted with an aldehyde and an acid catalyst produces a group of polymers known as vinyl acetal resins, and separately designated by type names, as polyvinyl butyral and polyvinyl formal. The polyvinyl alcohols are called Solvars, and the polyvinyl acetates are called Gelvas. The vinyl ethers range from vinyl methyl ether, $CH_2:CHOCH_3$, to vinyl ethylhexyl ether, from soft compounds to hard resins. Vinyl ether is a liquid that polymerizes, or that can be reacted with hydroxyl groups to form acetal resins. Alkyl vinyl ethers are made by reacting acetylene with an alcohol under pressure, producing methyl vinyl ether, ethyl vinyl ether, or butyl vinyl ether. They have reactive double bonds that can be used to copolymerize with other vinyls to give a variety of physical properties. The polyvinyl formals, Formvars, are used in molding compounds, wire coatings, and impregnating compounds. They are one of the toughest of the thermoplastics.

Plastisol

A plastisol is a vinyl resin dissolved in a plasticizer to make a pourable liquid without a volatile solvent for casting. The poured liquid is solidified by heating. Plastigels are plastisols to which a gelling agent has been added to increase viscosity. The polyvinyl acetals, Alvars, are used in lacquers, adhesives, and phonograph records. The transparent polyvinyl butyrals, Butvars, are used as interlayers in laminated glass. They are made by reacting polyvinyl alcohol with butyraldehyde, C_3H_7CHO. Vinal is a general name for vinyl butyral resin used for laminated glass.

Vinyl Acetate

Vinyl acetate is a water-white mobile liquid with boiling point 70°C, usually shipped with a copper salt to prevent polymerization in transit. The composition is $CH_3:COO:CH:CH_2$. It may be polymerized in benzene and marketed in solution, or in water solution for use as an extender for rubber, and for adhesives and coatings. The higher the polymerization of the resin, the higher the softening point of the resin. The formula for polyvinyl acetate resin is given as $(CH_2:CHOOCCH_3)_x$. It is a colorless, odorless thermoplastic with density of 1.189, unaffected by water, gasoline, or oils, but soluble in the lower alcohols, benzene, and chlorinated hydrocarbons. Polyvinyl acetate resins are stable to light, transparent to ultraviolet light, and are valued for lacquers and coatings because of their high adhesion, durability, and ease of compounding with gums and resins. Resins of low molecular weight are used for coatings, and those of high molecular weight for molding. Vinyl acetate will copolymerize with maleic acrylonitrile or acrylic esters. With ethylene it produces a copolymer latex of superior toughness and abrasion resistance for coatings.

Vinyl Benzoate

Vinyl benzoate is an oily liquid of the composition $CH_2:CHOOCC_6H_5$, which can be polymerized to form resins with higher softening points than those of polyvinyl acetate, but that are more brittle at low temperatures. These resins, copolymerized with vinyl acetate, are used for water-repellent coatings. Vinyl crotonate, $CH_2:CHOOCCH:CHCH_3$, is a liquid of specific gravity of 0.9434. Its copolymers are brittle resins, but it is used as a cross-linking agent for other resins to raise the softening point and to increase abrasion resistance. Vinyl formate, $CH_2:CHOOCH$, is a colorless liquid that polymerizes to form clear polyvinyl formate resins that are harder and more resistant to solvents than polyvinyl acetate. The monomer is also copolymerized with ethylene monomers to form resins for mixing in specialty rubbers. Methyl vinyl pyridine, $(CH_3)(CHCH_2)C_5H_3N$, is used in making resins, fibers, and oil-resistant rubbers. It is a colorless liquid boiling at 64.4°C. The active methyl groups give condensation reactions, and it will copolymerize with butadiene, styrene, or acrylonitrile. Polyvinyl carbazole, under the name of Luvican, is used as a mica substitute for high-frequency insulation. It is a brown resin, softening at 150°C.

Vinyl Chloride

Vinyl chloride, CH_2CHCl, also called ethenyl chloride and chloroethylene, produced by reacting ethylene with oxygen from the air and ethylene dichloride, is the basic material for the polyvinyl chloride resins. It is a gas. The plastic was produced originally for cable insulation and for tire tubes. The tensile strength of the plastic may vary from the flexible resins with about 20 MPa to the rigid resin with a tensile strength to 62 MPa and Shore hardness of 90. The dielectric strength is high, up to 52×10^6 V/m. It is resistant to acids and alkalies. Unplasticized polyvinyl chloride is used for rigid chemical-resistant pipe. Polyvinyl chloride sheet, unmodified, may have a tensile strength of 57 MPa, flexural strength 86 MPa, and a light transmission of 78%.

POLYVINYL CHLORIDE

Polyvinyl chloride (PVC) is a thermoplastic polymer formed by the polymerization of vinyl chloride. Resins of different properties can be made by variations in polymerization techniques. These resins can be compounded with plasticizers, color, mineral filler, etc. and processed into usable forms, varying widely in physical and electrical properties, chemical resistance, and processing versatility in coloring and design. Compared with other thermoplastics of comparable cost, articles produced from the vinyl chloride plastics have outstanding chemical, flame, and abrasion resistance, tensile properties, and resistance to heat distortion.

PVC homopolymer resins, the largest single type of vinyl chloride-containing plastics, are produced by several methods of polymerization:

1. *Suspension:* The largest-volume method that produces resins for general-purpose use, processed by calendering, injection molding, extrusion, etc.
2. *Mass or solution:* Produces fine particle size resins used principally for calendering and solution coating.
3. *Emulsion:* Produces extremely fine particle size resins used for the preparation of liquid plastisols or organosols for use in slush molding, coatings, and foam.

The two largest-volume members of the family of vinyl chloride polymers are the pure polyvinyl chloride or homopolymer resins and the vinyl chloride–vinyl acetate copolymers containing approximately 5 to 15% vinyl acetate.

TYPES OF VINYLS/PVCs

Rigid Vinyls

Products made from rigid vinyls are perhaps of most interest in the engineering field. Rigid materials can be prepared by calendering, extrusion, injection molding, transfer molding, and solution casting processes. Rigid PVC products are available in sheets, films, rods, pipes, profiles, valves, nuts and bolts, etc. The products can be machined easily with wood- and metalworking tools.

Sheets or other forms can be conventionally welded by hot-air guns, using extruded welding rods of essentially the same composition as that of the sheet. The welded joints have strength equal to that of the base material. Rigid sheets can be thermoformed into many intricate shapes by several different thermoforming techniques such as vacuum forming, ring and plug forming, etc. Rigid pipe can be threaded and joined like steel pipe or sealed with adhesives in a manner similar to the sweating of copper pipe. Vinyl pipe is being used increasingly in waterworks, the petroleum industry, in natural gas distribution, irrigation, hazardous chemical application, and food processing.

Rigid vinyl made from vinyl acetate–vinyl chloride copolymers is prominent in sheeting used for thermoforming for such items as maps, packaging, advertising displays, toys, etc. Rigids made from homopolymer vinyl chloride reins are used in heavier structural designs, for example: pipe, pipe valve, heavy panels, electrical ducting, window and door framing parts, architectural moldings, gutters, downspouts, automotive trim, etc. In these fields, rigid vinyls compete with aluminum and other metals. Rigid vinyl products made from homopolymer resins are available in two types: Type I, or unmodified PVC, is approximately 95% PVC and has outstanding chemical resistance but low impact strength; Type II PVC, containing 10 to 20% of a resinous or rubbery polymeric modifier, has improved impact strength but reduced chemical resistance.

Flexible Vinyls

Flexible vinyl products are produced by the same general methods used for rigid vinyl products. Flexibility is achieved by the incorporation of plasticizers (mainly high-boiling organic esters) with the vinyl polymer. By proper choice of plasticizer type, flexible products can be obtained that excel in certain specific properties such as gasoline and oil resistance, low temperature flexibility, flame resistance, etc. The flexible sheeting and film can be fabricated by heat sealing to itself or other substrates by induction or high-frequency methods, solvent sealing, sewing, etc. Flexible vinyl film and sheeting find application in upholstery, packaging, agriculture, etc. The corrosion resistance of vinyl sheeting makes it ideal for a pipe wrap to prevent corrosion of underground installations. Flexible extrusions in many different shapes and forms have application as insulating and jacketing on electrical wire and cable, refrigerator gaskets, weather stripping, upholstery, and shoe welting. Injection-molded flexible vinyl products are used as shoes, electrical plugs, and insulation of various sorts.

Abrasion and stain resistance, coupled with unlimited coloring and design possibilities, have made flexible vinyl flooring one of the largest items in the floor-covering field. The major revolution in vinyl flooring is the greater emphasis on the use of relatively low to very low molecular weight homopolymer resins in place of the more expensive vinyl chloride–vinyl acetate copolymers.

Coatings

Coatings based on PVC polymers and copolymers can be applied from solutions, latex, plastisols, or organosols. Plastisols are liquid dispersions of fine particle size emulsion PVC in plasticizers. Organosols are essentially the same as plastisols but contain a volatile liquid organic diluent to reduce viscosity and facilitate processing. In many coating operations, conventional paint-spraying equipment is used. In other techniques, articles can be dip-, knife-, or roller-coated. Plastisols and organosols are used extensively for dip coating of wire products such as household utensils and knife coating of fabrics and paper. Plastisol and organosol products can be varied from hard (rigidsols) to very soft (vinyl foam). Vinyl coatings are used for a variety of applications requiring corrosion and/or abrasion resistance.

Production coatings with vinyl plastics involve a fluidized-bed technique. A metallic object, heated to 204 to 260°C, is immersed in a bed of finely ground plastic which is "fluidized" by air entering the bottom of the container. Articles can be coated by this method to a thickness of 7 to 60 mils. Both rigid and plasticized vinyls can be applied by this method.

Vinyl–Metal Laminates

These products are made by direct lamination of preprocessed, embossed, and designed vinyl sheet to metal, or continuous plastisol coating of metal sheet followed by fusing or curing of the plastisol and subsequent embossing. In the former process, the vinyl can be laminated to both sides of the metal sheet. Steel, aluminum, magnesium, brass, and copper have been used. The hardness, elongation, general properties, and thickness of the vinyl can be modified within wide limits to meet particular needs.

These laminates are dimensionally stable below 100°C and combine the chemical and flame resistance, decorative and design possibilities of vinyl with the rigidity, strength, and fabricating attributes of metals. The vinyl–metal laminates can be worked without rupture by many of the metalworking techniques, such as deep-drawing, crimping, stamping, punching, shearing, and reverse-bending. Disadvantages are inability to spot-weld, and lack of covering of metal edges, which is necessary where severe exposure conditions are encountered. Vinyl–metal laminates find use in appliance cabinets, machine housings, lawn and office furniture, automotive parts, luggage, chemical tanks, etc. The cost of these laminates is comparable to some lacquered metal surfaces.

VINYLIDENE CHLORIDE PLASTICS

Vinylidene chloride plastics are derived from ethylene and chlorine polymerized to produce a thermoplastic with softening point of 116 to 138°C. The resins are noted for their toughness and resistance to water and chemicals. The molded resins have a specific gravity of 1.68 to

1.75, tensile strength 27 to 48 MPa, and flexural strength of 103 to 117 MPa. Saran is the name of a vinylidene chloride plastic, extruded in the form of tubes for handling chemicals, brines, and solvents to temperatures as high as 135°C. It is also extruded into strands and woven into a box-weave material as a substitute for rattan for seating. Saran latex, a water dispersion of the plastic, is used for coating and impregnating fabrics. For coating food-packaging papers, it is waterproof and greaseproof, odorless and tasteless, and gives the papers a high gloss. Saran is also produced as a strong transparent film for packaging. Saran bristles for brushes are made in diameters from 0.025 to 0.051 cm.

Applications

The two largest-volume applications are in upholstery made from monofilaments and film for food packaging. Other uses are in window screening (monofilaments), paper and other coatings, pipe and pipe linings, and staple fiber.

Saran pipe and saran-lined metal pipe are of interest to the engineering field. Saran-lined pipe is prepared by swaging an oversize metal pipe on an extruded saran tube. These products can be installed with ordinary piping tool. Fittings and valves lined with saran and flange joints with saran gaskets are available. Vinylidene chloride–acrylonitrile copolymer has applications as coatings for tank car and ship-hold linings. Lacquers of these polymers are used in cellophane coatings yielding a product with the low moisture vapor permeability of vinylidene chloride polymers plus the handling ease of cellophane. The lacquers are also used for paper coatings, dip coatings, and sprayed packaging. Vinylidene chloride copolymers in latex form are used in paper coatings and specialty paints.

Vinylidene Fluoride

Vinylidene fluoride, $CH_2:CF_2$, has a high molecular weight, about 500,000. It is a hard, white thermoplastic resin with a slippery surface and has a high resistance to chemicals. It resists temperatures to 343°C, and does not become brittle at low temperatures. It extrudes easily, and has been used for wire insulation, gaskets, seals, molded parts, and piping.

VOLATILE ORGANIC COMPOUND

Many efforts have been under way regarding the biological control of volatile organic compound (VOC) emissions.

Water-based and hot-melt adhesives and coatings have been developed and evaluated extensively, but they are not satisfactory for all applications and may require solvent-based cleaners and primers. Various developments have shown how biological treatment can reduce solvent levels in air emissions and focuses in particular on footwear production. It is, however, applicable to all industries using organic solvents.

For example, the U.K. Environmental Protection Act 1990 sets a 5 metric tons/year adhesive solvent usage threshold, above which processes are subject to local authority air pollution control. This regulation apparently affects manufacturing plants producing as few as 5000 pairs of shoes per week.

By June 1998 shoemakers and material suppliers were set to meet an emission limit of 50 mg/m^3 measured as carbon, or a stringent mass emission control regime of 20 g/pair for footwear; similar controls are anticipated throughout Europe.

Catalytic combustion has been shown to be technically effective, but involves high capital and running costs. For example, adsorption on activated carbon is an established technique, but apparently it is unsuitable for the mixed cocktail of solvents in footwear production. Biological treatment processes are expected to be less costly and more suitable for arresting emissions of moderate concentration at ambient temperature, as found in shoe factory exhausts.

Biological treatment has been shown suitable for halting VOC emissions from the manufacturing industry with an on-site biological treatment unit set up to demonstrate VOC abatement.

Process

The process biologically breaks down VOCs into biomass by mineralization and utilization of the carbon; the by-products of breakdown are carbon dioxide and water vapor. Within the system, microorganisms grow as a biofilm on selected media where they produce enzymes to break down the VOC contaminants. Liquid is continuously recirculated to solubilize the solvents from the contaminated inlet gas. Water is sprayed evenly where the microorganisms oxidize the solvents, and monitoring of the recirculation liquid for determinants, including pH, nitrogen, phosphorus, suspended solids, and so on apparently allows for tighter process control and greater removal efficiency.

In this system there is a large interface between gas and liquid phases so that the bed can capture the compounds of poor solubility that are present in the off-gases. Inputs to the process include mains water supply, electricity to run pumps, solenoid valves, and a programmable logic controller, as well as nutrients to maintain the biomass, including nitrogen, phosphate, sulfate, and trace elements that are added according to the carbon load to the reactor.

Although this process was used at a footwear manufacturing plant, biotechnology also has application within other industries producing low-concentration mixed solvent waste streams, where recovery is inappropriate and the capital and operating costs of thermal and catalytic systems can be prohibitively expensive. Examples include printing, painting, laminating, metal and leather finishing, and furniture coating.

In another example, the U.S. Navy is currently studying a new environmentally safe paint coating for use on its fleet of helicopters. The three-shade flat, haze gray coats were sprayed on a new Navy fleet combat support helicopter. What makes the coating environmentally safe is the absence of VOCs that contribute to air pollution. The new coating eliminates the use of chemicals targeted by the federal government for reduction or elimination.

Normal Environmental Protection Agency (EPA)-compliant aircraft coatings currently used bT-fy contain about 3.5 lb of VOC/gallon, whereas this new paint has zero VOC.

The zero-VOC paint was developed by Deft Coatings, Inc. This zero-VOC coating also offers a significant weight benefit compared to the current paint and is nonflammable.

The first zero-VOC-coated helicopter is currently undergoing evaluation by the Navy to ensure that the new coating meets stringent Military Standard requirements.

VULCANIZED FIBER

Vulcanized fiber is a pure, dense, cellulosic material with good electrical insulating properties and high mechanical strength. It is half the weight of aluminum, easily machined and formed, and is used for parts such as for barriers, abrasive-disk backing, high-strength bobbin heads, materials-handling equipment, railroad-track insulation, and athletic guards.

Forms

Most manufacturers provide vulcanized fiber in the form of sheets, coils, tubes, and rods. Sheets are made in a thickness of 0.06 to 50.8 mm, approximately 1.2×2 m in size, or in rolls and coils from 0.06 to 2.3 mm thick. Tubes are made in the outside diameter range of 4.7 to 111.5 mm, and rods are produced 2.3 to 50.4 mm in diameter.

Sheets can be machined and formed to produce a variety of useful shapes for insulating or shielding purposes. Sheets, tubes, and rods can be machined using standard practices for cutting, punching, tapping, milling, shaping, sanding, etc.

Properties

Vulcanized fiber possesses a versatile combination of properties, making it a useful material for practically all fields. It has outstanding arc resistance, high structural strength per unit area, and can be formed and machined. In thin sections it possesses high tear strength, smoothness, and flexibility. In heavier thicknesses it resists repeated impact and has high tensile, flexural, and compressive strength.

The material is unaffected by normal solvents, gasoline, and oils, and therefore is recommended for applications where a structural support is required in the presence of these materials.

Moisture absorption is high and dimensional stability is affected by conditions of humidity when not protected by moisture-resistant coatings.

Vulcanized fiber is produced in 13 basic grades and numerous special grades to meet specific application requirements.

Applications

Vulcanized fiber serves as the insulating material in a signal block for railroad track insulations. At the end of a signal station, the rails are completely insulated from the next adjoining section to form what is termed a *block*. The two meeting rails and the coupling fixtures are insulated with formed parts made from vulcanized fiber. This junction is effective while absorbing the repeated impact from trains under all weather conditions.

Vulcanized fiber offers durability, ease of fabricating, excellent wear characteristics, and lightness of weight for materials-handling and luggage applications. The materials-handling equipment resists scuffing, battering, denting, rusting, and other general wearing conditions and provides protection by its hardness and resilience.

Formed pieces of vulcanized fiber offer outstanding service as arc barriers in circuit breakers. The arc-resistant properties of vulcanized fiber prevent a breakdown when the circuit breaker is subjected to an overload. The formed barrier is tested to take higher electrical loads than the maximum that can be produced by the circuit and, since repeated circuit breaks will not affect the performance of vulcanized fiber, the need for replacement is negligible.

Peerless control tape is used for programming data, processing equipment, or automatic machining equipment. The special properties that give this material outstanding service life are high tensile strength, high tear strength, low stretch, and good abrasive resistance.

Flame-resistant vulcanized fiber gives designers a structural material that can be used in those applications requiring a nonburning material and reduces fire hazards by containing a fire at its source. Flame-retardant parts serve as barriers in electrical equipment, materials-handling equipment, and wastebaskets.

Chemistry

Vulcanized fiber is produced by the chemical action of zinc chloride solution on a saturating grade of absorbent paper when processed under heat and pressure. The action of the zinc chloride converts the cellulosic fibers to a dense, homogeneous structure producing a laminated material that is refined to a chemically pure form. Final processing consists of drying to the proper moisture content and applying the proper calender to give smoothness and uniformity of thickness.

Grades

The 13 basic grades are as follows:

Electrical insulation grade. Primarily intended for electrical applications and others involving difficult bending or forming operations. It is sometimes referred to as "fishpaper."

Commercial grade. Considered to be the general-purpose grade, sometimes referred to as "mechanical and electrical grade." It possesses good physical and electrical properties and fabricates well.

Bone grade. Characterized by greater hardness and stiffness associated with higher specific gravity. It machines smoother with less tendency to separate the plies in the machining operations.

Trunk and case grade. Conforms to the mechanical requirements of "commercial grade," but has better bending qualities and smoother surface.

Flexible grade. Made sufficiently soft by incorporating a plasticizer, it is suitable for gaskets, packings, and similar applications. It is not recommended for electrical use.

Abrasive grade. Designed as the supporting base for abrasive grit for both disk and drum sanders. It has exceptional tear resistance, ply adhesion, resilience, and toughness.

White tag grade. It has smooth clean surfaces and can be printed or written on without danger of ink feathering.

Bobbin grade. Used for the manufacture of textile bobbin heads. It punches well under proper conditions, but is firm enough to resist denting in use. It machines to a very smooth surface.

Railroad grade. Used as railroad track joint, switch rods, and other insulating applications for track circuits.

Hermetic grade. Used as electric-motor insulation in hermetically sealed refrigeration units. High purity and low methanol extractables are essential because it is immersed in the refrigerant.

White grade. Recommended for applications where whiteness and cleanliness are essential requirements.

Shuttle grade. Designed for gluing to wood shuttles to withstand the repeated pounding received in textile power looms.

Pattern grade. Made to provide maximum dimensional stability and minimum warpage for use as patterns in cutting cloth, leather, and similar materials.

W

WASH PRIMERS

Wash primers are a special group of corrosion-inhibitive coatings designed for use on clean metal surfaces. They are also known as "wash-coat primers," "metal conditioners," and "etch primers."

The most widely utilized primers consist of a two-part system that is prepared at the point of use by simple mixing of specified proportions. The base grind portion contains a corrosion-inhibiting pigment, basic zinc chromate (also known as zinc tetroxy chromate), and a small amount of talc extender ground in an alcohol solution of polyvinyl butyral resin. The reducer portion consists of phosphoric acid, alcohol, and water. When these are mixed, a slow chemical reaction ensues, resulting in partial reduction of the chromate pigment. The life of the mixed primer is usually 8 to 12 h. Single-package primers are now in use.

WATCH

A watch is a portable timepiece. Its operation may be described as mechanical, electromechanical, or electronic.

Mechanical and Electromechanical Watches

In the mechanical watch a mainspring in the barrel stores operating energy; the user retightens the spring daily by means of the winding stem. The wheel train advances at five increments per second under control by the escapement. From there the dial train turns the minute and hour hands across the watch face. The momentarily engageable setting feature enables the user to position the hands in accordance with a primary clock. The wheel train of four pairs, with an overall turns ratio of 1:40,000, reduces the high torque from the barrel to a low value controllable by the escapement, yet sufficient to drive the dial train.

Among the variety of features incorporated into modern watches are self-winding mechanism, substitution of an electrochemical cell for the mechanical mainspring, several forms of electromechanical escapement in place of the balance-and-hairspring mechanism, and instead of hands over a dial, marked disks viewed through windows for readout, extended to days of the week in calendar watches.

A significant improvement in accuracy of the electromechanical watch is provided by relocation of the time base mechanism from the output of the wheel train to the input. Instead of deriving power from a mainspring, this arrangement obtains power from a dry cell. In place of an escapement at the end of the wheel train to control its incremental advance, a tuning fork as a resonant element in an electronic oscillator reciprocates an index finger that rapidly ratchets against a fine-toothed index wheel at the input to the wheel train to initiate its advance. Through these actions the tuning fork with its drive circuit serves both as time base and as electrical-to-mechanical transducer, introducing power from the dry cell into the wheel train at an intermediate level of torque. In one style of this watch the tuning fork vibrates at 360 Hz; in a smaller style the fork resonates at 480 Hz.

Electronic Watches

When solid-state electronic integrated circuits became available in quantity from production stimulated by digital computers, the all-electronic watch became a commercial reality. In it, a chain of binary dividers triggered from a crystal oscillator develops a train of second pulses. These pulses drive a digital counter or scaler, which develops minute and hour pulses to activate the digital display. An externally switched fast/slow capacitor in the crystal oscillator enables the user to set the readout in accordance with a standard time signal. The frequency of the crystal oscillator, typically in the tens of kilohertz, is chosen so that successive divisions by 2 produce the desired 1-s pulse rate. For example, an oscillator with a 65.536-kHz crystal is followed by 16 binary dividers.

Power may be supplied by mercury or silver oxide cells, which are replaced annually, or lithium batteries, which operate for up to 5 years. To extend operating life, a solar cell may charge a nickel–cadmium power cell while the watch is illuminated.

One form of readout uses a light-emitting diode (LED). An assembly of these on a monolithic chip illuminates appropriate bars of a seven-segment display for each digit of the readout. The LED display is self-illuminating and can be read in the dark. Because illumination of the readout consumes most of the power in an electronic watch, a liquid crystal display (LCD) is used where low power consumption is a first consideration. The LCD readout depends for its indication on ambient illumination; the display is brighter in incident light. In it, glass plates confine a thin layer of liquid crystal. On the inside surface of the front plate a transparent metallic coating in the seven-segment pattern receives signals from the readout counter. A highly reflective metal coating on the inside surface of the back plate operates at ground potential. When a bipolar high-frequency pulse train energizes a segment, the electric field established through the liquid causes that region to become turbulent and thereby scatter incident light so that the segment appears diffusely illuminated against a specularly illuminated background.

WATER

Water is a chemical compound with two atoms of hydrogen and one atom of oxygen in each of its molecules. It is formed by the direct reaction (1):

$$2H_2 + O_2 \to 2H_2O \quad (1)$$

of hydrogen with oxygen. The other compound of hydrogen and oxygen, hydrogen peroxide, readily decomposes to form water, reaction (2):

$$2H_2O_2 \to 2H_2O + O_2 \quad (2)$$

Water also is formed in the combustion of hydrogen-containing compounds, in the pyrolysis of hydrates, and in animal metabolism. Some properties of water are given in Table W.1.

GASEOUS STATE

Water vapor consists of water molecules that move nearly independently of each other. The atoms are held together in the molecule by chemical bonds, which are very polar — the hydrogen end of each bond is electrically positive relative to the oxygen. When two molecules near each other are suitably oriented, the positive hydrogen of one molecule attracts the negative oxygen of the other, and while in this orientation, the repulsion of the like charges is comparatively small. The net attraction is strong enough to hold the molecules together in many circumstances and is called a hydrogen bond.

When heated above 1200°C, water vapor dissociates appreciably to form hydrogen atoms and hydroxyl free radicals, reaction (3):

$$H_2O \to H + OH \quad (3)$$

These products recombine completely to form water when the temperature is lowered. Water vapor also undergoes most of the chemical reactions of liquid water and, at very high concentrations, even shows some of the unusual solvent properties of liquid water. Above 374°C, water vapor may be compressed to any density without liquefying, and at a density as high as 0.4 g/cm^3, it can dissolve appreciable quantities of salt. These conditions of high temperature and pressure are found in efficient steam power plants.

TABLE W.1
Properties of Water

Property	Value
Freezing point	0°C
Density of ice, 0°C	0.92 g/cm^3
Density of water, 0°C	1.00 g/cm^3
Heat of fusion	80 cal/g (335 J/g)
Boiling point	100°C
Heat of vaporization	540 cal/g (2260 J/g)
Critical temperature	347°C
Critical pressure	217 atm (22.0 MPa)
Specific electrical conductivity at 25°C	$1 \times 10^{-7}/\Omega$-cm
Dielectric constant, 25°C	78

Source: McGraw-Hill Encyclopedia of Science and Technology, 8th ed., Vol. 19, McGraw-Hill, New York, , 579. With permission.

SOLID STATE

Ordinary ice consists of water molecules joined together by hydrogen bonds in a regular arrangement. This unusual feature is a result of the strong and directional hydrogen bonds taking precedence over all other intermolecular forces in determining the structure of the crystal. If the water molecules were rearranged to reduce the amount of empty space, their relative orientations would no longer be so well suited for hydrogen bonds. This rearrangement can be produced by compressing ice to pressures in excess of 14 MPa. Altogether, five different crystalline forms of solid water have been produced in this way, the form obtained depending upon the final pressure and temperature. They are all denser than water, and all revert to ordinary ice when the pressure is reduced.

LIQUID STATE

The molecules in liquid water also are held together by hydrogen bonds. When ice melts, many of the hydrogen bonds are broken, and those that remain are not numerous enough to keep the molecules in a regular arrangement. Many of the unusual properties of liquid water may be understood in terms of the hydrogen bonds that remain. As water is heated from 0°C, it contracts until 4°C is reached and then begins the expansion that is normally associated with increasing temperature. This phenomenon and the increase in density when ice melts both result from a breaking down of the open, hydrogen-bonded structure as the temperature is raised. The viscosity of water decreases tenfold as the temperature is raised from 0 to 100°C, and this also is associated with the decrease of icelike character of the water as the hydrogen bonds are disrupted by increasing thermal agitation. Even at 100°C, the hydrogen bonds influence the properties of water strongly, for it has a high boiling point and a high heat of vaporization compared with other substances of similar molecular weight.

PROPERTIES

Pure water, either solid or liquid, is blue if viewed through a thickness of more than 2 m. The other colors often observed are due to impurities.

Water is an excellent solvent for many substances, but particularly for those that dissociate to form ions. Its principal scientific and industrial use as a solvent is to furnish a medium for purifying such substances and for carrying out reactions between them.

Among the substances that dissolve in water with little or no ionization and that are very soluble are ethanol and ammonia. These are examples of molecules that are able to form hydrogen bonds with water molecules, although, except for the hydrogen of the OH group in ethanol, it is the hydrogen of the water that makes the hydrogen bond. On the other hand, substances that cannot interact strongly with water, either by ionization or by hydrogen bonding, are only sparingly soluble in it. Examples of such substances are benzene, mercury, and phosphorus.

Water is not a strong oxidizing agent, although it may enhance the oxidizing action of other oxidizing agents, notably oxygen. Examples of the oxidizing action of water itself are its reactions with the alkali and alkaline earth metals, even in the cold.

Water is an even poorer reducing agent than oxidizing agent. One of the few substances that it reduces rapidly is fluorine.

Water reacts with a variety of substances to form solid compounds in which the water molecule is intact, but in which it becomes a part of the structure of the solid. Such compounds are called hydrates, and are formed frequently with the evolution of considerable amounts of heat.

WATER-SOLUBLE PLASTICS

Within the plastics industry, water-soluble materials offer a variety of desirable physical properties, yet retain the advantages inherent in a water system. These advantages include ease of handling, negligible solvent costs, low toxicity, and low flammability.

There is no sharp dividing line between water-dispersible and water-soluble polymers. Many so-called water-soluble plastics form colloidal dispersions rather than true solutions. In this text, emulsions or dispersions of water-insoluble polymers are not discussed (such as acrylics, polyvinyl acetate, styrene butadiene, polyvinyl butyral, etc.).

The water-soluble plastics can be roughly divided into two general classes: thermoplastic resins and thermosetting resins.

THERMOPLASTIC RESINS

These plastics are usually synthesized by addition polymerization techniques. That is, small units (or monomers) are joined together to develop the final molecular weight and polymer configuration. Rarely do these polymers develop into long straight chains; considerable branching often occurs. The molecular weight, chemistry of side groups, and extent of branching all determine the properties that are obtained. These plastics are available as white or light-colored powders or in solution. Films, moldings, and extrusions are also available based on some of the thermoplastic resins.

Alkali-Soluble Polyvinyl Acetate Copolymers

Polyvinyl acetate itself is water insoluble. However, copolymers are available in which vinyl acetate is copolymerized with an acidic comonomer. Such products retain the organic solubility of polyvinyl acetate but are soluble in aqueous alkali. These polymers exhibit low viscosity in solution and deposit high gloss films that are water resistant, provided a volatile alkali such as ammonia is used. The use of a fixed alkali will result in a film with permanent water sensitivity. They generally possess good adhesion to cellulose and a wide variety of other surfaces.

Major uses include loom-finish warp sizes for dope dyed yarns, repulpable adhesives or sizes for paper and board, conditioning agents for masonry prior to painting, protective coatings for metals, and leveling agent and film former in self-polishing waxes.

Ethylene-Maleic Anhydride Copolymers

High-molecular-weight polymers have been prepared by copolymerizing ethylene and maleic anhydride. These resins are available either in "linear" form or cross-linked with either anhydride, free acid, or amide-ammonium salt side chains.

Major applications include general thickening and suspending in adhesives, agricultural chemicals, cleaning compounds, and ceramics. This resin is used as a thickener for latex and as a warp size for acetate filament.

Polyacrylates

Commercially important polymers are prepared by polymerizing either acrylic or methacrylic acid. Usually these products are neutralized with bases to the salt form. Solution viscosity increases during neutralization. Cast films are hard, transparent, colorless, and somewhat brittle.

Polyacrylic acid itself is used as a warp size for nylon. The neutralized polymers (polyacrylates) are used in various coating and binding applications (ceramics, grinding wheels, etc.). Because of interesting solution properties, the polyacrylates are used as thickeners, flocculants, and sometimes as dispersants in applications such as ore processing, drilling muds, and oil recovery.

Polyethers

Two different polymer types are covered under this heading: polyoxyethylene (includes polyethylene glycols) and polyvinyl methyl ether and copolymers.

Polyoxyethylene (polyethylene glycol). These resins are available over a wide molecular weight range. Low-molecular-weight members are slightly viscous liquids, whereas the medium-molecular-weight types (1000 to 20,000) are waxy solids. Polymers up through this molecular-weight level are known as polyethylene glycols. Extremely high molecular-weight (several thousand to several million) homologues are also available. All types are soluble in water and in some organic solvents. Applications for the liquid and waxy solids include lubricants (rubber molds, textile fibers, and metal), bases for cosmetic and pharmaceutical preparations, and chemical intermediates for further reaction. The very high molecular-weight types are useful principally as thickeners in many application areas.

Polyvinyl methyl ether. This unique family of vinyl polymers shows inverse solubility in that the resins precipitate out above 35°C. They do redissolve upon cooling and the addition of low-molecular-weight alcohols increases the solubility in water and raises the precipitation temperature. Higher homologues are available that are water insoluble and quite tacky. The products exhibit pressure-sensitive adhesiveness coupled with good cohesive strength and high wet tack. Copolymers are available that contain maleic anhydride to modify physical properties, particularly solubility or tolerance to water and organic solvents and ease of insolubilization. Major uses take advantage of properties such as pressure-sensitive characteristics (adhesives), tackiness (various latex systems), thickening active, heat sensitizing (latices for dip forming), and binding power (pigments).

Polyvinyl Alcohol

These resins are available commercially in a wide range of types, which vary in viscosity and chemical composition. Polyvinyl alcohol exhibits good water solubility, high resistance to organic solvents, oils, and greases, high tensile strength, adhesion, and flexibility. In addition, the polymer is resistant to oxidation and in film form is an excellent barrier for various gases. Certain types exhibit surface activity in solution and all types are soluble in both acid and alkaline media. The resins can be cross-linked by borax and numerous organic and inorganic agents to produce thickening or even insolubilization.

In adhesives, polyvinyl alcohol contributes machinability, viscosity control, specific adhesion, and in some cases remoistenability. Other major uses include paper coating and sizing (for increased strengths, ink hold-out, and grease resistance), textile sizing, wrinkle-resistant finishes (wash-and-wear fabrics in conjunction with thermosetting resins), polyvinyl acetate emulsion polymerization (protective colloid), binder (for nonwoven ribbons, filters, etc.), film (release agent in polyester and epoxy molding and water-soluble packaging), cement additive (for improved strength, toughness, and adhesion), and photosensitive coating (in the graphic arts industry).

Polyvinyl Pyrrolidone

Polyvinyl pyrrolidone (PVP) exhibits good solubility in both water and various organic solvents. A nontoxic material and tacky substance when wet, the polymer is a dispersant, suspending agent, and an adhesive component for bonding difficult surfaces.

Major uses include cosmetic preparations (hair sprays, etc.), tablet binding and coating, detoxifying of dyes, drug, and chemicals, beverage clarification, and specialty textile and paper applications involving sizing, dyeing, and printing.

Copolymers are also available (like PVP-vinyl acetate) that have some advantages over the homopolymers in heat sealability, pressure-sensitive adhesiveness, and other properties.

Polyacrylamide

These high-molecular-weight polymers are soluble in both cold and hot water and in selected organic solvents. The resin is an efficient thickener and by reaction can be changed in physical and chemical properties.

In addition to general uses for water-soluble resins, these resins have shown an outstanding ability to flocculate fines and increase the filtration rate of slurries. Consequently, polyacrylamide is used in ore processing and in other such systems where dispersed materials are encountered.

Styrene-Maleic Anhydride

Copolymers of these two monomers are soluble in some organic solvents and alkaline water. Styrene-maleic anhydride resins produce viscous and stable aqueous solutions. This resin is a strong polyelectrolyte. It is used as a textile warp size, paper coating, and static-electricity conductor. The polymer is also used in alkaline latex systems as a protective colloid, emulsifier, pigment dispersant, and filming aid.

Cellosic Derivatives

Various commercial derivatives are prepared from alpha cellulose, which is obtained from several plant sources. One class of derivatives is the water-soluble ethers. These products produce viscous aqueous solutions. All have some resistance to organic solvents, are hygroscopic, and are difficult to insolubilize.

Major industries that use these polymers as well as the water-soluble synthetic resins include food, pharmaceutical, cosmetic, textile, paper, petroleum (drilling muds), ceramic, paint, emulsion polymerization, and leather.

Hydroxyethylcellulose

This polymer is manufactured by reacting alkali cellulose with ethylene oxide. It can be water-soluble or only alkali-soluble depending on the extent of reaction. The alkali-soluble types possess the advantage of increased water resistance in deposited films. This polymer is somewhat intermediate in properties between methylcellulose and sodium carboxymethyl-cellulose. It is a protective colloid and relatively insensitive to the inclusion of multivalent ions in solution. The polymer is soluble in both hot and cold water, nonionic, but depolymerized by strong acids.

Water-soluble hydroxyethylcellulose is used in polyvinyl acetate emulsions as a stabilizer and in latex paints as a thickener and leveling agent.

Methylcellulose

Methylcellulose exhibits inverse water solubility in that it is more soluble at low temperatures than at high temperatures. It is nonionic in solution and is a very efficient thickener.

A major use is in latex paints (both polyvinyl acetate and acrylic types). Methylcellulose thickens the paint and contributes to good brushing characteristics as well. Other uses include bulking in laxatives, and binding and thickening in cosmetics and pharmaceuticals.

Sodium Carboxymethylcellulose

This polymer is soluble in both hot and cold water. It exhibits good thickening action and suspending ability for particulates. Since solutions are ionic in character, they are somewhat sensitive to pH shifts and salt additions. Major uses include soil suspension in synthetic detergents and viscosity control in oil-well drilling muds.

THERMOSETTING RESINS

A number of thermosetting resins are available in water solutions or in water-soluble form. These are principally the addition reaction products of formaldehyde with urea, phenolic, or melamine. Resorcinol and thiourea may also be reacted with formaldehyde to form water-soluble precondensates, although these have not attained the volume of the three main classes defined above.

These resins develop high molecular weight by a condensation reaction. The properties may change as the reaction proceeds. A-stage resins are those in which the degree of polymerization is minor. In some cases the degree of polymerization is such that only dimers, trimers, or similar small units are prepared. These products are water soluble or at least can tolerate the addition of significant amounts of water.

Should the reaction proceed further, the polymers enter an area roughly defined as B-stage, in which they will tolerate addition of only small amounts of water but are soluble in certain organic solvents.

Manufacturers of thermosetting resins carry the polymerization reaction to the A- or B-stage. Further reaction (to the C-stage) is carried out by consumers of these resins.

As the condensation reaction proceeds and the molecular weight builds to form a rigid, three-dimensional system, the polymers reach a point where they will not dissolve in organic solvents and are then termed *cured* (or C-staged). Further heating of the resin beyond this point may establish additional cross-links, but the physical properties do not change drastically. Once the resin has reached the C-stage, excessive heating leads to chemical breakdown of the material.

Thermosetting water-soluble polymers are treated with heat and/or catalyst to advance the cure after deposition on a particular surface or within a particular structure. Thus, the water-solubility feature is important in that it allows easy manipulation without the cost and hazards of organic systems. However, water is an essential ingredient when cross-linking cellulose with low-molecular-weight thermosetting resins. The general characteristics of a fully cured, water-soluble, thermosetting resin are similar to those obtained by curing B-stage varnishes or molding compounds.

In many areas, the water-soluble thermosetting resins compete with one another. Certain ones may be preferred in an industry or in a certain particular application because of specific properties or cost. Generally, these resins offer high-temperature stability and hardness coupled with water and solvent resistance. Resistance to either acids or bases can also be obtained.

Cyclic Thermosetting Resins

In this category are cyclic ethylene urea–formaldehyde resins and triazones, which can be obtained by cyclization of dimethylolurea with a primary amine (usually ethylamine) and then adding 2 mol of formaldehyde. These cyclic thermosetting resins were developed primarily for textile applications because they do not react or polymerize with themselves (as do other water-soluble thermosetting resins) but do react with the hydroxyls in cellulose through the methylol groups. These resins are called cellulose reactants and are by far the largest class of resins used in the wash-and-wear treatment of fabrics. Such resins impart crease resistance, wrinkle recovery, stiffness, tensile strength, water repellency, and good resistance to yellowing by chlorine-containing bleaches. In addition there are increases in resiliency, dimensional stability, and permanent texturizing.

Cyclic ethylene urea–formaldehyde resins have the advantage over triazone in better color stability, absence of odor, and scorch resistance. Triazone resins have gained prominence particularly because they are more resistant to the effects of chlorine and are somewhat cheaper. Triazones are used principally to develop wash-and-wear properties on white cotton.

Melamine-Formaldehyde

In general, melamine-formaldehyde (MF) resins are the most expensive of the water-soluble thermosetting types. They possess good color, lack of odor, high abrasion resistance, and high resistance to alkali.

The major use is in decorating laminants for surfacing of wood, paper, and other products. Other uses include binding of rock wool and glass wool for thermal insulation, finishing of nylon for stiffness and resilience, and imparting dimensional stability to wool and some cellulosics.

Phenol-Formaldehyde

This very popular class of water-soluble thermosetting resins is intermediate in cost between the MF and urea–formaldehyde (UF) types. They possess good water resistance, toughness, and acid resistance although they are somewhat poorer than the MF resins in color, odor, and flame resistance. In most other physical and chemical properties they are superior to the UF types.

Applications include laminates (including plywood and fabrics), grinding wheels, thermal insulation, battery separators, brake linings, and foundry uses.

Urea–Formaldehyde

These resins are the lowest cost as a class of the three general types but still cure to hard and somewhat brittle resins that have many desirable properties. The UF resins have an added plus in that they can cure at room temperature with suitable catalysts, whereas both the melamine-formaldehyde and phenol-formaldehyde types normally require temperatures in the neighborhood of 149°C to develop their full properties. The UF resins suffer somewhat in comparison with the other two types in poorer water resistance, less toughness, and poorer resistance to cyclical changes in temperature or water exposure.

Urea–formaldehyde resins are used in plywood because of case of handling and lower temperature of cure, on paper for increased wet strength in air filters, and on certain rayon fabrics for improved stabilization and water resistance. The UF resins are used as insolubilizers for hydroxyl-containing polymers and in many of the general application areas for water-soluble, thermosetting resins.

WAX

Wax is a general name for a variety of substances of animal and vegetable origin, which are fatty acids in combination with higher alcohols, instead of with glycerin as in fats and oils. They are usually harder than fats, less greasy, and more brittle, but when used alone do not mold as well. Chemically, the waxes differ from fats and oils in that they are composed of high-molecular-weight fatty acids with high-molecular-weight alcohols. The most familiar wax is beeswax from the honeybee, but commercial beeswax is usually greatly mixed or adulterated. Another animal wax is spermaceti from the sperm whale. Vegetable waxes include Japan wax, jojoba oil, candelilla, and carnauba wax.

Mineral waxes include paraffin wax from petroleum, ozokerite, ceresin, and montan wax. The mineral waxes differ from the true waxes and are mixtures of saturated hydrocarbons.

The animal and vegetable waxes are not plentiful materials, and are often blended with or replaced by hydrocarbon waxes or waxy synthetic resins. However, waxes can be made from common oils and fats by splitting off the glycerin and reesterifying selected mixtures of the fatty acids with higher alcohols.

Types and Uses

Some plastics have wax characteristics, and may be used in polishes and coatings or for blending with waxes. Polyethylene waxes are light-colored, odorless solids of low molecular weight, up to about 6000. Mixed in solid waxes to the extent of 50%, and in liquid waxes up to 20%, they add gloss and durability and increase toughness. In emulsions they add stability.

Waxes are employed in polishes, coatings, leather dressings, sizings, waterproofing for paper, candles, carbon paper, insulation, and varnishes. They are softer and have lower melting points than resins, are soluble in mineral spirits and in alcohol, and insoluble in water.

Synthetic waxes are used in liquid floor waxes, temporary corrosion protection, release agents, and as a melting point booster. There is a micronized polyethylene wax that is a processing and performance additive for adhesives, coatings, color concentrates, cosmetics, inks, lubricants, paints, plastics, and rubber. It can also be constituted from low-molecular-weight homopolymer, oxidized homopolymer, or as a copolymer. Another is a methylene polymer used to blend with vegetable or paraffin waxes to increase the melting point, strength, and hardness. This is a mixture of terphenyls. It is a light-buff, waxy solid, highly soluble in benzene, and with good resistance to heat, acids, and alkalies. It is used to blend with natural waxes in candles, coatings, and insulation. Waxes are not digestible, and the so-called edible waxes used as water-resistant coatings for cheese, meats, and dried fruits are not waxes, but are modified glycerides. One is a white, odorless, tasteless waxy solid melting at 40°C, and is an acetylated monoglyceride of fatty acids.

Microcrystalline waxes are used for the vacuum impregnation of inorganic-filled, organic-bonded electrical insulation and coatings for ceramic capacitors and other electronic components. The wax is chosen because of its low moisture permeability.

Wax emulsions have been widely used as binders for dry-press mixes and glaze suspensions.

A high-melting-point paraffin also will make an excellent binder for dry-press granules. The paraffin is melted, added to the body, and then thoroughly incorporated by means of a heated, muller-type mixer.

Ordinary paraffin also can be used to bond ceramic parts to steel plates for attachment to magnetic chucks during grinding.

WEAR-RESISTANT STEEL

Many types of steel have wear-resistant properties, but the term usually refers to high-carbon, high-alloy steels used for dies, tooling, and parts subject to abrasion and for wear-resistant castings. They are generally cast and ground to shape. They are mostly sold under trade names for specific purposes. The excess carbon of the steels is in spheroidal form rather than as graphite. One of the earlier materials of this kind for drawing and forming dies is Adamite. It is a chromium–nickel–iron alloy with up to 1.5% chromium, nickel equal to half that of the chromium, and from 1.5 to 3.5% carbon with silicon from 0.5 to 2%.

The Brinell hardness ranges from 185 to 475 as cast, with tensile strengths to 861 MPa. The softer grades can be machined and then hardened, but the hard grades are finished by grinding. Others have about 13% chromium, 1.5% carbon, 1.1% molybdenum, 0.70% cobalt, 0.55% silicon, 0.50% manganese, and 0.40% nickel. They are used for blanking dies, forming dies, and cams. T15 tool steel, for extreme abrasion resistance in cutting tools, is classed as a super-high-speed steel. It has 13.5% tungsten, 4.5% chromium, 5% cobalt, 4.75% vanadium, 0.50% molybdenum, and 1.5% carbon. Its great hardness comes from the hard vanadium carbide and the complex tungsten–chromium carbides, and it has full red-hardness. The property of abrasion or wear resistance in steels generally comes from the hard carbides, and is thus inherent with proper heat treatment in many types of steel.

WELDING ALLOYS

Welding alloys are usually in the form of rod, wire, or powder used for either electric or gas welding, for building up surfaces, or for hardfacing surfaces. In the small sizes in continuous lengths, welding alloys are called welding wire. Nonferrous rods used for welding bronzes are usually referred to as brazing rods, as the metal to be welded is not fused when using them. Welding rods may be standard metals or special alloys, coated with a fluxing material or uncoated, and are normally in diameters from 0.239 to 0.635 cm. Compositions of standard welding rods follow the specifications of the American Welding Society. Molded carbon, in sizes from 0.318 to 2.54 cm in diameter, is also used for arc welding. Low-carbon steel rods for welding cast iron and steel contain less than 0.18% carbon. High-carbon rods produce a hard deposit that requires annealing, but these are also used for producing a hard filler. High-carbon rods, with 0.85 to 1.10% carbon, will give deposits with an initial hardness of 575 Brinell, whereas high-manganese rod deposits will be below 200 Brinell but will work-harden to above 500 Brinell. For high-production automatic welding operations, carbon-steel wire may have a thin coating of copper to ease operation and prevent spattering. Stainless steel rods are marketed in various compositions. There are welding rods that comprise a range of stainless steels with either titania-lime or straight-lime coatings. Stainless C is an 18-8 type of stainless steel with 3.5% molybdenum. Aluminum-weld is a 5% silicon aluminum rod for welding silicon–aluminum alloys, and the Tungweld rods, for hard surfacing, are steel tubes containing fine particles of tungsten carbide. Kennametal KT-200 has a core of tungsten carbide and a sheathing of steel. It gives coatings with a hardness of Rockwell C63. Chromang, for welding high-alloy steels, is an 18-8 stainless steel modified with 2.5 to 4% manganese.

Welding rods with grades of high-manganese steel give hardnesses from 500 to 700 Brinell, and high-speed-steel rods are used for facing worn cutting tools; others are used for facing surfaces requiring extreme hardness and have the alloy granules in a soft steel tube. The welded deposit has a composition of 30% chromium, 8% cobalt, 8% molybdenum, 5% tungsten, 0.05% boron, and 0.20% carbon.

There is a group of welding alloys made especially for welding machines. They are, in general, sintered tungsten or molybdenum carbides, combined with copper or silver, and are electrodes for spot welding rather than welding rods. Tungsten electrodes may be pure tungsten, thoriated tungsten, or zirconium tungsten, the latter two used for direct-current welding. Thoriated tungsten gives high arc stability, and the thoria also increases the machinability of the tungsten. Zirconium tungsten provides adhesion between the solid electrode and the molten metal to give uniformity in the weld.

Thermit is a mixture of aluminum powder and iron oxide used for welding large sections of iron or steel or for filling large cavities. The process consists of the burning of the aluminum to react with the oxide, which frees the iron in molten form. To ignite the aluminum and start the reaction, a temperature of about 1538°C is required, which is reached with the aid of a gas torch or ignition powder, and the exothermic temperature is about 2538°C. Cast iron thermit, used for welding cast iron, is thermit with the addition of about 3% ferrosilicon and 20% steel punchings. Railroad thermit is thermit with additions of nickel, manganese, and steel.

The Stellite hardfacing rods are cobalt-based alloys that retain hardness at red heat and are very corrosion resistant. The grades have tensile strengths to 723 MPa and hardnesses to Rockwell C52. Inco-Weld A is welding wire for stainless steels and for overlays. It contains 70% nickel, 16% chromium, 8% iron, 2% manganese, 3% titanium, and not more than 0.07% carbon. The annealed weld has a tensile strength of 551 MPa with elongation of 12%. Nickel welding rod is much used for cast iron, and the operation is brazing, with the base metal not melted. Nickel silver for brazing cast iron contains 46.5% copper, 43.4% zinc, 10% nickel, 0.10% silicon, and 0.02% phosphorus. The deposit matches the color of the iron. Colmonoy 23A is a nickel alloy welding powder for welding cast iron and filling blow holes in iron casting by torch application. It has a composition of 2.3% silicon, 1.25% boron, 0.10% carbon, not over 1.5% iron, and the balance nickel, with a melting point of 1066°C. For welding on large structures where no heat treatment of the weldment is possible, the welding rods must have balanced compositions with no elements that form brittle compounds. Rockide rods are metal oxides for hard surfacing.

WETTING AGENTS

These are chemicals used in making solutions, emulsions, or compounded mixtures, such as paints, inks, cosmetics, starch pastes, oil emulsions, dentifrices, and detergents, to reduce the surface tension and give greater ease of mixing and stability to the solution. In the food industries chemical wetting agents are added to the solutions for washing fruits and vegetables to produce a cleaner and bacteria-free product. Wetting agents are described in general as chemicals having a large hydrophilic group associated with a smaller hydrophilic group. Some liquids naturally wet pigments, oils, or waxes, but others require a proportion of a wetting agent to give mordant or wetting properties. Pine oil is a common wetting agent, but many are complex chemicals. They should be powerful enough not to be precipitated out of solutions in the form of salts, and they should be free of odor or any characteristic that would affect the solution. Aerosol wetting agents are in the form of liquids, waxy pellets, or free-flowing powders. There are other free-flowing powders, basically modified polyacrylates, that are soluble in water and less so in alcohol. There are sodium or ammonium dispersions of modified rosin, with 90% of the particles below 1 μm in size. Also there is a sodium lignosulfonate produced from lignin waste liquor. It is used for dye and pigment dispersion, oil-well drilling mud, ore flotation, and boiler feedwater treatment.

Increasingly used by the ceramic industry is a popular type of poly-oxyethylene alkylate ether with a very high resistance to water hardness. Sulfonated types and carboxylates have moderate wetting properties and strong detergent and solubilizing tendencies.

WHISKERS

Whiskers are very fine single-crystal fibers that range from 3 to 10 μm in diameter and have length-to-diameter ratios of from 50 to 10,000. Since they are single crystals, their strengths approach the calculated theoretical strengths of the materials. Alumina whiskers, which have received the most attention, have tensile strengths up to 0.2 million MPa and a modulus of elasticity of 0.5 million MPa. Other whisker materials are silicon carbide, silicon nitride, magnesia, boron carbide, and beryllia.

WHITE

BRASS

White brass is a bearing metal that is actually outside of the range of the brasses, bronzes, or babbitt metals. It is used in various grades; the specification are tin, 65%; zinc, 28 to 30%; and copper, 3 to 6%. It is used for automobile bearings, and is close-grained, hard, and tough. It also casts well. A different alloy is known under the name of white brass in the cheap jewelry and novelty trade. It has no tin, small proportions of copper, and the remainder zinc. It is a high-zinc brass, and varies in color from silvery white to yellow, depending on the copper content.

White nickel brass is a grade of nickel silver. The white brass used for castings where a white color is desired may contain up to 30% nickel. The 60:20:20 alloy is used for white plaque castings for buildings. The high-nickel brasses do not cast well unless they also contain lead. Those with 15 to 20% nickel and 2% lead are used for casting hardware and valves. White nickel alloy is a copper–nickel alloy containing some aluminum. White copper is a name sometimes used for copper–nickel alloy or nickel brass. Nickel brasses known as German silver are copper–nickel–zinc white alloys used as a base metal for plated silverware, for springs and contacts in electrical equipment, and for corrosion-resistant parts. The alloys are graded according to the nickel content. Extra-white metal, the highest grade, contains 50% copper, 30% nickel, and 20% zinc. The lower grade, called fifths, for plated goods, has a yellowish color. It contains 57% copper, 7% nickel, and 36% zinc. All of the early German silvers contained up to 2% iron, which increased the strength, hardness, and whiteness, but is not desirable in the alloys used for electrical work. Some of the early English alloys also contained up to 2% tin, but tin embrittles alloys.

CAST IRON

White cast iron solidifies with all its carbon in the combined state, mostly as iron carbide, F_3C (cementite). White iron contains no free graphite as does gray iron, malleable iron, and ductile iron. White iron derives its name from the fact that it shows a bright white fracture on a freshly broken surface.

The main use for white iron is as an intermediate product in the manufacture of malleable iron. In addition to this, white iron is made as an end product to serve specific applications that require a hard, abrasion-resistant material. White iron is very hard and resistant to wear, has a very high compressive strength, but has low resistance to impact, and is very difficult to machine.

By the proper balancing of chemical composition and section size, an iron casting can be made to solidify completely white throughout its entire section. By modifying the balance and adjusting the cooling rate, the casting can be made to solidify with a layer of white iron at the surface backed up by a core of gray iron. Castings with such a duplex structure are called "chilled iron" castings.

Castings of white iron and chilled iron find their main use in resistance to wear and abrasion. Typical applications include parts for crushers and grinders, grinding balls, coke and cinder chutes, shot-blasting nozzles and blades, parts for slurry pumps, car wheels, metalworking rolls, and grinding rolls.

By using a fairly low silicon content, cast iron can be made to solidify white without the use of any additional alloy. Carbon contents are kept high (about 3.6%) when high hardness (575 Bhn) is desired. Such irons have very low toughness and a strength of about 240 MPa. For somewhat higher toughness and strength, at some sacrifice of hardness, the carbon content is lowered to about 2.8%. Unalloyed white and chilled irons have a structure composed of particles of massive iron carbide (Fe_3C) in a matrix of fine pearlite. For highest hardness, strength, and toughness, the white iron is alloyed to produce a martensitic matrix surrounding particles of massive carbide.

Gold

White gold is the name of a class of jewelers' white alloys used as substitutes for platinum. The name gives no idea of the relative value of the different grades, which vary widely. Gold and platinum may be alloyed together to make a white gold, but the usual alloys consist of from 20 to 50% nickel, with the balance gold. Nickel and zinc with gold may also be used for white golds. The best commercial grades of white gold are made by melting the gold with a white alloy prepared for this purpose. This alloy contains nickel, silver, palladium, and zinc. The 14-karat white gold contains 14 parts pure gold and 10 parts white alloy. A superior class of white gold is made of 90% gold and 10% palladium. High-strength white gold contains copper, nickel, and zinc with the gold. Such an alloy, containing 37.5% gold, 28% copper, 17.5% nickel, and 17% zinc, when aged by heat treatment, has a tensile strength of about 689 MPa and an elongation of 35%. It is used for making jewelry, has a fine, white color, and is easily worked into intricate shapes. White-gold solder is made in many grades containing up to 12% nickel, up to 15% zinc, with usually also copper and silver, and from 30 to 80% gold. The melting points of eight grades range from 695 to 845°C.

Metals

Although a great variety of combinations can be made with numerous metals to produce white or silvery alloys, the name usually refers to the lead–antimony–tin alloys employed for machine bearings, packings, and linings, to the low-melting-point alloys used for toys, ornaments, and fusible metals, and to the type metals. Slush castings, for ornamental articles and hollow parts, are made in a wide variety of soft white alloys, usually varying proportions of lead, tin, zinc, and antimony, depending on cost and the accuracy and finish desired. These castings are made by pouring the molten metal into a metal mold without a core, and immediately pouring the metal out, so that a thin shell of the alloy solidifies against the metal of the mold and forms a hollow product. A number of white metals are specified by the American Society for Testing and Materials for bearing use. These vary in a wide range from 2 to 91% tin, 4.5 to 15% antimony, up to 90% lead, and up to 8% copper. The alloy containing 75% tin, 12% antimony, 10% lead, and 3% copper melts at 184°C, is poured at about 375°C, and has an ultimate compressive strength of 111 MPa and a Brinell hardness of 24. The alloy containing 10% tin, 15% antimony, and 75% lead melts at 240°C, and has a compressive strength of 108 MPa and a Brinell hardness of 22. The first of these two alloys contains copper–tin crystals; the second contains tin–antimony crystals.

Society of Automotive Engineers (SAE) Alloy 18 is a cadmium–nickel alloy with also small amounts of silver, copper, tin, and zinc. A bismuth–lead alloy containing 58% bismuth and 42% lead melts at 123.5°C. It casts to exact size without shrinkage or expansion, and is used for master patterns and for sealing.

Various high-tin or reverse bronzes have been used as corrosion-resistant metals, especially before the advent of the chromium, nickel, and aluminum alloys for this purpose.

A white metal sheet now much used for making stamped and formed parts for costume jewelry and electronic parts is zinc with up to 1.5% copper and up to 0.5% titanium. The titanium with the copper prevents coarse-grain formation, raising the recrystallization temperature. The alloy weighs 22% less than copper, and it plates and solders easily.

WHITEWARE CERAMICS

Technical whitewares include clays, porcelains, china, white stoneware, and steatites. The modern oxide ceramics would also be in this group. For technical use, these whitewares are usually vitrified (nonporous) or very nearly so. Most commonly the pieces are glazed. To produce white-bodied ceramics, the raw materials must be of superior quality and selection. The range of materials available is comparatively limited. Although clays are widely distributed over the Earth, large, uniform deposits of white burning clays are not common.

WHITEWARE CERAMICS

Composition and Types

Most whiteware clays are basically kaolins and chemically are hydrous aluminum silicates. Ball clays are less pure than kaolins and contain free silica and small amounts of other contaminants. Kaolins are generally not highly plastic, whereas ball clays are very plastic. The plasticity derives from the physical form of the minute particles, which are colloidal in size. Ball clays are not as white-burning as kaolins and impart color to the fired body.

Feldspars are used as fluxes to provide the alkaline oxides for the glassy phase surrounding the mullite ($3Al_2O_3 \cdot 2\ SiO_2$) crystals forming the mass of the body.

Porcelains, china, and stoneware are composed of clays, silica, and feldspar. Steatites are composed of talc (magnesium silicate) and clay. Minor variations are made to enhance special properties.

The porcelains, china, and stoneware are composed of alumina (Al_2O_3), silica (SiO_2), sodium and potassium oxides (Na_2O and K_2O), as well as calcia (CaO), zinc oxide (ZnO), zirconia (ZrO_2), titania (TiO_2), barium oxide (BaO), magnesia (MgO), and phosphoric oxide (P_2O_5). Some other oxides may be present as traces. Iron oxide is usually present in small amounts as an undesirable impurity.

The oxides are supplied in kaolin, ball clays, quartz, feldspar, whiting, magnesia, and talc. Oxide ceramics that do not use clays may use only mixtures of refined oxides.

Compounding and Forming

The raw materials are intimately mixed, usually by ball milling, and then prepared for forming into ware by one of the methods listed below. For jiggering and simple mechanical pressing or extruding, the ball-milled slip is dewatered, filter pressed, deaired, and extruded into convenient size billets.

> For casting, the specific gravity is adjusted to give a good casting viscosity.
> For dry pressing, the body is dried, shredded, and powdered or spray dried into minute granules.

The pieces are formed as close to size before firing as practical, allowing for the shrinkage during firing. This shrinkage may run as high as 25% and must be very closely controlled to avoid loss in the firing of off-dimension pieces.

The ware is formed in a number of ways, some of them unchanged for centuries or even thousands of years and others unknown 60 years ago. Following are the primary methods used:

1. Throwing or jiggering on mechanical potters wheel from plastic clay body
2. Casting in plaster of paris mold from liquid slurry or slip
3. Pressing from plastic clay with simple mold
4. Pressing from dry powder in metal dies
5. Isostatic pressing from dry powder in rubber sack with hydraulic pressure
6. Hot pressing with heated clay blank and mold
7. Simple extrusion through die
8. Extrusion with thermoplastic resin to the body (injection molding)

Many pieces are formed by a combination of pressing or extrusion followed by mechanical shaping in lathes by special tools or dry grinding.

After forming, pieces are bisque-fired, which drives out moisture and water of crystallization. Porcelain is not vitrified during the bisque firing. The porcelain bisque is dipped or sprayed with glaze and, in the second firing, the body and glaze mature or vitrify together.

China is vitrified in the first firing. In china manufacture, the glaze is applied to the fired body and the glaze matures in the second firing, which is at a lower temperature than the initial firing. In some cases the glaze can be applied to an unfired piece and only one firing is needed.

Today, most kilns are fired with natural or manufactured gas, fuel oil, or electricity. The round beehive kiln fired periodically has been largely replaced with continuous tunnel kilns, which may be built with movable cars moving from one end to the other on a straight track or as a circular tunnel with a moving floor. Periodic kilns are often used in specialized work where the volume does not justify use of a tunnel kiln. They are also more versatile.

The firing of technical ceramics is performed under the most carefully controlled conditions possible. Both the temperature and the atmosphere must be known and controlled. The ware must be properly placed in the kilns or damage to the piece will result. Warpage, uneven firing, and cracking can easily occur.

Maximum firing temperatures for unglazed refractory porcelain are usually about 1760°C. Laboratory chemical porcelain that is glazed is fired to 1454°C. Hotel, sanitary china, and electrical porcelain is usually fired about 1260°C. Steatite bodies mature at around 1260 to 1316°C.

WIRE CLOTH

Stiff fabrics made of fine wire woven with plain loose weave, wire cloth is used for screens to protect windows, for guards, and for sieves and filters. Steel and iron wire may be used plain, painted, galvanized, or rustproofed, or various nonferrous metal wires are employed. It is usually put up in rolls in widths from 46 to 122 cm. Screen cloth is usually 12, 14, 16, and 18 mesh, but wire cloth in copper, brass, or Monel metal is made regularly in meshes from 4 to 100. The size of wire is usually from 0.023 to 0.165 cm in diameter. Wire cloth for fine filtering is made in very fine meshes. Mesh indicates the number of openings per inch, and has no reference to the diameter of wire. A 200-mesh cloth has 200 openings each way on a square inch, or 40,000 openings per square inch (6.4 cm^2). Wire cloth as fine as 400 mesh, or 160,000 openings per square inch (6.4 cm^2), is made by wedge-shaped weaving, although 250 wires of the size of 0.010 cm when placed parallel and in contact will fill the space of 2.5 cm. Very fine mesh wire cloth must be woven at an angle because the globular nature of most liquids will not permit passage of the liquid through microscopic square openings. One wire screen cloth, for filtering and screening, has elongated openings. One way the 0.0140-cm wire count is 200 per 2.5 cm, while the other way the 0.018-cm warp wire is 40 per 2.5 cm.

Wire fabrics for reentry parachutes are made of heat-resistant nickel–chromium alloys, and the wire is not larger than 0.013 cm in diameter to give flexibility to the cloth. Wire fabrics for ion engines to operate in cesium vapor at temperatures to 1316°C are made with tantalum, molybdenum, or tungsten wire, 0.008 to 0.015 cm in diameter, with a twill weave. Meshes to a fineness of 350 by 2300 can be obtained. Porosity uniformity is controlled by pressure calendering of the woven cloth, but for extremely fine meshes in wire cloth it is difficult to obtain the uniformity that can be obtained with porous sintered metals.

High-manganese steel wire is used for rock screens. For window screening in tropical climates or in corrosive atmospheres, plastic filaments are sometimes substituted for the standard copper or steel wire. For example, Lumite screen cloth is woven of vinylidene chloride monofilament, 0.038 cm in diameter in 18 and 20 mesh.

WOOD

For most purposes wood may be defined as the dense fibrous substance that makes up the greater part of a tree. It is found beneath the bark, and in the roots, stems, and branches of trees and shrubs. Of the three sources, the stem or trunk furnishes the bulk of raw material for lumber products.

Wood is a renewable resource. It is grown just about everywhere and can be produced in any reasonable quantity needed for future consumption. Wood products and the management of forested lands are changing to meet modern conditions; hence, trees were grown to meet modern production requirements for size, quality, and quantity. For example, with the developments in the modern technique of gluing, the former use of extremely wide, thick, and excessively long lumber is no longer necessary. Laminated lumber and plywood have generally taken the place of these large boards and timbers. Not only are the raw materials for such products easier to grow and more economical to obtain than are solid timbers of comparable size, but the products are generally improved by the use of modern methods of fabricating.

Although there are many species of wood, the commercially important types can be grouped into two categories of about 25 for each group.

ANATOMY

Wood is composed mostly of hollow, elongated, spindle-shaped cells that are arranged parallel to each other along the trunk of a tree. The characteristics of these fibrous cells and their arrangement affect strength properties, appearance, resistance to penetration by water and chemicals, resistance to decay, and many other properties.

The combined concentric bands of light and dark areas constitute annual growth rings. The age of a tree may be determined by counting these rings at the stump.

In temperate climates, trees often produce distinct growth layers. These increments are called growth rings or annual rings when associated with yearly growth; many tropical trees, however, lack growth rings. These rings vary in width according to environmental conditions. Where there is visible contrast within a single growth ring, the first-formed layer is called earlywood and the remainder latewood. The earlywood cells are usually larger and the cell walls thinner than the latewood cells. With the naked eye or a hand lens, earlywood is shown to be generally lighter in color than latewood.

Because of the extreme structural variations in wood, there are many possibilities for selecting a species for a specific purpose. Some species (for example, spruce) combine light weight with relatively high stiffness and bending strength. Very heavy woods (for example, lignum vitae) are extremely hard and resistant to abrasion. A very light wood (such as balsa) has high thermal insulation value; hickory has extremely high short resistance; mahogany has excellent dimensional stability.

Many mechanical properties of wood, such as bending strength, crushing strength, and hardness, depend on the density of wood; the heavier woods are generally stronger. Wood density is determined largely by the relative thickness of the cell wall and the proportions of thick- and thin-walled cells present.

HARDWOODS AND SOFTWOODS

The terminology used in the classification of trees is confusing, but because it has become general in usage, it is important for those who make or purchase products of wood to understand it.

The terms *hardwood* and *softwood* have no direct application to hardness or softness of the materials. Basswood is a softer domestic species, yet the yellow pines, which are classed as softwood, are often much harder. Even balsa, a foreign species that everyone knows, the lightest and softest wood used in commerce, is classed as hardwood. For practical purposes, the hardwoods have broad leaves, whereas the softwoods have needlelike leaves. Trees (hardwoods) with broad leaves usually shed them at some time during the year, while the conifers (softwoods) retain a covering of the needlelike foliage throughout the year. There are quite a few exceptions to these criteria, but as users gain familiarity with the various woods, they will soon learn by experience into which group a species falls.

Those who use wood must know something of the botanical classification because the lumber industry is also divided into two distinct groups. The methods of doing business and manufacturing and grading for quality differ from each other.

Generally, the hardwoods are used for the manufacture of factory-made products, such as tools, furniture, flooring, instrument cases, etc. The largest market for softwoods is in the home construction field or for other building purposes. But there is no line of demarcation that is reliable, for hardwoods and softwoods are often interchangeable in use.

The more important hardwoods are as follows:

Alder	Holly
Ash	Locust
Aspen	Magnolia
Basswood	Maple (hard and soft)
Beech	Oak, red
Birch	Oak, white
Cherry	Sweet or red gum
Chestnut	Sycamore
Cottonwood	Tupelo or black gum
Elm	Walnut, black
Hackberry	Willow, black
Hickory and pecan	Yellow poplar

The more important softwoods are the following:

Cedar (several species)	Pine, Ponderosa
Cypress	Pine, red
Douglas Fir (not a true fir)	Pine, southern yellow (several species)
Firs (eastern and western)	Pine, sugar
Hemlock (eastern and western)	Pine, Virginia
	Pine, western white
Larch	Redwood
Pine, eastern white	Spruce, eastern
Pine, jack	Spruce, Englemann
Pine, lodgepole	Spruce, Sitka
Pine, pitch	Tamarack

Hardwood

The horizontal plane of a block of hardwood (for example, oak or maple) corresponds to a minute portion of the top surface of a stump or end surface of a log. The vertical plane corresponds to a surface cut parallel to the radius and parallel to the wood rays. The vertical plane corresponds to a surface cut at right angles to the radius and the wood rays, or tangentially within the log. In hardwoods, these three major planes along which wood may be cut are known commonly as end-grain, quarter-sawed (edge-grain), and plain-sawed (flat-grain) surfaces.

Softwood

The rectangular units that make up the end grain of softwood are sections through long vertical cells called tracheids or fibers. Because softwoods do not contain vessel cells, the tracheids serve the dual function of transporting sap vertically and giving strength to the wood. Softwood fibers range from about 3 to 8 mm in length.

Cell Walls and Composition

The principal compound in mature wood cells is cellulose, a polysaccharide of repeating glucose molecules that may reach 4 µm in length. These cellulose molecules are arranged in an orderly manner into structures about 10 to 25 nm wide called microfibrils. This ordered arrangement in certain parts (micelles) gives the cell wall crystalline properties that can be observed in polarized light with a light microscope. The microfibrils wind together like strands in a cable to form macrofibrils that measure about 0.5 µm in width and may reach 4 µm in length. These cables are as strong as an equivalent thickness of steel.

Wood, regardless of the species, is composed of two principal materials: cellulose, which is about 70% of the volume, and lignin, nature's glue for holding the cells and fibers together, which is from 20 to 28%. Residues in the form of minerals, waxes, tannins, oils, etc. compose the remainder. The residues, although small in volume, often provide a species with unusual properties. The oils in cypress are responsible for its renown as a decay-resistant wood. Aromatic oils provide many of the cedars with distinctive odors that make them valuable for clothing storage chests. Other chemicals provide resistance to water absorption, which is useful for constructing light, high-speed boats that are relatively free from increase in weight due to water absorption.

Many of the chemical residues of wood can be removed by neutral solvents, such as water, alcohol, acetone, benzene, and ether. Some of them may be caused to migrate from one part of the wood to another.

STRUCTURE OF WOOD

The roots, stem, and branches of a tree increase in size by adding a new layer each year, just as the size of one's hand is increased by putting on a glove. The layer or growth ring will vary in thickness due to the age of the tree, growing conditions, amount of foliage, and other factors.

The growth ring is divided into two parts — spring wood and summer wood. The former is usually lighter in weight than the latter and is denser and stronger. Generally, a tree or portion of a tree with the most summer wood is stronger than one that has less.

The thickness of the growth ring and the relative amounts of spring and summer wood have great effect on the appearance of wood and are often a deciding factor in the choice of furniture materials. Excessively thick growth rings usually provide a rather coarse-textured material, whereas narrow growth rings provide fine texture. However, in hardwoods a thick ring may and usually does have more summer wood and is therefore the stronger of the two. The situation is somewhat different for the softwoods.

Grades of Lumber

Modern grading of lumber is the result of experience. Over a period of 90 years there has been a gradual evolution to meet the changes in industry. This trend will continue as long as wood products are used.

The grading of hardwoods and softwoods is entirely different; the former are used almost entirely as raw materials for manufactured products. For them, basic grading is on the number of usable cuttings or pieces that can be cut from a board. For some grades these cuttings need only be sound, but for others they must be clear at least on one face. The cuttings must also be of a certain size.

Hardwood Grading

The grades of hardwood lumber from the highest quality to the lowest are "Firsts," the top-quality grade; the next are known as "Seconds." These two grades are usually marketed as one and called "Firsts and Seconds"; the designation for the grade is FAS. The next lowest grade is "Selects," followed by "No. 1 Common," "No. 2 Common," "No. 3A Common," and "No. 3B Common." Sometimes a grade is further differentiated, such as "FAS One Face," which means that it is of a much higher grade on one face than on the other. A prefix "WHND" is also used sometimes. It means that wormholes are not to be considered as defects in evaluating cuttings nor as a reason for disqualifying them as might otherwise be done.

Softwood Grading

The theory of softwood lumber grading is probably somewhat more difficult for the layperson to understand than that used for the hardwoods. This is because many of the species are graded under separate association rulings, which are similar, but with some important differences.

Furthermore, softwoods are divided into three general classes of products, each of which is graded under a different set of rules.

1. *Yard lumber*, which is used for building construction and other ordinary uses, is obtainable in "Finish Grades," which are called "A," "B," "C," and "D"; "Common Boards," which are called "No. 1," "No. 2," "No. 3," "No. 4," and sometimes "No. 5"; and as "Common Dimension" in grades "No. 1," "No. 2," and "No. 3."

2. *Structural lumber* is a relatively modern concept in the field of lumber grading. It is an engineered product, intended for use where definite strength requirements are specified. The allowable stresses designated for a piece of structural lumber depend upon the size, number, and placement of the defects. The relative position of a defect is of great importance; therefore, if the maximum strength of the piece is to be developed it must be used in its entirety. It cannot be remanufactured for width, thickness, or length.

3. *Factory and shop lumber* is the third general category. These grades are similar to those for hardwoods in that the lumber is graded by the number of usable cuttings that can be taken from a board, but here the resemblance ceases, for both the grade descriptions and nomenclature are different. The term "Factory and Shop" is descriptive of the uses for which the product is designed. Much of it is used for general millwork products, patterns, models, etc., and wherever it is necessary to cut up softwood lumber for the production of factory-made items.

Wood Chemicals

These are chemicals obtained from wood. The practice was carried out in the past, and continues wherever technical utility and economic conditions have combined to make it feasible. Woody plants comprise the greatest part of the organic materials produced by photosynthesis on a renewable basis, and were the precursors of the fossil coal deposits. Future shortages of the fossil hydrocarbons from which most organic chemicals are derived may result in the economic feasibility of the production of these chemicals from wood.

Wood is a mixture of three natural polymers — cellulose, hemicelluloses, and lignin — in an approximate abundance of 50:25:25. In addition to these polymeric cell wall components, which make up the major portion of the wood, different species contain varying amounts and kinds of extraneous materials called extractives. Cellulose is a long-chain polymer of glucose that is embedded in an amorphous matrix of the hemicelluloses and lignin. Hemicelluloses are shorter or branched polymers of five- and six-carbon sugars other than glucose. Lignin is a three-dimensional polymer formed of phenylpropane units. Thus the nature of the chemicals derived from wood depends on the wood component involved.

Modern Processes

Chemicals derived from wood at present include bark products, cellulose, cellulose esters, cellulose ethers, charcoal, dimethyl sulfoxide, ethyl alcohol, fatty acids, furfural, hemicellulose extracts, kraft lignin, lignin sulfonates, pine oil, rayons, rosin, sugars, tall oil, turpentine, and vanillin.

Most of these are either direct products or byproducts of wood pulping, in which the lignin that cements the wood fibers together and stiffens them is dissolved away from the cellulose. High-purity chemical cellulose or dissolving pulp is the starting material for such polymeric cellulose derivatives as viscose rayon and cellophane (regenerated celluloses from the xanthate derivative in fiber or film form), cellulose esters such as the acetate and butyrate for fiber, film, and molding applications, and cellulose ethers such as carboxymethylcellulose, ethylcellulose, and hydroxyethylcellulose for use as gums.

Potential Chemicals

Considerable development effort has been devoted to the conversion of renewable biomass, of which wood is the major component, into the chemicals usually derived from petroleum. Processes for which technical feasibility has been demonstrated are shown in the illustration. Economic feasibility is influenced by fossil hydrocarbon cost and availability.

WOOD DEGRADATION

This refers to decay of the components of wood. Despite its highly integrated matrix of cellulose, hemicellulose, and lignin, which gives wood superior strength properties and a marked resistance to chemical and microbial attack, a variety of organisms and processes are capable of degrading wood. The decay process is a continuum, often involving a number of organisms over many years. Wood degrading agents are both biotic and abiotic, and include heat, strong acids and bases, organic chemicals, mechanical wear, and sunlight (ultraviolet degradation).

Engineering Design

This is the process of creating products, components, and structural systems with wood and wood-based materials. Wood engineering design applies concepts of engineering in the design of systems and products that must carry loads and perform in a safe and serviceable fashion. Common examples include structural systems such as buildings or electric power transmission structures, components such as trusses or prefabricated stressed-skin panels, and products such as furniture or pallets and containers. The design process considers the shape, size, physical and mechanical properties of the materials, type and size of the connections, and the type of system response needed to resist both stationary and moving (dynamic) loads, and function satisfactorily in the end-use environment.

Wood is used in both light frame or heavy timber structures. Light frame structures consist of many relatively small wood elements such as lumber covered with a sheathing material such as plywood. The lumber and sheathing are connected to act together as a system in resisting loads; an example is a residential house wood floor system where the plywood is nailed to lumber bending members or joists. In this system, no one joist is heavily loaded because

WOOD

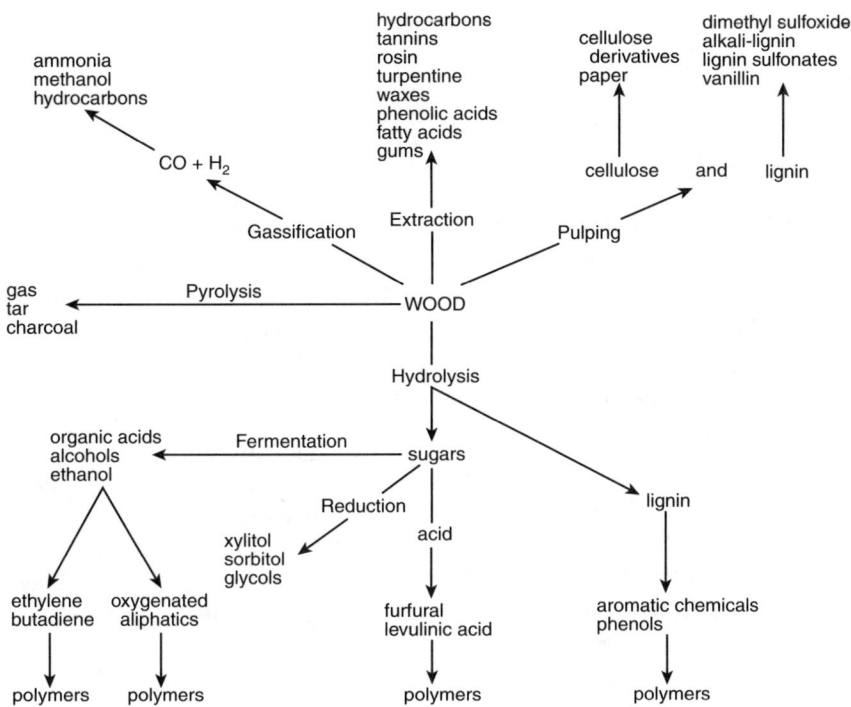

FIGURE W.1 Chemical pathways for obtaining chemicals from wood. (From *McGraw-Hill Encyclopedia of Science and Technology*, 8th ed., Vol. 19, McGraw-Hill, New York, 579. With permission.)

the sheathing spreads the load out over many joists. Service factors such as deflection or vibration often govern the design of floor systems rather than strength. Light frame systems are often designed as diaphragms or shear walls to resist lateral forces resulting from wind or earthquake.

In heavy timber construction, such as bridges or industrial buildings, there is less reliance on system action and, in general, large beams or columns carry more load transmitted through decking or panel assemblies. Strength, rather than deflection, often governs the selection of member size and connections. There are many variants of wood construction using poles, wood shells, folded plates, prefabricated panels, logs, and combinations with other materials.

Engineered Wood Composites

Wood composites are products composed of wood elements that have been glued together to make a different, more useful or more economical product than solid sawn wood. Plywood is a common example of a wood-composite sheathing panel product where layers of veneer are glued together. Plywood is used as a sheathing material in light frame wood buildings, wood pallets, and containers to distribute the applied forces to beams of lumber or other materials. Shear strength, bending strength, and stiffness are the most important properties for these applications and may be engineered into the panel by adjusting the species and quality of veneer used in the manufacture.

Processing

Processing involves peeling, slicing, sawing, and chemically altering hardwoods and softwoods to form finished products such as boards or veneer; particles or chips for making paper, particle, or fiber products; and fuel.

Most logs are converted to boards in a sawmill that consists of a large circular or band saw, a carriage that holds the log and moves past the saw, and small circular saws that remove excess bark and defects from the edges and ends of the boards. One method is to saw the log to boards with a single pass through several saw blades mounted on a single shaft (a gang saw). Sometimes, the outside of the log is converted to boards or chips until a rectangular center or cant remains. The cant is then processed to boards with a gang saw.

Other steps taken may be drying and machining, including veneer cutting. Wood is also ground to fibers for hardboard, medium-density fiberboard, and paper products. It is sliced and flaked for particle-board products, including wafer boards and oriented strand boards. Whether made from waste products (sawdust, planer shavings, slabs, edgings) or roundwood, the individual particles generally exhibit the anisotropy and hygroscopicity of larger pieces of wood. The negative effects of these properties are minimized to the degree that the three wood directions (longitudinal, tangential, and radial) are distributed more or less randomly.

Wood Products

Wood products are those products, such as veneer, plywood, laminate of products, particleboard, waferboard, pulp and paper, hardboard, and fiberboard, made from the stems and branches of coniferous (softwood) and deciduous (hardwood) tree species. The living portion of the tree is the region closest to the bark and is commonly referred to as sapwood; the dead portion of the tree is called heartwood. In many species, especially hardwoods, the heartwood changes color because of chemical changes in it. The heartwood of walnut, for example, is dark brown and the sapwood almost white.

Wood is one of the strongest natural materials for its weight. A microscopic view reveals thousands of hollow-tubed fibers held together with a chemical called lignin. These hollow-tubed fibers give wood its tremendous strength for its light weight. These fibers, after the lignin bonding material is removed, make paper. In addition to lignin, wood is composed of other chemicals, including cellulose and hemicellulose. This category includes lumber products, veneer, and structural plywood.

Lumber

Most small log sawmills try to maximize value and yield by automating as much of the manufacturing process as possible. The basic process of cutting lumber involves producing as many rectangular pieces of lumber as possible from a round tapered log. There are only a few sawing solutions, out of millions possible, that will yield the most lumber from any given log or larger timber.

Veneer and Structural Plywood

Structural plywood is constructed from individual sheets of veneer, often with the grain of the veneer in perpendicular directions in alternating plies. The most common construction is three-, four-, and five-ply panels. The alternating plies give superior strength and dimensional stability.

Composite Products

These include laminated products made from lumber, particleboard, waferboard, and oriented strand board.

Laminated Products

Laminated products are composite products, made from lumber, parallel laminated veneer, and sometimes plywood, particleboard, or other fiber product. The most common types are the laminated beam products composed of individual pieces of lumber glued together with a phenol resorcinol-type adhesive. Laminated beams are constructed by placing high-quality straight-grained pieces of lumber on the top and bottom, where tension and compression stresses are the greatest, and lower-quality lumber in the center section, where these stresses are lower.

Another form of composite beam is constructed from individual members made from parallel laminated veneer lumber. This type of lumber has the advantage that it can be made into any length, thus creating beams to span large sections. These beams can be made entirely of parallel laminated veneer lumber, or with the top and bottom flange made from parallel laminated veneer lumber, and the center web made from plywood or flakeboard. Some of these products use solid lumber for the flange material and resemble an I beam. All structural laminated products have the advantage over lumber in that much of the natural variation due to defects is removed, and wood structural members can be made much larger than the typical 51 × 305 mm lumber product.

Particleboard

Nonstructural particleboard is another type of composite product that is usually made from sawdust or planer shavings. It is sometimes made from flaked roundwood. This type of particleboard is one of the most widely used forms of wood product. Often hidden from view, it is used as a substrate under hardwood veneer or plastic laminates. It is commonly used in furniture, cabinets, shelving, and paneling. Particleboard is made by drying, screening, and sorting the sawdust and planer shavings into different size classifications.

Waferboard and Oriented Strand Board

Waferboard and oriented strand board are structural panels made from flakes or strands, and are usually created from very small trees. Unlike nonstructural particleboard, waferboard is designed for use in applications similar to those of plywood. Waferboard and oriented strand board can have flakes or strands oriented in the same direction, thus giving the board greater strength in the long axis. The most common type of waferboard is made from randomly oriented flakes.

FIBER PRODUCTS

The most common fiber products result from pulping processes that involve the chemical modification of wood chips, sawdust, and planer shavings. Such products include pulp and paper, hardboard, and fiberboard.

Pulp and Paper

Paper making begins with the pulping process. Pulp is made from wood chips created in the lumber manufacturing process, small roundwood that is chipped, and recycled paper. The fibers in the chips must be separated from each other by mechanically grinding the fibers or chemically dissolving the lignin from them. The most common chemical processes are sulfite and sulfate (kraft). Following the pulping process, the fibers are washed to remove pulping chemicals or impurities. In some processes (for example, writing papers), the fibers are bleached.

Dry, finished paper emerges from the end of this section, and it is placed in rolls for further manufacture into paper products.

Hardboard

Hardboard is a medium- to high-density wood fiber product made in sheets from 1.6 to 12.7 mm. Hardboard is used in furniture, cabinets, garage door panels, vinyl overlaid wall panels, and pegboard. It is made by either a wet or dry process.

Medium-Density Fiberboard

Medium-density fiberboard is used in many of the same applications where particleboard is used. It can be used in siding and is especially well suited to cabinet and door panels where edges are exposed. Unlike particleboard, which has a rough edge, medium-density fiberboard has a very fine edge that can be molded very well. Medium-density fiberboard is produced in much the same way that dry-processed hardboard is produced in its early stages. The chips are thermomechanically pulped or refined prior to forming into a dry mat. Following refining, medium-density fiberboard is produced in a fashion similar to that of particleboard. The dry pulp is sprayed with adhesive (usually urea-formaldehyde or phenol-formaldehyde) and formed into a dry mat prior to pressing in a multiopening hot press. After the panels have been formed, they are cooled and cut to smaller final product sizes prior to shipment.

OTHER ENGINEERED WOOD PRODUCTS

The list of processes and potential processes that are available for wood construction is almost endless. The laminating of wood can be extended to using species of one kind for the surfaces and another for the core. Also, other species having altogether different properties may be used in other places of the assembly. Decking for aircraft carriers is a good example: the surface of the decking must have high abrasion resistance, but great strength in bending is also required. The combination of several species in the same timber is therefore most effective.

Barrels and other cylindrical containers are made of laminated wood staves, plywood, or veneer. Hogsheads for tobacco export are of solid staves or plywood.

Wood can be treated to provide considerable fire resistance; it then can be destroyed by heat, but will not support combustion.

Natural wood is not adversely affected by extremes of cold such as are encountered in high latitudes and is therefore much in use for shelters, tools, sporting equipment, and other products where these conditions exist.

Wood is widely used for structures and products that must be nonmagnetic; hence, it is used for minesweepers and similar craft. Laminated wood has no peer for this type of product.

Special treatments have been developed for use in power-line construction. These treatment materials must act as nonconductors as well as preservatives.

Wood has excellent thermal insulating properties. It is used for refrigerated spaces, refrigerated delivery trucks, and in a great many other places where light weight and thermal insulation, combined with strength, are necessary. For example, specially treated milk containers are in widespread use in the dairy industry.

Special forms of timbers are manufactured for boxcar decking. Decking of this type provides a medium for fastening cargo as well as furnishing other needed functions.

In the textile industry, wood, both treated and in the natural form, is used for shuttles and bobbins, pulleys, and many other types of machinery.

Wood, in a natural form, treated, or laminated, is used for water and other liquid conduits, cooling towers, and chemical containers.

Preservation/Protection

There are several wood-preservation methods. The system used depends upon the service requirements. For example, a railroad tie, once in place, must withstand continuous exposure to the elements and almost every conceivable condition that promotes decay, until it is worn out some 20 to 30 years later. It is therefore necessary to provide this and similar items with maximum protection.

The two general methods of treating are surface applications where the wood is exposed to the chemicals by dipping, soaking, or brushing, and treatments wherein the chemicals are forced into the wood through pressure. These latter treatments are used for the most critical applications. Pressure-treating operations must be conducted in plants with considerable equipment. It is a specialized business, and a large industry has arisen to take care of the many needs for highly treated products.

The most effective way to prevent damage through exposure of untreated wood is to perform all major machining operations prior to treating. If this is impossible, the exposed parts should be given a surface treatment, by brushing or dipping. Unfortunately, neither of these treatments is as effective as the original pressure treating. All domestic species can be pressure-treated, but some of them require it less than others. The inherently decay-resistant species, such as cypress, redwood, some of the cedars, and white oak, need no treatment for most uses. But this applies only to the heartwood of any specie.

Of the various chemicals used for wood-preservative treatment, with the pressure system, creosote is one of the oldest and best. However, it has an odor that is sometimes offensive and is therefore not generally suitable for manufactured items that may be used in contact with the body, or adjacent to food or enclosed places where the fumes may become objectionable. It can also be somewhat of a fire hazard and is sometimes not used for this reason. A third objection to creosote is the difficulty of painting over it. However, creosote is very effective against decay fungi and insect damage.

There are several effective forms of copper salts. These can be painted over; they do not contribute to fire hazard and are odorless. They are extensively used in the manufacture of boats and other marine appliances.

A relatively new chemical for decay and insect prevention is pentatchloraphenol. This has most of the advantages of creosote, except that it is probably not as effective against termite damage and is more expensive, but it lacks most of the disadvantages and can be used in enclosed places.

For food containers preservative treatments are not generally recommended unless the conditions are closely examined and there is assurance that the chemicals are approved under the existing laws and are in no way harmful to health.

The second method of treatment is by one of the surface systems. Generally, the chemicals used for this purpose are almost the same as for pressure treating. However, they are often specially prepared and their carriers may differ to obtain greater natural penetration.

Of the various methods of application dipping is the most efficient, for in this way all of the surface is exposed to the chemicals and a maximum amount of usable chemical is deposited, which may not be the case with brush or spray treatments. Often, heating the chemicals and cooling the pieces during dipping will increase penetration. However, long periods of soaking are not usually of much advantage. It is better to dip the wood parts for several minutes (sufficient time to assure that surface exposure is complete) and then pile the pieces closely together soon after they are removed from the chemical bath. Several days of natural absorption under these conditions will often provide a surprising amount of penetration. The success of this method depends to a considerable extent on the type of chemical used and particularly the carrier.

In the millwork industry the surface method of application is of tremendous importance and is widely used for nearly all its products. Windows, storm sash, exterior trim, and most other products exposed to weather are effectively treated. In addition, the preservative chemicals are often combined with waxes or oils to provide a reasonable amount of dimensional stability.

PROPERTIES

The physical and mechanical characteristics of wood are controlled by specific anatomy, moisture content, and to a lesser extent mineral and extractive content. The properties are also influenced by the directional nature of wood, which results in markedly different properties in the longitudinal, tangential, and radial directions or axes. Wood properties within a species vary greatly from tree to tree and within a single axis. The physical properties (other than appearance) are moisture content, shrinkage, density, permeability, and thermal and electrical properties.

The mechanical properties of wood include elastic, strength, and vibration characteristics. These properties are dependent upon species, grain orientation, moisture content, loading rate, and size and location of natural characteristics such as knots.

Because wood is an orthotropic material, it has unique and independent mechanical properties in each of three mutually perpendicular axes — longitudinal, radial, and tangential. This orthotropic nature of wood is interrupted by naturally occurring characteristics such as knots that, depending on size and location, can decrease the stiffness and strength of the wood.

WOOD-BASED FIBER AND PARTICLE MATERIALS

Flat-formed board products can be classified into two groups: (1) Those primarily from fiber interfelted during manufacture with a predominantly natural bond, although extraneous material may be added to improve some property such as bond (or other strength property) and water resistance, and (2) those made from distinct fractions of wood with the primary bond produced by an added bonding material.

Moldings, based on a composition of wood-based fiber or particle and binding material similar to those used for boards, have found increasing use.

Production statistics for molded units of fiber and wood particle are difficult to obtain because the units are used in such widely different commodities as toilet seats, croquet balls, school desktops and seats, frames for luggage, and armrests, door panels, and other molded components for automobiles. The number of uses for these to moldings is increasing because they combine shape with adequate structural strength and durability for many uses. See **Wood**.

WOOD–METAL LAMINATES

Wood–metal laminates constitute a composite panel construction in which the core material is made up of a wood or wood-derivative slab to which metal facing sheets are adhesively bonded.

The core material is usually plywood or a composition board of wood fibers or chips compressed and bonded together to form a flat core slab. Balsa wood and insulation boards are frequently used as core materials where thermal insulation is a requirement.

The metal facing sheets may be steel, aluminum, stainless steel, porcelain enameled metals, rigidized metals, or metals with decorative finishes. Light-gauge metals are usually used, ranging in thickness from 0.25 to 1.5 mm depending upon the specific strength, stiffness, and service requirements.

Adhesives used in bonding the metal to the core material are selected to meet the desired service requirements. For most standard laminates the adhesive is water resistant and fungusproof so that the panels may be subjected to exterior as well its interior exposure. Continuous service temperatures may range from a minimum of –51°C to a maximum of 77°C, although specially prepared laminates are available for continuous use as high as 177°C.

Greater Stiffness

Wood–metal laminates are designed to utilize the best properties of each of the component parts providing a panel that is not only light in weight but also has good structural strength and high flexural rigidity. Because the metal facing sheets are supported by a core material of substantial thickness, smooth flat panels free from waves and buckles are obtained with light-gauge metals.

Wood–metal laminates are limited in size only by the size of the press equipment and commercial sizes of the metal sheets and core material available.

The panels may be sawed to exact size from stock size sheets. Frequently, however, the facing sheets are fabricated prior to bonding to the core material to provide special edge details.

Applications

Wood–metal laminates are used extensively in the architectural building and transportation fields. Curtain wall panels, column enclosures, partition panels, facia, and soffit panels are typical applications. Truck and trailer bodies, shipping containers, and railroad car partition panels and doors are common uses of the laminate. In general, wherever light weight, combined with high rigidity and structural strength, is prerequisite, wood–metal laminates may be used to good advantage.

WOODS, IMPORTED

There are over 100 different species of foreign woods that are imported into the United States. By far, the greatest number of these are used for decorative or nonengineering purposes — some are used for both.

Imported woods to be considered here are all defined as imported species, except those that are also native to the United States. Thus, all Canadian kinds are excluded because the United States has every species of wood that grows in Canada. Mexico, on the other hand, supplies some tropical woods not found in the United States. There are pines from the highlands of Mexico and Central America that are imported in fairly sizable volume at times, but the engineering aspects of their utilization coincide directly with the Southern pine found in the United States; consequently, no attempt is made to describe these. No other softwoods (conifers) are imported for use as engineering materials. Thus, all types discussed here are categorized as hardwood (broadleaf) species. Even balsa is a hardwood, because the terms "hardwood" and "softwood" in lumber-industry parlance allude to the botanical classification rather than the actual hardness or softness of the wood.

There are over 13 kinds of tropical hardwoods known to be used for engineering products or purposes. In certain parts of the world there are vast expanses of untouched tropical forests, which, no doubt, contain several other hardwoods of potential engineering value. However, the woods in current use are typical imported woods and applications:

1. Refrigeration: balsa
2. Wharves and docks: Greenheart, ekki, jarrah, ironbark, apitong, angelique
3. Boat construction: Philippine mahogany, Central and South American mahogany, African mahogany, balsa, apitong, teak, iroko, ironbark, jarrah, lignum vitae
4. Tanks and vats: Philippine mahogany, apitong
5. Building construction: mahogany (all types), apitong, balsa, greenheart
6. Poles, piling: greenheart, ekki
7. Machinery: lignum vitae
8. Aircraft and missiles: balsa
9. Vehicles: apitong

WROUGHT IRON

Wrought iron is commercially pure iron made by melting white cast iron and passing an oxidizing flame over it, leaving the iron in a porous condition, which is then rolled to unite it into one mass. As thus made, it has a fibrous structure, with fibers of slag through the iron in the direction of rolling. It is also made by the Aston process of shooting Bessemer iron into a ladle of molten slag. Modern wrought iron has a fine dispersion of silicate inclusions that interrupt the granular pattern and give it a fibrous nature.

Structurally, wrought iron is a composite material; the base metal and the slag are in physical association, in contrast to the chemical or alloy relationship that generally exists between the constituents of other metals.

The form and distribution of the iron silicate particles may be stringerlike, ribbonlike, or platelets. Practically, the physical effects of the incorporated iron silicate slag must be taken into consideration in bending and forming wrought iron pipe, plate, bars, and shapes, but when properly handled — cold or hot — fabrication is accomplished without difficulty.

MECHANICAL PROPERTIES

The value of wrought iron is in its corrosion resistance and ductility. It is used chiefly for rivets, staybolts, water pipes, tank plates, and forged work. Minimum specifications for ASTM wrought iron call for a tensile strength of 275 MPa, yield strength of 165 MPa, and elongation of 12%, with carbon not over 0.08%, but the physical properties are usually higher. Wrought iron 4D has only 0.02% carbon with 0.12% phosphorus, and the fine fibers are of a controlled composition of silicon, manganese, and phosphorus. This iron has a tensile strength of 330 MPa, elongation 14%, and Brinell hardness 105. Manganese wrought iron has 1% manganese for higher impact strength.

Ordinary wrought iron with slag may contain frequent slag cracks, and the quality grades are now made by controlled additions of silicate, and with controlled working to obtain uniformity. But for tanks and plate work, ingot iron is now usually substituted.

The Norway iron formerly much used for bolts and rivets was a Swedish charcoal iron brought to America in Norwegian ships. This iron, with as low as 0.02% carbon, and extremely low silicon, sulfur, and phosphorus, was valued for its great ductility and toughness and also for its permeability qualities for transformer cores. Commercial wrought iron is now usually ingot iron or fibered low-carbon steel.

The tensile properties of wrought iron are largely those of ferrite plus the strengthening effect of any phosphorus content, which adds approximately 6.8 MPa for each 0.01% above 0.10% of contained phosphorus. Strength, elasticity, and ductility are affected to some degree by small variations in the metalloid content and in even greater degree by the amount of the incorporated slag and the character of its distribution.

Nickel, molybdenum, copper, and phosphorus are added to wrought iron to increase yield and ultimate strengths without materially detracting from toughness as measured by elongation and reduction in area.

Fabrication

Forging

Wrought iron is an easy material to forge using any of the common methods. The temperature at which the best results are obtained lies in the range of 1149 to 1316°C. Ordinarily "flat and edge" working is essential for good results. Limited upsetting must be accomplished at "sweating to welding" temperatures.

Bending

Wrought-iron plates, bars, pipe, and structurals may be bent either hot or cold, depending on the severity of the operation, keeping in mind that bending involves the directional ductility of the material. Hot bending ordinarily is accomplished at a dull red heat (704 to 760°C) below the critical "red-short" range of wrought iron (871 to 927°C). The ductility available for hot bending is about twice that available for cold bending. Forming of flanged and dished heads is accomplished hot from special-forming, equal-property plate.

Welding

Wrought iron can be welded easily by any of the commonly used processes, such as forge welding, electric resistance welding, electric metallic arc welding, electric carbon arc welding, and gas or oxyacetylene welding. The iron silicate or slag included in wrought iron melts at a temperature below the fusion point of the iron-base metal, so that the melting of the slag gives the metal surface a greasy appearance. This should not be mistaken for actual fusion of the base metal; heating should be continued until the iron reaches the state of fusion. The siliceous slag content provides a self-fluxing action to the material during the welding operation.

Threading

The machinability or free-cutting characteristics of most ferrous metals are adversely influenced by either excessive hardness or softness. Wrought iron displays almost ideal hardness for good machinability, and the entrained silicate produces chips that crumble and clear the dies. Standard threading equipment that incorporates minor variations in lip angle, lead, and clearance is usually satisfactory with wrought iron.

Protective Coatings

Wrought iron lends itself readily to such cleaning operations as pickling and sandblasting for the application of protective coatings. Where protective coatings such as paint or hot-dipped metallic coatings are to be applied, the coatings are found to adhere more firmly to wrought iron and a thicker coat will be attained compared with other wrought ferrous metals. This is because the natural surface of wrought iron is microscopically rougher than other metals after cleaning, thus providing a better anchorage for coatings.

Corrosion Resistance

The resistance of wrought iron to corrosion has been demonstrated by long years of service life in many applications. Some have attributed successful performance to the purity of the iron base, the presence of a considerable quantity of phosphorus or copper, freedom from segregation, to the presence of the inert slag fibers disseminated throughout the metal, or to combinations of such attributes.

In actual service, the corrosion resistance of wrought iron has shown superior performance in such applications as radiant heating and snow-melting coils, skating-rink piping, condenser and heat-exchanger equipment, and other industrial and building piping services. Wrought iron has long been specified for steam condensate piping where dissolved oxygen and carbon dioxide present severe corrosion problems. Cooling water cycles of the once-through and open-recirculating variety are solved by the use of wrought-iron pipe.

Applications

> *Building Construction* — Hot and cold potable water, soil, waste, vent, and downspout piping; radiant heating, snow melting, air-conditioning cooling and chilled-water lines; gas, fire protection, and soap lines; condensate

and steam returns, ice-rink and swimming-pool piping; underground service lines and electrical conduit.

Industrial — Unfired heat exchangers, brine coils, condenser tubes, caustic soda, concentrated sulfuric acid, ammonia, and miscellaneous process lines; sprinkler systems, boiler feed and blowoff lines, condenser water piping, runner buckets, skimmer bars, smokestacks and standpipes, salt and water well pipe and casing.

Public Works/Infrastructure — Bridge railings, fenders, blast plates, drainage lines and troughs, traffic signal conduit, sludge digestor heating coils, aeration tank piping, sewer outfall lines, large outside diameter intake and discharge lines, trash racks, weir plates, dam gates, pier-protection plates, sludge tanks and lines, dredge pipe.

Railroad and Marine — Tie spacer bars, diesel exhaust- and air-brake piping, ballast and brine protection plates, brine, cargo and washdown lines on ships, hull and deck plating, rudders, fire screens, breechings, tanker heating coils, car retarder and yard piping, spring bands, car charging lines, nipples, pontoons, car and switch deicers.

Others — Gas collection hoods, staybolts, flue gas conductors, sulfur mining gut, air and transport lines, coal-handling equipment, chlorine, compressed air lines, distributor arms, cooling tower and spray pond piping.

Wrought iron is available in the form of plates, sheets, bars, structurals, forging blooms and billets, rivets, chain, and a wide range of tubular products, including pipe, tubing and casing, electrical conduit, cold-drawn tubing, and welded fittings.

X-RAY PHOTO-ELECTRON SPECTROSCOPY

Carbon black filled rubber parts may challenge many analytical methods because the compounded rubber may include up to 20 individual ingredients. Often, after mixing, it is difficult to extract ingredients from the polymer base because of their interactions with other ingredients such as carbon black.

Testing of bonded rubber-to-metal parts amplifies the problems twofold. The bonding adhesive is also a multicomponent polymeric blend, much like the rubber. Also involved are contaminants associated with the substrate. For example, a stamped steel insert may be contaminated with hydrocarbons. If the metal is porous, oil may bleed out of the metal grain over time and undermine a bond system. Or, if the oil is present initially, it may prevent adhesive wet-out, thus inhibiting adhesion to the substrate.

Disadvantages

Some techniques such as reflectants Fourier transform infrared spectrometer (FTIR) and energy dispersive x-ray (EDS) would not be adequate for testing rubber bonded to metal. EDS, using a windowless system, does not have sufficient resolution for lower-weight elements.

The Process

To gain the most definitive picture of failed bonded surfaces, scanning electron microscopy (SEM) and x-ray photo-electron spectroscopy (XPS or ESCA) methods have proved the most valuable.

XPS is an excellent method for determining chemical composition of a surface. This method penetrates the surface of the sample to a depth of only 50 Å. XPS measures the binding energies of the atoms present on the surface, yielding a very distinctive signature of the atomic species present. XPS is sensitive enough to generate signals that provide both a quantitative and qualitative picture of the surface composition over a wide elemental range. Computer analysis techniques convert these signals to atomic percentages to allow reconstruction of the various materials present.

Case Study

The following case study analyzes black rubber bonded to a steel part. The part was exposed to a hostile environment (hydrocarbon fluids at 150°C) for a relatively short period of time before failure occurred. Parts of this type are expected to last approximately 8 years; this part failed in under 6 months.

Initial visual analysis indicated failure primarily in an adhesive mode. However, no apparent cause of the failure was easily identifiable. Gaining a clear understanding of the cause of the adhesive failure at this point was difficult because the part was saturated with hydrocarbon oil.

The cross section of the remaining bonded area indicated that under normal conditions, the adhesive could not be distinguished from the rubber. This would indicate a high degree of commingling is necessary between rubber and adhesive to achieve a good bond.

At this point the unbond was believed to be caused by an overbake of the adhesive. In this situation, the poor bond is due to premature cross-linking of the adhesive prior to the application of rubber. The cross-linking of the adhesive causes mechanisms for adhesion to be compromised and inhibits physical interaction.

To confirm this, both a control part, with bond failure due to adhesive overbake, and the failed part were subjected to analysis by XPS. Only failed mating surfaces were examined by this method. Typical areas scanned by this method were 1 × 3 mm. It is not necessary to neutralize with the electron gun due to only minimal charging effects. Only a single sample was tested for each condition.

The XPS analysis of the mating surfaces of the defective part showed the surfaces to be very similar in elemental composition. Values from the failed part, when compared to the control part, indicated the hydrocarbon fluid did not affect the analysis. This is most likely because the high vacuum necessary in this testing technique removed the hydrocarbon fluid from the surface.

The trace elements present in analysis of both surfaces of both parts support the conclusion that the parts failed by the same mechanism. Test results support the idea that both samples failed in a boundary layer between the rubber and the adhesive. Trace elements, known not to be native to the rubber compound, indicate the source to be the adhesive.

From the presence of adhesive components on all mating surfaces tested, we can effectively eliminate contamination as the cause of the unbond. The presence of the adhesive and the pulled-out material present on the failed part would indicate the part may have been weakly bonded. Finally, the evident lack of adhesive and rubber interspersion would indicate one or both components had achieved sufficient viscosity to resist flow. The combined evidence using XPS tends to support our initial observation that all adhesive had prematurely cross-linked. These data allow the cause to be correctly identified and will help produce a more robust part and processing in the future.

X-RAY TECHNOLOGY

As with medical x-ray instruments there are analytical x-ray instruments that can produce images of internal structures of objects that are opaque to visible light. There are instruments that can determine the chemical elemental composition of an object, that can identify the crystalline phases of a mixture of solids, and others that determine the complete atomic and molecular structure of a single crystal. The determination of particle size and structural information for fibers and polymers, and the study of stress, texture, and thin films are x-ray applications that are growing in importance.

CHARACTERISTICS AND GENERATION OF X-RAYS

X-Ray Electromagnetic Spectrum

X-rays are a form of electromagnetic radiation and have a wavelength, λ, much shorter than visible light. The center of the visible light spectrum has a wavelength of about 0.56×10^{-6}m. The most commonly used methods for generating x-rays are the synchrotron and x-ray tubes.

Synchrotron Radiation

X-rays are produced when very energetic electrons traveling close to the speed of light are decelerated. In synchrotrons, electrons are accelerated with electromagnets while traveling along a linear path. Then they are inserted into a nearly circular path, which is maintained by bending magnets.

X-Ray Tubes

X-ray tubes are the most widely used source for the generation of x-rays. In these tubes, electrons are accelerated by a high electric potential (20 to 120 kV). These electrons strike the target (anode) of the tube and decelerate as they pass through the electron clouds of the atoms. This phenomenon produces a continuous spectrum similar, but much less intense, to that of the synchrotron. In addition, some high-energy electrons knock electrons out of the atomic orbitals of the atoms of the target material. When these orbitals are refilled by electrons, x-ray photons are generated. The resulting x-ray spectrum of intensity vs. wavelength has a series of peaks known as characteristic lines.

The materials that are used as targets in x-ray tubes depend on the application.

APPLICATIONS

X-ray applications can be placed into three categories:

1. X-ray radiography permits the imaging of the internal structure of an object (e.g., bones of a hand). It is based on the comparative observance of photons as they travel through the different materials making up the object.
2. X-ray fluorescence spectrometry consists of the measurement of the incoherent scattering of x-rays. It is used primarily to determine the elemental composition of a sample.
3. X-ray diffraction consists of the measurement of the coherent scattering of x-rays. X-ray diffraction is used to determine the identity of crystalline phases in a multiphase powder sample and the atomic and molecular structures of single crystals. It can also be used to determine structural details of polymers, fibers, thin films, and amorphous solids and to study stress, texture, and particle size.

X-Ray Fluorescence Spectrometry

X-ray fluorescence spectrometry is a technique for measuring the elemental composition of samples. The basis of the technique is the relationship between the wavelength or energy of the emitted incoherently scattered x-ray photons and the atomic number of the element. When an atom is bombarded with x-ray photons of sufficient energy, an inner-orbital electron may be displaced, leaving the atom in an excited state. The atom can return to the ground state by transference of a higher orbital electron into the vacancy (the resulting higher level vacancy is filled by an electron from a still higher level and so on). In so doing, the difference in energy between the electron ousted from the lower shell and the energy of the higher orbital electron is emitted as an x-ray photon. Each element produces a fluorescence spectrum of intensity vs. wavelength that is characteristic of that element.

X-Ray Radiography

X-ray imaging tests are widely used to examine interior regions of metal castings, fusion welds, composite structures, and brazed components. Radiographic tests are made on pipeline welds, pressure vessels, nuclear fuel rods, and other critical materials and components that may contain three-dimensional voids, inclusions, gaps, or cracks. Since penetrating radiation tests depend upon the absorption properties of materials on x-ray photons, the tests can reveal changes in thickness and density and the presence of inclusions in the material.

X-ray fluoroscopy is used for direct online examination. A fluorescent screen is used to convert x-ray photons into visible light photons. A television camera receives the visible image and displays it on a television screen. This type of system is used for security screening of carry-on luggage at airports.

As in medical x-ray imaging, computerized tomography (CT) can reveal the details of the internal structure of complex objects. Many detectors are used to measure the transmittance of x-rays along many lines through the object. A computer uses this information to produce an image of the internal structure of a slice of the object.

XYLENE

Xylenes and ethylbenzene (EB) are C_8 aromatic isomers with the molecular formula C_8H_{10}. The xylenes consist of three isomers: *o*-xylene (OX), *m*-xylene (MX), and *p*-xylene (PX). These differ in the positions of the two methyl groups on the benzene ring structures.

USES

The majority of xylenes, which are mostly produced by catalytic reforming or petroleum functions, are used in motor gasoline. The majority of the xylenes that are recovered for petrochemicals use are used to produce PX and OX. PX is the most important commercial isomer.

XYLYLENE POLYMERS

In a process capable of producing pinhole-free coatings of outstanding conformality and thickness uniformity through the unique chemistry of *p*-xylylene (PX), a substrate is exposed to a controlled atmosphere of pure gaseous monomer. The coating process is best described as a vapor deposition polymerization (VDP). The monomer molecule is thermally stable, but kinetically very reactive toward polymerization with other molecules of its kind. Although it is stable as a rarified gas, upon condensation it polymerizes spontaneously to produce a coating of high molecular weight, linear poly(*p*-xylylene) (PPX).

APPLICATIONS

Because the parylenes are generally insoluble in most solvents, even at elevated temperatures, they cannot be used as solvent-based coatings; neither can they be cast as films nor spun as fibers from solution.

The most important application of parylenes is as a conformal coating for printed wiring assemblies. These coatings provide excellent chemical resistance, and resistance to fungal attack. In addition, they exhibit stable dielectric properties over a wide range of temperatures.

The use of parylenes as a hybrid circuit coating is based on much the same rationale as its use in circuit boards. A significant distinction lies in obtaining adhesion to the ceramic substrate material, the success of which determines the eventual performance of the coated part. Adhesion to the substrate must be achieved using adhesion promoters, such as the organosilanes.

Parylenes are superior candidates for dielectrics in high-quality capacitors. Their dielectric constant and loss remain constant over a wide temperature range. The thermistor sensing probe of a disposable bathythermograph is coated with parylene. This instrument is used to chart the ocean water temperature as a function of depth.

Parylene is used in the manufacture of high-quality miniature stepping motors, such as those used in wristwatches, and as a coating for the ferrite cores of pulse transformers, magnetic tape-recording heads, and miniature inductors.

Use of parylene in the medical field is linked to electronics, for example, as a protective conformal coating on pacemaker circuitry.

As books age, the paper of their pages becomes brittle. A relatively thin coating of parylene can make these embrittled pages stronger.

By separating the coating from the substrate after deposition, the unique coating features of parylenes, especially continuity and thickness control and uniformity, can be imparted to a freestanding film. Applications include optical beam splitters, a window for a micrometeoroid detector, a detector cathode for an x-ray streak camera, and windows for x-ray proportional counters.

Parylenes can be used for contamination control, that is, securing small particles to prevent them from damaging a surface in a sealed unit; barrier coating; coating for corrosion control; and as dry lubricants.

Y

YARNS

Yarns are assemblages or bundles of fibers twisted or laid together to form continuous strands. They are produced with either filaments or staple fibers. Single strands of yarns can be twisted together to form ply or plied yarns, and ply yarns in turn can be twisted together to form cabled yarn or cord. Important yarn characteristics related to behavior are fineness (diameter or linear density) and number of twists per unit length. The measure of fineness is commonly referred to as yarn number. Yarn numbering systems are somewhat complex, and they are different for different types of fibers. Essentially, they provide a measure of fineness in terms of weight per unit or length per unit weight.

Cotton yarns are designated by numbers, or counts. The standard count of cotton is 840 yd/lb. Number 10 yarn is therefore 8400 yd/lb. A No. 80 sewing cotton is 80 × 840, or 67,200 yd/lb.

Linen yarns are designated by the lea of 300 yd. A 10-count linen yarn is 10 × 300, or 3,000 yd/lb.

The size or count of spun rayon yarns is on the same basis as cotton yarn. The size or count of rayon filament yarn is on the basis of the denier, the rayon denier being 450 m weighing 5 cg. If 450 m of yarn weighs 5 cg, it has a count of 1 denier. If it weighs 10 cg, it is No. 2 denier. Rayon yarns run from 15 denier, the finest, to 1200 denier, the coarsest.

Reeled silk yarn counts are designated in deniers. The international denier for reeled silk is 500 m of yarn weighing 0.05 g. If 500 m weighs 1 g, the denier is No. 20. Spun silk count under the English system is the same as the cotton count. Under the French system the count is designated by the number of skeins weighing 1 kg. The skein of silk is 1000 m.

A ply yarn is one that has two or more yarns twisted together. A two-ply yarn has two separate yarns twisted together. The separate yarns may be of different materials, such as cotton and rayon. A six-ply yarn has six separate yarns. A ply yarn may have the different plies of different twists to give different effects. Ply yarns are stronger than single yarns of the same diameter. Tightly twisted yarns make strong, hard fabrics. Linen yarns are not twisted as tightly as cotton because the flux fiber is longer, stronger, and not as fuzzy as the cotton. Filament rayon yarn is yarn made from long, continuous rayon fibers, and it requires only slight twist. Fabrics made from filament yarn are called twalle. Monofilament is fiber heavy enough to be used alone as yarn, usually more than 15 denier. Tow consists of multifilament reject strands suitable for cutting into staple lengths for spinning. Spun rayon yarn is yarn made from staple fiber, which is rayon filament cut into standard short lengths.

YTTRIUM AND ALLOYS

A chemical element, yttrium (symbol Y, atomic number 39, atomic weight 88.905) resembles the rare earth elements closely. The stable isotope yttrium-89 constitutes 100% of the natural element, which is always found associated with the rare earths and is frequently classified as one.

Yttrium forms a white oxide, Y_2O_3, which dissolves in acid to form trivalent yttrium salts. Yttrium has become commercially important since 1964. Yttrium forms the matrix for the europlumactivated yttrium phosphors. These phosphors, when excited by electrons, emit a brilliant, clear-red light. The television industry uses these phosphors in manufacturing television screens. It is claimed that this phosphor gives better color reproduction and a much brighter screen than did the older non-rare-earth red phosphor. Also, the yttrium iron garnets, $Y_3Fe_5O_{12}$, and other garnets have found important uses in radar and communication devices. They transmit shortwave energy with very small losses.

Yttrium metal absorbs hydrogen, and in alloys up to a composition of YH_2 they resemble metals very closely. In fact, in certain composition ranges, the alloy is a better conductor of electricity than the pure metal. The density of hydrogen near the YH_2 composition is greater per cubic centimeter than it is in water or liquid hydrogen; therefore, such alloys make excellent potential moderators for nuclear reactors. Also, these alloys can be heated to a white heat (about 1260°C) before the vapor pressure of hydrogen exceeds 1 atm (10^2 kPa), and therefore the moderator in the reactor can be operated at very high temperatures. Yttrium metal has a low nuclear cross section so it is also a potential structural material for reactors of the future.

Yttrium is used commercially in the metal industry for alloy purposes and as a "getter" to remove oxygen and nonmetallic impurities in other metals. Radioactive yttrium isotopes have been used in attempts at treating cancer.

Yttrium Aluminum Garnet

Yttrium aluminum garnet (YAG) has the formula, $Y_3Al_5O_{12}$; its crystals are capable of sustaining laser activity when doped with neodymium.

Yttrium Oxide

This oxide (Y_2O_3) has a melting point of 2685°C; a density of 5.03 g/cm^3; is soluble in acids, but only slightly soluble in water.

Yttrium is not a rare earth but always occurs with them in minerals because of similar general chemistry. Applications are in electrically conducting ceramics, refractories, insulators, phosphors, glass, special optical glasses, and other ceramics.

White powder has cubic crystal structure and small amounts of dysprosium oxide, gadolinium oxide, and terbium oxide as impurities.

Yttria can be compounded into polycrystalline as well as single-crystal garnets for use in microwave generation and detection devices. Such materials are important to microwave technology because they exhibit both good dielectric and magnetic properties, which can be controlled through compositional variations.

Yttria-stabilized zirconia can be used to produce a high-quality diamond substitute for jewelry or a rugged sensor for measuring oxygen in automotive exhaust, depending on the method of fabrication. Nd:YAG single crystal rods find many applications as lasers in industry and in research.

Y_2O_3 can be used (with scandium, lanthanum, and cesium oxides) with TiO_2 bodies for better control of properties than experienced with alkaline earths. In combination with europium oxide, yttria is used to make the red phosphor in color television picture tubes. Combined with ZrO_2, it makes good high-temperature refractories. It also is used in silicon nitride as a sintering aid.

Y-TZP

Yttria tetragonal zirconia polycrystal (Y-TZP) is a fine-grained ceramic used in special engineering applications that benefit from its high density, excellent wear resistance, and fine grain size, such as fiber-optic ferrules. High-purity fine reactive coprecipitated zirconia powders containing 3 mol% yttria are used to produce Y-TZP ceramics.

Z

ZINC AND ALLOYS

A bluish-white crystalline metal, zinc (symbol Zn) has a specific gravity of 7.13, melts at 420°C and boils at 906°C. The commercially pure metal has a tensile strength, cast, of about 62 MPa with elongation of 1%, and the rolled metal has a strength of 165 MPa with elongation of 35%. But small amounts of alloying elements harden and strengthen the metal, and it is seldom used alone.

Zinc is seldom used alone except as a coating. In addition to its metal and alloy forms, zinc also extends the life of other materials such as steel (by hot dipping or electrogalvanizing), rubber and plastics (as an aging inhibitor), and wood (in paints). Zinc is also used to make brass, bronze, and nickel silver; die-casting alloys in plate, strip, and coil; foundry alloys; superplastic zinc; and activators and stabilizers for plastics. Additionally, zinc is used for electric batteries; for die castings; and in alloyed sheets for flashings, gutters, and stamped and formed parts. The metal is harder than tin, and an electrodeposited plate has a Vickers hardness of about 45. Zinc is also used for many chemicals.

Production

The metallurgy of zinc is dominated by the fact that its oxide is not reduced by carbon below the boiling point of the metal.

A large fraction of the world's zinc is still produced from relatively small horizontal retorts with one furnace (or bank) containing hundreds of such units.

Other large, continuously operated vertical retorts have operated, with top charging of briquets of zinc oxide and bituminous coal, and metal tapping from an outside condenser.

Another continuous method involves electrothermic reduction, using a novel condenser in which the retort vapors are sucked through molten zinc.

Neither horizontal or vertical retorts, electrothermic units, nor blast furnaces normally produce zinc of the extreme high purity required by much of the total market for zinc. Since 1935, a redistillation process has been used as the thermal means of meeting this demand. The principles of fractional distillation are utilized and zinc of 99.99+% purity is made.

There is also an electrolytic method of producing metallic zinc. Because the selective flotation process made additional quantities of zinc concentrates available in localities where electric power is cheap, the production of electrolytic zinc was increased.

In the electrolytic process, the zinc content of the roasted ore is leached out with dilute sulfuric acid. The zinc-bearing solution is filtered and purified and the zinc content recovered from the solution by electrolysis, using lead alloy anodes and sheet aluminum cathodes. Current passing through the electrolytic cell, from anode to cathode, deposits the metallic zinc on the cathodes from which it is stripped at regular intervals, melted and cast into slabs. Zinc so produced is 99.9+ or 99.99+% pure depending on need and the process control exercised.

Compounds and Zinc Forms

Zinc is always divalent in its compounds, except for some of those with other metals, which are classed as zinc alloys. Most of the more important zinc compounds are inorganic, since they are much more widely used than the organic zinc compounds.

The old name *spelter*, often applied to slab zinc, came from the name spailter used by Dutch traders for the zinc brought from China.

Sterling spelter was 99.5% pure. Special high-grade zinc is distilled, with a purity of 99.99%, containing no more than 0.006% lead and 0.004 cadmium. High-grade zinc, used in alloys for die casting, is 99.9% pure, with 0.07 max lead. Brass special zinc is 99.10% pure, with 0.6 max lead and 0.5 max cadmium. Prime western zinc, used for galvanizing, contains 1.60 max lead and 0.08 max iron. Zinc crystals produced for electronic uses are 99.999% pure, metal.

On exposure to the air, zinc becomes coated with a film of carbonate and is then very corrosion-resistant. Zinc foil comes in thicknesses from 0.003 to 0.015 cm. It is produced by electrodeposition on an aluminum drum cathode and stripping off on a collecting reel. But most of the zinc sheet contains a small amount of alloying elements to increase the physical properties. Slight amounts of copper and titanium reduce grain size in sheet zinc. In cast zinc the hexagonal columnar grain extends from the mold face to the surface or to other grains growing from another mold face, and even very slight additions of iron can control this grain growth. Aluminum is also much used in alloying zinc. In zinc used for galvanizing, a small addition of aluminum prevents formation of brittle alloy layer, increases ductility of the coating, and gives a smoother surface. Small additions of tin give bright spangled coatings.

Zinc has 12 isotopes, but the natural material consists of 5 stable isotopes, of which nearly half is zinc-64. The stable isotope zinc–67, occurring to the extent of about 4% in natural zinc, is sensitive to tiny variations in transmitted energy, giving off electromagnetic radiations that permit high accuracy in measuring instruments. It measures gamma-ray vibrations with great sensitivity, and is used in the nuclear clock.

Zinc powder, or zinc dust, is a fine gray powder of 97% minimum purity usually in 325-mesh particle size. It is used in pyrotechnics, in paints, as a reducing agent and catalyst, in rubbers as a secondary dispersing agent and to increase flexing, and to produce Sherardized steel.

In paints, zinc powder is easily wetted by oils. It keeps the zinc oxide in suspension, and also hardens the film. Mossy zinc, used to obtain color effects on face brick, is a spangly zinc powder made by pouring the molten metal into water. Feathered zinc is a fine grade of mossy zinc. Photoengraving zinc for printing plates is made from pure zinc with only a small amount of iron to reduce grain size and is alloyed with not more than 0.2% each of cadmium, manganese, and magnesium. Cathodic zinc, used in the form of small bars or plates fastened to the hulls of ships or to underground pipelines to reduce electrolytic corrosion, is zinc of 99.99% purity with iron less than 0.0014% to prevent polarization.

Applications

For many years, the greatest use of zinc has been to protect iron and steel against atmospheric corrosion. Because of the relatively high electropotential of zinc, it is anodic to iron. If zinc and iron or steel are electrically connected and are jointly exposed in most corrosive media, the steel will be protected while the zinc will be attacked preferentially and sacrificially. This, along with the fact that zinc corrodes far less rapidly than iron in most environments, forms the basis for one of the great fields of use of zinc — in galvanizing (by hot dipping or electrolytically), metallizing, sherardizing, in zinc pigmented paint systems, and as anodes in systems for cathodic protection. The six techniques are described below.

Hot Dip Galvanizing

Zinc alloys readily with iron. Therefore, steel articles, suitably cleaned, will be wet by molten zinc and will acquire uniform coatings of zinc the thickness of which will vary with time, temperature, and rate of withdrawal. Such coats are continuous and reasonably ductile. Ductility is improved considerably by the restriction of immersion time and by the addition of small amounts of aluminum to the galvanizing bath.

Millions of tons of steel products are protected by zinc annually. The time before first rusting of the iron or steel base is proportional to the thickness of zinc coat which in turn is subject to control — depending on product and processing — within a range from thin wiped coats on some products to as much as 0.20 mm on certain low-alloy steels allowed to acquire a full natural coat.

The usefulness of zinc as a coating material comes from its dual ability to protect, first as a long-lasting sheath, and then sacrificially when the sheath finally is perforated.

Electrogalvanizing

Zinc may be electrolytically deposited on essentially all iron and steel products. Wire and strip are commonly so treated as are many fabricated parts. Electrodeposited coats are ductile and uniform but normally are thinner and therefore find application in less rigorous service.

Metallizing

Zinc wire or powder is melted and sprayed on suitably grit-blasted steel surfaces — a growing use. Its virtues are flexibility in application and substantial thicknesses that may be applied. The method is particularly useful for renewal of heavy coatings on areas exposed to particularly critical corrosive conditions and the coating of parts too large for hot dipping. Although metallized coats may be somewhat porous, the sacrificial nature of zinc nevertheless makes them protective. Suitable pore sealants may be used as a part of a metallizing system.

Sherardizing

Zinc powder is packed loosely around clean parts to be sherardized in an airtight container. When sealed, heated to temperature near but below the zinc melting point, then slowly rotated, the zinc alloys with the steel forming a thin, abrasion-resistant, and uniform protective coating (0.4 to 1.8 g/cm^2).

Sherardizing is used commonly to coat small items such as nuts, bolts, and screws; an exception is tubular electrical conduit. Sherardized coats receive varnish, paints, and lacquers particularly well.

Zinc-Pigmented Paints

Evidence has accumulated to demonstrate that paints heavily pigmented with zinc dust perform similarly to zinc coats otherwise applied. Electrical contact must exist between the steel and the zinc-dust particles; consequently, special vehicles must be used and the steel surface must be clean.

Zinc Anodes

High-purity zinc, normally alloyed with small additions of aluminum, with or without cadmium, is cast or rolled into anodes that, when electrically bonded to bare or painted steel, will protect large areas from the corrosive attack of such environments as seawater. The advantages of zinc in this application include self-regulation (no more current is generated than is required), a minimum generation of hydrogen, and long life. This is a growing application for the protection of ship hulls, cargo tanks in ballast, piers, pilings, etc.

General Comment

Reference has been made to the importance of coating thickness — the heavier the coat, the longer the time before first rusting. All evidence at hand indicates that the amount of zinc in a coat is the controlling factor and the method of application is of secondary importance. Uniformity of coat and adhesion must be good. No data are known to demonstrate that common zinc impurities normally present in amounts to or slightly above specification limits have any significantly deleterious or beneficial influence on the ability of zinc to protect iron or steel against atmospheric corrosion. Although any grade of zinc may be used for galvanizing, Prime Western is the one most commonly employed.

Die casting is a market for zinc that may soon become its largest market. These alloys melt readily, are highly fluid, and do not attack steel dies or equipment. When used under good temperature control and with good die design practices, casting surfaces are excellent and easily finished. Physical properties are good and dimensional stability excellent.

Alloy control within the specified limits ensures long life. Low aluminum results in decreased casting and mechanical properties and adversely affects the performance of plated coatings. High aluminum can lead to brittleness (an alloy eutectic forms at 5% aluminum). High copper content decreases dimensional stability. Iron as commonly encountered is not critical. Lead, tin, and cadmium, if present above specification limits, can lead to intercrystalline corrosion with objectionable growth and serious cracking or brittleness as a result. Magnesium minimizes the deleterious influence of lead, tin, or cadmium but at or near specification maximum decreases ductility and castability and can lead to objectionable hot shortness. Other impurities such as chromium and nickel, which may be encountered, are not critical.

Zinc die castings are used by the automotive, truck, and bus industry for functional, decorative–functional, or decorative purposes. A majority is plated with copper–nickel–chromium in a variety of plating systems especially adapted to withstand severe service conditions.

Other major outlets for zinc die casting include household appliances, business machines, machine tools, air-brake systems, and communication equipment.

Even such nonstructural materials as cardboard can be zinc-coated by low-temperature flame spraying. Other important uses of zinc are in brass and zinc die-casting alloys, in zinc sheet and strip, in electrical dry cells, in making certain zinc compounds, and as a reducing agent in chemical preparations.

A so-called tumble-plating process coats small metal parts by applying zinc powder to them with an adhesive, then tumbling them with glass beads to roll out the powder into a continuous coat of zinc. Rechargeable nickel-zinc batteries offer higher energy densities than conventional dry cells. Foamed zinc metal has been suggested for use in lightweight structures such as aircraft and spacecraft. Some other uses of zinc are in dry cells, roofing, lithographic plates, fuses, organ pipes, and wire coatings.

Zinc is believed to be needed for normal growth and development of all living species, including humans; actually, life without zinc would be impossible. Zinc is a common element that is present in virtually every type of human food, and zinc deficiency is therefore not considered to be a common problem in humans. Zinc is a trace element that is present in biological fluids at a concentration below 1 ppm (parts per million), and only a small amount (normally <25 mg) is required in the daily diet. (The recommended daily allowance for zinc is 15 mg/day for adults and 10 mg/day for growing children.) It is relatively nontoxic, without noticeable side effects at intake levels of up to ten times the normal daily requirement.

The adult human body contains approximately 2 g of zinc distributed in all cells but especially in bones and muscle. It occurs almost exclusively in association with other molecules, typically proteins, where it exists as the divalent on Zn^{2+}. Many of the zinc-containing proteins are enzymes, known as metalloenzymes, in which the zinc is bound to three amino acid side chains and a molecule of water. Zinc metalloenzymes are involved in most of the key steps of the replication, transcription, and translation of genetic material; hence, they are critical for growth and development. They also help catalyze all major pathways of metabolism, as well as many specialized reactions.

Many biological processes seem to have a relationship to zinc metabolism although they are undefined. The oxide has been used since the first century for treatment of skin rashes. Zinc has also been found to be helpful in other skin disorders such as burns, acne, and surgical wounds. It seems to modulate the immune system and consequently has been purported to have antiviral, antibacterial, and anticancer properties.

Zinc deficiency sometimes occurs due to an inadequate diet or one that includes a high content of phytic acid, fiber, phosphate, calcium, or copper, all of which diminish absorption.

Zinc telluride is a semiconductive material that has been found to become photorefractive when it is suitably doped with vanadium or with manganese and vanadium. The combination of photorefractivity and semiconductivity make this material attractive for use in a variety of applications, including optical power limiting (for shielding eyes or delicate sensors against intense illumination), holographic interferometry, providing reconfigurable optical interconnections for optical computing and optical communication, and correcting for optical distortions and combining laser powers via phase conjugation. In comparison with other important photorefractive materials based on III–V and II–VI binary compounds, ZnTe:V offers superior photorefractive performance at wavelengths from 0.6 to 1.3 µm.

Undoped or doped ZnTe can be grown by physical vapor transport in a closed ampoule.

Experiments were performed to investigate the utility of ZnTe:V:Mn for real-time resonant holographic interferometry. These experiments involved, variously, two- or four-wave mixing, using pulsed dye or continuous-wave helium-neon or diode lasers. Holographic image transfer and two-wavelength resonant holographic interferometry were demonstrated; in particular, a ZnTe:V:Mn crystal was used in a demonstration of resonant holographic interferometric spectroscopy, which is a technique for obtaining chemical-species-specific interferograms by recording two holograms simultaneously at two slightly different wavelengths near an absorption spectral peak of the species in question.

A vehicle (AT-1) was fitted with zinc air batteries and ran for 1043 miles at 20 to 25 mph. The batteries consist of 180 zinc-air cells weighing 1.9 kg and outputting 1.1 V each. On the record run they delivered 76 kW-h of power with 10% remaining, compared to the total capacity of the Saturn EV-1 of 16.2 kW-h from a lead-acid battery pack that weighs 50% more than that in the AT-1. A unique aspect of the batteries is that they are intended to be physically swapped from the vehicle in minutes instead of recharged onboard for hours.

Intended for emerging-market countries, the AT-1 is a five-passenger utility vehicle that can be configured in a variety of ways. Weighing 635.6 kg, it can carry a 590.2-kg load. Depending on motor, it has a top speed of 40 to 70 mph.

Although zinc-air batteries offer about four times the energy density of lead acid, their weakness is specific power. It should be noted that this could be addressed by combining zinc-air batteries with other batteries or an ultracapacitor. As vehicles based on crude oil lose their selling position, zinc-air fuel cells are expected to capture a substantial part of the market.

Alloys

Alloys of zinc are mostly used for die castings for decorative parts and for functional parts where the load-bearing and shock requirements are relatively low. Because the zinc alloys can be cast easily in high-speed machines, producing parts that weigh less than brass and have high accuracy and smooth surfaces that require minimum machining and finishing, they are widely used for such parts as handles, and for gears, levers, pawls, and other small parts. Zinc alloys for sheet contain only small amounts of alloying elements, with 92 to 98% zinc, and the sheet is generally referred to simply as zinc or by a trade name. The modified zinc sheet is used for stamped, drawn, or spun parts for costume jewelry and electronics, and contains up to 1.5% copper and 0.5% titanium. The titanium raises the recrystallization temperature, permitting heat treatment without coarse grain formation.

Hartzink had 5% iron and 2 to 3% lead, but iron forms various chemical compounds with zinc and the alloy is hard and brittle. Copper reduces the brittleness. Germania-bearing bronze contained 1% iron, 10% tin, about 5% each of copper and lead, and the balance zinc. The Fenton alloy had 14% tin, 6% copper, and 80% zinc, and the Ehrhard bearing metal contained 2.5% aluminum, 10% copper, 1% lead, and a small amount of tin to form copper–tin crystals. Binding metal, for wire-rope slings, has about 2.8% tin, 3.7% antimony, and the balance zinc. Pattern metal, for casting gates of small patterns, was almost any brass with more zinc and some lead added, but is now standard die-casting metal.

Zinc also is commonly used in varying degrees as an alloying component with other base metals, such as copper, aluminum, and magnesium. A familiar example of the latter is the association of varying amounts of zinc (up to 45%) with copper to produce brass.

Zinc alloys are commonly used for die castings, and the zinc used is high-purity zinc known as special high-grade zinc. The American Society for Testing and Materials (ASTM) AG40A (Society of Automotive Engineers, SAE 903) is the most widely used; others include AC41A (SAE 925), Alloy 7, and ILZRO 16. All typically contain about 4% aluminum, small amounts of copper, and very small amounts of magnesium. AG40A has a density of 6.6 Mg/m^3, an electrical conductivity 27% that of copper, a thermal conductivity of 113 W/m · K, an ultimate tensile strength of 283 MPa, and a hardness of Brinell 82. AC41A is stronger (331 MPa) and harder (Brinell 91), a trifle less electrically and heat conductive, and similar in density. The alloys have much greater unnotched Charpy impact strength than either die-cast aluminum or magnesium alloys, but are not especially heat resistant, losing about one third of their strength at temperatures above about 93°C. Both alloys have found wide use for auto and appliance parts, especially chromium-plated parts, as well as for office equipment parts, hardware, locks, toys, and novelties. Alloy 7 is noted primarily for its better castability and the smoother surface finish it provides. It is as strong as AG40A, although slightly less hard, and more ductile. ILZRO 16 is not nearly as strong (228 MPa), but more creep-resistant at room and elevated temperatures.

Casting Alloys

Zinc-casting alloys can be grouped into two general categories: standard zinc die-casting alloys, and the newer ZA (zinc–aluminum) casting alloys.

Standard Die-Casting Alloys

For pressure die casting, the established zinc alloys are the No. 3, 5, and 7 Zamak alloys. As die castings, they have good general-purpose tensile properties and can be cast in thin sections and with good dimensional accuracy. The alloys are often selected for plated or highly decorative applications because of their excellent finishing characteristics. Three major end-use areas for zinc die-cast components are automotive, building hardware, and electrical.

Zamak alloys contain approximately 4% aluminum with low percentages of magnesium, copper, and sometimes nickel. Impurities such as tin, lead, and cadmium are carefully controlled. These alloys are not recommended for gravity casting. They are cast by the hot-chamber die-casting process, which is different from, and more efficient than, the cold-chamber die-casting process commonly used for aluminum.

In addition, a specialized process is used for efficient production of miniature die-cast components, using these alloys as well as ZA-8.

Typical tolerances of zinc die-cast parts are ±0.0015 in./in. for the first inch with an additional ±0.002 in./in. for larger parts. New zinc-casting technology allows for thin walls down to 0.025 in. (0.6 mm), improved internal soundness, and surface finishes that range typically from 32 to 64 rms (root mean square).

Part dimensions change slightly when zinc die castings are aged. Zamak alloys No. 3 and 7 can shrink about 0.0007 in./in. after several weeks at room temperature. Alloy No. 5 responds similarly, but total shrinkage can be 0.0009 to 0.0024 in./in., followed by expansion of 0.0020 in./in. over a period of years. When it is necessary to bring these changes to completion within a short time after casting, a stabilizing treatment of 3 to 6 h at 100°C is recommended.

ZA Casting Alloys

Designated as ZA-8, ZA-12, and ZA-27 (the numerical suffix represents the approximate percent by weight of aluminum), the high-aluminum alloys differ radically from the standard Zamak alloys in composition, properties, and castability. Although the ZA alloys were first introduced for gravity casting (sand and permanent mold), they have expanded into pressure die castings as well as the new, precision graphite-mold process. *Important:* Alloys ZA-12 and ZA-27 must be cold-chamber die-cast; alloy ZA-8 is hot-chamber castable.

Gravity casting into low-cost graphite permanent molds provides high-quality ZA castings with excellent precision, eliminating much machining. It is particularly competitive for production quantities of 500 to 15,000 parts/year, where die casting or plastic injection molding would be prohibitive because of tooling costs.

ZA alloys combine high strength and hardness (up to 480 MPa and 120 Bhn), good machinability with good bearing properties, and wear resistance often superior to standard bronze alloys. ZA castings are now competing with cast iron, bronze, and aluminum because of various property and processing advantages.

Of the three alloys, ZA-12 is preferred for most applications, and particularly for gravity casting. However, ZA-27 offers the highest mechanical properties regardless of casting method. Both are excellent bearing materials. ZA-8, on the other hand, gives the best plating characteristics. Because of its hot-chamber die castability and high mechanical properties, ZA-8 is also used for high-performance applications where standard zinc alloys may be marginal. All ZA alloys offer superior creep resistance and performance at elevated temperatures compared with the Zamak alloys.

The most recent casting alloys discussed above are three high-aluminum zinc casting alloys for sand and permanent-mold casting: ZA-8, ZA-12, and ZA-27; the numerals in the designations indicate approximate aluminum content. They also contain more copper than AG40A and AC41A, from 0.5 to 1.2% in ZA-12 to 2 to 2.5% in ZA-27, and a bit less magnesium. As sand-cast, ultimate tensile strengths range from 248 to 276 MPa for ZA-8 and 400 to 441 MPa for ZA-27. Unlike the common die-casting alloys, the ZA alloys also exhibit clearly defined tensile yield strengths: from 193 MPa minimum for sand-cast ZA-8 to 365 MPa for sand-cast ZA-27. Tensile modulus is roughly 83×10^3 MPa. Also, because of their greater aluminum content, they are lighter in weight than the die-casting alloys.

Wrought Alloys

Wrought zinc alloys are available in rolled sheet, strip, foil, and as drawn rod or wire. With controlled rolling, zinc alloys can be tailored to meet a wide range of hardness, luster, and ductility requirements. Rolled zinc can be worked by common fabricating methods, and then polished, lacquered, painted, or plated.

When zinc alloys are formed in progressive presses, as in battery-shell manufacture, they are self-annealing. After successive forming operations in nonprogressive presses, however, the alloys work-harden and break. This can be overcome, in copper-free alloys, by intermediate annealing for 5 min in boiling water to which 20% glycerine has been added. Copper-bearing alloys should be heated 5 to 10 min at 177°C. The copper/titanium-containing alloy should be held at 199°C for about 15 min to bring about crystallization. Excessive exposure to higher temperatures should be avoided, however, to prevent grain growth, cleavage cracks, and property deterioration.

Highly workable and highly forgeable wrought-zinc alloys containing titanium, aluminum, lead, cadmium, copper, or iron in various quantities are easily machined. Forged or extruded parts are free from porosity and have good detail.

Other Alloys

Manganese–zinc alloys, with up to 25% manganese, for high-strength extrusions and forgings, are really 60–40 brass with part of the copper replaced by an equal amount of manganese, and are classed with manganese bronze. They have a bright white color and are corrosion resistant. Zam metal, for zinc-plating anodes, is zinc with small percentages of aluminum and mercury to stabilize against acid attack. Zinc solders are used for joining aluminum. The tin–zinc solders have 70 to 80% tin, about 1.5% aluminum, and the balance zinc. The working range is 260 to 310°C. Zinc–cadmium solder has about 60% zinc and 40% cadmium. The pasty range is between 266 and 315°C.

A group of wrought alloys, called superplastic zinc alloys, have elongations of up to 2500% in the annealed condition. These alloys contain about 22% aluminum. One grade can be annealed and air-cooled to a strength of 489 MPa. Parts of these alloys have been produced by vacuum forming and by a compression molding technique similar to forging but requiring lower pressures.

Chemicals

With the exception of the oxide, the quantities of zinc compounds consumed are not large compared with many other metals, but zinc chemicals have a very wide range of use; they are essential in almost all industries and for the maintenance of animal and vegetable life. Zinc is a complex element and can provide some unusual conditions in alloys and chemicals.

Zinc Oxide

Zinc oxide, ZnO, is a white, water-insoluble, refractory powder melting at about 1975°C, with a specific gravity of 5.66. It is much used as a pigment and accelerator in paints and rubbers. Its high refractive index, about 2.01, absorption of ultraviolet light, and fine particle size give high hiding power in paints, and make it also useful in such products as cosmetic creams to protect against sunburn. Commercial zinc oxide is always white, and in the paint industry is also called zinc white and Chinese white. But with a small excess of zinc atoms in the crystals, obtained by heat treatment, the color is brown to red.

In paints, zinc oxide is not as whitening as lithopone, but it resists the action of ultraviolet rays and is not affected by sulfur atmospheres, and is thus valued in outside paints. Leaded zinc oxide, consisting of zinc oxide and basic lead sulfate, is used in paints, but for use in rubber the oxide must be free of lead. The lead-free variety is also called French process zinc oxide. In insulating compounds zinc oxide improves electrical resistance. In paper coatings it gives opacity and improves the finish. Zinc-white paste for paint mixing usually has 90% oxide and 10% oil. Zinc oxide stabilizers, composed of zinc oxides and other chemicals, can be added to plastic molding compounds to reduce the deteriorating effects of sunlight and other types of degrading atmospheres.

The rubber industry is the largest consumer of zinc oxide, accounting for more than 50% of the market. Zinc oxide is most effective as an activator of accelerators in the vulcanization process.

The chemical industry has been opening new markets for zinc oxide. Examples are lubricating oil additives, water treatment, and catalysts. For photocopying, photoconductivity is a unique electronic property of zinc oxide.

In the ceramic industry, zinc oxide is used in the manufacture of glasses, glazes, frits, porcelain enamels, and magnetic ferrites. Here, the largest consuming plants are in the tile industry.

One ceramic grade of zinc oxide has these properties: specific gravity 5.6; apparent density 1201 kg/m^3; weight 5595.5 kg/m^3. Typical chemical analysis: 99.5% ZnO, 0.05% Pb, 0.02% Fe, 0.01% Cd, 0.02% S (total), 0.10% HCl (insoluble), 5 ppm magnetic iron.

In glass, zinc oxide reduces the coefficient of thermal expansion, thus making possible the production of glass products of high resistance to thermal shock. It imparts high brilliance of luster and high stability against deformation under stress. As a replacement flux for the more soluble alkali constituents, it provides a viscosity curve of lower slope. Specific heat is decreased and conductivity increased by the substitution of zinc oxide for BaO and PbO.

A 1% addition of zinc oxide to tank window glass lowers the devitrification temperature and improves chemical resistivity while maintaining good workability for drawing. It is used consistently in high-grade fluoride opal glass in which it greatly increases opacity, whiteness, and luster by inducing precipitation of fluoride crystals of optimum number and size. Apparently, zinc oxide makes its contribution to opacity through reduction of the primary opacifiers. It is used in optical glasses of high barium content to reduce their tendency to crystallize on cooling. The resistance of phosphate glasses to chemical attack is improved by the presence of zinc oxide. About 10% zinc oxide assists in the development of the characteristic color of cadmium sulfoselenide ruby glass, although its exact function is obscure.

Zinc oxide is used in many types of glazes, its function varying according to the particular composition in which it is included. In general, it provides fluxing power, reduction of expansion, prevention of crazing, greater gloss and whiteness, a favorable effect on elasticity, increased maturing range, increased brilliance of colors, and correction of eggshell finish. It is useful in preventing volatilization of lead by partial substitution for CaO, as high CaO tends to satisfy SiO_2, leaving PbO in a more volatile form.

In Bristol glazes for earthenware products, zinc oxide in combination with alumina produces both opacity and whiteness to a fair degree, provided the lime content is low. The use of zinc oxide in wall tile glazes is very general; the zinc oxide content of certain types is 10% or more. Small amounts are used in gloss or bright tile, whereas higher percentages are used where it is desired to develop a highly pleasing matte finish.

Zinc oxide is commonly used in dry-process cast iron porcelain enamels in amounts of 0.5 to 1% to 14%. In general, low lead content implies high zinc, and vice versa. Its specific functions are to increase fusibility, improve luster, contribute to opacity and whiteness, reduce expansion, and increase extensibility. It is probably a little stronger as a flux than is lime, but does not produce the sudden fluidity characteristic of lime.

Of great benefit to producers of cast iron enamels is the relative nontoxicity of zinc oxide.

A recent use for zinc oxide is its application to the manufacture of magnetic ferrites, which have been developed over the past 25 to 30 years. They usually are composed of ferric oxide in combination with zinc oxide (of high chemical purity) and any one or more of several other oxides of bivalent metals. The amount of zinc oxide used varies from 10 to 35%, depending on the characteristics desired in the finished magnetic ferrite. With their prime properties of high permeability and low hysteresis, they are used in the field of electronics for such devices as high-frequency transformer cores for television receivers.

Zinc oxide crystals are used for transducers and other piezoelectric devices. The crystals are hexagonal and are effective at elevated temperatures, as the crystal has no phase change up to its disassociation point. The resistivity range is 0.5 to 10 $\Omega \cdot$ cm.

Normally recognized as an n-type semiconductor, it has a resistivity less than 103 $\Omega \cdot$ cm. When doped with lithium, resistivity rises to 1012 $\Omega \cdot$ cm and it exhibits piezoelectricity about four times that of quartz.

Zinc oxide has luminescent and light-sensitive properties that are utilized in phosphors and ferrites. But the oxygen-dominated zinc phosphors used for radar and television are modifications of zinc sulfide phosphors.

The zinc sulfide phosphors, which produce luminescence by exposure to light, are made with zinc sulfide mixed with about 2% sodium chloride and 0.005% copper, manganese, or other activator, and fired in a nonoxidizing atmosphere. The cubic crystal structure of zinc sulfide changes to a stable hexagonal structure at 1020°C, but both forms have the phosphor properties. Thin films and crystals of zinc selenide with purities of 99.999% are used for photo or electroluminescent devices. Zinc sulfide, a white powder of the composition $ZnS \cdot H_2O$, is also used as a paint pigment, for whitening rubber, and for paper coating. Cryptone is zinc sulfide for pigment use in various grades, some grades containing barium sulfate, calcium sulfide, or titanium dioxide.

ZIRCON

Zircon, a mineral with the idealized composition $ZrSiO_4$, is one of the chief sources of the element zirconium. Trace amounts of uranium and thorium are often present and the mineral may then be partly or entirely metamict. The name *cyrtollite* is applied to an altered type of zircon. Structurally, zircon is a nesosilicate, with isolated SiO_4 groups.

Zircon is tetragonal in crystallization. It often occurs as well-formed crystals, which commonly are square prisms terminated by a low pyramid. The color is variable, usually brown to reddish brown, but also colorless, pale yellowish, green, or blue. The transparent colorless or tinted varieties are popular gemstones. Hardness is $7^1/_2$ on Mohs scale; specific gravity is 4.7, decreasing in metamict types.

Because of its chemical and physical stability, zircon resists weathering and accumulates in residual deposits and in beach and river sands.

Other properties include specific heat of 0.55 J/g/°C. It is chemically inert and stable to very high temperatures (liquidus >2205°C). Zircon has excellent thermal properties and its thermal conductivity is 14.5 Btu/ft²/hr/°F/in. and coefficient of thermal expansion is 1.4×10^{-6}.

The extremely high thermal conductivity and chilling action of zircon makes it very useful in controlling directional solidification and shrinkage in heavy metal sections.

USES

Zircon sand is used as refractory bedding material for heat-treating metal parts. It is used as a sealing medium for prevention of atmospheric leaks around doors and parts of heat-treating furnaces. Also, it is a high-qualilty, uniform sandblasting medium for metal preparation prior to plating, enameling, or buffing. The heavy, rounded grains give consistent peening without stray digs or gouges to mar the finish. The tough, resilient grains resist breakdown and loss.

ZIRCONIA (ZIRCONIUM OXIDE)

Zirconium oxide, ZrO_2, is a white crystalline powder with a specific gravity of 5.7, hardness 6.5, and refractive index 2.2. When pure, its melting point is about 2760°C, and it is one of the most refractory of the ceramics. It is produced by reacting zircon sand and dolomite at 1371°C and leaching out the silicates. The material is used as fused or sintered ceramics and for crucibles and furnace bricks. From 4.5 to 6% of CaO or other oxide is added to convert the unstable monoclinic crystal to the stable cubic form with a lowered melting point.

Zirconia is produced from the zirconium ores known as zircon and baddeleyite. The latter is a natural zirconium oxide. It is also called zirkite and Brazilite. Zircon is zirconium silicate, $ZrO_2 \cdot SiO_2$, and comes chiefly from beach sands. The sands are also called zirkelite and zirconite, or merely zircon sand. The white zircon sand has a zirconia content of 62%, and contains less than 1% iron.

Uses

Fused zirconia, used as a refractory ceramic, has a melting point of 2549°C and a usable temperature to 2454°C. The Zinnorite fused zirconia is a powder that contains less than 0.8% silica and has a melting point of 2704°C. A sintered zirconia can have a density of 5.4, a tensile strength of 82 MPa, compressive strength of 1378 MPa, and Knoop hardness of 1100. Zircoa B is stabilized cubic zirconia used for making ceramics. Zircoa A is the pure monoclinic zirconia used as a pigment, as a catalyst, in glass, and as an opacifier in ceramic coatings.

As an opacifier, zirconium compounds are used in glazes and porcelain enamels. Zirconium dioxide is an important constituent of ceramic colors and an important component of lead–zirconate–titanate electronic ceramics.

Pure zirconia also is used as an additive to enhance the properties of other oxide refractories. It is particularly advantageous when added to high-fired magnesia bodies and alumina bodies. It promotes sinterability and, with alumina, contributes to abrasive characteristics.

Zirconia brick for lining electric furnaces has no more than 94% zirconia, with up to 5% calcium oxide as a stabilizer, and some silica. It melts at about 2371°C, but softens at about 1982°C. The IBC 4200 brick is zirconia with calcium and hafnium oxides for stabilizing. It withstands temperatures to 2316°C in oxidizing atmospheres and to 1849°C in reducing atmospheres. Zirconia foam is marketed in bricks and shapes for thermal insulation. With a porosity of 75% it has a flexural strength above 3 MPa and a compressive strength above 0.7 MPa. For use in crucibles, zirconia is insoluble in most metals except the alkali metals and titanium. It is resistant to most oxides, but with silica it forms $ZrSiO_4$, and with titania it forms $ZrTiO_4$. Because structural disintegration of zirconia refractories comes from crystal alteration, the phase changes are important considerations. The monoclinic material, with a specific gravity of 5.7, is stable to 1010°C and then inverts to the tetragonal crystal with a specific gravity of 6.1 and volume change of 7%. It reverts when the temperature again drops below 1010°C. The cubic material, with a specific gravity of 5.55, is stable at all temperatures to the melting point, which is not above 2649°C because of the contained stabilizers. A lime-stabilized zirconia refractory with a tensile strength of 138 MPa has a tensile strength of 68 MPa at 1299°C.

Stabilized zirconia has a very low coefficient of expansion, and white-hot parts can be plunged into cold water without breaking. The thermal conductivity is only about one third that of magnesia. It is also resistant to acids and alkalies, and is a good electrical insulator.

To prepare useful formed products from zirconium oxide, stabilizing agents such as lime, yttrium, or magnesia must be added to the zirconia, preferably during fusion, to convert the zirconia to the cubic form. Most commercial stabilized zirconia powders or products contain calcium oxide as the stabilizing agent.

ZIRCONIA (ZIRCONIUM OXIDE)

The stabilized cubic form of zirconia undergoes no inversion during heating and cooling.

Stabilized zirconia refractories are used where extremely high temperatures are required. Above 1649°C, in contact with carbon, zirconia is converted to zirconium carbide.

Zirconia is of much interest as a construction material for nuclear energy applications because of its refractoriness, corrosion resistance, and low nuclear cross section. However, zirconia normally contains about 2% hafnia, which has a high nuclear cross section. The hafnia must be removed before the zirconia can be used in nuclear applications.

FORMS

Zirconia is available in several distinct types. The most widely used form is stabilized in cubic crystal form by a small lime addition. This variety is essential to the fabrication of shapes because the so-called unstabilized, monoclinic zirconia undergoes a crystalline inversion on heating, which is accompanied by a disruptive volume change.

Zirconia is not wetted by many metals and is therefore an excellent crucible material when slag is absent. It has been used very successfully for melting alloy steels and the noble metals. Zirconia refractories are rapidly finding application as setter plates for ferrite and titanate manufacture, and as matrix elements and wind tunnel liners for the aerospace industry.

OTHER TYPES

Toughening mechanisms, by which a crack in a ceramic can be arrested, complement processing techniques that seek to eliminate crack-initiating imperfections. Transformation toughening relies on a change in crystal structure (from tetragonal to monoclinic) that zirconia or zirconium dioxide (ZrO_2) grains undergo when they are subjected to stresses at a crack tip. Because the monoclinic grains have a slightly larger volume, they can "squeeze" a crack shut as they expand in the course of transformation. Because of the transformation toughening abilities of ZrO_2, which impart higher fracture toughness, research interest in engine applications has been high. In order for ZrO_2 to be used in high-temperature, structural applications, it must be stabilized or partially stabilized to prevent a monoclinic–tetragonal phase change. Stabilization involves the addition of calcia, magnesia, or yttria followed by some form of heat treatment. PSZ ceramic, the toughest known ceramic, has been investigated for diesel-engine applications.

PSZ is a transformation toughened material consisting of a cubic zirconia matrix with 20 to 50 vol% free tetragonal zirconia added in the matrix. The material is converted into the stabilized cubic crystal structure using oxide stabilizers (magnesia, calcia, yttria). The conversion is accomplished by sintering the doped zirconia at 1700°C. Magnesia-stabilized zirconia exhibits serrated plastic flow during compression at room temperature. The flow stress is strain rate sensitive. Several different grades are available for commercial use, and the properties of the material can be tailored to fit many applications.

One typical PSZ used for applications requiring maximum thermal shock resistance has a four-point bend strength of 600 MPa; PSZ is used experimentally as heat engine components, such as cylinder liners, piston caps, and valve seats. Vanadium impurities from fuel oil can cause zirconia destabilization, and sodium, magnesium, and sulfur impurities can cause yttria to dissociate from yttria-stabilized zirconia. Another area of interest for PSZ is in bioceramics, where it has use in surgical implants.

A new zirconia ceramic being developed, tetragonal zirconia polycrystal (TZP) doped with Y_2O_3, has the most impressive room-temperature mechanical properties of any zirconia ceramic. The commercial applications of TZP zirconia include scissors with TZP blades suitable for industrial use for cutting tough fiber fabrics, e.g., Kevlar, cables, and ceramic scalpels for surgical applications. One unique application is fish knives. The knife blades are Y-TZP and can be used when the delicate taste of raw fish would be tainted by slicing with knives with metal blades.

Another zirconia ceramic-developed material is zirconia-toughened alumina (ZTA). ZTA zirconia is a composite polycrystalline ceramic containing ZrO_2, as a dispersed phase (typically ~15 vol%). Close control of initial starting powder sizes and sintering schedules is thus necessary to attain the desired ZrO_2 particle dimensions in the finished ceramic. Hence, the mechanical properties of the composite ZTA ceramics limit current commercial applications to cutting tools and ceramic scissors.

PSZ is also finding application in the transformation toughening of metals used in the glass industry as orifices for glass fiber drawing. This material is termed zirconia grain-stabilized (ZGS) platinum.

Clear zircon crystals are valued as gemstones because the high refractive index gives great brilliance.

Zirconia fiber, used for high-temperature textiles, is produced from zirconia with about 5% lime for stabilization. The fiber is polycrystalline, has a melting point of 2593°C, and will withstand continuous temperatures above 1649°C. These fibers are as small as 3 to 10 μm and are made into fabrics for filter and fuel cell use. Zirconia fabrics are woven, knitted, or felted of short-length fibers and are flexible. Ultratemp adhesive, for high-heat applications, is zirconia powder in solution. At 593°C, it adheres strongly to metals and will withstand temperatures to 2427°C. Zircar is zirconia fiber compressed into sheets to a density of 320 kg/m³. It will withstand temperatures up to 2482°C and has low thermal conductivity. It is used for insulation and for high-temperature filtering.

ZIRCONIUM AND ALLOYS

A silvery-white metal, zirconium (symbol Zr), has a specific gravity of 6.5 and a melting point of about 1850°C. It is more abundant than nickel, but is difficult to reduce to metallic form as it combines easily with oxygen, nitrogen, carbon, and silicon. The metal is obtained from zircon sand by reacting with carbon and then converting to the tetrachloride, which is reduced to a sponge metal for the further production of shapes. The ordinary sponge zirconium contains about 2.5% hafnium, which is closely related and difficult to separate. The commercial metal usually contains hafnium, but reactor-grade zirconium, for use in atomic work, is hafnium-free.

Commercially pure zirconium is not a high-strength metal, with a tensile strength of about 220 MPa, elongation 40%, and Brinell hardness 30, or about the same physical properties as pure iron. But it is valued for atomic-construction purposes because of its low neutron-capture cross section, thermal stability, and corrosion resistance. It is employed mostly in the form of alloys but may be had in 99.99% pure single-crystal rods, sheets, foil, and wire for superconductors, surgical implants, and vacuum-tube parts. The neutron cross section of zirconium is 0.18 barn, compared with 2.4 for iron and 4.5 for nickel. The cold-worked metal, with 50% reduction, has a tensile strength of about 545 MPa, with elongation of 18% and hardness of Brinell 95. The unalloyed metal is difficult to roll, and is usually worked at temperatures to 482°C. Although nontoxic, the metal is pyrophoric because of its heat-generating reaction with oxygen, necessitating special precautions in handling powder and fine chips resulting from machining operations.

The metal has a close-packed hexagonal crystal structure, which changes at 862°C to a body-centered cubic structure that is stable to the melting point. At 300 to 400°C the metal absorbs hydrogen rapidly, and above 200°C it picks up oxygen. At about 400°C it picks up nitrogen, and at 800°C the absorption is rapid, increasing the volume and embrittling the metal.

The metal is not attacked by nitric (except red fuming nitric), sulfuric, or hydrochloric acids, but is dissolved by hydrofluoric acid. Zirconium powder is very reactive, and for making sintered metals it is usually marketed as zirconium hydride, ZrH_2, containing about 2% hydrogen, which is driven off when the powder is heated to 300°C. For making sintered parts, alloyed powders are also used. Zirconium copper (containing 35% zirconium), zirconium nickel (with 35 to 50% zirconium), and zirconium cobalt (with 50% zirconium), are marketed as powders of 200 to 300 mesh.

PROPERTIES

In addition to resisting HCl at all concentrations and at temperatures above the boiling temperature, zirconium and its alloys also have excellent resistance in sulfuric acid at temperatures above boiling and concentrations to 70%. Corrosion rate in nitric acid is less than 1 mil/year at temperatures above boiling and concentrations to 90%. The metals also resist most organics such as acetic acid and acetic anhydride as well as citric, lactic, tartaric, oxalic, tannic, and chlorinated organic acids.

Relatively few metals besides zirconium can be used in chemical processes requiring alternate contact with strong acids and alkalies. However, zirconium has no resistance to hydrofluoric acid and is rapidly attacked, even at very low concentrations.

Uses

Small amounts of zirconium are used in many steels. It is a powerful deoxidizer, removes the nitrogen, and combines with the sulfur, reducing hot-shortness and giving ductility. Zirconium steels with small amounts of residual zirconium have a fine grain, and are shock resistant and fatigue resistant. In amounts above 0.15% the zirconium forms zirconium sulfide and improves the cutting quality of the steel.

A noncrystalline metal that reportedly has twice the strength of steel and titanium, has been developed. The material, known as Vitrelloy, is an alloy composed of 61% zirconium, 12% titanium, 12% copper, 11% nickel, and 3% beryllium. Its yield strength is 1900 MPa, compared with 800 MPa for titanium alloy, Ti–6% Al–4% V, and 850 MPa for cast stainless steel.

Fracture toughness is said to be 55 MPa-$m^{1/2}$, the same as high-strength steel but half that of titanium. Its resistance to permanent deformation is said to be two to three times higher than that of conventional metals. The density of Vitrelloy is 6.1 g/cm^3 between cast titanium at 4.5 g/cm^3 and cast stainless steel at 7.8 g/cm^3. The material is particularly recommended for aerospace applications because of its surface hardness of 50 HRC. Cast titanium and steel are both tested at 30 HRC.

The beneficial properties of the alloy are ascribed to its noncrystalline structure. Because there are no patterns or grains within the structure, weak areas caused by grain boundaries are eliminated.

An advanced machinable ceramic that may be used to produce thermal shock-resistant components for aerospace, automotive, electrical, heat treating, metallurgical, petrochemical, and plastics applications up to 1550°C has been introduced. The new material (Aremcolox™ 502-1550) is based on the zirconium phosphate system ($Ba_{1+x}Zr_4P_{6-2x}Si_{2x}O_{24}$) and is especially unique because of its low coefficient of thermal expansion (CTE) of 0.5×10^{-6} in./in.°F. This characteristic sets the material apart from standard ceramic materials such as alumina and zirconia which have CTEs of 4.0×10^{-6} and 2.5×10^{-6}, respectively.

A low CTE ensures that as a component is thermally cycled the mechanical stress induced through expansion and contraction does not cause the part to crack. This feature enables engineers to adapt the material to high thermal shock applications, such as combustion and heater systems, that were not previously feasible.

Additional properties and applications of the machinable ceramic include their use as molds, optical stands, microwave housings, engine parts, and applications in which high mechanical strength, hardness, and low porosity are required. A low-density version of the material (502-1550 LD) is recommended for use as brazing fixtures, induction heating liners, rocket nozzles, and high-temperature gauges, tooling, and structures. The material is easily machined using carbide tooling and no postfiring is required.

Alloys

Zirconium alloys generally have only small amounts of alloying elements to add strength and resist hydrogen pickup. Zircoloy 2, for reactor structural parts, has 1.5% tin, 0.12% iron, 0.10% chromium, 0.05% nickel, and the balance zirconium. Tensile strength is 468 MPa, elongation 37%, and hardness Rockwell B89; at 316°C it retains a strength of 206 MPa.

Zirconium alloys can be machined by conventional methods, but they have a tendency to gall and work-harden during machining. Consequently, tools with higher than normal clearance angles are needed to penetrate previously work-hardened surfaces. Results can be satisfactory, however, with cemented carbide or high-speed steel tools. Carbide tools usually provide better finishes and higher productivity.

Mill products are available in four principal grades: 702, 704, 705, and 706. These metals can be formed, bent, and punched on standard shop equipment with a few modifications and special techniques. Grades 702 (unalloyed) and 704 (Zr–Sn–Cr–Fe alloy) sheet and strip can be bent on conventional press-brake or roll-forming equipment to a $5t$ bend radius at room temperature and to $3t$ at 200°C. Grades 705 and 706 (Zr–Cb alloys) can be bent to a $3t$ and $2.5t$ radius at room temperature and to about $1.5t$ at 200°C.

Small amounts of zirconium in copper give age-hardening and increase the tensile strength. Copper alloys containing even small amounts of zirconium are called zirconium bronze. They pour more easily than bronzes with titanium, and they have good electric conductivity. Zirconium–copper master alloy for adding zirconium to brasses and bronzes is marketed in grades with 12.5 and 35% zirconium. A nickel–zirconium master alloy has 40 to 50% nickel, 25 to 30% zirconium, 10% aluminum, and up to 10% silicon and 5% iron. Zirconium–ferrosilicon, for alloying with steel, contains 9 to 12% zirconium, 40 to 47% silicon, 40 to 45% iron, and 0.20% max carbon, but other compositions are available for special uses. SMZ alloy, for making high-strength cast irons without leaving residual zirconium in the iron, has about 75% silicon, 7% manganese, 7% zirconium, and the balance iron. A typical zirconium copper for electrical use is Amzirc. It is oxygen-free copper with only 0.15% of zirconium added. At 400°C it has a conductivity of 37% IACS, tensile strength of 358 MPa, and elongation of 9%. The softening temperature is 580°C.

Zirconium alloys with high zirconium content have few uses except for atomic applications. Zircoloy tubing is used to contain the uranium oxide fuel pellets in reactors because the zirconium does not have grain growth and deterioration from radiation. Zirconia ceramics are valued for electrical and high-temperature parts and refractory coatings. Zirconium oxide powder, for flame-sprayed coatings, comes in either hexagonal or cubic crystal forms. Zirconium silicate, $ZrSi_2$, comes as a tetragonal crystal powder. Its melting point is about 1649°C and hardness about 1000 Knoop.

Zirconium Beryllides

Intermetallic compounds, $ZrBe_{13}$ and Zr_2Be_{17}, have good strengths at elevated temperatures. $ZrBe_{13}$ is cubic, density 2.72 g/cm^3, melting point 1925°C; Zr_2Be_{17} is hexagonal, density 3.08 g/cm^3, melting point 1983°C; parts can be formed by all ceramic-forming methods plus flame and plasma-arc spraying. Materials are subject to safety requirements for all beryllium compounds.

These intermetallics, because of their greater densities (BeO = 1.85 g/cm^3), contain more beryllium atoms per unit volume than beryllia, a decided advantage for compact, beryllium-moderated nuclear reactors.

Zirconium Carbide

Zirconium carbide, ZrC_2, is produced by heating zirconia with carbon at about 2000°C. The cubic crystalline powder has a hardness of Knoop 2090, and a melting point of 3540°C. The powder is used as an abrasive and for hot-pressing into heat-resistant and abrasion-resistant parts.

Zirconium Diboride

Zirconium diboride (ZrB_2) has a density of 6.09 g/cm^3 and a hexagonal (AlB_2) crystal structure with a melting point of 3040°C.

Zirconium diboride is oxidation resistant at temperatures <1000°C and reacts slowly with nitric, hydrochloric, and hydrofluoric acids. It reacts with aqua regia and hot sulfuric acid, as well as with fused alkalies, carbonates, and bisulfates. Zirconium diboride has a typical room temperature electrical resistivity of 9.2×10^{-6} $\Omega \cdot$ cm and is superconductive at temperatures less than 2 K. Consolidation of ZrB_2 powder into parts is accomplished by hot pressing or pressureless sintering.

Similar to titanium diboride, ZrB_2 is wet by molten metals but is not attacked by them, making it a useful material for molten metal crucibles, free-formed nozzles, electric discharge machining electrodes, Hall–Heroult cell cathodes, and thermowell tubes for steel refining. This last use is one of the largest uses of zirconium diboride. Other uses for ZrB_2 include electrical devices, refractories, and applications where high oxidation resistance is required.

Others

Zirconium oxychloride, $ZrOCl_2 \cdot 8H_2O$, is a cream-colored powder soluble in water that is used as a catalyst, in the manufacture of color lakes, and in textile coatings. Zirconium-fused salt, used to refine aluminum and magnesium, is zirconium tetrachloride, a hygroscopic solid with 86% $ZrCl_4$. Zirconium sulfate, $Zr(SO_4)_2 \cdot 4H_2O$, comes in fine, white, water-soluble crystals. It is used in high-temperature lubricants, as a protein precipitant, and for tanning to produce white leathers. Soluble zirconium is sodium zirconium sulfate, used for the precipitation of proteins, as a stabilizer for pigments, and as an opacifier in paper. Zirconium carbonate is used in ointments for poison ivy, as the zirconium combines with the hydroxy groups of the urushiol poison and neutralizes it. Zirconium hydride has been used as a neutron moderator, although the energy moderation may be chiefly from the hydrogen.

Index

A

Abrasives, 1
  materials, 1
  wheels, 1
ABS plastics, 2–5, 527
  applications, 5
  fabrication, 4
  forms, 4–5
  properties, 2–4
Acetal plastics, 5–7, 605, 858
  copolymers, 7
    applications, 8
    properties, 7–8
  homopolymers, 7
    applications, 7
    properties, 7
  resins, 8–9, 858
    fabrication, 9
    processing, 8–9
Acetylene, 9
Acrylic plastics, 9–13, 780
  high-impact, 13
  standard, 9
    applications, 11–13
    fabrication, 11
    forms, 10
    properties, 10
Adhesives, 13–19, 513, 627, 729, 878
  applications, 19
  cements, 17–18
  characterization, 14
  classification, 14
    alloy, 16
    elastomeric, 16
    natural, 14–16
    thermoplastic, 16
    thermoset, 16
  forms, 14
  theory, 14
  types, 14, 627, 822
Alkyds, 19–22, 713
  applications, 22
  molding, 21
    characteristics, 22
    compounds, 21
  types, 21
    glass-fiber-reinforced, 21
    granular, 21
    putty, 21
Alloy, 22–28
  types, 23–28
    bearing, 23
    corrosion-resisting, 23
    dental, 23
    die-casting, 24
    eutectic, 24
    fusible, 24
    high-temperature, 24
    joining, 25
    light-metal, 25
    low-expansion, 25
    precious-metal, 26
    prosthetic, 27
    shape memory, 26
    superconducting, 28
Alloy structures, 28
Allylics, 29–31
  applications, 30
  molding compounds, 29
  types, 29–31
    diallyl phthalate (DAP), 29
    diallyl isophthalate (DAIP), 29
Alumina, 31–32, 488, 672
Aluminides (intermetallics), 32–38, 321, 645
  alloy types
    FeAl, 38, 322
    MoAl, 33, 323
    NiAl, 35–36, 322, 453, 646
    TiAl, 34, 322
  applications, 37
  future IM, 40, 324
  processing, 39
    gamma alloys (TiAl), 40, 322
Aluminum, 40
Aluminum alloys, 41–49
  aluminum, 42–49
    beryllium (AlBeMet), 47–48
    lithium, 45–46
    scandium, 48
  cast, 44
    permanent-mold, 45
    sand, 44–45
  mechanical alloying (MA), 47, 398
  powder metal (PM), 46, 615
  wrought
    heat-treatable, 41–44
    non-heat-treatable, 42–44
Amalgam, 49
Amine, 49
Amino, 49
Amorphous metals, 50
Anodic coatings, 50–54
  applications, 51–52
  production methods, 52
  properties, 51
  types of

aluminum, 50
    hard, 51
    magnesium, 53
    other, 51–53
    zinc, 54
Antimony, 54
Aramid (aromatic polyamide), 54
Arc discharge, 55
Argon, 55, 638
Aromatic hydrocarbon, 55
Arsenic, 56
Artificial intelligence, 56
Artificially layered structures, 56
Asbestos, 56–57
Asphalt, 57–58

B

Babbitt metal, 59
Barium, 59
Bearing materials, 60
Benzene, 61
Beryl, 61
Beryllium, 61–62
    applications, 62
    fabrication methods, 62
    forms, 61
Beryllium alloys, 63–65
    Be-Al, 47–48, 65
    Be-Cu, 63
    Be-Fe, 64
    Be-Ni, 64
    beryllides, 65, 772, 902
Beryllium oxide, 65–66, 488
Bismuth, 66–67
Bituminous coatings, 67–70
    applications, 70
    classes, 67–69
        cold-applied, 67
        hot-applied, 67
    types, 69
    uses, 69–70
Blow molding, 71–72, 531
Bonding, 72
Bonding materials, 72–73
    types, 72–73
        resins, 73
Boron, 73
    carbide (B_4C), 75
        whiskers, 75–76
    nitride (BN), 74, 465
Brass, 76–77, 177–178
Brazing, 77, 763
Brick, 77, 767
Bronze, 78, 178–179, 799
Buckyballs, 79
Butadiene-styrene resins, 79–80, 197
    properties, 79–80
Butyl rubber, 80–81, 468, 658

C

Cadmium, 83
Calcium, 84–85
Calendered sheet, 85, 535
Capacitor, 86–87
Carbides, 86–89
    refractory hard metals, 88, 644, 807, 818, 842
    silicon, 86–88, 645, 704
Carbon, 90–93
    applications, 93
        carbides, 92
        diamond, 91
    forms, 90
    properties, 92
    types
        fibers, 90, 911, see Graphite, 280, 281
Carbon–carbon composites (C–C), 94, 164
Carbon monoxide, 94
Carbon nitride (C–N), 94
    properties, 95
Carbon steels, 95–98
    production practices, 97–98
    types
        high-carbon, 97
        low-carbon, 96
        medium-carbon, 96
Case-hardening materials, 98–99
Castings, 99–102, 406
    metal-casting processes
        diecast, 100
        investment, 100, 259, 324–326
        lost foam, 101
        permanent mold, 100, 509
        replicast, 101
        sand, 100, 670
        shell mold, 101
        squeeze, 102
    rapid prototyping, 99, 634–635
Cast iron, 102–103, 106, 117
    types
        compacted graphite, 105
        ductible, 103, 195
        gray, 103, 281
        malleable, 105, 388
        white, 104, 865
Cast nonferrous alloys, 106–107
    heavy metals, 107
    light metals, 106, 373, 891
Cast plastics, 107–109, 534
    material types
        resins, 108–109
Cast steels, 109–111, 741
Cellulose, 110–112, 851, 860
Cellulose plastics, 112–115, 780
    acetates, 114–115
    nitrates, 115
Centrifugal castings, 115–118
    ferrous, 116
    nonferrous, 117
Ceramic fibers, 118–119

Index

alumina-silica, 118
  applications, 119
  silicon carbide, 119, 705, 706
Ceramic-matrix composite (CMC), 120, 164
  DIMOX process, 120–121, 193–194
  reinforcements, 120
Ceramics, 121–125, 300, 396, 898
  fabrication processes
    chemical, 123
    melt, 123
    powder, 123
    vapor
      CVD, 123
      CVI, 123
  properties, 122, 300
  types
    glass, 124, 204, 275
    metal oxide, 123, 488, 802, 810, 898
Cerium, 125
Cermets, 125–127
  applications, 127
  interactions, 126
  types, 125
Cesium, 127
Chemical milling process, 128–130
  characteristic, 129
  limitations, 129
  tolerances, 129
Chlorinated polyether, 130–131, 569
  fabrication, 130
  properties, 130
  uses, 131
Chlorinated polyethylene, 131
Chlorosulfonated polyethylene, 131
Chromium, 132–133
  alloys, 132
  biological aspects, 133
  uses, 132
Chromium alloys and steels, 133, 739
  copper, 133
  steels
    Cr, 133–134
    Cr-Mo, 133
    Cr-V, 134
Clad metals, 134–136
  applications, 135
  metal combinations, 135
  types, 135
Clay, 136, 753
  processing, 136
  properties, 137
  uses, 137
Coatings, 137–152, 421, 855
  conformal, 151
  conversion, 139–141, 515
  fluid, 149–151
  ionplating, 142–143
  metal, 138, 408
    aluminum, 138
    electroplated, 138, 211–214
    zinc, 138, 890

  plasma spraying, 145, 525, 777
  polymer, 146–149
  powder, 149, 412, 484, 528
  spray, 146, 410, 610, 776
  sputter, 141–142, 738, 838
  thermal spraying, 144, 776–778
  vacuum plasma, 143–144, 646
    PVD, 143
Cobalt, 152, 153
  applications, 153
  properties, 152
Cold-molded plastics, 153–154
Cold-rolled steel, 154, 802
Columbium and alloys, 155–158
Composite material, 156–165, 398, 734, 754, 784, 878
  applications, 162, 165
  classes, 159
    advanced composites
      reinforcements, 162
    carbon–carbon composites, 94, 164
    ceramic–matrix composites, 120, 164
    fiber–matrix composites, 160–161, 233
    metal–matrix composites, 164, 423
    organic–matrix composites, 163–164, 343, 580
  constituents, 158–159
  construction, 159
  fibers
    glass, 161, 234
    mineral, 161
  properties, 161
Compression and transfer molding, 165–167, 532
Concrete, 167–169, 754
  casting, 169
  flexible, 168
  polymer, 169
  rigid, 168
Conductive polymers and elastomers, 169, 729
Conductors, 170
Copolyesters, 170
Copolymer, 171
  block, 171, 791
  graft, 171
  random, 171
Copper and alloys, 171–179
  alloys, 173–179
    cast, 174
    powder, 179
    wrought, 174
  products
    cast, 172
    wrought, 173
  types and designations, 176–179
    brass, 177–178, 459
      aluminum, 177
      leaded, 178
      manganese, 177
      silicon, 177
      tin, 178
    bronze, 178–179, 459, 799
      aluminum, 178
      cadmium, 178

phosphor, 178, 518
silicon, 179, 703
cupronickels, 178
silvers
nickel, 178
Copper oxide, 179
Corrosion-resistant alloys, 179–180
cast, 179–180
Corundum, 180
Cryogenics, 181
Cutting tools, 181–183
PCNB, 181
PCD, 182

D

Dental materials, 185–186
orthodontic, 186
prosthodontic, 186
restorative, 185
Diamond, 186–188
applications, 187–188
CVD, 187
synthetic, 187
Diamond films, 188
Die castings, 188–190
applications, 189
characteristics, 189
materials, 190
production, 189
Die steels, 190
Dielectric materials, 190–192
industrial materials, 191
synthetic polymers, 191–192
Diffusion coatings, 192–193
calorized, 192
carbonitrided, 193
carburized, 192
chromized, 192
cyanided, 193
Ni–P, 193
nitrided, 193
siliconized, 193
Directed metal oxidation process (DIMOX), 193–194
Dispersion-strengthened metals, 194
Dolomite, 194
Dry ice, 195
Ductile iron, 195–196
applications, 196
forms, 196
joining, 195

E

Elastomers, 197–201, 219, 398, 786
definition, 197
types
natural rubber, 197, 201, 452, 656
butadiene, 197
GR–S, 197
synthetics

isoprene, 199, 768
neoprene, 198, 453, 659
nitrile (NBR), 199
polysulfide, 199, 596–597
silicone, 200, 660, 713–716
styrene-butadiene (SBR)-Bura S, 79, 197
Elastomeric linings, 201–202
applications, 202
materials
butyl rubber, 201
GR–S, 201
Hypalon, 202
natural rubber, 201
neoprene, 202
nitrile rubber, 202
polyvinyl chloride (PVC), 202
Electrical ceramics, 203–206
materials
electrical porcelain, 203
electronic ceramics, 205
glass, 204
bonded mica, 205
insulation, 203, 205
resistor ceramics, 206
semiconductor ceramics, 206, 680
thermal shock-resistant, 204
Electrical-contact materials, 206
Electrical-resistance metals and alloys, 207
Electrochemical process, 208–209
inorganic, 208
organic, 208
Electroformed parts
advantages, 209
sizes, 210
Electroless plating, 210
Electrolyte, 211
Electroplated coatings, 211–214
process, 212
properties, 214
specific metals
chromium, 212
copper, 213
gold, 213
nickel, 213
silver, 213
tin, 213
zinc, 214
Engineering films, 214–215
Epoxy resins, 215–217, 568, 581, 793
applications, 216
handling, 217
properties, 215–216
resin selection, 217
Etching materials, 218
Ether, 218, 583, 606
Ethylene, 218, 574, 845, 858
Ethylene glycol, 219
Ethylene-propylene elastomer, 219, 658
Extruded metals, 219–221
limitations, 221
shape, 221

Index

surface, 221
tolerances, 221
metal types, 221
process
    cold, 220
    hot, 219
Extruded plastics, 221–222, 531, 590
    materials, 222

F

Fabrics, nonwoven bonded, 223–224, 472
    applications, 224
    bonding agents, 224
    fiber types, 223
    production methods, 223
Fabrics, woven, 224–226, 775, 900
    textile constructions
        cordage, 225
        knit, 225
        woven, 225
Fasteners, 226–227, 654
    materials, 226
    types, 226
Feldspar, 227
Felts, 227–228
    applications, 228
    forms, 228
    properties, 228
Ferrite, 228, 729
    preparation, 229
    properties, 229
    types, 229
Ferrite devices, 229–230
    applications, 230
    chemistry, 229
    linear B–H, 230
    nonlinear B–H, 230
Ferroalloys, 230–232
    chromium, 231
    manganese, 231
    phosphorus, 232
    silicon, 232
    titanium, 232
Ferrous P/M, 232–233, 749
Fiber-reinforced glass, 233–236
    applications, 236
    properties, 234
    types, 234
Fiber-reinforced plastics, 236–240
    fibers, 236
    processing methods
        centrifugal casting, 239
        continuous laminating, 239
        filament winding, 238
        hand layup, 238
        injection molding, 238
        matched metal die molding, 238
        pultrusion, 239
        spray-up, 238
    properties, 239

resins, 237
Fibers, 239, 276, 705–706, 709, 775, 852, 887, 900
Fibrous glass, 240–242
Filament-wound reinforced plastics, 238, 242–243
    winding process
        accuracy, 243
        materials, 243
        range, 243
Flame-sprayed coatings, 243–245, 777–778
    finishing methods, 244–245
    types of guns
        plasma, 244, 777
        powder, 244, 777
        wire, 243, 777
Flint, 245
Fluidized-bed coatings, 245–247
    application, 245
    types of
        cellulosic, 245
        chlorinated polyether, 246
        epoxy, 246
        nylon, 246
        polyethylene, 246
Fluorocarbons, 247
Fluoroplastics, 247–251
    classes
        CFE, 248
        CTFE, 248
            properties, 250
        FEP, 247
        PFA, 247
        PTFE, 247
            properties, 248
        PVDF, 250
        PVF_2, 250
    melt processible class
        ECTFE, 251
        ETFE, 251
        PFA, 251
Fluorspar, 251
Flux, 252
Foam materials, 252–256, 533
    additives, 254
    applications, 254–256, 413
    cell, 254, 671
        density, 254
        structural, 254
    polymer, 253
        thermoplastic, 253
        thermoset, 253
    polyurethane, 253
    reinforcements, 254
    structural, 254, 751
Foil, 256
Forgings, 256–260
    advantages of, 258
    comparisons, 258–260
        CNC machining, 259
        composites, 260
        foundry casting, 259
        investment casting, 259

powder metallurgy, 260
reinforced plastics, 260
stampings, 258
weldments, 259
die types, 257
forgeability, 257
forging methods, 258–260
Fuel cells, 260–261
application, 261
types
proton exchange membrane (PEM), 261
solid oxide fuel cells (SOFC), 260
Fullerenes, 261–262
applications, 262
chemistry, 262
nanoparticles and tubes, 262
structures, 262
superconductivity, 262
Furfural, 263
Fusible alloys, 263

G

Gallium, 265–267
applications, 265
Galvanizing, 267
Garnet, 267
Gasket materials, 267
Gasoline, 268
Gel, 268–269, 730
gelation, 269
processing, 269
Germanium, 269–271
alloys, 269
uses, 270
infrared optical materials, 271
optoelectronics, 271
semiconductor, 270
Glass, 271–275, 397, 840
composition and structure, 271
processing, 271
products, 275
properties, 271
types
aluminosilicate, 273
boric oxide, 273
borosilicate, 272
colored, 274
conductive, 274
lead, 272
mirrors, 274
optical, 273
oxycarbide, 274
phosphate, 273
plate, 274
silica, 273
soda-lime, 271
Glass ceramics, 275–276
applications, 275
bearings, 276
heat exchanges, 276

precision instruments, 276
Glass fiber, 276–277
Glazing, 277–279
materials, 278
Gold, 279–280
alloys, 279
Graphite, 280–281, 300
properties, 281, see carbon and carbon composites, 92–94
types, 280
Gray iron, 103, 281–282
properties, 282
Gypsum, 282–283

H

Hafnium, 285
alloys, 285
properties, 285
Hard facings, 285–287
application methods, 286
materials, 287
Hard rubber, 287–288
fabrication, 287
properties, 287
uses, 288
Heat-resistant alloys (cast), 288–289
H-series, 289
properties, 288
Heat-resistant plastics (superpolymers), 290
Heavy alloy, 290
Helium, 290–292
applications, 291
properties, 291
High-energy rate forming (explosive forming), 292–293, 693
techniques of forming
bulkhead, 292
capacitor discharge, 293
cylinder, 292
Dynapak, 293
expanding gas, 293
free, 292
magnetic, 293, 693
High-speed steels, 293–295, 811
fabrication, 295
heat treatment, 294
properties, 294
uses, 295
High-strength hydrogen-resistant alloys, 296
High-temperature materials, 296–300
metallic materials, 296–297
design, 296
systems, 297
corrosion, 298
intermetallic compounds, 298, 321
oxidation, 298
refractory alloys, 297, 646
stainless alloys, 297, 738–741
superalloys, 297, 758–764
coatings, 300

Index

        directional solidification, 299
        oxide dispersion strengthening, 299
        powder metallurgy, 299
    nonmetallic materials, 300
Hot-dip coatings, 301–302
    aluminum, 302
    lead, 301
        application procedures, 301
        properties, 301
        uses, 301
    tin, 302, 801
        application procedures, 302
        properties, 302
        uses, 302
    zinc, 301, 890
        application procedures, 301
        properties, 301
        uses, 301
HSLA steels, 302
Hybrid materials, 302–303
    polymets, 303
        properties, 303
Hydrochloric acid, 304
Hydrogen, 304
    deuterium, 304
    heavy water, 304
    hydrides, 304
Hydrogen fluoride, 305
    properties, 305
    uses, 305

I

Immersion and chemical coatings, 307–308
    applications and properties, 307–308
        copper, 308
        electroless nickel, 307
        gold, 308
        silver, 308
        tin, 307
Impact extrusions, 308–309
    applications, 309
    design criteria, 309
    equipment, 309
    slugs, 309
Impregnating materials for coatings, 309–311
    materials, 310
        thermosetting plastics, 311
    processes, 310
Indium, 311–312
    applications, 312
    properties, 312
Industrial thermosetting laminates, 313
    properties, 313
    electrical, 313
    general, 313
    mechanical, 313
Injection molding, 313–317, 531
    limitations, 314
    molds, 315
    processes, 314–317
        electric molding technology (EMT), 316
        injection-compression molding (ICM), 315
Inorganic polymer, 317–319
    classification, 318
    properties, 317
    types
        boron, 318
        chalcogenide glasses, 318
        graphite, 318
        silicate, 318
        silicone, 318
Insulates, 319
Integrated circuits, 319
Interaction compounds, 319–320
    applications, 320
    properties, 320
Interface of phases, 320
Intermetallic compounds, 320–321
    family groups, 321
Intermetallic materials, 321–324
    ordered structures, 322–323
        iron aluminide, 322
        molybdenum disilicide, 323
        nickel aluminide, 322
        titanium aluminide, 322
    production
        primary processing, 323
        secondary processing, 324
Intermetallic matric composites (IMC), 324
Investment castings, 324–326
    applications, 326
    materials, 325
    process
        advantages, 326
        description, 324
Iodine, 327
Ion-implantation, 327
Iridium, 327–328
    uses, 327
Iron, 328–329, 879
    types, 328
Iron alloys, 329–334
    aluminum, 329
    carbon, 331
    chromium, 331
    cobalt, 332
    manganese, 332
    nickel, 333
    silicon, 333
    special types
        hard-facing, 334
        permanent-magnet, 334
        tungsten, 334
Isopolyester composites, 334
Isostatic pressing, 335–338
    cold (CIP), 335
        dry-bag, 336
        wet-bag, 336
    hot (HIP), 337
    warm (WIP), 337
Isothermal forging, 338

K

Kaolin (China clay), 339
Knitted and woven metals, 339–341

L

Laminated plastics, 343
Lanthanum, 343
  oxide, 343
Lanoxides, 344
Laser, 344–348
  advantages, 345
  applications, 346
  cooling, 346
  processes
    alloying, 344
    cutting, 346
      machines, 347
    inspection, 347
    lithography, 347
    peening, 348
  transitions, 344
Lava, 348
Lead, 348–352
  alloys, 350
    battery-grid, 351
    bearing metals, 351
    cable-sheathing, 350
    chemical-resistant, 351
    free-machining, 351
    solders, 351, 729
    type metals, 351
  forms, 349
    castings, 349
    cladding, 350
    coatings, 349
    extrusions
      foil, 349
    laminations, 349
    powder, 350
    sheet, 349
    shot, 350
    wool, 350
  uses, 351
Lead oxide, 352
Lead zirconate titanate, 352
Leather, 353, 354
  types, 353
Light-emitting diodes (LEDs), 354
Lime, 354–355
  limestone, 355
Liquid crystals, 355
Liquid crystal polymers (LCP), 355–356
  properties, 356
Lithia, 356–357
  uses, 356
Lithium, 357
  uses, 357
Low-alloy steels, 357–360
  carbon, 357
  fenitic, 359
  high-strength (ASLA), 302, 358–359
Low-expansion alloys, 360
Lubricants, 360
  types, 361
Lubricating grease, 362
  uses, 362

M

Machinability of metals, 363–364, 815
  process, 363
    fluid, 363
    speed, 363
    wear, 363
  properties, 363–364
    alloy systems, 364
    thermal conductivity, 364
    toughness, 363
Machining, 365–367
  operations
    automatic screw machines, 365
    boring, 365
    broaching, 365
    buffing, 367
    drilling, 365
    electric-discharge (EDM), 367
    grinding, 366
    honing, 366
    lapping, 367
    milling, 365
    nibbling, 366
    polishing, 367
    reaming, 365
    sawing, 365
    shaping, 365
    superfinishing, 367
    turning, 365
    ultrasonic, 367
Machining, high speed, 367–369
  data processing, 368
  tooling materials, 368
Magnesium, 369–375
  alloys, 372
    designation, 372
  applications, 374
  castings, 373
  characteristics, 369
  production, 371
  properties, 374
Magnesium oxide, 375, 488
Magnetic ferroelectrics, 375
Magnetic heating (UMH), 376–377
  applications, 377
  process, 377
    differences, 376
    principles, 376
Magnetic irons and steels, 377–379
  applications, 377
Magnetic materials, 380–387
  hard, 384–387

Index

soft, 381–383
Magnetic-shielding metals, 387
Magnetostriction, 387
Malleable irons, 388–389
   advantages, 388
   properties, 389
Manganese, 389–391
   alloys, 390
   uses, 390
Manganese dioxide, 391
Manganese modules, 391
Manganese steels, 391
Managing steels, 392–393, 821
Marble, 392
Materials handling, 393–394
   equipment, 393
Materials processing, 394–396
   synthesis processes
      combustion, 394
      reaction hot pressing, 396
      reaction sinter, 396
      sell-propagating high temperature (SHS), 394
      shock compression, 396
Materials science and engineering, 396–398
   ceramic materials, 121–125
      technical, 397
      traditional, 397
   composite materials, 156–165, 398
   elastomers, 197–201, 398
   glasses, 271–275, 397
   metallic materials, 396, 404–406
   polymeric materials, 397, 577–580
   thermoplastics, 397, 780, 783, 858
   thermosets, 398, 792–794, 860
Mechanical alloying, 398–400
   applications, 399
   process, 399
Mechanical finishes, 401–403
   finishes
      abrasives, 401, 402
         bonded, 402
         loose, 401
      methods, 401–402
         barrel, 402
         buffing, 401
         contact wheel, 401
         lapping, 402
         platen, 401
         pressure, 402
         slack belt, 401
   types of abrasives, 401
Melamine formaldehyde plastics, 403, 861
   applications, 403
Melamine resin, 403
Mercury, 404
Metal, 404–406, 734
   characteristics, 405
      directional bonds, 405
      metallic bonding, 405
   crystal structure, 405
   polycrystalline, 405

Metal casting, 406–408
   applications, 408
   cast to shape, 406, 407
   evaporative coding, 407
   ingot, 406
   zero gravity, 408
Metal coatings, 408–412
   anodic, 412
   cementation, 410
   ceramic, 411
   hot-dip, 409, 890–891
   immersion, 411
   powder, 149, 412
   sprayed, 410
   surface–conversion, 411
   vapors deposition, 411
   vitreous enamel, 411
Metal foams, 412–414
   applications, 413
   processing, 413
Metal forming, 414
Metal hydrides, 414–416
   covalent, 416
   ionic, 414
   transition, 414
Metal injection molding, 416
   applications, 417
Metal powders, 417–418
Metallic glasses, 418–419
   applications, 419
   properties
      electrical, 418
      magnetic, 418
      mechanical, 419
      superconductivity, 418
Metallic materials, 396, 419–420, 754
Metallic soap, 420
Metallizing, 421, 835–837, 891
   application methods, 422
   metallizing systems, 421
Metal-matrix composites, 164, 423–426
   applications, 426
   fabrication, 425
   production, 425
   properties, 424
   reinforcements, 423–424
      fibers, 423–424
      particles, 424
      whiskers, 424
Metal, mechanical properties of, 427–428
   creep, 428
   elongation, 427
   fatigue, 428
   fractural toughness, 428
   hardness, 428
   stress, 427
      rupture, 428
      states, 427
      strain curve, 427
   tensile strength, 427
   tension, 427

torsion, 427
yield strength, 427
Metal processing and mineral processing, 428–435
    extraction, 428
    occurrence, 428
    refining, 428
Methane, 428
Mica, 429
    types, 429
    uses, 429
Microelectromechanical systems (MEMS), 437
Microengineering, 437
Microgrinding, 438
Microinfiltrated macroliminated composites, 438–440
    applications, 440
    advantages, 439
    composition, 439
    fabrication, 440
Micromachining, 440
Microspheres, 441
Microwave heating, 441
Mold steel 441, 812
Molecular, 442–443
    materials, 442, 449
    nanotechnology, 443, 451
Molybdenum and alloys, 443–445
    alloys, 444
    applications, 444
    protective coatings, 444
Molybdenum disilicide, 445
Molybdenum steel, 446
Mullite, 446
Muntz metal, 446

N

Nanomaterials, 449
    applications, 449
Nanostructure, 450–451
    clusters, 450
    multilayers, 450
    reactivity, 451
Nanotechnology, 451
    principles, 451
Natural fiber, 451
Natural rubber, 452–453, 656–657, 768
    applications, 201, 452
    environmental effects, 452
    processing, 452
    vulcanization, 452, 656, 663
Neoprene rubber, 198, 453, 659
Nickel aluminide, 35, 322, 453, 646
Nickel and alloys, 453–458
    alloys
        binary, 455
        dispersion-strengthened, 454
        ferrous, 454
        magnetic, 454, 510
        nonferrous, 455
        specialty, 458
        superalloys, 458
        ternary, 457
Nickel-bonded titanium carbide, 458
Nickel brass, 459
Nickel bronze, 459
Nickel-chromium steel, 459
Nickel-molybdenum steel, 460
Nickel oxide, 460
Nickel silver, 460
Nickel steel, 460–461
Nickel sulfate, 461
Niobium and alloys, 462–465
    alloys, 462
    applications, 464
    properties, 462
Nitric acid, 465
Nitrides, 465–467, 645
    aluminum, 465, 645
    boron, 465
    silicon, 466
    titanium, 467, 645
    zirconium, 467
Nitriding steels, 467
    fabrication, 468
    processing, 467
        applications, 467
Nitrile rubber, 468–470, 659
    blends, 469
    composition, 469
    properties, 469
Nitrocarburizing, 470
    processing, 470
Nitrogen, 471
Nodular cast irons, 471
Nonmagnetic steels, 471
Nonwoven fabric, 472
Nylon plastics, 472–476
    applications, 475
    characteristics, 472
    processing, 474
    production, 472
    properties, 473
    types, 473

O

Olefin copolymers, 477, 787
Optical fibers, 478
    attenuation, 479
    designs, 478
Organic coatings, 479–484
    application and drying, 479
    other coatings
        dispersion, 483
        emulsion, 483
        hot melt, 484
        lining/sheeting, 484
        plastic powder, 484
        specialty finishes, 484
    types and systems
        coats, 480
            finish/top, 480

intermediate, 480
   fillers, 480
   sealers, 480
   surfacers, 480
  primer, 480
 composition, 480
 vehicles, 481–483
  drying oils, 481
  enamels, 482
  lacquer, 482
  paint, 483
  pigments, 481
  resins, 481
  varnishes, 483
Organometallic compounds, 485
Organosol coatings, 485–487
 application methods, 486
  dip, 486
  spray, 486
  spread, 486
  strand, 486
 properties, 486
 uses, 486
Osmium, 487
Osprey spray forming, 487
Oxide ceramics, 488–489, 734
 oxides
  mixed, 489
   cordierite, 489
   forsterite, 489
   steatite, 489
   zircon, 489
  single, 488, 489
   aluminum, 488
   beryllium, 488
   magnesium, 488
   thorium, 489
   zirconium, 488
Oxide coatings, 489–491
 heat-treatment methods, 490
  furnace/oven, 490
  molten salt bath, 490
 processes
  aqueous alkali-nitrate, 490
  browning, 491
  carbonia, 491
  nonferrous metals, 491
  steam, 491
Oxygen, 492–495
 production, 492
 uses, 492
 biological, 494
 cement/kiln, 495
 chemical synthesis, 493
 metallurgical, 492
 stone/clay/glass, 495

P

Paint, 497–501, 802
 coatings
  corrosion-resistant, 499
  high-temperature, 500
  immersion service, 500
  industrial, 499
 finishes
  air-dry, 499
  baking, 499
  commercial, 499
 removers, 501
 types of paints
  architectural, 497
  bituminous, 499
  enamel, 498
  house, 498
  solvent-thinned, 497
  water-thinned, 498
Palladium, 501, 541
 properties, 501
 uses, 501
Paper, 502–507
 applications/uses, 506
 papermaking, 504
 paper treatments, 505–506
  coatings, 505
  embossing, 506
  impregnation/saturation, 505
  laminations, 506
  mechanical, 506
  twisting, 506
 types, 503–505
  of paper
   cellulose fiber, 504
   inorganic fiber, 505
   synthetic organic fiber, 505
  of pulp, 503
   wood, 503
Particulates, 507–509
 characterization, 508
 particle size, 508
 processing, 507
 uses, 507
Pastes, 509
 types, 509
Permanent mold castings, 509
 advantages, 509
 casting alloys, 509
 disadvantages, 509
Permeability alloys, 510
 types, 510
 uses, 510
Pewter, 511, 800
Phenol, 511
 properties, 511
 uses, 511
Phenol-formaldehyde resin (phenolis), 511–515, 793, 861
 characteristics, 512
 uses, 512–514
  adhesives, 513
  bonding agents, 514
  fillers, 512

foundry shell molds, 514
molding compounds, 513
laminates, 513
Phosgene, 515
Phosphate conversion coatings, 515–518
applications, 516
bolt making, 517
coatings
plastisol, 516
vinyl, 516
extrusion, 518
wire drawing, 517
Phosphor bronze, 178, 518
Phosphor copper, 518
Phosphoric acid, 519
Phosphorus, 519–521
forms, 520
uses, 519
Photographic materials, 521–522
emulsions, 521
supports, 521
Piezoelectricity, 522
Pig iron, 522
Pigment material, 523–524, 657
colored, 523
special, 524
transparent, 523
white, 523
Plasma-arc coatings, 525–526, 777
characteristics, 525
fabrication, 525
materials, 525
Plaster mold castings, 526
Plaster of Paris, 526–527
applications, 526
Plastic alloys and blends, 527–528
ABS, 527
PPO, 528
PVC, 527, 603–604, 848
Plastic laminates, 528, 849
Plastic powder coatings, 528
Plastics, 529, 858
Plastics additives, 530–531
catalysts, 530
colorants, 530
fillers, 530
flame retardants, 530
plasticizers, 530
reinforced materials, 531
stabilizers, 530
Plastics processing, 531–535
blow molding, 531
calendering, 535
casting/encapsulation, 534
compression molding, 532
extrusion, 531
foam processes, 533
injection molding, 531
reinforced plastics/composites, 533
rotational molding, 532
thermoforming, 532

transfer molding, 532
Plastisol coatings, 535–536
application methods
dip, 536
spray, 536
spread, 535
properties, 536
uses, 536
Plastisols, 537–539, 847
molding methods
dip, 538
injection, 537
in-place, 538
pour, 537
rotational, 538
properties, 537
Platinum, 539
Platinum alloys, 539–540
cobalt, 540
gold, 540
iridium, 539
nickel, 540
rhenium, 540, 649–653
rhodium, 540, 653
ruthenium, 540, 667
tungsten, 540
Platinum group metals, 540, 541
iridium (Ir), 327, 541
osmium (Os), 541
palladium (Pd), 541
platinum (Pt), 541
rhodium (Rh), 541, 653
ruthenium (Ru), 541, 667
Plywood, 541–545
applications, 545
composition, 542
fabrication, 544
forms
grades, 542
properties, 543, 549
shapes, 545
sizes, 545
types, 542
Polonium, 545
Polyacrylate resin, 545, 858
Polyacrylic rubber, 546–547
curing, 546
forming, 546
Polyacrylonitrile resins, 547–548
Polyamides, 548, 859
Polyamide-imides, 548–549
properties, 549
Polyarylates, 549
Polyaryletherketones (PAEK), 549
Polybenzimidazole, 549
Polybutadiene rubber, 550–553
applications, 551
composition, 550
high-*cis*, 551
high-*trans*, 552
processing, 551

Index

uses, 553
Polycarbonates, 553–557
  applications, 556
  composition, 554
  fabrication, 554
  properties, 554
Polyester film, 557–559
  applications, 558
  fabrication, 558
  forms, 558
  properties, 557
Polyester plastics, 559–562
  composition, 559
  curing, 559
  fabrication techniques
    corrugated sheet molding, 560
    hand layup, 560
    matched die molding, 560
    premix molding, 561, 626
  forms, 561
  properties, 561
  reinforcement, 560
Polyester thermoplastic resins, 562–563, 787, 793, 825
  PBT, 562, 787
  PET, 562
Polyester thermosetting resins, 563
  laminates, 563
Polyetheretherketones (PEEK), 564–566
  applications, 565
  fabrication, 564
  properties, 564
  types, 564
  uses, 564
Polyetherimides (PEI), 566
  properties, 566
Polyether resins, 566–569
  epoxy, 568
  phenoxy, 568
  polyolefin oxide, 568
  polyoxymethylene, 569
  polyphenylene oxide (PPO), 568
Polyethylene, 569
  low-density, 571
  high-density, 572
  properties, 569
  UHMWPE, 572
Polyethylene glycols, 573–574
  applications, 573
  characteristics, 573
Polyfluorolefin resins, 574–575
  copolymers, 574
    FEP, 574
    HFP, 574
    CTFE, 574
  fluorinated elastomers, 575
  PVF$_2$, 574
  TFE, 574
Polyimides (PI), 575
  properties, 575
Polyketones, 575–577
  processing, 577

properties, 577
types
  PAEK/PEK, 577
  PEEK, 577
  PEKK, 577
Polymer, 577–580, 790, 858–859, 885
  additives, 578
  compounding, 578
  environmental stability, 578
  fabrication, 579
  history, 578
  polyblends, 579
  properties, 578
Polymer composite, 580–582
  mechanical properties, 580
  reinforced resins
    thermoplastic, 581, 790
    thermosetting, 581
Polymerization, 582
Polyolefin resins, 582–583, 780
Polyphenylene ether (PPE), 583–584
Polyphenylene oxide (PPO), 528, 583
Polyphenylene sulfide (PPS), 585–588
  applications, 585
  properties, 585
Polypropylene plastics (PP), 588–591
  fabrication
    compression molding, 590
    injection molding, 590
    extrusion, 590
  grades, 589
  molecular types, 588
  properties, 588
  uses, 590
Polystyrenes, 591–593, 780
  characteristics, 592
  properties, 592
  types
    chemical-resistant grades, 593
    general-purpose, 592
    impact grades, 592
    other special grades, 593
Polysulfide resins, 593–595
  applications, 593
Polysulfide rubber, 199, 596–597
  fabrication/applications, 596
  forms, 596
  polymers
    crudes/solid, 596
    liquid, 596
  properties, 596
Polysulfone resins, 597, 756
Polytetrafluoroethylene (PTFE), 597–598
  forms, 598
  uses, 598
Polyurethanes, 598–603, 790–791, 829–832
  applications, 602, 831
  fabrication, 602
    RIM, 602, 640
  foams, 599–602, 829
    flexible, 599, 829

integral-skin, 602
rigid, 599, 829
processing, 599
Polyvinyl chloride (PVC), 527, 603–604, 848
forms, 603, 605
processing, 603
properties, 603
Polyvinyl fluoride film, 604–605
applications, 604
properties, 604
Polyvinyl resins, 605–606
acetal, 605
acetate, 605, 847
alcohol, 605, 859
chloride, 605, 849
ether, 606
fluoride, 606
Porcelain, 606, 866–867
properties, 606
Porcelain enamels and ceramic coatings, 607–610
high-temperature coatings, 609
flame/arc-spraying, 610
low-temperature coatings, 607
chemical reactants, 607
inorganic colloids, 607
moderate temperature processes, 607
refractory enamels and costs, 609
Porous metals, 610
Potassium alloys and compounds, 611–615
alloys, 611
biological activity, 612
compounds, 611–615
carbonate, 611, 613
cholorate, 614
chloride, 611, 614
cyanide, 614
hydroxide, 611
nitrate, 611, 615
properties, 611
uses, 612
Powder metallurgy-parts (P/M), 615–618
applications, 617
characteristics, 617
processes
full-density powder, 617
powder injection molding, 616, 646
pressure/sintering, 616
safety/health considerations, 618
Powder plastics moldings, 616, 618–619
advantages, 618
materials used, 619
processing, 618
Precious metals, 619
Prefinished metals, 619–622
applications, 621
functional advantages, 621
types, 621
Preimpregnated decorative foil, 622–623
applications, 622
fabrication, 622
materials, 623

Preimpregnated materials for reinforced plastics, 623–626
advantages, 623
fabrication, 623
materials used
reinforcing, 624
molding, 625
resins, 625
Premix moldings, 561, 626–627
comparative benefits, 626
compounding, 627
molding, 626
Pressure-sensitive adhesives (PSA), 627–628
benefits, 628
materials, 628
Pyrolytic graphite, 628
Pyrolytic materials, 628–631
directionality, 629
forms, 629
materials, 629
properties, 629–631
anisotropic, 630
isotropic 630
carbides, 631
tungsten, 631

R

Radiant heating, 633–634
alloys, 634
characterization, 633
Rapid prototyping (RP), 99, 634–635
future, 634
tooling, 634
Rare earth elements/metals, 635–637
applications, 637
groups, 636
light, 636
heavy, 636
properties, 636
fabricability, 637
mechanical, 636
oxidation resistance, 637
Rare gases, 638–639
argon, 55, 638
helium, 290–292
krypton, 638
neon, 638
xenon, 638
Rare metals, 639
Rayon, 639
Reaction-injection molding (RIM), 640
Red brass, 640
Refractories specialty, 640–643
manufacture, 642
materials
aluminosilicate, 641
basic, 641
miscellaneous, 642
silica, 641
properties, 643

Index

Refractory cement, 643–644
Refractory hard metals (RHM), 644–646
  aluminides, 645
    nickel, 646
  borides, 645, 809
    chromium, 645
    molybdenum, 645
    zirconium, 645, 902
  carbides, 644, 704, 902
    tantalum/tungsten, 644, 772
    titanium, 644, 807–809
    tungsten, 644, 818
    tungsten-titanium (WTiC$_2$), 644
  disilicides, 645
    molybdenum (MoSi$_2$), 645
    tungsten (WSi$_2$), 645
  nitrides, 645, 708
    aluminum (AlN), 645
    nickel, 645
    titanium (TiN), 645, 809
  tribaloy intermetallic materials, 646
Refractory metals, 646–647
  processes
    low-pressure plasma spray (LLPS), 646
    plasma-activated sintering (PAS), 646
    powder injection molding (PIM), 616, 646
    small angle neutron scattering (SANS), 646
    vacuum plasma spray (VPS), 646
Reinforced concrete, 167–169, 647
Resin, 8, 545, 547, 566, 574, 593, 597, 605, 625, 647, 709, 825, 847
Resistance heating, 648–649
  applications, 649
  element forms, 648
  heating methods
    molten salt, 649
    materials, 648
Rhenium and alloys, 540, 649–653
  applications, 653
  fabrication, 652
  processing, 650
  properties, 651
Rhodium, 653
Rivet, 654
Robotics, 655–656
  degrees of freedom, 655
  joint actuators, 655
  manipulators, 655
  mechanisms, 655
  types of joints, 655
    articulated, 655
    prismatic, 655
Rubber, 546, 550–553, 596, 656–664, 673
  latex technology, 656, 663
  processing, 656
    curatives, 656
    pigments, 657
    vulcanization, 656, 663
  products, 662–663
    building, 662
    compounding, 662
    forming, 662
    mixing, 662
  specific types
    butyl, 658
    ethylene-propylene polymers (EPM, EPDM), 219, 658
    natural (NR), 452, 657
    neoprene (CR), 198, 453, 659
    nitrile (NBR), 468–470, 659
    polybutadiene (BR), 550, 658
    polyisoprene (IR), 659, 768
    specialty, 660
      elastomers, 661
    styrene-butadiene (SBR), 657, 673–674
Rust preventatives, 664–667
  selection of, 666
  types, 664–666
    emulsifiable, 666
    preservative oils, 665
    solid/semisolid waxes, 664
    solvent cutbacks, 665
  uses, 667
Ruthenium, 540–541, 667
Rutile, 667

S

Salt, 669
  classification, 669
Sand, 669–670
  characteristics, 669
  economic importance, 670
  origin, 669
  transport, 669
Sand castings, 100, 670–671
  process
    dry-sand molding, 670
    green-sand molding, 670
    pit molding, 671
Sandwich materials, 671–672
  advantages, 672
  construction, 671
  core materials, 671
    cellular, 671
    foam, 671
    solid, 671
  theory, 672
Sapphire, 672–673
  forms available, 673
  general natural, 672
  properties, 673
  uses, 673
SBR rubber, 657, 673–674
  compounding, 674
  properties, 673
  types, 673
Scandium, 674–676
  compounds, 676
    carbide, 676
    oxide, 676
    nitride, 676

fabrication, 675
preparation/purification, 675
properties, 675
uses, 675
Screw, 676
Screw-machine parts, 676–678
   design, 677
   dimensions, 678
   finish, 678
   materials, 677
Seals (ceramic-to-metal seals), 678
   types
      butt, 678
      external, 678
      glass-to-metal, 678
      internal, 678
      ram, 678
      taper, 678
Selenium, 679–680
   applications, 679
   forms, 679
   properties, 679
Semiconducting materials, 680–687
   applications, 685
   fabrication, 682
   Hall effect, 681
   materials, 206, 680, 683–685
      extrinsic, 680
      intrinsic, 680
   types, 682
Sensitizing compounds, 687
Shape memory alloys, 687–690
   applications, 688–690
   materials, 688
   properties, 688
   thermomechanical behavior, 688
   transformation characteristics, 687
Sheet metal forming, 690–692
   bending, 691
   deep drawing, 692
   miscellaneous processes, 692
   rubber forming, 692
   spinning, 692
   stretch forming, 691
Sheet metal parts, 693
   family of processes, 693
   materials, 693
   stamping, 693
   types of work, 693
Sheet molding compound (SMC), 694–695
   applications, 694
   developments, 694
Shell mold castings, 695
   advantages, 695
   process, 695
Silica, 696–700
   forms, 696–698
   properties, 696–697
   quartz, 696, 709
   uses, 697, 699
Silica minerals, 700

Silicides, 700–702
   availability, 701
   composition, 701
   fabrication, 701
   properties, 701
Silicon, 702–703
   fabrication, 702
   forms, 702
   hydrides, 703
   principal compounds, 703
   types, 702
   uses, 703
Silicon bronze, 179, 703
Silicon carbide, 704–707
   applications, 705
   fabrication, 704
   forms, 704–705
   properties, 704
   types, 704–705
      crystals, 705
      fibers, 705–706
      foam, 705
      platelets, 706
      whiskers, 706–707
Silicon cast iron, 707
Silicon copper, 707
Silicon halides, 707
Silicon manganese, 708
Silicon nitride, 708–709
   processing, 708
   types
      fibers, 709
      HIPSN, 708
      HPSN, 708
      RBSN, 708
      SSN, 708
Silicon oxides, 709
Silicon steel, 709
Silicone resins, 709–713
   composition, 710
   properties, 710
   types
      foaming powders, 711
      laminating resins, 710
      molding compounds, 711
      potting resins, 711
      varnishes, 712
   uses, 712
Silicone rubber, 200, 660, 713–716
   composition, 713
   fabrication, 715
   types, 713
   uses, 715
Silk, 716
Silver, 716–720
   alloys, 717, 719
   applications, 717–718
      compounds, 717–718
      classification, 716
      fabricability, 717
      joining, 717

Index

uses, 716–720
Single crystals, 720–721
   properties, 721
Sinter hardening, 721–723
   alloying, 722
   materials, 722
   processing, 722
Sizing agents, 723
Sodium, 724–727
   compounds, 725–726
   reactions
      inorganic, 724
      organic, 725
   uses, 727
Solder alloys, 727–730
   forms, 727
   types, 727
      cold, 729
      lead-free, 729
      polymer, 729
      silver, 728
Solder materials, 730
Sol-gel process, 730–731
   applications, 731
   glass formation, 731
      aging, 731
      densification, 731
      drying, 731
      gelation, 731
      hydrolysis/condensation, 731
      microstructural development, 731
Soybean oil, 731
Space processing, 732–734
   electronic/photonic materials, 732
      solution growth, 733
      vapor growth, 733
   metallic alloys/composites, 733
      dendrite growth, 733
      electrodeposition, 734
      liquid-phase sintering, 734
      monotectic alloys, 734
   protein crystal growth, 732
Spinel, 734
   applications, 735
Spray metal forming, 735–736
   advantages, 735
   applications, 735
   process, 735
   uses, 736
Spring steel, 736–737
   applications, 737
   forms, 737
   steels
      manganese, 737
      silicon, 737
Spun parts, 737–738
   process, 737
   shapes/tolerances, 738
Sputter texturing, 738
Stainless steel, 738–741
   fabrication, 739

      finishing, 741
      hot working, 739
      machining/forming, 740
      welding, 740
Stainless steel (cast), 740–741
   applications, 741
Stainless steel (wrought), 741–747
   family
      austenitic, 742
      ferritic, 746
      martensitic, 746
   specialty types, 747
Stainless steel products, 748
Steel, 748–749
Steel powder, 749
Steel wool, 749
Stereolithography, 749
   problems, 749
Strippable coatings, 750–751
   types
      acrylic, 750
      cellulosics, 750
      polyethylene, 750
      vinyl, 750
Strontium and alloys, 751
   compounds, 751
Structural foam, 751–753
   advantages, 752
   limitations, 753
   processing, 752
      high pressure, 752
      low pressure, 752
Structural materials, 753–755
   types
      building stones, 753
      clay products, 753
      composite materials, 754
      concrete, 754
      structural metals, 754
      wood, 754
Sulfone polymers, 755–756
   properties, 755
Sulfur, 756–757
   compounds, 757
   properties, 757
   uses, 757
Sulfuric acid, 757–758
   compounds, 757
   uses, 757
Superalloys, 758–764
   compositions, 760–761
   fabrication
      brazing, 763
      forging, 762
      forming, 763
      heat treatment, 762
      hot working, 762
      machining, 763
      welding, 763
   properties, 760
   strengthening mechanisms

carbide hardening, 759
deoxidizers, 759
malleabilizers, 759
precipitation hardening, 758
solid solution, 758
types
cobalt-base, 761
iron-base, 760
nickel-base, 760
Superconductivity, 764–765
applications, 765
processes, 764
properties, 764
superconductors, 764
Superplastic forming, 765–767
applications, 767
processes, 765–766
blow forming, 765
diffusion bonding, 766
movable-tool forming, 766
Superpolymers, 767
Surface pigments (for brick), 767–768
applications
dry, 768
wet, 768
compositions, 768
Synthetic natural rubber (isoprene), 768, 769
applications, 769
properties, 769

T

Talc, 771
applications, 771
Tantalum, 771, 772
alloys, 772
fabricability, 772
cold working, 772
hot working, 772
joining, 772
Tantalum beryllides/carbides, 772
Tape casting, 773–774
applications, 773
CAM-LEM process, 773–774
layered manufacturing, 773–774
Tellurium, 774–775
oxides, 775
Textile fibers, 775–776
spinning process, 775
textile, 775
Thallium, 776
applications, 776
toxicity, 776
Thermal spraying, 146, 410, 610, 776–779
coating, 777–779
characteristics, 778
selection, 778
types
detonation gun, 778
flame spray, 777
HVOF spray, 778

plasma spray, 777
wire arc spray, 777
Thermocouples, 779–780
Thermoformed plastic sheet, 780–783
materials, 780
processing
air blowing, 782
matched mold forming, 781
slip forming, 782
vacuum forming, 782
vacuum molds, 783
Thermoplastic polyesters, 562, 787–790
PBT, 562
PET, 562
properties, 559–561
Thermoplastic polypropylene, 588–591, 790
applications, 790
VHME, 790
Thermoplastic polyurethanes, 790–791, 829
properties, 791
TPU, 790–791
types, 253, 791
Thermoplastic styrenes, 591–593, 791
Thermoplastic urethanes, 792, 829
Thermoplastic vulcanizates, 792
properties, 792
uses, 792
Thermoset plastics, 253, 792–794, 860
composites, 792
properties, 794
resin systems, 793–794, 860
epoxies, 215–217, 568, 581, 793
phenolics, 511–515, 793
polyesters, 563, 793
silicones, 710, 794
Thixomag, 794
process, 794
Thixotropic processing, 794–796
Thermoplastic, 16, 253, 780, 783, 858
Thermoplastic continuous-fiber-reinforced materials (CFRTP), 784–786
applications, 786
basic, 784
molds, 785
processes, 784
compression molding, 785
Thermoplastic elastomers (TPE), 197–201, 786–787
applications/advancements, 786–787
drawbacks, 787
properties, 787
types, 786
Thermoplastic olefins, 582, 787
applications, 796
processing, 795
thixoforming, 795
thixotropic casting, 795
Thorium, 796
Thyristors, 796–797
Tile, 797
Tin, 797–801
alloys, 799–801

Index

babbitt/bearing metal, 800
bronzes, 799
die-cast tin base, 800
fusible tin, 800
pewters, 800
powders, 800, 801
soft solders, 728–729, 799
type metals, 800
applications, 798
forms, 798
properties, 798
Tinplate, 801
processing, 801
uses, 801
Titanates, 801–803
material types
barium, 802
calcium, 802
lead, 803
magnesium, 802
miscellaneous, 803
strontium, 802
Titanium, 803–807
alloys, 806–807
applications, 807
fabrication, 805
forms, 805
melting, 803
properties, 804
structure, 804
Titanium carbide (TiC), 807–809
grades, 808
properties, 807
uses, 808
Titanium diboride (TiB$_2$), 809
processing, 809
properties, 809
uses, 809
Titanium nitride (TiN), 809–810
whiskers, 809
Titanium oxide (TiO$_2$), 667, 802, 810–811
uses, 810
Tool steel, 811–815
coatings, 813
forms, 815
heat treatment, 814
properties, 814
types
cold-work, 812
high-speed, 811
hot-work, 812
mold, 441, 812
shock-resistant, 812
special-purpose, 813
water-hardening, 813
Tungsten, 815–818
alloys, 818
applications, 816–818
fabrication, 815
properties, 816
Tungsten carbide (WC), 818

Tungsten steel, 819

U

Ultrahigh-strength steels, 821
managing steels, 392–393, 821
Ultraviolet-curable hot-melt adhesives, 822
applications, 822
forms, 822
future, 822
properties, 822
types, 822
Uniform magnetic heating (UMH), 822–825
application, 823–825
induction, 823
Unsaturated polyester resin, 825–827
future, 827
gel-coat
developments/markets, 825–826
protection, 826
resistance to fire, 825
solid surfaces, 826
Uranium, 827–828
uses, 827
Urea, 828–829
types, 828
Urethanes, 792, 829–832
fabrication, 830
foams, 829
properties, 831
protection, 830
types
flexible, 829
rigid, 829
uses, 831

V

Vacuum assist mold processing, 783, 833
RTM, 833
VARI, 833
VARTM, 833
VRTM, 833
Vacuum carburizing, 834–835
advantages, 835
distortion, 834
process, 834
comparisons, 834
Vacuum coatings, 835–837
Vacuum processing, 837–841
applications, 838–841
data storage disks, 841
glass coatings, 840
optical coatings, 839
web coatings, 839
sources
cathode sputtering, 838
chemical vapor deposition (CVD), 839
electron beam evaporators, 838
thermal evaporators, 838
Vanadium and alloys, 841, 842

application, 842
fabrication
    cold working, 841
    hot working, 841
uses
    barides, 842
    carbides, 842
    oxides, 842
Vanadium steel, 842
Vapor-deposited coatings, 839, 843–845, 886
    applications, 843
    disadvantages, 844
Vinyl acetate ethylene (VAE), 845–846
    applications, 845
    future, 846
    processing, 845
Vinyl resins and plastics, 846, 859
    applications, 850
    products
        acetate, 848
        alcohol, 846
        benzoate, 847
        chloride, 847
            CH₂CHCl, 847–448
            PVC, 848, 859
        plastisol, 847
    types of vinyls/PVCs, 848–850, 859
        coatings, 849
        flexible, 848
        rigid, 848
        vinyl-metal laminates, 849
        vinylidene
            chloride plastics, 849
            fluoride, 850
Volatile organic compound, 850–851
    process, 851
Vulcanized fiber, 851–853
    applications, 852
    chemistry, 852
    forms, 851
    grades, 852
    properties, 851

W

Wash primers, 855
Watch, 855–856
    types
        electromechanical, 856
        electronic, 855
        mechanical, 856
Water, 856–858
    properties, 857
    states, 856–857
        gaseous, 856
        liquid, 857
        solid, 857
Water-soluble plastics, 858–862
    thermoplastic resins, 858–860
        alkali-soluble polyvinyl acetate copolymers, 858
        cellulosic derivatives, 860

ethylene-maleic anhydride copolymers, 858
hydroxyethylcellulose, 860
methylcellulose, 860
polyacrylamide, 859
polyacrylates, 545, 858
polyethers, 859
polyvinyl alcohol, 859
polyvinyl pyrrolidone, 859
sodium carboxymethyl-cellulose, 860
styrene-maleic anhydride, 860
thermosetting resins, 860–862
    cyclic, 861
    melamine-formaldehyde, 861
    phenol-formaldehyde, 861
    urea-formaldehyde, 862
Wax, 862–863
    types, 862
    uses, 862
Wear-resistant steel, 863
Welding alloys, 863–864
Wetting agents, 864
Whiskers, 864
White, 865–866
    brass, 865
    cast iron, 865
    gold, 866
    metal, 866
Whiteware ceramics, 866–868
    composition, 866
    compounding, 866
    forming, 866
    types, 866
Wire cloth, 868
Wood, 754, 868–877
    anatomy, 869
    chemicals, 871–872
        processing, 872–873
    composition, 870
    degradation, 872
    engineered composites, 873
    engineering design, 872
    grades of lumber, 871
        hardwood, 871
        softwood, 871
    preservation/protection, 876–877
    processing, 873
    products (composite), 874–875
        laminated, 874
        oriented strand board, 875
        particleboard, 875
        waferboard, 875
    products (fiber), 875
        hardboard, 875
        medium-density fiberboard, 875
        pulp and paper, 875
    products (wood), 874
        lumber, 874
        veneer/structural plywood, 874
    properties, 877
    types, 869–870
        hardwoods, 869–870

Index

softwoods, 869–870
Wood-based fiber and particle materials, 877
Wood-metal laminates, 878
    applications, 878
    stiffness, 878
Woods, imported, 878
Wrought iron, 879–881
    applications, 880–881
    fabrication
        bending, 880
        corrosion resistance, 880
        forging, 880
        protective coatings, 880
        welding, 880
    properties, 879

X

X-ray photo-to-electron spectroscopy, 883–884
    case study, 883
    disadvantages, 883
    process, 883
X-ray technology, 884–885
    applications, 885
        fluorescence spectrometry, 885
        radiography, 885
    characteristics/generation of x-rays, 884
        electromagnetic spectrum, 884
        synchrotron radiation, 884
        x-ray tubes, 884
Xylene, 885
Xylene polymers, 886
    applications, 886

Y

Yarns, 887
Yttrium and alloys, 887–888
    aluminum garnet, 888
    oxide, 888
Y-TZP, 888

Z

Zinc and alloys, 889–897
    alloys, 893–895
        casting, 891, 894
        other, 895
        wrought, 895
    applications, 890–897
    compounds, 889
    forms, 889
    oxide, 896–897
    processes, 890–891
        electrogalvanizing, 891
        hot dip galvanizing, 890
        metallizing, 891
        sherardizing, 891
        zinc anodes, 891
        zinc-pigmented paints, 891
    production, 889
Zircon, 897–898
    uses, 897
Zirconia (zirconium oxide) (ZrO_2), 898–900
    forms, 899
    types, 899–900
    uses, 898
Zirconium and alloys, 900–903
    alloys, 901–903
        beryllides, 902
        carbide, 902
        diboride, 902
        others, 903
    properties, 900
    uses, 901